T0137158

# Advances in Intelligent Systems and Computing

Volume 896

**Series editor**

Janusz Kacprzyk, Systems Research Institute, Polish Academy of Sciences,
Warsaw, Poland
e-mail: kacprzyk@ibspan.waw.pl

The series "Advances in Intelligent Systems and Computing" contains publications on theory, applications, and design methods of Intelligent Systems and Intelligent Computing. Virtually all disciplines such as engineering, natural sciences, computer and information science, ICT, economics, business, e-commerce, environment, healthcare, life science are covered. The list of topics spans all the areas of modern intelligent systems and computing such as: computational intelligence, soft computing including neural networks, fuzzy systems, evolutionary computing and the fusion of these paradigms, social intelligence, ambient intelligence, computational neuroscience, artificial life, virtual worlds and society, cognitive science and systems, Perception and Vision, DNA and immune based systems, self-organizing and adaptive systems, e-Learning and teaching, human-centered and human-centric computing, recommender systems, intelligent control, robotics and mechatronics including human-machine teaming, knowledge-based paradigms, learning paradigms, machine ethics, intelligent data analysis, knowledge management, intelligent agents, intelligent decision making and support, intelligent network security, trust management, interactive entertainment, Web intelligence and multimedia.

The publications within "Advances in Intelligent Systems and Computing" are primarily proceedings of important conferences, symposia and congresses. They cover significant recent developments in the field, both of a foundational and applicable character. An important characteristic feature of the series is the short publication time and world-wide distribution. This permits a rapid and broad dissemination of research results.

More information about this series at http://www.springer.com/series/11156

Rafik A. Aliev · Janusz Kacprzyk
Witold Pedrycz · Mo. Jamshidi
Fahreddin M. Sadikoglu
Editors

# 13th International Conference on Theory and Application of Fuzzy Systems and Soft Computing — ICAFS-2018

Springer

*Editors*
Rafik A. Aliev
Joint MBA Program, Azerbaijan State Oil
and Industry University
Baku, Azerbaijan

Janusz Kacprzyk
Systems Research Institute
Polish Academy of Sciences
Warsaw, Poland

Witold Pedrycz
Department of Electrical and Computer
Engineering
University of Alberta
Edmonton, AB, Canada

Mo. Jamshidi
Department of Electrical and Computer
Engineering
University of Texas at San Antonio
San Antonio, TX, USA

Fahreddin M. Sadikoglu
Department of Mechatronics
Near East University
Nicosia TRNC, Turkey

ISSN 2194-5357          ISSN 2194-5365   (electronic)
Advances in Intelligent Systems and Computing
ISBN 978-3-030-04163-2          ISBN 978-3-030-04164-9   (eBook)
https://doi.org/10.1007/978-3-030-04164-9

Library of Congress Control Number: 2018960997

This Springer imprint is published by the registered company Springer Nature Switzerland AG
The registered company address is: Gewerbestrasse 11, 6330 Cham, Switzerland

# Preface

The Thirteenth International Conference on Application of Fuzzy Systems and Soft Computing (ICAFS-2018) is the premier international conference organized by Azerbaijan Association of "Zadeh's Legacy and Artificial Intelligence" (Azerbaijan), Azerbaijan State Oil and Industry University (Azerbaijan), Berkeley Initiative in Soft Computing (BISC) (USA), Georgia State University (Atlanta, USA), Near East University (North Cyprus), TOBB Economics and Technology University (Turkey), University of Alberta (Alberta Canada), University of Siegen (Siegen, Germany), University of Texas, San Antonio (USA), University of Toronto (Toronto, Canada).

This volume presents an edited selection of the presentations from the Thirteenth International Conference on Application of Fuzzy Systems and Soft Computing (ICAFS-2018) which was held in Warsaw, Poland, August 27–28, 2018. ICAFS-2018 is held as a meeting for the communication of research on application of fuzzy logic, uncertain computation, Z-information processing, neuro-fuzzy approaches, and different constituent methodologies of soft computing applied in economics, business, industry, education, medicine, earth sciences, and other fields. The conference provided an opportunity to present and discuss state-of-the-art research in this expanding domain.

This volume will be a useful guide for academics, practitioners, and graduates in fuzzy logic and soft computing. It will allow for increasing of interest in development and applying of soft computing methods in various real-life fields.

August 2018

Rafik Aliev  
Chairman of ICAFS-2018

# Organization

## Chairman

R. A. Aliev, Azerbaijan

## Co-chairmen and Guest Editors

J. Kacprzyk, Poland
M. Jamshidi, USA
W. Pedrycz, Canada
F. S. Sadikoglu, North Cyprus

## International Program Committee

R. Abiyev, North Cyprus
B. Fazlollahi, USA
V. Nourani, Iran
I. G. Akperov, Russia
T. Fukuda, Japan
V. Novak, Czech Republic
R. R. Aliev, North Cyprus
M. Gupta, Canada
I. Perfilieva, Czech Republic
F. Aminzadeh, USA
R. Gurbanov, Azerbaijan
H. Prade, France
K. Atanassov, Bulgaria
O. Huseynov, Azerbaijan
H. Roth, Germany

N. Yusupbekov, Uzbekistan
Q. Imanov, Azerbaijan
F. Sadikoglu, North Cyprus
A. Averkin, Russia
C. Kahraman, Turkey
M. Salukvadze, Georgia
E. Babaei, Iran
O. Kaynak, Turkey
T. Takagi, Japan
M. Babanli, Azerbaijan
V. Kreinovich, USA
K. Takahashi, Japan
I. Batyrshin, Mexico
D. Kumar Jana, India
V. B. Tarasov, Russia

H. Berenji, USA
V. Loia, Italy
S. Ulyanov, Russia
K. Bonfig, Germany
P. Moog, Germany
H. Uzunboylu, North Cyprus
A. Danandeh Mehr, Iran

A. Musayev, Azerbaijan
A. Veliyev, Azerbaijan
D. Dubois, Japan
M. Nikravesh, USA
R. Yager, USA
D. Enke, USA
V. Niskanen, Finland

## Organizing Committee

### Chairman

U. Eberhardt, Germany

### Co-chairmen

T. Abdullayev, Azerbaijan
O. Huseynov, Azerbaijan
E. Tuncel, North Cyprus

### Members

L. Gardashova, Azerbaijan
A. Alizadeh, Azerbaijan
S. Uzelaltinbulat, North Cyprus
Gunay Sadikoglu, North Cyprus
B. Guirimov, Azerbaijan

A. Guliyev, Azerbaijan
K. Jabbarova, Azerbaijan
M. M. M. Elamin, North Cyprus
M. A. Salahli, Turkey
Şahin Akdağ, North Cyprus

## Conference Organizing Secretariat

Azadlig Ave. 20, AZ 1010 Baku, Azerbaijan
Phone: +99 412 493 45 38, Fax: +99 412 598 45 09
E-mail: raliev@asoa.edu.az

# Contents

# From Status Quo Bias to Innovative Multiagent Decisions Under Fuzzy Preferences and Fuzzy Majority

Janusz Kacprzyk[✉]

Polish Academy of Sciences, Ul. Newelska 6, 01-447 Warsaw, Poland
kacprzyk@ibspan.waw.pl

**Abstract.** Our point od departure is the general problem of multiagent group decision making under fuzzy preferences and a fuzzy majority. We assume a (finite, relatively small) set of options and a set of agents – human beings or software entities, even social groups, if homogeneous – who present their testimonies as fuzzy preference relations over a set of options. We wish to determine an option or a set options that is best acceptable to the group of agents as a whole as to their preferences. This model concerns mainly situations with a relatively small number of options and agents as in the case of human decisions in committees, juries, expert groupl, etc. Assuming Kacprzyk's concept of a fuzzy majority, equated with a fuzzy linguistic quantifiers, exemplified by "most", "almost all", "much more than a half", we use Zadeh's calculus of linguistically quantified propositions, Yager's OWA operators, etc. to derive some group decision solution, notably Kacprzyk's fuzzy Q-cores which is a fuzzy set of options which are not (strongly) defeated in pairwise comparisons (in the sense that other options are preferred over them) by the required fuzzy majority Q, e.g. "most", of (important) agents. All fuzzy preferences of agents are taken into account and we show that this can be viewed as a reflection of a so called status quo bias which, well known from psychology, which is an emotional bias with a preference for the current state of affairs, the principle of minimal change, etc. A group decision obtained in such a way may be viewed to be non-innovative, and we propose a new approach, with the innovation considered to be an opposite to the status quo. In the new approach, we derive first a social fuzzy preference relation by some aggregation of the individual fuzzy preference relations, that represents the group testimony, which can be viewed as representing some "consensus". Then, for each agent we find a distance of his/her individual fuzzy preference relations to that group (consensory) preference relation which is then used as an additional weight, in addition to the weights corresponding to the relevance of options and importance of individuals. We then argue that a solution (e.g. Q-core) that is best acceptable by, for instance, most agents whose preferences are close to the consensory group fuzzy preference relation, would rather reflect a conservative attitude, while that which is farer from consensory social fuzzy preference relation would rather reflect an innovative attitude. We present some examples of real innovative type decisions using this model.

© Springer Nature Switzerland AG 2019
R. A. Aliev et al. (Eds.): ICAFS-2018, AISC 896, p. 1, 2019.
https://doi.org/10.1007/978-3-030-04164-9_1

# Bimodal Information Clustering Methods

R. A. Aliev[1,2]([✉])

[1] Joint MBA Program, Georgia State University, Atlanta, USA
[2] Azerbaijan State Oil and Industry University,
20 Azadlig Ave., AZ1010 Baku, Azerbaijan
raliev@asoa.edu.az

**Abstract.** A large variety of clustering methods exist including the deterministic, probabilistic and fuzzy clustering methods. However, these methods are devoted to handling different types of uncertainty. No works exist on clustering taking into account a confluence of probabilistic and fuzzy information termed as a bimodal information. Indeed, real-world information is characterized by partial reliability of sources, uncertainty, incompleteness and imprecision of data.

In such cases, reliability of extracted knowledge is an important issue to be studied. Prof. Zadeh introduced the concept of Z-number, Z = (A, B), to formalize reliability of information under combination of fuzzy and probabilistic uncertainties. The first component, A, is a fuzzy restriction on the values of a random variable, X. The second component, B, is a fuzzy-valued measure of reliability of A.

In this study we suggest an approach to Z-number valued clustering of large data sets to describe reliability of data-driven knowledge. The clustering problem is formulated in terms of bimodal distribution. An objective function that compounds fuzzy clustering and probabilistic clustering criteria is used. A differential evolution optimization based method is used for solution of the formulated clustering problem.

A numerical example of Z-valued clustering of two-dimensional data is considered to compute reliability of knowledge extracted. The obtain results confirm validity of the proposed method.

# Maximum Likelihood Estimation from Interval-Valued Data. Application to Fuzzy Clustering

Hani Hamdan[✉]

Laboratoire des Signaux et Systèmes, CentraleSupélec, CNRS,
Université Paris-Sud, Université Paris-Saclay, Paris, France
Hani.Hamdan@centralesupelec.fr

**Abstract.** Interval-valued data are used in many applications where they represent data imprecision, measurement inaccuracy, or measurand variability. As a result of the increasing use of such data in data mining, many data analysis methods have been extended to interval data this last decade. The Expectation-Maximization (EM) algorithm has been widely used for maximum likelihood estimation of parameters in statistical models, where the model depends on unobserved latent variables. In our keynote talk, we will present the EM algorithm to interval-valued data. In this contribution, we provide an original likelihood expression for interval data. Then, we propose an original method to introduce the imprecision and the variability of data into the mathematical expectation of the EM algorithm. The maximization of the obtained expectation gives place to the EM algorithm for interval-valued data. We apply this EM algorithm to mixture model for maximum likelihood estimation of mixture model parameters from interval-valued data. A special attention is paid for the case of Gaussian mixture models. In order to show the usefulness of our approach, we apply it on real interval-valued data issued from a flaw diagnosis application using acoustic emission.

© Springer Nature Switzerland AG 2019
R. A. Aliev et al. (Eds.): ICAFS-2018, AISC 896, p. 3, 2019.
https://doi.org/10.1007/978-3-030-04164-9_3

# Theory and Practice of Material Development Under Imperfect Information

M. B. Babanli[✉]

Azerbaijan State University of Oil and Industry,
Azadlig Ave., 20, Baku AZ1010, Azerbaijan
mustafababanli@yahoo.com

**Abstract.** Material development is an important research problem in material science and engineering. Nowadays, computational approaches to these problems are used to alternate natural experiments. These approaches include data mining, machine learning and computational intelligence tools that rely on big data on material characteristics collected over long period experiments. One of the important issues in solving these problems is imperfect nature of information. In the present study we outline fuzzy logic and Z-number concept-based computational methodologies for material synthesis and selection to account for imprecision and partial reliability of relevant information. Several examples are provided to confirm validity of the study.

**Keywords:** Material synthesis · Material selection · Big data
Decision making · Fuzzy logic · Z-number

## 1 Introduction

Material selection is an important problem attracting theoretical and practical interest [4, 5]. Nowadays, a lot of materials and alloys are designed. In most alloys some properties are good and in compliance with the requirements, but some of them are not acceptable. Generally, for material selection methods a synergy of theoretical knowledge and practical experiences data is needed. Scientists used and developed some selection methods due to all of these.

Uncertainty of material properties requires to use fuzzy logic and soft computing methods to more adequately model and predict possible material behavior. This will help to deal with: imprecision of experimental data; partial reliability of experimental data, prediction results and expert opinions; uncertainty of material properties stemming from complex relationship between material components; a necessity to analyze, summarize, and reason with large amount of information of various types (numeric data, linguistic information, graphical information, geometric information etc.).

Z-number theory has a promising capability to account for fuzzy and partially reliable information due to ability to fuse fuzzy computation and probabilistic arithmetic. Indeed, variability of experimental conditions, complex content and structure of materials, imperfect expert knowledge demand to consider reliability of information on material behavior as restricted.

© Springer Nature Switzerland AG 2019
R. A. Aliev et al. (Eds.): ICAFS-2018, AISC 896, pp. 4–14, 2019.
https://doi.org/10.1007/978-3-030-04164-9_4

Fuzzy logic, Z-number theory and Soft computing may help to improve abilities of big data principles to deal with huge amount and variety of information. In this realm, fuzzy clustering, Neuro-fuzzy inference systems, intelligent databases, soft CBR, computational intelligence based KBs and information search algorithms provide bridge between complexity, imperfectness, qualitative nature of information and research techniques. Particularly, this may help to get intuitive general interpretation of material science results obtained by various techniques, and ways to get practical results would be then more evident.

In this paper the methodologies of material selection and synthesis under imperfect information are proposed. The methodologies are based on fuzzy set theory and computation with Z-numbers. Several examples for application of the proposed methodologies are provided.

## 2 Preliminaries

**Definition 1 *Fuzzy set.*** A fuzzy set A is a mapping

$$\mu_A : X \to [0, 1],$$

where $\mu_A(X)$ is the grade of x to the fuzzy set [22]. Fuzzy set A is described as $A = \{(x, \mu_A(x)), x \in X\}$.

**Definition 2 *Fuzzy number*** [22]. Fuzzy number is a fuzzy set of real numbers that satisfies normality, boundedness of support, continuity and unimodality.

**Definition 3 *Linguistic fuzzy models*** [2]. This type of models for objects with multi-input (n) and multi-output (m) objects (MIMO models) is expressed by means of fuzzy rules as:

$$\text{Rule k:} \begin{array}{l} IF\ x_1\ is\ A_{i1}\ AND\ x_2\ is\ A_{i2}\ AND \ldots x_n\ is\ A_{in}\ THEN \\ y_1\ is\ B_{i1}\ AND\ y_2\ is\ B_{i2}\ AND \ldots y_m\ is\ B_{im} \end{array}$$

where $x_i(i = \overline{1, n})$, $y_j(j = \overline{1, m})$ are linguistic variables of inputs and outputs, $A_{ij}, B_{ij}$ are fuzzy sets, (k = 1, N).

For single input and single output models (SISO-models) one has:

$$IF\ x\ is\ A_k\ THEN\ y\ is\ B_k\ (k = 1, \ldots, K)$$

The Z-number concept is proposed to describe partial reliability of information.

**Definition 4 Z-number** [21]. A Z-number is an ordered couple of fuzzy numbers, Z = (A, B), where A is a restriction on the values of random variable X and B is a value of probability measure of A. B is used to describe a level of reliability of information. A and B usually are described verbally.

The approach to computations over Z-number is proposed in [1].

# 3 Methodology of Material Selection Under Imperfect Information

## 3.1 Factors Affecting to Material Selection

Let us shortly outline a systematic consideration of material selection information. The structure of information includes various types of issues, such as required properties, costs, service requirements etc. The main questions for material selection are as follows [7].

1. What properties are required?
2. What is the availability of materials?
3. What is the cost?

Strength, Stiffness, Toughness are among the most important properties considered as selection criteria. Availability and cost of materials are feasibility conditions that form a sets of alternatives (materials) to choose from. The following factors are also under consideration as operational conditions for a material:

1. Environmental effects
2. Temperature factor
3. Chemical factor
4. Radiation
5. Time

In details, these factors are considered in [6, 7, 11, 19].

The other important issue is that real-world information related to choice criteria, feasibility conditions and other factors of material selection are characterized by imprecision and partial reliability. Below we formulate a problem of decision making on material selection under imprecision and partial reliability of information.

## 3.2 Statement of the Problem of Multiattribute Decision Making on Alloy Selection Under Z-Number Valued Information

Let a set of $n$ alloys (alternatives) be given, $F = \{f_1, f_2, \ldots, f_n\}$. Every alternative $f_i, i = 1, \ldots, n$ is characterized by $m$ criteria $C_j, j = 1, \ldots, m$ (mechanical properties, electrical properties etc.). The problem is to choose the best alloy. All the criteria evaluations and weights $W_j, j = 1, \ldots, m$ are described by Z-numbers. The problem is described by a decision matrix (Table 1):

$f_{ij} = (A_{ij}, B_{ij}), i = 1, \ldots, n, j = 1, \ldots, m$ are Z-number valued criteria evaluations. The considered problem of multiattribute choice is to determine the best alloy:

Find $f^* \in F$ such that $f^* \succ f_i, \forall f_i \in F$ where $\succ$ is a preference relation.

**Table 1.** Decision matrix

|       | $C_1$    | $C_2$    | $C_3$    | $\ldots$ | $C_m$    |
|-------|----------|----------|----------|----------|----------|
|       | $w_1$    | $w_2$    | $w_3$    |          | $w_m$    |
| $f_1$ | $f_{11}$ | $f_{12}$ | $f_{13}$ | $\ldots$ | $f_{1m}$ |
| $f_2$ | $f_{21}$ | $f_{22}$ | $f_{23}$ | $\ldots$ | $f_{2m}$ |
| $f_3$ | $f_{31}$ | $f_{32}$ | $f_{33}$ | $\ldots$ | $f_{3m}$ |
| $\vdots$ | $\vdots$ | $\vdots$ | $\vdots$ | $\vdots$ | $\vdots$ |
| $f_n$ | $f_{n1}$ | $f_{n2}$ | $f_{n3}$ | $\ldots$ | $f_{nm}$ |

# 4  Methodology of Material Synthesis Under Imperfect Information

## 4.1  Statement of Synthesis Problem and Its Solution

We consider fuzzy If-Then rules mining from experimental data which relate alloy composition to alloy properties [5]. In general, fuzzy rules are derived from data by using fuzzy clustering and other learning methods. Data-driven fuzzy If-Then rules based provide intuitively interpretable and mathematically consistent models of knowledge discovered from complex data.

Assume that big data on smart materials sourced from experiments is available. These big data describe relationship between alloy composition and its characteristics (Table 2):

**Table 2.** Big data of relationship between alloy composition and its characteristics

| Exp. | Alloy composition (in %) | | | Conditions | | | Alloy characteristics | | |
|------|------|------|------|------|------|------|------|------|------|
| #    | Metal1,$y_1$ | $\ldots$ | Metal n,$y_n$ | Cond.1 | $\ldots$ | Cond.$l$ | Char. 1, $z_1$ | $\ldots$ | Char. m, $z_m$ |
| 1    | $y_{11}$ | $\ldots$ | $y_{1n}$ | $T_{11}$ | $\ldots$ | $T_{1l}$ | $z_{11}$ | $\ldots$ | $z_{1m}$ |
| .    | . | | | | | | | | |
| .    | . | | | | | | | | |
| .    | . | | | | | | | | |
| s    | $y_{s1}$ | $\ldots$ | $y_{sn}$ | $T_{s1}$ | $\ldots$ | $T_{sl}$ | $z_{s1}$ | $\ldots$ | $z_{sm}$ |

The problem is to extract knowledge based model from considered data and to find an alloy composition which provides a predefined alloy characteristics. The problem is solved as follows [5].

First, fuzzy clustering of the big data is applied to determine fuzzy clusters $C_1, C_2, \ldots, C_K$.

Second, fuzzy IF-THEN rules based model is constructed from $C_1, C_2, \ldots, C_K$:

*IF $y_1$ is $A_{k1}$ and, ..., and $y_n$ is $A_{kn}$ THEN $z_1$ is $B_{k1}$ and, ..., and $z_m$ is $B_{km}$, k = 1, ..., K*

Third, fuzzy inference is implemented on the basis of the fuzzy IF-THEN rules to compute optimal values $B'_1, \ldots, B'_m$ of alloy characteristics $z_1, \ldots, z_m$. We propose to implement fuzzy inference by using linear interpolation [10, 23]. Optimal values $B'_1, \ldots, B'_m$ are found as close to the ideal vector of characteristics $B^* = (B^*_1, \ldots, B^*_m)$.

## 4.2 Fuzzy Approach to Estimation of Phase Diagram Under Uncertain Thermodynamic Data

Phase-equilibrium problems are important and interesting real-world optimization problems that provide a fundamental basis of material synthesis. The existing approaches to obtain a practically implementable solution include theoretical and computational approaches such as the CALPHAD, approaches based on field theory, statistical mechanics, electrochemical techniques and others [3, 12–16]. Let us mention that development of an adequate thermodynamic model is complicated due to uncertainty in thermodynamic data. More concretely, it is needed to take into account that boundaries between solidus and liquidus are uncertain. The main disadvantage of majority of the approaches is a low capability of dealing with real-world uncertainty of thermodynamic information. At the same time, one-dimensional problems are often considered. In order to deal with mutlidimensionality and real-world uncertainty, we consider an approach for solving two dimensional phase-equilibrium problems by using fuzzy logic and DE optimization.

The quantitative analysis of phase-equilibrium allows to construct a model of relationship between temperature, T, pressure, p, and mole fraction, x. This model describes equilibrium state of several homogeneous phases. The first necessary condition for a problem of equilibrium distribution of k components between two phases is the equality of the chemical potential:

$$\forall i \in \{1, \ldots, k\} : \mu_i^1 = \mu_i^2 \tag{1}$$

In notation $\mu_i^l$, upper index denotes phases and lower one denotes substances. The dependence of m on T, p, and x is described by a thermodynamic model. Thus, (1) can be described as:

$$\forall i \in \{1, \ldots, k\} : x_i^1 \varphi_i^1 = x_i^2 \varphi_i^2 \tag{2}$$

As a rule, concentrations $x_i^1$ and $x_i^2$ are found for fixed values of temperature and pressure respectively. The main problem is that such approach may generate trivial solutions $x_i^1 = x_i^2$ that do not have physical meaning (only a critical demixing point has meaning). In order to resolve this problem, it is needed to guess initial conditions for minimization that would not be too far away from the correct solutions.

In the considered problem, objective function is used [17] as a measure of the departure from equilibrium state between any two phases of one component:

$$f_1(x^1, x^2) = \sum_{i=1}^{3} \left| x_i^1 \varphi_i^1 - x_i^2 \varphi_i^2 \right| \tag{3}$$

Several useful formulations of the second objective exist. A general formulation based on the Euclidean norm of a vector of concentration differences is [17]

$$f_2(x^1, x^2) = \sqrt{2} - \left\| x^1 - x^2 \right\|_2 = \sqrt{2} - \sqrt{\sum_{i=1}^{3} \left( x_i^1 - x_i^2 \right)^2} \tag{4}$$

This can be easily extended for more components.

## 5  Examples

### 5.1  Synthesis of TiNiPd Alloys with Given Characteristics

A problem of computational synthesis of Ti-Ni-Pd alloy with predefined characteristics is considered. A big data fragment describing dependence alloy compo-sition and the corresponding characteristics is shown in Table 3:

**Table 3.**  A big data fragment on Ti-Ni-Pd alloy composition [8]

| Composition | | | Transformation temperatures | | | |
|---|---|---|---|---|---|---|
| $x_1$ (Ni, %) | $x_2$ (Ti, %) | $x_3$ (Pd, %) | $y_1$(marten. finish temp., K) | $y_2$(marten. start temp., K) | $y_4$(aust. finish temp., K) | $y_3$(aust. start temp., K) |
| 41 | 50 | 9 | 322.3 | 329.4 | 341.3 | 331.2 |
| 39 | 50 | 11 | 318.2 | 335.7 | 347.6 | 334.7 |
| 29 | 50 | 21 | 406.4 | 424.5 | 440.3 | 426.6 |
| 20 | 50 | 30 | 515.3 | 533.8 | 546.8 | 534.9 |

$B^* = (B_1^*, B_2^*, B_3^*, B_4^*) = (302.3, 323.3, 347.1, 331.3)$ can be considered as an ideal solution. In order to describe relationship between alloy composition and the characteristics values, the fuzzy IF-THEN rules were obtained by using FCM clustering of the considered big data:

*IF Ni is L and Pd is A2*

*THEN $M_f$ is A and $M_s$ is A and $A_f$ is A  and $A_s$ is A*

*IF Ni is A and Pd is A 1*

*THEN $M_f$ is L2 and $M_s$ is L 2 and $A_f$ is L 2 and $A_s$ is L2*

*IF $N_i$ is $H2$ and $P_d$ is $L1$*
*THEN $M_f$ is VL   and $M_s$ is VL and $A_f$ is L and $A_s$ is VL*'
*IF Ni is $H1$ and Pd is $L2$*
*THEN $M_f$ is $L1$  and $M_s$ is $L1$  and $A_f$ is $L1$  and $A_s$ is $L1$*
*IF Ni is VH and Pd is VH*
*THEN $M_f$ is H and $M_s$ is H  and $A_f$ is VH and $A_s$ is VH*

We have found that the output vector $B' = (B'_1, B'_2, B'_3, B'_4) = ((347.78), (364.86), (382.17), (375.22))$ induced by the input vector $A' = (A'_1, A'_2, A'_3) = (19.5, 50.5, 30)$ is is the closest one to the considered ideal fuzzy vector. The distance is $D(B', B^*) = 94$. The fuzzy model-based results for Ti-Ni-Pd shows the optimal alloy composition is: Ni is about 19%, Ti is about 51%, Pd is about 30%. The obtained characteristics: $M_f$ = about 347.78, $M_s$ = about 364.86, $A_f$ about = 382.17, and $A_s$ = 375.22.

## 5.2    Synthesis of TiNiHf Alloys with Given Characteristics

A fragment of the big data on relationship between composition of Ti-Ni-Hf alloy and transformation temperatures is available (Table 4):
The If-Then rules obtained by using FCM method:

**Table 4.** A fragment of the big data on Ti-Ni-Hf alloy composition [9]

| Composition | | | Transformation temperatures | | | |
|---|---|---|---|---|---|---|
| $x_1$ (Ni, %) | $x_2$ (Ti, %) | $x_3$ (Hf, %) | $y_1$(martensitic finish temperature, K) | $y_2$(martensitic start temperature, K) | $y_3$ (austenitic finish temperature, K) | $y_4$ (austenitic start temperature, K) |
| 49.8 | 46.2 | 4 | 325.5 | 358.8 | 406.8 | 368.5 |
| 49.8 | 44.2 | 6 | 329.0 | 363.9 | 421.4 | 381.3 |
| 49.8 | 35.2 | 15 | 451.0 | 480.0 | 530.9 | 512.0 |
| 49.8 | 30.2 | 20 | 546.4 | 573.1 | 610.6 | 595.8 |

If $x_2$ is Very High and $x_3$ is Very Low THEN $y_1$ is Very Low and $y_2$ is Very Low and $y_3$ is Very Low and $y_4$ is Very Low
If $x_2$ is Average and $x_3$ is Average THEN $y_1$ is Low and $y_2$ is Low and $y_3$ is Low and $y_4$ is Low
If $x_2$ is Low and $x_3$ is High THEN $y_1$ is Average and $y_2$ is Average and $y_3$ is Average and $y_4$ is Average
If $x_2$ is Very Low and $x_3$ is Very High THEN $y_1$ is Very High and $y_2$ is Very High and $y_3$ is Very High and $y_4$ is Very High
We have found that the fuzzy optimal output vector $B'$ induced by the fuzzy input vector $A' = (A'_1, A'_2, A'_3) = (49.8, 46.2, 4)$ is $B' = ((351.3), (384.9), (434.85),$

$(406.75))$. It is the closest vector to the considered ideal fuzzy vector $B^* = ((325.5), (358.8), (406.8), (368.5))$. The distance is $D(B', B^*) = 60$. The fuzzy model-based results for Ni-Ti-Hf shows that the optimal alloy composition is: Ni is 'about 49.8%', Ti is 'about 46.2%', Hf is 'about 5%'. The obtained characteristics: Mf is about 351.31, Ms is about 384.92, Af is about 434.85, and As is about 406.75.

## 5.3    Selection of an Alloy for Pressure Vessel by Using the VIKOR Method Under Z-Information

Assume that the following alternatives are considered for pressure vessel: $f_1$-alloy 825; $f_2$-alloy 59;$f_3$-alloy 625, $f_4$-alloy 718. Each alternative is evaluated by 4 criteria: $C_1$-PREN, $C_2$- yield strength, $C_3$- weldability, $C_3$-impact strength. The decision matrix of Z-valued scaled criteria evaluations of the alternatives is given in Table 5.

**Table 5.**  Z-number values decision matrix

|       | $C_1$ | $C_2$ |
|-------|-------|-------|
| $f_1$ | $(0.28; 0.31; 0.34)(0.5; 07; 1)$ | $(0.22; 0.24; 0.27)(0.5; 07; 1)$ |
| $f_2$ | $(0.66; 0.74; 0.81)(0.5; 0.7; 1)$ | $(0.28; 0.31; 0.34)(0.5; 07; 1)$ |
| $f_3$ | $(0.46; 0.51; 0.56)(0.3; 0.5; 0.7)$ | $(0.38; 0.42; 0.46)(0.3; 0.5; 0.7)$ |
| $f_4$ | $(0.29; 0.32; 0.35)(0.3; 0.5; 0.7)$ | $(0.73; 0.82; 0.9)(0.3; 0.5; 0.7)$ |
|       | $C_3$ | $C_4$ |
| $f_1$ | $(0.53; 0.59; 0.64)(0.5; 0.7; 1)$ | $(0.27; 0.31; 0.34)(0.5; 0.7; 1)$ |
| $f_2$ | $(0.42; 0.47; 0.51)(0.5; 0.7; 1)$ | $(0.74; 0.82; 0.91)(0.5; 0.7; 1)$ |
| $f_3$ | $(0.42; 0.47; 0.51)(0.3; 0.5; 0.7)$ | $(0.33; 0.37; 0.4)(0.3; 0.5; 0.7)$ |
| $f_4$ | $(0.42; 0.47; 0.51)(0.3; 0.5; 0.7)$ | $(0.27; 0.31; 0.34)(0.3; 0.5; 0.7)$ |

For simplicity, we consider numeric values of criteria weights: $w_1 = 0.47, w_2 = 0.12, w_3 = 0.23, w_4 = 0.18$. The ranking of alternatives with respect to regret, utility and VIKOR indexes $(R_i, S_i, Q_i)$ are shown in Table 6.

**Table 6.**  Ranking of alternatives with respect to $(R_i, S_i, Q_i)$

| $R_i$ | $S_i$ | $Q_i$ |
|-------|-------|-------|
| $f_1$ | $f_1$ | $f_2$ |
| $f_3, f_4$ | $f_3$ | $f_3$ |
| $f_2$ | $f_2$ | $f_1$ |
|       | $f_4$ | $f_4$ |

We have checked conditions C1 and C2 of VIKOR method [20]. According to these conditions, $f_2$ and $f_3$ are compromise solutions, but $f_2$ is preferable.

## 5.4    Estimation of Phase Diagram for the $UO_2 - BeO_2$ System Under Fuzzy Information

Consider estimation of phase diagram for the $UO_2 - BeO_2$ system under uncertainty by using the approach outlined in Sect. 4.2. The system is characterized by a complete solubility in the solid phase. The following formulas describe the mole fractions of liquidus ($x^{Liq}$) and solidus ($x^{Sol}$) for each fixed temperature [18]:

$$x^{Liq}_{UO_2 + Liq}(T) = 1 - \exp\left(\left(\frac{-\Delta H^M_{UO_2}}{RT}\right) \ln\left(\frac{T^M_{UO_2}}{T}\right)\right) \tag{5}$$

and

$$x^{Liq}_{Liq + BeO}(T) = \exp\left(\left(\frac{-\Delta H^M_{BeO}}{RT}\right) \ln\left(\frac{T^M_{BeO}}{T}\right)\right) \tag{6}$$

$R = 8.314 \ J/mol\,K$ is the gas constant.

In [18] for the first time uncertainty in phase diagram is taken into account in an interval-valued form. The authors consider computation of phase diagram by using one objective function. We will consider this problem on the basis of two objective functions under fuzzy information on values of $\Delta H^M_{UO_2}$, $T^M_{UO_2}$, $\Delta H^M_{BeO}$, and $T^M_{BeO}$. The corresponding fuzzy evaluations in form of triangular fuzzy numbers (TFNs) are shown in Table 7. Fuzziness of these values induces fuzzy uncertainty of information on values of the temperature and melting enthalpy $x^{Sol}$. In view of this, our purpose is to build an adequate model of by using fuzzy logic and DE optimization to fit uncertain experimental data.

**Table 7.** Fuzzy values of phase variables

| Variable | Units | Fuzzy value |
|---|---|---|
| $\Delta H^M_{UO_2}$ | kJ/mol | (40,80,165) |
| $\Delta H^M_{BeO}$ | kJ/mol | (42,85,165) |
| $T^M_{UO_2}$ | K | (3000,3100,3200) |
| $T^M_{BeO}$ | K | (2700,2800,2900) |

The fuzzy graphs describing the considered phase diagram under uncertainty is given in Fig. 1.

As one can see, a fuzzy eutectic point exists with the TFN-based coordinates:

$$T = (2300, 2500, 2700), x_{BeO} = (60\%, 68\%, 70\%).$$

The fuzziness of this point is the result of intersection of two fuzzy liquid lines describing real-world uncertainty intrinsic to thermodynamic data. In other words, this is a result of the fuzziness of the phase boundaries.

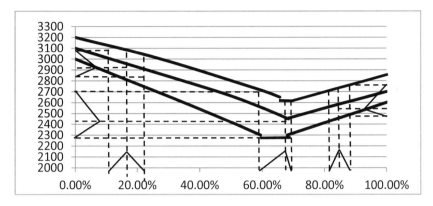

**Fig. 1.** The computed $UO_2 - BeO_2$ fuzzy phase diagram (the core values and bounds)

## 6    Conclusion

In this study we outline general approaches to material development. We have formulated the problem of multiattribute decision on material selection under Z-number valued information, the problem of material synthesis by using fuzzy If-Then rules extracted from big data on material characteristics, and the fuzzy set theory-based approach to construction of phase diagram under imprecise information. The proposed approaches to solving the considered problems allow to obtain intuitively interpretable solutions under imperfect information. Several practical examples such as selection of alloy for pressure vessel, estimations of phase diagram for the $UO_2 - BeO_2$ system and others are considered to illustrate validity of the proposed approaches.

## References

1. Aliev, R.A., Alizadeh, A.V., Huseynov, O.H.: The arithmetic of discrete Z-numbers. Inform. Sci. **290**, 134–155 (2015)
2. Aliev, R.A., Aliev, R.R.: Soft Computing and its Application. World Scientific, New Jersey (2001)
3. Averill, B.A., Eldredge, P.: Principles of General Chemistry. McGraw-Hill Education, New York City (2012)
4. Babanli, M.B., Huseynov, V.M.: Z-number-based alloy selection problem. Procedia Comput. Sci. **102**, 183–189 (2016)
5. Babanli, M.B.: Synthesis of new materials by using fuzzy and big data concepts. Procedia Comput Sci **120**, 104–111 (2017)
6. Amalgam, D.: A Scientific Review and Recommended Public Health Service Strategy for Research, Education and Regulation Final Report of the Subcommittee on Risk Management of the Committee to Coordinate Environmental Health and Related Programs Public Health Service. Department of Health and Human Services Public Health Service (1993) https://health.gov/environment/amalgam1/selection.htm
7. Factors Influencing Materials Selection. http://mechanical-materialstechnology.blogspot.com/2011/08/factors-influencing-materials-selection.html

8. Frenzel, J., Wieczorek, A., Opahle, I., Maa, B., Drautz, R., Eggeler, G.: On the effect of alloy composition on martensite start temperatures and latent heats in Ni–Ti-based shape memory alloys. Acta Mater. **90**, 213–231 (2015)
9. Hashimoto, K., Kimura, M., Mizuhara, Y.: Alloy design of gamma titanium aluminides based on phase diagrams. Intermetallics **6**(7–8), 667–672 (1998)
10. Kóczy, L.T.: Approximate reasoning by linear rule interpolation and general approximation. Int. J. Approx. Reason. **9**(3), 197–225 (1993)
11. Kosmač, A.: Factors affecting material selection for high temperature applications – review (2017). https://steelmehdipour.net/wp-content/uploads/2017/02/Factors-affecting-material-selection-for-high-temperature-applications.pdf
12. Laidler, K.J., Meiser, J.H.: Physical Chemistry. Oxford University Press, Oxford (1995)
13. Larson, E.: Thermoplastic Material Selection, a Practical Guide. William Andrew, London (2015)
14. Papon, P., Leblond, J., Meijer, P.H.E.: The Physics of Phase Transition: Concepts and Applications. Springer, Berlin (2002). https://doi.org/10.1007/3-540-33390-8
15. Petrucci, R.H., Harwood, W.S., Herring, F.G.: General Chemistry. Principles and Modern Applications. Prentice Hall, Upper Saddle River (2001)
16. Predel, B., Hoch, M., Pool, M.: Phase Diagrams and Heterogeneous Equilibria: A Practical Introduction. Springer, Berlin (2004). https://doi.org/10.1007/978-3-662-09276-7
17. Preuss, M., Wessing, S., Rudolph, G., Sadowski, G.: Solving phase equilibrium problems by means of avoidance-based multiobjectivization. In: Kacprzyk, J., Pedrycz, W. (eds.) Springer Handbook of Computational Intelligence. Springer, Heidelberg (2015). https://doi.org/10.1007/978-3-662-43505-2_58
18. Stan, M., Reardon, B.J.: A Bayesian approach to evaluating the uncertainty of thermodynamic data and phase diagrams. Comput. Coupling Phase Diagr. Thermochem. **27**(3), 319–323 (2003)
19. Welling, D.A.: A fuzzy logic material selection methodology for renewable ocean energy applications. Proquest, Umi Dissertation Publishing, 154 p. (2011)
20. Yazdani, M., Graeml, F.R.: VIKOR and its applications: a state-of-the-art survey. Int. J. Strat. Decis. Sci. **5**(2), 56–83 (2014)
21. Zadeh, L.A.: A note on Z-numbers. Inform. Sci. **181**, 2923–2932 (2011)
22. Zadeh, L.A.: Fuzzy Sets. Inform. Control **8**, 338–353 (1965)
23. Zadeh, L.A.: Interpolative reasoning in fuzzy logic and neural network theory. In: Proceedings of the First IEEE International Conference Fuzzy, San-Diego, CA, March 1992

# Forming and Quantifying Consensus in Distributed System Modeling and Group Decision-Making: A Perspective of Granular Computing

Witold Pedrycz[✉]

Department of Electrical and Computer Engineering, University of Alberta,
Edmonton AB T6R 2V4, Canada
wpedrycz@ualberta.ca

**Abstract.** In system modeling and decision-making, we encounter a variety of distributed sources of data that subsequently are transformed into a collection of sources of knowledge (viz. models and levels of preferences in case of group decision-making) producing an array of results. In general, these results differ from each other and therefore it becomes beneficial to build a consensus and quantify its quality. Furthermore the diversity of knowledge sources implies an inherent granular nature of the findings.

We advocate that the issues identified above can be formalized and supported by the formation of algorithmically sound solutions by engaging information granules (say, intervals, fuzzy sets, rough sets, etc.) and invoking the mechanisms of Granular Computing.

In this talk, we dwell upon the fundamental ideas and algorithms of Granular Computing such as the principle of justifiable granularity (and its generalizations) and an optimal allocation of information granularity. We show that the principle of justifiable granularity becomes instrumental in quantifying the diversity of individual sources of knowledge (say, in the form of fuzzy sets) and constructing information granules of type-2 (say, type-2 fuzzy sets). To endow the constructs with a required degree of flexibility, we admit some levels of information granularity across them by elevating their original numeric parameters to information granules.

© Springer Nature Switzerland AG 2019
R. A. Aliev et al. (Eds.): ICAFS-2018, AISC 896, p. 15, 2019.
https://doi.org/10.1007/978-3-030-04164-9_5

# Why Multidimensional Fuzzy Arithmetic?

Andrzej Piegat[1] and Marek Landowski[2(✉)]

[1] Faculty of Computer Science, West Pomeranian University of Technology,
Zolnierska 49, 71-210 Szczecin, Poland
`apiegat@wi.zut.edu.pl`
[2] Department of Mathematical Methods, Maritime University of Szczecin,
Waly Chrobrego 1-2, 70-500 Szczecin, Poland
`m.landowski@am.szczecin.pl`

**Abstract.** In the paper authors try to convince readers that application of multidimensional fuzzy arithmetic (MFAr) is useful because this arithmetic delivers more precise solutions of uncertain problems than low-dimensional fuzzy arithmetic, which is mostly used at present.

**Keywords:** Fuzzy arithmetic · RDM fuzzy arithmetic
Horizontal membership function · Granular computing
Multidimensional fuzzy arithmetic

## 1 Introduction

The subject of this paper is fuzzy arithmetic (FAr), However, because interval is special case of fuzzy set and interval arithmetic (IAr) is special case of FAr it also will be discussed in the paper. There exist many types of IAr. Examples of them are: standard IAr of Warmus, Sunaga and Moore, extended IAr of Kaucher, distributive IAr of Neumaier, constrained IAr of Lodwick, affine IAr of Figuiredo and Stolfi, complete IAr of Kulish, instantiation IAr of Dubois. There exist also a few types of FAr. Examples of them are: FAr based on Zadeh's extension principle, left-right FAr of Dubois and Prade, FAr of decomposed fuzzy numbers (FNs) based on $\alpha$-cuts and on standard interval arithmetic, advanced FAr based on transformation method of Hanss, constrained FAr of Klir, FAr of ordered FNs of Kosiński. Because any type of IAr can be used in FAr based on $\alpha$-cuts therefore the number of FAr-types is much higher than cited above. At present, according to observations of authors and Dymova [5] the mostly used IAr is the standard IAr (SIAr). It is non-complicated, easily understandable and intuitive. Below in Fig. 1 an example of addition of two intervals $A = [\underline{a}, \overline{a}] = [1, 3]$ and $B = [\underline{b}, \overline{b}] = [3, 5]$ with use of this arithmetic is given

One can easily notice that the operation result $X = A + B$ shown in Fig. 1 is the same mathematical object as the components $A$ and $B$: it also is interval, and the operation is realized in 1-dimensional space of real numbers. In the case of FAr the mostly used type of this arithmetic is the FAr based on $\alpha$-cuts and on standard IAr. It also is easily understandable and intuitive. Below, in Fig. 2 addition of two FNs, of $A = (1, 2, 3)$ and $B = (3, 4, 5)$ is shown.

© Springer Nature Switzerland AG 2019
R. A. Aliev et al. (Eds.): ICAFS-2018, AISC 896, pp. 16–23, 2019.
https://doi.org/10.1007/978-3-030-04164-9_6

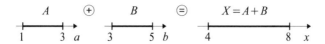

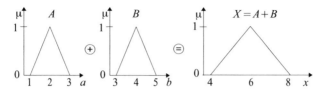

**Fig. 1.** Illustration of addition of two intervals $A$ and $B$ with use of standard interval arithmetic.

**Fig. 2.** Illustration of addition of two FNs, of $A = (1, 2, 3)$ and $B = (3, 4, 5)$ with use of FAr based on $\alpha$-cuts and on standard IAr.

Here, also the addition result is the same mathematical object, a 2-dimensional FN, similarly as the components $A$ and $B$, and the operation takes place in 2D-space $R \times \mu$, $R$-set of real numbers. All types of IAr and FAr known to authors operate in low-dimensional space. Because so many types of IAr and FAr exist one can ask: Why scientists have elaborated so many types of IAr and FAr? Is one type not sufficient?

The answer can be as follows: scientists have elaborated new types of IAr and FAr because already existing types are not ideal and require improvement. In what follows example of weaknesses of low-dimensional interval arithmetic will be presented in form of *Interval Crime Story*. Similar example for FAr can easily be constructed.

*Interval Crime Story*
A driver has to transport with his truck a load from city A to B. In city A the load had been weighed with a scales having error $\pm 1$ ton. From this measurement we know that the load $a \in A = [29, 31]$. During the travel the driver has stolen a part $x$ of the load and witnesses has seen it. After arrival to city B the load was weighed with a scales having also the error $\pm 1$ ton and from this weighing we know that the load $b \in B = [29, 31]$. How much load the driver has stolen?

It is not possible to determine the precise value $x$ of the stolen load. However, an approximate value can be determined. In terms of IAr Eq. (1) is to be solved.

$$A - X = B : [29, 31] - X = [29, 31] \tag{1}$$

But, what is $X$? Let us assume, similarly as it is done in the present IAr that the result $X$ is a 1-dimensional mathematical object: an interval, e.g. $X = [\underline{x}, \overline{x}]$. Under this assumption the Eq. (1) changes in Eq. (2).

$$[29, 31] - [\underline{x}, \overline{x}] = [29, 31] \tag{2}$$

Then with use of SIAr result (3) is achieved.

$$29 - \overline{x} = 29 : \overline{x} = 0; 31 - \underline{x} = 31 : \underline{x} = 31, X = [\underline{x}, \overline{x}] = [0, 0] \tag{3}$$

This result surprisingly informs that nothing has been stolen! But our common sense finds without difficulties that $X = (0, 2]$ tons, because the load in A could be equal to 31 tons and in B 29 tons. Conclusions from the *Interval Crime Story*: if we assume that the result $X$ is 1-dimensional object we achieve completely false solution of the story. Later, it will be shown that assuming multidimensional result allows for correct solving of this story. Standard IAr and FAr have many weak-points [10, 11]. A very important weak-point is phenomenon of multiple results described by L. Dymova in [5] and by M. Mazandarani in [6] as Unnatural Behavior Phenomenon (UBM). To understand this phenomenon let us analyze a system with two inputs $a$ and $b$ and output $x$ that is ruled by dependence $x = a + b$, where $a$, $b$, $x$ are precise values. The system law can be described not by only one equation $x = a + b$ but by 4 equivalent Eq. (4).

$$1.\, a + b = x; \quad 2.\, a = x - b; \quad 3.\, b = x - a; \quad 4.\, a + b - x = 0 \tag{4}$$

If we do not know precise values of $a$ and $b$ but only their approximate values and when our knowledge has form of fuzzy numbers $A = (1, 2, 3)$ and $B = (3, 4, 5)$ then we can make fuzzy extensions (5) of all 4 equivalent models (4).

$$\begin{aligned}
&1.\, A + B = X : (1, 2, 3) + (3, 4, 5) = X; X = (4, 6, 8) \\
&2.\, A = X - B : (1, 2, 3) = X - (3, 4, 5); X = (0, 3, 6) \\
&3.\, B = X - A : (3, 4, 5) = X - (1, 2, 3); X = (0, 3, 6) \\
&4.\, A + B - X = 0 : (1, 2, 3) + (3, 4, 5) - X = 0; X = (8, 6, 4)
\end{aligned} \tag{5}$$

From (5) we see that 3 different results have been achieved and no one of them satisfies all equivalent 4 forms (5) of the fuzzy model. It means that fuzzy number cannot be universal algebraic result of addition of fuzzy numbers. Next weak-points of 2-dimensional SFAr are non-existence of the inverse element $-X$ of addition that would satisfy equation $X - X = 0$, and of the inverse element $1/X$ of multiplication that would satisfy equation $X/X = 1$. In SFAr the distributive law $X(Y + Z) = XY + XZ$ does not hold. Lack of this law causes that equations cannot be transformed from one form to another to determine their solutions. In SFAr also cancellation law $XZ = YZ \Rightarrow (X = Y)$ does not hold. Lack of this law causes that fuzzy equations cannot be simplified for determining their solutions. In SFAr two ways of calculation of the difference $A - B$ officially are used: usual difference calculated as $X = A - B$ and Hukuhara difference $X^H$ calculated from equation $A = X^H + B$. Both ways deliver different results. Is it not strange? Summarizing numerous weak-points of low-dimensional FAr: the suspicion arises that uncertain fuzzy problems cannot be solved in the low-dimensional space and that their solution in multidimensional space should be tried. Hence, in Sect. 2 a few information pieces about multidimensional interval and fuzzy arithmetic (MIAr and MFAr) are given.

## 2  Methods

MIAr has been presented in about 40 papers. Its applications can e.g. be found in [1–3, 7, 8, 12]. Figure 3 presents basic notions of MIAr.

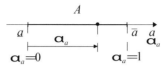

**Fig. 3.** Basic notions of multidimensional interval arithmetic: $\alpha_a$ – RDM variable specifying position of a point inside of interval.

Notation of MIAr is given by (6).

$$a \in B = [\underline{a}, \overline{a}], b \in [\underline{b}, \overline{b}]$$
$$a = \underline{a} + (\overline{a} - \underline{a})\alpha_a, \alpha_a \in [0, 1] \tag{6}$$
$$b = \underline{b} + (\overline{b} - \underline{b})\alpha_b, \alpha_b \in [0, 1]$$

Result $x^{gr}$ of arithmetic operation $* \in \{+, -, \times, /\}$ is given by (7).

$$A * B = X - \text{only symbolic notation}$$
$$a * b = x^{gr}, a \in A, b \in B - \text{correct notation}$$
$$x^{gr} = a * b = [\underline{a} + (\overline{a} - \underline{a})\alpha_a] * [\underline{b} + (\overline{b} - \underline{b})\alpha_b], \alpha_a, \alpha_b \in [0, 1] \tag{7}$$
$$span(x^{gr}) = \left[\min_{\alpha_a, \alpha_b} x^{gr}(\alpha_a, \alpha_b), \max_{\alpha_a, \alpha_b} x^{gr}(\alpha_a, \alpha_b)\right]$$

The result of arithmetic operations $a * b = x^{gr}(\alpha_a, \alpha_b)$ are information granules existing in 3D space $\alpha_a \times \alpha_b \times x$. Their simplified 1-dimensional indicators are *span* $(x^{gr})$, cardinality distribution *card* $x^{gr}$ and the center of gravity $COG(x^{gr})$ [9]. Figure 4 presents trapezoidal and triangular MFs.

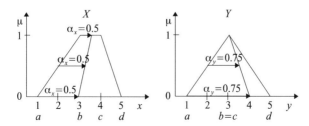

**Fig. 4.** Trapezoidal $X = (1, 3, 4, 5)$ and triangular $Y = (1, 3, 5)$ membership functions with denotation.

The trapezoidal MF from Fig. 4 can be expressed in form of the horizontal MF (HMF), see (8).

$$x = [a + (b-a)\mu] + [d-a-\mu(d-a+b-c)]\alpha_x, \alpha_x \in [0, 1] \tag{8}$$

The triangular MF can also be expressed in form of HMF (9).

$$y = [a + (b-a)\mu] + [d - a - \mu(d-a)]\alpha_y, \alpha_y \in [0, 1] \tag{9}$$

If * means an arithmetic operation $* \in \{+, -, \times, /\}$, then result $z^{gr}$ of the operation on $x(\mu, \alpha_x)$ and $y(\mu, \alpha_y)$ is expressed by (10).

$$x(\mu, \alpha_x) * y(\mu, \alpha_y) = z^{gr}(\mu, \alpha_x, \alpha_y), \mu, \alpha_x, \alpha_y \in [0, 1] \tag{10}$$

Below, example (11) of addition of two FNs is given.

$$
\begin{aligned}
&x + y = z = ? \\
&x \in X = (1, 3, 4, 5), y \in Y = (1, 3, 5) \\
&\text{Symbolic notation:} X + Y = Z \\
&\text{Horizontal MFs:} \\
&x = (1 + 2\mu) + (4 - 3\mu)\alpha_x, y = (1 + 2\mu) + 4(1 - \mu)\alpha_y, \mu, \alpha_x, \alpha_y \in [0, 1] \\
&z^{gr} = x + y = [(1 + 2\mu) + (4 - 3\mu)\alpha_x] + [(1 + 2\mu) + 4(1 - \mu)\alpha_y]
\end{aligned}
\tag{11}
$$

The result $z^{gr}(\mu, \alpha_x, \alpha_y)$ exists in 4D-space and cannot be seen. However, we can see its simplified low-dimensional indicators. The first indicator, the span can be calculated from (12). It is shown in Fig. 5.

$$
\begin{aligned}
span(z^{gr}) &= \left[\min_{\alpha_x, \alpha_y} z^{gr}(\mu, \alpha_x, \alpha_y), \max_{\alpha_x, \alpha_y} z^{gr}(\mu, \alpha_x, \alpha_y)\right] \\
&= [2 + 4\mu, 10 - 3\mu]
\end{aligned}
\tag{12}
$$

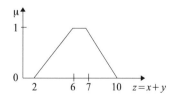

**Fig. 5.** Fuzzy span of the addition result being the first indicator of the 4D result-granule $z^{gr}(\mu, \alpha_x, \alpha_y)$.

In the general case it is not possible to exactly determine the span of the multidimensional result without this result itself. The second indicator of the direct result $z^{gr}(\mu, \alpha_x, \alpha_y)$ is its cardinality distribution [9] *card* $z^{gr}$ and position $z_{COG}$ of its center of gravity *COG*.

## 3 Results

To show effectiveness of MIAr first the problem of Interval Crime Story (ICS) will be solved. Its formulation is remembered by (13).

$$a \in A = [29, 31], a = 29 + 2\alpha_a, \alpha_a \in [0, 1]$$
$$b \in B = [29, 31], b = 29 + 2\alpha_b, \alpha_b \in [0, 1]$$
$$x = a - b : \alpha_a > \alpha_b$$
$$x^{gr} = a - b = (29 + 2\alpha_a) - (29 + 2\alpha_b) = 2(\alpha_a - \alpha_b), \alpha_a > \alpha_b$$
(13)

The multidimensional solution $x^{gr}(\alpha_a, \alpha_b)$ of ICS is shown in Fig. 6.

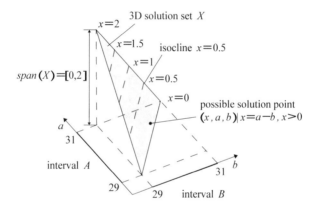

**Fig. 6.** Three-dimensional solution set $X = \{(x, a, b) | x = a - b, x > 0\}$ of the Interval Crime Story.

The solution span determined with (7) is equal to $span(x^{gr}) = (0, 2]$ tons. It can be seen in Fig. 6 that it is only one of a few measures of the 3D solution granule. This example has shown that problem which could not be solved with low-dimensional IAr has been solved with multidimensional IAr.

Second example presents solving fuzzy linear system (FLS) of equations taken from [4]. The FLS is given by (14) and $\mu$ means membership level.

$$
\begin{array}{llll}
2\tilde{x}_1 - \tilde{x}_2 + \tilde{x}_3 = \tilde{y}_1 & , & [\tilde{y}_1]_\mu = [\mu - 2, 2 - 3\mu] \\
-\tilde{x}_1 + \tilde{x}_2 - 2\tilde{x}_3 = \tilde{y}_2 & , & [\tilde{y}_2]_\mu = [1 + 2\mu, 7 - 4\mu] \\
\tilde{x}_1 - 3\tilde{x}_2 + \tilde{x}_3 = \tilde{y}_3 & , & [\tilde{y}_3]_\mu = [\mu - 3, -2\mu], \mu \in [0, 1]
\end{array}
$$
(14)

In [4] it has been solved with low-dimensional method called "interval inclusion FLS solving method". Solution (15) of FLS (14) delivered by this method consists of 3 triangular FNs.

$$[\tilde{x}_1]_\mu = [\mu - 2, 2 - 3\mu], \quad [\tilde{x}_2]_\mu = [1 + 2\mu, 7 - 4\mu], \quad [\tilde{x}_3]_\mu = [\mu - 3, -2\mu]. \quad (15)$$

One can easily check that the low-dimensional solution (15) is not the algebraic universal solution of FLS (14). However, such solution can be found with MFAr. To this aim, in the first step, RDM models (16) of $y_1$, $y_2$, $y_3$ are determined.

$$y_1 = (-2 + \mu) + 4(1 - \mu)\alpha_{y_1}, y_2 = (1 + 2\mu) + 6(1 - \mu)\alpha_{y_2},$$
$$y_3 = (-3 + \mu) + 3(1 - \mu)\alpha_{y_3}, \mu, \alpha_{y_1}, \alpha_{y_2}, \alpha_{y_3} \in [0, 1] \quad (16)$$

In the next step, with Cramer's rule 5D-solutions $x_1^{gr}$, $x_2^{gr}$, $x_3^{gr}$ are determined (17).

$$x_1^{gr} = [-5 + 8\mu + (1 - \mu)(20\alpha_{y_1} + 12\alpha_{y_2} - 3\alpha_{y_3}]/7$$
$$x_2^{gr} = [6 - 4\mu + (1 - \mu)(4\alpha_{y_1} - 6\alpha_{y_2} - 9\alpha_{y_3}]/7 \quad (17)$$
$$x_3^{gr} = [2 - 13\mu + (1 - \mu)(-8\alpha_{y_1} - 30\alpha_{y_2} - 3\alpha_{y_3}]/7$$

Spans of solutions (17) determined with (7) are given by (18).

$$span(x_1^{gr}) = [(-8 + 11\mu)/7, (27 - 24\mu)/7]$$
$$span(x_2^{gr}) = [(-9 + 11\mu)/7, (10 - 8\mu)/7] \quad (18)$$
$$span(x_3^{gr}) = [(-39 + 28\mu)/7, (2 - 13\mu)/7]$$

Figure 7 shows comparison of solution spans achieved with both methods.

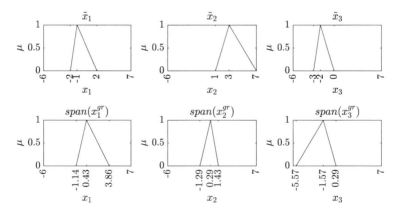

**Fig. 7.** Spans of solutions of FLS (14) achieved with low-dimensional "interval inclusion method" in [4] and with multidimensional FAr.

The spans $\tilde{x}_1$, $\tilde{x}_2$, $\tilde{x}_3$ are underestimated or overestimated: they are imprecise. Their precision can be checked. E.g. for $\alpha_{y1} = \alpha_{y2} = \alpha_{y3} = 1$ and $\mu = 0$ using (17) following solutions are achieved: $x_1 = 3.429, x_2 = -0.714, x_3 = -5.571$. These values satisfy

FLS (14). However, according to solution (15) from [4] they are not solutions. There exists an infinite number of similar examples.

## 4  Conclusions

To precisely determine solution of fuzzy arithmetic (mathematics) problem first its full, direct multidimensional solution (granule) has to be found. Only on its basis low-dimensional indicators (secondary, simplified "solutions") such as span, cardinality distribution, and center of gravity can precisely be determined. Multidimensional FAr enables this. Trials to find solution spans directly, only on the basis of low-dimensional versions of FAr are ineffective and deliver partly or fully incorrect spans. Sometimes the span cannot be found at all.

## References

1. Aliev, R.A.: Operations on Z-numbers with acceptable degree of specificity. Procedia Comput. Sci. **120**, 9–15 (2017)
2. Aliev, R.A., Huseynov, O.H., Aliyev, R.R.: A sum of a large number of Z-numbers. Procedia Comput. Sci. **120**, 16–22 (2017)
3. Aliev, R.A., Alizadeh, A.V., Huseynov, O.H.: An introduction to the arithmetic of Z-numbers by using horizontal membership functions. Procedia Comput. Sci. **120**, 349–356 (2017)
4. Allahviranloo, T., Ghanbari, M.: On the algebraic solution of fuzzy linear systems based on interval theory. Appl. Math. Model. **36**, 5360–5379 (2012)
5. Dymova, L.: Soft Computing in Economics and Finance. Springer, Heidelberg (2011). https://doi.org/10.1007/978-3-642-17719-4
6. Mazandarani, M., Pariz, N., Kamyad, A.V.: Granular differentiability of fuzzy-number-valued functions. IEEE Trans. Fuzzy Syst. **26**(1), 310–323 (2018)
7. Mazandarani, M., Zhao, Y.: Fuzzy bang-bang control problem under granular differentiability. J. Frankl. Inst.-Eng. Appl. Math. **355**(12), 4771–5350 (2018)
8. Najariyan, M., Zhao, Y.: Fuzzy fractional quadratic regulator problem under granular fuzzy fractional derivatives. IEEE Trans. Fuzzy Syst. **26**(4), 2273–2288 (2018)
9. Piegat, A., Landowski, M.: Two interpretations of multidimensional RDM interval arithmetic - multiplication and division. Int. J. Fuzzy Syst. **15**(4), 488–496 (2013)
10. Piegat, A., Plucinski, M.: Fuzzy number addition with application of horizontal membership functions. Sci. World J. **2015**, 16 (2015). https://doi.org/10.1155/2015/367214. Article no. ID367214
11. Piegat, A., Landowski, M.: Horizontal membership function and examples of its applications. Int. J. Fuzzy Syst. **17**(1), 22–30 (2015)
12. Zeinalova, L.M.: Application of RDM interval arithmetic in decision making problem under uncertainty. Procedia Comput. Sci. **120**, 788–796 (2017)

# On Using Fuzzy Sets in Healthcare Process Analysis

Uzay Kaymak[✉]

Information Systems Group, School of Industrial Engineering, Eindhoven
University of Technology, P.O. Box 513, 5600 MB Eindhoven, The Netherlands
u.kaymak@tue.nl, u.kaymak@ieee.org
http://is.ieis.tue.nl/staff/ukaymak/

**Abstract.** As the demand for health care services increases, healthcare organizations are seeking possibilities to optimize their care processes in order to
increase efficiency, while safeguarding the quality of the care. Process analytics
is an important input to the efforts for optimizing processes based on concrete
in-formation regarding process execution. Especially, process mining has
emerged recently as a promising methodology to discover process models based
on data from event logs. Until now, process analysis approaches have made little
use of soft computing and, in particular, fuzzy set-based techniques. Especially
processes that are characterized by a large complexity, much variability, flexibility and vagueness, such as healthcare processes, can gain much from the
applications of fuzzy set-based approaches in process analytics. In this paper, we
provide a systematic overview of the main approaches to applying fuzzy sets to
process analytics with a specific focus on the healthcare domain. In this way, we
aim to point to main directions for researchers in this area.

© Springer Nature Switzerland AG 2019
R. A. Aliev et al. (Eds.): ICAFS-2018, AISC 896, p. 24, 2019.
https://doi.org/10.1007/978-3-030-04164-9_7

# Optimization of Jobs in GIS by Coloring of Fuzzy Temporal Graph

Alexander Bozhenyuk[1(✉)], Stanislav Belyakov[1], and Janusz Kacprzyk[2]

[1] Southern Federal University, Nekrasovsky 44, 347922 Taganrog, Russia
{avb002, beliacov}@yandex.ru
[2] Systems Research Institute Polish Academy of Sciences, Newelska 6, 01-447 Warsaw, Poland
janusz.kacprzyk@ibspan.waw.pl

**Abstract.** The article proposes to consider the optimization of works in GIS as a task of coloring a fuzzy graph. The concept of fuzzy chromatic set of the second type is introduced and discussed in this paper as invariant fuzzy temporal graph. Fuzzy temporal graph is a graph in which the degree of connectivity of the vertices is changed in discrete time. Fuzzy chromatic set of the second type determines the greatest reparability degree of vertices of temporal fuzzy graph, when each of them can be assigned a specified number of colors at any discrete time. The example of finding the chromatic set of the second type is considered too.

**Keywords:** Fuzzy temporal graph · Invariant · Fuzzy subgraph
Graph coloring · Fuzzy chromatic set · Degree of reparability

## 1 Introduction

The application of geoinformation systems (GIS) for data analysis and decision making is practically important and has been studied for many years [1]. Areas of application of GIS are huge, therefore geoinformation technologies become leaders in the field of search, mapping, analytical tools and decision support [2, 3]. One of the main research problems is the construction of compact visual representations of the real world, which allow generating, evaluating and making reliable decisions in complex situations characterized by incompleteness and uncertainty of description [4].

The relevance and reliability of the spatial data used determines the quality of the solutions obtained with the help of GIS. Therefore, the process of updating data is important for GIS. One of the problems that arise with this is the joint updating of common areas of the electronic map. The update should be considered as a parallel flow of the processes of modification of local fragments of the electronic map. If two fragments intersect, then their simultaneous processing is impossible. To avoid losing updates, one of the processes must be blocked until the completion of the other. Because update transactions in GIS can take from several hours to several days, the update procedure should be planned in some special way in order to reduce the impact of such locks. This is possible due to the presence of layers of the electronic

© Springer Nature Switzerland AG 2019
R. A. Aliev et al. (Eds.): ICAFS-2018, AISC 896, pp. 25–32, 2019.
https://doi.org/10.1007/978-3-030-04164-9_8

map. Modification of objects on different layers is performed independently. Due to this, the spatial intersection of the map fragments does not lead to the blocking of the modification processes, if they are carried out on different layers. The more opportunities to process different layers the higher the likelihood of reducing the time to make changes [5].

To analyze the behavior of the described modification procedure, a model of a fuzzy temporal graph whose vertices correspond to the updating processes of individual fragments can be constructed, and fuzzy edges correspond to the degree of fuzzy relationship between them in a given discrete time. If the map layer to be associated with a certain color, then independent layer processing is reduced to the procedure for coloring the vertices of the temporal fuzzy graph. Thus, to plan the modification of the GIS map, the important question becomes: what possible number of colors can be associated with each vertex. The higher the number, the higher the probability that, with a lock threat, a couple of related processes will find different layers for processing. This will avoid blocking one or more processes.

In this paper, we propose a method for finding the greatest number of colors that can be assigned to each vertex of a fuzzy temporal graph with the greatest degree of separation between different colors.

In this paper we consider a fuzzy temporal graph in which fuzzy connections between the vertices of a graph change over discrete time. The vertices themselves remain unchanged.

## 2  Basic Concepts and Definitions

**Definition 1** [6, 7]. Let a fuzzy temporal graph $\widetilde{G} = (X, \{\widetilde{\Gamma}_t\}, T)$ be given, where set $X$ is a set of vertices ($|X| = n$), set of natural numbers $T$ defines discrete time and set $\{\widetilde{\Gamma}_t\}$ defines a family of sets, which display the vertices of $X$ into itself at time $t = \overline{1, T}$. In other words:

$$(\forall x \in X)(\forall t \in T)\left[\widetilde{\Gamma}_t(x) = \{ <\mu_t(y)/y > \}\right], y \in X, \mu_t \in [0, 1].$$

We consider the fuzzy subgraph $\widetilde{G}_t = (X, \widetilde{U}_t)$, in which the set of vertices $X$ is the same as in the original temporal fuzzy graph $\widetilde{G} = (X, \{\widetilde{\Gamma}_t\}, T)$; set $\widetilde{U}_t = \{\mu_t(x_i, x_j) | (x_i, x_j) \in X^2\}$ is a fuzzy set of edges at the discrete times $t = \overline{1, T}$, and a function $\mu_t$ is a membership function, which displays $X^2 \rightarrow [0, 1]$.

We color each vertex $x \in X$ of fuzzy subgraph $\widetilde{G}_t = (X, \widetilde{U}_t)$ by one of $k$ colors $(1 < k < n)$ and consider a subgraph $\widetilde{G}_i^{(t)} = (X_i, \widetilde{U}_i^{(t)})$. Here $X_i$ is a subset of vertices, which have $i$ color. Then the value $\alpha_i = 1 - \underset{x,y \in X_i}{\vee} \mu_G(x, y)$ defines the degree of internal stable of subset $X_i$ [8].

**Definition 2** [8]. Separation degree of fuzzy subgraph $\widetilde{G}_t$ with k colors is called a value:

$$L = \underset{i=\overline{1,k}}{\&} \alpha_i = \underset{i=\overline{1,k}}{\&} (1 - \underset{x,y \in X_i}{\vee} \mu_G(x,y)).$$

Separation degree $L$ of fuzzy subgraph $\tilde{G}_t$ depends on the number of colors k and concrete coloring of vertices.

We associate with fuzzy subgraph a family of fuzzy sets $\Re^{(t)} = \{\tilde{A}_G^{(t)}\}$, $\tilde{A}_G^{(t)} = \{<L_A^{(t)}(k)/k > | k = \overline{1,n}\}$. Here, the value $L_A^{(t)}(k)$ determines the separation degree of subgraph $\tilde{G}_t$ when coloring in particular $k$ colors.

**Definition 3.** A fuzzy set $\tilde{\gamma}^{(t)} = \{<L_\gamma^{(t)}(k)/k > | k = \overline{1,n}\}$ is called a fuzzy chromatic set of subgraph $\tilde{G}_t$ if the condition $\tilde{A}_G \subseteq \tilde{\gamma}$ is performed for any set $\tilde{A}_G^{(t)} \in \Re^{(t)}$, or else: $(\forall \tilde{A}_G^{(t)} \in \Re^{(t)})(\forall k = \overline{1,n})[L_A^{(t)}(k) \le L_\gamma^{(t)}(k)]$.

Chromatic set of subgraph $\tilde{G}_t$ determines the highest degree of separability in the color of it vertices by $1, 2, \ldots, n$ colors in time $t$.

**Definition 4** [8]. A fuzzy set $\tilde{\gamma} = \underset{t=\overline{1,T}}{\cap} \gamma^{(t)} = \{<a_1/1 >, <a_2/2 >, \ldots, <a_n/n > \}$ is called a fuzzy chromatic set fuzzy temporal graph $\tilde{G}$.

Chromatic set of fuzzy temporal graph $\tilde{G}$ determines the highest degree of separability in the color of it vertices by $1, 2, \ldots, n$ colors at time $t \in T$. However, each vertex is assigned only one color.

Now we consider the method of coloring a fuzzy temporal graph, when each vertex is assigned several colors. This method is based on the approach to the coloration of a fuzzy hypergraph [9, 10].

## 3 Fuzzy Coloring of the Second Type

Fuzzy coloring of the second type can be formulated as: to appropriate to every vertex of fuzzy temporal graph $\tilde{G}$ the maximum possible number of colors from $n$ possible so that they had the greatest separation degree on each color.

Let's attribute to each vertex $x \in X$ of the graph $k$ from n colors. Let $X_{K1}, X_{K2}, \ldots, X_{Kn}$ - subsets of vertices by which the first, the second,..., the $n$-th - colors are attributed with a degree of internal stability $\alpha_{K1}, \alpha_{K2}, \ldots, \alpha_{Kn}$ accordingly.

**Definition 5.** Value $L_K = \underset{i=\overline{1,n}}{\&} \alpha_{Ki}$ is called a separation degree of the second type of fuzzy temporal graph $\tilde{G}$ at assignment to its vertices k from n colors.

**Definition 6.** Fuzzy chromatic set of the second type is called set $\tilde{\gamma}_{II} = \{<L_{II}(k)/k | k = \overline{1,n}\}$ in which values $L_{II}(k)$ determine the greatest separation degrees of vertices the graph at assignment of each of them $k$ from $n$ colors.

Let's consider correlation between fuzzy chromatic set and fuzzy chromatic set of the second type. Let's denote through $t = ]\frac{n}{k}[$ - the whole from division of value $n$ on value $k$, then the following property is true:

**Property 1.** $(\forall k = \overline{1,n})(L_{II}(k) = L_I(t))$.

Differently, the separation degree of the second type at assignment to each vertex of the graph of $k$-colors coincides with a separation degree at coloring each vertex of the graph in one of $t = ]\frac{n}{k}[$ colors. The given property allows calculating fuzzy chromatic set of the second type on fuzzy chromatic set.

## 4  Method of Finding Fuzzy Chromatic Set of the Second Type

Consider the fuzzy subgraph $\widetilde{G}'_t = (X', \widetilde{U}'_t)$ of the temporal fuzzy graph $\widetilde{G}$, where $X' \subseteq X$ is the subset of vertices, and $\widetilde{U}'_t = \{\mu_t(x_i, x_j)|(x_i, x_j) \in X^2\}$ is the fuzzy set of edges over time $t$ with the membership function $\mu_t : X^2 \to [0,1]$. Denote by $\tau = \max\limits_{\forall x_i, x_j \in X'} \{\mu_t(x_i, x_j)\}$.

**Definition 7** [11]. A subset of vertices $X'$ is called an internally stable set for time $t$ with internal stability degree $\alpha(X') = 1 - \tau$.

**Definition 8** [11]. A subset of vertices $X' \subseteq X$ is called a maximal internally stable set at time $t$ with degree of internal stability $\alpha(X')$ if for any $X'' \supset X'$ the following inequality is fulfilled: $\alpha(X'') < \alpha(X')$.

Consider a property that establishes a relationship between a fuzzy chromatic set and internally stable sets.

**Property 2.** A fuzzy set $\tilde{\gamma}^{(t)} = \{<L_{\gamma}^{(t)}(k)/k > |k = \overline{1,n}\}$ is a fuzzy chromatic set if and only if it is not more then k maximal internally stable sets $\Psi_1, \Psi_2, \ldots \Psi_{k'}$, with the degrees of internal stability $\alpha_1, \alpha_2, \ldots \alpha_{k'}$, $(k' \leq k)$ and:

(1)  min $\{\alpha_1, \alpha_2, \ldots \alpha_{k'}\} = L_{\gamma}^{(t)}(k)$; $\cup_{j=\overline{1,k'}}\Psi_j = X$;

(2)  There is not another family $\{\Psi'_1, \Psi'_2, \ldots \Psi'_{k''}\}$, $k'' \leq k$ for which $\min\{\alpha'_1, \alpha'_2, \ldots, \alpha'_{k''}\} > \min\{\alpha_1, \alpha_2, \ldots, \alpha_{k'}\}$ and the condition 2 is true.

We will consider a method for determination of all maximal internally stable sets with the highest degree of internal stable. This method is a generalisation of the Maghout's method for fuzzy graphs [11, 12].

Let $\Psi$ be a certain maximal internally stable set with the degree of internal stable $\alpha(\Psi)$. For arbitrary vertices $x_i, x_j \in X$, one of the following cases may be realised: (a) $x_i \notin \Psi$; (b) $x_j \notin \Psi$; (c) $x_i \in \Psi$ and $x_j \in \Psi$. In the last case the degree $\alpha(\Psi) \leq 1 - \mu_U(x_i, x_j)$. In other words, the following expression is true:

$$(\forall x_i, x_j \in X)[x_i \notin \Psi \vee x_j \notin \Psi \vee (\alpha(\Psi) \leq 1 - \mu_U(x_i, x_j))]. \qquad (1)$$

We connect a Boolean variable $p_i$ taking 1 when $x_i \in \Psi$ and 0 when $x_i \notin \Psi$, with each vertex $x_i \in X$. We associate the expression $\alpha(\Psi) \leq 1 - \mu_U(x_i, x_j)$ with a fuzzy variable $\xi_{ij} = 1 - \mu_U(x_i, x_j)$.

Considering the expression (1) for all possible values i and j we obtain the truth of the following expression

$$\Phi_\Psi = \underset{i}{\&}\,\underset{j \neq i}{\&}\,(\bar{p}_i \vee \bar{p}_j \vee \xi_{ij}) = 1. \tag{2}$$

We open the parentheses and reduce the similar terms using the following rule

$$\xi' \,\&\, a \vee \xi'' \,\&\, a \,\&\, b = \xi' \,\&\, a, \text{ for } \xi' \geq \xi''. \tag{3}$$

Here, $a, b \in \{0, 1\}$ and $\xi', \xi'' \in [0, 1]$.

Then for each disjunctive term, the totality of all vertices corresponding to the variables missing in the totality, gives a maximal internal stable set with the obtained degree of internal stable.

We construct the matrix $F = \|f_{ij}\|$, $i = \overline{1, n}$, $j = \overline{1, p}$ where the lines correspond to the vertices of subgraph $\tilde{G}_t$ and columns correspond to the maximal internally stable sets. If $x_i \in \Psi_j$, then the value $f_{ij}$ has the value $\alpha_j$, if $x_i \notin \Psi_j$, then the value $f_{ij}$ has the value 0.

So, the task of finding fuzzy chromatic set $\tilde{\gamma}^{(t)}$ is the task of finding the covering of all lines by k columns ($k = \overline{1, n-1}$) with the maximum of the volume min $\{\alpha_{i_1}, \alpha_{i_2}, \ldots, \alpha_{i_k}\}$. Having calculated the fuzzy chromatic set, according to Property 1, we automatically define a fuzzy chromatic set of the second type $\tilde{\gamma}_{II}$.

**Example.** Consider the example of the temporal fuzzy graph $\tilde{G} = (X, \{\tilde{\Gamma}_t\}, T)$ [8], which is presented on Fig. 1:

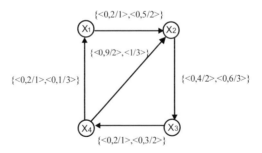

**Fig. 1.** Fuzzy temporal graph

Temporal fuzzy graph can be represented as the union of $T$ fuzzy subgraphs defined on the same set of vertices $X$. Since the graph $\tilde{G}$ is represented in the form $\tilde{G} = \bigcup_{t=1}^{3} \tilde{G}_t$, where fuzzy subgraphs $\tilde{G}_1$, $\tilde{G}_2$ and $\tilde{G}_3$ are shown on Fig. 2.

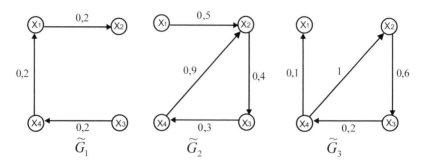

**Fig. 2.** Subgraphs at time $t$ = 1, 2, 3.

For fuzzy subgraph $\tilde{G}_t$, presented in Fig. 2, we find fuzzy chromatic set. To do this, we find the maximal internally stable sets for the fuzzy subgraphs $\tilde{G}_1$, $\tilde{G}_2$ and $\tilde{G}_3$. The corresponding expression (2) for this subgraph $\tilde{G}_1$ has the following form:

$$\Phi_\Psi = (\bar{p}_1 \vee \bar{p}_2 \vee 0.8) \& (\bar{p}_3 \vee \bar{p}_4 \vee 0.8) \& (\bar{p}_4 \vee \bar{p}_1 \vee 0.8) = 1.$$

Multiplying parenthesis 1 and 2, and using rules (3) we obtain:

$$\Phi_\Psi = (\bar{p}_1\bar{p}_3 \vee \bar{p}_1\bar{p}_4 \vee \bar{p}_2\bar{p}_3 \vee \bar{p}_2\bar{p}_4 \vee 0.8) \& (\bar{p}_4 \vee \bar{p}_1 \vee 0.8) = 1.$$

Completing the transformations of the fuzzy logical variables, we finally have:

$$\Phi_\Psi = \bar{p}_1\bar{p}_3 \vee \bar{p}_1\bar{p}_4 \vee \bar{p}_2\bar{p}_4 \vee 0.8 = 1.$$

It follows from the last expression that the considered fuzzy subgraph $\tilde{G}_1$ has 4 maximal internally stable sets: $\Psi_1 = \{x_2, x_4\}$, $\Psi_2 = \{x_2, x_3\}$, $\Psi_3 = \{x_1, x_3\}$, with the degrees of internal stability $\alpha_1 = \alpha_2 = \alpha_3 = 1$, and $\Psi_4 = \{x_1, x_2, x_3, x_4\}$ with the degree of internal stability $\alpha_4 = 0.8$.

For the fuzzy subgraph $\tilde{G}_1$ the matrix $F$ has view:

$$
F = \begin{array}{c} \\ x_1 \\ x_2 \\ x_3 \\ x_4 \end{array}
\begin{array}{cccc} \Psi_1 & \Psi_2 & \Psi_3 & \Psi_4 \\ \left| \begin{array}{cccc} 0 & 0 & 1 & 0.8 \\ 1 & 1 & 0 & 0.8 \\ 0 & 1 & 1 & 0.8 \\ 1 & 0 & 0 & 0.8 \end{array} \right| \end{array}.
$$

We write the expression (4) for finding of such covering:

$$\Phi_C = \underset{l=1,n}{\&} (r_{l1} \& \Psi_1 \vee r_{l2} \& \Psi_2 \vee \ldots \vee r_{lt} \& \Psi_t). \tag{4}$$

Here, $t$ is number of columns in the matrix $F$. We open the parentheses and reduce the similar terms using the rule (2). Then we can rewrite the expression (3) as:

$\Phi_C = \underset{i=\overline{1,m}}{\vee}(\Psi_{1i} \& \Psi_{2i} \& \cdots \& \Psi_{ki} \& r_i)$, $r_i \in [0,1]$, $\Psi_{ji} \in [\Psi_1, \Psi_2, \ldots, \Psi_l]$, $l \in \overline{1,n}$. So, for the fuzzy subgraph $\widetilde{G}_1$ the expression $\Phi_C$ is defined as:

$$\Phi_C = (\Psi_3 \vee 0.8\Psi_4) \& (\Psi_1 \vee \Psi_2 \vee 0.8\Psi_4) \& (\Psi_2 \vee \Psi_3 \vee 0.85\Psi_4) \& (\Psi_1 \vee 0.8\Psi_4).$$

Using the rule (2), we finally have: $\Phi_C = 0.8 \& \Psi_4 \vee \Psi_1 \& \Psi_3$.

So, for the fuzzy subgraph $\widetilde{G}_1$ presented in Fig. 2, the fuzzy chromatic set is:

$$\gamma^{(1)} = \{<0.8/1>, <1/2>, <1/3>, <1/4>\}.$$

Similarly, we define fuzzy chromatic sets of subgraphs $\widetilde{G}_2$ and $\widetilde{G}_3$ (Fig. 2):

$$\gamma^{(2)} = \{<0.1/1>, <0.6/2>, <0.7/3>, <1/4>\};$$
$$\gamma^{(3)} = \{<0/1>, <1/2>, <1/3>, <1/4>\}.$$

Hence the chromatic set of fuzzy temporal graph is defined as:

$$\tilde{\gamma} = \cap_{t=\overline{1,3}}\gamma^{(t)} = \{<0/1>, <0.6/2>, <0.7/3>, <1/4>\}.$$

Taking into account, the fuzzy chromatic set $\tilde{\gamma}$, we receive fuzzy chromatic set of the second type:

$$\tilde{\gamma}_{II} = \{<1/1>, <0.6/2>, <0/3>, <0/4>\}.$$

Thus, if 2 colors are assigned to each vertex of the fuzzy temporal graph $\widetilde{G}$, then the degree of separation is 0.6. When assigning 3 or 4 colors to each vertex, the degree of separation is 0.

## 5   Conclusions

In this paper, we proposed to consider the optimization of works in GIS as a task of coloring a fuzzy temporal graph. The concept of a fuzzy chromatic set of the second type as an invariant of the fuzzy temporal graph was introduced. The fuzzy chromatic set of the second type determines the highest degree of separability of the vertices of the temporal fuzzy graph, when they are assigned a specified number of colors at any time. The example of finding the chromatic set of the second type for fuzzy temporal graph was considered too.

**Acknowledgments.** This work has been supported by the Ministry of Education and Science of the Russian Federation under Project "Methods and means of decision making on base of dynamic geographic information models" (Project part, State task 2.918.2017), and the Russian Foundation for Basic Research, Project № 18-01-00023a.

# References

1. Malczewski, J.: GIS and Multicriteria Decision Analysis. Wiley, New York (1999)
2. Longley, P., Goodchild, M., Maguire, D., Rhind, D.: Geographic Information Systems and Science. Wiley, New York (2001)
3. Goodchild, M.: Modelling error in objects and fields. In: Goodchild, M.F., Gopal, S. (eds.) Accuracy of Spatial Databases, pp. 107–113. Taylor & Francis, Basingstoke (1989)
4. Zhang, J., Goodchild, M.: Uncertainty in Geographical Information. Taylor & Francis, New York (2002)
5. Belyakov, S., Belyakova, M., Bozhenyuk, A., Rozenberg, I.: Transformation of elements of geoinformation models in the synthesis of solutions. Adv. Intell. Syst. Comput. **679**, 526–535 (2018). https://doi.org/10.1007/978-3-319-68321-8_55
6. Bershtein, L., Bozhenyuk, A.: The using of temporal graphs as the models of complicity systems. Izvestiya UFY. Technicheskie nayuki. TTI UFY, Taganrog **4**(105), 198–203 (2010)
7. Bershtein, L., Bozhenyuk, A., Rozenberg, I.: Definition method of strong connectivity of fuzzy temporal graphs. Vestnik RGUPS, Rostov-on-Don **3**(43), 15–20 (2011)
8. Bozhenyuk, A., Belyakov, S., Rozenberg, I.: Coloring method of fuzzy temporal graph with the greatest separation degree. Adv. Intell. Syst. Comput. **450**, 331–338 (2016). https://doi.org/10.1007/978-3-319-33609-1_30
9. Monderson, J., Nair, P.: Fuzzy Graphs and Fuzzy Hypergraphs. Springer, Heidelberg (2000)
10. Bershtein, L., Bozhenyuk, A., Rozenberg, I.: Fuzzy coloring of fuzzy hypergraph. Adv. Soft Comput. **33**, 703–711 (2006). https://doi.org/10.1007/3-540-31182-3_65
11. Bozhenyuk, A., Belyakov, S., Knyazeva, M., Rozenberg, I.: Searching method of fuzzy internally stable set as fuzzy temporal graph invariant. Commun. Comput. Inf. Sci. **583**, 501–510 (2018). https://doi.org/10.1007/978-3-319-91473-2_43
12. Bershtein, L., Bozhenuk, A.: Maghout method for determination of fuzzy independent, dominating vertex sets and fuzzy graph kernels. Int. J. Gen. Syst. **1**(30), 45–52 (2001). https://doi.org/10.1080/03081070108960697

# Algebraic Properties of $Z$-Numbers Under Multiplicative Arithmetic Operations

R. A. Aliev[1](✉) and A. V. Alizadeh[2]

[1] Joint MBA Program, Azerbaijan State Oil and Industry University,
20 Azadlig Avenue, Baku AZ 1010, Azerbaijan
raliev@asoa.edu.az
[2] Department of Control and Systems Engineering, Azerbaijan State Oil and
Industry University, 20 Azadlig Avenue, Baku AZ 1010, Azerbaijan
akifoder@yahoo.com, a.alizade@asoiu.edu.az

**Abstract.** Prof. L.A. Zadeh introduced the concept of a Z-number for description of real-world information. A Z-number is an ordered pair $Z = (A, B)$ of fuzzy numbers $A$ and $B$ used to describe a value of a random variable $X$. $A$ is an imprecise estimation of a value of $X$ and $B$ is an imprecise estimation of reliability of $A$. A series of important works on computations with Z-numbers and applications were published. However, no study exists on properties of operations of Z-numbers. Such theoretical study is necessary to formulate the basics of the theory of Z-numbers. In this paper we prove that Z-numbers exhibit fundamental properties under multiplicative arithmetic operations.

**Keywords:** Fuzzy arithmetic · Probabilistic arithmetic · Associativity law
Distributivity law · Z-number

## 1 Introduction

Estimations of variables of interest in real-world problems are often characterized by fuzziness and partial reliability. Prof. Zadeh [12] introduced the concept of a Z-number as an ordered pair $Z = (A, B)$ of fuzzy numbers used to describe a value of a random variable $X$, where $A$ is a fuzzy constraint on values of $X$ and $B$ is a reliability considered as a value of probability measure of $A$. In [12] Zadeh suggested a general framework of computation with Z-numbers basis of the Zadeh's extension principle. The proposed approach is characterized by high complexity, it requires to deal with variational problems.

Kang et al. [4] proposed an approach is based on converting a Z-number to a fuzzy number on the basis of an expectation of a fuzzy set. However, this leads to loss of original Z-number-based information. The work of Zadeh [14] is devoted to computation with Z-numbers and several important practical problems. Aliev and colleagues [1–3] suggested a general and computationally effective approach to computation for arithmetic and algebraic operations and construction of typical functions.

However, existing approaches to computation with Z-numbers are based on classical fuzzy arithmetic. The main disadvantage of classical interval arithmetic and fuzzy arithmetic is that fundamental properties of arithmetic operations are lost [5–9]. This

© Springer Nature Switzerland AG 2019
R. A. Aliev et al. (Eds.): ICAFS-2018, AISC 896, pp. 33–41, 2019.
https://doi.org/10.1007/978-3-030-04164-9_9

creates problems with solving fuzzy equations, defining derivatives of fuzzy functions etc. In order to resolve this problem, in [7–10] they introduced a new approach to fuzzy arithmetic which relies on the use of so-called horizontal membership functions (HMFs).

Thus, there is a need for investigation of properties of operations over Z-numbers. This would provide a strong basis for the theory of Z-numbers and its further development. In this paper, we study properties of Z-numbers under multiplicative arithmetic operations. Validity of the study is illustrated in examples.

The paper is organized as follows. Section 2 includes basic concepts used in the paper. In Sect. 3 we prove basic laws of multiplicative arithmetic operations over Z-numbers and provide examples. Section 4 concludes.

## 2  Preliminaries

**Definition 1 Arithmetic operations of random variables [11].** Let $X_1$ and $X_2$ be two independent continuous random variables with probability distributions $p_1$ and $p_2$. A probability distribution $p_{12}$ of $X_{12} = X_1 * X_2$, $* \in \{+, -, \cdot, /\}$ is referred to as a convolution (a resulting probability distribution of an arithmetic operation) of $p_1$ and $p_2$ and is defined as follows.

$$X_{12} = X_1 + X_2 : p_{12}(x) = \int_{-\infty}^{\infty} p_1(x_1)p_2(x - x_1)dx_1,$$

$$X_{12} = X_1 - X_2 : p_{12}(x) = \int_{-\infty}^{\infty} p_1(x_1)p_2(x_1 - x)dx_1,$$

$$X_{12} = X_1 X_2 : p_{12}(x) = \int_{-\infty}^{\infty} p_1(x_1)p_2(x/x_1)\frac{1}{|x_1|}dx_1,$$

$$X_{12} = X_1/X_2 : p_{12}(x) = \int_{-\infty}^{\infty} |x_1|p_1(x_1)p_2(x_1/x)dx_1,$$

Let $X_1$ and $X_2$ be two independent discrete random variables with the corresponding outcome spaces $X_1 = \{x_{11}, \ldots, x_{1i}, \ldots, x_{1n_1}\}$ and $X_2 = \{x_{21}, \ldots, x_{2i}, \ldots, x_{2n_2}\}$ and the corresponding discrete probability distributions $p_1$ and $p_2$. The probability distribution of $X_1 * X_2$ is the convolution $p_{12} = p_1 \circ p_2$ of $p_1$ and $p_2$ which is determined as follows:

$$p_{12}(x) = \sum_{x=x_1*x_2} p_1(x_1)p_2(x_2),$$

for any $x \in \{x_1 * x_2 | x_1 \in X_1, x_2 \in X_2\}$, $x_1 \in X_1$, $x_2 \in X_2$.

**Definition 2 Probability measure of a continuous fuzzy number [13].** Let $X$ be a continuous random variable with pdf $p$. Let $A$ be a continuous fuzzy number describing a possibilistic restriction on values of $X$. A probability measure of $A$ denoted $P(A)$ is defined as

$$P(A) = \int_{\mathcal{R}} \mu_A(x)p(x)dx.$$

**Definition 3 A Z-number [12].** A Z-number is an ordered pair $Z = (A, B)$, where $A$ is a fuzzy number playing a role of a fuzzy constraint on values that a random variable $X$ may take:

$$X \ is \ A$$

and $B$ is a fuzzy number with a membership function $\mu_B : [0, 1] \to [0, 1]$, playing a role of a fuzzy constraint on the probability measure of $A$:

$$P(A) \ is \ B.$$

# 3   Algebraic Properties of Arithmetic Operations Under Z-Numbers

## 3.1   Commutative Law

Let us verify if the commutative law for product of Z-numbers is true:

$$Z_{12} = Z_1 \cdot Z_2 = Z_2 \cdot Z_1 = Z_{21}, \tag{1}$$

$$(A_1, B_1) \cdot (A_2, B_2) = (A_2, B_2) \cdot (A_1, B_1). \tag{2}$$

*Proof.* At first, we have to consider operation over fuzzy numbers:

$$A_{12} = A_1 \cdot A_2 = A_2 \cdot A_1 = A_{21}. \tag{3}$$

Commutativity (3) was proved in [7–9] in terms of HMFs. Thus, we need to prove equality of fuzzy restrictions over values of probability measures of $A_{12}$ and $A_{21}$:

$$B_{12} = B_{21}.$$

where the basic values are defined as

$$b_{12} = \int_R \mu_{A_{12}}(u)p_{12}(u)du, \tag{4}$$

$$b_{21} = \int_R \mu_{A_{21}}(u)p_{21}(u)du. \tag{5}$$

Taking into account that (3) holds, we have $\mu_{A_{12}} = \mu_{A_{21}}$. At the same time, it is known that commutativity law holds for convolution of sum $\circ_*$ of random variables:

$$p_{12} = p_1 \circ_* p_2 = p_2 \circ_* p_1 = p_{21}. \tag{6}$$

Then $B_{12} = B_{21}$. Thus, the commutativity law holds for Z-numbers:

$$Z_1 \cdot Z_2 = Z_2 \cdot Z_1$$

*Example.* Consider Z-numbers $Z_1 = (A_1, B_1)$ and $Z_2 = (A_2, B_2)$, the components of which are trapezoidal fuzzy numbers (TFNs):

$$A_1 = (10, 14, 14, 20), \ B_1 = (0.6, 0.8, 0.8, 0.9);$$
$$A_2 = (8, 16, 16, 18), \ B_2 = (0.4, 0.6, 0.6, 0.8).$$

Let us compute $Z_{12}$ and $Z_{21}$, and verify if (10) holds. At first, consider $Z_{12}$. We have obtained $A_{12}$:

$$A_{12} = (80, \ 224, \ 224, \ 360)$$

The fuzzy reliability value $B_{12}$ is obtained as

$$B_{12} = (0.3342, \ 0.6131, \ 0.6131, \ 0.7738).$$

For $Z_{21}$ we have obtained the following results:

$$A_{21} = (80, \ 224, \ 224, \ 360), \ B_{12} = (0.3342, \ 0.6131, \ 0.6131, \ 0.7738).$$

Thus, $A_{12} = A_{21}$ and $B_{12} = B_{21}$. Therefore, commutativity law holds:

$$Z_1 \cdot Z_2 = Z_2 \cdot Z_1.$$

## 3.2  Associative Law

Let us study whether associative law for sum of Z-numbers holds.

$$Z_{231} = Z_1(Z_2 Z_3) = (Z_1 Z_2)Z_3 = Z_{123}. \tag{7}$$

*Proof.* In [7–9] they proved that the associativity holds:

$$A_{231} = A_1(A_2 A_3) = (A_1 A_2)A_3 = A_{123}.$$

Let us now prove that

$$B_{231} = B_{123}, \tag{8}$$

$$b_{231} = \int_R \mu_{A_{231}}(u) \cdot p_{231}(u)du, \tag{9}$$

$$b_{123} = \int_R \mu_{A_{123}}(u) \cdot p_{123}(u)du. \tag{10}$$

Taking into account that associativity holds for fuzzy numbers, we have $\mu_{A_{231}} = \mu_{A_{123}}$. At the same time, it is known that associativity law holds for convolution of sum $\circ_*$ of random variables:

$$p_{231} = p_1 \circ_* p_{23} = p_1 \circ_* (p_2 \circ_* p_3) = (p_1 \circ_* p_2) \circ_* p_3 = p_{123}. \tag{11}$$

Then $B_{231} = B_{123}$. Thus, the associativity law holds for Z-numbers:

$$Z_1 \cdot (Z_2 \cdot Z_3) = (Z_1 \cdot Z_2) \cdot Z_3.$$

*Example.* Consider Z-numbers $Z_1 = (A_1, B_1)$, $Z_2 = (A_2, B_2)$, and $Z_3 = (A_3, B_3)$ the components of which are trapezoidal fuzzy numbers (TFNs):

$$A_1 = (10, 14, 14, 20), \ B_1 = (0.6, 0.8, 0.8, 0.9);$$
$$A_2 = (8, 16, 16, 18), \ B_2 = (0.4, 0.6, 0.6, 0.8);$$
$$A_3 = (16, 18, 18, 21), \ B_2 = (0.1, 0.3, 0.3, 0.6).$$

Let us verify if associativity (7) holds. Consider the right hand side of (7). For $A_{123}$ we have obtained:

$$A_{12} = (80, 224, 224, 360), \text{ and}$$
$$A_{123} = A_{12}A_3 = (1280, 4032, 4032, 7560).$$

The value of fuzzy reliability $B_{123}$ has been obtained as follows:

$$B_{12} = (0.3342, 0.6131, 0.6131, 0.7738), \text{ and}$$
$$B_{123} = (0.06867, 0.2461, 0.2461, 0.5376).$$

Consider the left hand side of (7). The fuzzy numbers are found as

$$A_{23} = (128, 288, 288, 378), \text{ and}$$
$$A_{231} = A_{23}A_1 = (1280, 4032, 4032, 7560).$$

For the value of fuzzy reliability $B_{231}$ we obtained:

$$B_{23} = (0.3824, 0.5175, 0.5175, 0.6892), \text{ and}$$
$$B_{231} = (0.1071, 0.2864, 0.2864, 0.5756).$$

Thus, $Z_1 \cdot (Z_2 \cdot Z_3) = (Z_1 \cdot Z_2) \cdot Z_3$.

### 3.3  Multiplicative Inverse Element $Z^{-1}$ of Element $Z$

Let us prove existence of an multiplicative inverse element of a Z-number:

$$Z_{11} = Z_1 \cdot (Z_1)^{-1} = (A_{11}, B_{11}); \tag{12}$$

$$Z_{11} = Z_1 \cdot (Z_1)^{-1} = 1. \tag{13}$$

*Proof.* The fuzzy $A_{11}$ is the singleton $A_{11} = 1$ as it is shown in [7–9]. Let us consider fuzzy reliability $B_{11}$. As $\mu_{A_{11}}(x) = 1$ iff x = 1, one has for any basic value

$$b_{11} = \int_R 1 \cdot p_{11}(u) du = 1; \tag{14}$$

So, $Z_{11} = \frac{Z_1(A_1, B_1)}{Z_1(A_1, B_1)} = Z(1, 1) = 1_Z$.

### 3.4  Cancellation Law

Let us investigate cancellation law for product of Z-numbers:

$$Z_3 Z_1 = Z_3 Z_2 \Rightarrow Z_1 = Z_2, \tag{15}$$

$$Z_3(A_3, B_3) \frac{1}{Z_3(A_3, B_3)} Z_1 = Z_3 \frac{1}{Z_3} Z_2. \tag{16}$$

*Proof.* In accordance with (14):

$$(1, 1) \cdot Z_1 = (1, 1) \cdot Z_2, \tag{17}$$

$$1_Z \cdot Z_1 = 1_Z \cdot Z_2. \tag{18}$$

So, $Z_1 = Z_2$.

## 3.5   Distributivity Law

We need to prove that

$$Z_1 \cdot (Z_2 + Z_3) = (Z_1 \cdot Z_2) + (Z_1 \cdot Z_3) \tag{19}$$

That is,

$$(A_1, B_1)((A_2, B_2) + (A_3, B_3)) = (A_1, B_1)(A_2, B_2) + (A_1, B_1)(A_3, B_3)$$

*Proof.* In [7–9] they proved that the distributivity holds for fuzzy numbers.

Let us now prove equality of fuzzy constraints $B$ of probability measures of $A$:

$$B_{23} = \int_R A_{23}(u)p_{23}(u)du, \tag{20}$$

$$B_1 = \int_R A_1(u)p_1(u)du, \tag{21}$$

$$B_{231} = \int_R A_{231}(u)p_{231}(u)du, \tag{22}$$

$$B_{12} = \int_R A_{12}(u)p_{12}(u)du, \tag{23}$$

$$B_{13} = \int_R A_{13}(u)p_{13}(u)du, \tag{24}$$

$$B_{123} = \int_R A_{123}(u)p_{123}(u)du = \int_R A_{123}(u)(p_{12} \circ_+ p_{13})(u)du, \tag{25}$$

$$B_{231} = \int_R A_{123}(u)p_{123}(u)du = \int_R A_{123}(u)(p_{23} \circ_* p_1)(u)du, \tag{26}$$

It is known that distributivity law holds for random variables:

$$p_{23} \circ_* p_1 = p_{12} \circ_+ p_{13}. \tag{27}$$

So, $B_{231} = B_{123}$.

*Example.* Consider Z-numbers $Z_1 = (A_1, B_1)$, $Z_2 = (A_2, B_2)$, and $Z_3 = (A_3, B_3)$ the components of which are trapezoidal fuzzy numbers (TFNs):

$$A_1 = (10, 14, 14, 20),\ B_1 = (0.6, 0.8, 0.8, 0.9);$$
$$A_2 = (8, 16, 16, 18),\ B_2 = (0.4, 0.6, 0.6, 0.8);$$
$$A_3 = (16, 18, 18, 21),\ B_2 = (0.1, 0.3, 0.3, 0.6).$$

At first, let us consider the left hand side of (7). For $A_{231}$ we have obtained:

$$A_{23} = (24, 34, 34, 39) \text{ and } A_{231} = A_1 A_{23} = (240, 476, 476, 780).$$

The value of fuzzy reliability $B_{231}$ has been obtained as follows:

$$B_{23} = (0.3824, 0.5175, 0.5175, 0.6892), \text{ and}$$
$$B_{231} = (0.2896, 0.495, 0.495, 0.6692).$$

Consider the right hand side of (7). The fuzzy numbers are found as

$$A_{12} = (80, 224, 224, 360),\ A_{13} = (160, 252, 252, 420), \text{ and}$$
$$A_{123} = (240, 476, 476, 780).$$

For the value of fuzzy reliability $B_{123}$ we obtained:

$$B_{12} = (0.3342, 0.6131, 0.6131, 0.7738),\ B_{13} = (0.1201, 0.3149, 0.3149, 0.6026) \text{ and}$$
$$B_{123} = (0.2896, 0.4950, 0.4950, 0.6692).$$

Thus, $Z_1 \cdot (Z_2 + Z_3) = Z_1 \cdot Z_2 + Z_1 \cdot Z_3$.

## 4  Conclusion

It is proved that the basic laws of multiplicative arithmetic operations hold for Z-numbers. The proofs are based on the analogous properties of fuzzy arithmetic and probabilistic arithmetic. The obtained results are necessary for solving important problems computation of Z-numbers, aggregation of Z-number valued information and other problems.

## References

1. Aliev, R.A., Alizadeh, A.V., Huseynov, O.H.: The arithmetic of discrete Z-numbers. Inf. Sci. **290**, 134–155 (2015)
2. Aliev, R.A., Alizadeh, A.V., Huseynov, O.H.: The arithmetic of continuous Z-numbers. Inf. Sci. **373**, 441–460 (2016)
3. Aliev, R.A., Alizadeh, A.V., Huseynov, O.H., Jabbarova, K.I.: Z-number based linear programming. Int. J. Intell. Syst. **30**, 563–589 (2015)

4. Kang, B., Wei, D., Li, Y., Deng, Y.: A method of converting Z-number to classical fuzzy number. J. Inf. Comput. Sci. **9**, 703–709 (2012)
5. Piegat, A., Plucinski, M.: Computing with words with the use of inverse RDM models of membership functions. Appl. Math. Comput. Sci. **25**(3), 675–688 (2015)
6. Piegat, A., Plucinski, M.: Fuzzy number addition with the application of horizontal membership functions. Sci. World J. **2015**, 16p. (2015). Article ID 367214
7. Piegat, A., Landowski, M.: Is the conventional interval-arithmetic correct? J. Theor. Appl. Comput. Sci. **6**(2), 27–44 (2012)
8. Piegat, A., Landowski, M.: Multidimensional approach to interval uncertainty calculations. In: Atanassov, K.T., et al. (eds.) New Trends in Fuzzy Sets, Intuitionistic: Fuzzy Sets, Generalized Nets and Related Topics, Volume II: Applications, pp. 137–151. IBS PAN -SRI PAS, Warsaw (2013)
9. Piegat, A., Landowski, M.: Two interpretations of multidimensional RDM interval arithmetic - multiplication and division. Int. J. Fuzzy Syst. **15**, 488–496 (2013)
10. Piegat, A., Plucinski, M.: Some advantages of the RDM-arithmetic of intervally-precisiated values. Int. J. Comput. Intell. Syst. **8**(6), 1192–1209 (2015)
11. Williamson, R.C., Downs, T.: Probabilistic arithmetic. I. Numerical methods for calculating convolutions and dependency bounds. Int. J. Approx. Reason. **4**(2), 89–158 (1990)
12. Zadeh, L.A.: A note on Z-numbers. Inf. Sci. **181**, 2923–2932 (2011)
13. Zadeh, L.A.: Probability measures of fuzzy events. J. Math. Anal. Appl. **23**(2), 421–427 (1968)
14. Zadeh, L.A.: Methods and systems for applications with Z-numbers, United States Patent, Patent No. US 8,311,973 B1, Date of Patent: 13 November 2012

# Z-Number Based TOPSIS Method in Multi-Criteria Decision Making

Latafat A. Gardashova[(⊠)]

Azerbaijan State Oil and Industry University, Azadlig 35, Nasimi, Baku,
Azerbaijan
latsham@yandex.ru

**Abstract.** In this paper, we propose a Z-number based TOPSIS method for multi-criteria decision making problem. Nowadays, a large diversity of approaches to multi-criteria decision making problems with uncertainty information and imperfect information exists. In view of this, Zadeh's Z-number theory is a very effective tool, but, up to day, there is no TOPSIS method which operates Z-number without conversion to fuzzy or crisp number. The existing Z-TOPSIS methods can't incorporate or only approximately reflect the advantage of Z-information and the properties of Z-number. Therefore, this article has offered a Z-TOPSIS method applying Z-numbers in a direct way. The presented method is applied to the vehicle choice problem. All the calculations are performed by using a Z-number software tool. The obtained results show applicability and validity of the proposed approach.

**Keywords:** Z-TOPSIS · Fuzzy number · Z-number · Vehicle choice problem

## 1 Introduction

Multi-criteria decision making (MCDM) methods concern a quite wide range of fuzzy approaches [1–4]. One of them is the TOPSIS (Technique for Order Preference by Similarity to Ideal Solution) method, developed by Hwang and Yoon [5] as an alternative to the ELECTRE method. In recent years, the TOPSIS method has had a lot of applications (in human resource management [6], planning, medicine [7], etc.) and has been widely discussed in the scientific literature. In [8], the uncertainty based TOPSIS for choosing sustainable energy alternatives is discussed, which is dealing with the linguistic estimation of the model uncertainty and inaccuracy. In [9], the comparative analysis of the AHP, ELECTRE, SAW and TOPSIS methods is performed, demonstrating the superiority of the TOPSIS method. In [10], the application of the TOPSIS method in the sustainable expansion issue of the waste management system is considered. In [11], the decision-making issue based on the interval type-2 TOPSIS method is considered. Yakoob and Gegov proposed a fuzzy rule-based approach for the selection of alternatives involving Z-numbers [12] and TOPSIS [13]. In [14], a fuzzy similarity based Z-TOPSIS method for performance assessment is considered. In [15], the TOPSIS method applying Z-numbers is proposed for vehicle choice and clothing evaluation problems, though the Z-numbers are converted to fuzzy numbers in the research.

© Springer Nature Switzerland AG 2019
R. A. Aliev et al. (Eds.): ICAFS-2018, AISC 896, pp. 42–50, 2019.
https://doi.org/10.1007/978-3-030-04164-9_10

Nowadays, real-world issues are commonly attempted to be described by linguistic information, which can be defined by several types of linguistic variables and terms. In this paper, we apply discrete Z-numbers (introduced by Zadeh) to define linguistic information, considering the Z-number based TOPSIS method. Most of the reported researches in decision making, if dealing with Z-numbers, apply the conversion procedure reducing Z-numbers to common fuzzy sets, which leads to the significant quantitative and qualitative information loss. In this paper, our main contribution is that we apply Z-numbers directly, without having to perform "from Z to fuzzy" transformation causing the loss of information.

The rest of the paper is organized as follows. In Sect. 2, we provide a brief description of discrete Z-numbers. In Sect. 3, we demonstrate the proposed Z-number based version of the TOPSIS method. Section 4 provides the experimental verification of the proposed method. Section 5 concludes the paper.

## 2  Preliminaries

In this section, we present some basic definitions of the Z-number theory.

**Definition 1 Discrete Z-number** [16]. A discrete Z-number is an ordered pair $Z = (A, B)$ of discrete fuzzy numbers $A$ and $B$. $A$ plays a role of the fuzzy constraint on the values that a random variable $X$ may take. $B$ is a discrete fuzzy number with a membership function $\mu_B : \{b_1, \ldots, b_n\} \to [0, 1], \{b_1, \ldots, b_n\} \subset [0, 1]$, playing a role of a fuzzy constraint on the probability measure of $A$, $P(A)$.

**Definition 2 Operations over discrete Z-numbers** [12, 17, 18]. Let $X_1$ and $X_2$ be discrete Z-numbers describing information about values of $X_1$ and $X_2$. Consider computation of $Z_{12} = Z_1 * Z_2, * \in \{+, -, \cdot, /\}$ and, square root, power. The first stage is the computation of $A_{12} = A_1 * A_2$.

The second stage involves construction of $B_{12}$. We realize that, in Z-numbers $Z_1$ and $Z_2$, the 'true' probability distributions $p_1$ and $p_2$ are not exactly known. In contrast, the fuzzy restrictions represented in terms of the membership functions are available:

$$\mu_{p_1}(p_1) = \mu_{B_1}\left(\sum_{k=1}^{n_1} \mu_{A_1}(x_{1k})p_1(x_{1k})\right), \mu_{p_2}(p_2) = \mu_{B_2}\left(\sum_{k=1}^{n_2} \mu_{A_2}(x_{2k})p_2(x_{2k})\right),$$

Probability distributions $p_{jl}(x_{jk}), k = 1, \ldots, n$ induce probabilistic uncertainty over $X_{12} = X_1 + X_2$. Given any possible pair $p_1$ and $p_2$, the convolution $p_{12} = p_1 \circ p_2$ is computed as follows:

$$p_{12}(x) = \sum_{x_1 + x_2 = x} p_1(x_1)p_2(x_2), \forall x \in X_{12}; x_1 \in X_1, x_2 \in X_2$$

Given $p_{12s}$, the value of probability measure of $A_{12}$ is computed as follows:

$$P(A_{12}) = \sum_{k=1}^{n} \mu_{A_{12}}(x_{12k})p_{12}(x_{12k}),$$

However, $p_1$ and $p_2$ are described by fuzzy restrictions which induce fuzzy set of convolutions:

$$\mu_{p_{12}}(p_{12}) = \max_{\{p_1,p_2:p_{12}=p_1 \circ p_2\}} \min\{\mu_{p_1}(p_1), \mu_{p_2}(p_2)\}$$

The fuzziness of information on $p_{12}$ induces fuzziness of $P(A_{12})$ as a discrete fuzzy number $B_{12}$. The membership function $\mu_{B_{12}}$ is defined as follows:

$$\mu_{B_{12}}(b_{12}) = \max \mu_{p_{12}}(p_{12}) \text{ subject to}$$

$$b_{12} = \sum_{i=1}^{n} \mu_{A_{12}}(x_i)p_{12}(x_i)$$

As a result, $Z_{12} = Z_1 * Z_2$ is obtained as $Z_{12} = (A_{12}, B_{12})$.

**Definition 3** [19] **A distance between Z-numbers.** The distance between Z-numbers $Z_1 = (A_1, B_1)$ and $Z_2 = (A_2, B_2)$ is defined as follows:

$$D(Z_1, Z_2) = \frac{1}{n+1} \sum_{k=1}^{n} \left\{ \left| a_{1\alpha_k}^{L} - a_{2\alpha_k}^{L} \right| \right\} + \left\{ \left| a_{1\alpha_k}^{R} - a_{2\alpha_k}^{R} \right| \right\} +$$
$$\frac{1}{m+1} \sum_{k=1}^{m} \left\{ \left| b_{1\beta_k}^{L} - b_{2\beta_k}^{L} \right| \right\} + \left\{ \left| b_{1\beta_k}^{R} - b_{2\beta_k}^{R} \right| \right\}$$

where $a_{ia_k}^{L} = \min A_i^{\alpha_k}$, $a_{ia_k}^{R} = \max A_i^{\alpha_k}$, $b_{i\beta_k}^{L} = \min B_i^{\beta_k}$, $b_{i\beta_k}^{R} = \max B_i^{\beta_k}$ and $A_i^{\alpha_k}$, $B_i^{\beta_k}$ are $k$-th $\alpha$-cuts of $A_i$ and $B_i$ respectively,

$$\alpha_k \in \{\alpha_1, \alpha_2, \ldots, \alpha_n\} \subset [0, 1], \ \beta_k \in \{\beta_1, \beta_2, \ldots, \beta_n\} \subset [0, 1], i = 1, 2$$

## 3   Z-TOPSIS

Consider an example of MCDM. Suppose that in MCDM a problem consists of m criteria and n alternatives. Decision matrix with Z-numbers is given in Table 1.

**Table 1.** Decision matrix with Z-number

|       | $C_{w1}: Z(A, B)$ | $C_{w2}: Z(A, B)$ |     | $C_{wm}: Z(A, B)$ |
|-------|-------------------|-------------------|-----|-------------------|
| $A_1$ | $Z_{11}(A_{11}, B_{11})$ | $Z_{12}(A_{12}, B_{12})$ |     | $Z_{1m}(A_{1m}, B_{1m})$ |
| $A_2$ | $Z_{21}(A_{21}, B_{21})$ | $Z_{22}(A_{22}, B_{22})$ |     | $Z_{2m}(A_{2m}, B_{2m})$ |
| ...   | ...               | ...               | ... | ...               |
| $A_n$ | $Z_{n1}(A_{n1}, B_{n1})$ | $Z_{n2}(A_{n2}, B_{n2})$ |     | $Z_{nm}(A_{nm}, B_{nm})$ |

In Table 1, $Z(A_{ij}, B_{ij})$ is a value of an alternative $A_i$ over $C_j$, and $C_{wj} : Z(A, B)$ is the weight of the criterion. $A_{ij}, B_{ij}$ are represented by trapezoidal-form fuzzy number:

$$f(x; a, b, c, d) = \max\left(\min\left(\frac{x-a}{b-a}, 1, \frac{d-x}{d-c}\right), 0\right)$$

In the formula shown above, $a$ and $d$ are the parameters locating the "feet", $b$ and $c$ – locating the "shoulders" of the trapezoid.

The goal is to define the best alternative by using Z-TOPSIS method, which consists of the following steps.

An element $r_{ij}$ of the normalized decision matrix $r(A_{ij}, B_{ij})$ can be calculated over Z-numbers as follows:

$$r_{ij}(A_{ij}, B_{ij}) = \frac{Z_{xij}(A_{ij}, B_{ij})}{\sum\limits_{i=1}^{n} Z_{xij}^2(A_{ij}, B_{ij})}, i = 1, \ldots m, j = 1, 2, \ldots n$$

Construct the weighted normalization decision matrix, obtaining elements by using

$$Z_{vi} = Z_{wi} r_{ij}(A_{ij}, B_{ij}), i = 1, \ldots m, j = 1, 2, \ldots n$$

We represent determined positive ideal and negative ideal Z-number valued solution in the following form:

$$A^+ = (r_{i1}^+(A_{ij}, B_{ij}), r_{i2}^+(A_{ij}, B_{ij}), \ldots, r_{im}^+(A_{ij}, B_{ij})),$$
$$r_{ij}^+(A_{ij}, B_{ij}) = \max_{1 \leq i \leq n}(r_{ij}(A_{ij}, B_{ij}), j = 1, 2, \ldots n$$

$$A^- = (r_{i1}^-(A_{ij}, B_{ij}), r_{i2}^-(A_{ij}, B_{ij}), \ldots, r_{im}^-(A_{ij}, B_{ij})),$$
$$r_{ij}^-(A_{ij}, B_{ij}) = \min_{1 \leq i \leq n}(r_{ij}(A_{ij}, B_{ij}), j = 1, 2, \ldots n$$

Calculate the distances from the positive and negative ideal solution of each alternative, between two Z-numbers according to the equation described in Definition 3.

For each alternative, calculate, the relative closeness to the ideal solution. The last step is to sort (rank) the alternatives by the relative closeness.

The proposed Z-TOPSIS method is claimed to perform decision making with no information loss, which is proved by the following example. In [19], an example is given about the following relation between Z-numbers and fuzzy numbers:

If $X_1$ is $(A_{11}, B_{11})$ and $X_2$ is $(A_{12}, B_{12})$ Then Y is $(A_1, B_1)$
If $X_1$ is $(A_{21}, B_{21})$ and $X_2$ is $(A_{22}, B_{22})$ Then Y is $(A_2, B_2)$

These rules provide a more adequate description of relation between random variables $X_1$, $X_2$ and Y under the combination of fuzzy and probabilistic uncertainties. Of course, computation with such rules is of a higher computational complexity as compared to that of pure fuzzy rules. In order to reduce the computational complexity of dealing with Z-numbers, in [20] the authors suggest converting a Z-number to a fuzzy number. However, the loss of information related to such conversion may lead to the following situation. Consider the following Z-number valued If-Then rules:

If X is ((26, 28, 30), (0.6, 0.64, 0.7)) Then Y is ((40, 50, 60), (0.7, 0.8, 0.9))
If X is ((34.7, 37.3, 40), (0.3, 0.36, 0.42)) Then Y is ((20, 30, 40), (0.5, 0.6, 0.7))

We have converted the Z-numbers in the considered If-Then rules into fuzzy numbers on the basis of the approach in [18]. As the result, the following fuzzy If-Then rules are obtained:

If X is (20.8, 22.4, 24) Then Y is (35.8, 44.7, 53.7)
If X is (20.8, 22.4, 24) Then Y is (15.5, 23.23, 31)

One can see that these rules are contradictory. Despite the fact that the inputs coincide, the outputs differ substantially. Thus, the loss of information related to the conversion of Z-information into the fuzzy one may lead to serious mistakes.

Let us now consider the conversion of Z-numbers into generalized fuzzy numbers.

If X is ((26, 28, 30), (0.6, 0.64, 0.7)) Then Y is ((40, 50, 60), (0.7, 0.8, 0.9))
If X is ((34.7, 37.3, 40), (0.3, 0.36, 0.42)) Then Y is ((20, 30, 40), (0.5, 0.6, 0.7))
The obtained generalized fuzzy If-Then rules are as follows:
If X is (26, 28, 30;0.64) Then Y is (40, 50, 60;0.8)
If X is ((34.7, 37.3, 40;0.36) Then Y is (20, 30, 40;0.6)

As one can see, in these rules the information on probabilistic uncertainty is lost, which means that the important fact of X and Y being random is disregarded.

## 4    Experimental Verification of the Z-TOPSIS Method

Consider MCDM problem, which consists of three criteria and three alternatives.

The **numerical example** [15] considers the vehicle selection for a journey (car, taxi or train). The criteria are price, journey time and comfort. Price and journey time are cost, comfort is a benefit criterion. The linguistic values VH, H and M stands for very high, high and medium, respectively. After the conversion of the linguistic values into Z-numbers, a new representation is in Table 2.

**Table 2.** Decision matrix described by Z-numbers expressed with linguistic values

| Alternative | Price(pounds) (VH, VH), C1 | Journey time (H, VH), C2 | Comfort (M, VH), C3 |
|---|---|---|---|
| (Car) | ((9, 9.9, 10.1, 12), VH)) | ((70, 98, 102, 120), VH)) | ((4, 4.8, 5, 6), H)) |
| (Taxi) | ((20, 23, 24, 25), H)) | ((60, 68, 70, 100), VH)) | ((7, 7.8, 8, 10), H)) |
| (Train) | ((15, 15.02, 15.03, 15.04), H)) | ((70, 78, 80, 90), H)) | ((1, 3.8, 4, 7), H)) |

In Table 3, the weights of the criteria C1, C2 and C3 are represented with Z-numbers:

**Table 3.** Decision matrix described by Z-numbers

|          | C1 | C2 | C3 |
|----------|----|----|----|
| $Z_{A1}$ | ((9, 9.9, 10.1, 12), (0.75, 0.9, 0.98, 1)) | ((70, 98, 102, 120), (0.75, 0.9, 0.98, 1)) | ((4, 4.8, 5, 6), (0.5, 0.72.0.75, 1)) |
| $Z_{A2}$ | ((20, 23, 24, 25), (0.5, 0.72, 0.75, 1)) | ((60, 68, 70, 100), (0.75, 0.9, 0.98, 1)) | ((7, 7.8, 8, 10), (0.5, 0.72, 0.75, 1)) |
| $Z_{A3}$ | ((15, 15.02, 15.03, 15.04), (0.5, 0.72, 0.75, 1)) | ((70, 78, 80, 90), (0.5, 0.72, 0.75, 1)) | ((1, 3.8, 4, 7), (0.5, 0.72, 0.75, 1)) |

w1 = ((0.75, 0.9, 0.98, 1), (0.75, 0.9, 0.98, 1));
w2 = ((05, 0.72, 0.75, 1), (0.75, 0.9, 0.98, 1));
w3 = ((0.25, 0.48, 0.5, 0.75), (0.75, 0.9, 0.98, 1)).

The calculated elements $Z_{rij}$ of the normalized decision matrix are given below:

Zr11(A, B) = ((0.00925, 0.01095, 0.01184, 0.01892), (0.198906, 0.410032, 0.480747, 0.879302))

Zr21(A, B) = ((0.020568, 0.02544501, 0.028146, 0.039414), (0.152487, 0.364515, 0.42126, 0.8793))

Zr31(A, B) = ((0.015436, 0.0166167, 0.017626, 0.0237), (0.15064, 0.364657, 0.420762, 0.879302))

Zr12(A, B) = ((0.0021535, 0.003594, 0.0050, 0.008955), (0.190528, 0.510952, 0.619738, 0.921207))

Zr22(A, B) = ((0.0018465, 0.003134, 0.003447, 0.007463), (0.190441, 0.509358, 0.6144, 0.894852))

Zr32(A, B) = ((0.0021538, 0.003594, 0.003939, 0.006716), (0.14609, 0.446239, 0.532967, 0.92121))

Zr13(A, B) = ((0.02162, 0.04571429, 0.050854, 0.090909), (0.128551, 0.314866, 0.355605, 0.9555))

Zr23(A, B) = ((0.037838, 0.0742857, 0.081367, 0.151515), (0.1305, 0.315092, 0.355544, 0.954774))

Zr33(A, B) = ((0.005405, 0.03619048, 0.04068, 0.10606), (0.130309, 0.314419, 0.355368, 0.95552))

By using the weights and elements of the normalized decision matrix, the weighted normalized decision matrix Z(V) is constructed, which elements are described as follows:

Zv11(A, B) = ((0.00694, 0.00985717, 0.011608, 0.018918), (0.19489, 0.402172, 0.4782, 0.879303));

Zv21(A, B) = ((0.015426, 0.0229005, 0.027583, 0.0394), (0.151089, 0.36085, 0.419945, 0.879303));

Zv31(A, B) = ((0.01158, 0.01495503, 0.01727, 0.0237), (0.150115, 0.361555, 0.419739, 0.879303));

… … ….

Zv33(A, B) = ((0.00135, 0.0173714, 0.020342, 0.07954), (0.12986, 0.312848, 0.355214, 0.955522)).

The determination of the positive ideal and negative ideal solutions from the matrix Z(V) is described below:

Z1+(A, B) = ((0.015426, 0.0229, 0.027583, 0.039414), (0.151089, 0.360847, 0.419945, 0.879303));

Z2+(A, B) = ((0.001077, 0.002588, 0.003767, 0.00896), (0.186716, 0.491486, 0.614477, 0.921208));

Z3+(AB) = ((0.009459, 0.035657, 0.04068, 0.113636), (0.130104, 0.31357, 0.355494, 0.954775));

Z1-(A, B) = ((0.00694, 0.009857, 0.011608, 0.018918), (0.194892, 0.402172, 0.4782, 0.879303));

Z2-(A, B) = ((0.000923, 0.002256, 0.002585, 0.00746), (0.186723, 0.4906527, 0.609222, 0.89485));

Z3-(A, B) = ((0.001351, 0.017371, 0.020342, 0.079545), (0.129864, 0.312848, 0.355214, 0.955522)).

Z(A+) indicates the ideal solution or the most preferable alternative:

Z(A+) = Z(A,    B) = ((0.009459459,    0.035657,    0.04068348,    0.113636), (0.130104303, 0.31357, 0.35549399, 0.954775))

Z(A-) indicates the negative-ideal solution or the least preferable alternative.

Z(A-) = Z(A,  B) = ((0.001351,  0.017371,  0.020342,  0.079545),  (0.129864, 0.312848, 0.355214, 0.955522))

The calculated separation measures are represented in the following form:

Z1+(A, B) = ((0.015426, 0.0229, 0.027583, 0.039414), (0.151089, 0.360847, 0.419945, 0.879303));

Z2+(A, B) = ((0.001077, 0.002588, 0.003767, 0.00895), (0.186716, 0.491486, 0.614477, 0.921208));

Z3+(A, B) = ((0.009459, 0.035657, 0.040683, 0.113636), (0.130104, 0.31357, 0.355494, 0.954775)).

Distance of each alternative to the positive ideal solution:

D(Z1+, ZV11) = 0.079329; D(Z1+, ZV21) = 0; D(Z1+, ZV31) = 0.016246; D(Z2+, ZV12) = 0; D(Z2+, ZV22) = 0.012974; D(Z2+, ZV32) = 0.072845; D(Z3+, ZV13) = 0.033731; D(Z3+, ZV23) = 0; D(Z3+, ZV33) = 0.034341.

$S_i+$ is the separation of each alternative from the positive ideal solution:

$S_i+$ = 0.336243; 0.113903; 0.351328
ZV11 = ((0.006942, 0.009857, 0.011608, 0.018918), (0.194892, 0.402172, 0.4782, 0.879303));
ZV21 = ((0.015426, 0.022901, 0.027583, 0.039414), (0.151089, 0.360847, 0.419945, 0.879303));

ZV31 = ((0.011577,   0.014955,   0.017274,   0.023711),   (0.150115,   0.361555, 0.419739, 0.879303));
Z2-(A,  B) = ((0.0009,  0.002256,  0.002585,  0.007463),  (0.186723,  0.490653, 0.609222, 0.8948530));
ZV12 = ((0.001077,   0.002588,   0.003767,   0.008955),   (0.186716,   0.491486, 0.614477, 0.9212080));
ZV22 = ((0.000923,   0.002256,   0.002585,   0.007463),   (0.186723,   0.490653, 0.609222, 0.894853));
ZV32 = ((0.001077,   0.002588,   0.002955,   0.006716),   (0.145431,   0.434026, 0.529128, 0.921208));
Z3-(AB) = ((0.001351,   0.017371,   0.020342,   0.079545),   (0.129864,   0.312848, 0.355214, 0.955522));
ZV13 = ((0.005405,   0.021943,   0.025427,   0.068182),   (0.128153,   0.313356, 0.355495, 0.955523));
ZV23 = ((0.009459,   0.035657,   0.040683,   0.113636),   (0.130104,   0.31357, 0.355494, 0.954775));
ZV33 = ((0.001351,   0.017371,   0.020342,   0.079545),   (0.129864,   0.312848, 0.355214, 0.955522));

The distance of each alternative to the negative ideal solution is shown below:

D($Z1^-$, ZV11) = 0; D(Z1-, ZV21) = 0.079329; D(Z1-, ZV31) = 0.063717;
D($Z2^-$, ZV12) = 0.012974; D(Z2-, ZV22) = 0; D(Z2-, ZV32) = 0.068284;
D(Z3-, ZV13) = 0.008991; D(Z3-, ZV23) = 0.034341; D(Z3-, ZV33) = 0,

Si- is the separation of each alternative from the negative ideal solution:
Si- = 0,148207; 0,337149; 0,363319,
*The relative closeness for alternatives is defined as follows:*

A1 = 0.305927829; A2 = 0.747473; A3 = 0.50838933.

The final ranking of the alternative is determined as max(A1, A2, A3) = 0.747473. The best alternative is A2 (selection of a taxi).

## 5   Discussion and Conclusions

Though a lot of approaches involving the TOPSIS method exists, there is almost no researches considering the TOPSIS method applying Z-numbers in the direct form, without transformation to fuzzy numbers. The approach proposed in this paper is based on the operations over Z-numbers (multiplication, division, distance, etc.), which have been performed without any "from Z to fuzzy" transformation.

The proposed approach has been verified on the vehicle choice problem. In [15], A1 = 0.2305; A2 = 0.1363; A3 = 0.1856 (the best alternative is A1-car). In our case: A1 = 0.305927829;   A2 = 0.747473;   A3 = 0.50838933   (the   best   alternative   is A2-Taxi). While converting Z-numbers to fuzzy numbers, the reliability information is omitted from the consideration. On the other hand, the fuzzy model becomes more strict and does not differentiate qualitatively distinct evaluation. The results of the performed experimental verification demonstrate the validity of the proposed approach.

# References

1. Zavadskas, E., Turskis, Z., Kildienė, S.: State of art surveys of overviews on MCDM/MADM methods. Technol. Econ. Dev. Econ. **20**, 165–179 (2014)
2. Løken, E.: Use of multicriteria decision analysis methods for energy planning problems. Renew. Sustain. Energy Rev. **11**, 1584–1595 (2007)
3. Lu, J., Zhang, G., Ruan, D., Wu, F.: Multi-Objective Group Decision Making – Methods, Software and Applications with Fuzzy Set Techniques. World Scientific, Singapore (2007)
4. Mardani, A., Jusoh, A., Nor, K., Khalifah, Z., Zakwan, N., Valipour, A.: Multiple criteria decision making techniques and their applications – a review of the literature from 2000 to 2014. Econ. Res.-Ekon. Istraživanja **28**, 516–571 (2015)
5. Hwang, C., Yoon, K.: Multiple Attribute Decision Making, 1st edn. Springer, Heidelberg (1981)
6. Saner, T., Gardashova, L., Alllahverdiyev, R., Eyupoglu, S.: Analysis of the job satisfaction index problem by using fuzzy inference. Procedia Comput. Sci. **102**, 45–50 (2016)
7. Aliev, B., Gardashova, L.: Selection of an optimal treatment method for acute pulpitis disease. Procedia Comput. Sci. **120**, 539–546 (2017)
8. Afsordegan, A., Sánchez, M., Agell, N., Zahedi, S., Cremades, L.: Decision making under uncertainty using a qualitative TOPSIS method for selecting sustainable energy alternatives. Int. J. Environ. Sci. Technol. **13**(6), 1419–1432 (2016)
9. Thor, J., Ding, S., Kamaruddin, S.: Comparison of multi criteria decision making methods from the maintenance alternative selection perspective. Int. J. Eng. Sci. **2**(6), 27–34 (2013)
10. Pires, A., Chang, N., Martinho, G.: An AHP-based fuzzy interval TOPSIS assessment for sustainable expansion of the solid waste management system in Setúbal Peninsula, Portugal. Resour. Conserv. Recycl. **56**(1), 7–21 (2011)
11. Chen, S., Lee, L.: Fuzzy multiple attributes group decision-making based on the interval type-2 TOPSIS method. Expert Syst. Appl. **37**(4), 2790–2798 (2010)
12. Zadeh, L.: A note on Z-numbers. Inf. Sci. **181**(14), 2923–2932 (2011)
13. Yaakob, A., Gegov, A.: Fuzzy rule based approach with z-numbers for selection of alternatives using TOPSIS. In: Proceedings of IEEE International Conference on Fuzzy Systems, Istanbul, pp. 1–8 (2015)
14. Khalif, K., Gegov, A., Bakar, A.: Z-TOPSIS approach for performance assessment using fuzzy similarity. In: Proceedings of IEEE International Conference on Fuzzy Systems, Naples, pp. 1–6 (2017)
15. Krohling, R., Pacheco, A., Santos, G.: TODIM and TOPSIS with Z-numbers. Front. Inf. Technol. Electron. Eng. (in press)
16. Aliev, R., Alizadeh, A., Huseynov, O.: The arithmetic of discrete Z-numbers. Inf. Sci. **290**, 134–155 (2015)
17. Aliev, R., Gardashova, L., Huseynov, O.: Z-numbers-based expected utility. In: Proceedings of Tenth International Conference on Application of Fuzzy Systems and Soft Computing, Lisbon, pp. 49–61 (2012)
18. Yager, R.: On Z-valuations using Zadeh's Z-numbers. Int. J. Intell. Syst. **27**, 259–278 (2012)
19. Aliev, R., Pedrycz, W., Huseynov, O., Eyupoglu, S.: Approximate reasoning on a basis of Z-number valued If-Then rules. IEEE Trans. Fuzzy Syst. **25**(6), 1589–1600 (2017)
20. Kang, B., Wei, D., Li, Y., Deng, Y.: Decision making using Z-numbers under uncertain environment. J. Comput. Inf. Syst. **8**(7), 2807–2814 (2012)

# Comparative Analysis of Artificial Intelligence Based Methods for Prediction of Precipitation. Case Study: North Cyprus

Selin Uzelaltinbulat[1(✉)], Fahreddin Sadikoglu[2], and Vahid Nourani[3,4]

[1] Department of Computer Engineering, Faculty of Engineering, Near East University, POBOX 99138, Nicosia, Mersin 10, TRNC, Turkey
selin.uzelaltinbulat@neu.edu.tr
[2] Department of Electrical and Electronic Engineering, Faculty of Engineering, Near East University, POBOX 99138, Nicosia, Mersin 10, TRNC, Turkey
[3] Department of Civil Engineering, Faculty of Engineering, Near East University, POBOX 99138, Nicosia, Mersin 10, TRNC, Turkey
[4] Department of Water Resources Engineering, Faculty of Civil Engineering, University of Tabriz, Tabriz, Iran

**Abstract.** Prediction of precipitation is important for design, management of water resources systems, planning, flood predicting and hydrological events. This study aimed to compare the performance of three different "Artificial Intelligence (AI)" techniques which are "Artificial Neural Network (ANN), Adaptive Neuro Fuzzy Inference System (ANFIS) and Least Square Support Vector Machine (LSSVM)" to estimate monthly rainfall in Kyrenia Station of Turkish Republic of Northern Cyprus (TRNC). The monthly data covering ten years' precipitation were used for the predictions. The comparative results showed that the LSSVM model can cause a bit more reliable performance in regard to ANN and ANFIS.

**Keywords:** Precipitation · ANFIS · LSSVM · ANN · Prediction
Rainfall station

## 1 Introduction

Precipitation is the most important environmental, natural and climatic process all around the world and plays crucial role in hydrological studies. The precipitation is defined as the quantity of water falling in drops in the atmosphere. Precipitation has negative and positive impacts on the agriculture, economy, tourism, ecosystem, drought and water resources. For these reasons it is important to develop reliable methods for prediction of precipitation. For prediction of precipitation as an extremely complex process some non-linear techniques have been already developed and employed bring out the complex patterns involved in the system.

Model of prediction of precipitation classified in two categories: physically based and black box models. Physically based model uses physical rules to modelling all of the proper physical processes that make contribution to the precipitation procedures. On the other hand, black box models use historically observed data to make further

R. A. Aliev et al. (Eds.): ICAFS-2018, AISC 896, pp. 51–64, 2019.
https://doi.org/10.1007/978-3-030-04164-9_11

estimations. Such black box methods are particularly developed on the basis of statistical and computational intelligence approaches. Recently, "Artificial Intelligence (AI)" methods as such black box methods showed great efficiency in modelling of the linear, non-linear and dynamic processes in presence the non-linearity, uncertainty and irregularity of data. Comparative researches have shown that AI based models may generate best results for precipitation prediction in consideration to physically based models [2]. The aim indicates the capability of some AI methods for prediction of monthly precipitation time series. "Least Square Support Vector Machine (LSSVM), Artificial Neural Network (ANN) and Adaptive Neuro Fuzzy Inference System (ANFIS)" are used as supervised techniques for data analysis and modelling. The nonlinearities, stochasticity and uncertainty of the precipitation process needs development of robust prediction methods based on AI algorithms.

One of the most commonly used AI methods for the precipitation modelling is ANN. In the recent decades, ANN has acquired increasing popularity to its flexibility and robustness to catch variety range of data [19]. For examples, [6] employed ANN technique for predicting of precipitation over 36 meteorological stations of India, for the last year's monsoon precipitation data to estimate monsoon precipitation of upcoming years. The model could catch input–output non-linear relationship and estimate the seasonal rainfall. [7] employed ANN for precipitation predicting and flood management in Bangkok, Thailand. Likewise, [2] have also worked ANN methods to predict precipitation over Queensland, Australia. More recently, [4, 8] employed ANN for predict of rainfall time series. However, sometimes ANN-based modelling can involve some deficiencies, like an over-fitting, converging to local minimum and slow training to achieve adequate efficiency when dealing with complex hydrological processes [5].

On the other hand, LSSVM was first proposed by [28] and is one of the most persuasive predicting methods as an alternative method of ANN. It was arising from SVM for non-linear classification problems, function estimation and density estimation [10]. SVM is machine learning algorithm to claim best model dealing with complicated classification problems. SVM is capable of predicting nonlinear, non-stationary and stochastic processes [27]. The SVM has been used for prediction of precipitation in the recent decades. [11] predicted monthly rainfall in a region of China using the SVM approach with different kernel functions. [9] introduced a novel SVM approach for predicting of precipitation using past observed data. [22] to show regard to a large set of predictive variables in the daily precipitation estimation and analyzed the importance of humidity and Equivalent Potential Temperature variables in the SVM based precipitation process involved high degree of uncertainty. Also, ANFIS model which serves the ability of both ANN and fuzzy set concepts within a unique framework can be considered as a robust model for precipitation prediction due to the ability of fuzzy concept to handle the uncertainty of the process. ANFIS is a Takagi-Sugeno-Kang (TSK) fuzzy based mapping algorithm. ANFIS is efficient tool to provide less overshoot, oscillation and minimal training time [1]. An ANFIS can simulate and analyse the mapping relation between the input and output data through a learning algorithm to optimize the parameters of a given "Fuzzy Inference System (FIS)" [26]. ANFIS training can use alternative algorithms to reduce the error of the training. Some previous investigations have indicated that ANFIS can be used as an efficient tool for rainfall modelling (e.g. see [25]).

## 2 Materials and Methods

### 2.1 Study Area and Data Gathering

Recently, ANN, ANFIS and LSSVM models have been frequently applied for prediction of the precipitation time series. However, each of these methods may give better precision for the particular condition of the process. In this study, individual calibration by each of the aforementioned technique was considered for providing minimum error prediction of precipitation. For training and validation of the models ten years' monthly data (during January 1, 2007, to December 31, 2016) of Kyrenia Station were used. The data were obtained from Meteorological Station of the Turkish Republic of Northern Cyprus. The 75% of this data set used for training and rest 25% for validation process. The monthly average gathered data were formatted by pre-processing based on normalization of the data into the interval [0, 1] by [3]:

$$x_{norm} = \frac{x_{(t)} - x_{\min(t)}}{x_{\max(t)} - x_{\min(t)}} \leq 1 \tag{1}$$

where $x_{\max(t)}$ *and* $x_{\min(t)}$ are maximum and minimum values of data.

The "Root Mean Square Error (RMSE)" and "Determination Coefficient (DC)" also used for evaluate the prediction efficiency of the models as [15]:

$$RMSE = \sqrt{\frac{\sum_{i=1}^{n}\left(o_{obs_i} - o_{com_i}\right)^2}{n}} \tag{2}$$

$$DC = 1 - \frac{\sum_{i=1}^{n}\left(o_{obs_i} - o_{com_i}\right)^2}{\sum_{i=1}^{n}\left(o_{obs_i} - \bar{o}_{obs}\right)^2} \tag{3}$$

where n is data number, $o_{obs_i}$ is the observed data, $o_{com_i}$ is the predicted data. DC ranges from $-\infty$ to 1 with a perfect score of 1.

### 2.2 Development of the Model

The conceptual model of the system involving above mentioned ANN, ANFIS and LSSVM models is shown in Fig. 1.

Where "$x(t-1), x(t-2)$ and $x(t-12)$" are previously monthly data corresponding to 1, 2 and 12 months ago; $x_{ANN(t)}$, $x_{ANFIS(t)}$ and $x_{LSSVM(t)}$ are results of prediction in current month by different models. Pre-processing stage was used to filter and normalize the gathered data. Argumentation of using as inputs "$x(t-1), x(t-2)$ and $x(t-12)$" for prediction of $x(t)$ are supported by the following:

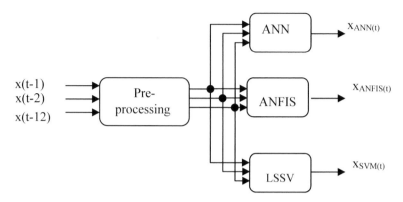

**Fig. 1.** Conceptual model of the system

(a) As shown by [16] and some others, in precipitation as a Marcovian (auto regres-
    sion) process, x(t) is more correlated with precipitation values at prior time spans as
    "x(t − 1) and x(t − 2)" and so on. For this reason, it is feasible to select previous
    time steps values as inputs for the AI based models as x(t − 1) and x(t − 2) (trial
    error showed not more correlation for delays more than (t − 2)).
(b) Selection of input x(t − 12) is related to the seasonality of the precipitation phe-
    nomenon. It means due to the seasonality of the process (i.e. periodicity), the
    precipitation value of current month has a strong relation (similarity) with the
    precipitation value at the same month in the previous year (i.e. x(t − 12) in
    monthly scale).

## 2.3   ANN

ANN is based on nonlinear algorithm that finds the relationship for the parameters of a
system [30]. ANN is mostly used in water resources and hydrological studies for an
estimation tool. In ANN, "Feed Forward (FF) Back Propagation (BP)" network models
are common which is a proof that BP model with three-layered, fulfilled for the
estimation and simulation [19]. Three-layered "Feed Forward Neural Network
(FFNN)" which were widely used for estimating hydrological time spans, provides a
framework for performing the nonlinear functional mapping between a set of input and
output, and linear combination of the input variables, which are transformed by a nonlinear activation function as expressed by Eq. (4). There is not any loop/cycle in this
network. The output value of a FFNN can be obtained through [18]:

$$\hat{y}_k = f_0 \left[ \sum_{j=1}^{M_N} W_{kj} . f_h \left( \sum_{i=1}^{N_N} W_{ji} X_i + W_{j0} \right) + W_{k0} \right] \tag{4}$$

where $w_{ji}$ is the applied weight to a neuron in hidden layer which connects *ith* neuron in
the input layer to the *jth* neuron in the hidden layer, $w_{jo}$ is the applied bias to the *jth*
neuron of hidden layer, $f_h$ denotes to the activation function of related hidden layer
neuron, $w_{kj}$ indicates the applied weight to a target neuron which connects *jth* hidden

neuron to the *kth* target neuron. $w_{k0}$ is the applied bias to the *kth* target neuron, $f_0$ stands for the activation function of the target neuron, $x_i$ is the *ith* input neuron and $y_k$ and $y$ are respectively the network output and observed values. $N_N$ and $M_N$ respectively show number of input neuron and hidden neurons [17]. Hidden and target layers' weights are different from each other and should be estimated during the training phase [13]. The developed ANN structure consists of 3-input neurons, 6-hidden neurons and 1-output neuron is shown in Fig. 2.

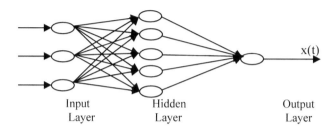

**Fig. 2.** Structure of a three layer FFNN

## 2.4 ANFIS

ANFIS is a type of ANN, depend on TSK FIS. ANFIS merges the ANN and FL concepts to benefits of both within a unique framework. Fuzzy systems need information to define fuzzy rules and tuning the membership functions parameters. However, ANFISs have more computational restrictions than ANNs [20]. In ANFIS, TSK type FIS is usually used. ANFIS used in this study has three inputs of "$x(t-1)$, $x(t-2)$, $x(t-12)$" and one output of $x(t)$ as shown in Fig. 3. The fuzzy system is combined by three main parts; fuzzification, database, defuzzification whereas the database part includes inference engine and fuzzy rules. Among different fuzzy inference systems which can be used for fuzzy operation, the TSK engine was employed in the current research.

The operation of ANFIS to create target function with 3 input vectors of "$x(t-1), x(t-2), x(t-12)$" and first order TSK applied to 2 fuzzy rules expressed as [25]:

Rule (1): if $\mu(x(t-1))$ is A1 and $\mu(x(t-2))$ is B1 and $\mu(x(t-12))$ is C1 then $f1 = p1(x(t-1)) + q1(x(t-2)) + t1(x(t-12)) + r1$
Rule (2): if $\mu(x(t-1))$ is A2 and $\mu(x(t-2))$ is B2 and $\mu(x(t-12))$ is C2 then $f1 = p2(x(t-1)) + q2(x(t-2)) + t2(x(t-12)) + r2$.

In which A1, A2, B1, B2 and C1, C2 show respectively the MFs of inputs $x(t-1)$, $x(t-2)$ and $x(t-12)$. p1, q1, t1, r1 and p2, q2, t2, r2 are the target function parameters [23]. The structure of the used ANFIS in this study is shown in Fig. 3.

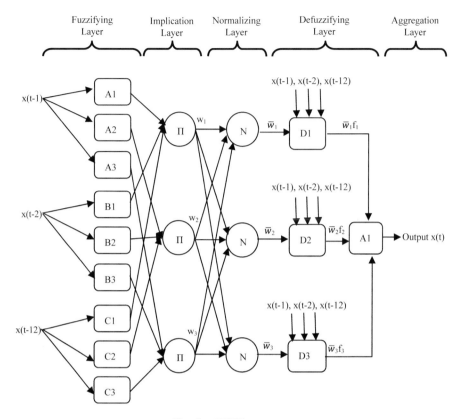

**Fig. 3.** ANFIS structure

*Layer 1 (Fuzzifying Layer):* Each node generates membership values of an input variable. The output of ith node in layer k is denoted as $Q_i^k$. For a generalized bell-function (gbellmf) with MF parameters of {ai, bi, ci}, the output $Q_i^1$ can be calculated as:

$$Q_i^1 = \mu_{A_i}(x) = \frac{1}{1 + \left(\frac{x - c_i}{a_i}\right)^{2b_i}} \tag{5}$$

*Layer 2 (Implication Layer):* The imposed signal to the layer is multiplied by each node of this layer as:

$$Q_i^2 = w_i = \mu_{A_i}(x(t-1)) \cdot \mu_{B_i}(x(t-2)) \cdot \mu_{C_i}(x(t-12)); \ i = 1, 2, 3 \tag{6}$$

*Layer 3 (Normalizing Layer):* Node i in this layer computes the normalized firing strength:

$$Q_i^3 = \bar{w}_i = \frac{w_i}{w_1 + w_2 + w_3}; \; i = 1, 2, 3 \tag{7}$$

*Layer 4 (Defuzzifying Layer):* The contribution of ith rule towards the target is determined where, $\bar{w}$ is the output of layer 3 and {pi, qi, ri} is the parameter set:

$$Q_i^4 = \bar{w}_i(p_i x(t-1) + q_i x(t-2) + t_i x(t-12) + r_i) = \bar{w}_i f_i \tag{8}$$

*Layer 5 (Aggregation Layer):* Finally, the output of the model is calculated by:

$$Q_i^5 = \bar{w}_i(p_i x(t-1) + q_i x(t-2) + t_i x(t-12) + r_i) = \sum_i \bar{w}_i f_i \tag{9}$$

To estimate the primary parameters set {ai, bi, ci} and consequence parameters set {pi, qi, ti, ri} of the ANFIS, the conjunction of the least squared and gradient descent methods are used as a hybrid calibration algorithm. The ANFIS models proposed in this study were trained using Gaussian and Generalizedbell MF, SugenoFuzzy model.

## 2.5 LSSVM

An SVM model is a representation of the set of supervised learning methodology that analyses and recognizes the examples and used them for classifying and regression [10]. "The Least Squares formulation of SVM" is called LSSVM. Thus, the solution in this method is obtained from solving a linear equations system. Efficient algorithms can be used in LSSVM [29]. According to the LSSVM method, a non-linear function can be expressed as [28]:

$$y = f(x) = w^T \varphi(x) + b \tag{10}$$

where $f$ indicates the relationship between the inputs and outputs, $w$ is the m-dimensional weight-vector, $\varphi$ is the mapping function that maps input vector x into the m-dimensional feature vector and b is the bias term [24]. The regression problem can be given as [28]:

$$\min J(w, b, e) = \frac{1}{2} w^T w + \frac{\gamma}{2} \sum_{i=1}^{m} e_i^2 \tag{11}$$

with has the following constraints:

$$y_i = w^T \varphi(X_i) + b + e_i \; (i = 1, 2, \ldots, m) \tag{12}$$

where $\gamma$ is the margin parameter and $e_i$ is the slack variable for $X_i$. To solve the optimization problem, the objective function can be obtained by changing the constraint problem into an unconstraint problem, according to the Lagrange multiplier $\alpha_i$ as [28]:

$$L(w, b, e, \alpha) = J(w, b, e) - \sum_{i=1}^{m} \alpha_i \{ w^T \varphi(X_i) + b + e_i - y_i \} \qquad (13)$$

Vector $w$ in Eq. (13) can be computed after solving the optimization problem as [11]:

$$w = \sum_{i=1}^{N} \alpha_i \varphi(x_i) \qquad (14)$$

So, the final form of LSSVM can be expressed as:

$$f(x, \alpha_i) = \sum_{i=1}^{N} \alpha_i K(x, x_i) + b \qquad (15)$$

where $K(x_i, x_j)$ is the kernel function performing the non-linear mapping into feature space. The commonly used kernel function used in LSSVM is Gaussian Radial Basis Function (RBF) as [17]:

$$K(x - x_i) = exp\left(-\gamma \|x - x_i\|^2 / \sigma^2\right) \qquad (16)$$

where $\gamma$ and $\sigma$ are the parameters of the kernel function (Fig. 4).

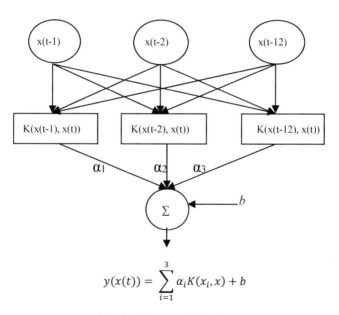

$$y(x(t)) = \sum_{i=1}^{3} \alpha_i K(x_i, x) + b$$

**Fig. 4.** Structure of LSSVM

**Table 1.** Results of monthly rainfall predictions using ANN, ANFIS and LSSVM methods

| Model | Epoch | Input variables | Network structure | DC | | RMSE | |
|---|---|---|---|---|---|---|---|
| | | | | Calibration | Verification | Calibration | Verification |
| ANN | 210 | x(t − 1), x(t − 2), x(t − 12) | 3-input 6-hidden, 1-output neurons | 0.82 | 0.80 | 0.070 | 0.094 |
| ANFIS | 210 | x(t − 1), x(t − 2), x(t − 12) | Gaussian MF | 0.83 | 0.78 | 0.067 | 0.099 |
| LSSVM | - | x(t − 1), x(t − 2), x(t − 12) | (20, 45) RBF-Kernel structure (γ, σ) | 0.88 | 0.84 | 0.056 | 0.085 |

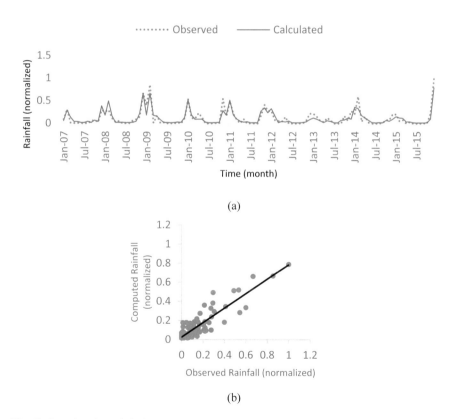

**Fig. 5.** Results of precipitation prediction using ANN - (a) Time series plot, (b) Scatter plot

## 3    Results and Discussion

For precipitation prediction of Kyrenia rainfall gauging station, monthly precipitation values were individually imposed into ANN, ANFIS and LSSVM models in order to predict one-month-ahead precipitation. For this purpose, the ANN, ANFIS and LSSVM models' architectures set depend on the priority of the precipitation process; however, monthly precipitation data obey from Markovian [13] and seasonality-based properties, so that at the current precipitation can be related with parameter values for prior data as well as twelfth months ago. Consequently, as the input values "x(t − 1), x (t − 2) and x(t − 12)" were precipitation values applied to the "LSSVM, ANN and ANFIS" models to monthly precipitation prediction x(t). The variables of RBF-kernel in LSSVM took into consideration, for each input combination to get highest efficiency. Therefore, different $\gamma$ and $\sigma$ were examined with grid method, also most precise values were choose. Then the trained LSSVM data used for verification of the precipitation prediction. The ANN model order for three layered model which are having input, output, and hidden layers. Also, the structure of ANFIS was set for this regard. The Gaussian MF (gaussmf) was used for ANFIS based precipitation prediction method. After suitable "ANN, LSSVM and ANFIS" architecture determination with

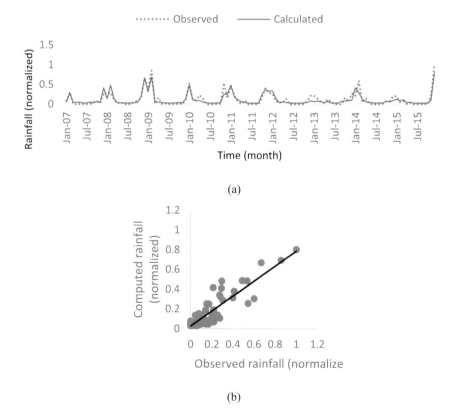

(a)

(b)

**Fig. 6.** Results of precipitation prediction using ANFIS - (a) Time series plot, (b) Scatter plot

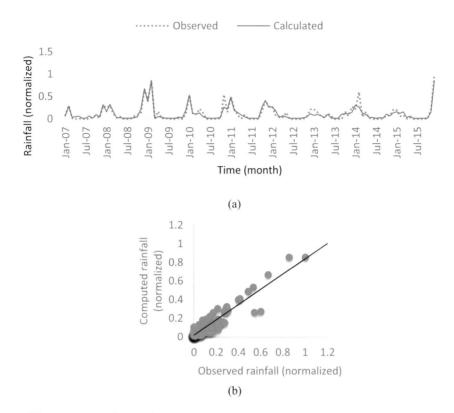

**Fig. 7.** Results of prediction using LSSVM - (a) Time series plot, (b) Scattered plot

regard to efficiency criteria, trained weights were saved for the apply to the verification purpose. The precipitation prediction results of Kyrenia station through individual proposed models showed in Table 1. According to the results, LSSVM could lead to more accurate results to comparison to "ANN and ANFIS" models.

The Figs. 5, 6 and 7 show time series and scatter plots of observed versus predicted values respectively using ANN, ANFIS and LSSVM methods. According to the obtained results see (Table 1), it is clear that LSSVM model could lead to a bit accurate results in each verification and calibration steps. Furthermore seems although, ANFIS model had similar performance to other methods in training phase, its complex structure including different values and MFs may cause a bit poor performance in the verification step. On the other hand, LSSVM could cause better performance in both calibration and verification steps mostly due to its ability to use data classification prior to fitting a non-linear regression as well as solving a Lagrangian based optimization procedure for calibration of the parameters.

It is clear that, DC and RMSE statistics show overall performance of the models, averaged over all samples. But in practice, it is also useful to see the model's performance at different time spans (intervals) since due to different climatic conditions of different seasons and months, the pattern of the time series may be different at different

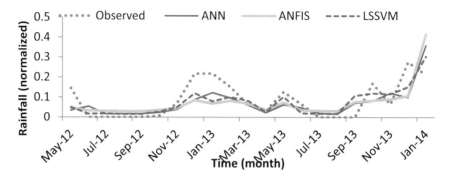

**Fig. 8.** An interval for ANN, ANFIS, LSSVM and observed rainfall.

spans. As shown in Fig. 8, in the interval of Jun-12, Oct-12, ANN and SVM provide the best fitting results to the observed values. But in interval of Apr-13, Jun-13, ANN and ANFIS models provide minimum errors. In interval of Dec-12, Feb-13 and in the interval Sep-13, Nov-13 all models could not lead to reliable prediction.

## 4   Conclusions

In this paper, performance of AI based ANN, ANFIS and LSSVM methods for prediction of monthly precipitation values for North Cyprus were obtained and examined results compared to together. The "ANN, ANFIS and SVM" predictors were developed using MATLAB codes. Calibration and verification of different methods were performed based on observed ten years' monthly data obtained from Metrological Agency of North Cyprus.

Analysis of the results regarding the obtained RMSE values for the considered methods show that all model causes to reliable results. However, in the different time intervals, these methods provide different performances. For this reason, as the plan for following study, it is advised to integrate (ensemble) the outputs of these methods to achieve more accurate predictions in all time spans.

## References

1. Akrami, S.A., Nourani, V., Hakim, S.J.S.: Development of nonlinear model based on wavelet-ANFIS for rainfall forecasting at Klang Gates Dam. Water Resour. Manag. **28**(10), 2999–3018 (2014)
2. Abbot, J., Marohasy, J.: Application of artificial neural networks to rainfall forecasting in Queensland, Australia. Adv. Atmos. Sci. **29**(4), 717–730 (2012)
3. Bisht, D., Joshi, M.C., Mehta, A.: Prediction of monthly rainfall of Nainital region using artificial neural network and support vector machine. Int. J. Adv. Res. Innov. Ideas Educ. **1**(3), 2395–4396 (2015)

4. Devi, S.R., Arulmozhivarman, P., Venkatesh, C.: ANN based rainfall prediction - a tool for developing a landslide early warning system. In: Advancing Culture of Living with Landslides - Workshop on World Landslide Forum, pp. 175–182 (2017)
5. Guo, J., Zhou, J., Qin, H., Zou, Q., Li, Q.: Monthly streamflow forecasting based on improved support vector machine model. Expert Syst. Appl. **38**(10), 13073–13081 (2011)
6. Guhathakurta, P.: Long lead monsoon rainfall prediction for meteorological sub-divisions of India using deterministic artificial neural network model. Meteorol. Atmos. Phys. **101**(2), 93–108 (2008)
7. Hung, N.Q., Babel, M.S., Weesakul, S., Tripathi, N.K.: An artificial neural network model for rainfall forecasting in Bangkok, Thailand. Hydrol. Earth Syst. Sci. **13**, 1413–1425 (2009)
8. Khalili, N., Khodashenas, S.R., Davary, K., Mousavi, B., Karimaldini, F.: Prediction of rainfall using artificial neural networks for synoptic station of Mashhad: a case study. Arab. J. Geosci. **9**, 624 (2016)
9. Kisi, O., Cimen, M.: Precipitation forecasting by using wavelet-support vector machine conjunction model. Eng. Appl. Artif. Intell. **25**(4), 783–792 (2012)
10. Kumar, M., Kar, I.N.: Non-linear HVAC computations using least square support vector machines. Energy Convers. Manag. **50**(6), 1411–1418 (2009)
11. Lu, K., Wang, L.: A novel nonlinear combination model based on support vector machine for rainfall prediction. In: Fourth International Joint Conference on Computational Sciences and Optimization, pp. 1343–1347. Computational Sciences and Optimization, China (2011)
12. Mokhtarzad, M., Eskandari, F., Vanjani, N.J., Arabasadi, A.: Drought forecasting by ANN, ANFIS and SVM and comparison of the models. Environ. Earth Sci. **76**(21), 729 (2017)
13. Murphy, R.A.: A predictive model using the Markov Property. http://arxiv.org/abs/1601.01700v1. Accessed 1 Nov 2016
14. Nourani, V.: An emotional ANN (EANN) approach to modelling rainfall-runoff process. J. Hydrol. **544**, 267–277 (2017)
15. Nourani, V., Andalib, G.: Daily and monthly suspended sediment load predictions using wavelet-based AI approaches. J. Mt. Sci. **12**(1), 85–100 (2015)
16. Nourani, V., Babakhani, A.: Integration of artificial neural networks with radial basis function interpolation in earthfill dam seepage modelling. J. Comput. Civ. Eng. **27**(2), 183–195 (2012)
17. Nourani, V., Komasi, M.: A geomorphology-based ANFIS model for multi-station modelling of rainfall–runoff process. J. Hydrol. **490**, 41–55 (2013)
18. Nourani, V., Mogaddam, A.A., Nadiri, A.O.: An ANN based model for spatiotemporal groundwater level forecasting. Hydrol. Process. **22**(26), 5054–5066 (2008)
19. Nourani, V., Parhizkar, M.: Conjunction of SOM-based feature extraction method and hybrid wavelet-ANN approach for rainfall–runoff modelling. J. Hydroinform. **15**(3), 829–848 (2013)
20. Nourani, V., Sharghi, E., Aminfar, M.H.: Integrated ANN model for earthfill dams seepage analysis: Sattarkhan Dam in Iran. Artif. Intell. Res. **1**(2), 22–37 (2012)
21. Ortiz-Garcia, E.G., Salcedo-Sanz, S., Casanova-Mateo, C.: Accurate precipitation prediction with support vector classifiers: a study including novel predictive variables and observational data. Atmos. Res. **139**, 128–136 (2014)
22. Sharifi, S.S., Delirhasannia, R., Nourani, V., Sadraddini, A.A., Ghorbani, A.: Using ANNs and ANFIS for modelling and sensitivity analysis of effective rainfall. In: Recent Advances in Continuum Mechanics, Hydrology and Ecology. ISBN 978-960-474-313-1
23. Cao, S.-G., Liu, Y.B., Wang, Y.P.: A forecasting and forewarning model for methane hazard in working face of coal mine based on LSSVM. J. China Univ. Min. Technol. **18**(2), 172–176 (2008)

24. Singh, V.K., Kumar, P., Singh, B.P., Malik, A.: A comparative study of adaptive neuro fuzzy inference system (ANFIS) and multiple linear regression (MLR) for rainfall-runoff modelling. Int. J. Sci. Nat. **7**(4), 714–723 (2016)
25. Sojitra, M.A., Purohit, R.C., Pandya, P.A.: Comparative study of daily rainfall forecasting models using ANFIS. Curr. World Environ. **10**(2), 529–536 (2015)
26. Solgi, A., Nourani, V., Pourhaghi, A.: Forecasting daily precipitation using hybrid model of wavelet-artificial neural network and comparison with adaptive neurofuzzy inference system. Adv. Civ. Eng. **2014**(3), 1–12 (2014)
27. Suykens, J.A.K., Vandewalle, J.: Least square support vector machine classifiers. Neural Process. Lett. **9**(3), 293–300 (1999)
28. Ye, J., Xiong, T.: SVM versus least squares SVM. In: Proceedings of the Eleventh International Conference on Artificial Intelligence and Statistics, no. 2, pp. 644–651 (2007)
29. Zhou, T., Wang, F., Yang, Z.: Comparative analysis of ANN and SVM models combined with wavelet pre-process for groundwater depth prediction. Water **9**(10), 781 (2017)

# Analysis and Processing of Information in Economic Problems. Crisp and Fuzzy Technologies

Araz R. Aliev[1,2(✉)], Vagif M. Mamedov[1], and Gasim G. Gasimov[1]

[1] Department of General and Applied Mathematics,
Azerbaijan State Oil and Industry University, Baku, Azerbaijan
alievaraz@yahoo.com, vaqifmammadoqlu@gmail.com,
q.qasim56@gmail.com
[2] Institute of Mathematics and Mechanics,
Azerbaijan National Academy of Sciences, Baku, Azerbaijan

**Abstract.** The existing mathematical methods of processing economic information from the perspective of the theory of measures are critically considered. Fuzzy and fuzzy-probabilistic measures that allow for processing of non-numerical economic data and decision making under uncertainty are proposed. Motivation to use fuzzy methods in economy is discussed.

**Keywords:** Fuzzy sets · Fuzzy measure · Data Mining · Probability
Reliability · Possibility

## 1 Introduction

A large variety of modern computer data analysis tools, more than a thousand software packages for Data Mining exists [7].

All these packages are subject to the following requirements for processing:

– Unlimited amount of information;
– The heterogeneity of information (quantitative, qualitative, textual, graphic, etc.);
– The results of the treatment must be specific and understandable to humans;
– The methods for processing raw materials (information) are simple and satisfy the Piatetsky-Shapiro condition, i.e. in the process of detection there are previously unknown, non-trivial and practically useful information.

It is clear that Data Mining methods and algorithms work well in conditions of redundant information. However, the majority of economic problems (forecasting problems, preventing unauthorized access to financial sources, etc.) have to be handled under uncertain, incomplete and vague initial information.

In the analysis and processing of information in economic problems, there is a need to classify them according to the type of uncertainty [2, 10–12]:

I. Information whose attributes are random in nature, i.e. if $x_1, x_2, \ldots, x_n$ are information units, then known (or unknown) $p(x_1), p(x_2), \ldots, p(x_n)$, where $p$ – is a probability.

© Springer Nature Switzerland AG 2019
R. A. Aliev et al. (Eds.): ICAFS-2018, AISC 896, pp. 65–72, 2019.
https://doi.org/10.1007/978-3-030-04164-9_12

II. Information, whose attributes are fuzzy, i.e. if $x_1, x_2, \ldots, x_n$ are information units, then information is described by membership functions (or fuzzy measures) $\mu(x_1), \mu(x_2), \ldots, \mu(x_n)$.

III. Information characterized by both fuzziness and randomness. Depending on the nature of the phenomena, this type is characterized by a fusion of fuzziness and randomness, and studied using the Lebesgue-Stieltjes integral if fuzzy and probabilistic measures are given.

## 2  Analysis and Processing of Random Information

The state of economic systems (from the position of analysis and synthesis) is determined by numerous and diverse complex factors, related to the internal properties of a particular component and the impact of external conditions of an objective and subjective nature. When the study and analysis of states of economic systems or phenomena (for example, the process of inflation, the change of economic systems in unexpected conditions, etc.) does not give judgments with certainty about the occurrence of events, then it is said that economic characteristics are of random or stochastic nature and probabilistic methods of research is the only effective device for solving the problems. For 350 years of existence of the theory of probability, three probability schools emerged and formed: odds counters (the ratio of the number of favorable events to the number of all possible outcomes); frequency bands (frequency limit of observed events); subjectivists (measure of uncertainty).

The main concepts and theorems of the theory of probability and mathematical statistics applied to the analysis and processing of economic information are described in detail in the monographs of well-known scientists, we only confine ourselves to relaxing the axioms of probability theory introduced by Kolmogorov [9]:

I. Probability P{A} of a random event $A \subseteq \Im$, where $\Im$ is a $\sigma$- algebra, is a non-negative number (the non-negativity axiom);

II. P{U} = 1 (the normality axiom); U is a sample space;

III. If the events A and B are mutually exclusive, i.e. $A \cap B = \emptyset$, then $P\{A \cup B\} = P\{A\} + P\{B\}$ (the additivity axiom).

The triple $\{U, \Im, P\}$ is referred to as a measure space and is widely used by subjectivists for modeling of uncertainty.

## 3  Probability and Possibility

Traditional models of imprecision and uncertainty are critically examined in [6] from the viewpoints of probability theory and error theory. Although the theory of probability - the classical apparatus is quite developed, the disputes about its interpretation (what kind of reality they want to express with the help of this mathematical model) still does not stop. Dubois and Prade note: "The probabilistic model is adapted to the processing of accurate, but distributed by realization information. As soon as there is an inaccuracy in an individual implementation, the model becomes unacceptable."

The following conditions (axioms) [6]:

$$\forall A, B \subseteq \Omega, \ g(A \cup B) \geq max\{g(A), g(B)\},$$

$$g(A \cap B) \leq \min\{g(A), g(B)\} \tag{1}$$

actually mean that we can neglect the restrictive condition III of probability theory, i.e. $A \cap B = \emptyset$, which is practically not satisfied in many economic problems.

The limiting case of measures of uncertainty is a set function $\Pi$ that satisfies

$$\forall A, B, \ \Pi(A \cup B) = \max[\Pi(A), \Pi(B)] \tag{2}$$

Real-valued function $\Pi$ is referred to as possibility measure of Zadeh. When is finite, then $\Pi$ is defined by the possibility distribution $(\omega) : \Omega \to [0, 1]$; $\Pi(A) = sup\{\pi(\omega)/\omega \in A\}$.

If $E \subseteq \Omega$ is a certain event, $\Pi$ satisfies axiom (2) and also satisfies

$$\Pi_E(A) = \begin{cases} 1, & if \ A \cap E \neq \emptyset, \\ 0, & otherwise, \end{cases}$$

then $\Pi_E(A) = 1$ means that event A is possible.

Another limiting case of measures of uncertainty is the so-called necessity measure [6] that satisfies an axiom dual to that of the possibility measure:

$$\forall A, B, N(A \cap B) = \min[N(A), N(B)] \tag{3}$$

is obtained by supposing

$$N(A) = \begin{cases} 1, & if \ E \subseteq A, \\ 0, & otherwise, \end{cases}$$

where $N(A) = 1$ indicates a certain event A. It is easy to prove that the measure of possibility and the measure of necessity are related to each other by the relation: $\forall A, \Pi(A) = 1 - N(\vec{A})$.

Further, the authors of [6] came to the conclusion that the class of probability measures $\rho$ is determined by the family:

$$\rho = \{P/\forall A, \ N(A) \leq P(A) \leq \Pi(A)\} \tag{4}$$

This allows to formulate rigorous definition of the concept of mathematical expectation within the framework of possibility measures. In the general case, De Soumen proved that a possible variable is a possible analogue of a random variable in probability theory.

However, formula (4) creates some difficulties for solving practical problems. To determine the bounds of truth, we need to construct two measures: the necessity measure (the lower bound), the possibility measure (the upper bound). Interval assessment is a desire to improve the mathematical results of measurements (or observations) of random variables or stochastic processes that are not subject to precise quantitative analysis and calculation, and at the same time an attempt to escape the responsibility of a posteriori consequences by blurring boundaries. This paradox has existed for many centuries in economy, defense, and other fields and completely not solved. Probabilistic assessment does not provide any guarantee. For example, assume that the probability of destroying a continental enemy rocket with a neutral zone as computed by using statistics is $P = 0.9$. Indeed, this is a "bad estimate" from the standpoint of reliability theory. Now, let's assume that according to the theory of possibility - "The possibility and the corresponding necessity computed on the basis of formula (4) is defined as [0.85, 0.95]". Surely, this is a more adequate estimate. However, it is not a guaranteed estimate in terms of safety and reliability.

From axioms (2), (3) we obtain:

$$\Pi(A \cup B) \leq \Pi(A) + \Pi(B), N(A \cap B) \leq N(A) + N(B).$$

The probability measure P also satisfies this condition:

$$P\{A \cup B\} \leq P\{A\} + P\{B\} \tag{5}$$

Consequently, the difficulties related to elementary of events are reduced to some extent by using the so-called "focal elements", and the problem remains generally unresolved.

## 4   Analysis and Processing of Fuzzy Information

A fuzzy set and its quasi-probabilistic interpretation were considered above. However, we intend to differentiate between fuzzy information and probabilistic information. In this regard, we introduce axioms [11] and a measure that satisfies this axioms will be referred to as a fuzzy measure. We note that the proposed fuzzy measure differs from Sugeno fuzzy measure, Clement fuzzy measure and fuzzy-valued Hele measure [16].

Denote X a universal set and denote $A_n$ a set of its subsets, that is, $\forall n, A_n \subset X$. Assume that a measure $g : A_n \to L$ exists that satisfies the following conditions:

$1^0$. $g_{A_n} \geq 0$ for $\forall A_n \subset X$ – non-negativity axiom;
$2^0$. $g_{\bigcup_{n \geq 1} A_n} = \max_n g_{A_n} \leq 1$ – boundedness axiom;
$3^0$. $g_{\bar{A}_n} = 1 - g_{A_n}$ – complement axiom.

As it can be seen from $1^0$–$3^0$, $A_i \cap A_j = \emptyset, \forall i \neq j$ condition is not necessary for the considered fuzzy measure, and g may take values not only within [0, 1] (when $L = [0, 1]$), but also within $R$ (when $L = R$). The case of a non-numerical scale $L$ is of an interest (in this work we don't consider this case).

Analogously to the classical theory of measure, let us prove that fuzzy measure is a monotonic function, i.e. $g_{A_i} \leq g_{A_j}$ for any $A_i \subset A_j$, $g_{\emptyset} = 0$, if $\lim_{n \to \infty} A_n = A$, then $\lim_{n \to \infty} g_{A_n} = g[\lim_{n \to \infty} A_n] = g_A$ and, for any $A_n \subset X (n = 1, 2, \ldots)$, $g_{\bigcup_{n=1}^{\infty} A_n} \leq \sum_{n=1}^{\infty} g_{A_n}$.

Let us also show that all the basic notions of the fuzzy set theory [8, 11] can be derived from axioms $1^0$–$3^0$ (see below).

I.  Equality. If $A_i = A_j$, $\forall i \neq j$, then $g_{A_i} = g_{A_j}$ and vice-versa, if $g_{A_i} = g_{A_j}$, then $A_i = A_j$, $\forall (A_i, A_j) \subset X$, $i \neq j$.

**Proof.** Necessity. Let $A_i = A_j$, $\forall i \neq j$. We prove that $g_{A_i} = g_{A_j}$. From condition $A_i = A_j$, $\forall i \neq j$ it follows that $A_i = A_i \cup A_j$ and $A_j = A_i \cup A_j$. According to axiom $2^0$ one has: $g_{A_i} = g_{A_i \cup A_j} = \max_{i,j}[g_{A_i}, g_{A_j}]$, $g_{A_j} = g_{A_i \cup A_j} = \max_{i,j}[g_{A_i}, g_{A_j}]$. As the right hand sides are equal, then the left hand sides are also equal, i.e. $g_{A_i} = g_{A_j}$.

Sufficiency. Let $g_{A_i} = g_{A_j}$. As $g$ is a fuzzy measure, then $g_{A_i} = g_{A_j}$ can be expressed as: $g_{A_i} = g_{A_j} = \max_{i,j}[g_{A_i}, g_{A_j}] = g_{A_i \cup A_j}$ which implies that $A_i = A_i \cup A_j$, $A_j = A_i \cup A_j$. This is true if and only if $A_i = A_j$, $\forall i \neq j$. Note that probability measure does not satisfy this condition, i.e. $P(A) = P(B)$ does not imply that $A = B$.

II.  Boundedness. $1^0$ and $2^0$ imply $0 \leq g_{A_n} \leq 1$. Axiom $2^0$ can be generalized: $g_{\bigcup_{n \geq 1} A_n} = \max_n g_{A_n} \leq M$, where $M \geq 1$ - any bounded whole number. Then, [0, M] scale can be considered instead of [0, 1] and we have $0 \leq g_{A_n} \leq M$.

III.  The measure of empty set. Let us show that if $\emptyset$ is the empty set then $g_{\emptyset} = 0$. Indeed, let $A_n$ be some non-empty subset of $X$. It is clear that $A_n = A_n \cup \emptyset$. Therefore, axiom $2^0$ implies $g_{A_n} = g_{A_n \cup \emptyset} = \max_n[g_{A_n}, g_{\emptyset}]$. However, this condition holds under axiom $1^0$ if and only if $g_{A_n} \geq g_{\emptyset}$. Two cases are possible: (1) $g_{A_n} = g_{\emptyset}$; (2) $g_{A_n} > g_{\emptyset}$.
According to property I and the Zadeh's definition, the $1^{st}$ case is equivalent to $A_n = \emptyset$. Thus, we get a contradiction. The $2^{nd}$ case implies $g_{\emptyset} = 0$ because $A_n$ is non-empty. Note, that $g_{\emptyset} = 0$ follows from the definition of the empty set.

IV.  Axiom of comparison (the inclusion concept). Let $A \subseteq B$. Then $B$ can be expressed as $B = A \cup C$, where $C$ is a common difference, i.e. $C = B \backslash A$. Taking into account axiom $2^0$ one has $g_A \leq g_{A \cup C} = g_B$. Thus, $g_A \leq g_B$.

V.  Intersection. Let us prove that given $\bigcap_{n \geq 1} A_n \neq \emptyset$ one has $g_{\bigcap_{n \geq 1} A_n} = \min_n g_{A_n}$.
According to De Morgan law (it also holds for fuzzy sets) $\overline{A \cap B} = \bar{A} \cup \bar{B}$. Then, $g_{\overline{A \cap B}} = g_{\bar{A} \cup \bar{B}}$. By applying axiom $2^0$, we get for the right hand side: $g_{\bar{A} \cup \bar{B}} = \max_{n=2}(g_{\bar{A}}, g_{\bar{B}})$. From axiom $3^0$ it follows that $g_{\overline{A \cup B}} = 1 - g_{A \cap B}$. Thus, we get $1 - g_{A \cap B} = \max_{n=2}(1 - g_A, 1 - g_B)$. This implies that the latter condition is true if and only if $g_{A \cap B} = \min_{n=2}(g_A, g_B)$ holds. The proof for $n > 2$ case is analogous. This important property significantly facilitates solving of various practical problems of reliability of complex control systems.

VI.  Limit of sequence. Let $A$ be a limit of a sequence of fuzzy sets $A_n$, i.e. $\lim_{n \to \infty} A_n = A$.    Then    $\lim_{n \to \infty} g_{A_n} = g_A$    and $g_A = \max_{A_n \subset X} \min_{\substack{A_k \subset X \\ k \geq n}} g_{A_k} = \min_{A_n \subset X} \max_{\substack{A_k \subset X \\ k \geq n}} g_{A_k}$. The proof for this result follows from the definition of a limit of sequence of fuzzy sets given below.

**Definition.** A fuzzy set $A$ is referred to as limit of sequence of fuzzy sets $A_n$, if $A = \bigcup_{n=1}^{\infty} \left( \bigcap_{k=n}^{\infty} A_k \right) = \bigcap_{n=1}^{\infty} \left( \bigcup_{k=n}^{\infty} A_k \right)$ holds.

According to the definition: $\lim_{n \to \infty} g_{A_n} = g_{\lim_{n \to \infty} A_n} = g_{\bigcup_{n=1}^{\infty} \left( \bigcap_{k=n}^{\infty} A_k \right)} = g_{\bigcap_{n=1}^{\infty} \left( \bigcup_{k=n}^{\infty} A_k \right)}$.

Taking into account axiom $2^0$ and properties I, V we get:

$$g_{\bigcup_{n=1}^{\infty} \left( \bigcap_{k=n}^{\infty} A_k \right)} = \max_{A_n \subset X} g_{\bigcap_{n=1}^{\infty} A_n} = \min_{A_n \subset X} \max_{\substack{A_k \subset X \\ k \geq n}} g_{A_k},$$

$$g_{\bigcap_{n=1}^{\infty} \left( \bigcup_{k=n}^{\infty} A_k \right)} = \min_{A_n \subset X} g_{\left( \bigcup_{n=1}^{\infty} \left( \bigcup_{k=n}^{\infty} A_k \right) \right)} = \min_{A_n \subset X} \max_{\substack{A_k \subset X \\ k \geq n}} g_{A_k}.$$

This completes the proof.

Let us prove that axiom $3^0$ can be omitted.

**Assertion.** If function $g_{A_n}$ satisfies axioms $1^0$, $2^0$ and $\exists A_k$ (among $A_n$), such that $\max_n g_{A_n} = g_{A_k} = 1$, then condition $3^0$ is true for complement of $A_n$ and vice-versa.

The proof follows from the equalities: $A \cup \bar{A} = E$, $A \cap \bar{A} = \emptyset$.

The proposed concept of fuzzy measure allows to formulate theorem of membership function representation [11].

**Theorem.** $g(\cdot)$ is a membership function if and only if it satisfies axioms $1^0$, $2^0$, $g_A : X \to L$ and $g_A$ is acceptable for experts.

Note that the last condition of the theorem is interactive and its mathematical proof is not possible. In essence, membership function is of a vectorial character, that is, in problems of reliability and security of information systems, any fuzzy set has its sense only as a collection its elements and the related membership degrees. Individual elements are of no sense. In contrast to the probability theory, introducing of additional elements is not effective and does not reduce an entropy. However, when construction of a strong mathematical model for an object is not possible, one can adopt a justified and an approximate operating model without a mathematical proof.

Examples of interactive conditions can be a curve of the aging process, a class of tall people etc. which has not been questioned and is unambiguously perceived by specialists.

Dubois and Prade defined information unit by the 4-tuple: object, sign, meaning, confidence. The latter is a reliability index of the information unit, and is formalized by fuzzy measure.

Fuzzy measures and integrals, the experimental definitions and the branches of Soft Computing, fuzzy modeling are described in monographs [2–5, 13–15].

## 5  Analysis and Processing of Biased Information

Let us consider processing of information characterized by fuzziness and randomness. There are two possible cases: (1) the event A is fuzzy and is defined in terms of a fuzzy measure $g(A)$, and its occurrence is random and is described by probability measure P; (2) the event A is random but its probability $P_A$ is fuzzy and is defined by using of possibility and necessity measures. These two cases are defined as [6, 10]:

1. $P_A = \int g_A(\cdot)dP$
2. $N(A) \leq P_A \leq \Pi(A)$.

## 6  Conclusion

Existing methods and methodologies for analyzing and processing information show that the nature of the information itself ("extracted", "inferred", etc.) from the point of view of human perception is very diverse, has a special significance and specificity in the field of decision-making under imprecision, fuzziness and uncertainty of the initial information. In most cases, instead of developing a mathematical model of forecasting, one has to deal with the prediction of the future state of the economic system (process) based on subjective judgment and knowledge, which in turn include large numbers of qualitative and quantitative facts. Here the information together with the consciousness act as a "certain memory" according to a certain mechanism (for example: fractals, trends, long-term memory, etc.) and due to the "hidden medium" (unknown for us) presents us with economic surprises. If the information about the structure and functional purpose of an object in a broad sense was not preserved and would not be transmitted, all the usual irrevocable would be violated, and there would be no economic growth. The predetermining role of information in the life of the economic society has been proven by states with advanced technology - bringing many billions in revenue [1]. Therefore, the study of information as a separate scientific and philosophical category, the development of non-traditional methods of processing, the creation of fuzzy technologies alongside with the classical technologies is of great interest in micro and macroeconomic problems.

## References

1. Ajemoglu, D., Robinson, J.A.: Why some countries are rich, and others are poor. In: The Origin of Power, Prosperity and Poverty. AST, Moscow (2015)
2. Aliev, R.A., Aliev, R.R.: Soft Computing and its Application. World Scientific, New Jersey, London, Singapore, Hong Kong (2001)
3. Aliev, R.A., Bonfig, K.W., Aliev, F.T.: Soft Computing. Technik Verlag, Berlin (2000)
4. Bocharnikov, V.P.: Fuzzy technology: mathematical foundations. In: The Practice of Modeling in Economics. Nauka, RAN, Saint Petersburg (2001)
5. Diligensky, N.V., Dymova, L.G., Sevastyanov, P.V.: Fuzzy modeling and multicriteria optimization of production systems under uncertainty. Machine building, Samara (2004)

6. Dubois, D., Prade, A.: Theory of possibilities. In: Application to Knowledge Representation in Computer Science. Radio and Communication, Moscow (1990)
7. Duke, V., Samoylenko, A.: Data Mining. Peter, Saint Petersburg (2001)
8. Kiyasbeyli, S.A., Mamedov, V.M.: Differences between fuzzy set and probability theories. Soviet J. Autom. Inf. Sci. **20**(3), 60–62 (1988)
9. Kolmogorov, A.N., Fomin, S.V.: Elements of the Theory of Functions and Functional Analysis. Nauka, Moscow (1981)
10. Mamedov, V.M.: F-reliability model. Trans. NAS Azerbaijan **22**(2–3), 3–9 (2002)
11. Mamedov, V.M.: Development of methods for assessing reliability and parameters of technical operation of complex control systems in conditions of uncertainty of the initial information. Ph.D. thesis. Riga Technical University (1982)
12. Mamedov, V.M.: Fuzzy and soft measurements in reliability problems of complex systems. In: Proceedings of the International Conference an Soft Computing and Measurements, Saint Petersburg, vol. 2, pp. 16–18 (2003)
13. Piegat, A.: Fuzzy Modeling and Control. Physica-Verlag, New York (2001)
14. Pospelov, D.A. (ed.): Fuzzy Sets in Control Models and Artificial Intelligence. Nauka, Moscow (1986)
15. Terano, T., Asai, K., Sugeno, M. (eds.): Applied Fuzzy Systems, 1st edn. Mir, Moscow (1993)
16. Yager, R.R.: Fuzzy Set and Possibility Theory: Recent Developments. Pergamon, New York (1982)

# Labeled Fuzzy Rough Sets in Multiple-Criteria Decision-Making

Alicja Mieszkowicz-Rolka [ID] and Leszek Rolka[✉] [ID]

The Faculty of Mechanical Engineering and Aeronautics,
Rzeszów University of Technology, 35-959 Rzeszów, Poland
{alicjamr, leszekr}@prz.edu.pl

**Abstract.** This paper presents a hybrid approach to constructing a decision-making process, basing on the concept of labeled fuzzy rough sets. We introduce a modified definition of fuzzy rough approximations which are necessary, when an increased value of the similarity threshold is chosen in the determination of fuzzy linguistic labels. Labeled fuzzy rough sets are used in the first stage of decision-making process for analyzing the decision system of an expert, who recommends objects by respecting the guidelines of a decision-maker. In the second stage, the linguistic labels obtained from the expert are applied for determining a final ranking of objects, according to preferences which are imposed by the decision-maker on fuzzy attributes and their linguistic values. A short example of analysis helps to elucidate the presented method.

**Keywords:** Fuzzy rough sets · Linguistic labels · Decision-making

## 1 Introduction

The problem of finding an optimal solution with respect to a set of criteria is an important task in the area of decision-making. Many sophisticated algorithms, such as SAW, TOPSIS, AHP, ELECTRE [1–3], were proposed for determining a compromise solution in the case of objective numerical criteria. Very often, however, the process of decision-making has to be performed under uncertainty, when several subjective criteria are evaluated by a skilled operator or a group of experts. Hence, the issue of knowledge representation and analysis of the human reasoning process must be considered. Decision-making was studied by many researchers active in the area of soft computing, who presented various methods basing on the concepts defined in the fuzzy set theory, such as classical, intuitionistic, hesitant, type-2 fuzzy sets, and others, see, e.g., [4–7]. In order to deal with this kind of problem, we propose a combined method based on the fuzzy set and the rough set theories. Although the standard fuzzy rough set approach is advantageous in the processing of imperfect knowledge, it requires a complex computation of fuzzy similarity matrices, and can also produce results which are not always easy interpretable. This is why we introduced the concept of fuzzy linguistic labels [8, 9], a mean of constructing a more human-oriented, intuitive, and straightforward fuzzy rough set model.

More recently, we demonstrated [10] that the idea of fuzzy linguistic labels can also be used for modeling the process of multiple-criteria decision-making performed by a

© Springer Nature Switzerland AG 2019
R. A. Aliev et al. (Eds.): ICAFS-2018, AISC 896, pp. 73–81, 2019.
https://doi.org/10.1007/978-3-030-04164-9_13

group of experts and a decision-maker. The experts do not compare the objects from a universe to each other, but rather evaluate the similarity of a perceived object to a group of ideal objects which possess distinctive characteristic features. The knowledge of the experts can be expressed with the help of fuzzy linguistic labels that correspond to those ideal objects. The decision-maker can impose a preference order on the fuzzy attributes of objects and their linguistic values. The preference degrees of the decision-maker are used for obtaining the final ranking of objects. It was assumed that the experts do not recommend particular objects, but only provide their expertise pertaining to the membership of objects in the linguistic values of condition attributes. Since it is a method of group decision-making, it requires detection of contradictions between experts and aggregation of their knowledge.

In this paper, we consider another special case of a hierarchical decision-making process. Contrary to the former approach, we also use the recommendation of the expert which can be taken into account by the decision-maker for producing the final ranking of objects. Furthermore, instead of basing on a decision matrix of a group of experts, we use a decision table of a single expert, who evaluates the membership of objects in the linguistic values of the condition attributes and the decision attribute, by respecting the preferences of the decision-maker. By applying a suitable set of weighting coefficients, the decision-maker is able to control the influence of the expert's recommendation on the final ranking of objects.

## 2  Labeled Fuzzy Rough Sets

We should start with a formal description of the decision process of the expert by defining a fuzzy decision system FDS, as a 4-tuple

$$\text{FDS} = \langle U, A, \mathbb{V}, f \rangle, \tag{1}$$

where:

$U$  – denotes a nonempty set of elements (objects), called the universe,

$A$  – is a sum of two disjoint finite sets of fuzzy attributes: $A = C \cup D$,
     $C$ denotes the set of condition attributes, $D$ is the set of decision attributes,

$\mathbb{V}$  – is a set of fuzzy (linguistic) values of attributes, $\mathbb{V} = \bigcup_{a \in A} \mathbb{V}_a$,
     $\mathbb{V}_a$ is the set of linguistic values of an attribute $a \in A$,

$f$  – is an information function, $f : U \times \mathbb{V} \to [0, 1]$,
     $f(x, V) \in [0, 1]$, for all $x \in U$, and $V \in \mathbb{V}$

We denote by $\mathbb{A}_i = \{A_{i1}, A_{i2}, \ldots, A_{in_i}\}$ the family of linguistic values of the fuzzy attribute $a_i \in A$, where $i = 1, 2, \ldots, n$. The membership degree of every element $x$ of the universe $U$, in particular linguistic values of all fuzzy attributes, should be assigned by an expert. Because we want to generalize the properties of a crisp decision system, in which a unique value of every attribute is assigned to any element $x \in U$, we assume that the following requirements are always respected by the expert:

$$\exists\, A_{ik}\left(A_{ik} \in \mathbb{A}_i,\ \mu_{A_{ik}}(x) \geq 0.5\right), \tag{2}$$

$$\text{power}(\mathbb{A}_i(x)) = \sum\nolimits_{k=1}^{n_i} \mu_{A_{ik}}(x) = 1. \tag{3}$$

The crucial point of our fuzzy rough set approach consists in discovering subsets of characteristic elements of the universe that can be described by the same tuple of linguistic values of attributes, in other words, by assigning a common fuzzy linguistic label to a subset of similar objects.

In order to determine what fuzzy labels are activated in the reasoning process of the expert, we should consider those linguistic values of attributes in which a selected element $x \in U$ has a higher degree of membership. For expressing formally the level of domination of a particular linguistic value, we use a similarity threshold $\beta$ which satisfies the inequality: $0.5 < \beta \leq 1$.

Depending on the value of the threshold $\beta$, we are able to divide the linguistic values of fuzzy attributes into three different categories. For a given fuzzy decision system FDS of an expert, we define with respect to any element $x \in U$, and a fuzzy attribute $a \in A$, the set $\widehat{\mathbb{V}}_a(x) \subseteq \mathbb{V}_a$ of dominating linguistic values

$$\widehat{\mathbb{V}}_a(x) = \{V \in \mathbb{V}_a : f(x, V) \geq \beta\}, \tag{4}$$

the set $\overline{\overline{\mathbb{V}}}_a(x) \subseteq \mathbb{V}_a$ of boundary linguistic values

$$\overline{\overline{\mathbb{V}}}_a(x) = \{V \in \mathbb{V}_a : 0.5 \leq f(x, V) < \beta\}, \tag{5}$$

and the set $\check{\mathbb{V}}_a(x) \subseteq \mathbb{V}_a$ of negative linguistic values

$$\check{\mathbb{V}}_a(x) = \{V \in \mathbb{V}_a : 0 \leq f(x, V) < 0.5\}. \tag{6}$$

Observe that the set $\widehat{\mathbb{V}}_a(x)$ can have at most one element $(\text{card}(\widehat{\mathbb{V}}_a(x)) \leq 1)$, because of the requirement (2). It can become empty, when the threshold $\beta$ is set to a higher value. When the dominating linguistic values of all attributes are taken into account for an element $x \in U$, then we obtain a linguistic label of that element. Given a subset of fuzzy attributes $P \subseteq A$, we define the set of linguistic labels $\widehat{\mathbb{L}}^P(x)$ of an element $x \in U$, as the Cartesian product of the sets of dominating linguistic values $\widehat{\mathbb{V}}_p(x)$, for an attribute $p \in P$

$$\widehat{\mathbb{L}}^P(x) = \prod\nolimits_{p \in P} \widehat{\mathbb{V}}_p(x). \tag{7}$$

For every element $x \in U$, it does hold: $\text{card}(\widehat{\mathbb{L}}^P(x)) \leq 1$.

We omit $\beta$ for in denotation for the sake of brevity. Furthermore, we express by:

- $L^P(x)$, the linguistic label which exists for an element $x \in U$,
- $\mathbb{L}^P$, the set of linguistic labels for all elements $x \in U$.
- $L^P$, a selected linguistic label from the set $\mathbb{L}^P$.

The elements of the universe $U$ that have a common linguistic label $L^P(x) \in \mathbb{L}^P$ can be easy found in the expert's decision table.

A remark must be added, concerning the possibility of setting higher values of the threshold $\beta$ in practical applications. It is possible that some elements $x \in U$ do not have a dominating linguistic value for one or more attributes, when we increase $\beta$. In such a case, we should discard those elements from further considerations, because they are not sufficiently similar to all linguistic values of some attributes, hence they do not have any linguistic label. Moreover, we should also admit the possibility for the decision-maker to intentionally exclude some objects. Therefore, we will use a restricted universe $U' \subseteq U$ in the next definitions.

Let us denote by $X_{L^P}$ the subset of elements of the universe $U'$ that correspond to a linguistic label $L^P \in \mathbb{L}^P$, obtained for a subset of fuzzy attributes $P \subseteq A$

$$X_{L^P} = \left\{ x \in U' : \; L^P(x) = L^P \right\}. \tag{8}$$

We call $X_{L^P}(e)$ the set of characteristic elements of the linguistic label $L^P \in \mathbb{L}^P$.

Using the membership degree in a linguistic label $L^P(x) \in \mathbb{L}^P$, for all elements $x \in U'$ (with $\mathrm{card}(U') = N$), we get a fuzzy similarity class denoted by $\tilde{L}^P(x)$

$$\tilde{L}^P(x) = \left\{ \mu_{L^P(x)}(x_1)/x_1, \mu_{L^P(x)}(x_2)/x_2, \ldots, \mu_{L^P(x)}(x_N)/x_N \right\}. \tag{9}$$

The resulting membership degree of $x \in U'$ in the linguistic label $L^P(x) \in \mathbb{L}^P$ will be obtained by aggregating the membership degrees of dominating linguistic values for all attributes $p \in P$. This is done by using an aggregation operator agr

$$\mu_{L^P(x)}(x) = \mathrm{aggr}\left( \mu_{\hat{V}_{p_1}}(x), \mu_{\hat{V}_{p_2}}(x), \ldots, \mu_{\hat{V}_{p_{|P|}}}(x) \right). \tag{10}$$

A suitable aggregation operator can be based on a distance measure between the element $x \in U'$ and an ideal element which corresponds to the linguistic label $L^P(x)$. Depending on the metric selected for expressing the distance, we can obtain different operators, such as min or ave (arithmetic mean).

The lower and upper approximations of a set constitute basic notions of the rough set theory. They are usually applied for approximating the similarity classes obtained with respect to decision attributes by the similarity classes generated with respect to decision attributes. We give a general definition of approximation for any fuzzy set $F$ on the domain $U'$. To this end, we will use a set denoted by $X_F$, containing characteristic elements of $F$, which is defined as follows

$$X_F = \{x \in U' : \mu_F(x) \geq \beta\}. \tag{11}$$

Lower approximation $\underline{\mathbb{L}}^P(F)$ of a fuzzy $F$ by the set of linguistic labels $\mathbb{L}^P$, obtained with respect to a subset of fuzzy attributes $P \subseteq A$, is defined as

$$\underline{\mathbb{L}}^P(F) = \bigcup\nolimits_{L^P \in \mathbb{L}^P} \widetilde{L}^P : X_{L^P} \subseteq X_F \tag{12}$$

Upper approximation $\overline{\mathbb{L}^P}(F)$ of a fuzzy set $F$ by the set of linguistic labels $\mathbb{L}^P$, which are obtained with respect to a subset of fuzzy attributes $P \subseteq A$, is defined as

$$\overline{\mathbb{L}^P}(F) = \bigcup\nolimits_{L^P \in \mathbb{L}^P} \widetilde{L}^P : X_{L^P} \cap X_F \neq \emptyset \tag{13}$$

Before we calculate the final ranking of objects, an explicit specification of the preferences must be given by the decision-maker. First of all, we need a vector of weights $[w(a_1), w(a_2), \ldots, w(a_n)]$ for representing the preferences for each fuzzy attribute, which satisfies the standard requirement: $\sum_{i=1}^n w(a_i) = 1$.

Secondly, a set of preference degrees for the linguistic of all attributes must be specified. We denote by $\text{pref}(A_{ik})$ the preference degree of the linguistic value $A_{ik}$ of an attribute $a_i \in A$, where $i = 1, 2, \ldots, n$, and $k = 1, 2, \ldots, n_i$. The dominating linguistic value of the attribute $a_i \in A$, which appears in the tuple describing the linguistic label $L^P$ for an element $x \in U'$, is denoted by $L^P(x, a_i)$.

The last question is how to calculate the ranking of all elements $x \in U'$. Although there are many possibilities of defining a ranking function, we propose here a straightforward way of calculating the rank of a given element $x \in U'$

$$\text{rank}(x) = \mu_{L^P(x)}(x) \times \sum\nolimits_{i=1}^n w(a_i) \times \text{pref}(L^P(x, a_i)). \tag{14}$$

With the help of the ranking procedure, a set of objects is obtained that best fit the preference of the decision-maker, basing on the knowledge of the expert.

## 3 Example

*Stage I: Evaluation of the expert's model of reasoning.*

We begin with the first stage of a small decision-making example, by analyzing the decision table of a single expert (Table 1). There are three condition attributes $C = \{c_1, c_2, c_3\}$, and one decision attribute $d_1$. All attributes have three linguistic values. The expert determines the membership degree of particular elements of the universe $U$ to linguistic values of attributes. The condition attributes express subjective criteria evaluated by the expert. The decision attribute represents the expert's recommendation for selecting the objects, as membership degrees in the linguistic values: $D_{11}$ – "Low", $D_{12}$ – "Medium", $D_{13}$ – "High".

The expert should determine his or her recommendation by respecting the preferences specified by the decision-maker. Let us assume the similarity threshold $\beta = 0.65$.

**Table 1.** Decision table of the expert.

| | $c_1$ | | | $c_2$ | | | $c_3$ | | | $d_1$ | | |
|---|---|---|---|---|---|---|---|---|---|---|---|---|
| | $C_{11}$ | $C_{12}$ | $C_{13}$ | $C_{21}$ | $C_{22}$ | $C_{23}$ | $C_{31}$ | $C_{32}$ | $C_{33}$ | $D_{11}$ | $D_{12}$ | $D_{13}$ |
| $x_1$ | 0.20 | **0.80** | 0.00 | 0.00 | 0.00 | **1.00** | 0.00 | 0.20 | **0.80** | 0.00 | 0.10 | **0.90** |
| $x_2$ | 0.00 | 0.15 | **0.85** | 0.00 | 0.10 | **0.90** | 0.00 | **1.00** | 0.00 | 0.25 | **0.75** | 0.00 |
| $x_3$ | **0.90** | 0.10 | 0.00 | 0.10 | **0.90** | 0.00 | 0.10 | **0.90** | 0.00 | 0.15 | **0.85** | 0.00 |
| $x_4$ | 0.30 | **0.70** | 0.00 | 0.00 | 0.10 | **0.90** | 0.00 | 0.25 | **0.75** | 0.00 | 0.15 | **0.85** |
| $x_5$ | 0.00 | 0.15 | **0.85** | 0.30 | **0.70** | 0.00 | **0.90** | 0.10 | 0.00 | **1.00** | 0.00 | 0.00 |
| $x_6$ | **0.85** | 0.15 | 0.00 | 0.00 | **0.90** | 0.10 | 0.00 | **1.00** | 0.00 | 0.00 | **1.00** | 0.00 |
| $x_7$ | 0.10 | **0.90** | 0.00 | 0.00 | 0.20 | **0.80** | 0.00 | 0.00 | **1.00** | 0.00 | 0.10 | **0.90** |
| $x_8$ | 0.00 | 0.25 | **0.75** | 0.00 | 0.15 | **0.85** | 0.00 | **0.90** | 0.10 | 0.20 | **0.80** | 0.00 |

We get the following linguistic labels for the condition attributes with corresponding sets of characteristic elements:

$$L_1^C = (C_{12}, C_{23}, C_{33}), \quad X_{L_1^C} = \{x_1, x_4, x_7\},$$
$$L_2^C = (C_{13}, C_{23}, C_{32}), \quad X_{L_2^C} = \{x_2, x_8\},$$
$$L_3^C = (C_{11}, C_{22}, C_{32}), \quad X_{L_3^C} = \{x_3, x_6\},$$
$$L_4^C = (C_{13}, C_{22}, C_{31}), \quad X_{L_4^C} = \{x_5\}.$$

The family of linguistic labels obtained with respect to the decision attribute includes:

$$L_1^D = (D_{11}), \quad X_{L_1^D} = \{x_5\},$$
$$L_2^D = (D_{12}), \quad X_{L_2^D} = \{x_2, x_3, x_6, x_8\},$$
$$L_3^D = (D_{13}), \quad X_{L_3^D} = \{x_1, x_4, x_7\}.$$

Observe that $X_{L_4^C} \subseteq X_{L_1^D}$, $X_{L_2^C} \subseteq X_{L_2^D}$, $X_{L_3^C} \subseteq X_{L_2^D}$, and $X_{L_1^C} \subseteq X_{L_3^D}$, hence, the expert applies the following decision rules, which are certain:

$$R_1 : (C_{13}, C_{22}, C_{31}) \rightarrow (D_{11}), \qquad R_2 : (C_{13}, C_{23}, C_{32}) \rightarrow (D_{12}),$$
$$R_3 : (C_{11}, C_{22}, C_{32}) \rightarrow (D_{12}), \qquad R_4 : (C_{12}, C_{23}, C_{33}) \rightarrow (D_{13}).$$

It can be concluded that the reasoning process of our expert is consistent.

*Stage II: Ranking of objects.*

Case 1. We select only the condition attributes from the expert's decision system. The resulting membership degree in the linguistic labels $L_1^C$, $L_2^C$, $L_3^C$, and $L_4^C$ is determined according to formula (10), for all elements $x \in U$. The aggregation operators min, and ave (arithmetic mean) are used, so, we obtain two sets of results (Table 2).

Preference degrees of linguistic values for particular condition attributes, given in Table 3, are provided by the decision-maker. The most preferable linguistic values (for

**Table 2.** Membership degree of elements $x \in U$ in the linguistic labels for condition attributes.

|       | $\mu_{L^c(x)}(x)$, for aggr = min | $\mu_{L^c(x)}(x)$, for aggr = ave |
|-------|------|-------|
| $x_1$ | 0.80 | 0.867 |
| $x_2$ | 0.85 | 0.917 |
| $x_3$ | 0.90 | 0.900 |
| $x_4$ | 0.70 | 0.783 |
| $x_5$ | 0.70 | 0.817 |
| $x_6$ | 0.85 | 0.917 |
| $x_7$ | 0.80 | 0.900 |
| $x_8$ | 0.75 | 0.833 |

**Table 3.** Preference degree of the linguistic values of condition attributes.

|            | $c_1$ | | | $c_2$ | | | $c_3$ | | |
|------------|----------|----------|----------|----------|----------|----------|----------|----------|----------|
| $C_{ik}$   | $C_{11}$ | $C_{12}$ | $C_{13}$ | $C_{21}$ | $C_{22}$ | $C_{23}$ | $C_{31}$ | $C_{32}$ | $C_{33}$ |
| pref($C_{ik}$) | 1.0 | 0.5 | 0.25 | 0.25 | 0.5 | 1.0 | 0.25 | 0.5 | 1.0 |

example $C_{11}$) have an assigned preference degree equal to 1.0. The decision-maker also chooses the preference weights for all condition attributes. The vector of weights is equal to $[0.25\ 0.35\ 0.40\ 0.0]$ in our case. As we can see, the attribute $c_3$ is the most important one for the decision-maker.

According to formula (14), we calculate value of the rank function and determine the ranking order for every element $x \in U$. The results are given in Table 4.

**Table 4.** Ranking obtained by the decision maker for condition attributes.

|       | aggr = min | | aggr = ave | |
|-------|--------------|----------|--------------|----------|
|       | rank($x$) | order($x$) | rank($x$) | order($x$) |
| $x_1$ | 0.7000 | 1 | 0.7583 | 2 |
| $x_2$ | 0.5206 | 6 | 0.5615 | 6 |
| $x_3$ | 0.5625 | 4 | 0.5625 | 5 |
| $x_4$ | 0.6125 | 3 | 0.6854 | 3 |
| $x_5$ | 0.2363 | 8 | 0.2756 | 8 |
| $x_6$ | 0.5313 | 5 | 0.5729 | 4 |
| $x_7$ | 0.7000 | 1 | 0.7875 | 1 |
| $x_8$ | 0.4594 | 7 | 0.5104 | 7 |

Case 2. We consider the condition as well the decision attributes. In this way the decision-maker also takes into account the recommendation of objects proposed by the expert. For the decision attribute $d_1$, we assume the same preference degree of the linguistic values as for the condition attributes $c_2$, $c_3$. The vector of weights for all attributes is equal to $[0.2\ 0.28\ 0.32\ 0.2]$. The final results are presented in Table 5.

**Table 5.** Ranking obtained by the decision maker for all attributes.

|        | aggr = min |          | aggr = ave |          |
|--------|------------|----------|------------|----------|
|        | rank($x$)  | order($x$) | rank($x$) | order($x$) |
| $x_1$  | 0.7200     | 1        | 0.7875     | 2        |
| $x_2$  | 0.4425     | 6        | 0.5163     | 6        |
| $x_3$  | 0.5100     | 4        | 0.5325     | 5        |
| $x_4$  | 0.6300     | 3        | 0.7200     | 3        |
| $x_5$  | 0.2240     | 8        | 0.2760     | 8        |
| $x_6$  | 0.5100     | 4        | 0.5625     | 4        |
| $x_7$  | 0.7200     | 1        | 0.8100     | 1        |
| $x_8$  | 0.4425     | 6        | 0.4868     | 7        |

Case 3. If the decision-maker sets the weights as $[0.0\ 0.0\ 0.0\ 1.0]$, the final ranking is equal to the ranking proposed by the expert.

The results of ranking obtained in all three cases are very similar. This is not surprising, because the expert's decision table was consistent. The objects $x_7$, $x_1$ turn out to be the best alternatives. Aggregation with the operator min produces the same results for these objects, due to a coarsening working of this operator. By using the operator ave, a refined unambiguous ranking can be obtained.

## 4   Conclusions

Labeled fuzzy rough sets can be included in multi-criteria decision-making, especially in the case of large information system, when subjective criteria play an important role. Different goals of the decision-maker can be achieved by setting a suitable preference order on attributes and their linguistic values. The final ranking is always based on the expert's assignment of objects to linguistic labels, obtained with respect to condition attributes, but the recommendation of the expert can be also taken into account. In the future work, the labeled fuzzy rough set approach will be combined with different types of fuzzy set and rough set concepts.

## References

1. Chou, S., Chang, Y., Shen, C.: A fuzzy simple additive weighting system under group decision-making for facility location selection with objective/subjective attributes. Eur. J. Oper. Res. **189**(1), 132–145 (2008)
2. Chen, C.: Extensions of the TOPSIS for group decision making under fuzzy environment. Fuzzy Sets Syst. **114**(1), 1–9 (2000)
3. Deni, W., Sudana, O., Sasmita, A.: Analysis and implementation fuzzy multi-attribute decision making SAW method for selection of high achieving students in faculty level. Int. J. Comput. Sci. **10**(1), 674–680 (2013)
4. Pedrycz, W., Ekel, P., Parreiras, R.: Fuzzy Multicriteria Decision-Making: Models, Methods and Applications. Wiley, Chichester (2010)

5. Szmidt, E., Kacprzyk, J.: Intuitionistic fuzzy sets in group decision making. Control Cybern. **31**(4), 1037–1053 (2002)
6. Atanassov, K., Pasi, G., Yager, R.: Intuitionistic fuzzy interpretations of multi-criteria multi-person and multi-measurement tool decision making. Int. J. Syst. Sci. **36**(14), 859–868 (2005)
7. Kabak, Ö., Ervural, B.: Multiple attribute group decision making: a generic conceptual framework and a classification scheme. Knowl.-Based Syst. **123**, 13–30 (2017)
8. Mieszkowicz-Rolka, A., Rolka, L.: A novel approach to fuzzy rough set-based analysis of information systems. In: Wilimowska, Z., et al. (eds.) Information Systems Architecture and Technology. Advances in Intelligent Systems and Computing, vol. 432, pp. 173–183. Springer International Publishing, Switzerland (2016)
9. Mieszkowicz-Rolka, A., Rolka, L.: Labeled fuzzy rough sets versus fuzzy flow graphs. In: Proceedings of the 8th International Joint Conference on Computational Intelligence (IJCCI 2016, FCTA, vol. 2, pp. 115–120. SCITEPRESS – Science and Technology Publications, Lda (2016)
10. Mieszkowicz-Rolka, A., Rolka, L.: Fuzzy linguistic labels in multi-expert decision making. In: Martin-Vide, C., et al. (eds.) Theory and Practice of Natural Computing, LNCS, vol. 10687, pp. 126–136. Springer, Heidelberg (2017)

# Comparing Image Distortion of LSB

Yucel Inan$^{(\boxtimes)}$

Near East University, Nicosia, TRNC, Mersin 10, Turkey
yucel.inan@neu.edu.tr

**Abstract.** Steganography plays an important role in computer science. With the development of technology, steganography technique is used for safe transmission of data in digital media. This technique hides the data in a symbol to prevent its comprehensibility and its solvability, so that it can be safely delivered to the receiver. Steganagrofiq method LSB was used in this study. The gray scaled jpg format was analyzed in the same images, visually unaware, that the data could be stored using the LSB method up to the 4th bit level. It worked on five same jpg but different pixel format images in $1024 \times 1024$, $512 \times 512$, $256 \times 256$, $128 \times 128$, $64 \times 64$ pixel sizes. From a visible standpoint, it seems logical in every five views. Histograms were plotted to visualize the differences between the original and the encoded. The rates of different images of the same images were calculated by embedding data, extracting data, comparing the byte and elapsed time, and calculating the distortion rates PSNR, SNR, MSE in the images.

**Keywords:** Image processing · Steganography · LSB method
Cover image · Hiding secret message · Stego-image · Extract cover image
PSNR · SNR · MSE

## 1 Introduction

Steganography, an important subdiscipline of information hiding, can be described as the concealment of a data within an object [1]. Johannes Trithemius is a German priest who lived between 1462–1516. Steganography is derived from a book called 'Steganographia' in his book on cryptology, astrology and number strings, and is derived from the Greek alphabet, which is called Steganography word roots "στεγανό-ς, γραφ-ειν". It is literally meaning "hidden writing", "cover writing" [2]. The goal in steganography is to hide the existence of the message. Techniques such as invisible ink usage, microdotting, text editing and encryption are used for this. As it can be understood from these explanations, steganography, whether numerical or not, is to ensure that any plain text with confidentiality, which is required to reach the target, is transmitted and received in a secure manner to the recipient via the media in such a way that third persons do not know [3]. Steganography has been widely used in various forms, especially military, diplomatic, personal and copyright applications, from ancient civilizations to this time. In short, steganography can be defined as the whole set of operations that are applied to conceal a message within an object that an observer can not notice, [4]. Steganography is not really an alternative to cryptography, but its complement. In cryptography, third parties know the existence of a secret

© Springer Nature Switzerland AG 2019
R. A. Aliev et al. (Eds.): ICAFS-2018, AISC 896, pp. 82–90, 2019.
https://doi.org/10.1007/978-3-030-04164-9_14

communication. However, the content of the confidential communication known to the sender and the recipient is that third parties can not resolve the content of the confidential communication. However, Steganography retains the existence of secret communication [5]. Steganography can be classified, Linguistics Steganography and Technical Steganography. Linguistics is steganography, the steganography of the carrier's text. Technical Steganography is involved in many aspects. Computer based methods: Data can be hidden by using text, audio, video, image picture files [6].

## 2 Structure of Steganography Technique

The structure of this technique creates a stego using a concealment function. The concealment function has three components: a cover for the hidden message and a secret message to be embedded, and a key code [7]. Below is a block diagram of the process. The steganography process shows the process of obtaining a secret image before and after embedding another secret image in the image. This structure of the design system is illustrated in Fig. 1. The resulting original image is first converted to a grayscale image and then scaled to $1024 \times 1024$, $512 \times 512$, $256 \times 256$, $128 \times 128$, $64 \times 64$ dimensions. In the grayscale Jpg pictures, data retention was tested with the LSB method, visually insignificant, up to the least significant 4.bit level. In general, the system consists of an algorithm that generates the key, encodes the message and decodes the message. Here the key is the extra secret key needed for the embedding opening operations which must be known by the sender and receiver. Steganography also has various methods of storing confidential information in the image [8]:

- Last Significant Bit Insertion (LSB)
- Masking and filtering
- Algorithms and transformations

In this study, the least significant bit insertion (LSB) method is examined.

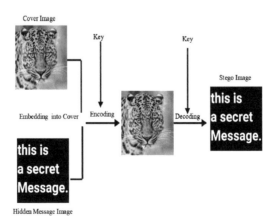

**Fig. 1.** Block diagram of the system.

## 3   Methodology

In the experiment, From the Web image was obtained Photo on Wallpapers Craft $1024 \times 1024^1$ was in database subset. This single image was digitized in 5 different sizes. Each image in a different pixel size was used for data storage by the LSB method, up to the least significant 4.bit plane. The system has been tested to see that it can be reasonably good to embed confidential data in each cover image at a different pixel size. These 5 different pixel size images were initially read using MATLAB in RGB format. RGB images were then converted to a grayscale image to reduce the computational cost and simplify the data embedding extraction process. These grayscale images were resized to fit the LSB method. It was used to examine the $1024 \times 1024$, $512 \times 512$, $256 \times 256$, $128 \times 128$, $64 \times 64$ pixel dimensions, which are appropriate in terms of cost of the calculation and which maintain adequate definition of image properties. In the next step the images were tested on 5 samples from each of the different pixel sizes of the images for the LSB. In each of these different sized images, a secret message was embedded and tested separately to calculate elapsed time and distortion rates in the images. The image shown in Fig. 2 below has been worked on JPG image in 5 different pixel sizes using the images.

**Fig. 2.** The image samples are increasing in size from left to right, the $1024 \times 1024$ image size can not be shown among the visuals you see because it covers a large area.

## 4   Image Steganography

5 JPG images of different pixel sizes of the same image have been worked on. JPG is a compression format that contains the true color value of large images. It is among the lossy formats due to compression by discarding details that are not very necessary for image perception. The missing details can not be brought in any way. Ideal for images that use 256 or more colors [9]. Figure 3 below shows the histograms for the original image and Stego object encrypted file. The histogram corresponds to the number of gray level pixels corresponding to each gray level in the image. It is a graph showing the number of each color value in a digital image. Based on this graph, histogram

---

[1] https://wallpaperscraft.com/download/leopard_color_spotted_predator_big_cat_53087/1024x1024.

analysis can be done about the state of the brightness or tones. The histograms of the image files can be seen visually, so that every 5 original covers and histograms can be seen in the stego image. Tables 1 and 2 also examine the hidden message JPG image file by comparing the changing values of the embedded file to the 5 different pixel JPG cover image objects.

Cover and Histogram of Transmission Stego Image, effect of changing the LSB's up fourth bit-plane

| Cover image1 64x64 | Cover image2 128x128 | Cover image3 256x256 | Cover image3 512x512 | Cover image3 1024x1024 |

**Fig. 3.** The histogram images of the original cover and the image of Stego embedded in the confidential data.

In Table 1, the file sizes, original cover file, byte differences of the embedded image files. There is no noticeable symbol on the stego image in this test. But there are hints that this view has changed.

**Table 1.** File sizes, original cover file, byte differences of embedded image files and embedding and elapsed times are examined.

| Image no | Image size | Size of cover (bytes) | Size of secret image (bytes) | Size of 4th bit stego (bytes) | Elapsed time is embed |
|---|---|---|---|---|---|
| Image 1 | 64 × 64 | 1.8 KiB (1,829 bytes) | 5.4 KiB (5,488 bytes) | 1.8 KiB (1,820 bytes) | 9.199033 s |
| Image 2 | 128 × 128 | 5.3 KiB (5,446 bytes) | 10.5 KiB (10,712 bytes) | 5.3 KiB (5,411 bytes) | 9.112614 s |
| Image 3 | 256 × 256 | 15.8 KiB (16,187 bytes) | 23.3 KiB (23,819 bytes) | 15.7 KiB (16,098 bytes) | 9.989000 s |
| Image 4 | 512 × 512 | 46.6 KiB (47,745 bytes) | 54.0 KiB (55,314 bytes) | 46.5 KiB (47,581 bytes) | 11.426355 s |
| Image 5 | 1024 × 1024 | 120.4 KiB (123,312 bytes) | 153.1 KiB (156,774 bytes) | 124.2 KiB (127,210 bytes) | 9.976279 s |

The byte size between the embedded image and the file sizes appears to be reduced. In fact, there is a loss of squeeze here. The rate of deterioration in lossy compression is important. When confrontation is made, the element that needs to be tested is not how it is compressed, but how it is quality compressed. The file size increases with the embedded image for Image 5 byte differences are 3,898. However, there is no noticeable symbol on the stego image. Image1, 2, 3 successful, quality has become a matter of compression. Although image 5 has a large file size, embedding time has been successful. In addition, there are variations within the histogram.

Table 2 shows the file sizes, original cover file, byte differences of embedded image files, This shows that the view changes. File size increased with embedded image. Although image 5 has a large file size, embedding time has been successful.

**Table 2.** File sizes, original cover file, byte differences of embedded files and hidden data extract and elapsed times are shown.

| Image no | Image size | Size of cover (bytes) | Size of secret image (bytes) | Size of 8th bit stego (bytes) | Elapsed time is extracted |
|---|---|---|---|---|---|
| Image 1 | 64 × 64 | 1.8 KiB (1,829 bytes) | 5.4 KiB (5,488 bytes) | 2.0 KiB (2,013 bytes) | 9.199033 s |
| Image 2 | 128 × 128 | 5.3 KiB (5,446 bytes) | 10.5 KiB (10,712 bytes) | 5.9 KiB (6,022 bytes) | 11.908559 s |
| Image 3 | 256 × 256 | 15.8 KiB (16,187bytes) | 23.3 KiB (23,819 bytes) | 17.5 KiB (17,958 bytes) | 12.247377 s |
| image 4 | 512 × 512 | 46.6 KiB (47,745 bytes) | 54.0 KiB (55,314 bytes) | 49.1 KiB (50,307 bytes) | 12.043904 s |
| image 5 | 1024 × 1024 | 120.4 KiB (123,312 bytes) | 153.1 KiB (156,774 bytes) | 128.3 KiB (131,417 bytes) | 12.880943 s |

## 5   Image Quality Assessment Methods

The changes made on the digital images are distorted afterwards. These distortions lead to visual differences in the picture. In steganographic applications, after the data is hidden, if it is too small in the picture, the distortions will change. Even though these changes are not visible to the human eye, they can be detected by analysis on digital media. There are basically two different approaches to measuring quality. The human vision system HSV (Human Vision System) is used to distinguish between original and distorted images [10]. Differences are revealed through this vision system. In the second approach, structural degradation is based on quality measurements. There are many algorithms in the literature to detect such changes. Applying the most commonly used methods in our studies will help to find out how healthy the image is against attack. Among these are the most known; MSE, SNR, PSNR [6].

### 5.1   Mean Squared Error (MSE)

The error is the average of the squares of the sum. The MSE is usually denoted as $\sigma^2$. Root Mean Squared Error (RMSE) is the square root of the MSE [3] (Figs. 4 and 5).

$$\sigma^2 = \frac{1}{N} \sum_{n=1}^{N} (x_n + y_n)^2 \tag{1}$$

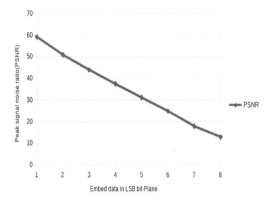

**Fig. 4.** $256 \times 256$ image size embedding data PSNR-based comparison from 1.bit plane to 8.bit plane.

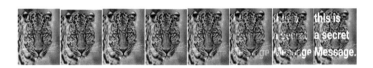

**Fig. 5.** On the comparison cover images, embedded secret message bits, from left to right, showed distortions of the images ranging from 1.bit plane to 8.bit plane.

### 5.2    PSNR (Peak Signal Noise Ratio)

PSNR is used to reveal similarities between images in digital images. After the information is embedded, the deterioration of the stego object is detected by the PSNR. The equation used for this calculation is given below. The SNR criterion is an estimate of the quality of the reconstructed image when compared to the original image. The PSNR is defined as [5] (Table 3):

$$\mathrm{PSNR(db)} = 10\log_{10}\frac{(x)^2_{\text{peak}}}{2\sigma^2} \qquad (2)$$

By examining the results in Table 4, the observations for the data are as follows: Cover images and similar areas of the image with confidential information, error measurements MSE, image quality indicator PSNR, SNR, and LSBs ratio between 1. bit plane to msb 4.bit plane can be accepted. Here, images of different size of 5 pixels were used to hide images with the same confidential information. The resulting stego images were analyzed by comparing MSE, SNR, PSNR with the original cover images.

Lsb 4. bit plane MSE, PSNR, SNR, measured values are given in the Fig. 6. Experimental study image dimensions were used in the blocks given in the Table 4. Error measurements show an increase in image5 in the MSE 4.bit plane. Image 1, image quality At the 4.bit level, PSNR improvement is observed to be greater than in other image images.

**Table 3.** PSNR, SNR, MSE, calculation values from 1.bit plane to 8.bit plane of 256x256 pixels size

| Cover image | Secret image | $1^{th}$ to $8^{th}$-bit plane LSB | MSE | PSNR | SNR |
|---|---|---|---|---|---|
| Cover Image3 with dimension $256 \times 256$ | Secret image with dimension $256 \times 256$ | 1.bit plane | 0.080 | 59.109 | 53.421 |
| | | 2.bit plane | 0.549 | 50.734 | 45.046 |
| | | 3.bit plane | 2.703 | 43.812 | 38.125 |
| | | 4.bit plane | 11.900 | 37.375 | 31.687 |
| | | 5.bit plane | 50.131 | 31.130 | 25.442 |
| | | 6.bit plane | 216.292 | 24.780 | 19.093 |
| | | 7.bit plane | 1064.714 | 17.858 | 12.171 |
| | | 8.bit plane | 3372.168 | 12.852 | 7.164 |

**Table 4.** PSNR, SNR, MSE-based comparison of the same image with different size cover image and hidden data embedded image dimensions.

| Image no | Cover image | Secret image | 4-bit plane LSB | MSE | PSNR | SNR |
|---|---|---|---|---|---|---|
| Image 1 | $64 \times 64$ | $64 \times 64$ | 4.bit plane | 10.909 | 37.753 | 32.066 |
| Image 2 | $128 \times 128$ | $128 \times 128$ | 4.bit plane | 11.572 | 37.497 | 31.812 |
| Image 3 | $256 \times 256$ | $256 \times 256$ | 4.bit plane | 11.900 | 37.375 | 31.687 |
| Image 4 | $512 \times 512$ | $512 \times 512$ | 4.bit plane | 12.124 | 37.294 | 31.605 |
| Image 5 | $1024 \times 12024$ | $1024 \times 1024$ | 4.bit plane | 12.125 | 37.294 | 31.604 |

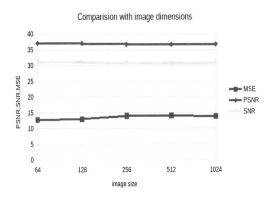

**Fig. 6.** Comparison chart of PSNR, SNR, MSE quality criterion values of 4.bit plane LSB method in different dimension of the same images.

When the SNR criterion is compared with the original image, an estimate of the image 1 reconstructed image quality between the other images has been examined (Fig. 7).

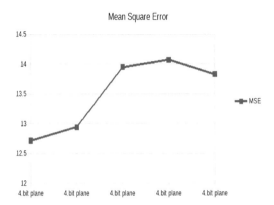

**Fig. 7.** For the LSB method, the MSE error criterion graph in the 4th bit plane of the different pixel image sizes 1024 × 1024, 512 × 512, 256 × 256, 128 × 128, 64 × 64.

## 6 Conclusions

The study was coded in the MATLAB programming language and tested on 5 different pixel size images of an image. By using the LSB method, the images were tested on the individual dimensions up to the 4th bit plane without visual difference, and the results were compared and the tables were also presented. Images of different sizes in the analysis made In the least significant bit method (LSB), similarity quality PSNR, error criterion MSE, reconstructed image quality SNR criterion values between the original data and the image containing the confidential data were examined. In addition, in this test, it was seen that the amount of distortion on the images increased in the data embedding and acquisition analysis from the 5.bit plane to the 8.bit plane.

## References

1. Haveliya, A.: A new approach for secret concealing in executable file. Int. J. Eng. Res. Appl. **2**(2), 1672–1674 (2012)
2. Agarval, M.: Text steganographic approaches: a comparison. Int. J. Netw. Secur. Its Appl. **5**(1) (2013)
3. Rosziati, I., Kuan, T.S.: Steganography algorithm to hide secret message inside an image. Comput. Technol. Appl. **2**, 102–108 (2011)
4. Seetha, D., Eswaran, P.: A study on steganography to hide secret message inside an image. Int. J. P2P Netw. Trends Technol. **3**(5), 33–37 (2013)

 5. Khan, M., Jamil, A., Farman, H., Zubair, M.: A novel image steganographic approach for hiding text in color images using HSI color model. Middle-East J. Sci. Res. **22**(5), 647–654 (2014)
 6. Rani, N., Chaudhary, J.: Text steganography techniques: a review. Int. J. Eng. Trends Technol. **4**(7), 3013–3015 (2013)
 7. Westfeld, A., Pfitzmann, A.: Attacks on steganographic systems. In: Pfitzmann, A. (ed.) Third International Workshop IH 1999, vol. 1768, pp. 61–76. Springer, Heidelberg (2000)
 8. Tiwari, R. K., Sahoo G.: Some new methodologies for image hiding using steganographic techniques. http://arxiv.org/abs/1211.0377. Accessed 13 Sept 2018
 9. Higgins, D.: Differences between file format. https://www.photoup.net/differences-between-fileformats-raw-dng-tiff-gif-png-jpeg/. Accessed 13 Sept 2018
10. Manjula, G.R., Danti, A.: Novel hash based least significant bit (2-3-3) image steganography in spatial domain. Int. J. Eng. Trends Technol. **4**(1), 11–20 (2015)

# An Effective Fuzzy Controlled Filter for Feature Extraction Method

Mohamad Alshahadat[✉], Bülent Bilgehan, and Cemal Kavalcıoğlu

Department of Electrical and Electronic Engineering, Faculty of Engineering,
Near East University, 99138 Nicosia, TRNC, Mersin 10, Turkey
{mohamad.alshahadat, bulent.bilgehan,
cemal.kavalcioglu}@neu.edu.tr

**Abstract.** Atomization of agricultural tasks such as disease removal is increasingly growing in European countries and thus accurate techniques are significantly required for efficient use of chemicals e.g. pesticides. In the present study, a computer vision-based technique is proposed which can be used for site specific spread of anti-fungal chemicals on strawberry leaves which alleviates yield's quality and quantity. The proposed technique mainly constitutes a band-pass filter for fungi-infection localization. The merit of this research work is taking into account human perception of fungi visual aspects to lower the computational load and ease the deploying technique on single chip processor for real-time application.

**Keywords:** Computer vision · Cypriot/mediterranean strawberry
Fungi-infection · Band-pass filter · Filter coefficients

## 1 Introduction

Cyprus is a country where more than 60% of the population relies on agriculture. Agriculture is the backbone of Cyprus economy. Strawberry farming contributes about 20% to overall agriculture income. Over 40% of Cyprus land area is cultivable. Chemical pesticides are commonly used to cure the infected strawberry diseases. This is a disadvantage because excessive use of pesticides for curing plant disease treatment increases costs and raises the danger of remaining toxicants on agricultural crops. The first check of the expertise would be to observe with the naked eye. However, such procedure requires continuous monitoring that leads to increased levels of cost. This can even be more costly when a remote located farm is considered. The farmer may need to travel to long distances to get in touch with an expert. The consultation fee may be too expensive for such applications. Considering a large field of a strawberry field, it would be a wise idea to introduce an automatic detection of plant diseases. The process is based on the machine vision that is to provide image based automatic inspection, process control and robot guidance. Comparatively, visual identification is labor intensive, less accurate and can be done only in small areas. The skill of detecting disease in earlier stage is very vital so that timely cure and control of such disease leads to decreasing dissatisfactory solutions. Hence, to overcome such type of traditional agriculture highly tech equipment's should be used. The general interest is to detect and

© Springer Nature Switzerland AG 2019
R. A. Aliev et al. (Eds.): ICAFS-2018, AISC 896, pp. 91–98, 2019.
https://doi.org/10.1007/978-3-030-04164-9_15

spry to cure the infected strawberry plant at early stages. This process offers the advantages to follow. Firstly, it avoids excessive amount of pesticide spraying. Secondly, it avoids direct contact of farmer with pesticide. Thirdly, it gives the proper diagnosis of the disease properly to avoid naked eye diagnosis, which is more time consuming and costly. Fourthly, it avoids farmers to travel to the farms, if they occur far from his place. Lastly, it avoids workload of farmers.

## 2  Literature Survey

Detection of disease, infection such a fungus while the plant is still in its early stages is a challenge for horticultural scholars (Boissard et al. 2008). Kim et al. 2009 have classified the grapefruit peel diseases using color texture features analysis. The texture features are calculated from the Spatial Gray-level Dependence Matrices (SGDM) and the classification is done using the squared distance technique.

El-Helly et al. 2003 developed a new method in which Hue Saturation Intensity (HIS) - transformation is applied to the input image, then it is segmented using Fuzzy C-mean algorithm. Feature extraction stage deals with the color, size and shape of the spot and finally classification are done using neural networks (El-Helly et al. 2003).

Al-Bashish et al. 2011 developed a fast and accurate method in which the leaf diseases are detected and classified using k-means based segmentation and neural networks based classification.

Sannakki et al. 2011 proposed an image processing based grading system using Fuzzy logic. The system built on Machine vision and fuzzy logic is useful to pathologist and it is better than manual grading.

Rishi et al. (2015) in this paper various image processing techniques are used as such as neural network, image segmentation, BP network, GRN network, fuzzy logic, SVM and many more are discussed in detail.

In this paper, detection and classification of leaf diseases have been proposed, this method is based on masking and removing of green pixels, applying a specific filter to extract the infected region and computing to evaluate the affected area. Plant diseases may be broadly classified into three types. They are bacterial, fungal and viral diseases. This work is based on the most important fungal disease because it is largely affecting the country at the present time.

## 3  Proposed Methodology

The infected image obtained from the camera and uploaded by frame. The target infected area is located in the center portion of the image and occupies the half (1/2) of the total image area. The traditional methods have some difficulties of extraction and identification of the fungus infected leaf. This paper is the first to introduce a mathematical model for such disease. Following through the statistical analysis, it is clear that RGB images include larger portion of G based resolution amongst the entire color space. The RGB color recombination process can be achieved by the combination of super green color $G_s$ and super-red color $R_s$ index as;

$$G_s = 2G - R - B \tag{1}$$

$$RGB = G_s - R_s \tag{2}$$

The step by step execution procedure listed in the block diagram (see Fig. 1)

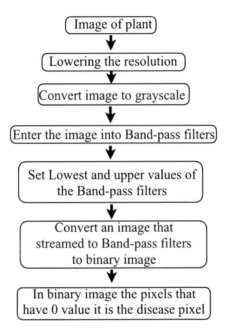

**Fig. 1.** Block diagram of proposed approach.

The images need to be pre-processed in order to remove the noise, reflections and mask the portions in the RGB image format. The second step is to lower the resolution of the image. The low resolution image converted from RGB to a grey scale image (see Fig. 2). The defected area as grey scale image next applied to three stage band-pass filters. The filters are designed to smooth the captured infected area on the leaf (see Fig. 3).

(a)                                           (b)

**Fig. 2.** Convert gray scale image to a binary image (a) Original image; (b) Gray scale image.

This work introduces a new model using parametric exponential bases. The mathematical model is applied to extract band-pass filter coefficients. The new model processed in geometric calculus and the parameters are evaluated from the original pixel values. The new model is can be applied to various engineering applications. The model is extremely applicable because large class of signals can be represented using the parametric exponential. The new model does not require a prior knowledge of the data source.

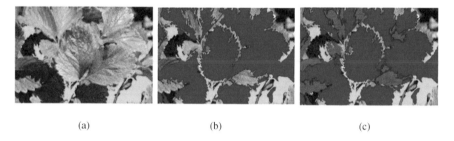

(a)                              (b)                              (c)

**Fig. 3.** Convert image to a binary image with lower and higher values of band-pass filters. (a) Result of first band-pass filter; (b) Result of second band-pass filter; (c) Result of third band-pass filter.

The band-pass filter design is executed using the algorithm known as the multiplicative least square method. The process of the multiplicative least square method was recently introduced in (Özyapıcı 2014). The multiplicative least square method can be implemented as:

$$S = \exp\left\{ \sum_{i=1}^{n} \ln\left( \frac{y_i}{f(b, x_i)} \right)^2 \right\} = \prod_{i=1}^{n} \exp\left( \frac{y_i}{b, x_i} \right)^2 \tag{3}$$

$$S = \prod_{i}^{n} \left[ \left( \frac{y_i}{f(b, x_i)} \right)^{\ln\left( \frac{y_i}{f(b, x_i)} \right)} \right] \tag{4}$$

Where $x_i$ represents the input pixel values and $y_i$ represents the corresponding output values. The process cannot be executed in the Newtonian calculus because the function is non-linear and does not produce an exact solution. However, the same function can be executed in geometric calculus because the exponential functions are linearly processed. The Eq. (4) includes some important parameter coefficients of vector $b$. Minimization method applied to Eq. (4) in order to determine the best representation of the coefficient values. The parameters are included in vector $b$ shown as $b_i$. Such process requires to minimize $S$ with respect to parameter $b_i$.

$$\left(\frac{\partial s}{\partial b_i}\right)^* = \exp\left\{\frac{\partial}{\partial b_i}\sum \ln(s_i)\right\} \tag{5}$$

$$\left(\frac{\partial s}{\partial bi}\right)^* = \exp\left\{-\sum_{i=1}^{n} 2\ln\left(\frac{y_i}{b,\,x_i}\right)\frac{1}{f(b,\,x_i}\frac{\partial f(b,x_i)}{\partial b_i}\right\} \tag{6}$$

The minimization process in the geometric calculus requires the Eq. (6) to be set to 1.

Hence,

$$\left(\frac{\partial s}{\partial bi}\right)^* = \exp\left\{-\sum_{i=1}^{n} 2\ln\left(\frac{y_i}{b,x_i}\right)\frac{1}{f(b,x_i}\frac{\partial f(b,x_i)}{\partial b_i}\right\} = 1 \tag{7}$$

*for all i*

Execution of the Eq. (7) produces as many equations as the subscript $i$. The main aim is to generate the same number of equations as the unknowns. This determines the parameter values to be applied in processing. The best suitable parameters will represent the infected region of the leaf. The parametric filter function is defined as:

$$f(c,x) = x^{c_1}e^{\left(c_2x^2 + c_3x + c_4\right)} \tag{8}$$

The parameter values $c_1, c_2, c_3$ and $c_4$ in the filter function allow a large degree of flexibility. It covers Gaussian, increasing/decreasing exponential, linearly varying data. The parameters values $c_1, c_2, c_3$ and $c_4$ have great impact to smooth out the area of interest on the image. Classical calculus faces difficulty to process and reveal the parameter values therefore we used the geometric calculus. The parameter values can be identified by the solutions of the following mathematical equations.

$$\left(\frac{\partial s}{\partial c_1}\right)^* = \exp\left\{-\sum_{i=1}^{n}\left(\ln(c_i)(\ln(y_i) - c_1\ln(x_i)c_2x_i^2 - c_3x_i - c_4)\right)\right\} \tag{9}$$

$$\left(\frac{\partial s}{\partial c_2}\right)^* = \exp\left\{-\sum_{i=1}^{n}\left(\ln(y_i) - c_1\ln(x_i) - c_2x_i^2 - c_3x_i - c_4\right)\right\} \tag{10}$$

$$\left(\frac{\partial s}{\partial c_3}\right)^* = \exp\left\{-\sum_{i=1}^{n}(x_i)\left(\ln(y_i) - c_1\ln(x_i) - c_2x_i^2 - c_3x_i - c_4\right)\right\} \tag{11}$$

$$\left(\frac{\partial s}{\partial c_4}\right)^* = \exp\left\{-\sum_{i=1}^{n}(1)\left(\ln(y_i) - c_1\ln(x_i) - c_2x_i^2 - c_3x_i - c_4\right)\right\} \tag{12}$$

The parameter values $c_1, c_2, c_3, c_4$ obtained from the solution of the Eqs. (9–12). The parametric values are substituted into the Eq. (8). The function in Eq. (8)

processed the same as the Gaussian, low pass, and high-pass filter function and produces more effective results.

The new type of band-pass filter function designed and applied in three stages for feature extraction. Each band-pass filter uses the present state of the pixel values to determine the new coefficient values. The parameter values are self-extracted from the pixel values. This enables almost exact fit to the application. This type of filter can detect the aimed pixels with high accuracy.

The three stage band-pass filters have different range of threshold values for detection. The ideal band-pass filter with a frequency range of $[D_L.....D_H]$ is defined as:

$$H(u, v) = \begin{cases} 1 & \text{if } D_L \leq D(u, v) \leq D_H \\ 0 & \text{otherwise} \end{cases}$$

Where $D_L$, $D_H$ denotes the lower, upper cutoff frequencies and $D(u, v)$ denotes the distance from the origin in spatial representation (see Fig. 4).

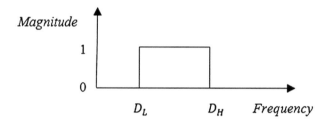

**Fig. 4.** Ideal band-pass filter. Lower ($D_L$), upper ($D_H$) values are to be calculated.

## 4   Result and Discussion

The extracted parameter values are normalized and applied as lower, higher cutoff values. Our produce range of values and run fuzzy optimization algorithm to select the optimum values. The optimal candidate for the first band-pass filter is indicated in grey color in Table 1 (2, 4, 91).

**Table 1.** Analysis of lower and higher value of the first band-pass filter

| Lower value of filter | Higher value of filter | Accuracy % |
|---|---|---|
| 1 | 6 | 83 |
| 5 | 7 | 65 |
| 6 | 10 | 48 |
| 7 | 10 | 30 |
| 2 | 4 | 91 |
| 4 | 6 | 75 |

The same process repeated for the second band-pass filter. The optimum values for the second band-pass filter are indicated in grey color in Table 2 (5, 8, 86).

**Table 2.** Analysis of lower and higher value of the second band-pass filter

| Lower value of filter | Higher value of filter | Accuracy % |
|---|---|---|
| 3 | 6 | 71 |
| 5 | 8 | 86 |
| 7 | 9 | 61 |
| 8 | 10 | 55 |
| 6 | 8 | 85 |
| 8 | 11 | 40 |

The third band pass filter produces the most optimal lower, higher and accuracy values of (9, 14, 89) as given in Table 3.

**Table 3.** Analysis of lower and higher value of the third band-pass filter

| Lower value of filter | Higher value of filter | Accuracy % |
|---|---|---|
| 10 | 12 | 69 |
| 9 | 15 | 85 |
| 9 | 14 | 89 |
| 8 | 11 | 54 |
| 10 | 12 | 76 |
| 14 | 16 | 10 |

The overall results for the three filters have an accuracy above 88%. The leaf with disease detected amongst a large background of obstacles with very much accepted accuracy. The work does not require neural network analysis that demands long time and large data samples for the learning period.

## 5   Conclusion

The method so far implemented is an effective and fast method of detection of fungus disease on strawberry crops. It is easy to use for early detection of fungus disease through leaf inspection. The leaf images captured are processed to determine the healthiness of each strawberry plant. The new filter design ensures to capture the fungi diseased area accurately. The accurate area detection enables to spry with the right dosage of chemicals. By using this recognition technique, it will identify the potential problem before serious damage whole field of strawberries. The proposed method avoids the use of harmful chemicals on plants and hence ensures a healthier environment. The method also lowers the production cost of the maintenance and produces a high quality of strawberries.

# References

Al-Bashish, D., Braik, M., Bani-Ahmad, S.: Detection and classification of leaf diseases using K-means-based segmentation and neural networks-based classification. Inf. Technol. J. **2**(3), 267–275 (2011)

Boissard, P., Martin, V., Moisan, S.: A cognitive vision approach to early pest detection in greenhouse crops. Comput. Electron. Agric. **60**, 81–93 (2008)

Durmus, H., Gunes, E.O., Kirci, M., Ustundag, B.B.: The design of general purpose autonomous agricultural mobile-robot. In: 4th International Conference on Agro-Geoinformatics, pp. 49–53. IEEE, Istanbul (2015)

El-Helly, M., Rafea, A., El-Gammal, S.: An integrated image processing system for leaf disease detection and diagnosis. In: IICAI 2003, pp. 1182–1195 (2003)

Pujari, J., Yakkundimath, R., Byadgi, A.: Identification and classification of fungal disease affected on agriculture/horticulture crops using image processing techniques. In: IEEE International Conference on Computational Intelligence and Computing Research, pp. 31–34. IEEE, Coimbatore (2014)

Kim, D.G., Burks, T.F., Qin, J., Bulanon, D.M.: Classification of grapefruit peel diseases using color texture feature analysis. Int. J. Agric. Biol. Eng. **35**, 41–50 (2009)

Rishi, N., Gill, J.S.: An overview on detection and classification of plant diseases in image processing. Int. J. Sci. Eng. Res. **3**(5), 3–6 (2015)

Özyapıcı, A.: On multiplicative and Volterra minimization methods. Numer. Algorithms **67**(3), 623–636 (2014)

Sannakki, S.S., Rajpurohit, V.S., Nargund, V.B., Yallur, P.S.: Leaf disease grading by machine vision and fuzzy logic. Int. J. Comput. Technol. Appl. **2**(5), 1709–1716 (2011)

# Analysis of Prediction Models for Wind Power Density, Case Study: Ercan Area, Northern Cyprus

Youssef Kassem[1,2(✉)], Hüseyin Gökçekuş[1], and Hüseyin Çamur[2]

[1] Faculty of Civil and Environmental Engineering, Near East University, 99138 Nicosia, North Cyprus, Turkey
{yousseuf.kassem,huseyin.gokcekus}@neu.edu.tr
[2] Faculty of Engineering, Mechanical Engineering Department, Near East University, 99138 Nicosia, North Cyprus, Turkey
huseyin.camur@neu.edu.tr

**Abstract.** This work focuses on the application of Multilayer Perceptron Neural Network (MLPNN), Radial Basis Function Neural Network (RPFNN) and Auto Regressive Integrated Moving Average (ARIMA) as predictive tools for the production of wind power density (WPD). The air temperature (AT), dew point (DP), atmospheric humidity (AH), pressure (P) and wind speed (WS) were used as the input variables for the models. Moreover, the performance of the models based on the R-squared value is presented. The results demonstrated that the MLPNN and ARIMA have the best accuracy for the prediction of WPD with the highest correlation coefficient of 0.99 compared to RPFNN. Consequently, it can be concluded that the MLPNN models developed in this study can be attractive for their incorporation in simulators.

**Keywords:** ARIMA · MLPNN · RPFNN · Wind power density

## 1 Introduction

Renewable energy sources such as wind energy have grown in popularity in recent years [1]. Wind energy is considered as a clean and environmental source of energy [2, 3]. The use of wind energy in many countries has reduced air pollution and fossil fuel consumption [4, 5]. Many studies have concluded that wind speed is one of the most significant parameters for wind power [6]. Wind power is an alternative source for generating electricity through wind turbines. Prediction of the effective wind power density is a significant aspect of wind power evaluation. Wind power density (WPD) is defined as [3, 7]

$$WPD = \frac{1}{2}\rho(h)v^3 \quad \left[\frac{w}{m^2}\right] \tag{1}$$

where, $v$ is wind speed in m/s, and $\rho(h)$ is air density at varying heights in kg/m$^3$, which can be calculated using Eq. (2) [7].

© Springer Nature Switzerland AG 2019
R. A. Aliev et al. (Eds.): ICAFS-2018, AISC 896, pp. 99–106, 2019.
https://doi.org/10.1007/978-3-030-04164-9_16

$$\rho(h) = \frac{P}{RT} exp\left(-\frac{gh}{RT}\right) \tag{2}$$

where; P is air pressure in hPa, R is a gas constant (R = 287.05 J/(K mol)), T is the air temperature in K, g is the gravitational constant (g = 9.81 m/s$^2$), and h is the considered height above sea level (in this study all measurement data are collected at a height of 10 m).

Several approaches have been proposed for predicting the wind power density of wind energy including ANN [8], ARIMA [9] and ANFIS [10].

In this study, evaluation, and comparison of the prediction and simulating efficiencies of Multilayer Perceptron Neural Network (MLPNN), Radial Basis Function Neural Network (RPFNN) and Auto Regressive Integrated Moving Average (ARIMA) have been conducted for modeling of the wind power density of the Ercan area in Northern Cyprus. The performance of the models is compared using statistical criteria like the determination coefficient (R$^2$).

## 2  Measurement Data

In the contents of this study, a meteorological model driven by historical weather data is used to predict the wind power density (WPD) using three different models: Multilayer Perceptron Neural Network (MLPNN), Radial Basis Function Neural Network (RPFNN) and Auto Regressive Integrated Moving Average (ARIMA). The air temperature (AT), dew point (DP), atmospheric humidity (AH), pressure (P) and wind speed (WS) data that were obtained over a period of 4 years from January 2013 to December 2016 are used in this study. The mean measurement data of the selected area including air density and wind speed values during the investigation period are shown in Figs. 1 and 2, respectively. It is observed from Fig. 1 that the mean wind speed values vary between 3 and 31 km/h. The maximum mean dailywind speed value of 31 km/h was recorded in April 2013 and July 2014 J, while the minimum value of 3 km/h was recorded in July 2013and January 2016. Moreover, it is observed that the air density values ranged between 1.13 and 1.29 kg/m$^3$, as shown in Fig. 2.

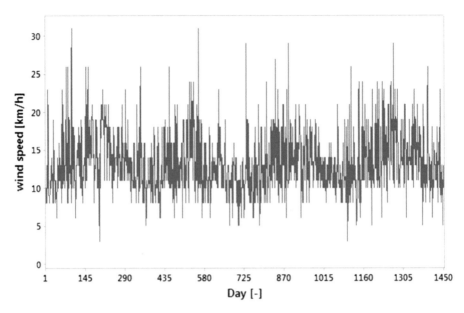

**Fig. 1.** Mean wind speed of the selected area (2013–2016).

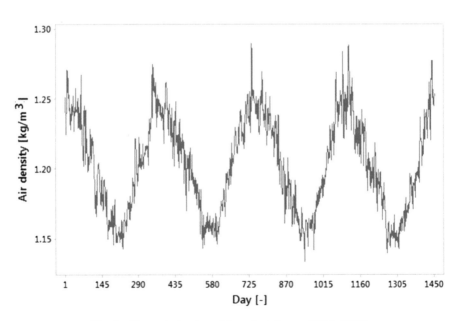

**Fig. 2.** Mean air density of the selected area (2013–2016).

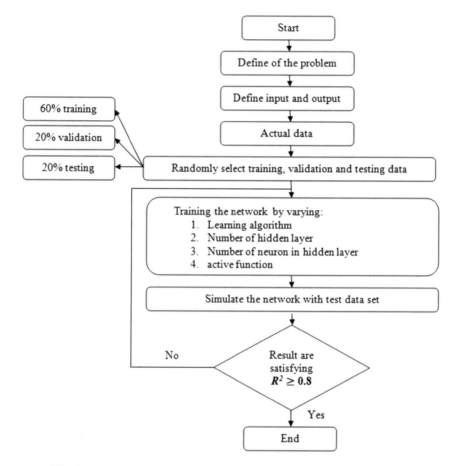

**Fig. 3.** Proposed flow chart of an MLPNN prediction model development.

## 3  Multilayer Perceptron Neural Network (MLPNN)

The creation of the MLPNN predictive model with PASW Statistics 18 for predicting the wind power density of the Ercan area in Northern Cyprus involves the following several stages. First stage: the input of the MLPNN model for wind power density prediction is day (D), air temperature (AT), atmospheric humidity (AH), pressure (P), dew point (DP) and wind speed (WS) values, as presented in Fig. 3. According to the calculation methodology in Fig. 3, the training results showed that the MLPNN with one hidden layer with three number units in the hidden layer (excluding the bias unit) has the best performance. Consequently, the best MLPNN architecture is shown in Fig. 4 as the final MLPNN model for the prediction of wind power density at the selected area.

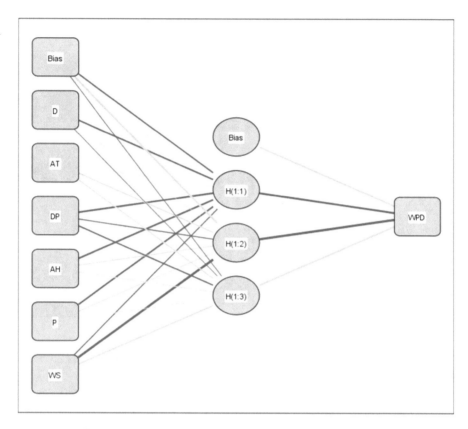

**Fig. 4.** MLPNN architecture for prediction wind power density.

## 4   Radial Basis Function Neural Network (RPFNN)

Radial Basis Function (RBF) networks form a class of ANNs (ANNs) that has some advantages compared to other types of ANNs. In this work, PASW Statistics 18 has been used for developing the RBFNN network implementation. Training of the network has been performed with a different number of RBF units. The developed network architecture with six input layers (D, AT, AH, DP, P, and WS) and one output (WPD) is shown in Fig. 5.

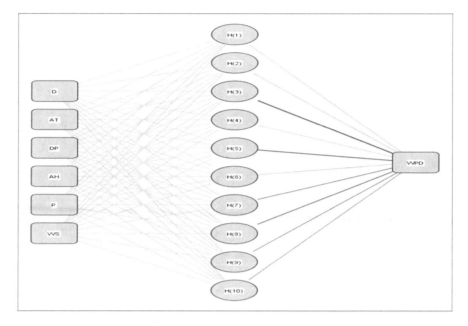

**Fig. 5.** RPFNN architecture for prediction wind power density.

## 5  ARIMA

The Auto-Regressive Integrated Moving Average (ARIMA) model is one of the most popular time series models used for predicting future data. ARIMA models are examined with different parameters to determine the best model that will give the best forecast. The representation of this model is done through ARIMA notation of order (p, d, q), where p is the number of seasonal auto-regressive terms, d is the number of seasonal differences (rarely should d > 2 be needed), and q is the number of seasonal media moving. In this study, the ARIMA model was performed with PASW Statistics 18. In this step, the WPD dependent variable and the independent variable are described, which are D, AT, AH, DP, P, and WS. The statistical calculations show that an ARIMA (1, 0, 2) is the best model obtained for the time series of wind power density.

## 6  Results and Discussion

In this study, the measurement data from 1,452 multiple local measurements in the Ercan area were supported by historical weather data to develop and validate the three different methods. The prediction results of the three methods are compared and discussed in this section. The test values obtained from the model's results were compared with actual values as shown in Fig. 6. As a result, the test values obtained from models were quite compatible with actual values. The accuracy of the models obtained from MLPNN, RBFNN, and ARIMA were examined by evaluating the *R-squared* values. The results (Fig. 7) showed that the two models, MLPNN and ARIMA gave good predictions due to the values of *R-squared* compared to RBFNN. However, MLPNN showed a clear lead over ARIMA because it had a higher *R-squared* value. It can be concluded that MLPNN and ARIMA were more effective than RPFNN in the modeling for predicting the wind power density.

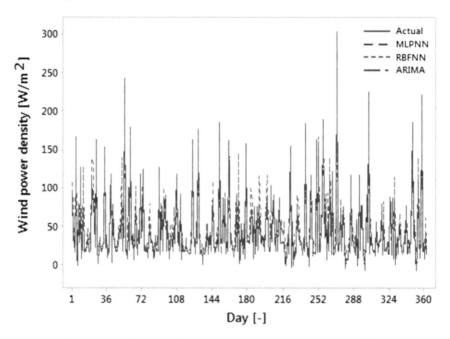

**Fig. 6.** Actual and predicting data of wind power density for 2016.

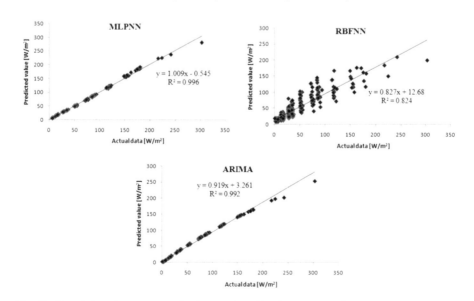

**Fig. 7.** Correlations between actual values and predicted WPD by different prediction methods.

## 7    Conclusions

In this paper, three models (MLPNN, RBFNN, and ARIMA)were developed to predict the wind power density from multiple local measurements, including air temperature, dew point, atmospheric humidity, pressure and wind speed. According to the results, it is concluded that both the MLPNN and ARIMA models can achieve good forecasts when applied to real-life problems and thus can be effectively engaged for wind power density prediction. Moreover, based on the results, the highest R-squared values are obtained from MLPNN models.

**Acknowledgments.**    The authors would like to thank the Faculty of Engineering, particularly the Civil Engineering Department and Mechanical Engineering Department of Near East University for their support and encouragement.

## References

1. Ozay, C., Celiktas, M.S.: Statistical analysis of wind speed using two-parameter Weibull distribution in Alaçatı region. Energy Convers. Manage. **121**, 49–54 (2016)
2. Shu, Z., Li, Q., Chan, P.: Statistical analysis of wind characteristics and wind energy potential in Hong Kong. Energy Convers. Manage. **101**, 644–657 (2015)
3. Fazelpour, F., Soltani, N., Rosen, M.A.: Wind resource assessment and wind power potential for the city of Ardabil, Iran. Int. J. Energy Environ. Eng. **6**(4), 431–438 (2014)
4. Jacovides, C.P., Theophilou, C., Tymvios, F.S., Pashiardes, S.: Wind statistics for coastal stations in Cyprus. Theoret. Appl. Climatol. **72**(3–4), 259–263 (2002)
5. Kumar, K.S., Gaddada, S.: Statistical scrutiny of Weibull parameters for wind energy potential appraisal in the area of northern Ethiopia. Renew.: Wind Water Solar **2**(1), 14 (2015)
6. Masseran, N., Razali, A.M., Ibrahim, K., Zaharim, A., Sopian, K.: The probability distribution model of wind speed over East Malaysia. Res. J. Appl. Sci. Eng. Technol. **6**(10), 1774–1779 (2013)
7. Olaofe, Z.O., Folly, K.A.: Wind energy analysis based on turbine and developed site power curves: a case-study of Darling City. Renewable Energy **53**, 306–318 (2013)
8. Singh, V.: Application of artificial neural networks for predicting generated wind power. Int. J. Adv. Comput. Sci. Appl. **7**(3), 250–253 (2016)
9. Alencar, D.B., Affonso, C.D., Oliveira, R.L., Rodríguez, J.M., Leite, J., Filho, J.R.: Different models for forecasting wind power generation: case study. Energies **10**(12), 1976–2003 (2017)
10. Hossain, M., Mekhilef, S., Afifi, F., Halabi, L.M., Olatomiwa, L., Seyedmahmoudian, M., Stojcevski, A.: Application of the hybrid ANFIS models for long term wind power density prediction with extrapolation capability. PLoS ONE **13**(4), e0193772 (2018)

# Evaluation of Image Representations for Player Detection in Field Sports Using Convolutional Neural Networks

Melike Şah[1(✉)] and Cem Direkoğlu[2]

[1] Department of Computer Engineering, Near East University, via Mersin 10, Nicosia, North Cyprus, Turkey
melike.sah@neu.edu.tr
[2] Department of Electrical and Electronics Engineering, Middle East Technical University - Northern Cyprus Campus, via Mersin 10, Kalkanli, Guzelyurt, North Cyprus, Turkey
cemdir@metu.edu.tr

**Abstract.** Player detection is an important task in sport video analysis. Once players are detected accurately, it can be used for player tracking, player activity/performance analysis as well as team activity recognition. Recently, convolutional Neural Networks (CNN) became the state-of-the-art in computer vision for object recognition. CNN based methods usually use gray or RGB images as an input. It is also possible to use other image representation techniques such as shape information image and polar transformed shape information image for player detection. In this paper, we evaluate various image representation techniques for player detection using CNN. In our evaluation, first the candidate image regions for players are determined using a sliding window technique. Then these regions are input to CNN for player detection. We examine four different types of image representations as an input to CNN: RGB, gray, shape information and polar transformed shape information image. Evaluation is conducted on a field hockey dataset. Results show that CNN based player detection is effective and different image representations yield different performances.

**Keywords:** Player detection · Field sports · Shape information image
Polar transformed shape information image · Convolutional neural networks

## 1 Introduction

In multimedia and computer vision, sport video analysis is a popular topic. Generally, as a first step of sport video analysis, player detection has to be applied, which can be challenging because of the resolution of images (players appear at a distance), changes of player appearances due to pose changes, illumination changes and weather conditions. On the other hand, there are two possible sources of sport videos: fixed camera setup around the playfield and TV broadcasting cameras. Since we use fixed cameras around the playground, we explain the related work in this area. Generally, fixed multi-camera systems cover all locations on the playground and thus all players are captured

© Springer Nature Switzerland AG 2019
R. A. Aliev et al. (Eds.): ICAFS-2018, AISC 896, pp. 107–115, 2019.
https://doi.org/10.1007/978-3-030-04164-9_17

simultaneously. With a fixed camera setup, background subtraction is a common method for player detection [1]. Background subtraction methods need to frequently update the background representation in order to overcome problems such as illumination changes, shadows and background objects. Han and Davis [2] use an adaptive background modelling method, however it works well for simple scenes with slow changes of illumination. After background subtraction for player detection, Carr et al. [3] generate shape-specific occupancy maps using the foreground regions. This increases the tolerance to shadows, but can only identify isolated players. Xu et al. [4] integrate the dominant color and geometry data of the playground to assist background subtraction. However, methods based on background subtraction cannot handle segmented regions that contain multiple players or when a single player is segmented into multiple regions. Direkoglu et al. [5] generate shape information images using a heat diffusion theory and applies it to player detection in field hockey games. They can handle occluded players and isolated players well.

On the other hand, in recent years, CNN became the state-of-the-art for object recognition. However, only a few approaches applied CNN for player detection. Lu et al. [6] proposes a light cascaded CNN for player detection in soccer and basketball games using broadcasting camera. They propose a CNN architecture with four classification branches and the input to the CNN is RGB images. Lehuger et al. [7] uses two convolutional layers and gray scale images for player detection. They tested their network on soccer games using broadcasting camera. Acuna [8] uses YOLO network for real-time player detection in basketball games. He focuses on system performance and input to the network is RGB images.

In this paper, we evaluate different image representation techniques for player detection using CNN. This is the first time a shape information image (SIM) and polar transformed shape information image (PSIM) is input to CNN for player detection. The SIM and PSIM is compared with RGB and gray scale images. For each image representation, we also find the optimal CNN architecture that yields the best performance in a field hockey dataset. Results show that CNN based player detection is effective, and that SIM and PSIM perform competitive results.

## 2   Player Detection Algorithm

To find players within a frame, first we determine candidate regions using a sliding window technique. Regions that contain a number of edge points above a predefined threshold are selected to be candidate regions. These candidate regions may have various image representations, where four of them are evaluated in this work. Finally, these images are classified using our CNN architecture.

### 2.1   Candidate Region Selection and Window Dimensions

The candidate region selection algorithm is adapted from [5]. First candidate regions on the image is found based on edge features derived from the image data and known playfield geometry. As a first step, we detect binary edges using the Canny method. In field sports, generally the playfield is homogeneous. Thus, the detected binary edges mostly belong to players, playground markings and noise on the playfield. Since outside the playground, there are audience and advertisements, these edges are also detected, which are removed using a geometry based playground mask. To accelerate the detection process, during the scanning (sliding window), we classify only the window regions which contain a significant number of edge points. Inside a window, if the total number of edge points, is greater than a threshold (*threshold = 10* in our experiments), we convert that image region to a image representation (i.e. gray, RGB, SIM or PSIM image) and give it to the CNN for classification.

The window dimensions are estimated based on known camera geometry and prior player information. We use the manually annotated players to estimate window dimensions that is adapted from [5]. Four pre-defined window dimensions are used in our work: $40 \times 25$, $48 \times 30$, $56 \times 35$ and $64 \times 40$. Finally, in CNN, all the input images must have the same size for classification. Therefore, window regions are normalized to a fixed size during the training and testing of CNN.

## 2.2 Image Representation Techniques

Generally, CNN approaches either use gray or RGB images for training and testing. In addition to these, we want to evaluate other image representation techniques such as SIM and PSIM images that are proposed by [5] for player detection. If any of these image representations together with CNN can improve player detection, then it would be alternative to the traditional gray and RGB image representations.

**Shape Information Image and Polar Transformed Shape Information Image.**
Recently, Direkoglu et al. [5] proposed a heat diffusion equation to create a shape information image (SIM) of an object using the binary edge points in a window. The diffusion equation fills inside the object shape while preserving the shape. Thus, remove appearance variations (color, texture, etc.) of the objects. They also presented how to convert SIM to the polar transformed shape information image (PSIM). Polar transformation removes the orientation variation of objects. Figure 1(a)–(d) shows different image representations for player (the first row) and nonplayer samples (the second row). Previously heat analogy has been also employed for feature extraction in computer vision [9–11]. In our work, SIM and PSIM images are used as an input to CNN and compared with RGB and gray scale representations.

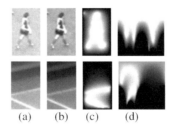

(a)     (b)     (c)     (d)

**Fig. 1.** Player/nonplayer image representations: (a) RGB image, (b) Gray image, (c) Shape information image and (d) Polar transformed shape information image

## 3    Player Detection Using Convolutional Neural Networks

### 3.1    CNN Architecture

We train CNNs to achieve accurate player detection using four types of image representations. For each image type, after several runs, the number of convolutional layers, filter sizes in each level and number of channels are experimentally determined to give the maximum F-score in the field hockey dataset.

We observe that SIM and PSIM perform better with the CNN architecture as shown in Fig. 2. We use three convolutional layers that is linked to a fully connected layer. SIM inputs are normalized to $56 \times 35$ image dimensions, while PSIM inputs are normalized to $56 \times 56$ image dimensions. The convolution layer one (convolution-one) utilizes $7 \times 7$ filters with 8 channels. After convolution-one, we use max pooling (with $5 \times 5$) for normalization. Convolution-two uses $5 \times 5$ filters with 16 feature maps followed by $3 \times 3$ max pooling. Convolution-three uses $3 \times 3$ filters with 32 feature maps followed by $2 \times 2$ max pooling. Increasing the number of convolutional layers did not increase the performance in our study. Overall, we can say that SIM and PSIM images works better with larger filters in the initial layers since they generate a smooth and homogeneous image after processing. On the other hand, for gray and RGB images, we observe that small size filters with $2 \times 2$ max pooling work better (see Fig. 3). In addition, RGB and gray scale images are normalized to $56 \times 35$ dimensions. For both architectures, we also use softmax layer. Finally, the input image is classified as player or nonplayer using the classification layer.

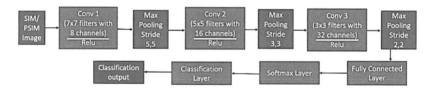

**Fig. 2.** CNN architecture for SIM + CNN and PSIM + CNN images

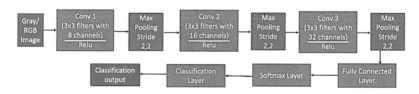

**Fig. 3.** CNN architecture for Gray + CNN and RGB + CNN images

## 3.2    CNN Training

In our approach, we use a field hockey dataset [5]; women hockey game between Ireland and Australia. The dataset uses fixed cameras mounted 20 m high around the field. The dataset is captured on a rainy weather, which also causes false detections. In this paper, we use a corner view camera. Figures 4 and 5 show sample frames, which has dimensions of 959 × 539. For the training of nonplayer image samples, frame edge regions are scanned after geometry-based masking to extract nonplayer samples. As a result, 13,420 nonplayer samples are used for training. In particular, the same 13,420 samples are converted to gray, RGB, SIM and PSIM images. We collect 1375 player samples from camera views with varying appearance, scale, rotation and pose. Similarly, we converted player samples to gray, RGB, SIM and PSIM images.

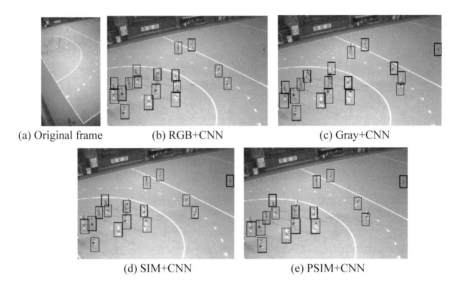

**Fig. 4.** Sample player detection on frame 9 of hockey dataset: (a) original frame, (b) RGB + CNN, (c) Gray + CNN, (d) SIM + CNN and (e) PSIM + CNN

Since, we have four image representations, we train four different CNNs; for Gray + CNN method, we use gray player and nonplayer training images; for RGB + CNN, we utilize RGB player and nonplayer training images and so forth. In CNN network training, initialization is random. Therefore, various networks give different performances. To find the best performing network for each input type, we run several CNN architectures with varying properties. As we explain above, gray and RGB images fit to the CNN architecture well shown in Fig. 3. On the other hand, SIM and PSIM images fit to the CNN architecture well illustrated in Fig. 2. While training, we use stochastic gradient descent as solver, with a learning rate of 0.01, epoch 12 for SIM/PSIM and epoch 4 for RGB/gray, since it gives the best results in this dataset.

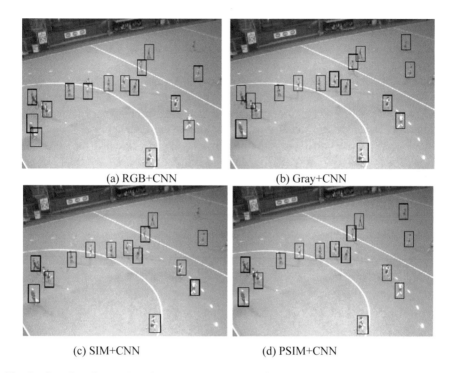

(a) RGB+CNN     (b) Gray+CNN

(c) SIM+CNN     (d) PSIM+CNN

**Fig. 5.** Sample player detection on frame 130 of hockey dataset: (a) RGB + CNN, (b) Gray + CNN, (c) SIM + CNN and (d) PSIM + CNN

## 4   Evaluations and Results

For evaluations, we prepare a dataset by manually labeling the ground truth Bounding Boxes (BBs). We manually annotated 4526 players in 301 consecutive frames. In Figs. 4 and 5, illustrative player detection results are shown for frame 9 and 130, using different CNN methods. Figure 4(a) shows the original frame. It can be seen that players appear at a distance with variations such as appearance, rotation and scale. Figures show that various CNN based method can handle player detection differently.

In evaluation, we calculate overlap ratio between the manually labelled BBs (ground truth) and the BBs determined by a method. For example, if the overlap ratio is greater than a predefined threshold (e.g. 0.25 or 0.5), then it is accepted as correct detection. In particular, we measure performances of methods using precision (P%), recall (R%) and f-score (F%) using Eqs. (1), (2) and (3) respectively:

$$P\% = (P_c/P_t) \times 100 \tag{1}$$

$$R = (R_c/R_t) \times 100 \tag{2}$$

$$F\% = 2 \times ((P\% \times R\%)/(P\% + R\%)) \tag{3}$$

where $P_c$ is the number of correctly predicted BB locations and $P_t$ is the total number of BBs predicted as player. $R_c$ is the total number of correctly predicted BB locations and $R_t$ is the total number of BB locations that manually labelled as player. F-score combines the precision and recall.

Results in Table 1 shows that when the overlap is greater than 0.25, Gray + CNN performs the best in terms of F-score, followed by SIM + CNN, PSIM + CNN and RGB + CNN. We observe that both SIM and PSIM perform similar to gray and RGB methods when the overlap is greater than 0.25. However, when the overlap is greater than 0.5, SIM and PSIM performs significantly better than both RGB and gray representations. SIM + CNN and PSIM + CNN achieve between 7–9% performance gain and you can observe the difference in Figs. 4 and 5 as well; SIM + CNN and PSIM + CNN detect the player regions better than Gray + CNN and RGB + CNN. RGB + CNN performs the worst for both 0.25 and 0.5 overlap compared to other methods. On the other hand, SIM + CNN and PSIM + CNN achieve the highest precision at both overlaps, as well as, both methods have significantly higher recall rates when the overlap is higher than 0.5.

**Table 1.** Precision (P%), Recall (R%) and F-score (F%) when the overlap between manually labelled BBs is greater than 0.25 and 0.5 (compared to 4526 manually labelled BBs).

| Method | P% (0.25) | R% (0.25) | F% (0.25) | P% (0.5) | R% (0.5) | F% (0.5) |
|---|---|---|---|---|---|---|
| RGB + CNN | 95.12 | 88.27 | 91.57 | 70.55 | 65.47 | 67.91 |
| Gray + CNN | 98.01 | 89.15 | 93.37 | 73.52 | 66.88 | 70.05 |
| SIM + CNN | 99.05 | 87.69 | 93.03 | 81.26 | 71.94 | 76.32 |
| PSIM + CNN | 98.28 | 87.07 | 92.34 | 80.40 | 71.23 | 75.54 |

Second, we compare various image representations with CNN under different occlusions: Heavy occlusion, partial occlusion and no occlusion. Results in Tables 2 and 3 show that when the overlap is greater than 0.25, Gray + CNN performs better especially for partial and heavy occlusions. We can say that SIM + CNN and PSIM + CNN performs similar results and they generally cannot handle occlusions as good as Gray + CNN at 0.25 overlap. When the overlap is greater than 0.5, SIM + CNN and PSIM + CNN achieve slightly better than Gray + CNN and RGB + CNN.

**Table 2.** Detection rates in occlusion cases when the overlap is greater than 0.25.

| Method | No occlusion | Partial occlusion | Heavy occlusion | Overall |
|---|---|---|---|---|
| RGB + CNN | 90.73 | 74.75 | 49.13 | 88.27 |
| Gray + CNN | 91.35 | 75.76 | 54.78 | 89.15 |
| SIM + CNN | 90.45 | 61.62 | 48.70 | 87.69 |
| PSIM + CNN | 90.09 | 62.63 | 42.61 | 87.07 |

**Table 3.** Detection rates in occlusion cases when the overlap is greater than 0.5.

| Method | No occlusion | Partial occlusion | Heavy occlusion | Overall |
|--------|--------------|-------------------|-----------------|---------|
| RGB + CNN | 67.45 | 61.62 | 30.87 | 65.47 |
| Gray + CNN | 69.17 | 47.47 | 33.48 | 66.88 |
| SIM + CNN | 74.34 | 48.48 | 38.29 | 71.94 |
| PSIM + CNN | 73.65 | 54.55 | 34.35 | 71.23 |

When the overlap is greater than 0.5, SIM + CNN can detect players under heavy occlusion well. Conversely, PSIM + CNN has a good performance under partial occlusion. RGB + CNN especially can handle partially occluded players for both 0.25 and 0.5 overlap.

## 5    Conclusions

We have presented a work that evaluates various image representations such as RGB, gray, shape information image (SIM) and polar transformed shape information image (PSIM) for player detection using convolutional neural networks (CNN). We also find optimal CNN architectures for each image representation. Results show that SIM and PSIM perform better when the overlap is greater than 0.5. When the overlap ratio is greater than 0.25, all of the methods achieve similar performances.

## References

1. Hamid, R., Kumar, R.K., Grundmann, M., Kihwan, K., Essa, I., Hodgins, J.: Player localization using multiple static cameras for sports visualization. In: IEEE Conference on Computer Vision and Pattern Recognition (CVPR), San Francisco, USA, pp. 731–738 (2010)
2. Han, B., Davis, L.S.: Density-based multifeature background subtraction with support vector machine. IEEE Trans. Pattern Anal. Mach. Intell. **34**(5), 1017–1023 (2012)
3. Carr, P., Sheikh, Y., Matthews, I.: Monocular object detection using 3D geometric primitives. In: European Conference on Computer Vision, Florence, Italy, vol. 7572, pp. 864–878 (2012)
4. Xu, M., Orwell, J., Lowey, L., Thirde, D.: Architecture and algorithms for tracking football players with multiple cameras. IEE Vis. Image Signal Process. **152**(2), 232–241 (2005)
5. Direkoglu, C., Sah, M., O'Connor, N.: Player detection in field sports. Mach. Vis. Appl. **29**(2), 187–206 (2018)
6. Lu, K., Chen, J., Little, J.J., He, H.: Light cascaded convolutional neural networks for accurate player detection. In: British Machine Vision Conference (BMVC), London, UK (2017)
7. Lehuger, A., Duffner, S., Garcia, C.: A robust method for automated player detection in sport videos. In: Compression et Representation des Signaux Audievisuels (2007)
8. Acuna, D.: Towards real-time detection and tracking of basketball players using deep neural networks. In: Conference on Neural Information Processing Systems, California, USA (2017)

9. Direkoglu, C., Nixon, M.S.: Moving edge detection via heat flow analogy. Pattern Recogn. Lett. **32**(2), 270–279 (2011)
10. Direkoglu, C., Nixon, M.S.: Image-based multiscale shape description using gaussian filter. In: IEEE Indian Conference on Computer Vision, Graphics and Image Processing, Bhubaneswar, India, pp. 673–678 (2008)
11. Direkoglu, C., Nixon, M.S.: Shape extraction via heat flow analogy. In: International Conference on Advanced Concepts for Intelligent Vision Systems, Delft, Netherlands, pp. 553–564 (2007)

# A Fuzzy Based Gaussian Weighted Moving Windowing for Denoising Electrocardiogram (ECG) Signals

Cemal Kavalcıoğlu$^{(\boxtimes)}$ and Bülent Bilgehan

Department of Electrical and Electronic Engineering, Near East University,
Near East Boulevard, P.O. Box: 99138, Nicosia, Mersin 10, TRNC, Turkey
cemal.kavalcioglu@neu.edu.tr

**Abstract.** The electrocardiogram is a graphical record of the biological signal that is thought to be susceptible to electrical activity of the heart and utilized in order to clinical diagnosis. Electrocardiogram signal is very responsive in nature, and even if there is small noise mixed with the original signal, assorted characteristics of signal change. ECG signal voltage level is as low as 0.5 to 5 mV and is sensitive to artifacts larger than this. Human electrocardiogram signal range frequency ingredients from 0.05 Hz to 100 Hz and are related to noise, muscle movements, network current, and ambient electromagnetic interference. Electrocardiogram is a very significant sign detects abnormal heart rhythms and examines cause of chest pain and widely utilized in cardiology. Most digital signals are infinitely large or too large to be manipulated as a whole. Because statistical calculations require that all points be present for analysis, it is difficult to statistically analyze sufficiently large signals. To avoid these problems, engineers characteristically analyze small subsets of the aggregate data with an operation named windowing. Fuzzy logic is a mathematical logic that attempts to solve problems with a clear, uncertain data spectrum that makes it possible to obtain a series of correct results. This manuscript suggests denoising method Gaussian Weighted Moving Windowing for denoising Electrocardiogram signals to remove random noise. This study is interpreted with the actual data set and confirmed according to Peak signal to noise ratio.

**Keywords:** Electrocardiogram · Windowing
Gaussian weighted moving windowing · Denoising · ECG signal processing
Fast fourier transform · Fuzzy logic

## 1 Introduction

The Electrocardiogram is the most known biological signal and non-invasive method; that is a graphical record of electrical activity of the heart over time. By signal extraction, ECG is widely utilized in detection of some diseases. Cardiovascular diseases and abnormalities change waveform of an electrocardiogram; each section of the electrocardiogram waveform carries information about the clinician to arrive at a suitable diagnosis [1].

© Springer Nature Switzerland AG 2019
R. A. Aliev et al. (Eds.): ICAFS-2018, AISC 896, pp. 116–128, 2019.
https://doi.org/10.1007/978-3-030-04164-9_18

Electrocardiogram signal from a patient is often impaired by external noise, so an appropriate noiseless electrocardiogram signal is required. A signal acquisition system includes several stages, consisting of: acquiring signals such as hardware and software instrumentation, processing sound and processing other features for filtering, and extracting information [2].

To detect intermittent disturbances in the heart rhythm, electrocardiographic signals are recorded on a long time scale. A standard schematic electrocardiogram illustrated in Fig. 1. For P, T, U and a QRS complicated, exact waveform is named the ECG labeled P, Q, R, S, and T, which state specific properties.

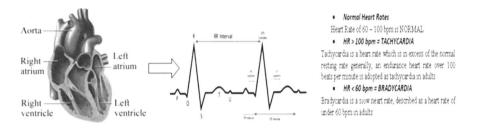

**Fig. 1.** Human heart and standard schematic electrocardiogram signal

Electrical activity of the heart is recorded by device named ECG. Electrical signals coming from heart characteristically precede normal mechanical function and the monitoring of this signal has major clinical presence. In cardiology Electrocardiogram is utilized in catheterization laboratories, coronary care units, and routine diagnostic implementations. The cardiologist easily interprets Electrocardiogram waveforms and classifies them as normal and abnormal patterns. Electrocardiogram signals includes P wave, QRS complex, T wave. The change in these parameters is indicative of heart disease that can occur for many reasons. In order to get a noise free signal, filter is used. Electrocardiography is the recording of electrical activity of heart for a period of time, which is sensed by electrodes attached to skin surface and externally viewed by an output appliance. Cardiac and cardiovascular state information that can be used for appropriate treatment and improve the quality of life is shown in ECG drawing. The record produced by this non-invasive method is called an electrocardiogram (ECG). An Electrocardiogram is utilized to measure the heart's electrical conductivity system. Electrically impulses formed by polarization and depolarization of the cardiac tissue are taken and converted into a waveform. This waveform is then utilized to measure the heart rate, size and location of the heart chambers, presence of any injury to the heart, and various medications or devices utilized to control or regulate cardiac function. It's like a heartbeat. At the same time, it may give information about balance of the nearby salts (electrolytes) (e.g. hypercarbia) or even reveal problems with sodium channels in Brugada syndrome (heart muscle cells). A heart attack (myocardial infarction) is, of course, one of the tests performed; The Electrocardiogram can determine, whether the heart muscle is hurted in particular areas even though the heart is not covered by all the zones. The Electrocardiogram cannot measure the ability to pump the heart, ultrasound-based (echocardiography), or reliably using nuclear medicine tests. The ECG signal from a patient is affected by external noise, so we need a convenient way to get a noiseless Electrocardiogram signal. The electrocardiogram signal is a combination of P, T, U wave and QRS complex. Full waveform is named the ECG labeled P, Q, R, S, and T, which define distinguishing features. The P wave originates from atrium depolarization and QRS complex originates from ventricles storage. T wave is caused by repolarization of ventricular muscle.

Fuzzy logic, in machine control, is widely utilized. The term "fuzzy" means that logic can deal with notions that cannot be expressed as "right" or "wrong", and is instead defined as "partially correct". Fuzzy logic has the advantage of solving the problem as human operators can understand, although it can perform many fuzzy logic besides alternate approaches like neural networks and genetic algorithms. The controller can be used in the design, which makes it easier for people to carry out tasks successfully accomplished [3]. Gaussian window is beneficial for time-frequency analysis because both the Fourier transform and the derivation of the Gaussian window are Gaussian functions.

This manuscript is organized as follows. Section 2 defines a denoising of ECG signal. Section 3 shows the Methodology of research. Section 4 gives the details of Gaussian Weighted Moving Windowing method. Section 5 demonstrates experimental outcomes of study done and discussion of these results and last part will be about conclusion.

## 2 Electrocardiogram Signal Denoising

For improving signal quality and minimize random error noise component, digital filtering techniques can be utilized [4]. The signal from the electrocardiograph is described as:

$$y(t) = x(t) + n(t) \tag{1}$$

The electrocardiogram signal measured at time "t" is represented by x(t), random noise affecting this signal is denoted by n(t) and y(t) is the signal of the electrocardiograph. A substantial problem, in low pass filtering is the normal overlap of the signal and noise spectra because without distorting x(t) it is not probable to subtract n(t) from y(t). Examine Gaussian Weighted Moving Windowing approach to analyze the ECG signal, is the aim of this manuscript.

## 3 Methodology of Research

The majority of electrocardiogram analysis and interpretation systems perform signal processing. Signal processing is utilized to remove some characteristic parameters. Nowadays, Biomedical Signal Processing aims at physiological systems quantitative or objective analysis and phenomena by signal analysis. Biomedical signal processing field or analysis has advanced to signal processing practical application and pattern analysis techniques in order to effective and advanced noninvasive diagnosis, critical patients rehabilitation and sensory support online rehabilitation in order to rehabilitation. Electrocardiogram is for signal processing and involves improving measurement accuracy and repeatability. Electrocardiogram analysis rests generate sets of basic algorithms to interpret electrocardiogram testing of stress, interpretation, ambulatory monitoring or intensive care monitoring, signal status associated with different noise and artifacts, heartbeat sensing and wave widths is based on electrocardiogram measurements, data compression in order to transmission or efficient storage.

Fundamental Electrocardiogram has frequency range between 0.5 Hz and 100 Hz. The elimination of artifacts plays a vital role in electrocardiogram signal processing. If specialists are present in the Electrocardiogram signal, it is difficult for specialists to diagnose the disease. Block diagram of the study done, is illustrated at Fig. 2 below.

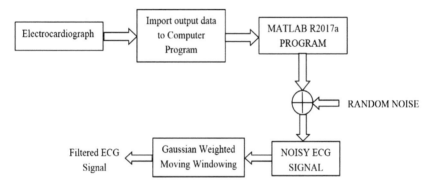

**Fig. 2.** Block diagram of the study done

The noisy Electrocardiogram signal was composed by adding convenient signal distributions with the reference signal. Gaussian Weighted Moving Windowing is tested with Random noise. The PSNR value between the original and denoised signals is calculated. Figure 3 illustrates a noiseless ECG signal and fast Fourier transform of the original signal that is utilized for this study. Noisy ECG signal with +10 dB random noise and discrete Fourier transform of the noisy signal was described in Fig. 4.

## 4   Filtering Techniques

In this section, Gaussian Weighted Moving Window filtering method applied for this study will be explained. In the process of designing digital filters, windowing techniques are utilized predominantly. The window activity is performed to convert infinite impulse response to the Finite Impulse Response filter design [5].

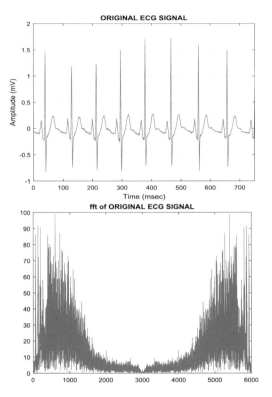

**Fig. 3.** Noiseless ECG signal with fft spectral density

Symmetric window arrays functions are for design of digital filter. These window functions are generally a strange length with a single maximum at the center. For spectral analysis [6], windows for Discrete Fourier Transform/Fast Fourier Transform is generated by taking the rightmost coefficient of a single-length, symmetrical window. The intermittent sequences are known periodically. When interrupted sequence is extended periodically, it can be restored with the deleted coefficient (a virtual copy of asymmetric leftmost coefficient). The window technique includes a function named window function; This returns a zero value outside of this range when a non-zero value and a time interval are selected at the end of the interval. The Gaussian window used to create the parabola is as follows; this can be used to estimate the exact quadrature interpolation frequency.

$$w(n) = e - \frac{1}{2}\left(\frac{n - \frac{N-1}{2}}{\sigma\frac{N-1}{2}}\right)^2, \sigma \leq 0.5 \tag{2}$$

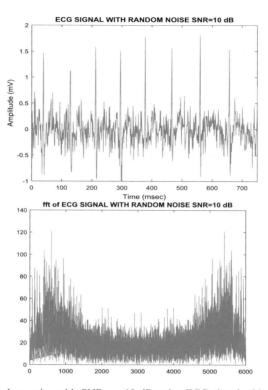

**Fig. 4.** Random noise with SNR = +10 dB noisy ECG signal with fft spectral

## 5 Classification

### 5.1 Fuzzy Logic

In 1965, by Lotfi A. Zadeh fuzzy logic was first introduced with fuzzy clusters concept as classical set theory extension specified by open clusters. Subsequently, an entire algebra that uses fuzzy sets to constructively account for classical logic as an extension of its correct operations was defined fuzzy logic [7]. In most status, fuzzy logic system is essentially a scalar output nonlinear map where an input data vector is defined as a linguistic expression whose computation is explicitly computed by the number.

Therefore, the fuzzy system is unique, the fuzzy logic system can handle numerical data and language knowledge. To many distinct associations, this logic, indicates that there are many probabilities that lead [8].

**Fuzzy Logic Preference Reasons**
Fuzzy logic observations general list, is given below:

- To uncertain data, fuzzy logic is tolerant.
- Fuzzy logic conceptual understanding is easy.
- Fuzzy logic is flexible.
- Traditional control techniques can be mixed with fuzzy logic.

- Fuzzy logic can model the complexity of random nonlinear functions.
- Fuzzy logic can be created on experience experts.
- Natural language is used for fuzzy logic.
- Fuzzy systems are not necessarily traditional control methods. Fuzzy systems simplify and facilitate their application, in many cases.

## 5.2    Sets of Fuzzy

Accession of an element in a set is either all or nothing in standard set theory. Therefore maps characteristic function of an element into either 1 (in the set) or 0 (not in the set) [27].

## 5.3    Functions of Membership (MF)

A Membership function is a curve that describes how each point in the input space is mapped to a membership value (or degree of membership) among 0 and 1. Input space is occasionally referred to as universe of discourse, a fancy name for a simple concept. Label is a name entitled for each MF that described. For instance, an input variable like, "Input_1" and "Input_2" might have two membership functions labeled as Filtered and Noisy signal in this study.

Chosen MF type is:

- **Triangle:** trimf is a name of Triangular MF. To form a triangle, it collects more than three points [28].

## 5.4    Variables of Linguistic

Whose values are natural words or phrases in place of numerical values called as the linguistic variables are system input or output variables. A linguistic variant is usually broken down into linguistic terms set, here we have Filtered and Noisy.

## 5.5    Universe of Discourse

Fuzzy set elements are taken from Universe in order to short or a discourse universe. Universe includes all elements that can take into account. Even universe depends on context [29].

## IF - THEN Rules

Operators and fuzzy sets of fuzzy are fuzzy logic verbs and subjects. To formulate conditional expressions containing fuzzy logic, these IF-THEN rule expressions are utilized.

A single fuzzy IF-THEN rule can be described as follows:

If x is "a" then y is "b".

Where "a" and "b" are variables of linguistic, fuzzy clusters are described in X and Y intervals (discourse universes), respectively. When if-part of rule "X is a" is named a priority or a premise, a part of rule "b with y" is called a result.

## 6    Experiment

Proposed study was implemented utilizing MATLAB (R2017a) software to reveal ECG data. Using the measurement data obtained with the help of doctors and nurses from Near East Hospital, the recommended method was utilized. The Electrocardiogram records the electrical activity resulting from depolarizations of the heart muscles, which allows electrical waves to propagate to the periphery. Although the amount of electricity is actually very small, it can be reliably taken with the Electrocardiogram electrodes attached to skin (data unit: microvolt, uV). Full Electrocardiogram setting includes at least four electrodes placed in accordance with classical nomenclature (RA = right arm; LA = left arm; RL = right leg; LL = left leg), chest or four limbs.

With help of generated data set, our proposed Gaussian Weighted Moving Window is interpreted and verified with PSNR. Study and measurement noise variance are initialized for Gaussian Weighted Moving Windowing as a window = 1 is setted.

## 7    Outcomes

The outcomes of this implementation is explained below; the Electrocardiogram signal measured at time "t" is generated, then Signal to Noise Ratio +10 dB random noise is added to ECG signal. Noisy Electrocardiogram signal is applied to the Gaussian Weighted Moving Window and compared to the resulting output. In this case, the true noiseless ECG signal, Peak Signal Noise Ratio is calculated. The results clearly demonstrate that the Gaussian Weighted Moving Windowing method can grab and combine the combined noise effects. The outcomes of denoising with Gaussian Weighted Moving Windowing is illustrated in Figs. 4, 5, 6 and Table 1.

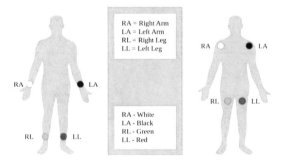

**Fig. 5.** The full ECG setup

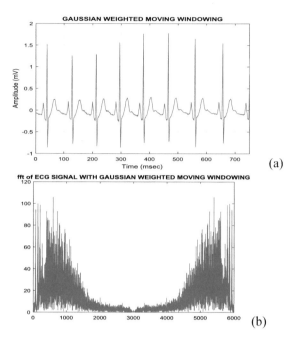

(a)

(b)

**Fig. 6.** (a) Noisy Electrocardiogram signal with SNR = 10 dB random noise denoised with Gaussian Weighted Moving Windowing (b) fft spectral

**Table 1.** Gaussian Weighted Moving Windowing PSNR values

| SNR (dB) | Order of Gaussian Weighted Moving Windowing | Noisy ECG signal | Gaussian Weighted Moving Windowing (PSNR VALUE) |
|---|---|---|---|
| SNR = +10 dB | 1 | PSNR = +71.568760 dB | PSNR = +94.002500 dB |
| | 2 | PSNR = +71.446902 dB | PSNR = +82.605035 dB |
| | 3 | PSNR = +71.605512 dB | PSNR = +78.066765 dB |
| | 4 | PSNR = +71.750399 dB | PSNR = +75.574702 dB |
| | 5 | PSNR = +71.481966 dB | PSNR = +74.097113 dB |

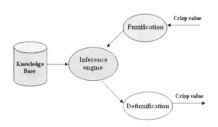

**Fig. 7.** Fuzzy inference system structure [28]

## 7.1    Classification Outcomes

Fuzzification module, Inference engine module, Knowledge base module, and Defuzzification modules are the four modules of Fuzzy Inference System (FIS). Direct methods and indirect methods are classified as fuzzy inference methods. The most commonly utilized direct methods are Mamdani's and Sugeno's methods. Indirect methods are more complicated. Most common fuzzy inference technique is Mamdani method. The Mamdani model is a knowledge-based prediction model. The Mamdani model works with clear data entries and with linguistic terms and intervals. An important advantage of this model is that the Mamdani model is a confidence measure in estimating the future value when true value is not known (Fig. 7).

## 7.2    Fuzzy Rules

The fuzzy rules depend on the number of input variables of the linguistic IF-THEN constructs in which general form is "IF a THEN b" and the Fuzzy Rules of "b" and MF. In fuzzy rule based selection model illustrated in Fig. 8, has 2 variables and 1 membership functions = 2 ^ 1 = 2 rules be obtained (Fig. 9).

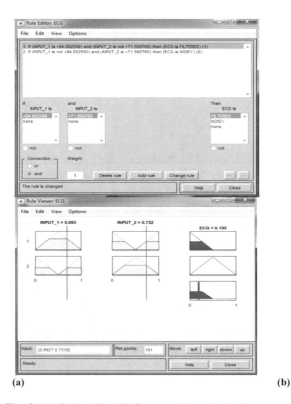

(a)                                                            (b)

**Fig. 8.** (a) Rule editor (b) Fuzzy rule based selection process

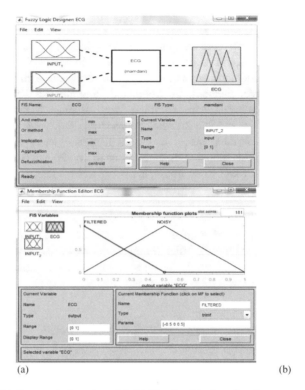

**Fig. 9.**  (a) Fuzzy rule based model (b) Membership function for output classifier

## 8   Conclusion

The electrocardiogram is a record of the bioelectrical potential that changes with the electrical activity of the heart. Various filters have been used to obtain a spurious signal. In communication technological advances and low power circuit design have led to safer development Electrocardiogram appliances with the latest diagnostic features. However, the sensitivity of electrocardiogram signal impairment to even a small noise makes Electrocardiogram filtering very important with various filters. For this reason, despite the technological advances, electrocardiogram signal acquisition and signal analysis remains a challenging task. Our proposed study is to classify electrocardiogram signals as filtered and noisy using fuzzy logic control with the Gaussian weighted moving windowing has verified its achievement in denosing ECG signal with actual data sets. In this manuscript, a solution that can be done in distinct types of faults and Electrocardiography in the ECG data was analyzed with random noise and results, the above given results was obtained. The main goal in the whole system is to get clear, better quality output signals for good consultation. A variety of other filtering methods, will be future study.

# References

1. Nayak, S., Soni, M.K., Bansal, D.: Filtering techniques for ECG signal processing. Int. J. Res. Eng. Appl. Sci., IJREAS **2**(2), 671 (2012). ISSN 2249-3905
2. Correia, S., Miranda, J., Silva, L., Barreto, A.: LABVIEW and MATLAB for ECG acquisition, filtering and processing. In: 3rd International Conference on Integrity, Reliability and Failure, Porto, Portugal, 20–24 July (2009)
3. Pedrycz, W.: Fuzzy Control and Fuzzy Systems, 2nd edn. Research Studies Press Ltd., Taunton (1993)
4. Anderson, B.D., Moore, J.B.: Optimal Filtering. Dover Publications Inc., Mineola (2005)
5. Subhadeep, C.: Advantages of Blackman window over hamming window method for designing FIR filter. Int. J. Comput. Sci. Eng. Technol. **4**(8), 1181–1189 (2013)
6. Nagarajan, T., Prasad, V.K., Murthy, H.A.: Minimum phase signal derived from root cepstrum. IEE Electron. Lett. **39**(12), 941–942 (2003)
7. Badiru, A.B., Cheung, J.Y.: Fuzzy Engineering Expert Systems with Neural Network Applications. Wiley, New York (2002)
8. Sandya, H.B., Hemanth Kumar, P., Himanshi, B., Susham, K.R.: Fuzzy rule based feature extraction and classification of time series signal. Int. J. Soft Comput. Eng. (IJSCE) **3**, 42–47 (2013)
9. Sadıkoğlu, F., Kavalcıoğlu, C., Dağman, B.: Electromyogram (EMG) signal detection, classification of EMG signals and diagnosis of neuropathy muscle disease. Procedia Comput. Sci. (2017). https://doi.org/10.1016/j.procs.2017.11.259
10. Kavalcıoğlu, C., Dağman, B.: Filtering maternal and fetal electrocardiogram (ECG) signals using Savitzky-Golay filter and adaptive least mean square (LMS) cancellation technique. Bull. Transylvania Univ. Bras.: Ser. III: Math., Inform., Phys. **9**(**58**)(2), 109–124 (2016)
11. Sadıkoğlu, F., Kavalcıoğlu, C.: Filtering continuous glucose monitoring signal using Savitzky-Golay filter and simple multivariate thresholding. Procedia Comput. Sci. (2016). https://doi.org/10.1016/j.procs.2016.09.410
12. Gomes, P.R., Soares, F.O., Correia, J.H.: ECG self diagnosis system at P- R interval. In: Proceedings of VIPIMAGE, pp. 287–290 (2007)
13. Pinheiro, E., Postolache, O., Pereira, J.M.D.: A practical approach concerning heart rate variability measurement and arrhythmia detection based on virtual instrumentation, pp. 112–115 (2007)
14. Sornmo, L., Laguna, P.: Electrocardiogram signal processing. In: Wiley Encyclopedia of Biomedical Engineering (2006)
15. Yatindra, K., Malik, G.K.: Performance analysis of different filters for power line interface reduction in ECG signal. Int. J. Comput. Appl. **3**(7), 1–6 (2010)
16. Joshi, P.J., Patkar, V.P., Pawar, A.B., Patil, P.B., Bagal, U.R., Mokal, B.D.: ECG denoising using MATLAB. Int. J. Sci. Eng. Res. **4**, 1401–1405 (2013)
17. Birle, A., Malviya, S., Mittal, D.: Noise removal in ECG signal using Savitzky - Golay filter. Int. J. Adv. Res. Electron. Commun. Eng. (IJARECE) **4**, 1331–1333 (2015)
18. Meireles, A.J.M.: ECG denoising based on adaptive signal processing technique. Master thesis, ISEP Instituto Superior de Engenharia do Porto (2011)
19. AlMahamdy, M., Riley, H.B.: Performance study of different denoising methods for ECG signals. In: The 4th International Conference on Current and Future Trends of Information and Communication Technologies in Healthcare (ICTH-2014) (2014)
20. Chandrika, B., Yadav, O.P., Chandra, V.K.: A survey of noise removal techniques for ECG signals. Int. J. Adv. Res. Comput. Commun. Eng. **2**, 1354–1357 (2013)

21. Islam, M.K., Haque, A.N.M.M., Tangim, G., Ahammad, T., Khondokar, H.: Study and analysis of ECG signal using MATLAB & LABVIEW as effective tools. Int. J. Comput. Electr. Eng. **4**, 404–408 (2012)
22. Kavitha, R., Christopher, T.: A study on ECG signal classification techniques. Int. J. Comput. Appl. **86**, 9–14 (2014)
23. Kumar, N., Ahmad, I., Rai, P.: Signal processing of ECG using Matlab. Int. J. Sci. Res. Publ. **2**, 1–6 (2012)
24. Nayak, S., Soni, K.M., Bansal, D.: Filtering techniques for ECG signal processing. IJREAS **2**, 2249–3905 (2012)
25. PubMed: The U. S. National Library of Medicine and the National Institutes of Health, A service of the U.S. (2008). http://www.pubmed.gov/
26. Kasar, S., Mishra, A., Joshi, M.: Performance of digital filters for noise removal from ECG signals in time domain. Int. J. Innov. Res. Electr., Electron., Instrum. Control. Eng. **2**, 1352–1355 (2014)
27. Jantzen, J.: Tutorial on fuzzy logic. Technical University of Denmark (2008)
28. Güler, I., Ubeyli, E.D.: Adaptive neuro-fuzzy inference system for classification of EEG signals using wavelet coefficients. J. Neurosci. Methods **148**, 113–121 (2005)
29. Matsuyama, A., Jonkman, M., de Boer, F.: Improved ECG signal analysis using wavelet and feature extraction. Methods Inf. Med. **46**, 227–230 (2007)

# Integrated Deep Learning Structures for Hand Gesture Recognition

Senol Korkmaz[⊠]

Applied Artificial Intelligence Research Centre,
Department of Computer Engineering, Near East University, Mersin-10, Nicosia,
North Cyprus, Turkey
senol.korkmaz@neu.edu.tr

**Abstract.** In this paper, object control with hand movements is proposed for controlling distant objects even when the user is far from the system. This method is based on finding hands, predicting the states and direction of the hand movement. This human-computer interface (HCI) is an assistive system for users at near/far from objects. The model is specifically designed for controlling computer mouse on big screens during the formal presentation. By moving the hand to the left, right, up and down moves the mouse pointer and sends mouse button command using hand states. Here, close hand triggers mouse button, until the same hand is opened. In this system, Single Shot Multi Box Detection (SSD) architecture is used for object detection and Convolutional Neural Network (CNN) is used for predicting hand states. This integrated system allows users to control the mouse from a distant position without using any hardware. The test results show that this system is robust and accurate. This invention uses a single camera and aids users who are far from the computer during a presentation to shuffle through the slides.

**Keywords:** Hand detection · Convolutional Neural Network · Computer vision
Deep learning

## 1 Introduction

Human-Computer Interaction should not be narrowed to just mouse and keyboard interactions. There are several sensory modes such as body expressions, speech and gestures where interaction between humans occurs. In many fields of human-computer interaction it is becoming highly important to naturally interact with the system. In human-computer interaction, gesture recognition has been an active area of study. As addressed in [1], the initial goal was to detect and recognize sign language. Gesture recognition has recently attracted high attention in research areas including; TV remote controlling [2], crisis management [3], direct interaction with computer system [4], augmented reality applications [5], gaming interfaces [6], interaction in hand-free car driving [7], fatigue detection in drivers [8] and car driving virtual training provision [9]. Computer system detecting gestures so as to understand user activities is another kind of case study. An example is robots confirmed to analyzing gestures so as to detect what jobs are being concluded in order to coherently take over with the next

© Springer Nature Switzerland AG 2019
R. A. Aliev et al. (Eds.): ICAFS-2018, AISC 896, pp. 129–136, 2019.
https://doi.org/10.1007/978-3-030-04164-9_19

stages/steps [10]. In the medical area, Robot nurses in medical field for instance, are conceived to detecting hand gestures of surgeons and to also aid with important instruments for surgery as depicted in [3]. For quality assurance purposes, documenting the observed gestures is sometimes very useful considering assembly lines scenario, to properly document the job [11]. Generally, major aim of detecting tasks of assembly line is a field of proactive research as seen in [12]. In the context of daily life logging activities, gesture recognition has been widely explored. For instance, as shown in [13], based on both smartphone and smart watch sensing, activity logging is utilized for detecting not eating or drinking too much coffee and finally in [14], the possibility of detecting eating habits via recognizing the gestures for drinking and eating was explored by the authors.

In this study, our major contribution to the field of gesture recognition is the implementation of a naturally interaction-oriented hand gestures control of objects from the distance. Our framework is composed based on; hand detection, predicting states and direction of the hand movement. This human computer interface (HCI) is an assistive system designed for users to operate the system from a distant position; an instance of controlling computer mouse on big screens during a formal presentation, by shuffling through the slides and moving objects from left, right, up and down via hand states and direction. Where, close hand triggers mouse button, until open again. In this system, Single Shot Multi Box Detection (SSD) architecture was used for object detection and Convolutional Neural Network (CNN) was used for predicting hand states. This integrated system allows users from the computer to control mouse pointer without using any hardware at a far distance.

To achieve hand gesture recognition, we conducted some literature surveys which demonstrate an insight to various implemented models. The conducted surveys aided us to understand various disadvantages and advantages found in the different algorithms. Cameras, hand belts and data gloves are some of the commonly explored methodologies for capturing input data from the users. Gesture recognition approach as implemented in [15] utilizes data gloves as an input data extraction method. The creative Senz3D camera was used in [16] by the researchers for capturing both the depth and colour information. To read hand movements, a gyroscope with a hand belt, bluetooth and an accelerometer was deployed as shown in [17]. In [18], the authors used a bumblebee2 stereo camera for model design. In [19], a monocular camera was used for data collection. As demonstrated in [20], cost-efficient model was explored and implemented by the authors using simple web cameras. In order to capture the color stream, method explored in [21] utilized kinect depth RGB camera. We see in [21] where the authors were able to obtain additional depth information for certain pixel by using depth cameras that is traditional images at a frame rate as well as depth images. By exploring colour space, most of the technologies allows robust extraction of the hand region. Background problem was not completely tackled by these methods. In [22], this background problem was addressed where the authors utilized monochrome glove (white and black pattern of augmented reality markers). So many methodologies have been used to pre-process images. These comprise techniques and algorithms for smoothening, edge detection, noise removal followed by other techniques of segmentation for extraction of boundary that is; extracting the background from foreground. To smoothen the contours after binarization researchers in [23] used the Gaussian filter. To eliminate noise, authors in [24] utilized the morphology algorithm to perform image dilation and image erosion. The depth map is computed through mapping right and left images with Sum of Absolute Differences (SAD) technique to perform segmentation as seen in [18]. Here, to find the contours, the authors explored Theo Pavildis algorithm which visits only the boundary pixels and the method was able to calculate the computational costs. In order to recognise 25 hand postures, the approach in [25] uses Euclidean distance based classifier. The largest contour was chosen in [26] to represent the hand palm contour after which the contour was simplified with the help of the polygonal approximation. Here, output prediction was seen as a procedure where individual objects were categorized with respect to similarity amongst the objects. In [27], Support Vector Machine (SVM) classifier is utilized for output prediction. To avoid generating depth information, inbuilt webcams requires less computing costs and for this reason, we used a webcam localized in our system without using hand gloves (hand markers) or any other additional cameras. Here, Single Shot Multibox Detection (SSD) architecture was used for object detection while Convolutional Neural Network (CNN) was used for predicting hand states.

The remaining part of the paper is organized as follows: Sect. 2 presents the proposed algorithm. Section 3 describes the simulation process of the system and Sect. 4 presents the conclusions of the paper.

## 2 Proposed Algorithm

### 2.1 Convolutional Neural Network (CNN)

Convolutional Neural Networks (CNN) is used in cases where the data can be expressed in "map" form. The proximity between the two data points here indicates how closely they are related to each other. The views are examples of maps in this case. This is why it is useful in image analysis. CNN takes an image that can be expressed as an array, and applies a set operation to that array and finally, returns the possibility of what object group an object in the image belongs.

CNN is a variation of multilayer perceptron that includes convolution layers, rectified linear units (ReLU layers), pooling layers, fully-connected layers. The convolution layers have a hierarchical structure and they are core building block of a CNN.

The first layer of CNN structure is always a Convolutional Layer. Convolutional layers apply a convolution operation to the input which is an $\underline{mxmxr}$ image where $m$ is the weight and height of the image and $r$ is the number of channels. The convolutional layer has $\underline{k}$ kernels of size $\underline{nxnxq}$ where $n$ is smaller than the dimension of image and $q$ can be same with $r$ or smaller. The size of filters used to produce locally connected structure which are each convolved with the image to k feature maps of size $m - n + 1$.

ReLU layers are used as an activation function for CNN structure and can be expressed as

$$f(x) = max(0, x) \tag{1}$$

The pooling layer aims to achieve shift-invariance by reducing the resolution of the feature maps. It is usually placed between two convolutional layers. Finally, fully-connected layer is applied for classification purpose.

## 2.2    Single Shot Multi Box Detection

Single Shot Multi Box Detector was released by Szegedy et al. in November 2016 [29] and reach new records performance and precision for object detection tasks, scoring over 74% mAP (mean average precision) at 59 frames per second (fps) on standard datasets such as PascalVOC.

The SSD architecture is based on the VGG-16 architecture, as shown in Fig. 1 above. However, it removes the layers that are completely connected and instead adds a series of convolutional layers. This allows for the extraction of features on multiple scales and the gradual reduction of the size of the input to each subsequent layer.

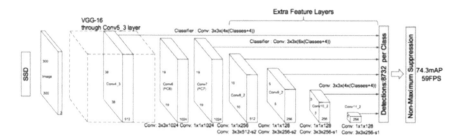

**Fig. 1.**  Structure of the SSD network

At the end, Multi Box only retains the top K predictions that have minimized both location and confidence losses.

## 3    Simulation of the System

### 3.1    The Structure of the System

The mouse control system using vision-based object detection and recognition is implemented using CNN. The first structure which is SSD MobileNet's input are images obtained from video camera located in front of the user. The video camera sends images to the system input. SSD uses the images and determines the possible position of hands. After that, mostly include hand position cropped and will be used as input for the second CNN structure for recognition purposes. We accept image center as "origin" and compute mouse signals from the difference of "origin - hand area center". The difference between this coordinates give us the expected mouse coordinate.

The second CNN model has three parameters, width, height and depth. The width and height are the sizes of the image which is cropped area, extracted by object detection system. The depth is the number of channels of input image. In our case, width, height and depth are 28, 28, 3 respectively.

The second CNN model is constructed as CONV1 => CONV2 => CONV3 => FLATTEN => FC LAYER. First Conv layer has 10 filters and (5 × 5) kernel size.

Then we applied ReLU activation function. Second Conv layer has 20 filters and (5 × 5) kernel size and then ReLU applied. Last Conv layer has 50 filters and same kernel size. We did not apply any pooling layer. In the Fully Connected layer, we have 1500 input neurons and class size output layer of "2" in our case. In the simulation, number of classes is two, "open" and "close" respectively to hand position.

## 3.2   Results

The CNN and its learning algorithm are used for "hand detection and classification" for assistive HCI system. Tensorflow [30] framework is used for this purpose. SSD MobileNet architecture is used for object detection. We have trained SSD with 5000 epoch and loss graph shown in Fig. 2.

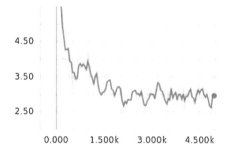

**Fig. 2.**  Mean squared error of SSD

Our CNN architecture is used for classification purposes. We defined 10 epochs, 1e-3 learning rate and 64 batch size. Images have been chosen from "ASL Finger Spelling Dataset" [31] "A" and "B" section. 2179 closed hand images and 2238 open hand images is used for training. 245 closed and 245 open hand images were used for training (Figs. 3 and 4).

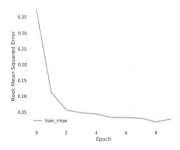

**Fig. 3.**  Root mean squared error of our CNN structure

**Fig. 4.** ASL dataset

## 4   Conclusion

The outcome of this research is an implementation of an integrated system that allows users at far distance from the computer to control mouse pointer without using any hardware. The implemented model is of high importance at formal presentation where the presenter could shuffle through slides and move objects from one point to the other even when at distant location from the system. Here, close hand triggers mouse button, until open again. This integrated system is made of Single Shot Multi Box Detection (SSD) architecture used for object detection and Convolutional Neural Network (CNN) used for predicting hand states. Inbuilt webcam was used input data collection medium. When we tested the obtained input dataset on our algorithm, 98.37% accuracy was achieved as depicted in Table 1. This high accuracy outperformed other author's models as demonstrated in the same table.

**Table 1.** Comparison results of SVM, FFNN and CNN

|             | HOG + SVM | FFNN  | CNN   |
|-------------|-----------|-------|-------|
| Accuracy    | 84.93     | 95.51 | 98.37 |
| Sensitivity | 79.67     | 99.59 | 99.59 |
| Specificity | 90.20     | 91.43 | 97.14 |

## References

1. Huang, C., Huang, W.: Sign language recognition using model-based tracking and a 3D Hopfield neural network. Mach. Vis. Appl. **10**, 292–307 (1998)
2. Ren, G., Li, C., O'Neill, E., Willis, P.: 3D freehand gestural navigation for interactive public displays. IEEE Comput. Graphics Appl. **33**, 47–55 (2013)
3. Wachs, J., Kölsch, M., Stern, H., Edan, Y.: Vision-based hand-gesture applications. Commun. ACM **54**, 60 (2011)
4. Kumar, P., Verma, J., Prasad, S.: Hand data glove: a wearable realtime device for human-computer interaction. Int. J. Adv. Sci. Technol. **43** (2012)

5. Kavakli, M., Taylor, M., Trapeznikov, A.: Designing in virtual reality (desire): a gesture-based interface. In: Proceedings of the 2nd International Conference on Digital Interactive Media in Entertainment and Arts, DIMEA, pp. 131–136 (2007)

6. Martins, T., Sommerer, C., Mignonneau, L., Correia, N.: Gauntlet: a wearable interface for ubiquitous gaming. In: MobileHCI: Proceedings of the 10th International Conference on Human Computer Interaction with Mobile Devices and Services, p. 367. ACM Request Permissions, New York (2008)

7. Molchanov, P., Gupta, S., Kim, K., Pulli, K.: Multi-sensor system for drivers hand-gesture recognition. In: IEEE Conference on Automatic Face and Gesture Recognition, pp. 1–8 (2015)

8. Lee, B., Lee, B., Chung, W.: Wristband-type driver vigilance monitoring system using smartwatch. IEEE Sens. J. **15**, 5624–5633 (2015)

9. Xu, D.: A neural network approach for hand gesture recognition in virtual reality driving training system of SPG. In: 18th International Conference on Pattern Recognition, vol. 3, pp. 519–522 (2006)

10. Roitberg, A., Somani, N., Perzylo, A., Rickert, M., Knoll, A.: Multimodal human activity recognition for industrial manufacturing processes in robotic workcells. In: Proceedings of the 2015 ACM on International Conference on Multimodal Interaction, pp. 259–266 (2015)

11. Stiefmeier, T., Roggen, D., Ogris, G., Lukowicz, P.: Wearable activity tracking in car manufacturing. Pervasive Comput. IEEE **7**(2), 42–50 (2008)

12. Ou, J., Shi, Y., Wong, J., Fussell, R., Yang, J.: Combining audio and video to predict helpers' focus of attention in multiparty remote collaboration on physical tasks. In: Proceedings of the 8th International Conference on Multimodal Interfaces, pp. 217–224 (2006)

13. Shoaib, M., Bosch, S., Scholten, H., Havinga, P., Incel. D.: Towards detection of bad habits by fusing smartphone and smartwatch sensors. In: IEEE International Conference on Pervasive Computing and Communication Workshops (PerCom Workshops), pp. 591–596 (2015)

14. Sen, S., Subbaraju, V., Misra, A., Balan, R., Lee, Y.: The case for smartwatch-based diet monitoring. In: IEEE International Conference on Pervasive Computing and Communication Workshops (PerCom Workshops), pp. 585–590 (2015)

15. Ji-Hwan, K., Nguyen, D., Tae-Seong, K.: 3-D hand motion tracking and gesture recognition using a data glove. In: IEEE International Symposium on Industrial Electronics, pp. 1013–1018 (2009)

16. She, Y., Wang, Q., Jia, Y., Gu, T., He, Q., Yang, B.: A real-time hand gesture recognition approach based on motion features of feature points. In: IEEE 17th International Conference on Computational Science and Engineering (CSE), pp. 1096–1102 (2014)

17. Hung, C.H., Bai, Y., Wu, H.: Home appliance control by a hand gesture recognition belt in LED array lamp case. In: IEEE 4th Global Conference on Consumer Electronics (GCCE), pp. 599–600 (2015)

18. Lee, D., Hong, K.: A hand gesture recognition system based on difference image entropy. In: 6th International Conference on Advanced Information Management and Service (IMS), pp. 410–413 (2010)

19. Dulayatrakul, J., Prasertsakul, P., Kondo, T., Nilkhamhang, I.: Robust implementation of hand gesture recognition for remote human-machine interaction. In: 7th International Conference on Information Technology and Electrical Engineering (ICITEE), pp. 247–252 (2015)

20. Hussain, I., Talukdar, A., Sarma, K.: Hand gesture recognition system with real-time palm tracking. In: Annual IEEE India Conference (INDICON), pp. 1–6 (2014)

21. Wang, C., Liu, Z., Chan, S.: Superpixel-based hand gesture recognition with kinect depth camera. IEEE Trans. Multimedia **17**, 29–39 (2015)
22. Ishiyama, H., Kurabayashi, S.: Monochrome glove: a robust real-time hand gesture recognition method by using a fabric glove with design of structured markers. In: IEEE Virtual Reality (VR), pp. 187–188 (2016)
23. Huong, T.N., Huu, T., Le, T.: Static hand gesture recognition for vietnamese sign language (VSL) using principal components analysis. In: International Conference on Communications, Management and Telecommunications (ComManTel), pp. 138–141 (2015)
24. Suriya, R., Vijayachamundeeswari, V.: A survey on hand gesture recognition for simple mouse control. In: International Conference on Information Communication and Embedded Systems (ICICES), pp. 1–5 (2014)
25. Luzhnica, G., Simon, J., Lex, E., Pammer, V.: A sliding window approach to natural hand gesture recognition using a custom data glove. In: IEEE Symposium on 3D User Interfaces (3DUI), pp. 81–90 (2016)
26. Chen, W., Wu, C., Lin, C.: Depth-based hand gesture recognition using hand movements and defects. In: International Symposium on Next-Generation Electronics (ISNE), pp. 1–4 (2015)
27. Chen, Y., Ding, Z., Chen, Y., Wu, X.: Rapid recognition of dynamic hand gestures using leap motion. In: IEEE International Conference on Information and Automation, pp. 1419–1424 (2015)
28. Abiyev, R., Arslan, M., Gunsel, I., Cagman, A.: Robot pathfinding using vision based obstacle detection (2017)
29. Liu, W., Anguelov, D., Erhan, D., Szegedy, C., Reed, S., Fu, Y., Berg, A.: SSD: single shot multibox detector. In: European Conference on Computer Vision, pp. 21–37. Springer (2016)
30. Abadi, M.: TensorFlow: learning functions at scale. In: ACM SIGPLAN Notices, vol. 51, p. 1 (2016)
31. Pugeault, N., Bowden, R.: Spelling it out: real-time ASL fingerspelling recognition. In: Proceedings of the 1st IEEE Workshop on Consumer Depth Cameras for Computer Vision, jointly with ICCV. (2011)

# Rule Based Intelligent Diabetes Diagnosis System

Elbrus Imanov[1]([⊠]), Hamit Altıparmak[1], and Gunay E. Imanova[2]

[1] Department of Computer Engineering, Near East University, via Mersin 10,
Nicosia, North Cyprus, Turkey
{elbrus.imanov,hamit.altiparmak}@neu.edu.tr
[2] Department of Business Administration, Near East University, via Mersin 10,
Nicosia, North Cyprus, Turkey
gunel-gunay@hotmail.com

**Abstract.** It is known that expert systems have been used for many years. By courtesy of the advanced technology and the recent studies made on expert systems, this field of study has been gained popularity and many successful progresses have been made over time. As an evidence of this improvement we can discuss about the results shown by the expert system that gave us very close and sometimes exact values as human decision making.

The purpose of this study is to design diabetes diagnosis system. Acquiring right data is needed for the application of rules to this design. These rules determine whether a person is healthy or diabetes patient, along with its types such as type1 diabetes, type2 diabetes, gestational diabetes, and at risk. VP-Expert rule based system was used to design this diabetes diagnosis system, and this system passed many tests with success. System was tested on 15 patients and able to achieve exact results as doctors. System that we have designed can be used effectively and efficiently to determine diagnoses for diabetes especially in undeveloped and crowded countries where the number of doctors is not enough compared to the population. Due to the annual increasing number of patients, rule based intelligent system targets to reduce the dependence on doctors, and therefore it will help both doctors and patients to make more correct and quicker decisions.

**Keywords:** Expert system (ES) · Diabetes mellitus (DM) · VP expert
Artificial intelligent (AI) · Diabetes diagnose expert system (DDES)
Certainty factors (CF)

## 1 Introduction

This paper aimed at developing a diabetes diagnosis system, the developed system can be used efficiently for diagnosis of diabetes where the number of patients is increasing daily. Hence the designed system will help to reduce the dependency on doctors, and save time for both doctors and patients. An Expert System as the part of artificial intelligence is a computer system that is able to imitate or mimic the tasks of a human being's intelligence as a result of effective decision making similar to human. Expert system is utilized for the purpose of solving complicated problems by acquiring

© Springer Nature Switzerland AG 2019
R. A. Aliev et al. (Eds.): ICAFS-2018, AISC 896, pp. 137–145, 2019.
https://doi.org/10.1007/978-3-030-04164-9_20

knowledge that is often in terms of IF-THEN production rules. Expert system is a section of the broad phenomenon of artificial intelligence (AI). Currently expert systems represent one of the main areas of use for medicine, chemistry and social sciences such as economics, and political science, which areas need decision making practices [1]. In the real world, almost no perfect or ideal information is offered in any section of the decision making process to be used in determination of condition of nature, probabilities and outcomes. Transition techniques that consist of describing one object to others are associated with the intelligent system that comprises knowledge concepts [2]. Knowledge base is the main part of an expert system. Heuristic knowledge, scientific ideologies and computational algorithms are used to solve engineering problems. Critical knowledge of an expert system is congregated inside the knowledge base, which plays the crucial role in the success of a system [3]. Knowledge base involves 2 types of knowledge; they are stationary knowledge which comprises facts about situation, objects and events, and dynamic knowledge which is related to sequence of action. For the representation of knowledge, mainly IF-THEN production rules are used. Moreover, 'if' stands for condition, while 'then' aims recommendation. Materials such as decision tables, and block diagrams assist in representation of the acquired knowledge. Diabetes diagnosis expert system (DDES) which is a knowledge base expert system that comprises acknowledge base, inference engine, and a user face (basic module of an expert system), targets to help patients that suffer diabetics with a systematic diagnosis treatment advice, which is also helpful for doctors [3]. Direct questioning (survey) with the expert doctors and medicine dealers related with diabetes disease, provides us with the acquisition of knowledge to develop a rule based expert system [4]. Data stored in the knowledge base is the key determinant of the success of an expert system with the level of superiority, comprehensiveness and certainty of that data.

The authors brought forward a rule based process to evaluate the diabetes diagnosis system. First of all, important information is collected and placed in the system of diagnosis, and the diagnosis is performed as a second step and finally database software of this system is obtained in the third step. VP-Expert which is a rule based expert system shell comprises the inference engine, the user interface, and other important modules needed for a working expert system. A shell is an expert system with an empty knowledge base, however, designing a knowledge base for a significant domain, creates an expert system special for that domain [3]. Hence VP-Expert works if there is only knowledge base representation.

## 2    Description of Rule Based System and Inference Algorithm

Production rule based If-Then system comprises knowledge that can be described variously. Logical calculus and structured models are example techniques used for the representation of the knowledge. This study is based on knowledge representation of the rule based system, which is composed of three elemental parts:

1. Collection of rules.
2. Active database, termed as working memory.
3. Control interpreter that is in interpreting database by collecting rules [5].

Inference engine is the driving software of knowledge based systems, and it implements and operates on the knowledge form the source of knowledge which is presented in the data bases, for the solutions of the problems that arise [6]. VP-Expert operates based on the backward reasoning for inference. Forward chain involves a logical motion that is from a rule's conclusion towards another rule's premise. However backward chain involves the motion forms the conclusion to the facts. Chaining backward is realized when the rule is noticed and by obtaining the proof of the certainty of each rule's premises (with same attitude) we try to prove it [7]. The tool has an inference engine for checking the knowledge base to reply queries, an editor for coding rules of the knowledge base, and a user interfaces for handling the queries, asking questions to the patient, and offering suggestions and clarifications, where desirable. The inference engine chooses the kind of search to be used to solve the problem. In fact, the inference engine runs the expert system, defining which rule is useful, executing the rules and defining when a suitable solution is attained. The expert systems are called as systems rather than programs, because an expert system involves different elements such as facts, goals, rules and inference engine for decision making action [8]. The Structure of rule based system is represented in Fig. 1.

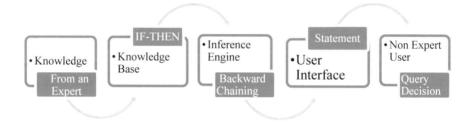

**Fig. 1.** Structure of rule based system.

In VP-Expert consultation screen there are 3 windows. The communication windows where data or information is inputted by the user and results are revealed here. The rules window permits one to see the action of the VP-Expert's inference engine. The values window records the middle and last resultant values throughout the path of the consultation. The values are shown as variable = value CF. Expert replied to the question that asks about how much certainty they can propose for this conclusion, by joining CF to the rules [9]. Patients might also offer confidence in the information they enter as answer to queries [10]. If system needed to input Panadol to treatment with a confidence of 90, this may print, "I recommend Panadol CF 90 for your symptoms". Since it wasn't probable to show the confidence in an extra useful method, your system would possibly create obvious to the patients precisely what the CF 90 is all about.

## 3   Statement of the Problem

Expert system can be started to develop as soon as the proper tools related to knowledge are selected. As a first step a plan of a hierarchical flowchart, decision table and other plans that will be used in setting and understanding of the knowledge are designed. As a result, translation of the knowledge into IF-THEN production rules is realized. Second step involves producing a sample of a part of the system (after acquiring the basic design). If the system seems to be going to perform element well and correctly, sample into the final system can be increased as required [11]. Rule based expert system is used in determination of diagnosis and treatment in endocrinology branch which has a lot and hard to determine parameters. Processes that determine the diagnosis of disease have been studied principally in some stages. Necessary information from the expert is acquired, then translation of the knowledge to rule based expert system is realized and, the database software is created.

As the composed framework is a rule based master framework, for information speaking to a few principles have been utilized. Structure of these standards is IF THEN. In the event that is exhibiting the circumstances and afterward, demonstrates the proposal [12]. For changing specialists information to these standards, there are phases that ought to be regarded, involving decision tables. In Table 1, the essential test reports including: Pregnancy, Fasting Blood Sugar 1 (FBS1), Fasting Blood Sugar 2 (FBS2), HBA1C lastly the choice about the state of the patient have been illustrated.

The second table is a choice table of analysis which demonstrates different blends of patient's circumstance, patient's age, symptoms, effective factors and tests and investigating them gives an ultimate conclusion of determination. It is clear that every line of this table demonstrates a rule of the analysis choice. Diabetes diagnosing Expert system was coded via the use of a VP-Expert shell, the shell is a precise tool.

Diabetes diagnosing Expert system was coded via the use of a VP-Expert shell, the shell is a precise tool for designing the expert systems thus only expert systems' developers are acquainted with it. The production rules of this expert system include 7 attribute questions which serve as the input of this system (Table 2).

fbs1 result equal or higher than 126?, fbs1 result is equal or higher than 126?, hba1c result is equal or higher than 6?, Please enter your patient situation, your age is equal or higher than 20 years, what is your symptoms result, what is your effective factors' result?.

The sample of this expert system rules is being demonstrated below for proper explanation.

IF FBS1 result not equal or higher than 126 AND, FBS2 result equal or higher than 126 AND HBA1C result equal or higher than 6 AND, Your patient situation male AND Your age is not equal or higher than 20 years AND, Symptoms result is diabetes AND, Effective factory test result is healthy THEN Treatment = Diabetes Type1.

IF FBS1 = no, AND FBS2 = yes, AND HBA1C = yes, AND Patient Situation = male, AND Age = no, AND symptoms = diabetes, AND Effective Factor = healthy, THEN treatment = diabetes type1.

Pattern of building up the framework has prepared, all rules, ways and connections between traits have been tested and the vital changes have been mended. The outlined framework has been assessed by the internists and diabetes experts of Near East Hospital [14]. After validation and approval of the framework, the last outlined framework has been provided.

**Table 1.** Decision making table of tests [13].

| Pregnancy | FBS 1 | FBS 2 | HBA 1C | Tests |
|---|---|---|---|---|
| Female and NoPreg. | More/equal 126 | More/equal 126 | – | Unhealthful |
| Female and Pregn. | More/equal 126 | More/equal 126 | More/equal 6% | Unhealthful |
| Female and Pregn. | More/equal 126 | More/equal 126 | Less 6% | Unhealthy-Gestational |
| Female and NoPreg. | Less126 | More/equal 126 | Less 6% | Healthful |
| Female and Preg. | Less 126 | More/equal 126 | Less 6% | More Consideration |
| Female and NoPreg. | More/equal 126 | Less 126 | Less 6% | Healthful |
| Female and Preg. | More/equal 126 | Less 126 | Less 6% | More Consideration |
| None | More/equal 126 | More/equal 126 | – | Unhealthy |
| None | Less 126 | More/equal 126 | Less 6% | Healthful |
| None | More/equal 126 | Less 126 | Less 6% | Healthful |
| – | Less 126 | More/equal 126 | More/equal 6% | Unhealthful |
| – | Less 126 | Less 126 | Less 6% | Healthful |
| – | More/equal 126 | Less 126 | More/equal 6% | Unhealthful |
| – | Less126 | Less 126 | More/equal 6% | More Consideration |

## 4    Computer Simulation

Before presenting a sample running of the expert sys-tem, it is essential to illustrate the Use-case diagram for better understanding on how the system runs. The Fig. 2 use-case diagram demonstrates the interaction between the user and the expert system. The Figs. 3 and 4 demonstrate the execution of the VP-Expert that is the user interface. On the execution, the user interrelates with system through the user inter-face. The user interface has 3 windows which are question windows where the questions should be asked systematically to the users, and their alternatives are illustrated. Rules window represents the rules by indicating how the users have to reply. The Facts window displays the users' answers, which may determine the systems final decision. The main advantage of facts and rules windows is vivid clarification for the system's decision-making. After the system has received the facts, the system represents the final answer or decision.

**Table 2.** Decision making table of diagnosis [13].

| Test | Patient state | Age | State | Factors | Diagnoses |
|------|---------------|-----|-------|---------|-----------|
| Healthful | – | – | Healthful | Healthful | Healthy |
| Healthful | – | – | Healthful | At. Risk | At. Risk |
| Healthful | – | – | Diabetes | At. Risk | At. Risk |
| Healthful | – | – | Diabetes | Healthful | More Consideration |
| Unhealthful | Male | Less 20 | Diabetes | – | Type 1 |
| Unhealthful | Male | More/equal 20 | Diabetes | – | Type 2 |
| Unhealthful | Female and No Preg. | Less 20 | Diabetes | – | Type 1 |
| Unhealthful Gestational | Female and Pregnant | – | Diabetes | – | Gestational Diabet. |
| Unhealthful | Female and Pregnant | More/equal 20 | Diabetes | – | Type 2 |
| Unhealthful | Female and Pregnant | Less 20 | – | – | More Consideration |
| Unhealthful | – | – | Healthful | – | More Consideration |
| Unhealthful | Female and No Preg. | More/equal 20 | Diabetes | – | Type 2 |
| Unhealthy. Gestational | Female and Pregnant | – | Healthful | – | More Consideration |
| More Cons. | – | – | – | – | More Consideration |

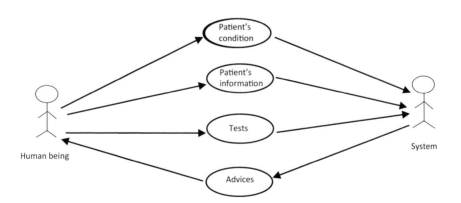

**Fig. 2.** Treatment advices use-case diagram.

We tested the developed expert system by 15 patients. It is shown that 10 patients out of 15 have diabetes 2 which concludes that we have been able to achieve the same results with the doctors. However, it is scientifically known that for each of the type 2

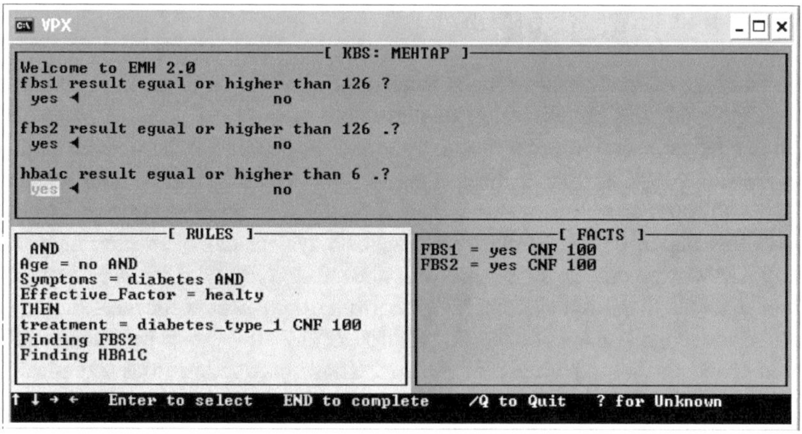

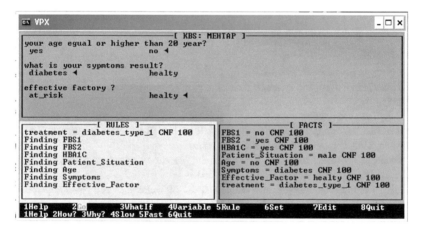

**Fig. 3.** Fragment of computer simulation (blood test results).

**Fig. 4.** Fragment of computer simulation (system diagnosis sample).

diabetes patients there is 10% chance or probability of being MODY, which is a genetic disorder. Although this condition also affects type1 diabetes patient, our system was able to get accurate diagnoses for one type1 diabetes patient out of 15 (1/15), patients, as the doctors. Accuracy is accepted as 90% because our expert system does not check MODY test results. Furthermore, the developed expert system diagnosed 3 patients as healthy and 1 patient as at risk, same as the doctors again. By using this developed expert system, our purpose is to help patients where consulting to doctor is limited.

## 5  Conclusion

People can take advantage of artificial intelligence and use it as an information guidance system with the consideration of automated counseling administration. Expert system can be exploited in order to diagnose and determine the therapeutic finding, or use as instructive guide. Application of AI to more specific field of interest has been resulted with more crucial progresses than applying it to more generic problems.

With this paper, design of diabetes diagnosis system has been presented. Acquisition of knowledge and its stages are described following with knowledge representation of diabetes diagnosis system. IF-THEN rules are selected for the solution of the system. Knowledge needed for the preparation of 280 If – Then production rules, is acquired based on the experience of doctors. These production rules include 7 input attributes and 5 diagnoses as the output of production rules. Also this article focuses on analyzing type1, type2 diabetes together.

This diabetes diagnosis expert system is designed by using VP expert system shell and tested on 15 patients in Near East Hospital. As a conclusion obtained results had been monitored and reviewed by experienced medical doctors, and efficiency of our diabetes diagnosis expert system has been satisfied by these obtained results.

## References

1. Mishkoff, H.: Understanding artificial intelligence. Instrument learning Centre, Dallas, Texas (1985)
2. Zadeh, L.A.: Computing with words and perceptions a paradigm shift. In: 2009 Proceedings of the IEEE International Conference on Information Reuse and Integration, Las Vegas, Nevada, USA, pp. 450–452. IEEE Press (2009)
3. Sayedah, T., Tawfik, S., Zaki, Y.: Developing an expert system for diabetics' treatment advices. Int. J. Hosp. Res. **2**(3), 155–162 (2013)
4. Patel, T.: Knowledge models, knowledge acquisition techniques and developments. Orient. J. Comput. Sci. Technol. **6**(4), 467–472 (2013)
5. Aliev, R.A., Fazlollahi, B., Aliev, R.R.: Soft Computing and its Application in Business and Economics, p. 446. Springer, Heidelberg (2004)
6. Karray, F.O., de Silva, C.: Soft Computing and Intelligent Systems Design Pearson Education Limited, London, pp. 4–13, 223–224. British library (2004)
7. Griffin, N.L., Lewis, F.D.: A rule-based inference engine which is optimal and VLSI implementable. Technical report, Department of Computer Science, University of Kentucky (1989)
8. Dennis, M.: Building Expert System in Prolog. Azmi Inc. Publishers, Lebanon (1989). An On-Line Edition
9. Jackson, P.: Introduction to Expert System Pearson Wesley Longman Limited, Harlow, pp. 2–9. Harlow Essex CM20 2Je, England (1999)
10. Jose, A., Prasad, A.: Design and Development of Rule Base Expert System for AACR: A Study of the Application of Artificial Intelligence Techniques in Library and Information Field. Saarbrucken VDM Verlag Publishers, Saarbrücken (2011)
11. Nilsson, N.J.: Principles of Artificial Intelligence. Narosa Publishing House, New Delhi (1998)

12. Akter, M., Uddin, M.S., Hague, A.: Diagnosis and management of diabetes through knowledge based system. In: 2009 13th International Conference on Biomedical Engineering, pp. 1000–1003. Springer, Heidelberg (2009)
13. Zeki, T.S., Malakooti, M.V., Ataeipoor, Y., Tabibi, S.T.: An expert system for diabetes diagnosis. Am. Acad. Sch. Res. J. **4**(5), 1 (2012)
14. Altıparmak, H.: Diabetes diagnose system by using VP expert thesis, Nicosia (2016)

# Intensive Investigation in Differential Diagnosis of Erythemato-Squamous Diseases

Idoko John Bush$^{(\boxtimes)}$, Murat Arslan, and Rahib Abiyev

Department of Computer Engineering, Applied Artificial Intelligence Research Centre, Near East University, North Cyprus, Mersin-10, Turkey
john.bush@neu.edu.tr

**Abstract.** Research in the field of dermatology shows that differential diagnosis of erythemato-squamous diseases is one of the challenges seeking attention and to contribute to this problem, we designed four novel machine learning models exploring; Random Forest (RF), Multilayer Perceptron (MLP), Support Vector Machines (SVM) and Fuzzy Neural Network (FNN) techniques to accurately recommend the best model to dermatologists when diagnosing patients with erythemato-squamous diseases. At the design stage, we considered a dataset characterizing the six classes of the disease. To reduce the training time, the input data was normalized and scaled in interval; 0–1. Furthermore, we implored 10-fold cross-validation where the original sample was randomly segmented into 10 equal sized subsamples. These 10 outcomes from the folds are then averagely computed and produce a single prediction. Total performance of each of the models as depicted in table one shows that FNN outperformed the other 3 models hence, recommended for the differential diagnoses of these six classes of the disease.

**Keywords:** Random forest · Support vector machine · Multilayer Perceptron
Fuzzy Neural Network · Dermatology · Erythemato-squamous disease

## 1 Introduction

In dermatology, the differential diagnosis of erythemato-squamous diseases is a difficult problem. The different classes of the disease share the clinical features of scaling and erythema, with very few differences. The set of erythemato-squamous diseases to be classified are pityriasis rubra pilaris, seboreic dermatitis, psoriasis, lichen planus, chronic dermatitis and pityriasis rosea. Some diagnostic considerations for differential erythemato-squamous diseases include; degree of scaling and erythema, with or without defined lesions borders, the presence of itching and koebner phenomenon, papules formation, whether there is a family history or not, whether the oral mucosa, knees, elbows and the scalp are involved or not.

Acquisition of the implored dataset involves the evaluation of 22 histopathological features of skin samples [1]. At the beginning stage, a disease may show the histopathological features of another disease and may have the characteristic features at the following stages and this is another difficulty faced in differential diagnosis. Here, some of the samples might display typical histopathological variables of the illness

© Springer Nature Switzerland AG 2019
R. A. Aliev et al. (Eds.): ICAFS-2018, AISC 896, pp. 146–153, 2019.
https://doi.org/10.1007/978-3-030-04164-9_21

while some does not. The explored dataset consist of feature of family history that has a value of 1 if any of these diseases is been observed in the family otherwise, it has a zero value. Another feature identified in the domain is age and this represents the age of the diagnosed patient. The other features; histopathological and clinical were given a degree ranging from 0–3. Where 0 depicts absent of the feature, 1 and 2 depicts the relative intermediate values while 3 indicate the largest possible amount possible.

Basically, the database was created to determine and differentiate Erythemato-Squamous Disease types. The domain contains 34 attributes out of which 33 are linear valued and one of which is noted to be nominal. The dataset explored in this study is known to having problem of missing data. In tackling this menace, we implored minimum distance method to generate the missing values. The method includes two stages; determining cluster space according to output signals and finding the minimum value of the sum of distances.

Recently, several machine learning approaches have been applied in medical domains to rightly predict outcomes. For instance; in [2] we see that by adapting the rule-base system behavior to the most recent information available about a patient, the case-based BOLERO system is able to learn both goals and plans states to improve performance of a rule-based system. To determine the diagnostic value of clinical data, DIAGAID is used as connectionist method as seen in [3]. Others include k-NNFP [4] and the VFI representation based on Feature Projections as seen in [5]. Several fuzzy and neural topologies have been explored to solve different classification and feature extraction problems as seen in [6–12]. In [6, 7], adaptive neuro-fuzzy inference system (ANFIS) was applied for feature extraction. [8] used neuro-fuzzy system for Crohn's disease classification. [9] used Haralick features and backpropagation neural networks for glaucoma classification. [10] applied neural networks to a distinctive medical problem; one-year survival prediction of myocardial infarction. In [11], genetic algorithm is applied to design multi-input and single output neuro- fuzzy system. A renowned Adaptive Neuro-Fuzzy Inference System (ANFIS) algorithm was utilized to optimize the chiller loading in [12].

## 2   Proposed Machine Learning Techniques

### 2.1   Random Forest (RF)

Random Forest (RF) algorithm which is an ensemble learning method, used mostly for regression and classification purposes, was created by Tin Kam Ho in 1995 [13] using the feature bagging method [14]. RF can be used to process both classical and numerical data: most machine learning algorithms are useful in classification problems or useful for numerical applications.

RF, uses the "Classification and Regression Tree" (CART) algorithm to generate trees without pruning [15–18]. In the CART algorithm, division is performed by applying a certain criterion in a node. Selection is performed on these divisions. Knots with homogeneous class distribution are preferred for splitting operations. In the measurement of node homogeneity; Gini Index is used. The GINI can be expressed by the following formula:

$$\sum\sum_{j\neq i}(f(C_i,T)/|T|)(f(C_i,T)/|T|) \tag{1}$$

Here, T is the training set, $C_i$ is the randomly selected feature of the class $C_i$ and $f(C_i,T)/|T|$, indicates the selected sample belongs to class $C_i$ (Fig. 1).

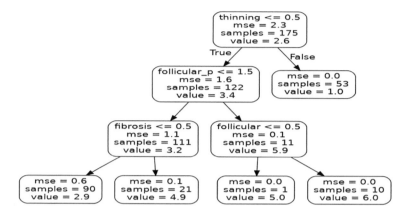

**Fig. 1.** Extracted random tree from the dataset

## 2.2   Support Vector Machines (SVM)

Support vector machine (SVM) tries to find out a hyperplane that has best separation which can be achieved by largest distance to the nearest training data point of any class [19]. SVM separates two groups by drawing a border between the two groups in a plane for classification. Here, SVM determines how this boundary is drawn as shown in Fig. 2. In order to do this, two near and two parallel border lines are drawn on the two groups and these boundary lines are drawn closer together to produce a common boundary line.

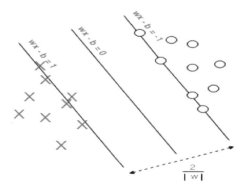

**Fig. 2.** SVM boundaries

From the above figure, two groups are shown on a two-dimensional plane. It is possible to think of these planes and dimensions as properties. In other words, a feature extraction is performed on each input that enters the system in a simple sense, resulting in a different point showing each input on this two-dimensional plane. The classification of these points is the classification of inputs according to the properties that have been extracted. The definition of each point in this plane can be made by the following notation:

$$D = \{(X_i, C_i)|X_i \in R^p, C_i \in \{-1, 1\}\}_{i=1}^{n} \tag{2}$$

For every x, c, the vector X is a point in our space and c is the value indicating that this point is −1 or +1. This set of points goes up i = 1 to n. This representation is on an extreme plane (hyperplane). Here, every point in this display is expressed by:

$$wx - b = 0 \tag{3}$$

Where w is the normal vector with perpendicular overstrain, x is the variable of the dot and b is the shear rate. It is possible to compare this equation to the equation for calculating wx + b. According to the above equation, $\frac{b}{||w||}$ gives us the distance difference between the two groups.

## 2.3    Multilayer Perceptron (MLP)

Artificial neural networks can be described as "parallel, interrelated nets of simple elements and their hierarchical arrangements that aim to interact with real-life objects, such as in the biological nervous system" [20]. Multilayer Perceptron (MLP) is a kind of feedforward artificial neural network which has at least three layers of nodes as shown in Fig. 3. First layer is the input layer where the information is transmitted to the hidden layer. In the output layer, the output values for each input are determined against the information received from the hidden layer, and the process elements are interlinked between all the layers.

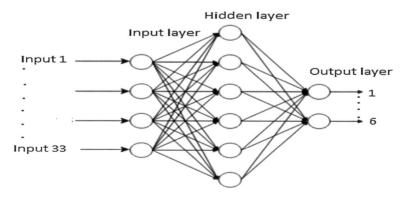

**Fig. 3.** Explored MLP architecture

In our MLP model, we calculate the square of the difference between the expected value and the real value of the output value, $E$, using the expressed:

$$E = \frac{1}{2}\sum_i (y_{dj} - y_j)^2 \tag{4}$$

Here $y_{dj}$ is the expected value of the output neuron with indices $j$ and $y_i$ is the real value of the neuron. MLP calculates the error of the output and reorganize the neuron weights to reduce the error. $\triangle w_{ij}$ is added to the weight for each $w_{ij}$ be reorganized.

## 2.4  Fuzzy Neural Network (FNN)

Due to much similarities and ambiguities in the input space of the dataset, we implored Fuzzy Neural System to first simplify the impreciseness then accurately classifies the six classes of the disease (erythemato-squamous disease). In the paper, the design of the FNN classifier is considered using data sets characterizing the six classes. The initial design of the classifier consumed a lot of time and to solve this problem, we normalized and scaled the input data in interval 0–1 as demonstrated in [21–32]. The structure of the system used for this classification is given in Fig. 4.

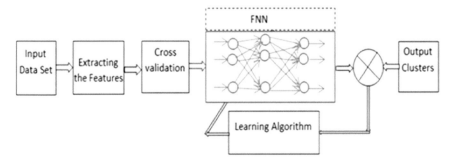

**Fig. 4.** FNN structure

We implored TSK type fuzzy rules in this paper to design the fuzzy neural system model. Fuzzy inference systems linguistic rule interpretation and neural networks learning capabilities are combined in our proposed fuzzy neural system. We designed the proposed FNS using fuzzy rules having the IF-THEN structure. By dint of optimal definition of the consequent and premise parts of the fuzzy IF-THEN rules, we were able to accomplish the classification system through the neural networks learning capability. This type of fuzzy system approximates linear systems with nonlinear system and takes the following form:

$$\text{If } x_1 \text{ is } A_{1j} \text{ and } x_2 \text{ is } A_{2j} \text{ and}\ldots\text{and } x_m \text{ is } A_{mj} \text{ then } y_j \text{ is } \sum_{i=1}^{m} a_{ij}x_i + b_j \tag{5}$$

Here, $x_i$ and $y_j$ represent the system's input and output signals respectively. The number of input signals is represented by $i = 1, \ldots, m$ while the number of rules is represented by $j = 1\ldots r$. We denoted the input fuzzy sets by $A_{ij}$ while the coefficients are represented by $b_j$ and $a_{ij}$ respectively. The FNS have six layers and the $x_i$ ($i = 1, \ldots, m$) input signals are distributed in the first layer. We incorporated membership functions in the second layer to map linguistic term to each node, and the Gaussian membership function is utilized to properly evaluate the linguistic terms. Finally, we calculate FNS output signals using the illustration:

$$u_k = \frac{\sum_{j=1}^{r} w_{jk} y_j}{\sum_{j=1}^{r} \mu_j(x)} \tag{6}$$

From the above illustration, the FNS output signals represent the $u_k$. Between layers 5 and 6, we used ($k = 1, \ldots, n$), $w_{jk}$ for weight coefficients of connections. The training of the network begins immediately after calculating the output signal.

## 3   Results

After obtaining our results from the above illustrations, we compared 4 different techniques and in the first technique, we applied random forest algorithm to differentiate the six classes of erythemato-squamous diseases, where we got 97.02 accuracy, which is pretty good. We further applied Support Vector Machines to same data set and accuracy of 97.54 was obtained. After this, MLP Classifier which uses "relu" activation function, 16 batch size, 32 => 16 => 6 neurons with "adam" solver was explored and 97.90 accuracy was obtained. Finally, intense experiment was carried out on same domain with fuzzy neural system having 32 neurons (rules) at 100 epochs and 98.37 accuracy was achieved as depicted in Table 1.

**Table 1.** Experimental results comparison

| Techniques | RMSE values | Accuracy values |
|------------|-------------|-----------------|
| RF         | 0.4373      | 97.02           |
| SVM        | 0.5989      | 97.54           |
| MLP        | 0.4777      | 97.90           |
| FNN        | 0.2136      | 98.37           |

## 4  Conclusion

In this paper, we designed 4 models using machine learning approaches such as Random Forest (RF), Support Vector Machines (SVM), Multilayer Perceptron (MLP) and Fuzzy Neural Network (FNN) to accurately investigate the model to be recommended to dermatologists when diagnosing patients with erythemato-squamous diseases. From the evaluations of the entire experiments carried out on same domain, FNN classifier having the highest accuracy and lowest root mean square error is therefore recommended for differential diagnoses of erytheamo-squamous diseases. Our further engagement would be the deployment of the recommended classifier (FNN) to hospitals, beginning with the prestigious Near East University Hospital for the differential diagnoses of erytheamo-squamous diseases, where unseeing datasets will be used.

## References

1. Güvenir, H., Demiröz, G., İlter, N.: Learning differential diagnosis of erythemato-squamous diseases using voting feature intervals. Artif. Intell. Med. **13**, 147–165 (1998)
2. López, B., Plaza, E.: Case-based learning of plans and goal states in medical diagnosis. Artif. Intell. Med. **9**, 29–60 (1997)
3. Forsström, J., Eklund, P., Virtanen, H., Waxlax, J., Lähdevirta, J.: DIAGAID: a connectionist approach to determine the diagnostic value of clinical data. Artif. Intell. Med. **3**, 193–201 (1991)
4. Akkus¸ A., Guvenir, H.A.: K nearest neighbor classification on feature projections. In: Proceedings of ICML 1996, pp. 12–19 (1995)
5. Guvenir, H., Sirin, I.: Classification by feature partitioning. Mach. Learn. **23**, 47–67 (1996)
6. Subhi Al-batah, M., Mat Isa, N., Klaib, M., Al-Betar, M.: Multiple adaptive neuro-fuzzy inference system with automatic features extraction algorithm for cervical cancer recognition. Comput. Math. Methods Med. **2014**, 1–12 (2014)
7. Wang, D., He, T., Li, Z., Cao, L., Dey, N., Ashour, A., Balas, V., McCauley, P., Lin, Y., Xu, J., Shi, F.: Image feature-based affective retrieval employing improved parameter and structure identification of adaptive neuro-fuzzy inference system. Neural Comput. Appl. **29**, 1087–1102 (2016)
8. Ahmed, S., Dey, N., Ashour, A., Sifaki-Pistolla, D., Bălas-Timar, D., Balas, V., Tavares, J.: Effect of fuzzy partitioning in Crohn's disease classification: a neuro-fuzzy-based approach. Med. Biol. Eng. Comput. **55**, 101–115 (2016)
9. Samanta, S., Ahmed, S.S., Salem, M., Nath, S., Dey, N., Chowdhury, S.S.: Haralick features based automated glaucoma classification using back propagation neural network. In: Proceedings of the 3rd International Conference on Frontiers of Intelligent Computing (FICTA), pp. 351–358 (2014)
10. Helwan, A., Uzun, D., Abiyev, R., Bush, J.: One-year survival prediction of myocardial infarction. Int. J. Adv. Comput. Sci. Appl. **8**, 173–178 (2017)
11. Dey, N., Ashour, A., Beagum, S., Pistola, D., Gospodinov, M., Gospodinova, E., Tavares, J.: Parameter optimization for local polynomial approximation based intersection confidence interval filter using genetic algorithm: an application for brain MRI image de-noising. J. Imaging **1**, 60–84 (2015)

12. Lu, J., Chang, Y., Ho, C.: The optimization of chiller loading by adaptive neuro-fuzzy inference system and genetic algorithms. Math. Probl. Eng. **2015**, 1–10 (2015)
13. Ho, T.K.: Random decision forests. In: Proceedings of the 3rd International Conference on Document Analysis and Recognition, Montreal, QC, pp. 278–282 (1995)
14. Tin, K.H.: The random subspace method for constructing decision forests. IEEE Trans. Pattern Anal. Mach. Intell. **20**, 832–844 (1998)
15. Breiman, L.: Random forests. Mach. Learn. **45**(1), 5–32 (2001)
16. Archer, K., Kimes, R.: Empirical characterization of random forest variable importance measures. Comput. Stat. Data Anal. **52**, 2249–2260 (2008)
17. Breiman, L., Cutler, A.: Random forest (2005)
18. Horning, N.: Random forests: an algorithm for image classification and generation of continuous field data sets. In: International Conference on Geoinformatics for Spatial Infrastructure Development in Earth and Allied Sciences (GIS-IDEAS) 9–11 (2010)
19. Abiyev, R., Arslan, M., Gunsel, I., Cagman, A.: Robot pathfinding using vision based obstacle detection (2017)
20. Kohonen, T.: State of the art in neural computing. In: IEEE First International Conference on Neural Networks, vol. 1, pp. 79–90 (1987)
21. Idoko, J.B., Rahib, H.A., Mohammad, K.M.: Intelligent machine learning algorithms for colour segmentation. WSEAS Trans. Signal Process. **13**, 232–240 (2017)
22. Bush, I., Abiyev, R., Sallam Ma'aitah, M., Altıparmak, H.: Integrated artificial intelligence algorithm for skin detection. In: ITM Web of Conferences, vol. 16, p. 02004 (2018)
23. Khaleel, M., Abiyev, R., John, I.: Intelligent classification of liver disorder using fuzzy neural system. Int. J. Adv. Comput. Sci. Appl. **8**, 25–31 (2017)
24. Rahib, A., Mohammad, M.K.S.: Deep convolutional neural networks for chest diseases detection. J. Healthc. Eng. **2018** (2018)
25. Abiyev, R., Altunkaya, K.: Neural network based biometric personal identification with fast iris segmentation. Int. J. Control Autom. Syst. **7**, 17–23 (2009)
26. Abiyev, R., Abizade, S.: Diagnosing Parkinson's diseases using fuzzy neural system. Comput. Math. Methods Med. **2016**, 1–9 (2016)
27. Rahib, H.A., Kemal, K.: Adaptive Iris segmentation. In: Lecture Notes in Computer Sciences. Springer, CS Press (2009)
28. Rahib, A., Koray, A.: Personal iris recognition using neural networks. Int. J. Secur. Its Appl. **2**(2), 41–50 (2008)
29. Rahib, A., Koray, A.: Neural network based biometric personal identification. LNCS, Springer, CS press (2007)
30. Kamil, D., Idoko, J.B.: Automated classification of fruits: pawpaw fruit as a case study. In: International Conference on Man–Machine Interactions, pp. 365–374. Springer, Cham (2017)
31. Bush, I., Dimililer, K.: Static and dynamic pedestrian detection algorithm for visual based driver assistive system. In: ITM Web of Conferences, vol. 9, p. 03002 (2017)
32. Helwan, A., Idoko, J., Abiyev, R.: Machine learning techniques for classification of breast tissue. Procedia Comput. Sci. **120**, 402–410 (2017)

# A Computational-Intelligence Based Approach to Diagnosis of Diabetes Mellitus Disease

Elif Dogu$^{(\boxtimes)}$ and Y. Esra Albayrak

Galatasaray University, Besiktas, 34349 Istanbul, Turkey
{edogu, ealbayrak}@gsu.edu.tr

**Abstract.** Diabetes Mellitus (DM) is a disease that occurs when the pancreas cannot produce enough insulin or when insulin that it produces cannot be used effectively. High frequency of urination and hunger and thirst are general symptoms of high levels of blood glucose. Global estimates of 2015 claims that 415 million people are living with diabetes and 90% of them belongs to Type 2 DM.

DM have equal rates for men and woman, and a rate of 8.3% in total adults. Diagnosis of the disease is not challenging however, it requires blood glucose measurements in different times. In emergency cases where the patient is unconscious, the possibility to overlook the disease is high. In this study, fuzzy c-means clustering algorithm, in which each variable can belong to more than one class, is used to classify the two groups of patients with and without diabetes through other blood test data and demographic factors. In the first application with 100 patients of a hospital, the algorithm correctly classified 81% of patients.

**Keywords:** Fuzzy c-means clustering · Unsupervised learning
Medical decision support system

## 1 Introduction

Diabetes is a serious public health problem with its increasing number of cases and prevalence over the past decades. In April 2016, World Health Organization (WHO) published the Global report on diabetes, which calls the world leaders for action to diminish exposure to the risk factors and to develop accessibility and quality of health care for people who have diabetes mellitus.

108 million adults were living with diabetes in 1980s and in 2014; this number is increased to 422 million adults, worldwide. Age-standardized global prevalence augmented from 4.7% to 8.5% in adults. In 2012, 1.5 million deaths are recorded as caused by DM. High levels of blood sugar, by augmenting other cardiovascular diseases' risk, triggered another 2.2 million deaths. Which totally means 3.7 million deaths that occurs before the age of 70. The percentage of deaths is higher in low- and middle-income countries [1].

Diagnosis of diabetes is not difficult however; its process needs time and consciousness of the patient. When an unconscious patient is brought to emergency for a reason that is not related to DM, it is highly problematic to understand if he/she had

© Springer Nature Switzerland AG 2019
R. A. Aliev et al. (Eds.): ICAFS-2018, AISC 896, pp. 154–159, 2019.
https://doi.org/10.1007/978-3-030-04164-9_22

DM as a comorbid disease. This information is vital because it might change the treatment given in an emergency.

The demographics, examination and blood test results are the fastest information gathered on a patient. In this study, fuzzy c-means clustering algorithm is performed to classify the patients' DM status using this information. Fuzzy c-means clustering is a machine-learning algorithm, which is widely used as a decision aid in medical problems. In the last five years, Dutta et al. used weighted fuzzy c-means algorithm to identify the target class thresholds for the classification of diabetic retinopathy images [2], Prakash et al. implemented multidimensional thresholding, region-growing, fuzzy c-means and neural network algorithms for automatic segmentation of brown adipose tissue which normalizes metabolic disorders in diabetes [3], Oliveira et al. integrated deformable models with fuzzy c-means for retinal vessel segmentation that help to predict cardiovascular related diseases as diabetes and hypertension [4], Iliyasu et al. used possibilistic fuzzy c-means with evidence accumulation clustering for the diagnosis of hepatitis, breast cancer and diabetes [5], Tasgaonkar and Khambete integrated Mahalanobis metric classification and fuzzy c-means for exudate detection in color fundus imaging for diabetic patients [6], Mahendran and Dhanasekaran used fuzzy c-means for detection and localization of retinal exudates for diabetic retinopathy [7], Hassanien et al. used bee colony swarm optimization, pattern search and fuzzy c-means for retinal blood vessel localization [8] and Ozsen and Ceylan compared artificial immune system and fuzzy c-means on the classification of breast cancer and diabetes [9].

The remainder of the study is structured as follows: Sect. 2 explains fuzzy c-means clustering in detail. Numerical application is given in Sect. 3 and the paper is concluded in Sect. 4.

## 2 Fuzzy c-Means Clustering

Fuzzy c-means is an extension of conventional k-means clustering method using the fuzzy sets. To develop the method in clustering, a family of fuzzy sets is defined a fuzzy c-partition on a data points' universe. Since fuzzy sets accept for membership degrees, the crisp classification idea can be extended into a fuzzy classification concept. Then, membership degrees can be assigned to the several data points of each fuzzy set (fuzzy class, fuzzy cluster). Hence, a single point can have different membership degrees for many classes [10]. Fuzzy c-means clustering was developed by J.C. Dunn in 1973 [11], and improved by J.C. Bezdek in 1981 [12]. Fuzzy c-means is grounded on the objective function's minimization:

$$J_m = sum_{i=1}^{D}(sum_{i=1}^{N}(\mu_{ij}^m\|x_i - c_j\|^2)) \tag{1}$$

where D is the number of data points, N is the number of clusters, m is fuzzy partition matrix exponent for controlling the degree of fuzzy overlap (with m > 1, fuzzy overlap refers to how fuzzy the boundaries between clusters are, that is the number of data points that have significant membership in more than one cluster.), $x_i$ is the i'th data point, $c_j$ is the center of the j'th cluster and $m_{ij}$ is the degree of membership of $x_i$ in the j'th cluster. For a given data point, $x_i$, the sum of the membership values for all clusters is one.

Fuzzy c-means algorithm performs 5 steps during clustering.

1. The cluster membership values, $m_{ij}$ are initialized randomly.
2. The cluster centers, $c_j$ are calculated using Eq. (2)

$$c_j = \left( \text{sum}_{i=1}^{D} \mu_{ij}^m \cdot x_i \right) \Big/ \left( \text{sum}_{i=1}^{D} \mu_{ij}^m \right) \tag{2}$$

3. Update $m_{ij}$ according to Eq. (3)

$$\mu_{ij} = 1 \Big/ \left( \text{sum}_{k=1}^{N} \left( \|x_i - c_j\| / \|x_i - c_k\| \right) \right)^{2/(m-1)} \tag{3}$$

4. The objective function, $J_m$ is calculated.
5. Steps 2–4 are repeated until $J_m$ improves by less than a pre-defined minimum value or until after a number of iterations.

## 3   Numerical Application for a State Hospital in Istanbul

For the numerical application, patient data is gathered in a full-fledged state hospital in Istanbul. A sample of 100 patients admitted to hospital with diverse complaints is prepared. After an interview with the physicians, mostly demographic factors and blood test results are included in application for a faster diagnosis. 23 types of patient data are observed: Age, gender, height, weight, glucose, urea, creatinine, total protein, ALB, CRP, ESR, WBC, HGB, HCT, PLT, AST, ALT, PH, CO2, O2, SPO2, breaths per minute (BPM) and number of pulses per minute (PPM).

Numerical application is performed with MATLAB Software Fuzzy Logic Toolbox. First, the data matrix (100×23) is prepared as in Table 1.

**Table 1.** Sample patient data

| Patient No | Age | Gender | Glucose | ... | SPO2 | BPM | PPM |
|------------|-----|--------|---------|-----|------|-----|-----|
| 1 | 73 | 1 | 91 | ... | 89 | 20 | 105 |
| 2 | 65 | 1 | 215 | ... | 85 | 25 | 90 |
| ... | ... | ... | ... | ... | ... | ... | ... |
| 99 | 75 | 1 | 201 | ... | 76 | 18 | 100 |
| 100 | 70 | 0 | 425 | ... | 91 | 15 | 110 |

Membership values are randomly initialized and for the first iteration, the value of the objective function is calculated as 899932896162.19. Stopping condition of the algorithm is set as 50 iterations. The objective function converged after 20 iterations as shown in Fig. 1.

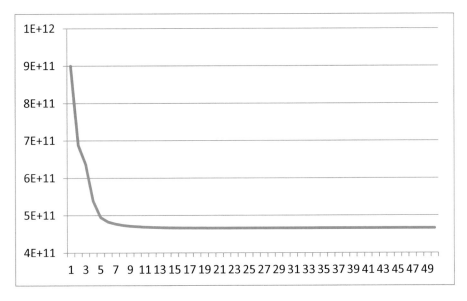

**Fig. 1.** Convergence of the objective function

Membership degrees calculated by the algorithm for each patient and the patient's DM status are given in Table 2.

**Table 2.** Fuzzy c-means results and DM status of the patients

| P. 1–25 | DM stat. | P. 25–50 | DM stat. | P. 51–75 | DM stat. | P. 76–100 | DM stat. |
|---------|----------|----------|----------|----------|----------|-----------|----------|
| 0.007   | 0        | 0.992    | 1        | 0.000    | 0        | 0.062*    | 1        |
| 0.450*  | 1        | 0.005    | 0        | 0.003    | 0        | 0.603     | 1        |
| 0.028   | 0        | 0.010    | 0        | 0.533*   | 0        | 0.371*    | 1        |
| 0.908   | 1        | 0.175    | 0        | 0.022    | 0        | 0.529     | 1        |
| 0.044   | 0        | 0.717    | 1        | 0.298    | 0        | 0.168     | 0        |
| 0.215   | 0        | 0.013    | 0        | 0.283    | 0        | 0.774     | 1        |
| 0.800   | 1        | 0.211    | 0        | 0.028    | 0        | 0.688     | 1        |
| 0.050   | 0        | 0.583*   | 0        | 0.065    | 0        | 0.008     | 0        |
| 0.010   | 0        | 0.676*   | 0        | 0.402*   | 1        | 0.010*    | 1        |
| 0.716   | 1        | 0.424    | 0        | 0.443    | 0        | 0.004     | 0        |
| 0.740   | 1        | 0.492    | 0        | 0.381    | 0        | 0.003     | 0        |
| 0.908   | 1        | 0.015    | 0        | 0.837    | 1        | 0.842     | 1        |
| 0.604   | 1        | 0.357    | 0        | 0.075    | 0        | 0.020*    | 1        |
| 0.003   | 0        | 0.929    | 1        | 0.028*   | 1        | 0.098*    | 1        |
| 0.114*  | 1        | 0.001    | 0        | 0.905    | 1        | 0.003     | 0        |
| 0.012   | 0        | 0.000    | 0        | 0.014    | 0        | 0.625     | 1        |
| 0.765   | 1        | 0.388*   | 1        | 0.410*   | 1        | 0.617     | 1        |

*(continued)*

**Table 2.**  (*continued*)

| P. 1–25 | DM stat. | P. 25–50 | DM stat. | P. 51–75 | DM stat. | P. 76–100 | DM stat. |
|---------|----------|----------|----------|----------|----------|-----------|----------|
| 0.710 | 1 | 0.260 | 0 | 0.040 | 0 | 0.058* | 1 |
| 0.116* | 1 | 0.015 | 0 | 0.703 | 1 | 0.044 | 0 |
| 0.035* | 1 | 0.346 | 0 | 0.607 | 1 | 0.838 | 1 |
| 0.870 | 1 | 0.007* | 1 | 0.023 | 0 | 0.015 | 0 |
| 0.071* | 1 | 0.624 | 1 | 0.290 | 0 | 0.094 | 0 |
| 0.630 | 1 | 0.132 | 0 | 0.037 | 1 | 0.978 | 1 |
| 0.522 | 1 | 0.703 | 1 | 0.514 | 1 | 0.708 | 1 |
| 0.044 | 0 | 0.013 | 0 | 0.702 | 1 | 0.555 | 1 |

# 4   Conclusions

Many factors affect the performance and accuracy of a segmentation algorithm. This study is about proposing fuzzy c-means algorithm for faster diagnosis of diabetes via other patient information in an emergency. A sample of 100 patients from a state hospital in Istanbul is evaluated. The results of the numerical application are interpreted by the physicians. When the threshold value is chosen as 0.5 between the two classes, 19 out of 100 patients were incorrectly classified, which are marked in Table 2. Fuzzy c-means reached an accuracy of 81%. In order to obtain better results, the data set might be extended and additional information on the patients might be used in the model. As a medical decision support model, 81% accuracy is interpreted as useful by the physicians. The model can be integrated with preprocessing steps as data reduction or factor analysis. As further research directions, other clustering algorithms as logistic regression, neural networks etc. can be implemented to the same data for comparison and integrated models can be proposed for better accuracy.

**Acknowledgments.**  Authors would like to thank to Assoc. Prof. Esin Tuncay MD and Özlem Yılmaz Ünlü MD for their contribution. This study is financially supported by Galatasaray University Research Fund, project 18.402.004.

# References

1. WHO - World Health Organization: Global report on diabetes (2016)
2. Dutta, S., Manideep, B.C.S., Basha, S.M., Caytiles, R.D., Iyengar, N.: Classification of diabetic retinopathy images by using deep learning models. J. Grid Distr. Comput. **11**(1), 89–106 (2018)
3. Prakash, K.N.B., Srour, H., Velan, S.S., Chuang, K.H.A.: Method for the automatic segmentation of brown adipose tissue. Magn. Reson. Mater. Phys. Biol. Med. **29**(2), 287–299 (2016)
4. Oliveira, W.S., Teixeira, J.V., Ren, T.I., Cavalcanti, G.D.C., Sijbers, J.: Unsupervised retinal vessel segmentation using combined filters. Plos one **11**(2) (2016)

5. Iliyasu, A.M., Fatichah, C., Abuhasel, K.A.: Evidence accumulation clustering with possibilitic fuzzy C-means base clustering approach to disease diagnosis. Automatika **57**(3), 822–835 (2016)
6. Tasgaonkar, M., Khambete, M.: Integrating fuzzy c-means and Mahalanobis metric classification for exudate detection in color fundus imaging. J. Mech. Med. Biol. **15** (5) (2015)
7. Mahendran, G., Dhanasekaran, R.: Detection and localization of retinal exudates for diabetic retinopathy. J. Biol. Syst. **23**(2), 195–212 (2015)
8. Hassanien, A.E., Emary, E., Zawbaa, H.M.: Retinal blood vessel localization approach based on bee colony swarm optimization, fuzzy c-means and pattern search. J. Vis. Commun. Image Represent. **31**, 186–196 (2015)
9. Ozsen, S., Ceylan, R.: Comparison of AIS and fuzzy c-means clustering methods on the classification of breast cancer and diabetes datasets. Turk. J. Electr. Eng. Comput. Sci. **22**(5), 1241–1254 (2014)
10. Ross, T.J.: Fuzzy Logic with Engineering Applications, 3rd edn. John Wiley, Chichester (2010)
11. Dunn, J.C.: A fuzzy relative of the ISODATA process and its use in detecting compact well-separated clusters. J. Cybern. **3**(3), 32–57 (1973)
12. Bezdek, J.C.: Pattern Recognition with Fuzzy Objective Function Algorithms, pp. 43–93. Plenum Press, New York (1981)

# Fuzzy Expert System for Rectal Cancer

Yusif R. Aliyarov[1], Latafat A. Gardashova[2(✉)],
and Shamil A. Ahmadov[3]

[1] National Center of Oncology, 79B, H. Zardabi Street, Baku, Azerbaijan
yusifaliyarov@yahoo.com
[2] Azerbaijan State Oil and Industry University, Azadlig 20, Nasimi, Baku,
Azerbaijan
latsham@yandex.ru
[3] French-Azerbaijani University, 183 Nizami Street, Baku, Azerbaijan
shamilahmadov@yandex.ru

**Abstract.** Colorectal cancer is the second most common cancer in women and the third most common cancer in men. The goal of this paper is to design a fuzzy rule based medical expert system for colorectal cancer. In the initial stage of the conceptual modeling the designation parameter of a colorectal cancer was performed using clinical data. The goal of the next stage consists of the soft computing based evaluation of the factors. At the third stages is given a possibility measure based fuzzy inference algorithm and examples. In developing knowledge-base of the offered system 2 years case data of 70 persons (patient) of the National Center of Oncology are used. Veracity of 20 diagnoses of patients was checked, and 15 from them were defined as correct.

**Keywords:** Colorectal cancer · IF-THEN rules
Possibility measure based inference system · Tumor response

## 1 Introduction

In the world, intestinal infections as colorectal cancer (CRC) are widely spread among humans. Colorectal cancer statistics is discussed in [1] and shown that research is needed to explain causes for increasing CRC in young adults.

Research work [2] is devoted epidemiology of colorectal cancer and a rectum, including the represented risk factors for its development. Authors provide with pack short discussion of problems of medical care as they concern colorectal cancer. In this work is shown that the risk of growing rectal cancer changes depend on age, but the general role of gender remains obscure. It demands the exact information analysis about colorectal cancer.

Data and operation over them for analysing the survival of patients with colorectal cancer is discussed in [3, 4]. Authors of papers are described 2 approaches of analyzing given data. Patient's pathological data was analyzed for 403 persons. Determined results from logistic regression analysis is approximately 66%, and using a neural network based approach is defined 78%. The obtained results claims that the neural network is better than the regression analysis. Neural network have several advantages, such as processing large amount of data, reducing likelihood of error relevant

© Springer Nature Switzerland AG 2019
R. A. Aliev et al. (Eds.): ICAFS-2018, AISC 896, pp. 160–166, 2019.
https://doi.org/10.1007/978-3-030-04164-9_23

information and decrease of diagnosis time. In [5] discussed expert systems in medicine. The qualities of these systems are characterized. Nowadays all existing systems are based on a crisp or fuzzy information and result is defining by using fuzzy inference methods, which is characterized by loss information. Therefore possibility measure based fuzzy inference method is effective [6, 7]. This measure based algorithm is kernel of information processing of software system ESPLAN.

The purpose of this study is to design a fuzzy rule-based expert system for diagnosis of Colorectal cancer based on possibility measure. The rest of the paper is organized as follows: Sect. 2 discusses representation of production type fuzzy if-then rules and inference engine. Statement of the problem and its solution is given in Sect. 3. Finally, Sect. 4 concludes the paper.

## 2   Representation of Fuzzy if-then Rules and Fuzzy Inference Algorithm

Representation of the knowledgebase is based on fuzzy interpretation in production rules [6]:

$$R^k : IF \ x_1 \ is \ \tilde{A}_{k1} \ and \ x_2 \ is \ \tilde{A}_{k2} \ and \ ...and \ x_m \ is \ \tilde{A}_{km} \ THEN$$

$$y_1 \ is \ \tilde{B}_{k1} \ and \ y_2 \ is \ \tilde{B}_{k2} \ and \ ...and \ y_{kl} \ is \ \tilde{B}_{kl}, k = \overline{1, K}$$

where $x_i, \ i = \overline{1, m}$ and $y_j, j = \overline{1, l}$ are inputs and output variables, $\tilde{A}_{kj}, \tilde{B}_{kj}$ are described by fuzzy sets, and $k$ is the amount of production rules.

The main stages of the possibility measure based algorithm are given below:

1. Representation of the knowledgebase, by using trapezoidal number
2. Calculation of the truth level of the production rules based on possibility measure:
   If the sign is " $=$ " and $r_k = (1 - Poss(\tilde{v}_k | \tilde{a}_{jk})) \cdot cf_k$, then $r_{jk} = (1 - Poss(\tilde{v}_k | \tilde{a}_{jk})) \cdot cf_k$

   If the sign is " $\neq$ " then $Poss$ is determined as $Poss(\tilde{v} | \tilde{a}) = \max_y \min(\mu_{\tilde{v}}(y),$

   $(\mu_{\tilde{a}}(y)) \in [0, 1], \tau_j = \min(r_{jk})$
   First the objects are evaluated, i.e. every $w_i$ object has appropriate linguistic value defined as $(v_i, cf_i)$, where $v_i$ is linguistic value, $c_{f_i} \in ]0, 100]$ is confidence degree of the value $v_i$. $v_k$- linguistic value of the rule object, $a_{jk}$- current linguistic value (j is index of the rule, k is index of relation) value.
3. To determine for each rule:

$$R_j = (\min_j r_{jk}) * CF_j / 100$$

   where CF is the confidence degree of the rule.
4. To define the firing level ($\pi$) and to check $R_j \geq \pi$. If the condition holds true, then the consequent part of rule is calculated.

5. The evaluated $w_i$ objects have $S_i$ value: $w_i, (v_i^1, cf_i^1), \ldots, \ldots, (v_i^{S_i}, cf_i^{S_i})$ $S_i$ is the number of the rules in fuzzy inference process
6. Calculation of resulting value by using the fuzzy average value as follows:

$$\bar{v}_i = \frac{\displaystyle\sum_{n=1}^{S_i} v_i^n \cdot cf_i^n}{\displaystyle\sum_{n=1}^{S_i} cf_i^n}$$

IF $x_1 = \tilde{a}_1^j$ AND $x_2 = \tilde{a}_2^j$ AND... THEN $y_1 = \tilde{b}_1^j$ AND $y_2 = \tilde{b}_2^j$ AND...
IF... THEN $Y_1 = AVR(y_1)$ AND $Y_2 = AVR(y_2)$ AND...
This model has a built-in function AVRG which calculates the average value.

## 3   Statement of the Problem and Its Solution

Colorectal cancer (CRC) is the second most common cancer in women and the third most common one in men worldwide. Approximately in 1/3 of case tumor is localised in rectum- the most distal part of gastrointestinal tract. Surgical treatment of rectal cancer affects quality of life of patients very strongly, causing dysfunctions such as fecal incontinence, urine disorders and sexual dysfunctions. The last 20 years the main aim of treatment of rectal cancer patients was to increase the rate of sphincter pre-serving procedures, now it is directed towards organ preserving strategy.

The management of rectal cancer patients is complex, using in approximately 75–80% of cases neoadjuvant radio- and chemotherapy to improve local control and overall survival. But only 20–30% of patients who underwent neoadjuvant treatment achieve complete pathological response, others 70–80% have poor or no effect of neoadjuvant treatment. The main question now is how can we predict tumor response in rectal cancer patients.

This variation is thought to depend on tumor size, tumor height from anal verge, depth of invasion, differentiation of tumor and of course of molecular factors such as KRAS, BRAF mutation. Creating expert system to predict tumor response value after neoadjuvant chemoradiation is goal of this paper. Defining predict value tumor response after neoadjuvant chemoradiation is a very important problem. The basic problem is to evaluate tumor response by using related parameters. For determining tumor response value, we use fuzzy rules. The tumor response value after neoadjuvant chemoradiation denoted R is a compound index built fromfive parameters each of which is assessed by an expert judgement. The five components are: V-age, LO-localization of tumors, I-infestation rate, N-state of lymph nodes, G-mutation in the genes, R-predict value of tumor response.

Using the above mentioned parameters, the tumor response value model can be expressed as (21 rules):

   *1. If V = about 80 and LO = about 2 and T = about 2 and N = about 1 and G = about 0 THEN R = about 0;*

2. *If V = about 40 and LO = about 2 and T = about 2 and N = about 3 and G = about 0 THEN R = about 0;*

*... ... ... ...*

21. *If V = about 20 and LO = about 92 and T = about 2 and N = about 3 and G = about 1 THEN R = about 100*

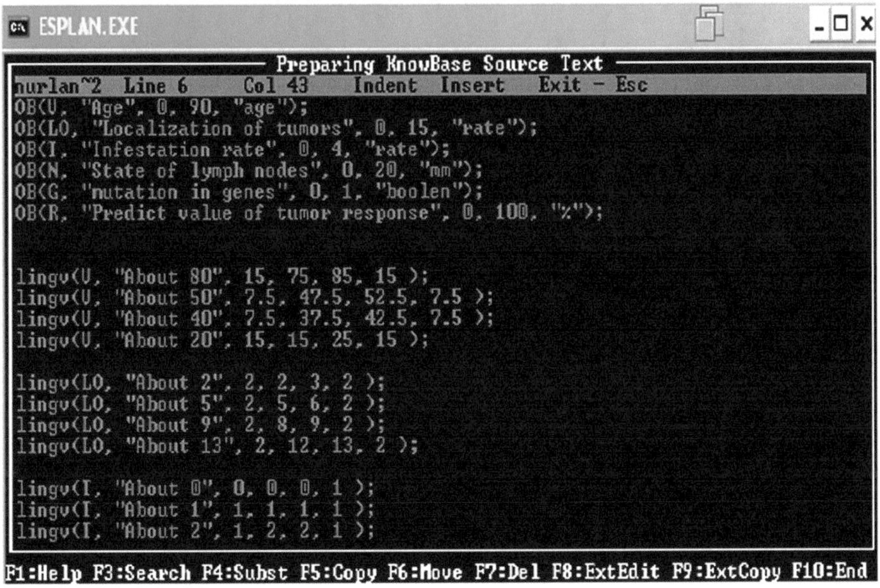

**Fig. 1.** Objects and linguistic terms for "tumor response"

Our aim is to define the level of the tumor response value using five parameters represented by fuzzy linguistic terms. Representation objects, linguistic terms by using ESPLAN is given in Fig. 1.

The trapezoidal fuzzy numbers describing the used linguistic terms are given below:

less than K: (0, M, K − D,D);

approximately K: (D, K,K, D);

more than a: (D, K + D, F, 0);

neutral: (D, M + 2 * D, M + 3*D, D);

much: (D, F − D, F, 0)

where M and F- respectively minimum and maximum value of universe, D = (F − M)/5.

Representation of linguistic terms of tumor response is given in Fig. 2.

For instance,

object = "localization of tumors",

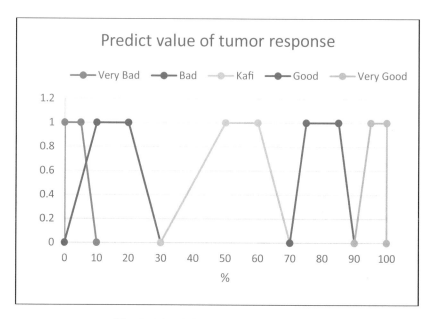

**Fig. 2.** Linguistic terms of tumor response

M = minimum = 0, F = maximum = 15,
linguistic term = "about 5": About 5 = (D, M + 2*D, M + 3*D, D)

These rules have been extracted from experts' knowledge based on interviews conducted by us (Fig. 3). The above mentioned model is implemented by using the

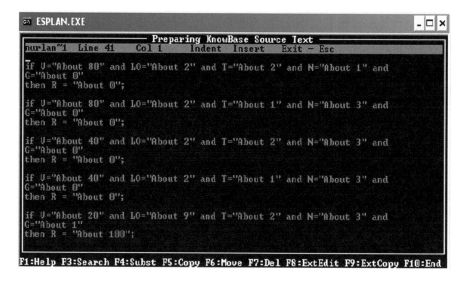

**Fig. 3.** Fragment of the rules of expert system

fuzzy expert system ESPLAN and different tests are performed. Different current information in tests is used.

*Test 1:  If age = about 38 and localization of tumors = about 4 and infestation rate = about 2 and state of lymph nodes = about1 and mutation in the genes = about 0 THEN predict value of tumor response = ?*
FOR TEST1. ANSWER:
EXPERT system shell ESPLAN's decision is "predict value of tumor response is from 0 to 2".
The results of tests is shown in Fig. 4.

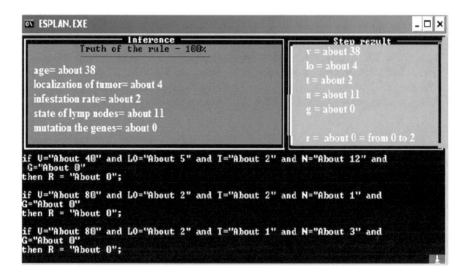

**Fig. 4.** Fragment of results

## 4   Conclusion

In this paper for the evaluation of value of tumor response a possibility measure based method is used. The fuzzy rules obtained from experts were performed in the expert system shell ESPLAN. Various tests were performed and the results were compared to real data about patients.

Description of created system and examples of using it in medicine show, that this system possesses a wide potential of capacities in decision-making on the basis of fuzzy knowledge in conditions of uncertainty. Experimental results show efficiency of the offered expert system.

# References

1. Siegel, R.L., Miller, K.D., Fedewa, S.A., Ahnen, D.J., Meester, R.G.S., Barzi, A., Jemal, A.: Colorectal cancer statistics. CA Cancer J. Clin. **67**(3), 177–193 (2017)
2. Farin Amersi, M.D., Michelle Agustin, M.P.H., Ko, C.Y.: Colorectal cancer: epidemiology, risk factors, and health services. Clin. Colon. Rectal. Surg. **18**(3), 133–140 (2005)
3. Grumett, S., Snow, P., Kerr, D.: neural networks in the prediction of survival in patients with colorectal cancer. Clin. Color. Cancer **2**(4), 239–244 (2003)
4. Amato, F., López, A., Peña-Méndez, E.M., Vaňhara, P., Hampl, A., Havel, J.: Artificial neural networks in medical diagnosis. J. Appl. Biomed. **11**, 47–58 (2013)
5. Rivas Echeverría, F., Rivas Echeverría, C.: Application of expert systems in medicine. In: Proceedings of the 2006 Conference on Artificial Intelligence Research and Development, pp. 3–4. IOS Press, Amsterdam (2006)
6. Aliev, R.A., Aliev, R.R.: Soft Computing and its Application, p. 444. World Scientific, New Jersey, London, Singapore, Hong Kong (2001)
7. Abdullayev, T.S., Gardashova, L.A., Aliev, B.F., Aliev, A.G., Ismailov, B.I.: Fuzzy expert system ESPLAN and its application in business, medicine and technics. In: Seventh International Conference on Application of Fuzzy Systems and Soft Computing, ICAFS-2006, Germany, 13–14 September 2006, pp. 205–215 (2006)

# Dynamics and Control of HIV/AIDS in Cyprus Using Real Data

Evren Hincal[1], Tamer Sanlidag[3,4(✉)], Farouk Tijjani Saad[1],
Kaya Suer[5], Isa Abdullahi Baba[1], Murat Sayan[2,4],
Bilgen Kaymakamzade[1], and Nazife Sultanoglu[6]

[1] Department of Mathematics, Near East University, 99138 Nicosia, Cyprus
[2] Faculty of Medicine, Clinical Laboratory, PCR Unit, Kocaeli University,
41780 Kocaeli, Turkey
[3] Faculty of Medicine, Department of Medical Microbiology,
Celal Bayar University, 45010 Manisa, Turkey
[4] Research Center of Experimental Health Sciences, Near East University,
99138 Nicosia, Cyprus
tamer.sanlidag@neu.edu.tr
[5] Faculty of Medicine, Department of Infectious Diseases and Clinical
Microbiology, Near East University, 99138 Nicosia, Cyprus
[6] Faculty of Medicine, Department of Medical Microbiology and Clinical
Microbiology, Near East University, 99138 Nicosia, Cyprus

**Abstract.** This article presents a mathematical model that studies the dynamics of HIV in North and South Cyprus. The global stability of the two equilibrium points involved are disease-free and endemic, and are performed using Lyapunov function. We have showed that the stability is dependent on the magnitude of the basic reproduction number $R_0$. If $R_0 < 1$, the disease free equilibrium point is globally asymptotically stable and the disease vanishes, whereas if $R_0 \geq 1$, the endemic equilibrium point is globally asymptotically stable and epidemics will occur. Real data obtained from the Turkish Republic of Northern Cyprus Ministry of Health is used to examine and predict the progress of HIV in North Cyprus, as well as comparing our results with South Cyprus using their published data. Reported HIV positive cases of only Turkish and Greek Cypriots were included from the data obtained from Turkish Republic of Northern Cyprus Ministry of Health and South Cyprus data, respectively. The results showed that, the basic reproduction number of North and South Cyprus are 0.00012 and 0.00034 respectively; which are less than one; hence, this indicates that there is currently no epidemic in the country. Furthermore, the number of HIV positive individuals in North Cyprus is likely to increase by almost 50%, whereas for South Cyprus an increase of 100% of the initial value (of 2017) is estimated in the next 20 years. Thus, the authorities should take the necessary actions and strategic measures for controlling the spread of the disease.

**Keywords:** HIV · Basic reproduction number · Cyprus · Lyapunov function
Stability

© Springer Nature Switzerland AG 2019
R. A. Aliev et al. (Eds.): ICAFS-2018, AISC 896, pp. 167–177, 2019.
https://doi.org/10.1007/978-3-030-04164-9_24

# 1   Introduction

HIV is a major health problem in many developed and developing countries, where it is often the most common disease for all ages combined. It is currently a disease of greater demographic diversity, affecting all ages, sexes, races and involving multiple transmission risk behaviors [1]. At least 50000 new HIV infections will continue to be added each year; however, one-fifth of the newly infected individuals may not know their status, and a substantial proportion of those who know they are infected are not engaged in HIV care [2].

Since the recognition of AIDS, it has been estimated that more than 70 million people have encountered HIV and approximately 35 million have died due to the infection. By the end of 2017, 36.9 million people were living with HIV and 940000 HIV-related deaths occurred worldwide. Of the 36.7 million people infected with HIV in 2016, 34.5 million are adults; 17.8 million are women and 16.7 million are men and 2.1 million are children. HIV continues to be a major global public health issue because 1.8 million people became newly infected in 2017 [3, 4].

Mainly two communities exist in Cyprus. These are namely, Turkish and Greek Cypriots which are ruled by different governments since 1974. The Turkish Cypriots live in the Northern part, whereas Greek Cypriots live in the Southern part of the island. In South Cyprus (SC), the preponderance of HIV was historically less until 2005. The disease however, started growing since then with new HIV infected individuals emerging and the rate of infection increased to 6.5% per 100000 in 2014 [5]. Pineda-Peña et al. presented a statistical study of the transmission of HIV-1 in SC using a "densely sampled transmission cohort that included 85% of HIV-1-infected individuals linked to clinical care between 1986 and 2012 based on detailed clinical, epidemiological, behavioral and HIV-1 genetic information" [6]. Their results explored a greater understanding of the dynamics of HIV-1 in SC and may have a direct impact on the evolution and implementation of prevention strategies.

The purpose of this study is to mathematically analyze the information of HIV infections in Cyprus. This is the first study of its kind for NC; which has a population of 286257. The country was deemed potentially interesting from an epidemiological perspective. Moreover, the inter-communal strife and immigrants from Turkey and other third world countries after 1974, and the fighting that NC experienced in the last 40 years might be expected to have had adverse effects on HIV infection cases [7].

We have reviewed the HIV incidence information obtained from Turkish Republic of Northern Cyprus Ministry of Health from 1997–2017 for NC and compared the results with published data from Pineda-Peña et al. for SC [6]. Reported HIV positive cases of only Turkish and Greek Cypriots were included in the study. The foreigners diagnosed as HIV positive in the country were excluded for the purpose of this study. The present study gives also, the initial findings of the NC HIV Registry (NCHR). This is an official population-based HIV registry, set up in 1997, based at the Dr. Burhan Nalbantoglu Hospital, which is the only unit in NC where, HIV drugs may be prescribed. Thus, the NCHR is the main source of information for NC. For the purposes of the present analyses, only the HIV cases registered and diagnosed from 1997 to end of 2017 were considered. People who did not reside in NC for more than six months prior

to their HIV diagnosis were excluded from the study. All information received were treated as "confidential" and ethically approved by the local authorities.

### 1.1   Quality of Data

To ensure continuous improvement quality of the data, the following measures were taken:

1. Since fluctuations would be expected in the number of new HIV cases from year to year, consistency checks were used to detect gross differences over time mainly with respect to site, gender and age distribution.
2. Before analysis, hospital discharge data was organized in order to eliminate possible repeated admissions. Only first admission was retained in each case. Patients in other countries were excluded from the data analyzed.

## 2   Construction of the Model

The model consists of the dynamics of three mutually exclusive populations: the susceptible $S(t)$, the exposed $E(t)$, and the infected $I(t)$. Let the total population at any time t, be denoted as $N(t)$, then $N(t) = S(t) + E(t) + I(t)$.

### 2.1   Dynamics of the Population of Susceptible

$$\dot{S}(t) = \Lambda - \beta_1 SE - \beta_2 SI - \mu S. \tag{1}$$

Here, $\dot{S}(t)$ represents the change in the population of susceptible with time. $\Lambda$ is the recruitment rate into susceptible class. $\beta_1 SE$ and $\beta_2 SI$ are the effective contact rates of susceptible population with exposed and infective populations, respectively. $\mu$ is the natural death rate in the susceptible population.

### 2.2   Dynamics of the Population of Exposed Individuals

$$\dot{E}(t) = \beta_1 SE + \beta_2 SI - (\mu + \delta)E. \tag{2}$$

Here, $\dot{E}(t)$ represents the change in the population of exposed individuals with time. The exposed individuals included in this model are those who are HIV positive however; either did not disclose their status or refuse to seek for medical help or services elsewhere. $\beta_1 SE$ is the effective contact rate as a result of susceptible individuals coming in contact with exposed population. $\beta_2 SI$ is the effective contact rate as a result of susceptible individuals coming in contact with infective population. $\delta$ is the rate at which people in the exposed population are diagnosed and they subsequently know their HIV positive status. $\mu$ is the natural death rate in the exposed population.

## 2.3    Dynamics of the Population of Infective Individuals

$$\dot{I}(t) = \delta E - \mu I. \tag{3}$$

Here, $\dot{I}(t)$ represents the change in the population of infective individuals with time. $\delta$ is the rate at which people in the exposed population know their status and consequently moved to infective class. $\mu$ is the natural death rate in the infective population.

Finally, the interaction between susceptible, exposed and infective individuals leads to the following system of nonlinear ordinary differential equations.

$$\begin{aligned}
\dot{S}(t) &= \Lambda - \beta_1 SE - \beta_2 SI - \mu S \\
\dot{E}(t) &= \beta_1 SE + \beta_2 SI - (\mu + \delta)E \\
\dot{I}(t) &= \delta E - \mu I,
\end{aligned} \tag{4}$$

where,

$$S_0 > 0, E_0 \geq 0, \text{ and } I_0 \geq 0,$$

and

$$N(t) = S(t) + E(t) + I(t).$$

## 2.4    Boundedness

**Theorem 1.** The solutions of system (1) are bounded.

**Proof.** Let $\dot{N}(t) = \dot{S}(t) + \dot{E}(t) + \dot{I}(t)$, then we have,

$$\begin{aligned}
\dot{N}(t) &= \Lambda - \beta_1 SE - \beta_2 SI - \mu S + \beta_1 SE + \beta_2 SI - (\mu + \delta)E + \delta E - \mu I \\
&= \Lambda - \mu(S + E + I).
\end{aligned} \tag{5}$$

Then, $\dot{N}(t) + \mu N(t) = \Lambda$; which gives a linear ordinary differential equation. Solving the equation, we obtain,

$$N(t) = \frac{\Lambda}{\mu} + ce^{-\mu t},$$

$$\lim_{t \to \infty} N(t) = \frac{\Lambda}{\mu}. \tag{6}$$

## 2.5 Existence of Equilibrium Points

Two equilibrium points are found; disease free equilibrium

$$U_0 = (S_0, E_0, I_0) = \left(\frac{\Lambda}{\mu}, 0, 0\right),$$

and endemic equilibrium

$$U_1 = (S_1, E_1, I_1)$$
$$= \left(\frac{\mu(\mu+\delta)}{\mu\beta_1 + \beta_2\delta}, \frac{\Lambda\mu\beta_1 + \Lambda\beta_2\delta - \mu^3 - \mu^2\delta}{\beta_1\mu^2 + \beta_2\mu\delta + \beta_1\mu\delta + \beta_2\delta^2}, \frac{\delta(\Lambda\mu\beta_1 + \Lambda\beta_2\delta - \mu^3 - \mu^2\delta)}{(\beta_1\mu^2 + \beta_2\mu\delta + \beta_1\mu\delta + \beta_2\delta^2)\mu}\right).$$

$$(7)$$

Since both $S_0, E_0, I_0, S_1 \geq 0$ they exist and are biologically meaningful. $E_1, I_1$ also exist if, $\Lambda\mu\beta_1 + \Lambda\beta_2\delta - \mu^3 - \mu^2\delta \geq 0$, that is if,

$$\frac{\Lambda(\mu\beta_1 + \beta_2\delta)}{\mu^2(\mu+\delta)} \geq 1. \tag{8}$$

## 2.6 Basic Reproduction Ratio

It is defined as the number of secondary infections caused by an infectious individual in a completely susceptible population. In this paper, for the basic reproduction ratio calculation, the next generation matrix method will be used. Define

$$F = \begin{bmatrix} \beta_1 SE + \beta_2 SI \\ 0 \end{bmatrix}, V = \begin{bmatrix} (\mu+\delta)E \\ \mu I - \delta E \end{bmatrix}. \tag{9}$$

Then the Jacobian matrices corresponding to F and V from (9) are,

$$\partial F = \begin{bmatrix} \beta_1 S & \beta_2 S \\ 0 & 0 \end{bmatrix} \partial V = \begin{bmatrix} \mu+\delta & 0 \\ -\delta & \mu \end{bmatrix}. \tag{10}$$

Then, the next generation matrix (NGM) is given by

$$\partial F(\partial V)^{-1}(U_0) = \begin{bmatrix} \frac{\Lambda(\beta_1\mu + \beta_2\delta)}{\mu^2(\mu+\delta)} & \frac{\beta_2\Lambda}{\mu^2} \\ 0 & 0 \end{bmatrix}. \tag{11}$$

Thus, the eigenvalues of the NGM are,

$$\lambda_1 = \frac{\Lambda(\beta_1\mu + \beta_2\delta)}{\mu^2(\mu+\delta)} \text{ and } \lambda_2 = 0. \tag{12}$$

Thence, the basic reproduction ratio, which is defined as the dominant eigenvalue is

$$R_0 = \frac{\Lambda(\beta_1 \mu + \beta_2 \delta)}{\mu^2(\mu + \delta)}. \tag{13}$$

## 3   Stability Analysis

In this chapter we study the stability analysis of both disease free and endemic equilibrium points.

**Theorem 2.** The disease free equilibrium $(U_0)$ is globally asymptotically stable if (17) holds.

**Proof.** Define Lyapunov function

$$V(S,E,I) = (S - S_0 lnS) + E + I. \tag{14}$$

It is clear $V(S,E,I) > 0$ and $V(U_0) = 0$. It remains to show $\dot{V}(S,E,I) \leq 0$.

$$\dot{V}(S,E,I) = \left(1 - \frac{S_0}{S}\right)\dot{S} + \dot{E} + \dot{I} \tag{15}$$

$$= \left(1 - \frac{S_0}{S}\right)[\Lambda - \beta_1 SE - \beta_2 SI - \mu S] + \beta_1 SE + \beta_2 SI - (\mu + \delta)E + \delta E - \mu I$$

$$= \Lambda - \beta_1 SE - \beta_2 SI - \mu S - \frac{\Lambda S_0}{S} + \beta_1 S_0 E + \beta_2 S_0 I + \mu S_0 + \beta_1 SE + \beta_2 SI - \mu E - \mu I$$

$$= 2\mu S_0 - \frac{\Lambda S_0}{S} - \mu S - E(\mu - \beta_1 S_0) - I(\mu - \beta_2 S_0)$$

$$= \mu S_0\left(2 - \frac{S_0}{S} - \frac{S}{S_0}\right) - E(\mu - \beta_1 S_0) - I(\mu - \beta_2 S_0). \tag{16}$$

Since $2 - \frac{S_0}{S} - \frac{S}{S_0} < 0$, by the relation between arithmetic and geometric mean, this implies $\dot{V}(S,E,I) \leq 0$ if

$$max\left\{\frac{\beta_1 S_0}{\mu}, \frac{\beta_2 S_0}{\mu}\right\} \leq 1. \tag{17}$$

**Theorem 3.** The endemic equilibrium $(U_0)$ is globally asymptotically stable if $R_0 \geq 1$.

**Proof.** Define Lyapunov function

$$V(S,E,I) = S + (E - E_0 lnE) + I. \tag{18}$$

It is clear $V(S,E,I) > 0$ and $V(U_1) = 0$. It remains to show $\dot{V}(S,E,I) \leq 0$.

$$\dot{V}(S,E,I) = \left(1 - \frac{E_1}{E}\right)\dot{E} + \dot{S} + \dot{I}$$

$$= \Lambda - \beta_1 SE - \beta_2 SI - \mu S + \left(1 - \frac{E_1}{E}\right)[\beta_1 SE + \beta_2 SI - (\mu+\delta)E] + \delta E - \mu I$$

$$= \mu(S_1 - S - E - I) - E_0\left[\beta_1 S + \beta_2 S\frac{I}{E} - (\mu+\delta)\right] \qquad (19)$$

$$= -\mu(N - S_1) - E_1\left[\beta_1 S + \beta_2 S\frac{I}{E} - (\mu+\delta)\right].$$

Since, $N - S_1 \geq 0$ and $\beta_1 S + \beta_2 S\frac{I}{E} - (\mu+\delta) \geq 0$, then $\dot{V}(S,E,I) \leq 0$ if $E_1 \geq 0$. But $E_1 \geq 0$ if,

$$R_0 = \frac{\Lambda(\mu\beta_1 + \beta_2\delta)}{\mu^2(\mu+\delta)} \geq 1. \qquad (20)$$

## 4  Numerical Simulations

There are five parameters in model (1): $\Lambda, \mu, \beta_1, \beta_2$, *and* $\delta$. According to the data obtained from Turkish Republic of Northern Cyprus Ministry of Health and Pineda-Peña et al. [6], these parameter values are given in Tables 1 and 2, respectively. Some variables/parameters are not easily obtainable; hence, their estimate values are given. For example, $E = [1.2285 I_0]$. We use MATLAB version 2017b to perform the simulation of our model using parameter values from Tables 1 and 2. The prediction of what is likely to happen in the next 40 years in the number of infective individuals is given in

**Table 1.** List of variables needed for North Cyprus

| Variable/Parameter | Meaning | Value |
|---|---|---|
| N | Total population | 286,257 |
| S | Number of susceptible individuals | 286,117 |
| I | Number of infected individuals | 63 |
| E | Number of exposed individuals | 77 |
| $\Lambda$ | Birth rate | 1.3 |
| $\mu$ | Natural death rate | 0.004 |
| $\beta_1$ | Contact rate between susceptible and exposed individuals | $1.25 \times 10^{-13}$ |
| $\beta_2$ | Contact rate between susceptible and infected individuals | $1.7 \times 10^{-10}$ |

Fig. 1. Hence, according Fig. 1, the estimated number of HIV infected individuals in NC will almost increase to 50% of its initial value (63 HIV infected individuals) in the next 20 years.

**Table 2.** List of variables needed for South Cyprus

| Variable/Parameter | Meaning | Value |
|---|---|---|
| N | Total population | 854800 |
| S | Number of susceptible individuals | 854048 |
| I | Number of infected individuals | 233 |
| E | Number of exposed individuals | 286 |
| $\Lambda$ | Birth rate | 1.37 |
| $\mu$ | Natural death rate | 0.0026 |
| $\beta_1$ | Contact rate between susceptible and exposed individuals | $1.25 \times 10^{-12}$ |
| $\beta_2$ | Contact rate between susceptible and infected individuals | $1.7 \times 10^{-9}$ |

**Table 3.** Basic reproduction numbers of North and South Cyprus

| North Cyprus ($R_0$) | South Cyprus ($R_0$) |
|---|---|
| 0.000012 | 0.00034 |

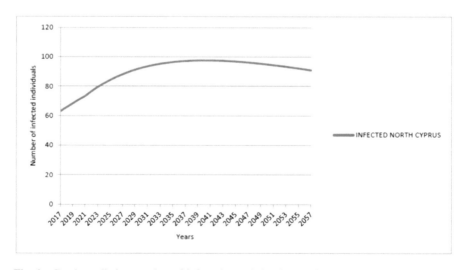

**Fig. 1.** Graph predicting number of infected population in North Cyprus in the next 40 years

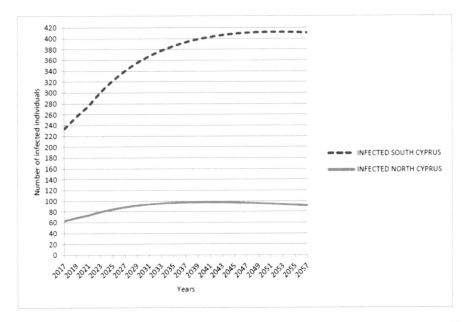

**Fig. 2.** Comparison of infected individuals between North and South Cyprus in the next 40 years

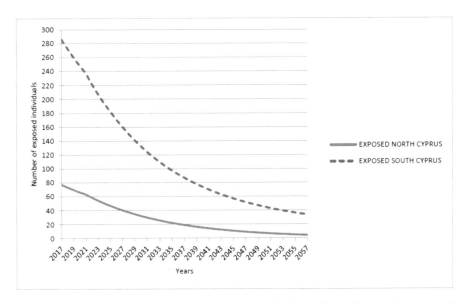

**Fig. 3.** Comparison of exposed individuals between North and South Cyprus in the next 40 years

## 5  Discussions and Conclusion

We have formulated a mathematical model that studied the dynamics of HIV in Cyprus. Two equilibrium points were obtained; disease free and endemic equilibrium. The global stability analysis of the equilibria was performed using Lyapunov function, and the stability of the equilibrium points depends on the magnitude of the basic reproduction number $R_0$. If the basic reproduction number is less than one, then no epidemic will occur, and that implies the disease will be eradicated. However, if the basic reproduction number is bigger than or equal to one, then epidemic will occur.

We have used the data obtained from Turkish Republic of Northern Cyprus Ministry of Health and Pineda-Peña et al. [6] for NC and SC respectively, to simulate our model. Solely, Turkish and Greek Cypriots cases of HIV positive were included, and foreigners were excluded from the study. From the parameters in Table 1, the basic reproduction number $R_0$ of NC is approximately 0.000012; which is less than one. This implies that, presently there is no HIV epidemic in NC. Similarly, the basic reproduction number for SC using the parameters from Table 2 is approximately 0.00034. Hence, there is also no epidemic in SC because $R_0$ is less than one. It can be observed from Table 3 that, the basic reproduction number of SC is bigger than that of NC and is much closer to one; therefore, SC is more likely to experience an epidemic first when compared to NC.

In addition, we have simulated our results and made a prediction of what is likely to happen in subsequent years. The numerical simulations revealed that, the number of HIV positive individuals in NC would almost increase by 50% of its initial value in the next 20 years and this is given in Fig. 1.

It can be observed from Fig. 2 that, the number of HIV infected individuals in SC and NC are both increasing in the following 20 years. It is also important to note that, the increase in SC is more rapid than in NC. Therefore, the authorities in SC need to significantly improve their control strategy so to avoid this robust increase observed.

On the other hand, the number of exposed individuals in both SC and NC will decrease in the next 40 years (Fig. 3). In the same figure, it can be also observed that, the decrease in SC is more rigorous than in NC. However, this might be due to the fact that, the population of SC is bigger compared to NC.

In conclusion, the mathematical model formulated in this article indicated that the $R_0 < 1$ for both NC and SC, hence, there is currently no HIV epidemic for both communities in the island.

## References

1. Kaymakamzade, B., Şanlıdağ, T., Hıncal, E., Sayan, M., Saad, F.T., Baba İ.B.: Role of awareness in controlling HIV/AIDS: a mathematical model. Qual Quant (2017) https://doi.org/10.1007/s11135-017-0640-2
2. UNAIDS, 12 May 2013. http://www.unaids.org/en/media/unaids/contentassets/documents/epidemiology/2013/gr2013/UNAIDS_Global_Report_2013_en.pdf. Accessed 13 May 2014
3. WHO HIV/AIDS Global Health Observatory (GHO) data. http://www.who.int/gho/hiv/en/. Accessed 18 July 2016

4. WHO HIV/AIDS Key facts. http://www.who.int/en/news-room/fact-sheets/detail/hiv-aids. Accessed 18 July 2016
5. European Centre for Disease Prevention and Control – ECDC - & WHO Regional Office for Europe. Technical mission: HIV in Cyprus, 15–17 October 2014. ECDC, Stockholm (2015)
6. Pineda-Pena, A.C., Theys, K., Demetriades, I., Abecasis, A.B., Kostrikis, L.G.: HIV-1 Infection in Cyprus, the Eastern Mediterranean European Frontier: A Densely Sampled Transmission Dynamics Analysis from 1986 to 2012, Scientific Reports, vol. 8, no. 1702, pp. 1–15 (2018). https://doi.org/10.1038/s41598–017-19080-5
7. Centers for Disease Control and Prevention (CDC) Estimates of new HIV infections in the United States. http://www.cdc.gov/hiv/topics/surveillance/resources/factsheets/incidence.htm. Accessed 28 Feb 2010

# Intuitionistic Fuzzy Sets for Estimating the Parameters of Distributive Task

Alexander Bozhenyuk[(✉)], Margarita Knyazeva, and Olesiya Kosenko

Southern Federal University, Nekrasovsky 44, 347922 Taganrog, Russia
avb002@yandex.ru, margarita.knyazeva@gmail.com,
o_kosenko@mail.ru

**Abstract.** This article proposes an approach to assessing factors that affect the solution of distribution problems. Distribution tasks are widely used at present. The system principle of investigating the objects of the distribution system corresponds to the understanding that when studying them it is necessary to start from internal connections and multilateral interdependencies between a large number of elements. The increase of the system parameters allows to optimize complex resource allocation problems and to take into account a greater number of factors affecting the final result. One of the important parameters of the distribution system is demand. A correct definition of the magnitude of demand affects the solution of several problems: planning and organization of production procedure; calculation of optimal levels of orders for resources, as well as the determination of volumes and the rational functioning of the transport subsystem. Since the total number of factors influencing the level of demand is very high, an expert needs a tool to distinguish groups of such factors. In order to solve this problem it is proposed to use intuitionistic fuzzy sets, which allow to take into consideration the influence degree of factors on the controlled parameter. This approach allows a large number of unordered factors to be converted into a small number of significant and agreed factors, which can provide the basis for a visual and informative analysis.

**Keywords:** Distribution of resources · Fuzzy parameters · Factors of influence
Intuitionistic fuzzy set · Measure of similarity

## 1 Introduction

The control of the work of enterprises, workshops, equipment lines requires the application of methods that ensure the systematization of a large number of controlled parameters. In this case the solutions found should ensure the optimization of the system as a whole [1]. When optimizing the functioning of large systems distribution methods are used successfully [2]. Mathematical apparatus for modelling distribution problems are usually based on applied researches in the design of systems using system approaches. A common feature of such problems is the allocation of resources between producers and consumers of these resources in such a way that a group of pre-selected criteria is met [3].

© Springer Nature Switzerland AG 2019
R. A. Aliev et al. (Eds.): ICAFS-2018, AISC 896, pp. 178–184, 2019.
https://doi.org/10.1007/978-3-030-04164-9_25

The allocation of resources is an important part of the production activities of enterprises. However, the role of the defining function in the organization of enterprises, which it acquired relatively recently [4], was called logistics. n logistics, allocation refers to the physical movement of resources to meet the subsequent demand for products [4]. The main task of distribution is to optimize the process of allocating resources in specified places with minimal costs [4–7]. The tasks of distribution optimization find solutions in different fields of science and technology, therefore the meaning of distribution and optimality can be different depending on the specific task.

The creation of a new transport system that is stable against external disturbances leads to the need to solve a number of specific problems, among which [8, 9]:

– estimation and forecasting of the demand for resources;
– planning and organization of production of resources;
– calculation of optimal levels of orders for resources and associated definition of volumes and organization of the rational functioning of the transport subsystem.

The magnitude of demand is an important parameter in the solution of the above tasks. In [10–12] some methods were proposed, they aimed at optimizing resource allocation. The disadvantage of the presented methods is that the problem of efficient allocation is achieved in a deterministic setting. Non-deterministic demand can significantly affect the outcome of the solution of the problem.

In [13], a solution to this problem was proposed, taking into account the stochasticity of demand. Accounting for demand parameters in the form of fuzzy intervals was considered in [14–16].

## 2   Basic Concepts and Definitions of Demand Parameters

Uncertain exact parameters of the problem are defined in the form of intervals, the boundaries of which correspond to certain values. These boundaries can be set with a sufficient degree of reliability according to statistical data, or by the accumulated experience and intuitive assumptions of experts. If we determine the boundaries of interval estimates accurately, then there is a possibility that the boundaries can be either overvalued or underestimated, which will cause doubt in the results of calculations.

Setting the parameters in the form of a fuzzy interval will be both overestimated and underestimated, and the carrier of the fuzzy interval will be chosen so that the kernel contains the most plausible values, and it will be guaranteed to find the parameter in question within the required limits.

The fuzzy interval is defined on the set $Q$ by four parameters [16] - $\tilde{Q} = (\alpha, \beta, A, B)$, where $\alpha$ is the left coefficient of fuzziness; $\beta$ is the right coefficient of fuzziness; A is lower modal value; B is the upper modal value of the fuzzy interval. Parameters $A$ and $B$ define the core of the fuzzy interval. Figure 1 shows a fuzzy interval of $(L\text{-}R)$ type [17].

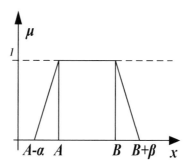

**Fig. 1.** Fuzzy interval of (L-R) type.

The membership function of (L-R)-type fuzzy interval is specified as a composition of some $L$ and $R$ functions [18, 19]:

$$\mu_Q(x) = \begin{cases} L\left(\frac{A-x}{\alpha}\right), x \leq A \\ 1, \quad A \leq x \leq B \\ R\left(\frac{x-B}{\beta}\right), x \geq B. \end{cases}$$

It is proposed to implement the problem of determining the location of intermediate centers depending on the parameters of demand (Fig. 2). Setting the parameters of demand in the problem of determining the location of intermediate centers in the form of fuzzy intervals allows to obtain adequate results in the absence of the possibility of clearly indicating the magnitude of demand, and to determine the existence of tolerances for the change in the corresponding quantities [15].

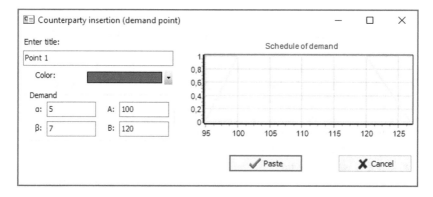

**Fig. 2.** Specifying demand parameters in the form of a fuzzy interval.

Determination of demand parameters in the problem of allocating centers in the form of fuzzy intervals is a more general way of formalizing these parameters, allowing taking into account the existing a priori uncertainty. The proposed fuzzy-interval technique corresponds to the intuitive views of experts about the predicted parameters. At the same time, experts should take into account a variety of factors, both direct and indirect, that affect, including demand. In quantifying the range of demand values, the expert should be guided by the degree of influence of the relevant factors.

## 3   Fuzzy Factors Estimation Based on Intuitionistic Fuzzy Sets

The task of selecting the most influential factors precedes the solution of other problems, for example, optimization problems.

The intuitionistic approach is proposed to be used to assess the degree of influence of the factor on the magnitude of demand. Therefore, the value of the sought-for factor will be represented by an intuitionistic fuzzy set [18, 19]:

$$\Phi = \{ <x, \mu_\Phi(x), v_\Phi(x) > | x \in X \},$$

where $\mu_\Phi(x):X \to [0,1]$ is the membership function of the intuitionistic fuzzy set; $v_\Phi(x): X \to [0,1]$ is a function of non-membership of the intuitionistic fuzzy set. In addition, the functions $\mu_\Phi$ and $v_\Phi$ such that: $(\forall x \in X)[0 \leq \mu_\Phi(x) + v_\Phi(x)]$. For each intuitionistic set, in addition to membership functions and non-membership functions, an intuitionistic fuzzy index is determined in accordance with the following formula:

$$\pi_\Phi(x) = 1 - \mu_\Phi(x) - v_\Phi(x).$$

Let $n$ factors $(\Phi_n)$ be given, which, in the opinion of experts, can influence the demand value when solving transport-distribution problems. According to the theory of intuitionistic fuzzy sets, it is necessary to choose a creeps-element, to determine the so-called distance, which is an auxiliary tool for calculating the similarity of a fuzzy set. As a creeps-element, we choose a conditional factor with the following values of membership, non-membership and intuitionistic fuzziness:

$$\Phi_0 = \{x_0, \mu_\Phi(x_0), v_\Phi(x_0)\} = \{x_0, 1, 0\}.$$

That is, about factor $\Phi_0$, it is possible to tell, that it unequivocally influences size of demand. We write down the values of the factors, in accordance with the assumptions of the experts in the form of an intuitionistic fuzzy sets $(\Phi_i = \{x_i, \mu_\Phi(x_i), v_\Phi(x_i)\}$, $i = \overline{1,5})$: $\Phi_0 = \{x_0, 1, 0\}$, $\Phi_1 = \{x_1, 0.5, 0.2\}$, $\Phi_2 = \{x_2, 0.3, 0.6\}$, $\Phi_3 = \{x_3, 0.7, 0,2\}$, $\Phi_4 = \{x_4, 0.4, 0.4\}$, $\Phi_5 = \{x_5, 0.1, 0.4\}$.

To calculate the distance between intuitionistic fuzzy sets, we use the Hausdorff formula, which is defined as follows:

$$L(\Phi_0, \Phi_i) = \max(|\mu_{\Phi_0}(x) - \mu_{\Phi_i}(x)|, |v_{\Phi_0}(x) - v_{\Phi_i}(x)|, |\pi_{\Phi_0}(x) - \pi_{\Phi_i}(x)|),$$

and we use the Hamming formula to estimate the single-valued of the proposed approach:

$$L(\Phi_0, \Phi_i) = \frac{1}{2n} \Sigma(|\mu_{\Phi_0}(x) - \mu_{\Phi_i}(x)| + |v_{\Phi_0}(x) - v_{\Phi_i}(x)| + |\pi_{\Phi_0}(x) - \pi_{\Phi_i}(x)|),$$

where $n$ is the power of intuitionistic fuzzy sets $\Phi_0, \Phi_i$.

*The following* Table 1 *gives* the results of calculated distances:

**Table 1.** Computing distance with intuitionistic index of fuzziness.

| $\Phi_0, \Phi_i$ | $L(\Phi_0, \Phi_i)$ by Hausdorff's formula | $L(\Phi_0, \Phi_i)$ by Hamming formula |
|---|---|---|
| $\Phi_0, \Phi_1$ | 0.5 | 0.5 |
| $\Phi_0, \Phi_2$ | 0.7 | 0.7 |
| $\Phi_0, \Phi_3$ | 0.3 | 0.3 |
| $\Phi_0, \Phi_4$ | 0.6 | 0.6 |
| $\Phi_0, \Phi_5$ | 0.9 | 0.9 |

When calculating distances of a fuzzy set without taking into account the intuitionistic index of fuzziness, the values will be the following (Table 2):

**Table 2.** Computing the distance without taking into account the intuitionistic index of fuzziness.

| $\Phi_0, \Phi_i$ | $L(\Phi_0, \Phi_i)$ by Hausdorff's formula | $L(\Phi_0, \Phi_i)$ by Hamming formula |
|---|---|---|
| $\Phi_0, \Phi_1$ | 0.5 | 0.35 |
| $\Phi_0, \Phi_2$ | 0.7 | 0.65 |
| $\Phi_0, \Phi_3$ | 0.3 | 0.25 |
| $\Phi_0, \Phi_4$ | 0.6 | 0.5 |
| $\Phi_0, \Phi_5$ | 0.9 | 0.65 |

Thus, the values of distances calculated without taking into account the intuitionistic fuzzy index according to the Hamming and Hausdorff formulas do not coincide, in contrast to the values taking into account this index. It can be concluded that calculating distances using only two parameters of a fuzzy set leads to inconsistent results. At the same time, taking into account the membership function, the non-membership function and the intuitionistic fuzzy index give the same results.

We calculate the measure of similarity, which depends on the quantitative value of the distances between the considered sets. There are various ways to calculate the

**Table 3.** The value of the similarity measure of intuitionistic fuzzy sets.

|          | $\Phi_0$ | $\Phi_1$ | $\Phi_2$ | $\Phi_3$ | $\Phi_4$ | $\Phi_5$ |
|----------|------|------|------|------|------|------|
| $\Phi_0$ | 1    | 0.5  | 0.3  | 0.7  | 0.4  | 0.1  |
| $\Phi_1$ | 0.5  | 1    | 0.6  | 0.8  | 0.8  | 0.6  |
| $\Phi_2$ | 0.3  | 0.6  | 1    | 0.6  | 0.8  | 0.6  |
| $\Phi_3$ | 0.7  | 0.8  | 0.6  | 1    | 0.7  | 0.4  |
| $\Phi_4$ | 0.4  | 0.8  | 0.8  | 0.7  | 1    | 0.7  |
| $\Phi_5$ | 0.1  | 0.6  | 0.6  | 0.4  | 0.7  | 1    |

degree of similarity between two sets. In particular, in [20, 21] the following way of calculating the degree of similarity between two intuitionistic fuzzy sets was proposed:

$$Sim(\Phi_j, \Phi_i) = \frac{1}{n}\Sigma(1 - \frac{|\mu_{\Phi_j}(x) - \mu_{\Phi_i}(x)|}{2} - \frac{|\nu_{\Phi_j}(x) - \nu_{\Phi_i}(x)|}{2} - \frac{|\pi_{\Phi_j}(x) - \pi_{\Phi_i}(x)|}{2})$$
$$= 1 - \frac{1}{2n}\Sigma(1 - |\mu_{\Phi_j}(x) - \mu_{\Phi_i}(x)| - |\nu_{\Phi_j}(x) - \nu_{\Phi_i}(x)| - |\pi_{\Phi_j}(x) - \pi_{\Phi_i}(x)| = 1 - L(\Phi_j, \Phi_i),$$

here $L(\Phi_j, \Phi_i)$ is the distance between intuitionistic fuzzy sets $\Phi_j$, $\Phi_i$.

The results of calculations of the similarity measure are presented in Table 3:

## 4    Conclusions

This paper examines similarity measures for intuitionistic fuzzy sets. It allows to consider a problem of similar factors determination and to reduce the total number of factors which influence the level of demand in allocation problem.

Based on the results of calculations it is possible to identify the most similar sets (factors) and to determine the application of the most similar decisions to them. This approach has a wide range of applications: when choosing preferences, in diagnosing, in solving various social problems, where the number of factors is large and it makes sense to determine similar, thereby simplifying the analysis of the effect of multiple parameters on the final result of the solution.

**Acknowledgments.** This work has been supported by the Russian Foundation for Basic Research, Project № 18-01-00023a.

## References

1. Schenk, M., Tolujew, J., Reggelin, T.: A mesoscopic approach to the simulation of logistics systems. In: Advanced Manufacturing and Sustainable Logistics, pp. 15–25. Springer, Heidelberg (2010)
2. Giraud, L., Bavière, R., Vallée, M., Paulus, C.: Recent advances in modelling, simulation and operational optimization of DH systems. Euroheat and Power (English Edition) **13**(4), 12–15 (2016)
3. Muckstadt, J., Sapra, A.: Principles of Inventory Management. When You Are Down to Four Order More. Springer Series in Operations Research and Financial Engineering. Springer, New York (2010)

4. Brandimarte, P., Zotteri, G.: Introduction to Distribution Logistics. Wiley, Hoboken (2007)
5. Gudehus, T., Kotzab, H.: Comprehensive Logistics. Springer, Heidelberg (2012)
6. Wardlow, D., Wood, D., Johnson, P.: Modern logistic. Murphy. Trudged., Publ. house Williams (2002)
7. Rushton, A., Croucher, P., Baker, P.: The Handbook of Logistics and Distribution Management: Understanding the Supply Chain. Kogan Page, London (2014)
8. Du, D.-Z., Ko, K.I., Hu, X.: Design and Analysis of Approximation Algorithms. Springer, Heidelberg (2012)
9. Kuzmin, E.: Uncertainty and Certainty in Management of Organizational-Economic Systems. LAP LAMBERT Academic Publishing, Munich (2012)
10. Mac Queen, J.: Some methods for classification and analysis of multivariate observations. In: Proceedings of the Fifth Berkeley Symposium on Mathematical Statistics and Probability, pp. 281–297 (1967)
11. Lambert, D., Stock, J., Ellram, L.: Fundamentals of Logistics Management. McGraw-Hill/Irwin, New York (1997)
12. Ross, D.: Introduction to Supply Chain Management Technologies. CRC Press, Boca Raton (2010)
13. Seraya, O.: Mnogomernye modeli logistiki v usloviyah neopredelennosti. FOP Stecenko I. I., Kharkiv (2010)
14. Kosenko, O., Sinyavskaya, E., Shestova, E., Kosenko, E., Chemes, O.: Method for solution of the multi-index transportation problems with fuzzy parameters. In: XIX IEEE International Conference on Soft Computing and Measurements (SCM), pp. 179–182 (2016)
15. Kosenko, O., Shestova, E., Sinyavskaya, E., Kosenko, E., Nomerchuk, A., Bozhenyuk, A.: Development of information support for the rational placement of intermediate distribution centers of fuel and energy resources under conditions of partial uncertainty. In: XX IEEE International Conference on Soft Computing and Measurements (SCM), pp. 224–227 (2017)
16. Dubois, D., Prade, H.: Fuzzy Sets and Systems. Academic Press, New York (1980)
17. Raskin, L., Seraya O.: Nechetkaya matematika. Osnovy teorii. Prilozheniya. Parus, Kharkiv (2008)
18. Atanassov, K.: On Intuitionistic Fuzzy Sets Theory. Springer, New York (2012)
19. Atanassov, K.: New operations defined over the intuitionistic fuzzy sets. Fuzzy Sets Syst. **61** (2), 137–142 (1994)
20. Shabir, M., Khan, A.: Intuitionistic fuzzy filters of ordered semigroups. J. Appl. Math. Inform. **26**(5–6), 213–220 (2008)
21. Pagurova, V.: A limiting multidimensional distribution of intermediate order statistics. Mosc. Univ. Comput. Math. Cybern. **41**(3), 130–133 (2017)

# Fuzzy Logic Based Modelling of Decision Buying Process

Gunay Sadikoglu[1(✉)] and Tulen Saner[2]

[1] Faculty of Business Administration, Near East University, PO Box: 99138, Nicosia, Mersin 10, North Cyprus, Turkey
gunay.sadikoglu@neu.edu.tr
[2] School of Tourism and Hotel Management, Near East University, PO Box: 99138, Nicosia, Mersin 10, North Cyprus, Turkey
tulen.saner@neu.edu.tr

**Abstract.** Understanding of buyer behaviour plays a significant role in marketing. Modelling of buyer behaviour provides an ability to segment the market effectively and develop marketing-mix usefully. Existing models of buyer decision-making are not well suited to uncertain, imprecise real-life situation and there is a need to develop a new conceptual and quantitative model.

In this paper, we suggest fuzzy logic based on 2-level hierarchical decision-making model which includes main factors affecting consumer buying behaviour such as environmental, situational, psychological, social and other that are inherently uncertain and imprecise.

In the first level of the model impact of shopping environment and time pressure on psychological variables such as shopping motivation and emotion of consumers are considered. In the second level, the relationship between shopping motivation and emotion of consumers' buying intensity is established. Decision processes are performed by using fuzzy aggregation methods and fuzzy reasoning on the bases of obtained fuzzy "If-Then" rules.

Finally, the numerical example is provided in order to demonstrate the validity of the proposed model.

**Keywords:** Decision-making · Hierarchical model · Emotion
Shopping motivation · Consumer buying intensity
Horizontal membership function

## 1 Introduction

The consumer behaviour has always been of great interest in marketing research [1], the consumer buying behaviour has become an integral part of the strategic marketing planning and includes mental, emotional and physical activities. The consumer behaviour and decision-making process are usually subject to uncertainties related to the influences of marketing, psychological, socio-cultural, environmental and personal factors [2, 3].

Motivation and emotion are usually viewed as two psychological variables that energize behaviour [4].

© Springer Nature Switzerland AG 2019
R. A. Aliev et al. (Eds.): ICAFS-2018, AISC 896, pp. 185–194, 2019.
https://doi.org/10.1007/978-3-030-04164-9_26

Motivation has been defined in various ways over the years, but a common component of the different definitions is that motivation is a force that energizes, activates and directs behaviour [5].

Emotion refers to the conscious and subjective experience that is characterized by mental states, biological reactions and psychological or physiologic expressions.

Time pressure that is important for airport retail environment can be considered as a situational variable affecting consumers' decision making [6]. Shopping motivation, emotion and time pressure are the key factors affecting passengers' shopping behaviours within the retailing environment [7].

Importance of applications of fuzzy logic in humanity, social sciences, life sciences, medicine, and so on was stated in the many papers [8–14].

## 2  Preliminaries

*Linguistic IF-THEN rules.* Fuzzy sets and linguistic variables are widely used for modeling nonlinear uncertain process and for approximating functions [15–18]. The idea of fuzzy modeling consists in substituting the precise mathematical relation between object parameters by some qualitative relations expressed via linguistic IF-THEN rules.

The structure of linguistic IF-THEN rules for objects or systems with multiple inputs (n) and outputs (m) (MIMO models) is expressed as follows:

$$IF\ x_1\ is\ A_{11}\ AND\ x_2\ is\ A_{12}\ AND\ \ldots x_n\ is\ A_{1n}\quad THEN$$
$$IF\ y_1\ is\ B_{11}\ AND\ y_2\ is\ B_{12}\ AND\ \ldots y_m\ is\ B_{1m}$$

*ALSO*

$$IF\ x_1\ is\ A_{21}\ AND\ x_2\ is\ A_{22}\ AND\ \ldots x_n\ is\ A_{2n}\quad THEN$$
$$IF\ y_1\ is\ B_{21}\ AND\ y_2\ is\ B_{22}\ AND\ \ldots y_m\ is\ B_{2m}$$

*ALSO*

$$IF\ x_1\ is\ A_{r1}\ AND\ x_2\ is\ A_{r2}\ AND\ \ldots x_n\ is\ A_{r\,n}\quad THEN$$
$$IF\ y_1\ is\ B_{r1}\ AND\ y_2\ is\ B_{r2}\ AND\ \ldots y_m\ is\ B_{r\,m}$$

Here $x_i(i = \overline{1,n})$, $y_j(j = \overline{1,m})$ are input and output linguistic variables of the fuzzy system $A_{ij}, B_{ij}$ are fuzzy sets. Each rule can be represented by IF-THEN fuzzy relation (linguistic implication).

The conceptual model of the buyer behaviour decision process is shown in Fig. 1.

Main factors affecting emotion and shopping motivation of a buyer are shopping environment $(X_1)$ and time pressure $(X_2)$. The shopping environment and time pressure are defined by subfactors, such as "The shopping area is well arranged", "Shopping environment is quitter" and "I feel pressured to complete my shopping quickly" etc.

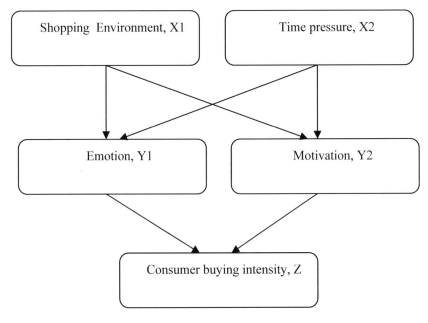

**Fig. 1.** Conceptual model

Calculation of emotion $(Y_1)$ and shopping motivation $(Y_2)$ measures provide an ability to define consumer (buyer) behaviour index (Z).

Collection of relevant data is needed in order to construct mathematical description model of the suggested conceptual model of consumer behaviour. To do so, first, we have to identify all factors and subfactors affecting the shopping environment, time pressure, emotional states, etc.

## 3   Methodology

Measurement variables for the buyer decision-making process were obtained from the questionnaire given in Table 1. Shopping environment $(X_1)$, is defined by 8 items (X11, X18), Time Pressure $(X_2)$, - by 3 items (X21, X22, X23), Emotion $(Y_1)$ - by 7 items (Y11,…Y17), Shopping motivation $(Y_2)$ - by 14 items (Y21,…Y214) and Consumer Buying Intensity (Z) - by 5 items (Z11,…Z14). The reliability and validity of the questionnaires were proven in [7, 19, 20].

Decision relevant information in the considered problem is characterized by uncertainty and imprecision. Because of this, factors and subfactors are expressed linguistically by fuzzy numbers.

**Table 1.** Measurement variables

| Shopping environment, $X_1$ | |
|---|---|
| X11 | The shopping environment is clean and tidy |
| X12 | The colours used create adds excitement to the store environment |
| X13 | The signs and labels used in the shop environment helped me to find my way and do my shopping |
| X14 | Shopping environment is quieter |
| X15 | Lighting environment is quieter |
| X16 | The airport provided sufficient number of comfortable seats |
| X17 | The shopping area is well arranged |
| X18 | Shop assistants were friendly and helpful |
| **Time Pressure, $X_2$** | |
| X21 | I must hurry if I am to complete my shopping trip on time |
| X22 | I feel pressured to complete my shopping quickly |
| X23 | I do not have enough time to shop |
| **Emotional states, $Y_1$** | |
| Y31 | Shopping environment is pleasant |
| Y32 | Shopping environment is attractive |
| Y33 | Shopping environment is boring |
| Y34 | Shopping environment is unsatisfactory |
| Y35 | Shopping environment is active |
| Y36 | Shopping environment is exciting |
| Y37 | Shopping environment is lively |
| **Shopping Motivation, $Y_2$** | |
| Y41 | In the airport, to what extent do the following reasons make you wish to buy something? |
| Y42 | There are many duty-free products |
| Y43 | There are good exchange rates when buying products |
| Y44 | There are significant discounts on the prices of products |
| Y45 | Products are of high quality |
| Y46 | There are newly released products |
| Y47 | There are products with local characteristics |
| Y48 | There are souvenirs |
| Y49 | The staff answer customers questions in a nice way |
| Y410 | The customers can trust the staff |
| Y411 | The staff speak various languages |
| Y412 | There is good access |
| Y413 | One feels comfortable |
| Y414 | The goods are like those in foreign counties |
| **Behavioural Intention, $Z$** | |
| Z51 | I would like to come back to this store in the future |
| Z52 | I enjoyed Shopping at this store |
| Z53 | I would like to stay longer than I planned at this store |
| Z54 | I am willing to spend more than I planned at this store |

The codebook of Fuzzy numbers is shown in Table 2.

**Table 2.** Codebook for fuzzy numbers

| 1 | Strongly disagree (SD) | (1 1 2) |
|---|---|---|
| 2 | Disagree (D) | (1 2 3) |
| 3 | Neither agree nor disagree (NAD) | (2 3 4) |
| 4 | Agree (A) | (3 4 5) |
| 5 | Strongly agree (SA) | (4 5 5) |

## 4 Solution of the Problem

The solution of the consumer behaviour problem described above is given below step by step using a horizontal membership function approach [21] and hierarchical aggregation decision procedure [16, 22].

Step 1. Compute fuzzy valued evaluation of shopping environment as follows

$$x_1^j = \frac{\sum_{i=1}^{8} x_{1i}^j}{8}, \tag{1}$$

Here $j = 1, 2, \ldots, n$, n-number of experts, $i = 1, 2, \ldots, 8$.

Step 2. Compute fuzzy valued evaluation of time pressure as follows.

$$x_2^j = \frac{\sum_{i=1}^{3} x_{2i}^j}{3}, \tag{2}$$

Step 3. Compute fuzzy valued evaluation of emotional states of a buyer $Y_1$ as follows.

$$Y_1^j = \frac{\sum_{i=1}^{7} Y_{1i}^j}{7}, \tag{3}$$

Step 4. Compute fuzzy valued evaluation of shopping motivation of a buyer as follows.

$$Y_2^j = \frac{\sum_{i=1}^{14} Y_{2i}^j}{14}, \tag{4}$$

Step 5. Compute fuzzy valued evaluation of a buyer behavioural intention as follows.

$$Z^j = \frac{\sum Z_i^j}{4} \qquad (5)$$

Step 6. Construct fuzzy If-Then rule-based models that describe the relationship between shopping environment $(X_1)$, time pressure $(X_2)$ and emotional states and shopping motivations as follows.

(a)  Fuzzy rule-based model for shopping environment

$$\text{If } X_1^j \text{ is } A_1^j \text{ and } X_2^j \text{ is } A_2^j \text{ Then } Y_1^j \text{ is } B_1^j \qquad (6)$$

where j = 1,2,...,n

(b)  Fuzzy rule-based model for time pressure

$$\text{If } X_1^j \text{ is } A_1^j \text{ and } X_2^j \text{ is } A_2^j \text{ Then } Y_2^j \text{ is } B_2^j \qquad (7)$$

Step 7. Construct the fuzzy rule-based model for evaluation of a buyer behavioural intention (Z)

$$\text{If } Y_1^j \text{ is } C_1^j \text{ and } Y_2^j \text{ is } C_2^j \text{ Then } Z^j \text{ is } D^j \qquad (8)$$

## 5   Numerical Results

Aggregation results for estimates given by the 1st survey participant:
Please note that the first paragraph of a section or subsection is not indented. The first paragraphs that follows a table, figure, equation etc. does not have an indent, either.

Subsequent paragraphs, however, are indented.

$$X_1^1 = \frac{SA + A + SA + A + SA + NAD + A + SA}{8} = \frac{(4 \quad 5 \quad 5) + (3 \quad 4 \quad 5)}{8}$$

$$\frac{+ (4 \quad 5 \quad 5) + (3 \quad 4 \quad 5) + (4 \quad 5 \quad 5) + (2 \quad 3 \quad 4) + (3 \quad 4 \quad 5) + (4 \quad 5 \quad 5)}{8}$$

$$= \frac{(32.30 \quad 35 \quad 36.81)}{8} = (4.04 \quad 4.38 \quad 4.60)$$

$$X_2^1 = \frac{A + D + D}{3} = \frac{(3 \quad 4 \quad 5) + (1 \quad 2 \quad 3) + (1 \quad 2 \quad 3)}{3} = \frac{(7 \quad 8 \quad 9)}{3} = (2.33 \quad 2.67 \quad 3)$$

$$X_2^1 = \frac{A+D+D}{3} = \frac{(3 \quad 4 \quad 5)+(1 \quad 2 \quad 3)+(1 \quad 2 \quad 3)}{3} = \frac{(7 \quad 8 \quad 9)}{3} = (2.33 \quad 2.67 \quad 3)$$

$$Y_1^1 = \frac{A+A+D+SD+A+A+A}{7} = \frac{(3 \quad 4 \quad 5)+(3 \quad 4 \quad 5)+(1 \quad 2 \quad 3)}{7}$$

$$\frac{+(1 \quad 1 \quad 2)+(3 \quad 4 \quad 5)+(3 \quad 4 \quad 5)+(3 \quad 4 \quad 5)}{7} = \frac{(20.95 \quad 23 \quad 25.29)}{7}$$

$$=(2.99 \quad 3.29 \quad 3.61)$$

$$Y_2^1 = \frac{A+SA+SD+SD+SA+A+NAD+A+D+D+NAD+A+D+NAD}{14}$$

$$= \frac{(3 \quad 4 \quad 5)+(4 \quad 5 \quad 5)+(1 \quad 1 \quad 2)+(1 \quad 1 \quad 2)+(4 \quad 5 \quad 5)+(3 \quad 4 \quad 5)}{14}$$

$$\frac{+(2 \quad 3 \quad 4)+(3 \quad 4 \quad 5)+(1 \quad 2 \quad 3)+(1 \quad 2 \quad 3)+(2 \quad 3 \quad 4)+(3 \quad 4 \quad 5)}{14}$$

$$\frac{+(1 \quad 2 \quad 3)+(2 \quad 3 \quad 4)}{14} = \frac{(20.95 \quad 23 \quad 25.29)}{14} = (1.50 \quad 1.64 \quad 1.80)$$

$$Z^1 = \frac{A+A+NAD+NAD}{4} = \frac{(3 \quad 4 \quad 5)+(3 \quad 4 \quad 5)+(2 \quad 3 \quad 4)+(2 \quad 3 \quad 4)}{4}$$

$$= \frac{(13 \quad 14 \quad 15)}{4} = (3.25 \quad 3.5 \quad 3.75)$$

The IF-THEN rules between the variables are:

IF $X_1^1$ is A and $X_2^1$ is NAD  THEN $Y_1^1$ is NAD

IF $X_1^1$ is A and $X_2^1$ is NAD  THEN $Y_2^1$ is D

IF $Y_1^1$ is NAD and $Y_2^1$ is D  THEN $Z^1$ is NAD

For n − 1 experts, fuzzy rules are constructed analogously.

## 5.1   Example

Given two rules:

IF $Y_1$ is NAD and $Y_2$ is A  THEN Z is A

IF $Y_1$ is A and $Y_2$ is NAD  THEN Z is NAD

and a current input

$$Y_1 \ is \ (3.15 \quad 3.5 \quad 3.85), \quad Y_2 \ is \ (2.85 \quad 3.2 \quad 3.55)$$

We need to compute behavioural intention Z. Let us compute Z by using linear interpolation method. First, we compute the similarity of current input with the inputs of the rules:

$$S((3.15 \quad 3.5 \quad 3.85), NAD) = 0.38$$

$$S((2.85 \quad 3.2 \quad 3.55), A) = 0.29$$

$$S((3.15 \quad 3.5 \quad 3.85), A) = 0.38$$

$$S((2.85 \quad 3.2 \quad 3.55), NAD) = 0.52$$

Next, we computed firing degrees of the rules:

$$S_1 = \min(0.38, 0.29) = 0.29, \ S_2 = \min(0.38, 0.52) = 0.38$$

On the base of the firing degrees we computed the coefficients of linear interpolation:

$$w_1 = \frac{S_1}{S_1 + S_2} = 0.43, \quad w_2 = \frac{S_2}{S_1 + S_2} = 0.57.$$

The resulting output is computed as follows:

$$Z = 0.43 * A + 0.57 * NAD = 0.43 * (3 \ 4 \ 5) + 0.57 * (2 \ 3 \ 4) = (2.6, \ 3.6, \ 4.2).$$

According to the codebook, this output can be labelled as "Agree".

## 6   Conclusion

In this paper, the hierarchical decision-making system and its application on marketing research were considered. The marketing data is gathered by a survey through which questionnaires were used. Imprecision and uncertainty marketing data requires human-like judgment that can be performed by using the fuzzy logic. Application of HMFs allows to simplify computation procedure and increases the precision of results.

This study is a significant contribution to quantitative marketing research. A proposed methodology based on fuzzy logic can be used to solve many other problems in marketing research. The adequacy and usability of the model are proven through the computer simulation by using real data obtained from the passengers who visited the North Cyprus Ercan Airport retail environment.

# References

1. Engel, F.J.: Consumer Behaviour. Delmar Cengage Learning, Boston (2005)
2. Sadikoglu, G.: Modelling of Consumer Buying Behaviour Using Z-Number Concept Intelligent Automation and Soft Computing (2017). https://doi.org/10.1080/10798587.2017.1327159
3. Sadikoglu, G.: Modeling of the travelers' shopping motivation and their buying behaviour using fuzzy logic. Procedia Comput. Sci. **120**, 805–811 (2017)
4. Sarah, M. S.: Motivation and emotion are usually viewed as two psychological features that seemingly share cause-and-effect relationship. https://explorable.com/motivation-and-emotion. Accessed 23 Oct 2017
5. Infantolino, Z.P., Miller, G.A., Crocker, L.D., Heller, W., Waren, S.L., O'Hare, A.J.: Relationships among cognition, emotion, and motivation: implications for Intervention and neuroplasticity in psychopathology. Front. Hum. Neurosci. **7**, 261 (2013)
6. Tomaszewski, K.: The application of horizontal membership functions to fuzzy arithmetic operations. J. Theor. Appl. Comput. Sci. **8**(2), 3–10 (2014)
7. Yi-Hsin, L., Ching-Fu, Ch.: Passengers' shopping motivations and commercial activities at airports - The moderating effects of time pressure and impulse buying tendency. Tour. Manag. **36**, 426–434 (2013)
8. Li, S.: Development of a hybrid intelligent system for developing marketing strategy. Decis. Support Syst. **27**, 395–409 (2000)
9. Casillas, J., Martínez-López, F.J.: Mining uncertain data with multiobjective genetic fuzzy systems to be applied in consumer behaviour modelling. Expert Syst. Appl. **36**(2, part 1), 1645–1659 (2009)
10. Enache, I.C.: Marketing a web-site using a fuzzy logic approach Marketing. In: Proceeding of the International Conference on RISK MANAGEMENT, ASSESSMENT and MITIGATION (2010). www.wseas.us/e-library/conferences/2010/Bucharest/RIMA/RIMA-59.pdf. Accessed 25 Apr 2017
11. Martínez-López, F.J., Casillas, J.: Artificial intelligence-based systems applied in industrial marketing: an historical overview, current and future insights. Ind. Mark. Manag. **42**, 489–495 (2013)
12. Sadikoglu, F., Huseynov, O., Memmedova, K.: Z-regression analysis in psychlogical and educational researches. Procedia Comput. Sci. **102**, 385–389 (2016)
13. Moreno, J.: Trading strategies modelling in Colombian power market using artificial intelligence techniques. Energy Policy **37**, 836–843 (2009)
14. Neshat, M., Baghi, A., Pourahmad, A., Sepidnam, G., Sargolzaei, M., Masoumi, A.: FHESMM: fuzzy hybrid expert system for marketing mix. Int. J. Comput. Sci. Issues (IJCSI) **8**, 126(2011)
15. Zadeh, L.A.: The concept of a linguistic variable and its application to approximate reasoning. Inf. Sci. **8**(3), 199–249 (1975)
16. Aliev, R.A.: Uncertain computation based on decision theory. World Scientific Publishing, Singapore (2017)
17. Dubois, D., Prade, H.: Fuzzy Set and Systems: Theory and Applications. Academic Press, New York (1980)
18. Pedrycz, W.: Fuzzy Control and Fuzzy Systems, 2nd edn. Wiley, New York (1993)
19. Donovan, R.J., Rossiter, J.R.: Store atmosphere: an environmental psychology approach. J. Retail. **58**(1), 34–57 (1982)
20. Bohl, J.: The impact of airport shopping environments and dwell time on consumer spending, VEZETÉSTUDOMÁNY, XLV. ÉVF. 2014. 11. SZÁM (2014)

21. Piegat, A., Plucinski, M.: Computing with words with the use of inverse RDM models of membership functions. Appl. Math. Comput. Sci. **25**(3), 675–688 (2015)
22. Mendel, J.M., Wu, D.: Computing with words for hierarchical decision making applied to evaluating a weapon. IEEE Trans. Fuzzy Syst. **18**, 441–460 (2010)

# An Integrated Fuzzy Decision Framework for Neuromarketing Technology Selection Problem

Mehtap Dursun[✉] and Nazli Goker

Industrial Engineering Department, Galatasaray University, Istanbul, Turkey
{mdursun, nagoker}@gsu.edu.tr

**Abstract.** Companies that want to increase profitability try to have a deeper understanding of consumers' complex purchasing habits. This complexity has forced academics and companies to seek ways beyond traditional marketing research methods. Neuroscience, together with developing medical technologies, reveals new, evolving and synthesized findings about the functioning of the human brain. This finding, which emerges with technological developments, helps consumers to investigate how consumers react consciously and subconsciously to brands, advertisements and products. Neuromarketing can offer complementary alternatives to researchers in areas that traditional marketing methods cannot account for. This study introduces a 2-tuple linguistic representation modeling based fuzzy multi-criteria decision making (MCDM) framework to determine the best performing neuromarketing technology. The proposed decision framework identifies the most suitable neuromarketing technology while enabling experts to cope with the information loss problem.

**Keywords:** Neuromarketing technology selection · Linguistic hierarchies
2-Tuple linguistic representation · MCDM

## 1 Introduction

Determining consumer needs and offering low-cost dynamic solutions to these needs has become one of the most fundamental issues for companies in a market where globalization and information access affect purchasing behavior. When consumers make purchasing decisions, they go through subconscious emotional processes and their behaviors are based on emotional processes rather than cognitive processes. It is important that these emotional processes are analyzed correctly by marketing managers. For this reason, traditional market research and data collection methods such as questionnaires, observations, and experiments are used. But new data emerging about changing consumer profiles and the workings of the human brain have propelled marketing managers into deeper knowledge.

The development of technology has revealed significant developments in medicine as well as in many areas of life. In the last decade, the cognitive developments have begun to present detailed and previously unobtainable findings on the human brain and to give insights into how the brain works. Neuroscience is a general investigation of the

© Springer Nature Switzerland AG 2019
R. A. Aliev et al. (Eds.): ICAFS-2018, AISC 896, pp. 195–200, 2019.
https://doi.org/10.1007/978-3-030-04164-9_27

nervous system, taking advantage of these developing technologies. It has been understood that studies on neuroscience and human brain and behavior can shed light on changing consumer behaviors. Thus, the field of neuroscience combined with the field of psychology led to neuromarketing, which could be called consumer neuroscience [1]. The information obtained in the neuromuscular spectrum can be used effectively in creating value for companies, better positioning against competitors and finding effective ways to influence consumer decisions. In areas where traditional marketing methods need to be supported, deeper analyzes can be done using neuromarketing techniques and used in the direction of the company's goals [2].

Marketing management and neurology are trying to predict human behaviors with different methods. In neuromarketing, the relationship between nervous system and behavior is examined by combining common marketing research techniques such as observation, questionnaire and focus group research and nervous system findings analyzed by psychological, sociological and physiological factors of neurology [3].

The main purpose of neuromarketing is to better understand consumers by solving the nervous link between consumer brains and behavior and presenting it with predictable models and data. In behavioral experiments with traditional marketing research methods, it has been discovered that people sometimes express their preferences, sometimes they do not, and sometimes they are not even cognitively aware of their preferences. With this awareness, a deeper knowledge of the decision-making mechanisms of the people and the causes of their behavior has arisen. Neuromarketing has offered solutions that can measure what the consumer reacts to under the consciousness with various technical methods. In order to employ neuromarketing methods, companies utilize brain imaging techniques that can be called as "neuromarketing technologies" in this work. Throughout the medical literature, there are a lot of neuromarketing technologies namely functional magnetic resonance imaging (fMRI), electroencephalography (EEG), magnetoencephalography (MEG), transcranial magnetic stimulation (TMS), eye tracking, galvanic skin response, electrocardiography, electromyography, analysis of pupil dilation, blush, blinking, heartbeat, or breathing [2, 4]. fMRI, EEG, MEG and TMS are defined as medical diagnostic devices, which are considered as the most frequently used neuromarketing technologies [5]. fMRI is the most widely used brain imaging technology in the world [2, 5]. EEG utilizes electrodes that are placed on the head of a person to measure variations in the electrical area of the brain region underneath [2, 6]. MEG, being an expensive version of EEG, is applied to measure the variations in the magnetic area induced by neuronal activity. TMS creates a magnetic field for inducing electrical currents in underlying neurons by using an iron core, which is placed on one's head [6].

The objective of this work is to introduce a fuzzy MCDM methodology for determining the most appropriate neuromarketing technology alternative. A fuzzy MCDM method is thought to be appropriate because of the complexity of the decision framework, the presence of conflicting criteria, which influence neuromarketing technology selection problem, and lack of crisp data. For determining the most appropriate technology, subjective assessment is employed by utilizing expert knowledge. The rest of the study is organized as follows. The methodology and the decision framework employed in the study are given in Sects. 2 and 3, respectively. The case study is given in Sect. 4. Conclusion part is provided in Sect. 5.

## 2  Methodology

This study employs 2-tuple linguistic representation model introduced by Herrera and Martínez [7], and linguistic hierarchies proposed by Cordon et al. [8]. 2-tuple linguistic representation model is employed for describing the linguistic information as a 2-tuple that is made up of a linguistic variable and a number. It can be represented as $(s_i, \alpha)$ where $s_i$ denotes the linguistic label of the described linguistic term set $S_T$, and $\alpha$ is a numerical value.

**Definition 1 [9]:** Define $L = (\gamma_0, \gamma_1, \ldots, \gamma_g)$ as a fuzzy set in $S_T$. A transformation function $\chi$ that converts $L$ into a numerical value in $S_T, [0, g]$ is given as

$$\chi : F(S_T) \rightarrow [0, g],$$

$$\chi(F(S_T)) = \chi(\{(s_j, \gamma_j), j = 0, 1, \ldots, g\}) = \frac{\sum\limits_{j=0}^{g} j\gamma_j}{\sum\limits_{j=0}^{g} \gamma_j} = \beta \tag{1}$$

**Definition 2 [7]:** Define $S = \{s_0, s_1, \ldots, s_g\}$ as a linguistic term set and $\beta \in [0, g]$ a numerical value, the 2-tuple that expresses the equal information to $\beta$ is obtained as

$$\Delta : [0, g] \rightarrow S \times [-0.5, 0.5),$$

$$\Delta(\beta) = \begin{cases} s_i, & i = round(\beta) \\ \alpha = \beta - i, & \alpha \in [-0.5, 0.5) \end{cases} \tag{2}$$

**Proposition 1 [7]:** Define $S = \{s_0, s_1, \ldots, s_g\}$ as a linguistic term set and $(s_i, \alpha)$ as a 2-tuple. $\Delta^{-1}$ function, that compute the equal numerical value $\beta \in [0, g] \subset \Re$. of a 2-tuple is described as

$$\Delta^{-1} : S \times [-0.5, 0.5) \rightarrow [0, g],$$

$$\Delta^{-1}(s_i, \alpha) = i + \alpha = \beta \tag{3}$$

**Definition 3 [10]:** Define $x = \{(s_1, \alpha_1), \ldots, (s_n, \alpha_n)\}$ as a set of linguistic 2-tuples and $W = \{(w_1, \alpha_1^w), \ldots, (w_n, \alpha_n^w)\}$ as their weights. The 2-tuple linguistic weighted average $\bar{x}_l^w$ is calculated using the following function:

$$\bar{x}_l^w \left( [(s_1, \alpha_1), (w_1, \alpha_1^w)], \ldots, [(s_n, \alpha_n), (w_n, \alpha_n^w)] \right) = \Delta \left( \frac{\sum\limits_{i=1}^{n} \beta_i \cdot \beta_{w_i}}{\sum\limits_{i=1}^{n} \beta_{w_i}} \right) \tag{4}$$

Linguistic hierarchies are used to resolve hierarchical systems of linguistic rules. They are employed to prevent the information loss problem that arises in the fusion phase of multigranular linguistic information. Each level belonging to a linguistic hierarchy is denoted as $l(t, n(t))$, where $t$ indicates the level of the hierarchy, and $n(t)$ is the granularity of the linguistic term set of the level $t$ [8]. A linguistic hierarchy, $LH$, can be described as the fusion of all levels $t$ as $LH = \bigcup_t l(t, n(t))$.

The linguistic term set of level $t + 1$ is achieved as [8]

$$L(t, n(t)) \rightarrow L(t + 1, 2.n(t) - 1) \tag{5}$$

The transformation function is as follows [8].

$$TF_{t'}^t : l(t, n(t)) \rightarrow l(t', n(t'))$$

$$TF_{t'}^t \left( s_i^{n(t)}, \alpha^{n(t)} \right) = \Delta \left( \frac{\Delta^{-1} \left( s_i^{n(t)}, \alpha^{n(t)} \right) (n(t') - 1)}{n(t) - 1} \right) \tag{6}$$

## 3   Proposed Decision Framework

The explication of the employed decision methodology is as follows.

**Step 1.** Alternatives and evaluation criteria are determined through the literature.
**Step 2.** Experts are provided their opinions on the evaluation of alternatives regarding to criteria using the linguistic hierarchy $LH = \bigcup_t l(1, 3)$, given in Table 1.

**Table 1.** The linguistic hierarchy [11].

| $LH_1$ | $s_0^3$ | $s_1^3$ | $s_2^3$ | | | | | | |
|--------|---------|---------|---------|---------|---------|---------|---------|---------|---------|
| $LH_2$ | $s_0^5$ | $s_1^5$ | $s_2^5$ | $s_3^5$ | $s_4^5$ | | | | |
| $LH_3$ | $s_0^9$ | $s_1^9$ | $s_2^9$ | $s_3^9$ | $s_4^9$ | $s_5^9$ | $s_6^9$ | $s_7^9$ | $s_8^9$ |

**Step 3.** A linguistic terms set is determined to unify the multi-granular information given by the experts, and the evaluation provided by the experts are unified using Eq. (6).
**Step 4.** The 2-tuple linguistic weighted average given in Eq. (4) is employed to compute the ranking index of alternatives.

## 4   Case Study

This section presents the illustration of the decision framework for neuromarketing technology selection problem. The example problem involves 5 neuromarketing technology alternatives that are fMRI, EEG, MEG, TMS and PET, and 5 evaluation criteria namely "cost", "spatial resolution", "temporal resolution", "complexity", and "prejudgments of participant", which are cost-related criteria. The evaluation is done by a committee of three experts ($E_1$, $E_2$, $E_3$). $E_1$ announced his preferences in $l(2, 5)$, $E_2$ utilized $l(1, 3)$, and $E_3$ used $l(3, 9)$. Data related to neuromarketing technology evaluation problem are reported in Table 2.

**Table 2.** Data related to neuromarketing technology evaluation problem.

|        | $C_1$ | $C_2$ | $C_3$ | $C_4$ | $C_5$ |
|--------|-------|-------|-------|-------|-------|
| $A_1$ | $(s_4^5, s_2^3, s_8^9)$ | $(s_2^5, s_1^3, s_5^9)$ | $(s_3^5, s_1^3, s_5^9)$ | $(s_1^5, s_2^3, s_2^9)$ | $(s_1^5, s_0^3, s_1^9)$ |
| $A_2$ | $(s_1^5, s_0^3, s_1^9)$ | $(s_2^5, s_3^3, s_6^9)$ | $(s_0^5, s_1^3, s_0^9)$ | $(s_2^5, s_3^3, s_3^9)$ | $(s_2^5, s_1^3, s_6^9)$ |
| $A_3$ | $(s_3^5, s_1^3, s_6^9)$ | $(s_3^5, s_1^3, s_5^9)$ | $(s_0^5, s_0^3, s_1^9)$ | $(s_2^5, s_0^3, s_7^9)$ | $(s_3^5, s_1^3, s_3^9)$ |
| $A_4$ | $(s_2^5, s_0^3, s_3^9)$ | $(s_4^5, s_2^3, s_7^9)$ | $(s_1^5, s_0^3, s_0^9)$ | $(s_2^5, s_0^3, s_4^9)$ | $(s_2^5, s_0^3, s_4^9)$ |
| $A_5$ | $(s_3^5, s_2^3, s_7^9)$ | $(s_2^5, s_1^3, s_3^9)$ | $(s_4^5, s_2^3, s_7^9)$ | $(s_3^5, s_1^3, s_7^9)$ | $(s_2^5, s_1^3, s_7^9)$ |
| Weight | $(s_4^5, s_1^3, s_7^9)$ | $(s_3^5, s_2^3, s_6^9)$ | $(s_3^5, s_2^3, s_7^9)$ | $(s_1^5, s_1^3, s_3^9)$ | $(s_3^5, s_1^3, s_8^9)$ |

$l(2, 5)$ is utilized to unify the heterogenous information given by the experts. Equal weights are appointed to the experts. Hence, the unified evaluations of experts are aggregated by incorporating 2-tuple mean operator, and the aggregated data are shown in Table 3.

**Table 3.** Aggregated data related to neuromarketing technology evaluation problem.

|        | $C_1$ | $C_2$ | $C_3$ | $C_4$ | $C_5$ |
|--------|-------|-------|-------|-------|-------|
| $A_1$ | $(s_4^5, 0)$ | $(s_2^5, 0.17)$ | $(s_3^5, -0.5)$ | $(s_2^5, 0)$ | $(s_1^5, -0.5)$ |
| $A_2$ | $(s_1^5, -0.5)$ | $(s_3^5, 0.5)$ | $(s_1^5, -0.33)$ | $(s_3^5, -0.5)$ | $(s_2^5, 0.33)$ |
| $A_3$ | $(s_3^5, -0.33)$ | $(s_3^5, -0.5)$ | $(s_0^5, 0.17)$ | $(s_2^5, -0.17)$ | $(s_2^5, 0.17)$ |
| $A_4$ | $(s_1^5, 0.17)$ | $(s_4^5, -0.17)$ | $(s_0^5, 0.33)$ | $(s_1^5, 0.33)$ | $(s_1^5, 0.33)$ |
| $A_5$ | $(s_4^5, -0.5)$ | $(s_2^5, -0.17)$ | $(s_4^5, -0.17)$ | $(s_3^5, -0.17)$ | $(s_3^5, -0.5)$ |
| Weight | $(s_3^5, 0.17)$ | $(s_3^5, 0.33)$ | $(s_4^5, -0.5)$ | $(s_2^5, -0.5)$ | $(s_3^5, 0)$ |

By employing Eq. (4), the final ranking index of each alternative is determined as $(s_2^5, 0.29)$, $(s_2^5, -0.30)$, $(s_2^5, -0.16)$, $(s_2^5, -0.37)$, $(s_3^5, -0.08)$. Then, using the process of comparison among linguistic 2-tuples, the ranking of alternatives is obtained as $A_4 \succ A_2 \succ A_3 \succ A_1 \succ A_5$.

# 5 Conclusions

In this work, a fuzzy MCDM approach, which combines 2-tuple fuzzy linguistic modeling and linguistic hierarchies is presented. The developed approach is adequate to cope with multi-granular linguistic information. Furthermore, it enables experts to utilize different semantic types, and deal with loss of information, which may be occur due to the classical MCDM methods.

The contributions of this paper to marketing science can be listed as follows. The proposed approach obtains the most suitable neuromarketing technology when the crisp data are not available, helps marketers utilize the best performing neuromarketing technology for their market research studies, and provides the novelty by being the first study, which uses a MCDM method in neuromarketing field. Through a numerical illustration, which is given by collecting data from a neurology specialist, TMS is considered as the most appropriate neuromarketing technology alternative. Future research will focus on constructing a decision framework, which considers consumer requirements along with expert knowledge.

**Acknowledgements.** This work is supported by Galatasaray University Research Fund Project 18.402.007.

# References

1. Gordon, W.: The darkroom of the mind: what does neuropsychology now tell us about brands? J. Consum. Behav. **1**, 280–292 (2002)
2. Ariely, D., Berns, G.S.: Neuromarketing: the hope and hype of neuroimaging in business. Nat. Rev. Neurosci. **11**, 284–292 (2010)
3. Hubert, M., Kenning, P.: A current overview of consumer neuroscience. J. Consum. Behav. **7**(4–5), 272–292 (2008)
4. Fisher, C.E., Chin, L., Klitzman, R.: Defining neuromarketing: Practices and professional challenges. Harv. Rev. Psychiatry **18**, 230–237 (2010)
5. Ruanguttamanun, C.: Neuromarketing: i put myself into a fMRI scanner and realized that I love Louis Vuitton ads. Procedia – Soc. Behav. Sci. **148**, 211–218 (2014)
6. Burgos-Campero, A.A., Vargas-Hernandez, J.G.: Analitical approach to neuromarketing as a business strategy. Procedia – Soc. Behav. Sci. **99**, 517–525 (2013)
7. Herrera, F., Martínez, L.: A 2-tuple fuzzy linguistic representation model for computing with words. IEEE Trans. Fuzzy Syst. **8**(6), 746–752 (2000)
8. Cordon, O., Herrera, F., Zwir, I.: Linguistic modeling by hierarchical systems of linguistic rules. IEEE Trans. Fuzzy Syst. **10**(1), 2–20 (2002)
9. Herrera, F., Martínez, L.: An approach for combining linguistic and numerical information based on 2-tuple fuzzy representation model in decision-making. Int. J. Uncertain., Fuzziness Knowl.-Based Syst. **8**(5), 539–562 (2000)
10. Herrera-Viedma, E., Herrera, F., Martínez, L., Herrera, J.C., López, A.G.: Incorporating filtering techniques in a fuzzy linguistic multi-agent model for information gathering on the web. Fuzzy Sets Syst. **148**(1), 61–83 (2004)
11. Huynh, V.N., Nakamori, Y.: A satisfactory-oriented approach to multiexpert decision-making with linguistic assessments. IEEE Trans. Syst. Man Cybern. Part B-Cybern. **35**(2), 184–196 (2005)

# Performance Indicators Evaluation of Business Process Outsourcing Employing Fuzzy Cognitive Map

Nazli Goker[(✉)], Y. Esra Albayrak, and Mehtap Dursun

Galatasaray University, 34349 Istanbul, Turkey
nagoker@gsu.edu.tr

**Abstract.** The aim of this work is evaluating and analyzing the performance indicators of business process outsourcing. The criteria influencing the performance of business process outsourcing are indicated through a large literature survey and experts' opinions, and a multi-criteria decision model is thought to be appropriate because of the complexity of the problem. Fuzzy cognitive map methodology is a suitable tool due to the presence of causalities, positive as well as negative directions of relationships among criteria, and the difficulty of expressing the interrelations with crisp numbers. The proposed methodology provides an evaluation for clients to assess their service providers, a self-evaluation for service providers. Hence, this work proposes a mutual assessment.

**Keywords:** Outsourcing · Performance evaluation · Business processes
Fuzzy cognitive map

## 1 Introduction

Outsourcing is a strategy of operations that influences the supply chain performance and a managerial constituent of operations administrations. Outsourcing enables companies to decrease costs of assets, manufacturing costs, administrative costs while providing agility. Moreover, customers can directly concentrate on their core competencies by transferring their peripheral issues to outsourcing providers [1]. Companies can employ outsourcing when they require managing the activities in a more effective manner. Thereby, effective use of the sources affects positively the profitability of firms. Hence, financial ratios as well as flexibility are improved by outsourcing peripheral activities to the service providers.

Making use of the information of the outsourcing provider allows firms to obtain fast growth rate. Companies have know-how capability, however, they are not capable technically for starting and then continuing the process. Thus, they prefer to outsource technical activities to an expert to survive in competitive environments and to grow rapidly [2].

In recent years, researchers have focused on fuzzy cognitive map (FCM) method and employed in a few research fields and industries. Papageorgiou et al. [3] and [4] employed FCM in crop management to forecast the yield obtained from cotton and

© Springer Nature Switzerland AG 2019
R. A. Aliev et al. (Eds.): ICAFS-2018, AISC 896, pp. 201–208, 2019.
https://doi.org/10.1007/978-3-030-04164-9_28

apples. Büyüközkan and Vardaloğlu [5] utilized FCM for indicating and analyzing the factors that provide the combination of collaborative planning estimation and replenishment strategy in retail sector. Zhao et al. [6] proposed a decision framework for a flexible operating mechanism in wind power industry employing FCM technique. Baykasoğlu and Gölcük [7] used FCM to determine the causal relations of a SWOT-based strategy evaluation problem for an industrial engineering department. Ahmadi et al. [8] applied FCM to the forecasting of readiness of an organization to adapt a new enterprise resource planning (ERP) system. Büyükavcu et al. [9] assessed breast cancer risk factors by combining FCM and fuzzy inference system, they employed several scenario analyses. Bagdatli et al. [10] used FCM to model a complex system constructed to provide a cost-benefit analysis of highway projects.

This work is to assess and analyze the performance evaluation of business process outsourcing. The criteria influencing the performance of business process outsourcing are indicated through a large literature survey and experts' opinions, and a multi-criteria decision model is thought to be appropriate because of the complexity of the problem. Fuzzy cognitive map methodology (FCM) is a suitable tool due to the presence of causalities, positive as well as negative directions of relationships among criteria, and the difficulty of expressing the interrelations with crisp numbers.

The rest of the study is organized as follows. Section 2 explains fuzzy cognitive map methodology. Section 3 summarizes the application procedure of the proposed approach. Subsequent section provides the case study. Concluding remarks and future research directions are delineated in the last section.

## 2    Fuzzy Cognitive Map Methodology

Cognitive maps (CMs) were originally proposed by Axelrod [11] as a tool to model decision support systems in political and social sciences. CMs comprise directed edges which provide modeling causalities and interrelationships among factors.

A crisp CM, which is indeterminate, can be solved by providing a numerical weighting, however, it requires computational and conceptual efforts [12]. FCM, enabling to construct complex decision frameworks, is a cause-based technique which is arised from the combination of neural networks and fuzzy logic [12]. Afterwards, Kosko [12] elaborated the methodology and incorporated fuzzy numbers or linguistic terms for indicating the causalities among factors in FCM. These factors represent a state, an entity, a variable or a system's behavior, a characteristic of a cause-based framework is stated by factors in FCM. Concept (factors) nodes and weighted arcs are the components of FCM which can be represented with feedback. Signed edges are to determine the direction of causal links: whether the causal relationship is positive, negative or null, and connect the concept nodes through which causal links among concepts are generated [9].

The sign of $w_{ji}$ indicates the direction of causal links between concepts. If $w_{ji} > 0$, then there is a positive causal relation, if $w_{ji} < 0$, then there is a negative causal relation between concepts $C_j$ and $C_i$. Besides, if $w_{ji} = 0$, then there is no causality between associated concepts. In addition, the direction of causal links represents if concept $C_j$ causes concept $C_i$, or vice versa. In order to determine the power of these causal relations, a value has to be assigned to weight $w_{ji}$. Each concept's value is computed by taking into account the influence of the other concepts on the evaluating concept, and by running the iterative formulation as follows:

$$A_i^{(k+1)} = f\left(A_i^{(k)} + \sum_{\substack{j \neq i \\ j=1}}^{N} A_j^{(k)} w_{ji}\right) \qquad (1)$$

where $A_i^{(k)}$ refers to the value of concept $C_i$ at $k^{\text{th}}$ iteration, $w_{ji}$ is the weight of the causal link from $C_j$ to $C_i$, and $f$ represents threshold function.

The levels of activation for the concepts are synchronously updated in FCM, which represents a discrete time system. Hence, the system is to be updated in a simultaneous way. The activation level of concept $C_i$ is denoted by $A_i^t$, $t$ is the time step. The vector $A^t = [A_1^t, A_2^t, \ldots, A_n^t]$ indicates the entity of the FCM at time step $t$, $n$ denotes the concept number. Each concept has an initial and a final vector, which indicate a state for the system at the initial and the last time step, respectively. The objective of FCM method is to identify the final vector, which provides determining the value of each concept [9].

## 3   Proposed Decision Framework

The application steps of the proposed approach are as [9].

**Table 1.** Performance criteria for business process outsourcing

| Label | Concept |
|-------|---------|
| $C_1$ | Cost reduction |
| $C_2$ | Firm size |
| $C_3$ | Age of relationship |
| $C_4$ | Operational efficiency |
| $C_5$ | Overall financial position |
| $C_6$ | Contract schedule |
| $C_7$ | Responsiveness |
| $C_8$ | Employee productivity |
| $C_9$ | Reliability |
| $C_{10}$ | IT capability |
| $C_{11}$ | Task complexity |
| $C_{12}$ | Communication capability |
| $C_{13}$ | Flexibility |

**Step 1:** Performance criteria are decided through expert opinions, 13 factors are listed in Table 1.

**Step 2:** Decision-makers indicate the direction of causal links in three classes: positive, negative, null.

**Step 3:** Decision-makers indicate the degree of causal links by using linguistic terms; subsequently linguistic terms are converted into triangular fuzzy numbers. The corresponding triangular fuzzy numbers for these linguistic terms are listed in Table 2.

**Table 2.** Scale of triangular fuzzy numbers [9]

| Linguistic variable | Triangular fuzzy number |
|---|---|
| nvs (negatively very strong) | $(-1, -1, -0.75)$ |
| ns (negatively strong) | $(-1, -0.75, -0.5)$ |
| nm (negatively medium) | $(-0.75, -0.5, -0.25)$ |
| nw (negatively weak) | $(-0.5, -0.25, 0)$ |
| z (zero) | $(-0.25, 0, 0.25)$ |
| pw (positively weak) | $(0, 0.25, 0.5)$ |
| pm (positively medium) | $(0.25, 0.5, 0.75)$ |
| ps (positively strong) | $(0.5, 0.75, 1)$ |
| pvs (positively very strong) | $(0.75, 1, 1)$ |

**Step 4:** Using MAX method, linguistic terms belonging to each causal relationship are converted into a single fuzzy number, then, this triangular fuzzy number that is in the interval $[-1,1]$ is converted to a crisp number by employing Centre of Gravity (COG) technique and is transformed into a crisp value.

**Step 5:** The initial vector starts the FCM process.

**Step 6:** Updating the values of the initial vector is provided by employing Formulation (1) and a threshold function. $tanh(x)$ is a suitable threshold function for placing the values of concepts in the interval $[-1,1]$.

**Step 7:** Computing concepts' values is completed by running the iterative formulation (1).

# 4   Case Study

In this section of the study, a FCM approach is employed in order to provide a performance assessment for business process outsourcing. FCM methodology is preferred to be applied due to the lack of crisp numbers and consequently the requirement of the use of linguistic variables or fuzzy numbers, and the presence of cause-effect relationships among concepts in performance assessment of business process outsourcing.

Performance criteria for business process outsourcing are listed in order to provide a decision framework to be evaluated. Academic studies which are related to outsourcing performance evaluation are reviewed, and the performance criteria that are appropriate as well as the most selected for business process outsourcing performance assessment are identified. Thereafter, selected criteria are sent to three different decision makers, whose jobs are associated with business process outsourcing in large companies, which are located in Turkey. Experts' opinions are incorporated in order to obtain the final performance criteria and 13 factors mentioned in the previous section are included to the study.

The performance criteria, which are indicated through a literature survey and experts' knowledge, are sent to the three decision makers whose job description is directly related to business process outsourcing, and who are from different large companies, which are located in Turkey. The decision makers have large knowledge about outsourcing processes, background and market experience. Hence, each decision maker is able to indicate the effect of one concept on another.

At the initial step, they determine whether there is a causal relationship between each pair of concepts, or not. If there is no relation, they skip the associated pair of concepts, but if there is a causal link, they indicate the direction (sign) of the relation such as positive or negative. Experts then indicate the power of causal relations by using linguistic terms; subsequently linguistic terms are converted to triangular fuzzy numbers. Power of causality matrix of the expert 1 is given in Table 3.

**Table 3.** The matrix of sign according to the Expert 1

|          | $C_1$ | $C_2$ | $C_3$ | $C_4$ | $C_5$ | $C_6$ | $C_7$ | $C_8$ | $C_9$ | $C_{10}$ | $C_{11}$ | $C_{12}$ | $C_{13}$ |
|----------|-----|-----|-----|-----|-----|-----|-----|-----|-----|------|------|------|------|
| $C_1$    | z   | z   | z   | z   | z   | z   | z   | z   | ps  | z    | z    | z    | z    |
| $C_2$    | pw  | z   | z   | pw  | pm  | nw  | nw  | z   | pw  | ps   | z    | z    | nw   |
| $C_3$    | pm  | z   | z   | z   | z   | nvs | pm  | z   | pm  | z    | z    | pm   | pm   |
| $C_4$    | pvs | ps  | z   | z   | ps  | z   | z   | z   | ps  | z    | z    | z    | z    |
| $C_5$    | z   | pvs | z   | z   | z   | z   | z   | z   | pw  | z    | z    | z    | nm   |
| $C_6$    | pw  | z   | z   | pm  | z   | z   | pm  | pw  | z   | z    | z    | z    | nm   |
| $C_7$    | z   | z   | z   | z   | z   | nw  | z   | z   | pw  | z    | z    | z    | z    |
| $C_8$    | pm  | z   | z   | pvs | ps  | z   | z   | z   | ps  | z    | z    | z    | z    |
| $C_9$    | z   | z   | z   | z   | z   | nm  | z   | z   | z   | z    | z    | z    | z    |
| $C_{10}$ | ps  | pm  | z   | ps  | pm  | z   | z   | pw  | pw  | z    | z    | z    | z    |
| $C_{11}$ | ns  | z   | z   | nm  | z   | ps  | nm  | nw  | z   | z    | z    | z    | z    |
| $C_{12}$ | z   | z   | z   | z   | z   | nw  | pm  | pw  | pm  | z    | z    | z    | z    |
| $C_{13}$ | z   | z   | z   | z   | z   | nw  | pw  | pw  | z   | z    | z    | pm   | z    |

Experts initially determine each interrelationship, and then the causal links for the same interrelation are indicated, the fuzzy numbers from the three decision makers for the associated interrelation are transformed into a single fuzzy set via MAX method, which is coded in MATLAB Fuzzy Toolbox.

The single fuzzy set, which is obtained from MAX aggregation method, is converted into numerical value, $w_{ji}$, with the defuzzification method of COG, which is coded in MATLAB Fuzzy Toolbox. The formulation of this method is given in the following formulation [13].

$$z^* = \frac{\int \mu_{\tilde{A}}(z) \cdot z \quad dz}{\int \mu_{\tilde{A}}(z) \, dz} \tag{2}$$

**Table 4.** The weight matrix

|        | $C_1$ | $C_2$ | $C_3$ | $C_4$ | $C_5$ | $C_6$ | $C_7$ | $C_8$ | $C_9$ | $C_{10}$ | $C_{11}$ | $C_{12}$ | $C_{13}$ |
|--------|-------|-------|-------|-------|-------|-------|-------|-------|-------|----------|----------|----------|----------|
| $C_1$    | 0.00  | 0.00  | 0.00  | 0.00  | 0.00  | −0.13 | 0.00  | 0.00  | 0.50  | 0.00     | 0.00     | 0.00     | 0.00     |
| $C_2$    | 0.13  | 0.00  | 0.00  | 0.50  | 0.63  | −0.13 | −0.25 | 0.00  | 0.50  | 0.80     | 0.00     | 0.00     | −0.50    |
| $C_3$    | 0.38  | 0.00  | 0.00  | 0.00  | 0.00  | −0.65 | 0.63  | 0.00  | 0.67  | 0.00     | 0.00     | 0.67     | 0.63     |
| $C_4$    | 0.49  | 0.50  | 0.00  | 0.00  | 0.50  | 0.00  | 0.13  | 0.00  | 0.50  | 0.00     | 0.00     | 0.00     | 0.13     |
| $C_5$    | 0.00  | 0.67  | 0.00  | 0.25  | 0.00  | −0.13 | 0.00  | 0.00  | 0.50  | 0.00     | 0.00     | 0.00     | −0.13    |
| $C_6$    | 0.13  | 0.00  | 0.00  | 0.25  | 0.00  | 0.00  | 0.38  | 0.13  | 0.00  | 0.00     | 0.00     | 0.13     | −0.63    |
| $C_7$    | 0.00  | 0.00  | 0.00  | 0.00  | 0.00  | −0.13 | 0.00  | 0.00  | 0.50  | 0.00     | 0.00     | 0.40     | 0.25     |
| $C_8$    | 0.25  | 0.00  | 0.00  | 0.67  | 0.50  | 0.00  | 0.25  | 0.00  | 0.50  | 0.00     | 0.00     | 0.00     | 0.00     |
| $C_9$    | 0.00  | 0.00  | 0.00  | 0.00  | 0.00  | −0.38 | 0.00  | 0.00  | 0.00  | 0.00     | 0.00     | 0.00     | 0.00     |
| $C_{10}$ | 0.63  | 0.25  | 0.00  | 0.50  | 0.25  | 0.00  | 0.00  | 0.25  | 0.13  | 0.00     | 0.00     | 0.00     | 0.00     |
| $C_{11}$ | −0.80 | 0.00  | 0.00  | −0.38 | 0.00  | 0.50  | −0.38 | −0.38 | 0.00  | 0.13     | 0.00     | 0.00     | 0.13     |
| $C_{12}$ | 0.00  | 0.00  | 0.00  | 0.00  | 0.00  | −0.13 | 0.38  | 0.13  | 0.38  | 0.00     | 0.00     | 0.00     | −0.13    |
| $C_{13}$ | 0.00  | 0.00  | 0.00  | 0.25  | 0.00  | −0.38 | 0.38  | 0.13  | 0.00  | 0.00     | 0.00     | 0.25     | 0.00     |

The weight matrix for performance indicators of business process outsourcing is generated by employing aggregation and defuzzification processes for every relation between each pair of connected concepts, and it is given in Table 4.

For obtaining concepts' values, Formulation (1) starts to be activated with the initial vector $A^0 = [1, 1, \ldots, 1]$. The values of this vector finalized by utilizing formulation (1) and a transform ld function. $f(x) = \tanh(x)$ is the suitable threshold function since the values of $A_i$ can be negative and the interval that these values belong to is $[-1, 1]$. The new vector, which is obtained by running the iterative formulation with this threshold function, is considered as the initial vector for the next iteration. These vectors are updated by using Formulation (1) until positive as well as negative interrelations between the concepts have acquired equilibrium. In other words, the process continues till $|vector(t) - vector(t+1)| \leq \varepsilon$, where $\varepsilon > 0$, and small enough [9]. After a lot of iterations, the system reaches the equilibrium and the stabilization is provided.

The concepts' values of business process outsourcing performance are listed in Table 5, which indicates that the age of relationship and task complexity are not powerful factors. Apart from these three criteria, the others are quite powerful on performance assessment of business process outsourcing. In addition, contract schedule is negatively very important performance indicator.

**Table 5.** The concepts' values of business process outsourcing performance

| Label | Concept | Concept's value |
|-------|---------|-----------------|
| $C_1$ | Cost reduction | 0.97140 |
| $C_2$ | Firm size | 0.98301 |
| $C_3$ | Age of relationship | 0.26164 |
| $C_4$ | Operational efficiency | 0.98850 |
| $C_5$ | Overall financial position | 0.99124 |
| $C_6$ | Contract schedule | −0.98035 |
| $C_7$ | Responsiveness | 0.85893 |
| $C_8$ | Employee productivity | 0.75700 |
| $C_9$ | Reliability | 0.99971 |
| $C_{10}$ | IT capability | 0.94257 |
| $C_{11}$ | Task complexity | 0.26164 |
| $C_{12}$ | Communication capability | 0.90945 |
| $C_{13}$ | Flexibility | 0.85696 |

# 5  Concluding Remarks

In this study, performance criteria of business process outsourcing were initially determined by reviewing the literature and using experts' opinions. Afterwards, thirteen factors were indicated and sent to three different decision makers, whose job description is directly related to business process outsourcing. Experts determined firstly whether there is causality between each pair of concept, and the sign of the relationship if there exists. Then, the power of causal links for each relationship was determined by utilizing linguistic terms. These linguistic terms are transformed into triangular fuzzy numbers according to the associated membership function. By means of MATLAB Fuzzy Toolbox, obtained fuzzy numbers from three decision makers were aggregated and then defuzzified by using MAX and center of gravity methods, respectively. The final weight matrix and FCM were constructed; outdegree, indegree and centrality values were calculated. By running the iterative formulation of FCM, concepts' values were computed.

Future research will focus on incorporating a performance domain as a criterion into the assessment of business process outsourcing, beside this adding outsourcing provider alternatives to the decision framework and selecting the most appropriate service provider by applying TOPSIS (technique for order preference by similarity to ideal solution) methodology or proposing an integrated model called as FCM-TOPSIS.

**Acknowledgements.** This work has been financially supported by Galatasaray University Research Fund 18.402.006.

# References

1. Gotzamani, K., Longinidis, P., Vouzas, F.: The logistics services outsourcing dilemma quality management and financial performance perspectives. Supply Chain Manag.: Int. J. **15**(6), 438–453 (2010)
2. Apak, S., Gümüş, S., Kurban, Z.: Strategic dimension of outsourcing in the information technologies intensified businesses. Procedia – Soc. Behav. Sci. **58**, 783–791 (2012)
3. Papageorgiou, E.I., Markinos, A.T., Gemtos, T.A.: Fuzzy cognitive map based approach for predicting yield in cotton crop production as a basis for decision support system in precision agriculture application. Appl. Soft Comput. **11**, 3643–3657 (2011)
4. Papageorgiou, E.I., Aggelopoulou, K.D., Gemtos, T.A., Nanos, G.D.: Yield prediction in apples using fuzzy cognitive map learning approach. Comput. Electron. Agric. **91**, 19–29 (2013)
5. Büyüközkan, G., Vardaloğlu, Z.: Analyzing of CPFR success factors using fuzzy cognitive maps in retail industry. Expert Syst. Appl. **39**(12), 10438–10455 (2012)
6. Zhao, Z.Y., Zhu, J., Zuo, J.: Sustainable development of the wind power industry in a complex environment: a flexibility study. Energy Policy **75**, 392–397 (2014)
7. Baykasoğlu, A., Gölcük, I.: Development of a novel multiple-attribute decision making model via fuzzy cognitive maps and hierarchical fuzzy TOPSIS. Inf. Sci. **301**, 75–98 (2015)
8. Ahmadi, S., Yeh, C.H., Papageorgiou, E.I., Martin, R.: An FCM-FAHP approach for managing readiness-relevant activities for ERP implementation. Comput. Ind. Eng. **88**, 501–517 (2015)
9. Büyükavcu, A., Albayrak, Y.E., Göker, N.: A fuzzy information-based approach for breast cancer risk factors assessment. Appl. Soft Comput. **38**, 437–452 (2016)
10. Bagdatli, M.E.C., Akbiyikli, R., Papageorgiou, E.I.: A fuzzy cognitive map approach applied in cost-benefit analysis for highway projects. Int. J. Fuzzy Syst. **19**(5), 1512–1527 (2017)
11. Axelrod, R.: Structure of Decision. Princeton University Press, Princeton (1976)
12. Kosko, B.: Fuzzy cognitive maps. Int. J. Man-Mach. Stud. **24**, 65–75 (1986)
13. Ross, T.J.: Fuzzy Logic with Engineering Applications, 3rd edn. Wiley, Hoboken (2010)

# Evaluating Lung Cancer Treatment Techniques Using Fuzzy PROMETHEE Approach

Mordecai Maisaini[1(✉)], Berna Uzun[2], Ilker Ozsahin[1],
and Dilber Uzun[1,3]

[1] Department of Biomedical Engineering, Near East University,
Nicosia, North Cyprus, Turkey
{mordecai.maisaini,dilber.uzunozsahin}@neu.edu.tr
[2] Department of Mathematics, Near East University,
Nicosia, North Cyprus, Turkey
[3] Radiology Massachusetts General Hospital and Harvard Medical School,
Gordon Center for Medical Imaging, Boston, USA

**Abstract.** Lung cancer also known as lung carcinoma is a disease caused by an uncontrolled growth of cells in the lung(s). This uncontrolled growth is influenced by a mutation of the DNA that causes growth of the cells in the lungs, and the mutation is caused by various factors which include inhalation of radon gas whose products ionize genetic material, asbestos, genetic makeup of an individual and also tobacco smoke products. Early diagnosis and therapy is very important in order to increase the chances of survival from the disease. The aim of this study is to analyse the most common therapeutic techniques of lung cancer such as surgery, chemotherapy, radiation therapy, immunotherapy, and hormone therapy as they affect the patient and the hospital. In this project, Fuzzy *PROMETHEE (preference ranking organization method for enrichment of evaluations)* a decision making process that uses a multi-criteria method was used to analyse the therapeutic techniques based on factors such as radiation dose, cost of treatment, treatment time, chances of survival, side effects and cost of the method for the hospital. Evaluation results showed that the surgery among other techniques showed a great performance on lung cancer treatment based on the criteria, importance, and weights we have selected. Fuzzy PROMETHEE also shows that one can easily modify the method by adding more criteria and change their importance and weights depending on the specific application.

**Keywords:** Lung cancer · Therapeutic techniques · Fuzzy PROMETHEE

## 1 Introduction

Lung cancer is the foremost cause of cancer related deaths, almost 36000 deaths recorded in the United Kingdom in 2016 [1] and a projected 154050 expected deaths in the United States in 2018, according to the American Cancer Society [2]. To increase the chances of survival, the importance of early diagnosis and therapy cannot be overemphasized. Various factors contribute to the increase in risk of lung cancer, such

© Springer Nature Switzerland AG 2019
R. A. Aliev et al. (Eds.): ICAFS-2018, AISC 896, pp. 209–215, 2019.
https://doi.org/10.1007/978-3-030-04164-9_29

as inhalation of radon gas, whose products ionize genetic material, asbestos, genetic makeup of an individual and tobacco related smoking. Lung cancer can either be primary where cancerous cells originate from the lungs or secondary in which cancerous cells metastasize from other parts of the body to the lungs.

The treatment method of lung cancer depends on both the health condition of the patient and the stage of the cancer. The stage of the cancer is a reference to the extent to which the cancer has spread, either to the other lung or to parts of the body. The most common treatment methods are chemotherapy (drug use), surgery (surgical removal of the malignant tissue), and radiotherapy (using x-rays or other forms of radiation). Certain factors such as the chances of survival, treatment time, radiation dose, and the cost of treatment are important contributing factors in determining the most suitable treatment technique to pursue. In this study, we propose using fuzzy PROMETHEE for the analysis of lung cancer treatment options corresponding to their parameters.

## 1.1  Lung Cancer Treatment Techniques

**Surgery.** Surgery generally increases the chances of getting rid of cancerous cells. Pulmonary function tests are carried out prior to the surgical procedure to ensure that the patient would still have sufficient healthy tissues afterwards [3]. Other tests on the heart function and other vital organs are also carried out to verify that the patient is in a healthy enough state to tolerate the operation.

**Immunotherapy.** It is the use of specific therapeutic agents to stimulate a person's immune system to recognize cancer cells and destroy them more effectively. Cancerous cells use the same checkpoints as those used by healthy cells to prevent them from being recognized by the immune system. Drugs called immune checkpoint inhibitors have been developed to prevent the cancerous cells from using this advantage [4].

**Chemotherapy.** This treatment technique involves the use of drugs to eliminate cancerous cells. Administration of chemotherapy drugs is done either orally or intravenously. Chemotherapy is administered in cycles, with a single cycle lasting for 3 to 4 weeks and may require daily dosage of drugs, followed by a period of rest to allow the body to rest and healthy cells to recover. Most often chemotherapy is done with a combination of two or more drugs. However, some patients may require just a single chemotherapy drug, as they might be unlikely to cope with combination chemotherapy due to poor health or age factor [5].

**Radiation Therapy.** It is the utilization of particles or rays with high energy level to eliminate cancerous cells. Radiation therapy may be used as the main treatment technique (sometimes alongside chemotherapy), after surgery to eliminate residual cancer cells, before surgery to shrink the cancer cells, to treat a single area of cancerous cells or to expel cells from blocking large airways of the lungs.

There are two principal types of radiation therapy, external beam radiation therapy (EBRT) and internal radiation therapy (brachytherapy). In external beam radiation therapy, an external radiation source is focused on the cancerous cells. Brachytherapy uses radioactive substances encapsulated in delivery systems such as catheters, which are placed close to or directly on the cancerous cells [6].

**Hormone Therapy.** This technique lessens or inhibits the growth of cancerous cells by blocking the body's ability to produce estrogen or progesterone receptor hormones that the cancerous cells are sensitive to, thereby disrupting their growth.

This technique can be used in combination with chemotherapy or as a standalone treatment alternative. It can also be utilized prior to surgery or post-surgery to prevent reoccurrence of cancerous cells.

## 2    Materials and Methods

Obtaining crisp data that accurately captures a problem, makes analysis and arrives at an optimal solution is often a challenging feat. Fuzzy PROMETHEE is a technique that combines the concept of fuzzy logic and that of PROMETHEE method. Fuzzy logic enables the user to define the problem using vague but realistic conditions and convert the linguistic variables to mathematical variables. Using PROMETHEE, the decision maker can compare different fuzzy values [7]. In their 2017 study compared different nuclear medicine imaging devices using fuzzy PROMETHEE and discussed fuzzy PROMETHEE technique in depth [8]. Applied fuzzy PROMETHEE to evaluate and rank energy exploitation schemes of a low temperature geothermal field [9]. In their study proposed a method that combined modified DELPHI method, AHP (Analytical Hierarchy Process) and PROMETHEE for curbing the difficulties experienced in the selection process of appropriate tools and machines that affect a manufacturing operations' effectiveness. In other studies, [10, 11] made analysis evaluating breast cancer treatment techniques as well as evaluation of different cancer treatment alternatives using fuzzy PROMETHEE [12]. Evaluated different cancer treatment alternatives, [13] made analysis of x-ray based medical imaging devices and effective analysis of image reconstruction algorithms in nuclear medicine all using fuzzy PROMETHEE.

Table 1 shows the importance of the parameters on a linguistic scale, using a triangular fuzzy scale. The decision for the weights of these parameters were arrived upon based on the opinion of a specialist, but these weights can be changed or more parameters can be added, based on the desires of the decision maker. Yager index was used to defuzzify the triangular fuzzy numbers and obtain the weight for each criterion.

**Table 1.** Linguistic scale of importance.

| Linguistic scale for evaluation | Triangular fuzzy scale | Importance ratings of criteria |
|---|---|---|
| Very high (VH) | (0.75, 1, 1) | |
| Important (H) | (0.50, 0.75, 1) | Survival percentage |
| Medium (M) | (0.25, 0.50, 0.75) | Cost of treatment, radiation dose, side effects |
| Low (L) | (0, 0.25, 0.50) | Cost of machine, treatment time |
| Very low (VL) | (0, 0, 0.25) | |

After all the parameters of the lung cancer treatment techniques were collected, Gaussian preference function was applied for each criterion using visual PRO-METHEE decision lab program (Table 2).

**Table 2.** Visual PROMETHEE application for lung cancer treatment alternatives.

| Criteria | Cost of Treatment | Cost of machine | Radiation dose | Treatment time | Survival percentage | Side effects |
|---|---|---|---|---|---|---|
| Unit | $ | $ | Gy | min. | % | |
| Preferences | | | | | | |
| (min/max) | Min | min | min | max | Max | min |
| Weight | 0,50 | 0,25 | 0,50 | 0,25 | 0,75 | 0,50 |
| PreferencFn. | Gaussian | Gaussian | Gaussian | Gaussian | Gaussian | Gaussian |
| Evaluations | | | | | | |
| Radiation therapy | 0,41 | 0,94 | 0,53 | 0,25 | 0,50 | 0,75 |
| Chemotherapy | 0,16 | 0,06 | 0,47 | 0,75 | 0,75 | 0,92 |
| Immunotherapy | 0,18 | 0,00 | 0,00 | 0,75 | 0,25 | 0,50 |
| Hormone therapy | 0,05 | 0,00 | 0,00 | 0,92 | 0,08 | 0,50 |

## 3   Results and Discussion

The results show that with the high percentage of survival, minimum treatment time, radiation dose, side effect and cost, surgery is the best treatment alternative while chemotherapy and the immunotherapy are second and third respectively in terms of the patients (Table 3).

**Table 3.** Complete ranking of lung cancer treatment alternatives in terms of patients.

| Complete ranking | Alternative | Positive outranking flow | Negative outranking flow | Net flow |
|---|---|---|---|---|
| 1 | Surgery | 0,0116 | 0,0001 | 0,0115 |
| 2 | Chemotherapy | 0,0034 | 0,0052 | −0,0019 |
| 3 | Immunotherapy | 0,0024 | 0,0043 | −0,0020 |
| 4 | Radiation therapy | 0,0024 | 0,0052 | −0,0028 |
| 5 | Hormone therapy | 0,0025 | 0,0074 | −0,0049 |

The advantages and the disadvantages of lung cancer treatment alternatives can be seen in Fig. 1 in terms of patients and in Fig. 2 in terms of hospitals.

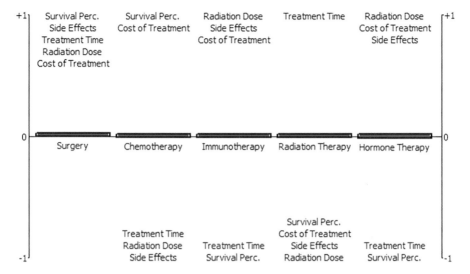

**Fig. 1.** PROMETHEE evaluation results for patients.

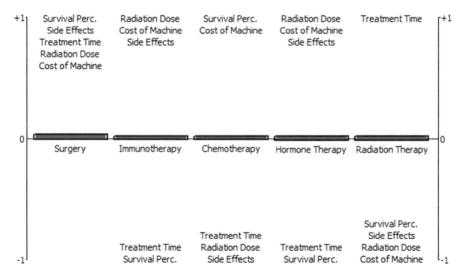

**Fig. 2.** PROMETHEE evaluation results for hospitals

Table 4 shows that surgery is also the best treatment alternative in terms of hospital while immunotherapy and chemotherapy are second and third respectively with a small difference.

During analysis in terms of patients, the cost of machine was deactivated as this parameter only affects the hospitals, while the cost of treatment was deactivated for analysis in terms of hospitals since the parameter relating to the cost of treatment relates only to patients.

**Table 4.** Complete ranking of lung cancer treatment alternatives in terms of hospital.

| Complete ranking | Alternative | Positive outranking flow | Negative outranking flow | Net flow |
|---|---|---|---|---|
| 1 | Surgery | 0,0141 | 0,0000 | 0,0141 |
| 2 | Immunotherapy | 0,0038 | 0,0048 | −0,0010 |
| 3 | Chemotherapy | 0,0047 | 0,0058 | −0,0011 |
| 4 | Hormone therapy | 0,0036 | 0,0082 | −0,0046 |
| 5 | Radiation therapy | 0,0026 | 0,0101 | −0,0074 |

# 4   Conclusion

Using Fuzzy PROMETHEE in this study, we were able to achieve good decision results by utilizing fuzzy input data. The results of this study provide the decision maker the ability to choose the best alternative in order to achieve desirable results from a treatment technique for lung cancer. This study shows that the proposed method of fuzzy PROMETHEE is an effective method to employ when solving decision-making problems related to healthcare.

# References

1. Lung cancer statistics. https://www.cancerresearchuk.org/health-professional/cancer-statistics/statistics-by-cancer-type/lung-cancer. Accessed 21 June 2018
2. Key Statistics for Lung Cancer. https://www.cancer.org/cancer/non-small-cell-lung-cancer/about/key-statistics.html. Accessed 04 July 2018
3. Keenan, R., Landreneau, R., Maley, R., Singh, D., Macherey, R., Bartley, S., Santucci, T.: Segmental resection spares pulmonary function in patients with stage I lung cancer. Ann. Thorac. Surg. **78**, 228–233 (2004)
4. Dine, J., Gordon, R., Shames, Y., Kasler, M., Barton-Burke, M.: Immune checkpoint inhibitors: An innovation in immunotherapy for the treatment and management of patients with cancer. Asia-Pac. J. Oncol. Nurs. **4**(2), 127 (2017)
5. Chemotherapy for Non-Small Cell Lung Cancer. https://www.cancer.org/cancer/non-small-cell-lung-cancer/treating/chemotherapy.html. Accessed 21 June 2018
6. Types of Radiation Therapy SEER Training. https://training.seer.cancer.gov/treatment/radiation/types.html. Accessed 21 June 2018
7. Ozsahin, D., Uzun, B., Musa, M., Şentürk, N., Nurçin, F., Ozsahin, I.: Evaluating nuclear medicine imaging devices using fuzzy PROMETHEE method. Procedia Comput. Sci. **120**, 699–705 (2017)
8. Goumas, M., Lygerou, V.: An extension of the PROMETHEE method for decision making in fuzzy environment: ranking of alternative energy exploitation projects. Eur. J. Oper. Res. **123**(3), 606–613 (2000)
9. Özgen, A., Tuzkaya, G., Tuzkaya, U., Özgen, D.: A multi-criteria decision making approach for machine tool selection problem in a fuzzy environment. Int. J. Comput. Intell. Syst. **4**(4), 431–445 (2011)
10. Ozsahin, D., Ozsahin, I.: A fuzzy PROMETHEE approach for breast cancer treatment techniques. Int. J. Med. Res. Health Sci. **7**(5), 29–32 (2018)

11. Uzun, D., Uzun, B., Sani, M., Helwan, A., Nwekwo, C., Veysel, F., Sentürka, N., Ozsahin, I.: Evaluating cancer treatment alternatives using fuzzy PROMETHEE method. Int. J. Adv. Comput. Sci. Appl. **8**(10), 177–182 (2018)

12. Uzun, D., Uzun, B., Sani, M., Ozsahin, I.: Evaluating X-Ray based medical imaging devices with fuzzy preference ranking organization method for enrichment evaluations. Int. J. Adv. Comput. Sci. Appl. **9**(3), 7–10 (2018)

13. Ozsahin, D., Isa, N., Uzun, B., Ozsahin, I.: Effective analysis of image reconstruction algorithms in nuclear medicine using fuzzy PROMETHEE. In: 2018 Advances in Science and Engineering Technology International Conferences (ASET) (2018)

# A Comparative Study on the Economic Development Level of the Countries by Fuzzy DEA Methodologies

Mujde Erol Genevois and Michele Cedolin[(✉)]

Industrial Engineering Department, Galatasaray University, Istanbul, Turkey
mcedolin@gsu.edu.tr

**Abstract.** In this work, we aim to compare the economic development level of the countries, regarding to their macroeconomic indicators, also considering their financial service accessibility, including as factor the number of bank branches and the automated teller machines. In this context, we took into account the data of the sixteen European countries, obtained by the consensus of financial experts and we applied different Data Envelopment Analysis (DEA) models for finding the most efficient countries in the selected context. We observed that although the models differ, we obtained similar results for each model.

**Keywords:** Fuzzy decision making · Data envelopment analysis
Performance evaluation

## 1 Introduction

Financial performance evaluation of the countries is a widely investigated topic in the sector and in the literature. There are many special institutions that rates countries considering different parameters, such as gross domestic product (GDP), GDP per capita, consumer price index (CPI), unemployment rate, inflation rate etc. and their ratings seriously affects the economy and the investments of the investigated country.

These ratings models differ from each other and there is no evidence of the unique correct form of it. Basing on that idea, in this work, sixteen European countries are selected and investigated according to predefined five outputs and two inputs. Our main focus is to integrate the number of the financial services tools into the model, so that economic performance efficiencies are calculated by using classical DEA model, however, due to the poor discriminating power of this pioneer model, we employed also a well-known common weight DEA model which yield to only two efficient decision making units (DMUs).

The rest of the paper is organized as follows. The second section introduces the DEA methodology, providing the two employed models and the last section provides the application of the fuzzy DEA models.

© Springer Nature Switzerland AG 2019
R. A. Aliev et al. (Eds.): ICAFS-2018, AISC 896, pp. 216–222, 2019.
https://doi.org/10.1007/978-3-030-04164-9_30

# 2   Methodology

Data Envelopment Analysis (DEA) is a non-parametric linear programming-based technique employed as a decision making technique in comparing the efficiency of decision making units (DMUs) such as health services, local authority departments, education departments, factories, banks and also a decision help for selection problems [1, 2]. DEA is first proposed by Charnes et al. (1978) and they based their study on productive efficiency [3]. The efficiency measure obtained as the maximum of a ratio weighted outputs to weighted inputs subject to the condition that the similar ratios for every DMU be less than or equal to unity [4, 5].

DEA mainly considers the crisp data. However, in real-life problems such performance evaluation, the decision makers confront with vagueness and uncertainty while evaluating the candidates and they prefer to use linguistic terms [6]. Fuzzy DEA, is an extension of DEA which incorporates imprecision in DEA [7–9]. In the present study, some of the alternative fuzzy DEA methodologies are provided.

## 2.1   Optimistic and Pessimistic Models

In this section the model that is first employed to evaluate the performance of the selected countries are provided [10].

Let $(E_{J_0})^U$ and $(E_{J_0})^L$ denote the upper and lower bounds of the $\alpha$-cut of the membership function of the efficiency score for the evaluated DMU ($j_0$) the general optimistic scenario DEA model incorporating crisp and fuzzy data can be written as:

$$\max(E_{J_0})^U = \sum_{r \in C_R} u_r y_{rj_0} + \sum_{r \in F_R} u_r y_{rj_0c} - \mu_r(y_{rj_0c} - y_{rj_0b}) \tag{1}$$

subject to

$$\sum_{i \in C_I} v_i x_{ij_0} + \sum_{i \in F_I} v_i x_{ij_0a} + w_i(x_{ij_0b} - x_{ij_0a}) = 1$$

$$\sum_{r \in C_R} u_r y_{rj_0} + \sum_{r \in F_R} u_r y_{rj_0c} - \mu_r(y_{rj_0c} - y_{rj_0b}) - \sum_{i \in C_I} v_i x_{ij_0} - \sum_{i \in F_I} v_i x_{ij_0a} + w_i(x_{ij_0b} - x_{ij_0a}) \le 0$$

$$\sum_{r \in C_R} u_r y_{rj} + \sum_{r \in F_R} u_r y_{rja} + \mu_r(y_{rjb} - y_{rja}) - \sum_{i \in C_I} v_i x_{ij} - \sum_{i \in F_I} v_i x_{ijc} - w_i(x_{ijc} - x_{ijb}) \le 0 \, j = 1, 2 \dots, n; j \neq j_0$$

$$\mu_r - u_r \le 0, r \in F_R, \quad w_i - v_i \le 0, i \in F_I \, \mu_r \ge 0, r \in F_R, w_i \ge 0, i \in F_I$$

$$u_r \ge \varepsilon \ge 0, r \in C_R, r \in F_R \quad v_i \ge \varepsilon \ge 0, i \in C_I, i \in F_I$$

where $F_R$ and $F_I$ respectively represent the subset of fuzzy outputs $(F_R \subseteq R)$ and the subset of fuzzy inputs $(F_I \subseteq I)$ where R denotes the set of outputs $(C_R \cup F_R = R)$ and I represents the set of inputs $(C_I \cup F_I = I)$. The pessimistic scenario model is quite similar to the optimistic model and the interested reader may find it at [10].

## 2.2   Ranking Model

In this part, fuzzy DEA model proposed by Saati et al. [11] is employed due to the need to consider qualitative as well as quantitative data. The concept of $\alpha$-cut is used to develop the following model.

$$\max E = \bar{y}_p \tag{2}$$

subject to

$$\bar{x}_p = 1$$
$$\bar{y}_j - \bar{x}_j \leq 0$$
$$v(\alpha x_j^m + (1-\alpha)x_j^l) \leq \bar{x}_j \leq v(\alpha x_j^m + (1-\alpha)x_j^u) \qquad \forall j,$$
$$u(\alpha y_j^m + (1-\alpha)y_j^l) \leq \bar{y}_j \leq u(\alpha y_j^m + (1-\alpha)y_j^u) \qquad \forall j, \qquad u,v \geq 0$$

As a result of standard DEA models, efficient DMUs have an efficiency score equals to 1. Thus, it is not possible to rank these DMUs. The objective of performance evaluation and selection processes is to obtain a comparison among the units, so that the same score for more than one unit is not desirable. The following model enables to give a ranking for efficient DMUs:

$$\min z = \theta \tag{3}$$

subject to

$$\theta(\alpha x_p^m + (1-\alpha)x_p^l) \geq \sum_{j=1}^{n} \lambda_j(\alpha x_j^m + (1-\alpha)x_j^u)$$
$$\alpha y_{rp}^m + (1-\alpha)y_{rp}^u \leq \sum_{j=1}^{n} \lambda_j(\alpha y_{rj}^m + (1-\alpha)y_{rj}^l) \qquad \forall r, \lambda_j \geq 0 \qquad \forall j.$$

## 3   Application

In this part of the study, we employed the data belonging to the sixteen European countries. Table 1 provides the inputs and outputs of the study; Fig. 1 is for the adapted fuzzy scale, while the Table 2 contains all the corresponding data that is utilized.

**Table 1.** The examined factors of the countries

|     | Factor |
| --- | --- |
|     | ATM numbers |
| O1  | ATM numbers |
| O2  | Account ownership (% of population ages 15+) |
| O3  | Adjusted net national income per capita (US$) |
| O4  | Bank branches |
| O5  | GDP per capita (current US$) |
| I1  | Consumer price index (CPI) |
| I2  | Unemployment, total (% of total labor force) |

(DL: (0, 0, 0.16), VL: (0, 0.16, 0.33), L: (0.16, 0.33, 0.50), M: (0.33, 0.50, 0.66), H: (0.50, 0.66, 0.83), VH: (0.66, 0.83, 1), DH: (0.83, 1, 1)).

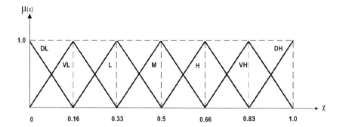

**Fig. 1.** Membership functions for linguistic variables

**Table 2.** The employed data set

|          | O1 | O2 | O3 | O4 | O5 | I1 | I2 |
|----------|----|----|----|----|----|----|----|
| Austria     | VH | VH | H  | L  | H  | H  | M  |
| Belgium     | H  | VH | H  | VL | H  | H  | M  |
| Bulgaria    | H  | L  | L  | H  | VL | M  | M  |
| Denmark     | VL | VH | H  | M  | VH | M  | M  |
| France      | H  | H  | H  | M  | M  | M  | H  |
| Germany     | DH | VH | H  | L  | H  | M  | DL |
| Greece      | L  | M  | M  | M  | L  | L  | VH |
| Hungary     | L  | L  | VL | L  | L  | H  | L  |
| Italy       | H  | H  | M  | H  | M  | M  | H  |
| Netherlands | VL | DH | H  | L  | H  | M  | M  |
| Norway      | VL | VH | VH | VL | VH | H  | VL |
| Poland      | M  | M  | M  | M  | L  | M  | M  |
| Portugal    | VH | H  | M  | H  | L  | M  | H  |
| Spain       | H  | H  | M  | VH | M  | M  | VH |
| Switzerland | H  | VH | VH | H  | VH | L  | L  |
| Turkey      | M  | L  | L  | L  | L  | VH | H  |

The models that are represented in the previous section, are applied with $\varepsilon = 0.001$ for different alpha-cut scenarios with step size 0.1 The results for the optimistic model yielded to an efficient score of 1 for each decision making unit (DMU), except for Turkey and Hungary, when $\alpha = 1$ (the fuzziness of the system is eliminated). Therefore, we conclude that optimistic scenario is lack of the discriminating power. However, when we applied the pessimistic scenario in the same data set, we had the chance to compare the efficiency results of the countries for different alpha-cuts. In Table 3 we provide the results, and in Fig. 2, the evaluation of the efficiencies can be examined (Table 4).

It is worth noting that, this model yielded to a better discrimination, comparing both with the optimistic and pessimistic model. For the first four alpha levels, only one DMU is efficient, while for $\alpha = 1$ and the fuzziness of the system is eliminated, only three countries remain efficient. On the other hand both models, shows similarly that, DMU8 and DMU16 have the poorest performance.

**Table 3.** The pessimistic scenario results

| DMU/α | 0 | 0.1 | 0.2 | 0.3 | 0.4 | 0.5 | 0.6 | 0.7 | 0.8 | 0.9 | 1 |
|---|---|---|---|---|---|---|---|---|---|---|---|
| Austria | 0.436 | 0.417 | 0.4 | 0.382 | 0.366 | 0.35 | 0.334 | 0.319 | 0.304 | 0.29 | 0.276 |
| Belgium | 0.436 | 0.417 | 0.4 | 0.382 | 0.366 | 0.35 | 0.334 | 0.319 | 0.304 | 0.29 | 0.276 |
| Bulgaria | 0.333 | 0.317 | 0.301 | 0.285 | 0.27 | 0.256 | 0.242 | 0.229 | 0.216 | 0.203 | 0.191 |
| Denmark | 0.402 | 0.385 | 0.369 | 0.353 | 0.338 | 0.323 | 0.309 | 0.295 | 0.281 | 0.268 | 0.255 |
| France | 0.361 | 0.343 | 0.325 | 0.309 | 0.293 | 0.277 | 0.262 | 0.248 | 0.234 | 0.22 | 0.207 |
| Germany | 0.506 | 0.487 | 0.469 | 0.452 | 0.435 | 0.419 | 0.403 | 0.387 | 0.372 | 0.358 | 0.344 |
| Greece | 0.33 | 0.314 | 0.299 | 0.284 | 0.269 | 0.255 | 0.242 | 0.229 | 0.216 | 0.204 | 0.192 |
| Hungary | 0.123 | 0.108 | 0.093 | 0.079 | 0.066 | 0.052 | 0.04 | 0.028 | 0.016 | 0.005 | 0 |
| Italy | 0.362 | 0.343 | 0.325 | 0.309 | 0.293 | 0.277 | 0.262 | 0.248 | 0.234 | 0.22 | 0.207 |
| Netherlands | 0.338 | 0.321 | 0.305 | 0.289 | 0.274 | 0.259 | 0.245 | 0.232 | 0.218 | 0.206 | 0.193 |
| Norway | 0.436 | 0.417 | 0.4 | 0.382 | 0.366 | 0.35 | 0.334 | 0.319 | 0.304 | 0.29 | 0.276 |
| Poland | 0.227 | 0.211 | 0.196 | 0.181 | 0.167 | 0.154 | 0.141 | 0.128 | 0.116 | 0.104 | 0.092 |
| Portugal | 0.436 | 0.417 | 0.4 | 0.382 | 0.366 | 0.35 | 0.334 | 0.319 | 0.304 | 0.29 | 0.276 |
| Spain | 0.436 | 0.417 | 0.4 | 0.382 | 0.366 | 0.35 | 0.334 | 0.319 | 0.304 | 0.29 | 0.276 |
| Switzerland | 0.325 | 0.311 | 0.298 | 0.285 | 0.273 | 0.261 | 0.249 | 0.238 | 0.227 | 0.217 | 0.206 |
| Turkey | 0.234 | 0.218 | 0.203 | 0.188 | 0.174 | 0.16 | 0.147 | 0.134 | 0.121 | 0.109 | 0.097 |

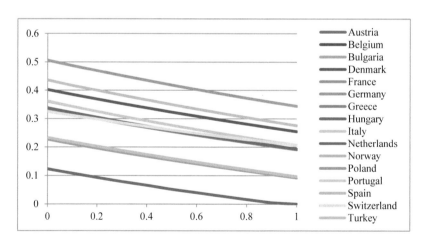

**Fig. 2.** The evolution of the efficiencies according to the pessimistic scenario

In order to enrich our study, we also employed the ranking model for the same data set, and we obtained the results given in Table 5.

As can easily be observed form the Fig. 3 also, when the alpha levels increase, the efficient scores decrease. However, between the models the efficiency are not changing. In both models, Germany and Norway show the best performances in addition to that the worst performances are not changing too. This is quite insightful, for commenting that the models mathematical background move in the same manner. On further studies, we will try to solve the same data set with different fuzzy DEA models and

**Table 4.** Saati model results

| DMU/α | 0 | 0.1 | 0.2 | 0.3 | 0.4 | 0.5 | 0.6 | 0.7 | 0.8 | 0.9 | 1 |
|---|---|---|---|---|---|---|---|---|---|---|---|
| Austria | 1 | 1 | 1 | 1 | 1 | 1 | 0.949 | 0.857 | 0.773 | 0.698 | 0.629 |
| Belgium | 1 | 1 | 1 | 1 | 1 | 0.97 | 0.866 | 0.773 | 0.689 | 0.613 | 0.545 |
| Bulgaria | 1 | 1 | 1 | 1 | 1 | 1 | 1 | 0.986 | 0.863 | 0.755 | 0.66 |
| Denmark | 1 | 1 | 1 | 1 | 1 | 1 | 1 | 0.986 | 0.848 | 0.748 | 0.66 |
| France | 1 | 1 | 1 | 1 | 1 | 1 | 1 | 0.963 | 0.849 | 0.748 | 0.66 |
| Germany | 1 | 1 | 1 | 1 | 1 | 1 | 1 | 1 | 1 | 1 | 1 |
| Greece | 1 | 1 | 1 | 1 | 1 | 1 | 1 | 1 | 1 | 0.923 | 0.795 |
| Hungary | 1 | 1 | 1 | 1 | 0.935 | 0.795 | 0.676 | 0.575 | 0.489 | 0.416 | 0.352 |
| Italy | 1 | 1 | 1 | 1 | 1 | 1 | 1 | 0.986 | 0.863 | 0.755 | 0.66 |
| Netherlands | 1 | 1 | 1 | 1 | 1 | 1 | 1 | 0.986 | 0.863 | 0.755 | 0.66 |
| Norway | 1 | 1 | 1 | 1 | 1 | 1 | 1 | 1 | 1 | 1 | 1 |
| Poland | 1 | 1 | 1 | 1 | 1 | 1 | 0.872 | 0.76 | 0.662 | 0.576 | 0.5 |
| Portugal | 1 | 1 | 1 | 1 | 1 | 1 | 1 | 1 | 1 | 0.938 | 0.83 |
| Spain | 1 | 1 | 1 | 1 | 1 | 1 | 1 | 1 | 1 | 0.945 | 0.83 |
| Switzerland | 1 | 1 | 1 | 1 | 1 | 1 | 1 | 1 | 1 | 1 | 1 |
| Turkey | 0.976 | 0.854 | 0.748 | 0.668 | 0.596 | 0.532 | 0.475 | 0.424 | 0.379 | 0.338 | 0.301 |

**Table 5.** Saati ranking model results

| DMU/α | 0 | 0.1 | 0.2 | 0.3 | 0.4 | 0.5 | 0.6 | 0.7 | 0.8 | 0.9 | 1 |
|---|---|---|---|---|---|---|---|---|---|---|---|
| Austria | 0.436 | 0.417 | 0.4 | 0.382 | 0.366 | 0.35 | 0.334 | 0.319 | 0.304 | 0.29 | 0.276 |
| Belgium | 0.436 | 0.417 | 0.4 | 0.382 | 0.366 | 0.35 | 0.334 | 0.319 | 0.304 | 0.29 | 0.276 |
| Bulgaria | 0.333 | 0.317 | 0.301 | 0.285 | 0.27 | 0.256 | 0.242 | 0.229 | 0.216 | 0.203 | 0.191 |
| Denmark | 0.402 | 0.385 | 0.369 | 0.353 | 0.338 | 0.323 | 0.309 | 0.295 | 0.281 | 0.268 | 0.255 |
| France | 0.361 | 0.343 | 0.325 | 0.309 | 0.293 | 0.277 | 0.262 | 0.248 | 0.234 | 0.22 | 0.207 |
| Germany | 0.506 | 0.487 | 0.469 | 0.452 | 0.435 | 0.419 | 0.403 | 0.387 | 0.372 | 0.358 | 0.344 |
| Greece | 0.33 | 0.314 | 0.299 | 0.284 | 0.269 | 0.255 | 0.242 | 0.229 | 0.216 | 0.204 | 0.192 |
| Hungary | 0.123 | 0.108 | 0.093 | 0.079 | 0.066 | 0.052 | 0.04 | 0.028 | 0.016 | 0.005 | 0 |
| Italy | 0.362 | 0.343 | 0.325 | 0.309 | 0.293 | 0.277 | 0.262 | 0.248 | 0.234 | 0.22 | 0.207 |
| Netherlands | 0.338 | 0.321 | 0.305 | 0.289 | 0.274 | 0.259 | 0.245 | 0.232 | 0.218 | 0.206 | 0.193 |
| Norway | 0.436 | 0.417 | 0.4 | 0.382 | 0.366 | 0.35 | 0.334 | 0.319 | 0.304 | 0.29 | 0.276 |
| Poland | 0.227 | 0.211 | 0.196 | 0.181 | 0.167 | 0.154 | 0.141 | 0.128 | 0.116 | 0.104 | 0.092 |
| Portugal | 0.436 | 0.417 | 0.4 | 0.382 | 0.366 | 0.35 | 0.334 | 0.319 | 0.304 | 0.29 | 0.276 |
| Spain | 0.436 | 0.417 | 0.4 | 0.382 | 0.366 | 0.35 | 0.334 | 0.319 | 0.304 | 0.29 | 0.276 |
| Switzerland | 0.325 | 0.311 | 0.298 | 0.285 | 0.273 | 0.261 | 0.249 | 0.238 | 0.227 | 0.217 | 0.206 |
| Turkey | 0.234 | 0.218 | 0.203 | 0.188 | 0.174 | 0.16 | 0.147 | 0.134 | 0.121 | 0.109 | 0.097 |

compare the results. Another focus will be enlarging data set, as input-output amount, besides as the decision making unit numbers. Ultimately, we will try to integrate DEA models with different decision making tools.

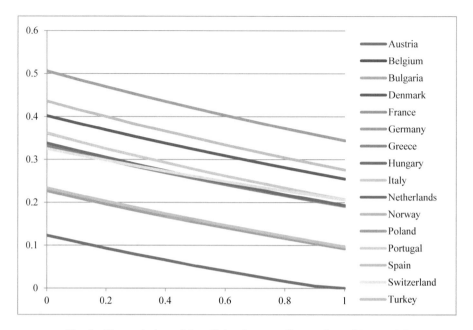

**Fig. 3.** The evolution of the efficiencies according to the ranking model

**Acknowledgments.** This work has been financially supported by Galatasaray University Research Fund 18.402.011.

# References

1. Hatefi, S.M., Torabi, S.A.: A common weight MCDA–DEA approach to construct composite indicators. Ecol. Econ. **70**(1), 114–120 (2010)
2. Doyle, J., Green, R.: Efficiency and cross-efficiency in DEA: derivations, meanings and uses. J. Oper. Res. Soc. **45**, 567–578 (1994)
3. Farrell, M.J.: The measurement of productive efficiency. J. R. Stat. Soc. 253–290 (1957)
4. Charnes, A., Cooper, W.W., Rhodes, E.: Measuring the efficiency of decision making units. Eur. J. Oper. Res. 429–444 (1978)
5. Li, X.-B., Reeves, G.R.: A multiple criteria approach to data envelopment analysis. Eur. J. Oper. Res. **115**, 507–517 (1999)
6. Zadeh, L.A.: Fuzzy sets. Inf. Control **8**, 338–353 (1965)
7. Sengupta, J.K.: A fuzzy systems approach in data envelopment analysis. Comput. Math. Appl. **24**, 259–266 (1992)
8. Triantis, K., Girod, O.: A mathematical programming approach for measuring technical efficiency in a fuzzy environment. J. Prod. Anal. **10**, 85–102 (1998)
9. Hatami-Marbini, A., Emrouznejad, A., Tavana, M.: A taxonomy and review of the fuzzy data envelopment analysis literature: two decades in the making. Eur. J. Oper. Res. **214**, 457–472 (2011)
10. Karsak, E.E.: Using data envelopment analysis for evaluating flexible manufacturing systems in the presence of imprecise data. Int. J. Adv. Manuf. Technol. **35**, 867–874 (2008)
11. Saati, M.S., Memariani, A., Jahanshahloo, G.R.: Efficiency analysis and raking of DMUs with fuzzy data. Fuzzy Optim. Decis. Making **1**(3), 255–267 (2002)

# Fuzzy Analysis of Macroeconomic Stability

Gorkhmaz Imanov[1]($\boxtimes$), Yadulla Hasanli[2], and Malahat Murtuzaeva[1]

[1] Control Systems of Azerbaijan National Academy of Science,
B. Vahabzadeh Street 9, AZ1102 Baku, Azerbaijan
korkmazi2000@gmail.com, malaxat-55@rambler.ru
[2] Scientific Research Institute of Economic Studies, Azerbaijan State University
of Economics, Istiglaliyyat Street 6 (I Building), AZ1001 Baku, Azerbaijan
yadulla59@mail.ru

**Abstract.** In the proposed paper, we investigate macroeconomic stability using instruments of intuitionistic linguistic fuzzy set. For calculating the level of macroeconomic stability were used the indicators proposed by World Bank and Maastricht Treaty. By using statistical data of Azerbaijan Republic, the level of macroeconomic stability of the country is defined for 2010–2016 years.

**Keywords:** Macroeconomic stability · Intuitionistic linguistic fuzzy number Aggregate index of macroeconomic stability

## 1 Introduction

World Bank describes macroeconomic stability as follows: when the inflation rate is low and predictable, real interest rates are appropriate, the real exchange rate is competitive and predictable, public sector saving rates are compatible with the resource mobilization requirements of the program, and the balance of payments situation is perceived as variable [1].

According to the Maastricht Treaty [2] macroeconomic stability is measured through five variables:

– Low and stable inflation (the Maastricht criteria capped at 3%);
– Low long-term interest rate (the Maastricht criteria restricted to the range of 9%);
– Low debt to Gross Domestic Product ratio (the Maastricht criteria capped at 60% of GDP);
– Low deficit (the Maastricht criteria capped at 3% of GDP);
– Monetary stability (the Maastricht criteria permitted fluctuation of at most 2.5%).

In order to calculate the level of macroeconomic stability, econometric models are mainly used. By means of linguistic intuitionistic fuzzy number, we calculate aggregate index of macroeconomic stability of Azerbaijan for the period 2010–2016 years [3]. With this purpose, we use the following macroeconomic indicators:

1. Growth rate of Gross Domestic Product - **GGD;**
2. Inflation % - **INF**;
3. Interest rate % - **INR**;
4. National debt relative to GDP % - **NAD;**

© Springer Nature Switzerland AG 2019
R. A. Aliev et al. (Eds.): ICAFS-2018, AISC 896, pp. 223–229, 2019.
https://doi.org/10.1007/978-3-030-04164-9_31

5. Budget deficit (% of GDP) - **DEF**;
6. Exchange rate - **EXR**;
7. Current account balance (% of GDP) **- CAB;**
8. Unemployment rate % - **UNE**;
9. Growth rate of Foreign Investment - **FEI**

## 2   Calculating Parameters of Linguistic Intuitionistic Fuzzy Set

On the basis of Atanassov intuitionistic fuzzy set (IFS) [4] Wang and Li proposed a linguistic intuitionistic fuzzy set [5].

$$A = \left\{ \left\langle x, \left[ S_{\theta(x)}, \mu_A(x), v_A(x) \right] \right\rangle | x \in X \right\} \tag{1}$$

where $S_{\theta(x)} \in S$, $\mu_A : X \to [0, 1]$ and $v_A : X \to [0, 1]$, that satisfies the condition $\mu_A(x) + v_A(x) \leq 1$, $\mu_A(x)$ and $v_A(x)$, represent the membership and non-membership degrees, respectively, of elements $x$ to the linguistic value $S_{\theta(x)}$.

For each intuitionistic linguistic set $A = \left\{ \left\langle x, \left[ S_{\theta(x)}, \mu_A(x), v_A(x) \right] \right\rangle | x \in X \right\}$, there is $\pi_A(x) = 1 - \mu_A(x) - v_A(x)$, which is called the fuzzy intuitionistic index of the element $x$ of the linguistic variable $S_{\theta(x)}$.

For the intuitionistic linguistic set $A = \left\{ \left\langle x, \left[ S_{\theta(x)}, \mu_A(x), v_A(x) \right] \right\rangle | x \in X \right\}$, $\left( S_{\theta(x)}, \left( \mu_A(x), v_A(x) \right) \right)$ triple is called an intuitionistic linguistic fuzzy number (IFN).

By using thresholds recommended by Maastricht treatment and Alert Mechanism European Commission [6] for indicators of macroeconomic stability, linguistic variable $S_{\theta(x)}$ is defined (Table 1):

In order to define membership and non-membership values, Attanassov's function [7] is used:

$$\mu_{\tilde{x}}(x) = \begin{cases} \frac{u_{\tilde{x}}\left(x - \underline{t}\right)}{t - \underline{t}} & \text{if } \underline{t} \leq x < t \\ u_{\tilde{x}} & \text{if } x = t \\ \frac{u_{\tilde{x}}(\bar{t} - x)}{\bar{t} - t} & \text{if } t < x \leq \bar{t} \\ 0 & \text{if } x < \underline{t} \text{ or } x > \bar{t} \end{cases} \tag{2}$$

and

$$v_{\tilde{x}}(x) = \begin{cases} \frac{\left[ t - x + w_{\tilde{x}}\left(x - \underline{t}\right) \right]}{t - \underline{t}} & \text{if } \underline{t} \leq x < t \\ w_{\tilde{x}} & \text{if } x = t \\ \left[ x - t + \frac{w_{\tilde{x}}(\bar{t} - x)}{\bar{t} - t} \right] & \text{if } t < x \leq \bar{t} \\ 1 & \text{if } x < \underline{t} \text{ or } x > \bar{t} \end{cases} \tag{3}$$

For calculating membership and non-membership function are used reduction coefficients $(u_{\tilde{x}}, w_{\tilde{x}})$, which take into account accuracy of statistical information. The

**Table 1.** Linguistic variables of macroeconomic stability indicators.

| Indicators | Unstable-$S_0$ | | | Low stable-$S_1$ | | |
|---|---|---|---|---|---|---|
| GGD | $-\infty$ | -1.05 | 0.1 | 0 | 1.25 | 2.5 |
| INF | 9 | 9.5 | $+\infty$ | 3 | 6.25 | 9.5 |
| NAD | 55 | 65 | $+\infty$ | 25 | 42.5 | 60 |
| INR | 9 | 9.5 | $+\infty$ | 3.5 | 6.5 | 9.5 |
| DEF | 9 | 9.5 | $+\infty$ | 3.5 | 6.75 | 10 |
| EXR | $-\infty$ | -50 | -32 | -30 | -11 | -10 |
| CAB | $-\infty$ | 2 | 2.5 | -3 | -2.5 | -2 |
| UNE | 11 | 11.5 | $+\infty$ | 7.5 | 9.75 | 12 |
| FDI | $-\infty$ | -3 | -2.5 | -3 | -2.25 | -1.5 |

**Table 1** (continued)

| Indicators | Stable-$S_2$ | | | High stable-$S_3$ | | |
|---|---|---|---|---|---|---|
| GGD | 2 | 2.5 | 3 | 2.5 | 4.25 | 7 |
| INF | 2.5 | 3.25 | 4 | 0 | 1.5 | 3 |
| NAD | 10 | 20 | 30 | 0 | 7.5 | 15 |
| INR | 1 | 2.5 | 4 | -3 | 0 | 3 |
| DEF | 1.5 | 2.75 | 4 | -1.5 | 0.25 | 2 |
| EXR | -11 | 0 | 11 | -30 | -30 | -30 |
| CAB | -4 | 1 | 6 | 5 | 17.5 | 30 |
| UNE | 5.5 | 6.75 | 8 | 4 | 5 | 6 |
| FDI | -2 | 0.65 | 3.3 | 3 | 10 | $+\infty$ |

calculations' results of membership, non-membership degree and linguistic indices are presented in Table 2.

Then, the weights of k-th macroeconomic indicators in t-years are obtained by applying the following formula, which introduced in [8] (Table 3):

$$\lambda_k = \frac{\left(\mu_k + \pi_k\left(\frac{\mu_k}{v_k}\right)\right)}{\sum_{k=1}^{l}\left(\mu_k + \pi_k\left(\frac{\mu_k}{v_k}\right)\right)} \tag{4}$$

and $\sum_{k=1}^{l} \lambda_k = 1$

In order to calculate Aggregate Index of Macroeconomic Stability (**AIMS**) for each year, intuitionistic linguistic weighted average (**ILWA**) formula developed by Wang and Li [5] is used:

$$\textbf{AIMS} = \textbf{ILWA} = \left\langle S_{\sum_{k=1}^{t} \lambda_k \theta(a_{ij}^k)}, \left(1 - \prod_{k=1}^{t}\left(1 - \mu\left(a_{ij}^k\right)\right)^{\lambda_k}\right), \prod_{k=1}^{t}\left(v\left(a_{ij}^k\right)\right)^{\lambda_k}\right\rangle \tag{5}$$

**Table 2.** IFN of macroeconomic stability indicators.

| Indicators | 2010 | | | 2011 | | | 2012 | | |
|---|---|---|---|---|---|---|---|---|---|
| | $S_\theta$ | $\mu$ | $\nu$ | $S_\theta$ | $\mu$ | $\nu$ | $S_\theta$ | $\mu$ | $\nu$ |
| GGD | $S_3$ | 0.8 | 0.16 | $S_1$ | 0.07 | 0.92 | $S_2$ | 0.36 | 0.62 |
| INF | $S_1$ | 0.67 | 0.25 | $S_1$ | 0.39 | 0.56 | $S_3$ | 0.53 | 0.4 |
| NAD | $S_1$ | 0.78 | 0.18 | $S_3$ | 0.024 | 0.97 | $S_0$ | 0.9 | 0.05 |
| INR | $S_3$ | 0.74 | 0.22 | $S_3$ | 0.56 | 0.41 | $S_3$ | 0.66 | 0.3 |
| DEF | $S_3$ | 0.53 | 0.43 | $S_3$ | 0.68 | 0.28 | $S_3$ | 0.83 | 0.13 |
| EXR | $S_2$ | 0.83 | 0.08 | $S_2$ | 0.73 | 0.19 | $S_2$ | 0.81 | 0.09 |
| CAB | $S_2$ | 0.13 | 0.86 | $S_3$ | 0.22 | 0.75 | $S_3$ | 0.64 | 0.28 |
| UNE | $S_2$ | 0.34 | 0.62 | $S_3$ | 0.51 | 0.43 | $S_3$ | 0.24 | 0.68 |
| FDI | $S_2$ | 0.51 | 0.42 | $S_2$ | 0.66 | 0.25 | $S_2$ | 0.62 | 0.31 |

**Table 2** (continued)

| Indicators | 2013 | | | 2014 | | | 2015 | | | 2016 | | |
|---|---|---|---|---|---|---|---|---|---|---|---|---|
| | $S_\theta$ | $\mu$ | $\nu$ | $S_\theta$ | $\mu$ | $\nu$ | $S_\theta$ | $\mu$ | $\nu$ | $S_\theta$ | $\mu$ | $\nu$ |
| GGD | $S_3$ | 0.48 | 0.49 | $S_2$ | 0.36 | 0.62 | $S_1$ | 0.79 | 0.16 | $S_0$ | 0.9 | 0.05 |
| INF | $S_3$ | 0.32 | 0.64 | $S_3$ | 0.75 | 0.04 | $S_1$ | 0.25 | 0.72 | $S_0$ | 0.8 | 0.1 |
| NAD | $S_0$ | 0.9 | 0.05 | $S_0$ | 0.9 | 0.05 | $S_0$ | 0.9 | 0.05 | $S_3$ | 0.44 | 0.54 |
| INR | $S_3$ | 0.53 | 0.44 | $S_3$ | 0.26 | 0.72 | $S_2$ | 0.33 | 0.45 | $S_1$ | 0.78 | 0.17 |
| DEF | $S_3$ | 0.68 | 0.28 | $S_3$ | 0.49 | 0.49 | $S_3$ | 0.15 | 0.85 | $S_3$ | 0.53 | 0.43 |
| EXR | $S_2$ | 0.84 | 0.06 | $S_2$ | 0.85 | 0.05 | $S_1$ | 0.17 | 0.81 | $S_0$ | 0.85 | 0.05 |
| CAB | $S_3$ | 0.71 | 0.2 | $S_3$ | 0.53 | 0.40 | $S_2$ | 0.58 | 0.35 | $S_2$ | 0.13 | 0.86 |
| UNE | $S_3$ | 0.85 | 0.05 | $S_3$ | 0.76 | 0.15 | $S_3$ | 0.85 | 0.05 | $S_3$ | 0.76 | 0.15 |
| FDI | $S_2$ | 0.67 | 0.24 | $S_2$ | 0.65 | 0.28 | $S_2$ | 0.71 | 0.2 | $S_2$ | 0.7 | 0.21 |

**Table 3.** Weights of macroeconomic indicators.

| Indicators | 2010 | 2011 | 2012 | 2013 | 2014 | 2015 | 2016 |
|---|---|---|---|---|---|---|---|
| GGD | 0.15 | 0.02 | 0.06 | 0.08 | 0.06 | 0.16 | 0.15 |
| INF | 0.13 | 0.10 | 0.10 | 0.05 | 0.16 | 0.05 | 0.14 |
| NAD | 0.14 | 0.01 | 0.16 | 0.15 | 0.16 | 0.18 | 0.07 |
| INR | 0.14 | 0.14 | 0.12 | 0.09 | 0.04 | 0.08 | 0.13 |
| DEF | 0.10 | 0.17 | 0.15 | 0.11 | 0.08 | 0.03 | 0.09 |
| EXR | 0.16 | 0.19 | 0.15 | 0.15 | 0.16 | 0.03 | 0.15 |
| CAB | 0.02 | 0.06 | 0.12 | 0.12 | 0.09 | 0.12 | 0.02 |
| UNE | 0.06 | 0.13 | 0.04 | 0.15 | 0.14 | 0.18 | 0.13 |
| FDI | 0.10 | 0.17 | 0.11 | 0.11 | 0.12 | 0.15 | 0.12 |
| | 1 | 1 | 1 | 1 | 1 | 1 | 1 |

The obtained results of calculations are given below:

$$\textbf{AIMS (2010)} = \langle S_{2.1}(0.71, 0.22) \rangle$$

$$\textbf{AIMS (2011)} = \langle S_{2.4}(0.60, 0.33) \rangle$$

$$\textbf{AIMS (2012)} = \langle S_{2.2}(0.74, 0.19) \rangle$$

$$\textbf{AIMS (2013)} = \langle S_{2.3}(0.76, 0.14) \rangle$$

$$\textbf{AIMS (2014)} = \langle S_{2.2}(0.74, 0.13) \rangle$$

$$\textbf{AIMS (2015)} = \langle S_{1.6}(0.75, 0.15) \rangle$$

$$\textbf{AIMS (2016)} = \langle S_{1.3}(0.78, 0.13) \rangle$$

As can be seen from result of calculation, macroeconomic stability was satisfying in 2010–2014, but in 2015–2016 the level of macroeconomic stability decreased and became low.

As shown in Table 2, fluctuation and decrease of GGD from high stability (S3) in 2010 to instability level (S0) in 2016 in dynamics of GGD trio can be mainly associated with price changes in oil sector due to global financial crisis. The change in oil price in the world market had its impact on GDP growth, as oil sector has the large share in GDP of Azerbaijan. Thus, the sharp decline in oil prices since the end of 2014 has led to a decline of the oil volume in GDP and the fact, that a devaluation has not been observed with a noticeable increase in the non-oil sector in a short time, led to insta-bility in GDP.

Since large oil revenues in the country led to increase in the volume of currency reserves, fluctuations in inflation can be mainly related to monetary policy governed by Central Bank. In order to ensure and diversify economic stability in the country, monetary policy regulating inflation rate was implemented. In 2013–2014, as a con-sequence of implemented policy, high stability in the rate of inflation was provided. In subsequent years, the financial crisis occurred in the world resulted in decline in the national currency of the country. As a result, since long-term economic stability could not be achieved with a regulated monetary policy, the transition to floating exchange rate was started, which led to a change in inflation rate. Thus, fluctuations toward inflation instability were started.

The main factor of economic growth in the country during oil boom, which lasted until 2015, was oil revenues.

Loans were mainly directed to households (44% of credits in 2014), trade (15%), which are mostly non-commercial sectors that depend heavily on oil revenues, and construction (14%). The sum of share of industrial sector in credit portfolio of banks was 10%. Thus, the role of interest rate (INR) in economic growth this period was low and it is wrong to link high economic growth to the interest rate. The fall of interest rate from S3 to S1 in 2015–2016 is associated with the reduction of role of oil factor in this period.

The main reason of transition of national debt to GDP ratio (NAD) from low stability level (S1) in 2010 to high stability level in 2011 is increase in oil production, foreign currency flow to the country and relative increase of national currency (exchange rate - EXR). However, fall to instability level (S0) started from 2012 and continued up to 2016 was linked to decline in oil production on a regular basis. The high level of stability in 2016 can be associated with a downturn in indebtedness and an increase in gas production.

The main reason of high level stability (S3) in budget deficit (DEF) was at the expense of transfers to State Budget by Oil Fund.

Macroeconomic stability level (S2) of exchange rate (EXR) in 2010–2014 became the main factor for keeping exchange rate of national currency (manat) stable during those period. The transition to low stability level (S1) and instability level (S0) can be explained with a sharp decline of oil price in the world market and decrease in exchange rate of manat.

The rise of current account balance (CAB) from medium stability level (S2) in 2010 to high stability level (S3) in 2011 was related to increase in positive saldo of CAB - from 15.0 blln. U.S. dollars to 17.1 blln. dollars. Due to the substitution of positive saldo with negative one in 2015–2016 (0.2 blln. dollars and 1.4 blln. dollars relatively) related with decline in crude oil price in the world market for more than two times, its stability level decreased from high stability (S3) in 2014 to medium stability (S2) in 2015.

Unemployment rate (UNE) was almost 5% and remained in high stability level (S3) during 2010–2016. State programs directed to ensuring social-economic development of regions, creating new workplaces, developing non-oil sector and etc. have a certain role in maintaining high stability level observed in unemployment rate.

The stability level of foreign investment growth rate (FDI) remained stable (medium stability level-S2) during 2010–2016 years. It is associated with high level and dynamic growth of foreign direct investments. It was 3.5 blln. dollars in 2010, 4.4 blln. dollars in 2011, 5.3 blln. dollars in 2012, 6.3 blln. dollars in 2013, 7.5 blln. dollars in 2014, 7.5 blln. dollars in 2015 and 7.4 blln. dollars in 2016.

# 3   Conclusion

Proposed approach to the analysis of macroeconomic stability gives us a possibility to define weak and strong sides of macroeconomic process in the country. It enables optimal control over macroeconomic processes. By using the result of investigation, in the future, we can forecast the direction of development of macroeconomic state of the country.

# References

1. The East Asian economic miracle: economic growth and public policy. World Bank (1993). http://documents.worldbank.org/curated/en/975081468244550798/Main-report
2. Afxentiou, P.C.: Convergence, the Maastricht criteria and their benefits. Brown J. World Aff. **1**(7), 245–254 (2000)
3. State Statistical Committee of Azerbaijan Republic. https://www.stat.gov.az/
4. Atanassov, K.: Intuitionistic fuzzy sets. Fuzzy Set Syst. **1**(20), 87–96 (1986)
5. Wang, J.Q., Li, H.B.: Multi-criteria decision-making method based on aggregation operators for intuitionistic linguistic fuzzy numbers. Control Decis. **10**(25), 1571–1574 (2010)
6. Alert Mechanism Report 2018. To the European Parliament, the Council, the European Central bank and the European Economic and Social Committee European Commission, p. 48. Brussels (2017)
7. Atanassov, K.: Intuitionistic Fuzzy Sets. Springer Physica-Verlag, Heidelberg (1999)
8. Boran, F.E., Genc, S., Kurt, M., Akay, D.: A multi-criteria intuitionistic fuzzy group decision making for supplier with TOPSIS method. Expert Syst. Appl. **36**(8), 11363–11368 (2009)

# Wind Speed Prediction of Four Regions in Northern Cyprus Prediction Using ARIMA and Artificial Neural Networks Models: A Comparison Study

Youssef Kassem[1,2]($\boxtimes$), Hüseyin Gökçekuş[1], and Hüseyin Çamur[2]

[1] Faculty of Civil and Environmental Engineering, Near East University, 99138 Nicosia, North Cyprus, Turkey
{yousseuf.kassem, huseyin.gokcekus}@neu.edu.tr
[2] Faculty of Engineering, Mechanical Engineering Department, Near East University, 99138 Nicosia, North Cyprus, Turkey
huseyin.camur@neu.edu.tr

**Abstract.** Wind speed data is one of the most critical factors affecting the operation of wind power farm systems. This paper examines the forecasting performance of Auto-Regressive Integrated Moving Average (ARIMA) and Artificial Neural Networks (ANN) models for predicting wind speeds in four regions in Northern Cyprus: Lefkoşa, Girne, Salamis, and Boğaz. For the application of the methodology, the meteorological measurements including wind speed, air temperature, humidity, sunshine duration, global solar radiation and rainfall values, from 1 January 2013 to 31 December 2016, were used. The obtained results demonstrated that the ANN model realizes the best accuracy for the prediction of the wind speeds with the highest R-squared value.

**Keywords:** ARIMA · ANN · Northern cyprus · Wind speed

## 1 Introduction

Recently, wind power has been growing at an unprecedented rate globally. The increasing importance of wind power integration into the power system has encouraged researchers to develop more reliable techniques to forecast wind power. Wind speed prediction methods using various models have been studied by several researchers [1–6]. In general, prediction methods can be divided into two categories including statistical methods such as Autoregressive Integrated Moving Average (ARIMA) and soft computing techniques like Artificial Neural Networks (ANN).

The Box-Jenkins model/ARIMA model is method that is widely used in analysis and forecasting [7], particularly for time series data like wind speed data.

Artificial neural networks (ANNs) as a soft computing technique are the most accurate and widely used as forecasting models in many areas including science and engineering [8]. Furthermore, ANNs are frequently used, as a network can efficiently approximate a continuous function to the desired level of accuracy. ANNs have been

R. A. Aliev et al. (Eds.): ICAFS-2018, AISC 896, pp. 230–238, 2019.
https://doi.org/10.1007/978-3-030-04164-9_32

found to be very efficient in solving nonlinear problems including those in the real world [9].

The objective of this paper is to predict wind speeds using Auto Regressive Integrated Moving Average (ARIMA) and Artificial Neural Networks (ANN) models. The proposed models were applied to a case study in four regions in Northern Cyprus: Lefkoşa, Girne, Salamis and Boğaz, and results are according to reality. In this paper, the performance of the ANN and ARIMA models is studied and compared.

## 2  Measurement Data at the Selected Sites

Cyprus is situated at a latitude of 35° North and a longitude of 33° East, surrounded by the Eastern Mediterranean Sea. The surface winds over Cyprus are controlled by local surface effects, such as the temperature contrast between the land and open sea (land and sea breezes), the differential heating of the land (anabatic and catabatic winds) and the constraints imposed by topography. The data used in this present study consists of wind speed (W), air temperature (AT), humidity (H), sunshine duration (SD), global solar radiation (GSR) and rainfall (RF) values recorded at four stations that are located in different parts of the island. Data (from 1 January 2013 to 31 December 2016) were collected from the Meteorological service in Cyprus. The coordinates, the period of the records of each region and geographic information of the selected areas are presented in Table 1.

**Table 1.** Details of regions used in the analysis

| Region name | Coordinates | | Period records | Characteristics of the region |
|---|---|---|---|---|
| | Latitude [°N] | Longitude [°E] | | |
| Lefkoşa | 35.19253 | 33.35984 | 4 years | Urban |
| Girne | 35.32371 | 33.31494 | 4 years | Coastal |
| Salamis | 35.13117 | 33.92595 | 4 years | Coastal |
| Boğaz | 35.31677 | 33.95410 | 4 years | Coastal |

The monthly wind speed patterns of four regions in Northern Cyprus are shown in Fig. 1. It is observed that the mean monthly wind speed values for all areas are ranging between 1.5 and 5 m/s. In addition, it is noticed that Boğaz has the maximum value of mean monthly wind speeds. The monthly average measurement data in the selected areas, including wind speed (W), air temperature (AT), humidity (H), sunshine duration (SD), global solar radiation (GSR) and rainfall (RF) values during the period from 2013 to 2016 are tabulated in Table 2. It can be seen that Boğaz had the maximum values for mean monthly wind speed and global solar radiation. Additionally, it can be observed that sunshine duration level at all regions peaked during July at approximately 12 day/h. During the period (2013–2016), the average monthly humidity values of the four selected areas varied between 43 and 80% and Salamis had the highest humidity values compared to the other regions. The maximum value of the mean monthly air temperature was recorded in July for the years 2013–2015 in Girne and in July 2016 for Lefkoşa. Moreover, Boğaz and Lefkoşa have the highest mean monthly rainfall values compared to Girne and Salamis.

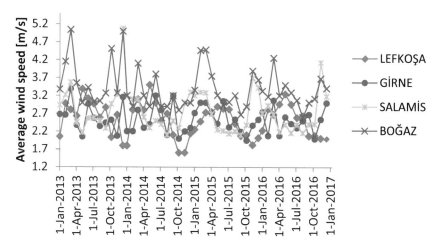

**Fig. 1.** Average wind speed of the four selected regions (2013–2016).

## 3    Auto-Regressive Integrated Moving Average (ARIMA)

Stochastic modeling and forecasting of a time series require adequate knowledge about mathematical techniques for identifying patterns in time series data and for expressing the physical process regarding the mathematical models. However, the physical process is very complex, the patterns of time series data are unclear, and individual data points involve significant errors. Box and Jenkins (1976) developed an Auto-Regressive Integrated Moving Average (ARIMA) model and successfully demonstrated their applications in forecasting of the physical process. The ARIMA modeling is an inherently powerful technique and contains excellent flexibility. Figure 2 illustrates the proposed algorithm for the ARIMA model. The representation of this model is done through ARIMA notation of order (p, d, q), where: p is the number of seasonal auto-regressive terms; d is the number of seasonal differences (rarely should d > 2 be needed), and q is the number of seasonal media moving. The wind speed values are forecasted by using ARIMA to investigate multiple local meteorological data effects. The wind speed values are not only predicted by historical wind speed values but also air temperature, humidity, sunshine duration, global solar radiation and rainfall values, as shown in Fig. 2. ARIMA models are examined with different parameters to determine the model that will give the best forecast, as indicated in Table 3. The actual wind speed and predicted values during 2016 are presented in Table 4. It is observed that the

**Table 2.** Mean monthly measurement data, W (m/s), GSR (Cal/cm$^2$.day), SD (day/h), AT (°C), H (%) and RF (mm), of four regions in Northern Cyprus (2013–2016).

| M | Lefkoşa | | | | | | Girne | | | | | |
|---|---|---|---|---|---|---|---|---|---|---|---|---|
| | W | GSR | SD | AT | H | RF | W | GSR | SD | AT | H | RF |
| J | 2.1 | 233.1 | 5.3 | 9.9 | 74.1 | 36.9 | 2.6 | 219.4 | 4.9 | 13.4 | 64.4 | 75.8 |
| F | 2.8 | 328.3 | 6.7 | 11.4 | 72.9 | 21.1 | 2.5 | 311.0 | 5.9 | 14.4 | 66.0 | 67.3 |
| M | 2.8 | 434.8 | 7.8 | 13.9 | 62.7 | 33.9 | 3.1 | 419.8 | 7.3 | 16.0 | 61.7 | 20.4 |
| A | 2.8 | 547.1 | 9.2 | 17.8 | 55.9 | 26.1 | 2.4 | 551.9 | 9.3 | 18.4 | 62.4 | 18.5 |
| M | 3.3 | 592.0 | 9.7 | 21.9 | 53.5 | 49.7 | 2.4 | 594.6 | 9.7 | 21.9 | 64.5 | 27.7 |
| J | 3.0 | 674.3 | 11.2 | 26.2 | 48.5 | 0.0 | 2.9 | 680.2 | 11.4 | 25.7 | 60.1 | 5.3 |
| J | 2.7 | 676.8 | 11.5 | 29.0 | 47.8 | 0.1 | 2.5 | 673.1 | 11.6 | 28.7 | 58.7 | 0.0 |
| A | 2.6 | 609.3 | 10.7 | 29.4 | 49.3 | 0.2 | 2.4 | 602.9 | 10.8 | 29.5 | 58.7 | 0.0 |
| S | 2.3 | 497.5 | 9.1 | 26.0 | 53.3 | 9.8 | 2.6 | 498.1 | 9.5 | 27.1 | 56.9 | 2.1 |
| O | 2.0 | 389.0 | 8.0 | 21.3 | 49.7 | 23.9 | 2.1 | 358.1 | 7.3 | 23.2 | 56.2 | 21.5 |
| N | 2.0 | 280.1 | 6.5 | 15.9 | 59.7 | 16.6 | 2.3 | 270.7 | 6.4 | 19.1 | 57.6 | 35.3 |
| D | 2.0 | 211.5 | 5.1 | 11.1 | 65.3 | 55.7 | 2.8 | 208.2 | 5.1 | 14.5 | 61.9 | 142.1 |
| M | Salamis | | | | | | Boğaz | | | | | |
| | W | GSR | SD | AT | H | RF | W | GSR | SD | AT | H | RF |
| J | 3.0 | 225.7 | 5.0 | 11.4 | 75.1 | 38.3 | 3.3 | 229.0 | 5.0 | 10.2 | 74.3 | 79.2 |
| F | 3.1 | 309.4 | 6.3 | 12.6 | 74.9 | 21.3 | 3.7 | 325.3 | 6.1 | 11.6 | 72.3 | 75.3 |
| M | 3.2 | 424.7 | 7.2 | 14.5 | 68.1 | 24.0 | 4.5 | 419.4 | 7.6 | 13.3 | 65.1 | 47.2 |
| A | 2.5 | 549.8 | 9.0 | 17.3 | 68.0 | 18.2 | 3.4 | 539.1 | 9.4 | 16.8 | 58.8 | 19.5 |
| M | 2.3 | 592.5 | 9.3 | 21.2 | 68.4 | 33.0 | 3.2 | 589.0 | 9.7 | 20.4 | 60.2 | 42.3 |
| J | 2.4 | 679.7 | 10.8 | 25.3 | 61.8 | 0.6 | 3.4 | 669.9 | 11.1 | 24.4 | 54.1 | 1.1 |
| J | 2.4 | 675.7 | 11.2 | 28.3 | 62.5 | 0.0 | 3.0 | 658.0 | 11.5 | 27.0 | 55.4 | 0.0 |
| A | 2.2 | 607.6 | 10.5 | 28.8 | 63.0 | 0.0 | 2.9 | 592.6 | 10.7 | 27.5 | 57.6 | 0.0 |
| S | 2.3 | 506.3 | 8.4 | 26.1 | 61.6 | 16.7 | 3.0 | 471.3 | 8.8 | 24.5 | 59.6 | 3.4 |
| O | 2.5 | 387.6 | 7.7 | 21.3 | 64.3 | 36.7 | 3.3 | 387.5 | 7.7 | 20.6 | 59.0 | 19.0 |
| N | 3.5 | 268.6 | 6.1 | 17.4 | 66.3 | 18.3 | 3.5 | 274.5 | 6.4 | 16.2 | 63.7 | 48.7 |
| D | 3.8 | 204.4 | 4.9 | 12.7 | 71.7 | 69.6 | 3.8 | 202.7 | 5.2 | 11.3 | 70.4 | 168.1 |

forecast error (Eq. (1)) is quite low and impressive as the predicted values are close to the actual values.

$$Forecast\ error = \frac{Actual - Predicted}{Actual} \tag{1}$$

# 4 Artificial Neural Network

The ANN model has been developed using Matlab 2015a for predicting the wind speed data for the selected regions and involves the following steps

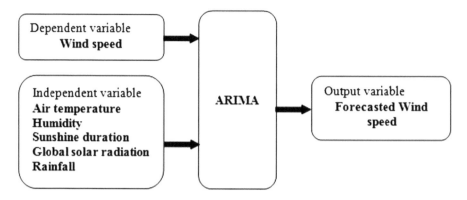

**Fig. 2.** Block diagram of the ARIMA model.

**Table 3.** ARIMA model for four regions.

| Forecasting | Magnitude | Regions | ARIMA model (p, q, d) |
|---|---|---|---|
| Long-term | Month-year | Lefkoşa | (10, 0, 5) (1, 0, 1) |
| | | Girne | (9, 0, 6) (1, 0, 1) |
| | | Salamis | (11, 0, 5) (1, 0, 1) |
| | | Boğaz | (9, 0, 4) (1, 0, 1) |

i. ***Creating the network topology.*** This consists of the selection of the number of inputs (in this case 5 inputs, GSR, SD, AT, H and RF), the number of hidden layers, the number of hidden neurons in the hidden layer, and the number of outputs (one, in this case, W). The inputs of the ANN model for wind speed prediction are air temperature (AT), humidity (H), sunshine duration (SD), global solar radiation (GSR) and rainfall (RF) values, as presented in Fig. 3. The database was randomly divided into three groups with 60% to training, 20% to testing and 20% to validation. Since the input and output variables on the artificial neural network have different magnitudes, normalization is required.

ii. ***Training the network.*** In this study, a feed-forward back-propagation algorithm was selected as the network training algorithm, TRAINGDM, LEARNGDM, and TANSIG were chosen (by a trial and error) and selected as training function adaptation learning function and transfer function, respectively. Mean squared error (MSE) was used for evaluating the performance of the function.

By a process of trial and error, the best model, i.e., 5:5:1, (see Fig. 3) which has ten neurons as the optimum, was chosen due to the *R-squared* values of both the training and testing sets. Hence, the actual wind speed and predicted values during 2016 are presented in Table 5. It can be seen that the ANN error values showed that the actual data for wind speed has an excellent correlation with the models for all regions.

**Table 4.** A sample of empirical results of ARIMA model of wind speed during the period 2016.

| Month | Lefkoşa | | | Girne | | |
|---|---|---|---|---|---|---|
| | Actual | Predicted | Forecast error | Actual | Predicted | Forecast error |
| Jan-16 | 2.23 | 2.28 | 0.024 | 2.7 | 2.64 | −0.039 |
| Feb-16 | 2.83 | 2.64 | −0.066 | 2.1 | 2.18 | 0.048 |
| Mar-16 | 2.68 | 2.58 | −0.035 | 3.1 | 3.17 | 0.032 |
| Apr-16 | 2.97 | 3.16 | 0.065 | 2.1 | 2.16 | 0.038 |
| May-16 | 3.25 | 3.34 | 0.029 | 2.6 | 2.64 | 0.013 |
| Jun-16 | 2.96 | 3.03 | 0.022 | 2.4 | 2.72 | 0.124 |
| Jul-16 | 2.59 | 2.68 | 0.035 | 2.3 | 2.4 | 0.042 |
| Aug-16 | 2.64 | 2.75 | 0.041 | 2.5 | 2.53 | 0.013 |
| Sep-16 | 2.07 | 2.34 | 0.132 | 2.7 | 2.78 | 0.039 |
| Oct-16 | 2.09 | 1.93 | −0.075 | 2.0 | 2.08 | 0.043 |
| Nov-16 | 2.01 | 2.13 | 0.064 | 2.5 | 2.35 | −0.075 |
| Dec-16 | 2.00 | 2.06 | 0.031 | 3.0 | 3.01 | 0.003 |
| Month | Salamis | | | Boğaz | | |
| | Actual | Predicted | Forecast error | Actual | Predicted | Forecast error |
| Jan-16 | 2.7 | 2.99 | 0.105 | 3.3 | 3.37 | 0.011 |
| Feb-16 | 2.9 | 2.68 | −0.085 | 3.2 | 3.45 | 0.086 |
| Mar-16 | 2.7 | 2.88 | 0.074 | 4.3 | 4.12 | −0.032 |
| Apr-16 | 2.2 | 2.32 | 0.058 | 3.2 | 3.53 | 0.102 |
| May-16 | 2.4 | 2.44 | 0.019 | 3.5 | 3.59 | 0.028 |
| Jun-16 | 2.2 | 2.34 | 0.083 | 3.3 | 3.41 | 0.036 |
| Jul-16 | 2.4 | 2.37 | −0.010 | 3.1 | 3.02 | −0.019 |
| Aug-16 | 2.2 | 2.33 | 0.080 | 2.6 | 2.79 | 0.070 |
| Sep-16 | 2.4 | 2.24 | −0.067 | 3.0 | 2.99 | −0.007 |
| Oct-16 | 2.4 | 2.5 | 0.040 | 3.1 | 2.98 | −0.044 |
| Nov-16 | 4.1 | 3.84 | −0.069 | 3.7 | 3.57 | −0.030 |
| Dec-16 | 3.2 | 3.25 | 0.021 | 3.4 | 3.41 | 0.001 |

## 5 Comparison Between ARIMA and ANN Models

The accuracy of the models obtained from ARIMA and ANN were examined by evaluating the values of *R-squared*. The results (Table 6) showed that the two optimization tools gave good predictions due to the values of *R-squared*. However, ANN showed a clear lead over ARIMA because the *R-squared* value was higher. Additionally, data fitting of the models was tested, and ANN demonstrated a better fit than ARIMA (see Tables 4 and 5). ANN was better than ARIMA in terms of the modeling for predicting the wind speed.

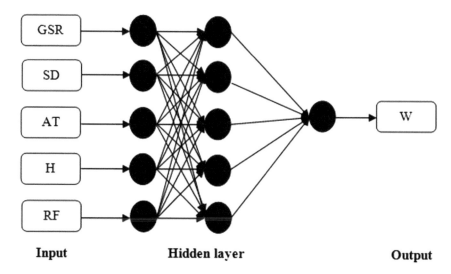

**Fig. 3.** ANN architecture for the wind speed prediction.

**Table 5.** A sample of empirical results of ANN model of wind speed during the period 2016.

| Month | Lefkoşa | | | Girne | | |
|---|---|---|---|---|---|---|
| | Actual | Predicted | ANN error | Actual | Predicted | ANN error |
| 16-Jan | 2.23 | 2.2 | −0.012 | 2.75 | 2.73 | −0.007 |
| 16-Feb | 2.83 | 2.85 | 0.009 | 2.08 | 2.11 | 0.014 |
| 16-Mar | 2.68 | 2.72 | 0.016 | 3.07 | 3.28 | 0.067 |
| 16-Apr | 2.97 | 3.02 | 0.018 | 2.08 | 2.23 | 0.073 |
| 16-May | 3.25 | 3.32 | 0.023 | 2.61 | 2.58 | −0.01 |
| 16-Jun | 2.96 | 3.14 | 0.06 | 2.42 | 2.48 | 0.025 |
| 16-Jul | 2.59 | 2.64 | 0.019 | 2.3 | 2.35 | 0.02 |
| 16-Aug | 2.64 | 2.72 | 0.03 | 2.5 | 2.62 | 0.048 |
| 16-Sep | 2.07 | 2.1 | 0.014 | 2.68 | 2.77 | 0.035 |
| 16-Oct | 2.09 | 2.08 | −0.001 | 1.99 | 2.15 | 0.079 |
| 16-Nov | 2.01 | 2.15 | 0.072 | 2.54 | 2.43 | −0.043 |
| 16-Dec | 2 | 2.23 | 0.114 | 3 | 3.11 | 0.037 |
| Month | Salamis | | | Boğaz | | |
| | Actual | Predicted | ANN error | Actual | Predicted | ANN error |
| Jan-16 | 2.71 | 2.84 | 0.049 | 3.33 | 3.29 | −0.013 |
| Feb-16 | 2.93 | 2.84 | −0.030 | 3.18 | 3.35 | 0.055 |
| Mar-16 | 2.68 | 2.95 | 0.100 | 4.25 | 4.02 | −0.055 |
| Apr-16 | 2.19 | 2.46 | 0.121 | 3.20 | 3.35 | 0.046 |
| May-16 | 2.39 | 2.59 | 0.081 | 3.49 | 3.54 | 0.013 |
| Jun-16 | 2.16 | 2.31 | 0.069 | 3.29 | 3.31 | 0.006 |
| Jul-16 | 2.39 | 2.35 | −0.018 | 3.08 | 3.17 | 0.030 |

*(continued)*

**Table 5.** (*continued*)

| Month | Salamis | | | Boğaz | | |
|---|---|---|---|---|---|---|
| | Actual | Predicted | ANN error | Actual | Predicted | ANN error |
| Aug-16 | 2.16 | 2.26 | 0.047 | 2.61 | 2.56 | −0.018 |
| Sep-16 | 2.40 | 2.37 | −0.011 | 3.01 | 3.14 | 0.043 |
| Oct-16 | 2.40 | 2.55 | 0.061 | 3.12 | 3.13 | 0.004 |
| Nov-16 | 4.13 | 4.07 | −0.014 | 3.68 | 3.75 | 0.019 |
| Dec-16 | 3.18 | 3.24 | 0.018 | 3.41 | 3.45 | 0.013 |

**Table 6.** R-squared generated by ARIMA and ANN models.

| Regions | ARIMA | ANN |
|---|---|---|
| Lefkoşa | 0.908 | 0.970 |
| Girne | 0.882 | 0.938 |
| Salamis | 0.904 | 0.956 |
| Boğaz | 0.852 | 0.932 |

## 6   Conclusions

Wind speed prediction is significant for wind farms studies in all locations. The performance of ANN and ARIMA models were compared by using a determination coefficient ($R^2$). According to the results, it is concluded that both models can provide good forecasts when applied to real-life problems and thus can be effectively engaged for wind speed prediction. Moreover, based on the results, the highest R-squared values are obtained from ANN models.

**Acknowledgments.** The authors would like to thank the Faculty of Civil and Environmental Engineering especially the Civil Engineering Department.

## References

1. Patlakas, P., Drakaki, E., Galanis, G., Spyrou, C., Kallos, G.: Wind gust estimation by combining a numerical weather prediction model and statistical post-processing. Energy Procedia **125**, 190–198 (2017)
2. Kusiak, A., Zheng, H., Song, Z.: Wind farm power prediction: a data-mining approach. Wind Energy **12**(3), 275–293 (2009)
3. Alencar, D.B., Affonso, C.D., Oliveira, R.L., Rodríguez, J.M., Leite, J., Filho, J.R.: Different models for forecasting wind power generation: case study. Energies **10**(12), 1976 (2017)
4. Kulkarni, M.A., Patil, S., Rama, G.V., Sen, P.N.: Wind speed prediction using statistical regression and neural network. J. Earth Syst. Sci. **117**(4), 457–463 (2008)
5. Deligiorgi, D., Philippopoulos, K., Kouroupetroglou, G.: Artificial neural network based methodologies for the estimation of wind speed. In: Assessment and Simulation Tools for Sustainable Energy Systems Green Energy and Technology, pp. 247–266 (2013)

6. Aquino, R.R., Gouveia, H.T., Lira, M.M., Ferreira, A.A., Neto, O.N., Carvalho, M.A.: Models based on neural networks and neuro-fuzzy systems for wind power prediction using wavelet transform as data preprocessing method. In: Engineering Applications of Neural Networks Communications in Computer and Information Science, pp. 272–281 (2012)
7. Stankovic, J.: Tools and Techniques for Economic Decision Analysis. IGI Global, Business Science Reference, Hershey (2017)
8. Litta, A.J., Idicula, S.M., Mohanty, U.C.: Artificial neural network model in prediction of meteorological parameters during premonsoon thunderstorms. Int. J. Atmos. Sci. **2013**, 1–14 (2013)
9. Khashei, M., Bijari, M.: An artificial neural network (p, d, q) model for time-series forecasting. Expert Syst. Appl. **37**(1), 479–489 (2010)

# Head Movement Mouse Control Using Convolutional Neural Network for People with Disabilities

Murat Arslan[✉], Idoko John Bush, and Rahib H. Abiyev

Applied Artificial Intelligence Research Centre,
Department of Computer Engineering, Near East University,
Mersin-10, North Cyprus, Turkey
murat.arslan@neu.edu.tr

**Abstract.** In this paper, computer mouse control with head movements and eye blinks was proposed for people with impaired spinal cord injury. The head mouse control is based on finding and predicting eye states and direction of the head. This human computer interface (HCI) is an assistant system for people with physical disabilities who are suffering from motor neuron diseases or severe cerebral palsy. By moving the head to the right, left, up and down moves the mouse pointer and sends mouse button commands using eye blinks. Here, left eye-blink triggers left mouse button, right eye-blink triggers right mouse button and double-eyed sends "holds" command. In this system, eye blink and head movement used same Convolutional Neural Network (CNN) architecture with different number of classes (output). In head movement part, CNN has 5 outputs, (forward, up, down, left and right), in eye-blink part CNN has 2 output either opened or closed. This combined system allows people with down to neck paralyzed, to control computer using head movement and eye blinking. The test results reveal that this system is robust and accurate. This invention allows people with disabilities to use computer with head movements and eye blinks without using any extra hardware.

**Keywords:** Computer mouse · People with disabilities · Convolutional Neural Network · Computer vision · Deep learning · Haar cascade

## 1 Introduction

The necessity of computer systems has made them inevitable items in human lives. Unfortunately, disabled people that impedes movement of the upper limbs finds it impossible to correctly utilize the standard interface of a computer component such as the mouse. Computer mouse has become even more important considering Microsoft Windows interfaces (Windows 98 and Windows NT) popularity. Hence, to aid disabled people operate their computers, it is of high importance to invent a simple mouse system. Paralyzed people as a result of spinal cord injuries (SCIs) have increasingly applied electronic assistive devices to improve their ability of performing certain essential functions. Powered wheelchairs, communication and daily activity devices are some of the electronic equipments modified to benefit people with disabilities.

© Springer Nature Switzerland AG 2019
R. A. Aliev et al. (Eds.): ICAFS-2018, AISC 896, pp. 239–248, 2019.
https://doi.org/10.1007/978-3-030-04164-9_33

These wide rang interfaces can be an enlarged mouse or a complex system that aid the impede people to control a movement with the help of infrared or ultrasound-controlled mouse system, electroencephalogram (EEG) signals and eye imaged input system [1–3]. The head mouse system is designed using computers and control objects basically for people with disabilities. This system predicts the position of the head and converts it into mouse position. The system also detects and analyzes the eye states and uses those information to control mouse buttons. This module needs a video camera for capturing images and also high-speed computer. The aim of this module is basically to track the head position and eye states. The eye state modules analyzes the video images using process techniques and then detects the eyes and classify them as either "opened" or "closed" state in the video frames. The performances of these modules are very important for detecting profile region, predicting the head position, eyes and their states. This procedures controls head movements and eye states in a short time.

This paper proposes head pose estimation, eye detection and classification. To initialize the profile region and eye regions, we used dlib library [4]. Here, we extracted profile region and eye regions. After this 3 layered same structured except output layers, Convolutional Neural Networks (CNN) was applied to classify whether eyes are opened or closed and head directions. CNN is deep learning structure inspired by the natural perception mechanism of the living creatures. A number of research papers are already published on deep neural structures of which CNN is one of the structures. CNN is a special variation of multilayer perceptrons (MLPs) which have one or more convolution and max-pooling layers. Its architecture was proposed by LeCun et al. in [5], comprising seven-level convolutional network called "LeNet-5" that classifies hand-written numbers in $32 \times 32$ pixel images. The network was trained by the back propagation algorithm. The system could recognize the images directly from the image pixels [5, 6]. In [7], CNN architecture that shows improvements in image recognition is presented. The reference [7] designed deep structure called AlexNet which was similar to LeNet with a deeper structure. Many researches have been performed to improve the performances of the systems and to overcome the difficulties related to training. These include; ZFNet [8], VGGNet [9], GoogleNet [10] and ResNet [11]. These works are extended deeper to better approximate complex nonlinear function by applying non-linear activation function and to have better feature representations. But increasing the number of layers leads the increase of complexity of the network and also difficulties of parameter learning. To this end, different methodologies are developed to deal with these problems. Several CNN structures have been developed to solve classification, pattern recognition and image processing problems. It was applied in [12] for face recognition, in [13] for handwritten character classification, in [14] for visual document analysis, in [15] for face sketch synthesis, in [16] for micro aneurysm detection, in [17] for fingerprint enhancement, in [18] for segmentation of glioma tumors in brain, in [19] for handwritten recognition, in [20] for granite tiles classification and in [21] for detection of chest diseases.

In this paper, the design of CNN, head pose estimation for vision-based mouse control for people with disabilities were considered. CNN can constrain the architecture in more intelligent way and require minimal preprocessing steps compared to other image classification methods [22]. The design of mouse control with head movements system from low-quality images captured from a monocular camera mounted on computer based on head pose estimation using CNN is presented. The real-time head tracking and eye state system that uses video camera have some challenging threads. While controlling mouse, light conditions can be changed, people may have different facial shapes and also eyes may be in different shape and size according to their origin. In this paper, we designed a novel algorithm that improves the performance of detecting head position and eye states by integrating some image processing techniques and CNNs.

The remaining part of the paper is organized as follows: Sect. 2 presents CNNs used for eye and head pose classification where the architecture of CNN is demonstrated. Section 3 describes the simulation process and the results obtained from the simulations. Section 4 presents the conclusions of the paper.

## 2    Learning Algorithms

### 2.1    Haar Cascade

Haar cascade classifiers designed for object detection proposed by Viola and Jones [23] is an effective object detection method in machine learning. Cascade function is trained from positive and negative images and then used to detect objects in other images. The features searched for by the sensing framework are the sum of the pixels in the rectangles. These algorithms are more complex than they can withstand multiple rectangle fields. The value of any feature of the inputs is the sum of the pixels in the clear rectangles and the sum of the pixels extracted from the pointed rectangles as depicted in Fig. 1.

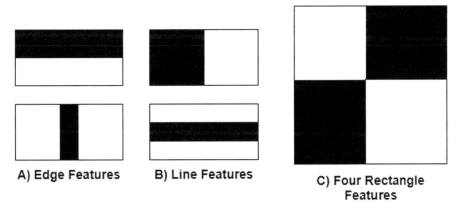

**Fig. 1.** Haar features

All human faces have similar properties, and those properties can be used as Haar features. For example, eye regions are darker than the upper-cheeks. To define the face, location and size of eyes, mouth, nose etc. and value of oriented gradients of pixels intensities are needed. Here, we calculate the value thus:

$$Value = \Sigma(black\,area\,pixels) - \Sigma(white\,area\,pixels) \qquad (1)$$

The integral image evaluates rectangular features. Each rectangular feature is adjacent to at least one other rectangle. For primitive image analyses, a standard $24 \times 24$ pixel image has more than 160000 features and it needs too much time to analyze. Because of this reason, selecting useful features from image is important. For this purpose, a variant of Adaboost is used. For each feature, algorithm finds the suitable threshold to classify the faces. Each image has equal weights at the beginning. Each step weights and errors is calculated and updated until suitable accuracy and error rate is reached.

## 2.2 Convolutional Neural Network

Convolutional Neural Networks was first proposed by LeCun in [5]. CNN is multilayer neural network structure that includes convolution layers, pooling layers, full-connected or feedforward layers. Convolution layers are basic building blocks of the CNN. Figure 2 depicts the simple structure of CNN. The input layer of CNN is convolutional layer. The input for the convolutional layer is the image and the output is the feature map which is an input for the next convolutional layer. The input and output of next convolutional layers are feature maps of input space. Depending on the considered problem, the programmer can select the number of convolutional layers. After processing input space, the convolutional layers produces feature maps of input image. These feature maps are input for the pooling layer. In pooling layer, activation function is applied for transformation of feature map. For this purpose, we applied ReLU function to retain relevant features. The determined features are rearranged as one-dimensional array and is called feature vector. This feature vector is input for fully connected array and is used for classification purpose.

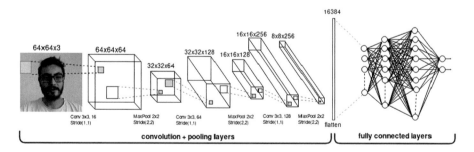

**Fig. 2.** The structure of CNN

After determining output signals of the CNN, the training of the network begins. It includes updating unknown parameters of the network. Training of CNN is performed by minimizing the loss function. As a result of training, the accurate values of the parameters are determined. Adam optimizer (adaptive moment estimation) is used for updating and finding the network parameters [24].

# 3  Simulation of the System

Many intelligent systems have been designed for solving different engineering problems [27–37]. In this section using CNN and Haar Cascades the design of vision-based head and eye tracking and prediction system is implemented. The input of the systems is the images of the head profile obtained from video camera located in front of the user. Haar cascade classifier finds and extracts user's head-shoulder profile and eyes from the whole image. Based on the CNN's output, the control of the mouse pointer and buttons is performed.

CNN models have four parameters; width, height, depth and number of classes. The width and height are the sizes of the profile picture and eye regions from the input image. The depth is the number of channels of our input images and 3-dimensional depth value for standard RGB is used. Human profile region input is $64 \times 64 \times 3$ and eye profile input is $24 \times 24 \times 3$. Here, width of image is 64 or 24, the height of image is 64 or 24 and width is 3 respective channels; RGB.

The CNN models are constructed sequentially; CONV1 => RELU => POOL1 => CONV2 => RELU => POOL2 => etc. as demonstrated in Fig. 2. The first convolutional layer learns 64 convolutional filters with each input having $5 \times 5$ dimensions. Then rectified linear unit (ReLU) activation function has output 0 if the input is less than 0 and output 1 otherwise. Furthermore, $2 \times 2$ max pooling in both x and y direction is applied. In the second and third convolutional layer, the same structures are used but this time, 128 and 256 convolutional filters are used and in the third layer which has 256 convolutional filters, input is downgraded in $3 \times 3$.

As explained, to prevent overfitting, ReLU activation layers are used in between convolutional layers. After convolutional layers, flatten layer is applied to get output from last convolutional layer and flatten into a single vector. In the first model, after flatten layer, 1000 dense fully connected layer and 5 output dense layer (for the 5 classes; forward, up, down, left and right) are applied for classification purpose. In the second structure, same dense layers are applied but output is 2 since classification is binary; either opened or closed.

The CNN based classifier system is used for "head mouse control system" for people with disabilities. Tensorflow [25] framework is used with these parameters shown in Fig. 3. In the system, 50 training epochs, 1e−3 learning rate and 64 batch size are defined. For the head pose prediction, our handcrafted dataset were used. The dataset has 853 images from 7 persons in our university, the Near East University Cyprus. These 7 persons are of different nationalities, gender and age with 5 classes; up, down, left, right and forward as depicted in Fig. 4. The datasets is divided into 90% for training and 10% for testing. This also helps to reduce overfitting and boosting the test performance. For the eye classification, CEW Dataset [26] having 2385 closed and 2463 opened in $24 \times 24$ pixels of eye images is used.

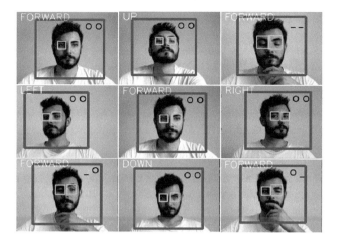

**Fig. 3.** System simulation results

**Fig. 4.** Head pose dataset from 7 people and 5 classes

In the data augmentation stage, some generator techniques which include crops, shifts, random rotations etc. for getting more reliable results from small datasets were applied. Training and validation accuracy and loss are shown in Figs. 5 and 6.

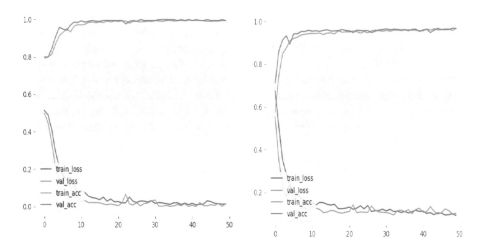

**Fig. 5.** Head CNN training and validation results  **Fig. 6.** Eye CNN training and validation results

**Table 1.** Performances of the CNN models

| Model | Training | | | Validation | | | Testing | | | | | | |
|---|---|---|---|---|---|---|---|---|---|---|---|---|---|
| | RMSE | AUC | ACC | RMSE | AUC | ACC | TP | FP | TN | FN | SPE | SENS | ACC |
| Eye | 0.058 | 0.992 | 0.960 | 0.059 | 0.989 | 0.952 | 2345 | 85 | 2377 | 40 | 0.965 | 0.983 | 0.974 |
| Head | 0.019 | 0.999 | 0.977 | 0.009 | 1.000 | 0.988 | – | – | – | – | – | – | 0.991 |

Both of the CNN models are trained at 50 epochs. The performances of the system are depicted for train, evaluation and testing stages. During simulation, the values of RMSE, ACC, AUC, TP, FP, TN, FN, SPE and SENS are calculated. Table 1 shows the results obtained from each of the models for training, evaluation and testing stages.

## 4 Conclusion

The outcome of this study is an implementation of a vision-based computer mouse controlled by head movements and eye blinks which aid people with impaired spinal cord injury to operate the computer. Originally, sensor based solutions is used for head mouse system but they are expensive, uncomfortable and serves limited data for future computing. To overcome these challenges and also let people with disabilities operate computer, we invented this system that could be used without wearing any hardware parts. Experimental results depicts that our implemented system is robust and effective. One of the advantages of CNN discovered during the design stage is low preprocessing steps. Our invention can automatically detect head direction and eye states from camera image and sends the detected signal to the computer to control the mouse without using any extra hardware. In conclusion, this invention aids people with disabilities to use computer mouse and we intend applying Z-number fuzzy system for further investigations.

# References

1. Vasa, J.J.: Electronic aids for the disabled and the elderly. Med. Instrument. **16**(5), 263–264 (1982)
2. Chen, Y.L., Tang, F.T., Chang, W.H., Wong, M.K., Shih, Y.Y., Kuo, T.S.: The new design of an infrared-controlled human computer interface for the disabled. IEEE Trans. Rehab. Eng. **7**, 474–481 (1999)
3. Computer access solutions for the motor impaired, 12 April 1999. http://www.orin.com/access/
4. King, D.E.: Dlib-ml: a machine learning toolkit. JMLR **10**, 1755–1758 (2009)
5. Lecun, Y., Bottou, L., Bengio, Y., Haffner, P.: Gradient-based learning applied to document recognition. Proc. IEEE **86**(11), 2278–2324 (1998). https://doi.org/10.1109/5.726791
6. Hecht-Nielsen, R.: Theory of the backpropagation neural network. Neural Netw. **1**, 445 (1988). https://doi.org/10.1016/0893-6080(88)90469-8
7. Russakovsky, O., Deng, J., Su, H., Krause, J., Satheesh, S., Ma, S., et al.: ImageNet large scale visual recognition challenge. Int. J. Comput. Vis. **115**(3), 211–252 (2015). https://doi.org/10.1007/s11263-015-0816-y
8. Zeiler, M.D., Fergus, R.: Visualizing and understanding convolutional networks. In: Fleet, D., Pajdla, T., Schiele, B., Tuytelaars, T. (eds.) Computer Vision – ECCV 2014, ECCV 2014. Lecture Notes in Computer Science, vol. 8689, pp. 818–833. Springer, Cham (2014). https://doi.org/10.1007/978-3-319-10590-1_53
9. Simonyan, K., Zisserman, A.: Very deep convolutional networks for large-scale image recognition. In: Proceedings of ICLR 2015, pp. 1–14 (2014)
10. Szegedy, C., Liu, W., Jia, Y., Sermanet, P., Reed, S., Anguelov, D., Erhan, D., Vanhoucke, V., Rabinovich, A.: Going deeper with convolutions. In: Proceedings of the IEEE Conference on Computer Vision and Pattern Recognition (CVPR), pp. 1–9 (2015). https://doi.org/10.1109/cvpr.2015.7298594
11. He, K., Zhang, X., Ren, S., Sun, J.: Deep residual learning for image recognition. In: Proceedings of the IEEE Conference on Computer Vision and Pattern Recognition (CVPR), pp. 770–778 (2016). https://doi.org/10.1109/cvpr.2016.90
12. Lawrence, S., Giles, C., Tsoi, A.C., Back, A.: Face recognition: a convolutional neural-network approach. IEEE Trans. Neural Netw. **8**(1), 98–113 (1997). https://doi.org/10.1109/72.554195
13. Ciresan, D., Meier, U., Gambardella, L., Schmidhuber, J.: Convolutional neural network committees for handwritten character classification. In: Proceedings of 2011 International Conference on Document Analysis and Recognition, pp. 3207–3220 (2011). https://doi.org/10.1109/icdar.2011.229

14. Simard, P., Steinkraus, D., Platt, J.: Best practices for convolutional neural networks applied to visual document analysis. In: Proceedings of Seventh International Conference on Document Analysis and Recognition, pp. 958–963. https://doi.org/10.1109/icdar.2003. 1227801

15. Jiao, L., Zhang, S., Li, L., Liu, F., Ma, W.: A modified convolutional neural network for face sketch synthesis. Pattern Recogn. **76**, 125–136 (2018). https://doi.org/10.1016/j.patcog. 2017.10.025

16. Chudzik, P., Majumdar, S., Calivá, F., Al-Diri, B., Hunter, A.: Microaneurysm detection using fully convolutional neural networks. Comput. Methods Programs Biomed. **158**, 185– 192 (2018). https://doi.org/10.1016/j.cmpb.2018.02.016

17. Li, J., Feng, J., Kuo, C.: Deep convolutional neural network for latent fingerprint enhancement. Sig. Process. Image Commun. **60**, 52–63 (2018). https://doi.org/10.1016/j. image.2017.08.010

18. Hussain, S., Anwar, S., Majid, M.: Segmentation of glioma tumors in brain using deep convolutional neural network. Neurocomputing **282**, 248–261 (2018). https://doi.org/10. 1016/j.neucom.2017.12.032

19. Baldominos, A., Saez, Y., Isasi, P.: Evolutionary convolutional neural networks: an application to handwriting recognition. Neurocomputing **283**, 38–52 (2018). https://doi.org/ 10.1016/j.neucom.2017.12.049

20. Ferreira, A., Giraldi, G.: Convolutional neural network approaches to granite tiles classification. Expert Syst. Appl. **84**, 1–11 (2017). https://doi.org/10.1016/j.eswa.2017.04. 053

21. Abiyev, R., Ma'aitah, M.K.S.: Deep convolutional neural networks for chest diseases detection. J. Healthc. Eng. **2018** (2018)

22. Wachinger, C., Reuter, M., Klein, T.: DeepNAT: deep convolutional neural network for segmenting neuroanatomy. Neuroimage **170**, 434–445 (2018). https://doi.org/10.1016/j. neuroimage.2017.02.035

23. Viola, P., Jones, M.: Robust real-time face detection. Int. J. Comput. Vis. **57**(2), 137–154 (2004)

24. Kingma, D.P., Ba, J.L.: ADAM: a method for stochastic optimization. In: ICLR 2015 (2015)

25. TensorFlow. https://www.tensorflow.org/. Accessed 19 June 2018

26. Song, F., Tan, X., Liu, X., Chen, S.: Eyes closeness detection from still images with multi-scale histograms of principal oriented gradients. Pattern Recogn. **47**(9), 2825–2838 (2014)

27. Abiyev, R.H., Altunkaya, K.: Neural network based biometric personal identification using fast iris segmentation. Int. J. Control Autom. Syst. **7**(1), 17–23 (2009)

28. Abiyev, R.H., Abizade, S.: Diagnosing Parkinson's diseases using fuzzy neural system. Comput. Math. Methods Med. **2016** (2016). Article ID 1267919

29. Abiyev, R.H., Kilic, K.: Adaptive Iris Segmentation. Lecture Notes in Computer Sciences. Springer/CS Press, Heidelberg (2009)

30. Abiyev, R., Altunkaya, K.: Personal iris recognition using neural networks. Int. J. Secur. Appl. **2**(2), 41–50 (2008)

31. Abiyev, R., Altunkaya, K.: Neural Network Based Biometric Personal Identification. Lecture Notes in Computer Sciences. Springer/CS Press, Heidelberg (2007)

32. Helwan, A., Ozsahin, D.U., Abiyev, R., John, B.: One-year survival prediction of myocardial infarction. (IJACSA). Int. J. Adv. Comput. Sci. Appl. **8**, 173–178 (2017)

33. Idoko, J.B., Abiyev, R.H., Ma'aitah, M.K.S.: Intelligent machine learning algorithms for colour segmentation. WSEAS Trans. Sig. Process. **13**, 232–240 (2017)

34. Idoko, J.B., Abiyev, R., Ma'aitah, M.K.S., Hamit, A.: Integrated artificial intelligence algorithm for skin detection. In: International Conference Applied Mathematics, Computational Science and Systems Engineering (AMCSE 2017), ITM Web of Conferences, vol. 16, p. 02004 (2018)

35. Ma'aitah, M.K.S., Abiyev, R., Idoko, J.B.: Intelligent classification of liver disorder using fuzzy neural system. Int. J. Adv. Comput. Sci. Appl. (IJACSA) **8**, 25–31 (2017)

36. Helwan, A., Idoko, J.B., Abiyev, R.H.: Machine learning techniques for classification of breast tissue. In: 9th International Conference on Theory and Application of Soft Computing, Computing with Words and Perception, ICSCCW 2017. Procedia Computer Science, vol. 120, pp. 402–410 (2017)

37. Abiyev, R.H., Arslan, M., Gunsel, I., Cagman, A.: Robot pathfinding using vision based obstacle detection. In: IEEE International Conference on Cybernetics, pp. 1–6 (2017)

# Data Coding and Neural Network Arbitration for Feasibility Prediction of Car Marketing

Adnan Khashman[1,2(✉)] and Gunay Sadikoglu[3]

[1] Final International University, Kyrenia, Mersin 10, Turkey
adnan.khashman@final.edu.tr
[2] European Centre for Research and Academic Affairs (ECRAA),
Nicosia, Mersin 10, Turkey
[3] Department of Marketing, Near East University, Nicosia, Mersin 10, Turkey
gunay.sadikoglu@neu.edu.tr

**Abstract.** In this paper, we investigate and develop an intelligent system for predicting the marketing feasibility of cars based on using selected decisive car features and a supervised neural network prediction model. The novelty in our work is firstly using simple yet important car feature indicators that consumers often rely on when making a purchase decision. Secondly, designing and implementing a neural prediction model using a large freely available dataset with 1728 car evaluation instances. Our obtained experimental results suggest that using neural networks can be effectively used in predicting the feasibility of car selling or marketing based on its offered basic vehicle features.

**Keywords:** Neural networks · Prediction · Marketing feasibility
Car sales

## 1 Introduction

Currently, there is an overwhelming number of options and choices available to car buyers. Manufacturers and dealers continue to explore different methods and use impressive advertising approaches to convince potential consumers to buy their cars. To make the matter even more complex, we, the consumers, are also offered products with emerging technologies such as electric and hybrid cars; which makes our decision making task even more difficult. This brings us to the question of what do consumers and car buyers really focus on when purchasing a new or used car?

In our hypothesis, we believe that the average consumer is more inclined to consider the basic features of a car and thus tend to make a purchase based on such simple criteria. For example, price and comfort are two major and general categories that consumers examine. The technical specifications of a car play also a major role in decision making by the majority of consumers. These main basic categories have been overlooked in many recent car marketing evaluation or feasibility studies.

For example, in [1] the authors explore the place and position of consumer neuroscience for effective marketing management realization in a car selling enterprise. In [2] the authors focus on electric and hybrid vehicles while analyzing and comparing different policy scenarios to learn willingness to pay data from the French new-car market

© Springer Nature Switzerland AG 2019
R. A. Aliev et al. (Eds.): ICAFS-2018, AISC 896, pp. 249–255, 2019.
https://doi.org/10.1007/978-3-030-04164-9_34

in 2014. In [3] the work focuses on the subjective knowledge of first time buyers and how it influences their preference attributes in the emerging Chinese car market. In [4] the authors propose a price evaluation model using a back propagation neural network for price prediction of second-hand cars. In [5] this work investigates Chinese car dealing records to predict prices based on car make and series. In [6] investigates ways to integrate electric vehicles into the market focusing on the energy aspect. In [7] evaluation of regional vehicle marketing performance is investigated. In [8] the authors examine the impact upon used vehicle resale prices in the USA from the annual reliability evaluations done by Consumer Reports magazine. In [9] the work investigates car's country-of-origin effect on new car prices using hedonic price analysis. All these recent works have shown that there is a demand for applying computational aids and artificial intelligence in order to improve their analysis method of marketing cars in particular. Of course each work follows different method and have specific objectives, but all agree on providing efficient and easier to use marketing analysis methods.

In this paper, we differ by considering only price, comfort and technical specifications as our categories when buying a car, and we propose the use of artificial neural networks as the prediction tool. Neural models have been widely used over the past two decade and have been shown to provide efficient solutions to a various range of applications [10–20].

Our proposed intelligent feasibility prediction system uses car features, belonging to price, comfort and technical specifications categories, as attributes for training the neural network to make a decision whether the related car has a potential to sell or not; in other words, is marketing this car feasible? To train the neural model we use an online car evaluation dataset that has 1728 instances [21]. We choose the back propagation neural network as it is easy to train and implement, and has also been successfully used in various application.

The subsequent sections in the paper describe the car evaluation database and data set coding, the neural network design and implementation, the results and conclusions.

## 2  Database and Data Coding

Most machine learning algorithms rely in their arbitration on large datasets. These are often available in databases containing instances of examples of input attributes and their corresponding yielded output or classes. In this work, we use a publicly-available dataset [21] that has 1728 instances on car evaluation. The input attributes represent the basic important categories that car buyers often consider when planning to buy a car. The comfort and cost (or price) of a car have a major impact on a customer's buying decision. Table 1 shows the input and output categories and their indicative features and classes. Table 2 provides an explanation of the terms used in the dataset description in Table 1. Since there are six input attributes and considering the maximum number of classes for each input attribute is four, then the total number of input attributes' classes is 24 which will be the number of input layer neurons in our proposed neural network model. However, the dataset classes require numerical coding prior to presentation to the neural network input layer. The data coding involves firstly, indicating the attribute class in each attribute and in each instance by numbers 1, 2, 3 or 4. Secondly by replacing each of these numerical values with a 4-digit binary code as follows:

| 1: 1 0 0 0 | 2: 0 1 0 0 | 3: 0 0 1 0 | 4: 0 0 0 1 |

The feasibility output is coded in a similar manner with 4-digit binary code for each instance. Tables 3, 4, and 5 demonstrate the data coding process for the first five instances in the dataset.

**Table 1.** Car evaluation dataset categories and attributes [21].

| Input | | | | | | | Output |
|---|---|---|---|---|---|---|---|
| Category | Price | | Technical specifications | | | | Feasibility |
| Attributes | Buying | Maint | Comfort doors | | | Safety | |
| Classes | vhigh | vhigh | 2 | 2 | small | low | unacc |
| | high | high | 3 | 4 | med | med | acc |
| | med | med | 4 | more | big | high | good |
| | low | low | 5-more | – | – | – | v-good |

**Table 2.** Explanation of dataset terminology.

| Term | Explanation | Classes |
|---|---|---|
| Buying | Buying price | Very high, high, medium, low |
| Maint | Maintenance cost | Very high, high, medium, low |
| Doors | Number of doors | 2, 3, 4, 5 or more |
| Persons | Passenger carrying capacity | 2, 4, more |
| Lug_Boot | Size of luggage boot | Small, medium, big |
| Safety | Estimated safety of the car | Low, medium, high |
| Feasibility | Marketing potential | Unacceptable, acceptable, good, very good |

**Table 3.** Dataset original values prior to coding-showing first five instances.

| Attributes | Buying | Maint | Doors | Persons | Lug_Boot | Safety | Feasibility |
|---|---|---|---|---|---|---|---|
| Car 01 | vhigh | vhigh | 2 | 2 | small | low | unacc |
| Car 02 | vhigh | vhigh | 2 | 2 | small | med | unacc |
| Car 03 | vhigh | vhigh | 2 | 2 | small | high | unacc |
| Car 04 | vhigh | vhigh | 2 | 2 | med | low | unacc |
| Car 05 | vhigh | vhigh | 2 | 2 | med | med | unacc |

**Table 4.** Dataset numerical coding-showing first five instances.

| Attributes | Buying | Maint | Doors | Persons | Lug_Boot | Safety | Feasibility |
|---|---|---|---|---|---|---|---|
| Car 01 | 1 | 1 | 1 | 1 | 1 | 1 | 1 |
| Car 02 | 1 | 1 | 1 | 1 | 1 | 2 | 1 |
| Car 03 | 1 | 1 | 1 | 1 | 1 | 3 | 1 |
| Car 04 | 1 | 1 | 1 | 1 | 2 | 1 | 1 |
| Car 05 | 1 | 1 | 1 | 1 | 2 | 2 | 1 |

**Table 5.**  Dataset binary coding-showing first five instances.

| Attributes | Buying | Maint | Doors | Persons | Lug_Boot | Safety | Feasibility |
|---|---|---|---|---|---|---|---|
| Car 01 | 1,0,0,0 | 1,0,0,0 | 1,0,0,0 | 1,0,0,0 | 1,0,0,0 | 1,0,0,0 | 1,0,0,0 |
| Car 02 | 1,0,0,0 | 1,0,0,0 | 1,0,0,0 | 1,0,0,0 | 1,0,0,0 | 0,1,0,0 | 1,0,0,0 |
| Car 03 | 1,0,0,0 | 1,0,0,0 | 1,0,0,0 | 1,0,0,0 | 1,0,0,0 | 0,0,1,0 | 1,0,0,0 |
| Car 04 | 1,0,0,0 | 1,0,0,0 | 1,0,0,0 | 1,0,0,0 | 0,1,0,0 | 1,0,0,0 | 1,0,0,0 |
| Car 05 | 1,0,0,0 | 1,0,0,0 | 1,0,0,0 | 1,0,0,0 | 0,1,0,0 | 0,1,0,0 | 1,0,0,0 |

## 3   Neural Network Implementation

In this work, a supervised neural network based on the backpropagation algorithm method (BPNN) is designed and implemented for the task of predicting car marketing feasibility. The BPNN has 24 input neurons receiving binary-coded values of the car attributes (see example in Table 5). The output layer of this model has four neurons according to the number of feasibility output classes. The number of hidden layer neurons is usually determined during training; and in this work it was adjusted until optimum performance was achieved based on a predefined minimum error value of 0.007. The activation function of the processing neurons in our neural networks is the sigmoid activation function. The tunable parameters in supervised learning are the learning coefficient ($\eta$) and the momentum rate ($\alpha$). These parameters indicate how well a neural network learns, and are usually adjusted during learning to achieve minimal error and higher recognition rates. The topological design of the BPNN model is shown in Fig. 1, while the final training parameters are shown in Table 6.

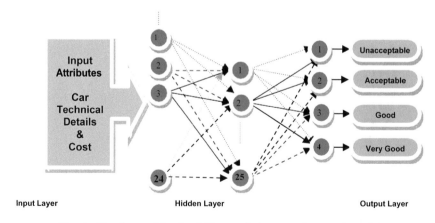

**Fig. 1.**  Topology of the feasibility prediction neural network model

**Table 6.** BPNN model final training parameters and prediction results

| Neural network model | BPNN |
|---|---|
| Training instances | 864 |
| Testing instances | 864 |
| Input neurons | 119 |
| Hidden neurons | 25 |
| Output neurons | 4 |
| Learning rate | 0.009 |
| Momentum rate | 0.800 |
| Minimum error | 0.0070 |
| Obtained error | 0.0051 |
| Maximum iterations | 5000 |
| Training time (s)[a] | 0.017 |
| CPR[b] – training data | (864/864) 100% |
| CPR[b] – testing data | (606/864) 70.14% |
| **CPR[b] – combined** | **(1470/1728) 85.07%** |

[a]Using a 3.20 GHz PC with 4 GB of RAM, Windows 10 OS and Dev-C++ compiler.
[b]CPR: Correct Prediction Rate.

## 4   Experimental Results

The implementation of the neural network model in this work was carried out using a 3.20 GHz PC with 4 GB of RAM, Windows 10 OS and Dev-C++ compiler. Training the BPNN model was carried out following a learning scheme where (50:50) training-to-validation data ratio was employed, i.e. 50% of the dataset was used for training; and the remaining 50% of the dataset used for testing or validation. Table 6 shows a detailed listing of the experimental results. A combined correct prediction rate (CPR) of 85.07% was achieved and is considered as sufficient since this prediction task relies on a relatively small number of input attributes. The time costs are minimal with training time of 0.017 s; which is considered very fast for such a large number of training instances (864 instances). In summary, we consider the speed of learning (training time) and the CPR values when evaluating the performance of the feasibility prediction model. Therefore, based on our performance criteria and the obtained experimental results, this model has been successful in performing this feasibility prediction task.

## 5   Conclusions

In this paper, we presented a practical application for artificial neural networks, where we developed a simple; yet efficient prediction model for marketing feasibility of different cars. The dataset used in this work is a freely available online benchmark database containing 1728 instances for car evaluation.

To simplify the design and implementation we focus on basic features that in our hypothesis most car buyers do look for when making a purchase. These include the price, some technical specifications and comfort. We, therefore, avoid the overwhelming car-related information that sales persons often use to secure a deal. The output classification or the prediction results provide four classes for each car in the data set; namely unacceptable, acceptable, good, and very good.

The neural network prediction model, which is based on the back propagation neural network was implemented and its parameters optimized yielding a fast training time of 0.017 s and a correct combined prediction rate (CPR) of 85.07%, Thus, providing a practical solution to this feasibility prediction task. Future work will involve applying different neural network learning scheme to investigate the possibility of improving the prediction rates.

# References

1. Bercik, J., Horska, H., Viragh, R., Sulaj, A.: Advanced mapping and evaluation of consumer perception and preferences on the car market based on eyetracking. Polish J. Manag. Stud. **16**(2), 28–39 (2017). https://doi.org/10.17512/pjms.2017.16.2.03
2. Fernandez-Antolin, A., de Lapparent, M., Bierlaire, M.: Modeling purchases of new cars: an analysis of the 2014 French market. Theory Decis. **84**(2), 277–303 (2018). https://doi.org/10.1007/s11238-017-9631-y
3. Qian, L., Soopramanien, D., Daryanto, A.: First-time buyers' subjective knowledge and the attribute preferences of Chinese car buyers. J. Retail. Consum. Serv. **36**, 189–196 (2017). https://doi.org/10.1016/j.jretconser
4. Sun, N., Bai, H.X., Geng, Y.X., Shi, H.: Price evaluation model in second-hand car system based on BP neural network theory. In: Hochin, T., Hirata, H., Hiroki, N. (eds) Software Engineering, Artificial Intelligence, Networking and Parallel/Distributed Computing (SNDP 2017), pp. 431–436 (2017)
5. Chen, C.C., Hao, L.L., Xu, C.: Comparative analysis of used car price evaluation models. In: You, Z., Xiao, J., Tan, Z. (eds) Material Science, Energy Technology and Power Engineering I (AIP 2017), vol. 1839 (2017). https://doi.org/10.1063/1.4982530
6. Illing, B., Warweg, O.: Business cases evaluation for electric vehicle market integration. In: IEEE International Symposium on Smart Electric Distribution Systems and Technologies (EDST 2015), pp. 499–504 (2015)
7. Cui, L.X., Zhang, J.L.: Evaluation indicator system of regional vehicle marketing team performance with Delphi. In: IEEE International Conference on Information Management, Innovation Management and Industrial Engineering, vol. 1, pp. 319–323 (2009). https://doi.org/10.1109/iciii.2009.84
8. Hollenbacher, A., Yerger, D.B.: Third party evaluations and resale prices in the US used vehicle market. Appl. Econ. Lett. **8**(6), 415–418 (2001). https://doi.org/10.1080/13504850175023788
9. Saridakis, C., Baltas, G.: Modeling price-related consequences of the brand origin cue: an empirical examination of the automobile market. Mark. Lett. **27**(1), 77–87 (2016). https://doi.org/10.1007/s11002-014-9304-3

10. Khashman, A., Sekeroglu, B.: Multi-banknote Identification using a single neural network. In: Blanc-Talon, J., Philips, W., Popescu, D., Scheunders, P. (eds.) Advanced Concepts for Intelligent Vision Systems (ACIVS 2005). LNCS, vol. 3708, pp. 123–129. Springer, Heidelberg (2005)
11. Khashman, A., Sekeroglu, B.: A novel thresholding method for text separation and document enhancement. In: 11th Panhellenic Conference on Informatics, Greece, pp. 323–330 (2007)
12. Khashman, A.: Intelligent face recognition: local versus global pattern averaging. In: Australasian Joint Conference on Artificial Intelligence, pp. 956–961 (2006)
13. Khashman, A., Dimililer, K.: Medical radiographs compression using neural networks and haar wavelet. In: EUROCON 2009, pp. 1448–1453. IEEE (2009)
14. Olaniyi, E.O., Khashman, A.: Onset diabetes diagnosis using artificial neural network. Int. J. Sci. Eng. Res. **5**(10), 754–759 (2014)
15. Khashman, A.: An emotional system with application to blood cell type identification. Trans. Inst. Meas. Control **34**(2–3), 125–147 (2012)
16. Oyedotun, O.K., Khashman, A.: Document segmentation using textural features summarization and feedforward neural network. Appl. Intell. **45**(1), 198–212 (2016)
17. Khashman, Z., Khashman, A.: Anticipation of political party voting using artificial intelligence. Procedia Comput. Sci. **102**, 611–616 (2016)
18. Khashman, A.: Investigation of different neural models for blood cell type identification. Neural Comput. Appl. **21**(6), 1177–1183 (2012)
19. Khashman, Z., Khashman, A.: Modeling people's anticipation for Cyprus peace mediation outcome using a neural model. Procedia Comput. Sci. **120**, 734–741 (2017)
20. Olaniyi, E.O., Oyedotun, O.K., Khashman, A.: Heart diseases diagnosis using neural networks arbitration. Int. J. Intell. Syst. Appl. **7**(12), 72 (2015)
21. Lichman, M.: UCI Machine Learning Repository. University of California, School of Information and Computer Science, Irvine, CA (2013). http://archive.ics.uci.edu/ml

# Signature Recognition Using Backpropagation Neural Network

Yucel Inan[1(✉)] and Boran Sekeroglu[2]

[1] Department of Computer Engineering, Near East University,
Nicosia, Mersin 10, Turkey
yucel.inan@neu.edu.tr
[2] Department of Information Systems Engineering, Near East University,
Nicosia, Mersin 10, Turkey

**Abstract.** Imitation or the fake signatures is the global fraud that cause the waste of financial sources, time and human effort. For this reason, signature recognition is the most widely used biometrics system for security and personal identification. Signatures are the most complex human patterns which are used to identify and approve the authorized persons. They can be varied according to the paper and pen influences, and human psychology and characteristics at the signature moment. Therefore, effective recognition of signatures is required in order to minimize the fraud. The usage of neural networks in biometrics, yet signature recognition, provides more steady and accurate identification thus authorization of person. This paper presents the preliminary results of developed offline signature recognition system using backpropagation neural network. Signature database is created by collecting the multiple signatures of 27 persons and the accuracy of the system is tested under artificially created conditions. System achieved 86% of highest recognition rate.

**Keywords:** Signature recognition · Neural networks · Backpropagation

## 1 Introduction

Biometric systems are security systems that enable people to access the system by themselves, without having any information or object. Biometric systems are basically divided into two groups: physical (passive) and behavioral (active) systems [1]. Physical biometric systems includes fingerprint, hand geometry, face, sound, iris, retina and behavioral biometric systems includes signature, dynamics, and lip movements during speech and walking pattern.

Signature recognition uses the signature, which can be defined as the handwriting of the person, in many areas of social life. A signature under any document indicates that the signer has read, wrote or approved the document. Behavioral characteristics related to the signing process, such as the duration of signing, speed, acceleration, intensity of printing etc. are considered for the recognition of signature, thus the identification of the person. Even the imitation of signatures is difficult, it is possible to sign any document similar to real user. Because signature pattern of any user can not be

© Springer Nature Switzerland AG 2019
R. A. Aliev et al. (Eds.): ICAFS-2018, AISC 896, pp. 256–261, 2019.
https://doi.org/10.1007/978-3-030-04164-9_35

same as the previous pattern that can be affected by the user's mental state, health, and urgency [2].

Signatures form a special class that legible text or words cannot be exhibited. They carry hidden meanings for real estate, legal authority, banking and other high security areas. It involves rules that are difficult to formulate. It requires carefully performed empirical analysis.

The signature identification system can be classified as online or offline [3, 4]. An online systems use electronic pen and consider spring and dynamic speed character- istics such as speed of release, pressure applied number of strokes to recognize sig- nature. In offline signature identification system, it is sufficient to throw the signatures onto paper.

Neural networks (NN) are effective tool for any kind of real life applications require the improvement of humans' flexible decision making by simulating the perception of them. One of the most widely used and effective learning algorithm is backpropagation (BP) learning algorithm. It is based on gradient and propagating the error values to the previous layers to update the weights until the learning criteria is achieved.

Backpropagation Neural Network (BPNN) was implemented to solve several and varied real life applications such as the workload prediction for cloud computing [5], land change for remotely sensed images [6], leaves recognition [7], microRNA analysis [8], threshold estimation [9] etc.

Several researches were conducted to distinguish real signatures from fake ones [10–12]. However, the complexity and performances of these methods are based solely on visual control and resulting in huge amounts of money being lost. Specialists focused on neural based systems to reduce the time and financial costs during the recognition of the signatures [13].

Karoni et al. [14] implemented offline signature recognition system using NN to recognize 100 different types of signatures and similarly, Arora and Choubey [1] proposed offline signature verification and recognition system using NN to recognize 100 different types of signatures. Bhatia [15] developed another offline handwritten signature verification using NN to recognize 50 different people.

In this study, offline signature recognition system based on BPNN is developed. Signatures which are obtained from 27 persons with 10 different samples are acquired by a digital camera and images are processed, trained and tested under artificially created conditions.

The rest of the paper organized as follows; Sect. 2 explains the general structure of the system and Sect. 3 presents obtained results. Section 4 concludes the results obtained in this research.

## 2   Structure of the System

In this section, the structure of the system is presented. System consists 2 phases; image preprocessing and learning phases.

## 2.1    Image Preprocessing Phase

Image preprocessing phases are applied to minimize the computational cost and to improve the learning ability of the neural systems. Various data preparation methods can be applied in order to provide robust patterns for NN.

The first step is the acquisition of the signature images which is the initial step for all offline recognition systems. Digitized images are converted to grayscale to reduce the 3D color range to 1D, and then images are sized to 50 × 50 to minimize the inputs of NN thus training time of the BPNN. In this phase, 3 types of additional noise; Salt and Pepper, Gaussian and Poison noise is added to the resized images of each person which generates different input data to test the efficiency of the system under different and various conditions. Figure 1 demonstrates the noisy images of original signatures.

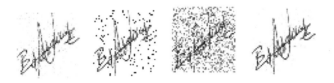

**Fig. 1.**  Sample noisy images

Finally, gray level values of signature images are normalized to be sent to the input layer of BPNN. General block diagram of image preprocessing phase can be seen in Fig. 2.

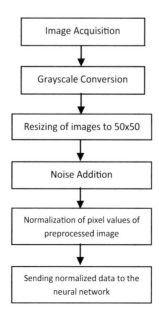

**Fig. 2.**  Block diagram of image preprocessing phase.

## 2.2   Learning Phase

Multi-hidden layered BPNN is used to train the signature images, and normalized pixel values of preprocessed images directly fed to input layer of network which creates totally 2500 input neurons. After several experiments, 2 hidden layers with 150 neurons for both, is determined to be used in the system. 27 output neurons which represents unique person are used. Table 1 and Fig. 3 shows the final parameters and BPNN architecture respectively.

**Table 1.** Neural network parameters

| Parameter | Value |
|---|---|
| Number of input neurons | 2500 |
| Number of hidden neurons in hidden layer 1 | 150 |
| Number of hidden neurons in hidden layer 2 | 150 |
| Number of output neurons | 27 |
| Maximum epochs | 1500 |
| Target error | $5 \times 10^{-7}$ |
| Learning rate | 0.05 |
| Momentum | 0.9 |

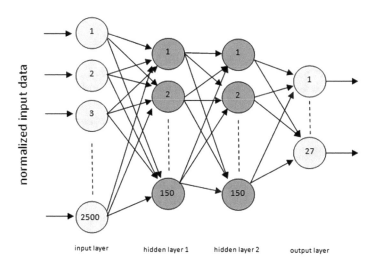

**Fig. 3.** Architecture of backpropagation neural network.

BPNN was trained using 9 patterns of each signature and totally 243 training patterns were used during the training. Rest of the signature images which are 108, are used to test the efficiency of the system.

## 3  Results and Discussions

As it is mentioned above, training of the BPNN was performed using 2500 input neurons, two hidden layers with 150 neurons and 27 outputs with 9 patterns of signature. Network reached to maximum epoch number in 38 s and the recognition rate of trained data which is not included into the test data was obtained 100% as expected.

Testing was applied by dividing the untrained images into 3 sets according to the noise type with 4 different levels of noise density and total 108 images for each. BPNN achieved 86% of recognition rate in Salt and Pepper noise added test set which is the highest rate in this research. In Test Set II, which the Gaussian noise is added, BPNN recognized 83.9% of images. BPNN recognized only 73.2% of Poison noise added images in Test Set III. Table 2 shows the testing results of the Test set I, Test set II, Test set III.

**Table 2.** Results for all Test sets

| Test set | Noise type | Recognition rate |
|---|---|---|
| Test set I | Salt and Pepper | 86.3% |
| Test set II | Gaussian | 83.9% |
| Test set III | Poison | 73.2% |

## 4  Conclusions

Biometric recognition systems are developed to identify the user by recognizing the physical and behavioral characteristics of the user. These systems work in a similar way to the human brain's recognition and differentiation methods. Determining whether the signature, which is perhaps the most commonly used method of authentication, has been signed by that person, is an important problem.

In this study, preliminary results of offline signature recognition and verification system using backpropagation neural networks were presented. Signature database was created by collecting signatures from 27 persons and adding 3 noise types; Salt and Pepper, Gaussian and Poison, to create artificially varied extreme conditions, totally 351 images. 70% of signature database was used in training and 30% in testing. Testing was performed in three test sets and BPNN achieved 86.3% of recognition rate in set 1 which consists Salt and Pepper noise within the images.

Future work will include the training and testing of system by collecting more signature samples from more people on different pen and paper types. Image preprocessing methods will also be investigated to increase the recognition ability of the system and the comparison will be conducted with recent researches.

# References

1. Arora, A., Choubey, A.S.: Offline signature and recognition using neural network. Int. J. Sci. Res. **2**(8), 196–200 (2013)
2. Prabhakar, S., Pankanti, S., Jain, A.K.: Biometric recognition: security and privacy concerns. IEEE Secur. Priv. **99**(2), 33–42 (2003)
3. Maiorana, E., Campisi, P., Fierrez, J., Ortega-Garcia, J., Neri, A.: Cancelable templates for sequence-based biometrics with application to on-line signature recognition. IEEE Trans. Syst. Man Cybern.-Part A: Syst. Hum. **40**(3), 525–538 (2010)
4. Mlaba, A., Gwetu, M., Viriri, S.: Online signature verification using hybrid transform features. In: 2018 Conference on Information Communications Technology and Society (ICTAS), pp. 1–5. IEEE, South Africa (2018)
5. Kumar, J., Singh, A.K.: Workload prediction in cloud using artificial neural network and adaptive differential evolution. Future Gener. Comput. Syst. **81**, 41–52 (2018)
6. Wu, K., Zhong, Y., Wang, X., Sun, W.: A novel approach to subpixel land-cover change detection based on a supervised back-propagation neural network for remotely sensed images with different resolutions. IEEE Geosci. Remote Sens. Lett. **14**(10), 1750–1754 (2017)
7. Sekeroglu, B., Inan, Y.: Leaves recognition using neural networks. Procedia Comput. Sci. **102**(C), 578–582 (2016)
8. Adali, T., Sekeroglu, B.: Analysis of microRNAs by neural network for early detection of cancer. Procedia Technol. **1**, 449–452 (2012)
9. Khashman, A., Sekeroglu, B.: Document image binarisation using a supervised neural network. Int. J. Neural Syst. **18**, 405–418 (2008)
10. Nasser, A.T., Dogru, N.: Signature recognition by using SIFT and SURF with SVM basic on RBF for voting online. In: 2017 International Conference on Engineering and Technology (ICET), pp. 1–5. IEEE, Antalya (2017)
11. Patil, P., Almeida, B., Chettiar, N., Babu, J.: Offline signature recognition system using histogram of oriented gradients. In: 2017 International Conference on Advances in Computing, Communication and Control (ICAC3), pp. 1–5. IEEE, Mumbai (2017)
12. Hiryanto, L., Yohannis, A.R., Handhayani, T.: Hand signature and handwriting recognition as identification of the writer using gray level cooccurrence matrix and bootstrap. In: 2017 Intelligent Systems Conference (IntelliSys), pp. 1103–1110. IEEE, London (2017)
13. Demir, Z., Çikoğlu, S., Temurtaş, F., Yumuşak, N.: Signature recognition using a neural network model. SAU Fen Bilimleri Enstitüsü Dergisi **7**(2) (2003)
14. Karoni, A., Daya, B., Bahlak, S.: Offline signature using neural networks approach. Procedia Comput. Sci. **3**, 155–161 (2011)
15. Bhatia, M.: Hand-written signature verification using neural network. Int. J. Appl. Inf. Syst. (IJAIS) **1**(2), 44–49 (2012)

# Topological Ordering on Interval Type-2 Fuzzy Graph

Margarita Knyazeva[1(✉)], Stanislav Belyakov[1(✉)],
and Janusz Kacprzyk[2(✉)]

[1] Southern Federal University, Nekrasovsky 44, Taganrog, Russia
margarita.knyazeva@gmail.com, beliacov@yandex.ru
[2] Systems Research Institute, Polish Academy of Sciences, Newelska 6,
Warsaw, Poland
kacprzyk@ibspan.waw.pl

**Abstract.** The topological ordering for graphs has many practical applications where the nature of the stated problem requires sequential processing. It might be linking and loading problems, planning and scheduling algorithms, assembly line processing and many other practical applications where precedence constraints are met. Sequencing of vertices execution usually depends on problem and can be represented by directed acyclic graph structure with expert estimations of uncertain variables one can come across. Difficulties in ordering and scheduling vertices of such fuzzy-estimated weighted graph are investigated in this paper. An algorithm for topological ordering and directed minimum spanning tree problem of interval type-2 fuzzy graph for scheduling problem is developed.

**Keywords:** Fuzzy graph · Scheduling · Decision-making
Type-2 fuzzy numbers

## 1 Introduction

### 1.1 Topological Ordering for Graphs-Based Problems

Topological sorting or ordering for Directed Acyclic Graph $G = (V, E)$, where $V$ is a finite non-empty set of data elements (vertices), and $E$ represents the set of relations between elements (edges) $e_{ij}$, and vertex $v_i$ comes before $v_j$ in the ordering. Usually it happens that there can be more than one topological sorting for a directed acyclic graph (DAG) depending on graph context. Context may consider resource requirements, time and precedence constraints, and multiple execution modes for vertices. Topological ordering is mainly used for scheduling jobs from the given dependencies among them. In computer science applications of this type arise in instruction scheduling, ordering of formula cell evaluation when recomputing formula values in spreadsheets, logic synthesis, determining the order of compilation tasks to perform in makefiles, data serialization, and resolving symbol dependencies in linkers.

Depending on information to be ordered within graph presentation model one can decide which algorithm need to be chosen as a basic to solve ordering and scheduling

© Springer Nature Switzerland AG 2019
R. A. Aliev et al. (Eds.): ICAFS-2018, AISC 896, pp. 262–269, 2019.
https://doi.org/10.1007/978-3-030-04164-9_36

problem. The approaches differ depending on whether ordering starts from source vertex or sink one (forward pass and backward pass propagation when calculating critical path length and schedules that show the planned start/finish dates for activities, schedule duration, and float values for the individual activities on the project). Various topological sorting problems take into consideration some relative weights available between the vertices, relative precedence relations as well as their relative weight, which creates the need of checking through all possible topological ordering.

A number of approaches were suggested to solve classical topological ordering problem including Kahn's algorithm, greedy algorithms such as Kruskal's algorithm for Minimum Spanning Tree, Prim's Undirected Minimum Spanning Tree (MST), Depth First Traversal (or Search). Algorithm starts with an empty spanning tree. The idea is to consider two sets of vertices simultaneously. The first set contains the vertices already included in the Minimum Spanning Tree, the second set contains the vertices not yet scheduled. At every step it considers all the edges that connect the two sets, and picks the minimum weight edge from these edges. After picking the edge, it moves the other endpoint of the edge to the set containing MST.

A group of edges that connects two set of vertices in a graph is called *cut* according to graph theory and at every step of Prim's algorithm, one can find a cut (of two sets, one contains the vertices already included in MST and other contains rest of the vertices), pick the minimum weight edge from the cut and include this vertex to MST Set (the set that contains already scheduled vertices).

Let's consider basic concepts and definitions that are needed to precede algorithm searching process on graph.

## 1.2  Basic Concepts and Definitions for Fuzzy Graph Topological Ordering

In terms of graph theory let's consider two graphs $G_1 = (V_1, E_1)$ and $G_2 = (V_2, E_2)$. If $V_2 \subseteq V_1$ and $E_2 \subseteq E_1$, then $G_2$ is a *sub-graph* of $G_1$, that is $G_1$ contains $G_2$. Subgraphs usually needed while procession ordering.

**Definition 1.** If the two graphs $G_1 = (V_1, E_1)$ and $G_2 = (V_2, E_2)$ have the same topological structure, they are considered to be *isomorphic*; that is, there exists a corresponding relationship between $V_1$ and $V_2$ such that each edge in $E_1$ corresponds to only one edge in $E_2$ and vice-versa. Sub-graph isomorphic means that there exist sub-graphs of $G_1$ and $G_2$ that are isomorphic.

As long as graph structure represents a particular relationship between elements of a set $V$ it may give an idea about the way that relationship between any two elements of $V$ should be formalized. In most of the situations the relationships between elements are uncertain in nature and fuzzy relation can deal with the situation in appropriate way. As an example, if $V$ represents certain events or activities to be scheduled and they presuppose certain precedence relations, then their flexible execution durations can be fuzzy. Durations can be treated to some extent using expert estimations as fuzzy relations. Thus, fuzzy graph models are more appropriate and realistic in modelling real-life situations.

Rosenfeld [1] first considered fuzzy relations on fuzzy sets and developed fuzzy graph theory, introduced fuzzy analogs of several basic graph-theoretic concepts. In book [2] authors consider dynamic fuzzy graph representation for fuzzy machine learning procedure and assume topological ordering on fuzzy graph. Authors in [3, 4] study a timely overview of fuzzy graph theory and applications in a broad range areas.

**Definition 2.** *A fuzzy graph* $\tilde{G} = (V, \sigma, \mu)$ is a triple consisting of a nonempty set $V$ together with a pair of functions $\sigma: V \to [0, 1]$ and $\mu: E \to [0, 1]$ such that for all $x$, $y \in V$, $\mu(xy) \leq \sigma(x) \wedge \sigma(y)$ [4].

**Definition 3.** The fuzzy set $\sigma$ is considered to be *fuzzy vertex set* of $\tilde{G}$ and $\mu$ the *fuzzy edge set* of $\tilde{G}$. So $\mu$ here is a *fuzzy relation* on $\sigma$.

We consider $\tilde{G}$ or $(\sigma, \mu)$ as fuzzy graph representation $\tilde{G} = (V, \sigma, \mu)$. Analogously $\sigma*$ and $\mu*$ represent the supports of $\sigma$ and $\mu$, and denoted by *Supp*($\sigma$) and *Supp*($\mu$) respectively.

**Definition 4.** Let $\tilde{G} = (V, \sigma, \mu)$ be a fuzzy graph. Then a fuzzy graph $H = (V, \tau, v)$ is called a *partial fuzzy subgraph* of $\tilde{G}$ if $\tau \subseteq \sigma$ and $v \subseteq \mu$. Similarly, the fuzzy graph $H = (P, \tau, v)$ is called a *fuzzy subgraph* of $G$ induced by $P$ if $P \subseteq V$, $\tau(x) = \sigma(x)$ for all $x \in P$ and $v(xy) = \mu(xy)$ for all $x, y \in P$.

**Definition 5.** Let $\tilde{G} = (\sigma, \mu)$ be a fuzzy graph. Then a partial fuzzy subgraph $(\tau, v)$ of $\tilde{G}$ is called to *span* $\tilde{G}$ if $\sigma = \tau$. In this case, we call $(\tau, v)$ *a spanning fuzzy subgraph* of $(\sigma, \mu)$.

Let's consider a connected graph $\tilde{G} = (\sigma, \mu)$, then *a spanning tree* of the graph $\tilde{G} = (\sigma, \mu)$ is a tree that *spans* $\tilde{G}$ (that is, it includes every vertex of $\tilde{G}$) and is a subgraph of $\tilde{G}$ (every edge in the tree belongs to $\tilde{G}$).

The cost of the spanning tree is the sum of the weights of all the edges in the tree. Usually when solving optimization graph problems such as scheduling and planning one use "Activity-on-Node" (AoN) representation, where cost or weights of edges represents costs of resources needed to perform activity, activity durations or time-lags between them or so-called mode. Mode can combine several profile elements (costs) to schedule activity. Thus there can be many spanning trees. Minimum spanning tree (MST) is the spanning tree where the cost is minimum among all the spanning trees.

In this paper we suggest to use the following notation (see Fig. 1).

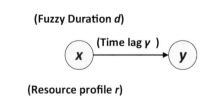

**Fig. 1.** An Activity-on-Node presentation of fuzzy precedence graph.

A triple $\{\tilde{d}, r, \gamma\}$ forms a weighted graph presentation that can be performed within modified algorithm to obtain Directed MST that minimizes the weight of the edges under certain criteria. In this paper we assume that only time-dependent variables are presented as interval type-2 fuzzy numbers while other criteria are crisp.

**Definition 6.** A *Time Lag* between start time of activity $x$: $\tilde{s}_x$ and start time of activity $y$: $\tilde{s}_y (x \neq y)$, is stated as follows: Activity-on-Node (AoN) graph $N = (V, E, Lag)$ with arc weighted function $\gamma$ and arc $\langle x, y \rangle \epsilon E$ indicates time lag between earliest starting times $\tilde{s}_x$ and $\tilde{s}_y$ with forward propagation method. Arc weight function $\gamma$ assigns to each arc $\langle x, y \rangle \epsilon E$ a $|M_x| \times |M_y|$- matrix $\gamma_{xy}$ of arc weight [5].

## 2 Advanced Techniques for Topological Ordering and Directed Minimum Spanning Tree with Fuzzy Interval Type-2 Structures

### 2.1 Type-2 Fuzzy Sets Estimations of Time-Dependent Variables

Basically in decision-making problems there is a necessity to consider several criteria and find optimal or near-optimal decision. Operations research study and graph theory investigate such solutions to complex problems. For example, in scheduling and allocation problems the goal is to construct special plan for activity performance with respect to precedence relations, durations, resources profiles and objective function to optimize; or to construct special allocation or transportation plan.

Allocation of resources, variables estimation, job sequencing and other activities are usually made by experts and in practice decision-makers are required to choose the best alternative among several. They are able to evaluate each alternative under several criteria (time, resource usage) and choose the best or eligible alternative (profile) for activity to be scheduled. This type of problem can also be related to multi-criteria decision making problem (MCDM).

As a result, with the increasing complexity of decision-making environment and the limitation of knowledge in some specific projects, expressing preference attitude and variable estimation process can't be performed under exact crisp numbers.

Fuzzy numbers and their different types are usually associated with imprecise or approximate variables, which can stay for a measure of preference or attitude of decision-maker and thus reflect flexibilities.

In this paper we propose an approach based on Interval Type-2 Fuzzy Numbers to capture and illustrate preprocessing uncertainty of initial data and variables estimation. Lack of information about some untypical projects and variation amongst the individual opinions of experts, their attitude towards variables dispersion, all this degrees of uncertainty and its merging into one can be managed with Interval Type-2 Fuzzy Numbers. Here we refer to classical terminology and definition of type-2 fuzzy sets [6].

**Definition 7.** A type-2 fuzzy set (T2FS) is defined on the universe of disclosure X and can be characterized by its membership function $\mu_{\tilde{A}}(x, u)$ and represented by Eq. (1):

$$\tilde{A} = \left\{\left((x, u), \mu_{\tilde{A}}(x, u)\right)/\forall x \in X, \forall u \in J_x \subseteq [0, 1]\right\} \tag{1}$$

where $0 \le \mu_{\tilde{A}}(x, u) \le 1$, the subinterval $J_x$ is called *primary membership function* of $x$, and $\mu_{\tilde{A}}(x, u)$ is called *secondary membership function* that defines the possibilities of primary membership function. Uncertainty in $J_x$ of type-2 fuzzy number $\tilde{A}$ is usually called footprint of uncertainty (FOU) and defined as the union of all primary memberships.

**Definition 8.** An Interval Type-2 Fuzzy Set (IT2FS) is defined on the universe of disclosure X and can be characterized by upper membership function (UMF) and lower membership function (LMF) and denoted as follows:

$$\tilde{A} = \left\{\langle \mu_{A^U}(x), \mu_{A^L}(x)\rangle/\forall x \in X\right\} \tag{2}$$

where $A^U$ is an upper T1FS and its membership function is $\mu_{A^U}(x) = max\{J_x\}$ for $\forall x \in X$, and $A^L$ is a lower T1FS whose membership function is equal to $\mu_{A^L}(x) = min\{J_x\}$ for $\forall x \in X$.

Let $\tilde{A} = \langle A^U, A^L \rangle = \langle a_1^U, a_2^U, a_3^U, a_4^U, h(A^U); a_1^L, a_2^L, a_3^L, a_4^L, h(A^L) \rangle$ be a IT2FS on the set of real numbers R. Then its upper and lower membership function can be defines as follows:

$$\mu_{A^U}(x) = \begin{cases} h(A^U) * \frac{x - a_1^U}{a_2^U - a_1^U}, a_1^U \le x < a_2^U \\ h(A^U), a_2^U \le x \le a_3^U \\ h(A^U) * \frac{a_4^U - x}{a_4^U - a_3^U}, a_3^U < x \le a_4^U \\ 0, otherwise \end{cases} \tag{3}$$

$$\mu_{A^L}(x) = \begin{cases} h(A^L) * \frac{x - a_1^L}{a_2^L - a_1^L}, a_1^L \le x < a_2^L \\ h(A^L), a_2^L \le x \le a_3^L \\ h(A^L) * \frac{a_4^L - x}{a_4^L - a_3^L}, a_3^L < x \le a_4^L \\ 0, otherwise \end{cases} \tag{4}$$

Figure 2a illustrates interval type-2 fuzzy trapezoidal job duration (as well as ready time $\tilde{b}_j$ and due date $\tilde{e}_j$ according to expert estimation profile). A footprint of uncertainty (FOU) here is the value of the membership function at each point on its two-dimensional domain, which reflects a decision-maker satisfaction/attitude towards optimistic and pessimistic profile.

Figure 2b illustrates IT2FS and FOU with respect to decision-maker attitude towards fuzzy variable. Each upper and lower point $a^L$ and $a^U$ shows optimistic and pessimistic evaluation of the same variable within a fuzzy number representation.

In this case expert data estimations is usually preprocessed with preference matrix $K = (k_{ij})_{n \times n}$, where $(k_{ij}) = (x_{ij}, u_{ij})$. Data processing using type-2 fuzzy sets from interval-valued data was discussed in paper [7]. This paper supposes interval type-2

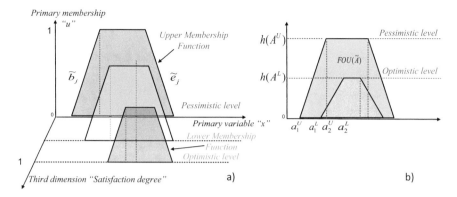

**Fig. 2.** Interval type-2 fuzzy variable estimation

fuzzy numbers and $\alpha$-cuts or zSlices for general type-2 fuzzy numbers which are interval type-2 fuzzy numbers with a secondary membership grade of $z_i$.

## 2.2 Greedy Algorithm for Fuzzy Topological Sorting Based on Minimum Spanning Tree

Basically in literature one can use several optimization graph algorithms to order undirected weighted graph so that it will obtain the Minimum Spanning Tree. Usually among them are Prim's or Kruskal's greedy algorithms for Minimum Spanning Tree [8, 9]. For directed graphs, the minimum spanning tree problem is called the *Arborescence problem* and can be solved in quadratic time using the Chu–Liu/ Edmonds algorithm or its modifications [9]. Chu–Liu/Edmonds' algorithm is an algorithm that allows to find a spanning arborescence of minimum weight (or optimum branching). It is the directed version of the minimum spanning tree problem, which was solved for undirected crisp graph ordering problem.

Minimum Spanning Tree in directed graphs is based on the following assumptions:

(a) a spanning tree rooted at source 0;
(b) a set of $n - 1$ edges contains paths from 0 to every vertex of minimum total edge so that optimum branching here supposes that $\sum\limits_{(x,y)\in V} C(x, y) \rightarrow min$.

An algorithm starts choosing the Minimum Weight Arc (MWA) from all the vertices including a root vertex (source), and gets rid of the vertices in which cycle occurs after sorting them in an ascending order.

Let's consider the formal description of algorithm based on Edmonds algorithm modification [10]. The algorithm introduced in this paper takes as input a directed fuzzy graph $\tilde{G} = (V, \sigma, \mu)$ *or* $\tilde{G} = (V, E)$ on each zSlice based on Interval Type-2 fuzzy number representation and distinguish root (source vertex 0) and a real-valued weight triple $\{\tilde{d}, r, \gamma\} = \{w_d(e), w_r(e), w_\gamma(e)\}$ for each edge $e \in E$. It returns a spanning *arborescence* $A$ rooted at 0 of minimum weight, where the weight of an arborescence is defined to be the sum of its edge weights:

$$w(A) = \sum_{e \in A} w(e) \tag{5}$$

Let $f(D, r, w_i)$ denote the function which returns a spanning arborescence rooted at root 0 of minimum weight [11].

**1 Step:** Remove any edge from set $E$ whose destination is 0.Then replace any set of parallel edges by a single edge with weight equal to the minimum of the weights of these parallel edges.

**2 Step:** For each vertex $v$ other than the root 0, find the edge incoming to $v$ of lowest weight (with ties broken arbitrarily). Denote the source of this edge by $p(y)$.

If the set of edges $P = \{(p(y), y), y \in V \backslash r\}$ does not contain cycles, then $f(D, r, w_i) = P$. Otherwise, $P$ contains cycle and algorithm arbitrarily chooses one cycle and calls it $C$.

Cycles on directed graph in scheduling problems correspond to the fact that activity can be performed several times. In order to guarantee, that precedence relations are performed, one needs to break all the possible cycles when considering scheduling graph.

**3 Step:** Define a new weighted directed fuzzy graph $G' = (V', \sigma', \mu')$ in which the cycle $C$ is "contracted" into one node as follows: the vertices $V'$ are the vertices of $V$ not included in $C$ plus a new vertex denoted by $v_c$.

If $(x, y)$ is an edge in $E$ with $x \notin C$ and $y \in C$, then include in $E'$ a new edge $e = (x, y_c)$ and define $w_i'(e) = w_i(x, y) - w_i(p(y), y)$. The case edge comes into the cycle.

If $(x, y)$ is an edge in $E$ with $y \notin C$ and $x \in C$, then include in $E'$ a new edge $e = (y_c, y)$ and define $w_i'(e) = w_i(x, y)$. The case edge goes out the cycle.

If $(x, y)$ is an edge in $E$ with $y \notin C$ and $x \notin C$, then include in $E$ a new edge $e = (x, y)$ and define $w_i'(e) = w_i(x, y)$. The case edge is unrelated to the cycle.

For each edge in set $E'$ fix which edge in set $E$ it corresponds to.

**4 Step:** Find a minimum spanning arborescence $A'$ of $D'$ using function $f(D, r, w_i)$.

If $A'$ is a spanning arborescence, then each vertex has exactly one incoming edge. Let $(x, y_c)$ be the only incoming edge to $y_c$ in $A'$, then this edge corresponds to an edge $(x, y) \in C$ with vertex $y \in C$. Remove the edge $p(y)$, $y)$ from $C$, break the cycle. Mark each remaining edge in $C$. For each edge in $A'$, mark its corresponding edge in $E$. Define $f(D, r, w_i)$ to be the set of marked edges, which form a minimum spanning arborescence. Observe that $f(D, r, w_i)$ is defined in terms of $f(D', r, w_{i'})$, with $D'$ having strictly fewer vertices than $D$.

**Algorithm Termination Condition:** If one find $f(D, r, w_i)$ for a single-vertex graph as $D$ itself, then the recursive algorithm is guaranteed to terminate.

The running time of this algorithm is $O(EV)$. This running time corresponds to Prim's classical algorithm for searching undirected minimum spanning tree.

# 3   Conclusion

This paper investigates ordering procedure on fuzzy graph and fuzzy relations between vertices. The algorithm was introduced, and it takes as input a directed fuzzy graph on each zSlice based on Interval Type-2 fuzzy number representation and uses weight triple of criteria for each edge on graph. Algorithm returns a spanning *arborescence A* of minimum weight and ordered graph structure. Thus given connected graph with positive edge weights, we can find a min weight set of edges that connect all the vertices and solve directed Minimum Spanning Tree (MST) problem.

**Acknowledgments.**  This work has been supported by the Ministry of Education and Science of the Russian Federation under Project "Methods and means of decision making on base of dynamic geographic information models" (Project part, State task 2.918.2017), and the Russian Foundation for Basic Research, Project № 18-01-00023a.

# References

1. Rosenfeld, A.: Fuzzy graphs. In: Zadeh, L.A., Fu, K.S., Shimura, M. (eds.) Fuzzy Sets and Their Applications, pp. 77–95. Academic Press, New York (1975)
2. Li, F., Zang, L., Zang, Z.: Dynamic Fuzzy Machine Learning, pp. 251–252. Walter De Gruyter GmbH, Berlin/Boston (2018)
3. Mordeson, J.N., et al. (eds.): Fuzzy Graphs and Fuzzy Hypergraphs. Springer, Heidelberg (2000)
4. Mathew, S., Mordeson, J.N., Malik, D.S.: Fuzzy Graph Theory. Studies in Fuzziness and Soft Computing, pp. 19–20. Springer, Cham (2018)
5. Knyazeva, M., Bozhenyuk, A., Rozenberg, I.: Scheduling alternatives with respect to fuzzy and preference modeling on time parameters. In: Proceedings of the 10th Conference of the European Society for Fuzzy Logic and Technology (EUSFLAT 2017), Warsaw, Poland, 11–15 September 2017. Advances in Intelligent Systems and Computing Book Series (AISC), vol. 642, pp. 358–369. Springer, Cham (2017)
6. Mendel, J.M., John, R.I.: Type-2 fuzzy sets made simple. IEEE Trans. Fuzzy Syst. **10**(2), 117–127 (2002)
7. Miller, S., Wagner, C., Garibaldi, J.M., Appleby, S.: Constructing general Type-2 fuzzy sets from interval-valued data. In: Proceedings of WCCI 2012 IEEE World Congress on Computational Intelligence (2012)
8. Choi, Myeong-Bok, Lee, Sang-Un: An efficient implementation of Kruskal's and reverse-delete minimum spanning tree algorithm. J. Inst. Internet Broadcast. Commun. **13**(2), 103–114 (2013)
9. Choi, Myeong-Bok, Lee, Sang-Un: A prim minimum spanning tree algorithm for directed graph. J. Inst. Internet Broadcast. Commun. **12**(3), 51–61 (2012)
10. Edmonds, J.: Optimum branchings. J. Res. Nat. Bur. Stand. **71B**, 233–240 (1967)
11. An open source implementation of Edmonds's algorithm written in C ++ and licensed under the MIT License: http://edmonds-alg.sourceforge.net/

# Z-Number Clustering Based on General Type-2 Fuzzy Sets

Rafik Aliev[1](✉) and Babek Guirimov[2]

[1] Azerbaijan State Oil and Industry University, Baku, Azerbaijan
raliev@asoa.edu.az
[2] SOCAR Midstream Operations, Baku, Azerbaijan
guirimov@hotmail.com

**Abstract.** Clustering is a convenient tool to extract or summarize information from large data sets. Data sets collected by modern information systems are constantly increasing in size. These data sets may include imprecise and partially reliable information. Usually this uncertainty is both probabilistic and fuzzy. Unfortunately, up-to-date there is almost no research on clustering which takes into account a synergy of both probability and fuzziness in produced information. In this paper, we first suggest an approach to clustering large data sets with probabilistic and fuzzy uncertainties. This approach is based on relationship of general Type-2 Fuzzy and Z-number concepts. A numerical example is considered to demonstrate the validity of the proposed method.

**Keywords:** Z-number · C-means clustering · Z-clustering · Type-2 fuzzy set

## 1 Introduction

Clustering is a very convenient and widely used tool for data mining and knowledge discovery from large data sets produced by contemporary information systems. Data sets from real life applications substantially increase in size and contain partially precise and reliable entries.

There exist different clustering algorithms for various applications [1–3]. All of them can be divided into hard clustering (bimodal) and soft clustering (fuzzy) methods. A well-known hard clustering method is K-means algorithm [4]. The method provides partition of data vectors into disjoint sets (crisp clusters). Every data vectors belongs to any cluster with bivalent membership degree $\{0, 1\}$.

Unlike K-means, a soft clustering method, such as Fuzzy C-means [5], divides data into a number of soft sets (fuzzy clusters). Any data vector belongs to these clusters with certain membership degrees in the range $[0, 1)$. The centroid of a fuzzy set, forming a cluster, defines that cluster's center vector. Fuzzy C-means algorithm has a manually defined numerical (crisp) parameter m from the range $(1, +\infty)$, called a fuzzifier that defines the softness of the cluster partitions. When m is close to 1, C-means result becomes the same as K-means. For m $\rightarrow +\infty$, the C-means algorithm provides the fuzziest possible result (with same membership for all data vectors). The most convenient and frequently used value for m is 2, however, specific application experts testify some data sets may give more desired partitioning with other values of

© Springer Nature Switzerland AG 2019
R. A. Aliev et al. (Eds.): ICAFS-2018, AISC 896, pp. 270–278, 2019.
https://doi.org/10.1007/978-3-030-04164-9_37

fuzzifier [6–8]. Some experts recommend to use data for m from the range [1.5, 4.5] and avoid extreme values [9].

To further improve quality of FCM clustering m is allowed to take a set of values. Then a generalized clustering result for the whole set could be produced. To this end Interval Type-2 Fuzzy (IT2FC) and General Type-2 Fuzzy (GT2FC) clustering methods are suggested. In IT2FC m takes interval values while in GT2FC, the fuzzifier's values are fuzzy numbers. Work [3] provides the detailed description of the mentioned improved clustering algorithms.

Sometimes, especially in cases with large data sets and for the values of fuzzifier striving to its extremal values, the iterative clustering algorithms (e.g. standard FCM) fail to converge to produce a quality result. In such cases, evolutionary optimization methods, such as Differential Evolution (DE) [10], can be of great help [11, 12]. Hence an adequate evolutionary algorithm is used to directly minimize the FCM's objective function and produce optimal cluster centers, which can be considered as virtual data vectors with membership values equal to 1 to one of clusters and 0 to others. The memberships of data vectors to clusters are updated after the valid cluster centers are found.

Unfortunately, up-to-date there is almost no research on clustering which takes into account a synergy of both probability and fuzziness in produced information, which can be done by Z-numbers [13, 14]. Relying on the relationship between Type-2 fuzzy sets and Z-numbers, we transform Type-2 fuzzy data to cluster memberships into Z-number based data partitioning [12].

The authors propose to use a DE based clustering algorithm to minimize FCM objective function with the value of m considered as fuzzy set to produce clusters with Type-2 Fuzzy memberships of data vectors to clusters. Then, the obtained set of Type-2 Fuzzy sets (piecewise-linearized fuzzy membership functions) are used to produce Z-number based cluster descriptions. Finally, we demonstrate how to use the obtained clusters as Z-number based knowledge terms (summaries) to construct Z-number based IF-THEN rules, which can be potentially exploited in knowledge discovery systems.

The paper is structured as follows. This section is Introduction. The next section, Preliminaries, considers required preliminary information, existing methods and definitions. The third section is the Statement of Problem. The fourth section, considers the proposed solution of the stated problem. An example problem is considered in section number five. The final section is Conclusions.

## 2 Preliminaries

### 2.1 Z-Number

A Z-number is a high level data construct (A, B) consisting of two fuzzy terms (fuzzy sets or numbers), the first of which (A) defines the constraint over the values of represented data element, and the second (B) defines the constraint over the reliability measure of A [13, 14]. Terms A and B are usually normalized and convex fuzzy sets (fuzzy numbers). Part B sometimes is treated as fuzzy probability, which implies existence of a fuzzy set (generally infinite) of probability density distributions $p_A$ over the elements of set A.

An example of a Z-number based linguistic term to define the height of a tall person is Z = "approximately 180 cm or higher (part A), most likely (part B)".

## 2.2  Differential Evolution

Differential Evolution is a very convenient and efficient population based evolutionary algorithm designed specifically for fast numerical optimization. The brilliant idea of the authors of algorithm [10] to use, in addition to random factors, also the difference of two vectors from population to create new potential offspring, provides the algorithm for inexhaustible search directions.

DE is widely used in any application where numerical function optimization is required including operations research, pattern recognition, and clustering, as well as Soft Computing based applications for training of artificial neural networks, adjusting rule-based systems and others [11].

## 2.3  Fuzzy (Type-1) Number

Fuzzy (more exactly, Type-1 Fuzzy) number A is a (usually continuous) fuzzy set whose elements are numerical and memberships are normalized and convex.

$A = \{x_i/\mu(x_i)\},\ x_i \in \mathbb{R},\ \mu(x_i) \in [0, 1]$
$\exists x_j \in A/\mu(x_j) = 1$
$\forall x_i, x_j \in A, t \in [0, 1]/\mu(tx_i + (1 - t)x_j) \leq t\mu(tx_i) + (1 - t)\mu(tx_j)$

## 2.4  Type-2 Fuzzy Number

A formal well-known definition is as follows [15]:

A (General) Type-2 (GT2) fuzzy set $\tilde{A}$, is characterized by a Type-2 Membership function $\mu_{\tilde{A}}(x, u)$, where $x \in X$ and $u \in J_x \subseteq [0, 1]$:

$$\tilde{A} = \{((x, u), \mu_{\tilde{A}}(x, u))/\forall x \in X, \forall u \in J_x \subseteq [0, 1]\}, 0 \leq \mu_{\tilde{A}}(x, u) \leq 1.$$

This also can be expressed by using unions of all admissible $x$ and $u$ as:

$$\tilde{A} = \int_{x \in X} \int_{u \in J_x} \mu_{\tilde{A}}(x, u)/(x, u),\ J_x \subseteq [0, 1]$$

It can be said also that a Type-2 fuzzy set is a fuzzy set A, the elements of which $a_i$ are also fuzzy sets.

$A = \{m_i/v(m_i)\}/m_i = \{a_j/\mu(a_j)\},$
$\mu(.) \in [0, 1]$ is the primary membership function and
$v(.) \in [0, 1]$ is the secondary membership function.

The memberships of elements $m_i$ are called secondary memberships. The footprint of uncertainty (FOU) of a Type-2 fuzzy set is the fuzzy set formed of elements from all $a_i$ whose secondary memberships is greater than 0. FOU(A) is upper and lower bounded by $\overline{FOU}(A)$ and $\underline{FOU}(A)$, respectively [15]. For the purpose of this research, we require for a Type-2 fuzzy number that its $\overline{FOU}(A)$ be a Type-1 fuzzy number.

### 2.5   Relationship Between Type-2 Fuzzy Set and Z-Number

**Representation Theorem [16].** Every monotonic type-2 fuzzy set $d(y, \mu)$ can be represented as a result of applying an appropriate data processing algorithm $y = f(x_1, \ldots, x_n)$ to some Z-numbers $X_1, \ldots, X_n$.

Proof of this theorem is given in [16].

## 3   Statement of Problem

Consider, we have a data set consisting of N vectors of dimension D:

$$X = \{x_i\}, \quad i = 1, \ldots, N$$

We need to partition the data set X into C clusters using DE based Type-2 fuzzy clustering algorithm with FCM objective (cost) function. Using the existing relationship between type-2 fuzzy numbers and Z-numbers, the produced Type-2 data-to-cluster membership degrees are transformed into discrete Z-numbers. Then, the generated Z numbers are used to formulate Z-number based IF-THEN rules.

## 4   Solution Method

Refer to the detailed solution method to [12]. The algorithm in brief is as follows:

1.  Apply DE for minimization of FCM objective function for a fuzzy fuzzifier

$$m = \{ m_1/g_1, \ m_2/g_2, \ m_3/g_3, \ \ldots, \ m_K/g_K \}$$

2.  Obtain a set of cluster centers:

$$v = \{ v_1/g_1, \ v_2/g_2, \ v_3/g_3, \ \ldots, \ v_K/g_K \}$$

3.  Obtain a Type-2 set of data-to-cluster memberships:

$$\left\{ \left\{ \left[ \underline{u}_{1ij}, \overline{u}_{1ij} \right] \right\}/\alpha_1, \ \left\{ \left[ \underline{u}_{2ij}, \overline{u}_{2ij} \right] \right\}/\alpha_2, \ \left\{ \left[ \underline{u}_{3ij}, \overline{u}_{3ij} \right] \right\}/\alpha_3, \ldots, \ \left\{ \left[ \underline{u}_{Lij}, \overline{u}_{Lij} \right] \right\}/\alpha_L \right\},$$

$i = 1, \ldots N; \ j = 1, \ldots C.$

Here $\alpha_i$ (i = 1,...L) are formed from $g_j$ (j = 1,..., K), assuring that $\alpha_1 < \alpha_2 < \ldots < \alpha_L$ and

$$\left[\underline{u}_{1ij}, \overline{u}_{1ij}\right] \supseteq \left[\underline{u}_{2ij}, \overline{u}_{2ij}\right] \supseteq \left[\underline{u}_{3ij}, \overline{u}_{3ij}\right] \supseteq \ldots \supseteq \left[\underline{u}_{Lij}, \overline{u}_{Lij}\right].$$

4. Compose/select a representational set of data-to-cluster membership functions $u_K(x)$ by piecewise linearizing membership data in a form:

$$\{u_1(x)/g_1\}, \ \{u_2(x)/g_2\}, \ \{u_3(x)/g_3\}, \ \ldots, \ \{u_K(x)/g_K\}$$

5. Type-reduce the Type-2 set of data-to-cluster memberships to produce a fuzzy set representing "A" part of cluster's Z-number representation:

$$\mu_A(x) = \frac{\sum_k u_{g_k}(x) \cdot g_k}{\sum_k g_k}$$

6. Construct the fuzzy set of probability density functions (PDFs) on the basis of all representational membership functions $u_k(x)$. For simplicity, use a non-symmetric Gaussian-like functions (area under whose are equal to unity) and find PDFs by curve fitting to meet the compatibility constraints:

$$G_{c, w_l, w_r}(x) = \begin{cases} \frac{1}{\frac{\sqrt{2\pi}}{2}(w_l + w_r)} Exp\left\{-\frac{1}{2}\left(\frac{x-c}{w_l}\right)^2\right\}, & \text{if } x < c \\ \frac{1}{\frac{\sqrt{2\pi}}{2}(w_l + w_r)} Exp\left\{-\frac{1}{2}\left(\frac{x-c}{w_r}\right)^2\right\}, & \text{otherwise} \end{cases}$$

7. Find the cluster probabilities:

$$P_j(\alpha_k) = \int_{x \in X} \mu_j(x) p_{j, g_k}(x) dx \approx \sum_{i=2}^{N} \mu_{ij} p_{j, g_k}(x_i)(x_i - x_{i-1})$$

8. Having probabilities and memberships $g_k$ construct the "B" part of cluster's Z-number representation

## 5   Numerical Example

For our experiment we used 81 two-dimensional data vectors. A fragment of data set is demonstrated in Table 1.

**Table 1.** Table captions should be placed above the tables.

| $x_{i,1}$ | $x_{i,2}$ |
|---|---|
| ... | ... |
| 1.05 | 0.774906 |
| 1.10 | 0.822311 |
| 1.15 | 0.874949 |
| 1.20 | 0.933029 |
| 1.25 | 0.996711 |
| 1.30 | 1.066098 |
| 1.35 | 1.141221 |
| 1.40 | 1.22203 |
| 1.45 | 1.308371 |
| 1.50 | 1.399982 |
| 1.55 | 1.496474 |
| ... | ... |
| 4 | 7.386384 |

To produce Type-2 Fuzzy clusters we used a fuzzy fuzzifier as shown in Fig. 1.

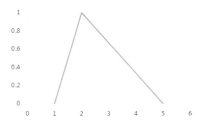

**Fig. 1.** The fuzzifier's membership function.

In Fig. 2 we demonstrated Type-2 membership functions for both dimensions of one of the clusters obtained by DE based GT2FC.

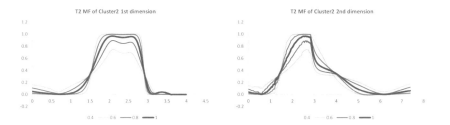

**Fig. 2.** Type-2 data-to-cluster membership function produced for one of clusters (cluster 2).

A-parts of Z-numbers are obtained as centroids of the above functions (Fig. 3)

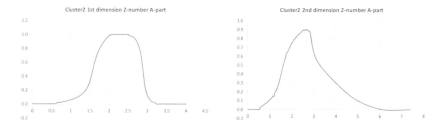

**Fig. 3.** A-parts of for each dimension of cluster 2.

Now we need to compute B part of Z-valued clusters. First, for each cluster we construct probability density functions $p_{j,l,k}$, $l = 1, 2$ by using the Type-1 MFs $\mu_{A_{j,l,k}}$. For convenience, we consider $p_{j,l,k}$ as a class of Gaussian-like functions (Fig. 4).

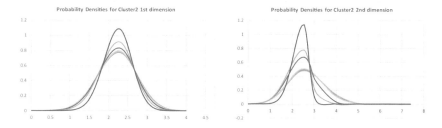

**Fig. 4.** Probability density functions.

Finally, we compute the probability measure for $A_{j,l}$, $P(A_{j,l}) = \sum_{i=1}^{n-1} \mu_{A_{j,l}}(x_{i,l})p_{j,l,k}(x_{i,l})$ and construct the "B" parts for both dimensions (Fig. 5).

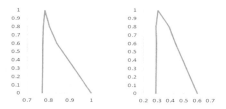

**Fig. 5.** B-parts of Z-numbers: first and second dimensions.

As can be seen, one of possible IF-THEN rules might be:

IF  Z1  THEN Z2,

where Z1 is a Z-number whose A-part is represented by the first MF in Fig. 3 and B-part is represented by first MF in Fig. 5 and Z2 is a Z-number whose A-part is the second MF in Fig. 3 and B-part is second MF in Fig. 5.

## 6   Conclusion

The paper suggests a method of clustering data sets with probabilistic and fuzzy uncertainties. The approach is based on relationship of general Type-2 Fuzzy and Z-number concepts suggested by authors. A numerical example has demonstrated the validity of the proposed method.

## References

1. Borgelt, C.: Fuzzy and Probabilistic Clustering (2015). http://www.cost-ic0702.org/summercourse/files/clustering.pdf
2. Pedrycz, W.: Fuzzy equalization in the construction of fuzzy sets. Fuzzy Sets Syst. **119**(2), 329–335 (2001)
3. Linda, O., Manic, M.: General type-2 fuzzy c-means algorithm for uncertain fuzzy clustering. IEEE Trans. Fuzzy Syst. **20**(5), 883–897 (2012)
4. MacQueen, J.: Some methods for classification and analysis of multivariate observations. In: LeCam, L.M., Neyman, J. (eds.) Proceedings of the Fifth Berkeley Symposium on Mathematical Statistics and Probability, vol. 1, pp. 281–297. University of California Press, Berkeley (1967)
5. Bezdek, J.C., Harris, J.: Fuzzy partitions and relations. Fuzzy Sets Syst. **1**(2), 111–127 (1978)
6. Pal, N.R., Bezdek, J.C.: On cluster validity for the fuzzy c-means model. IEEE Trans. Fuzzy Syst. **3**(3), 370–379 (1995)
7. Dembele, D., Kastner, P.: Fuzzy c-means method for clustering microarray data. Bioinformatics **19**(8), 973–980 (2003)
8. Schwämmle, V., Jensen, N.O.: A simple and fastmethod to determine the parameters for fuzzy c-means cluster analysis. Bioinformatics **26**, 2841–2848 (2010)
9. Ozkan, I., Turksen, I.B.: Upper and lower values for the level of fuzziness in FCM. Inf. Sci. **177**(23), 5143–5152 (2007)
10. Price, K., Storm, R., Lampinen, J.: Differential Evolution – A Practical Approach to Global Optimization. Springer, Berlin (2005)
11. Aliev, R.A., Pedrycz, W., Guirimov, B.G., Aliev, R.R., Ilhan, U., Babagil, M., Mammadli, S.: Type-2 fuzzy neural network with fuzzy clustering and differential evolution optimization. Inf. Sci. **181**(9), 1591–1608 (2011)
12. Aliev, R.A., Guirimov, B.G., Huseynov,O.H.: Z-number based clustering for knowledge discovery with reliability measure of results. In: Proceedings of the International Conference on Information Society and Smart Cities (ISC 2018), Fitzwilliam College, University of Cambridge, Cambridge, UK (2018). ISBN: 978-1-912532-02-5

13. Zadeh, L.A.: A note on Z-numbers. Inf. Sci. **181**, 2923–2932 (2010)
14. Aliev, R.A., Huseynov, O.H., Aliyev, R.R., Alizadeh, A.V.: The Arithmetic of Z-numbers, Theory and Applications. World Scientific, Singapore (2015)
15. Mendel, J.M., Liu, F., Zhai, D.: $\alpha$-plane representation for type-2 fuzzy sets: theory and applications. IEEE Trans. Fuzzy Syst. **17**(5), 1189–1207 (2009)
16. Aliev, R.A., Kreinovich, V.: Z-Numbers and Type-2 Fuzzy Sets: A Representation Result, Intelligent Automation and Soft Computing. Taylor & Francis, Abingdon (2017). https://doi.org/10.1080/10798587.2017.1330310

# One Approach to Multi-criteria Evaluation of Alternatives in the Logical Basis of Neural Networks

Misir Mardanov and Ramin Rzayev$^{(\boxtimes)}$

Institute of Mathematics and Mechanics of ANAS, Baku AZ1141, Azerbaijan
raminrza@yahoo.com

**Abstract.** An approach to a multicriteria evaluation of alternatives in conditions of insufficient information is proposed. By means of an adequate translation of the external notion of expert conclusions based weighted summary estimates of alternatives in effective internal knowledge is compiled in the logical basis of a multi-layer feedforward neural network.

**Keywords:** Knowledge compilation · Fuzzy set · Neural network

## 1 Introduction

Multi-criteria selection of one or several alternatives from a certain set is the essence of the solution of the decision-making problem, which today is one of the most widespread in any subject area. At the same time, the choice of the method of solution of such problem depends on the quantity and quality of the relevant information. However, analytic models, including econometric models and expert systems, which are used for multifactorial evaluation of alternatives often suffer from the complexity of information support and the shortcomings in quantitative descriptions of the majority of factors necessary for analyzing the influences they exert on the desired result. Moreover, the models and methods used for multi-criteria evaluation of alternatives have both their advantages and disadvantages. Therefore, optimal is approach that combines the best aspects of each of the assessment methods, which provides a more balanced and objective measurement and interpretation of the results.

In the absence of a sufficient amount of quantitative data that form a numerical representation of the evaluated alternatives, information on evaluation criteria, information on preferences, and environmental information required for multi-criteria evaluation of alternatives, the expert systems are applied by heuristic knowledge as the main resource. However, expert systems are constantly criticized, because they are not able to identify internal cause-effect relationships that characterize the validity of choosing the best alternative. The latter, in particular, does not allow providing a gradation of final estimates of alternatives from the existing set, i.e. to determine the appropriate characteristics for their classification. Therefore, on these premises the importance and actuality of further research on the methods of multi-criteria evaluation of alternatives under uncertainty becomes apparent.

© Springer Nature Switzerland AG 2019
R. A. Aliev et al. (Eds.): ICAFS-2018, AISC 896, pp. 279–287, 2019.
https://doi.org/10.1007/978-3-030-04164-9_38

## 2  Problem Definition

Let $X = \{x_1, x_2, \ldots, x_n\}$ ise the set of factors that influence the state (behavior) of alternatives from a certain set $A = \{a_1, a_2, \ldots, a_s\}$. For the coordinated selection of the factors $x_i$ ($i = 1 \div n$), there are mobilized $m$ experts from among the most experienced profile specialists, each of which forms a rank estimate of the selected $i$-th factor ($i = 1 \div n$) in the form of $r_{ij}$, and the corresponding normalized value of the weight estimate in the form of $w_{ij}$, so that for each $j = 1 \div m$ the equality $w_{1j} + w_{2j} + \ldots + w_{nj} = 1$ is satisfied. Then the assessment of the $x_i$-factors is carried out on the base of two methods of expertise: a comparative qualitative assessment by a simple ranking method (or by the expert preference method) and a quantitative evaluation by the method of setting the appropriate weights (normalized values). On the assumption of these prerequisites, it is necessary to determine the measure of agreement between expert evaluations regarding the priority of the $x_i$-factors, their generalized weights, and to initiate the deduction of a weighted total index, theoretically located within the interval [0; 100].

For consistent compilation of cause-and-effect relations between influences and total indexes of comparison of alternatives, or, in other words, for the purpose of knowledge compilation by transfering an external representation of knowledge about the weighted composite scores of alternatives, which are obtained on the base of expert conclusions, in the effective (adequate) internal representation, it is necessary to develop an analytical model in the logical basis of a multi-layer feedforward neural network. The testing of the proposed approach will be carried out on a hypothetical example of the finite set of alternatives $A$, which are estimated by results of the relative influence of factors from the bounded set $X$.

## 3  Ranking of Alternatives on the Base of Expert Assessments of Influences

Each alternative $a_r$ ($r = 1 \div s$) from the set $A$ will be considered as a multifactor category, i.e. as category characterized by a combined system of various (including by nature) influences. Let us perform the ranking of alternatives from the claimed set $A$ by quantitative estimates of the $x_i$-factors ($i = 1 \div n$), which have a significant influence on them. At the same time, the consolidated indicator of the evaluation of each alternative aggregates the relative influence of the considered number of factors (variables) $x_i$ by weighted summation of their expert estimates. The ranking of alternatives includes: (1) selection of $x_i$-factors, which have a significant influence on indicators of alternative estimates; (2) identification of the weights of the selected $x_i$-factors on the base of their relative influence on the indicator of the alternative estimation; (3) expert evaluation of $x_i$-factors using the expert scale, for example, a five-point rating system; (4) definition of indexes reflecting weighted quantitative estimates of alternatives.

## 3.1   Ranking of Influences in the Orders of Experts' Preferences

Suppose that by independent questionnaire of 15-th profile specialists, expert assessments of the importance degrees of the $x_i$-factors ($i = 1 \div 5$) are determined, which have a relative impact on the estimates of the declared alternatives. Each expert was suggested to arrange the variable $x_i$ according to the following principle: the most important variable is designated by the number "1", the next, less important – by the number "2" and further in descending order of importance. The rank estimates thus obtained are summarized in Table 1. Further, to establish the degree of agreement between the expert opinions on the priority of the influences, the Kendall concordance coefficient is applied, demonstrating a multiple rank correlation of expert opinions. According to [1], this coefficient is calculated by formula:

**Table 1.** Expert ranging of influences.

| Expert | Estimated factors and their ranks ($r_{ij}$) | | | | | Expert | Estimated factors and their ranks ($r_{ij}$) | | | | |
|---|---|---|---|---|---|---|---|---|---|---|---|
| | $x_1$ | $x_2$ | $x_3$ | $x_4$ | $x_5$ | | $x_1$ | $x_2$ | $x_3$ | $x_4$ | $x_5$ |
| 01 | 2 | 1 | 4 | 3 | 5 | 09 | 1 | 3 | 2 | 4 | 5 |
| 02 | 1 | 3 | 2 | 4 | 5 | 10 | 1 | 3 | 2 | 5 | 4 |
| 03 | 2 | 1 | 5 | 4 | 3 | 11 | 1 | 3 | 4 | 2 | 5 |
| 04 | 1 | 2 | 4 | 5 | 3 | 12 | 1 | 2 | 3 | 5 | 4 |
| 05 | 2 | 1 | 3 | 4 | 5 | 13 | 2 | 1 | 4 | 3 | 5 |
| 06 | 1 | 2 | 4 | 3 | 5 | 14 | 3 | 1 | 2 | 4 | 5 |
| 07 | 2 | 1 | 4 | 3 | 5 | 15 | 1 | 2 | 5 | 4 | 3 |
| 08 | 1 | 2 | 3 | 5 | 4 | $\sum r_{ij}$ | 22 | 28 | 51 | 58 | 66 |

$$W = 12 \times S / \left[ m^2 \left( n^3 - n \right) \right] \qquad (1)$$

where $m$ is the number of experts; $n$ is the number of influences, and $S$ is the deviation of expert conclusions from the average ranking value of $x_i$-factors ($i = 1 \div 5$), which is calculated as:

$$S = \sum_{i=1}^{n} \left[ \sum_{j=1}^{m} r_{ij} - \frac{m(n+1)}{2} \right]^2, \qquad (2)$$

where $r_{ij} \in \{1; 2; \ldots; 5\}$ is the rank of the $i$-th influence established by the $j$-th expert.

According to (1)–(2) and to data from Table 1 we have following values: $S = 1464$, $W = 12 \cdot 1464 / [15^2 (5^3 - 5)] = 0.6507$, which indicates a sufficiently *strong* agreement of expert opinions on the importance of $x_i$-factors ($i = 1 \div 5$).

## 3.2    Identification of the Weights of Influences

Now suppose that at the preliminary stage of independent annotation each expert was also instructed to set the values of the normalized weights for the $x_i$-factors ($i = 1 \div 5$). As a result of the questionnaire survey, experts' values are summarized in Table 2.

**Table 2.**  Expert ranging of influences.

| Expert | Normalized weights for factors ($\alpha_{ij}$) | | | | | Expert | Normalized weights for factors ($\alpha_{ij}$) | | | | |
|---|---|---|---|---|---|---|---|---|---|---|---|
| | $x_1$ | $x_2$ | $x_3$ | $x_4$ | $x_5$ | | $x_1$ | $x_2$ | $x_3$ | $x_4$ | $x_5$ |
| 01 | 0.250 | 0.3 | 0.150 | 0.225 | 0.075 | 09 | 0.275 | 0.175 | 0.200 | 0.100 | 0.250 |
| 02 | 0.350 | 0.175 | 0.200 | 0.150 | 0.125 | 10 | 0.300 | 0.200 | 0.250 | 0.100 | 0.150 |
| 03 | 0.225 | 0.250 | 0.150 | 0.175 | 0.200 | 11 | 0.300 | 0.175 | 0.150 | 0.250 | 0.125 |
| 04 | 0.275 | 0.250 | 0.175 | 0.100 | 0.200 | 12 | 0.300 | 0.250 | 0.150 | 0.100 | 0.200 |
| 05 | 0.250 | 0.275 | 0.200 | 0.175 | 0.100 | 13 | 0.225 | 0.250 | 0.175 | 0.200 | 0.150 |
| 06 | 0.300 | 0.250 | 0.150 | 0.200 | 0.100 | 14 | 0.200 | 0.300 | 0.250 | 0.150 | 0.100 |
| 07 | 0.200 | 0.375 | 0.150 | 0.175 | 0.100 | 15 | 0.300 | 0.250 | 0.125 | 0.150 | 0.175 |
| 08 | 0.325 | 0.300 | 0.150 | 0.025 | 0.200 | $\sum r_{ij}$ | 4.075 | 3.775 | 2.625 | 2.275 | 2.250 |

Starting from the data of Table 2, it is necessary to make preliminary calculations for the subsequent identification of the weights of the $x_i$-factors. Let us define group estimates of $x_i$ and numerical characteristics (levels) of competence of each of experts. In particular, to calculate the average value $\alpha_i$ of $i$-th group of normalized estimates of $x_i$-factors ($i = 1 \div 5$) one can use the following equation:

$$\alpha_i(t+1) = \sum_{j=1}^{m} w_j(t)\alpha_{ij}, \qquad (3)$$

where $w_j(t)$ is the weight characterizing the competence of the $j$-th expert ($j = 1 \div m$) at time $t$. In this case, the finding process of the group estimates of the normalized values has an iterative character. It ends after fulfillment of condition:

$$\max_i\{|a_i(t+1) - a_i(t)|\} \leq \varepsilon, \qquad (4)$$

where $\varepsilon$ is the feasible accuracy of calculations, which is set in advance. In our case, let it be $\varepsilon = 0.0001$. At the initial stage $t = 0$ we will assume that experts have the same levels of competence. Then, setting $w(0) = 1/m$ as initial value of the competence level of each expert, the average value for the $i$-th group of normalized estimates of the $x_i$-weights in the first approximation is obtained from the equality: $\alpha_i(1) = \sum_j w_j(0)\alpha_{ij} = \sum_j \alpha_{ij}/m$. Then the average group estimates of the $x_i$-factors ($i = 1 \div 5$) in the first approximation are the following numbers: $\{\alpha_1(1); \alpha_2(1); \alpha_3(1); \alpha_4(1); \alpha_5(1)\} = \{0.2717; 0.2517; 0.1750; 0.1517; 0.1500\}$. It is not difficult to define that the requirement (4) for the first approximation is not satisfied. Therefore, before passing to the next stage of the iteration, it is necessary to calculate the normalizing coefficient as following:    $\eta(1) = \sum_i(\sum_j \alpha_i(1)\alpha_{ij}) = 0.2717 \cdot 4.075 + 0.2517 \cdot 3.775 + 0.175 \cdot 2.625 +$

$0.1517 \cdot 2.275 + 0.1500 \times 2.250 = 3.199$. In this case, the competence levels of experts can already be calculated using the following equations:

$$w_j(1) = \frac{1}{\eta(1)} \sum_{i=1}^{5} \alpha_i(1) \cdot \alpha_{ij} (j = \overline{1, 14}), \ w_{15}(1) = 1 - \sum_{j=1}^{14} w_j(1), \ \sum_{j=1}^{15} w_j(1)$$
$$= 1,$$

(5)

where $w_{15}(1)$ is the competence indicator of the 15-th expert in the $1^{st}$ approximation. On the base of expressions (5) the indicators of expert competence in the $1^{st}$ approximation were obtained as following: $\{w_1(1); w_2(1); w_3(1); w_4(1); w_5(1); w_6(1); w_7(1); w_8(1); w_9(1); w_{10}(1); w_{11}(1); w_{12}(1); w_{13}(1); w_{14}(1); w_{15}(1)\} = \{0.0672; 0.0674; 0.0647; 0.0667; 0.0668; 0.0675; 0.0677; 0.0700; 0.0645; 0.0667; 0.0652; 0.0675; 0.0649; 0.0661; 0.0673\}$. In addition to this, let us calculate the average value of the group estimate of the $x_i$ in the $2^{nd}$ approximation by particular expression of the formula (3): $\alpha_i(2) = \sum_j w_j(1)\alpha_{ij}$. In this case, the average group estimates of the $x_i$-factors in the $2^{nd}$ approximation are the following numbers: $\{\alpha_1(2); \alpha_2(2); \alpha_3(2); \alpha_4(2); \alpha_5(2)\} = \{0.2720; 0.2522; 0.1748; 0.1511; 0.1498\}$. Checking these values for condition (4) and making sure that it is not fulfilled again: $\max\{|\alpha_i(2) - \alpha_i(1)|\} = \max\{|0.2720 - 0.2717|; \ |0.2522 - 0.2517|; \ |0.1748 - 0.1750|; \ |0.1511 - 0.1517|; \ |0.1498 - 0.1500|\} = 0.0005 > \varepsilon$, it is need to calculate the normalizing coefficient $\eta(2)$ as:

$\eta(2) = \sum_i(\sum_j \alpha_i(2)\alpha_{ij}) = 0.272 \cdot 4.075 + 0.2522 \cdot 3.775 + 0.1748 \cdot 2.625 + 0.1511$ $2.275 + 0.1498 \times 2.250 = 3.199$. Then according to (5), the experts' competence in the $2^{nd}$ approximation $w_j(2)$ $(j = 1 \div 15)$ will be the corresponding numbers: $\{w_1(2); w_2(2); w_3(2); w_4(2); w_5(2); w_6(2); w_7(2); w_8(2); w_9(2); w_{10}(2); w_{11}(2); w_{12}(2); w_{13}(2); w_{14}(2); w_{15}(2)\} = \{0.0672; 0.0674; 0.0646; 0.0667; 0.0668; 0.0675; 0.0677; 0.0700; 0.0645; 0.0667; 0.0651; 0.0675; 0.0649; 0.0661; 0.0673\}$. Applying particular case of formula (3) the average group estimates of the $x_i$-factors in the $3^{rd}$ approximation was obtained as corresponding numbers: $\{\alpha_1(3); \alpha_2(3); \alpha_3(3); \alpha_4(3); \alpha_5(3)\} = \{0.27205; 0.25218; 0.17481; 0.15113; 0.14982\}$. In this case, the accuracy of the group estimates of $x_i$-factors in the $3^{rd}$ approximation: $\max\{|\alpha_i(3) - \alpha_i(2)|\} = \max\{|0.27205 - 0.27205|; |0.25218 - 0.25218|; |0.17781 - 0.17482|; |0.15113 - 0.15114|; |0.14982-0.14982|\} = 0.00001 < \varepsilon$, already satisfies condition (4), which is the basis for stopping the computations. Then the values of the group estimates of the $x_i$-factors in the $3^{rd}$ approximation, i.e. the numbers $\{\alpha_i(3)\}$ $(i = 1 \div 5)$ will be considered their final (consolidated) weights.

## 3.3 Determination of Weighted Indexes of Alternatives on the Base of Expert Assessments of Influences

The method of expert evaluations assumes discussion of the $x_i$-factors by a group of experts invited from the context subject area. Each of them is offered to evaluate individually the degree of influence of $x_i$-factors on the index values of acceptable alternatives from the finite set $A = \{a_1, a_2, \ldots, a_s\}$, for example, by following five-point scale: 5 – TOO STRONG; 4 – SIGNIFICANTLY STRONG; 3 – STRONG; 2 – WEAK; 1 –

INSIGNIFICANT; $0$ – TOO WEAK. Such expert evaluations are analyzed for coordination according to the rule: the maximum acceptable difference between two expert opinions for any $x_i$-factor should not exceed 3. This rule allows filter out unacceptable deviations in expert assessments of alternatives for each particular influence. The computation of the total index, theoretically located in the range from 0 to 100, can be carried out by the following criteria:

$$C = \frac{\sum_{i=1}^{5} \alpha_i e_i}{\max_i \sum_{i=1}^{5} \alpha_i e_i} \times 100, \qquad (6)$$

where $\alpha_i$ is the weight of importance of the $x_i$-factors ($i = 1 \div 5$), $e_i$ is the expert evaluation of the alternative in terms of the influence of the $i$-th factor by five-point scale. Herewith, the maximum index denotes the consolidation of too strong influence of all factors. Table 3 presents various scenarios for the formation of indexes of 30 alternatives with the application of criterion (6).

**Table 3.**  Scenarios for the formation of indexes of alternatives.

| Alt. | Weights of influences | | | | | Index | Alt. | Weights of influences | | | | | Index |
|---|---|---|---|---|---|---|---|---|---|---|---|---|---|
| | $\alpha_1$ | $\alpha_2$ | $\alpha_3$ | $\alpha_4$ | $\alpha_5$ | | | $\alpha_1$ | $\alpha_2$ | $\alpha_3$ | $\alpha_4$ | $\alpha_5$ | |
| 01 | 5 | 5 | 5 | 5 | 5 | 100.00 | 16 | 2 | 1 | 3 | 2 | 1 | 35.63 |
| 02 | 4 | 4 | 4 | 4 | 4 | 80.00 | 17 | 5 | 3 | 4 | 3 | 3 | 74.19 |
| 03 | 3 | 3 | 3 | 3 | 3 | 60.00 | 18 | 4 | 3 | 5 | 5 | 5 | 81.36 |
| 04 | 2 | 2 | 2 | 2 | 2 | 40.00 | 19 | 2 | 4 | 3 | 2 | 3 | 57.10 |
| 05 | 1 | 1 | 1 | 1 | 1 | 20.00 | 20 | 2 | 3 | 3 | 3 | 1 | 50.18 |
| 06 | 0 | 0 | 0 | 0 | 0 | 0.00 | 21 | 5 | 3 | 4 | 4 | 3 | 76.14 |
| 07 | 5 | 4 | 4 | 3 | 3 | 82.29 | 22 | 3 | 2 | 1 | 1 | 3 | 39.53 |
| 08 | 3 | 1 | 4 | 2 | 4 | 53.37 | 23 | 5 | 3 | 4 | 3 | 5 | 77.08 |
| 09 | 2 | 3 | 3 | 1 | 2 | 41.11 | 24 | 4 | 4 | 3 | 2 | 3 | 70.02 |
| 10 | 2 | 3 | 2 | 4 | 3 | 51.95 | 25 | 4 | 3 | 5 | 5 | 5 | 84.72 |
| 11 | 3 | 4 | 2 | 2 | 4 | 57.74 | 26 | 3 | 1 | 2 | 2 | 3 | 42.75 |
| 12 | 2 | 3 | 1 | 2 | 3 | 41.84 | 27 | 3 | 1 | 1 | 2 | 3 | 38.91 |
| 13 | 4 | 3 | 4 | 3 | 2 | 66.42 | 28 | 2 | 2 | 0 | 1 | 0 | 24.97 |
| 14 | 1 | 3 | 2 | 3 | 1 | 41.61 | 29 | 2 | 2 | 0 | 1 | 0 | 24.53 |
| 15 | 4 | 2 | 3 | 2 | 2 | 58.46 | 30 | 3 | 4 | 5 | 5 | 5 | 82.91 |

## 4   Compilation of Expert-Quantitative Estimates of Alternatives in the Logical Basis of the Neural Networks

Expert systems are reasonably criticized that they do not transpose internal cause-effect relations. The quantitative approach to the evaluation of alternatives allows to compare alternatives using a single numerical factor that summarizes the relative influence of the certain number of factors by multi-factor function $R = R(x_1, x_2, \ldots, x_n)$. Nevertheless, the econometric models of type of $R$ used in decision making suffer from the difficulty of providing the current data sources of $x_i$-factors, most of which are weakly structured. Therefore, it is advisable to represent the working model in the form of a "*black box*", whose inputs and outputs are described by fuzzy sets [2].

In our case, external knowledge of 30 alternatives is represented by the information model of the form $\{(x_{1j}, x_{2j}, x_{3j}, x_{4j}, x_{5j}) \rightarrow y_j\}_j$ for each $j = 1 \div 30$ (see Table 3), where $y_j$ is the quantitative index of the $j$-th alternative, calculated according to (6), which weighted sums up the consolidated expert the relative influence of the $x_i$-factors. In the simplest case, when quantitative estimates of the influence of factors on alternatives are not in doubt, the function $R$ can be approximated by a three-layer feedforward neural network that induces outputs as $z_j = \sum_{k=1}^{m} c_k \phi[w_{ki} x_{ij}) - \theta_k]$, $(i = 1 \div 5; j = 1 \div 30)$, where $m$ is the number of nonlinear neurons in the hidden layer selected by the user during the simulation; $w_{ki}$ and $c_i$ are the weights of the input and output synoptic connections, respectively; $\theta_i$ is threshold of the $k$-th nonlinear neuron of the hidden layer; $\varphi(\cdot)$ is the activation function of the nonlinear neuron of the hidden layer, for example, of the sigmoidal type $\varphi(t) = 1/(1 + e^{-t})$. However, the task and its solution are much more difficult, because in the general case the nature of influences and the determination of their relative weights are the main problem of the quantitative estimation method.

Taking into account all the difficulties, with which expert systems and econometric models collide, special mathematical core based scoring system equally freely operating objective (quantitative) and subjective (qualitative) values can be solution of a multifactorial evaluation of alternatives. To form such core, we use the Fuzzy Inference System (FIS) in the logical basis of the feedforward neural network, assuming that the influences $x_i$ ($i = 1 \div 5$) as linguistic variables (LV) are well known, unambiguously understood and adequately represent the final vision of a multifactorial evaluation, providing, thus, the principle of the unity of measurements. Therefore, accepting scenarios for the formation of indexes of alternatives as the training set (see Table 3), we will load it into the Sugano-type MATLAB\ANFIS-editor (see Fig. 1). Next, we will generate the structure of the Sugeno-type MATLAB\FIS, for which we activate the "bell-typed" membership functions for fuzzy description of the terms of inputs $x_i$ ($i = 1 \div 5$) and output $y$. As a result, the structure of the fuzzy inference system in the logical basis of the neural network is visualized as shown in Fig. 1.

For the structural and parametric optimization of the generated FIS, the hybrid training method was used, by which, in particular, the membership functions of the fuzzy set describing the terms of the input linguistic variables are optimized [3]. In particular, the optimized membership functions of fuzzy sets describing the terms of the input linguistic variables $x_1$ and $x_5$ are shown in Fig. 2. Thus, in the process of simulating a multi-criteria evaluation of alternatives by MATLAB\ANFIS-editor it was possible to obtain an adequate cause-and-effect relations between the influences factors $x_i$ ($i = 1 \div 5$) and the summary indexes of alternatives. The graphical interface for viewing the optimized rule set of the generated FIS is shown in Fig. 2, where, in particular, the expert's estimate of the 10-th alternative as 51.9 is shown (see Table 3). Simulation of

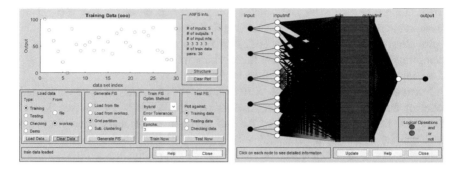

**Fig. 1.**  Generating the MATLAB\Fuzzy inference system.

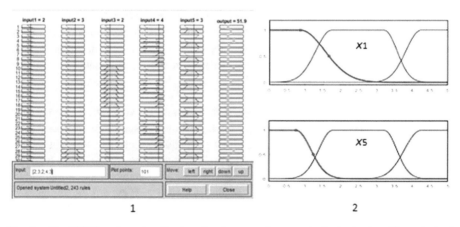

**Fig. 2.** (1) ANFIS rules visualization in the neural network logic basis; (2) optimized membership functions describing the terms of the LVs $x_1$ and $x_5$.

ANFIS in the logical basis of the five-layer neural network showed that it is able to adequately evaluate the alternatives not only from the declared list, but also outside it.

## 5   Conclusion

The proposed approach to the evaluation of alternatives can be quickly and easily adapted to the problem of multi criteria evaluation of alternatives from different areas and for any number of influences. The main advantage of this approach is its ability to identify the internal regularity of the process of evaluating alternatives and make them transparent. However, it is necessary to collect sufficient statistics of consolidated expert assessments of typical alternatives from the context of the research area.

    This work was supported by the Science Development Foundation under the President of the Republic of Azerbaijan – **Grant №EİF/MQM/Elm-Tehsil-1-2016-1 (26)-71/15/5)**.

# References

1. Lin, A.S., Wu, W.: Statistical Tools for Measuring Agreement. Springer, New York (2012). https://doi.org/10.1007/978-1-4614-0562-7
2. Zadeh, L.A.: The concept of a linguistic variable and its application to approximate reasoning. Inf. Sci. **8**(3), 199–249 (1965)
3. Lin, C.T., George Lee, C.S.: Supervised and unsupervised learning with fuzzy similarity for neural network-based fuzzy logic control systems. In: Yager, R.R., Zadeh, L.A., Reinhold, V. N. (eds.) Fuzzy Sets, Neural Network and Soft Computing, New York (1994)

# Self-confidence Preference Based Decision Making in Personnel Selection

S. Z. Eyupoglu[(⊠)] and G. E. Imanova

Department of Business Administration,
Near East University, Mersin 10, North Cyprus, TRNC, Turkey
`serife.eyupoglu@neu.edu.tr, gunel-gunay@hotmail.com`

**Abstract.** The efficiency of the decision making process is significantly related with preference relations of the decision maker. In the existing literature there are different types of preference relations, mainly fuzzy preference relations, linguistic preference relations, etc. In the existing literature preference representation formats that consider the self-confidence of the decision maker is neglected. In this paper we investigate the personnel selection decision making problem, specifically choosing the best candidate for a vacant faculty post taking into account the self confidence level of the decision maker. A numerical example with 5 criteria and 5 alternatives is provided.

**Keywords:** Fuzzy preference relations · Linguistic preference relations
Self-confidence level · Linear programming

## 1 Introduction

Decision making in personnel selection is very important in the fields of human resource management, organizational behaviour, industrial engineering, etc. Personnel selection and evaluation by using multi-criteria decision making approach is considered in [1]. It is shown that the appropriate solution of this problem can significantly affect the performance of the organization. In [2] the authors give an extensive review of personnel selection approaches. The aim of the study was to make use of different decision making methods. In [3] TOPSIS decision tool is used to solve personnel selection problem. The authors considered the selection of the most effective personnel in the automotive sector in Turkey. Multi-criteria decision making on personnel selection was considered in [4]. The problem was to choose the best candidate among 6 alternatives based on 5 criteria. For the solution compositional rule of inference was used.

However, in these studies the self-confidence level of the decision maker was not taken into account. In this paper we consider the personnel selection problem by using fuzzy preference relation with self-confidence. The paper is structured as follows; Sect. 2 introduces some basic information on fuzzy preference relation, linguistic variable, preference with self-confidence. In Sect. 3 the statement of the personnel selection problem is given. In Sect. 4 solution of the stated problem is given and Sect. 5 presents the conclusion.

© Springer Nature Switzerland AG 2019
R. A. Aliev et al. (Eds.): ICAFS-2018, AISC 896, pp. 288–292, 2019.
https://doi.org/10.1007/978-3-030-04164-9_39

## 2  Preliminaries

*Fuzzy Preference [4]*
Fuzzy preference relation on set of alternatives X is a relation on $X \times X$, $\mu_p$ : $X \times X \rightarrow [0, 1]$ $P_{ij}$ is the degree of preference of $x_i$ over $x_j$.

*Linguistic Variable for Self-confidence [4]*
In this paper self-confidence levels are described by linguistic terms, for example as:

$$S = \{low, medium, \ldots, very\ high\} \tag{1}$$

*Fuzzy Preference with Self-confidence [5]*
This index is presented as

$$P = \left(P_{ij},\ S_{ij}\right) \tag{2}$$

Here $P_{ij}$ is degree of preference $x_i$ over $x_j$, $S_{ij}$ is self-confidence level on the $P_{ij}$.

## 3  Statement of the Problem

The problem is to choose the best candidate for a vacant faculty post. The following criteria are used for decision making.

- Publication results (C1);
- Industrial experience (C2);
- Teaching quality (C3);
- Grant taking ability (C4);
- Intelligence level (C5).

Assume that there are 5 alternatives (candidates)

$$A = \{a_1,\ a_2,\ a_3,\ a_4, a_5\}$$

It is necessary to find priority vector of fuzzy preference with self-confidence w = {w1, w2, w3, w4, w5}, which minimizes total information deviation between the DM preference and vector w. The ranking of {w1, w2,…,w5} enables finding the best alternative.

## 4  Solution

The fuzzy self-confidence related preference relation is given in Table 1.

**Table 1.** The fuzzy self-confidence related preference relation

|       | $A_1$      | $A_2$      | $A_3$      | $A_4$      | $A_5$      |
|-------|------------|------------|------------|------------|------------|
| $A_1$ | (0.5, VH)  | (0.4, VH)  | (0.6, H)   | (0.8, VH)  | (0.7, H)   |
| $A_2$ | (0.6, VH)  | (0.5, VH)  | (0.7, VH)  | (0.9, VH)  | (0.8, VH)  |
| $A_3$ | (0.4, H)   | (0.3, VH)  | (0.5, VH)  | (0.7, VH)  | (0.6, H)   |
| $A_4$ | (0.2, VH)  | (0.1, VH)  | (0.3, VH)  | (0.5, VH)  | (0.4, VH)  |
| $A_5$ | (0.3, H)   | (0.2, VH)  | (0.4, H)   | (0.6, VH)  | (0.5, VH)  |

Here VH and H are linguistic terms of self-confidence level and are desired by triangular fuzzy numbers

$$VH = \text{Very High} = \{1.1, 0.9\}$$

$$H = \text{High} = \{0.7, 0.8, 0.9\}$$

To find priority vector of P we formulate the following linear programming model [5].

Objective function.

$$z = z_{12} + z_{13} + z_{14} + z_{15} + z_{23} + z_{24} + z_{25} + z_{34} + z_{35} + z_{45} \text{ min} \qquad (3)$$

subject to

$$0.5w_1 - 0.5w_2 - y_{12} = (0.4 - 0.5) = -0.1$$

$$0.5w_1 - 0.5w_3 - y_{13} = (0.6 - 0.5) = 0.1$$

$$0.5w_1 - 0.5w_4 - y_{14} = (0.8 - 0.5) = 0.3$$

$$0.5w_1 - 0.5w_5 - y_{15} = (0.7 - 0.5) = 0.2$$

$$0.5w_2 - 0.5w_3 - y_{23} = (0.7 - 0.5) = 0.2$$

$$0.5w_2 - 0.5w_4 - y_{24} = (0.9 - 0.5) = 0.4$$

$$0.5w_2 - 0.5w_5 - y_{25} = (0.8 - 0.5) = 0.3$$

$$0.5w_3 - 0.5w_4 - y_{34} = (0.7 - 0.5) = 0.2$$

$$0.5w_3 - 0.5w_5 - y_{35} = (0.6 - 0.5) = 0.1$$

$$0.5w_4 - 0.5w_5 - y_{45} = (0.4 - 0.5) = -0.1$$

$$Z_{12} - 5 * Y_{12} \geq 0$$

$$Z_{12} + 5 * Y_{12} \geq 0$$

$$Z_{13} - 3 * Y_{13} \geq 0$$

$$Z_{13} + 3 * Y_{13} \geq 0$$

$$Z_{14} - 5 * Y_{14} \geq 0$$

$$Z_{14} + 5 * Y_{14} \geq 0$$

$$Z_{15} - 3 * Y_{15} \geq 0$$

$$Z_{15} + 3 * Y_{15} \geq 0$$

$$Z_{23} - 5 * Y_{23} \geq 0$$

$$Z_{23} + 5 * Y_{23} \geq 0$$

$$Z_{24} - 5 * Y_{24} \geq 0$$

$$Z_{24} + 5 * Y_{24} \geq 0$$

$$Z_{25} - 5 * Y_{25} \geq 0$$

$$Z_{25} + 5 * Y_{25} \geq 0$$

$$Z_{34} - 5 * Y_{34} \geq 0$$

$$Z_{34} + 5 * Y_{34} \geq 0$$

$$Z_{35} - 3 * Y_{35} \geq 0$$

$$Z_{35} + 3 * Y_{35} \geq 0$$

$$Z_{45} - 5 * Y_{45} \geq 0$$

$$Z_{45} + 5 * Y_{45} \geq 0$$

$$w_1 + w_2 + w_3 + w_4 + w_5 = 1$$

$$w_i \geq 0, i = \overline{1,5}$$

$$z_{ij} \geq 0, i,j = \overline{1,5}$$

Solution of ((3)) gives

$$w = \{0.33, 0.53, 0.13, 0, 0\}$$

So best candidate is $A_2$, z = 2.9.

## 5   Conclusion

In this paper we have formulated a decision model on personnel selection taking into account the self-confidence level of the decision maker presented by linguistic terms. By using linear programming tool the priority vector of fuzzy preference relation with self-confidence is obtained. Priority vector enables the decision maker to find the best alternative. The numerical example for decision making to choose the best candidate for a vacant faculty post proved validity of the considered approach.

## References

1. Bogdanovic, D., Miletic, S.: Personnel evaluation and selection by multicriteria decision making method. Econ. Comput. Econ. Cybern. Stud. Res./Acad. Econ. Stud. **48**(3), 179–196 (2014)
2. Hudson, I., Reinerman, J.-L., Teo, G.: A review of personnel selection approaches for the skill of decision making. In: Augmented Cognition. Enhancing Cognition and Behavior in Complex Human Environments, pp. 474–485 (2017)
3. Senel, B., Senel, M., Aydemir, G.: Multi criteria decision making method topsis with personnel selection. Int. Ref. J. Res. Econ. Manag., 19–70 (2017)
4. Aliev, R.A., Aliev, R.R.: Soft Computing and Its Applications. Word Scientific, Singapore (2001)
5. Liu, W., Dong, Y., Chiclana, F., Herrera, E.-V., Cabrerizo, J.F.: A new type of preference relations: Fuzzy preference relations with self-confidence. In: International Conference on Fuzzy Systems (FUZZ-IEEE), pp. 1677–1684 (2016)

# The Incrementality Issue in the Wu-Mendel Approach for Linguistic Summarization Using IF-THEN Rules

Vuqar E. Mirzakhanov$^{(\boxtimes)}$ and Latafat A. Gardashova

Azerbaijan State Oil and Industry University,
Azadlig 20, Nasimi, Baku, Azerbaijan
mirzakhanv@inbox.ru, latsham@yandex.ru

**Abstract.** The incrementality issue in the Wu-Mendel approach for linguistic summarization of datasets is considered in this paper. An incremental version of the Wu-Mendel approach is proposed in order to increase the efficiency of the approach. Since the proposed and implemented modification is the analytical one, obtained in the equation form and derived from the original Wu-Mendel approach itself, the increase in efficiency is not followed by any decrease in the effectiveness of the approach. The proposed modification has been successfully put through the experimental verification process. Moreover, a MATLAB toolbox for linguistic summarization using IF-THEN rules has been designed and presented in this paper. The designed toolbox is made publicly available and allows a user to perform non-incremental/incremental summarization of datasets by applying the Wu-Mendel approach.

**Keywords:** Linguistic summarization · IF-THEN rules · Wu-Mendel approach
Incrementality

## 1 Introduction

Linguistic summarization (LS) is a Data Mining branch performing the knowledge extraction from data [1]. Almost all recently introduced LS techniques are based on the concepts of Zadeh's fuzzy logic (FL) [2]. The implementation of FL in LS was initially introduced by Yager [3–6] and then further developed by Kacprzyk et al. [7, 8]. Since its introduction, LS has had a lot of various applications, and most of the reported researches fall under the one of the following categories: LS of time series [9–11] and LS of datasets/databases [12–14].

This paper considers the Wu-Mendel approach for LS [14], which performs the generation of dataset-based summaries consisting of IF-THEN rules. This approach has several specific properties, which presence allows it to stand out in the current diversity of LS approaches. First, the Wu-Mendel approach applies five manifold measures to evaluate the quality of linguistic summaries, while most of the existing LS approaches applies only one quality measure, evaluating the degree of truth [1, 14]. Second, the Wu-Mendel approach can incorporate different types of fuzzy logic: the implementation of type-1 and interval type-2 fuzzy logic has been reported in [14]. Third, the

© Springer Nature Switzerland AG 2019
R. A. Aliev et al. (Eds.): ICAFS-2018, AISC 896, pp. 293–300, 2019.
https://doi.org/10.1007/978-3-030-04164-9_40

Wu-Mendel approach generates a linguistic summary consisting of IF-THEN rules, which makes the approach implementable in the design of rule-based systems, being one of the main types of knowledge-based systems [14].

The incrementality issue in the Wu-Mendel approach has been considered in this paper, and the approach has been modified in order to become incremental. The modified Wu-Mendel approach has been implemented in the form of the MATLAB toolbox, which has been made publicly available.

The rest of the paper is organized as follows. Section 2 briefly reviews the Wu-Mendel approach. Section 3 considers the incrementality issue in the Wu-Mendel approach. In Sect. 4, an incremental version of the Wu-Mendel approach is proposed, and the corresponding designed MATLAB toolbox is briefly introduced. In Sect. 5, the proposed incremental version of the Wu-Mendel approach is experimentally verified. Finally, Sect. 6 concludes the paper.

## 2  A Brief Review of the Wu-Mendel Approach

Let's define the dataset to be put through LS as follows:

$$D = \left\{ d_j^i \middle| i = 1, \ldots, I; j = 1, \ldots, J \right\} \tag{1}$$

where $d_j^i$ is the value of the $j$-th attribute in the $i$-th record, $I$ and $J$ are the numbers of the dataset records and attributes, respectively.

The Wu-Mendel approach processes $D$ and generates its linguistic summary consisting of all possible rules (by using the exhaustive search technique) being in the following form:

$$\textbf{IF } A_1 \text{ is } S_1 \& \ldots \& A_m \text{ is } S_m, \textbf{ THEN } A_{m+1} \text{ is } S_{m+1} \& \ldots \& A_{m+n} \text{ is } S_{m+n} [Q] \tag{2}$$

where $A_j (j = 1, \ldots, m+n)$ is an attribute of $D$, $S_j (j = 1, \ldots, m+n)$ is a linguistic term of $A_j$ user defined by type-1 (T1) or interval type-2 (IT2) fuzzy set, and $Q \in [0, 1]$ is a quality measure of the rule.

It should be mentioned that, in [14], the authors persistently claim that not T1 but IT2 fuzzy sets should be used in LS. However, the Wu-Mendel approach is not actually confined by any type of fuzzy sets, which is demonstrated in [14] itself. In this paper, the proposed incrementality modification is not related to any specific type of fuzzy sets either; so, for the sake of plainness and intelligibility, only T1 fuzzy sets are used in this paper.

The main quality measure $Q$ in the Wu-Mendel approach is represented by the degree of reliability ($R$), which is a function of two less complex quality measures: the degree of truth ($T$) and the degree of sufficient coverage ($C$). Moreover, there are two additional quality measures proposed in [14]: the degree of outlier ($O$) and the degree of simplicity ($S$).

The degree of truth defines the inference accuracy of a rule (the greater $T$ is, the greater the accuracy is) and is computed as follows:

$$T = \frac{\sum_{i=1}^{l} \min\left(\mu_{S_1}\left(d_1^i\right), \ldots, \mu_{S_{m+n}}\left(d_{m+n}^i\right)\right)}{\sum_{i=1}^{l} \min\left(\mu_{S_1}\left(d_1^i\right), \ldots, \mu_{S_m}\left(d_m^i\right)\right)} = \frac{c_D(S_1, \ldots, S_{m+n})}{c_D(S_1, \ldots, S_m)} \tag{3}$$

The degree of sufficient coverage defines the data support of a rule (the greater $C$ is, the greater the support is) and is computed as follows:

$$C = \begin{cases} 0, & r_c \le r_c^{min} \\ 2\left(\frac{r_c - r_c^{min}}{r_c^{max} - r_c^{min}}\right)^2, & r_c^{min} < r_c < \frac{r_c^{min} + r_c^{max}}{2} \\ 1 - 2\left(\frac{r_c^{max} - r_c}{r_c^{max} - r_c^{min}}\right)^2, & \frac{r_c^{min} + r_c^{max}}{2} \le r_c < r_c^{max} \\ 1, & r_c \ge r_c^{max} \end{cases} \tag{4}$$

where $r_c$ is the relative number of dataset records covered by a rule, $r_c^{max}$ is a user-defined fully satisfactory value of $r_c$ (if $r_c \ge r_c^{max}$, then $C = 1$) set equal to 0.15 in [14], and $r_c^{min}$ is a user-defined critically low value of $r_c$ (if $r_c \le r_c^{min}$, then $C = 0$) set equal to 0.02 in [14].

$r_c$ is computed as follows:

$$r_c = \frac{\sum_{i=1}^{l} t_i}{l} \tag{5}$$

where $t_i$ is computed as follows:

$$t_i = \begin{cases} 1, & \mu_{S_1}\left(d_1^i\right) > 0 \ and \ldots and \ \mu_{S_{m+n}}\left(d_{m+n}^i\right) > 0 \\ 0, & otherwise \end{cases} \tag{6}$$

The *degree of reliability* defines the trustworthiness of a rule (the greater $R$ is, the greater the trustworthiness is) and is computed as follows:

$$R = \min(T, C) \tag{7}$$

The *degree of outlier* defines the unexpectedness of the inference made a rule (the greater $O$ is, the greater the unexpectedness is) and is computed as follows:

$$O = \begin{cases} \min(\max(T, 1 - T), 1 - C), & T > 0 \\ 0, & T = 0 \end{cases} \tag{8}$$

The *degree of simplicity* defines the complexity (length) of a rule (the lesser $S$ is, the greater the complexity is) and is computed as follows:

$$S = 2^{2-l} \tag{9}$$

where $l$ is the total number of rule antecedents and consequents ($l = m + n$).

## 3    The Incrementality Issue in the Wu-Mendel Approach

The Wu-Mendel approach, despite having multiple advantages, has the great lack of efficiency, mainly, for the following reasons:

- The approach consecutively processes all available records in the initial dataset.
- The approach generates the combinatorial number of IF-THEN rules, which means that the resulting summary contains all possible rules.
- If some data is added to the initial dataset, LS has to be performed again on the whole modified dataset, even if the added data amount is exiguous.

In [14], the authors stated that the proposed LS approach has to be modified in order to become incremental. In case of the Wu-Mendel approach, the incrementality means the ability to update quality measures of the summary, when some new data comes in, only by processing the new data and avoiding the processing of the whole boosted dataset.

The incrementality issue is not too significant when dealing with small datasets, but becomes of great importance when dealing with complex datasets which summarization can take days/weeks/months. So, the incrementality is the issue closely related to the computational cost of LS, and the corresponding solution will make the Wu-Mendel approach more applicable in the summarization of enlargeable complex data.

## 4    An Incremental Version of the Wu-Mendel Approach and the Corresponding MATLAB Toolbox

As stated in Sect. 3, an incremental version of the Wu-Mendel approach has to be able to incrementally modify the quality measures $(T, C, R, O, S)$ of the summary. The quality measure $S$ doesn't need to be incrementally modified, since it depends only on the rule length, which is the same for a certain rule in the summary and doesn't change when new data comes in. The quality measures $R$ and $O$ are the functions of $T$ and $C$, so they can be computed directly by using the incrementally modified values of the degree of truth and the degree of sufficient coverage. Thus, an incremental version of the Wu-Mendel approach demands the incremental update of only two quality measures: $T$ and $C$.

Let's consider the initial dataset $D$ as a conjunction of two subsequent subsets $D_1$ and $D_2$ containing $I_1$ and $I_2$ records[1], respectively. The LS approach is considered to be incremental if, after LS of $D_1$, it can perform LS of $(D_1 + D_2)$ without having to process $D_1$ again.

The incrementality of $T$ assumes the reconsideration of Formula (3). According to (3), $T$ in case of LS of $D_1$ and in case of LS of $(D_1 + D_2)$ is computed as follows:

---

[1] $I_1 + I_2 = I$.

$$T(D_1) = \frac{\sum_{i=1}^{l_1} \min\left(\mu_{S_1}\left(d_1^i\right), \ldots, \mu_{S_{m+n}}\left(d_{m+n}^i\right)\right)}{\sum_{i=1}^{l_1} \min\left(\mu_{S_1}\left(d_1^i\right), \ldots, \mu_{S_m}\left(d_m^i\right)\right)} = \frac{c_{D_1}(S_1, \ldots, S_{m+n})}{c_{D_1}(S_1, \ldots, S_m)} \tag{10}$$

$$T(D_1 + D_2) = \frac{\sum_{i=1}^{l_1} \min\left(\mu_{S_1}\left(d_1^i\right), \ldots, \mu_{S_{m+n}}\left(d_{m+n}^i\right)\right) + \sum_{i=l_1+1}^{l_1+l_2} \min\left(\mu_{S_1}\left(d_1^i\right), \ldots, \mu_{S_{m+n}}\left(d_{m+n}^i\right)\right)}{\sum_{i=1}^{l_1} \min\left(\mu_{S_1}\left(d_1^i\right), \ldots, \mu_{S_m}\left(d_m^i\right)\right) + \sum_{i=l_1+1}^{l_1+l_2} \min\left(\mu_{S_1}\left(d_1^i\right), \ldots, \mu_{S_m}\left(d_m^i\right)\right)}$$

$$= \frac{c_{D_1}(S_1, \ldots, S_{m+n}) + c_{D_2}(S_1, \ldots, S_{m+n})}{c_{D_1}(S_1, \ldots, S_m) + c_{D_2}(S_1, \ldots, S_m)}$$

$$\tag{11}$$

As seen from (10) and (11), $T$ can be calculated incrementally as follows:

1. In case of LS of $D_1$, $T$ is calculated and added to the linguistic summary not alone but with $c_{D_1}(S_1, \ldots, S_{m+n})$ and $c_{D_1}(S_1, \ldots, S_m)$.
2. In case of the consequent LS of $(D_1 + D_2)$, $c_{D_2}(S_1, \ldots, S_{m+n})$ and $c_{D_2}(S_1, \ldots, S_m)$ are calculated, and $T$ is computed by using the calculated $c_{D_2}(S_1, \ldots, S_{m+n})$ and $c_{D_2}(S_1, \ldots, S_m)$ with the previously saved $c_{D_1}(S_1, \ldots, S_{m+n})$ and $c_{D_1}(S_1, \ldots, S_m)$.

The incrementality of $C$ assumes the reconsideration of Formula (5). According to (5), $r_c$ in case of LS of $D_1$ and in case of LS of $(D_1 + D_2)$ is computed as follows:

$$r_c(D_1) = \frac{\sum_{i=1}^{l_1} t_i}{l_1} \tag{12}$$

$$r_c(D_1 + D_2) = \frac{\sum_{i=1}^{l_1} t_i + \sum_{i=l_1+1}^{l_1+l_2} t_i}{l_1 + l_2} \tag{13}$$

As seen from (12) and (13), $C$ can be calculated incrementally as follows:

1. In case of LS of $D_1$, $r_c$ and $C$ are calculated, and $C$ is added to the linguistic summary not alone but with $\sum_{i=1}^{l_1} t_i$ and $l_1$.
2. In case of the consequent LS of $(D_1 + D_2)$, $\sum_{i=l_1+1}^{l_1+l_2} t_i$ is calculated, and $r_c$ with $C$ are computed by using the calculated $\sum_{i=l_1+1}^{l_1+l_2} t_i$ and $l_2$ with the previously saved $\sum_{i=1}^{l_1} t_i$ and $l_1$.

The aforesaid considerations and proposals are implemented in the form of the MATLAB toolbox performing non-incremental/incremental LS of datasets. The designed LS toolbox is made publicly available [15], consists of 9 scripts with 4 functions, and is provided with the tutorial giving the comprehensive explanation of how to handle the toolbox.

## 5   Experimental Verification of the Proposed Incremental Version of the Wu-Mendel Approach

The proposed incremental version of the Wu-Mendel approach for LS is experimentally verified on the "Skin Segmentation" dataset [16], having the dimension 245057 × 4 and being a collection of RGB color values belonging to the one of two classes: skin color and non-skin color. The attribute information on the "Skin Segmentation" dataset is provided in Table 1. The attributes of the dataset are user defined as linguistic variables depicted in Fig. 1.

**Table 1.** The attribute information on the "Skin Segmentation" dataset

| № | Name | Type[a] | Range | Description |
|---|------|---------|-------|-------------|
| 1 | Blue | Antecedent | [0, 255] | B (Blue) in RGB |
| 2 | Green | Antecedent | [0, 255] | G (Green) in RGB |
| 3 | Red | Antecedent | [0, 255] | R (Red) in RGB |
| 4 | Class | Consequent | {1, 2} | 1 – skin color, 2 – non-skin color |

[a]The rules to be mined during LS in this section are in the form *Red&Green&Blue* → *Class*. having 1 consequent and 3 antecedent attributes.

The "Skin Segmentation" dataset, for verification purposes, is randomly divided into 10 equal-size subsets. The subsets will be consecutively conjoined and put through the iterative LS procedures of two types: the original non-incremental LS and the proposed incremental one. For example, at the first iteration, 10% (subset 1) of the "Skin Segmentation" dataset is put through LS of two types; at the second iteration, 20% (subsets 1 & 2) of the "Skin Segmentation" dataset is put through LS of two types; etc. At the end of each iteration, the obtained two linguistic summaries are compared in terms of LS identity and LS time[2]. LS identity, in our case, means the equality of the obtained linguistic summaries and can take 2 values: equal (E) and unequal (UE). LS time, in this paper, is the amount of time spent by a computer (in our case, the *Dell Latitude E6420* running MS Windows 8.1 and MATLAB 2016a) to perform non-incremental/incremental LS.

As seen from Fig. 2, the incremental version of the Wu-Mendel approach, in comparison with the original non-incremental one, provides the same performance (measured by LS identity) accompanied by the lower computational cost (measured by LS time), which means that the results of LS are the same in case of both incremental and non-incremental approaches but are achieved at different costs: the incremental version is more efficient.

---

[2] Note that each LS time measurement in this section is taken 10 times, and the mean values are used in the corresponding figure. The RSD (relative standard deviation) of each taken LS time measurement is less than 5%.

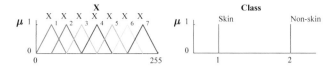

**Fig. 1.** The user-defined linguistic variables $X(X = Red, Green, Blue)$ and *Class*

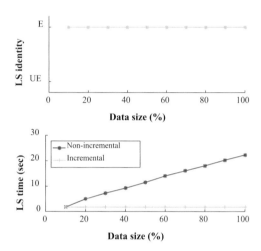

**Fig. 2.** The comparison of the original Wu-Mendel approach with its incremental version

## 6  Discussion and Conclusions

This paper has presented an incremental version of the Wu-Mendel approach for LS using IF-THEN rules. The incrementality of the Wu-Mendel approach makes it applicable in the up-to-date summarization of complex enlargeable datasets. The proposed incremental version of the Wu-Mendel approach has been compared with the original non-incremental one and has demonstrated the superior efficiency (measured by LS time) accompanied by the same effectiveness (measured by LS identity).

The obvious advantage of the incrementality modification proposed and considered in this paper is that it is the fully analytical one, which means that it is obtained in the equation form from the original Wu-Mendel approach itself. So, the results (effectiveness) of the proposed incremental version of the Wu-Mendel approach, in any case, cannot differ from the ones obtained by applying the non-incremental LS.

Moreover, the MATLAB toolbox for LS using IF-THEN rules has been presented in this paper and has been made publicly available.

# References

1. Boran, F., Akay, D., Yager, R.: An overview of methods for linguistic summarization with fuzzy sets. Expert Syst. Appl. **61**, 356–377 (2016)
2. Zadeh, L.: Fuzzy sets. Inf. Control **8**(3), 338–353 (1965)
3. Yager, R.: A new approach to the summarization of data. Inf. Sci. **28**(1), 69–86 (1982)
4. Yager, R.: On linguistic summaries of data. In: Knowledge Discovery in Databases, pp. 347–366. MIT Press, Cambridge (1991)
5. Yager, R.: Linguistic summaries as a tool for database discovery. In: Proceedings of Workshop on Fuzzy Database Systems and Information Retrieval at FUZZ-IEEE/IFES, Yokohama, pp. 79–82 (1995)
6. Yager, R.: Database discovery using fuzzy sets. Int. J. Intell. Syst. **11**(9), 691–712 (1996)
7. Kacprzyk, J., Yager, R., Zadrożny, S.: A fuzzy logic based approach to linguistic summaries of databases. Int. J. Appl. Math. Comput. Sci. **10**(4), 813–834 (2000)
8. Kacprzyk, J., Yager, R.: Linguistic summaries of data using fuzzy logic. Int. J. Gen Syst **30**(2), 133–154 (2001)
9. Wilbik, A., Kacprzyk, J.: On the evaluation of the linguistic summarization of temporally focused time series using a measure of informativeness. In: Proceedings of the International Multiconference on Computer Science and Information Technology, Wisla, pp. 155–162 (2010)
10. Castillo-Ortega, R., Marín, N., Sánchez, D., Tettamanzi, A.: Linguistic summarization of time series data using genetic algorithms. In: Proceedings of the 7th conference of the European Society for Fuzzy Logic and Technology, Aix-les-Bains, pp. 416–423 (2011)
11. Degtiarev, K., Remnev, N.: Linguistic resumes in software engineering: the case of trend summarization in mobile crash reporting systems. Procedia Comput. Sci. **102**, 121–128 (2016)
12. Pilarski, D.: Linguistic summarization of databases with quantirius: a reduction algorithm for generated summaries. Int. J. Uncertainty, Fuzziness Knowl.-Based Syst. **18**(3), 305–331 (2010)
13. Niewiadomski, A., Superson, I.: On multi-subjectivity in linguistic summarization of relational databases. J. Theor. Appl. Comput. Sci. **8**(1), 15–34 (2014)
14. Wu, D., Mendel, J.: Linguistic summarization using IF-THEN rules and interval type-2 fuzzy sets. IEEE Trans. Fuzzy Syst. **19**(1), 136–151 (2011)
15. The toolbox for the Wu-Mendel linguistic summarization approach. https://www.mathworks.com/matlabcentral/fileexchange/66837. Accessed 09 Apr 2018
16. Skin Segmentation Dataset. https://archive.ics.uci.edu/ml/datasets/Skin+Segmentation. Accessed 05 Apr 2018

# A Genetic Programming Approach to Forecast Daily Electricity Demand

Ali Danandeh Mehr[1], Farzaneh Bagheri[2], and Rifat Reşatoğlu[3(✉)]

[1] Civil Engineering Department, Antalya Bilim University, Antalya, Turkey
[2] Department of Electrical and Electronic Engineering, Eastern Mediterranean University, Famagusta, Mersin 10, Turkey
[3] Faculty of Civil and Environmental Engineering, Near East University, P.O. Box: 99138, Nicosia, Mersin 10, Turkey
rifat.resatoglu@neu.edu.tr

**Abstract.** A number of recent researches have compared machine learning techniques to find more reliable approaches to solve variety of engineering problems. In the present study, capability of canonical genetic programming (GP) technique to model daily electrical energy consumption (ED) as an alternative for electrical demand prediction was investigated. For this aim, using the most recent ED data recorded at northern part of Nicosia, Cyprus, we put forward two daily prediction scenarios subjected to train and validate by GPdotNET, an open source GP software. Minimizing root mean square error between the modeled and observed data as the objective function, the best prediction model at each scenario has been presented for the city. The results indicated the promising role of GP for daily ED prediction in Nicosia, however it suffers from lagged prediction that must be considered in practical application.

**Keywords:** Genetic programming · Electricity demand · Time series analysis

## 1 Introduction

Accurate forecast of electricity demand (ED) is one of the important tasks for a number of issues in energy industry including (but not limited to) sustainable energy supply, power plant operation, electricity market, industrialization etc. So far, various models/approaches have been recommended to forecast ED either through time series analysis of historical energy consumption or regression analysis between ED and germane forcing parameters/factors. To this end, variety of conventional statistical models (such as linear regression (LR), auto regressive (AR) or autoregressive moving average) or soft computing techniques (such as support vector regression (SVM), artificial neural networks (ANN), genetic programming (GP) and fuzzy logic) have been used. The current literature shows have been enunciated the superiority of soft computing techniques to conventional statistical models in ED forecasting (e.g., Fan and Chen 2006; Ekonomou 2010; Naghbe et al. 2018; Al-Musaylh et al. 2018; Hrnjica and Danandeh Mehr 2019). For instance, ANN, AR, and decision tree methods were compared by Tso and Yau (2007) and the results showed that ANNs canforecast ED more accurate than the counterparts. In a similar study by Ekonomou (2010), ANN was used to long-term

© Springer Nature Switzerland AG 2019
R. A. Aliev et al. (Eds.): ICAFS-2018, AISC 896, pp. 301–308, 2019.
https://doi.org/10.1007/978-3-030-04164-9_41

forecast of ED in Greece. The author proved that ANN is more accurate method than either LR or SVR models. Because of the non-stationary feature of ED series, particularly in developing cites, the commonly used standalone soft computing approaches may not be accurate enough to forecast ED signals. In such circumstances, hybrid soft computing techniques have been used to model the signals provided a set of observed ED is at hand. Azadeh et al. (2008) demonstrated that hybrid soft computing techniques can significantly improve the accuracy of standalone ANN model techniques. By enforcing economic indicators in the country, it integrated the genetic algorithm (GA) optimization with ANN to regress the ED in the Iranian agricultural ED. The results indicated that the hybrid GA-ANN outperforms ad hoc ANN and conventional LR. More recently, owing to the advances in metaheuristic optimization, variety of hybrid models like particle swarm optimization-ANN (Amjadi et al. 2010) developed for ED estimation. Although such hybrid models can provide promising forecasts, these are implicit models, and so difficult to apply in practice. It was well-documented that ANN-based methods create complex matrix of weights and bias vector that makes them often being criticized as "ultimate black boxes" techniques. Thus, the modeler may not distilling knowledge about the physics of the process (Danandeh Mehr and Nourani 2018). To overcome such problems, GP, which is relatively a new soft computing technique,has beensuggested and applied in the recent studies. Examples of such applications include Huo et al. (2007); Karabulut et al. (2008); Çunkaş and Taşkıran (2011); Forouzanfar et al. (2012). Referring to the literature, the first application of GP for ED prediction may belongs to Bhattacharya et al. (2002) who applied linear genome GP(LGP) to forecast ED in Australia. They showed that the LGP can produce straightforward and accurate models. In another outstanding research, Bakhshaii and Stull (2012) used gene expression programming (GEP) to regress ED series to the meteorological predictors in Canada. The model was found efficient and accurate enough to use in practice. More recently, Çunkaş and Taşkıran (2011) and Mousavi et al. (2014) reported the advantages of GP and GEP in ED forecasting.

Following the methodologies presented in the above-mentioned studies, we have investigated the capability of monolithic GP for one-day and two-day ahead prediction of ED in Cyprus for the first time in this article. To this end, GPdotNETv5 (Hrnjica and Danandeh Mehr 2019), an open source GP tool is used and introduced to GP users' community for the first time. To model ED process, we approach to the problem through time series forecasting analysis. Thus, the paper puts forward two univariate forecasting scenarios that use antecedent ED records to forecast future values. In each scenario, different setups are used to create simple and complex solutions. As a result, the efficiency of all developed models is compared according to different evaluation criteria, which are automatically calculated by GPdotNETv5.

## 2   Study Area and Data

Cyprus is the third largest island of the Mediterranean Sea with an area of 9251 km$^2$. Since 1974, the island covers two separated zone of Republic of Southern Cyprus and Turkish Republic of Northern Cyprus. It has typical Mediterranean climate with annual average electricity consumption about 4000 MWh on the northern part and about 6000

MWH on the southern part. On the basis of historical consumptions, the rapid increasing trend is observed since 2013. The total ED in the island is typically met through the consumption of the imported oil and fuel products. Like the island, Nicosia, the capital city, has been divided into two northern and southern parts. The present study uses the historical electricity consumption of northern Nicosia, also called Lefkoşa. Based upon the census reported in 2011 under the auspices of the UN, the Turkish Republic of Northern Cyprus has about 300,000 inhabitants living in Nicosia. (State Planning Organization 2011). The generation and distribution of electrical energy in Turkish Republic of Northern Cyprus is tracked by the Turkish Electricity Authority of Cyprus locally called KIB-TEK. The total production capacity of Turkish Electricity Authority is around 400 MW and is dependent on oil and fuel products. Figure 1 shows the daily ED of Lefkoşa during 2011–2016 years. As it is clear in the figure, the ED of Northern Cyprus is regularly increasing, and such a growing trend in the amount of ED is generally relevant to the nonstop grows of population and urbanization in the city. The statistical properties of the observed ED series used in this study are shown in Table 1.

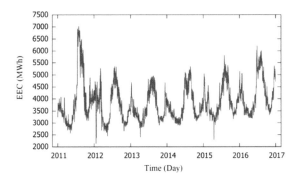

**Fig. 1.** Daily ED records innorthern part of Nicosia for the 2011–2016 period.

**Table 1.** Statistical properties of implemented data shown in Fig. 1

|  | Number | Mean | Median | Min | Max | Std. Dev. |
|---|---|---|---|---|---|---|
| All data | 2190 | 3875 | 3663 | 2130 | 7032 | 784.8 |
| Train set | 1534 | 3781 | 3585 | 2130 | 7032 | 783.4 |
| Test set | 656 | 4096 | 3911 | 2293 | 6215 | 745.5 |

## 3   GPdotNET for ED Forecasting

The selected ED prediction scenarios in this study include two scenarios of a one-day and two-day forecast, both of which are a function of the current and previous values that delayed receiving the ED value in the future. To obtain the minimum number of lags required to accurate forecast, first, the relatively large number of lags up to seven days was considered in this study as suggested by Danandeh Mehr et al. (2018). Then, the principal of the survival of the best gene in GP was applied to reach the optimum number of lags that effectively contribute in the prediction of future ED.

$$ED_{t+1} = f(ED_t, ED_{t-1}, \ldots, ED_{t-i}, \ldots, ED_{t-n}) + \varepsilon(t) \tag{1}$$

$$ED_{t+2} = f(ED_t, ED_{t-1}, \ldots, ED_{t-i}, \ldots, ED_{t-n}) + \varepsilon(t) \tag{2}$$

where, EDt denotes electricity demand at the day t. The index i represent slag values that varies from one up to six. The term $\varepsilon(t)$ is small bias that is considered as random noise in GP models.

Detailed description of different GP variants can be found in Hrnjica and Danandeh Mehr (2019). Therefore, a brief introduction of GP methodology is provided here. Monolithic GP is an automotive programming approach that uses evolutionary algorithms to generate computer programs (chromosomes) that can solve a specific problem. As a forecasting tool, GP is often used for self-structuring model to solve time series and symbolic regression problems. (e.g., Kumar et al. 2014). However, it can be applied as a binary or multiclass classifier (Danandeh Mehr et al. 2017). The major evolutionary operations that use to breed a chromosome in an initial population of chromosomes (i.e., random programs) to a set of preferred solutions are: reproduction, crossover, and mutation. Reproduction is direct transfer of the best chromosome into the new population without alteration. Crossover is substituting sub-chromosomes between ideal parents to yield better solution called offspring, and mutation is changing a randomly selected node (function or terminal) from a chromosome with another node/chromosomes from the initial population (Safari and Danandeh Mehr 2018).

To create GP solutions at each forecasting scenario, separate run was accomplished by GPdotNETv5.0. The first 70% of the observed data were chosen for training aim and the remaining 30% of ED signals was used to test the GP-induced models. All the input/output data were normalized using min-max normalization method in order to reach dimensionless programs. Thus, dimensionally correct ED forecasts can be easy obtained via de-normalizing the model induced values. All these issues can be automatically done by GPdotNET toolbox automatically. Thus, the users need to only import the raw inputs to GPdotNET. In the second step, the modeler may determine the sets of functions and terminals. The latter are typically the input variables and (optionally) some random constants. As the input and output data were normalized to have values within the range [0–1], the same range for random constant may be a wise selection. As previously mentioned, the present study considers two setups. In the first setup four basic arithmetic operators were considered and maximum tree depth was restricted to 8 levels. In the second setup, the aim was to reach more accurate solution neglecting their complexity. Thus, to achieve more efficient (and possibly more complicated) models, trigonometric and exponential functions were added to the function set. The model was permitted to be created by up to the 10 genes. Other run parameters include: population size = 500 program, fitness function = root mean square error (RMSE), initialization method = ramped Half and Half, and selection approach tournament. The over fitting problem perhaps happens when a small number of data samples in training set or a large number of generations during the training are used. To avoid over fitting problem in this study, we limited the maximum tree depth and generations to 10 and 500, respectively. In addition, we monitored efficiency of the evolved models at each generation with respect to the validation data. Thus the best model was selected based on its efficiency in both training and validation periods. For both the abovementioned forecasting scenarios, the configuration was found good enough to reach a suitable solution for the problem at hand.

## 4  Results and Discussion

To create the best forecasting models, GPdotNETv5.0 was used in this study. For each scenario different runs were done and after the various experiments the best evolved models were chosen. AS examples of solutions, the best simple GP models are shown in Fig. 2. The efficiency results of the best simple an complex models in terms of RMSE, Nash-Sutcliffe (NSE), percentage bias (PB), and correlation coefficient (CC) were given in Table 2. The parameters x1, x2,..., and x7 given in the figures represent the normalized values of $ED_{t-6}$, $ED_{t-5}$,..., and $ED_t$. The parameter $r$ is the random constants produced by GPdotNETv5.0. Regarding the evaluation criteria given in Table 2, it can be concluded that the models are precise enough to anticipate ED of Nicosia. To control the models' efficiency a giants timing error, we have visually examined the ED graphs of the best obtained models for simple cases in the test period, as shown in Fig. 3.

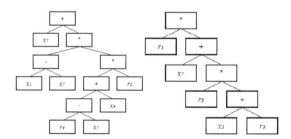

**Fig. 2.** The best evolved simple models at one-day (left) and two-day (right) ahead ED forecasting scenarios

In order to investigate in detail the timing errors of the GP models, the ED sequences observed during one of the highest ED month during the test period (i.e., examples 460–490) are shown in Fig. 3 for both estimation scenarios. While the timing difference between observed and modelled data is scarcely distinguishable in Fig. 3 (left), the right hand figure clearly shows that the models suffer from timing error in which the forecasts were shifted forward, i.e. they are advanced in time. Such error may be due to the high autocorrelation feature of ED signals, which causes GP give more chances of appearance for the most recent input.

To improve timing accuracy of the induced models, variety of data pre-processing procedures (with changeable performance and purposes) such as season algorithm,

**Table 2.** Summary of the model evaluation measures at training and testing period

| | Simple model | | Complex model | |
|---|---|---|---|---|
| | Train | Test | Train | Test |
| **One-day ahead scenario** | | | | |
| RMSE | 234.4 | 216.9 | 230.4 | 218.7 |
| NSE | 0.911 | 0.915 | 0.914 | 0.914 |
| PB | 0.042 | 0.038 | 0.041 | 0.038 |
| CC | 0.955 | 0.957 | 0.956 | 0.956 |
| **Two-day ahead scenario** | | | | |
| RMSE | 302.7 | 268.7 | 300.1 | 266.7 |
| NSE | 0.851 | 0.870 | 0.854 | 0.872 |
| PB | 0.055 | 0.048 | 0.055 | 0.048 |
| CC | 0.923 | 0.933 | 0.924 | 0.934 |

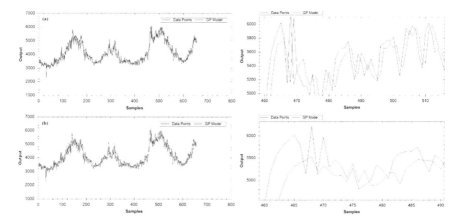

**Fig. 3.** Measured and modeledEDsignals at testing period (Left) and during a peak consumption (right): (a) one-day ahead scenario and (b) two-day ahead scenario.

firefly optimization, wavelet denoising, moving average filtering, singular spectrum analysis, were suggested in the literature. For instance, the hybrid season algorithm GP model developed by Danandeh Mehr and Nourani (2018) significantly cleaned timing error of the associated forecasts. Thus, the same methodology can be considered as the topic for further study in order to decrease timing error of ED forecasts.

## 5    Conclusion

In this study, based upon historical electricity demand patterns on Nicosia city, Cyprus, we investigated the ability of GP (as an explicit AI technique) as well as GPdotNET software to model daily ED. In order to present explicit GP trees at each scenario, we limited our GP instructions to the use of only mathematical functions. The models' efficiency was shown via square error, RMSE, Nash-Sutcliff efficiency, percentage bias, correlation coefficient, and determination coefficient. Our results demonstrated that the GP can handle the daily ED process. With respect to both scenarios examined here, GP resulted in high performance prediction models. It is also obtained that both simple and complex modelling setups resulted in more or less same efficiency. The results revealed the fact that increasing in lead time decreases the GP efficiency. Regarding timing accuracy, our results showed lagged prediction in the models. The reason is the evolutionary optimization feature of GP that lets the model to give more chance to the appearance of the latest input in the modelled output vector. Evidently, a wisdom methodology is required to remove this draw back. Owing to the restricted data in our study region, we used only the historical ED values as inputs. In the future studies, it may be instructive to include other parameters such as air temperature, economic factors which are effective in ED. Furthermore, it is worth mentioning that there is a tendency to get energy from renewable sources these days. Renewable energy sources are increasing steadily as current standard resources are exhausted and they cause damage to the environment with carbon dioxide emissions. North Cyprus needs to benefit from renewable energy sources such as sun, wind and even wave power. The electricity demand in Northern Cyprus can be financially affordable from the sources mentioned above.

# References

Al-Musaylh, M.S., Deo, R.C., et al.: Short-term electricity demand forecasting with MARS, SVR and ARIMA models using aggregated demand data in Queensland, Australia. Adv. Eng. Inform. **35**, 1–16 (2018)

Amjadi, M.H., Nezamabadi-pour, H., Farsangi, M.M.: Estimation of electricity demand of Iran using two heuristic algorithms. Energy Convers. Manag. **51**(3), 493 (2010)

Azadeh, A., Ghaderi, S.F., Sohrabkhani, S.: Annual electricity consumption forecasting by neural network in high energy consuming industrial sectors. Energy Convers. Manag. **49**, 2272–2278 (2008)

Bakhshaii, A., Stull, R.: Electric load forecasting for Western Canada: a comparison of two non-linear methods. Atmos.-Ocean. **50**(3), 352–363 (2012)

Bhattacharya, M., Abraham, A., Nath, B.: A linear genetic programming approach for modeling electricity demand prediction in Victoria. In: Abraham, A., Köppen, M. (eds.) Hybrid Information Systems (Advances in Soft Computing), p. 734. Springer, Heidelberg (2002). https://doi.org/10.1007/978-3-7908-1782-9_28

Çunkaş, M., Taşkıran, U.: Turkey's electricity consumption forecasting using genetic programming. Energy Sources Part B: Econ. Plan. Policy **6**(4), 406–416 (2011)

Danandeh Mehr, A., Nourani, V., Hrnjica, B., Molajou, A.: A binary genetic programing model for teleconnection identification between global sea surface temperature and local maximum monthly rainfall events. J. Hydrol. **555**, 397–506 (2017)

Danandeh Mehr, A., Nourani, V.: Season algorithm-multigene genetic programming: a new approach for rainfall-runoff modelling. Water Resour. Manag. **32**(8), 2665–2679 (2018)

Danandeh Mehr, A., Nourani, V., Khosrowshahi, V.K., Ghorbani, M.A.: A hybrid support vector regression–firefly model for monthly rainfall forecasting. Int. J. Environ. Sci. Technol., 1–12 (2018)

Ekonomou, L.: Greek long-term energy consumption prediction using artificial neural networks. Energy **35**, 512–517 (2010)

Fan, S., Chen, L.: Short-term load forecasting based on an adaptive hybrid method. IEEE Trans. Power Syst. **21**, 392–401 (2006)

Forouzanfar, M., Doustmohammadi, A., Hasanzadeh, S., Shakouri, H.G.: Transport energy demand forecast using multi-level genetic programming. Appl. Energy **91**(1), 496–503 (2012)

Hrnjica, B., Danandeh Mehr, A.: Optimized Genetic Programming Applications: Emerging Research and Opportunities. IGI-Global (2019). ISBN 13:9781522560050, https://doi.org/10.4018/978-1-5225-6005-0

Huo, L., Fan, X., Xie, Y., Yin, J.: Short-term load forecasting based 440 on the method of genetic programming. IEEE International Conference on Mechatronics and Automation, Harbin, China, 5–8 August, pp. 839–843 (2007)

Karabulut, K., Alkan, A., Yılmaz, A.S.: Long term energy consumption forecasting using genetic programming. Math. Comput. Appl. **13**, 71–80 (2008)

Kumar, B., Jha, A., Deshpande, V., Sreenivasulu, G.: Regression model for sediment transport problems using multi-gene symbolic genetic programming. Comput. Electron. Agric. **103**, 82–90 (2014)

Mousavi, S.M., Mostafavi, E.S., Hosseinpour, F.: Gene expression programming as a basis for new generation of electricity demand prediction models. Comput. Ind. Eng. **74**, 120–128 (2014)

Nagbe, K., Cugliari, J., Jacques, J.: Electricity Demand Forecasting Using a Functional State Space Model (2018)

State Planning Organization, General population and housing unit census. Statistical Yearbook, 2015. Nicosia, Turkish Republic of Northern Cyprus (2011)

Safari, M.J.S., Danandeh Mehr, A.: Multigene genetic programming for sediment transport modeling in sewers for conditions of non-deposition with a bed deposit. Int. J. Sedim. Res. **33**(3), 262–270 (2018)

Tso, G.K.F., Yau, K.K.W.: Predicting electricity energy consumption: a comparison of regression analysis, decision tree and neural network. Energy **32**(9), 1761–1768 (2007)

# Teaching Fuzzy Control Using an Embedded Processor

Dogan Ibrahim$^{(\boxtimes)}$

Near East University, Nicosia, Mersin 10, Turkey
dogan.ibrahim@neu.edu.tr

**Abstract.** Fuzzy control has become one of the most popular topics in automatic control engineering. The classical and intelligent fuzzy based control theory is taught in most undergraduate and postgraduate control engineering courses. In addition, the theory is backed-up by real control experiments carried out in the laboratories. One of the problems with real laboratory experiments is that they are usually very expensive and are beyond the budgets of many institutions. This paper describes the design of a low-cost fuzzy temperature control experiment using a microcontroller as an embedded processor.

**Keywords:** Fuzzy control · Fuzzy temperature control · Fuzzy education

## 1 Introduction

The automatic control theory is one of the important engineering fields as it continues to grow both in theory and in practical applications. The theory is based on highly complex mathematical derivations and equations. In general, as suggested by [1], the present day control theory was developed in three distinct stages:

In the first stage we see the simple on/off type control with very simple algorithms where the actuator that controls the plant is turned on if the plant output is below the desired set-point, and it is turned off should the plant output is equal to or is above the desired set-point. On/off type control requires no mathematical model of the plant under control. Although this type of control is extremely simple, it has several disadvantages in practice. Firstly, the actuator can be damaged as a result of the rapid requests to turn on and off. Secondly, the plant output becomes oscillatory and does not settle to a constant value making it undesirable in applications where the output accuracy is a critical factor. Thirdly, not all plants can be controlled with on/off type control.

The second stage of the control theory started with the development of complex mathematical control algorithms such as the PID. Here, the controller output is derived from a signal which is proportional to the error signal, its integral, and its derivative. One important advantage of the PID type control is that very accurate results can be obtained from the controller if an accurate model of the plant is available. But unfortunately it is very difficult or impossible to derive the mathematical models of complex systems such as non-linear systems. Another advantage of the PID controller is that, as suggested by [2], accurate control is achieved even in the presence of external disturbances.

© Springer Nature Switzerland AG 2019
R. A. Aliev et al. (Eds.): ICAFS-2018, AISC 896, pp. 309–314, 2019.
https://doi.org/10.1007/978-3-030-04164-9_42

The final stage of the control theory started with the development of the intelligent controllers as an alternative to the conventional control theory. Perhaps the most important topic in this stage is the Fuzzy Logic Control (FLC), [3], which is based on the design of linguistic and rule-based controller algorithms not requiring complex mathematical models, [4].

With the availability of low-cost embedded processors such as microcontrollers the popularity of fuzzy control has increased over the last decade and nowadays there are many applications of fuzzy controllers in many diverse fields. Some application areas of fuzzy control are: water purification, automatic train operation systems, water level control, quadcopter attitude control, motor speed control, battery charge control, and many others. Non-linear processes have complex behaviours and it is generally very difficult if not impossible to drive an accurate mathematical model of such a process. One important advantage of FLC is that it can be used to control complex non-linear processes where a model of the plant is not available. One of the problems in teaching fuzzy control is that the laboratory experimentation kits are expensive and are usually beyond the budgets of many educational institutions. This paper describes the design and development of a low-cost fuzzy control based microcontroller driven temperature control system for educational purposes.

## 2   The Designed System

Figure 1 shows the block diagram of the designed fuzzy temperature control system. Temperature is read by the sensor and is sent to the microcontroller which compares this reading with the set-point value and generates the error signal as the difference between the set-point and the measured value. This error signal goes through the fuzzy controller which generates a crisp controller signal that controls the heater to increase or decrease the temperature so that the desired set-point temperature is achieved.

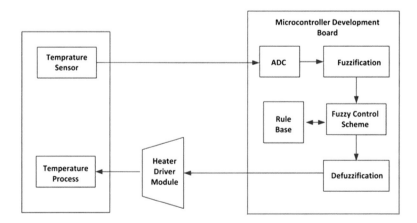

**Fig. 1.** Block diagram of the designed fuzzy temperature control system

## 2.1    The Hardware

Figure 2 shows a simplified hardware setup of the designed system. Here, the aim has been to keep the cost as low as possible by using standard off the shelf components wherever possible.

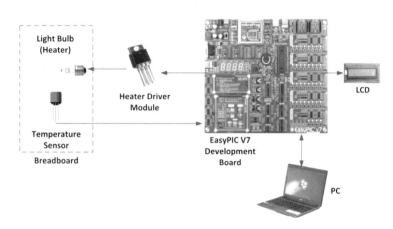

**Fig. 2.** Hardware of the designed temperature control system

**The Temperature Sensor.** In this design a TMP36DZ type analog temperature sensor, chip is used which can measure the temperature in the range −40 °C to +150 °C with a linearity of ±0.5 °C and an accuracy of ±2 °C, operating with +2.7 V to +5.5 V supply voltage. The output voltage of this sensor is directly proportional to the measured voltage. The output of the temperature sensor is fed to the Analog-to-Digital converter (ADC) input channel of the microcontroller where it is converted into digital format so that it can be processed by the microcontroller.

**The Heater.** A small 12 V incandescent light bulb was used in the design as the heat generator and the temperature sensor chip was placed very close to the light bulb. It was measured experimentally that the surface temperature of the light bulb changes with the applied voltage and it can be in excess of +100 °C when full 12 V is applied.

**The Microcontroller Development System.** In this design a microcontroller with a development system was used as an embedded processor. Any type of microcontroller development system could be used in the design as long as it has one ADC channel, LCD interface, and a PWM module. In this design the EasyPIC V7 microcontroller development system was used which is based on the medium performance PIC18F45K22 8-bit microcontroller.

**The Heater Driver Module.** The heater driver module controls the voltage supplied to the heater (the light bulb in this case). In this project the output of the heater driver module is a PWM signal. The duty cycle as a percentage is an important parameter of a PWM signal.

# 3 The Fuzzyfication Process

## 3.1 Fuzzyfication of the Inputs

Triangular and trapezoidal membership functions are frequently used to fuzzify the inputs and it is necessary to determine the range of fuzzy variables related to the crisp inputs. Here, as shown in Fig. 3, the following fuzzy variables are used to define the error input: SNEG = small negative, NEG = negative, ZERO = zero, SPOZ = small positive, POZ = positive, and LPOZ = large positive.

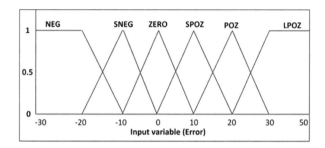

**Fig. 3.** Graphical representation of the input error linguistic variables

## 3.2 Fuzzification of the Outputs

The following fuzzy membership values were assigned for the output variable (Fig. 4): ZE = zero, SM = small, ME = medium, SH = small high, HI = high, VH = very high.

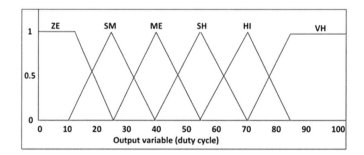

**Fig. 4.** Graphical representation of the output linguistic variables

## 3.3 The Rule Base

The rule base consists of the fuzzy rules in the form of IF…ELSE blocks where the IF part represents the situation and the THEN part describes the response of the system. Table 1 shows the fuzzy rules for the system.

**Table 1.** Rule base

| IF the input error is... | **THEN** the duty cycle is... |
| --- | --- |
| NEG | VH |
| SNEG | HI |
| ZERO | SH |
| SPOZ | ME |
| POZ | SM |
| LPOZ | ZE |

### 3.4  Defuzzification

The process of defuzzification gives a numeric value for the duty cycle to be used to power the heater, There are many ways to carry out the defuzzification process [5]. One of the popular ones used in this paper is the weighted average defuzzification (or the center of mass defuzzification) technique.

## 4  Results

The set-point temperature was displayed as a constant on the first row of the LCD, while the measured temperature was displayed on the second row. Figure 5 shows a typical result where the ambient temperature was 30 °C and the set-point was at 50 °C. It is clear from the graph that the temperature settled to the desired set-point without any overshoots. Experiments with different set-point values gave similar successful results.

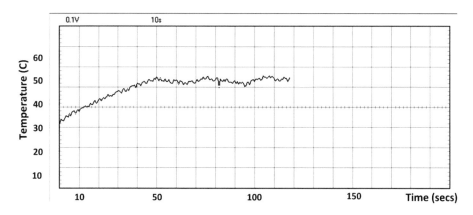

**Fig. 5.** A typical output

## 5   Conclusions

The design of a low-cost educational fuzzy temperature control system has been described. Perhaps the most attractive feature of this system is that it is extremely low-cost, consisting of a small lamp, a temperature sensor chip, a power transistor, and a microcontroller development system. The system is easy to construct and use, it is safe (uses low voltage), thus making it a suitable laboratory experiment in undergraduate teaching of feedback control systems and related fields.

Although the temperature control system described in this paper uses the well-known and established fuzzy control algorithms, there is no reason why it cannot be used as a laboratory experiment in teaching the practical aspects of classical feedback control. For example, the system can be used in experiments to teach the practical aspects of various classical controller design algorithms such as PI, PID, and also the design of other fuzzy logic based control algorithms.

## References

1. Alla, M.E., Taha, W.I.M: PID and fuzzy logic in temperature control system. In: International Conference on Computing, Electrical and Electronic Engineering (ICCEEE), pp. 172–177 (2013)
2. Johnson, C.D.: Process Control Instrumentation Technology, 8th edn. Pearson Education Ltd, Essex (2014)
3. Bolton, W.: Instrumentation and Control Systems. Newnes, Oxford (2006)
4. Zadeh, L.A.: A rationale for fuzzy control. Trans. ASME J. Dynam. Syst. Measur. Control **94**, 3–4 (1972)
5. Patil, A.R., et al.: Embedded fuzzy module for battery charger control. Int. J. Adv. Res. Electr. Electr. Instr. Eng. **2**, 4072–4078 (2013)

# PSO Algorithm Applied to Enhance Power Quality of Multilevel Inverter

Yahya Naderi[1], Fahreddin Sadikoglu[2(✉)],
and Seyed Hossein Hosseini[1,2]

[1] Faculty of Electrical and Computer Engineering, Tabriz University,
Tabriz, Iran
[2] Engineering Faculty, Near East University,
99138 Nicosia, North Cyprus, Mersin 10, Turkey
fahreddin.sadikoglu@neu.edu.tr

**Abstract.** In this paper, a multilevel inverters' output voltage THD is the object of optimizationdu. Since the higher number of voltage levels in output of a multilevel leads to high quality output voltage and current, it is desired to have more levels in output, and it needs more switches to handle this issue. As the number of equations increase, it becomes complicated to find optimum switching angles that lead to least THD value, So particle swarm optimization method is used to find optimum switching angles in this paper. Lower THD will improve the power quality of output voltage, and it can be achieved by selected harmonic elimination method, that has been used to eliminate desired harmonics. The topology used in this paper is an optimized topology of multilevel inverter, cascaded H-bridge with unequal DC voltage sources. The novelty of this work is in its very low output THD and the optimized configuration calculations of the presented cascaded multilevel inverter.

**Keywords:** Multilevel inverters · Selective harmonic elimination
Particle swarm optimization (PSO) · THD

## 1 Introduction

Multilevel inverters was introduced in Early 1970s and since then it has attracted attention of researchers in medium-voltage and high-power applications such as motor drivers [1–3], static AVR compensators (STATCOMs) [4] and renewable energy systems [5] because of its unique characteristics. A multilevel inverter combines several DC voltage sources to have an output voltage waveform which is nearly sinusoidal. The advantages of multilevel converter in comparison with traditional converters are its lower harmonic components, higher power quality, lower switching losses, needlessness of transformer at distribution voltage level and better EMI due to its smaller voltage steps, however, the most important drawback of multilevel inverter is the high number of DC energy sources and other power electronic devices.

Multilevel inverters are classified into three categories with respect to their configuration; diode clamping, flying capacitors and cascaded H-bridge multilevel inverters [6]. Between the variety of topologies for multilevel inverters, there is a

© Springer Nature Switzerland AG 2019
R. A. Aliev et al. (Eds.): ICAFS-2018, AISC 896, pp. 315–324, 2019.
https://doi.org/10.1007/978-3-030-04164-9_43

specific attention for cascaded multilevel inverter due to its modularity and simplicity of control [7, 8]. The output voltage and elimination of undesired harmonics can be controlled by different modulation techniques. Sinusoidal pulse width modulation (SPWM) and space vector PWM (SVPWM) are the modulation techniques which are discussed on [9, 10]. Due to the fact that PWM techniques can not eliminate lower-order harmonics completely, selective harmonic elimination (SHE) or PWM techniques are presented in [11, 12]. The principle of SHE method is to get an arithmetic solution of nonlinear transcendental equations which has trigonometric terms and naturally will generate multiple solutions. There are some methods that can be used for solving the mentioned equations. The first one is numerical iterative techniques Newton-Raphson. A proper initial guess is needed for Newton-Raphson method to work properly, however, it is not available for SHE problem due to the fact that, the search space of SHE is unknown. The second strategy is a symmetric approach which is based on resultant theory and it is proposed in [13, 14]. In this method the transcendental equations are converted to polynomial equation and the algorithm is applied on the equations to find all possible solutions for the study case. The drawback of this method lies in the fact that it is not possible to solve the equations under the circumstance of large degrees of polynomial equation. Other strategies for solving these equations are modern stochastic search techniques such as GA and PSO.

## 2   Cascaded Multilevel H-Bridge Topology

The Cascaded H-bridge multilevel shown in Fig. 1-a is used as multilevel inverter in this paper, as it can be seen, it is made up of series connection of n basic units that are followed by an H-bridge to output positive and negative voltage levels and zero also. Each basic unit is built up of two unidirectional switches (an IGBT and antiparallel Diode as shown in the Fig. 1-a) with an isolated DC source, it can produce both positive and zero polarity voltage in the output, it should be noted that, these switches work in a complementary mode. The output voltage of this inverter before and after connecting the H-bridge is shown in Fig. 1-b, as it can be seen maximum output voltage is sum of all series DC sources, $V'_o = 2^n - 1$. By adding the single phase H-bridge to the inverter, sub multilevel is formed and output voltage will alternate between positive and negative values. Extended multilevel topology shown in Fig. 2 is formed by cascaded connection of sub multilevel units.

To have less THD and higher number of voltage steps, it is better to use unequal DC sources, so an optimized method for determining DC source magnitudes is used. In this algorithm, number of output levels is larger in comparison with other conventional methods. Voltage magnitude of DC sources is calculated based on Binary basis, so that value of voltage DC source for the first basic unit and $j_{th}$ voltage DC source is computed by Eq. (1).

$$V_{1,j} = 2^{j-1} V_{dc} \qquad \text{for} \quad j = 1, 2, \cdots, m \qquad (1)$$

for the first basic unit of second sub-multilevel it is calculated by Eq. (2)

$$V_{2,1} = V_{dc} + 2 \sum_{i=1}^{m} V_{1,i} = V_{dc}[2^{m+1} - 1] \tag{2}$$

Finally for the $j_{th}$ unit of second sub-multilevel $V_{2,j}$ is calculated as follows:

$$V_{2,j} = (2^{m+1} - 1) \times (2^{j-1}) \; V_{dc} \qquad \text{for} \quad j = 2, 3, \cdots, m \tag{3}$$

In this equation "m" is the number of voltage DC source in each sub-multilevel, for the $n_{th}$ sub-multilevel unit and n is the number of all sub-multilevel units.
For the nth sub-multilevel, the $V_{n,1}$ is calculated as:

$$V_{n,1} = V_{dc} + 2 \left( \sum_{i=1}^{n-1} \sum_{k=1}^{m} V_{i,k} \right) = V_{dc}(2^{m+1} - 1)^{n-1} \tag{4}$$

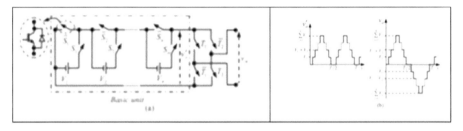

**Fig. 1.** (a) Sub-multilevel inverter with its components, (b) Output voltage of basic unit and sub-multilevel inverter

The number of steps and variety of DC voltage source values is calculated in (5, 6)

$$N_{step} = (2^{m+1} - 1)^{n} \tag{5}$$

$$N_{variety} = nm \tag{6}$$

Maximum of output voltage is presented in (7).

$$V_{o,max} = \sum_{j=1}^{n} \sum_{k=1}^{m} V_{k,j} = (2^{m} - 1) \sum_{j=1}^{n} V_{j,1} = V_{dc} \left[ \frac{(2^{m+1} - 1)^{n} - 1}{2} \right] \tag{7}$$

For optimal structure the number of DC voltage sources must be equal in each unit [15–18]. Therefore for 15-level multilevel inverter we can compute the number of DC voltage source for one basic unit by (9).

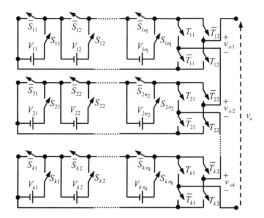

**Fig. 2.** Extended multilevel converter topology

$$m_1 = m_2 = m_3 = \ldots = m_n \tag{8}$$

$$\left(2^{m+1} - 1\right)^n = 15 \tag{9}$$

In this paper, one basic unit is used to be simulated and optimized, then the number of DC voltage sources is $m = 3$, $n = 1$ for $n_{step} = 15$.

According to the Fig. 1, the number of switches used for this topology can be obtained from (10, 11)

$$N_{switch} = N_{switch,1} + N_{switch,2} + \cdots + N_{switch,n} \tag{10}$$

$$N_{switch,j} = 2\,(m+2) \tag{11}$$

Therefore 10 switches are needed for this structure. Figure 3 shows a quarter cycle of a typical stepped waveform of the phase voltage for 15-level inverter with binary based DC source values. Values of each DC voltage is calculated as,

$$V_{1,1} = V_{base} = V_{dc}, \quad V_{1,2} = 2\,V_{dc}, \quad V_{1,3} = 4V_{dc} \tag{12}$$

Table 1 shows the switching table for mentioned structure. The maximum possible value of the fundamental component is obtained from. To determine the exact value for $V_{dc}$, the peak of voltage must be considered equal to grid voltage ($220 \times \sqrt{2}$). Therefore it can be computed, such that:

$$V_{o,Max} = 220\sqrt{2} = 311.127 \Rightarrow 7V_{dc} = 311.127 \tag{13}$$

Then,

$$V_{1,1} = V_{dc} = 44.447; \quad V_{1,2} = 2\,V_{dc} = 88.88; \quad V_{1,3} = 4V_{dc} = 176.76 \qquad (14)$$

## 3  Harmonic Minimization Issue of Multilevel Inverters

To find the optimum switching angles of a multilevel inverter and overcoming harmonic problems it is needed to solve the cost function that leads to the minimum THD value. The Fourier transform of the following expression is as follows;

$$V_n = \left\{ \sum_{n=1}^{\infty} \frac{4}{n\pi} V_{dc} \sin(n\pi)\{\cos n\theta_1 + \cos n\theta_2 + \cos n\theta_3 \\ + \cos n\theta_4 + \cos n\theta_5 + \cos n\theta_6 + \cos n\theta_7\} \right\} \qquad (15)$$

Because the quarter-wave is symmetric in the waveform, it only contains odd-order harmonics. Normalizing the fundamental components based on the maximum magnitude and presenting it in per unit (P.U).

$$V_{pu} = \frac{V_n}{V_{o,Max}} = \frac{1}{7V_{dc}} \sum_{n=1}^{\infty} \left\{ \frac{4}{n\pi} V_{dc} \sin(n\pi)\{\cos n\theta_1 + \cos n\theta_2 + \cos n\theta_3 + \\ \cos n\theta_4 + \cos n\theta_5 + \cos n\theta_6 + \cos n\theta_7\} \right\} \qquad (16)$$

In which $\theta_i$ is the switching angle for the presented waveform in Fig. 3. The switching angles $\theta_1, \theta_2, \cdots, \theta_7$ must satisfy the following limits,

$$0 \le \theta_1 < \theta_2 < \cdots < \theta_7 \le \frac{\pi}{2} \qquad (17)$$

To have the desired value of the fundamental harmonic while minimizing the other harmonics, (18) must be solved using minimization algorithm for different M values. Table 1 presents the switching table for the proposed multilevel inverter.

## 4  Particle Swarm Optimization (PSO)

PSO algorithm which stands for the particle swarm optimization is a stochastic search algorithm based on population, which could be an alternative solution for non-linear complex optimization problems. It has been introduced on the basic idea of social behaviour development of the animas, such as bird migration, fish schooling and etc. based on the natural group knowledge which could be transferred to other group members, if a member of a group finds the best position, the other members are able to follow the member to reach the best position and it is called the social knowledge of a group member [19, 20]. This is the basic idea behind PSO, so that all the group members will start to go to the best position if a member reaches there.

## 5 Implementation of PSO Algorithm to Harmonic Minimization Problem

In three phase systems the triple harmonics are automatically emitted from line-line voltage, so by solving the equation set in (18) the harmonics with the order of $6k \pm 1^{th}$ will be minimized.

$$\begin{cases} (\cos \theta_1 + \cos \theta_2 + \cos \theta_3 + \cos \theta_4 + \cos \theta_5 + \cos \theta_6 + \cos \theta_7) = 2\pi M \\ (\cos 5\theta_1 + \cos 5\theta_2 + \cos 5\theta_3 + \cos 5\theta_4 + \cos 5\theta_5 + \cos 5\theta_6 + \cos 5\theta_7) = 0 \\ (\cos 7\theta_1 + \cos 7\theta_2 + \cos 7\theta_3 + \cos 7\theta_4 + \cos 7\theta_5 + \cos 7\theta_6 + \cos 7\theta_7) = 0 \\ \qquad\qquad\qquad \cdots \\ (\cos 19\theta_1 + \cos 19\theta_2 + \cos 19\theta_3 + \cos 19\theta_4 + \cos 19\theta_5 + \cos 19\theta_6 + \cos 5\theta_7) = 0 \end{cases}$$

$$(18)$$

**Table 1.** The switching table for presented structure

| $V_o$ | $S_{11}$ | $S_{12}$ | $S_{13}$ | $T_{11}$ | $T_{12}$ |
|---|---|---|---|---|---|
| $-7\,V_{dc}$ | 1 | 1 | 1 | 1 | 1 |
| $-6\,V_{dc}$ | 0 | 1 | 1 | 1 | 1 |
| $-5\,V_{dc}$ | 1 | 0 | 1 | 1 | 1 |
| $-4\,V_{dc}$ | 0 | 0 | 1 | 1 | 1 |
| $-3\,V_{dc}$ | 1 | 1 | 0 | 1 | 1 |
| $-2\,V_{dc}$ | 0 | 1 | 0 | 1 | 1 |
| $-1\,V_{dc}$ | 1 | 0 | 0 | 1 | 1 |
| 0 | 0 | 0 | 0 | 1 | 1 |
| $1\,V_{dc}$ | 1 | 0 | 0 | 0 | 0 |
| $2\,V_{dc}$ | 0 | 1 | 0 | 0 | 0 |
| $3\,V_{dc}$ | 1 | 1 | 0 | 0 | 0 |
| $4\,V_{dc}$ | 0 | 0 | 1 | 0 | 0 |
| $5\,V_{dc}$ | 1 | 0 | 1 | 0 | 0 |
| $6\,V_{dc}$ | 0 | 1 | 1 | 0 | 0 |
| $7\,V_{dc}$ | 1 | 1 | 1 | 0 | 0 |

Since there are 7 DOF (Degrees of Freedom) in solving this equation, up to 6 undesired harmonic values could be omitted, which means up to $19^{th}$ harmonic component could be minimized using this method.

As a result the cost function includes both undesired harmonics and fundamental harmonic. PSO optimization algorithm is applied on THD function using MATLAB programming environment. The equations are solved and optimum switching angles for different modulation factors are calculated. As an example, Table 2 show the results of this optimization process for M = 1.

**Table 2.** Optimum switching angles to minimize output voltage THD based on PSO for M = 1

| $\theta_1$ | $\theta_2$ | $\theta_3$ | $\theta_4$ | $\theta_5$ | $\theta_6$ | $\theta_7$ |
|------|-------|-------|-------|-------|-------|------|
| 2.59 | 12.38 | 20.36 | 30.54 | 39.60 | 50.10 | 68.5 |

**Fig. 3.** Quarter cycle of a typical stepped waveform of the phase voltage for 15-level inverter

## 6   Minimization of THD

An important index that shows the power quality of a voltage or current waveform is the total harmonic distortion (THD), which could be calculated based on the Fourier transform of the voltage or current. Another way to improve the power quality of a waveform is to minimize the THD for that output. In this paper, up to 31 harmonics are used in calculation of THD, and the cost function is written as,

$$F(\theta_1, \theta_2, \cdots, \theta_7) = \left( \frac{|v_5| + |v_7| + \cdots + |v_{31}|}{7v} \right) \tag{19}$$

It should be note that this function is on the basis that, power system can eliminate harmonics that are multiples of 3, and also even harmonics are eliminated because of symmetric output of inverter, so the residual harmonics are of $6K \pm 1^{th}$ degrees. Output Voltage THD is the objective function which is minimized by choosing proper switch angles so that the optimization cost function is as, $F(\theta_1, \theta_2, \cdots, \theta_7)$ Dependant on, $0 \le \theta_1 < \theta_2 < \cdots < \theta_7 \le \frac{\pi}{2}$.

## 7   Simulation Results

The over mentioned structure is simulated by PSCAD/EMTDC software and output voltage and current THD is computed also. For equal switching angles output voltage THD is calculated about 16.75% however, with proper switching angles which are obtained from PSO optimization process it decreases to 3.32%. THD for output current for R-L load of ($R = 5\,\Omega$, $L = 0.75\,\text{mH}$) is about 2.50%, that shows the filtering process of R-L load. Figure 4 shows the simulation voltage and current output results for 15-level inverter based on equal switching angles and Fig. 5 shows the results for 15-level inverter based on optimized switching angles. As it can be seen the phase difference of current and voltage waveforms are obvious. Figures 6 and 7 show the output voltage harmonic spectrum in ordinary and zoomed stance. Finally to show the other results, the calculated THD values for different modulation factors are calculated and shown in Fig. 8.

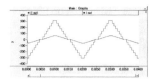

**Fig. 4.** Output voltage and output current waveforms of a 15-level multilevel inverter based on equal switching angles

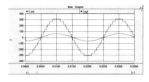

**Fig. 5.** Output voltage and Current waveforms of a 15-level inverter based on optimum switching angles

**Fig. 6.** Harmonic spectrum of output voltage up to $31^{th}$ harmonic

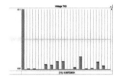

**Fig. 7.** Zoomed harmonic spectrum of output voltage up to $31^{th}$ harmonic

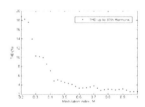

**Fig. 8.** Calculated THD values for different modulation factors

# 8 Conclusion

The PSO technique has been used to find optimum switching angles for minimum output voltage THD of a multilevel inverter. The structure used in this paper is cascaded H-bridge multilevel inverter with unequal DC sources that is also an optimized structure from the view point of number of DC sources and switches Vs number of output levels. The optimization and simulation results are presented for 15-level cascaded H-bridge inverter. PSCAD software is used for simulation and MATLAB programming is utilized for optimization process.

# References

1. Hosseini, S.H., Varesi, K., Ardashir, J.F., Gandomi, A.A., Saeidabadi, S.: An attempt to improve output voltage quality of developed multi-level inverter topology by increasing the number of levels. In: 2015 9th International Conference on Electrical and Electronics Engineering (ELECO), 26 November 2015, pp. 665–669. IEEE (2015)
2. Oskuee, M.R., Karimi, M., Naderi, Y., Ravadanegh, S.N., Hosseini, S.H.: A new multilevel voltage source inverter configuration with minimum number of circuit elements. J. Central S. Univ. **24**(4), 912–920 (2017)
3. Hosseini, S.H., Ravadanegh, S.N., Karimi, M., Naderi, Y., Oskuee, M.R.: A new scheme of symmetric multilevel inverter with reduced number of circuit devices. In: 2015 9th International Conference on Electrical and Electronics Engineering (ELECO), 26 November 2015, pp. 1072–1078. IEEE (2015)
4. Tolbert, L.M., Peng, F.Z., Habetler, T.G.: Multilevel converter for large electric drives. IEEE Trans. Ind. Appl. **35**(1), 36–44 (1999)
5. Babaei, E.: A cascade multilevel converter topology with reduced number of switches. IEEE Trans. Power Electron. **23**(6), 2657–2664 (2008)
6. Banaei, M.R., Oskuee, M.R., Khounjahan, H.: Reconfiguration of semi-cascaded multilevel inverter to improve systems performance parameters. IET Power Electron. **7**(5), 1106–1112 (2014)
7. Cecati, C., Ciancetta, F., Siano, P.: A multilevel inverter for PV systems with fuzzy logic control. IEEE Trans. Ind. Electron. **57**(12), 4115–4125 (2010)
8. Saeedifard, M., Iravani, R., Pou, J.: Analysis and control of DC-capacitor-voltage-drift phenomenon of a passive front-end five-levelconverter. IEEE Trans. Ind. Electron. **vol. 54**, no. pp. 3255–3266 (2007)
9. Kaviani, A.K., Fathi, S.H., Farokhnia, N., Ardakani, A.J.: PSO, an effective tool for harmonics elimination and optimization in multi-level inverters. In: Proceedings of 4th IEEE ICIEA, 25–27 May 2009, pp. 2902–2907 (2009)
10. Yousefpoor, N., Fathi, S.H., Farokhnia, N., Sadeghi, S.H.: Application of OHSW technique in cascaded multi-level inverter with adjustable DC sources. In: Proceedings of IEEE International Conference on EPECS, pp. 1–6 (2009)
11. Taghizadeh, H., Hagh, M.T.: Harmonic elimination of cascade multilevel inverters with non-equal DC sources using particle swarm optimization. IEEE Trans. Ind. Electron. **57**(11), 3678–3684 (2010)
12. Naderi, Y., Hosseini, S.H., Mahari, A., Naderi, R.: A new strategy for harmonic minimization based on triple switching of multilevel converters. In: 2013 21st Iranian Conference on Electrical Engineering (ICEE), 14 May 2013, pp. 1–6 (2013)
13. Chiasson, J.N., Tolbert, L.M., McKenzie, K.J., Zhong, D.: Elimination of harmonics in a multilevel converter using the theory of symmetric polynomials and resultants. IEEE Trans. Control Syst. Technol. Mar. **13**(2), 216–223 (2005)
14. Ozpineci, B., Tolbert, L.M., Chiasson, J.N.: Harmonic optimization of multilevel converters using genetic algorithms. IEEE Power Electron. Lett. **3**(3), 92–95 (2005)
15. Ebrahimi, J., Babaei, E., Gharehpetian, G.B.: A new topology of cascaded multilevel converters with reduced number of components for high-voltage applications. IEEE Trans. Power Electron. **26**(11), 3109–3118 (2011)
16. Zarnaghi, Y.N., Hosseini, S.H., Zadeh, S.G., Mohammadi-Ivatloo, B., Quintero, J.C., Guerrero, J.M.: Distributed Power Quality Improvement in Residential Microgrids. In: Eleco 2017 10th International IEEE Conference on Electrical and Electronics Engineering 2017. IEEE (2017)

17. Sadeghian, H., Athari, M.H., Wang, Z.: Optimized solar photovoltaic generation in a real local distribution network. In: 2017 IEEE 2017 Power & Energy Society Innovative Smart Grid Technologies Conference (ISGT), pp. 1–5. IEEE, 23 April 2017

18. Nouri, T., Vosoughi, N., Hosseini, S.H., Sabahi, M.: A novel interleaved nonisolated ultrahigh-step-up DC–DC converter with ZVS performance. IEEE Trans. Ind. Electron. **64** (5), 3650–3661 (2017)

19. Kennedy, J.: Particle swarm optimization. In: Sammut, C., Webb, G.I. (eds.) Encyclopedia of Machine Learning. Springer, Boston (2011). https://doi.org/10.1007/978-0-387-30164-8

20. Shi, Y.: Particle swarm optimization: developments, applications and resources. In: Proceedings of the 2001 Congress on 2001 evolutionary computation, vol. 1, pp. 81–86. IEEE (2001)

# Development of Instruments of Fuzzy Identification of Extended Objects Based on the Results of Satellite Monitoring

Imran G. Akperov$^{(\boxtimes)}$ and Vladimir V. Khramov

Southern University (IMBL), M. Nagibin Ave. 33a/47,
Rostov-on-Don 344068, Russia
maoovo@yandex.ru

**Abstract.** The approaches and methods of both the actual identification of objects of the earth's surface and improving the accuracy of determining their position and location in the local coordinate system of the target extended object, in terms of uncertainty and fuzziness of the source information, by taking into account its location relative to its neighboring objects, were considered. The task appears in monitoring of the state of agricultural facilities, infrastructure of railway and vehicle transport, the system of pipelines for various purposes.

**Keywords:** Satellite monitoring · Identification · Informativity
Fuzziness · TIN (Triangulated Irregular Network) model

## 1 Introduction

At the stage of design and construction of large infrastructure facilities, enterprises, railways and vehicle roads it is necessary to have relevant and objective information about the underlying surface. The modern method of collecting and analyzing such information is remote sensing of the Earth (further - ERS).

It is also possible to assess the suitability of the area for this type of construction on the basis of data obtained by remote sensing of the Earth.

Remote sensing methods are based on measuring the characteristics of the underlying surface at a distance, i.e. these ones do not require the direct presence of experts at the place of research and construction.

A special place is occupied by the economic validity of the use of ERS. The expenses are reduced by this fact that it is not required to provide ground personnel, there is no need to organize field work, expeditions. The efficiency of data acquisition gives effective control over the state of the region. The scale of one image can cover an area of tens of thousands of square kilometers. As a result, the total cost of work with the use of remote sensing methods is much lower than the cost with the use of traditional methods of study of the underlying surface.

R. A. Aliev et al. (Eds.): ICAFS-2018, AISC 896, pp. 325–332, 2019.
https://doi.org/10.1007/978-3-030-04164-9_44

## 2  Basic Information

The equipment of artificial earth satellites has the ability to use both optical and acoustic, and microwave signals [1]. It is important to register exactly optical signals at the stage of railway design. However, the frequency and speed of data transmission to ground services are not critical. The main requirement is the maximum spatial resolution. On this basis we will analyze the existing satellite survey equipment for the suitability of monitoring of operation of existing and the design of new, including high-speed, railways.

Modern optoelectronic ERS allow to receive panchromatic, multispectral and hyperspectral images. Panchromatic images are obtained in the visible range of the electromagnetic spectrum and are black and white images of higher resolution than multispectral images of the same satellites [1, 2]. Multispectral imaging systems receive few separate images for spectral zone from visible to infrared radiation. At the present, the most practical interest is represented by multispectral data from new generation of spacecraft, such as RapidEye (5 spectral zones) and WorldView-2 (8 spectral zones).

One of the main parameters of satellite images is the spatial resolution – value, characterized the size of the smallest objects that can be distinguished in the image. The spatial resolution is influenced by the parameters of the radar or optoelectronic system and altitude of the satellite orbit, that is, the distance from the satellite to the subject. The best spatial resolution is achieved at shooting in nadir, but resolution deteriorates at the deviation from nadir. Space images can have a low (more than 10 m), medium (10 to 2.5 m), high (2.5 to 1 m), and ultra-high (less than 1 m) resolution.

Radar space survey realizes another approach in area research. This survey is realized in the ultra-short-wave (i.e., ultra-high-frequency) area of radio waves. Radar satellites have several survey modes, characterized by different parameters of spatial resolution and the width of survey band.

At the present, none of the satellites can provide the customer with all the necessary information, because none of the sensors is not able to carry out all the necessary measurements of the underlying surface.

Data obtained from various ERS are used in pre-project researches. One of the methods of receiving such information is the use of groups of satellites that rotate in the same orbit with some lagging behind each other [3, 4]. This fact minimizes the impact of temporary changes, which greatly simplifies the analysis of the data. Prospective groups are considered KADMC, consisting of satellites of different countries, and the group of German mini-satellites "Rapideye".

Information obtained from satellites is a multispectral image with a geographical reference. Depending on the resolution of the scanner, modern satellites can distinguish objects up to 0.5 m in size.

The location of natural and artificial structures in the study area is of primary interest.

One of the most effective methods of object recognition is the sliding window method [1]. The essence of the method is that the brightness of each pixel is processed depending on the brightness values of the pixels adjacent to it.

Herewith, the coefficients of the window, as well as its size, may vary depending on the image. The contours of the objects in the window can be used to obtain information about the boundaries of the territories in the study area.

The Sobel method, which works on the same principle of sliding window, allows to detect the faults of the railway track in time, which, in turn, plays an important role in the safety of high-speed highways. $3 \times 3$ lattice increases image accuracy, but does not filter the resulting noise.

The water component of the soil, of course, has a great impact on the complex of works during construction [2]. Analysis of water resources and vegetation as potential barriers during construction may be carried out using vegetation indices and research of the red part of the visible and near infrared spectra image of the area. Places of high humidity and the presence of groundwater requires strengthening of rocks and construction of additional structures to ensure an appropriate level of traffic safety and reliability of railway electrification [3].

For identification of extended objects as initial data we use two-dimensional graphic images of these objects received with use of means of supervision of an optical range.

The identification procedure includes:

- the compression process of data about the object (conversion to binary mind, the formation of a binary matrix image, filtering);
- the selection of pattern of the contour of the body of the object and patterns of the contour of signs of objects board number using a chain code of Freeman;
- decomposition of parametric descriptions of object contours into a number of orthogonal exponential functions (further - OEF);
- formation of a set of informative features of identification based on the coefficients of the expansion of the OEF;
- definition of multiple classes;
- description of classes in the language of signs;
- the actual identification of the object.

## 2.1 Selection and Study of the Method of Preliminary Processing of Basic Data

In the process of analysis of graphic images obtained by photographing the board numbers from the objects of rolling stock (railway cars, tanks, platforms), it was found that they have different kinds of noise. The sources of noise are additional inscriptions, fuel oil stains, rust, poor quality drawing of the number, sun glare, etc. The applied method of preparing the image for the identification process should take into account the real situation, respond adequately in these conditions and be able to form the output data in accordance with the accepted concept of identification and the features used for this purpose. The procedure for evaluation of inference influence by introducing random variation of feature values sometimes leads to errors that are difficult to account. For disposal of such errors, it is advisable to work with images obtained from real objects in conditions as close as possible to the conditions of the functioning of developed system.

Separate pixels are used to perform the filtering process. The quality of filtration is determined by the quality factor $\rho$:

$$\rho = \frac{M_{\varPi}}{M_{\varPhi}},$$

where $M_{\varPi}$ - the number of points on the distorted by interference image that do not coincide with the corresponding points on the standard (the original image);

$M_{\varPhi}$ - the number of points on the filtered image that do not match the corresponding points on the standard.

At present anisotropic filtering is used for spatial filtering of image. In a discrete interpretation it can be represented as follows:

$$\tilde{a}_{ij} = A \left[ \sum_{x=-N_a/2}^{N_a/2} \sum_{y=-N_a/2}^{N_a/2} a_{i+x;j+y}\omega_{xy} - \eta \right],$$

where $\tilde{a}_{ij}$ - the matrix element of the filtered image (i – column, j – row of the matrix); $\eta$- the filtering threshold; $A$ - threshold function;

$$A(x) = \begin{cases} 0, & \text{if } x = \eta; \\ 1, & \text{if } x = \eta; \end{cases} *$$

$\omega_{xy}$ - the aperture control (a matrix size of $N_a \times N_a$), located at the intersection of column x and row y;

$a_{i+x;j+y}$ - the matrix element of the distorted image ($i + x$ – column, $j + y$ – matrix row); we convert them to black and white for simplify the process of image processing.

To filter the image matrix, fill in 1 and 0. The size of the matrix is determined by the size of the image. In our case, if the pixel is "black", then the matrix cell is filled with 1, if "white" $-0$.

Then, we determine the size of the aperture. With increase of matrix $N_a$ the quality of filtering and the time required for image processing increases proportionally. Relying on the existing experience in the field of identification [1], we assume a size equal to $5 \times 5$.

The elements of the aperture $\omega_{xy}$ are determined from the normal uncorrelated two-dimensional circular distribution. The weights of the elements increase, approaching the center of the matrix. This distribution is characterized by mean-square deviation $\sigma_a$, it is truncated with the rationing of the weights. The sum of the weights of the aperture should be set to 1. Moreover, the smaller the value, the greater the weight of the central element of the matrix. If $W_1 = 1$, that the peripheral weights $W_2$, $W_3$ will be 0. If the threshold at the this $\eta \leq 1$, that image will remain the same during the filtering process. At $\sigma_a < 1$, the aperture is called "narrow", at $\sigma_a \geq 1$ – "wide".

In the course of research [2] of the anisotropic filtering method it was found that at $\eta \leq 0.35$ and multiple start procedure the image fills constantly by 1, as a result, it leads to complete painting of the image. It was also found that at $\eta \geq 0.71$ the first time you run the filtering procedure, the image is completely erased. Thus, we have determined the lower and upper thresholds. In addition, it was found that at $0.35 < \eta \leq 0.5$ our procedure works in saturation mode "black". In this mode, the noise that is inside the object is well filtered. If you use the threshold filter in the range $0.5 < \eta \leq 0.71$, the filtration procedure is focused on the removal of external noise, i.e. noise that are outside the body of the object.

To get rid of both internal and external noise, it is logical to assume that it is necessary to use a filtering procedure with dynamically changing filtering thresholds.

## 2.2   The Realization of Allocation Pattern of the Object Contour

Thus, at the end of the filtering process, the image took a clear (contrast) shape and has a closed smoothed contour. The pattern of the object contour will be selected using the chain code algorithm. The algorithm is brought to software implementation [2].

It is possible to store information about the object in a compressed form, having the coordinates of the first point of the circuit and the chain code.

Thus, at the current step of the object identification process, there is all the information necessary to calculate the integral features (numerical coefficients). An example of processing real source data is described in [2].

In general, the information properties of the landscape and, accordingly, its features, allowing to assess its features and suitability for the construction of railway facilities, can be integrated into a single multidimensional image [4]. The invariants of this image form an information model suitable for object design.

## 3   Method of Identification of Extended Objects of the Earth's Surface

The proposed method relates to the methods of research and monitoring of extended objects on the Earth's surface during remote sensing and can be used both for the operational identification and improvement of the accuracy of the location of the target objects and for the assessment of the characteristics of the spacecraft location.

The method is aimed both at solving the problems of proper identification, and to improve the location exactness and location in the local coordinate system of the target object by taking into account its location relative to its neighboring objects.

The task arises in monitoring the state of agricultural objects, infrastructure of railway and vehicle transport, the network of pipelines for various purposes, as well as to solve the navigation problem on spacecraft, etc. The results of its solution relate to computer technology and can be used in vision systems to identify extended objects of the earth's surface in satellite images.

Taking into account the fuzziness of the initial information, the following information granulation structure was used within the TIN model:

$$GR = \langle X, G, C, M, T \rangle,$$

where $X$ – field of reasoning,
  $G$ – class of information granules;

$C$ – the set of generalized limits;
(each type of limits defines the requirements for the choice of the method of granulation)
$M$ – the set of formal methods of granulation;
$T$ – the set of transitions between levels of granulation of (transformations of the granules).

The method provides for preliminary reduction of the image of the object entered into the computer to a single view for this method - changing the scale, turning to the desired position, centering, inscribing into a rectangle of the required size, converting the image of the object into an image made in gradations - different degrees of brightness - one color, which is sequentially, alternately superimposed images stored in the computer memory templates. In this case, the object recognition program can step by step combine normalized images of recognized objects, centered and inscribed in the same size of the table cells and templates, centered and inscribed in similar cells of the template table, with a step equal to the height of the row with cells or the width of the column of table cells, and in each of the columns or in each of the rows of the template table, the number of columns or rows in the table of recognized objects, there is a complete set of templates. The absence of an account of invariance to the affine transformations of the object rotation in the method, and also the necessity for a full search of large patterns in the recognition could be significant drawbacks. Then the selection of contours (Fig. 1) all objects in the space image invariant to turn are performed, the recognition signs, invariant to rotation, are formed, are compared in turn with the standards stored in the computer memory, which are stored in vector model, are compared them by neuronet, besides the comparison is made by analyzing the features of the shape of the contour of each of the objects of the earth's surface on the image, comparison for each feature is made, the decision about the coincidence of the vector models of the target object of the image and the reference objects is accepted.

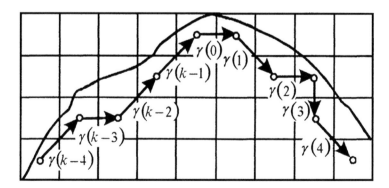

**Fig. 1.** Selection of object outlines

In addition to the target, the neighboring objects are identified, the centers of gravity of the target and its neighboring contours of objects are determined, and graphs are built (Fig. 2) and the matrix of connectivity of the set of extended objects, which are used as additional signs of identification, are built, their comparison is made and the decision on the identification of the target object is accepted.

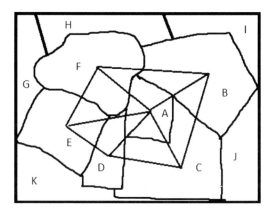

**Fig. 2.** The centers of gravity of the target and adjacent contours of objects.

Significant feature of the proposed method is that the space image of the earth's surface is fixed, and the digital raster images are processed. The contours of objects are distinguished, using Freeman chain code in connectivity 4. Then the starting points of the bypass contour are selected, which are the end: top, right, left and bottom. Then the bypass is performed, with respect to each of the initial points, the external contour of the object. During the bypass, the signs are obtained (numerical coefficients are calculated) according to the formulas (1) and (2):

$$U_{xy} = \frac{\sum\limits_{x=0}^{S} F_x \left[ l_x \right] e^{wx}}{\sum\limits_{y=0}^{S} F_y \left[ l_y \right] e^{wy}}, \tag{1}$$

$$U_{yx} = \frac{\sum\limits_{y=0}^{S} F_y \left[ l_y \right] e^{wy}}{\sum\limits_{x=0}^{S} F_x \left[ l_x \right] e^{wx}}, \tag{2}$$

where $F_x$, $F_y$ - Laplace transform of parametric description of the contour of a recognized object; $l_x$ and $l_y$ - the values of elementary vectors at the step $x$ and $y$, respectively, when moving along the contour from the first (0) to the last (S) point; $w$ – weight coefficient.

The calculated coefficients are consistently used as input parameters of the neural network. The pre-trained three-layer neural network consistently receives a two-dimensional array of shape features. Taking into account information from all sources, a decision is made to identify the object.

To improve the accuracy of the target object identification and its location accuracy, the principles of quorum reservation are used [10].

## 4    Conclusion

It is obviously, the storage and support of such a constantly developing model of identification of extended objects of the earth's surface requires appropriate information resources available at this stage within the concept of data storage.

The form of geodatabase organization that combines digital territory and relational databases is the most common at this point. However, the complexity of such an organization caused by a set of instruments for creating and supporting a data topology creates certain problems.

The method of identification considered in this article solves the problems of application of the algorithm and methods provided by fuzzy triangulation [9, 11].

In the framework of the TIN model for individual components of the identification system in the process of forming a digital plan diagram for the designers of transport corridors, it is necessary to obtain additional information about the objects included in the resulting information product.

## References

1. Author, F.: Article title. Journal **2**(5), 99–110 (2016)
2. Author, F., Author, S.: Title of a proceedings paper. In: Editor, F., Editor, S. (eds.) Conference 2016, LNCS, vol. 9999, pp. 1–13. Springer, Heidelberg (2016)
3. Author, F., Author, S., Author, T.: Book title, 2nd edn. Publisher, Location (1999)
4. Author, F.: Contribution title. In: 9th International Proceedings on Proceedings, pp. 1–2. Publisher, Location (2010)
5. LNCS Homepage. http://www.springer.com/lncs. Accessed 21 Nov 2016

# Optimal Placement of Capacitor Using Particle Swarm Optimization

Farzad Mohammadzadeh Shahir[1], Ebrahim Babaei[1,2],
and Fahreddin Sadikoglu[2(✉)]

[1] Faculty of Electrical and Computer Engineering, University of Tabriz,
Tabriz, Iran
[2] Engineering Faculty, Near East University, 99138 Nicosia, North Cyprus,
Mersin 10, Turkey
fahreddin.sadikoglu@neu.edu.tr

**Abstract.** In this paper, a research regarding finding an optimal place for installing the capacitor under particle swarm optimization (PSO) has been presented in radial networks by considering the size and price of different capacitors. In this paper, the calculations and results for a typical radial network have been calculated by considering harmonic distortion calculation (HDF) and its results have been stated. Based on this, all the losses have been calculated and compared too. The results of this study have been calculated by MATLAB software.

**Keywords:** Capacitor · Placing capacitor · PSO algorithm · Radial network

## 1 Introduction

Nowadays compensation and stability are important problem in power system designing and researchers have been tried to introduce new methods for improve them [1, 2]. Capacitors are usually used for compensation of the losses of radioactive power of systems of distribution and creation of flat voltage profile [3]. Selection of an optimal place for placement of capacitor is a good research issue [4–6]. Different kinds of methods of placement are available as theory or software all of which try to optimize reference parameters. Some also formulate this use as limitation of optimization and consider the voltage constraints as well. The placement way, size and place of capacitor must be exactly investigated otherwise the capacitors enhance harmonic currents and voltages which results from aggravation in one or many harmonic frequencies. Capacitor measures are usually considered as continuous variables whose costs are determined in accordance with capacitor sizes in previous papers [4–6]. Regarding the issue of process of capacitors mentioning this point will be of benefit that the price of the capacitors is discrete from the measures of its reactive power (KVAR) and they don't have a linear relationship with their nominal measures and depend on the type and material and manufacturing companies; thus, if the methods of previous papers [4–6] are used for selection of the capacitor size by the method of the continuous variable method, it may not lead to an optimal solution and it may even lead to creation of an adverse harmonic and non-harmonic aggravation phenomenon and lead the system to

© Springer Nature Switzerland AG 2019
R. A. Aliev et al. (Eds.): ICAFS-2018, AISC 896, pp. 333–343, 2019.
https://doi.org/10.1007/978-3-030-04164-9_45

the border of instability [7]. Thus, selection of a method for optimization also requires great studies in this field that should be exerted in the system of changes such that the system stability has been ensured and become better than previous states. In this paper, using the formative algorithms and specifically PSO algorithm the selection of the parameters is taken place.

## 2    Formulation of the Problem

### 2.1    Assumption

In order to simplify this analysis merely fixed capacitors have been used for investigation of the presented method. The fixed capacitors assumed that: (1) balanced conditions (2) regardless of interface transmission lines and its voltage drop (3) invariant loads (4) harmonic production.

### 2.2    Power System

Figure 1 clearly shows an m bus radial distribution system in which an $i$ line includes a load and a parallel capacitor. In a fixed frequency, we have:

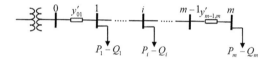

**Fig. 1.** Single line diagram for model of radial systems.

$$P_i = \left|V_i^1\right|^2 G_{ii} + \sum_{j=1, j \neq i}^{m} \left|V_i^1 V_j^1 V_{ij}^1\right| \cos(\theta_{ij}^1 + \delta_j^1 + \delta_i^1) \quad i = 1, 2, 3, \ldots, m \quad (1)$$

$$Q_i = -\left|V_i^1\right|^2 B_{ii} + \sum_{j=1, j \neq i}^{m} \left|V_i^1 V_j^1 V_{ij}^1\right| \sin(\theta_{ij}^1 + \delta_j^1 + \delta_i^1) \quad i = 1, 2, 3, \ldots, m \quad (2)$$

$$P_i = P_{1i} + P_{ni} \quad (3)$$

$$Q_i = Q_{1i} + Q_{ni} \quad (4)$$

And in Eqs. (1) to (4) the following relations are established:

$$\theta_{ij}^1 = -y_{ij}^1 \ if \ i \neq j \ and \ -y_{i-1,i}^1 + y_{i+1,i}^1 + y_{ci}^1 \ if \ i = j \quad (5)$$

$$Y_{ii} = G_{ii} + jB_{ii} \quad (6)$$

In the basic frequency, the active power losses between the $i$ and $i+1$ lines are:

$$P^1_{loss(i,i+1)} = R_{i,i+1}\left(\left|V^1_{i+1} - V^1_i\right|\left|Y^1_{i,i+1}\right|\right)^2 \tag{7}$$

$$P_{loss} = \sum_{n=1}^{N}\left(\sum_{i=0}^{m-1} P^n_{loss(i,i+1)}\right) \tag{8}$$

The sum of application of the annual cost resulting from capacitor placement is:

$$f = A_l A_P P_{loss} + \sum_{j=1}^{m} Q_{cj} A_{cj} \tag{9}$$

In which $i = 1, 2, \ldots, m$ shows the sizes of the capacitor. The reactive power equals:

$$Q_{cj} = j \times A_s \tag{10}$$

Also, Eq. (1) has the following limitations:

$$V_{\min} \leq |V_i| \leq V_{\max} i = 1, 2, 3, \ldots, m \tag{11}$$

$$IDF_i \leq HDF_{\max} \; i = 1, 2, 3, \ldots, m \tag{12}$$

## 3 Proposed Algorithm

### 3.1 Produced Harmonic Currents

The whole load in higher frequencies can be modeled by means of a combination of harmonic current resources and passive and non-linear elements [7]. Disregarding the skin effect, the n[th] harmonic frequency of parallel load and capacitor and the feeding system will be respectively:

$$Y^n_{li} = \frac{P_{li}}{|V^1_i|^2} - j\frac{Q_{li}}{n|V^1_i|^2} \tag{13}$$

$$Y^n_{li} = nY^1_{ci} \tag{14}$$

$$Y^n_{i,i+1} = \frac{1}{R_{i,i+1} + jnX_{i,i+1}} \tag{15}$$

$$I^1_i = \left[\frac{P_{ni} + jQ_{ni}}{V^1_i}\right]^* \tag{16}$$

$$I_i^n = C(n)I_i^1 \tag{17}$$

The produced harmonics voltages, effective voltage and harmonic are obtained as:

$$Y^n V^n = I^n \tag{18}$$

$$|V_i| = \sqrt{\sum_{n=1}^{N} |V_i^n|^2} \tag{19}$$

$$HDF_i(\%) = \frac{\sqrt{\sum_{n=1}^{N} |V_i^n|^2}}{V_i^1} \times 100\% \tag{20}$$

### 3.2    Selection of Optimal Location for the Capacitor

Firstly the PSO considers an optimal size of capacitor according to the capacitor locations. Subsequently, by placement of other locations in the process of repetition of the $i$ degree the size and locations of the $i - 1$ process of repetition will be used as assumption. If the new results are more optimal than the previous state, the new results and solutions are referred to as pre-assumption and the repetition will continue unless other optimal solutions and results are obtained such that the results of the $i$ repetition don't have any difference with the $i - 1$ process. When the repetition is more, the answers obtained are convergent, minimal (or optimal). The best situation that has been found by all particles us shown in the form of $x_{g,best}$ that is selected by comparison of $f_{i,best}$ measures for all particles and among $x_{i,best}$. The function rate of $x_{g,best}$ is shown as $f_{g,best}$. If the number of available particles in the population is n, then the following relations can be written:

$$x_{i,best}(t) = \arg\min(x_i[T]) = \arg\min\{f(x_{i,best}, [T], F(x_{i,best}[t-1])\} \ t \leq T \tag{21}$$

$$f_{i,best}[t] = f(x_{i,best}y[t]) = \min f_i[t] = \min\{f_i[T] = f_{i,best}[t-1]\} \ t \leq T \tag{22}$$

$$x_{g,best}[t] = \arg\min f(x_{i,best}[t]) \tag{23}$$

$$f_{g,best}[t] = f(x_{i,best}[t]) = \min f_{i,best}[t] \ \ i = 1, \ldots, n \tag{24}$$

Particles are created by random situations and speeds. Common kinds of capacitors with their power and price have been shown in Table 5. In this paper, the maximum size of the capacitor, as considered in Table 5, $4050kVar$ and the kind is 27. The best kind of capacitor based on PSO algorithm is placed in the shown test circuit and evaluation under different methods of load distribution such as Newton-Raphson, Gauss, Saydl. In every stage a new generation of the optimal states of the past generation is obtained and the results are obtained which are stated as:

$$f_i = (f_{\max} - f_a)/f_{\max} \tag{25}$$

$$f_a = A_1 A_P P_{loss} + \sum_{j=1}^{m} Q_{cj} A_{cj} \tag{26}$$

Figure 2(a) presents the main flowchart in this paper. The processes of selection of optimal place for capacitor and calculation of PSO have been shown in Fig. 2(b) and (c). Figure 3(a) presents the flowchart of subprograms presented in Fig. 2(b) and (c).

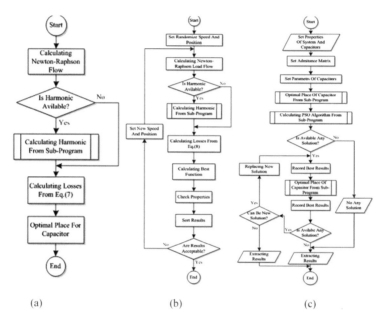

(a)                              (b)                              (c)

**Fig. 2.** (a) Flowchart of subprogram of the calculation of optimal place of capacitor from Fig. 1; (b) Flowchart of subprogram of calculation of PSO; (c) Main flowchart of the solution.

## 4    Calculation and Simulation Results

This experiment distribution system has been shown in Fig. 3(b). This feeding system has 9 transmission lines with 23KV voltages and 100MW power. The Tables 1 and 2 show the fixed numbers and the transmission line. Non-linear loads that are produced by every subscriber have also been replaced by harmonic current resources that are depicted in Table 3. The minimum and maximum voltages have been considered as 0.9$p.u.$ and 1$p.u.$. In this paper $A_P$ has been regarded as 168$/KW. Table 4 has presented the commercial size and price of the capacitors under study. In Table 5 the possible measures for selection of size and price of capacitors have been stated. In the study case 1, the results for the assumed system in Fig. 1 have been obtained without PSO algorithm. These results have been presented in Table 6. The Study cases 2 3 have

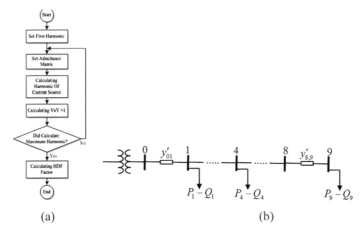

(a)                                        (b)

**Fig. 3.** (a) Flowchart of subprogram of harmonic calculation from Fig. 2(a); (b) The test system including 9 buses.

been calculated by considering the PSO algorithm and respectively without considering the harmonic and by considering the harmonic and the results have been presented in Tables 7 and 8. The HDF measures calculated in each bus and in each study case have been presented in related tables. The highest HDF of cases 1 and 2 and 3 are 6.25%, 1.65% and 1.37, respectively. The improvement percentage of HDF of the case 3 in ratio to the case 1 is 78.08%. Improvement percentage of HDF of case 2 according to case 1 is calculated as 16.69%. Improvement percentage of HDF of case 3 according to case 2 is obtained 73.6%. Based on results, HDF is improved on case study 2 and 3 rather than case study 1. According to the results of stimulation it can also be stated that the best place for installing the bus capacitor is 9. Other results of stimulation have been presented in Table 9.

**Table 1.** The load data in the test system.

| Bus | 1 | 2 | 3 | 4 | 5 | 6 | 7 | 8 | 9 |
|---|---|---|---|---|---|---|---|---|---|
| P(kW) | 1750 | 8750 | 1695 | 1450 | 1550 | 690 | 1080 | 890 | 1580 |
| Q(kW) | 450 | 323 | 435 | 1795 | 601 | 106 | 70 | 125 | 195 |
| Non-linear% | 53.7 | 19.1 | 93.2 | 5.7 | 2.1 | 39.3 | 4.6 | 6.4 | 5.3 |

**Table 2.** Information feeder in the test system.

| From $i$ | From $j$ | $R_{i,i+1}(\Omega)$ | $X_{i,i+1}(\Omega)$ | From $i$ | From $j$ | $R_{i,i+1}(\Omega)$ | $X_{i,i+1}(\Omega)$ |
|---|---|---|---|---|---|---|---|
| 0 | 1 | 0.1233 | 0.4127 | 1 | 2 | 0.0140 | 0.06051 |
| 2 | 3 | 0.7463 | 1.2050 | 3 | 4 | 0.6984 | 0.6084 |
| 4 | 5 | 1.9831 | 1.7276 | 5 | 6 | 0.9053 | 0.7889 |
| 6 | 7 | 2.0552 | 1.1640 | 7 | 8 | 4.7953 | 2.7160 |
| 8 | 9 | 5.3434 | 3.0264 | | | | |

**Table 3.** Harmonic of the current resource.

| Bus-Harmonic | 5 | 7 | 11 | 13 | 17 | 19 | 23 | 25 |
|---|---|---|---|---|---|---|---|---|
| 1 | 0 | 0 | 0 | 0 | 0 | 0 | 0 | 0 |
| 2 | 8.99 | 4.97 | 1.17 | 1.01 | 0.65 | 0.62 | 0.45 | 0.35 |
| 3 | 3 | 1.65 | 0.55 | 0.37 | 0.25 | 0.22 | 0.17 | 0.98 |
| 4 | 5.8 | 3.45 | 1.22 | 0.74 | 0.51 | 0.35 | 0.27 | 0.18 |
| 5 | 15.4 | 3 | 3.89 | 8.3 | 4.87 | 2.8 | 2.88 | 0 |
| 6 | 0 | 0 | 9.7 | 5.33 | 0 | 0 | 3.55 | 2.55 |
| 7 | 0.28 | 0 | 0 | 0 | 0 | 0 | 0 | 0 |
| 8 | 0.78 | 0.45 | 0.25 | 0 | 0 | 0 | 0 | 0 |
| 9 | 0.88 | 8.7 | 2.98 | 1.77 | 1.17 | 0.9 | 0.55 | 0.45 |

**Table 4.** Measure and prices of available capacitor for KVAr and $.

| Size | 150 | 300 | 450 | 600 | 900 | 1200 | Price | 750 | 975 | 1140 | 1320 | 1650 | 2040 |
|---|---|---|---|---|---|---|---|---|---|---|---|---|---|

**Table 5.** The measures possible in selection of size and price of capacitors.

| J | 1 | 2 | 3 | 4 | 5 | 6 | 7 | 8 | 9 |
|---|---|---|---|---|---|---|---|---|---|
| $Q_{cj}$ | 150 | 300 | 450 | 600 | 750 | 900 | 1050 | 1200 | 1350 |
| $A_{cj}$ | 0.5 | 0.35 | 0.253 | 0.22 | 0.276 | 0.183 | 0.228 | 0.17 | 0.207 |
| J | 10 | 11 | 12 | 13 | 14 | 15 | 16 | 17 | 18 |
| $Q_{cj}$ | 1500 | 1650 | 1800 | 1950 | 2100 | 2250 | 2400 | 2550 | 2700 |
| $A_{cj}$ | 0.201 | 0.193 | 0.187 | 0.211 | 0.176 | 0.197 | 0.17 | 0.189 | 0.179 |
| J | 19 | 20 | 21 | 22 | 23 | 24 | 25 | 26 | 27 |
| $Q_{cj}$ | 2850 | 3000 | 3150 | 3300 | 3450 | 3600 | 3750 | 3900 | 4050 |
| $A_{cj}$ | 0.183 | 0.18 | 0.195 | 0.174 | 0.188 | 0.17 | 0.183 | 0.182 | 0.179 |

**Table 6.** Results in case study 1.

| Bus-Har. | $1^{st}$ | $5^{th} \times 10^{-2}$ | $7^{th} \times 10^{-3}$ | $11^{th} \times 10^{-3}$ | $13^{th} \times 10^{-3}$ | $17^{th} \times 10^{-4}$ | $19^{th} \times 10^{-4}$ | $V_{rms}$ | $HDF\%$ |
|---|---|---|---|---|---|---|---|---|---|
| 1 | 0.986 | 5.12 | 3.81 | 2.54 | 1.87 | 9.44 | 7.95 | 0.989 | 5.88 |
| 2 | 0.982 | 5.01 | 3.63 | 2.42 | 1.72 | 9.11 | 7.45 | 0.985 | 5.95 |
| 3 | 0.971 | 4.95 | 3.54 | 2.35 | 1.63 | 8.56 | 7.18 | 0.971 | 6.01 |
| 4 | 0.962 | 4.83 | 3.42 | 2.18 | 1.51 | 8.33 | 6.96 | 0.935 | 6.25 |
| 5 | 0.945 | 4.75 | 3.24 | 1.98 | 1.47 | 7.89 | 6.74 | 0.913 | 5.97 |
| 6 | 0.923 | 4.68 | 3.13 | 1.62 | 1.28 | 7.72 | 6.35 | 0.901 | 5.84 |
| 7 | 0.901 | 4.32 | 3.01 | 1.47 | 1.17 | 7.35 | 6.17 | 0.892 | 5.76 |
| 8 | 0.881 | 4.15 | 2.97 | 1.41 | 1.06 | 6.95 | 6.01 | 0884 | 5.72 |
| 9 | 0.871 | 4.01 | 2.87 | 1.35 | 0.99 | 6.84 | 5.95 | 0.872 | 5.62 |

**Table 7.** Results in case study 2.

| Bus-Har. | $1^{st}$ | $5^{th} \times 10^{-2}$ | $7^{th} \times 10^{-3}$ | $11^{th} \times 10^{-3}$ | $13^{th} \times 10^{-3}$ | $17^{th} \times 10^{-4}$ | $19^{th} \times 10^{-4}$ | $V_{rms}$ | $HDF\%$ |
|---|---|---|---|---|---|---|---|---|---|
| 1 | 0.995 | 1.32 | 5.13 | 2.11 | 1.55 | 8.45 | 7.56 | 0.998 | 1.65 |
| 2 | 0.982 | 1.27 | 4.89 | 2.04 | 1.31 | 7.95 | 6.07 | 0.992 | 1.49 |
| 3 | 0.979 | 1.21 | 4.71 | 1.95 | 1.25 | 7.17 | 5.47 | 0.985 | 1.44 |
| 4 | 0.964 | 1.18 | 4.62 | 1.81 | 1.01 | 6.86 | 4.12 | 0.965 | 1.38 |
| 5 | 0.945 | 1.06 | 4.49 | 1.75 | 0.91 | 6.13 | 3.26 | 0.942 | 1.24 |
| 6 | 0.913 | 0.991 | 4.35 | 1.61 | 0.82 | 5.72 | 2.79 | 0.918 | 1.12 |
| 7 | 0.908 | 0.891 | 4.25 | 1.35 | 0.65 | 4.31 | 1.89 | 0.901 | 1.02 |
| 8 | 0.904 | 0.861 | 4.17 | 1.15 | 0.54 | 3.34 | 1.16 | 0895 | 0.98 |
| 9 | 0.899 | 0.795 | 3.89 | 0.991 | 0.45 | 2.95 | 0.72 | 0.889 | 0.87 |

**Table 8.** Results in case study 3.

| Bus-Har. | $1^{st}$ | $5^{th} \times 10^{-2}$ | $7^{th} \times 10^{-3}$ | $11^{th} \times 10^{-3}$ | $13^{th} \times 10^{-3}$ | $17^{th} \times 10^{-4}$ | $19^{th} \times 10^{-4}$ | $V_{rms}$ | $HDF\%$ |
|---|---|---|---|---|---|---|---|---|---|
| 1 | 0.999 | 1.17 | 5.17 | 2.03 | 1.37 | 8.23 | 7.44 | 0.998 | 1.37 |
| 2 | 1.000 | 1.12 | 4.69 | 1.99 | 1.27 | 7.81 | 5.95 | 0.999 | 1.25 |
| 3 | 0.998 | 1.07 | 4.26 | 1.82 | 1.15 | 7.02 | 5.32 | 0.987 | 1.12 |
| 4 | 0.989 | 0.972 | 3.84 | 1.67 | 0.951 | 6.71 | 4.01 | 0.958 | 1.08 |
| 5 | 0.987 | 0.917 | 3.27 | 1.54 | 0.843 | 6.23 | 2.91 | 0.935 | 1.02 |
| 6 | 0.968 | 0.889 | 2.69 | 1.44 | 0.795 | 5.61 | 2.24 | 0.928 | 0.99 |
| 7 | 0.959 | 0.815 | 2.45 | 1.17 | 0.615 | 4.08 | 1.42 | 0.924 | 0.92 |
| 8 | 0.946 | 0.782 | 2.18 | 0.984 | 0.471 | 3.12 | 0.95 | 0.917 | 0.91 |
| 9 | 0.935 | 0.691 | 2.02 | 0.881 | 0.335 | 2.85 | 0.68 | 0.908 | 0.82 |

**Table 9.** Final results of PSO algorithm.

| Per unit | Max. voltage | Min. voltage | Rate of losses | Max. HDF% |
|---|---|---|---|---|
| Case study 1 | 0.989 | 0.872 | 0.00613 | 6.25 |
| Case study 2 | 0.998 | 0.889 | 0.00595 | 1.65 |
| Case study 3 | 0.999 | 0.908 | 0.00569 | 1.37 |

# 5 Conclusion

In this paper, a formative algorithm by the name of PSO was presented for searching the size and place of optima parallel capacitor with harmonic considerations. The function of cost or adjustment is constrained by voltage and harmonic wave's agent (HDF). The optimal results have been evaluated based on comparison of the obtained results under the mentioned algorithm. In PSO algorithm by selection of the number of proper particles in addition to acceleration in the process of repetition the minimum cost can be obtained as well and the voltage profile and HDF% improve considerably and in comparison with other intelligent methods precision and selection of the optimal place and the investigated parameters of the introduced algorithm is revealed more than before. Based on the results presented in the tables, using the PSO algorithm in optimal displacement of capacitors, the best measures of power system can be attained.

# References

1. Shahir, F.M., Babaei, E.: Evaluating the dynamic stability of power system using UPFC based on indirect matrix converter. J. Autom. Control Eng. **1**(4), 279–284 (2013)
2. Shahir, F.M., Babaei, E.: Dynamic modeling of UPFC by two shunt voltage-source converters and a series capacitor. Int. J. Comput. Electr. Eng. **5**(5), 476–481 (2013)
3. Marti, L.: Effects of series compensation capacitors on geomagnetically induced currents. IEEE Trans. Power Deliv. **29**(4), 2032–2035 (2014)
4. Rezaei, P., Vakilian, M.: Distribution system efficiency improvement by reconfiguration and capacitor placement using a modified particle swarm optimization algorithm. In: Proceedings MEPS, pp. 1–6 (2010)
5. Sayed, A.G., Youssef, H.K.M.: Optimal sizing of fixed capacitor banks placed on a distorted interconnected distribution networks by genetic algorithms. In: Proceedings Computational Technologies in Electrical and Electronics Engineering, pp. 180–185 (2008)
6. Young Choi, J., Swaminthan, M.: Decoupling capacitor placement in power delivery networks using MFEM. IEEE Trans. Compon. Packag. Manuf. Technol. **1**(10), 1651–1661 (2011)
7. Pedder, D.A.G., Brown, A.D., Ross, J.N., Williams, A.C.: A parallel-connected active filter for the reduction of supply current distortion. IEEE Trans. Ind. Electron. **47**(5), 1108–1117 (2000)

# Predicting Fresh Water of Single Slope Solar Still Using a Fuzzy Inference System

Lida Ebrahimi Vafaei[1(✉)] and Melike Sah[2]

[1] Department of Mechanical Engineering, Near East University, Nicosia,
Northern Cyprus, Turkey
lida.ebrahimi@neu.edu.tr
[2] Department of Computer Engineering, Near East University, Nicosia,
Northern Cyprus, Turkey
melike.sah@neu.edu.tr

**Abstract.** According to the Cyprus Demographic Report, about 55% of households in Northern Cyprus do not have access to treated water. For treating water, distillation is utilized. Although there are many distillation methods, they are either energy intensive or contribute to environment. This research we use experimental data of a solar distillation system that produces potable water for drinking purposes from sea or salty water. The solar energy heats the water in the tank the water temperature will increase then evaporate and condensate on the glass cover, this condensate drops down the fresh water collector at the bottom, which settles the mass of salt in the bottom of distillation basin. The results showed that the thermal performance, improvements can be made to achieve still water production rates. The water (polluted sea water and salty well water) laboratory test results show that the distillation process eliminated the bacteria, being appropriated for human use. In this work, we developed a fuzzy inference system (FIS) to predict the fresh water of single slope solar still distillation. In the used FIS, we only utilize angle, water temperature, surface temperature and amount of fresh water of the solar still distillation. Evaluations show that predicted values are correlating with the experimental data with 29.40, 4.75, root mean square error and average forecasting error respectively. Therefore, the soft computing approach can be very useful for predicting fresh water of single slope solar still distillation with speed and simplicity.

**Keywords:** Basin still · Experimental performance · Fresh water
Fuzzy inference system

## 1 Introduction

Consumption of energy is increasing every year in the TRNC where there is an abundance of solar energy resources. TRNC has no other important natural sources of energy and for this reason solar energy is of particular importance. There are four primary sources of energy namely petroleum, natural gas, coal and wood. Most of the world's energy sources are derived from these fossil fuels which are called conventional sources [1]. The two likely sources are nuclear and solar energy which are most significant. Nuclear energy requires advanced technology and costly means for its safe.

© Springer Nature Switzerland AG 2019
R. A. Aliev et al. (Eds.): ICAFS-2018, AISC 896, pp. 344–351, 2019.
https://doi.org/10.1007/978-3-030-04164-9_46

But solar energy does not require high technology and there are no polluting effects. An experimental discovery was observed of how efficient a solar still from various inclination angles which include 15°, 25°, 35°, 45° et 55° react. They noticed that the most extreme water creation is acquired with an inclination point of 35°. The trial comes about acquired, show that the water saltiness influences the creation of the distillate, even at a low fixation [2]. Only few approaches apply soft computing methods for water distillation. In the work of Mashaly and Alazba [3] a neuro-fuzzy inference system (ANFIS) is utilized for the prediction of the SSP required by designers, operators, and beneficiaries of solar stills. On the other hand, for water quality index estimation, Yaseen et al. [4] use a hybrid adaptive neuro-fuzzy models.

The point of this work is to look at the impact of water temperature, the climatic conditions and ambient temperature thereby suggest solution to the problem water impurities in the town of Nicosia -Northern part of Cyprus (located 35° 10' N in scope and 33° 22' E longitude) on the working characteristic of the work. The acquired outcomes enable us to initiate the development of various temperatures capable of solving the problem of no dependable water, the amount of distillate versus time and the physio-substance examination of the distillation. We developed a fuzzy inference system (FIS) to predict the amount of fresh water. The predicted values were highly correlating with the experimental data with 29.40 and 4.75, root mean square error (RMSE) and average forecasting error respectively. To the best of our knowledge, there have not been previous studies in Cyprus that combine experimental data obtained from fresh water of single slope solar still distillation in Cyprus and apply this data to a FIS for predicting amount of fresh water.

## 2   Slope Solar Still Distillation and Experimental Data Gathering

We obtained experimental data using a slope solar still distillation in Nicosia, North Cyprus. The following equipment is used for measuring the amount of fresh water: flat plate collector, solenoid valve, solar still and mercury thermometer. The process of getting pure water by distillation using solar thermal energy is known as solar distillation. For this process solar stills are used and the experiments that had been done in our research work at a varied time using different angles, and temperatures, to obtain the quantity of distilled water from salt water. To calculate the total heat losses from the design it should be known all the temperatures: Surrounding temperature, Outer surface temperature of the glass, Inner surface temperature of the glass, Inside temperature, Water temperature, Inner surface steel wall temperature, Outer surface steel wall temperature, Insulation temperature. When angle is equal to 5° and time of experiment from 10:00 am to 03:00 pm ($\Delta t = 5$ h), at this angle of time the temperature varied from 29 °C to 31 °C at surrounding temperature while the inside temperature varied from 31 °C to 53 °C that way the water temperature had a gradual increasing from 27 °C to 56 °C. That's leads to evaporating of water and the water vapor condensed at the glass surface from that the quantity of water condensed from evaporated was 300 mL (Fig. 1).

**Fig. 1.** Experimental setup for the solar still distillation system.

# 3   The Fuzzy Inference System (FIS)

In our work, we use a Mamdani [5] FIS, since both inputs and outputs are represented with fuzzy sets. The used FIS is described in the following subsections as follows.

## 3.1   Input/Output Fuzzy Sets

During the experimental analysis of solar still distillation, we gathered various data regarding to the solar still (explained in Sect. 3); angle, $\Delta$Tout, $\Delta$T water and amount of fresh water. We observed that prediction accuracy is affected by angle, $\Delta$Tout, $\Delta$T water and amount of fresh water experimental data. Thus, we only use angle, $\Delta$Tout and $\Delta$T water temperatures as the inputs of the FIS. As an output of the FIS, the measured amount of fresh water from the real experimental data is utilized.

First, the universe of discourse (U) for angle, $\Delta$Tout, $\Delta$T water and fresh water is determined. For input and output fuzzy sets, trapezoid membership function (MF) is utilized since it gives the minimum RMSE. U angle = [50, 360], where the universe is divided into three unequal intervals that corresponds to trapezoid MFs Low, Medium and High (Fig. 2(a)). U $\Delta$Tout = [27.5, 31], where the universe is divided into two intervals according to our experimental data (Fig. 2(b)), represented by trapezoid MFs that corresponds to Very Low and Low fuzzy linguistic values. U $\Delta$water = [39.5, 42.5], where the universe is divided into three intervals (Fig. 2(c)) and represented by trapezoid MFs of Low, Medium and High. Finally, for the output fuzzy set, U fresh water = [300, 740], where the universe is divided into five intervals (Fig. 2(d)) and represented by trapezoid MFs of Very Low, Low, Medium, High and Very High. In our work, MFs and the number of fuzzy sets are experimentally determined that give the minimum RMSE. Experimental data is fuzzified based on the MFs shown in Fig. 2

(a–c); for an input value, a membership degree (ranging between 0 and 1) to each fuzzy linguistic value is calculated [6].

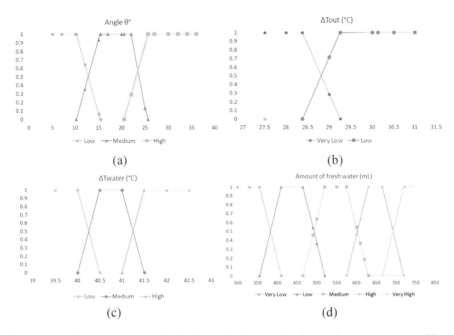

**Fig. 2.** Input fuzzy sets (a) Angle, (b) Tout, (c) Twater. (d) Output fuzzy set amount of fresh water

## 3.2   Fuzzy Reasoning Rules

Fuzzy inference rules are determined based on the values of angle, $\Delta T_{\text{out}}$ and $\Delta T_{\text{water}}$, which are obtained from the experimental data as shown in Table 2. By examining the experimental data, we have seen that when angle increases the amount of fresh water increases until $12°$. When the angle increases from $12°$, the amount of fresh water decreases. In other words, when we check at different angle $(\theta°)$ amount of fresh water is not constant and varied, maximum data at $12°$ and minimum at $5°$. According to the experimental data, we generated 12 fuzzy inference rules as shown in Table 1.

## 3.3   Prediction

To combine the outputs of fuzzy inference rules, we used the minimum implication operator and maximum aggregation operator in our FIS that gives the lowest RMSE. In particular, fuzzy inference rules that do not contain empty antecedents are fired. A rule conclusion is obtained by applying the minimum operator. In Table 2, the fired rules are shown according to the fuzzified input data. For instance, following the rule name, in parenthesis, the fuzzy output membership degree is represented (i.e. after applying the minimum operator). Finally in the defuzzification, outputs of the fired inference

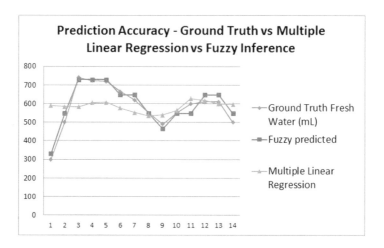

**Fig. 3.** Prediction accuracy – multiple linear regression versus fuzzy inference.

**Table 1.** Fuzzy inference rules.

| Rule | Angle | Tout | T water | Fresh water |
|------|-------|------|---------|-------------|
| Rule1 | Low | Low | High | Very low |
| Rule2 | Low | Low | Medium | Medium |
| Rule3 | Low | Low | Low | Very high |
| Rule4 | Low | Low | Medium | Very high |
| Rule5 | Medium | Low | High | Very high |
| Rule6 | Medium | Low | Medium | High |
| Rule7 | Medium | Very low | Medium | High |
| Rule8 | Medium | Very low | Low | Medium |
| Rule9 | High | Very low | Medium | Low |
| Rule10 | High | Low | Low | Medium |
| Rule11 | High | Low | High | High |
| Rule12 | High | Low | Medium | High |

rules are combined and a numeric prediction value is created. In particular, the fuzzy inference rule with the highest conclusion is chosen as the output by utilizing the maximum operator. To generate a numeric value, we use the centroid function on the output of amount of fresh water MFs. Various linguistic values of Amount of Fresh Water output fuzzy set are defuzzified using the centroid function (middle of the membership value of 1.0) as follows: *Very Low* is represented as 330 numeric value, *Low* is represented as 465 numeric value, *Medium* is represented as 548.3 numeric value, *High* is represented as 647.5 numeric value and *Very High* is represented as 730 numeric value.

We explain the rule firing process and prediction for the data (angle = 15°, $T_{out} = 30$ and $T_{water} = 41.5$) in Table 2. R1 (with 0.05 membership degree) and R5

**Table 2.** Amount of fresh water distillation prediction using fuzzy inference rules.

| Angle | Tout | T water | Rules fired | Predicted fuzzy output | Predicted value using fuzzy deffuzification | Ground truth fresh water (experimental) |
|-------|------|---------|-------------|------------------------|---------------------------------------------|------------------------------------------|
| 5 | 30 | 41.5 | R1(1) | Very low | 330 | 300 |
| 7 | 30 | 41 | R2(1) | Medium | 548.3 | 500 |
| 10 | 30.5 | 40 | R3(1) | Very high | 730 | 740 |
| 12 | 31 | 41 | R4(1) | Very high | 730 | 725 |
| 15 | 30 | 41.5 | R1 (0.06), R5 (0.94) | Very high | 730 | 720 |
| 17 | 29 | 41 | R6 (0.71), R7 (0.29) | High | 647.5 | 665 |
| 20 | 28 | 41 | R7(1) | High | 647.5 | 620 |
| 22 | 28 | 39.5 | R8(1) | Medium | 548.5 | 550 |
| 25 | 27.5 | 40.5 | R9(1) | Low | 465 | 490 |
| 27 | 29 | 40 | R10(1) | Medium | 548.5 | 550 |
| 30 | 31 | 42 | R11(1) | Medium | 548.5 | 600 |
| 32 | 31 | 41 | R12(1) | High | 647.5 | 610 |
| 34 | 30 | 41 | R12(1) | High | 647.5 | 610 |
| 36 | 30.5 | 40 | R10(1) | Medium | 548.5 | 500 |

(with 0.94 membership degree) are fired. R1 and R5 conclusions are calculated by applying the minimum implication operator. Then, the rule with the highest weight is selected as the output of the rule, where in this case R5 (0.94) is selected. This means that the fuzzy output is predicted as Very High. Finally, predicted Very High output is defuzzified by taking the centroid of *Very High* MF of the amount of fresh water shown in Fig. 2(d), which is 730. In Table 2, the prediction process for the whole experimental data is presented.

## 4   Evaluations

We compared the FIS with the results of multiple linear regression using the comparison metrics of RMSE (Eq. 1), average forecasting error (Eq. 2) and Sum of Squared Error (SSE – Eq. 3). As shown Table 3 and Figs. 3 and 4, we achieve RMSE of 29.40 and average forecasting error of 4.75 in comparison to the RMSE of 105.80 and average forecasting error of 16.65 of multiple linear regression. This shows that the FIS can better predict the amount of fresh water that is more inline with the experimental data. Prediction accuracy and the SSE graphics also shows the efficiency of the FIS (Fig. 3).

**Table 3.** Comparison of results.

|  | Multiple linear regression | Fuzzy inference |
|---|---|---|
| RMSE | 105.80 | 29.40 |
| Average forecasting error | 16.65 | 4.75 |

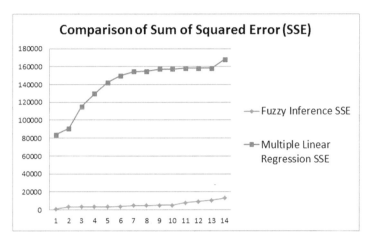

**Fig. 4.** Sum of Squared Error (SSE) comparison – multiple linear regression versus fuzzy inference.

$$RMSE = \sqrt{\frac{\sum_{j=1}^{n}\left(actual\ efficiency_j - predicted\ efficiency_j\right)^2}{n}} \qquad (1)$$

$$average\ forecasting\ error = \frac{\left|actual\ efficiency - predicted\ efficiency\right|}{actual\ efficiency} \times 100 \qquad (2)$$

$$SSE = \sum_{j=1}^{n}\left(actual\ efficiency_j - predicted\ efficiency_j\right)^2 \qquad (3)$$

## 5 Conclusions

A single inclined solar still was produced and tested under the thermal radiation. We use the experimental data obtained from a single slope solar still distillation in the fuzzy inference system (FIS). With 29.40 root mean square error and 4.75 average forecasting error, the predicted values are in close agreement with the experimental counterparts.

# References

1. Howell, J.R., Bannerot, R.B., Vliet, G.C.: Solar Thermal Energy Systems Analysis and Design. McGraw Hill, New York City (1985)
2. Ghosh, G.K.: Solar Energy: The Infinite Source. South Asia Books, Columbia (1992)
3. Mashaly, A.F., Alazba, A.A.: Membership function comparative investigation on productivity forecasting of solar still using adaptive neuro-fuzzy inference system approach. Environ. Prog. Sustain. Energy **15**(5), 555–578 (2017)
4. Yaseen, Z.M., Ramal, M.M., Diop, L., Jaafer, O., Demir, V., Kisi, O.: Hybrid adaptive neuro-fuzzy models for water quality index estimation. Water Resour. Manag. **32**(7), 2227–2245 (2018)
5. Zadeh, L.A.: Fuzzy sets. Inf. Control **8**, 338–353 (1965)
6. Mamdani, E.H.: Application of fuzzy logic to approximate reasoning using linguistic synthesis. IEEE Trans. Comput. **26**(12), 1182–1191 (1977)

# Evaluation of the Impact of State's Administrative Efforts on Tax Potential Using Sugeno-Type Fuzzy Inference Method

Samir Rustamov[1(✉)], Akif Musayev[2,3], and Shahzada Madatova[4]

[1] ADA University, Institute of Control Systems of ANAS, Baku, Azerbaijan
srustamov@ada.edu.az
[2] Institute of Economics, ANAS, Baku, Azerbaijan
akif.musayev@gmail.com
[3] Near East University, Nicosia, TRNC, Turkey
[4] Azerbaijan State University of Economics (UNEC), Baku, Azerbaijan
shahzada.madatova@unec.edu.az

**Abstract.** Evaluation of the impact of state's administrative efforts on tax potential via Sugeno-type fuzzy inference method has been investigated in the article. For this purpose, input data of the model has been fuzzified on the base of expert knowledge via different membership functions, and the output function has been evaluated on the base of the determined rules. Effective model-specific parameters have been selected in order to calculate the output function. The results obtained by Sugeno-type fuzzy inference method have been compared with the results evaluated via the Mamdani-type fuzzy inference method.

**Keywords:** Tax potential · Sugeno-type fuzzy inference method
Membership functions

## 1 Introduction

Taxation theory has two types of tax relations: material and organizational. Material tax relations reflect the movement of the money flowing from the taxpayer to the state, that is, in economic terms, these relations imply taxation or payment, however, in a legal term, implementation of tax liabilities.

Organizational tax relations occur related to the formation and functioning of the system of tax agencies, determination and application of state tax activity procedures (for example, the formation of tax service agencies, determination of their authorities and operating rules). Some authors call organizational tax relations as "tax administrative relations."

Tax legislation involves such issues as application of tax forms, the methods of calculating separate types of tax, tax deductions, tax rates, taxation object, definition of taxation base and so on. However, tax administration includes such issues as organization of taxpayer service, organization of mobile and cameral tax inspection, effective tax control, organization of the work on obtaining unreported declarations and tax

© Springer Nature Switzerland AG 2019
R. A. Aliev et al. (Eds.): ICAFS-2018, AISC 896, pp. 352–360, 2019.
https://doi.org/10.1007/978-3-030-04164-9_47

debts, state registration of commercial bodies in tax agencies, installation of POS-terminals, the rules of paid funds accounting, etc. [1].

Tax administration is a system of dynamically developing management of tax relations. This system coordinates the activities of tax agencies in the market economy condition. Tax administration ensures implementation of the state's tax policy. This policy is implemented through the tax mechanism. Tax mechanism is a set of organizational-legal norms and methods of taxation management. The role that taxes play in the life of the state and society is manifested through tax policy and tax mechanism. Therefore, the task of tax administration is to increase the effectiveness and improvement of the tax relations management system in the country. Effective tax administration will allow to finance the measures aimed at minimizing the impact of the crisis on the world economy and to make success in achieving long-term economic and social goals [2, 3].

There is an immediate connection between strong economic growth and tax potential framed by it. Such factors as tax legislation, administration, population's knowledge on economy and taxes, and so on. also have serious impact on this dependency. Therefore, from the point of taxation optimization, the main goal is not to increase the tax burden of the economy, but it should correctly estimate the tax potential formed by the economic system and maximize its collection level [4–6].

In the research work, Sugeno-type fuzzy inference method was used to assess the influence of the changes made in tax administration on the tax potential. The input data of the model was fuzzified on the base of expert knowledge via different membership functions, and the output function was evaluated on the base of the determined rules. The model-specific parameters were selected in order to calculate the output function. The results obtained by Sugeno-type fuzzy inference method were compared with the results evaluated via the Mamdani-type fuzzy inference method.

## 2  Fuzzy Inference System

The Fuzzy Inference System (FIS) uses expert knowledge and experience to construct a system for managing the process that is being studied. The connection between the input and output of this process is expressed via *IF − THEN* rules. In the fuzzy inference model, two major types of information are processed. First, membership functions of the input and output variables are determined. The precise selection of these functions in the construction of the model is one of the most critical steps. The second type of information is related to rule-based information that processes fuzzy output data from fuzzy input data. FIS generally consists of 3 blocks [7]. The first is fuzzification, that is, converting valuable crisp input data into fuzzy data via membership functions on the base of expert knowledge. Being inference block, the second one is an issue of assessing the membership degree of output variables to the fuzzy set from input data by using fuzzy rules. Finally, the fuzzy output turns into a crisp value in the defuzzification block. The inference block is the key block of FIS and produces a similar process to human decision-making, to obtain a management strategy by performing an approximate result (Fig. 1).

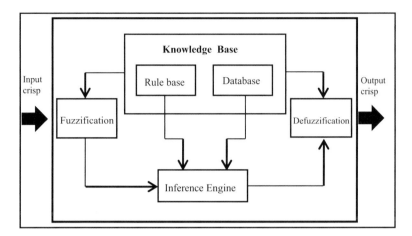

**Fig. 1.** General structure of fuzzy expert system [8].

In the fuzzy inference model, the judgement process is implemented via *IF −THEN* rules, which is based on the expert knowledge. Expert knowledge is included in the fuzzy control system and described as a set of rules:

"*If x is A then Y is B*", or, more generally, "*If $X_1$ is $A_1$ and* … *and $X_n$ is $A_n$ − then Y is B*".

where $A, A_n, B$ are fuzzy sets. Popular used FIS model was suggested by Mamdani. However, the number of rules in the Mamdani-type FIS is growing due to the number of variables in the condition. As the number of the rules is increased, combination of rules becomes heavier and sometimes it is difficult to cover the relationship between all the conditions and results.

Sugeno-type fuzzy inference method was introduced by M.Sugeno in 1985 and the first two steps are identical to the Mamdani method. These steps are: fuzzifying input data and application of fuzzy operator. Being linear or constant of the Sugeno output membership function is the primary distinction between Mamdani and Sugeno strategies. The Mamdani-type fuzzy inference method requires to find the center of a two-dimensional area formed by a continuous function, and this process is not considered effective in terms of calculation. The Sugeno-type method has fuzzy output having fuzzy input and linear input combination, which is effective in terms of calculation. The Sugeno method is a systematic approach producing fuzzy rules from a set of input-output data [9].

The fuzzy rule is defined as follows in the Sugeno-type fuzzy model:

$$If\ x\ is\ A\ and\ y\ is\ B\ then\ z = f(x, y)$$

Where $A$ and $B$ are fuzzy sets in *IF − THEN* rule, and $z = f(x, y)$ is the crisp function expressing the consequent of the rule. In many issues, the function $f(x, y)$ is expressed in the polynomial of input variables $x$ and $y$. If $f(x, y)$ is a fisrt-order polynomial, the fuzzy inference system is called the first order Sugeno fuzzy inference model, for example, $z = ax + by + c$. If $f$ is constant, the model is called zero-order

Sugeno fuzzy inference model $(a = b = 0)$ and this is the particular case of the Mamdani fuzzy inference model. Here, the consequent of each rule is determined by fuzzy singleton. Let's note that, special mam2sug function was developed in Matlab in order to convert Mamdani model into a zero-order Sugeno model.

The output of each $z_i$ rule is strengthened via $w_i$ weight. For example, the weight function for the rule $AND$ with first input x and the second input y is defined as follows:

$$w_i = And\,Method(F_1(x), F_2(y))$$

Where $F_1(x)$ and $F_2(y)$ are membership functions for the first and second input data.

As the weight of all rules, the final output of the system is calculated in the following formula [7]:

$$final\,output = \frac{\sum_{i=1}^{N} w_i z_i}{\sum_{i=1}^{N} w_i} \tag{1}$$

where $N$ is the number of rules. In comparison with Mamdani system, Sugeno model is expressed more compactly, and is effective in terms of calculation. Therefore, this model is used as an adaptive method in the construction of fuzzy models. Such adaptive methods allow to adaptive fuzzy systems to the data by specifying membership functions.

The main advantages of the Sugeno model are its efficiency in terms of calculation, its successful application in linear models, its use in adaptive methods and their optimization, its guarantee continuity of the output surface, as well as its efficiency in mathematical analysis. The primal precedencies of the Mamdani strategy are to be instinctive, having broad acceptance and appropriateness to human input.

## 3   Application of Sugeno-Type FIS

The changes made in tax administration and their impact on tax potential were evaluated by 20 different economist experts on 0-100 points scale. In order to increase the quality of the evaluation, it was checked by other experts and accuracy weight coefficient was included in this evaluation. In order to access the input data of the system, first, the weight coefficients were initially evaluated, they were multiplied by numbers and then the average value was found for each input data (Table 1). Let's indicate the impact of the changes and additions made in tax administration on the tax potential by y.

Five rules were determined by experts for evaluation of impact of the changes and additions made in tax administration on tax potential (Fig. 2).

The initial phase of FIS is the determinationn of input variables and their membership functions. As input variables $x_1$ is a creating new departments, $x_2$ is an applying ASAN signature, $x_3$ is an applying MOBILE signature, $x_4$ is an applying electronic audit, $x_5$ is an applying a unit standard in the system. At the next stage, membership functions, the appropriate parameters of these functions and linguistic variables are

**Table 1.** Average value of the expert estimation of tax reform in tax administration.

| Sign | The reforms in tax administration | Average value |
|------|-----------------------------------|---------------|
| $x_1$ | Some departments have been established on the basis 2, 11, 12 of the Regional Tax Offices | 9,85 |
| $x_2$ | "ASAN" signature application | 59,4 |
| $x_3$ | "MOBILE" signature application | 59,4 |
| $x_4$ | E-audit application | 58,9 |
| $x_5$ | The application of unique standards of services to taxpayers | 49,6 |

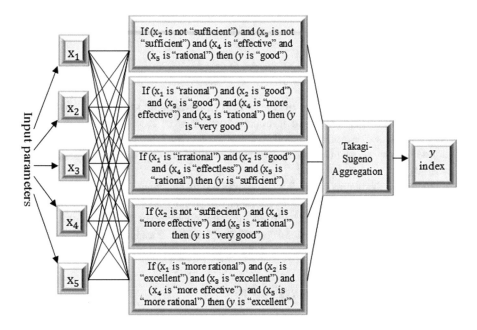

**Fig. 2.** Architecture of a fuzzy expert system.

defined by experts to fuzzify the input data (Table 2). After defining variables and membership functions, $IF - THEN$ fuzzy rules are defined. The determination of these rules is based on various combinations of input data fuzzified through the fuzzy operators [11–14]. We define five $IF - THEN$ rules for this problem (Fig. 2).

Following the fuzzification of the inputs, the degree which each part of the antecedent is satisfied for each rule. In case anterior of the provided rule possesses more than one part, the fuzzy operator is employed to acquire the single number demonstrating the outcome of anterior for that rule [7]. Aftermath, the acquired number is employed to the output function. The input to the fuzzy operator is two or more membership values from fuzzified input variables. The output is a single truth value.

**Table 2.** Membership functions of inputs and their appropriate parameters.

| Input variables | Linguistic variables | | | Membership functions |
|---|---|---|---|---|
| $x_1$ | "irrational"<br>$a = -39,7$<br>$b = 0,27$<br>$c = 35,3$ | "rational"<br>$a = 10$<br>$b = 50$<br>$c = 90$ | "more rational"<br>$a = 60$<br>$b = 100$<br>$c = 140$ | Triangle MF<br><br>$f(x; a, b, c) = \begin{cases} 0, & x \leq a \\ \frac{x-a}{b-a}, & a \leq x \leq b \\ \frac{c-x}{c-b}, & b \leq x \leq c \\ 0, & c \leq x \end{cases}$ |
| $x_2$ | "sufficient"<br>$a = -36$<br>$b = -4$<br>$c = 12,57$<br>$d = 36$ | "good"<br>$a = 14$<br>$b = 37,7$<br>$c = 50,9$<br>$d = 86$ | "excellent"<br>$a = 64$<br>$b = 88,5$<br>$c = 104$<br>$d = 136$ | Trapezoid MF<br><br>$f(x; a, b, c, d) = \begin{cases} 0, & x \leq a \\ \frac{x-a}{b-a}, & a \leq x \leq b \\ 1, & b \leq x \leq c \\ \frac{d-x}{d-c}, & c \leq x \leq d \\ 0, & d \leq x \end{cases}$ |
| $x_3$ | "sufficient"<br>$a = -36$<br>$b = -4$<br>$c = 13,62$<br>$d = 36$ | "good"<br>$a = 14$<br>$b = 47$<br>$c = 62,8$<br>$d = 86$ | "excellent"<br>$a = 64$<br>$b = 89,6$<br>$c = 104$<br>$d = 136$ | Trapezoid MF<br><br>$f(x; a, b, c, d) = \begin{cases} 0, & x \leq a \\ \frac{x-a}{b-a}, & a \leq x \leq b \\ 1, & b \leq x \leq c \\ \frac{d-x}{d-c}, & c \leq x \leq d \\ 0, & d \leq x \end{cases}$ |
| $x_4$ | "effect less"<br>$\sigma_1 = 10,4$<br>$c_1 = -7,5$<br>$\sigma_2 = 65,9$<br>$c_2 = 0,916$ | "effective"<br>$\sigma_1 = 12,2$<br>$c_1 = 41,1$<br>$\sigma_2 = 12,1$<br>$c_2 = 58,9$ | "more effective"<br>$\sigma_1 = 17,75$<br>$c_1 = 99,6$<br>$\sigma_2 = 136$<br>$c_2 = 104$ | Gaussian-2 MF<br><br>$f(x; \sigma, c) = e^{-\frac{(x-c)^2}{2\sigma^2}},$ |
| $x_5$ | "irrational"<br>$a = 0,2$<br>$b = 2,5$<br>$c = 0$ | "rational"<br>$a = 0,2$<br>$b = 2,5$<br>$c = 0,5$ | "more rational"<br>$a = 0,2$<br>$b = 2,5$<br>$c = 1$ | Bell-shaped MF<br>$f(x; a, b, c) = \frac{1}{1 + \left\lvert \frac{x-c}{a} \right\rvert^{2b}},$ |

We deployed operators such as "AND" and "OR" in our system. Being probabilistic, "OR" operator is specified as follows:

$$probor(a, b) = a + b - ab$$

4 linguistic variables were determined for Sugeno-type fuzzy inference system ($excellent - z_1, very\ good - z_2, good - z_3, sufficient - z_4$).

Constant function was used as output function in our experiment with parameters: $z_1 = 1, z_2 = 0.75, z_3 = 0.5, z_4 = 0.25$. The output of the system is calculated by the (1) formula and received 51.7 (Fig. 3).

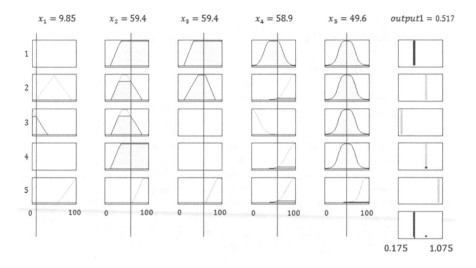

**Fig. 3.** Architecture of a fuzzy expert system.

As another experiment, linear functions with following coefficients were taken as a system output:

$$z_1 = 0.1x_1 + 0.5x_2 + 0.2x_3 + 0.5x_4 + 1$$
$$z_2 = 0.1x_1 + 0.4x_3 + 0.3x_4 + 0.2x_5 + 0.5$$
$$z_3 = 0.1x_1 + 0.4x_3 + 0.3x_4 + 0.2x_5 + 0.5$$
$$z_4 = 0.1x_1 + 0.5x_2 + 0.3x_3 + 0.2x_4 + 0.3x_5 + 0.25$$

In this experiment the output of the system is $y = 56.3$. This value is different for other parameters of the functions. For this reason, more effective parameters require adaptive learning of input and output data according to the expert's knowledge.

## 4   Conclusion

The article investigates the possibility of applying the Sugeno-type fuzzy inference method for studying the impact of government's administrative efforts on tax potential. As an example, the changes and additions made in the tax administration of the Republic of Azerbaijan in 2013 and their impact on tax potential were evaluated. As the input data of the system, the changes made in the tax administration were taken, and these data were fuzzified by using triangular, trapesoid, Gaussian, and Bell membership functions. The parameters of membership functions were evaluated for fuzzy linguistic variables.

5 $IF - THEN$ rules determined by the expert were used in Sugeno-type fuzzy inference system. Constant and linear functions were used as the system output. Sugeno-type fuzzy inference method was realized in the Matlab software package for

the problem being studied. According to the results, value 51.7 was obtained on the output of the fuzzy inference system using a constant output function, so that this is similar to the result of the Mamdani-type fuzzy control system presented in [15, 16]. An exemplary experiment was carried out for the Sugeno-type FIS system using a linear output function. In this case, in order to find out more effective combinations of the system parameters, it is necessary to use expert knowledge or adaptive teaching on the base of input and output data.

# References

1. Tax reforms in EU Member States, Tax policy challenges for economic growth and fiscal sustainability. European Economy 008/2015
2. Jean-François, B., Maïmouna, D.: Tax potential and tax effort: an empirical estimation for non-resource tax revenue and VAT's revenue. 2016.10
3. Slobodchikov, D.N.: Dissertation. Tax potential in the system of inter-budgetary relations (Code, HAC- 08.00.10) (2010)
4. Musayev, A.F.: Tax potential and its assessment methods. Tax magazine of Azerbaijan N5 (119) (2014)
5. Musayev, A.F.: Innovation Economics and Tax Stimulation, p. 184. The University of Azerbaijan, Baku (2014)
6. Musayev, A., Gahramanov, A.: Introduction to Econometrics, p. 173. The University of Azerbaijan, Baku (2011)
7. Mathwork. Fuzzy Inference Process. http://www.mathworks.com/
8. Fausto, C.: A Takagi-Sugeno fuzzy inference system for developing a sustainability index of biomass. Sustainability **7** (2015)
9. Shleeg, A.A., Ellabib, I.M.: Comparison of Mamdani and Sugeno fuzzy interference systems for the breast cancer risk. Int. J. Comput., Electr. Autom. Control. Inf. Eng. **7**(10), 387–391 (2013)
10. Kamyar, M.: Takagi-Sugeno fuzzy modeling for process control. Industrial Automation, Robotics and Artificial Intelligence (2008)
11. Sugeno, M.: Industrial Applications of Fuzzy Control. Elsevier, Amsterdam (1985)
12. Yen, J., Langari, R.: Fuzzy Logic. Pearson Education, London (2004)
13. Ross, T.J.: Fuzzy Logic with Engineering Applications. Wiley, Hoboken (2010)
14. Koop, G.: Analysis of Economic Data. Wiley, Chichester (2000)
15. Musayev, A., Madatova, S., Rustamov, S.: Evaluation of the impact of the tax legislation reforms on the tax potential by fuzzy inference method. Procedia Comput. Sci. **102**, 507–514 (2016)
16. Musayev, A., Madatova, S., Rustamov, S.: Mamdani-type fuzzy inference system for evaluation of tax potential. In: Recent Developments and the New Direction in Soft-Computing Foundations and Applications. Studies in Fuzziness and Soft Computing, vol. 361, pp. 511–523. Springer (2018)
17. Kamil, A., Rustamov, S., Clements, M.A., Mustafayev, E.: Adaptive neuro-fuzzy inference system for classification of texts. In: Recent Developments and the New Direction in Soft-Computing Foundations and Applications. Studies in Fuzziness and Soft Computing, vol. 361, pp. 63–70. Springer (2018)
18. Rustamov, S.: A hybrid system for subjectivity analysis. Adv. Fuzzy Syst. (2018)

19. Rustamov, S., Mustafayev, E., Clements, M.A.: Context analysis of customer requests using a hybrid adaptive neuro fuzzy inference system and hidden Markov models in the natural language call routing problem. Open Eng. **8**(1), 61–68 (2018)
20. Rustamov, S.S.: An application of neuro-fuzzy model for text and speech understanding systems. In: PCI 2012, The IV International Conference "Problems of Cybernetics and Informatics", Baku, Azerbaijan, vol. I, pp. 213–217 (2012)

# Analysing the Economic Impacts of the Euro for the Three Largest Economies in the Emu (European Monetary Union) and the Place of the Euro in Global Economics

Hüseyin Özdeşer$^{(\boxtimes)}$

Near East University, Nicosia, North Cyprus, Turkey
huseyin.ozdeser@neu.edu.tr

**Abstract.** Before the monetary integration process in the EU, there were also other attempts establish monetary integrations such as the Latin, Scandinavian and Germany monetary unions. For the European Union, achieving monetary unification can be accepted as the main factor for a stable unification between the member countries. Application of the common economic policies under the umbrella of monetary union also has significant importance for the EU in order to establish political unification. It is undeniable that the introduction of the euro has had significant importance for global politics and economics. The inability of the EU to act as one body, as in the United States, is an important factor that has led to the euro being a more uncertain currency compared to the dollar. The likelihood of Brexit and similar events raises the risk of the collapse of the EU, which leads to uncertainty about the future of the euro, and undoubtedly makes it a more risky investment instrument against the dollar. In this study, the economic impacts of the euro on the three largest economies in the EMU (European Monetary Union) member countries is analysed the efficiency of the euro created in the optimum currency area and the place of the euro in the global economy compared with the dollar will be investigated.

**Keywords:** Euro · Monetary integration · European Union
European Monetary Union · Risk · Uncertainty

## 1 Introduction

In any study which covers the EU, it is important that the structure of the union is comprehensively understood. Only in this way will the studies that investigate the EU be more effective. It is an unquestionable reality that the EU, which has 27 members, has a significantly important role in global economics and politics. After the Second World War, a concept emerged where it was perceived to be more beneficial to act with common goals rather than countries competing with each other. During the period of the Cold War, a new world order was formed by Western Europe and North America. During the Cold War, the countries discovered that they could not neglect the security of their citizens. The tension between countries caused a number of international

© Springer Nature Switzerland AG 2019
R. A. Aliev et al. (Eds.): ICAFS-2018, AISC 896, pp. 361–370, 2019.
https://doi.org/10.1007/978-3-030-04164-9_48

institutions to develop rapidly after the Second World War, one of the most important of which was the United Nations (McCormick, 2005).

The most important instruments in the EU that have power are the national governments, the EU Commission, the EU Court of Justice and the European Central Bank. The other important elements for the EU are political parties, non-governmental institutions and also public opinion (Hancock, Castle, Conradt et al., 2007).

In order to strengthen the economic integration in the EU and also to improve the collaboration between the EU member countries, the European Coal and Steel Union was established (Dyker, 1999). In reality, after the Second World War, many new developments occurred. It is possible to say that, among these new developments, the establishment of the EU was the most important. Although the EU currently has 27 members, it is not determined how many nations will ultimately become members. Undoubtedly, the importance of the EU for global economics and politics will continue to increase in the future. The EU plays a significantly important role in the globalization of world economics and politics. As it is known, attempts have been made to minimize the economic and political restrictions between the member countries of the EU as much as possible. As a result of the increased mobility of the population caused by migration, as well as countries' reduced ability to satisfy the needs of their citizens, the rapid developments in technology, science and transportation and the increasingly complex global world, the formation of international institutions has rapidly accelerated (McCormick, 2005).

## 2   Monetary Integration in the EU

For the formation of monetary integration between the countries within the monetary union, it is necessary for them to adopt a single currency and the exchange rate of the member countries that will be used in the monetary union should be stabilized in an inevitable way against each other. The presence of three main elements are also necessary for the monetary union, which are: the harmonization of monetary policies between the member countries, the formation of a common pool for foreign exchange reserves, and formation of a single Central Bank. The existence of these three elements is particularly important for the member countries that participate in the monetary union. If the member countries that form the monetary union have their own exchange reserves and apply their own monetary and fiscal policies, in the scenario where there is a fall in exchange reserves, the countries will have to apply restrictionary monetary and fiscal policies. The restrictionary policies or any failures in the applied policies could create stress on the exchange rate (EL-Agraa, 2007). The greatest risk for euro is the degree of the economic collapse risk of the EMU member countries in a similar manner as Greece, which certainly lowers the reliability of the euro.

## 3   The Place of the Euro in the Globalized World

The euro lost ground to the dollar in terms of global payment transactions in 2014 compared with 2013. Because of financial activities, the apparent power of the dollar is continuing. Approximately 80% of the financial activities which are realized in global markets use the dollar. In 2015, the dollar was the most frequently used currency as a global payment agent, accounting for 43% of all transactions.

In 2014, the demand of the euro in foreign markets was very high. Because of the political instability in the regions outside the euro area, the demand for the euro in foreign markets in 2014 was three times more than that in 2013. It is not known to what extent the euro is used outside the euro area. According to the data from the end of December 2014, it is thought that approximately 175 billion euro existed outside the euro area. For the same period, this amount was equal to 18% of the total euro in the euro area (Draghi, 2015) (Table 1).

**Table 1.**  Comparing Euro Area with United States

| United States | 2015 | Euro area | 2015 |
|---|---|---|---|
| Population (million) | 321 | Population (million) | 333 |
| GDP (billion dollar) | 17,947 | GDP (billion euro) | 10,400 |
| Per-capita income | 55,868 | Per-capita income | 31,213 |
| Economic growth (GDP annual %) | 2.4 | Economic growth (GDP annual %) | 1.5 |
| Unemployment rate(%) | 5.3 | Unemployment rate | 11.6 |

Source:http://www.focus-economics.com/countries/united-states
http://www.focus-economics.com/countries/eurozone

## 4   Analysis of the Effects of the Euro on the European Monetary Union Member Countries

The formation of a new currency unit and the formation of a new central bank which cooperates with the central banks of the countries which participate in the monetary union, thus forming a new monetary authority called the Euro system for the EU, enabled the EU to realize its monetary integration target. The most important target of the Euro system is to achieve price stability in the monetary integration area (Hitiris, 2003).

## 5   Methodology of the Study

The aim of the study is to compare the per capita income, inflation rate, and real interest rates for Germany, France, and Italy (GDP level-current US dollar in 2016 values GDP in Germany 3.478 trillion, GDP in France 2.465 trillion, GDP in Italy 1.859 trillion-data.worldbank.org), which have the largest economies in the EMU, for the 14-year period before 2002 and the 14-year period after 2002. In this manner, it will be possible to investigate whether the euro has created the expected positive effects on the per capita income, inflation rate and real interest rates for the three largest economies in the EMU. In the study, averages for per capita income, inflation rate, and real interest rate for Germany, France and Italy have been calculated for the 14-year period before the countries adopted the euro as a national currency and for the 14-year period after the countries adopted the euro as a national currency. After calculating the averages, the changes are measured.

## 6    The Effects of the Euro on Germany, France and Italy Economies

### 6.1    The Effects of the Euro on German Economy

See (Tables 2, 3 and 4).

**Table 2.** Per capita income in Germany

| Per – Capita income | Per – Capita income |
|---|---|
| 1998 – 2001 | 2002 – 2015 |
| 88 – 16,000 | 2002 – 25,200 |
| 89 – 17,000 | 2003 – 30,400 |
| 90 – 22,200 | 04 – 34,200 |
| 91 – 23,300 | 05 – 34,700 |
| 92 – 26,300 | 06 – 36,400 |
| 93 – 25,500 | 07 – 41,800 |
| 94 – 27,100 | 08 – 45,700 |
| 95 – 31,700 | 09 – 41,700 |
| 96 – 30,600 | 10 – 41,800 |
| 97 – 27,000 | 11 – 45,900 |
| 98 – 27,300 | 12 – 44,000 |
| 99 – 26,800 | 12 – 44,000 |
| 2000 – 23,700 | 14 – 47,800 |
| 2001 – 23,700 | 15 – 41,200 |

348,200/14 = 24,871.43
554,800/14 = 39,628.57
On average, per capita income has increased by 59.33%.
(39,628.57 − 24,871.43/24,871.43 * 100 = 59.33%)
Source: data.worldbank.org

**Table 3.** GDP deflator for Germany

| 1988 – 2001 | 2002 – 2015 |
|---|---|
| 1988 – 1.69 | 2002 – 1.35 |
| 1989 – 2.88 | 2003 – 1.4 |
| 1990 – 3.4 | 2004 – 1.2 |
| 1991 – 3.09 | 2005 – 1.1 |
| 1992 – 5.29 | 2006 – 0.6 |
| 1993 – 4.14 | 2007 – 0.3 |
| 1994 – 2.16 | 2008 – 1.7 |
| 1995 – 1.98 | 2009 – 0.8 |
| 1996 – 0.62 | 2010 – 1.8 |
| 1997 – 0.26 | 2011 – 0.8 |
| 1998 – 0.60 | 2012 – 1.1 |
| 1999 – 0.31 | 2013 – 1.5 |
| 2000 – 0.45 | 2014 – 2.1 |
| 2001 – 1.28 | 2015 – 2.06 |

26.65/14 = 1.90
17.81/14 = 1.27
On average, GDP deflator has decreased by 35.26%
(1.27 − 1.90/1.27 * 100 = − 35.26%)
Source: data.worldbank.org

**Table 4.** Real interest for Germany

| 1988 – 2001(%) | 2001 – 2015(%) |
|---|---|
| 1988 – 6.53 | 2002 – 8.24 |
| 1989 – 6.86 | 2003 – 4.27 |
| 1990 – 7.93 | 2004 – 3.98 |
| 1991 – 9.1 | 2005 – 4.51 |
| 1992 – 7.89 | 2006 – 5.08 |
| 1993 – 9.37 | 2007 – 4.19 |
| 1994 – 9.11 | 2008 – 5.09 |
| 1995 – 8.79 | 2009 – 3.15 |
| 1996 – 9.34 | 2010 – 3.09 |
| 1997 – 8.84 | 2011 – 2.58 |
| 1998 – 8.36 | 2012 – 2.41 |
| 1999 – 8.46 | 2013 – 3.14 |
| 2000 – 10.13 | 2014 – 3.06 |
| 2001 – 8.62 | 2015 – 3.05 |

119,33/14 = 8.52% 55.84/14 = 3.98%
On average real interest has decreased
by 53.28%
(3.98% − 8.52%/8.52% * 100 = −53.28%)
Source: Master Big Data Solution http://ychats.com/indicators

## 6.2    The Effects of the Euro on French Economy

See (Tables 5, 6 and 7).

**Table 5.** Per capita income for France

| 1988 – 2001 | 2001 – 2015 |
|---|---|
| 1988 – 17,700 | 2002 – 24,300 |
| 1989 – 17,700 | 2003 – 29,700 |
| 1990 – 21,800 | 2004 – 33,900 |
| 1991 – 21,800 | 2005 – 34,900 |
| 1992 – 23,900 | 2006 – 36,500 |
| 1993 – 22,500 | 2007 – 41,600 |
| 1994 – 23,600 | 2008 – 45,400 |
| 1995 – 27,00 | 2009 – 41,600 |
| 1996 – 27,000 | 2010 – 40,700 |
| 1997 – 24,400 | 2011 – 43,800 |
| 1998 – 25,100 | 2012 – 40,800 |
| 1999 – 24,800 | 2013 – 42,600 |
| 2000 – 22,500 | 2014 – 42,500 |
| 2001 – 22,500 | 2015 – 36,200 |

322,300/14 = 23,021.42
534,500/14 = 38,178.57
On average, per capita income has increased by 65.83% (38,178.57−
23,021.42/23,021.42 * 100 = 65.83%)
Source: data.worldbank.org

**Table 6.** GDP deflator for France

| 1988 – 2001 | 2001 – 2015 |
|---|---|
| 1988 – 3.21 | 2002 – 2.07 |
| 1989 – 3.3 | 2003 – 1.87 |
| 1990 – 2.67 | 2004 – 1.65 |
| 1991 – 2.57 | 2005 – 1.94 |
| 1992 – 1.99 | 2006 – 2.16 |
| 1993 – 1.64 | 2007 – 2.56 |
| 1994 – 0.93 | 2008 – 2.38 |
| 1995 – 1.15 | 2009 – 0.097 |
| 1996 – 1.37 | 2010 – 1.08 |
| 1997 – 0.88 | 2011 – 0.94 |
| 1998 – 0.95 | 2012 – 1.16 |
| 1999 – 0.22 | 2013 – 0.77 |
| 2000 – 1.54 | 2014 – 0.55 |
| 2001 – 2 | 2015 – 1.23 |

24.42/14 = 1.74
20.45/14 = 1.46
On average, it has declined by 16.09%.
(1.46 − 1.74/1.74 * 100 = − 16.09%)
Source: data.worldbank.org

**Table 7.** Real interest rate for France

| 1988 – 2001(%) | 2001 – 2015(%) |
|---|---|
| 1988 – 6.02 | 2002 – 4.44 |
| 1989 – 6.49 | 2003 – 4.64 |
| 1990 – 7.69 | 2004 – 4.87 |
| 1991 – 7.46 | 2005 – 2.85 |
| 1992 – 7.85 | 2006 – 3.73 |
| 1993 – 7.14 | 2007 – 4.6 |
| 1994 – 6.89 | 2008 – 5.62 |
| 1995 – 6.89 | 2009 – 7.36 |
| 1996 – 5.33 | 2010 – 5.52 |
| 1997 – 5.41 | 2011 – 5.05 |
| 1998 – 5.54 | 2012 – 5.03 |
| 1999 – 6.13 | 2013 – 5 |
| 2000 – 5.08 | 2014 – 4.46 |
| 2001 – 4.88 | 2015 – 4.91 |

88.8/14 = 6.34%
68.08/14 = 4.86%
On average real interest decreased by 23.34% (4.86% − 6.34%/
6.34% * 100 = − 23.34%)
Source: Master Big Data Solution http://ychats.com/indicators

### 6.3   The Effects of the Euro on the Italian Economy

See (Tables 8, 9 and 10).

**Table 8.** Per capita income for Italy

| 1988 – 2001 | 2001 – 2015 |
|---|---|
| 1988 – 15,700 | 2002 – 22,200 |
| 1989 – 16,300 | 2003 – 27,400 |
| 1990 – 20,800 | 2004 – 31,200 |
| 1991 – 21,900 | 2005 – 32,200 |
| 1992 – 23,200 | 2006 – 33,400 |
| 1993 – 18,700 | 2007 – 37,700 |
| 1994 – 19,300 | 2008 – 40,600 |
| 1995 – 20,600 | 2009 – 37,000 |
| 1996 – 23,000 | 2010 – 35,900 |
| 1997 – 21,800 | 2011 – 38,300 |
| 1998 – 22,300 | 2012 – 41,700 |
| 1999 – 21,900 | 2013 – 42,800 |
| 2000 – 20,100 | 2014 – 44,000 |
| 2001 – 20,400 | 2015 – 40,400 |

286,000/14 = 20,428.57
504,800/14 = 36,057.14
On average, per capita income has increased by 76.50% (36,057.14−
20,428.57/20,428.57 * 100 = 76.50%)
Source: data.worldbank.org

**Table 9.** GDP deflator for Italy

| 1988 – 2001 | 2001 – 2015 |
|---|---|
| 1988 – 6.65 | 2002 – 3.35 |
| 1989 – 6.2 | 2003 – 3.18 |
| 1990 – 8.91 | 2004 – 2.52 |
| 1991 – 7.58 | 2005 – 1.89 |
| 1992 – 4.37 | 2006 – 1.9 |
| 1993 – 3.89 | 2007 – 2.43 |
| 1994 – 3.54 | 2008 – 2.48 |
| 1995 – 4.93 | 2009 – 1.96 |
| 1996 – 4.55 | 2010 – 0.31 |
| 1997 – 2.6 | 2011 – 1.47 |
| 1998 – 2.53 | 2012 – 1.38 |
| 1999 – 1.62 | 2013 – 1.22 |
| 2000 – 1.97 | 2014 – 0.80 |
| 2001 – 2.99 | 2015 – 0.75 |

62.33/14 = 4.45

25.64/14 = 1.83

On average GDP deflator has decreased by 58.8%.

(1.83 − 4.45/4.45 * 100 = − 58.8%).

Source: data.worldbank.org

**Table 10.** Real interest rate for Italy

| 1988 – 2001(%) | 2001 – 2015(%) | |
|---|---|---|
| 1988 – 6.49 | 2002 – 3.08 | |
| 1989 – 7.55 | 2003 – 2.57 | |
| 1990 – 5.45 | 2004 – 2.91 | |
| 1991 – 6.59 | 2005 – 3.36 | |
| 1992 – 11.65 | 2006 – 3.65 | |
| 1993 – 10.34 | 2007 – 3.81 | |
| 1994 – 8.15 | 2008 –  4.25 | |
| 1995 – 7.92 | 2009 – 2.75 | |
| 1996 – 7.9 | 2010 – 3.7 | |
| 1997 – 7.91 | 2011 – 3.09 | |
| 1998 – 5.96 | 2012 – 3.79 | |
| 1999 – 4.65 | 2013 – 3.89 | |
| 2000 – 4.96 | 2014 – 3.87 | |
| 2001 – 4.18 | 2015 – 3.44 | |

99.7/14 = 7.12%

48.16/14 = 3.44%

On average, real interest rate has declined by 51.68%

(3.44% − 7.12%/7,12% * 100 = −51.68%)

Source: Master Big Data Solution http://ychats.com/indicators

# 7 Results

When the effects of the euro on Germany, France and Italy economy are analysed, it is observed that the euro has created positive impacts on the per-capita income, inflation levels, and also real interest rates for the three countries as expected when comparing the 14 years before the euro was adopted as the national currency and the 14 years after the euro was adopted as the national currency.

# 8 Conclusion

The EU member countries signed the Maastricht Treaty in 1992, which supported the formation of the Economic and Monetary Union in the EU. In March of 2002, the euro was adopted as the national currency of the EU (Zestos, 2006).

By this study it is possible to see the positive economic impacts of euro that was expected to be seen before the introduction of euro. This shows that the euro is the right decision for the economic integration process of the EU. This is because as the euro was able to achieve the positive effects on the per capita income, inflation rate and at the same time on the interest rate within 14 years, this means this positive movement of the euro will continue. In the formation of economic integration, there are five main phases, which are: free trade areas, customs union, common market, economic and monetary union and political union. The establishment of a political union, for example, which is concerned with matters such as foreign policy and security, indicates that there is a need for an authority above the national governments (Nello, 2005). The greatest cost for the countries joining the EMU is the loss of monetary sovereignty and the application of a common monetary policy without inflation (Bainbridge, Burkitt, Whyman, 2000).

One of the most important conditions for the formation of monetary integration is political integration (Mintz: 1970, Paolo from 1977). The application of similar policies between the member countries is beneficial for the creation of a monetary area. It is a natural result to perceive the euro as a stable and powerful currency in global economics. The euro started to be maintained as a reserve currency with the dollar in many countries (Özdeşer, 2015).

# References

Bainbridge, M., Burkitt, B., Whyman, P.: The Impact of The Euro, 1st edn. Macmillan Press LTD, London (2000)

Data.worldbank.org. Accessed 18 Aug 2016

Draghi, M.: Eurosystem, 14th edn. The International Role of the Euro. Euuropean Central Bank, Frankfurt (2015)

Dyker, A.: The European Economy, 2nd edn. Addison Wesley Longman Limited, England (1999)

EL-Agraa, A.: The European Union, Economics and Politics, 8th edn. Cambridge University Press, Cambridge (2007)

FocusEconomics. https://www.focus-economics.com/countries/united-states. Accessed 26 Sept 2016

FocusEconomics. https://www.focus-economics.com/countries/eurozone. Accessed 2 Sep 2016

Hancock, D., Carman, C., Castle, M., Conradt, D., et al.: Politics in Europe, 4th edn. CQ Press, Washington (2007)

Hitiris, T.: European Union Economics, 5th edn. Prentice Hall, England (2003)

Master Big Data Solution http://ychats.com/indicators. (2018/05/20)

McCormick, J.: Understanding The European Union, 3rd edn. Palgrave Macmillan, New York (2005)

Mintz, N.: Monetary Union and Economic Integration, 1st edn. New York University, New York (1970)

Nello, S.: The European Union economics, policies and History, 1st edn. McGrawHill Education, London (2005)

Özdeşer, H.: Avrupa Merkez Bankası'nın (AMB) Euro Yönetimindeki Etkinlik Sorununun Analizi. Yönetim ve Ekonomi **2**(22), 467–482 (2015)

Zestos, K.: European Monetary Integration The Euro, 1st edn. Thomson South-Western, America (2006)

# Evaluation of Tourism Sector Based on the Internal Environment by Using a Fuzzy Approach

İhsan Yüksel[1(✉)], Metin Dağdeviren[2], and Gülsüm Alicioğlu[2]

[1] Kırıkkale University, 71450 Kırıkkale, Turkey
yuksel@kku.edu.tr
[2] Gazi University, 06570 Maltepe, Turkey
{metindag,gulsumalicioglu}@gazi.edu.tr

**Abstract.** Organizations have been trying to achieve their goals considering the internal and external environment conditions, parameters and variables. For this reason, the management activities have been performed by a strategic approach at the organizations. Different analyses have been made in the strategic management process first stage. Main reason of this is difficulties of the explanation completely and detailed of strategic analysis stage in the organization. Primary purpose of this study is to carry out the strategic analysis based on the internal environment. In this study strategic analysis was conducted on sectoral basis and by an integrated approach. Analysis unit of the study is Turkey tourism sector. Content of this study is limited by the internal environment of the Turkey tourism sector. Firstly in this study the vision's components of Turkey tourism sector were determined by the content analysis. Then, the weights of the components were calculated. Weights of the internal environment factors that are strengths and weaknesses were determined according to the components of vision. In addition to these analyses the present state of each internal factors were evaluated by the scale that was consisted with fuzzy numbers. Finally the present level of each internal environment factor was calculated as separated and whole. In the result of the study tourism sector were evaluated by the strategic and integrated approach with fuzzy numbers.

**Keywords:** Fuzzy approach · Turkey tourism sector
Internal environment factors

## 1 Introduction

Organizations have been trying to achieve their goals considering the internal and external environment conditions, parameters and variables. Nowadays organizations have carried out their functions and activities in the dynamic and competitive conditions [1]. At the same time organizations have been trying the performance in the unstable environment [2]. Consequently organizations have seen the issue by the strategic management approach. In the literature strategic management and process have been explained from the different arguments and approaches [3–5].

© Springer Nature Switzerland AG 2019
R. A. Aliev et al. (Eds.): ICAFS-2018, AISC 896, pp. 371–377, 2019.
https://doi.org/10.1007/978-3-030-04164-9_49

Strategic management process has different stage [6, 7]. But strategic management has been begun with the analysis stage [8]. An analysis of the environment of organization or unit of analysis has been performed in the strategic analysis stage. Environment of organization or unit of analysis has been consisted of internal and external environment. Internal environment is an environment that is under control of organization. Key elements of organization's internal environment are resources and capabilities. And resources and capabilities determine the strengths and weaknesses of organization [6, 7]. Strengths and weaknesses of organization have affected their activities and competitive advantage of organization [3, 9].

On the other side organization have been located an external environment. External environment have opportunities and threats. Threats have hindered the activities and functions of organization and goals. On the contrary opportunities are situation and conditions that make easer to perform the goals of organization [6, 7]. Due to these reasons organizations have to consider the internal and external environment. Because environment of organization can affect the all activities and functions of organization, therefore organizations have made the analysis of internal and external [8]. In the literature it has been seen that strategic analysis process has been studied with different approaches and methods and level [9–12].

In strategic analysis, the vision of organization must be determined and known because activities of organization must be toward the vision. In this way resources and capabilities of organization can be canalized to perform of the vision or other goals of organization. Due to these reasons an organization must have a vision that indicated their goals. And organization must carry out their activities based on the vision because strategic management concept has required this thinking. In the related literature there have been studies that are contain strategic analysis based on the internal environment issue [3, 9]. However when reviewed the literature, it were not encountered with strategic analysis studies according to the internal environment based on the vision as integrated approach and fuzzy approach. In this study, internal factors' present states were tried to evaluate by the fuzzy approach considering vision.

In this study fuzzy numbers were used because calculated the measurement of the internal factors' present states by the crisp is difficult. Indeed in the literature that study with fuzzy numbers [13] has been seen to use the fuzzy numbers more calculable in decision making and evaluating. For this reason in this study internal environment factors were evaluated with fuzzy numbers.

## 2   Method

This study was made devoted to Turkey tourism sector. Turkey is located in the most beautiful geographic of earth. All the same it is known that numerous civilization and culture were experienced and lived in Turkey geography. Because of this Turkey has high potential level in tourism. Tourism sector has important weight in Turkey economy. Also Turkey has given particular importance to tourism sector. And long range plans have been made to develop and to sustain tourism sector. One of those is Tourism Strategy of Turkey-2023 [14]. It is most important study for tourism sector development that was prepared by a strategic approach.

In today's world tourism activity has been conducted by the strategic approach is a requirement. In this study Analytic Hierarchy Process (AHP) and Technique for Order Performance by Similarity to Ideal Solution (TOPSIS) that are the multi criteria decision making techniques and fuzzy numbers were used to determine present states of the internal environments factors. In this study weights of the vision components were calculated by the AHP [15]. AHP technique applications were used in different studies that are related with performance and evaluation [16, 17]. Weights of the internal factors were determined by the TOPSIS technique [18, 19].

In this study TOPSIS technique was preferred because alternatives can be evaluated based on the two or more criteria and other reason is that TOPSIS is not limited by the alternative numbers. In this study internal environment factors' present state were evaluated by the fuzzy numbers. Main reason of this is that internal environment factors' present state cannot be measured by the crisp number. In this study internal environment factors were evaluated by the fuzzy scale that is used in the literature [20] that was given as Table 1. As seen Table 1, internal environment factors were evaluated by the five categories.

**Table 1.** Linguistic mean of fuzzy numbers.

| Importance of strengths and weaknesses | Triangular fuzzy scale |
| --- | --- |
| Very low (VL) | (1, 1, 3) |
| Low (L) | (1, 3, 5) |
| Medium (M) | (3, 5, 7) |
| High (H) | (5, 7, 9) |
| Very high (VH) | (7, 9, 9) |

Membership functions of triangular fuzzy numbers were given in Fig. 1. In this study data that it was taken that were evaluated by the scale that is Table 1 that indicates internal environment factors' present states were converted to the crisp numbers by the algorithm [21, 22].

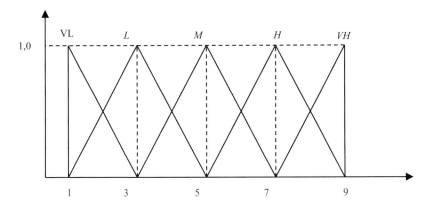

**Fig. 1.** Membership functions of linguistic values for internal environment factors

In this study based on the vision evaluation of Turkey tourism sector internal environment factors were made as follow steps:

- Determination internal environment factors.
- Determination the criteria.
- Determination weights of criteria.
- Calculation weights of internal environment factors.
- Evaluation of the tourism sector.

## 3   Results

Step 1. Determination internal environment factors. In the first step Turkey tourism sector's internal environment factors were identified. Internal factors contain the strengths and weaknesses factors. Strengths and weaknesses factors were taken from the Tenth Development Plan Specialization Commission Report [23] that was prepared by the Republic of Turkey Ministry of Development. Strengths factor as follows:

- Climatic conditions (S01)
- History and cultural heritage (S02)
- Intangible cultural heritage (S03)
- Turkish hospitality (S04)
- Dynamic population structure (S05)
- Genuine socio-cultural features (S06)
- Famousness in the regular market (S07)
- Undiscovered destination (S08)
- Activation in the domestic tourism (S09)
- Geographic situation and transport facilities (S010)
- Attract great attention by the investors (S11)
- Conformance to the tourism demand(S12)
- Winter tourism (S13)
- Health and thermal tourism (S14)
- Yacht tourism (S15).

  Weaknesses factors as follows:

- The infrastructure, service quality, and support sectors (W01)
- Renewable energy potential (W02)
- The disparity of large scale hotels' with rural area (W03)
- An authority complexity derived from legislative in planning (W04)
- Distribution of the harbors and marinas for yacht tourism (W05)
- Different standard on the tourism operation licenses (W06)
- Insufficient preservation of the authentic characteristics of the natural, history and intangible cultural heritage (W07)
- Coordination in resource utilization for promotion (W08)
- Ineffective usage of social media's (W09)
- Insufficiency on certification of professional qualifications' (W10)

- Insufficiency of sustainable environment management (W11)
- Deficiency of the townscape concept (W12)
- Insufficiency recycling of the municipal solid wastes in the seaside (W13)
- Non-functional tourism councils (W14)
- Lack of the holistic approach of the tourism development (W15)
- Not to attend of the local people and officials to the decision-making mechanisms (W16).

Step 2. Determination the criteria. Evaluation of the internal environment factors were made according to the vision components of Turkey tourism sector. For this in this study Tourism Strategy of Turkey-2023 and Action Plan (2017-2013) [14] that were prepared and published were analyzed. As a result of the context analysis sustainable tourism and international brand were identified as component vision.

Step 3. Determination weight of criteria. In this step of the study vision components' weights were calculated by the AHP. Pairwise comparisons of the criteria were made by the Saaty [15] scale 1–9. Pairwise comparison and weights of the vision components were given in Table 2.

**Table 2.** Pairwise comparison matrix and weights of the vision components

| Vision components | ST | IB | Weights |
|---|---|---|---|
| Sustainable tourism (ST) | 1 | 2 | 0.333 |
| International brand (IB) | 1/2 | 1 | 0.667 |

Step 4. Calculation weights of internal environment factors. In this study weights of the internal factors were calculated by the TOPSIS technique. Each strengths and weaknesses factors were evaluated based on the components of vision by the scale that is Likert type. Scale range is from 1 to 5. One is minimum value and five is maximum value for the evaluation of the internal factors. Results of TOPSIS analysis were given Table 3 that are weights of strengths in the second column and weights of weaknesses in the seventh column.

Step 5. Evaluation of the tourism sector. In this step present state of tourism sector were evaluated according to the internal factors. Strengths and weaknesses factors' results were given Table 3. In Table 3, first column is indicated the code of the strengths factors. In second column weights of the strengths internal environment factors were given that were calculated by the TOPSIS. In third column strengths factors' linguistic value were given by the Table 1. In fourth column triangular fuzzy numbers for linguistic value were given. And in the fifth column each strengths factors' present state level were given as percentage. In Table 3, sixth column the code of the weaknesses factors were given. In seventh column weights of the weaknesses factors were given that were calculated by the TOPSIS. In the eighth column weaknesses factors' linguistic value were given by the Table 1. In ninth column triangular fuzzy numbers for linguistic value were given. And in the tenth column each weaknesses factors' present state level were given as percentage.

**Table 3.** Results of the strengths and weaknesses factors

| Strengths | | | | | Weaknesses | | | | |
|---|---|---|---|---|---|---|---|---|---|
| Factors | Weights | Linguistic value | Fuzzy value | Present state | Factors | Weights | Linguistic value | Fuzzy value | Present state |
| S01 | 0.0876 | H | (5.7.9) | 0.133 | W01 | 0.0786 | VH | (7.9.9) | 0.071 |
| S02 | 0.0657 | L | (1.3.5) | 0.043 | W02 | 0.0180 | H | (5.7.9) | 0.014 |
| S03 | 0.0805 | VL | (1.1.3) | 0.029 | W03 | 0.0541 | VH | (7.9.9) | 0.049 |
| S04 | 0.0438 | VL | (1.1.3) | 0.016 | W04 | 0.0541 | H | (5.7.9) | 0.041 |
| S05 | 0.0438 | VL | (1.1.3) | 0.016 | W05 | 0.0656 | H | (5.7.9) | 0.050 |
| S06 | 0.0728 | L | (1.3.5) | 0.047 | W06 | 0.0836 | H | (5.7.9) | 0.063 |
| S07 | 0.1095 | M | (3.5.7) | 0.119 | W07 | 0.0721 | VH | (7.9.9) | 0.065 |
| S08 | 0.0728 | L | (1.3.5) | 0.047 | W08 | 0.0476 | H | (5.7.9) | 0.036 |
| S09 | 0.0290 | L | (1.3.5) | 0.019 | W09 | 0.0771 | H | (5.7.9) | 0.058 |
| S10 | 0.0657 | M | (3.5.7) | 0.071 | W10 | 0.0656 | H | (5.7.9) | 0.050 |
| S11 | 0.0805 | VL | (1.1.3) | 0.029 | W11 | 0.0426 | VH | (7.9.9) | 0.038 |
| S12 | 0.0947 | L | (1.3.5) | 0.062 | W12 | 0.0656 | VH | (7.9.9) | 0.059 |
| S13 | 0.0438 | VL | (1.1.3) | 0.016 | W13 | 0.0721 | VH | (7.9.9) | 0.065 |
| S14 | 0.0438 | VL | (1.1.3) | 0.016 | W14 | 0.0541 | H | (5.7.9) | 0.041 |
| S15 | 0.0657 | VL | (1.1.3) | 0.024 | W15 | 0.0836 | H | (5.7.9) | 0.063 |
| | | | | | W16 | 0.0656 | VH | (7.9.9) | 0.059 |
| Overall evaluation of factors | | | | 0.687 | | | | | 0.822 |

In Table 3, in the last row, total level of strengths and weakness factors were given as percentage. As it is seen in Table 3, the present state of the strengths factors is 68.7%. This level has indicated that tourism sector has not used all potential. Level of the present state of the weaknesses is 82.2%. It can be said that weaknesses factors is in unfavorable state for tourism sector.

## 4  Conclusion

In this study Turkey tourism sector were analyzed based on the internal environment factors that are strengths and weakness of the tourism sector. Tourism sector were evaluated by the strategic approach using the multi criteria decision techniques and fuzzy numbers in study. Tourism sector's internal environment factors were analyzed based on the vision components of the tourism sector with integrated approach. Present situation of strengths and weaknesses factors that are internal environment factors were evaluated with fuzzy numbers. According to the results of the study each strengths and weaknesses factors' present situation level could be calculated. In this way each internal factor's potential degree in the tourism sector was identified. Another a result, strengths and weaknesses factors' general degree can be calculated as a whole. In this way potential of tourism sector's internal factors to what extent used have been seen. Study is limited with the internal environment factors of Turkey tourism sector. In future study external environment factors can be considered for evaluating the tourism sector analysis.

# References

1. Misankova, M., Kocisova, K.: Strategic implementation as a part of strategic management. Procedia-Soc. Behav. Sci. **110**, 861–870 (2014)
2. Brews, P., Purohit, D.: Strategic planning in unstable environments. Long Range Plan. **40**, 64–83 (2007)
3. Duncan, W.J., Ginter, P.M., Swayne, L.E.: Competitive advantage and internal organizational assessment. Acad. Manag. Exec. **12**(3), 6–15 (1998)
4. Lynch, R.: Strategic Management. Pearson Education Limited, England (2009)
5. Pop, Z.C., Borza, A.: New perspectives on strategic management process. Ann. Fac. Econ. Univ. Oradea Fac. Econ. **1**(1), 1573–1580 (2013)
6. Dinçer, Ö.: Stratejik yönetim ve işletme politikası. Beta Yayınları, İstanbul (2004)
7. Ülgen, H., Mirze, S.K.: İşletmelerde Stratejik Yönetim. Arıkan Yayınları, İstanbul (2007)
8. Me, A.: Analysis of the impact of strategic management on the business performance of SMES in Nigeria. Acad. Strat. Manag. J. **17**(1), 1–20 (2018)
9. Brownlie, D.: Scanning the internal environment: impossible precept or neglected art? J. Mark. Manag. **4**(3), 300–329 (1989)
10. Glaister, K.W., Falshaw, R.: Strategic planning: still going strong? Long Range Plan. **32**(1), 107–116 (1999)
11. Yüksel, İ., Dağdeviren, M.: Using the analytic network process (ANP) in a SWOT analysis-a case study for a textile firm. Inf. Sci. **177**(6), 3364–3382 (2007)
12. Gökdeniz, İ., Kartal, C., Kömürcü, K.: Strategic assessment based on 7S McKinsey model for a business by using analytic network process (ANP). Int. J. Acad. Res. Bus. Soc. Sci. **7**(6), 342–353 (2017)
13. Lin, C., Hsieh, P.-J.: A fuzzy decision support system for strategic portfolio management. Decis. Support Syst. **38**, 383–398 (2004)
14. Ministry of Culture and Tourism (MCT): Tourism Strategy of Turkey-2023. T.R. Ministry of Culture and Tourism Publications Number: 3090, Ankara (2007)
15. Saaty, T.L.: The Analytic Hierarchy Process. McGraw-Hill International Book Company, New York City (1980)
16. Yüksel, M.: Evaluating the Effectiveness of the chemistry education by using the analytic hierarchy process. Int. Educ. Stud. **5**(5), 79–91 (2012)
17. Yüksel, M., Geban, Ö.: Evaluation of teacher performance according to the special area competencies of chemistry teachers. H. U. J. Educ. **30**(1), 299–312 (2015)
18. Hwang, C.L., Yoon, K.: Multiple Attribute Decision Making: Methods and Applications A State of the Art Survey. Springer, New York City (1981)
19. Shih, H.-S., Shyur, H.-J., Lee, E.S.: An extension of TOPSIS for group decision making. Math. Comput. Model. **45**, 801–813 (2007)
20. Yüksel, İ. Dağdeviren, M., Kurt, M.: Aristo ve Bulanık Mantık Temelinde Üst Yönetim Stratejilerinin Değerlendirilmesi. In: 14 Ulusal Yönetim ve Organizasyon Kongresi, pp. 340–345, Atatürk Üniversitesi, Erzurum (2006)
21. Cheng, A.-C., Chen, C.-J., Chen, C.-Y.: A fuzzy multiple criteria comparison of technology forecasting methods for predicting the new materials development. Techonological Forecast. Soc. Chang. **75**(1), 131–141 (2008)
22. Dağdeviren, M.: Performans Değerlendirme Sürecinin Çok Ölçütlü Karar Verme Yöntemleri İle Bütünleşik Modellenmesi. Yayınlanmamış Doktora Tezi Gazi Üniversitesi Fen Bilimleri Enstitüsü, Ankara (2005)
23. Republic of Turkey Ministry of Development (RTMD): Tenth Development Plan Specialization Commission Report. Publications Number: 2859, Ankara (2014)

# Evaluation of the Impact of the Changing the Term of Tax Liability Performance on Tax Receipts by Minmax Composition Method

A. A. Musayeva[1], T. M. Musayev[2], and M. Kh. Gazanfarli[3(✉)]

[1] The Azerbaijan University, J. Hajibeyli Street, 71, AZ1007 Baku, Azerbaijan
aygun.musayeva@gmail.com
[2] ANAS Institute of Control Systems, B. Vahabzadeh Street, 9, AZ1141 Baku, Azerbaijan
turac.musayev@gmail.com
[3] ANAS Institute of Economy, H. Javid Avenue, 115, AZ1143 Baku, Azerbaijan
m.qezenferli@gmail.com

**Abstract.** This paper investigates the application of methods that enable the selection of the most suitable one of the many possible options based on the theory of fuzzy logic in the process of changing tax obligation period. Applied methods analyze factors reflecting the activities of tax authorities or taxpayers in uncertainty, choosing the most appropriate decision to reduce the loss of state budget and the risks.

**Keywords:** Fuzzy sets · Decision-making · Multi-criteria selection
Tax liability · Tax policy

## 1 Introduction

Lack or inaccuracy of information has created difficulties in the decision-making process of people and finding ways to eliminate these problems has become one of the most important issues from the earliest times of human history. The decision-making problem under uncertainty has found its solution more efficiently by methods based on the theory of "Fuzzy logic" proposed by Zadeh [1] in 1965. One of the most widely-used and known branch of decision-making is multi-criteria selection or multi-criteria decision-making. Multi-criteria decision-making (MCDM) or multi-criteria decision analysis (MCDA) enables to evaluate many contradictory criteria explicitly in this process. There are different classifications of MCDM problems and solution methods of them. However, the main qualifying is based on finding solution explicitly and implicitly that was proposed by lots of authors (one of them is Zimmermann 1996) [2] and divided into two subdivision: multiple-criteria evaluation problems and multiple-criteria design problems.

The first type of MCDM problems consists of a finite number of alternatives, which are defined from the beginning of the solution process and each of them is presented by several criteria that reflect its performance. The problem can be solved by choosing the best alternative or set of alternatives for the decision maker.

© Springer Nature Switzerland AG 2019
R. A. Aliev et al. (Eds.): ICAFS-2018, AISC 896, pp. 378–385, 2019.
https://doi.org/10.1007/978-3-030-04164-9_50

In the second one, alternatives are unknown and can be found by solving the applied mathematical model.

There are numerous methods for solving MCDM problems [2]: The weighted sum model (WSM) is one of the most widely used and simple methods that proposed by Fishburn [3] in 1967. This method chooses the best solution by evaluating the alternatives based on the weight of the criterion (the degree of belonging to the elements of fuzzy logic) . The analytic hierarchy process (AHP) is a process structured based on mathematics and psychology to analyze and organize complicated decisions. It has been designed by Saaty in the 1970s and extensively studied and improved in subsequent years [4]. The source of Decision expert (DEX) method is coming from the idea that was presented by Efstathion and Rajkovic in 1979. Their proposal was to use words instead of figures and tables instead of functions. This method has been developed by a research team of Bohanec, Bratko and Rajkovics [5, 6] in 1983. Best – Worst method (BWM) was proposed by Rezaei [7, 8] in 2015. This method was developed basing on a systematic double-comparison of criteria, it means, identifying the membership degree of each criterion and selecting two critical criteria that one of them plays the most important role and another one has not any role in decision-making process. So, selecting the best alternative, the final points are obtained by aggregation of weights of different pairs. Zak [9] compares alternative formula inputs with results, it means, with supplier ratings, expressing comparable analysis of suppliers in various industries as a multi-criteria decision-making problem. The final result is obtained from the implementation of expert assessment and AHP methods of MCDM. A trade-off ranking method was developed for ranking alternatives which are given as contradictory criteria in presented article by Jaini and Utyuzhnikov [10]. The proposed method is used to sort alternatives by comparing two known method - (TOPSIS) and (VIKOR) of MCDM.

## 1.1   Statement of the Problem

In our paper the solved problem as follows: Tax policy and its efficient organization is one of the most essential factors for forming of the state budget and economic development of each country. For efficient tax policy implementation, considerable experience has been accumulated in international taxation and incentive forms of tax liability have been compiled [11]. Tax administration and legislation have been investigated by numerous scientific researchers all over the world. Some of them have been proposed by Musayev, Rustamov and Madatova and in these papers, the effects of changes and additions of tax administration and legislation to the tax potential were assessed by the Mamdani method [12, 13]. The tax legislation of Azerbaijan Republic is also widely used by stimulants. According to so called 'Terms of tax obligations fulfillment and alteration of such terms' Article 85 [14, 15] of the Tax Code of Azerbaijan Republic, the terms of fulfillment of tax obligations may be extended for the period 1–9 months within tax year, if the taxpayer is damaged as a result of natural disaster or any other force-majeure circumstances or in the case of threat of bankruptcy of the tax payer, if the taxpayer is damaged when paying one-off tax and if the relevant collateral or guarantee is given. Factors that reflect the financial and economic activity of each taxpayer are analyzed for applying concession to the most appropriate taxpayer defined by legislation, also reducing risks and budget losses when allowance is applied and this analysis plays an important role in decision making in uncertainty.

In this issue, the choice of a suitable taxpayer in uncertainty is expressed as a MCDM problem based on the theory of fuzzy sets and its solution is sought by using the minmax composition. At the same time, alternatives are evaluated with another known method of MCDM and the results obtained in both methods are compared.

Each taxpayer who applying to the tax authority to extend the term of the tax liability is considered as objects (alternatives) and the indicators reflecting these objects are criteria. The method of selecting the most suitable object (alternative), based on given criteria, is presented in the next section.

## 2   Solution of Problem with Multi-criteria Decision-Making

Before mentioning these methods, let us look through a represented formula for summarizing sub-criteria of each criterion (for appropriate objects):

$$C^k = \left(c_{ij}^k\right); k = 1\ldots K; i = 1\ldots nk; \ j = 1\ldots m \tag{1}$$

$C^k$ – are criteria given for appropriate objects (alternatives) $A_j$ ($j = 1\ldots m$), $A_j$ – are objects (alternatives), $j = 1\ldots m$ is number of objects; $n_k$ – are sub-criteria (subsets) of each i criterion. In this case, we can express each $c_{ij}^k$ matrices that consisting of subcriteria as follows:

If $k = K$, $i = 1\ldots n_k$, $j = 1\ldots m$ then matrix $c_{ij}^K$ is:

$$c_{ij}^K = \begin{pmatrix} c_{1,1}^K & \cdots & c_{1,m}^K \\ \vdots & \ddots & \vdots \\ c_{nk,1}^K & \cdots & c_{nk,m}^K \end{pmatrix}$$

Using the above, we can estimate the average of criteria for objects by the following formula:

$$\overline{c}_J^k = \sqrt[n_k]{\prod_{i=1}^{n_k} c_{ij}^k} \ k = 1\ldots K; i = 1\ldots n_k; j = 1\ldots m \tag{2}$$

### 2.1   Minmax Composition Method

Now let's solve the problem of choosing the most appropriate object (solution) in accordance to each of the average criterian obtained by using MDCM based on fuzzy logic. The membership rates of each medium criterion obtained for the given objects are defined by experts and membership functions are established. For each of the observed criteria, fuzzy polynomials are discretized as follows:

$$\mu_{c_j^{-1}} = c_1^1/\mu_1^1 + c_2^1/\mu_2^1 + \ldots + c_m^1/\mu_m^1$$
$$\mu_{c_j^{-2}} = c_1^2/\mu_1^2 + c_2^2/\mu_2^2 + \ldots + c_m^2/\mu_m^2$$

$$\ldots\ldots\ldots\ldots\ldots\ldots\ldots\ldots\ldots\ldots\ldots$$ 
$$\ldots\ldots\ldots\ldots\ldots\ldots\ldots\ldots\ldots\ldots\ldots$$

$$\mu_{c_j^{-K}} = c_1^K/\mu_1^K + c_2^K/\mu_2^K + \ldots + c_m^K/\mu_m^K$$

$$(3)$$

Utilizing the information obtained, we get the following decision-making matrix:

$$D = \begin{pmatrix} \mu_1^1 & \cdots & \mu_1^K \\ \vdots & \ddots & \vdots \\ \mu_m^1 & \cdots & \mu_m^K \end{pmatrix} \tag{4}$$

Via the support of minmax composition method, the object (alternative) A* considered as the best solution is found as below:

$$A^* = min\left[max_j\left\{\mu_j^k\right\}\right] \tag{5}$$

On the other hand, since combination of fuzzy sets brought to the maximization process, (5) can be written as follows:

$$B^k = \mu_1^k \cup \mu_2^k \cup \ldots \cup \mu_m^k \, ; \, A^* = min_k B_k \tag{6}$$

Thus, according to (5) or (6), we can make a decision on applying concession to the lowest indices object.

## 2.2  The Weighted Sum Model (WSM)

The general expression of this method is as follows:

$$A_j = \sum_{k=1}^{K} \bar{c}_j^{-k} w_j^k \tag{7}$$

Where, $\bar{c}_j^{-k}$, k = 1...K; j = 1...m - are the average evaluation of criteria found with (2) expression; $A_{j,}$ j = 1...m - are objects (alternatives); $w_j^k$, k = 1...K; j = 1...m - are the weights of generalized criteria that have found as a result of (2) and where $\sum_{k=1}^{K} w_j^k = 1$, $w_j^k \geq 0$; k = 1...K; j = 1...m . Each alternative found via (7) is compared to one another and selects the most appropriate alternative.

**Note:** Determination of weights is one of the most important issues of the MCDM methods. The weight of the criteria reflects the individual choice of the decision maker. If the significance of the criteria is the same, weights are equal. If different, they are determined for their importance by paying the given terms [16].

## 3  Application of Minmax Composition and WSM of MCDM in Problem Solving Process

### 3.1  Problem Description

In this problem, taxpayers are objects and each of the indicators reflecting their financial position consists of subsets. These indicators are evaluated based on the forms compiled by each entity and are listed in the following table (Tables 1 and 2):

**Table 1.**  Indicators of objects financial condition assessment

| Criteria (and sub criteria of each of them) | Object | | | | | |
|---|---|---|---|---|---|---|
| | $A_1$ | $A_2$ | $A_3$ | $A_4$ | $A_5$ | $A_6$ |
| 1. Property Estimation of Object | | | | | | |
| Active part of main resources | 0 | 0.63 | 0.46 | 1 | 0.6688 | 0.971 |
| Etching coefficient of main resources | 0 | 0 | 0.17 | 0.12 | 0.2091 | 0 |
| 2. Estimation of liquidity | | | | | | |
| Maneuvered skill of objects resource | 0.008 | 0.096 | 0.46 | 0.000 | $129*10^{-5}$ | 1.183 |
| General covering coefficient | 0.973 | 3.187 | 0.93 | 0.091 | 0.627 | 4.483 |
| Speedy liquidity coefficient | 0.631 | 2.295 | 0.41 | 0.048 | 0.0015 | 4.461 |
| Absolute liquidity coefficient | 0.008 | 0.210 | 0.03 | 0.000 | 4.851 | 4.123 |
| Circulating resources of objects actives | 0.968 | 0.389 | 0.72 | 0.082 | 0.3558 | 0.452 |
| 3. Financial soundness assessment | | | | | | |
| Concentration coefficient of objects capital | 0.005 | 0.877 | 0.23 | 0.038 | 0.4325 | 0.898 |
| Financial depending coefficient | 171.9 | 1.139 | 4.43 | 26.28 | 2.3121 | 1.112 |
| Maneuvered coefficient of objects capital | 4.462 | 0.304 | 0.25 | 24.60 | 0.4895 | 0.391 |
| Concentration coefficient of involved capital | 0.994 | 0.122 | 0.77 | 0.962 | 0.5675 | 0.101 |
| 4. Estimation of business activity | | | | | | |
| Fund capacity | 1 | 0.665 | 1.81 | 1.199 | 0.9808 | 1 |
| Coefficient of the funds in settlements | 0.441 | 7.145 | 3.42 | 0.872 | 821.2 | 232.1 |
| Circulating of industrial resources | 0.776 | 6.057 | 1.06 | 0.009 | 1.5337 | 3189.0 |
| Maneuvered skill of objects capital | 47.11 | 0.868 | 2.07 | 1.006 | 1.6276 | 8.787 |
| Circulating skill of main capital | 0.273 | 0.762 | 0.47 | 0.038 | 0.7034 | 7.899 |
| Sustainability coefficient of economic growth | 1 | 1 | 0.01 | 1 | 1 | 1 |
| 5. Estimation of profitability | | | | | | |
| Profitability of product | 1.173 | 0.033 | 0.02 | 81.35 | 0.0605 | 0.045 |
| Profitability of main activity | 1.13 | 0.038 | 0.02 | 81.35 | 0.0782 | 0.045 |
| Profitability of main capital | 1 | 0.0014 | 1 | 1 | 0.0186 | 0.236 |
| Profitability of capital of object | 1 | 0.0016 | 1 | 1 | 0.0431 | 0.2626 |

Where, Aj, j = 1...6 – are objects (alternatives); $C^k$, k = 1...5 - are main criteria and each of them consists of subsets. Using the formula (2), let's estimate the main criteria consisting of subcriteria:

**Table 2.** Average evaluation of main criteria

| Criteria | Objects | | | | | |
|---|---|---|---|---|---|---|
| | $A_1$ | $A_2$ | $A_3$ | $A_4$ | $A_5$ | $A_6$ |
| $c_J^{-1}$ | 0 | 0 | 0.28 | 0.35 | 0.37 | 0 |
| $c_J^{-2}$ | 0.13 | 0.56 | 0.33 | 0 | 0.07 | 2.13 |
| $c_J^{-3}$ | 1.39 | 0.44 | 0.67 | 2.20 | 0.73 | 0.45 |
| $c_J^{-4}$ | 1.29 | 1.63 | 0.92 | 0.27 | 3.35 | 19.2 |
| $c_J^{-5}$ | 1.07 | 0.01 | 0.45 | 9.01 | 0.044 | 0.105 |

## 3.2   Evaluation of Alternatives via Minmax Composition Method

Determined membership rates for each average criterion (for each object) based on expert knowledge and discrete form of fuzzy sets define by (3)

$$D = \begin{pmatrix} 0 & 0,16 & 0,62 & 0,5 & 0,5 \\ 0 & 0,41 & 0,36 & 0,52 & 0,01 \\ 0,2 & 0,23 & 0,47 & 0,49 & 0,3 \\ 0,26 & 0 & 0,9 & 0,12 & 1 \\ 0,3 & 0,1 & 0,5 & 1 & 0,02 \\ 0 & 0,9 & 0,39 & 1 & 0,14 \end{pmatrix}$$

Choice the most appropriate alternative via minmax composition method by using the decision-making matrix:

$$A^* = min\left[max_j\left\{\mu_j^k\right\}\right]; A^* = min[0.62, 0.52, 0.49, 1, 1, 1]$$

$A_3$, $A_2$, $A_1$, $A_4 = A_5 = A_6$ according to this order, we can give an opportunity to object $A_3$ for extending term of tax liability (Fig. 1).

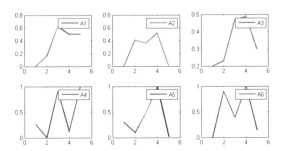

**Fig. 1.**  Membership function of alternatives

### 3.3  Evaluation of Alternatives via WSM

Importance rates of criteria are different, so weights: $w_1 = 0.1$; $w_2 = 0.2$; $w_3 = 0.3$ $w_4 = 0.2$; $w_5 = 0.2$

$$\sum_{k=1}^{K} w_j^k = 1, \ w_j^k \geq 0; A_j = \sum_{k=1}^{K} \bar{c}_j^{\,k} \, w_j^k; \text{where}, k = 1\ldots5, \ j = 1\ldots6;$$
$$(A_1, \ A_2, \ A_3, \ A_4, \ A_5, \ A_6) = (0.915, 0.572, 0.569, 1.969, 0.9488, 4.422)$$

$A_3, A_2, A_1, A_5, A_4, A_6$ -in this order, we make a decision about applying the concession (Table 3).

**Table 3.** The consequences of alternatives by two method

| | |
|---|---|
| The result obtained by the Minmax method | $A_3, A_2, A_1, A_4 = A_5 = A_6$ |
| The result obtained by the WSM method | $A_3, A_2, A_1, A_5, A_4, A_6$ |

## 4  Conclusion

As a result, in this paper, it is proposed new formulation for summarizing criteria that consists of finite numbers of subsets, also to use minmax composition method for selecting the most appropriate alternative. Furthermore, the problem has been solved with the other known method of MCDM (WSM) and alternatives have been evaluated.

The sequences are the same in the first three lines in both methods. However, comparing the 4.5th and 6th enterprises, we find that the results obtained on the basis of the minmax composition method are more accurate. And implementing this method, we get the most appropriate order of taxpayers (enterprises) for applying them concession in the process of changing term of the tax liability fulfillment that expressed as a MCDM problem.

## References

1. Zadeh, L.A.: Fuzzy sets. Inf. Control **8**(3), 338–353 (1965)
2. Triantaphyllou, E.: Multi-criteria Decision Making Methods. A Comparative Study (2000)
3. Fishburn, P.C.: Additive Utilities with Incomplete Product Set: Applications to Priorities and Assignments. Operations Research Society of America (ORSA), Baltimore (1967)
4. Saaty, T.L., Peniwati, K.: Group Decision Making: Drawing out and Reconciling Differences. RWS Publications, Pittsburgh (2008)
5. Bohanec, M., Bratko, I., Rajkovič, V.: An expert system for decision-making. In: Sol, H.G. (ed.) Processes and Tools for Decision Making, pp 235–248. North-Holland, Amsterdam (1983)
6. Bohanec, M.: DEXi: A Program for Multi-Attribute Decision Making (2012)
7. Rezaei, J.: Best-worst multi-criteria decision-making method. Omega **53**, 49–57 (2015)

8. Rezaei, J.: Best-worst multi-criteria decision-making method: some properties and a linear model. Omega **64**, 126–130 (2016)
9. Jacek, Z.: Comparative analysis of multiple criteria evaluations of suppliers in different industries. In: 18th Euro working Group on Transportation, EWGT 2015, Delft, The Netherlands (2015)
10. Jaini, N.I., Utyuzhnikov, S.V.: A fuzzy trade-off ranking method for multi-criteria decision-making. Axioms **7**(1), 1 (2017)
11. Musayev, A.F.: Economic problems of tax policy. CBS "Poliqraphic Production", Baku (2004)
12. Musayev, A.F., Madatova, Sh.G., Rustamov, S.S.: Evolution of the impact of the tax legislation reforms on the tax potential by fuzzy inference method. In: 12th International Conference on Application of Fuzzy Systems and Soft Computing (ICAFS), Vienna, pp. 507–514. Elsevier (2016)
13. Musayev, A.F., Madatova, Sh.G., Rustamov, S.S.: Mamdani-type fuzzy inference system for evaluation of tax potential. In: Sixth World Conference on Computing dedicated to 50th Anniversary of Fuzzy Logic and Applications and 95th Birthday Anniversary of Lotfi A. Zadeh, Berkley, USA (2016)
14. The Tax Code of Azerbaijan Republic (2017). http://www.taxes.gov.az
15. Musayev, A.F., Kalbiyev, Y.A., Huseynov, A.A.: The tax system of Azerbaijan Republic: reforms and results, Baku (2002)
16. Roszkowska, E.: Rank ordering criteria weighting methods – a comparative overview. Optimum. Studio Ekonomicze **5**(65), 14–33 (2013)

# Evaluating US Dollar Index Movements Using Markov Chains-Fuzzy States Approach

Berna Uzun[1(✉)] and Ersin Kıral[2]

[1] Near East University, 99138 Nicosia, TRNC Mersin-10, Turkey
berna.uzun@neu.edu.tr
[2] Cukurova University, 01330 Adana, Turkey

**Abstract.** The U.S. dollar (USD) is one of the most used currencies in the world and also the most commonly used currency in international payments. The U.S. dollar index (USDX) is a measure of the value of the USD relative to the value of a basket of currencies of the majority of the U.S.'s most significant trading partners. Therefore, dollar index gives to market players and regulators more valuable information about the dollar value rather than the regional value among the currencies. Since it can be counted as a key indicator for the direction of the USD, the Central Banks are also closely monitoring the USDX. In recent years, large fluctuations in dollar value have caused US price instability to increase. The aim of this study is to classify the USDX with triangular fuzzy sets and evaluate the USDX movements using Markov Chain of the Fuzzy States method. The data used in this study consist of the monthly changes rate of the USDX over the January 2003 to May 2018 period. The movements of the monthly USDX have been analysed with the probabilistic transition matrix of the fuzzy states, then the steady condition of the changes rate of the USDX has been presented. These outcomes give significant information to the decision makers about the USDX movements. With this model, we are able to evaluate USDX movements and estimate the expected USDX for long and short term without missing any movements between boundaries of the states.

**Keywords:** USD index · Fuzzy sets · Markov chains of the fuzzy states

## 1 Introduction

The USDX measures the U.S. dollar's value regarding majority of its most prominent trading partner's currencies including Euro (57.6%), Japanese Yen (13.6%), UK Pound (11.9%), Canadian Dollar (9.1%), Swedish Krona (4.2%) and Swiss Franc (3.6%). The start of USDX was in March 1973, right after the dismantling of the Bretton Woods system. At its beginning, the U.S. Dollar Index value was 100.000. Since it has traded, it has reached the highest level with 164.7200 in February 1985, with its lowest point in March 16, 2008 at 70.698. Apart from showing the strength of the U.S. dollar, it is also used as an exchange-trading unit. Therefore, there have been many studies carried out by the researchers in order to analyze USDX movements and relations of the USD with other financial instruments. Landefeld et al. analyzed the correlation between US-GDP and US dollar index [1]. Manning and Dimitri applied the method of co-integration to

© Springer Nature Switzerland AG 2019
R. A. Aliev et al. (Eds.): ICAFS-2018, AISC 896, pp. 386–391, 2019.
https://doi.org/10.1007/978-3-030-04164-9_51

show the dollar movements and inflation [2]. Plat evaluated the relationship between major currencies with USD [3]. Cretien presented the relation between US dollar index, euro-dollars, gold and silver [4]. Kim analyzed the US inflation and the dollar exchange rate using a vector error correction technique [5]. In this study, we propose Markov chains of the fuzzy states for the analysis of the direction of the USDX movements.

Zadeh has defined fuzzy logic in order to model the vague conditions and linguistic information's mathematically [7]. Using fuzzy logic, we can classify objects to the fuzzy sets, which has uncertain boundaries by using membership functions. In recent years, the applications of fuzzy logic in many different areas have increased significantly.

A Markov chain is a stochastic model that describes the sequence of possible events in which the likelihood of each event depends only on the situation obtained in the previous event. In the theory of probability and in related fields the Markov process, which has named with Russian mathematician Andrey Markov, has been applied successfully in many fields for the analysis of the dynamic systems. This model uses the crisp boundaries, which depend on the classical set for defining the situations of the systems.

Fuzzy Markov chains was first defined by Kruse et al., as a perception of the classical Markov chains based on fuzzy probabilities [7]. Then Yoshida constructed a fuzzy Markov process with finite states space and transition matrix [8]. Zadeh showed that we can apply to fuzzy Markov chains using fuzzy Markov algorithms [9]. All of these studies have made significant improvements in making optimal decisions in dynamic systems under uncertain conditions. In the study conducted by Pardo and Fuente in 2010 the Markov chain of the fuzzy states technique was used for calculation of the publicity decisions for queuing system and discussed in detail [10]. Kıral and Uzun used the fuzzy states- Markov chain technique in order to analyse daily Borsa Istanbul Index movements [11] and monthly gold price movements [12]. Kıral also used this technique in monthly Brent oil prices direction's evaluation [13]. In this study, the same method is used to analyse the USDX movements.

## 2 Materials and Methods

The data of this study consist of monthly percentage change rates of the USDX between January 2003 and May 2018, which was obtained from "investing.com" [14]. Monthly percentage change rates of the USDX ($R_t$) are used for the analysis of the USDX movements between the fuzzy states of $R_t$. The conditional probability of the fuzzy event $\tilde{A}_j$ given the fuzzy event $\tilde{A}_i : i, j \in \{1, \ldots, n\}$ can be calculated as a linear combination of probabilities $P(\tilde{A}_j|m) : m \in \{0, \ldots, N\}$ as the following function:

$$P\left(\tilde{A}_j|\tilde{A}_i\right) = P\left\{\tilde{X}1 = \tilde{A}_j|\tilde{X}_0 = \tilde{A}_i\right\} = \sum_{m=0}^{N} P\left(\tilde{A}_j|m\right) \frac{P\left\{\tilde{X}_0 = m\right\} \mu \tilde{A}_i(m)}{P\left(\tilde{A}_i\right)}$$

This shows the one step transition probability matrix of fuzzy states (from the fuzzy initial state to the fuzzy final state) [10].

The average of the USDX percentage change rates data is $-0.02\%$ for the given period while standard deviation is $2.38\%$, which is 119 times higher than the average return.

In this analysis, probabilistic transition matrix of the $R_t$ categorized as 21 fuzzy states: from $S_{-10}$ (high decrease) to $S_{10}$ (high increase). In order to classify $(R_t)$ to its fuzzy states we calculate the monthly membership degree of the $(R_t)$ using triangular fuzzy sets as seen in Fig. 1.

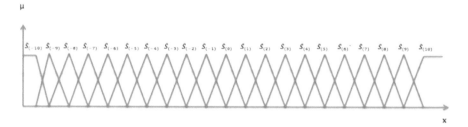

**Fig. 1.** Fuzzy states of $R_t$

First, we denote the $\tilde{S}_i$ and $\tilde{S}_{i+1}$ as the fuzzy state components of the $R_t$ and then we calculate the membership degree of the $R_t$ with the following function:

$$\text{If } -2.25\% < R_t < 2.25\% \text{ then } i = \left[\!\left[\frac{R_t}{0.225}\right]\!\right] \text{ and}$$
$$\tilde{S}_i = \frac{(i+1)0.225 - R_t}{0.225}, \tilde{S}_{i+1} = 1 - \tilde{S}_i.$$

If $R_t \leq -2.25\%$ or $R_t \geq 2.25\%$ then $\tilde{S}_{-10} = 1$ or $\tilde{S}_{10} = 1$ respectively while $[\![x]\!]$ denotes the greatest integer function. Therefore, each $R_t$ has been transformed to its fuzzy state components for the selected period as shown in Table 1.

**Table 1.** Transformed fuzzy states of the monthly $R_t$ for some months

| Date | $R_t$ | $\tilde{S}_{-10}$ | $\tilde{S}_{-9}$ | $\tilde{S}_{-8}$ | $\tilde{S}_{-7}$ | $\tilde{S}_{-6}$ | $\tilde{S}_{-5}$ | $\tilde{S}_{-4}$ | $\tilde{S}_{-3}$ | $\tilde{S}_{-2}$ | $\tilde{S}_{-1}$ | $\tilde{S}_0$ | $\tilde{S}_1$ | $\tilde{S}_2$ | $\tilde{S}_3$ | $\tilde{S}_4$ | $\tilde{S}_5$ | $\tilde{S}_6$ | $\tilde{S}_7$ | $\tilde{S}_8$ | $\tilde{S}_9$ | $\tilde{S}_{10}$ |
|---|---|---|---|---|---|---|---|---|---|---|---|---|---|---|---|---|---|---|---|---|---|---|
| May. 2018 | 2.33% | 0 | 0 | 0 | 0 | 0 | 0 | 0 | 0 | 0 | 0 | 0 | 0 | 0 | 0 | 0 | 0 | 0 | 0 | 0 | 0 | 1 |
| Apr. 2018 | 2.08% | 0 | 0 | 0 | 0 | 0 | 0 | 0 | 0 | 0 | 0 | 0 | 0 | 0 | 0 | 0 | 0 | 0 | 0 | 0 | .76 | .24 |
| Marc. 2018 | -0.71% | 0 | 0 | 0 | 0 | 0 | 0 | .16 | .84 | 0 | 0 | 0 | 0 | 0 | 0 | 0 | 0 | 0 | 0 | 0 | 0 | 0 |
| Feb. 2018 | 1.66% | 0 | 0 | 0 | 0 | 0 | 0 | 0 | 0 | 0 | 0 | 0 | 0 | 0 | 0 | 0 | 0 | 0 | .62 | .38 | 0 | 0 |

After we obtained the fuzzy states of the $R_t$'s we calculated the transition probabilities of the USDX via conditional probability of the $\tilde{S}_j$, given the fuzzy state $\tilde{S}_i$.

## 3 Results and Discussion

Transition probability matrix (P) shows that the transition probability of the USDX passing to the positive states from $\tilde{S}_9$ (where $1.80\% \leq R_t \leq 2.25\%$) is 62% and 61% from $\tilde{S}_{-3}$ (where $-0.90\% \leq R_t \leq -0.45\%$) while it is only %9 from $\tilde{S}_6$ (where $1.125\% \leq R_t \leq 1.575\%$) and 19% from $\tilde{S}_{-5}$ (where $-1.35\% \leq R_t \leq -0.90\%$) (Table 2).

**Table 2.** Transition probability matrix (P) of $R_t$.

| | $\tilde{S}_{-10}$ | $\tilde{S}_{-9}$ | $\tilde{S}_{-8}$ | $\tilde{S}_{-7}$ | $\tilde{S}_{-6}$ | $\tilde{S}_{-5}$ | $\tilde{S}_{-4}$ | $\tilde{S}_{-3}$ | $\tilde{S}_{-2}$ | $\tilde{S}_{-1}$ | $\tilde{S}_0$ | $\tilde{S}_1$ | $\tilde{S}_2$ | $\tilde{S}_3$ | $\tilde{S}_4$ | $\tilde{S}_5$ | $\tilde{S}_6$ | $\tilde{S}_7$ | $\tilde{S}_8$ | $\tilde{S}_9$ | $\tilde{S}_{10}$ |
|---|---|---|---|---|---|---|---|---|---|---|---|---|---|---|---|---|---|---|---|---|---|
| $\tilde{S}_{-10}$ | .08 | .01 | .06 | .13 | .03 | .03 | .02 | .02 | 0 | .10 | .05 | .02 | .02 | 0 | .05 | .07 | .04 | .09 | .02 | .04 | .15 |
| $\tilde{S}_{-9}$ | .16 | .08 | .03 | 0 | .06 | .01 | 0 | 0 | .08 | .06 | .01 | 0 | .09 | .10 | .07 | 0 | 0 | 0 | 0 | .02 | .24 |
| $\tilde{S}_{-8}$ | .21 | .02 | .01 | .07 | .01 | .03 | .01 | 0 | .05 | .10 | .02 | 0 | .02 | .02 | .04 | .10 | .03 | 0 | .08 | .03 | .15 |
| $\tilde{S}_{-7}$ | .04 | 0 | 0 | .03 | 0 | .10 | .07 | 0 | .08 | .06 | .12 | 0 | .09 | .10 | .06 | .06 | .02 | 0 | .03 | .04 | .09 |
| $\tilde{S}_{-6}$ | .47 | .05 | 0 | 0 | 0 | 0 | .03 | .04 | 0 | 0 | .12 | .01 | .01 | 0 | .01 | .01 | 0 | 0 | .10 | .09 | .04 |
| $\tilde{S}_{-5}$ | .14 | .02 | .18 | .02 | .15 | .01 | .06 | .24 | 0 | 0 | 0 | 0 | 0 | 0 | 0 | 0 | 0 | 0 | .03 | .02 | .14 |
| $\tilde{S}_{-4}$ | .36 | 0 | .09 | .01 | .11 | .02 | 0 | .01 | 0 | 0 | 0 | .01 | .01 | 0 | 0 | .12 | 0 | .12 | .02 | .06 | .06 |
| $\tilde{S}_{-3}$ | .18 | .04 | .10 | .01 | .05 | .01 | 0 | .01 | 0 | 0 | 0 | .07 | .11 | .02 | .02 | .08 | 0 | .03 | 0 | .10 | .18 |
| $\tilde{S}_{-2}$ | .21 | .10 | .03 | .02 | .08 | 0 | .01 | .13 | .06 | 0 | 0 | .05 | .04 | .09 | .04 | .06 | 0 | 0 | 0 | 0 | .08 |
| $\tilde{S}_{-1}$ | .11 | .18 | .05 | 0 | 0 | 0 | 0 | .12 | .01 | 0 | 0 | .02 | .14 | .03 | .02 | .10 | .09 | 0 | 0 | 0 | .11 |
| $\tilde{S}_0$ | .27 | .08 | 0 | 0 | 0 | 0 | 0 | .02 | .04 | .09 | 0 | .17 | .17 | .12 | 0 | .01 | .04 | 0 | 0 | 0 | 0 |
| $\tilde{S}_1$ | .02 | .11 | .03 | 0 | 0 | 0 | 0 | .13 | .18 | .12 | .04 | .11 | .05 | 0 | 0 | 0 | 0 | .01 | 0 | 0 | .21 |
| $\tilde{S}_2$ | 0 | .02 | 0 | 0 | 0 | 0 | .06 | .17 | .12 | .04 | .05 | .04 | .02 | .05 | 0 | 0 | 0 | .13 | .02 | 0 | .30 |
| $\tilde{S}_3$ | .20 | .09 | .03 | 0 | 0 | 0 | .08 | .11 | .03 | 0 | .05 | .04 | .02 | .05 | .01 | .01 | 0 | .06 | .04 | 0 | .18 |
| $\tilde{S}_4$ | .17 | .02 | .01 | .03 | .10 | .03 | .01 | .02 | .06 | 0 | 0 | 0 | .08 | .03 | .13 | .15 | 0 | 0 | 0 | 0 | .15 |
| $\tilde{S}_5$ | .19 | 0 | .03 | .03 | .06 | .06 | .05 | .13 | .08 | 0 | 0 | 0 | .03 | .02 | .10 | .02 | 0 | .06 | .04 | 0 | .11 |
| $\tilde{S}_6$ | .23 | 0 | .14 | .05 | .01 | .25 | 0 | .01 | .20 | .03 | 0 | .04 | .01 | 0 | .01 | 0 | 0 | .01 | .01 | 0 | .01 |
| $\tilde{S}_7$ | 0 | .09 | .12 | .13 | 0 | 0 | .05 | .13 | .07 | 0 | 0 | .08 | .02 | 0 | 0 | 0 | 0 | 0 | 0 | .03 | .28 |
| $\tilde{S}_8$ | 0 | .03 | .01 | .10 | .01 | 0 | .04 | .13 | .12 | .15 | .09 | .01 | .19 | 0 | 0 | 0 | 0 | 0 | 0 | .05 | .08 |
| $\tilde{S}_9$ | .19 | 0 | 0 | .13 | .01 | 0 | 0 | 0 | 0 | .03 | .02 | 0 | .08 | .02 | 0 | .17 | 0 | .12 | 0 | 0 | .23 |
| $\tilde{S}_{10}$ | .14 | 0 | .04 | .08 | .05 | .04 | 0 | .03 | .04 | 0 | 0 | .03 | .03 | .07 | .02 | .07 | .03 | .07 | 0 | .03 | .21 |

USDX transition probability matrix has been reached to its stable condition in 5 sessions as shown in Table 3.

**Table 3.** Steady states of the $R_t$ ($P^5$)

| | $\tilde{S}_{-10}$ | $\tilde{S}_{-9}$ | $\tilde{S}_{-8}$ | $\tilde{S}_{-7}$ | $\tilde{S}_{-6}$ | $\tilde{S}_{-5}$ | $\tilde{S}_{-4}$ | $\tilde{S}_{-3}$ | $\tilde{S}_{-2}$ | $\tilde{S}_{-1}$ | $\tilde{S}_0$ | $\tilde{S}_1$ | $\tilde{S}_2$ | $\tilde{S}_3$ | $\tilde{S}_4$ | $\tilde{S}_5$ | $\tilde{S}_6$ | $\tilde{S}_7$ | $\tilde{S}_8$ | $\tilde{S}_9$ | $\tilde{S}_{10}$ |
|---|---|---|---|---|---|---|---|---|---|---|---|---|---|---|---|---|---|---|---|---|---|
| $\tilde{S}_{-10}$ | .14 | .04 | .04 | .05 | .03 | .03 | .02 | .06 | .05 | .04 | .03 | .03 | .05 | .04 | .03 | .05 | .02 | .04 | .02 | .03 | .16 |
| $\tilde{S}_{-9}$ | .14 | .04 | .04 | .05 | .03 | .03 | .02 | .06 | .05 | .04 | .03 | .03 | .05 | .04 | .03 | .05 | .02 | .04 | .02 | .03 | .16 |
| $\tilde{S}_{-8}$ | .14 | .04 | .04 | .05 | .03 | .03 | .02 | .06 | .05 | .04 | .03 | .03 | .05 | .04 | .03 | .05 | .02 | .04 | .02 | .03 | .16 |
| $\tilde{S}_{-7}$ | .14 | .04 | .04 | .05 | .03 | .03 | .02 | .06 | .05 | .04 | .03 | .03 | .05 | .04 | .03 | .05 | .02 | .04 | .02 | .03 | .16 |
| $\tilde{S}_{-6}$ | .14 | .04 | .04 | .05 | .03 | .03 | .02 | .06 | .05 | .04 | .03 | .03 | .05 | .04 | .03 | .05 | .02 | .04 | .02 | .03 | .16 |
| $\tilde{S}_{-5}$ | .14 | .04 | .04 | .05 | .03 | .03 | .02 | .06 | .05 | .04 | .03 | .03 | .05 | .04 | .03 | .05 | .02 | .04 | .02 | .03 | .16 |
| $\tilde{S}_{-4}$ | .14 | .04 | .04 | .05 | .03 | .03 | .02 | .06 | .05 | .04 | .03 | .03 | .05 | .04 | .03 | .05 | .02 | .04 | .02 | .03 | .16 |
| $\tilde{S}_{-3}$ | .14 | .04 | .04 | .05 | .03 | .03 | .02 | .06 | .05 | .04 | .03 | .03 | .05 | .04 | .03 | .05 | .02 | .04 | .02 | .03 | .16 |
| $\tilde{S}_{-2}$ | .14 | .04 | .04 | .05 | .03 | .03 | .02 | .06 | .05 | .04 | .03 | .03 | .05 | .04 | .03 | .05 | .02 | .04 | .02 | .03 | .16 |
| $\tilde{S}_{-1}$ | .14 | .04 | .04 | .05 | .03 | .03 | .02 | .06 | .05 | .04 | .03 | .03 | .05 | .04 | .03 | .05 | .02 | .04 | .02 | .03 | .16 |
| $\tilde{S}_0$ | .14 | .04 | .04 | .05 | .03 | .03 | .02 | .06 | .05 | .04 | .03 | .03 | .05 | .04 | .03 | .05 | .02 | .04 | .02 | .03 | .16 |
| $\tilde{S}_1$ | .14 | .04 | .04 | .05 | .03 | .03 | .02 | .06 | .05 | .04 | .03 | .03 | .05 | .04 | .03 | .05 | .02 | .04 | .02 | .03 | .16 |
| $\tilde{S}_2$ | .14 | .04 | .04 | .05 | .03 | .03 | .02 | .06 | .05 | .04 | .03 | .03 | .05 | .04 | .03 | .05 | .02 | .04 | .02 | .03 | .16 |
| $\tilde{S}_3$ | .14 | .04 | .04 | .05 | .03 | .03 | .02 | .06 | .05 | .04 | .03 | .03 | .05 | .04 | .03 | .05 | .02 | .04 | .02 | .03 | .16 |
| $\tilde{S}_4$ | .14 | .04 | .04 | .05 | .03 | .03 | .02 | .06 | .05 | .04 | .03 | .03 | .05 | .04 | .03 | .05 | .02 | .04 | .02 | .03 | .16 |
| $\tilde{S}_5$ | .14 | .04 | .04 | .05 | .03 | .03 | .02 | .06 | .05 | .04 | .03 | .03 | .05 | .04 | .03 | .05 | .02 | .04 | .02 | .03 | .16 |
| $\tilde{S}_6$ | .14 | .04 | .04 | .05 | .03 | .03 | .02 | .06 | .05 | .04 | .03 | .03 | .05 | .04 | .03 | .05 | .02 | .04 | .02 | .03 | .16 |
| $\tilde{S}_7$ | .14 | .04 | .04 | .05 | .03 | .03 | .02 | .06 | .05 | .04 | .03 | .03 | .05 | .04 | .03 | .05 | .02 | .04 | .02 | .03 | .16 |
| $\tilde{S}_8$ | .14 | .04 | .04 | .05 | .03 | .03 | .02 | .06 | .05 | .04 | .03 | .03 | .05 | .04 | .03 | .05 | .02 | .04 | .02 | .03 | .16 |
| $\tilde{S}_9$ | .14 | .04 | .04 | .05 | .03 | .03 | .02 | .06 | .05 | .04 | .03 | .03 | .05 | .04 | .03 | .05 | .02 | .04 | .02 | .03 | .16 |
| $\tilde{S}_{10}$ | .14 | .04 | .04 | .05 | .03 | .03 | .02 | .06 | .05 | .04 | .03 | .03 | .05 | .04 | .03 | .05 | .02 | .04 | .02 | .03 | .16 |

Table 3 can be represented with $\pi$ as shown in Table 4.

**Table 4.** Steady condition of $R_t$ as $\pi$ .

| | $\tilde{S}_{-10}$ | $\tilde{S}_{-9}$ | $\tilde{S}_{-8}$ | $\tilde{S}_{-7}$ | $\tilde{S}_{-6}$ | $\tilde{S}_{-5}$ | $\tilde{S}_{-4}$ | $\tilde{S}_{-3}$ | $\tilde{S}_{-2}$ | $\tilde{S}_{-1}$ | $\tilde{S}_0$ | $\tilde{S}_1$ | $\tilde{S}_2$ | $\tilde{S}_3$ | $\tilde{S}_4$ | $\tilde{S}_5$ | $\tilde{S}_6$ | $\tilde{S}_7$ | $\tilde{S}_8$ | $\tilde{S}_9$ | $\tilde{S}_{10}$ |
|---|---|---|---|---|---|---|---|---|---|---|---|---|---|---|---|---|---|---|---|---|---|
| $\tilde{S}_i$ | .14 | .04 | .04 | .05 | .03 | .03 | .02 | .06 | .05 | .04 | .03 | .03 | .05 | .04 | .03 | .05 | .02 | .04 | .02 | .03 | .16 |

This shows us the probability distribution of the USDX movements in the long run. According to the steady condition of the USDX transition matrix, it can be seen that the number of decreases at USDX is expected to be 3% higher than the number of increases in 5 months. And no matter in which state USDX for any starting time; in five months it has the highest transition probabilities of passing to $\tilde{S}_{-10}$ with 14% (where $R_t \leq -2.25\%$) and $\tilde{S}_{10}$ (where $R_t \geq 2.25\%$) with 16%.

## 4 Conclusion

The result of this study gives valuable signals in some states, for the short-term and long-term return rate of the USDX. With the monthly transition probability matrix of the USDX we can predict the monthly movement directions of the USDX. Related to the steady condition of the USDX, we can say the movements of USDX either increases with a high rate or decreases with a high rate. In this case, it can be said that the USDX investment is risky but can provide high profit opportunity to the investor also.

Using Markov chains of the fuzzy states in this study, we were able to achieve good decision results by fuzzifying the states of the USDX instead of using state with crisp boundaries. Analysing the transition matrix result for each state gives sensitive and valuable information for the USDX direction. This study can be expanded for the evaluation of other investment instruments. Using different fuzzy sets can also give more information about the system's future. The results of this study provide the decision maker with the ability to choose the best investment option. This study shows that the proposed method of Markov chains of the fuzzy states is an effective method to employ when analyzing the dynamic systems.

## References

1. Landefeld, J.S., Moulton, B.R., Vojtech, C.M.: Chained-dollar indexes. J. Surv. Curr. Bus. **11**, 8–16 (2003)
2. Manning, L., Andrianacos, D.: Dollar movements and inflation: a vector error correction model. Appl. Econ. **25**(12), 1483–1488 (2004)
3. Platt, G.: Dollar stays strong, but yen tumbles. J. Glob. Finance **4**, 68–69 (2009)
4. Cretien, P.D.: Currencies, eurodollars, silver and gold: not your average. Futur. Mag. **9**, 40–43 (2009)
5. Kim, K.: US inflation inflation and the dollar exchange rate: a vector error correction model. Appl. Econ. **30**(5), 613–619 (2010)
6. Zadeh, L.A.: Fuzzy sets. Inf. Control **8**, 338–353 (1965)
7. Kruse, R., Buck-Emden, R., Cordes, R.: Processor power considerations - an application of fuzzy Markov chains. Fuzzy Sets Syst. **21**, 289–299 (1987)
8. Yoshida, Y.: Markov chains with a transition possibility measure and fuzzy dynamic programming. Fuzzy Sets Syst. **66**, 39–57 (1994)
9. Zadeh, L.A.: Maksimizing sets and fuzzy Markoff algorithms. IEEE Trans. Syst. Man Cybern. – Part C: Appl. Rev. **28**, 9–15 (1998)
10. Pardo, M.J., Fuente, D.: Fuzzy Markovian decision processes: application to queueing systems. Comput. Math Appl. **60**, 2526–2535 (2010)
11. Kıral, E., Uzun, B.: Forecasting closing returns of Borsa istanbul index with Markov chain process of the fuzzy states. J. Econ. Finance Account. **4**(1), 15–23 (2017)
12. Uzun, B., Kıral, E.: Application of Markov chains-fuzzy states to gold price. Procedia Comput. Sci. **120**, 365–371 (2017)
13. Kıral, E.: Modeling brent oil price with Markov chain process of the fuzzy states. J. Econ. Finance Account. **5**(1), 79–83 (2018)
14. Investing.com. https://www.investing.com. Accessed 07 Jan 2018

# The Statistical Reasoning of the Occupational Attitudes of Students. Case Study: Gastronomy and Culinary Arts Program

Serdar Oktay[✉] and Saide Sadıkoğlu

School of Tourism and Hotel Management, Near East University, 98010 Nicosia,
North Cyprus, Turkey
serdar.oktay@neu.edu.tr

**Abstract.** In this study, it was aimed to determine the occupational attitudes of
students studying in the Gastronomy and Culinary Arts Program at the undergraduate level at the university. For this purpose, a questionnaire consisting of
positive and negative questions was applied to students studying in the
Department of Gastronomy and Culinary Arts at Near East Vocational School of
Tourism and Hotel Management. The t-test analysis was performed at the end of
the questionnaire. As a result of this analysis, it was determined that students
consider specializing in an area of the culinary profession, the culinary profession will satisfy them financially and morally, they believe that they will be
more connected to the profession as their experiences in the culinary profession
increase and that they have voluntarily selected the culinary program they are
currently studying in. Furthermore, it was determined that there was a significant
difference between the gender profile and attitudes towards the culinary profession of the students who participated in the study.

**Keywords:** Cookery · Occupational attitude
Undergraduate students studying in the department of gastronomy

## 1 Introduction

Cookery means making food ready to eat by various methods. The most basic method
of cookery is cooking. There are reasons to cook foods. Some foods cannot be eaten
raw. Some parasites or bacteria that can be found in foods die or lose their effects for a
while under the effect of cooking. Cooked foods facilitate digestion [1]. For these
reasons, it is a necessity to study cookery. In the 18th century, the Scottish writer James
Boswell drew attention to the fact that only humans among living beings are able to
cook and lay an appetizing table [2]. Nowadays, like the Turkish cuisine, other cuisines
in the world have become known, and different styles of cooking have begun to blend.

It is important for kitchen staff to put their knowledge about nutrition into practice
[3]. In this context, qualified kitchen staff should know best what, why, and how to do
[4]. Besides having a good culinary art skill, it is very important to catch up with the
developing technology [5]. Schools that provide culinary education have also enabled
cooking to be considered as a profession in the eyes of the public [6]. The development
of positive attitudes towards the profession allows students to be more effective in

© Springer Nature Switzerland AG 2019
R. A. Aliev et al. (Eds.): ICAFS-2018, AISC 896, pp. 392–398, 2019.
https://doi.org/10.1007/978-3-030-04164-9_52

practicing their profession [7]. Cookery is a profession that requires a high-level skill and planned practice. The need for qualified personnel in the profession has become a very important issue in the development of occupational attitudes of students who receive culinary education. Furthermore, the opportunities of institutions providing culinary education have become highly important in cook candidates' preparing themselves for their profession [8]. The culinary profession was first recognized as a branch of art in England [9]. The first restaurant in the modern sense was opened in Paris in 1765. Most of the palace cooks who were unemployed after the French Revolution started to open restaurants in Paris and other big cities of the country. Like many other professions until those periods, cooks learned the profession by the master-apprentice method [10]. Cookery is also considered as a branch of art, which includes the stages of purchasing, storing, and presenting materials as a final product. Culinary education constitutes the research subject of gastronomy with many educational titles such as the selection of quality materials, hygiene, material management, production management, and human resources management [11]. Schools that provide culinary education not only trained quality kitchen staff but also allowed cooking to be considered as a profession in the eyes of the public [6]. However, the culinary profession is considered as an area of the profession in which there is a lack of qualified personnel [12].

## 2 Method

A questionnaire was applied to students studying in the Department of Gastronomy and Culinary Arts at the Near East University Vocational School of Tourism and Hotel Management. The questionnaire consists of three sections. In the first section, demographic questions were answered. In the second section, students' participation in 27 positive statements was analyzed to determine their attitudes towards the culinary profession. In the third section of the questionnaire, students' participation in 8 negative statements was analyzed. As a result of these analyses, information about students' choices was obtained [13]. As it is, the questionnaire consists of 35 statements. In this questionnaire, the attitudes of culinary students towards the profession were attempted to be determined by adapting the Attitude Scale Towards the Teaching Profession, which was developed by Üstüner [14] and determined to be valid and reliable. In the scale used in the questionnaire, the KMO value calculated by Üstüner [14] was determined as 0.91 and the internal consistency coefficient as 0.93. In the study, the reliability test was applied to the scale used, and the Cronbach's alpha was found to be 0.854. The fact that these values are greater than 0.80 means that the reliability of this scale prepared is high [15]. The distribution, frequency, and (t) test analysis of the data obtained from the questionnaire was performed.

## 3 Findings

Of the students participating in the study, 22% are females, and 78% are males, and 60% constitute the age range of 20–21 years. The ratio of the participants in the age range of 18–19 years is 40%. Upon examining the grade point averages, 64% of the sample has 1.76–2.50 GDP, and 16% have 2.51–3.25 GDP. 84% of the students participating in the study are university students (Table 1).

**Table 1.** Demographic characteristics of the students participating in the study

| Gender | n | % |
|---|---|---|
| Male | 39 | 78% |
| Female | 11 | 22% |
| **Age** | **n** | **%** |
| 15–19 | 20 | 40% |
| 20–24 | 30 | 60% |
| **Grade point average** | **n** | **%** |
| Between 1.75–2.50 | 32 | 64% |
| Between 2.51–3.25 | 8 | 16% |
| **Grade** | **n** | **%** |
| 1st Grade | 25 | 50% |
| 2nd Grade | 25 | 50% |
| **Educational status** | **n** | **%** |
| Associate Program | 8 | 16% |
| Undergraduate Program | 42 | 84% |

Upon examining the participation of the students in the positive statements aimed at determining their attitudes towards the culinary profession, the following conclusions are made; the ratio of those who consider specializing in an area of the culinary profession (cold, hot, pastry, etc.) the most is 54%. The ratio of those who believe that the culinary profession will satisfy them financially and morally is 42%. The ratio of those who believe that they will become more connected to their profession as their experiences in the culinary profession increase is 58%. The ratio of those who have expressed they have voluntarily chosen the culinary program they are currently studying in is 44%. At the same time, the ratio of students who think that the idea of being able to make people taste something they do not know makes them happy is 48%. The ratio of those who think they can overcome the difficulties they will encounter in the culinary profession is 48% and those who think they will be knowledgeable and competent cooks in the future is 42%.

The ratio of those who lean towards the statement that cookery gives opportunities to produce and create something is 46%. The ratio of students who have given a positive answer to the statement that there are many things that I can do as a cook is 38%. The ratio of students who have given a positive answer to the statement that I can professionally practice the culinary profession is 44% (Table 2).

The participation status of the students in the negative statements aimed at determining their attitudes towards the culinary profession is as follows; the ratio of those who think that the culinary profession is boring is 2% and those who think it is not boring is 58%. While the ratio of those who have stated they do not regret choosing this profession is 58%, the ratio of those who have given a negative answer to this

**Table 2.** The participation status of the students in the positive statements for determining their attitudes towards the culinary profession

|   |   | 1% | 2% | 3% | 4% | 5% |
|---|---|---|---|---|---|---|
| 1 | Even the thought of being a cook is attractive to me | 6% | 4% | 18% | 24% | 48% |
| 2 | If I were to choose a profession again, I would still choose cookery | 8% | 12% | 16% | 34% | 30% |
| 3 | I consider specializing in an area of the culinary profession (Cold, Hot, Pastry, etc.) | 6% | 6% | 10% | 24% | 54% |
| 4 | I believe I can achieve international success in cookery | 8% | 16% | 32% | 12% | 32% |
| 5 | I am pleased that I have chosen this department related to the culinary profession | 8% | 8% | 14% | 32% | 38% |
| 6 | I believe I can overcome the difficulties I will encounter in the culinary profession | 8% | 2% | 10% | 32% | 48% |
| 7 | I think there are professional cooks that I can take as an example | 6% | 6% | 8% | 38% | 42% |
| 8 | I have confidence in the requirements of the culinary profession | 8% | 4% | 10% | 38% | 40% |
| 9 | I am of the opinion that I have a special talent for cookery | 8% | 8% | 14% | 32% | 38% |
| 10 | I think cookery will provide me with opportunities for producing and creating | 10% | 6% | 10% | 28% | 46% |
| 11 | I believe I can professionally practice the culinary profession | 10% | 4% | 6% | 36% | 44% |
| 12 | I will be able to make people taste something they don't know makes me happy | 10% | 6% | 8% | 28% | 48% |
| 13 | I sympathize with people working as cooks | 8% | 2% | 26% | 28% | 36% |
| 14 | I think there are plenty of things that I can do when I become a cook | 6% | 4% | 12% | 40% | 38% |
| 15 | The working conditions of cookery attract me | 6% | 12% | 24% | 22% | 34% |
| 16 | I care about being successful in the courses of professional culinary knowledge | 8% | 8% | 4% | 36% | 44% |
| 17 | I like to have a talk with cooks | 8% | 12% | 6% | 22% | 42% |
| 18 | I discuss and talk about the topics of cooking, foods, kitchen, and food culture | 6% | 12% | 18% | 22% | 40% |
| 19 | I believe I will become a knowledgeable and competent cook | 6% | 4% | 6% | 34% | 42% |
| 20 | I believe I will gain prestige in the society with cookery | 4% | 6% | 18% | 36% | 36% |
| 21 | I have voluntarily chosen the culinary program I am currently studying in | 8% | 8% | 12% | 22% | 44% |

*(continued)*

**Table 2.**  (*continued*)

|  |  | 1% | 2% | 3% | 4% | 5% |
|---|---|---|---|---|---|---|
| 22 | It makes me proud to shape the tastes of people with the food I will cook as a cook | 8% | 10% | 14% | 22% | 44% |
| 23 | I believe the society will appreciate me enough when I become a cook | 6% | 2% | 22% | 32% | 38% |
| 24 | The continuity of the culinary profession gives me confidence | 8% | 12% | 12% | 26% | 42% |
| 25 | I can be a cook for a lifetime | 12% | 4% | 24% | 20% | 40% |
| 26 | I believe I will become more connected to the profession as my experiences increase | 10% | 6% | 8% | 18% | 58% |
| 27 | I believe the culinary profession will satisfy me financially and morally | 8% | 10% | 12% | 28% | 42% |

1 = Strongly Disagree, 2 = Disagree, 3 = Neutral, 4 = Agree, 5 = Strongly Agree

statement is 6%. While the ratio of those who have stated that cookery is suitable for their personality and lifestyle is 58.09%, no students have given a negative answer to this statement. The ratio of those who have stated that the culinary profession does not cause distress is 44%. The ratio of those stating that they like to talk about cooking, foods, kitchen, and food culture is 62%. Students have given a positive answer to the negative statement "I am intimidated by the thought of being a cook" by 70%. While 62% of the students have given a positive answer to the statement "I do not like to talk about cooking, foods, kitchen, and food culture", 2% have given a negative answer to this statement (Table 3).

**Table 3.**  The participation status of the students in the negative statements for determining their attitudes towards the culinary profession

|  |  | 1% | 2% | 3% | 4% | 5% |
|---|---|---|---|---|---|---|
| 1 | The culinary profession is boring | 58% | 26% | 12% | 2% | 2% |
| 2 | I regret choosing the culinary profession | 58% | 26% | 6% | 4% | 6% |
| 3 | I think cookery is not suitable for my personality and lifestyle | 58% | 22% | 14% | 6% | |
| 4 | The attitudes and behaviors of kitchen staff make my interest decrease | 48% | 24% | 22% | 4% | 2% |
| 5 | Thinking about being a cook intimidates me | 70% | 18% | 10% | 2% | 2% |
| 6 | I will not recommend cookery to those who are choosing a profession | 58% | 26% | 6% | | |
| 7 | I am worried that the culinary profession will cause distress | 44% | 24% | 12% | 2% | 2% |
| 8 | I do not like to talk about cooking, foods, kitchen, and food culture | 62% | 30% | 4% | 2% | 2% |

1 = Strongly Disagree, 2 = Disagree, 3 = Neutral, 4 = Agree, 5 = Strongly Agree

# 4   Conclusion

In the historical process, the food takes an important place in all societies. For example, it is encountered in almost all areas of the society, such as in strengthening the family, friendship, and fellowship bonds, in banquets, for recreational purposes, and as a status symbol when appropriate. This is where the culinary profession steps in. The employment area of cooks in our country is mostly in tourism and accommodation enterprises.

Nowadays, due to the need for qualified personnel, departments of culinary arts are opened in schools. In this study, it was aimed to determine the occupational attitudes of students studying at the undergraduate level of the departments in question. Besides, the variation of students' occupational attitudes in terms of the variables such as gender, attitude, age, grade point average, and grade was examined. It is observed that the majority of the participants are male, between the age of 20–21 years, with the grade point average of between 1.76–2.50, and the first-grade students. It is observed that students have voluntarily chosen the department of culinary arts, consider specializing in an area of the culinary profession, and believe that the culinary profession will satisfy them financially and morally. However, the students stated that they do not regret choosing the culinary profession, the profession is not boring, and the idea of being a cook does not intimidate them; on the contrary, they will become knowledgeable and competent cooks, and they are proud of the idea that the foods they will cook will shape the tastes of people. In the light of the findings and hypotheses obtained from the study, it can be said that the attitudes of the culinary program students towards the profession are generally positive.

Cookery is a contemporary profession. Students should be provided with the necessary knowledge and skills to enable them to specialize in a specific area of the profession [16]. Besides, platforms on which students can discuss cooking, foods, kitchen, and meals should be created, and their knowledge and skills should be measured. Therefore, the difficulties that students may encounter in the culinary profession will be more easily overcome. Furthermore, the increased student experiences will make their attitudes towards the profession more positive and increase their commitment to the profession.

It is thought that knowing the attitudes of the cook candidates who will work in the service sector towards the profession will guide on what kind of training should be given to culinary students in the future. Moreover, the development of positive attitudes towards the profession by cook candidates will ensure that they become more effective and productive when practicing the culinary profession.

# References

1. Aslan, S., Çakıroğlu, P.: Aşçıların Besin Güvenliği Konusundaki Bilgileri ve Bu Konuda Verilecek Eğitimin Bilgi Düzeylerine Etkisinin İncelenmesi. Gazi Üniversitesi Mesl. Eğitim Derg. **6**(11), 133–150 (2004)
2. Barrows, A., Shapleigh, B.E.: An Outline on the History of Cookery. Kessinger Publishing, Whitefish (1915)

3. Erdinç, B.Ş., Kahraman, S.: Turizm Mesleğini Seçme Nedenlerinin İncelenmesi. VI Lisansüstü Turizm Öğrencileri Araştırma Kongresi Kemer Antalya, 12 Nisan, pp. 229–237 (2012)

4. Gömeç, İ.: Otel İşletmelerinin Beklentilerine Göre Otelcilik Okulu Mutfak Bölümü Öğrencilerinin Mesleki Eğitimi. Yayınlanmamış Yüksek Lisans Tezi Gazi Üniversitesi Sosyal Bilimler Enstitüsü Turizm İşletmeciliği Ana Bilim Dalı Ankara (1995)

5. Harbalıoğlu, M., Ünal, İ.: Aşçılık Programı Öğrencilerinin Mesleki Tutumlarının Belirlenmesi, Ön Lisans Düzeyinde Bir Uygulama. Turiz. Akad. Derg. 01, 55–65 (2004)

6. Hughes, M.H.: Culinary Professional Training - Measurement of Nutrition Knowledge among Culinary Students Enrolled in a Southeastern Culinary Arts Institute. Unpublished Doctoral Dissertation Alabama University (2003)

7. Karahan, C.: Aşçıların Beslenme Bilgi Düzeyleri. Yayınlanmamış Yüksek Lisans Tezi Gazi Üniversitesi Eğitim Bilimleri Enstitüsü Aile Ekonomisi Ve Beslenme Eğitimi Anabilim Dalı Ankara (2010)

8. Kurnaz, A., Kurnaz, H.A., Kılıç, B.: Önlisans Düzeyinde Eğitim Alan Aşçılık Programı Öğrencilerinin Mesleki Tutumlarının Belirlenmesi. Muğla Sıtkı Koçman Üniversitesi Sos. Bilim. Enstitüsü Derg. 32, 41–61 (2014)

9. MEB: Yiyecek İçecek Hizmetleri – Aşçı. Ankara MEB (2007)

10. Özdamar, K.: Paket programlar ve istatistiksel veri analizi. Kaan Yayınları, Eskişehir (2001)

11. Öztürk, Y., Görkem, O.: Ulusal Aşçılık Meslek Standardı Çerçevesinde Öğrenci Yeterliklerine İlişkin Üç Boyutlu Değerlendirme. İşletme Araştırmaları Derg. 3(2), 69–89 (2011)

12. Robinson, R.N.S., Barron, P.E.: Developing a Framework for Understanding the Impact of Deskilling and Standardization on The Turnover and Attrition of Chefs. Hosp. Manag. 26, 913–926 (2007)

13. Terzi, A.R., Tezci, E.: Necatibey Eğitim Fakültesi Öğrencilerinin Öğretmenlik Mesleğine İlişkin Tutumları. Kuram ve Uygul. Eğitim Yönetimi Derg. 52, 593–614 (2007)

14. Santich, B.: Hospitality and gastronomy: natural allies. In: Hospitality: A Social Lens, pp. 47–59. Elsevier (2007)

15. Üstüner, M.: Öğretmenlik Mesleğine Yönelik Tutum Ölçeğinin Geçerlik ve Güvenirlik Çalışması. Kuram ve Uygul. Eğitim Yönetimi 45, 109–127 (2006)

16. Womersley, D.: The Life of Samuel Johnson. Penguin Classics, London (2008)

# Statistical Reasoning of Education Managers Opinions on Institutional Strategic Planning

Gülsün Başarı[✉], Ali Aktepebaşı, Ediz Tuncel, Emete Yağcı,
and Şahin Akdağ

Near East University, 99138 Nicosia, North Cyprus, Turkey
gulsun.basari@neu.edu.tr

**Abstract.** This research was carried out in order to investigate the opinions of the education administrators working in the Turkish Republic of Northern Cyprus on strategic planning and implementation.

The participants of research are 118 education administrators (principals, assistant principals and acting principal teachers) who worked in 2016-2017 academic year. The survey model was prepared in order to investigate the opinions of directors on strategic planning and implementation in education through using a five-level Likert-scale. Participants in the questionnaire were asked to mark the most appropriate choices for each statement varying from "Never Participate", "Participate", "Partially Participate", "Participate", "Totally Participate".

The results of this research have revealed that the educational administrators have positive opinions on strategic planning and that the strategic planning is necessary and beneficial. Educational administrators believe that they and the teachers need to be trained on strategic planning. The results of the research revealed that the education administrators had a positive opinion on strategic planning and thought that strategic planning was necessary and beneficial.

**Keywords:** Strategic planning · Education · Administrator · Implementation

## 1 Introduction

Strategic planning is a long-term plan that takes into account the external environment outside the control of the organization, which is concerned with issues affecting the development of the institution. Strategic planning is a holistic approach. A holistic approach arises from both the weaknesses and strengths of the organization itself, the opportunities and threats that result in due to external factors and the need for a planning process, as well as the full support of those who work at all levels in the organization. The stakeholders need to take an active role at all stages of the planning process. These stages include all the processes from the determination of the needs and expectations to the evaluation. At this point, the importance of communication in the planning process comes to the forefront.

"It is a systematic approach that involves the vision of the organization, its mission, its values, its needs (strengths - weaknesses, opportunities and threats), its objectives, its strategies, its action plans and its plans." According to Genç [1], although strategic

© Springer Nature Switzerland AG 2019
R. A. Aliev et al. (Eds.): ICAFS-2018, AISC 896, pp. 399–403, 2019.
https://doi.org/10.1007/978-3-030-04164-9_53

plans and decisions are shown as a way of planning future activities, they are more concerned with today's problems than the opportunities of the future. If strategic plans are to be considered within the framework of these definitions, the following points are noteworthy: The planning of the results is focused on the results, not the inputs [2].

Without a strategic plan, the institution can not know where it is going, at least whether it will reach its desired spot. What is important in strategic planning for the success of the institution is to know that everybody must work as required to fulfill the objectives of the institution. The success of the institution depends on its functioning as a single body. Strategic planning has three main elements: (1) Developing the Plan (2) Implementing the Plan (3) Evaluating the Plan [3].

Together with the social and economic changes, the roles of education and educational institutions are also changing. There is a two-way interaction between education and change. First, education has to reorganize itself in accordance with the changes in society. Secondly, education should lead to the reform of the society. This is because the educational institutions have the opportunity to influence other systems through their output. The basic building blocks that help to fulfill the objectives of the education are the schools and it is necessary to start the planned applications first at the schools. Education, which is in constant interaction with the economic, cultural and social structures, is influenced by the innovations, developments and rapid changes at the same time. Any change or developmental training in any one of them is affected and changed equally. For this reason, schools need to improve themselves in accordance with the needs of the era and innovations in order to make the necessary arrangements [4].

The most important feature that distinguishes the strategic plans from other plans is its long-term scope and its consideration of the external factor throughout the planning process. Like the other institutions, the schools should be able to see the strengths and weaknesses of these changes and developments, and, at the same time, they should take the threats that may arise into consideration carefully. What will ensure this is the strategic plan. It will be very difficult to avoid the setbacks that arise when strategic plan is not conducted and implemented. Because, as it is known, the raw material of the education is human being like it is in other service areas, and it may be very difficult to correct the mistakes caused by the human beings. At this point, both schools need to be able to fulfill their duties exactly as they are expected to, and make plans to avoid mistakes that can not be rectified.

## 2 Method

This research is based on a survey aiming to investigate the opinions of the stakeholders. The survey is a process of data collection to identify specific features that a group possesses. An important feature of the survey model is that it provides the researcher with a lot of information from a relatively high number of samples. In other words, in survey studies a wide range of views and features can be described [5]. In this respect, the population of the research consists of the managers who are working in the state and private schools affiliated to the Ministry of National Education in 2016–2017 Academic Year. Within this randomly selected population, a certain number of samples from preschool, primary and secondary education were involved. A questionnaire focusing on "Opinions of Education Managers on Strategic Planning and Implementation" was completed by 118 education managers. A Likert type scale was used to obtain the opinions of the participants. Participants in the questionnaire were asked to mark the most appropriate choice of the scale ranging from "Never Participate", "Participate", "Partially Participate", "Participate", "Entirely Solid". The data obtained have been processed and analyzed through "IBM SPSS Statistics 21" in order to find out the frequencies, percentages, mean and standard deviation of the outcomes.

## 2.1 Participants

In total, 118 participants including 42 female, 76 male education managers were involved in the research.

### Gender

Table 1 shows the gender distribution of education administrators. There are a total of 118 participants, of which 64% are male (f = 76) and 36% (f = 42) are female.

**Table 1.** Gender distribution.

| Gender | F | % |
|--------|-----|--------|
| Male | 76 | 64.00 |
| Female | 42 | 36.00 |
| Total | 118 | 100.00 |

### Position

In Table 2 shows that 72% (f = 84) of the training administrators are school principals, 20% (f = 24) vice principals and 8% (f = 10) are the acting principals.

**Table 2.** Distribution of education managers positions.

| Groups | f | % |
|--------|-----|--------|
| Principal | 84 | 72.00 |
| Vice Principal | 24 | 20.00 |
| Acting Principal | 10 | 8.00 |
| Total | 118 | 100.00 |

### School Types

In Table 3, the types of school where education managers are in charge are presented; 46% (f = 54) of education administrators are in primary education, 36% (f = 42) are in secondary education and 19% (f = 22) are in preschool education.

**Table 3.** School types where education managers work.

| Groups | $f$ | $\%$ |
|---|---|---|
| Elementary school | 54 | 46.00 |
| Secondary school | 42 | 36.00 |
| Preschool | 22 | 19.00 |
| Total | 118 | 100.0 |

## 2.2   Instruments

The questionnaire used in the research was developed by the researcher and consisted of 15 questions prepared to find out the opinions of the education managers regarding strategic planning and implementation. Crobach's Alpha value of the questionnaire is 0.765 which indicates that the findings of the questionnaire were reliable.

## 2.3   Data Analysis

The data of the study were analyzed to find out the descriptive statistics including mean, standard deviation and frequency levels of the data through the SPSS program The mean and standard deviation values of the questions regarding the opinions about strategic planning and implementation were included.

# 3   Results and Discussion

Table 4 shows the average and standard deviation values regarding the opinions education managers on strategic planning and implementation.

**Table 4.** Educational managers' opinions on strategic planning and implementation

| Statements | Mean | SD |
|---|---|---|
| I have sufficient information on strategic planning | 3.92 | .81 |
| I believe that strategic planning is important at schools | 3.66 | .95 |
| In my opinion all teaching staff of school must have training on strategic planning | 3.85 | .90 |
| Strategic planning improves cooperation among teaching staff at school | 3.87 | .86 |
| Strategic planning helps to take precautions against possible problems at school | 3.88 | .68 |
| I believe that our school will be successful through strategic planning | 3.67 | .70 |
| School teaching staff are aware of the vision and mission of the school | 3.16 | .96 |
| Implementation of strategic planning is very difficult | 3.41 | .72 |
| Strategic plan prepared at our school is easily implemented | 3.66 | .58 |
| In my opinion our school will be successful through strategic planning | 3.63 | .55 |
| School teaching staff know about the vision and mission of the school | 3.51 | .59 |
| School teaching staff are aware of the vision and mission of the school | 3.53 | .60 |
| I evaluate the strategic plan throughout the academic year | 3.66 | .64 |
| I obtain support from teachers while preparing the strategic plan | 3.64 | .69 |
| I believe that the school will benefit from the strategic plan in the long run | 3.60 | .78 |

As shown in Table 4, the education managers responded to all statements at the level of "I agree". The standard deviation figures also indicate that the responses of the participants were consistent within themselves.

## 4  Conclusions

According to the results of the research, it has been revealed that education managers have a positive opinion about strategic planning and that the strategic planning is necessary and useful. Education managers believe that they and their teachers should be trained on strategic planning. In addition, many education managers receive support from teachers when preparing the strategic plans. Despite the fact that education managers think that they have sufficient knowledge, it is understood that managers need more information and training on the subject. They underlined the need for strategic planning training in their research.

The researchers who will work on the topic should investigate the opinions of the teachers in more depth. Determining teachers' knowledge, attitudes and studies on the subject is important in terms of preparing and implementing the strategic plans.

## References

1. Genç, N.: Management and Organization, pp. 91–92. Seckin Publications, Ankara (2005)
2. Başaran, İ.E.: Management. Feryal Printing Press, Ankara (2000)
3. Özdemir, S.: Institutional Development in Education. Pegem Publications, Ankara (1998)
4. Büyüköztürk, Ş., Cakmak, E.K., Akgün, Ö.E., Karadeniz, Ş., Demirel, F.: Scientific Research Methods. Pegem Akademi, Ankara (2010)
5. Yelken, T.Y., Üredi, L., Tanrıseven, I., Kılıç, F.: Opinions of primary education inspectors about the constructive program and the level of teachers constructivist learning environment. Çukurova Univ., J. Soc. Sci. Inst. **19**(2) (2010)

# The Impact of Using Social Media on University Students Socialization: Statistical Reasoning

Sahin Akdag[✉], Huseyin Bicen, Gulsun Basari, and Ahmet Savasan

Near East University, POBOX 99138 Near East Boulevard, Nicosia, TRNC,
Mersin 10, Turkey
sahin.akdag@neu.edu.tr

**Abstract.** In today's communication age the role of the internet has brought about rapid changes in both individual and community life. The websites (Facebook, Twitter, Youtube, Instagram etc.), with millions of members, have become so powerful with no boundaries. The social media shapes an individual's educational, social, cultural, and political life. They communicate on the internet by sharing information and this is a clear indication that the social media is not a tool for fun and entertainment only. This study was carried out to specify the impact of social media and the reason for using it among university students, who use the internet more that anyone else because of their ages and education levels. Questionnaires, to specify the effects of social media on students, were conducted with 228 students, 97 male and 131 female, studying at Near East University.

**Keywords:** Social media · Technology · Sociology · Society

## 1 Introduction

Today, social media applications for various purposes by many differ from person to person and this has become almost an inevitable part of life. People use the social websites to reach the large mass sharing experiences or following the shared and while doing so, their level of media use is affected. Several facilities have been provided for sharing information or communication through internet technology [5]. The users began to share their thoughts, interest areas, and information on the internet [6]. Recently, social media was introduced into educational fields and is contributing positively to sharing and communication between teaches and their learners. However, it should always be taken into consideration that once the aim and length of using the media is out of control of the users, it may end in negative results [7].

When the development processes of the internet is looked at, it can be noted that technology and the internet developed incredibly between the 1970s and to the 1900s [2] and its use has begin to affect every phase of our lives. Social media can be defined as the applications developed and based on technology to fulfill social interaction and collaborative planning [1].

© Springer Nature Switzerland AG 2019
R. A. Aliev et al. (Eds.): ICAFS-2018, AISC 896, pp. 404–409, 2019.
https://doi.org/10.1007/978-3-030-04164-9_54

Besides the widespread application and easy access as well as its being costless, the social media has become very popular [3]. However, it has been found out in recent studies that students used the social media to meet new people and create an imaginary communication environment rather than using it for education and searching for new resources [4].

With this reality in mind, the ongoing and increasing sociological effects of social media on university students were the target investigation of this research.

## 2 Methodology

227 students, 96 male and 131 females, studying at Near East University, composed the participants of this research. The data obtained by questionnaires were analysed through IBM SPSS 23. The 5-likert type questions on the questionnaire were arranged as "Strongly disagree", "Disagree", "Undecisive", "Agree", and "Strongly agree". The participants were asked to pick one choice. The study aimed at gathering independent variable information in terms of gender and age.

At this stage of the research, the frequency and percentage distribution of the responses by the participants were specified and the values were shown in Tables.

### 2.1 Gender

Gender distribution of the students, %57.50 (f = 131) female and %42.50 (f = 97) male, is presented in Table 1.

**Table 1.** Gender distribution.

| Gender | f | % |
|--------|-----|--------|
| Female | 131 | 57.00 |
| Male | 97 | 43.00 |
| Total | 288 | 100.00 |

### 2.2 Age

As it is shown in Table 2, %44.30 (f = 101) of the students are 19, %27.20 (f = 62) are 18, %44.30 (f = 44) 20, %17.50 (f = 17) 21, and %1.80 (f = 4) are 22 years old.

**Table 2.** Age distribution.

| Age | f | % |
|-------|-----|--------|
| 18 | 63 | 27.00 |
| 19 | 101 | 44.00 |
| 20 | 44 | 19.00 |
| 21 | 17 | 8.00 |
| 22 | 4 | 2.00 |
| Total | 228 | 100.00 |

## 2.3   Instruments

The questionnaire, with 33 questions seeking answers to the students habits in using of the social media and websites, was developed by the researcher. The Cronbach alpha value of the questionnaire was found as 0.784.

## 2.4   Data Analysis

The data was analysed through SPSS program and frequencies, percentages, and averages were specified.

# 3   Results and Discussions

As it can be seen in Table 3, %93.40 (f = 213) of the students responded as "YES" and %6.60 (f = 15) of them responded as "NO" to the question, "Do you approve the use of the internet sociologically?" (Table 4).

**Table 3.** The appropriateness of the use of the internet sociologically

| Response | f | % |
| --- | --- | --- |
| Yes | 213 | 93.00 |
| No | 15 | 7.00 |
| Total | 218 | 100.00 |

**Table 4.** The benefits of the use of the internet in our daily life

| Response | f | % |
| --- | --- | --- |
| Yes | 204 | 90.00 |
| No | 24 | 10.00 |
| Total | 228 | 100.00 |

To the question, "Do you benefit from the internet in your daily life?", %89.50 (f = 204) of the students responded as "YES" and %10.50 (f = 24) responded as "NO".

As Table 5 reveals, %44.30 (f = 101) of the students admitted that they spared more than four hours, %39.90 (f = 91) spared 1–2 h, %11.40 (f = 26) spared 30 min–1 h, and %4.40 (10) spared less than 30 min for daily use of the social media.

**Table 5.** Daily time allocated to social media

| Time | f | % |
|---|---|---|
| Less than 30 min | 10 | 4.00 |
| Between 30 min–one hour | 26 | 11.00 |
| 1–2 h | 91 | 40.00 |
| More than four hours | 101 | 44.00 |
| Total | 228 | 100.00 |

**Table 6.** Students' priority in the use of social websites

| Time | f | % |
|---|---|---|
| Facebook | 13 | 6.00 |
| Twitter | 29 | 13.00 |
| Youtube | 44 | 19.00 |
| Instagram | 129 | 57.00 |
| Others | 13 | 6.00 |
| Total | 228 | 100.00 |

As it is revealed in Table 6, %56.60 (f = 129) said their priority was "Instagram", %19.30 (f = 44) preferred "Youtube", %12.70 (f = 29) used "Twitter" the most, %5.70 (f = 13) used "Facebook", and %5.70 (f = 13) preferred others (Table 7).

**Table 7.** Average and standard deviation results of student views

| Items | Mean | SD |
|---|---|---|
| 1. I approve meeting new friends on social media | 3.08 | 1.04 |
| 2. I use the social media for fun and relaxation | 3.30 | 1.19 |
| 3. I follow the news | 3.25 | 1.27 |
| 4. I reach daily information | 2.99 | 1.35 |
| 5. I can follow the activities of political parties | 3.20 | 1.04 |
| 6. I can reach charity organizations | 3.21 | 1.08 |
| 7. I can know the weather forecast | 3.00 | 1.26 |
| 8. I can reach travelling agencies | 3.01 | 1.23 |
| 9. I can know about holiday programs | 2.95 | 1.24 |
| 10. I can make holiday programs | 2.89 | 1.23 |
| 11. I can book travelling tickets | 2.68 | 1.34 |
| 12. I can have ticket options | 2.72 | 1.34 |
| 13. I can contact my old friends | 2.62 | 1.42 |
| 14. I can make new friends | 3.08 | 1.15 |
| 15. I use the social media to keep me busy when I'm free | 3.10 | 1.21 |
| 16. I can share information with my friends | 2.94 | 1.40 |

(*continued*)

**Table 7.**  (*continued*)

| Items | Mean | SD |
|---|---|---|
| 17. I can study with friends | 3.01 | 1.25 |
| 18. I can shop online | 2.92 | 1.27 |
| 19. I can buy new products in a short time | 2.96 | 1.27 |
| 20. I can follow prices and do shopping | 2.96 | 1.27 |
| 21. I can buy tickets | 2.82 | 1.32 |
| 22. I can sell something I want to sell | 2.96 | 1.14 |
| 23. I don't trust news on the social media | 2.93 | 0.97 |
| 24. Friends made on the social media aren't forever | 3.03 | 1.19 |
| 25. It isn't reliable for official affairs | 2.78 | 1.2 |
| 26. It isn't wise to shop online | 2.86 | 1.1 |
| 27. It isn't wise sharing privacy | 2.76 | 1.4 |
| 28. Buying tickets online isn't reliable | 2.88 | 1.1 |
| 29. It is a waste of time | 2.76 | 1.4 |
| 30. I don't approve improper sharings | 2.81 | 1.3 |
| 31. It's sociplogically harmful | 3.01 | 1.1 |
| 32. I go sleepless | 3.00 | 1.1 |
| 33. I don't approve celebrations on the social media | 2.80 | 1.1 |

When the average and standard deviation results are examined, the highest average is seen to be (3.10) which reflects the participants response as "*I use the social media to keep myself busy when I am free*". On the other hand, the lowest average is (2.62) and refers to "*I use the social media to reach my old friends*". In general, the Table reveals that the students are mostly "undecisive".

## 4   Conclusion and Recommendations

The findings obtained from this research reveal that the use of the internet is approved sociologically. It was found out that the longest time spared is more than four hours daily and Instagram is the web most frequently used. While university students seemed undecisive in meeting friends on the internet, they admit that they often use it for fun and entertainment as well as to follow the news. They also pointed out that they referred to the internet to learn about travelling programs. The students also added that they were not affected either positively or negatively by the internet in their daily lives. They stated that they were undecisive about online shopping. In the light of these findings, it is recommended that more studies are done to lead students in using the internet effectively and for beneficial purposes.

# References

1. Akar, E.: Sanal toplulukların bir türü olarak sosyal ağ siteleri-bir pazarlama iletişimi kanalı olarak işleyişi (2010)
2. Dilmen, N.E.: Yeni medya kavramı çerçevesinde internet günlükleri-bloglar ve gazeteciliğe yansımaları. Marmara İletişim Dergisi. sayı:12 (2007)
3. Gonzales, L., Vodicka, D.: Top ten internet resources for educators. Leadership **39**(3), 8–37 (2010)
4. Özmen, F., Aküzüm, C., Sünkür, M.: Sosyal ağ sitelerinin eğitsel ortamlardaki işlevselliği. Educ. Sci. **7**(2), 496–506 (2012)
5. Neumann, M., O'Murchu, I., Breslin, J., Decker, S., Hogan, D., MacDonaill, C.: Semantic social network portal for collaborative online communities. J. Eur. Ind. Train. **29**(6), 472–487 (2005)
6. Sayımer, İ.: Halkla İlişkilerde Hedef Kitlelerle Çift Yönlü Simetrik İletişim Kurmak Amacıyla Web Siteleri Kullanımı. Yeni İletişim Ortamları ve Etkileşim Uluslararası Konferansı Kitabı, pp. 1–3 (2008)
7. Yang, S.C., Tung, C.J.: Comparison of internet addicts and non-addicts in taiwanese high school. Comput. Hum. Behav. **23**(1), 79–96 (2007)

# Statistical and Structural Equation Modelling the Relationship Between Creativity and Material Design Self-efficacy Beliefs of Preschool Woman Preservice Teachers

Konul Memmedova[1(⊠)], Dervise Amca Toklu[2],
and Saide Sadikoglu[1,2]

[1] Department of Psychological Counselling and Guidance, Near East University,
TRNC, Mersin 10, Turkey
konul.memmedova@neu.edu.tr
[2] Tourism and Hotel Management Vocational School, Near East University,
TRNC, Mersin 10, Turkey

**Abstract.** The aim of this research is to analyse the relationship between creativity and the self-efficacy beliefs of preservice woman teachers in material design. The study group involves 219 women preservice teachers who obtain education in preschool teaching through the 2017–2018 school year. The creativity scale aiming to find out "How creative are you" and developed by Çoban [1] was used. The "Material design self-efficacy beliefs scale" developed by Bakaç and Özen [2] was used to determine material design oriented self-efficacy. The collected data were processed by using the statistical package for the social sciences (SPSS 24.0) and structural equation modelling software AMOS 24.0. Normality of distribution of preservice teachers' data created from the questionnaires of "How creative are you" and "Material design self-efficacy beliefs scales" were processed and examined through Kolmogorov-Smirnov's, Shapiro-Wilks tests, QQ plot and coefficient of Skewness-Kurtosis. Correlations between creativity and material design self-efficacy beliefs scales were examined with Pearson correlation analysis.

The AMOS allows to determine the creativity effect towards material design self-efficacy beliefs scale by using the structural equation modelling. The positive relationship between the creativity of preschool teaching preservice teachers and their self-efficacy belief level was determined.

**Keywords:** Preschool teaching · Preservice teachers · Creativity
Material design · Self-efficacy belief

## 1 Introduction

Material design is the most important tool that speeds up learning intriguingly and provides continuity of effective teaching. According to Yalın [3], a teaching material must be qualified in the form of presented information with different tools, messages and supplies. From another point of view, Yanpar [4] argued that teaching materials vary from simple to complex, for instance; teaching materials started with chalk and

© Springer Nature Switzerland AG 2019
R. A. Aliev et al. (Eds.): ICAFS-2018, AISC 896, pp. 410–416, 2019.
https://doi.org/10.1007/978-3-030-04164-9_55

paper and continued with a computer and related technologies at the present time. Yanpar argued that intramural and out-of-school tools are needed in order to compose materials. Thus, students can use their own creativity to prepare teaching materials instead of taking pre-prepared materials. During this process, they will learn many concepts and processing methods while taking the pleasure of producing and revealing their own products.

Material utilization provides a multi-agent learning environment, helps to supply the individual needs of students, draws attention by ensuring easily understandable concept, makes remembering easier, materialises abstract stuff, provides opportunities to use the time well and to observe [1]. The main goal of the visual design is to draw the attention of the audience and to organize communication media to easily understand information [5].

When research results are analyzed, it is observed that students get involved in the learning process by using more than one sense organ during material used educations. Therefore, more lasting experiences are gained [3, 6].

Role of a teacher in material design is in the forefront. In order to lead students, a teacher should be sufficiently qualified on the subject of material design. For this reason, according to Yelken [7], teachers should have qualifications such as; recognizing materials and knowing how to use, having knowledge about education design, knowing material design principles and use, leading students to design materials, knowing material types and learning principles, knowing development psychology and recognise learner, using appropriate material for target objective and achievement level, knowing learning material preparation principles and application procedures for lecture material development, observing well family features of students and environmental aspects, being tolerant and patient, having skill of material preparation and having prepared to learners, and having good communication skills. Also, Kopcha and Sullivan [8] emphasized that the teachers use the internet activity in order to design learning materials or evaluate education achievement level by using learning materials. Material design skill gives rise to increasing of self-efficacy belief level, too.

Self-efficacy is a belief of individual about his or her capability to achieve goals. The self-efficacy concept was first defined by Bandura in Social Learning theory [9]. Bandura defines self-efficacy as a personal judgment of how one can execute courses of action required to deal with prospective situations. Bandura also stated that self-efficacy is one of the key determinants for human behaviors and behavioral changes [9, 10].

It is thought that candidate teachers with positive self-efficacy beliefs will be more successful in instructional technology and material design lectures and perform unique designs [11]. Candidate teachers should pay sufficient attention to material design because they have to design and use appropriate materials for their own field [2]. According to Demirel [12], creative ability development of students is closely associated with the education program in school, education method and techniques.

Creativity is characterised by the human existence and means evolution, improvement. Creative thinking is a factor that should be energised and improved in the information age. Creativity is a process, it can be developed through education. Creativity should be accepted as a breathing activity during the education period. Namely, raising creative generations is possible with teachers who can organise required environments [13].

The purpose of this research is to determine the relationship between creativity and material design self-efficacy beliefs of preservice female preschool teachers. In accordance with this purpose, answers are sought for the questions below:

1. What is the creativity level of female teacher candidates who obtain education in preschool teaching?
2. What is material design self-efficacy belief level of female teacher candidates who obtain education on preschool teaching?
3. Do creativity levels of female teacher candidates who receive education on preschool teaching cause an influence on material design self-efficacy belief?

## 2  Method

### 2.1  Model of the Research

This research is descriptive research, and the relational screening model is used to determine the relationship between creativity and material preparation self-efficacy of female candidate teachers.

### 2.2  Population and Sample

The population of this research involves 219 female candidate teachers who obtain education through the 2017–2018 academic year at the Department of Preschool Teaching.

### 2.3  Data Collection Tools

With the intention to determine creativity level of candidate teachers, the scale of "How Creative Are You" as a creativity scale including 50 statements was developed by Raudsepp [14] and translated into Turkish by Çoban [2], and was used by taking into consideration of the individual's behaviours, values, interests, motivations, personal characteristics and many more variables. The scale has the following options; strongly agree (−2), agree (−1), indecisive (0), disagree (1) and strongly disagree (2). Each of the statements was addressed with one of the above-mentioned values. Later, creativity points were obtained through the responses. Scores between 100 and 80 have a high level of creativity, scores between 79 and 60 have above average level of creativity, scores between 59 and 40 have a medium level of creativity, scores between 39 and −20 have under the level of creativity and the ones which received scores between −19 and −100 have no creativity. The Cronbach's Alpha value was found as 0.94 in this research scale. Similarly, the reliability coefficient of the scale of Cronbach's Alpha was found to be 0.95 in Aydın's (2009) research.

The efficacy material design self-belief scale, which was developed by Bakaç and Özen [2] is used to determine the material design self-efficacy level of teachers. The scale is prepared by using the 5-point Likert scale and includes 25 statements. The construct validity of the scale was tested by using the exploratory and confirmatory factor analysis and a three-factor structure scale which accounts for 48% of total variance was obtained. The Cronbach's Alpha reliability coefficient of the scale was calculated as 0.92. High scores show that material design directed self-efficacy is high. Cronbach's Alpha value is found 0.96 in this research scale.

## 3  Findings

### 3.1  Descriptive Statistics of Data

SPSS (Statistical package for social sciences) and IBM AMOS 24.0 data analysis programs were used in the statistical analysis.

Within the scope of research, questions focusing on the theme of "How Creative Are You?" were asked to preservice teachers and descriptive statistics were processed such as average, standard deviation, lower and upper values in line with the Material Design Self-Efficacy Beliefs Scale points.

Normal distribution adaptation of preservice teacher's responses and Material Design Self-Efficacy Beliefs Scales were examined with Kolmogorov-Smirnov test, Shapiro-Wilk test, QQ plot and coefficient of Skewness-Kurtosis and it was determined that the data set is compatible with the normal distribution.

The correlations between points related to "How creative are you" and "Material design self-efficacy beliefs scales" were examined with Pearson correlation analysis. Structural equation modelling is used to determine the effect on the level of creativity of students related to material design self-efficacy beliefs scale.

Table 1 shows that students' mean of creativity is $\bar{x} = 42.50$, with standard deviation material design self-efficacy, scales $\pm 29.80$. This shows that the creativity level of survey participants is at the medium level.

**Table 1.** Descriptive statistics

|  | n | x | s | Min | Max |
|---|---|---|---|---|---|
| **How creative are you scale** | 219 | 42.50 | 29.80 | −44 | 100 |
| Material preparation on computer | 219 | 51.17 | 7.02 | 26 | 60 |
| Three-dimensional material preparation | 219 | 30.26 | 4.12 | 17 | 35 |
| Two-dimensional material preparation | 219 | 25.84 | 3.47 | 15 | 30 |
| **General of material design self-efficacy belief scale** | 219 | 107.27 | 13.71 | 58 | 125 |

Analysis of Material design self-efficacy scale sub-dimensions sows that for material preparation on computer sub-dimension $\bar{x} = 51.17 \pm 7.02$, three-dimensional material preparation sub-dimension has $\bar{x} = 30.26 \pm 4.12$ and two-dimensional material preparation sub-dimension has $\bar{x} = 25.84 \pm 3.47$. Thus, material preparation regarding the computer belief is higher than other dimensions.

The total mean of material design self-efficacy belief scale is $\bar{x} = 107.27 \pm 13.71$. This result indicates that material design self-efficacy beliefs scale is positive and sufficient.

Table 2 presents the results of correlation analysis between "How creative are you" and "Material design self-efficacy belief scale" of students. Analysis of the results shows that there are statistically significant correlations between the material design self-efficacy belief scale and creativity (p < 0.05). These correlations are positive and with the increase of creativity of students, the material design self-efficacy belief scale's sub-dimensions have also increased.

**Table 2.** Correlations between creativity self-efficacy belief scale points

|  | How creative are you scale | |
| --- | --- | --- |
|  | r | p |
| Material preparation on computer | 0.272 | 0.000* |
| 3-D material preparation | 0.260 | 0.000* |
| 2-D material preparation | 0.293 | 0.000* |
| **General of material design self-efficacy scale** | 0.291 | 0.000* |

*p < 0.05*

### 3.2    Statistical Equation Modelling (SME)

The Fig. 1 presents path diagram of the impact of creativity on material design self-efficacy belief scale generated by AMOS software.

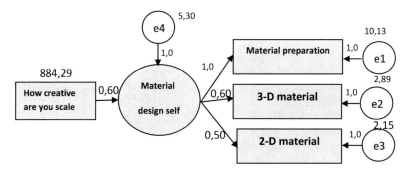

**Fig. 1.** I. SEM shown influence of creativity to material design self-efficacy belief

As shown in Table 3 the goodness of fit results is $\chi^2/dof = 0.402$. According to Klein (2005), $\chi^2/dof$ value below 3.0 provides good model fit and values $\chi^2/dof = 3 - 5$ show acceptable fit model.

Fair fit model is determined between the range of $\chi^2/dof = 0.05 - 0.08$.

The value of Root Mean Square Error of Approximation (RMSEA) below of 0.05 indicates good model fit.

**Table 3.** Goodness of fit index of the model

| Goodness of fit index | Value |
|---|---|
| $\chi^2$/dof (Chi-square/degrees of freedom) | 0.402 |
| The Root Mean Square Error of Approximation (RMSEA) | 0.001 |
| Normalized Fit Index (NFI) | 0.999 |
| Comparative Fit Index (CFI) | 1.000 |
| Goodness Fit Index (GFI) | 0.998 |

The normalised fit index assesses the model ranges between 0 and 1. NFI values greater than 0.95 indicate a good fit and NFI values between the range of 0.90 and 0.95 indicate acceptable fit. Hereunder, the scale from the point of NFI value with 0.999 corresponds to goodness fit model.

The limit value for CFI is 0.90. Values higher than 0.90 indicates acceptable fit and values higher than 0.95 indicate good fit.

GFI value greater than 0.90 indicates a well-fitting model. The model has 0.998 GFI value and good model fit.

According to this, it is observed that the established Structural Equation Modeling for determination of "How creative are you" scale has positive influences on material design self-efficacy belief scale and this indicates a good model fit.

It is observed that (Table 4) creativity of students (How creative are you scale-HCAYS) has a significant influence on the general material design self efficacy belief scale (MDSES). This influence is positive and increasing of the creativity level yields to increasing of material design self-efficacy of students.

**Table 4.** Regression coefficient in reference to the model

|  |  | Estimate | SE | CR | p |
|---|---|---|---|---|---|
| **MDSES** | **HCAYS** | 0.064 | 0.014 | 4.48 | *** |
| Material preparation on computer | MDSES | 1.0 |  |  |  |
| 3-D material preparation | MDSES | 0.6 | 0.03 | 19.6 | *** |
| 2-D material preparation | MDSES | 0.502 | 0.02 | 19.4 | *** |

## 4   Conclusion

The relation between creativity and material design self-efficacy belief levels of pre-school teaching female candidate teachers was analysed. As a result of the study, it is determined that female teacher candidates have medium creativity and sufficient material design self-efficacy beliefs. It is determined that the material preparation on computer belief on sub-dimension of material design self-efficacy belief scale is higher than the other sub-dimensions. Therefore, technology, pedagogy and field knowledge take an important place in the material preparation process of female candidate teachers. As a result of these findings, it is also determined that students' creativity has a positive influence on material design self-efficacy belief. This was also found in a similar study which was conducted by Bakaç and Özen [11]. According to the results obtained from this study, it is thought that the more the creativity level of female candidate teachers increases, the more their self-efficacy belief levels increase too.

The human sense of self-efficacy is related to a belief of efficacy level more than the real efficacy level. In addition to self-efficacy beliefs, they need other factors. Processing materials regarding creativity are also emphasised. Torrance [15] defined creativity as "a process of becoming sensitive to problems, deficiencies, gaps in knowledge, missing elements, disharmonies; identifying the difficulty; searching for solutions, making guesses or formulating hypotheses: testing and retesting these hypotheses and possible modifying an retesting them and finally communicating the results."

As a result, high creativity level necessity for material design skill is determined with this study, too. It is thought that gaining advanced creativity behaviours is possible with positive motivation and self-conception of an individual.

# References

1. Çoban, S.: The relation between creativity levels of directors and leadership styles. Unpublished postgraduate thesis, İstanbul: İstanbul University, Institute of social sciences, Turkey (1999)
2. Bakaç, E., Özan, R.: Material design self-efficacy belief scale: validity and reliability study1. Int. J. Hum. Sci. (JHS) (2015). ISSN 2458-9489
3. Yalın, H.İ.: Instructional Technologies and Material Development. Nobel Dağıtım, Ankara (2015)
4. Yanpar, T.: Instructional Technologies and Material Design. PegemA Yayıncılık, Ankara (2006)
5. Szabo, M., Kanuka, H.: Effects of violating screen design principles of balance, unity, and focus on recall learning, study time, and completion rate. J. Educ. Multimedia Hypermedia **8** (1), 23–42 (1998)
6. Seferoğlu, S.: Instructional Technologies and Material Design. Pegem Akademi Yayınları, Ankara (2011)
7. Yelken, Y.T.: Instructional Technologies and Material Design. Anı Yayıncılık, Ankara (2007, 2012)
8. Kopcha, T.J., Sullivan, H.: Learner preferences and prior knowledge in learner-controlled computer-based instruction. Educ. Technol. Res. Dev. **56**(3), 265–286 (2008)
9. Bandura, A.: Social Learning Theory. Prentice Hall, Englewood Cliffs (1977)
10. Özerkan, E.: The relation between self-efficacy sense of teachers and social studies self-conception sense of students. Postgraduate thesis. Trakya Üniversitesi Sosyal Bilimler Enstitüsü, Turkey (2007)
11. Bakaç, E., Özen, R.: Examining preservice teachers' material design self-efficacy beliefs based on their technological pedagogical content knowledge competent. Ahi Evran Univ. Kırşehir Faculty Educ. J. **18**(2), 613–632 (2017)
12. Demirel, Ö., Bandura, A.: Exercise of personal and collective efficacy in changing societies. In: Bandura, A. (ed.) Self-efficacy in Changing Societies, pp. 1–45. Cambridge University Press, Cambridge (1995)
13. New orientations in education. Pegem A Yayıncılık, Ankara
14. Senemoğlu, N.: Creativity and teacher qualifications. Creativity and Education Panel. Kara Harp Okulu, Ankara (1996)
15. Raudsepp, E.: 12 Vital characteristics of the creative supervisor. Supervision **45**, 14–15 (1977)

# Directions of Complex Regional Socio-Economic Development Based on Cognitive Modeling of Realization of Investment Strategy

Imran Akperov, Nataliya Brukhanova[✉], and Dmitriy Dynnik

Southern University (IMBL), M. Nagibin Ave. 33a/47,
Rostov-on-Don 344068, Russia
natalia@iubip.ru

**Abstract.** The main problems of the investment strategy of the region, deterring support its complex socio-economic development, are considered in the article. The methodology of cognitive modeling is considered as one of the directions of development of the theory of fuzzy sets. Scenario analysis is carried out to determine the priorities of regional development based on the developed cognitive model of the mechanism of implementation of the investment strategy of the regional socio-economic space. As a result, it is proposed to create a regional project office that performs the function of supporting investment projects.

**Keywords:** Region · Socio-economic development · Cognitive model
Scenario analysis · Project office · Investment strategy · Impulsive modeling

## 1 Introduction

The activation of investment activity is one of the main conditions for the stable development of the region, which in turn is aimed at attracting financial and material resources of investors, and also their rational use in the most priority sectors of the economy. The volume of investments involved in the economy of the region is one of the criteria for the efficiency of its functioning. At the same time, there is a huge gap between the level of investment in the economy of the Russian regions and the level necessary to ensure their sustainable development.

## 2 Analysis of Problems

In The National rating of the investment climate in the subjects of the Russian Federation at the end of 2017, the Rostov region took 17th place. In common, Rostov Region has a great potential to increase its investment attractiveness for both Russian and foreign investors.

The indicators, which are characterized the current state of the investment sector are presented in Table 1.

© Springer Nature Switzerland AG 2019
R. A. Aliev et al. (Eds.): ICAFS-2018, AISC 896, pp. 417–425, 2019.
https://doi.org/10.1007/978-3-030-04164-9_56

**Table 1.** Dynamics of key indicators of the investment sector of the Rostov region in 2011–2017 [3].

| | 2011 | 2012 | 2013 | 2014 | 2015 | 2016 | 2017 |
|---|---|---|---|---|---|---|---|
| *Investments in basic capital, billion rubles* | | | | | | | |
| Rostov Region (RR) | 166.0 | 207.9 | 253.6 | 264.1 | 309.4 | 287.4 | 319.3 |
| Share of RR in Russian Federation (RF), % | 1.5 | 1.7 | 1.9 | 1.9 | 2.2 | 2.0 | - |
| Place of RR in RF | 8 | 13 | 11 | 11 | 10 | 11 | - |
| *Foreign direct investment (according to the balance of payments of the Russian Federation), million dollars USA* | | | | | | | |
| RR | 705.0 | 679.0 | 507.0 | 801.0 | 257.0 | 192.0 | 198.0 |
| Share of RR in Russian Federation (RF), % | 0.07 | 0.06 | 0.04 | 0.07 | 0.03 | 0.02 | - |
| Place of RR in RF | 22 | 23 | 26 | 19 | 36 | 22 | - |

At the end of 2016, the volume of investments in basic capital in the Rostov region amounted to 287.4 billion rubles, which corresponds to 11th place among the Russian regions (the share in the Russian Federation was 2.0%). During the period under review, this indicator in absolute value terms increased by 121.4 billion rubles.

Private investments in the region are gradually growing – at the end of 2016, their volume amounted to 151.8 billion rubles, which is more than in 2012 by 49.0 billion rubles. At the same time, the share of private investments in the structure of investments in basic capital by forms of ownership has changed: in 2012 it was 49.4% of the total volume of investments in basic capital, while in 2016 it was 52.8%.

The largest investment activity is demonstrated by large organizations – they account for 86.1% of the total volume of investments in fixed assets in 2016.

In the context of economic activities (for large and medium-sized organizations) at the end of 2016, the largest volume of investments in fixed assets (based on the total volume of investments for large and medium-sized organizations) shared on transport and communications (30.5%), production and distribution of electricity, gas and water (17.6%) and manufacturing (15.2%).

The volume of foreign direct investment in the Rostov region at the end of 2016 amounted to 192.0 million dollars, that corresponds to 39 places in the Russian Federation, and 513.0 million dollars of USA is less than in 2011.

In the context of the partner countries at the end of 2016, the largest volumes of foreign direct investment in the Rostov region share on the following States:

– Cyprus - 25%;
– Switzerland - 14.1%;
– Luxembourg - 11.5%;
– Turkey - 7.2%.

The key problems deterring the effective realization of the investment strategy of the Rostov region are the following:

1. High cost and long terms of technological connection to engineering infrastructure, due to the following factors:

   - insufficient level of development of engineering infrastructure in some areas of the Rostov region. As a result, the potential of using land plots for use in economic activities is significantly limited, the cost and terms of investment projects are increased;
   - administrative barriers to access to infrastructure, high duration and complexity of licensing procedures;
   - high tariffs for electricity connection and consumption.

2. Low availability of credit (loan) funds for investors, which is mainly due to:

   - high interest rates on loans;
   - high requirements for the size of the collateral base.

   As a result of the survey of entrepreneurs, the following problems were highlighted as key ones when applying to a bank for a loan: high interest rates (67.4% of the number of respondents); strict conditions for the loan (37.4%); long term of the application (21.9%) collateral base.

   Also, as the factors hindering access to financial resources, we can highlight the underdevelopment of project financing of the regional financial market.

3. Imperfection of the legal framework. The legal and regulatory framework forms the institutional environment conducive to the growth of investment activity in the region. Its imperfection and non-compliance with modern trends in investment legislation negatively affects the volume of investment in the economy. The need to correct regional legal acts in a short time is largely dictated by the high rate of development of the Federal regulatory legal framework.

4. Lack of specialists for high-tech investment projects.

   - the availability of the necessary labor resources in the region is one of the most important factors in the evaluation of the project (especially high-tech project) and affects to the investor's decision on the feasibility of its implementation.
   - lack of "innovative" education (education as part of the innovation system that meets the modern requirements of distribution and development of innovations), lack of educational programs that meet the modern needs of the market.

5. The mechanism of attracting and further support of investors is not sufficiently effective. Complex support both at the initial stage of discussion and at the stage of project implementation increases the investment attractiveness of the region for potential investors.

   According to the results of the National rating of the investment climate in the subjects of the Russian Federation, the Rostov region in 2016 worsened its positions on the following factors:

   - "the effectiveness of institutional mechanisms of support of business";
   - "the efficiency of business registration procedures";
   - "the effectiveness of the procedures for issuance of construction permits";

- "the effectiveness of the procedures for issuance of licenses";
- "administrative pressure on business".

The problems described above are characterized by a large number of heterogeneous in their economic nature of the elements and the relationships between them, the presence of uncertainty and the risk of functioning, the description at the qualitative level, the ambiguity of the consequences of management decisions. Such systems are called complex and difficult to present in the form of traditional formal quantitative models, the most common of which are regression analysis and time series analysis.

Cognitive analysis and modeling technologies are used to work with such systems [1].

## 3  Methodology

In 2006, the National Science Foundation, under the auspices of which the lion's share of scientific research in the United States is conducted, together with the U.S. Department of Commerce issued a report predicting the development of science for 50 years. The report was named NBIC – an abbreviation of the first letters of the names of four megatechnologies that define our near future: nanotechnology, biotechnology, information technology and cognitive technology.

Cognitive technologies are software and hardware that mimic the work of the human brain. The Kurchatov Institute, known for its ability to convert new trends in science into a solid infusion of public money, is one of the first in Russia, which caught fashionable theme "NBIC". Thus, within the framework of the Kurchatov Institute TheInstitute of cognitive research is formed. But the Institute of management problems of the Russian Academy of Sciences, where the laboratory № 51 was created (V. I. Maksimov, N. A. Abramov and many others), was the first in Russia, which began research in the direction of "Cognitive analysis and management of situations" since 1991. Later, cognitive modeling technologies were developed by other Russian scientists, in particular, at the Taganrog Radio-Engineering University.

Let us consider the method of cognitive modeling of complex socio-economic development of the region.

Stage 1 - Building of a cognitive map of the investment strategy of realization in the regional socio-economic space. This stage includes the collection, systematization, analysis of existing statistical and qualitative information on the state of socio-economic development of the region, the results of the investment strategy, which includes the number of investment projects implemented in the region, the amount of funding, as well as the number of jobs created in the region.

Such information may include basic data on the economy and socio-economic systems, theoretical material on investment activities, statistical material located on the website of the Federal state statistics service, information collected by the Ministry of economic development on the implementation of investment projects in the region and also the results of sociological surveys of investment project managers, heads of infrastructure organizations and representatives of regional authorities in charge of regional investment activities.

Determination of cause-and-effect relationships between factors, which will identify the main directions of influence of factors; positive impact (positive "+", negative "−"). Building of cognitive map by the person, who makes decision, based on the study of documents by experts, based on a survey of a group of experts or through open sample surveys. As a result, we have a cognitive map:

G = <V, E>, where
$V_i$, i = 1, 2, …, k - tops;
$E_i$, i = 1, 2, …, k - curves represent the relationship between the factors.

Stage II - development of cognitive model of interaction of investment sphere with the external environment in the form of a functional graphs.

For building of a cognitive model, it is important to determine the relationship between the factors in the form of the degree of influence of the vertices, i.e. to ascribe weight $w_{ij}$ to the curves or set of the functional dependencies between the tops.

As a result, a cognitive model in the form of a functional graph is constructed:

M = <<V, E>, X, F, θ>, where
G = <V, E> - cognitive map, X - many of the parameters of the tops, F = F(X, E) - functional relationship between $V_i V_j$, θ - parameter space of tops. Herewith (a) If the indicators are qualitative, the weight coefficient $w_{ij}$ set by experts from a certain interval, for example from [−10; +10], where

– very much increases (9;10);
– greatly increases (7; 8);
– significantly increases (5; 6);
– mildly increases (3;4);
– very little increases (1; 2);
– very weak decreases (−1; −2);
– mildly decreases (−3; −4);
– significantly decreases (−5; −6);
– greatly decreases (−7; −8);
– very much decreases (−9; −10) [2].

Stage 2 - cognitive modeling. This stage includes impulse modeling, scenario analysis and perturbation stability analysis. Scenario analysis is carried out on the models of impulse processes. It makes it possible to identify the factors that cause significant changes in the behavior of the simulated system. Scenario analysis is carried out in stages:

– the definition of initial trends characterizing the development of situations at this stage is necessary, which increases the confidence in the results;
– setting the desired target areas (increase, decrease) and strength (weak, strong) changes in the trends of processes in the situation;
– selection of a set of measures (set of control actions), determination of their strength and focus on the situation;

- the choice of the observed factors (indicators) characterizing the development and compliance of the situation with the desired result is carried out depending on the analysis objectives and the user's desire;
- carrying out pulse modeling considering certain initial trends, set the desired target areas, the selected set of activities;
- expert "qualitative" analysis of possible scenarios of the system under study to identify the worst and best possible scenarios of development and the choice of the desired scenario. Moreover, the choice of the desired scenario can be carried out according to various criteria and rules of decision-making, depending on the objectives of the study of the system.

To build a cognitive map of the mechanism of investment strategy implementation in the regional socio-economic space, the basic data on the economy and socio-economic systems, theoretical material on the subject area, as well as the results of sociological surveys of the Ministry of economic development of the Rostov region were used.

The main factors affecting the investment attractiveness of the region are presented in Table 2.

**Table 2.** Factors influencing investment attractiveness (control factors)

| № | Full name | Abbreviated name |
|---|---|---|
| 0 | Investment attractiveness | $V_0$ |
| 1 | State of the region infrastructure | $V_1$ |
| 2 | Market size | $V_2$ |
| 3 | Political stability | $V_3$ |
| 4 | Availability of tax breaks and benefits, subsidies and other support from the regional budget in the region | $V_4$ |
| 5 | Business risk | $V_5$ |
| 6 | Corruption | $V_6$ |
| 7 | Crime rate | $V_7$ |
| 8 | Legal and regulatory framework | $V_8$ |
| 9 | Interregional and foreign economic relations | $V_9$ |
| 10 | Natural resource potential | $V_{10}$ |
| 11 | Economic and geographical location | $V_{11}$ |
| 12 | Income level of the population | $V_{12}$ |
| 13 | Regional authority | $V_{13}$ |

In turn, a favorable investment climate and the implementation of the investment strategy contribute to the creation of jobs in the region, the increase in tax revenues to the budget of the region, the growth of gross regional product, etc. Dependent (controlled) factors are presented in Table 3.

**Table 3.** Dependent (controlled) on the implementation of the strategy factors.

| № п/п | Full name | Abbreviated name |
|---|---|---|
| 14 | Employment | $V_{14}$ |
| 15 | Budget revenues | $V_{15}$ |
| 16 | Living standard | $V_{16}$ |
| 17 | Population life quality | $V_{17}$ |
| 18 | Gross regional product | $V_{18}$ |
| 19 | Volume of investment resources | $V_{19}$ |
| 20 | Real sector of the economy | $V_{20}$ |

Table 4 presents the results of the expert survey on the impact of the factors of the model on each other.

Based on information (statistical and expert) on the state of the investment climate in the Rostov region and with the help of the software "Software system of cognitive modeling" (PSCM) was built integrated a cognitive model of implementation of the investment strategy in the regional socio-economic space (Fig. 1).

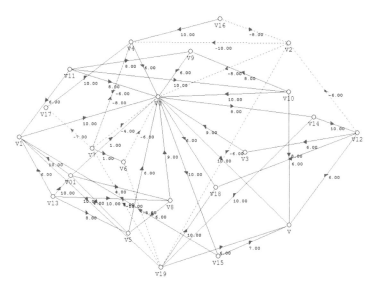

**Fig. 1.** Enlarged cognitive model of the investment strategy implementation in the regional socio-economic space, built in the software environment of the PSCM.

## 4   Results

The model allows using the technology of cognitive analysis to study the interaction of factors. The developed model is close to the assessment of the real mechanism of investment strategy implementation in the regional socio-economic space (due to the linearity of the relationships between the tops of the model).

Stage 3 is a scenario analysis.

Scenario analysis, conducted on the developed cognitive model, showed that significant changes in the socio-economic development of the region begin after the fourth cycle of modeling.

The figure shows that after the fourth year of implementation of the strategy, there is a significant growth of all the targets of the strategy: together with the investment attractiveness, the standard of living of the population is growing, the gross regional product is growing, there is a growth of the real sector of the economy. At the same time, the implementation of this scenario in modern conditions will not allow the region even to maintain its existing positions in the rating of the regions of the Russian Federation, not to mention the breakthrough development. Therefore, new approaches in regional management are becoming popular and relevant, which allow to solve the problems of investment development of the region at a new qualitative level and as soon as possible. In this situation, the project office can become an effective mechanism of investment activity activation, which can give a new vector of successful investment development.

Let us consider the forecast of the investment strategy implementation if we introduce an additional element – the project office (Fig. 2).

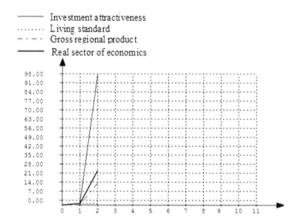

**Fig. 2.** The first step of simulation modeling after the introduction of the project office

The figure shows that after the first year of the strategy implementation, all the targets are growing. Thus, it should be noted that the basis for the successful implementation of any strategy is the continuous monitoring of initiatives and the use of a number of institutional mechanisms. At the same time, the key success factors are the support of project management based on the project office as the main center of responsibility for project activities. In solving business tasks and achieving strategic goals of project management, the project office should focus not only on their implementation, but primarily be a feedback tool and provide support to management representing opposite points of view and interests, and also plans for different periods.

As the practice of project offices functioning in Russia shows, this direction of development in project activities is a new area in organizational management, attracting special attention of the professional community.

# References

1. Volkova, V., Denisov, A.: Fundamentals of Systems Theory and System Analysis: Manual. Publishing of SPbGTU, Saint-Petersburg (1999)
2. Gorelova, G., Zakharova, E., Ginis, L.: Cognitive analysis and modeling of sustainable development of socio-economic systems. Publishing of RGU (2005)

# Expert-Analytical Support for the Document Marking Process Using a Fuzzy Analysis of Data Confidentiality

Ramin Rzayev[1(✉)], Elkhan Aliyev[1], and Afet Suleymanova[2]

[1] Institute of Control Systems of ANAS, Baku AZ1141, Azerbaijan
raminrza@yahoo.com
[2] Baku State University, Baku AZ1148, Azerbaijan

**Abstract.** It is proposed to use the fuzzy inference for the preliminary classi-fication and processing of expert opinions from remote sources regarding the degree of confidentiality of data to be included in the future document and use them for subsequent aggregation in order to identify a consolidated estimate the level of secrecy of a particular document.

**Keywords:** Confidentiality factor · Expert conclusion · Fuzzy set

## 1 Introduction

Existing approaches to the procedure for establishing the degree of confidentiality of data to be included in future documents (DDC) are based on the application of heuristic knowledge (in particular, the system of preferences) of the chief DC service officer (DCO) responsible for assigning the security label. However, the procedure for iden-tifying contextual knowledge, the system of preferences of DCO are very complex and, therefore, it requires to invite a consultant in the process of evaluating, synthesizing and choosing a solution from the variety of alternative options.

Another method of analytical support for the decision-making process for the choosing of reasoned decision can be a fuzzy analysis of the confidentiality factors (CF) of data using the expert survey by the method of scoring. During the decision-making process relative to the degree of data confidentiality to be included in the evaluated document, preliminary expert assessments of the confidentiality of docu-ments are usually expressed in terms that can be described by appropriate membership functions (MF) of corresponding fuzzy sets (FS). To order the CFs estimated by the experts and, therefore, the FS reflecting their degree of preference, method of fuzzy inference have been used.

## 2 The Concept of Expert Support System of the Document Marking Process

The identification of the data, knowledge and preference scheme of DCO for the assignment the security label of the data to be included in future documents (SLD) is carried out by expert survey. The amount of expert information relative to confidentiality

R. A. Aliev et al. (Eds.): ICAFS-2018, AISC 896, pp. 426–434, 2019.
https://doi.org/10.1007/978-3-030-04164-9_57

of data is greater the higher the dimension of the set of assessment criterions and the amount of data to be identified for their confidentiality. In addition, the development of universal questionnaires covering the circulation of documents in economic, technical, social, managerial, and other types of activities is impossible. Therefore, the permanent participation of a consultant directing the sequence of the DCO's arguments in the process of collecting of expert information is required. It disturbs the principle of confidentiality and the necessary of information hardness. Thus, the resolution of a problem of assigning the SLD in the familiar mode does not allow organizing the collection and processing of the initial information, even from several experts. Therefore, the development of expert support system of the marking of documents (ESSDM) from different areas allow to significantly accelerate the implementation of this procedure and, thereby, to improve its quality.

When developing the ESSDM it is necessary to take into account the following requirements for the operational characteristics of the interactive system (IS) [1]: (1) adaptation of DCO to the IS of the ESSDM should occur in natural language, which provides for the presence the elements of the artificial intelligence; (2) the uniformity of computing and survey procedures, and terminology regardless of the areas of documents circulation; (3) determination of the reporting sequence and the possibility of obtaining, as required, in-depth information; (4) the opportunity to adaptation as far as expand of the DCO's competence, heuristic knowledge and accumulated experience in operation. These requirements provide both the adaptation of DCO to the interactive system, and the invers process of system adaptation, taking into account the level of the increased DCO's competence. As result, the mental burden of DCO is minimized by reducing the number of questions, reducing their wording, etc., and, thereby, the system can provide the essentially great degree of objectivity of assigned SLD.

## 3  The Traditional Approach

The DDC is traditionally determined by comparing the expert level of confidentiality with the given maximum level in the applied system of assessment criterions of confidentiality. The corresponding index can be calculated by following formula [2]:

$$DC = [\sum_{i=1}^{K} \alpha_i \frac{c_{ri}}{c_{di}}] / [\max_i \{\sum_{i=1}^{K} \alpha_i \frac{c_{ri}}{c_{di}}\}], \tag{1}$$

where $K$ is the number of assessment criterions of confidentiality; $\alpha_i$ is the weight of the $i$-th assessment criterion of confidentiality, which determines the degree of its importance; $c_{di}$ is the maximum DDC according to the $i$-th assessment criterion; $c_{ri}$ – is the DDC explored by the expert according to the $i$-th assessment criterion of confidentiality. Determination of the weights $\alpha_i$ and estimation of the ratio $c_i = c_{ri}/c_{di}$ are carried out using an expert survey by scoring scale. The overall index of the consolidated opinion of all experts relative to $c_i$ should satisfy to follows [2]:

$$\sum_{i=1}^{n} w_i c_i \rightarrow \max, \quad \sum_{i=1}^{n} w_i = 1, \tag{2}$$

where $w_i$ is the value of the weight of the function $c_i$. Then the resultant value of the ratio of the DDC explored by the expert to the specified maximal level in the applied system of assessment criterions of confidentiality ($c_i$) is determined in the form of averaging: $c_i = \sum_{j=1}^{m} c_{ij}/m$, where $m$ is the number of experts; $c_{ij}$ is the estimate of DDC given by $j$-th expert. At the same time, the measure of agreement between experts ($W$) as a whole on the totality of all FC is defined as [2]:

$$W = \frac{12}{K^2 - K} \sum_{i=1}^{K} (c_i - \frac{K+1}{2})^2. \tag{3}$$

The expert opinions concerning the values of $\alpha_i$, which determine the relative importance of the $i$-th assessment criterion of confidentiality, are generalized similarly.

## 4  Classification of Assessments of the FC Influences on the SLD

Estimation of the DDC is a multi-criterion procedure. It implies the use of the composite rating aggregation rule for each FC $x_i$ ($i = 1 \div 6$). Therefore, to evaluate the FC influence five evaluative concepts are selected as follows: $u_1$ – "INSIGNIFICANT"; $u_2$ – "APPRECIABLE"; $u_3$ – "SIGNIFICANT"; $u_4$ – "CONSIDERABLE"; $u_5$ – "VERY LARGE". Simply put, $C = (u_1, u_2, u_3, u_4, u_5)$ is the set of the attributes by which the degrees of FC influence are classified. Then, to classify assessments of the FC influences on the SLD the following non-contradictory reasoning can be chosen as the base, though these arguments can and must be agreed with the DCO.

$r_1$ – "If the document contains FCs $x_1$, $x_2$ and $x_3$, then their cumulative SLD effect is significant";

$r_2$ – "If, in addition to these FCs, $x_6$ also takes place, then their cumulative SLD effect is more than significant";

$r_3$ – "If, in addition to the conditions of $r_2$, the factor $x_5$ takes place, then their cumulative SLD effect is very significant";

$r_4$ – "If the document contains all FCs: $x_1 \div x_6$, then their cumulative SLD effect is too significant";

$r_5$ – "If the document contains FCs $x_1$, $x_2$, $x_3$ and $x_4$, and the rest: $x_5$ and $x_6$ are absent, then cumulative SLD effect is significant";

$r_6$ – "If in the document there are no FCs $x_1$, $x_2$, $x_4$ and $x_5$, then SLD effect is insignificant".

In the above statements, the inputs are considered the presence (or absence) of the FC $x_k$ ($k = 1 \div 6$), and the output $y$ is the degree of their cumulative SLD effect. Then, defining the appropriate terms of the linguistic variables (LV) $x_k$ ($k = 1 \div 6$) and $y$, the fuzzy implicative rules can be constructed on the base of statements $r_1 \div r_6$ as follows:

$r_1$: "If $x_1$ = PRESENT and $x_2$ = PRESENT and $x_3$ = PRESENT, then $y$ = SIGNIFICANT";

$r_2$: "If $x_1$ = PRESENT and $x_2$ = PRESENT and $x_3$ = PRESENT and $x_6$ = PRESENT, then $y$ = MORE THAN SIGNIFICANT";

$r_3$: "If $x_1$ = PRESENT and $x_2$ = PRESENT and $x_3$ = PRESENT and $x_5$ = PRESENT and $x_6$ = PRESENT, then $y$ = VERY SIGNIFICANT";

$r_4$: "If $x_1$ = PRESENT and $x_2$ = PRESENT and $x_3$ = PRESENT and $x_4$ = PRESENT and $x_5$ = PRESENT and $x_6$ = PRESENT, then $y$ = TOO SIGNIFICANT";

$r_5$: "If $x_1$ = PRESENT and $x_2$ = PRESENT and $x_3$ = PRESENT and $x_4$ = PRESENT and $x_5$ = ABSENT and $x_6$ = ABSENT, then $y$ = SIGNIFICANT";

$r_6$: "If $x_1$ = ABSENT and $x_2$ = ABSENT and $x_4$ = ABSENT and $x_5$ = ABSENT, then $y$ = INSIGNIFICANT".

Let $y$ is defined on the discrete set $J$ = {0; 0.1; …; 1}. Then $\forall j \in J$ the terms from the right-hand parts of the rules $r_1 \div r_6$ can be described by FS with the following appropriate MF [2]: $Y_1$ = SIGNIFICANT: $\mu_{Y1}(j) = j$; $Y_2$ = MORE THAN SIGNIFICANT: $\mu_{Y2}(j) = j^{1/2}$; $Y_3$ = VERY SIGNIFICANT: $\mu_{Y3}(j) = j^2$; $Y_4$ = TOO SIGNIFICANT: $\mu_{Y4}(j) = 1$, if $j = 1$ and $\mu_{Y4}(j) = 0$, if $j < 1$; $Y_0$ = INSIGNIFICANT: $\mu_{Y0}(j) = 1 - j$. According to the approach described in [2], the fuzzification of terms in the left-hand parts of these rules can be realized by Gaussian MF: $\mu(u_i) = \exp\{-(u_i - 1)^2/\sigma_j^2\}$ ($j = 1 \div 5$), which restore the fuzzy subsets of the discrete universe ($u_1$, $u_2$, $u_3$, $u_4$, $u_5$), where $u_i = (d_{i-1} + d_i)/2$ ($i = 1 \div 5$). In this case, the values for the parameters $\sigma_j$ are chosen on the base of the importance of the FCs. The segment $[d_0; d_5]$ can always be easily transformed into the interval [0; 1] by simple transformation: $x = d_0 + t(d_5 - d_0)$ ($t \in$ [0; 1]). Therefore, estimating the presence of FC from the point of view of their significance, graded in the scale [0; 1], where $d_i = 0.2i$ ($i = 0 \div 5$), all terms from the left-hand parts of the rules $r_1 \div r_6$ can be written in the form of following FSs:

BE PRESENT ($x_1$): $A$ = {0.0013/$u_1$; 0.0183/$u_2$; 0.1299/$u_3$; 0.4797/$u_4$; 0.9216/$u_5$};
BE PRESENT ($x_2$): $B$ = {0.0001/$u_1$; 0.0043/$u_2$; 0.0622/$u_3$; 0.3679/$u_4$; 0.8948/$u_5$};
BE PRESENT ($x_3$): $C$ = {0.0063/$u_1$; 0.0468/$u_2$; 0.2096/$u_3$; 0.5698/$u_4$; 0.9394/$u_5$};
BE PRESENT ($x_4$): $D$ = {0.0183/$u_1$; 0.0889/$u_2$; 0.2910/$u_3$; 0.6412/$u_4$; 0.9518/$u_5$};
BE PRESENT ($x_5$): $E$ = {0.0392/$u_1$; 0.1409/$u_2$; 0.3679/$u_3$; 0.6977/$u_4$; 0.9608/$u_5$};
BE PRESENT ($x_6$): $F$ = {0.0687/$u_1$; 0.1979/$u_2$; 0.4376/$u_3$; 0.7427/$u_4$; 0.9675/$u_5$}.

Then the rules $r_1 \div r_6$ can be rewritten as:

$r_1$: "If $x_1$ = $A$ and $x_2$ = $B$ and $x_3$ = $C$, then $y$ = $Y_1$";
$r_2$: "If $x_1$ = $A$ and $x_2$ = $B$ and $x_3$ = $C$ and $x_6$ = $F$, then $y$ = $Y_2$";
$r_3$: "If $x_1$ = $A$ and $x_2$ = $B$ and $x_3$ = $C$ and $x_5$ = $E$ and $x_6$ = $F$, then $y$ = $Y_3$";
$r_4$: "If $x_1$ = $A$ and $x_2$ = $B$ and $x_3$ = $C$ and $x_4$ = $D$ and $x_5$ = $E$ and $x_6$ = $F$, then $y$ = $Y_4$";
$r_5$: "If $x_1$ = $A$ and $x_2$ = $B$ and $x_3$ = $C$ and $x_4$ = $D$ and $x_5$ = $\neg E$ and $x_6$ = $\neg F$, then $y$ = $Y_1$";
$r_6$: "If $x_1$ = $\neg A$ and $x_2$ = $\neg B$ and $x_4$ = $\neg D$ and $x_5$ = $\neg E$, then $y$ = $Y_0$".

For the left-hand parts of these rules the cumulative FSs $M_i$ ($i = 1 \div 6$) are defined by finding the minimum of the values of the corresponding MFs as follows:

$$M_1 = \{0.0001/u_1; \ 0.0043/u_2; \ 0.0622/u_3; \ 0.3679/u_4; \ 0.8948/u_5\};$$
$$M_2 = \{0.0001/u_1; \ 0.0043/u_2; \ 0.0622/u_3; \ 0.3679/u_4; \ 0.8948/u_5\};$$
$$M_3 = \{0.0001/u_1; \ 0.0043/u_2; \ 0.0622/u_3; \ 0.3679/u_4; \ 0.8948/u_5\};$$
$$M_4 = \{0.0001/u_1; \ 0.0043/u_2; \ 0.0622/u_3; \ 0.3679/u_4; \ 0.8948/u_5\};$$
$$M_5 = \{0.0001/u_1; \ 0.0043/u_2; \ 0.0622/u_3; \ 0.2573/u_4; \ 0.0325/u_5\};$$
$$M_6 = \{0.9608/u_1; \ 0.8591/u_2; \ 0.6321/u_3; \ 0.3023/u_4; \ 0.0392/u_5\}.$$

In this case, the rules $r_1 \div r_6$ will be written in even more compact form, namely as:

$r_1$: "If $x = M_1$, then $Y = Y_1$"; $r_2$: "If $x = M_2$, then $Y = Y_2$"; $r_3$: "If $x = M_3$, then $Y = Y_3$";
$r_4$: "If $x = M_4$, then $Y = Y_4$"; $r_5$: "If $x = M_5$, then $Y = Y_1$"; $r_6$: "If $x = M_6$, then $Y = Y_0$".

Transformations of these rules by Lukashevich's implication $\mu_D(u, j) = \min\{1; 1 - u + j\}$ on $D = U \times J$ for each pair $(u, j) \in D$ allowed to obtain fuzzy relations in the form of matrices $R_1, R_2, \ldots, R_6$. As a result, it is succeed to obtain the following general functional solution $R$ by intersection of these matrices using the operation "min":

$$R = \begin{bmatrix} & 0 & 0.1 & 0.2 & 0.3 & 0.4 & 0.5 & 0.6 & 0.7 & 0.8 & 0.9 & 1 \\ u_1 & 0.9999 & 0.9999 & 0.9992 & 0.9492 & 0.8792 & 0.7892 & 0.6792 & 0.5492 & 0.3992 & 0.2292 & 0.0392 \\ u_2 & 0.9957 & 0.9957 & 0.9957 & 0.9957 & 0.9809 & 0.8909 & 0.7809 & 0.6509 & 0.5009 & 0.3309 & 0.1409 \\ u_3 & 0.9378 & 0.9378 & 0.9378 & 0.9378 & 0.9378 & 0.9378 & 0.9378 & 0.8779 & 0.7279 & 0.5579 & 0.3679 \\ u_4 & 0.6321 & 0.6321 & 0.6321 & 0.6321 & 0.6321 & 0.6321 & 0.6321 & 0.6321 & 0.6321 & 0.6321 & 0.6977 \\ u_5 & 0.1052 & 0.1052 & 0.1052 & 0.1052 & 0.1052 & 0.1052 & 0.1052 & 0.1052 & 0.1052 & 0.1052 & 0.9608 \end{bmatrix}$$

According to [2], the fuzzy conclusion concerning the $k$-th SLD ($k = 1 \div 5$) depending on the corresponding FC is reflected in the form of FS $E_k$ with the corresponding values of the MF from the $k$-th row of the matrix $R$. In particular, the fuzzy conclusion concerning the presence of the first-level of FC is interpreted as FS $E_1 = \{0.9999/0; \ 0.9999/0.1; \ 0.9999/0.2; \ 0.9492/0.3; \ 0.8792/0.4; \ 0.7892/0.5; 0.6792/0.6; \ 0.5492/0.7; \ 0.3992/0.8; \ 0.2292/0.9; \ 0.0392/1\}$. Then, setting the $\alpha$-level sets $E_{1\alpha}$ and calculating the corresponding cardinality $M(E_{1\alpha})$ by formula: $M(E_\alpha) = \sum_j^n i_j/n, \ i \in E_\alpha$, we have:

- for $0 < \alpha < 0.0392$: $\Delta\alpha = 0.0392$, $E_{1\alpha} = \{0; 0.1; 0.2; \ldots; 0.9; 1\}$, $M(E_{1\alpha}) = 0.5$;
- for $0.0392 < \alpha < 0.2292$: $\Delta\alpha = 0.19$, $E_{1\alpha} = \{0; \ 0.1; \ 0.2; \ \ldots; \ 0.8; \ 0.9\}$, $M(E_{1\alpha}) = 0.45$;
- for $0.2292 < \alpha < 0.3992$: $\Delta\alpha = 0.17$, $E_{1\alpha} = \{0; \ 0.1; \ 0.2; \ \ldots; \ 0.7; \ 0.8\}$, $M(E_{1\alpha}) = 0.40$;
- for $0.3992 < \alpha < 0.5492$: $\Delta\alpha = 0.15$, $E_{1\alpha} = \{0; \ 0.1; \ 0.2; \ \ldots; \ 0.6; \ 0.7\}$, $M(E_{1\alpha}) = 0.35$;
- for $0.5492 < \alpha < 0.6792$: $\Delta\alpha = 0.13$, $E_{1\alpha} = \{0; \ 0.1; \ 0.2; \ \ldots; \ 0.5; \ 0.6\}$, $M(E_{1\alpha}) = 0.30$;

- for $0.6792 < \alpha < 0.7892$:  $\Delta\alpha = 0.11$, $E_{1\alpha} = \{0;\ 0.1;\ 0.2;\ 0.3;\ 0.4;\ 0.5\}$, $M(E_{1\alpha}) = 0.25$;
- for $0.7892 < \alpha < 0.8792$: $\Delta\alpha = 0.09$, $E_{1\alpha} = \{0; 0.1; 0.2; 0.3; 0.4\}$, $M(E_{1\alpha}) = 0.20$;
- for $0.8792 < \alpha < 0.9492$: $\Delta\alpha = 0.07$, $E_{1\alpha} = \{0; 0.1; 0.2; 0.3\}$, $M(E_{1\alpha}) = 0.15$;
- for $0.9492 < \alpha < 0.9992$: $\Delta\alpha = 0.05$, $E_{1\alpha} = \{0; 0.1; 0.2\}$, $M(E_{1\alpha}) = 0.10$;
- for $0.9992 < \alpha < 0.9999$: $\Delta\alpha = 0.0007$, $E_{1\alpha} = \{0; 0.1\}$, $M(E_{1\alpha}) = 0.05$.

Then, according to [2] the numerical estimate of the fuzzy output $E_1$ is obtained as:

$$F(E_1) = \frac{1}{\alpha_{max}} \int_0^{\alpha_{max}} M(E_{1\alpha})d\alpha = [0.5 \cdot 0.0392 + 0.45 \cdot 0.19 + 0.4 \cdot 0.17 + 0.35 \cdot 0.15 + 0.3 \cdot 0.13$$
$$+ 0.25 \cdot 0.11 + 0.2 \cdot 0.09 + 0.15 \cdot 0.07 + 0.1 \cdot 0.05 + 0.05 \cdot 0.0007]/0.9999 = 0.3257.$$

The numerical estimates for others fuzzy conclusions are obtained by similar actions: at the level of the estimated concept $u_2 - F(E_2) = 0.3647$; $u_3 - F(E_3) = 0.4350$; $u_4 - F(E_4) = 0.5470$; $u_5 - F(E_5) = 0.9453$. In this case, the value of $F(E_1) = 0.3257$ is the upper limit of the interval, within which the evaluation of the FC is characterized as "HAS INSIGNIFICANT INFLUENCE". Similarly, the deffuzified output $F(E_2) = 0.3647$ is the upper limit for assessing the FC as "HAS SIGNIFICANT INFLUENCE"; $F(E_3) = 0.4350$ is the limit for assessing the FC as "HAS MORE THAN SIGNIFICANT INFLUENCE"; $F(E_4) = 0.5470$ is the limit for assessing the FC as "HAS VERY SIGNIFICANT INFLUENCE"; $F(E_5) = 0.9453$ is the limit for assessing the FC as "HAS TOO SIGNIFICANT INFLUENCE". Then, the final estimation of SLD is obtained by following criterion:

$$E = 100 \times F(E_k)/F_{max}, k = 1 \div 5 \tag{4}$$

where $F(E_k)$ is the estimate of the $k$-th level of availability and influence of the FC (to wide extent and any other estimate); $F_{max} = F(E_5) = 0.9453$. Then, in the accepted assumptions we get a valid scale for assessing the level of SLD at the scale of the segment $[0; 100]$, namely: $A = (57.86; 100]$; $B = (46.01; 57.86]$; $C = (38.58; 46.01]$; $D = (34.45; 38.58]$; $E = [0.00; 34.45]$.

## 5   Estimation of SLD by Fuzzy Inference

Now, suppose that the expert community is offered to test 10 documents $a_k$ ($k = 1 \div 10$) by five-point system relative to degree of influence of the FC ($i = 1 \div 6$) on the level of their secrecy. Assume that the documents $a_k$ ($k = 1 \div 10$) are estimated by equality (4) on the base of cumulative expert conclusions, which satisfied to requirements (2) and agreed according to formula (3) [3]. Obtained results are summarized in Table 1.

**Table 1.** Classification of $a_k$ relative to confidence degree on the base of expert conclusions.

| Document | Weight of estimation FC $x_i$ ($i = 1 \div 6$) | | | | | | Final estimate (%) | SLD |
|---|---|---|---|---|---|---|---|---|
| | $\alpha_1$ | $\alpha_2$ | $\alpha_3$ | $\alpha_4$ | $\alpha_5$ | $\alpha_6$ | | |
| | 0.261 | 0.207 | 0.177 | 0.159 | 0.122 | 0.074 | | |
| $a_1$ | 4.67 | 4.67 | 4.80 | 4.53 | 4.47 | 4.67 | 93.27 | A |
| $a_2$ | 3.87 | 4.47 | 4.47 | 4.20 | 4.27 | 4.47 | 85.19 | B |
| $a_3$ | 3.47 | 3.80 | 4.13 | 4.07 | 4.07 | 4.13 | 77.22 | C |
| $a_4$ | 2.80 | 2.80 | 2.87 | 3.00 | 3.40 | 3.00 | 58.27 | D |
| $a_5$ | 2.00 | 2.40 | 2.73 | 2.27 | 2.73 | 2.47 | 47.90 | E |
| $a_6$ | 4.47 | 4.60 | 4.67 | 4.73 | 4.47 | 4.53 | 91.62 | A |
| $a_7$ | 3.87 | 4.20 | 4.07 | 4.60 | 4.00 | 4.27 | 81.91 | B |
| $a_8$ | 1.67 | 1.87 | 2.07 | 2.07 | 1.87 | 2.07 | 37.78 | E |
| $a_9$ | 3.20 | 3.60 | 3.13 | 3.80 | 3.60 | 3.60 | 68.09 | C |
| $a_{10}$ | 2.53 | 2.33 | 2.67 | 2.47 | 2.60 | 2.80 | 50.84 | D |

In order to construct a system of fuzzy inference relative to estimates of SLD of the documents $a_k$ ($k = 1 \div 10$), which examined by the five-point rating system for the effect of the FC $x_i$ ($i = 1 \div 6$) (see Table 1), it possible to use the construction of the verbal model, formulated above by the statements $r_1 \div r_6$. In this case, for the terms from their left-hand parts the fuzzification procedure is applied somewhat differently, namely: each term is reflected as the fuzzy subset of the finite set of evaluated alternatives (or documents) $\{a_1, a_2, ..., a_{10}\}$ in the following form: $A_i = \{\mu_{Ai}(a_1)/a_1; \mu_{Ai}(a_2)/a_2; ...; \mu_{Ai}(a_{10})/a_{10}\}$, where $\mu_{Ai}(a_k)$ ($k = 1 \div 10$) is the value of the MF of the FS $A_i$, which establishes how much the document $a_k$ corresponds to the evaluated criterion $A_i$. As such function the following Gaussian membership function $\mu_{Ai}(a_k) = \exp\{-[e_i(a_k) - 5]^2/\sigma_i^2\}$ was chosen, where $e_i(a_k)$ is the averaged expert judgment by five-point scale relative to the influence degree of $i$- th FC on the SLD of the document $a_k$ ($k = 1 \div 10$); $\sigma_i^2$ is the density of the location of the nearest elements, which was chosen as 4 for all cases of the fuzzification. Then, assuming the influence levels of the $x_i$ ($i = 1 \div 6$) as LV, their most sensitive terms, in particular, "VERY LARGE" can be described in the form of the appropriate fuzzy subsets $A_i$ of the discrete universe $U = \{a_1, a_2, ..., a_{10}\}$ as follows:

- $A_1 = \{0.9731/a_1; 0.7267/a_2; 0.5570/a_3; 0.2982/a_4; 0.1054/a_5; 0.9322/a_6; 0.7267/a_7; 0.0625/a_8; 0.4449/a_9; 0.2176/a_{10}\}$;
- $A_2 = \{0.9731/a_1; 0.9322/a_2; 0.6977/a_3; 0.2982/a_4; 0.1845/a_5; 0.9608/a_6; 0.8521/a_7; 0.0864/a_8; 0.6126/a_9; 0.1683/a_{10}\}$;
- $A_3 = \{0.9900/a_1; 0.9322/a_2; 0.8276/a_3; 0.3217/a_4; 0.2758/a_5; 0.9731/a_6; 0.8056/a_7; 0.1169/a_8; 0.4172/a_9; 0.2574/a_{10}\}$;
- $A_4 = \{0.9463/a_1; 0.8521/a_2; 0.8056/a_3; 0.3679/a_4; 0.1552/a_5; 0.9819/a_6; 0.9608/a_7; 0.1169/a_8; 0.6977/a_9; 0.2019/a_{10}\}$;
- $A_5 = \{0.9322/a_1; 0.8753/a_2; 0.8056/a_3; 0.5273/a_4; 0.2758/a_5; 0.9322/a_6; 0.7788/a_7; 0.0864/a_8; 0.6126/a_9; 0.2369/a_{10}\}$;

- $A_6 = \{0.9731/a_1; 0.9322/a_2; 0.8276/a_3; 0.3679/a_4; 0.2019/a_5; 0.9463/a_6; 0.8753/a_7; 0.1169/a_8; 0.6126/a_9; 0.2982/a_{10}\}$.

Then, the basis fuzzy estimation model can be written in the following form:
$r_1$: "If $x_1 = A_1$ and $x_2 = A_2$ and $x_3 = A_3$, then $y = Y_1$";
$r_2$: "If $x_1 = A_1$ and $x_2 = A_2$ and $x_3 = A_3$ and $x_6 = A_6$, then $y = Y_2$";
$r_3$: "If $x_1 = A_1$ and $x_2 = A_2$ and $x_3 = A_3$ and $x_5 = A_5$ and $x_6 = A_6$, then $y = Y_3$";
$r_4$: "If $x_1 = A_1$ and $x_2 = A_2$ and $x_3 = A_3$ and $x_4 = A_4$ and $x_5 = A_5$ and $x_6 = A_6$, then $y = Y_4$";
$r_5$: "If $x_1 = A_1$ and $x_2 = A_2$ and $x_3 = A_3$ and $x_4 = A_4$ and $x_5 = \neg A_5$ and $x_6 = \neg A_6$, then $y = Y_1$";
$r_6$: "If $x_1 = \neg A_1$ and $x_2 = \neg A_2$ and $x_4 = \neg A_4$ and $x_5 = \neg A_5$, then $y = Y_0$".

Transformation of these rules in the familiar manner induces the final solution $R$, which reflects the cause-effect relations between the consolidated expert assessments of documents by FC $x_i$ ($i = 1 \div 6$) and the corresponding levels of confidentiality.

$$R = \begin{array}{c} \\ a_1 \\ a_2 \\ a_3 \\ a_4 \\ a_5 \\ a_6 \\ a_7 \\ a_8 \\ a_9 \\ a_{10} \end{array}$$

|     | 0 | 0.1 | 0.2 | 0.3 | 0.4 | 0.5 | 0.6 | 0.7 | 0.8 | 0.9 | 1 |
|-----|-----|-----|-----|-----|-----|-----|-----|-----|-----|-----|-----|
| $a_1$ | 0.0269 | 0.0678 | 0.0678 | 0.0678 | 0.0678 | 0.0678 | 0.0678 | 0.0678 | 0.0678 | 0.0678 | 0.9731 |
| $a_2$ | 0.2733 | 0.2733 | 0.2733 | 0.2733 | 0.2733 | 0.2733 | 0.2733 | 0.2733 | 0.2733 | 0.2733 | 0.9322 |
| $a_3$ | 0.4430 | 0.4430 | 0.4430 | 0.4430 | 0.4430 | 0.4430 | 0.4430 | 0.4430 | 0.4430 | 0.4430 | 0.8056 |
| $a_4$ | 0.7018 | 0.7018 | 0.7018 | 0.7018 | 0.7018 | 0.7018 | 0.7018 | 0.7018 | 0.7018 | 0.6273 | 0.5273 |
| $a_5$ | 0.8946 | 0.8946 | 0.8946 | 0.8946 | 0.8758 | 0.7758 | 0.6758 | 0.5758 | 0.4758 | 0.3758 | 0.2758 |
| $a_6$ | 0.0678 | 0.0678 | 0.0678 | 0.0678 | 0.0678 | 0.0678 | 0.0678 | 0.0678 | 0.0678 | 0.0678 | 0.9819 |
| $a_7$ | 0.2733 | 0.2733 | 0.2733 | 0.2733 | 0.2733 | 0.2733 | 0.2733 | 0.2733 | 0.2733 | 0.2733 | 0.9608 |
| $a_8$ | 0.9378 | 0.9378 | 0.9169 | 0.8169 | 0.7169 | 0.6169 | 0.5169 | 0.4169 | 0.3169 | 0.2169 | 0.1169 |
| $a_9$ | 0.5828 | 0.5828 | 0.5828 | 0.5828 | 0.5828 | 0.5828 | 0.5828 | 0.5828 | 0.5828 | 0.5828 | 0.6977 |
| $a_{10}$ | 0.8317 | 0.8317 | 0.8317 | 0.8317 | 0.8317 | 0.7369 | 0.6369 | 0.5369 | 0.4369 | 0.3369 | 0.2369 |

According to the above reasoning, the $k$-th row of the matrix $R$ is the fuzzy conclusion relative to the aggregated level of confidentiality for the $k$-th document. To numerically interpret each of these fuzzy conclusions it is necessary to apply the defuzzification. So, for fuzzy conclusion $E_1$ relative to the SLD of the 1st document, i.e. for fuzzy set $E_1 = \{0.0269/0;\ 0.0678/0.1;\ 0.0678/0.2;\ 0.0678/0.3;\ 0.0678/0.4;\ 0.0678/0.5;\ 0.0678/0.6;\ 0.0678/0.7;\ 0.0678/0.8;\ 0.0678/0.9;\ 0.0678/1\}$ we have:

- for $0 < \alpha < 0.0269$: $\Delta\alpha = 0.0269$; $E_{1\alpha} = \{0;\ 0.1;\ 0.2;\ \dots;\ 0.9;\ 1\}$, $M(E_{1\alpha}) = 0.50$;
- for $0.0269 < \alpha < 0.0678$: $\Delta\alpha = 0.0410$; $E_{1\alpha} = \{0.1;\ 0.2;\ \dots;\ 0.9;\ 1\}$, $M(E_{1\alpha}) = 0.55$;
- for $0.0678 < \alpha < 0.9731$: $\Delta\alpha = = 0.9053$; $E_{1\alpha} = \{1\}$, $M(E_{1\alpha}) = 1.00$.

Then numerical estimate of fuzzy conclusion $E_1$ will be following number:

$$F(E_1) = \frac{1}{0.9731} \int_0^{0.9731} M(E_{1\alpha})d\alpha$$
$$= [0.5 \cdot 0.0269 + 0.55 \cdot 0.0410 + 1 \cdot 0.9053]/0.9731 = 0.9673.$$

The numerical estimates of fuzzy conclusions relative to the SLD of others documents are obtained by similar actions: $a_2 - F(E_2) = 0.8534$; $a_3 - F(E_3) = 0.7250$; $a_4 - F(E_4) = 0.4823$; $a_5 - F(E_5) = 0.3753$; $a_6 - F(E_6) = 0.9655$; $a_7 - F(E_7) = 0.8578$;

$a_8 - F(E_8) = 0.2981$; $a_9 - F(E_9) = 0.5823$; $a_{10} - F(E_{10}) = 0.3756$. As a result, the ratios of the resulting estimates are obtained by multiplying above values by 100. These results are summarized in Table 2 in the scale of the segment [0; 100].

**Table 2.** Comparing estimates of confidentiality levels.

| Document | Weight-estimating technique | | | Fuzzy inference | | |
|---|---|---|---|---|---|---|
| | Estimate | SLD | Order | Estimate | SLD | Order |
| $a_1$ | 93.27 | A | 1 | 96.73 | A | 1 |
| $a_2$ | 85.19 | B | 3 | 85.34 | A | 4 |
| $a_3$ | 77.22 | C | 5 | 72.50 | A | 5 |
| $a_4$ | 58.27 | D | 7 | 48.23 | B | 7 |
| $a_5$ | 47.90 | E | 9 | 37.53 | B | 9 |
| $a_6$ | 91.62 | A | 2 | 96.55 | A | 2 |
| $a_7$ | 81.91 | B | 4 | 85.78 | A | 3 |
| $a_8$ | 37.78 | E | 10 | 29.81 | E | 10 |
| $a_9$ | 68.09 | C | 6 | 58.23 | A | 6 |
| $a_{10}$ | 50.84 | D | 8 | 37.56 | D | 8 |

## 6   Conclusion

Comparison of the results obtained by both methods for estimating the SLD of alternative documents $a_k$ ($k = 1 \div 10$) is presented in Table 2. As it is not difficult to notice, that the estimates of the SLD do not always coincide, especially when comparing the SLD by their markers. It is explained by different approaches to the formation of the grading scale. Nevertheless, the ranking of documents by fuzzy estimation of their SLD provides a higher level of trust, since in this case, the cause-effect relations between the FC influences and SLD are traced. As for the order of the ratios of the final ratings, they are quite similar, except for the results obtained for alternatives $a_2$ and $a_7$.

## References

1. Alexentsev, A.I.: About the classification of confidential information by types of secrets. Inf. Technol. Secur. **3**, 65–71 (1999). (in Russian)
2. Rzayev, R.R.: Analytical Support for Decision-Making in Organizational Systems. Palmerium Academic Publishing, Saarbruchen (2016). (in Russian)
3. Suleymanova, A.N.: About one approach to the establishment of the document security label subject to the weighted coefficients of the estimated confidentiality factors. Math. Mach. Syst. **1**, 160–167 (2018). (in Russian)

# An Interval Number Based Input-Output Analysis for a Regional Economy

Behiye Cavusoglu[1(✉)], Serife Z. Eyupoglu[2], Pinar Sharghi[3], and Tulen Saner[4]

[1] Economics Department, Near East University,
Nicosia, Mersin 10, North Cyprus, Turkey
behiye.cavusoglu@neu.edu.tr
[2] Business Administration Department, Near East University,
Nicosia, Mersin 10, North Cyprus, TRNC, Turkey
serife.eyupoglu@neu.edu.tr
[3] Maritime Management Department, University of Kyrenia,
Kyrenia, Mersin 10, North Cyprus, Turkey
pinar.sharghi@kyrenia.edu.tr
[4] School of Tourism and Hotel Management, Near East University,
Nicosia, Mersin 10, North Cyprus, Turkey
tulen.saner@neu.edu.tr

**Abstract.** Assessment of major impacts of economic events is based primarily on the Input-Output (I/O) model. However, in all of the existing I/O models exact information is used. Yet, in real life, it is difficult to obtain such exact information for problems, in particular related to the future. This paper presents an interval number based I/O analysis using the most resent information available for a national economy. The analysis will provide an important source of information in order to understand the inter-relations existing among the different sectors of an economy. This paper focuses on the North Cyprus economy.

**Keywords:** Input-output analysis · Interval numbers · Input-output table
North cyprus

## 1 Introduction

According to Turco and Kelsey [1] the net economic change in a region that arises because of expenditure attributed to an activity or event is known as economic impact. Variations in the amount as well as circulation of regional employment, sales, income and wealth are frequently the aim of analysts' from the perspective of local planning [2]. An economic impact analysis should be able to identify whether the economic development effects of an event/activity/project are net gains to the national economy, or are simply transfers of activity that would have occurred elsewhere.

The foremost device for the measurement of economic impact is what is known as the multiplier. The multiplier is the multiple of a preliminary exogenous variation in income (the multiplicand) in an economic system through which total income in the

© Springer Nature Switzerland AG 2019
R. A. Aliev et al. (Eds.): ICAFS-2018, AISC 896, pp. 435–443, 2019.
https://doi.org/10.1007/978-3-030-04164-9_58

system varies [3]. There is, though, a variety of multiplier concepts, namely Keynesian, Economic-Base, and Input-Output.

The Input-Output (I/O) analytical model was put forth by Wassily Leontief [4] towards the end of the 1930s. The I/O model is also considered as the Leontief model [5]. In developing the I/O analytical framework Leontief's aim was to merge economic facts (empirical content) and theory (the formulation of problems in a mathematical state) in order to verify the inter-industry relationships of a specific region. I/O models provide us with an immense amount of information about the economic transactions that exist in an economic region and as a result provide understanding as to how impacts that originate in one sector are transmitted throughout the whole economy of that particular region.

This paper is significant in that it conducts an interval number based I/O analysis for a national region, namely North Cyprus.

## 2  Preliminary Information

This study will utilize an interval based approach to the analysis therefore it is necessary to provide the preliminaries of internal numbers.

**Definition 1:** Addition of Interval numbers [6]

$$A = [a_1, a_2]; B = [b_1, b_2]$$

If

$$X \in [a_1, a_2] \text{ and } y \in [b_1, b_2] \tag{2.1}$$

Then

$$x + y \in [a_1, b_1, a_2 + b_2] \tag{2.2}$$

This can symbolically be expressed as

$$A + B = [a_1, a_2] + [b_1, b_2] = [a_1 + b_1, a_2 + b_2] \tag{2.3}$$

The image of A. If $x \in [a_1, a_2]$ then $-x \in [a_1, a_2]$. Symbolically this is conveyed as

$$-A = [-a_2, -a_1]$$

Considering the result of $A + (-A)$ and in accordance to the calculations shown above:

$$A + (-A) == [a_1, a_2] + [-a_2, -a_1] = [-a_1 - a_2, a_2 - a_1]. \text{ Note that } A + (-A) \neq 0. \tag{2.4}$$

**Definition 2:** Subtraction of Interval numbers [6].
If $x \in [a_1, a_2] y \in [b_1, b_2]$ then

$$A - B = [a_1, a_2] - [b_1, b_2] = [a_1 - b_2, a_2 + b_1]. \tag{2.5}$$

**Definition 3:** Multiplication of Interval numbers [5].
The production of intervals $A, B \subset R$ is defined as follows:

$$A . B[\min(a_1 . b_2, a_2 . b_1, a_2.b_2), \max(a_1 . b_1, a_1 . b_1, a_2 . b_1,)]. \tag{2.6}$$

for the case, $A \ B \subset R_+$ the result is obtained as

$$A . B = [a_1, a_2] [b_1, b_2] = [a_1 . b_1, a_2 . b_2]. \tag{2.7}$$

The multiplication of an interval $A$ by a real number $k \in R$ is defined as:

$$\text{If } k > 0 \text{ then } k . A = k . [a_1, a_2] = [ka_1, ka_2],$$
$$\text{if } k < 0 \text{ then } k . A = k . [a_1, a_2] = [ka_2, ka_1].$$

**Definition 4:** Division of Interval numbers [6].
Assuming that the dividing interval does not contain 0 and $A \ B, \subset R_+$ one has

$$A : B[a_1, a_2] : [b_1, b_2][a_1/b_2, a_2/b_1]. \tag{2.8}$$

According to (2.8), the inverse of A is defined as:
If $x \in [a_1, a_2]$ then

$$\frac{1}{x} \in \left[\frac{1}{a_2}, \frac{1}{a_1}\right]$$

and

$$A^{-1} = [a_1, a_2]^{-1} = [1/a_2, \ 1/a_1]. \tag{2.9}$$

Assuming a general case, the ratio of A and B can be as indicated below:

$$[a_1, a_2] : [b_1, b_2]$$
$$= [a_1, a_2] . [1/b_2, \ 1/b_1] \tag{2.10}$$
$$[\min\{a_1/b_1, \ a_2/b_2, \ a_2/b_1, \ a_2/b_2\}, \max\{a_1/b_1, \ a_1/b_2, a_2/b_1, a_2/b_2\}]$$

Note that,

$$A . A^{-1} = [a_1/a_2, a_2/a_1] \neq 1.$$

The division by a number $k > 0$ is equivalent to multiplication by a number $1/k$.

## 3  Interval Based Input-Output Analysis

### 3.1  Input-Output Model

Input-Output analysis tracks the interactions between industrial sectors within an economy. It helps predict how a change in the output in one industrial sector affects other industrial sectors and the overall economy. The essential of I/O analysis is the I/O table [7]. For each industrial sector, the I/O table displays the allocation of the inputs purchased and the outputs sold, and by using I/O tables, predictions are made on how spending that starts in one industrial sector or originates from the government will have effects on other industrial sectors from initial rounds of spending and then through subsequent rounds generated by the initial round. Therefore, input-output analysis is the representation and study of the production structure of an economy. An economy's production processes are interdependent with the products of one process being utilized in another while the product of that process may in turn be used in many others [7].

As seen in Table 1 below, an I/O table is separated into four different parts, namely intermediate demand, final demand, primary inputs to industries, and primary inputs to direct consumption.

National I/O tables and macroeconomic indicators are the two main sources of data used in this study. I/O tables not only exhibit data on final output of sectors/industries but also provide their intermediate demand and present the association amongst different sectors/industries of the economy. The row sum of the sectors gives total intermediate demand. Total final demand counted with the total intermediate demand and the total of private and government consumption, gross fixed capital formation and export. On the other hand column sum of the sectors gives total inter industry input. The total input for the economy counted as a sum of inter industry input and total GDP arising.

The I/O analysis aims to get equality in calculation of total input and total output and also equality in total intermediate demand and total inter-industry input. In Table 1;

$X_{1...n}$ = represents total input/output per sector
$x_{1...n}$ = represents inter industry input and output flow with interval numbers
$Y_{1...n}$ = total final demand per sector

Furthermore the total output per sector is symbolized by X1, X2, X3 and so on respectively. The final demands for these sectors are represented by Y1, Y2, Y3 and so forth, whereas the internal flows within the economy is represented with x11, x12, x13 and so forth. The data required for the construction of the I/O table for North Cyprus were collected from the North Cyprus Prime Ministry, State Planning Organization [8] and initial information is presented with interval numbers.

### 3.2  Technical Coefficients

Considering the I/O table presented in Table 1, the technical coefficients for each sector are deliberated using the interval numbers and formula below:

$$aij = \frac{[xij]}{\sum_{i=1}^{n} [Xj]} \tag{3.1}$$

Where;

$[x_{ij}]$ = flow of sector $i$ with interval numbers

$\sum_{j=1}^{n} [Xj]$ = column sum of the sector with interval numbers

n = number of rows in column

i = represents the number of row

j = represents the number of column

**Table 1.** Fragment of input-output table

| Output→ Input↓ | $X_1$ | $X_2$ | $X_3$ | $X_4$ | $X_5$ | $X_6$ | $X_7$ | $X_8$ | $X_9$ | $X_{10}$ | Total Final Demand | Total Output |
|---|---|---|---|---|---|---|---|---|---|---|---|---|
| $X_1$ | $[x_{11}]$ | $[x_{12}]$ | $[x_{13}]$ | $[x_{14}]$ | $[x_{15}]$ | $[x_{16}]$ | $[x_{17}]$ | $[x_{18}]$ | $[x_{19}]$ | $[x_{110}]$ | $[Y_1]$ | $[X_1]$ |
| $X_2$ | $[x_{21}]$ | $[x_{22}]$ | $[x_{23}]$ | $[x_{24}]$ | $[x_{25}]$ | $[x_{26}]$ | $[x_{27}]$ | $[x_{28}]$ | $[x_{29}]$ | $[x_{210}]$ | $[Y_2]$ | $[X_2]$ |
| $X_3$ | $[x_{31}]$ | $[x_{32}]$ | $[x_{23}]$ | $[x_{34}]$ | $[x_{35}]$ | $[x_{36}]$ | $[x_{37}]$ | $[x_{38}]$ | $[x_{39}]$ | $[x_{310}]$ | $[Y_3]$ | $[X_3]$ |
| $X_4$ | $[x_{41}]$ | $[x_{42}]$ | $[x_{43}]$ | $[x_{44}]$ | $[x_{45}]$ | $[x_{46}]$ | $[x_{47}]$ | $[x_{48}]$ | $[x_{49}]$ | $[x_{410}]$ | $[Y_4]$ | $[X_4]$ |
| $X_5$ | $[x_{51}]$ | $[x_{52}]$ | $[x_{53}]$ | $[x_{54}]$ | $[x_{55}]$ | $[x_{56}]$ | $[x_{57}]$ | $[x_{58}]$ | $[x_{59}]$ | $[x_{510}]$ | $[Y_5]$ | $[X_5]$ |
| $X_6$ | $[x_{61}]$ | $[x_{62}]$ | $[x_{63}]$ | $[x_{64}]$ | $[x_{65}]$ | $[x_{66}]$ | $[x_{67}]$ | $[x_{68}]$ | $[x_{69}]$ | $[x_{610}]$ | $[Y_6]$ | $[X_6]$ |
| $X_7$ | $[x_{71}]$ | $[x_{72}]$ | $[x_{73}]$ | $[x_{74}]$ | $[x_{75}]$ | $[x_{76}]$ | $[x_{77}]$ | $[x_{78}]$ | $[x_{79}]$ | $[x_{710}]$ | $[Y_7]$ | $[X_7]$ |
| $X_8$ | $[x_{81}]$ | $[x_{82}]$ | $[x_{83}]$ | $[x_{84}]$ | $[x_{85}]$ | $[x_{86}]$ | $[x_{87}]$ | $[x_{88}]$ | $[x_{89}]$ | $[x_{810}]$ | $[Y_8]$ | $[X_8]$ |
| $X_9$ | $[x_{91}]$ | $[x_{92}]$ | $[x_{93}]$ | $[x_{94}]$ | $[x_{95}]$ | $[x_{96}]$ | $[x_{97}]$ | $[x_{98}]$ | $[x_{99}]$ | $[x_{910}]$ | $[Y]$ | $[X_9]$ |
| $X_{10}$ | $[x_{101}]$ | $[x_{102}]$ | $[x_{103}]$ | $[x_{104}]$ | $[x_{105}]$ | $[x_{106}]$ | $[x_{107}]$ | $[x_{108}]$ | $[x_{109}]$ | $[x_{1010}]$ | $[Y_{10}]$ | $[X_{10}]$ |
| Total input | $[X_1]$ | $[X_2]$ | $[X_3]$ | $[X_4]$ | $[X_5]$ | $[X_6]$ | $[X_7]$ | $[X_8]$ | $[X_9]$ | $[X_{10}]$ | - | - |

In order to determine the interval based technical coefficient for a given sector, it is necessary to calculate the column sum of the given sector with interval numbers first and then divide each row of the given sector with the column sum of it;

$$\sum_{j=1}^{n} Xj\,min = 5523586 + 7342211 + 1917381 + 168 + 1785263 + 1796 + 712874 + 741428 + 261458 = 18286165$$

$$\sum_{j=1}^{n} Xj\,max = 105016 + 8115075 + 2119211 + 1973185 + 1986 + 787914 + 819472 + 288980 = 20211025$$

$$\left[\sum_{j=1}^{n} Xj\right] = [18286165 - 20211025] \tag{3.2}$$

$$aij = \frac{[5523589 - 6105016]}{[18286165 - 20211025]} = [5,5 - 6,1] \tag{3.3}$$

As a result of the above calculation procedure, all of the technical coefficients have been calculated.

All inter-industry technical coefficients are calculated by dividing the given sector output by the corresponding column totals in terms of interval. Technical coefficients show the immediate influence of change in final demand for a given sector. The indirect effect of changes in final demand for a given sector is measured through the interdependence of coefficients.

The various inter-industry flows can be shown by the following system of linear equations. Each inter-industry technical coefficient is calculated through the division of the given sector's output by the corresponding column totals.

$$
\begin{aligned}
[X_1] &= [x_{11}] + [x_{12}] + [x_{13}] + [x_{14}] + [x_{15}] + [x_{16}] + [x_{17}] + [x_{18}] + [x_{19}] + [x_{110}] + [Y_1] \\
[X_2] &= [x_{21}] + [x_{22}] + [x_{23}] + [x_{24}] + [x_{25}] + [x_{26}] + [x_{27}] + [x_{28}] + [x_{29}] + [x_{210}] + [Y_2] \\
[X_3] &= [x_{31}] + [x_{32}] + [x_{33}] + [x_{34}] + [x_{35}] + [x_{36}] + [x_{37}] + [x_{38}] + [x_{39}] + [x_{310}] + [Y_3] \\
[X_4] &= [x_{41}] + [x_{42}] + [x_{43}] + [x_{44}] + [x_{45}] + [x_{46}] + [x_{47}] + [x_{48}] + [x_{49}] + [x_{410}] + [Y_4] \\
[X_5] &= [x_{51}] + [x_{52}] + [x_{53}] + [x_{54}] + [x_{55}] + [x_{56}] + [x_{57}] + [x_{58}] + [x_{59}] + [x_{510}] + Y_5] \\
[X_6] &= [x_{61}] + [x_{62}] + [x_{63}] + [x_{64}] + [x_{65}] + [x_{66}] + [x_{67}] + [x_{68}] + [x_{69}] + [x_{610}] + [Y_6] \\
[X_7] &= [x_{71}] + [x_{72}] + [x_{73}] + [x_{74}] + [x_{75}] + [x_{76}] + [x_{77}] + [x_{78}] + [x_{79}] + [x_{710}] + [Y_7] \\
[X_8] &= [x_{81}] + [x_{82}] + [x_{83}] + [x_{84}] + [x_{85}] + [x_{86}] + [x_{87}] + [x_{88}] + [x_{89}] + [x_{810}] + [Y_8] \\
[X_9] &= [x_{91}] + [x_{92}] + [x_{93}] + [x_{94}] + [x_{95}] + [x_{96}] + [x_{97}] + [x_{98}] + [x_{99}] + [x_{910}] + [Y_9] \\
[X_{10}] &= [x_{101}] + [x_{102}] + [x_{103}] + [x_{104}] + [x_{105}]_+ [x_{106}] + [x_{107}] + [x_{108}] + [x_{109}] + [x_{1010}] + [Y_{10}]
\end{aligned}
\tag{3.4}
$$

$$
[x_{ij}] = [a_{ij}][X_j]
\tag{3.5}
$$

$a_{ij}$. is the technical coefficient of the $i^{th}$ row and $j^{th}$ column. Technical coefficients for each sector are represented by $a_{ij}$ and determined by the procedure below.

$$
\begin{aligned}
[X_1] &= [a_{11}]X_1 + [a_{12}]X_2 + [a_{13}]X_3 + a_{14}]X_4 + a_{15}]X_5 + [a_{16}]X_6 + [a_{17}]X_7 + [a_{18}]X_8 + [a_{19}]X_9 + [a_{110}]X_{10} + [Y_1] \\
[X_2] &= [a_{21}]X_1 + [a_{22}]X_2 + [a_{23}]X_3 + [a_{24}]X_4 + [a_{25}]X_5 + [a_{26}]X_6 + [a_{27}]X_7 + [a_{28}]X_8 + [a_{29}]X_9 + [a_{210}]X_{10} + [Y_2] \\
[X_3] &= [a_{31}]X_1 + [a_{32}]X_2 + [a_{33}]X_3 + [a_{34}]X_4 + [a_{35}]X_5 + [a_{36}]X_6 + [a_{37}]X_7 + [a_{38}]X_8 + [a_{39}]X_9 + [a_{310}]X_{10} + [Y_3] \\
[X_4] &= [a_{41}]X_1 + [a_{42}]X_2 + [a_{43}]X_3 + [a_{44}]X_4 + [a_{45}]X_5 + [a_{46}]X_6 + [a_{47}]X_7 + [a_{48}]X_8 + [a_{49}]X_9 + [a_{410}]X_{10} + [Y_4] \\
[X_5] &= [a_{51}]X_1 + [a_{52}]X_2 + [a_{53}]X_3 + [a_{54}]X_4 + [a_{55}]X_5 + [a_{56}]X_6 + [a_{57}]X_7 + [a_{58}]X_8 + [a_{59}]X_9 + [a_{510}]X_{10} + [Y_5] \\
[X_6] &= [a_{61}]X_1 + [a_{62}]X_2 + [a_{63}]X_3 + [a_{64}]X_4 + [a_{65}]X_5 + [a_6]_6 X_6 + [a_{67}]X_7 + [a_{68}]X_8 + [a_{69}]X_9 + [a_{610}]X_{10} + [Y_6] \\
[X_7] &= [a_{71}]X_1 + [a_{72}]X_2 + [a_{73}]X_3 + [a_{74}]X_4 + [a_{75}]X_5 + [a_{76}]X_6 + [a_{77}]X_7 + [a_{78}]X_8 + [a_{79}]X_9 + [a_{710}]X_{10} + [Y_7] \\
[X_8] &= [a_{81}]X_1 + [a_{82}]X_2 + [a_{83}]X_3 + [a_{84}]X_4 + [a_{85}]X_5 + [a_{86}]X_6 + [a_{87}]X_7 + [a_{88}]X_8 + [a_{89}]X_9 + [a_{810}]X_{10} + [Y_8] \\
[X_9] &= [a_{91}]X_1 + [a_{92}]X_2 + [a_{93}]X_3 + [a_{94}]X_4 + [a_{95}]X_5 + [a_{96}]X_6 + [a_{97}]X_7 + [a_{98}]X_8 + [a_{99}]X_9 + [a_{910}]X_{10} + [Y_9] \\
[X_{10}] &= [a_{101}]X_1 + [a_{102}]X_2 + [a_{103}]X_3 + [a_{104}]X_4 + [a_{105}]X_5 + [a_{106}]X_6 + [a_{107}]X_7 + [a_{108}]X_8 + [a_{109}]X_9 + [a_{1010}]X_{10} + [Y_{10}]
\end{aligned}
\tag{3.6}
$$

The inter-industry technical coefficients shown in Eq. (3.6) are usually referred to as the A matrix. Transferring all the X's to the left hand side and re-writing the equations we get:

$$(1 - [a_{11}])X_1 - [a_{12}]X_2 - [a_{13}]X_3 - [a_{14}]X_4 - [a_{15}]X_5 - [a_{16}]X_6 - [a_{17}]X_7 - [a_{18}]X_8 - [a_{19}]X_9 - [a_{110}]X_{10} = Y_1$$
$$-[a_{21}]X_1 + (1 - [a_{22}])X_2 - [a_{23}]X_3 - [a_{24}]X_4 - [a_{25}]X_5 - [a_{26}]X_6 - [a_{27}]X_7 - [a_{28}]X_8 - [a_{29}]X_9 - [a_{210}]X_{10} = Y_2$$
$$-[a_{31}]X_1 - [a_{32}]X_2 + (1 - [a_{33}])X_3 - [a_{34}]X_4 - [a_{35}]X_5 - [a_{36}]X_6 - [a_3]_7X_7 - [a_{38}]X_8 - [a_{39}]X_9 - [a_{310}]X_{10} = Y_3$$
$$- a_{41}]X_1 - [a_{42}]X_2 - [a_{43}]X_3 + (1 - [a_{44}])X_4 - [a_{45}]X_5 - [a_{46}]X_6 - [a_{47}]X_7 - [a_{48}]X_8 - [a_{49}]X_9 - [a_{410}]X_{10} = Y_4$$
$$- a_{51}]X_1 - [a_{52}]X_2 - [a_{53}]X_3 - [a_{54}]X_4 + (1 - [a_{55}])X_5 - [a_{56}]X_6 - [a_{57}]X_7 - [a_{58}]X_8 - [a_{59}]X_9 - [a_{510}]X_{10} = Y_5$$
$$-[a_{61}]X_1 - [a_{62}]X_2 - [a_{63}]X_3 - [a_{64}]X_4 - [a_{65}]X_5 + (1 - [a_{66}])X_6 - [a_{67}]X_7 - [a_{68}]X_8 - [a_{69}]X_9 - [a_{610}]X_{10} = Y_6$$
$$-[a_{71}]X_1 - [a_{72}]X_2 - [a_{73}]X_3 - [a_{74}]X_4 - [a_{75}]X_5 - [a_{76}]X_6 + (1 - [a_{77}])X_7 - [a_{78}]X_8 - [a_{79}]X_9 - [a_{710}]X_{10} = Y_7$$
$$-[a_{81}]X_1 - [a_{82}]X_2 - a_{83}]X_3 - [a_{84}]X_4 - [a_{85}]X_5 - [a_{86}]X_6 - [a_{87}]X_7 + (1 - [a_{88}])X_8 - [a_{89}]X_9 - [a_{810}]X_{10} = Y_8$$
$$-[a_{91}]X_1 - [a_{92}]X_2 - [a_{93}]X_3 - [a_{94}]X_4 - [a_{95}]X_5 - [a_{96}]X_6 - [a_{97}]X_7 - [a_{98}]X_8 + (1 - [a_{99}])X_9 - [a_{910}]X_{10} = Y_9$$
$$-[a_{101}]X_1 - [a_{102}]X_2 - [a_{103}]X_3 - [a_{104}]X_4 - [a_{105}]X_5 - [a_{106}]X_6 - [a_{107}]X_7 - [a_{108}]X_8 - [a_{109}]X_9 + (1 - a_{1010}])X_{10} = Y_{10}$$

$$(3.7)$$

## 3.3    Interdependence Coefficients

After some mathematical calculations the following equation arises:

$$X_1 = [z_1]Y_1 + [z_2]Y_2 + [z_3]sY_3 + [z_4]Y_4 + [z_5]Y_5$$

$$+ [z_6]Y_6 + [z_7]Y_7 + [z_8]Y_8 + [z_9]Y_9 + [z_{10}]Y_{10} \qquad (3.8)$$

Where $z_1, z_2, z_3, \ldots z_{10}$ are the interdependence coefficients in interval form. The interdependence coefficients are depicted by;

$$(I - A)X = Y$$
$$X = (I - A)^{-1}Y \qquad (3.9)$$

Substituting the technical coefficients from Eq. (3.6) for the equation above we obtain;

$$[5, 5 - 6, 1]X_1 - [7, 4 - 8, 1]X_2 - [0, 0001 - 0, 0002]X_3 - [0, 0]X_4 - [0, 0]X_5 - [0, 13 - 0, 15]X_6 -$$
$$[0, 0009 - 0, 001]X_7 - [0, 0003 - 0, 0004] \ X_8 - [0, 0036 - 0, 0037]X_9 - [0, 004 - 0, 006]X_{10} = Y_1$$
$$[7, 3 - 8, 1]X_1 + [7, 07 - 7, 81]X_2 - [0, 67 - 0, 82]X_3 - [0, 66 - 0, 81]X_4 - [0, 074 - 0, 081]X_5 -$$
$$[0, 4 - 0, 5]X_6 - [0, 36 - 0, 44]X_7 - [0, 22 - 0, 27]X_8 - [0, 34 - 0, 42]X_9 - [0, 52 - 0, 64]X_{10} = Y_2$$
$$[1, 9 - 2, 1]X_1 + [2, 4 - 2, 6]X_2 - [0, 12 - 0, 13] \ X_3 - [0, 0012 - 0, 0013]X_4 - [0, 009 -$$
$$0, 01]X5 - [0, 071 - 0, 087]X_6 - [0, 007 - 0, 0084]X_7 - [0, 04 - 0, 05]X_8 - [0, 062 - 0, 074]X_9 - [0, 015 - 0, 018]X_{10} = Y_3$$
$$[0, 00016 - 0, 00018]X_1 + [0, 0]X_2 - [0, 0009 - 0, 001]X_3 - [0, 0]X_4 - [0, 0016 - 0, 0018]X_5$$
$$-[0, 04 - 0, 055]X_6 - \ [0, 0002 - 0, 00022]X_7 - [0, 021 - 0, 025]X_8 - [0, 001 - 0, 002]X_9 - [0, 002 - 0, 003]X_{10} = Y_4$$
$$[1, 7 - 1, 9]X_1 + [2, 73 - 3, 02]X_2 - [0, 064 - 0, 078]X_3 - [0, 12 - 0, 15]X_4 - [0, 49 - 0, 51]X_5 -$$
$$[0, 15 - 0, 19]X_6 - [0, 062 - 0, 075]X_7 - [0, 062 - 0, 082]X_8 - [0, 09 - 0, 12]X_9 - [0, 11 - 0, 141]X_{10} = Y_5$$
$$[0, 0017 - 0, 0019]X_1 + [0, 056 - 0, 068]X_2 - [0, 001 - 0, 002]X_3 - [0, 003 - 0, 004]X_4 -$$
$$[0, 005 - 0, 006]X5 - [0, 006 - 0, 007]X_6 - [0, 11 - 0, 12]X_7 - [0, 045 - 0, 055]X_8 - [0, 014 - 0, 024]X_9 - [0, 002 - 0, 003]X_{10} = Y_6$$
$$[0, 71 - 0, 78]X_1 + [0, 04 - 0, 05]X_2 - [0, 029 - 0, 033]X_3 - [0, 031 - 0, 037]X_4 - [0, 025 -$$
$$0, 031]X5 - [0, 04 - 0, 05]X_6 - [0, 22 - 0, 26]X_7 - [0, 18 - 0, 22]X_8 - [0, 085 - 0, 104]X_9 - [0, 0132 - 0, 039]X_{10} = Y_7$$
$$[0, 74 - 0, 82]X_1 + [0, 003 - 0, 004]X_2 - [0, 024 - 0, 029]X_3 - [0, 075 - 0, 092]X_4 - [0, 34 -$$
$$0, 37]X5 - [0, 05 - 0, 067]X_6 - [0, 15 - 0, 18]X_7 - [0, 33 - 0, 41]X_8 - [0, 29 - 0, 37]X_9 - [0, 18 - 0, 22]X_{10} = Y_8$$
$$[0 - 0]X_1 + [0 - 0]X_2 - [0 - 0]X_3 - [0 - 0]X_4 - [0, 001 - 0, 02]X_5 - [0 - 0]X_6 - [0, 0034 -$$
$$0, 004]X_7 - [0, 0003 - 0, 0004]X_8 - [0 - 0]X_9 - [0 - 0]X_{10} = Y_9$$
$$[0, 26 - 0, 28]X_1 + [0, 015 - 0, 016]X_2 - [0 - 0]X_3 - [0 - 0]X_4 - [0, 0007 - 0, 0008]X_5 -$$
$$[0, 0007 - 0, 0009X_6 - [0, 001 - 0, 0012]X_7 - [0, 007 - 0, 008]X_8 - [0, 006 - 0, 008]X_9 - [0, 0004 - 0, 0006]X_{10} = Y_{10}$$

$$(3.10)$$

The system of interval Eq. (3.10) is solved through the Differential Evolutionary Optimization Method [9].

The equation:

$$X_1 = z_1 Y_1 + z_2 Y_2 + z_3 Y_3 + z_4 Y_4 + z_5 Y_5 + z_6 Y_6 + z_7 Y_7 + z_8 Y_8 + z_9 Y_9 + z_{10} Y_{10}$$
$$(3.11)$$

Can be interpreted as

$$X_1 = f(Y_1, \ Y_2, \ Y_3, \ Y_4, \ Y_5, \ Y_6, \ Y_7, \ Y_8, \ Y_9, \ Y_{10}) \qquad (3.12)$$

From the equations above it can be seen that the output of sector 1 is contingent on the final demand for the goods of sector 1, sector 2, and sector 3 and so on. The interdependence coefficients show the relationship between the outputs of the sectors (agriculture, industry, electricity water, etc.) and the final demand for the goods of the sectors.

Consider the following interdependence coefficients;

$$X_1 = [-42.430, \ -42.431]Y_1 + [20.218, \ 20.225]Y_2 + [-0.591, \ -0.271]Y_3 +$$
$$[22.848, \ 22.893]Y_4 + [38.955, \ 43.476]Y_5 + [39.278, \ 39.282]Y_6 + [8.222, \ 8.251]Y_7 +$$
$$[8.870, \ 8.960]Y_8 + [3.855, \ 3.968]Y_9 + [53.899, \ 54.094]Y_{10}$$
$$(3.13)$$

For each unit of final demand for goods of sector 1, the total output of sector 1 changes between $[-42.430, \ -42.431]$ and the total output of sector 2 changes between $[20.218, \ 20.225]$. As can be seen from the $z_i$'s given with interval numbers, the direct as well as the indirect effects of the variations in final demand calculated with interval numbers.

The interdependence coefficients illustrate the direct and indirect influences of increasing final demand by a single unit of value for any sector/industry. If final demand for a specific good increases in the inter-industry demand, then this will results in an rise in the output of the other sectors. Interval number based analysis enabled the observation of the effects of final demand on inter-industries within a certain interval instead of an exact number with the consideration of all possible risks.

As a result of the analysis conducted the question "How is output influenced through changes in final demand?" can be countered. To attain specific objectives national authorities can use this technique for policy purposes.

## 4   Conclusion

In this paper an interval number based regional Input-Output model is constructed for the North Cyprus economy. The outcome of this study indicated the validity of the proposed model which was applied to North Cyprus. The limitations to this study are the available data used for the analysis which is from 1998. Therefore, this study's simulation results are only an initial indicator for policy makers and the availability of up-to-date data would provide for more reliable calculations and judgments

**Acknowledgements.** The research has been conducted with the support of the Near East University, Nicosia, North Cyprus and Azerbaijan State Oil and Industry University, Baku, Azerbaijan.

# References

1. Turco, D.M., Kelsey, C.W.: Conducting Economic Impact Studies of Recreation and Parks Special Events. National Recreation and Park Association, Washington, D.C. (1992)
2. Shaffer, R.E.: Community Economics, Economic Structure and Change in Smaller Commnities. Iowa State University Press, Ames (1989)
3. Lewis, A.J.: Economic impact analysis: a UK literature survey and bibliography. Plan. Prog. **30**(3), 157–209 (1988)
4. Leontief, W.W.: The Structure of American Economy: An Empirical Application of Equilibrium Analysis. Harvard University Press, Cambridge (1941)
5. Miller, R., Blair, P.D.: Input-Output Analysis: Foundations and Extensions. Prentice-Hall, Englewood Cliffs (1985)
6. Aliev, R.A., Huseynov, O.H., Aliev, R.R., Alizadeh, A.A.: The Arithmetic of Z-Numbers: Theory and Applications. World Scientific Publications, London (2015)
7. Surugiu, C.: The economic impact of tourism, an input-output analysis. Rom. J. Econ. **29**, 142–161 (2009)
8. State Planning Organization: Statistical Yearbook, TRNC, Prime Ministry, Nicosia (2015)
9. Aliev, R.A., Guirimov, B.G.: Type-2 Fuzzy Neural Networks and Their Applications. Springer, Heidelberg (2014)

# Fuzzy Stability Analysis of an Isothermal First-Order Reaction

Mahsati Babanli[(⊠)]

Azerbaijan State Oil and Industry University, Azadlig 20,
Nasimi, Baku, Azerbaijan
mehseti.babanli@gmail.com

**Abstract.** In this paper we are focusing on an investigation of fuzzy solution of the dimensionless first-order isothermic reaction. Stability analysis of chemical kinetics, described by second-order fuzzy differential equation is considered. Sensitivity analysis of the fuzzy solution of the considered isothermic first-order reaction is provided, that gives ability to user to find desirable stable behavior of designed chemical reactor.

**Keywords:** Fuzzy stability · Isothermic reaction · Fuzzy differential equation

## 1 Introduction

In kinetic modeling of a chemical reactor usually it is needed to find solutions to the rate equations to determine the time-varying concentration of all species in reaction. This equations in existing kinetic modeling study are presented as mathematically intractable differential equations of different order depending of types of reactions. Such equations are of fundamental importance in the design of chemical reactors. In real model practice parameters and variables of these equations are uncertain and should be described as interval probabilistic or fuzzy values.

In [1] statement and solution of fuzzy isothermal equation is given. Solution of the reaction by using generalized differentiability is presented.

Steady state approximation approach to solving the rate equations is presented in [2]. In accordance with this approach coupled differential equations is converted into a system of algebraic equations, one for each species in the reactions.

Boundary value problem solution for chemical reaction is considered in [3]. Modeling of chemical kinetic processes is discussed in [4]. The mass, energy and component balance equation is considered.

The paper [5] is devoted to investigation of the steady- state equation of exothermic reactions. The paper is focused on stability analysis of this type of reactions. Randomly perturbations in the input variables are within $\pm 3\%$ of the study-state values. In this paper fuzzy flow control of a reactor vessel also is considered.

Analyzing state-of-the art of the considered in this paper problem we can conclude, that research results on modeling of chemical reactions with uncertain parameters and variables for designing proper chemical reactors are very scarce. To fill in this gap in

© Springer Nature Switzerland AG 2019
R. A. Aliev et al. (Eds.): ICAFS-2018, AISC 896, pp. 444–450, 2019.
https://doi.org/10.1007/978-3-030-04164-9_59

this paper we try to investigate stability analysis of isothermic reaction described by fuzzy differential equations. Fuzzy stability theory is applied for solving the problem.

The paper is structured as follows. In Sect. 2, we present some prerequisite material on fuzzy number, fuzzy function, fuzzy stability etc. In Sect. 3 statement of the problem is given. In Sect. 4, we describe solution of the fuzzy differential equation described dynamics of investigated isothermic reaction. Section 5 offers some conclusions.

## 2   Preliminaries

**Definition 1. Fuzzy number [6].** A fuzzy number is a fuzzy set $A$ on $R$ which possesses the following properties: (a) $A$ is a normal fuzzy set; (b) $A$ is a convex fuzzy set; (c) $\alpha$-cut of $A$, $A^{\alpha}$ is a closed interval for every $\alpha \in (0, 1]$; (d) the support of $A$, $A^{+0}$ is bounded.

**Definition 2. Fuzzy function [6].** A fuzzy function $f$ from a set $X$ into a set $Y$ assigns to each x in $X$ a fuzzy subset $f(x)$ of $Y$. We denote it by $f : X \to Y$. We can identify $f$ with a fuzzy subset $G_j$ of $X \times Y$ and $f(x)(y) = G_j(x, y)$.

If $A$ is a fuzzy subset of $X$, then the fuzzy set $f(A)$ in $Y$ is defined by

$$f(A)(y) = \sup_{x \in X} [G_j(x, y) \wedge A(x)]$$

**Definition 3. Fuzzy differential equation.** Fuzzy differential system is:

$$x = f(t, x) \tag{2}$$

where $f$ in (2) is continuous and has continuous partial derivatives $\frac{\partial f}{\partial x}$ on $R_+ \times E^n_{,\text{i.e.}} f \in C^1[R_+ \times E^n, E^n]$
   and
   $x(t_0) = y_0 \in E^n, t \geq t_0, t_0 \in R_+$.

**Definition 4. Zadeh-Aliev stability criteria:** The solution $x(t, t_0, y_0)$ of the system (2) is said to be fuzzy Lipschitz stable with respect to the solution $x(t, t_0, x_0)$ of the system (2) for $t \geq t_0$, where $x(t, t_0, x_0)$ is any solution of the system (2), if there exists a fuzzy number $M = M(t_0) > 0$, such that

$$\|x(t, t_0, x_0) - {}_h x(t, t_0, x_0)\|_{fH} \leq M(t_0)\|y_0 - {}_h x_0\|_{fH} \tag{3}$$

If $M$ is independent on $t_0$, then the solution $x(t, t_0, x_0)$ of the system (2) is said to be uniformly fuzzy Lipschitz stable with respect to the solution $x(t, t_0, x_0)$.

## 3  Statement of the Problem

The dimensionless an isothermic first-order reaction A → B is described by following fuzzy differential Eq. [1].

$$E_a \frac{d^2 C_A}{dx^2} - \vartheta \frac{dC_A}{dx} - kC_A = 0 \tag{4}$$

or

$$\frac{d^2 y}{dz^2} - N \frac{dy}{dz} - NRy = 0 \tag{5}$$

Here

$$y = \frac{C_A}{C_{A0}}, \quad z = \frac{x}{L}, \quad N = \frac{\vartheta L}{E_a}, \quad R = \frac{kL}{\vartheta} \tag{6}$$

$C_A$ is concentration of species A, k is rate of reaction, $C_A$ is concentration of the entering A, $\vartheta$ is the axial velocity of A, L is length of the tube.
Boundary condition is

$$\frac{dy(1)}{dz} = 0 \tag{7}$$

Equation (5) with (7) represents boundary value problem.

## 4  Solution of the Problem

We introduce following notation

$$z = 1 - s, y = 1 - fz = 1 - s$$

Then (5) is transformed to

$$f''(s) = -f'(s) + f(s) - 1 \tag{8}$$

where $f(s) \in E^1$.
First N = 1 and R = 1 is assumed.
To solve fuzzy differential Eq. (8) we used α- cut approach, well known from fuzzy calculus [7].
α- cut of (8) is presented as

$$f''(s, ) = -f'(s, \alpha) + f(s, \alpha) - 1$$

and can be written as

$$[f_l''s, \alpha), f_r''s, \alpha)] = [f_l(s, \alpha), f_r(s, \alpha)] - [f_l'(s, \alpha), f_r'(s, \alpha)] - 1$$

Then

$$f_l''(s, \alpha) = f_l(s, \alpha) - f_l'(s, \alpha) - 1, f_r''s, \alpha) = f_r(s, \alpha) - f_r'(s, \alpha) - 1$$

If substituted $f_l(s, \alpha) = x_1$, $f_l'(s, \alpha) = x_2$, then

$$\dot{x}_1 = x_2,$$
$$\dot{x}_2 = x_1 - x_2 - 1.$$
$$A_l = \begin{pmatrix} 0 & 1 \\ 1 & -1 \end{pmatrix}, X_l = \begin{pmatrix} x_1 \\ x_2 \end{pmatrix}, \dot{X}_l = \begin{pmatrix} \dot{x}_1 \\ \dot{x}_2 \end{pmatrix}, U_l = \begin{pmatrix} 0 \\ -1 \end{pmatrix},$$
$$\dot{X}_l = A_l * X_l + U_l$$

If substituted $f_r(s, \alpha) = x_3$, $f_r'(s, \alpha) = x_4$, then

$$\dot{x}_3 = x_4$$
$$\dot{x}_4 = x_3 - x_4 - 1.$$
$$A_r = \begin{pmatrix} 0 & 1 \\ 1 & -1 \end{pmatrix}, X_r = \begin{pmatrix} x_3 \\ x_4 \end{pmatrix}, \dot{X}_r = \begin{pmatrix} \dot{x}_3 \\ \dot{x}_4 \end{pmatrix}, U_r = \begin{pmatrix} 0 \\ -1 \end{pmatrix},$$
$$\dot{X}_r = A_r * X_r + U_r$$
$$Al = \begin{pmatrix} 0 & 1 \\ 1 & -1 \end{pmatrix}; Ar = \begin{pmatrix} 0 & 1 \\ 1 & -1 \end{pmatrix};$$
$$Ul = \begin{pmatrix} 0 \\ -1 \end{pmatrix}; Ur = \begin{pmatrix} 0 \\ -1 \end{pmatrix}.$$
$$X0l = \begin{pmatrix} -0.53 \\ 1.5 \end{pmatrix}, X0r = \begin{pmatrix} 0.53 \\ -0.2 \end{pmatrix}.$$

Fl = MatrixExp[Al * t]=

$$\{\{\frac{1}{10} (5e^{\frac{1}{2}(-1-\sqrt{5})t} - \sqrt{5}e^{\frac{1}{2}(-1-\sqrt{5})t} + 5e^{\frac{1}{2}(-1+\sqrt{5})t} + \sqrt{5}e^{\frac{1}{2}(-1+\sqrt{5})t}),$$
$$-\frac{e^{\frac{1}{2}(-1-\sqrt{5})t} - e^{\frac{1}{2}(-1+\sqrt{5})t}}{\sqrt{5}}\}, \{-\frac{e^{\frac{1}{2}(-1-\sqrt{5})t} - e^{\frac{1}{2}(-1+\sqrt{5})t}}{\sqrt{5}},$$
$$\frac{1}{10} (5e^{\frac{1}{2}(-1-\sqrt{5})t} + \sqrt{5}e^{\frac{1}{2}(-1-\sqrt{5})t} + 5e^{\frac{1}{2}(-1+\sqrt{5})t} - \sqrt{5}e^{\frac{1}{2}(-1+\sqrt{5})t})\}\}$$

Fr = MatrixExp[Ar * t]=

$$\{\{\frac{1}{10}(5e^{\frac{1}{2}(-1-\sqrt{5})t} - \sqrt{5}e^{\frac{1}{2}(-1-\sqrt{5})t} + 5e^{\frac{1}{2}(-1+\sqrt{5})t} + \sqrt{5}e^{\frac{1}{2}(-1+\sqrt{5})t}),$$

$$-\frac{e^{\frac{1}{2}(-1-\sqrt{5})t} - e^{\frac{1}{2}(-1+\sqrt{5})t}}{\sqrt{5}}\} : \{-\frac{e^{\frac{1}{2}(-1-\sqrt{5})t} - e^{\frac{1}{2}(-1+\sqrt{5})t}}{\sqrt{5}},$$

$$\frac{1}{10}(5e^{\frac{1}{2}(-1-\sqrt{5})t} + \sqrt{5}e^{\frac{1}{2}(-1-\sqrt{5})t} + 5e^{\frac{1}{2}(-1+\sqrt{5})t} - \sqrt{5}e^{\frac{1}{2}(-1+\sqrt{5})t})\}\}$$

Fls = MatrixExp[Al * s]/.s → t − s

$$\{\{\frac{1}{10}(5e^{\frac{1}{2}(-1-\sqrt{5})(-s+t)} - \sqrt{5}e^{\frac{1}{2}(-1-\sqrt{5})(-s+t)} + 5e^{\frac{1}{2}(-1+\sqrt{5})(-s+t)} + \sqrt{5}e^{\frac{1}{2}(-1+\sqrt{5})(-s+t)}),$$

$$-\frac{e^{\frac{1}{2}(-1-\sqrt{5})(-s+t)} - e^{\frac{1}{2}(-1+\sqrt{5})(-s+t)}}{\sqrt{5}}\}, \{-\frac{e^{\frac{1}{2}(-1-\sqrt{5})(-s+t)} - e^{\frac{1}{2}(-1+\sqrt{5})(-s+t)}}{\sqrt{5}},$$

$$\frac{1}{10}(5e^{\frac{1}{2}(-1-\sqrt{5})(-s+t)} + \sqrt{5}e^{\frac{1}{2}(-1-\sqrt{5})(-s+t)} + 5e^{\frac{1}{2}(-1+\sqrt{5})(-s+t)} - \sqrt{5}e^{\frac{1}{2}(-1+\sqrt{5})(-s+t)})\}\}$$

Frs = MatrixExp[Ar * s]/.s → t − s

$$\{\{\frac{1}{10}(5e^{\frac{1}{2}(-1-\sqrt{5})(-s+t)} - \sqrt{5}e^{\frac{1}{2}(-1-\sqrt{5})(-s+t)} + 5e^{\frac{1}{2}(-1+\sqrt{5})(-s+t)} + \sqrt{5}e^{\frac{1}{2}(-1+\sqrt{5})(-s+t)}),$$

$$-\frac{e^{\frac{1}{2}(-1-\sqrt{5})(-s+t)} - e^{\frac{1}{2}(-1+\sqrt{5})(-s+t)}}{\sqrt{5}}\}, \{-\frac{e^{\frac{1}{2}(-1-\sqrt{5})(-s+t)} - e^{\frac{1}{2}(-1+\sqrt{5})(-s+t)}}{\sqrt{5}},$$

$$\frac{1}{10}(5e^{\frac{1}{2}(-1-\sqrt{5})(-s+t)} + \sqrt{5}e^{\frac{1}{2}(-1-\sqrt{5})(-s+t)} + 5e^{\frac{1}{2}(-1+\sqrt{5})(-s+t)} - \sqrt{5}e^{\frac{1}{2}(-1+\sqrt{5})(-s+t)})\}\}$$

Sl = Fl.X0l + Integrate[Fls.Ul, {s, 0, t}]=

$$\{\{1. - 0.4362980073075309e^{\frac{1}{2}(-1+\sqrt{5})t} - 1.0937019926924691e^{-\frac{1}{2}(1+\sqrt{5})t}\},$$

$$\{e^{-\frac{1}{2}(1+\sqrt{5})t}(1.769646997739904 - 0.26964699773990397e^{\sqrt{5}t})\}\}$$

Sr = Fr.X0r + Integrate[Frs.Ur, {s, 0, t}]=

$$\{\{0.9999999999999999 - 0.42953791404248176e^{\frac{1}{2}(-1+\sqrt{5})t}$$

$$- 0.040462085957518296e^{-\frac{1}{2}(1+\sqrt{5})t}\},$$

$$\{e^{-\frac{1}{2}(1+\sqrt{5})t}(0.06546903033498441 - 0.2654690303349845e^{\sqrt{5}t})\}\}$$

Geometrical representation of solution is shown in Fig. 1.

As fundamental matrix solution Fl and Fr are bounded the reaction system is Lipschitz stable.

Now assume that N and R changed as

$R_L = 0.5$, $R_r = 1.5$, $N_L = 0.5$, $N_r = 1.5$

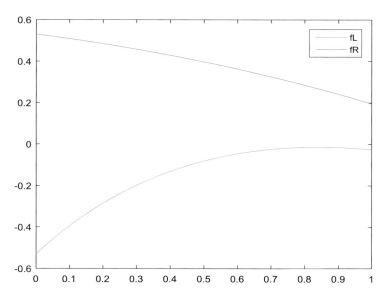

**Fig. 1.** Solution of dynamics of the reaction in accordance with (8)

Solution shows that dynamics of the reaction is unstable. Figure 2 illustrates unstable behavior of dynamics of considered isothermic reaction.

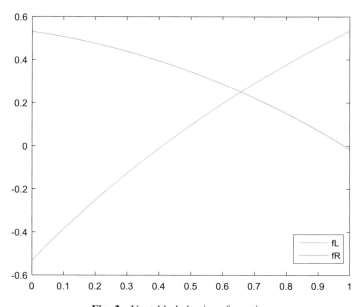

**Fig. 2.** Unstable behavior of reaction

## 5  Conclusion

Research results on modeling chemical reactions with uncertain parameters and variables are very scarce. For first time in this paper we suggest the approach to investigate dynamics of first-order isothermic reactions and its stability, described by fuzzy differential equation. Computer simulations have proved validity of the suggested approach.

## References

1. Can, E., Bayrak, M.A.: A new method for solution of fuzzy reaction equation. Commun. Math. Comput. Chem. **73**, 649–661 (2015). Claire Vallance. Reaction kinetics. 31 p. http://vallance.chem.ox.ac.uk/pdfs/KineticsLectureNotes.pdf
2. Erturk, V.: Differential transform method for solving a boundary value problem arising in chemical reactor theory. In: Conference Mathematical Modelling and Analysis (2017). inga.vgtu.lt/ ~ art/konf/programme_detailed.php
3. Modeling of CSTR process, 16 p. http://shodhganga.inflibnet.ac.in/bitstream/10603/142921/10/10_chapter%202.pdf
4. Ross, T.J., Booker, J.M., Parkinson, W.J.: Fuzzy Logic and Probability Applications: Bridging the Gap. Series: ASA-SIAM Series on Statistics and Applied Mathematics, pp. xxiii + 409 (2002)
5. Aliev, R.A.: Fundamentals of the Fuzzy Logic-Based Generalized Theory of Decisions, 322 p. Springer, Heidelberg (2013)
6. Beg, I.: Fuzzy multivalued functions (2011). https://wenku.baidu.com/view/71a84c136c175f0e7cd1372d.html
7. Aliev, R.: Uncertain Computation-Based Decision Theory, 521 p. World Scientific Publishing (2018)

# Prioritization of Mosque Facility Site Selection Criteria Under Fuzzy Environment

Abdurrahman Yagmur Toprakli[1]([⊠]), Mehmet Kabak[2],
Eren Özceylan[3], and Aylin Adem[4]

[1] Architecture Department, Gazi University, 06570 Ankara, Turkey
toprakli@gazi.edu.tr
[2] IE Department, Gazi University, 06570 Ankara, Turkey
[3] IE Department, Gaziantep University, 27310 Gaziantep, Turkey
[4] Department of Administration and Organization, Gazi University,
Ankara, Turkey

**Abstract.** Mosques are important buildings to conduct daily worship together with the community, which is precious to the belief in Islam. Simultaneously, architectural standing of the mosques is important to embellish the city landscape. The place of the mosques, which is very important both in terms of religious and public use, is an important decision that must be specified carefully. The location of the mosque should correspond to the needs of the public and at the same time should not distort the architectural structure of the city. The needs of the city can be identified as both physical and social needs. Accordingly, this paper deals with the prioritization of the criteria affecting the selection of the mosque location. Criteria affecting the choice of Mosque location was adapted from Abu Dhabi Mosque Development Regulations Planning section as foremost developed instance for Turkey. Prioritization of the selection criteria of the mosque was carried out under fuzzy environment. In decision matrices where, triangular fuzzy numbers were employed; criteria weights were determined by geometric mean and alpha cut methods.

**Keywords:** Site selection · Mosque · Prioritization
Analytic hierarchy process · Fuzzy logic · Multi-criteria

## 1 Introduction

Mosques are significant buildings to conduct daily worship together with the community, which is central to the belief in Islam. It is known that many people use the mosque to perform prayer worship five times a day in Muslim communities. Criteria affecting the location of the mosque can be expressed in a wide range from people's access to mosque to physical requirements. Simultaneously, architectural standing of mosques is highly regarded in Muslim communities in terms of the city scenery. The place of the mosques, which is very important both in terms of religious and public use, is an important decision that must be prearranged prudently. The location of the mosque should correspond to the needs of the city and at the same time should not misrepresent the architectural assembly of the city. The needs of the city can be

© Springer Nature Switzerland AG 2019
R. A. Aliev et al. (Eds.): ICAFS-2018, AISC 896, pp. 451–457, 2019.
https://doi.org/10.1007/978-3-030-04164-9_60

classified as both physical and social needs. Criteria affecting the choice of Mosque location was adapted from Abu Dhabi Mosque Development Regulations [1] Planning section as the most developed instance and adapted for Turkey. In the relevant regulatory report, the construction stages of a mosque are covered under three headings (Planning-Design-Operation). Location selection criteria determined in this study were adapted from the planning part of the related document. Determined criteria were given as follows: site suitability for construction, accessibility, environmental issues, site cost, suitability in terms of social life and finally physical requirements. Sub-criteria of these criteria have been determined and detailed explanations have been made in the application part of the paper. This issue has been chosen because the selection of the mosque location is important in many aspects and it is desired to use it for a new mosque construction decision for Ankara, the capital of Turkey. Therefore, this study deals with the prioritization of the criteria affecting the selection of the mosque location. The problem of site selection in the multi criteria decision-making (MCDM) literature has been studied frequently and is a popular subject in academic terms [2–5]. In this study, prioritization of the selection criteria of the mosque was carried out under fuzzy environment. In decision matrices where, triangular fuzzy numbers were employed; criteria weights were determined by geometric mean and alpha cut methods. Detailed information of the method implementation was given in the related section.

The paper is organized as follows; after the introduction, in the second part steps of methods, which are used in this study, are explained. In Sect. 3 the proposed approach is given while in the 4th section results of the study is showed. Finally, in Sect. 4 the conclusions of the study are interpreted.

## 2   The Fuzzy Set Theory and Analytic Hierarchy Process

Individuals feel more comfortable to express their evaluation in linguistic terms instead of exact numeric values when their evaluation or assessment are required. Also, human judgment on qualitative attributes is always idiosyncratic and imprecise. Therefore, the fuzzy set theory is generally used in human judgements to solve decision-making

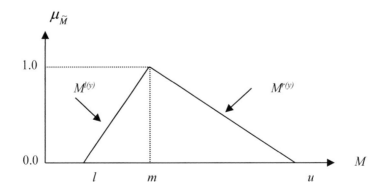

**Fig. 1.**  A triangular fuzzy number, $\tilde{M}$

problems. Fuzzy numbers expand on the idea of the confidence interval and are defined over a fuzzy subset of real numbers. A triangular fuzzy number (TFN) shown in Fig. 1 is a type of fuzzy number [6]. Detailed information for fuzzy set theory could be easily find in different papers by researchers and is not explained in this paper because of page restraint [7–10].

The Analytic Hierarchy Process (AHP) was developed by Thomas L. Saaty in 1970s and it is a practical MCDM method to calculate weights/priorities. The AHP can be simply applied for decision-making problems by researchers and does not require complex information. It allows individual or group decision-making process also, quantitative and qualitative values can be used in a pairwise comparison and criteria weights are calculated by pairwise comparison of decision makers [11, 12]. The readers can find details about the steps of AHP in different papers [13–15]. In this study, the fuzzy AHP approach is applied to determine criteria weights for mosque facility site selection. Please see Table 1 for the linguistic scale used in the current study.

**Table 1.** Linguistic scale

| Linguistic scale for importance | Triangular fuzzy scale |
|---|---|
| Just equal | (1, 1, 1) |
| Equally important (EI) | (1/2, 1, 3/2) |
| Weakly more important (WMI) | (1, 3/2, 2) |
| Strongly more important (SMI) | (3/2, 2, 5/2) |
| Very strongly more important (VSMI) | (2, 5/2, 3) |
| Absolutely more important (AMI) | (5/2, 3, 7/2) |

## 3 Application

In this part of the study, selection criteria of the mosque location were prioritized to gain more accurate results for the new mosque construction process in Ankara. Figure 2 shows the adapted criteria from Abu Dhabi Mosque Development Guidelines to Turkish circumstances.

Determined main criteria and their meaning are given as follows:

- Site Suitability for Construction: This criterion refers to the substructure and formal suitability for construction of the selected area for mosque construction. For example, the maximum height allowed for that region.
- Accessibility: What is meant by accessibility is that the mosques are easily reachable for pedestrians and public transport users. At the same time, this criterion also means to reach believers in a safe way to the mosque.
- Environmental Issues: Environmental Concerns is mostly related to city landscape and urban architecture.
- Site Cost: This criterion is mostly being relevant that some locations are more expensive because they are in the city's preferred places.

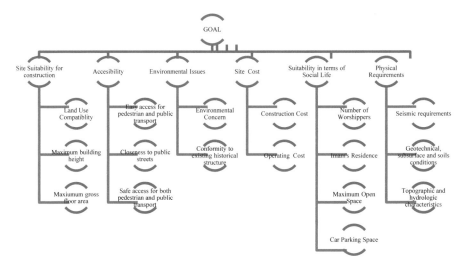

**Fig. 2.** The hierarchical decision model of site selection for mosques

- Suitability in terms of Social Life: Another criterion that influences the choice of mosque location is that the region concerned is in harmony with social life. For example; the number of worshippers, car parking requirements etc. In addition to this, mosque should be in a distance to alcohol consumed places such as restaurants.
- Physical Requirements: Physical Requirements generally refer to the seismic attitude of soil, geotechnical aspects etc.

Figure 2 shows that the hierarchical decision model of site selection for mosques. Similarly, Table 2 shows linguistic evaluations taken from the decision maker. After

**Table 2.** Experts' pairwise comparison matrix for main criteria

|    | C1 | C2 | C3 | C4 | C5 | C6 |
|----|------|-----------|-----------|-----------|-----------|-----------|
| C1 | Just equal | Equally important | Strongly more important | Just equal | Weakly more important | Strongly more important |
| C2 |    | Just equal | Strongly more important | Equally important | Strongly more important | Strongly more important |
| C3 |    |    | Just equal | Very strongly more important | Just equal | Just equal |
| C4 |    |    |    | Just equal | Very strongly more important | Just equal |
| C5 |    |    |    |    | Just equal | Just equal |
| C6 |    |    |    |    |    | Just equal |

the linguistic evaluations were taken from decision maker, the main criteria weights were obtained by the geometric mean method and the alpha cut method (please see Table 3). Similar calculations were performed for sub-criteria groups, and then global weights of sub-criteria were found (Please see, Table 3).

**Table 3.** Main criteria weights

| Main Criteria | Site Suitability For Construction | | | Accesibility | | | Environmental Issues | | | Site Cost | | | Suitability in terms of social life | | | Physical Requirements | | | Weights |
|---|---|---|---|---|---|---|---|---|---|---|---|---|---|---|---|---|---|---|---|
| Site Suitability For Construction | 1.00 | 1.00 | 1.00 | 0.50 | 1.00 | 1.50 | 1.50 | 2.00 | 2.50 | 1.00 | 1.00 | 1.00 | 1.00 | 1.50 | 2.00 | 1.50 | 2.00 | 2.50 | 0.215 |
| Accesibility | 0.67 | 1.00 | 2.00 | 1.00 | 1.00 | 1.00 | 1.50 | 2.00 | 2.50 | 0.50 | 1.00 | 1.50 | 1.50 | 2.00 | 2.50 | 1.50 | 2.00 | 2.50 | 0.231 |
| Environmental Issues | 0.40 | 0.50 | 0.67 | 0.40 | 0.50 | 0.67 | 1.00 | 1.00 | 1.00 | 2.00 | 2.50 | 3.00 | 1.00 | 1.00 | 1.00 | 1.00 | 1.00 | 1.00 | 0.148 |
| Site Cost | 1.00 | 1.00 | 1.00 | 0.67 | 1.00 | 2.0 | 0.33 | 0.40 | 0.50 | 1.00 | 1.00 | 1.00 | 2.00 | 2.50 | 3.00 | 1.00 | 1.00 | 1.00 | 0.162 |
| Suitability in terms of social life | 0.50 | 0.67 | 1.00 | 0.40 | 0.50 | 0.67 | 1.00 | 1.00 | 1.00 | 0.30 | 0.40 | 0.50 | 1.00 | 1.00 | 1.00 | 1.00 | 1.00 | 1.00 | 0.115 |
| Physical Requirements | 0.40 | 0.50 | 0.67 | 0.40 | 0.50 | 0.67 | 1.00 | 1.00 | 1.00 | 1.00 | 1.00 | 1.00 | 1.00 | 1.00 | 1.00 | 1.00 | 1.00 | 1.00 | 0.127 |

According to Table 3, the most important criterion, which effects mosque location selection, is accessibility. Criteria's order of importance is given as follows: accessibility, site suitability for construction, site cost, environmental issues, physical requirements, suitability in terms of social life.

As seen in Table 4, the most important first three sub-criteria determined as construction cost, land use compatibility, closeness to the public street. The least important three sub-criteria are determined as Seismic requirements; Geotechnical, subsurface, and soils conditions; Topographic and hydrologic characteristics.

**Table 4.** Sub-criteria's local and global weights

| | Sub-criteria local weight | Main criteria weight | Sub-criteria global weight |
|---|---|---|---|
| C11 | 0.412 | 0.215 | 0.089 |
| C12 | 0.261 | | 0.056 |
| C13 | 0.327 | | 0.070 |
| C21 | 0.330 | 0.231 | 0.076 |
| C22 | 0.377 | | 0.087 |
| C23 | 0.292 | | 0.068 |
| C31 | 0.500 | 0.148 | 0.074 |
| C32 | 0.500 | | 0.074 |
| C41 | 0.595 | 0.162 | 0.097 |
| C42 | 0.405 | | 0.066 |
| C51 | 0.378 | 0.115 | 0.044 |
| C52 | 0.240 | | 0.028 |
| C53 | 0.191 | | 0.022 |
| C54 | 0.191 | | 0.022 |
| C61 | 0.323 | 0.127 | 0.041 |
| C62 | 0.438 | | 0.056 |
| C63 | 0.239 | | 0.030 |

## 4   Results and Conclusion

Mosques are central buildings of social life in Muslim communities and to conduct worship together with the community is highly valued in the Islamic belief. Mosques are also highly respected structures for Islamic city. It is a difficult procedure to decide the location of the mosque both in terms of physical and social needs.

In this study, criteria, which provide the all-embracing direction for the planning of mosques in Ankara, are studied in terms of order of importance to inform the decision-making process throughout the planning, design and assessment of a sample development. As a result of our study, the easy access to the mosque, which many people need to perform their worship at five times of the day, has been identified as the most important criterion affecting the selection of the mosque site. Thus, Mosques are to be located to ensure they are easily accessible by worshippers travelling on foot or by public transport. The cost of construction work has also been shown to have a very significant place in terms of sub-criteria, and it has been identified as the most important sub-criterion. The results obtained from the study of prioritizing the mosque location selection criteria, which is a very frequently studied field in the literature, shed light on the decision makers to determine the location of the new mosques.

## References

1. Abu Dhabi Mosque Development Guidelines. https://www.upc.gov.ae/en/publications/manuals-and-guidelines/abu-dhabi-mosque-development-regulations. Accessed 24 July 2018
2. Darani, S.K., Eslami, A.A., Jabbari, M., Asefi, H.: Parking lot site selection using a fuzzy AHP-TOPSIS framework in Tuyserkan, Iran. J. Urban Plan. Dev. **144**(3) (2018)
3. Deveci, M., Canıtez, F., Gökaşar, I.: WASPAS and TOPSIS based interval type-2 fuzzy MCDM method for a selection of a car sharing station. Sustain. Cities Soc. **41**, 777–791 (2018)
4. Wu, Y., Liu, L., Gao, J., Chu, H., Xu, C.: An extended VIKOR-based approach for pumped hydro energy storage plant site selection with heterogeneous information. Information **8**(3), 106 (2017)
5. Villacreses, G., Goana, G., Martinez-Gomez, J., Jjion, D.J.: Wind farms suitability location using geographical information system (GIS), based on multi-criteria decision making (MCDM) methods: The case of continental Ecuador. Renew. Energy **109**, 275–286 (2017)
6. Facchinetti, G., Ricci, R.G.: A characterization of a general class of ranking functions on triangular fuzzy numbers. Fuzzy Sets Syst. **146**(2), 297–312 (2004)
7. Zimmermann, H.J.: Fuzzy set theory. Wiley Interdisc. Rev.: Comput. Stat. **2**(3), 317–333 (2010)
8. Kabak, M., Dağdeviren, M., Burmaoğlu, S.: A hybrid SWOT-FANP model for energy policy making in Turkey. Energy Sources Part B-Econ. Plan. Policy **11**, 6 (2016)
9. Taylan, O., Kaya, D., Demirbas, A.: An integrated multi attribute decision model for energy efficiency processes in petrochemical industry applying fuzzy set theory. Energy Convers. Manag. **117**, 501–512 (2016)
10. Ebrahimnejad, A., Verdegay, J. L.: Fuzzy set theory. In: Fuzzy Sets-Based Methods and Techniques for Modern Analytics. Springer, Cham (2018)

11. Saaty, T.L.: A scaling method for priorities in hierarchical structures. J. Math. Psychol. **15** (3), 234–281 (1977)
12. Saaty, T.L.: The Analytic Hierarchy Process. McGraw-Hill International, New York (1980)
13. Saaty, T. L.: Analytic hierarchy process. In: Encyclopedia of Operations Research and Management Science. Springer, Boston (2013)
14. Vargas, L.G.: An overview of the analytic hierarchy process and its applications. Eur. J. Oper. Res. **48**(1), 2–8 (1990)
15. Govindan, K., Kaliyan, M., Kannan, D., Haq, A.N.: Barriers analysis for green supply chain management implementation in Indian industries using analytic hierarchy process. Int. J. Prod. Econ. **147**, 555–568 (2014)

# Statistical Computation of the Effect of Using Mobile Applications as a Travel Information Tool

Huseyin Bicen[✉] and Ahmet Arnavut

Near East University, Nicosia, Cyprus
{huseyin.bicen,ahmet.arnavut}@neu.edu.tr

**Abstract.** Today, mobile users have many advantages in terms of communication and many other aspects, and they are able to make plans through applications on their mobile devices. Consequently, many applications have been developed for travel purposes. In particular, applications called peer-to-peer accommodation rentals such as Airbnb allow people to rent their property to tourists who intend to visit their country. Through the UBER application, it is possible to communicate through a smart device with the nearest available driver using GPS and subsequently obtain services. By using social media applications, individuals are now also influenced by the travel plans of the people they follow. In this study, the aim is to examine student opinions about how mobile applications will have an impact if used in terms of information and travel planning. Even though these devices play an important role in the lives of the 123 students participating in the study, they are hesitant in using the mobile applications related to travelling. This means that if users are trained in the use of secure mobile applications, they will have the opportunity to take advantage of these applications more easily. The study was a quantitative study and data analysis was performed with the SPSS program. The data were analyzed and interpreted by tabulating frequency, percentages, and Anova tests. Future studies will provide training for students on the advantages and disadvantages of applications used for travel and how they can be used safely.

**Keywords:** Statistical computation · Mobile applications · Information

## 1 Introduction

Today, mobile technologies can guide us in many aspects of our lives. Mobile technologies, which have rapidly gained momentum and provide convenience to our lives every day, are increasingly being used as a means of entertainment and information. [1]. Around the world, the age range of people that use the most mobile technologies has been determined by studies as 18–29 years old. [8]. Commonly known as "Mobile Apps", mobile applications are popularized and developed according to many mobile device operating systems. [4]. By 2022, smartphone usage will have increased further and an estimated 5.5 billion people are expected to be using such devices. [11]. The intensive use of mobile technologies and smartphones has also led to new trends in the tourism sector and many mobile applications have been developed accordingly.

© Springer Nature Switzerland AG 2019
R. A. Aliev et al. (Eds.): ICAFS-2018, AISC 896, pp. 458–463, 2019.
https://doi.org/10.1007/978-3-030-04164-9_61

Peer-to-peer accommodation, which is also known as airplane renting of properties by people with the application of Airbnb application tourism has caused a new direction. [3, 6, 9]. Additionally, the mobile taxi booking application UBER allows passengers to reach the nearest driver in their surroundings through GPS and mobile applications. [2, 12] The effects that social media have on people and the photos they share in their Instagram accounts can also influence many people's travel plans. [7]. In addition to these applications, many airlines allow their customers to earn points by enabling their members to participate in Airline loyalty programs through their own mobile applications, allowing them to plan for their next flight. [5]. In this study, students' opinions were taken regarding mobile travel applications.

## 2   Aim of the Research

In this study, the aim was obtaining student opinions about how mobile applications will have an impact if used in terms of information and travel planning.

### 2.1   Participants

The participants in the research consisted of 123 volunteer students who were taking History, Turkish and English lessons at the Near East University Distance Education Center via the distance education method using the Moodle system. This study was conducted in the spring term of the 2017–2018 academic year.

**Gender**
Table 1 shows the gender distribution of the students. As seen in Table 1, 46% of the students (f = 57) were male students while 54% (f = 66) were female students.

**Table 1.**  Gender distribution

| Gender | F | % |
|--------|-----|-----|
| Female | 66 | 54 |
| Male | 57 | 46 |
| Total | 123 | 100 |

**Age**
Table 2 shows the age distribution of students. As shown in Table 2, 10% (f = 12) of the students were 18 years old, 21% (f = 26) were 19 years old, 25% (f = 31) were 20 years old, 14% (f = 17) were 21 years old, 11% (f = 13) were 22 years old, 8% (f = 10) were 23 years old, 4% (f = 5) were 24 years old and 7% (f = 9) were students who were 25 years or older.

**Table 2.** Age distribution

| Age | F | % |
|-----|----|----|
| 18 | 12 | 10 |
| 19 | 26 | 21 |
| 20 | 31 | 25 |
| 21 | 17 | 14 |
| 22 | 13 | 11 |
| 23 | 10 | 8 |
| 24 | 5 | 4 |
| 25+ | 9 | 7 |

## 2.2  Instruments

The survey used in the research consists of 20 positive phrases about the impact of the use of mobile applications as a travel information tool developed by the researchers. The Cronbach's alpha of the questionnaire prepared according to the 5-point Likert type was 0.93.

## 2.3  Data Analysis

The data of the study were analyzed by SPSS program and the data were tabulated and interpreted as frequency, percentage and average.

# 3  Results and Discussion

This section contains the average values of the opinions about then students' daily usage habits of smart devices and how they will have a positive effect when used in terms of travel planning for mobile applications in order to shed light on future travel possibilities.

**Daily Mobile Device Usage Times**

Table 3 shows the daily usage hours of mobile devices used by the students. As shown in Table 3, while 65% of students use mobile devices more than 8 h a day (f = 80), 27% use them between 4–8 h (f = 33) and 8% between 1–3 h (10).

**Table 3.** Daily mobile device usage times

| Hours | f | % |
|-------|----|----|
| 1–3 h | 10 | 8 |
| 4–8 h | 33 | 27 |
| More than 8 h | 80 | 65 |

**Student Opinions on Mobile Applications as a Travel Information Tool**
Table 4 reviews the average and standard deviation of student opinions for mobile applications as a travel information tool.

**Table 4.** Student opinions on mobile applications as a travel information tool

| Items | Mean | StD. |
|---|---|---|
| 1. With the Airbnb application, I can find the cheapest place to stay. | 2.86 | 1.04 |
| 2. With the Airbnb application, I find it safe to make reservations. | 2.86 | 1.04 |
| 3. With the Airbnb application, I am make suggestions on where to stay for others | 2.70 | 1.07 |
| 4. With the Airbnb application I can open my own property for reservations | 2.66 | 1.06 |
| 5. I believe that the Airbnb application is safer than other applications | 2.82 | .97 |
| 6. I prefer to use Ubers rather than taxis in foreign countries I visit | 2.90 | 1.04 |
| 7. I prefer Ubers rather than taxis because it is safer | 2.91 | 1.03 |
| 8. I use Ubers because it is cheaper than taxis | 3.00 | 1.11 |
| 9. I recommend the Uber application to my friends | 3.03 | 1.09 |
| 10. The photos my friends share on Instagram grasp my attention | 2.93 | 1.21 |
| 11. My friends' posts about their travels play a big role in my travel plans | 2.91 | 1.21 |
| 12. I follow their Instagram stories before I visit a country | 3.03 | 1.30 |
| 13. The comments on Instagram of those who visited those places while traveling are important to me. | 2.92 | 1.17 |
| 14. I like to explore the places I am going to visit from tags | 3.00 | 1.15 |
| 15. Thanks to Instagram, I can easily meet other travelers in the country | 3.21 | 1.15 |
| 16. I trust Instagram posts about travel | 2.78 | 1.12 |
| 17. I use travel agents in order to travel more economically | 3.04 | 1.22 |
| 18. I participate in flight loyalty programs to get cheap air tickets | 2.94 | 1.18 |
| 19. I prefer to use the same apps to get more flight points | 3.01 | 1.31 |
| 20. I prefer to earn more points on the same mobile application on my hotel reservations | 3.08 | 1.26 |

According to these findings, the student's responses are as follows: With the Airbnb application, I can find the cheapest place to stay. (M = 2.86, SD = 1.04); With the Airbnb application, I find it safe to make reservations, (M = 2.86, SD = 1.04); With the Airbnb application, I can make suggestions on where to stay for others, (M = 2.70, SD = 1.07); With the Airbnb application, I can open my own property for reservations (M = 2.66, SD = 1.06); I believe that the Airbnb application is safer than other applications (M = 2.82, SD = .97); I prefer to use Ubers rather than taxis in foreign countries I visit (M = 2.90, SD = 1.04); I use Ubers because it is cheaper than taxis (M = 3.00, SD = 1.11), recommending uber services to their friends (M = 3.03, SD = 1.09); The photos my friends share on Instagram grasp my attention (M = 2.93 SD = 1.21); My friends' posts about their travels play a big role in my travel plans (M = 2.91, SD = 1.21); I follow their Instagram stories before I visit a country (M = 3.03, SD = 1.30); The comments on Instagram of those who visited those places while traveling are important to me. (M = 2.92, SD = 1.17); I like to explore the places I am going to visits from tags (M = 3.00, SD = 1.15); Thanks to Instagram, I can easily meet other travelers in the country (M = 3.21, SD = 1.15); I trust Instagram posts about travel (M = 2.78, SD = 1.12); I use travel agents in order to travel more economically (M = 3.04, SD = 1.22); I participate in flight loyalty programs to get cheap air tickets (M = 2.94, SD = 1.18); I prefer to use the same apps to get more flight points (M = 3.01, SD = 1.31) and I prefer to earn more points on the same mobile application on my hotel reservations is rated as (M = 3.08, SD = 1.26), which shows that the students are "neutral". [10] noted that the barrier in restricting Peer to Peer accommodation platforms is the problem of trust. According to the findings of this study, students who were against Airbnb, a Peer to Peer accommodation platform, also stated that they are hesitant about cheap accommodation or the availability of reservations.

## 4   Conclusions

Even though these devices hold an important place in the lives of the students participating in the study, they are hesitant in using the mobile applications about travelling. This means that if users are trained in the use of secure mobile applications, they will have the opportunity to take advantage of these applications more easily. In particular Airbnb, an application for peer-to-peer accommodation, has some uncertainty in terms of whether travelers will stay at cheap or safe accommodation. Users are also hesitant to use UBER as a taxi service in the countries to which they travel. Participants who stated that the people they follow in their Instagram accounts did not have a negative impact on their travel plans also expressed their hesitancy about participating in flight loyalty programs to obtain cheap airline tickets. These results reveal that while mobile users tend to get services through applications, they sometimes feel confident that they can make their travel plans according to the news they read or other user comments on their own travel plans, but they sometimes avoid using it. Future studies will provide training for students on the advantages and disadvantages of applications used for travel and how they can be used safely.

## References

1. Cavus, N.: Investigating mobile devices and LMS integration in higher education: Student perspectives. Procedia Comput. Sci. **3**, 1469–1474 (2011). https://doi.org/10.1016/j.procs. 2011.01.033
2. Chan, J.W., Chang, V.L., Lau, W.K., Law, L.K., Lei, C.J.: Taxi app market analysis in Hong Kong. J. Econ. Bus. Manage. **4**(3), 239–242 (2016). https://doi.org/10.7763/joebm.2016.v4. 397
3. Cheng, M.: Sharing economy: a review and agenda for future research. Int. J. Hosp. Manage. **57**, 60–70 (2016)
4. Garg, R., Telang, R.: Inferring app demand from publicly available data. MIS Q. (2012)

5. Jong, G.D., Behrens, C., Ommeren, J.V.: Airline loyalty (programs) across borders: a geographic discontinuity approach. Int. J. Ind. Organ. (2018). https://doi.org/10.1016/j.ijindorg.2018.02.005

6. Martin-Fuentes, E., Fernandez, C., Mateu, C., Marine-Roig, E.: Modelling a grading scheme for peer-to-peer accommodation: Stars for Airbnb. Int. J. Hosp. Manage. **69**, 75–83 (2018). https://doi.org/10.1016/j.ijhm.2017.10.016

7. Mukhina, K.D., Rakitin, S.V., Visheratin, A.A.: Detection of tourists attraction points using Instagram profiles. Procedia Comput. Sci. **108**, 2378–2382 (2017). https://doi.org/10.1016/j.procs.2017.05.131

8. Pew: Mobile fact sheet (2017). http://www.pewinternet.org/fact-sheet/mobile/. Accessed May 2018 from Pew Research Center

9. Sigala, M.: Collaborative commerce in tourism: implications for research and industry. Curr. Issues Tour. **20**(4), 346–355 (2015). https://doi.org/10.1080/13683500.2014.982522

10. Tussyadiah, I.P., Pesonen, J.: Drivers and barriers of peer-to-peer accommodation stay – an exploratory study with American and Finnish travellers. Curr. Issues Tour. **21**(6), 703–720 (2016). https://doi.org/10.1080/13683500.2016.1141180

11. WARC: 5.5bn people will use mobile devices by 2022, WARC (2017). https://www.warc.com/NewsAndOpinion/News/5.5bn_people_will_use_mobile_devices_by_2022/39021. Accessed 30 May 2018

12. Weng, G.S., Zailani, S., Iranmanesh, M., Hyun, S.S.: Mobile taxi booking application service's continuance usage intention by users. Transp. Res. Part D: Transp. Environ. **57**, 207–216 (2017). https://doi.org/10.1016/j.trd.2017.07.023

# Improvement in Strength of Radio Wave Propagation Outside the Coverage Area of the Mobile Towers for Cellular Mobile WiFi

Jamal Fathi$^{(\boxtimes)}$

Near East University, Nicosia, Turkish Republic of Northern Cyprus, Turkey
jamalfathi2004@gmail.com

**Abstract.** Base station of any mobile-phone is designed to provide coverage for more than one geographical area called cells. While the network, is made up of several base stations that are operating in conjunction with the adjacent base stations. Moreover, base stations have to be carefully analyzed and efficiently located in order to minimize the interference between cells with better signal qualities. Where, the most important point in cellular mobiles is the dropped in calls while downloading data. This research provides a new recommendation for repositioning the tower's places in order to provide more easily interface to replace the traditional methods for controlling the level of the signals. Utilizing the college buildings at Near East University (NEU) campus in the Northern part of Cyprus, the loss in WiFi propagation is studied outside the coverage areas. These buildings serve as good experimental settings because they exemplify typical signal dead spots, locations where little to no WiFi signal is available. In this study, we researched several ways of path loss propagation that spread between the base stations; we recognized and arranged these issues. We at that point applied our path loss propagation algorithmic model to demonstrate that signal quality is fundamentally enhanced and no loss in the signal. At last, we demonstrated the proficiency of the proposed positions and clarify the specifics of our framework.

**Keywords:** Lossy wifi · Path loss propagation · Signal strength
Radio wave · Towers

## 1 Introduction

Repeated delay in the movement of the signals for each one, by concentrating the beam of radial signals through the WiFi in the line of sight and the unique path. In Fig. 1 [1] and [4].

A directional radio wire, in a regular manner assigns the wire signals through the single drop within the same width of the beam signals [3].

All these depends on the power of the transmitter's signal which is $P_{tx}$ watts (W) in combination of the gain of the antenna Gt dBi, this gives the sum of the Effective Isotropic Radiated Power (EIRP) the power $P_{tx} \times G_{tx}$. Simply, EIRP and the log

© Springer Nature Switzerland AG 2019
R. A. Aliev et al. (Eds.): ICAFS-2018, AISC 896, pp. 464–471, 2019.
https://doi.org/10.1007/978-3-030-04164-9_62

**Fig. 1.** Library scene [1–3].

domain are mainly described in dB$_m$ P$_{tx}$ + G$_{tx}$. All this system, it can be represented by one equation that combines the log-domain and the radio signal by Eq. (1):

$$P_{rx} = P_{tx} + G_{tx} + G_{rx} - P_L \tag{1}$$

where, P$_{tx}$ and G$_{rx}$ are the received power at the receiver and the gain of the antenna in the transmitter's direction, while PL is the path loss.

This paper is organized in five sections as: first section represented radio wave propagation and its performance. Secondly, related research. While in section three, the proposed model is described. The forth section represents the obtained results. Finally, section five presents the obtained conclusion.

## 2 Related Research

The progress of downloading and uploading of the data through the WiFi is the important point of view, which deserves all the studies, so that there are some places there will be a loss in the WiFi signal. Moreover, increasing the number of base stations also has a widely disadvantage in the quality of the signal, so that there will be overlapping. The importance of security in exchanging the information in our daily life is widely increasing. Wireless Local Area Network (WLAN) is considered as the faster network, available, this gives the promotion to re-evaluate the uploaded and down-loaded information within the radio wave propagation in case of indoor through the Line of Sight (LOS) and the (NLOS). This progress of exchanging the information through radio waves using the WiFi between the transmitter and the receiver [5]. [6] Examined the overall collection of promulgation mensuration within 3.5 GHz in order to support the pertinence of the country case, residential case, and urbanized areas. As much as, rearranging the positions of the base stations is required in order to avoid overlapping process [7]. In [8], suggested a new model which delicate foretell the outdoor to indoor propagation loss; which relies on subordinate's angle of the losses that transfuse all indoor area. These approaches can be treated separately. Where, the channel loses between both base and mobile stations. [9] Upgraded and advanced the prediction of the losses while diffusion; the calculations are aimed on the improvement of the communication through the WiFis, allocation of wireless, and the internal loops.

[10] predestined the losses in the diffusions as the angles between the height and way, and then spotted the cyst within the assessment in order to the desired power with different rates of transmission in the IMT-2000. [11] Suggested a method which concentrates on the path between both transmitter and receiver in order to design wireless networks, and called it the prevailing method. As much as did comparison between propagations in different cases, one as WLAN in urbanized city-centers as multi floor structures, and the immediate shaft in the indoors. [12] Offered a new tactic with guaranteed results to rectify the networks in both urbanized and indoor as 3GPP Long Term Evolution (LTE), as much as predestined the strength of the signal's capacity in MIMO. [13] Adequate the deliberated saved data through the normal diffusion losses, managed the mensuration frequencies at 203.25 MHz and 583.25 MHz using several tracks in Ilorin City, as much as used minimal methods in order to scrutinized the attitude of the broadcasted TV signals within several floors buildings. [14] Offered diffusions within 5.725–5.825 GHz through the Unlicensed National Information Infrastructure (UNII) used in the US. Moreover, deliberated the propagation-losses in the domestically areas with frequency 5.8 GHZ, also unattached the information to LOS and NLOS, finally, gained remarkable conclusions ever after propagation that intended for both mobiles and PCs with lower-frequencies within narrow band channels. [15] Offered using Error Control Coding (ECC) within the WSNs in order to decide the efficiency of the energy used in WSNs. [16] Offered a frequency with 900 MHz with ultra-low power radio frequency to be used in WSNs, then expound these frequencies in order to carry them over 16 m using 20 kps. [17] Exercised an adjusted back-off strategy to attain completely steady chunk head allocation at the network. [18] Outlined the obtained works in the experiments in the field of attenuation of radio waves in the forest environments. [19] Theorized Silicon-on-insulator (SOI) visual wave-guides which underwrite the custody of the higher magnetic fields, without concentrating on the permeation of the harshness of the side wall while diffusion process. [20] Summarized all obtained results from the fineness of the direct ray, also the screed of the single path-loss, as much as the convenient of the Seidel Rappaport propagation which is obtained as a results of several analysis of 2.4 GHz, 802.11 g used in the Georgia Institute of Technology. [21] Used least square method, offered the obtained results by studying high frequency VHF 92.1 MHz radio wave propagation loss within the forest areas. [22] Investigated the differences in the obtained results in Stanford University Interim SUI, Okumara, Hata Cost-231, and Ericsson 9999 models, as much as got the more suitable model which is suitable for calculating the radio wave propagation for LTE. [23] Communicated the way path-loss in simple in order to utilize recipes as the aggregate of a far off ward way as a scourge factor, a floating capture, and a shadowing factor that limits the mean square mistake fit to the experimental information.

# 3    Proposed Model

The model utilized for this examination is an immediate beam, single way that considered as a change to [1–4, 24 and 25]. This model is reasonable for the considered models in all the studied models, as much as it's suitable for inside the lifts. This model, which computes the way in view of the transmitter to collector partition separately, and the number and sort of obstructions crossing the straight line between the transmitter and recipient. Moreover, it's reasonable for every climate condition. In this paper, overlapping points of the signal way as a special case inside lifts are incorporated as the proliferation influencing factors of the displayed condition. The condition for the signal way (in dB) utilized as a part of this investigation is then given in Eq. (2).

$$Y = 24.5 + 33.8log(d) + 4.0K_{floor} - 16.6S_{win} - 9.8G_{G/1} - 0.25A_{Elv} \qquad (2)$$

Taking into account the transmitter to beneficiary way appeared in Fig. 3, for this way, the demonstrated signal crosses the library working of the University of Bridgeport in [1] and [3] with its 7 stories with the base floor frameworks, and 2 foliage limits as appeared in Fig. 1.

$P_L$ characterizes the measure of signal quality with the lost among proliferation from transmitter to beneficiary. Free space is assorted on recurrence and separation. Equation (1) is utilized in order to describe the calculations of the path loss (Fig. 2).

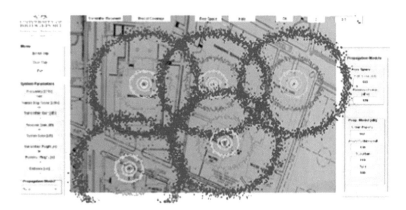

**Fig. 2.** Results

The used parameters are classified in Table 1.

**Table 1.** Used parameters [1–3]

| Parameter | Value |
|---|---|
| Frequency | 1800 MHz |
| Transmitted power | 46 dBm |
| Transmitted gain | 16 dBi |
| System loss | 3 dB |
| Transmitter height | 30 m |
| Received height | 1.5 m |
| Distance | 15 km |

The arranged parameters were utilized to obtain Fig. 3.

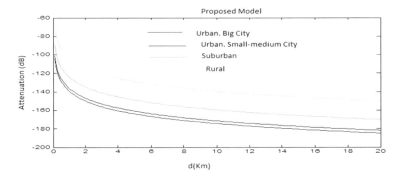

**Fig. 3.** Obtained outcome

## 4 Simulation Results

Figure 4 demonstrates the areas of every single estimated point incorporated into this examination. As beforehand said, every estimation area may contain estimated values, and subsequently, 810 aggregate way path-loss estimations have been incorporated into the examination from 35 base stations.

**Fig. 4.** Strenuous base-stations

Management of a foreordained number of relationship by any base-station consequently in regions of fame, this requires extra radio wires as a portion of the time added in order to transmit and get more calls and extra services for mobile-stations, in some cases, its required to nominate base-station. This infers incalculable stations are required to empower extra clients especially moving stations. This will offer clients more coverage and extra number of clients. In any case, this paper is presenting the overlapping caused by adding extra stations, so that the proposed model gives us the new arrangements of the stations in order to eliminate the overlapping process. This process of elimination is done by adding splitters for each station, which will save time instead of cancelling all working plans of the available base-stations, and the additional of the extension zone so that there will be no dis-accessibility as showed up in Fig. 5 and included by yellow shading (Fig. 6).

**Fig. 5.** Base station scan in/out the campus

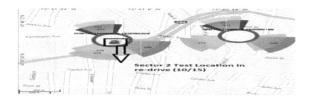

**Fig. 6.** Available location of the base stations in/out the campus

In mobile system, the reception apparatus destinations should be closer, where individuals request to utilize in order to get continuous connection while uploading and downloading process. Because requiring the changes in the available system, sounds impossible to be re-arranged, and the growth of the technology is considered as Incredible, as much as this growth, the needs of uploading and downloading the required information from the cloud driver are proportionally increasing. The proposed system solved more than what was required for a permanent of time. This is depicted in Fig. 7.

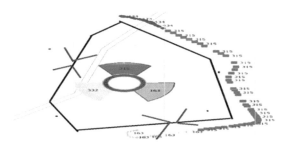

**Fig. 7.** Proposed base station location

## 5   Conclusion

Similar examination demonstrated earlier using multipath and NLOS condition in Urban area, while all models encounters higher way base-station emerge from rural basic scopes. Moreover, no any single model that might be prescribed for all conditions together.

The previous models were spotted mostly in urbanized areas, Ericsson model obtained 163.6277 dB using 10 m height station comparing with other models.

According to the obtained results in this paper, ECC-33 used antenna with 3 m height 348.9756 dB, as much as SUI system got a variable range of results as 137.6485–131.8531 dB, while Ericsson system used variable mobile antenna height ranges between 6–10 m.

COST-HATA displayed a result of 148.2541 dB using high signal in LOS mode with 10 m antenna height, while SUI 153.4329 dB and Ericsson had 148.2541 dB. Moreover, while using antenna with 3 m height, the results were as: SUI had 162.7613 dB, and in 6 m height was 159.9675 dB Additionally, COST-HATA in 3 m height had 167.9407 dB, and in 6 m height was 158.9635 dB, more effectively results than in [1–3] are obtained.

On the loss of the signal, the proposed model can serve more conditions than all previous models, so that the proposed model is considering inside the building and even inside the elevators as an extra condition. Even though, the obtained results are more accurate than what is discussed in [1–3].

## References

1. Nossire, Z., Dichter, J., Fathi, J.: A new algorithm to enhance radio wave propagation strength in dead spots for cellular mobile WiFi downloads (2014). 978-1-4577-1343-9/12/ ©2014 IEEE
2. Nossire, Z., Dichter, J., Fathi, J.: A new selected points to enhance radio wave propagation strength outside the coverage area of the mobile towers in the dead spots of cellular mobile WiFi downloads. In: Long Island Section Systems, Applications and Technology Conference. IEEE (2015)
3. Nossire, Z.F., Ditcher, J., Gupta, N., AlKhawamleh, R.: New mathematical model for wireless signal path loss inside building. In: 14th International Symposium on Pervasive Systems, Algorithms and Networks, Long Island, USA (2017)
4. Abu Hasna, J.F.: Estimating coverage of radio transmission into and within buildings for line of sight visibility between two points in terrain by linear prediction filter. In: Third Mosharaka International Conference on Communications, Signals and Coding, MIC-CSC, pp. 1–5 (2009)
5. Japertas, S., Orzekauskas, E.: Investigation of WI-Fi indoor signals under LOS and NLOS conditions. Int. J. Dig. Inf. Wirel. Commun. (IJDIWC) 2(1), 26–32 (2012). ISSN 2225-658X
6. Abhayawardhana, V.S., Wassell, I.J., Crosby, D., Sellars, M.P.: Comparison of empirical propagation path loss models for fixed wireless access systems, pp. 1–5. BT Mobility Research Unit, Ipswich (2004)
7. Bose, A., Foh, C.H.: A practical path loss model for indoor WiFi positioning enhancement (2007). 1-4244-0983-7/07, 2007 IEEE

8. Miura, Y., Oda, Y., Taga, T.: Outdoor-to-indoor propagation modeling with the identification of path passing through wall openings. Wireless Laboratories, NTT DoCoMo, Inc., Yokosuka-shi (2002). 0-7803-7589-0/02, 2002 IEEE

9. Durgin, G., Rappaport, T.S., Xu, H.: Measurements and models for radio path loss and penetration loss in and around homes and trees at 5.85 GHz. IEEE Trans. Commun. **46**(11), 1484–1496 (1998)

10. Iskandar, Shimamoto, S.: Prediction of propagation path loss for stratospheric platforms mobile communications in urban site LOS/NLOS environment, pp. 5643–5648 (2006). 1-4244-0355-3/06, 2006 IEEE

11. Wolfle, G., Wahl, R., Wildbolz, P., Wertz, P.: Dominant path prediction model for indoor and urban scenarios. AWE Communications GmbH, Böblingen. www.awe-communications. com

12. Stabler, O., Hoppe, R., Wölfle, G., Hager, T., Herrmann, T.: Consideration of MIMO in the planning of LTE networks in urban and indoor scenarios. AWE Communications GmbH, Böblingen (2011)

13. Faruk, N., Ayeni, A.A., Adediran, Y.A.: Characterization of propagation path loss at VHF/UHF bands for Ilorin City, Nigeria. Niger. J. Technol. (NIJOTECH) **32**(2), 253–265 (2013). ISSN 1115-8443. www.nijotech.com

14. Schwengler, Th., Gilbert, M.: Propagation models at 5.8 GHz–path loss & building penetration, US WEST Advanced Technologies, Boulder

15. Howard, S.L., Schlegel, C., Iniewski, K.: Error control coding in low-power wireless sensor networks: when is ECC energy-efficient. Department of Electrical & Computer Engineering, University of Alberta, Edmonton

16. Molnar, A., Lu, B., Lanzisera, S., Cook, B.W., Pister, K.S.J.: An ultra-low power 900 MHz RF transceiver for wireless sensor networks. In: IEEE 2004 Custom Integrated Circuits Conference (2004). 0-7803-8495-4/04, 02004 IEEE

17. Wang, J., Cao, Y.-T., Xie, J.-Y., Chen, S.-F.: Energy efficient backoff hierarchical clustering algorithms for multi-hop wireless sensor networks. J. Comput. Sci. Technol. **26**(2), 283–291 (2011). https://doi.org/10.1007/s11390011-1131-x

18. Meng, Y.S., Lee, Y.H., Ng, B.C.: Study of propagation loss prediction in forest environment. Prog. Electromagnet. Res. B **17**, 117–133 (2009)

19. Grillot, F., Vivien, L., Laval, S., Cassan, E.: Propagation loss in single-mode ultrasmall square silicon-on-insulator optical waveguides. J. Lightwave Technol. **24**(2), 891–896 (2006)

20. Liechty, L.C.: Path loss measurements and model analysis of a 2.4 GHz wireless network in an outdoor environment. A Thesis Presented to the Academic Faculty. Georgia Institute of Technology (2007)

21. Michael, A.O.: Further investigation into VHF radio wave propagation loss over long forest channel. Int. J. Adv. Res. Electr. Electron. Instrum. Eng. **2** (1), 705–710 (2013). www.ijareeie.com

22. Shabbir, N., Sadiq, M.T., Kashif, H., Ullah, R.: Comparison of radio propagation models for long term evolution (LTE) network. Int. J. Next-Gener. Netw. (IJNGN) **3**(3), 27–41 (2011)

23. MacCartney, Jr., G.R., Zhang, J., Nie, S., George, R., Rappaport, T.S.: Path loss models for 5G millimeter wave propagation channels in urban microcells. In: IEEE Global Communications Conference, Exhibition & Industry Forum, pp. 1–6 (2013)

# Application of Interval Approximation Method of a Fuzzy Number to the Supplier Selection

Kamala Aliyeva[✉]

Department of Instrument Making Engineering,
Azerbaijan State Oil and Industry University,
20 Azadlig Avenue, AZ1010 Baku, Azerbaijan
`kamalann@gmail.com`

**Abstract.** In this paper the idea of an approximation interval method of a fuzzy number is used for supplier selection. This is the interval that fulfils two stipulations. at the beginning, its distance is the same as the distance of a fuzzy number existing approximated. Next, the Hamming approximation among this distance and the approximated data is minimum. We obtain formula of defining the approximation distance for a fuzzy data presented in a basic kind as well as for a fuzzy number of L-R type. In practice, fuzzy intervals are frequently used to represent uncertain or imperfect information.

**Keywords:** Weighted interval approximation method of fuzzy numbers
Supplier selection · Fuzzy numbers · Multiple criteria decision making

## 1 Introduction

Trapezoidal type of fuzzy distances are often utilized in praxis. Sapid problem is to approximate basic fuzzy intervals by trapezoidal ones, to facilitate estimations. The exploration in this sphere determined by authors [1, 2] that suggested the symmetric triangular distance. In reality, symmetric triangular approximation is a specific type of the trapezoidal approximation that was considered by different researchers [6, 7].

In this article, are presented the problem of weighted interval approximation method of fuzzy numbers. An essential task is to approximate basic fuzzy intervals by interval triangular, and trapezoidal fuzzy numbers, so as to make easy accounting. Lately, a lot of researches explored these approximations of fuzzy numbers. The represented approximation distance is contrasted with the anticipated distance of a fuzzy data informed in [4, 5].

## 2 Preliminaries

The estimated distance of a fuzzy data has the identical distance as this data (Fig. 1).

© Springer Nature Switzerland AG 2019
R. A. Aliev et al. (Eds.): ICAFS-2018, AISC 896, pp. 472–477, 2019.
https://doi.org/10.1007/978-3-030-04164-9_63

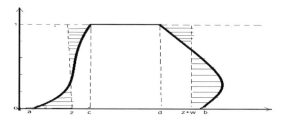

**Fig. 1.** The Hamming distance between interval $I_{z_0}$ and fuzzy number A.

**Definition 1.** If A is fuzzy data of L-R we apply of the following formula [2]:

$$w_A = \bar{a} - \underline{a} + \alpha_a * \lambda + \beta_a * \rho \tag{1}$$

$z_0$ value, when $\lambda = \rho$ [2]:

$$z_0 = \underline{a} - \alpha_A \lambda \tag{2}$$

**Definition 2.** The best interval approximation of A [2]:

$$I_{z_0} = [z_0, z_0 + W_A] \tag{3}$$

**Definition 3.** In this paper, we consider that

$$\int_{-\infty}^{\infty} \mu_A(x)dx = W_A < \infty \tag{4}$$

and call value $W_A$ the width of the fuzzy number A.

**Definition 4.** Advantage of intervals

$$d(I,J) = \begin{cases} 1, & \\ \dfrac{\bar{I} - \bar{J}}{\left|(\bar{I} - \bar{J}) + (\underline{I} - \underline{J})\right|}, & \bar{I} > \bar{J}, \underline{I} > \underline{J} \\ & \bar{I} = \bar{J}, \underline{I} > \underline{J} \text{ or } \bar{I} > \bar{J}, \underline{I} > \underline{J} \text{ or } \bar{I} = \bar{J}, \underline{I} = \underline{J} \\ 1 - d(J,I), & \text{otherwise} \end{cases}$$

is attached for comparing and evaluating superiority of $d(I,J)$ intervals.

## 3   Statement of the Problem

In this article is represented supplier choosing problem which there are 3 suppliers $A_1$, $A_2$, $A_3$, examined by the size for the criteria $C_1$, $C_2$, $C_3$, $C_4$ and $C_5$ defined as $C_1$ - Process and design capabilities; $C_2$ - Quality; $C_3$ - Reliability; $C_4$ - Supplier background; $C_5$ - Support services.

Taking solution on these suppliers decision-making presents the linguistic definition in expression of Type-2 fuzzy numbers. Our linguistic performance definitions are
shown in Table 1.

**Table 1.** The linguistic performance rating.

| Attribute | Supplier A1 | Supplier $A_2$ | Supplier $A_3$ |
|---|---|---|---|
| $C_1$ (Quality) | Very good | Good | Average |
| $C_2$ (Delivery) | Good | Good | Very good |
| $C_3$ (Flexibility) | Average` | Very good | Good |
| $C_4$ (Supplier background) | Good | Average | Good |
| $C_5$ (Support services) | Average | Good | Very good |

Priority of decision-making relatively importance of factors $C_1$, $C_2$, $C_3$, and $C_4$ is
presented as weight vector W = {0.3, 0.2, 0.2, 0.1, 0.2}.
Good = <0.6, 0.7, 0.8, 0.9>.
Very good = <0.8, 0.9, 1, 1>.
Average = <0.4, 0.5, 0.6, 0.7> (Fig. 2).

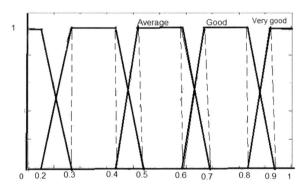

**Fig. 2.** Interval-valued approximation to fuzzy number.

We determined the need of the use of a fuzzy utility function for appropriate
estimating submission of uncertain precedence's. The problem under review (choosing
suitable supplier) is to define on alternative A, i = 1, 2, 3 (supplier 1, supplier 2,
supplier 3) with maximum number of the utility function U:

$$U(A^*) = \max U(A_i), \ A_1 \in A \tag{5}$$

## 4  Solution of the Problem

$$L(\frac{a.-x}{\alpha}) = 1 - (\frac{a.-x}{\alpha}); \ R(\frac{x-\bar{a}}{\beta}) = 1 - (\frac{x-\bar{a}}{\beta})$$

$$\int_{-\infty}^{\infty} \mu_A(x)dx = W_A < \infty$$

For average - $L = 1 - \frac{0.5-x}{0.5-0.4};$   $R = 1 - \frac{x-0.6}{0.7-0.6}$

For good - $L = 1 - \frac{0.7-x}{0.7-0.6};$   $R = 1 - \frac{x-0.8}{0.9-0.8}$

For very good - $L = 1 - \frac{0.9-x}{0.9-0.8};$   $R = 1 - \frac{x-1}{1-1}$

In second step we calculate value $W_A$ the distance of the fuzzy number A and $z_0$.

$$W_A = \bar{a} - \underline{a} + \alpha * \lambda + \beta * \rho$$

$$\lambda = \int_0^1 L(t)dt \int_0^1 (1-t)dt = 1 - \frac{1}{2} - 0 = 0.5$$

$$\rho = \int_0^1 R(t)dt \int_0^1 (1-t)dt = 0.5$$

For average

$$W_A = 0.6 - 0.5 + 0.1 * 0.5 + 0.1 * 0.5 = 0.1 + 0.1 = 0.2$$

$$z_0 = \underline{a} - \alpha * \lambda = 0.5 - 0.1 * 0.5 = 0.5 - 0.05 = 0.45$$

$$[z_0, z_0 + W_A] = [0.45, 0.45 + 0.2] = [0.45, 0.65]$$

For good

$$W_A = 0.8 - 0.7 + 0.1 * 0.5 + 0.1 * 0.5 = 0.1 + 0.1 = 0.2$$

$$z_0 = \underline{a} - \alpha * \lambda = 0.7 - 0.1 * 0.5 = 0.65$$

$$[z_0, z_0 + W_A] = [0.65, 0.65 + 0.2] = [0.65, 0.85]$$

For very good

$$W_A = 1 - 0.9 + 0.1 * 0.5 = 0.1 + 0.05 = 0.15$$

$$z_0 = \underline{a} - \alpha * \lambda = 0.9 - 0.1 * 0.5 = 0.85$$

$$[z_0, z_0 + W_A] = [0.85, 0.85 + 0.15] = [0.85, 1]$$

In last step we calculate utility functions for all 3 alternatives by using RDM: W = {0.3, 0.2, 0.2, 0.1, 0.2}

$$U_{A_1} = d_{11} * C_1 + d_{12} * C_2 + d_{13} * C_3 + d_{14} * C_4 + d_{15} * C_5 = [0.255, 0.3]$$
$$+ [0.13, 0.17] + [0.09, 0.13] + [0.06, 0.08] + [0.09, 0.13] = [0.6679, 0.7674]$$

$$U_{A_2} = d_{21} * C_1 + d_{22} * C_2 + d_{23} * C_3 + d_{24} * C_4 + d_{25} * C_5 = [0.19, 0.25]$$
$$+ [0.13, 0.17] + [0.17, 0.2] + [0.04, 0.06] + [0.06, 0.17] = [0.6433, 0.7938]$$

$$U_{A_3} = d_{31} * C_1 + d_{32} * C_2 + d_{33} * C_3 + d_{34} * C_4 + d_{35} * C_5 = [0.13, 0.19]$$
$$+ [0.17, 0.2] + [0.13, 0.17] + [0.06, 0.08] + [0.17, 0.2] = [0.5213, 0.6110]$$

In result by using robust decision making method we get

$$U_{A_1} = [0.6679, 0.7674]$$
$$U_{A_2} = [0.6433, 0.7938]$$
$$U_{A_3} = [0.5213, 0.6110]$$

Then we benchmark the defined intervals on the basis of Definition 4

$$d(I, J) = \frac{(0.6679 - 0.6433)}{|(0.6679 - 0.6433) + (0.7674 - 7938)|} = 13.66$$

(Comparing $U_{A_1}$ and $U_{A_2}$)

$$d(I, J) = 1 (\text{Comparing } U_{A_2} \text{ and } U_{A_3})$$
$$d(I, J) = 1 (\text{Comparing } U_{A_1} \text{ and } U_{A_3})$$

Amount of $d(I, J)$ is the largest in the result of benchmarking $U_{A_1}$ and $U_{A_2}$ It is evident that best alternative (supplier) first.

## 5   Conclusion

In this article we have proposed an extension of the interval approximation of fuzzy numbers approach to supplier selection problem. Distance approximation instruction, which is ideal one with respect to a determined measure of interval between different fuzzy numbers. 5-criteria supplier selection problem were considered with three alternatives. Decision was made on the base of ranging the distances.

# References

1. Buckley, J.J.: Ranking alternatives using fuzzy numbers. Fuzzy Sets Syst. **15**, 21–31 (1985)
2. Buckley, J.J., Chanas, S.: A fast method of ranking alternatives using fuzzy numbers. Fuzzy Sets Syst. **30**(1989), 337–338 (1989)
3. Dubois, D., Prade, H.: Operations on fuzzy numbers. Int. J. Syst. Sci. **9**(6), 613–626 (1978)
4. Dubois, D., Prade, H.: Fuzzy Sets Syst. **24**, 279–300 (1987)
5. Heilpern, S.: The expected value of a fuzzy number. Fuzzy Sets Syst. **47**(1), 81–86 (1992)
6. Jimenez, M.: Ranking fuzzy intervals through the comparison of its expected intervals. Int. J. Uncertainty Fuzziness Knowl.-Based Syst. **4**, 379–388 (1996)
7. Liou, T.S., Wang, M.J.: Ranking fuzzy numbers with integral value. Fuzzy Sets Syst. **50**, 247–255 (1992)

# Multifactor Personnel Selection by the Fuzzy TOPSIS Method

Kamala Aliyeva[(⊠)]

Department of Instrument Making Engineering,
Azerbaijan State Oil and Industry University,
20 Azadlig Avenue, AZ1010 Baku, Azerbaijan
kamalann@gmail.com

**Abstract.** Generally managers are withstanding with different factors to be regarded until any decision is making. This is the event of multi factor decision making, where basic aim to determine the all precedence between the accessible choices. In same time, where the characteristics are uncertain the fuzzy logic method is used. One of the very required method in multifactor decision making is the technique for order preference by similarity method. The basic goal of this article is to apply the fuzzy TOPSIS method represented by fuzzy data in solution of multifactor personnel chousing problems.

**Keywords:** Fuzzy TOPSIS · Personal selection · Fuzzy numbers
Multiple criteria decision making

## 1 Introduction

Various businesses depend on the inspection of their human resources department to make significant employing decisions on behalf of the firm. Selecting the correct applicants to respond a job stance within a company can be calling, but some general criteria's exist that impact the HR selection method. For this aim a fuzzy multi factor decision making method, which is a part of the fuzzy technique for order preference by analogy to Ideal Solution method is applied. TOPSIS was offered in 1981 [1]. The essential conception is that the most leading alternative represents the shortest interval from the positive ideal solution and the longest distance from the negative ideal solution [2]. Some scientists offered [3], positive ideal personal is the one that maximizes the advantage factor and minimizes the cost factor, where the negative ideal personal functions in the contrary way. As contrary to the real employing of TOPSIS where the weight of the factor and the ratings of alternatives are known precisely, more decision problems are opposed with uncertain, imperfect and non-permissible information [4] that make accurate decision impossible. This is when fuzzy TOPSIS is used where the factor weights and option ratings are linguistic variables, expressed by fuzzy numbers. In 2000, Chen [5] had used an algorithm of a class multi-factor decision making. From our results, the appropriate experience, education, salary requirements and relocation are the most significant in personnel selection. Fuzzy TOPSIS, that used by this paper represent a explanation for decision makers when working with practical

© Springer Nature Switzerland AG 2019
R. A. Aliev et al. (Eds.): ICAFS-2018, AISC 896, pp. 478–483, 2019.
https://doi.org/10.1007/978-3-030-04164-9_64

data that are multi factors and include a composite decision making solution. This article represents personal selection for company, by applying this method.

## 2 Preliminaries

**Definition 1.** A fuzzy set $\tilde{A}$ of $X$ is characterized by a membership function $\mu\tilde{A}$ $(x)$ which associate with each component $x$ in $X$ in the interval $[0, 1]$. The value $\mu\tilde{A}(x)$ is named the rank of membership of $x$ in $\tilde{A}$ [6].

**Definition 2.** An element $r_{ij}$ - of the normalized decision matrix R can be calculated as follows [5]:

$$r_{ij} = \frac{x_{ij}}{\sqrt{\sum_{i=1}^{M} x_{ij}^2}} \tag{1}$$

**Definition 3.** A group of weights $W = (w_1, w_2, w_3, ..., w_N)$, (where: $\sum w_i = 1$) represented by the decision maker is placed to the decision matrix to created the weighted normalized matrix V as follows [5]:

$$V = \begin{bmatrix} w_1 r_{11} & w_2 r_{12} & w_3 r_{13} & \cdots & w_N r_{1N} \\ w_1 r_{21} & w_2 r_{22} & w_3 r_{23} & \cdots & w_N r_{2N} \\ \cdot & & & & \cdot \\ \cdot & & & & \cdot \\ \cdot & & & & \cdot \\ w_1 r_{M1} & w_2 r_{M2} & w_3 r_{M3} & \cdots & w_N r_{MN} \end{bmatrix} \tag{2}$$

**Definition 4.** The positive A* and the negative $A^-$ solutions are determined as follows:

$$A^* = \left\{ (\max_i v_{ij} | j \in J), (\min_i v_{ij} | j \in J) | i = 1, 2, 3, ..., M| \right\} = \{v_{1^*}, v_{2^*}, ..., v_{N^*}\}$$
$$A^- = \left\{ (\min_i v_{ij} | j \in J), (\max_i v_{ij} | j \in J) | i = 1, 2, 3, ..., M| \right\} = \{v_{1-}, v_{2-}, ..., v_{N-}\}$$

$$\tag{3}$$

For this aim a fuzzy multi-attribute decision making method, which is an raising of the fuzzy technique for order preference by analogy to ideal solution (FTOPSIS) approach is used.

**Definition 5.** The Euclidean distance method is next used to gauge the parting distances of each alternative to the positive ideal solution and negative-ideal solution [5].

$$S_{i^*} = \left(\sum (v_{ij} - v_{j^*})^2\right)^{1/2}, i = 1, 2, 3, \ldots, M. \tag{4}$$

**Definition 6.** The relative nearness of an alternative $A_i$ with respect to the ideal solution $A^*$ is defined as follows [5]:

$$C_{i^*} = S_{i\bullet}/(S_{i^*} + S_{i\bullet}), \ 0 \le C_{i^*} \le 1, \ i = 1, 2, 3, \ldots, M. \tag{5}$$

Apparently, $C_{i^*} = 1$ **if** $A_i = A^*$, **and** $C_{i-} = 1$, **if** $A_i = A^-$.

## 3 Statement of the Problem

In this article is represented, multifactor decision making method problem which consists of 4 criteria - $C_1, C_2, C_3, C_4$ and 4 alternatives - $A_1, A_2, A_3, A_3$.

$C_1$ - Relevant experience; $C_2$ - Education; $C_3$ - Technical skills; $C_4$- Relocation. We determine relative weights for this four factors.

Taking solution on these suppliers decision-making presents the linguistic definition in expression of Type-2 fuzzy numbers. Our linguistic performance definitions are shown in Table 1. The codebook of the linguistic terms is given in Fig. 1.

**Table 1.** Linguistic performance rating.

| Attribute | Candidate $A_1$ | Candidate $A_2$ | Candidate $A_3$ | Candidate $A_4$ |
|---|---|---|---|---|
| $C_1$-(relevant experience) | Very high | High | Average | High |
| $C_2$-(education) | High | Very high | High | Average |
| $C_3$-(technical skills) | High | High | High | Average |
| $C_4$-(relocation) | Average | Average | Very high | Very high |

Very high = <0.6, 0.7, 0.8, 0.9>
High = <0.8, 0.9, 1, 1>
Average = <0.4, 0.5, 0.6, 0.7>

Priority of decision-making relatively importance of factors $C_1, C_2, C_3$, and $C_4$ is presented as weight vector W = {0.35, 0.30, 0.25, 0.1}.

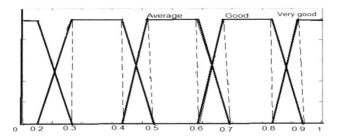

**Fig. 1.** Interval-valued approximation to fuzzy number.

# 4 Solution of the Problem

**Step 1:** We state decision matrix for fuzzy numbers of feasible options with respect to attribute and the weights of criteria (Table 2).

**Table 2.** Fuzzy decision matrix and fuzzy weights of four candidates.

|       | $C_1$ | $C_2$ | $C_3$ | $C_4$ |
|-------|-------|-------|-------|-------|
| W     | 0.35  | 0.30  | 0.25  | 0.1   |
| $A_1$ | 0.6, 0.7, 0.8, 0.9 | 0.8, 0.9, 1, 1 | 0.8, 0.9, 1, 1 | 0.4, 0.5, 0.6, 0.7 |
| $A_2$ | 0.8, 0.9, 1, 1 | 0.6, 0.7, 0.8, 0.9 | 0.8, 0.9, 1, 1 | 0.4, 0.5, 0.6, 0.7 |
| $A_3$ | 0.6, 0.7, 0.8, 0.9 | 0.8, 0.9, 1, 1 | 0.8, 0.9, 1, 1 | 0.6, 0.7, 0.8, 0.9 |
| $A_4$ | 0.8, 0.9, 1, 1 | 0.6, 0.7, 0.8, 0.9 | 0.6, 0.7, 0.8, 0.9 | 0.6, 0.7, 0.8, 0.9 |

**Step 2:** In this step we **establish the normalized decision matrix for our data.** A complex of weights $W = (w_1, w_2, w_3, \ldots, w_N)$, (where: $\sum w_i = 1$) determined by the managers generate the weighted normalized matrix V as follows (Table 3):

$$
V = \begin{bmatrix}
w_1 r_{11} & w_2 r_{12} & w_3 r_{13} & \cdots & w_N r_{1N} \\
w_1 r_{21} & w_2 r_{22} & w_3 r_{23} & \cdots & w_N r_{2N} \\
\cdot & & & & \cdot \\
\cdot & & & & \cdot \\
\cdot & & & & \cdot \\
w_1 r_{M1} & w_2 r_{M2} & w_3 r_{M3} & \cdots & w_N r_{MN}
\end{bmatrix}
$$

**Table 3.** Fuzzy normalized weighted decision matrix of four candidates.

|       | $C_1$ | $C_2$ | $C_3$ | $C_4$ |
|-------|-------|-------|-------|-------|
| $A_1$ | 0.28, 0.31, 0.35, 0.35 | 0.21, 0.27, 0.30, 0.30 | 0.10, 0.12, 15, 17 | 0.04, 0.05, 0.06, 0.07 |
| $A_2$ | 0.24, 0.31, 35, 35 | 0.24, 0.27, 0.3, 0.3 | 0.17, 0.22, 0.25, 0.25 | 0.04, 0.05, 0.06, 0.07 |
| $A_3$ | 0.14, 0.17, 0.21, 0.24 | 0.21, 0.27, 0.30, 0.30 | 0.17, 0.22, 0.25, 0.25 | 0.08, 0.09, 0.1, 0.1 |
| $A_4$ | 0.24, 0.31, 0.35, 0.35 | 0.12, 0.15, 0.18, 0.21 | 0.1, 0.12, 0.15, 0.17 | 0.08, 0.09, 0.1, 0.1 |

**Step 3:** In this **step we define the ideal and the negative-ideal points**. $A^*$ and $A^-$ are determined by next functions (Table 4):

$$
A^* = \left\{ \left( \max_i v_{ij} \middle| j \in J \right), \left( \min_i v_{ij} \middle| j \in J \right) \middle| i = 1, 2, 3, \ldots, M \right\} = \{ v_{1^*}, v_{2^*}, \ldots, v_{N^*} \}
$$

$$
A^- = \left\{ \left( \min_i v_{ij} \middle| j \in J \right), \left( \max_i v_{ij} \middle| j \in J \right) \middle| i = 1, 2, 3, \ldots, M \right\} = \{ v_{1-}, v_{2-}, \ldots, v_{N-} \}
$$

Table 4. The ideal and the negative-ideal points

|  | $C_1$ | $C_2$ | $C_3$ | $C_4$ |
|---|---|---|---|---|
| $A^*$ | 0.28, 0.31, 0.35, 0.35 | 0.24, 0.27, 0.3, 0.3 | 0.17, 022, 0.25, 0.25 | 0.08, 0.09, 0.01, 1 |
| $A^-$ | 0.14, 0.17, 0.21, 0.24 | 0.12, 0.15, 0.18, 0.21 | 0.1, 0.12, 0.15, 0.17 | 0.04, 0.05, 0.06, 0.7 |

**Step 4:** The Euclidean distance method is used to calculate intervals of different alternatives of positive and negative-ideal solutions (Table 5).

$$S_{i^*} = \left( \sum (v_{ij} - v_{j^*})^2 \right)^{1/2}, \ i = 1, 2, 3, \ldots, M.$$

**Table 5.** Delimitation distances of different alternatives to the positive and negative-ideal solutions.

|  | $A_1$ | $A_2$ | $A_3$ | $A_4$ | $A_1$ | $A_2$ | $A_3$ | $A_4$ |
|---|---|---|---|---|---|---|---|---|
|  | $C_1$ | | | | $C_2$ | | | |
| $A^*$ | 0 | 0.04 | 0.26 | 0.05 | 0.03 | 0 | 0.03 | 0.21 |
| $A^-$ | 0.26 | 0.28 | 0 | | 0.26 | 0.35 | 0.35 | 0.35 | 0 |
|  | $C_3$ | | | | $C_4$ | | | |
| $A^*$ | 0.03 | 0 | 0 | 0.17 | 0.07 | 0.07 | 0 | 0 |
| $A^-$ | 0 | 0.35 | 0.36 | 0 | 0 | 0 | 0.22 | 0.22 |

## Step5: <u>In this step we define the relative nearness to the ideal solution.</u>

$$C_{i^*} = S_{i\bullet}/(S_{i^*} + S_{i\bullet}), \quad 0 \leq C_{i^*} \leq 1, \quad i = 1, 2, 3, \ldots, M.$$

## For example

$$C_{1^*} = S_{1-}/(S_{1^*} + S_{1-}) = [(0.26/0.26 + 0) + 0.35/(0.35 + 0.03) + 0 + 0)]/4 = 0.48$$

Other $C_{1^*}$ are calculated analogously and shown below

$$C_{i^*} = (0.48, \ 0.71, \ 0.73, \ 0.45)$$

**Step 6:** The best alternative can be defined by ranking the predominance sequence of $C_{1^*}$. The best choice will be alternative that has smallest approximation to the $C_{1^*}$. The relation of alternatives represent that, alternative with smallest approximation to the $C_{1^*}$ is allow to highest approximation to the negative-ideal solution (Table 6).

Table 6. Rank the preference order

|  | $A_1$ | $A_2$ | $A_3$ | $A_4$ |
|---|---|---|---|---|
| $C_{1^*} =$ | 0.48 | 0.71 | 0.73 | 0.45 |

$$A_3 > A_2 > A_1 > A_4$$

By applying fuzzy TOPSIS rule the sequence ranking of workforce selection are as $A_3 > A_2 > A_1 > A_4$. The result represents that candidate $A_3$ is the ideal candidate and candidate $A_4$ is undesirable candidate in personnel selection.

## 5  Conclusion

In our article, fuzzy TOPSIS was used in the choosing of the best candidate according to four factors for personnel selection. First criteria are relevant experience, second criteria is education, third criteria is technical skills and fourth criteria is relocation. Results defined from the relative nearness to the ideal decision utilized to grade the predominance order in the selection of candidate for personnel selection. Evidently, the applying of fuzzy setting in relation with TOPSIS is efficacious in determining more real definition to the solving problem in workforce selection.

## References

1. Hwang, C.L., Yoon, K.: Multiple Attributes Decision Making Methods and Applications. Springer, Heidelberg (1981)
2. Opricovick, S., Tzeng, G.H.: Compromise solution by MCDM methods: a comparative analysis of VIKOR and TOPSIS. Eur. J. Oper. Res. **156**, 445–455 (2004)
3. Wang, Y.M., Elhag, T.M.S.: Fuzzy TOPSIS method based on alpha level sets with an application to bridge risk assessment. Expert Syst. Appl. **31**, 309–319 (2006)
4. Olcer, A.I., Odabasi, A.Y.: A new fuzzy multiple attributive group decision making methodology and its application to population/maneuvering system selection problem. Eur. J. Oper. Res. **166**, 93–114 (2005)
5. Chen, C.-T.: Extension of the TOPSIS for group decision-making under fuzzy environment. Fuzzy Sets Syst. **114**, 1–9 (2000)
6. Zadeh, L.A.: Fuzzy sets. Inf. Control **8**, 338–353 (1965)

# Fractal Analysis of Chaotic Fluctuations in Oil Production

E. E. Ramazanova[1], A. A. Abbasov[2], H. Kh. Malikov[1(✉)],
and A. A. Suleymanov[1]

[1] Scientific Research Institute "Geotechnological Problems of Oil, Gas and
Chemistry", 227 D. Aliyeva str., AZ1010 Baku, Azerbaijan
{e.ramazan, h.malikov}@gpogc.az
[2] SOCAR, Oil and Gas Reservoirs and Reserves Management Department,
121 H. Aliyev Ave., AZ1029 Baku, Azerbaijan

**Abstract.** Diagnosis criteria of dynamic series chaotic data are described in this
paper.

The chaotic data can conventionally divide in rectifiable and non-rectifiable in
fractal surface.

A simplified approach of fractal dimension evaluation of dynamic series has
been developed.

Developed a coefficient for calculating of processes with neighbor fractal
dimensions.

Offered nonparametric criteria for non-rectifiable on fractal surface chaotic
data. The criteria are can recognize changing in dynamical system state.

The offered methods have been validated on modeling and oil field cases.
Diagnosis of oil well performance has been realized as a case study of practical
using.

Evaluation of the offered methods is easily carried out which is valuable in
practical computations.

**Keywords:** Chaotic fluctuations · Fractal analysis · Oil production

## 1 Introduction

Dynamical systems complexities are associated with invariable parameters of its
conditions under impact of inner and exterior actions [1–3].

Most of technological data has fluctuating character, and therefore it is important
applying methods, which allow assessing the state, level of nonequilibrium and self-
organization of dynamic processes using vibrations analysis, and diagnosing of fluid
flow and decision-making on its management [3, 4]. Different indices of variations of
time-series are often used in solution of this kind of problems, such as variance,
coefficient of variation, normalized deviate, Theil criteria, etc. [4]. It is worth mentioning that these indices are parametric and their use is only correct in case if the
analyzed time-series corresponds to normal distribution.

© Springer Nature Switzerland AG 2019
R. A. Aliev et al. (Eds.): ICAFS-2018, AISC 896, pp. 484–490, 2019.
https://doi.org/10.1007/978-3-030-04164-9_65

Meanwhile, it is known that most of the natural processes do not submit to normal distribution [5, 6]. Therefore, the application of parametric criteria, developed for normal distribution, during analysis of most natural processes is incorrect [5].

Various nonparametric criteria of dynamic systems condition diagnosis on the basis of technological parameters fluctuation analysis is offered in this work. The fluctuations of dynamic-series can be conventionally divided in rectifiable and non-rectifiable on fractal surface.

## 2    Diagnosing of the Fluctuations Which Are Rectifiable on Fractal Surface

Fractal theory [7, 8], which is being widely used nowadays in many areas [3, 9], allows to state that the characteristics of many natural processes submit to fractal distribution of Pareto [6]. So, the processes, which submit to fractal distribution, excluding the particular case of normal distribution, have "infinite variance" and have no average value, characterizing the whole selection.

Nowadays, the dynamic-series fractal dimension is generally calculated by methods of covering [8]. This method match for the continuous curved lines, but it is inaccurate for time-series. It is known, that measuring of technical data is discontinuous. The measurements points, not the straight-line segment connect them for visualization, should be allow for evaluating fractal dimension (FD) for dynamic series [10, 11].

Fractal dimension evaluation method of dynamic series chaotic data are described in this paper.

### 2.1    Method for Fractal Dimension Calculation of Time Series

The method of FD evaluation is as follows. In period of time $T$ there is $n$ number of measurements of technological parameter $y_i$ (measuring time step-interval $\Delta t$).

Form a sequence of samples from the existing set of data as follows:

First sample contains all data period of time $T$: $y_1, y_2, y_3, \ldots, y_{n-1}, y_n$;

Next sample consists data with $2\Delta t$: $y_1, y_3, y_5, \ldots, y_{n-2}, y_n$ etc.

Length of dynamic series for sample is sum of $k$ magnitude of differences between neighboring values, ande $k = \mathrm{int}[(n-1)/m]$.

Therefore for first sample

$$|y_1 - y_2| + |y_2 - y_3| + \ldots + |y_{n-1} - y_n|, \tag{1}$$

next selection is determined in the same way $|y_1 - y_3| + |y_3 - y_5| + \ldots + |y_{n-2} - y_n|$.

The length of a curved line is evaluated in depending on s $\Delta t_m = m\Delta t$.

Because of the limited variations of the measured parameters of the curved line length $L$ at small $\Delta t_m$ is well characterized relationship $L \sim \Delta t_m^{1-D}$.

Considering that $\Delta t_m$ is inversely proportional to $k = (n-1)/m$ can be written:

$$L \sim k^{D-1}, \tag{2}$$

here $k = n-1, \text{int}\left[\frac{n-1}{2}\right], \text{int}\left[\frac{n-1}{3}\right], \ldots, \text{int}\left[\frac{n-1}{m}\right]$.

Reconstruct relationship (2) in $\log L - \log k$ coordinates. For large values $k$, this dependence rests on a straight line along the slope of which the value is determined $D$.

Modeling researches suggest that this method of FD evaluating has a greater accuracy compared with covering techniques.

In addition, this method enables the calculation of FD to evade measurement units not matching the measured values and the time [5].

Let us consider the approbation of this method on a model example of values of the Weierstrass-Mandelbrot fractal function [8, 12] with a given FD

$$C(t) = \sum_{N=-\infty}^{\infty} \frac{(1 - \cos(b^N t))}{b^{(2-D)N}} \tag{3}$$

here $1 < D < 2$.

Figure 1 shows case of calculating the FD with $D = 1.6$ and the number of measurements $n = 2500$. The valuation of the FD evaluated by the proposed algorithm is 1.597.

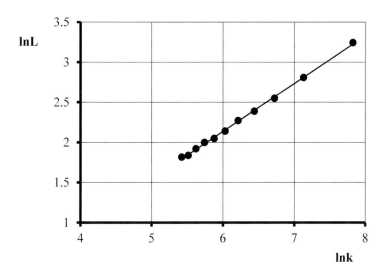

**Fig. 1.** Calculation of fractal dimension.

## 2.2   Improvement of the Computational Algorithm

The accuracy of evaluating the FD can be improved by considering all available values. For example, if $\Delta t_2 = 2\Delta t$, dependency on the choice of the starting point, you can calculate 2 $L$ and accordingly 2 values $k$:

$$L_2^1 = |y_1 - y_3| + |y_3 - y_5| + |y_7 - y_5| + \ldots,$$

$$L_2^2 = |y_2 - y_4| + |y_6 - y_4| + |y_8 - y_6| + \ldots.$$

Thus, the FD can be determined from the arithmetic mean $L$ and $k$ from the relationship

$$L_{av} \sim k_{av}^{D-1} \tag{4}$$

$$L_{av} = \frac{L_m^1 + L_m^2 + \ldots + L_m^m}{m}, k_{av} = \frac{k_m^1 + k_m^2 + \ldots + k_m^m}{m}.$$

Comparing shows that the relationship (4) is much better rectifying by averaging values $L$ and $k$.

## 3   Diagnosing of the Fluctuations Which Are Non-rectifiable on Fractal Surface

A reliable calculation of the FD for dynamic-series, in connection with the fineness of the time interval between measurements, is fraught with difficulties: such as requirement for long-term studies to obtain a large data quantity, the variability of the state of system during measuring etc.

The limitation of applicability of fluctuation examination for temporal process is related to the case that a lot of the processes characterized by fluctuation process parameters on fractal surface is non-rectifiable (e.g., multifractal data [8, 12]). Therefore, it is not always possible to use the Hurst index [6, 12]. It should also be noted that it is not correct to use this indicator to analyze fluctuations with the trend [6, 13].

The algorithm for estimating behavior of complex dynamic processes is as follows:

For analysis of fluctuation processes, a nonparametric criterion for the variation of dynamic-series is offered in this paper:

$$l = \frac{\sum_{1}^{n-1} |y_{i-1} - y_i|}{n-1} = \frac{|y_1 - y_2| + |y_2 - y_3| + \ldots + |y_{n-1} - y_n|}{n-1} \tag{5}$$

$y_i$ - the values of technological data measurements; $n$ - is the number of data.

The following modifications of non-parametric criterion (5) can Be used for assessment of dynamic process condition:

$$l_1 = \frac{n}{2(n-1)} \frac{\sum\limits_{1}^{n-1} |y_{i-1} - y_i|}{\sum\limits_{1}^{n} |y_i|} = \frac{n}{2(n-1)} \frac{|y_1 - y_2| + |y_2 - y_3| + \ldots + |y_{n-1} - y_n|}{|y_1| + |y_2| + \ldots + |y_n|} \quad (6)$$

or

$$l_2 = \frac{\ln\left(\sum\limits_{1}^{n-1} |y_{i-1} - y_i|\right)}{\ln(n-1)} = \frac{\ln(|y_1 - y_2| + |y_2 - y_3| + \ldots + |y_{n-1} - y_n|)}{\ln(n-1)} \quad (7)$$

Using the proposed coefficients, it is possible to diagnose qualitative changing in the dynamic process performance.

## 4   Application of the Suggested Criteria in Oil Production

In oil production, the results of hydrodynamic tests are used to diagnose the state of the reservoir-well system, to evaluate the filtration properties of the bottomhole zone and to determine the technological parameters of the well operation. Methods of hydrodynamic studies are related to the study of the reaction of the reservoir-well system to the change in the operating performance of the wells [14].

However, realization of hydrodynamic tests is associated with some technical and technological difficulties, material and financial costs, undesirable shutdowns of wells, etc.

The foregoing predetermines the significance of the diagnostic methods proposed in this work, which allow estimating both the state and characteristics of the reservoir systems from the data of normal well operation (well flow, wellhead and bottomhole pressure, temperature, etc.).

Consider the application of the proposed indicators to a retrospective analysis of wells of one of the oil fields in Azerbaijan.

For the calculations were used measures of wellhead pressure recorded in Dec.04 and Jan.05.

Let's consider application of suggested criteria to retrospective analysis of well performance.

The FD calculated from the trend slopes (Fig. 2) is close: 1.83 and 1.85. Though, coefficients A for this lines are 0.26 and 0.35 respectively.

Analysis of the further performance of the well showed that the change in the character of the oscillations in this case was associated with the beginning of the watering of the well, which compliant with the non-equilibrium filtration process for multiphase systems.

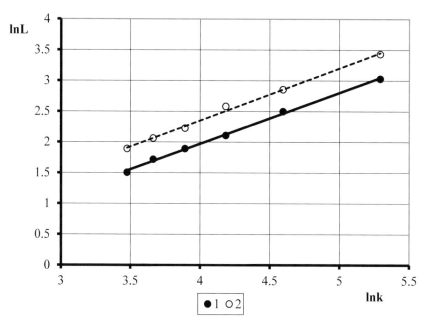

**Fig. 2.** Evaluation of wellhead pressure measurements: 1 – Dec.04; 2 – Jan.05.

## 5   Conclusions

A simplified approach of fractal dimension evaluation of dynamic series has been developed.

Proposed a coefficient for calculating of processes with neighbor fractal dimensions.

Offered nonparametric criteria for non-rectifiable on fractal surface chaotic data. The criteria are can recognize changing in dynamical system state.

The advantage of introduced fractal approach is that it can be used for data transformed by different trend removal methods.

The suggested algorithm for calculating the fractal dimension has greater accuracy and faster convergence, on a limited number of measurements, compared to the covering methods.

The offered methods have been validated on modeling and oil field cases. Diagnosis of oil well performance has been realized as a case study of practical using.

Evaluation of the offered methods is easily carried out which is valuable in practical computations.

# References

1. Haken, H.: Synergetics: Introduction and Advanced Topics. Springer, Berlin (2004)
2. Mirzajanzadeh, A., Shahverdiyev, A.: Dynamic processes in oil and gas production: system analysis, diagnosis, prognosis. Nauka, Moscow (1997)
3. Mirzajanzadeh, A., Hasanov, M., Bahtizin, R.: Modeling of oil and gas production processes. ICR, Moscow (2004)
4. Mirzajanzadeh, A., Aliev, N., Yusifzade, Kh.: Fragments on development of offshore oil and gas fields. Elm, Baku, Azerbaijan (1997)
5. Mandelbrot, B.: Fractals, Hasard et Finance. Flammarion, Paris (1997)
6. Peters, E.: Chaos and Order in the Capital Markets. Wiley, New York (1996)
7. Mandelbrot, B.: Statistical self-similarity and fractional dimension. Science **156**, 636–638 (1967)
8. Mandelbrot, B.: The Fractal Geometry of Nature. Freeman, New York (1982)
9. Hardy, H., Beier, R.: Fractals in Reservoir Engineering. World Scientific, London (1994)
10. Abbasov, A., Suleymanov, A., Ismaylov, A.: Determination of fractal dimension of time series. Azerbaijan Oil Ind. **6**, 8–11 (2000)
11. Suleymanov, A., Abbasov, A., Ismaylov, A.: Application of fractal analysis of time series in oil and gas production. Pet. Sci. Technol. **27**, 915–922 (2009)
12. Feder, E.: Fractals. Plenum Press, New York (1988)
13. Peters, E.: Fractal Market Analysis. Wiley, New York (2003)
14. Dake, L.: The Practice of Reservoir Engineering. Elsevier, New York (2001)

# Non-parametric Criteria of Chaotic Data Analysis in Oil Production

T. Sh. Salavatov[1], A. A. Abbasov[2], H. Kh. Malikov[1],
D. F. Guseynova[3], and A. A. Suleymanov[1(✉)]

[1] Scientific Research Institute "Geotechnological Problems of Oil,
Gas and Chemistry", 227 D. Aliyeva Str., Baku AZ1010, Azerbaijan
petrotech@asoiu.az
[2] Oil and Gas Reservoirs and Reserves Management Department, SOCAR,
121 H.Aliyev Ave., Baku AZ1010, Azerbaijan
[3] Oil and Gas Research and Design Institute, SOCAR, 88a Zardabi Str.,
Baku AZ1012, Azerbaijan

**Abstract.** This paper presents distribution analysis of oil production data chaotic fluctuations. Use of distribution analysis allows giving numerical characterization to fluctuation processes. This enables prediction of certain problems in well life based on the change of this numerical characterization.

In particular, water break-through prediction is considered in this paper with application of distribution analysis.

The paper suggests non-parametric criteria for analysis of production data chaotic fluctuations.

The suggested methods enable analysis changing of technological process with data distribution skewness, and also if using of other method is not proper or not to purpose.

The offered non-parametric method criteria enable simplifying of processes' analysis, which are characterized by multi-fractal, chaotic data, and their evaluation procedure can be simply implemented.

Validity of diagnosis methods has been confirmed in modeling and practical examples.

**Keywords:** Chaotic fluctuations · Distribution · Non-parametric criteria

## 1 Introduction

Oil and gas production data is normally observed in the form of dynamic and continuous fluctuations. The shape of these fluctuations changes once the dynamic system is disturbed by certain processes such as water breakthrough, sanding or other types of problems. The qualitative analysis of the production history trends allows correlating certain changes in trend of data fluctuations with such processes. However, it does not provide any quantitative characteristics of these processes. Forecasting and predicting of these problems requires their numerical characterization.

© Springer Nature Switzerland AG 2019
R. A. Aliev et al. (Eds.): ICAFS-2018, AISC 896, pp. 491–498, 2019.
https://doi.org/10.1007/978-3-030-04164-9_66

The complexity of dynamic systems and processes is related to infinite variation of its condition under outer and inner impacts [1–4]. These processes require permanent control of main technological data and timely reaction to variations in their condition.

With development of IT technologies, diagnostic methods were involved in the solution of such tasks, which help to effectively manage technological processes.

Procedures of analysis technological dynamic processes can be separated into two major groups: test and functional diagnosing methods.

However, the using of test diagnosing methods can be associated with technological complexities, extra finance expenditures, etc.

The abovementioned pre-defines the necessity of using the functional diagnostics methods, allowing assessing state and properties of the dynamic systems both condition and characteristics of the dynamic systems founded on the production data [3, 4].

One of such methods is energetic method, which is based on measuring the power or amplitude of the controlled signal [5]. Temperature, pressure, noise, vibration and many other parameters may be used as a diagnostic signal. Technology is built on measuring the degree of the signals in the controlled points and comparing them with threshold values.

Its further development is amplitude-frequency technology [5], offering allocation of measured signal constituents in a given frequency ranges and allowing assessment of the state, level of non-equilibrium and self-organization of reservoir system based on specific features of fluctuations, and diagnosing fluid flow and decision making regarding their management [3, 6].

In statistics, two types of criteria are used: parametric, built on the basis of statistical parameters of given sampling (such as dispersion, variation coefficient, normalized deviate, Theil criterion, etc.); and non-parametric, which are functions, dependant directly on values of studied data aggregates with their frequencies. Parametric criteria are used for verification of hypothesis about aggregates parameters, which are distributed according to normal law, while non-parametric criteria are used for verification of hypothesis independent of aggregates distribution shape [7].

It is worth mentioning that using parametric criteria [3, 7] in analyzing majority of natural processes is not always correct. Their use is valid only in case if the studied time series submits normal distribution. However, it is often difficult to talk about analyzed data subjugation to a definite distribution law [8–10].

Present work suggests non-parametric criteria of diagnosing the state of dynamic systems based on technological parameters fluctuation analysis.

## 2    Non-parametric Criterion of Identification Chaotic Data Distribution

The principles of assessment of dynamic systems variation is based on the following.

Let's assume that some dynamic process is being analysed, and is represented by $n$ time series of some $y$ parameter values.

Initial data are ranked in increasing order. Then ratio is determined of the square of obtained curve deviation from distribution straight line, which connects points $(1, y_{min})$ and $(n, y_{max})$, to the square of *abc* triangle, with coordinates of $a(1, y_{min})$, $b(n, y_{max})$ and $c(n, y_{min})$:

$$S = \frac{s_1 + s_2}{s_{\Delta abc}} = \frac{s_1 + s_2}{0.5 \cdot (y_{max} - y_{min}) \cdot (n - 1)} = \frac{2(s_1 + s_2)}{(y_{max} - y_{min}) \cdot (n - 1)}, \qquad (1)$$

where $y_{max}$ - maximal value of $y$, $y_{min}$ - minimal value of $y$, $n$ - number of values, $s_1$ and $s_2$ - are magnitudes of squares between ranked values of studied parameter and even distribution straight line, $s_{\Delta abc}$ - is the square of *abc* triangle (see Fig. 1).

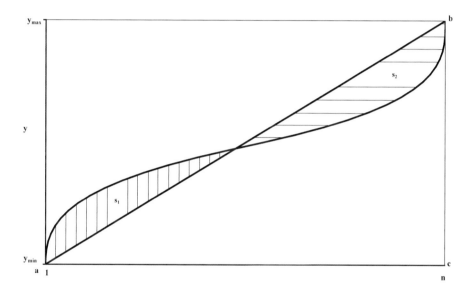

**Fig. 1.** Ranked data curve.

The value of $S$ varies in interval from 0 to 1.

Several dynamic processes comparison with various numbers of measurements allows to use the following calculation algorithm of $S$ criterion.

Initial data are ranked in increasing order and are normalized relative to maximal and minimal values of $Y_i = \frac{y_i - y_{min}}{y_{max} - y_{min}}$, where $i = 1, 2, \ldots, n$. The numbers of ranked values of $N_i$ are normalized in similar way. $Y_i$ and $N_i$ values vary from 0 to 1 (Fig. 2).

Such transformation (normalization) of initial data does not impact their distribution shape, and in the meantime allows visually presenting variations in data distribution character with different number of measurements and different maximum and minimum values.

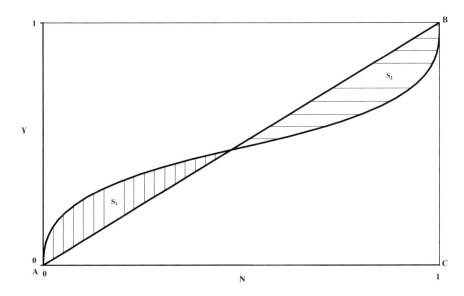

**Fig. 2.** Normalized ranked data curve.

Then the ratio is determined of the square of obtained curve deviation from distribution straight line, which connects points (0, 0) and (1, 1), to the square of $ABC$ triangle, with coordinates of $A(0, 0)$, $B(1, 1)$ and $C(0, 1)$:

$$S = \frac{S_1 + S_2}{S_{\triangle ABC}} = 2(S_1 + S_2),\tag{2}$$

where $S_1$ and $S_2$ - are magnitudes of squares between normalized ranked values of studies parameter and even distribution straight line, $S_{\triangle ABC}$ - is $ABC$ triangle square (Fig. 2).

One can judge the variation of dynamic system condition based on $S$ value variation.

As example of suggested criterion application, let us consider three types of data distribution: normal, uniform and Cauchy distributions.

Figure 3 shows ranked data, which subjugate to normal, uniform and Cauchy distribution functions. $S$ criterion values for them are 0.215, 0.015 and 0.485 respectively. It is obvious from the given example that the suggested $S$ criterion responds very well to the variations in distribution types.

The advantage of this approach is that it can be used for data, which are transformed by different trend removal methods.

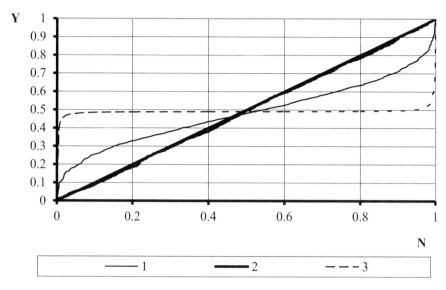

**Fig. 3.** Ranked normalized data: 1 - Normal distribution; 2 - Uniform distribution; 3 - Cauchy distribution.

## 3  Evaluation of Data Distribution Skewness

Important advantage of this approach in comparison with others methods [11] is that it can be used in case of analysing data with skew distribution.

Skewness degree can be evaluated using equation:

$$A_s = \frac{S_2 - S_1}{S_1 + S_2},$$
(3)

where $S_1$ and $S_2$ are magnitudes of the squares between ranked normalized values of studies parameter and even distribution straight line (Fig. 2).

$A_s$ value varies from $-1$ to 1. As example, let us consider cases of non-symmetrical distributions with right- and left-side skewness.

$A_s$ coefficient values for presented data equal 0.608 and $-0.608$ respectively, and S parameter values equal in both cases 0.283.

## 4  Oil Well Performance Diagnostics

Consider the application of the offered criteria to analysis of dynamic of watercut of oil production well.

The problems caused by significant water breakthrough in production wells are well known [12]. The content of water in oil has a negative influence on oil production of oil and gas. Early time prediction of watercut allows managing of oil well performance. There are methods of well testing which help to evaluate the motion of water front to oil well, but their realization costs are very expensive and are not always reasonable.

Considering the importance of creating a technique to accurately predict water breakthrough timing based on the ongoing measured data analysis, it was decided to apply the distribution analysis to a well where water breakthrough has occurred.

Let's consider application of suggested indices to retrospective analysis of oil well performance. The well data is taken from Caspian Sea region was selected.

Figure 4 measurements of wellhead pressures, taken on an oil well in July 2005 and in July 2006.

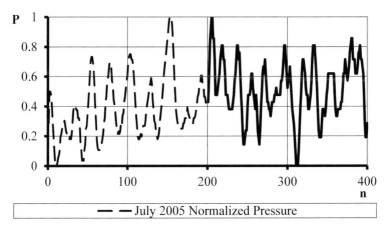

**Fig. 4.** Wellhead pressure trend (normalized) of an oil well in July 2005 and July 2006.

The values of suggested criterion $S$ for the given data considerably differ and equal 0.261 and 0.189 respectively (Fig. 5). Skewness $A_s$ equals 0.877 and $-0.187$ respectively.

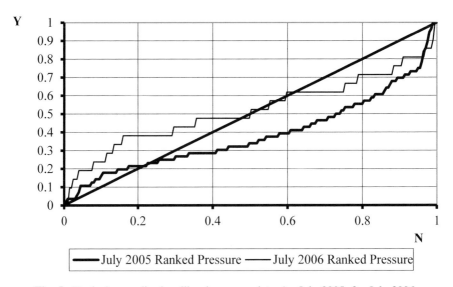

**Fig. 5.** Ranked normalized wellhead pressure data: 1 - July 2005; 2 - July 2006.

The analysis of obtained results allowed assuming that qualitative changes are taking place in the reservoir-well system's behaviour. Further production from well indicated that changing fluctuations profile was related to start of water breakthrough.

Hence, on the basis of application of suggested methods analysis of fluctuations of technological data it is possible diagnosing the changes in well performance.

## 5    Conclusions

The paper suggests non-parametric criteria of diagnosis changes in processes' condition based on analysis of production data chaotic fluctuations.

The suggested methods enable analysis changing of technological process with data distribution skewness, and also if using of other method is not proper or not to purpose.

The offered non-parametric method criteria enable simplifying of processes' analysis, which are characterized by multi-fractal, chaotic data, and their evaluation procedure can be simply implemented.

Validity of diagnosis methods has been confirmed in modelling and practical examples.

The simplified logic sequence for the approach used in the paper can be described as following:

- Review historic production data prior to and following the water break-through event.
- Perform retrospective analysis of this data using the distribution analysis parameters. Define the numerical characteristics of data series fluctuations.
- Identify the changing parameters and correlate the change with water breakthrough.
- The identified change in parameters can be applied in monitoring the ongoing trends in other wells.

Distribution analysis is applied in retrospective analysis of an offshore well in Caspian Sea region, where water breakthrough occurred. The numeric characterization of the flowing wellhead pressure data fluctuations has identified a pattern behavior of the data. Numerical value of suggested statistical criteria change is observed before water breakthrough takes place.

## References

1. Haken, H.: Synergetics: Introduction and Advanced Topics. Springer, Berlin (2004)
2. Feder, E.: Fractals. Plenum Press, New York (1988)
3. Mirzajanzadeh, A., Aliev, N., Yusifzade, K.: Fragments on Development of Offshore Oil and Gas Fields. Elm, Baku (1997)
4. Mirzajanzadeh, A., Hasanov, M., Bahtizin, R.: Modeling of Oil and Gas Production Processes. ICR, Moscow (2004)
5. Bendat, J., Piersol, A.: Random Data: Analysis and Measurements Procedures. Wiley, New York (1971)

6. Mirzajanzadeh, A., Sultanov, Ch.: Reservoir Oil Recovery Process Diacoptics. APC, Baku (1995)
7. Jensen, J., Lake, L., Corbett, P., Goggin, D.: Statistics for Petroleum Engineers and Geoscientists. Elsevier, Amsterdam (2000)
8. Mandelbrot, B.: Fractals, hasard et finance, 246 p. Flammarion, Paris (1997)
9. Belfield, W.C.: Incorporating spatial distribution into stochastic modeling of fractures: multifractals and levy-stable statistics. J. Struct. Geol. **20**(4), 473–486 (1998)
10. Aguilera, R.F., Ramirez, J.F., Ortega, C., Aguilera, R.: A variable shape distribution model for characterization of pore throat radii, drill cuttings, fracture apertures and petrophysical properties in tight, shale and conventional reservoirs. In: SPE Asia Pacific Oil and Gas Conference 2012, SPE 158808 (2012)
11. Klikushin, Y.: Method of fractal classification of compound signals. Radioelectroniks **4**, 1–11 (2000)
12. Dake, L.: The Practice of Reservoir Engineering. Elsevier, New York (2001)

# An Application of the VIKOR Method to Decision Making in Investment Problem Under Z-Valued Information

Konul Jabbarova[1(✉)] and Naila Hasanova[2,3]

[1] Department of Computer Engineering, Azerbaijan State Oil and Industry University, 20 Azadlig Avenue, AZ1010 Baku, Azerbaijan
konul.jabbarova@mail.ru
[2] Azerbaijan State Oil and Industry University,
20 Azadlig Avenue, AZ1010 Baku, Azerbaijan
naila.hasanova@gmail.com
[3] Georgia State University, Atlanta, USA

**Abstract.** The multi-criteria decision making is one of the important activities in real-world problems. Real-world decision problems are often characterized by imperfect information. In this paper we consider application of the VIKOR method to solving real world investment problem with Z-information.

**Keywords:** Z-number · VIKOR index · Utility measure · Regret measure
Investment problem

## 1 Introduction

Multi-criteria decision-making (MCDM) is one of the most important activities in various spheres. The main methods of MCDM are analytic hierarchy process (AHP), the weighted sum method (WSM), technique for order preference by similarity to ideal solution (TOPSIS), compromise ranking method (VIKOR) etc. The VIKOR method has become popular due to its computational simplicity and efficiency [1]. This method allows a decision maker to determine the best or compromise solutions under conflicting criteria.

Real-world decision problems are often characterized by imperfect information. In such cases, it is not adequate to use a numerical description for alternatives and criteria for decision analysis. Therefore, some researchers consider the use of VIKOR method under fuzzy information. In [1] the performance of original VIKOR method and its five variants is analyzed. Two demonstrative examples for application of interval VIKOR method and fuzzy VIKOR method are considered.

In [2] researchers used fuzzy numbers to deal with vague data related to criteria evaluations of alternatives. In this framework, TOPSIS and VIKOR multi-criteria methods are compared. The obtained results show superiority of the VIKOR method.

In [3] they consider solving an MCDM problem on the basis of neutrosophic hesitant fuzzy VIKOR method. The result shows the validity of the proposed approach.

© Springer Nature Switzerland AG 2019
R. A. Aliev et al. (Eds.): ICAFS-2018, AISC 896, pp. 499–506, 2019.
https://doi.org/10.1007/978-3-030-04164-9_67

In [4] they applied fuzzy set theory and a VIKOR method for calculation of the service quality of online auction. Service quality is a composite of various attributes, many of which are rather qualitative than quantitative. Due to this reason, the authors applied fuzzy set theory to the measurement of service quality.

In [5], for solving fuzzy multicriteria problem with conflicting and non-commensurable criteria, the fuzzy VIKOR method has been developed, where both criteria values and weights are described by fuzzy sets. The developed fuzzy VIKOR method relies on fuzzy arithmetic operations and procedures for ranking fuzzy numbers. An application of the method to water resources planning problem is considered.

An interval type-2 fuzzy VIKOR method based on a ranking method for interval type-2 fuzzy sets is considered in [6]. In this work, interval type-2 fuzzy numbers are used to describe the weights of criteria. Applicability of the proposed method is illustrated on a typical MCDM problem.

However, real-world problems are characterized by both fuzziness and partial reliability of information. Unfortunately, this is not taken into account in the existing works on the MCDM problems. In order to deal with fuzzy and partially reliable information, Prof. Zadeh suggested the concept of a Z-number. A Z-number is a pair of fuzzy numbers Z = (A, B), where A is a soft constraint on a value of a variable of interest, and B is a soft constraint on a value of a probability measure of A, playing a role of reliability of A. In this paper we consider application of the VIKOR method for solving an investment problem a Z-number-valued information.

The paper is organized as follows. In Sect. 2 we provide a prerequisite material used in the sequel. In Sect. 3, we consider an application of the VIKOR method under fuzziness and partial reliability of information. Section 4 is conclusion.

## 2    Preliminaries

**Definition 1. A Discrete Z-Number [7–10].** A discrete Z-number is an ordered pair $Z = (A, B)$ where $A$ is a discrete fuzzy number playing a role of a fuzzy constraint on values of a random variable $X$: $X$ is $A$. Is a discrete fuzzy number with a membership function $\mu_B : \{b_1, \ldots, b_n\} \to [0, 1], \{b_1, \ldots, b_n\} \subset [0, 1]$, playing a role of a fuzzy constraint on the probability measure of $A$: $P(A) = \sum_{i=1}^{n} \mu_A(x_i)p(x_i)$ is $B$.

**Definition 2. Operations over Discrete Z-Numbers [7–10]:** Let $X_1$ and $X_2$ be discrete Z-numbers describing information about values of $X_1$ and $X_2$. Consider computation of $Z_{12} = Z_1 * Z_2$, $* \in \{+, -, \cdot, /\}$. The first stage is computation of $A_{12} = A_1 * A_2$.

The second stage involves construction of $B_{12}$. We realize that in Z-numbers $Z_1$ and $Z_2$, the 'true' probability distributions $p_1$ and $p_2$ are not exactly known. In contrast, fuzzy restrictions represented in terms of the membership functions are available

$$\mu_{p_1}(p_1) = \mu_{B_1}\left(\sum_{k=1}^{n_1} \mu_{A_1}(x_{1k})p_1(x_{1k})\right), \quad \mu_{p_2}(p_2) = \mu_{B_2}\left(\sum_{k=1}^{n_2} \mu_{A_1}(x_{2k})p_2(x_{2k})\right).$$

Probability distributions $p_{jl}(x_{jk}), k = 1, .., n$ induce probabilistic uncertainty over $X_{12} = X_1 + X_2$. Given any possible pair $p_1, p_2$, the convolution $p_{12} = p_1 \circ p_2$ is computed as $p_{12}(x) = \sum\limits_{x_1 + x_2 = x} p_1(x_1) p_2(x_2), \forall x \in X_{12}; x_1 \in X_1, x_2 \in X_2$.

Given $p_{12s}$, the value of probability measure of $A_{12}$, is computed:

$$P(A_{12}) = \sum_{k=1}^{n} \mu_{A_{12}}(x_{12k}) p_{12}(x_{12k}).$$

However, $p_1$ and $p_2$ are described by fuzzy restrictions which induce fuzzy set of convolutions:

$$\mu_{p_{12}}(p_{12}) = \max{}_{\{p_1, p_2 : p_{12} = p_1 \circ p_2\}} \min\left\{ \mu_{p_1}(p_1), \mu_{p_2}(p_2) \right\} \qquad (1)$$

Fuzziness of information on $p_{12}$ induces fuzziness of $P_{12}(A_{12})$ as a discrete fuzzy number $B_{12}$. The membership function $\mu_{B_{12}}$ is defined as

$$\mu_{B_{12}}(b_{12}) = \max \mu_{p_{12}}(p_{12}) \qquad (2)$$

$$\text{subject to} \quad b_{12} = \sum_{i=1}^{n} \mu_{A_{12}}(x_i) p_{12}(x_i) \qquad (3)$$

As a result, $Z_{12} = Z_1 * Z_2$ is obtained as $Z_{12} = (A_{12}, B_{12})$.

A scalar multiplication $Z = \lambda Z_1, \; \lambda \in R$ is a determined as $Z = (\lambda A_1, B_1)$.

**Definition 3. Fuzzy Pareto Optimality (FPO) Principle Based Comparison of Z-Numbers [11].** Fuzzy Pareto optimality (FPO) principle allows to determine degrees of Pareto Optimality of multiattribute alternatives. We apply this principle to compare Z-numbers as multiattribute alternatives – one attribute measures value of a variable, the other one measures the associated reliability. According to this approach, by directly comparing Z-numbers $Z_1 = (A_1, B_1)$ and $Z_2 = (A_2, B_2)$ one arrives at total degrees of optimality of Z-numbers: $do(Z_1)$ and $do(Z_2)$. These degrees are determined on the basis of a number of components (the minimum is 0, the maximum is 2) with respect to which one Z-numbers dominates another one. $Z_1$ is considered higher than $Z_2$ if $do(Z_1) > do(Z_2)$.

Let us consider a MCDM problem under Z-valued information.

**Definition 4. [13] A distance between Z-numbers.** The distance between Z-numbers $Z_1 = (A_1, B_1)$ and $Z_2 = (A_2, B_2)$ is defined as

$$D(Z_1, Z_2) = \frac{1}{n+1} \sum_{k=1}^{n} \left\{ \left| a_{1\alpha_k}^L - a_{2\alpha_k}^L \right| + \left| a_{1\alpha_k}^R - a_{2\alpha_k}^R \right| \right\} + \frac{1}{m+1} \sum_{k=1}^{m} \left\{ \left| b_{1\alpha_k}^L - b_{2\alpha_k}^L \right| + \left| b_{1\alpha_k}^R - b_{2\alpha_k}^R \right| \right\}$$

where $a_\alpha^L = \min A^\alpha$, $a_\alpha^R = \max A^\alpha$, $b_\alpha^L = \min B^\alpha$, $b_\alpha^R = \max B^\alpha$.

## 3   Statement of the Problem and a Solution Method

Let us consider an MCDM problem under Z-number-valued information. Assume that a company is planning to invest into three areas: $f_1$-development of small business; $f_2$-tourism sector; $f_3$-transportation [12]. Each alternative is evaluated by 4 criteria: $C_1$-quality of product, $C_2$-degree of risk, $C_3$-quality of service, $C_3$-volume of income. Due to impression and partially readability of relevant information, the criteria evaluations are described by Z-numbers $f_{ij} = (A_{ij}, B_{ij})$ (Table 1). For simplicity, we consider numeric importance weights of criteria: $w_1 = 0.25$, $w_2 = 0.2$, $w_3 = 0.25$, $w_4 = 0.3$.

**Table 1.**  Z-number-valued decision matrix

|       | $f_1$ | $f_2$ | $f_3$ |
|-------|-------|-------|-------|
| $C_1$ | {0/0.42, 1/0.53, 0/0.64} | {0/0.56, 1/0.67, 0/0.78} | {0/0.66  1/0.79  0/0.92} |
|       | {0/0.25, 1/0.31, 0/0.37} | {0/0.25  1/0.31  0/0.37} | {0/0.25  1/0.31  0/0.37} |
| $C_2$ | {0/0.62, 1/0.74, 0/0.86} | {0/0.48  1/0.50  0/0.72} | {0/0.72  1/0.83  0/0.94} |
|       | {0/0.62, 1/0.75, 0/0.88} | {0/0.62  1/0.75  0/0.88} | {0/0.62  1/0.75  0/0.88} |
| $C_3$ | {0/0.72, 1/0.83, 0/0.94} | {0/0.66  1/0.77  0/0.88} | {0/0.46  1/0.57  0/0.68} |
|       | {0/0.13, 1/0.44, 0/0.75} | {0/0.13  1/0.44  0/0.75} | {0/0.13  1/0.44  0/0.75} |
| $C_4$ | {0/0.72  1/0.83  0/0.94} | {0/0.62  1/0.73  0/0.84} | {0/0.66  1/0.73  0/0.88} |
|       | {0/0.38  1/0.5  0/0.62} | {0/0.38  1/0.5  0/0.62} | {0/0.38  1/0.5  0/0.62} |

Let us solve this problem by using the algorithm of the VIKOR method under Z-number-valued information.

At the *first stage*, we define ideal point and negative ideal point for every criteria. As all the criteria values are Z-numbers, the ideal point and the negative ideal point will also be defined as Z-numbers. A parts of these endpoints are fuzzy lower and upper bounds of [0, 1], the corresponding B parts are the fuzzy upper bound of [0, 1]: $f_j^+ = (0.98 \quad 0.99 \quad 1)(0.8 \quad 0.9 \quad 1)$, $f_j^- = (0 \quad 0.01 \quad 0.02)(0.8 \quad 0.9 \quad 1)$ (almost completely reliable points are used).

At the *second stage* we need to calculate the values of regret measure $R_i = (A_{R_i}, B_{R_i})$ for each alternative:

$$R_i = \max \left[ w_j \frac{\left( f_j^+ - f_{ij} \right)}{\left( f_j^+ - f_j^- \right)} \right] \quad i = 1, \ldots, n. \tag{4}$$

The operations of subtraction, multiplication and division of Z-numbers in (4) and all the next formulas are performed by using Definition 2.

In order to compute $R_i = (A_{R_i}, B_{R_i})$ for all $f_i, i = 1, \ldots, n$, we obtained the following results:

$$\left(f_1^+ - f_{11}\right) = (0.98\ 0.99\ 1)(0.8\ 0.9\ 1) - (0.42\ 0.53\ 0.64)(0.25\ 0.31\ 0.37) =$$
$$= (0.34\ 0.46\ 0.58)(0.22\ 0.29\ 0.37);$$
$$\left(f_1^+ - f_1^-\right) = (0.98\ 0.99\ 1)(0.8\ 0.9\ 1) - (0\ 0.01\ 0.02)(0.8\ 0.9\ 1) =$$
$$= (0.96\ 0.98\ 1)(0.68\ 0.8\ 0.97);$$
$$\frac{\left(f_1^+ - f_{11}\right)}{\left(f_1^+ - f_1^-\right)} = \frac{(0.34\ 0.46\ 0.58)(0.22\ 0.29\ 0.37)}{(0.96\ 0.98\ 1)(0.68\ 0.8\ 0.97)} = (0.34\ 0.47\ 0.60)(0.18\ 0.25\ 0.36);$$
$$w_1 \frac{\left(f_1^+ - f_{11}\right)}{\left(f_1^+ - f_1^-\right)} = 0.25 \cdot (0.34\ 0.47\ 0.60)(0.18\ 0.25\ 0.36) = (0.09\ 0.12\ 0.15)(0.18\ 0.25\ 0.36);$$
$$\left(f_2^+ - f_{21}\right) = (0.98\ 0.99\ 1)(0.8\ 0.9\ 1) - (0.62\ 0.74\ 0.86)(0.62\ 0.75\ 0.88) =$$
$$= (0.12\ 0.25\ 0.38)(0.53\ 0.69\ 0.86);$$
$$\left(f_2^+ - f_2^-\right) = (0.98\ 0.99\ 1)(0.8\ 0.9\ 1) - (0\ 0.01\ 0.02)(0.8\ 0.9\ 1) = (0.96\ 0.98\ 1)(0.68\ 0.8\ 0.97);$$
$$\frac{\left(f_2^+ - f_{21}\right)}{\left(f_2^+ - f_2^-\right)} = \frac{(0.12\ 0.25\ 0.38)(0.53\ 0.69\ 0.86)}{(0.96\ 0.98\ 1)(0.68\ 0.8\ 0.97)} = (0.12\ 0.26\ 0.4)(0.41\ 0.58\ 0.84);$$
$$w_2 \frac{\left(f_2^+ - f_{21}\right)}{\left(f_2^+ - f_2^-\right)} = 0.2 \cdot (0.12\ 0.26\ 0.4)(0.41\ 0.58\ 0.84) = (0.02\ 0.05\ 0.08)(0.41\ 0.58\ 0.84);$$
$$\left(f_3^+ - f_{31}\right) = (0.98\ 0.99\ 1)(0.8\ 0.9\ 1) - (0.72\ 0.83\ 0.94)(0.13\ 0.44\ 0.75) =$$
$$= (0.04\ 0.16\ 0.28)(0.12\ 0.41\ 0.74);$$
$$\left(f_3^+ - f_3^-\right) = (0.98\ 0.99\ 1)(0.8\ 0.9\ 1) - (0\ 0.01\ 0.02)(0.8\ 0.9\ 1) = (0.96\ 0.98\ 1)(0.68\ 0.8\ 0.97);$$
$$\frac{\left(f_3^+ - f_{31}\right)}{\left(f_3^+ - f_3^-\right)} = \frac{(0.04\ 0.16\ 0.28)(0.12\ 0.41\ 0.74)}{(0.96\ 0.98\ 1)(0.68\ 0.8\ 0.97)} = (0.04\ 0.16\ 0.29)(0.1\ 0.36\ 0.72);$$
$$w_3 \frac{\left(f_3^+ - f_{31}\right)}{\left(f_3^+ - f_3^-\right)} = 0.25 \cdot (0.04\ 0.16\ 0.29)(0.1\ 0.36\ 0.72) = (0.01\ 0.04\ 0.07)(0.1\ 0.36\ 0.72);$$
$$\left(f_4^+ - f_{41}\right) = (0.98\ 0.99\ 1)(0.8\ 0.9\ 1) - (0.72\ 0.83\ 0.94)(0.38\ 0.5\ 0.62) =$$
$$= (0.04\ 0.16\ 0.28)(0.33\ 0.47\ 0.61);$$
$$\left(f_3^+ - f_3^-\right) = (0.98\ 0.99\ 1)(0.8\ 0.9\ 1) - (0\ 0.01\ 0.02)(0.8\ 0.9\ 1) = (0.96\ 0.98\ 1)(0.68\ 0.8\ 0.97);$$
$$\frac{\left(f_4^+ - f_{41}\right)}{\left(f_4^+ - f_4^-\right)} = \frac{(0.04\ 0.16\ 0.28)(0.33\ 0.47\ 0.61)}{(0.96\ 0.98\ 1)(0.68\ 0.8\ 0.97)} = (0.04\ 0.16\ 0.29)(0.1\ 0.36\ 0.72);$$
$$w_4 \frac{\left(f_4^+ - f_{41}\right)}{\left(f_4^+ - f_4^-\right)} = 0.3 \cdot (0.04\ 0.16\ 0.29)(0.26\ 0.41\ 0.6) = (0.01\ 0.05\ 0.09)(0.26\ 0.41\ 0.6).$$

Thus, we computed value of $R_1$ as follows:

$$R_1 = \max(((0.09\ 0.12\ 0.15)(0.18\ 0.25\ 0.36)), (0.02\ 0.05\ 0.08)(0.41\ 0.58\ 0.84)),$$
$$((0.01\ 0.04\ 0.07)(0.1\ 0.36\ 0.72)), ((0.01\ 0.05\ 0.09)(0.26\ 0.41\ 0.6))) =$$
$$= (0.29\ 0.48\ 0.8)(0.1\ 0.27\ 0.67).$$

Analogously, we computed the regret measures for the other alternatives:

$$R_2 = (0.05\ 0.1\ 0.11)(0.41\ 0.58\ 0.84);$$
$$R_3 = (0.08\ 0.11\ 0.14)(0.1\ 0.36\ 0.72).$$

At the *third stage*, we calculate the values of utility measures $S_i = (A_{S_i}, B_{S_i})$:

$$S_i = \sum_{j=1}^{n} \left[ w_j \frac{\left( f_{ij}^+ - f_{ij}^+ \right)}{\left( f_{ij}^+ - f_{ij}^- \right)} \right] \quad i = 1, \ldots, n \tag{5}$$

The obtained results:

$S_1 = (((0.09\ 0.12\ 0.15)(0.18\ 0.25\ 0.36)) + (0.02\ 0.05\ 0.08)(0.41\ 0.58\ 0.84)) +$
$\quad + ((0.01\ 0.04\ 0.07)(0.1\ 0.36\ 0.72)) + ((0.01\ 0.05\ 0.09)(0.26\ 0.41\ 0.6)) =$
$\quad = (0.13\ 0.26\ 0.39)(0.04\ 0.12\ 0.28);$

Analogously, we computed the utility measures for the other alternatives:

$$S_2 = (0.17\ 0.32\ 0.44)(0.16\ 0.18\ 0.29);$$
$$S_3 = (0.14\ 0.27\ 0.4)(0.02\ 0.07\ 0.21).$$

At the *fourth stage*, we need to compute VIKOR index $Q_i = (A_{Q_i}, B_{Q_i})$ for all alternatives:

$$Q_i = \left[ v \frac{(S_i - S^-)}{(S^+ - S^-)} + (1 - v) \frac{(R_i - R^-)}{(R^+ - R^-)} \right]. \tag{6}$$

$S^-, S^+, R^-, R^+$ are to be obtained as the highest and lowest values by comparing the Z-numbers on the basis of the FPO principle (Definition 3):

$$S^- = \min_i S_i; S^+ = \max_i S_i; R^- = \min_i R_i; R^+ = \max_i R_i.$$

The computed $S^-$, $S^+$, $R^-$, $R^+$ are as follows:

$R^+ = (0.05, 0.1, 0.11)(0.42, 0.58, 0.84), \quad R^- = (0.09, 0.12, 0.15)(0.18, 0.25, 0.36)$

and

$S^+ = (0.17, 0.32, 0.44)(0.16, 0.18, 0.29), \quad S^- = (0.14, 0.27, 0.4)(0.02, 0.07, 0.21).$

The values of VIKOR index are obtained as

$Q_1 = (12.1 \quad -0.1 \quad 12.4)(0.04 \quad 0.06 \quad 0.08); Q_2 = (-24.25 \quad 1 \quad 22.95)(0.14 \quad 0.16 \quad 0.21);$
$Q_3 = (-13.5 \quad 0.25 \quad 13.5)(0.008 \quad 0.07 \quad 0.12).$

At the *fifth stage*, we compare the values of $R, S, Q$ measures by using Definition 3 (Table 2).

The results of ordering of the alternatives with respect to $R, S, Q$ measures are shown in Table 3:

At the *sixth stage*, according to the VIKOR algorithm, we need to analyze the results of this ordering by verification of the conditions C1 and C2 [1]. For brevity, we don't describe them here (they are described in details in [1]). We have found out that these conditions are satisfied. Now, we need to evaluate the difference between values of $Q_1$ and $Q_2$. For computational efficiency we will do this by calculating distance $D(Q_1, Q_2)$ (Definition 4). The obtained result is $D(Q^1, Q^2) = 1.2 > 0.5$. This implies that the alternative $f_1$ is a compromise solution.

**Table 2.** Ranking of $R, S, Q$ measures

| Ranking values of regret measure $R$ | Ranking values of utility measure $S$ | Ranking values of VIKOR index $Q$ |
|---|---|---|
| Alternative $f_1$ vs. Alternative $f_2$: $do(R_1) = 0.54, do(R_2) = 1$ | Alternative $f_1$ vs. Alternative $f_2$: $do(R_1) = 0., do(R_2) = 1,$ | Alternative $f_1$ vs. Alternative $f_2$: $do(R_1) = 0.68, do(R_2) = 1$ |
| Alternative $f_3$ vs. Alternative $f_2$: $do(R_3) = 0.65, do(R_2) = 1$ | Alternative $f_3$ vs. Alternative $f_2$: $do(R_3) = 0, do(R_2) = 1,$ | Alternative $f_3$ vs. Alternative $f_2$: $do(R_3) = 0.02, do(R_2) = 1$ |
| Alternative $f_3$ vs. Alternative $f_1$: $do(R_3) = 1, do(R_1) = 0.37$ | Alternative $f_3$ vs. Alternative $f_1$: $do(R_3) = 0.15, do(R_1) = 1$ | Alternative $f_3$ vs. Alternative $f_1$: $do(R_3) = 1, do(R_1) = 0$ |

**Table 3.** The alternatives sorted with respect to $R, S, Q$ measures

| $R$ | $S$ | $Q$ |
|---|---|---|
| $f_1$ | $f_3$ | $f_1$ |
| $f_3$ | $f_1$ | $f_3$ |
| $f_2$ | $f_2$ | $f_2$ |

## 4 Conclusion

In this paper we consider an application of VIKOR method to decision making in investment problem when all criteria values are described by Z-numbers. We propose a way to calculate the Z-values of utility measure, regret measure and VIKOR index. A fuzzy Pareto optimality principle-based comparison of Z-numbers is applied for ranking of the alternatives with respect to these measures. The obtained results coincide with human-like decision making.

## References

1. Chatterjeea, P., Chakraborty, S.: Comparative analysis of VIKOR method and its variants. Decis. Sci. Lett. **5**(4), 469–486 (2016)
2. Chowdhury, S., Poet, R., Mackenzie, L.: Multicriteria optimization to select images as passwords in recognition based graphical authentication systems. In: Conference 2013, HAS, pp. 13–22. Springer (2013)

3. Liu, P., Zhang, L.: The extended VIKOR method for multiple criteria decision making problem based on neutrosophic hesitant fuzzy set. http://fs.unm.edu/TheExtended VIKORMethod.pdf
4. Wang, C.-H., Pang, C.-T.: Using VIKOR method for evaluating service quality of online auction under fuzzy environment. Int. J. Comput. Sci. Eng. Technol. **1**(6), 307–314 (2011)
5. Opricovic, S.: Fuzzy VIKOR with an application to water resources planning. Expert Syst. Appl. **38**, 12983–12990 (2011)
6. Yazici, I., Kahraman, C.: VIKOR method using interval type two fuzzy sets. J. Intell. Fuzzy Syst. **29**(1), 411–421 (2015)
7. Aliev, R.A., Alizadeh, A.V., Huseynov, O.H.: The arithmetic of discrete Z-numbers. Inf. Sci. **290**, 134–155 (2015)
8. Aliev, R.A., Alizadeh, A.V., Huseynov, O.H., Jabbarova, K.I.: Z-number based linear programming. Int. J. Intell. Syst. **30**(5), 563–589 (2015)
9. Aliev, R.A., Huseynov, O.H.: Decision Theory with Imperfect Information. World Scientific, Singapore (2014)
10. Aliev, R.A., Huseynov, O.H., Aliyev, R.R., Alizadeh, A.V.: The Arithmetic of Z-Numbers: Theory and Applications. World Scientific, Singapore (2015)
11. Aliev, R.A., Huseynov, O.H., Serdaroglu, R.: Ranking of Z-numbers and its application in decision making. Int. J. Inf. Technol. Decis. Making **15**(6), 1503–1519 (2016)
12. Jabbarova, A.I.: Solution for the investment decision making problem through interval probabilities. Procedia Comput. Sci. **102**, 465–468 (2016)
13. Aliev, R.A., Pedrycz, W., Huseynov, O.H., Eyupoglu, S.Z.: Approximate reasoning on a basis of Z-number-valued if–then rules. IEEE Trans. Fuzzy Syst. **25**(6), 1589–1600 (2017)

# Energy Consumption Prediction of Residential Buildings Using Fuzzy Neural Networks

Sanan Abizada[1] and Esmira Abiyeva[2]([✉])

[1] Electrical and Electronic Engineering Department, Near East University,
Nicosia, North Cyprus, Turkey
[2] Economics Department, Near East University, Nicosia, North Cyprus, Turkey
esmira.abiyeva@neu.edu.tr

**Abstract.** This paper presents an energy consumption prediction model of residential buildings using fuzzy neural networks (FNN). The design of FNN prediction model has been performed using clustering and gradient descent algorithms. A cross-validation procedure is used for the training of the FNN. The descriptions of the training algorithms have been given. The statistical data is applied to design FNN. The obtained simulation results prove the effectiveness of using FNN in the energy consumption prediction of residential buildings. Based on prediction results of the energy consumption, the efficient ventilation system of the buildings can be planned. As a result, the energy waste can be decreased considerably.

**Keywords:** Energy consumption · Fuzzy neural networks · Prediction

## 1 Introduction

The design of energy efficiency buildings needs the prediction of heating and cooling loads of the energy consumption of the buildings. The efficient controlling of energy consumption allows the decrease of the energy waste and its negative effect on the environment. The energy consumption prediction of buildings has been investigated by many research works [1, 2]. For the controlling of heating and cooling loads, implementation of the efficient building design needs to be carried out. For this purpose, different building energy simulation tools are designed. These simulation programs are DOE-2 given in [3], ESP-r [4], Energy Plus [5], and Designer's Simulation Toolkit DeST [6] are used to compute energy consumption of buildings. These systems can evaluate the impact of design alternatives to the energy consumption and can give reliable solutions. But these systems have complicated structure and time-consuming. Sometimes the correctness of the results may vary in different simulation software [7]. Recently, artificial intelligence systems are applied to analyze the effect of different building parameters on energy consumption. Statistical methods [8], different machine learning techniques are applied to forecast building energy consumption to obtaining required accuracy. Machine learning techniques can present an accurate, flexible and robust solution for the data sets having incomplete and inaccurate data sets. The references [9], [10] use support vector machines (SVM), [2] uses decision trees, [11] and [12] use neural networks for predicting the energy consumption. The effects of

© Springer Nature Switzerland AG 2019
R. A. Aliev et al. (Eds.): ICAFS-2018, AISC 896, pp. 507–515, 2019.
https://doi.org/10.1007/978-3-030-04164-9_68

different building parameters on energy consumption were studied in [13], [14] and [15]. Heating and cooling loads are estimated using different building variables such as relative compactness, wall area, roof area, surface area, climate, orientation and glazing [6, 15]. It was found that these parameters have an adverse impact on energy consumption and also heating and cooling load respectively.

One of an effective way for prediction of building energy consumption is the use of soft computing elements such as fuzzy sets, artificial neural networks and evolutionary computation. The reference [16] used the neuro-fuzzy system, [17, 18] used fuzzy wavelet networks for energy consumption prediction. The systems based on fuzzy logic can make classifications using vague, imprecise, noisy, or missing input information. In the literature, different fuzzy and neural systems are developed for solving various problems. FNN structure is designed for stock-price prediction [19], for classification of brain signals [20], diagnosing of parkinson diseases [21], for credit scoring [22], for channel equalization [23]. In the paper, the fuzzy neural network model is designed for building energy consumption. A multi-input and multi-output FNN based on TSK rules are developed for prediction of building energy consumption in particularly for prediction of heating and cooling parameters.

## 2   Fuzzy Neural Networks for Building Energy Consumption

The construction of the FNN is considered for prediction of the heating and cooling loads. The inputs for the system are eight parameters; relative compactness, roof area, wall area, surface area, orientation, overall height, glazing area and glazing area distribution. The heating and cooling loads are outputs of the system. The design of FNN includes the development of the fuzzy IF-THEN rules. In the paper, the Takagi-Sugeno-Kang (TSK) types of IF-THEN rules are used.

$$\text{If } x_1 \text{ is } A_{1j} \text{ and } x_2 \text{ is } A_{2j} \text{ and}\dots\text{and } x_m \text{ is } A_{mj} \text{ Then } y_j \text{ is } \sum_{i=1}^{m} a_{ij}x_i + b_j \quad (1)$$

Here $x_i$ and $y_j$ are input and output signals correspondingly. $A_{ij}$ are input fuzzy sets, $b_j$ and $a_{ij}$ are coefficients, $i = 1 ,..., m$ and $j = 1 \dots r$, $m$ is the number of input signals, $r$ is the number of rules.

The FNN structure presented in [21] for prediction purpose (Fig. 1) is used. The input layer is used to distribute the $x_i$ $(i = 1 ,..., m)$ signals. The FNN includes

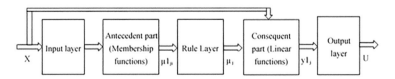

**Fig. 1.** FNN based identifier

membership functions each one of which represents one linguistic term. The linguistic terms are represented using Gaussian membership functions.

$$\mu 1_j(x_i) = e^{-\frac{(x_i - c_{ij})^2}{\sigma_{ij}^2}}, \quad i = 1..m, j = 1..r \tag{2}$$

where $c_{ij}$ and $\sigma_{ij}$ are centers and widths of the membership functions (2) correspondingly. $m$ is the number of input signals, $r$ is the number of fuzzy rules.

After calculating membership grades $\mu 1_j(x_i)$ using t-norm min (AND) operation the outputs of rule layer is calculated.

$$\mu_j(x) = \prod_i \mu 1_j(x_i) \tag{3}$$

where $i = 1$, $m$, $j = 1$,..., $r$. $\prod$ is the min operation.

The obtained $\mu_j(x)$ signals are inputs for the consequent part. Consequent part includes n linear functions that are determined as:

$$y_j = \sum_{i=1}^{m} x_i w_{ij} + b_j, \quad j = 1,\ldots,r \tag{4}$$

In the consequent layer $\mu_j(x)$ signals are multiplied by the linear functions. The output of this layer is computed as $y1_j = \mu_j(x) \cdot y_j$

The outputs of the FNN are determined as:

$$u_k = \sum_{j=1}^{r} w_{jk} y1_j / \sum_{j=1}^{r} \mu_j(x) \tag{5}$$

Here $y1_j = \mu_j(x) \cdot y_j$, $u_k$ are the output signals of FNN, ($k = 1, .., n$). After computing the output signals, the learning of the network starts.

## 3   Parameter Updates

In the design of FNN model, the basic problem is the finding of the unknown parameters of networks, which are the parameters of the fuzzy If-Then rules (1), $c_{ij}$ and $\sigma_{ij}$ ($i = 1, .., m, j = 1, .., r$) and $w_{jk}$, $a_{ij}$, $b_j$ ($i = 1, .., m, j = 1, .., r, k = 1, .., n$). In the paper, clustering and gradient algorithms are used for the parameter update. Learning of FNN starts with the update of the parameters of the antecedent part. At first Fuzzy C-means (FCM) clustering is used to design antecedent part of the fuzzy rules. In the result of clustering, the cluster centers are determined. These centers correspond to the centers of the membership functions used in the input layer of the FNN. The widths of the membership functions are determined using the distances between the cluster centers.

The design of the consequent part of the fuzzy rules is implemented using gradient descent algorithm. The momentum term is also used to improve the speed of convergence of learning. In the learning of FNN, 10 fold cross validation is applied for separating the data into training and testing set.

The error on the output of the network is determined as:

$$E = \frac{1}{2}\sum_{k=1}^{n}\left(u_k^d - u_k\right)^2 \qquad (6)$$

where $u_k$ and $u_k^d$ are current and desired output signals of the FNN, correspondingly. $n$ is the number of output signals, The $w_{jk}$, $a_{ij}$, $b_j$, $(i = 1, .., m, j = 1, .., r, k = 1, .., n)$ parameters in consequent part and $c_{ij}$ and $\sigma_{ij}(i = 1, .., m, j = 1, .., r)$ parameters in the premise part are updated as:

$$w_{jk}(t+1) = w_{jk}(t) - \gamma\frac{\partial E}{\partial w_{jk}} + \rho\left(w_{jk}(t) - w_{jk}(t-1)\right) \qquad (7)$$

$$a_{ij}(t+1) = a_{ij}(t) - \gamma\frac{\partial E}{\partial a_{ij}} + \rho\left(a_{ij}(t) - a_{ij}(t-1)\right) \qquad (8)$$

$$b_i(t+1) = b_j(t) - \gamma\frac{\partial E}{\partial b_j} + \rho\left(b_j(t) - b_j(t-1)\right) \qquad (9)$$

$$c_{ij}(t+1) = c_{ij}(t) - \gamma\frac{\partial E}{\partial c_{ij}} + \rho\left(c_{ij}(t) - c_{ij}(t-1)\right) \qquad (10)$$

$$\sigma_{ij}(t+1) = \sigma_{ij}(t) - \gamma\frac{\partial E}{\partial \sigma_{ij}} + \rho\left(\sigma_{ij}(t) - \sigma_{ij}(t-1)\right) \qquad (11)$$

Here $m$ and n are the numbers of input and output signals of the FNN correspondingly. $r$ is the number of hidden neurons. $\gamma$ is the learning rate and $\lambda$ is the momentum. Using Eqs. (7–11) the learning of the FNN is carried out.

## 4  Simulations

The above described FNN is used for prediction of the building energy consumption. For this purpose, the real data set that includes the heating and cooling loads requirements of buildings is used. The energy data set includes 768 data instances for 768 different buildings. The data set includes eight input and two output parameters. The input data set of the FNN are the relative compactness, wall area, roof area, surface area, orientation, overall height, glazing area and glazing area distribution. The output data set is heating and cooling loads. Fragment of energy efficiency dataset is shown in Table 1.

The training pair for the prediction system is the eight-dimensional input vector and the two-dimensional output vector. Based on input-output data vectors the structure of FNN is generated with the 8 input and two output neurons. 10-fold cross-validation that partitions the original data samples into 10 groups of equal size is used. In each subsample, one set is used for validation and the other remaining sets are used for training. The results achieved from the folds are then averaged in order to produce a

**Table 1.** Fragment of energy consumption data set

| X1 | X2 | X3 | X4 | X5 | X6 | X7 | X8 | Y1 | Y2 |
|----|----|----|----|----|----|----|----|----|----|
| 0.98 | 514.5 | 294.0 | 110.25 | 7.0 | 2 | 0.0 | 0 | 15.55 | 21.33 |
| 0.98 | 514.5 | 294.0 | 110.25 | 7.0 | 3 | 0.0 | 0 | 15.55 | 21.33 |
| 0.98 | 514.5 | 294.0 | 110.25 | 7.0 | 4 | 0.0 | 0 | 15.55 | 21.33 |
| ... | ... | ... | ... | ... | ... | ... | ... | ... | ... |
| 0.98 | 514.5 | 294.0 | 110.25 | 7.0 | 3 | 0.1 | 1 | 24.63 | 26.37 |

single estimation. During simulation training, evaluation and testing results are recorded. The mean squared error (MSE) and root mean squared error (RMSE) are used to measure the performance of the system.

$$MSE = \frac{1}{N}\sum_{i=1}^{N}\left(Y_i^d - Y_i\right)^2; \quad RMSE = \sqrt{\sum_{i=1}^{N}\left(Y_i^d - Y_i\right)^2} \qquad (12)$$

where N is the number of data items in training or testing subsets.

Simulation has been done using a different number of rules. During learning FCM is used to find the parameters of the membership functions, The gradient descent algorithm is used to find the parameters of the linear functions. 100 epochs are used for training. As a result of training, the parameters of FNN prediction system are found. The simulation is performed using eight hidden neurons. MSE and (RMSE) are used for estimation of the model performance. Figure 2 describes the plot of RMSE values. Figure 3 describes the plot of errors. During simulation, the RMSE value for training data was obtained as 0.068550, for evaluation as 0.071671. As a result of learning the RMSE for the test data was obtained as 0.070848. The simulation results of building

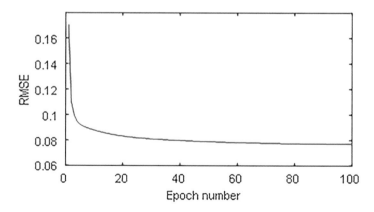

**Fig. 2.** Plot of RMSE

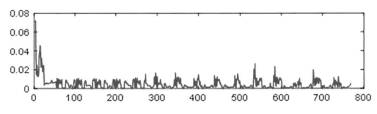

**Fig. 3.** Plot of errors

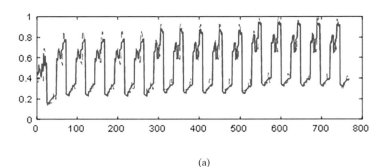

(a)

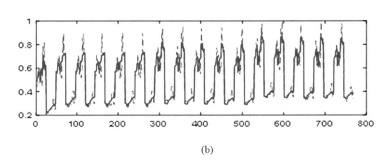

(b)

**Fig. 4.** Plots of output heating (a) and cooling (b) signals.

energy consumption prediction system were obtained using eight and sixteen hidden neurons. Figure 4(a) and (b) describe the plots of output signals after training.

We fix the performance of the prediction system for each output using 10 fold cross-validation and 100 training epochs. MSE and RMSE values for each output signal for train, evaluation and test data are given in Table 2. The simulation results are

**Table 2.** Simulation results obtained with 10 fold cross-validation and 100 epochs

| Number of rules | MSE | | | RMSE | | |
|---|---|---|---|---|---|---|
| | Train | Evaluation | Test | Train | Evaluation | Test |
| 8 | 0.006046 | 0.0060055 | 0.0059019 | 0.077755 | 0.077495 | 0.076824 |
| 16 | 0.004699 | 0.0051367 | 0.0050194 | 0.068550 | 0.071671 | 0.070848 |

**Table 3.** MSE and RMSE values obtained for each output

| 100it | | Train | | Evolution | | Test | |
|---|---|---|---|---|---|---|---|
| | | MSE | RMSE | MSE | RMSE | MSE | RMSE |
| 8 rakes | Y1 | 5 | 2.236 | 5 | 2.236 | 4.9 | 2.2159 |
| | Y2 | 8 | 2.828 | 8 | 2.828 | 7.5 | 2.7386 |
| 16 rules | Y1 | 4 | 2 | 4 | 2 | 4.3 | 2.0736 |
| | Y2 | 6 | 2.4495 | 7 | 2.645 | 6.2 | 2.4899 |

obtained using 8 and 16 rules. Table 3 describes MSE and RMSE values of each output for original data.

The results are obtained using averaged results of 10 simulations. For comparative analysis, the building energy consumption prediction system has been simulated using neural networks. The results of comparisons of neuro-fuzzy and neural networks models are given in Table 4. The comparison has been done at the same initial conditions using 16 hidden neurons. The table also includes the comparison of test results of the iteratively reweighted least squares (IRLS) [7]. The simulation results demonstrate that the FNN based system has achieved small RMSE value in comparison with

**Table 4.** Comparative results

| 100it | | Train | | Evolution | | Test | |
|---|---|---|---|---|---|---|---|
| | | MSE | RMSE | MSE | RMSE | MSE | RMSE |
| Fuzzy neural networks | Y1 | 4 | 2 | 4 | 2 | 4.3 | 2.0736 |
| | Y2 | 6 | 2.4495 | 7 | 2.645 | 6.2 | 2.4899 |
| Neural networks | Y1 | 5 | 2.236 | 5.5 | 2.345 | 4.9 | 2.2136 |
| | Y2 | 7 | 2.645 | 7.7 | 2.774 | 7.5 | 2.7386 |
| IRLS [7] | Y1 | - | - | - | - | 9.87 | - |
| | Y2 | - | - | - | - | 11.46 | - |

the other models. The comparative results demonstrate the effectiveness of using FNN in building energy consumption.

## 5    Conclusions

The design of the energy consumption prediction model of residential buildings using fuzzy neural networks is performed. The structure of FNN and its learning algorithm is presented. The design of FNN is performed using classification, cross-validation and gradient descent techniques. FCM classification is used for the selection of parameters of antecedent part of FNN. In next stage, the parameter update of the consequent part is performed using cross-validation and gradient descent algorithm. In the result of learning the synthesis of FNN model for energy consumption prediction of residential buildings has been performed. The prediction model is tested using statistical data.

# References

1. Yu, Z., Haghigrat, F., Fung, B.C.M., Yoshimo, H.: A decision tree method for building energy demand modeling. Energy Build. **42**, 1637–1646 (2010)
2. Perez-Lombard, L., Ortiz, J., Pout, C.: A review on buildings energy consumption information. Energy Build. **40**(3), 394–398 (2008)
3. The Home of DOE2 based Building Energy Use and Cost Analysis Software (1998). http://www.doe2.com/
4. Strachan, P.A., Kokogiannakis, G., Macdonald, I.A.: History and development of validation with the ESP-r simulation program. Build. Environ. **43**(4), 601–609 (2008)
5. Crawley, D.B., Lawrie, L.K., Winkelmann, F.C., Buhl, W.F., Huang, Y.J., Pedersen, C.O., Strand, R.K., Liesen, R.J., Fisher, D.E., Witte, M.J., Glazer, J.: EnergyPlus: creating anew-generation building energy simulation program. Energy Build. **33**(4), 319–331 (2001)
6. Yan, D., Xia, J., Tang, W., Song, F., Zhang, X., Jiang, Y.: DeST—an integrated building simulation toolkit part I: fundamentals. Build. Simul. **1**(2), 95–110 (2008)
7. Tsanas, A., Xifara, A.: Accurate quantitative estimation of energy performance of residential buildings using statistical machine learning tools'. Energy Build. **49**, 560–567 (2012)
8. Platt, G., Li, J., Li, R., Poulton, G., James, G., Wal, J.: Adaptive HVAC zone modelling for sustainable buildings. Energy Build. **42**, 412–421 (2010)
9. Dong, B., Cao, C., Lee, S.E.: Applying support vector machines to predict building energy consumption in tropical region. Energy Build. **37**, 545–553 (2005)
10. Li, Q., Meng, Q., Cai, J., Yoshino, H., Mochida, A.: Applying support vector machine to predict hourly cooling load in the building. Appl. Energy **86**, 2249–2256 (2009)
11. Yezioro, A., Dong, B., Leite, F.: An applied artificial intelligence approach towards assessing building performance simulation tools. Energy Build. **40**, 612–620 (2008)
12. Zhang, J., Haghighat, F.: Development of artificial neural network based heat convection for thermal simulation of large rectangular cross-sectional area earth-to-earth heat exchanges. Energy Build. **42**(4), 435–440 (2010)
13. Schiavon, S., Lee, K.H., Bauman, F., Webster, T.: Influence of raised floor on zone design cooling load in commercial buildings. Energy Build. **42**(8), 1182–1191 (2010)
14. Wan, K.K.W., Li, H.W., Liu, D., Lam, J.C.: Future trends of building heating and cooling loads and energy consumption in different climates. Build. Environ. **46**, 223–234 (2011)
15. Tsanas, A., Goulermas, J.Y., Vartela, V., Tsiapras, D., Theodorakis, G., Fisher, A.C., Sfirakis, P.: The Windkessel model revisited: a qualitative analysis of the circulatory system. Med. Eng. Phys. **31**, 581–588 (2009)
16. Abiyev, R., Abiyev, V.H., Ardil, C.: Electricity consumption prediction model using neuro-fuzzy system. Int. J. Comput. Inf. Eng. **3**(12), 2963–2966 (2009)
17. Abiyev, R.H.: Fuzzy wavelet neural network for prediction of electricity consumption. AIEDAM: Artif. Intell. Eng. Des. Anal. Manuf. **23**(2), 109–118 (2009)
18. Abiyev, R.H.: Fuzzy wavelet neural network based on fuzzy clustering and gradient techniques for time series prediction. Neural Comput. Appl. **20**(2), 249–259 (2011)
19. Abiyev, R.H., Abiyev, V.H.: Differential evaluation learning of fuzzy wavelet neural networks for stock price prediction. J. Inf. Comput. Sci. **7**(2), 121–130 (2012). ISSN 1746-7659, England, UK
20. Abiyev, R.H., Akkaya, N., Aytac, E., Günsel, I., Çağman, A.: Brain-computer interface for control of wheelchair using fuzzy neural networks. Biomed Res. Int. **2016**, 1–9 (2016)
21. Abiyev, R.H., Abizade, S.: Diagnosing Parkinson's diseases using fuzzy neural system. Comput. Math. Methods Med. **2016**, 1–9 (2016)

22. Abiyev, R.H.: Credit rating using type-2 fuzzy neural networks. Math. Probl. Eng. **2014**, 1–8 (2014)
23. Abiyev R.H., Alshanableh T.: Neuro-fuzzy network for adaptive channel equalization. In: 5th Mexican International Conference on Artificial Intelligence, MICAI – 2006, pp. 13–17. IEEE CS press, Apizaco (2006)

# Classification of Diseases on Chest X-Rays Using Deep Learning

Sertan Kaymak, Khaled Almezhghwi[(✉)], and Almaki A. S. Shelag

Near East University, North Cyprus, Mersin-10, Turkey
sertan.kaymak@neu.edu.tr, khaldalmezghwi84@gmail.com,
almakkishilag16@gmail.com

**Abstract.** Doctors and radiologists are still using manual and visual manners in ordert to diagnose the chest radiographs. Thus, there is a need for an intelligent and automatic system that has the capability of diagnosing the chest X-rays. This thesis aims to employ a deep neural network named as stacked auto-encoder for the classification of chest X-rays into normal and abnormal images. The stacked auto-encoder is trained and tested on chest X-rays obtained for public databases which contain normal and abnormal radiographs. A performance based comparison is carried out between two networks where the first one uses input chest X-rays without processing or enhancement and the other one uses input images that are processed and enhanced using histogram equalization.

Experimentally, it is concluded that the Stacked auto-encoder achieved a good generalization power in diagnosing the unseen chest X-rays into normal or abnormal. Moreover, it is seen that the enhancement of images using histogram equalization helps in improving the learning and performance of network due to the rise in the accuracy achieved when image are enhanced.

**Keywords:** Deep network · Stacked auto-encoder · Radiographs
Classification · Generalization · Intelligent

## 1 Introduction

Medical X-rays are images are generally used to diagnose some sensitive human body parts such as bones, chest, teeth, skull, etc. Medical experts have used this technique for several decades to explore and visualize fractures or abnormalities in body organs [1]. Chest diseases can be shown in CXR images in the form of cavitations, consolidations, infiltrates, blunted costophrenic angles, and small broadly distributed nodules. The interpretation of a chest X-ray can diagnose many conditions and diseases such as pleurisy, effusion, pneumonia, bronchitis, infiltration, nodule, atelectasis, pericarditis, cardiomegaly, pneumothorax, fractures and many others [2].

Classifying the chest x-ray abnormalities is considered a tough task for radiologists. Hence, over the past decades, computer aided diagnosis (CAD) systems have been developed to extract useful information from X-rays to help doctors in having a quantitative insight about an X-ray [3, 4].

© Springer Nature Switzerland AG 2019
R. A. Aliev et al. (Eds.): ICAFS-2018, AISC 896, pp. 516–523, 2019.
https://doi.org/10.1007/978-3-030-04164-9_69

Recently, accurate images classification has been achieved by deep learning based systems. Those deep networks showed superhuman accuracies in performing such tasks. This success motivated the researchers to apply those networks on medical images for diseases classification tasks and the results showed that deep networks can efficiently extract useful features that distinguish different images classes [5, 6].

In this work, deep learning based networks are employed to classify most common thoracic diseases. Two stacked auto-encoder are examined in this study to classify the chest X-rays into two common classes: normal and abnormal which may have different types of diseases that may be found in chest X-ray, i.e., Atelectasis, Cardiomegaly, Effusion, Infiltration, Mass, Nodule, Pneumonia, Pneumothorax, Consolidation, Edema, Emphysema, Fibrosis. In this work, we aim to train the deep network on the same number of chest X-ray images and evaluate their performances in classifying different chest X-rays. The data used in obtained from two public databases which are the Shenzhen Hospital X-ray Set/China data set: X-ray images in this data set [7], and the Montgomery County X-ray Set [7].

## 2    Methodology

This study presents an original research for the diagnosis of chest X-rays using deep learning. A deep network named as stacked auto-encoder (SAE) is selected to be used as the brain this work. This selection came from the few researches that were conducted for the chest X-rays classification using this kind of networks. Thus, there is a need to investigate the effectiveness and performance of stacked auto-encoder [8] in classifying the chest X-rays and detecting whether a radiograph has a disease or it is normal (healthy).

The proposed network is trained to classify chest images into normal which have no abnormalities or diseased images regardless of the type of the disease. A sample of the database normal and abnormal chest X-rays is shown in Fig. 1.

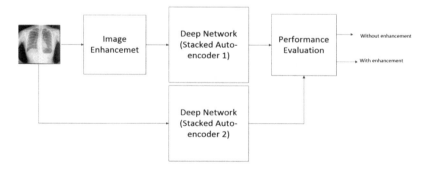

**Fig. 1.** Flowchart of the proposed methodology.

Note that in this work, two deep models are employed. Both models are stacked auto-encoders with the same learning parameters, however, for the first model, which we call SAE1, the chest X-rays are fed directly into network, without processing and enhancement. The second network model, which is called SAE2, was trained on images that are processed and enhanced before being fed into network. The aim of the use of two models is to investigate the effects of processing and image enhancement on the auto-encoder training and testing performance.

Figure 1 shows the workflow of the proposed methodology. As seen, the network model is trained first on the chest images without enhancement and the network is then tested and the performance is evaluated. Same network is then trained and tested on same images but here they are enhanced using image Histogram equalization and similarly, the network is also evaluated and tested in order to investigate the one that outperforms in terms of accuracy and less error achieved.

### 2.1   Database

A deep network an intelligent classifier that is hungry for data. The more data it is trained on the more intelligent it will be. Therefore, there is need for a good database that has good number of normal and abnormal images to train and test the developed network. Therefore, the images in this work are all obtained from two public and well-known databases. The first one is Shenzhen Hospital X-ray Set/China data set: X-ray images in this data set [7], while the other database is the Montgomery County X-ray Set [7]. The first database contains chest X-rays of both normal and abnormal cases and they were acquired as part of the routine care at Shenzhen Hospital. The images are of JPEG format and there 340 normal x-rays and 275 abnormal x-rays showing various aspects of tuberculosis. The second database contains 58 abnormal x-rays and 80 normal images.

### 2.2   Training the Deep Models

In this section the training of the two deep models which are SAE1 and SAE 2 is discussed. Note that the SAE1 is the stacked auto-encoder network that uses X-ray images without enhancement while SAE2 is the same network but with enhanced images as inputs. It is important to mention that SAE1 and SAE2 are both trained on the same number of images which is 470 images; among them 240 are normal and 230 are abnormal.

For output classes coding it was considered as the following:

Abnormal output class [1 0],
Normal output class [0 1].

Note that the networks are first pre-trained as they are deep networks. Pre-training means that the networks are first trained layer by layer using Greedy-layer wise training (Hinton, 2006). In this phase there is no output labeling because the network is trained here to reconstruct its input from the extracted features in the hidden layer, which is why the number of output neurons is equal to the number of input neurons which is 4096.

**SAE1 Training.** SAE1 is a stacked auto-encoder that is trained on 470 images that are fed into it without any processing or enhancement technique. This deep model is composed of one input layer of 4096 neurons since the input images size is 64 * 64 pixels; two hidden layers of 100 and 65 neurons, respectively. Also, it has an output layer of two neurons as the output classes are only two. Figure 2 shows the architecture of the SAE1. Table 2 shows the values of the learning parameters of the SAE1 when it is trained on 200 images.

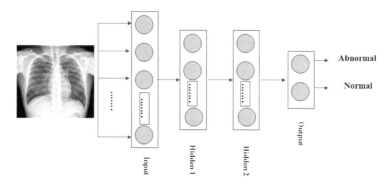

**Fig. 2.** The deep network model 1 (SAE1) structure.

The Fig. 3 depicts the learning of the network SAE1 during fine-tuning. It is also seen that the network error is diminishing sharply and it reaches a very small error of 0.0009 at iteration 323 which indicates a good learning results of the network during this stage.

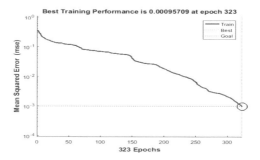

**Fig. 3.** Learning curve of SAE1 during fine-tuning.

Table 1 summarizes the performance of the SAE1 during pre-training and fine-tuning. It is seen that the network SAE1 performed very good during fine-tuning where it achieved a high recognition rate of 100% with a very short time (38 s) and with small number of iterations (323). Moreover, it is seen that the network reached a very small error during fine-tuning (0.0009). However, the network couldn't achieve similarly in the pre-training since it achieved a low recognition rate with long time and high error margins. This is not an issue because the network performance is evaluated in the fine-tuning stage where it is trained to classify however, in the pre-training the network is just trying to get the good and right weights that can be used on the fine-tuning.

**Table 1.** Training network performance.

| Learning results | Pre-training | Fine-tuning |
|---|---|---|
| Number of training images | 470 | 470 |
| Training recognition rate | 87% | 100% |
| Minimum square error achieved (MSE) | 0.0402 | 0.0009 |
| Iterations required | 500 | 323 |
| Training time | 250 s | 38 s |

**SAE2 Training.** SAE2 is the stacked auto-encoder network that is trained on the same chest X-rays, however, the images are processed and enhanced using histogram equalization. This image enhancement technique is an image processing tool that is used to enhance the contrast of the image by mapping or transforming the values of the intensities of pixels in image into different ranges so that the histogram becomes flat. This results in better quality images where the pixels are sharper and brighter. The result of histogram equalization is shown in Fig. 4.

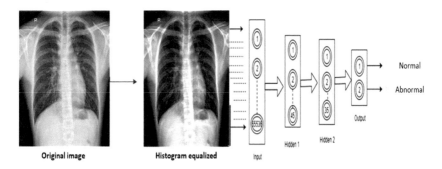

**Fig. 4.** SAE2 network architecture.

From the Fig. 4, it is seen that the histogram helped in brightening the image which may be good in highlighting the important features that distinguish the abnormality of the chest X-rays.

Figure 5 shows the learning curve of the SAE2 network during fine-tuning. It is seen that the network achieved a very small error of 0.0008 within a short time (139 s) and small number of iterations (195).

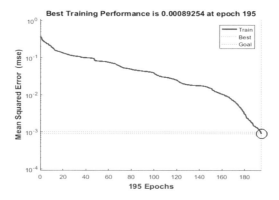

**Fig. 5.** Learning curve of SAE2 during fine-tuning

The Table 2 shows the comparison of the network performance of SAE2 which uses image that are enhanced using histogram equalization. It is noticeable that this image processing technique has contributed to enhancing the network performance in both pre-training and fine-tuning. This is seen in the recognition ration during pre-training which is 90%, greater that of SAE1. Moreover, this network has achieved a smaller error rates (0.0376 and 0.008) than that obtained when histogram equalization is not used. Nevertheless, the use of histogram equalization has resulted in longer training time of the network (524 and 139 s).

**Table 2.** Learning results of SAE2

| Learning results | Pre-training | Fine-tuning |
|---|---|---|
| Number of training images | 470 | 470 |
| Training recognition rate | 90% | 100% |
| Minimum square error achieved (MSE) | 0.0376 | 0.0008 |
| Iterations required | 491 | 195 |
| Training time | 524 s | 139 s |

## 2.3   Testing the Deep Models

Once trained, both networks (SAE1 and SAE2) are tested using 283 chest X-rays. Among them, 180 are normal and the rest are abnormal. Figure 7 shows a sample of the chest images used for testing the networks. Table 3 shows the classification rates of networks during testing.

From Table 3, it can be seen that the stacked auto-encoder which was trained on enhanced images has outperformed the one that used unenhanced images (SAE1) in terms of classification rates.

**Table 3.** Classification rates of both networks during testing

| Deep networks | Number of testing images | Classification rates |
|---|---|---|
| SAE1 | 283 | 89.5% |
| SAE2 | 283 | 93% |

## 3   Results Discussion

In this work, images are first fed into network without enhancement and the network was trained and tested. Then, same images are enhanced using histogram equalization and they are fed into network which is also trained and tested on these processed images. The aim of this is to study if the use of image enhancement technique in particular, histogram equalization affects the learning and performance of the network. Table 4 shows a comparison of the deep networks performances with and without enhancement. Moreover, this table shows the results of the backpropagation neural network (BPNN) when it is trained and tested on images with enhancement.

From Table 4, it can be seen that the network that was trained on images which were enhanced has achieved a 93% recognition rate which is higher than that obtained by the network that uses images without enhancement (89%). Moreover, the SAE2 has achieved a smaller error of 0.0008 smaller than that achieved by SAE1 (0.0009). It is also important to mention that the SAE has achieved such high recognition rate with smaller number of iterations however, with a longer training time. Furthermore, it is noted that the BPNN achieved lower accuracies and higher error rates compared to the other employed deep networks.

**Table 4.** Performance comparison with and without enhancement.

| Performance parameters | SAE1 | SAE2 | BPNN |
|---|---|---|---|
| Number of training images | 470 | 470 | 470 |
| Number of testing images | 283 | 283 | 283 |
| Training recognition rates | 100% | 100% | 99% |
| Testing recognition rates | 89% | 93% | 84.2% |
| Minimum square error achieved (MSE) | 0.0009 | 0.0008 | 0.052 |
| Iterations required | 323 | 195 | 1230 |
| Training time | 38 s | 139 s | 123 s |

Overall, it can be concluded that the enhancement of images using image histogram equalization helps in achieving a higher recognition rate and error; however it requires longer training time which can be due to the process which needs more time.

# 4   Conclusion

A stacked auto-encoder is employed in this work and it was trained and tested on 470 and 283 images, respectively. The network is first trained on images taken directly from the database, without processing and enhancing. Then same network was tested and performance was evaluated in terms of training time, error, and accuracy. The same network was trained again on the same images but here the images were enhanced using histogram equalization. Also, this network was tested and an evaluation of its performance was carried out.

The performance of both networks was discussed and a comparison of the two network performance was shown, in terms of accuracy, error reached, training time, and number of iterations needed. After this comparison it was seen that the network that uses enhanced images outperformed the one that used unprocessed images as it achieved a higher recognition rate during testing.

In conclusion, the testing of stacked auto-encoder showed that it gained a good capability of diagnosing the new unseen chest X-rays and correctly classifying them into normal or abnormal images. Thus, it can be stated that the SAE can be a good classifier for the chest X-rays classification with a small margin of errors. Moreover, it is seen that the enhancement of chest X-rays using histogram equalization has a good role in improving the learning of the stacked auto-encoder, which results in a better accuracy during the testing of the network.

# References

1. Er, O., Yumusak, N., Temurtas, F.: Chest diseases diagnosis using artificial neural networks. Expert Syst. Appl. **37**(12), 7648–7655 (2010)
2. Jaeger, S., Karargyris, A., Candemir, S., Folio, L., Siegelman, J., Callaghan, F., Thoma, G.: Automatic tuberculosis screening using chest radiographs. IEEE Trans. Med. Imaging **33**(2), 233–245 (2014)
3. Helwan, A., Abiyev, R.: Shape and texture features for the identification of breast cancer. In: Proceedings of the 24th World Congress on Engineering and Computer Science, pp. 19–21. Newswood Limited, Hong Kong (2016)
4. Helwan, A., Tantua, D.P.: IKRAI: intelligent knee rheumatoid arthritis identification. Int. J. Intell. Syst. Appl. **8**(1), 18–24 (2016)
5. Dong, Y., Pan, Y., Zhang, J., Xu, W.: Learning to read chest x-ray images from 16000+ examples using CNN. In: Proceedings of the 2nd IEEE/ACM International Conference on Connected Health: Applications, Systems and Engineering Technologies, pp. 51–57. IEEE Computer Society, Washington (2017)
6. Kaymak, S., Helwan, A., Uzun, D.: Breast cancer image classification using artificial neural networks. Procedia Comput. Sci. **120**, 126–131 (2017)
7. Jaeger, S., Candemir, S., Antani, S., Wáng, Y.X., Lu, P.X., Thoma, G.: Two public chest X-ray datasets for computer-aided screening of pulmonary diseases. Quant. Imaging Med. Surg. **4**(6), 475–477 (2017)
8. LeCun, Y., Bengio, Y., Hinton, G.: Deep learning. Nature **521**, 436–444 (2015)

# The Ecosystem of ICT Based Innovation in the Public Sector. The Role of Public Procurement

Ani Matei[(⊠)] [iD], Carmen Săvulescu,
and Corina-Georgiana Antonovici

National University of Political Studies and Public Administration,
Bucharest, Romania
amatei@snspa.ro

**Abstract.** The intrinsic link between public procurement and social innovation is moreover present both in the field literature and policies, documents of the European Union or other international organizations or institutions.

The priority topics refer both to public procurement as driver of innovation in the public sector and to the "smart guide" on innovation for services. The 'smart' feature reflects ICT role and importance as support of innovation in the public services and in general, in social innovation.

In Europe, as in other states/regions of the world, the actual trends describe a systemic behaviour of social innovation, stimulated by substantial measures and tools for its promotion through continuous improvement of its informational and technological support.

The paper aims to substantiate and describe the ecosystem of ICT based innovation in the public sector.

**Keywords:** Social innovation · Public procurement · Ecosystem

## 1 Introduction

For most European states the social innovation has become an important concern, gaining often the characteristics of a public policy or even a strategy.

The modalities and tools for promoting social innovation are moreover diverse, their interaction fostering the premises of an ecosystem. The term covers multiple significations related to the scientific field. Originating from physical or biological sciences, the concept of ecosystem has developed and it has gained new contents, the socio-ecological system being closer to the approach in the current paper, as it is at the crossroad between social and natural sciences.

## 2 Theoretical Framework of the Ecosystem of ICT Based Social Innovation

Characterised by complexity, dynamics and adaptability, the socio-ecological system provides a relevant support for better understanding in a systemic way the evolution of the social innovation processes.

© Springer Nature Switzerland AG 2019
R. A. Aliev et al. (Eds.): ICAFS-2018, AISC 896, pp. 524–532, 2019.
https://doi.org/10.1007/978-3-030-04164-9_70

We find arguments in favour of such an approach in several studies, theoretical and empirical researches. Thus, we mention: "the field of social innovation remains fragmented and there is a need for more developed networks as well as innovation intermediaries for brokering the connection needed to nurture and scale up social innovations" [1].

Pulford [2] identifies three relevant areas concerning the creation of a social innovation ecosystem: learning and training; support structures at a policy level; networks and innovating as a global community. Pulford's final conclusion refers to the fact that "an ecosystem is a set of complex, interdependent relationships that function best through effective networks and communities".

Other researches leading to the use of complex systems have been developed, concerning the European Union administrative system [3] or models for cost – benefit analysis of the anticorruption strategies impact [4].

Social complexity proved to be an enabling framework for social innovation while its modelling is better highlighted through ICT. Approaching digital social innovation, European Commission [5] concludes: "digital technologies are particularly well suited to helping civic action: mobilizing large communities, sharing resources and spreading power".

Conceptualizing and operationalizing the emergence between social innovation and digital technologies, the report defines "DSI Ecosystem", its characteristics deriving from the contemporary developments of new technologies, such as: open access; awareness networks; collaborative economy; new ways of making; open democracy; acceleration and incubation.

Hansson et al. [6] provide an interesting description concerning a strategic agenda for a social innovation ecosystem, which includes:

- "Actors who drive the supply of a social innovations ecosystem;
- Actors who drive the demand for social innovations, and
- The intermediaries who act as brokers between the supply and demand for social innovations by supporting networks and meeting places" [6].

## 3  Public Procurement and Social Innovation

Public procurement proves to be an important support and enabler for social innovation. "Public procurement is at centre of recent demand-side innovation of policy initiatives" [7]. Revealing important concerns about the role of public procurement in the innovation processes, Rolfstam [8] presents analyses from various states (Canada, China, New Zealand, India or Japan), and Edquist et al. [9] for Austria, Greece, France and Sweden.

For the European Union, "public procurement was identified as an important tool for reaching the innovation targets drawn up in the wake of the Lisbon Agenda goals, set to increase competitive advantage in a global economy" [8].

It is also worth to mention Edler and Georghiou's [10] contributions, who consider public procurement as one of the key elements of a demand-oriented innovation policy.

Uyarra et al. [11] approach the barriers for innovation and public procurement, remarking: "the main barriers reported by suppliers refer to the lack of interaction with procuring organizations, the use of over-specified tenders as opposed to outcome based specifications, low competences of procurers and a poor management of risk during the procurement process".

Motivating a good understanding of the mutual impact of innovation and public procurement, [12] accomplish an interesting synthesis, revealed by Table 1.

**Table 1.** Interactions between innovation and public procurement (Source: Adapted after [11])

| Procurement type | Adapted procurement | Technological procurement | Experimental procurement | Efficient procurement |
|---|---|---|---|---|
| Innovation type | Market niche | Architectural | Radical | Regular |

# 4   A Socio-Statistical Analysis of E-Procurement Impact on Social Innovation

The below analysis aims to exemplify the interaction between the e-procurement processes and social innovation in the framework of the social innovation ecosystem in Romania.

The socio-statistical research was accomplished in the framework of FP7 project: Learning from Innovation in Public Sector Environments (LIPSE), while their substantiation and analysis have been provided by Nasi et al. [13] report.

The research was structured for the eight development regions of Romania, which are purely statistical entities.

## 4.1   Descriptive Statistics on the Institutional Context of e-Procurement

The county and regional institutions in Romania have already preoccupations and consecrated methods in view to promote and adopt e-procurement.

Benefiting of an adequate legal framework, which is continuously up dated, e-procurement has become, for the analysed institutions, the main driver and modality to access the resources necessary for institutional development and optimum administrative operation.

The quantitative analysis highlights the fact that 100% of the respondent organizations have adopted and use currently e-procurement.

During time, the attitude concerning the use of e-procurement procedures and especially their generalization has been different related to the local context, resources, training level etc. The evolution has been positive, attaining the uniformization of the opinions and unanimous acceptance.

Concerning the appreciation of the level for each organization related to the e-procurement process, most options reveal the position of "Follower" and "Late adopter".

Analysing the regional level, the averages are as shown by Table 2.

**Table 2.** Type of adopters clustered for geographical area (percentage values) (Source: the authors)

| Type of adopter | Bucharest Ilfov | Center | North East | North West | South | South East | South West | West |
|---|---|---|---|---|---|---|---|---|
| Pioneer | - | - | 14.3% | - | - | 16.6% | - | - |
| Follower | 12.5% | 66.7% | 71.4% | 40% | 42.8% | 66.7% | 40% | - |
| Late adopter | 87.5% | 33.3% | 14.3% | 60% | 57.2% | 16.7% | 60% | 100% |
| Laggards | - | - | - | - | - | - | - | - |
| Non-adopters | - | - | - | - | - | - | - | - |

## 4.2 Descriptive Statistics on Determinants and Barriers of the Outer/Inner Context

The appreciations concerning the determinants and barriers of the outer and inner context are often at the level of the average of the scale of appreciation. They take into consideration the inter-institutional dynamics, economic, social, political, demographical and technological factors in the outer context.

On regions, however, relevant differences emerge both regarding the role of each external or internal factor and its specificity (Table 3).

**Table 3.** Determinants and barriers of the outer and inner context (Source: the authors)

| Factor | Bucharest Ilfov | Center | North East | North West | South | South East | South West | West | Average |
|---|---|---|---|---|---|---|---|---|---|
| Average outer context | 3.54 | 3.41 | 3.27 | 3.38 | 3.39 | 3.47 | 2.89 | 3.29 | 3.33 |
| Average inner context | 3.54 | 3.49 | 3.53 | 3.51 | 3.35 | 3.39 | 3.40 | 3.60 | 3.47 |

A hierarchy of the determinants and barriers of the outer context on regions highlights the differences above mentioned.

Analysing the averages of appreciations concerning the determinants and barriers of the outer context, their positions are situated in totality above the arithmetic average of 2.5, with specific differences on regions, as it results from Table 4.

Correlating the averages of determinants and barriers of the outer context with the type of adopters clustered for geographical areas, we obtain possible influences of these factors on the level of e-procurement process.

It is worth to remark that the variable "Costs for introducing e-procurement in public sector organizations" becomes the less important barrier for adopting e-procurement in most development regions (Table 5).

**Table 4.** Hierarchy of the determinants and barriers of the outer context on development regions (Source: the authors)

| Factor outer context region | Most important determinant and barrier | Less important determinant and barrier |
|---|---|---|
| Bucharest Ilfov (BI) | Territorial ICT infrastructures | Resistance of key external stakeholders |
| Center (C) | Providers' familiarity with ICTs | The imitation of other e-procurement experiences<br>Political vision for e-procurement<br>Conflicts among public institutions and/or political parties<br>The fluent relationship between public sector organizations and external stakeholders<br>Resistance of key external stakeholders |
| North East (NE) | The imitation of other e-procurement experiences | Political vision for e-procurement |
| North West (NW) | Territorial ICT infrastructures | Political vision for e-procurement |
| South (S) | Territorial ICT infrastructures | Conflicts among public institutions and/or political parties<br>The fluent relationship between public sector organizations and external stakeholders |
| South East (SE) | The imitation of other e-procurement experiences | Regional budget constraints |
| South West (SW) | Legislative obligations and standards | Resistance of key external stakeholders |
| West (W) | Law frustrates the adoption of e-procurement in public sector organizations | Resistance of key external stakeholders |

We mention relative influence as in appreciating the type of adopter, one should take also into consideration other determinants such as those concerning the year of implementation, corruption, political will etc.

**Table 5.** Hierarchy of determinants and barriers in inner context (Source: the authors)

| Factor inner context region | Most important determinant and barrier | Less important determinant and barrier |
|---|---|---|
| Bucharest Ilfov (BI) | Managers' professionalism and capabilities | The costs for introducing e-procurement in public sector organizations |
| Center (C) | Consulting, training and other supporting activities | The costs for introducing e-procurement in public sector organizations |
| North East (NE) | Consulting, training and other supporting activities Managers' professionalism and capabilities | The costs for introducing e-procurement in public sector organizations |
| North West (NW) | Consulting, training and other supporting activities Employees' technological skills The quality and quantity of ICT | The costs for introducing e-procurement in public sector organizations |
| South (S) | The lack of interoperability | Personal characteristics of employees |
| South East (SE) | Consulting, training and other supporting activities The quality and quantity of ICT | The costs for introducing e-procurement in public sector organizations |
| South West (SW) | Consulting, training and other supporting activities Employees' technological skills | Personal characteristics of employees The lack of interoperability |
| West (W) | Managers' professionalism and capabilities | Organizational slack resources |

The data and conclusions of Tables 6, 7 and 8 reflect a specificity of the determinants and barriers depending on the type of adopter,

**Table 6.** Correlation type of adopter and average of determinants and barriers of outer and inner context (Source: the authors)

| Region | Type of adopters | Average of determinants and barriers of outer context | Average of determinants and barriers of inner context |
|---|---|---|---|
| Bucharest-Ilfov (BI) | Late adopter | 3.54 | 3.54 |
| Center (C) | Follower | 3.41 | 3.49 |
| North East (NE) | Follower | 3.27 | 3.53 |
| North West (NW) | Late adopter | 3.38 | 3.51 |
| South (S) | Late adopter | 3.39 | 3.35 |
| South East (SE) | Follower | 3.47 | 3.39 |
| South West (SW) | Late adopter | 2.89 | 3.40 |
| West (W) | Late adopter | 3.29 | 3.60 |

They are synthetized in Table 7 (outer context) and Table 8 (inner context).

**Table 7.** Correlation type of adopter/determinants and barriers in outer context (Source: the authors)

| Type of adopter | Determinants and barriers | |
|---|---|---|
| | Most important | Less important |
| Late adopter | Territorial ICT infrastructures | Political vision for e-procurement<br>Resistance of key external stakeholders |
| Follower | The imitation of other e-procurement experiences | Regional budget constraints |

**Table 8.** Correlation type of adopter/determinants and barriers in inner context (Source: the authors)

| Type of adopter | Determinants and barriers | |
|---|---|---|
| | Most important | Less important |
| Late adopter | Managers' professionalism and capabilities | The costs for introducing e-procurement in public sector organizations |
| Follower | Consulting, training and other supporting activities | The costs for introducing e-procurement in public sector organizations |

# 5   Conclusion

The quantitative analysis presents in details and quantifies in a more profound manner the internal and external mechanisms determining the characteristics of the e-procurement processes.

We should remark the fact that although the statistic deviation of the variables is often very small, the results reveal a plausible hierarchy of the role and importance of determinants and barriers in influencing the evolution of e-procurement processes.

A simple analysis of linear regression establishes the direct dependency of type of adopter (T) related to the determinants and barriers of the outer context (OC) and indirect one, but very low, related to the determinants and barriers of the inner context (IC).

Thus, we obtain: $T = 1.906 + 0.236 * OC - 0.057 * IC$.

The regression process reveals standard errors, higher than 1.295 for constant, 0.238 for the coefficient of OC and 0.300 for the coefficient of IC, triggering the necessity of a more detailed analysis for separating the determinants and barriers, so that we could decide the use of the direct variable or its inverse one.

Such modelling could lead to results closer to reality.

Relevant conclusions could be extracted in view of creating an agenda for enhancing the social innovation ecosystem in Romania and improving the impact of public procurement. Thus a main conclusion reveals the **powerful dependency on the financial resources** allocated to the development and modernisation of the activities of e-procurement, as well as on the **social and political environment**.

In fact analysing the determinants and barriers related to regions holding the higher/less relevance, we find out that the **ICT infrastructure represents the most important determinant in outer context.**

For Romania, in spite of the fact that the introduction of procedures for e-procurement belongs to the latest 7–8 years, the development of ICT infrastructure had an important evolution in the context of promoting and implementing the concept of informational, knowledge-based society.

**At the other extreme, there are ranked "the political vision for e-procurement", and "conflicts among public institutions and/or political parties".**

Speaking about these two barriers, we refer both to conflicts in the decisional processes concerning e-procurement and delays in the processes for awarding the contracts due to local or party interests.

**In inner context**, as we do believe that it is specific to several states, **the most important determinant** is the "**managers' professionalism and capabilities**", followed by "consulting, training and other supporting activities".

This finding reveals the constant effort of the Romanian public authorities both for conceiving and operationalizing specific training programmes concerning management and operationalization of e-procurement, and for cooperation with similar entities in the European Union in view of the transfer of best practices in the activities specific for e-procurement.

The quantitative analysis highlights also the **differences of opinions, expressed at regional level related to various determinants and barriers**.

Herewith, we refer to the fact, that for example, interoperability emerges in contrary hypostases in the regional appreciations, fact which could be explained, in our opinion through misunderstanding of the concept and lack of a conclusive vision.

# References

1. TEPSIE Homepage Report, Building the Social Innovation Ecosystem in Europe, p. 4 (2014). http://www.tepsie.eu/images/documents/d73final.pdf. Accessed 21 May 2018
2. Pulford, L.: The global ecosystem for social innovation, social innovation mosaic, pp. 112–113 (2011). https://centres.smu.edu.sg. Accessed 31 May 2018
3. Matei, L., Matei, A.: The administrative system of the European union - from concept to reality. Transylv. Rev. Adm. Sci. **33E**, 170–196 (2011)
4. Matei, L., Matei, A.: Corruption in the public organizations. towards a model of cost-benefit analysis for the anticorruption strategies. Transylv. Rev. Adm. Sci. **27E**, 145–171 (2009)
5. NESTA Homepage, Growing a Digital Social Innovation Ecosystem for Europe, p. 4 (2015). https://www.nesta.org.uk/sites/default/files/dsireport.pdf. Accessed 21 May 2018
6. Hansson, J., Björk, F., Lundborg, D., Olofsson, L.-E. (eds.): An ecosystem for social innovation in Sweden. A strategic research and innovation agenda, p. 10. Lund University, Lund (2014)
7. OECD: Demand Side Innovation Policies. OECD Publishing, Paris (2011)

8. Rolfstam, M.: Public Procurement and Innovation. The Role of Institutions. Edward Elgar, USA (2013)

9. Edquist, Ch., Hommen, L., Tsipouri, L.: Public Technology Procurement and Innovation. Economics of Science, Technology and Innovation, vol. 16. Springer, New York (2000). https://www.springer.com/gp/book/9780792386858. Accessed 19 July 2018

10. Edler, J., Georghiou, L.: Public procurement and innovation – resurrecting the demand side. Res. Policy **36**, 949–963 (2007). https://doi.org/10.1016/j.respol.2007.03.003

11. Uyarra, E., Edler, J., Garcia-Estevez, J., Georghiou, L., Yeow, J.: Barriers to innovation through public procurement: a supplier perspective. Technovation **10**(34), 631–645 (2014). https://doi.org/10.1016/j.technovation.2014.04.003

12. Uyarra, E., Kieron, F.: Understanding the innovation impacts of public procurement. Manchester Business School working paper, No. 574, p. 22 (2009)

13. Nasi, G., Cucciniello, M., Mele, V., Valoti, G., Bazurli, R., De Vries, H., Bekkers, V., Tummers, L., Gascó, M., Ysa, T., Fernández, C., Albareda, A., Matei, A., Savulescu, C., Nemec, J., Svidroňová, M., Mikusova Merickova, B., Oviska, M., De Froidcourt, V., Eymeri - Douzans, M., Monthubert, E.M.: LIPSE research report, determinants and barriers of adoption, diffusion and upscaling of ICT-driven social innovation in the public sector: a comparative study across 6 EU countries (2015). LIPSE: http://lipse.org/. Accessed 28 May 2018

# Fuzzy Set-Theory Analysis of Deregulations Effects on the National Taxable Capacity

Rovshan Akbarov[1]([✉]), Musa Aghamaliyev[2], and Emin Garibli[1]

[1] Azerbaijan State University of Economics, Baku AZ1101, Azerbaijan
*rovshanakperov@yahoo.com*
[2] Baku Eurasian University, Baku AZ1073, Azerbaijan

**Abstract.** Two approaches to assessing the impact of changes in legislation on the tax potential of the country, implemented on the base of the weighted coefficients of influence factors, and verbal model based fuzzy inference are considered. On arbitrary examples of legislative initiatives, a comparative analysis of the results obtained with the application of both methods was carried out.

**Keywords:** Legislative initiative · Taxable capacity · Fuzzy set

## 1 Introduction

Adequate management of the taxation system (TS) is a dynamically complex problem, and the tax management is a weakly structured and, accordingly, difficulty formalizable procedure. At the current stage of the development of information and related technologies, any TS cannot exist in isolation from human. Therefore, its evaluation has not only an objective, but also a subjective component, because in the final stage the analysis of the influence of existing legislation and state administration on the level of TS development is carried out by the person himself. This is a quite important factor, which specifies the use of qualitative categories to evaluation the current level of TS, i.e. terms of linguistic variables (LV), which are the main structural units of the natural language of the agent of management [1]. Just this paradigm explains the necessity of using elements of fuzzy logic.

The basic leads towards optimal taxation are the foresight of the taxable capacity (TC), prevention, localization and elimination of damage from unbalanced decisions. At the same time, the estimation of the TC-level is always relative, and the desire to assign a numerical value to it is unacceptable from the point of view of further interpretation of complex results. TS and TC are complex concepts and cannot be viewed as a simple aggregate of their interrelated and/or interdependent components, since each of them is critically important. During integrated assessment of TC, the numerical description of assessment criterion is not possible. Therefore, it is necessary to develop such a model for integrated assessment of TC-level, so that it can unify approaches to the formation of TS.

© Springer Nature Switzerland AG 2019
R. A. Aliev et al. (Eds.): ICAFS-2018, AISC 896, pp. 533–541, 2019.
https://doi.org/10.1007/978-3-030-04164-9_71

## 2   Expert Estimation of Deregulations Effects on the TC

Realized in the state budget revenues TC depends on a large number of factors, i.e. within the framework of the state TS the forming of the TC is a multi-factor process. There are various ways of classifying these factors. Let us chose one generalized internal and seven external factors of influence (FI) [2]: $x_1$ – *internal*, including regional legislative, socio-demographic, resource-source, infrastructure, innovation, environmental, investment factors, as well as the factors of the "shadow economy", the smoothing of TC and tax culture; $x_2$ – *legislative*, providing direct and/or indirect influence to TC; $x_3$ – *the state form of government*, which directly influences to the formation of revenues of the state budget system; $x_4$ – *macroeconomic*, characterizing the state of the global economy and foreign-economic activity; $x_5$ – *political*, characterizing the political situation and implemented through relations between budgets of different levels; $x_6$ – *inflationary* caused by rise in prices; $x_7$ – *geographical*, defining the perspectives for the development of resource environment; $x_8$ – *natural and climatic conditions*, specified by climatic conditions, availability and composition of natural resources, etc. Suppose that this list of FIs is agreed by the expert community, and its preliminary analysis is carried out on the base of two methods of examination: the comparative qualitative assessment of FIs by the method of experts' preferences, and quantitative estimation of FI-parameters by setting the weights (or normalized values). Further, determining the goodness of fit of expert judgements and the generalized weights of FIs, a decision is made on the base of expert opinions to form weighted assessments of the effect of certain changes in the Tax Code on the TC.

Now, suppose that priority estimates $r_{ij}$ ($j = 1 \div 15$) for the declared FI $x_i$ ($i = 1 \div 8$) were obtained by independent questionnaire of 15 experts in the field of taxation. Each expert was individually suggested to rank the FI $x_i$ by the principle: the most important factor should be designated by the number "1", the next less important one is the number "2" and then the descending order of the expert's preferences. The rank estimates obtained in this way are summarized in the form of Table 1.

To define the coordination degree of expert conclusions it is necessary to apply the Kendall's concordance coefficient, which demonstrates the rank correlation of expert opinions. According to [3], this coefficient is calculated as following:

$$W = 12 \cdot S/[m^2(n^3 - n)], \tag{1}$$

where $m$ is the number of experts; $n$ is the number of FIs; $S$ is the deviation of expert conclusions from the average value of the FIs-ranking, which is calculated as follows:

$$S = \sum_{i=1}^{n} [\sum_{j=1}^{m} r_{ij} - \frac{1}{2}m(n+1)]^2, \tag{2}$$

where $r_{ij} \in \{1, 2, \ldots, 8\}$ is the rank of the $i$-th FI, which is defined by the $j$-th expert ($j = 1 \div m$). On the base of data from Table 1 and formula (2) we have: S = 7424. Then, according to (1), the rank correlation of FI is characterized by the number $W = 12 \cdot 7424/[15^2(8^3 - 8)] = 0.7856$, which indicates a rather high coordination degree of expert conclusions relative to FIs $x_i$ ($i = 1 \div 8$).

**Table 1.** Ranking of FIs subject to priority of experts.

| Expert | Expert rang estimates of FI $x_i$ ($i = 1 \div 8$) | | | | | | | |
|--------|-------|-------|-------|-------|-------|-------|-------|-------|
|        | $x_1$ | $x_2$ | $x_3$ | $x_4$ | $x_5$ | $x_6$ | $x_7$ | $x_8$ |
| 01 | 1 | 3 | 2 | 4 | 5 | 6 | 7 | 8 |
| 02 | 1 | 2 | 4 | 3 | 7 | 6 | 5 | 8 |
| 03 | 2 | 1 | 3 | 5 | 7 | 4 | 6 | 8 |
| 04 | 1 | 3 | 2 | 4 | 7 | 6 | 8 | 5 |
| 05 | 3 | 1 | 2 | 5 | 4 | 7 | 6 | 8 |
| 06 | 1 | 3 | 4 | 2 | 5 | 8 | 6 | 7 |
| 07 | 2 | 3 | 1 | 4 | 6 | 5 | 7 | 8 |
| 08 | 1 | 4 | 2 | 3 | 5 | 6 | 7 | 8 |
| 09 | 2 | 1 | 5 | 4 | 3 | 7 | 6 | 8 |
| 10 | 1 | 2 | 4 | 5 | 3 | 8 | 7 | 6 |
| 11 | 4 | 1 | 2 | 3 | 6 | 5 | 8 | 7 |
| 12 | 1 | 2 | 5 | 4 | 3 | 7 | 6 | 8 |
| 13 | 2 | 1 | 5 | 3 | 4 | 6 | 7 | 8 |
| 14 | 1 | 2 | 3 | 4 | 5 | 7 | 6 | 8 |
| 15 | 2 | 3 | 4 | 1 | 5 | 6 | 8 | 7 |
| $\sum r_{ij}$ | 25 | 32 | 48 | 54 | 75 | 94 | 100 | 112 |

Suppose that at the preliminary stage of independent questioning, each of the experts invited was also instructed to define the values of the normalized estimates of the FIs $x_i$ ($i = 1 \div 8$). The results of this questionnaire survey are summarized in Table 2.

**Table 2.** Expert values of normalized estimates of the FIs $x_i$ ($i = 1 \div 8$).

| Expert | FI $x_i$ ($i = 1 \div 8$) and their normalized estimates ($\alpha_{ij}$) | | | | | | | |
|--------|-------|-------|-------|-------|-------|-------|-------|-------|
|        | $x_1$ | $x_2$ | $x_3$ | $x_4$ | $x_5$ | $x_6$ | $x_7$ | $x_8$ |
| 01 | 0.250 | 0.150 | 0.200 | 0.125 | 0.100 | 0.085 | 0.065 | 0.025 |
| 02 | 0.250 | 0.200 | 0.125 | 0.185 | 0.050 | 0.065 | 0.100 | 0.025 |
| 03 | 0.175 | 0.250 | 0.150 | 0.100 | 0.065 | 0.125 | 0.085 | 0.050 |
| 04 | 0.250 | 0.150 | 0.200 | 0.125 | 0.060 | 0.075 | 0.040 | 0.100 |
| 05 | 0.175 | 0.225 | 0.200 | 0.100 | 0.150 | 0.050 | 0.075 | 0.025 |
| 06 | 0.250 | 0.175 | 0.125 | 0.200 | 0.100 | 0.025 | 0.075 | 0.050 |
| 07 | 0.200 | 0.175 | 0.250 | 0.125 | 0.075 | 0.100 | 0.050 | 0.025 |
| 08 | 0.250 | 0.125 | 0.200 | 0.175 | 0.100 | 0.075 | 0.050 | 0.025 |
| 09 | 0.200 | 0.250 | 0.100 | 0.125 | 0.175 | 0.050 | 0.075 | 0.025 |
| 10 | 0.250 | 0.200 | 0.125 | 0.100 | 0.175 | 0.025 | 0.050 | 0.075 |
| 11 | 0.125 | 0.250 | 0.200 | 0.175 | 0.075 | 0.100 | 0.025 | 0.050 |
| 12 | 0.250 | 0.200 | 0.100 | 0.125 | 0.175 | 0.050 | 0.075 | 0.025 |
| 13 | 0.200 | 0.250 | 0.100 | 0.175 | 0.125 | 0.075 | 0.050 | 0.025 |
| 14 | 0.250 | 0.200 | 0.175 | 0.125 | 0.100 | 0.050 | 0.075 | 0.025 |
| 15 | 0.200 | 0.175 | 0.125 | 0.250 | 0.100 | 0.075 | 0.025 | 0.050 |
| $\sum \alpha_{ij}$ | 3.275 | 2.975 | 2.375 | 2.210 | 1.625 | 1.025 | 0.915 | 0.600 |

To identify FI-weights, as the first step, it is necessary to determine the group estimates for all FIs on the base of data from Table 2. So, the average $\alpha_i$ for the $i$-th group of normalized estimations of PVs is calculated by following equation:

$$\alpha_i(t+1) = \sum_{j=1}^{m} w_j(t)\alpha_{ij}, \tag{3}$$

where $w_j(t)$ is the value characterizing the competence of the $j$-th expert ($j = 1 \div m$) at time t. Averaging of group estimates of normalized values is an iterative process that completes under condition:

$$\max_i \{\alpha_i(t+1) - \alpha_i(t)\} \leq \varepsilon, \tag{4}$$

where $\varepsilon$ is the feasible accuracy, which we choose as $\varepsilon = 0.001$. At the initial stage ($t = 0$) it is assumed that all experts have the same levels of competence as: $w_j(0) = 1/m$ ($j = 1 \div 15$). Then, in the 1st approximation averaging for the $i$-th group of normalized estimates of the FIs are calculated as $\alpha_i(1) = \sum_{j=1}^{15} w_j(0)\alpha_{ij} = \frac{1}{15}\sum_{j=1}^{15} \alpha_{ij}$, and these will be following values: $\alpha_1(1) = 0.2183$; $\alpha_2(1) = 0.1983$; $\alpha_3(1) = 0.1583$; $\alpha_4(1) = 0.1473$; $\alpha_5(1) = 0.1083$; $\alpha_6(1) = 0.0683$; $\alpha_7(1) = 0.0610$; $\alpha_8(1) = 0.0400$. It is not difficult to verify that condition (4) does not fulfillment for these values. Therefore, to search next generation of average values it is necessary to define the normalizing factor as following: $\eta(1) = \sum_{i=1}^{8}\sum_{j=1}^{15}\alpha_i(1)\alpha_{ij} = 0.2183 \cdot 3.275 + 0.1983 \cdot 2.975 + 0.1583 \cdot 2.375 + 0.1473 \cdot 2.210 + 0.1083 \cdot 1.625 + 0.0683 \cdot 1.025 + 0.0610 \cdot 0.9150 + 0.04 \cdot 0.600 = 2.3326$.

Then, the experts' competence indicators are updated by following expressions:

$$\begin{cases} w_j(1) = \dfrac{1}{\eta(1)}\sum_{i=1}^{8}\alpha_i(1)\cdot\alpha_{ij}, j = 1 \div 14; \\ w_{15}(1) = 1 - \sum_{j=1}^{14} w_j(1); \quad \sum_{j=1}^{15} w_j(1) = 1, \end{cases} \tag{5}$$

where $w_{15}(1)$ is the competency indicator of the 15-th expert.

According to (5), the appropriate expert competence indicators will be: $w_1(1) = 0.0669$; $w_2(1) = 0.0678$; $w_3(1) = 0.0639$; $w_4(1) = 0.0654$; $w_5(1) = 0.0662$; $w_6(1) = 0.0676$; $w_7(1) = 0.0666$; $w_8(1) = 0.0672$; $w_9(1) = 0.0666$; $w_{10}(1) = 0.0667$; $w_{11}(1) = 0.0655$; $w_{12}(1) = 0.0671$; $w_{13}(1) = 0.0676$; $w_{14}(1) = 0.0687$; $w_{15}(1) = 0.0662$. Then, in the 2nd approximation averaging for the groups of normalized values of the FIs relevant to the partial expression of formula (3) will be the following numbers: $\alpha_1(2) = 0.5194$; $\alpha_2(2) = 0.4289$; $\alpha_3(2) = 0.3423$; $\alpha_4(2) = 0.4016$; $\alpha_5(2) = 0.2693$; $\alpha_6(2) = 0.1590$; $\alpha_7(2) = 0.1986$; $\alpha_8(2) = 0.1540$. As can be seen from the following excretion: $\max\{\alpha_i(2) - \alpha_i(1)\} = \max\{|0.5194 - 0.2183|,$ $|0.4289 - 0.1983|,$ $|0.3423 - 0.1583|,$ $|0.4016 - 0.1473|,$ $|0.2693 - 0.1083|,$ $|0.1590 - 0.0683|, |0.1986 - 0.0610|, |0.1540 - 0.0400|\} = 0.3011 > \varepsilon$, condition (4) does not fulfillment for these values. Therefore, to search third generation of average values let us define the normalizing multiplier as following: $\eta(2) = \sum_{i=1}^{8} \sum_{j=1}^{15}\alpha_i(2)\alpha_{ij} = 0.51944 \cdot 3.275 + 0.4289 \cdot 2.975 + 0.3423 \cdot 2.37 + 0.4016 \cdot 2.21$

$+ 0.2693 \cdot 1.625 + 0.159 \cdot 1.025 + 0.1986 \cdot 0.9150 + 0.1540 \cdot 0.600 = 5.5523$. According to (5), the appropriate expert competence indicators will be: $w_1(2) = 0.0666$; $w_2(2) = 0.0685$; $w_3(2) = 0.0633$; $w_4(2) = 0.0656$; $w_5(2) = 0.0654$; $w_6(2) = 0.0687$; $w_7(2) = 0.0657$; $w_8(2) = 0.0675$; $w_9(2) = 0.0665$; $w_{10}(2) = 0.0668$; $w_{11}(2) = 0.0648$; $w_{12}(2) = 0.0673$; $w_{13}(2) = 0.0675$; $w_{14}(2) = 0.0683$; $w_{15}(2) = 0.0673$. Then average estimates for the FIs-groups in the third approximation $\alpha_i(3) = \sum_{j=1}^{15} w_j(2)\alpha_{ij}$ will be the follows: $\alpha_1(3) = 0.5204$; $\alpha_2(3) = 0.4297$; $\alpha_3(3) = 0.3430$; $\alpha_4(3) = 0.4027$; $\alpha_5(3) = 0.2703$; $\alpha_6(3) = 0.1599$; $\alpha_7(3) = 0.1997$; $\alpha_8(3) = 0.1551$. As can be seen from: $\max\{\alpha_i(3) - \alpha_i(2)\} = \max\{|0.5204 - 0.5194|, |0.4297 - 0.4289|, |0.3430 - 0.3423|, |0.4027 - 0.4016|, |0.2703 - 0.2693|, |0.1599 - 0.1590|, |0.1997 - 0.1986|, |0.1551 - 0.1540|\} = 0.0012 > \varepsilon$, and for these values the condition (4) is not satisfied again. Therefore, it is necessary to continue the search for group average estimates of the FIs in the 4-th approximation. So, assuming that:

- the normalizing multiplier is $\eta(3) = 0.5204 \cdot 3.2750 + 0.4297 \cdot 2.975 + 0.343 \cdot 2.375 + 0.4027 \cdot 2.21 + 0.2703 \cdot 1.625 + 0.1599 \cdot 1.025 + 0.1997 \cdot 0.9150 + 0.1551 \cdot 0.6 = 5.5666$;
- expert competence indicators in third approximation are: $w_1(3) = 0.0668$; $w_2(3) = 0.0687$; $w_3(3) = 0.0635$; $w_4(3) = 0.0658$; $w_5(3) = 0.0656$; $w_6(3) = 0.0689$; $w_7(3) = 0.0658$; $w_8(3) = 0.0677$; $w_9(3) = 0.0667$; $w_{10}(3) = 0.0670$; $w_{11}(3) = 0.0649$; $w_{12}(3) = 0.0675$; $w_{13}(3) = 0.0677$; $w_{14}(3) = 0.0685$; $w_{15}(3) = 0.0649$,

and according to (3) one can to obtain the following average estimates of the FI-groups: $\alpha_1(4) = 0.5206$; $\alpha_2(4) = 0.4299$; $\alpha_3(4) = 0.3433$; $\alpha_4(4) = 0.4026$; $\alpha_5(4) = 0.2705$; $\alpha_6(4) = 0.1601$; $\alpha_7(4) = 0.1999$; $\alpha_8(4) = 0.1553\}$. As can be seen from: $\max\{\alpha_i(4) - \alpha_i(3)\} = \max\{|0.5206 - 0.5204|, |0.4299 - 0.4297|, |0.3433 - 0.3430|, |0.4026 - 0.4027|, |0.2705 - 0.2703|, |0.1601 - 0.1599|, |0.1999 - 0.1997|, |0.1553 - 0.1551|\} = 3 \cdot 10^{-4} < \varepsilon$, condition (4) is satisfied. In this case, group averaging of the FIs in the 4-th approximation: $\{\alpha_1(4); \alpha_2(4); \alpha_3(4); \alpha_4(4); \alpha_5(4); \alpha_6(4); \alpha_7(4); \alpha_8(4)\}$, can be considered as final appropriate weights of corresponding FIs.

The method of expert evaluations provides for discussion of legislative initiatives (LIs) that affect the TC-level. Experts are invited to make an independent assessment of deregulations effects on the TC in the basis of the FIs $x_i$ ($i = 1 \div 8$) on the base of hundred-point scale: 100 – STRONG; 75 – SIGNIFICANT; 50 – VISIBLE; 25 – WEAK; 0 – INSIGNIFICANT. Expert judgements are subjected to preliminary analysis for their consistency according to the rule: the maximum permissible difference between two expert opinions on any LI for compliance with the FI should not exceed 50 points. This filters unacceptable deviations in expert judgements. The value of the final weighted assessment of LI for its compliance with the FIs $x_i$ ($i = 1 \div 8$) can be calculated by criterion: $C = 100 \times \sum_{i=1}^{8} \alpha_i e_i / (\max_i \sum_{i=1}^{8} \alpha_i e_i)$, where $\alpha_i$ is the weight of the $i$-th FI; $e_i$ is expert estimate of LI for its compliance with the $i$-th FI by the hundred-point scale. Then, based on the average estimates of the expert opinions and the evaluation criterion $C$, the final estimates of deregulations effects on the TC are obtained. Their values are presented in Table 3.

**Table 3.** Comparison of the final estimates of deregulations effects on TC.

| LI | WE | Rang | FI | Rang | LI | WE | Rang | FI | Rang |
|----|------|---|--------|---|----------|-------|----|--------|----|
| $a_1$ | 36.34 | 7 | 0.3168 | 9 | $a_6$ | 28.92 | 9 | 0.3984 | 7 |
| $a_2$ | 61.44 | 3 | 0.4895 | 5 | $a_7$ | 20.34 | 10 | 0.2265 | 10 |
| $a_3$ | 58.38 | 4 | 0.4989 | 3 | $a_8$ | 31.94 | 8 | 0.4260 | 6 |
| $a_4$ | 44.98 | 6 | 0.4973 | 4 | $a_9$ | 52.80 | 5 | 0.3744 | 8 |
| $a_5$ | 73.91 | 2 | 0.5006 | 2 | $a_{10}$ | 85.97 | 1 | 0.7832 | 1 |

# 3  Estimation the Deregulations Effects on the TC by Fuzzy Inference

For the aggregation of expert conclusions relative to LI in the basis of the FIs $x_i$ ($i = 1 \div 8$) the following logically consistent reasoning are chosen as the base:

$d_1$:   "If the LI has the specifiers of following FIs: $x_1$, $x_2$, $x_4$ and $x_5$, then its consolidated influence (CI) on the TC-volume is strong";

$d_2$:   "If in addition to this the LI also has the specifiers of FIs: $x_6$ and $x_8$, then its CI on the TC-volume is more than strong";

$d_3$:   "If the LI has specifiers of all FI, then its CI on the TC-volume is too strong";

$d_4$:   "If the LI has the specifiers of following FIs: $x_1$, $x_2$, $x_3$, $x_4$, $x_5$ and $x_6$, then its CI on the TC-volume is very strong";

$d_5$:   "If the LI has the specifiers of following FIs: $x_1$, $x_2$, $x_4$, $x_5$ and $x_6$, but it does not reflect the specifiers of $x_7$ and $x_8$, then its CI on the TC-volume is strong";

$d_6$:   "If the LI does not reflect the specifiers of following FIs: $x_1$, $x_2$, $x_4$ and $x_6$, then its CI on the TC-volume is weak".

These reasoning can be interpreted in the form of cause-and-effect relations, where inputs are the availability (and/or absence) of specifiers of the FIs $x_i$ ($i = 1 \div 8$) presented by their terms: $X_i$ = OCCUR and/or $\neg X_i$ = NO OCCUR, and output is a linguistic variable (LV) $y$ presented by its terms: $S$ = STRONG, $MS$ = MORE THAN STRONG, $VS$ = VERY STRONG, $TS$ = TOO STRONG and $US$ = WEAK, which reflect the quantity of the consolidated influence. Fuzzification of the terms of the input LP $x$ is carried out on the base of the discrete universe $\{a_1, a_2, ..., a_{10}\}$ by Gaussian membership function: $\mu(u) = \exp[-(u - 100)^2/\sigma^2]$, where $\sigma^2 = 1225$ is the density chosen for all cases. Then the influences of the LI on the TC-volume are reflected by following fuzzy sets (FS):

- *internal* as: $X_1$ = $\{0.0041/a_1; 0.2369/a_2; 0.0032/a_3; 0.1299/a_4; 0.9798/a_5; 0.0061/a_6; 0.0005/a_7; 0.0007/a_8; 0.0090/a_9; 0.9216/a_{10}\}$;
- *legislative* as: $X_2$ = $\{0.0036/a_1; 0.8114/a_2; 0.9710/a_3; 0.9216/a_4; 0.8322/a_5; 0.4111/a_6; 0.0027/a_7; 0.0145/a_8; 0.1525/a_9; 0.6004/a_{10}\}$;
- *the state form of government* as: $X_3$ = $\{0.1525/a_1; 0.9870/a_2; 0.1100/a_3; 0.0705/a_4; 0.5273/a_5; 0.0007/a_6; 0.0010/a_7; 0.1778/a_8; 0.5759/a_9; 0.8322/a_{10}\}$;
- *macroeconomic* as: $X_4$ = $\{0.0018/a_1; 0.0392/a_2; 0.4797/a_3; 0.0041/a_4; 0.8711/a_5; 0.0004/a_6; 0.0004/a_7; 0.0012/a_8; 0.1409/a_9; 0.1409/a_{10}\}$;

- *political* as: $X_5 = \{0.0529/a_1; 0.9059/a_2; 0.4335/a_3; 0.0256/a_4; 0.0318/a_5; 0.0007/a_6; 0.7448/a_7; 0.6004/a_8; 0.9798/a_9; 0.9216/a_{10}\}$;
- *inflationary* as: $X_6 = \{0.0925/a_1; 0.1648/a_2; 0.8114/a_3; 0.0145/a_4; 0.0013/a_5; 0.4111/a_6; 0.0013/a_7; 0.5033/a_8; 0.1100/a_9; 0.7214/a_{10}\}$;
- *geographical* as: $X_7 = \{0.9059/a_1; 0.0016/a_2; 0.0090/a_3; 0.0021/a_4; 0.9927/a_5; 0.0529/a_6; 0.0090/a_7; 0.0018/a_8; 0.0846/a_9; 0.9798/a_{10}\}$;
- *natural and climatic* as: $X_8 = \{0.9608/a_1; 0.0061/a_2; 0.9927/a_3; 0.2535/a_4; 0.1100/a_5; 0.2059/a_6; 0.4111/a_7; 0.1010/a_8; 0.1100/a_9; 0.9798/a_{10}\}$.

Above terms of LP $y$ can be described by fuzzy subset of a discrete universe $J = \{0; 0.1; \ldots; 1\}$ with following appropriate membership functions (MF) [1, 4]: $\mu_S(j) = j$; $\mu_{MS}(j) = j^{1/2}$; $\mu_{VS}(j) = j^2$; $\mu_{TS}(j) = 1$, if $j = 1$ and $\mu_{TS}(j) = 0$, if $j < 1$; $\mu_{US}(j) = 1 - j$, where $j \in J$. Then, taking into account the formalisms introduced, the statements $d_1 \div d_6$ can be written as:

$d_1$: "If $x_1 = X_1$ and $x_2 = X_2$ and $x_4 = X_4$ and $x_5 = X_5$, then $y = S$";
$d_2$: "If $x_1 = X_1$ and $x_2 = X_2$ and $x_4 = X_4$ and $x_5 = X_5$ and $x_6 = X_6$ and $x_8 = X_8$, then $y = MS$";
$d_3$: "If $x_1 = X_1$ and $x_2 = X_2$ and $\ldots$ and $x_8 = X_8$, then $y = TS$";
$d_4$: "If $x_1 = X_1$ and $x_2 = X_2$ and $x_3 = X_3$ and $x_4 = X_4$ and $x_5 = X_5$ and $x_6 = X_6$, then $y = VS$";
$d_5$: "If $x_1 = X_1$ and $\ldots$ and $x_6 = X_6$ and $x_7 = \neg X_7$ and $x_8 = \neg X_8$, then $y = S$";
$d_6$: "If $x_1 = \neg X_1$ and $x_2 = \neg X_2$ and $x_4 = \neg X_4$ and $x_6 = \neg X_6$, then $y = US$".

Realizing the logical operation "AND" for the left-hand parts of the rules by intersection of FS, namely [1]:

- $\mu_{M1}(u) = \min\{\mu_{X1}(a), \mu_{X2}(a), \mu_{X4}(a), \mu_{X5}(a)\}$, $M_1 = \{0.002/a_1; 0.039/a_2; 0.003/a_3; 0.004/a_4; 0.032/a_5; 0.0004/a_6; 0.0004/a_7; 0.0007/a_8; 0.009/a_9; 0.6004/a_{10}\}$;
- $\mu_{M2}(u) = \min\{\mu_{X1}(a), \mu_{X2}(a), \mu_{X4}(a), \mu_{X5}(a), \mu_{X6}(a), \mu_{X8}(a)\}$, $M_2 = \{0.002/a_1; 0.006/a_2; 0.003/a_3; 0.004/a_4; 0.001/a_5; 0.0004/a_6; 0.0004/a_7; 0.0007/a_8; 0.009/a_9; 0.600/a_{10}\}$;
- $\mu_{M3}(u) = \min\{\mu_{X1}(a), \mu_{X2}(a), \ldots, \mu_{X8}(a)\}$, $M_3 = \{0.002/a_1; 0.002/a_2; 0.003/a_3; 0.002/a_4; 0.001/a_5; 0.0004/a_6; 0.0004/a_7; 0.0007/a_8; 0.009/a_9; 0.6004/a_{10}\}$;
- $\mu_{M4}(u) = \min\{\mu_{X1}(a), \mu_{X2}(a), \ldots, \mu_{X6}(a)\}$, $M_4 = \{0.002/a_1; 0.039/a_2; 0.003/a_3; 0.004/a_4; 0.001/a_5; 0.0004/a_6; 0.0004/a_7; 0.0007/a_8; 0.009/a_9; 0.6004/a_{10}\}$;
- $\mu_{M5}(u) = \min\{\mu_{X1}(a), \mu_{X2}(a), \mu_{X4}(a), \mu_{X5}(a), \mu_{X6}(a), 1-\mu_{X7}(a), 1 - \mu_{X8}(a)\}$, $M_5 = \{0.002/a_1; 0.039/a_2; 0.007/a_3; 0.004/a_4; 0.001/a_5; 0.0004/a_6; 0.0004/a_7; 0.001/a_8; 0.110/a_9; 0.0202/a_{10}\}$;
- $\mu_{M6}(u) = \min\{1 - \mu_{X1}(a), 1 - \mu_{X2}(a), 1 - \mu_{X4}(a), 1 - \mu_{X6}(a)\}$, $M_6 = \{0.908/a_1; 0.189/a_2; 0.029/a_3; 0.078/a_4; 0.020/a_5; 0.589/a_6; 0.997/a_7; 0.497/a_8; 0.848/a_9; 0.078/a_{10}\}$,

and Lukasiewicz's implication: $\mu_{X \times Y}(u, j) = \min\{1; 1 - \mu_X(u) + \mu_Y(j)\}$, fuzzy relations in the form of matrices $R_1 \div R_6$ were formed for each pair $(u, j) \in X \times Y$. As a result, following required functional solution is obtained by intersection of these relations:

$$R = \begin{array}{c|ccccccccccc} & 0 & 0.1 & 0.2 & 0.3 & 0.4 & 0.5 & 0.6 & 0.7 & 0.8 & 0.9 & 1 \\ \hline a_1 & 0.9982 & 0.9925 & 0.8925 & 0.7925 & 0.6925 & 0.5925 & 0.4925 & 0.3925 & 0.2925 & 0.1925 & 0.0925 \\ a_2 & 0.9608 & 0.9708 & 0.9984 & 0.9984 & 0.9984 & 0.9984 & 0.9984 & 0.9984 & 0.9984 & 0.9114 & 0.8114 \\ a_2 & 0.9927 & 0.9968 & 0.9968 & 0.9968 & 0.9968 & 0.9968 & 0.9968 & 0.9968 & 0.9968 & 0.9968 & 0.9710 \\ a_4 & 0.9959 & 0.9979 & 0.9979 & 0.9979 & 0.9979 & 0.9979 & 0.9979 & 0.9979 & 0.9979 & 0.9979 & 0.9216 \\ a_5 & 0.9682 & 0.9987 & 0.9987 & 0.9987 & 0.9987 & 0.9987 & 0.9987 & 0.9987 & 0.9987 & 0.9987 & 0.9798 \\ a_6 & 0.9996 & 0.9996 & 0.9996 & 0.9996 & 0.9996 & 0.9111 & 0.8111 & 0.7111 & 0.6111 & 0.5111 & 0.4111 \\ a_7 & 0.9996 & 0.9027 & 0.8027 & 0.7027 & 0.6027 & 0.5027 & 0.4027 & 0.3027 & 0.2027 & 0.1027 & 0.0027 \\ a_8 & 0.9988 & 0.9993 & 0.9993 & 0.9993 & 0.9993 & 0.9993 & 0.9033 & 0.8033 & 0.7033 & 0.6033 & 0.5033 \\ a_9 & 0.8900 & 0.9900 & 0.9525 & 0.8525 & 0.7525 & 0.6525 & 0.5525 & 0.4525 & 0.3525 & 0.2525 & 0.1525 \\ a_{10} & 0.3996 & 0.3996 & 0.3996 & 0.3996 & 0.3996 & 0.3996 & 0.3996 & 0.3996 & 0.3996 & 0.3996 & 0.9216 \end{array}$$

According to [4], the fuzzy conclusion relative to final estimation of the $k$-th LI-influence on the TC is reflected as a fuzzy subset $E_k$ of the universe $J$ with the corresponding values of the MF from the $k$-th row of the matrix $R$. In particular, the conclusion concerning the influence of the 1-st LI is reflected as FS: $E_1 = \{0.9982/0;$ $0.9925/0.1;\ 0.8925/0.2;\ 0.7925/0.3;\ 0.6925/0.4;\ 0.5925/0.5;\ 0.4925/0.6;\ 0.3925/0.7;$ $0.2925/0.8;\ 0.1925/0.9;\ 0.0925/1\}$. Defining its level sets $(E_{1\alpha})$ and calculating the corresponding cardinal numbers as $M(E_{1\alpha}) = \sum(i_r/n)$, where $i_r \in E_{1\alpha}$, $r = 1 \div n$ [4], namely:

- for $0 < \alpha < 0.0925$: $\Delta\alpha = 0.0925$, $E_{1\alpha} = \{0;\ 0.1;\ 0.2;\ \ldots;\ 0.9;\ 1\}$. $M(E_{1\alpha}) = 0.50$;
- for $0.0925 < \alpha < 0.1925$: $\Delta\alpha = 0.1$, $E_{1\alpha} = \{0;\ 0.1;\ \ldots;\ 0.8;\ 0.9\}$, $M(E_{1\alpha}) = 0.45$;
- for $0.1925 < \alpha < 0.2925$: $\Delta\alpha = 0.1$, $E_{1\alpha} = \{0;\ 0.1;\ 0.2;\ \ldots;\ 0.7;\ 0.8\}$, $M(E_{1\alpha}) = 0.40$;
- for $0.2925 < \alpha < 0.3925$: $\Delta\alpha = 0.1$, $E_{1\alpha} = \{0;\ 0.1;\ 0.2;\ \ldots;\ 0.6;\ 0.7\}$, $M(E_{1\alpha}) = 0.35$;
- for $0.3925 < \alpha < 0.4925$: $\Delta\alpha = 0.1$, $E_{1\alpha} = \{0;\ 0.1;\ 0.2;\ \ldots;\ 0.5;\ 0.6\}$, $M(E_{1\alpha}) = 0.30$;
- for $0.4925 < \alpha < 0.5925$: $\Delta\alpha = 0.1$, $E_{1\alpha} = \{0;\ 0.1;\ 0.2;\ 0.3,\ 0.4,\ 0.5\}$, $M(E_{1\alpha}) = 0.25$;
- for $0.5925 < \alpha < 0.6925$: $\Delta\alpha = 0.1$, $E_{1\alpha} = \{0;\ 0.1;\ 0.2;\ 0.3,\ 0.4\}$, $M(E_{1\alpha}) = 0.20$;
- for $0.6925 < \alpha < 0.7925$: $\Delta\alpha = 0.1$, $E_{1\alpha} = \{0;\ 0.1;\ 0.2;\ 0.3\}$, $M(E_{1\alpha}) = 0.15$;
- for $0.7925 < \alpha < 0.8925$: $\Delta\alpha = 0.1$, $E_{1\alpha} = \{0;\ 0.1;\ 0.2\}$, $M(E_{1\alpha}) = 0.10$;
- for $0.8925 < \alpha < 0.9925$: $\Delta\alpha = 0.1$, $E_{1\alpha} = \{0;\ 0.1\}$, $M(E_{1\alpha}) = 0.05$;
- for $0.9925 < \alpha < 0.9982$: $\Delta\alpha = 0.0057$, $E_{1\alpha} = \{0\}$, $M(E_{1\alpha}) = 0.00$,

defuzzified estimate is obtained as following [4]: $F(E_1) = \frac{1}{\alpha_{max}} \int_0^{\alpha_{max}} d\alpha = \frac{1}{0.9982} [0.0925 \cdot 0.5 + 0.1 \cdot 0.45 + 0.1 \cdot 0.4 + 0.1 \cdot 0.35\ +\ 0.1 \cdot 0.3\ +\ 0.1 \cdot 0.25\ + 0.1 \cdot\ 0.2 +\ + 0.1 \cdot 0.15\ + 0.1 \cdot 0.1\ + 0.1 \cdot 0.05 + 0.0057 \cdot 0] = 0.3168$. Defuzzified estimates for other conclusions were found by similar actions, which are summarized in Table 3.

## 4   Conclusion

On the base of averaged preliminary expert judgements of deregulations effects on TC, aggregated estimates of ten LI were obtained by method of weighted estimation (WE) and fuzzy inference (FI). The obtained results are placed in Table 3.

**Acknowledgement.** The authors consider necessary to express their appreciation to Professor R. R. Rzayev for the help that he rendered during the process of writing and preparing this article.

# References

1. Zadeh, L.A.: The concept of a linguistic variable and its application to approximate reasoning. Inf. Sci. **8**(3), 199–249 (1965)
2. Filatova, E.A.: Analysis of factors affecting the volume of the region tax potential. Bull. Econ. Law Sociol. **1**, 112–116 (2013). (in Russian)
3. Lin, A.S., Wu, W.: Statistical Tools for Measuring Agreement. Springer, New York (2012). https://doi.org/10.1007/978-1-4614-0562-7
4. Rzayev, R.R.: Analytical Support for Decision-Making in Organizational Systems. Palmerium Academic Publishing, Saarbruchen (2016). (in Russian)

# Ranking of Universities on the Base of Fuzzy Evaluation Procedure of Their Index of Competitiveness

Zeynal Jamalov[1](✉), Alla Khudadova[2], Inara Abasova[3], and Narmina Aliyeva[3]

[1] Institute of Control Systems of ANAS, Baku AZ1141, Azerbaijan
zjamalov@mail.ru
[2] Baku State University, Baku AZ1148, Azerbaijan
[3] Azerbaijan State Oil and Industrial University, Baku AZ1010, Azerbaijan

**Abstract.** Fuzzy models for estimating the marketing space of educational services and the competitive position of universities in the labor market are developed and tested. In the context of this study, indexes of competitiveness characterized and randomly selected universities are ranked.

**Keywords:** Competitiveness · Marketing space · Fuzzy set

## 1 Introduction

First, any university is a complex, open and dynamically developing system. Moreover, its distinctive feature is the ability to timely adaptation to environmental conditions, which is expressed by a system of indicators that characterizes the competitiveness of university. In terms of economists, the competitiveness of the university is a cumulative characteristic of the education process, which reflects the level of its compliance with concrete social utility. In other words, under the competitive advantage of the university we will understand how it can gain an edges in the market of educational services by consolidating and effectively organizing the available resources.

The existing tendencies of the globalization of education (for example, within the framework of the Bologna Education System) dictate an urgent and actual problem for any university, related to its survival and prospects in the market of educational services.

The question naturally arises: how to achieve a competitive advantage and keep it in the rapidly changing competitive environment? Unfortunately, majority of universities in the post-Soviet space do not have the necessary competitive advantages to compete not only with leading, but also even with the "middling" from the world rating list of universities.

Any university can gain a competitive advantage in the framework of the solution of the issue of its comprehensive assessment and effective management. Therefore, at the initial stage, a methodology for assessing the competitiveness of the national university is needed, which adapted to the task solution of multi-criteria assessment of the behavior of a humanistic type system, characterized by weakly structured indicators.

© Springer Nature Switzerland AG 2019
R. A. Aliev et al. (Eds.): ICAFS-2018, AISC 896, pp. 542–550, 2019.
https://doi.org/10.1007/978-3-030-04164-9_72

## 2  Indicators of the Marketing Environment

Each university is obliged to correctly estimate the current conjuncture in the national market of educational services in order to offer adequate means of competition subject to its own characteristics and capabilities. The first and, perhaps, the most difficult problem here is to investigate the cause-effect relationships for achievement of competitive advantages. The result of such reasoning should be the university strategy selection in the market of educational services, which is formed on the base of research of the marketing environment (ME) and evaluation of the competitive position of educational services in the labor market.

Therefore, starting from this paradigm, let us consider the cause-effect relationships that characterize the current market environment in order to evaluate the competitiveness of the educational service of a particular university in support of developing the appropriate behavior strategy in the market of educational services.

The review of characteristics of subjects and factors of the ME made it possible to identify 13 indicators of the ME subject to activities of other universities [1]:

- government support ($x_1$);
- competition of applicants ($x_2$);
- relationships with the secondary schools ($x_3$);
- popularization by media ($x_4$);
- state of the national economy ($x_5$);
- socio-demographic dynamic ($x_6$);
- socio-cultural situation of the society ($x_7$);
- political and legal support of the society ($x_8$);
- target orientation ($x_9$);
- applied education technology ($x_{10}$);
- organizational structure of the university ($x_{11}$);
- level of faculty and training of support staff ($x_{12}$);
- number of universities-competitors ($x_{13}$).

Obviously, for carrying out analytical studies, i.e. for the knowledge compilation that requires the transfer of accumulated external knowledge of the problem into internal ones, it is necessary to conduct the statistical analysis of the ME for each of these criteria. This is the labor-intensive process, which requires the reliability of the relevant information.

We proceed from the assumption that statistical analysis is the subject of separate consideration. Our task is to compile the expert knowledge about the ME in terms of arbitrary universities characterized by their external and internal factors of competitiveness influences. Therefore, let us assume that expert judgments of the ME indicators $x_i$ ($i = 1 \div 13$) for hypothetical five universities $a_1$, $a_2$, $a_3$, $a_4$ and $a_5$ were obtained by independent questionnaire of specialists in the subject area. Consolidated average expert estimates are obtained on the base of ten-point scale and are summarized in Table 1.

**Table 1.** Consolidated average expert estimates of ME.

| Factors | Indicators of the ME | Consolidated average expert estimates | | | | |
|---------|----------------------|-------|-------|-------|-------|-------|
| | | $a_1$ | $a_2$ | $a_3$ | $a_4$ | $a_5$ |
| $x_1$ | Government support | 7.23 | 2.91 | 9.02 | 2.48 | 2.13 |
| $x_2$ | Competition of applicants | 7.08 | 9.13 | 6.30 | 8.50 | 7.56 |
| $x_3$ | Relationships with the secondary schools | 8.95 | 3.96 | 6.23 | 6.30 | 4.34 |
| $x_4$ | Popularization by media | 3.19 | 2.60 | 7.69 | 3.91 | 6.24 |
| $x_5$ | State of the national economy | 6.42 | 9.36 | 7.66 | 2.63 | 4.98 |
| $x_6$ | Socio-demographic dynamic | 6.38 | 6.65 | 5.99 | 5.47 | 2.19 |
| $x_7$ | Socio-cultural situation of the society | 6.54 | 6.28 | 3.11 | 9.78 | 4.08 |
| $x_8$ | Political and legal support of the society | 5.11 | 8.53 | 9.42 | 2.49 | 7.61 |
| $x_9$ | Target orientation | 5.36 | 6.23 | 9.47 | 2.70 | 9.36 |
| $x_{10}$ | Applied education technology | 4.37 | 6.85 | 4.88 | 4.16 | 5.43 |
| $x_{11}$ | Organizational structure of the university | 8.14 | 3.71 | 8.44 | 6.99 | 8.05 |
| $x_{12}$ | Level of faculty and training of staff | 9.28 | 4.56 | 6.95 | 6.68 | 9.33 |
| $x_{13}$ | The number of universities-competitors | 7.12 | 8.09 | 5.33 | 7.95 | 7.09 |

## 3   The Choice of the Most Competitive University with the Application of Fuzzy Inference

In most cases, the values of the indicators $x_i$ ($i = 1 \div 13$) are, in fact, qualitative criteria for assessing the ME of universities (MEU), which are best reflected in the terms of the appropriate linguistic variables (LV). These terms are inherently weakly structured evaluation concepts of the type: "SUFFICIENT", "HIGH", "ACCEPTABLE", etc. Such concepts are best described by fuzzy sets [2]. Therefore, to estimate the MEU the following reasonable and logically consistent judgments can be chosen:

$d_1$:   "If the government support is sufficient, the current state of the national economy is stable, there is a growth in the socio-demographic dynamics, the target orientation of the university is reasonable and the level of faculty and training of staff is high, then the MEU is favorable";

$d_2$:   "If, in addition to the above requirements, the competition of applicants is high, the current socio-cultural situation of the society is high, the organizational structure of the university is balanced and the number of universities-competitors is large, then the MEU is more than favorable";

$d_3$:   "If, in addition to the conditions stipulated in $d_2$, the relationships of university with the secondary schools are close, the popularization of the university by media is continuous, the political and legal support of the society is appropriate, the university applies the modern education technologies, then its ME is perfect";

$d_4$:   "If the MEU is characterized by everything stipulated in $d_3$, except for information about the popularization by media and, the target orientation of the university is reasonable, then the MEU is very favorable";

$d_5$:    "If the government support is sufficient, the competition of applicants is high, the current state of the national economy is stable, there is a growth in the socio-demographic dynamics, the current socio-cultural situation of the society is high, the political and legal support of the society is appropriate, the target orientation of the university is reasonable, the university applies the modern education technologies, the level of faculty and training of staff is high, the number of universities-competitors is large, but the relationships of university with the secondary schools are not close, the university do not popularize by media and its organizational structure is unbalanced, then the MEU is still favorable";

$d_6$:    "If the competition of applicants is not high and the university do not apply the modern education technologies, its organizational structure is unbalanced and the level of faculty and training of staff is not high, then the MEU is unfavorable".

Analysis of the given statements in the form of cause-effect relations allows us to identify 13 inputs (values of ME indicators $x_k$) and one output ($y$ – favorability of the ME) as corresponding evaluation concepts. Then the model for evaluating the MEU can be interpreted in the form of the following implicative rules:

$d_1$:    "If $x_1$ = SUFFICIENT and $x_5$ = STABLE and $x_6$ = GROWING and $x_9$ = REASONABLE and $x_{12}$ = HIGH, then $y$ = FAVORABLE";

$d_2$:    "If $x_1$ = SUFFICIENT and $x_2$ = HIGH and $x_5$ = STABLE and $x_6$ = GROWING and $x_7$ = HIGH and $x_9$ = REASONABLE and $x_{11}$ = BALANCED and $x_{12}$ = HIGH and $x_{13}$ = LARGE, then $y$ = MORE THAN FAVORABLE";

$d_3$:    "If $x_1$ = SUFFICIENT and $x_2$ = HIGH and $x_3$ = CLOSE and $x_4$ = CONTINUOUS and $x_5$ = STABLE and $x_6$ = GROWING and $x_7$ = HIGH and $x_8$ = APPROPRIATE and $x_9$ = REA-SONABLE and $x_{10}$ = MODERN and $x_{11}$ = BALANCED and $x_{12}$ = HIGH and $x_{13}$ = LARGE, then $y$ = PERFECT";

$d_4$:    "If $x_1$ = SUFFICIENT and $x_2$ = HIGH and $x_3$ = CLOSE and $x_5$ = STABLE and $x_6$ = GROW-ING and $x_7$ = HIGH and $x_8$ = APPROPRIATE and $x_{10}$ = MODERN and $x_{11}$ = BALANCED and $x_{12}$ = HIGH and $x_{13}$ = LARGE, then $y$ = VERY FAVORABLE";

$d_5$:    "If $x_1$ = SUFFICIENT and $x_2$ = HIGH and $x_3$ = NOT CLOSE and $x_4$ = CONTINUOUS and $x_5$ = STABLE and $x_6$ = GROWING and $x_7$ = HIGH and $x_8$ = APPROPRIATE and $x_9$ = REA-SONABLE and $x_{10}$ = MODERN and $x_{11}$ = BALANCED and $x_{12}$ = HIGH and $x_{13}$ = LARGE, then $y$ = FAVORABLE";

$d_6$:    "If $x_{12}$ = NOT HIGH and $x_{10}$ = NOT MODERN and $x_{11}$ = UNBALANCED and $x_{12}$ = NOT HIGH, then $y$ = UNFAVORABLE".

The solution set is characterized by the set of criteria – the evaluated concepts of the indicators $x_i$ ($i = 1 \div 13$), which can be describe by fuzzy subsets of the discrete universe $U = \{a_1, a_2, a_3, a_4, a_5\}$. According to [3], as the membership function of fuzzy sets a Gaussian function is chosen in the type of $\mu(u) = \exp[-(u - 10)^2/\sigma^2]$, where $\sigma^2$ is the density of the distribution of neighboring elements, which determines the width of the function. Then, assuming $\sigma = 3.1$ it is possible to describe the evaluated concepts in the left-hand parts of the rules $d_1 \div d_6$ in the form of the following fuzzy sets:

- SUFFICIENT (government support): $X_1 = \{0.4498/a_1;\ 0.0053/a_2;\ 0.9052/a_3;\ 0.0028/a_4;\ 0.0016/a_5\}$,
- HIGH (competition of applicants): $X_2 = \{0.4111/a_1;\ 0.9244/a_2;\ 0.2406/a_3;\ 0.7923/a_4;\ 0.5378/a_5\}$,
- CLOSE (relationships with the secondary schools): $X_3 = \{0.8924/a_1;\ 0.0224/a_2;\ 0.2287/a_3;\ 0.2414/a_4;\ 0.0356/a_5\}$,
- CONTINUES (popularization by media): $X_4 = \{0.0080/a_1;\ 0.0034/a_2;\ 0.5729/a_3;\ 0.0211/a_4;\ 0.2292/a_5\}$,
- STABLE (state of the national economy): $X_5 = \{0.2645/a_1;\ 0.9587/a_2;\ 0.5669/a_3;\ 0.0035/a_4;\ 0.0728/a_5\}$,
- GROWING (socio-demographic dynamic): $X_6 = \{0.2551/a_1;\ 0.3121/a_2;\ 0.1871/a_3;\ 0.1183/a_4;\ 0.0018/a_5\}$,
- HIGH (socio-cultural situation of the society): $X_7 = \{0.2873/a_1;\ 0.2373/a_2;\ 0.0071/a_3;\ 0.9948/a_4;\ 0.0262/a_5\}$,
- APPROPRIATE (political and legal support of the society): $X_8 = \{0.0835/a_1;\ 0.7981/a_2;\ 0.9655/a_3;\ 0.0028/a_4;\ 0.5519/a_5\}$,
- REASONABLE (target orientation): $X_9 = \{0.1063/a_1;\ 0.2280/a_2;\ 0.9709/a_3;\ 0.0039/a_4;\ 0.9588/a_5\}$,
- MODERN (applied education technology): $X_{10} = \{0.0368/a_1;\ 0.3570/a_2;\ 0.0655/a_3;\ 0.0289/a_4;\ 0.1137/a_5\}$,
- BALANCED (organizational structure of the university): $X_{11} = \{0.6973/a_1;\ 0.0163/a_2;\ 0.7767/a_3;\ 0.3890/a_4;\ 0.6742/a_5\}$,
- HIGH (level of faculty and training of staff): $X_{12} = \{0.9473/a_1;\ 0.0460/a_2;\ 0.3807/a_3;\ 0.3181/a_4;\ 0.9543/a_5\}$,
- LARGE (the number of universities-competitors): $X_{13} = \{0.4219/a_1;\ 0.6837/a_2;\ 0.1032/a_3;\ 0.6449/a_4;\ 0.4139/a_5\}$.

To describe terms from the right-hand parts of the rules $d_1 \div d_6$ the discrete set $J = \{0;\ 0.1;\ 0.2;\ \ldots;\ 1\}$ is chosen as a universe. Then $\forall j \in J$ their appropriate membership functions are chosen as follows [2, 3]: $F$ = FAVORABLE: $\mu_F(j) = j$; $MF$ = MORE THAN FAVORABLE: $\mu_{MF}(j) = j^{1/2}$; $VF$ = VERY FAVORABLE: $\mu_{Y3}(j) = j^2$; $P$ = PERFECT: $\mu_P(j) = 1$, if $j = 1$ and $\mu_P(j) = 0$, if $j < 1$; $UF$ = UNFAVORABLE: $\mu_{UF}(j) = 1 - j$.

Thus, taking into account the introduced formalisms for evaluation concepts, the rules $d_1 \div d_6$ can be rewritten in the following form:

$d_1$:  "If $x_1 = X_1$ and $x_5 = X_5$ and $x_6 = X_6$ and $x_9 = X_9$ and $x_{12} = X_{12}$, then $y = F$";

$d_2$:  "If $x_1 = X_1$ and $x_2 = X_2$ and $x_5 = X_5$ and $x_6 = X_6$ and $x_7 = X_7$ and $x_9 = X_9$ and $x_{11} = X_{11}$ and $x_{12} = X_{12}$ and $x_{13} = X_{13}$, then $y = MF$";

$d_3$:  "If $x_1 = X_1$ and $x_2 = X_2$ and $x_3 = X_3$ and $x_4 = X_4$ and $x_5 = X_5$ and $x_6 = X_6$ and $x_7 = X_7$ and $x_8 = X_8$ and $x_9 = X_9$ and $x_{10} = X_{10}$ and $x_{11} = X_{11}$ and $x_{12} = X_{12}$ and $x_{13} = X_{13}$, then $y = P$";

$d_4$:  "If $x_1 = X_1$ and $x_2 = X_2$ and $x_3 = X_3$ and $x_5 = X_5$ and $x_6 = X_6$ and $x_7 = X_7$ and $x_8 = X_8$ and $x_{10} = X_{10}$ and $x_{11} = X_{11}$ and $x_{12} = X_{12}$ and $x_{13} = X_{13}$, then $y = VF$";

$d_5$:   "If $x_1 = X_1$ and $x_2 = X_2$ and $x_3 = \neg X_3$ and $x_4 = \neg X_4$ and $x_5 = X_5$ and $x_6 = X_6$ and $x_7 = X_7$ and $x_8 = X_8$ and $x_9 = X_9$ and $x_{10} = X_{10}$ and $x_{11} = \neg X_{11}$ and $x_{12} = X_{12}$ and $x_{13} = X_{13}$, then $y = F$";

$d_6$:   "If $x_2 = \neg X_2$ and $x_{10} = \neg X_{10}$ and $x_{11} = \neg X_{11}$ and $x_{12} = \neg X_{12}$, then $y = UF$".

Fuzzy logical operation "AND" in the left-hand parts of the rules $d_1 \div d_6$ are realized by crossing the corresponding fuzzy sets, i.e. by finding the minimum of the values of appropriate membership functions, namely:

$d_1$:   $\mu_{M1}(u) = \min\{\mu_{X1}(u),\ \mu_{X5}(u),\ \mu_{X6}(u),\ \mu_{X9}(u),\ \mu_{X12}(u)\}$, where $M_1 = \{0.1063/a_1;\ 0.0053/a_2;\ 0.1871/a_3;\ 0.0028/a_4;\ 0.0016/a_5\}$;

$d_2$:   $\mu_{M2}(u) = \min\{\mu_{X1}(u),\ \mu_{X2}(u),\ \mu_{X5}(u),\ \mu_{X6}(u),\ \mu_{X7}(u),\ \mu_{X9}(u),\ \mu_{X11}(u),\ \mu_{X12}(u)\}$, where $M_2 = \{0.1063/a_1;\ 0.0053/a_2;\ 0.0071/a_3;\ 0.0028/a_4;\ 0.0016/a_5\}$;

$d_3$:   $\mu_{M3}(u) = \min\{\mu_{X1}(u),\ \mu_{X2}(u),\ \ldots,\ \mu_{X13}(u)\}$, where $M_3 = \{0.0080/a_1;\ 0.0034/a_2;\ 0.0071/a_3;\ 0.0028/a_4;\ 0.0016/a_5\}$;

$d_4$:   $\mu_{M4}(u) = \min\{\mu_{X1}(u),\ \mu_{X2}(u),\ \mu_{X3}(u),\ \mu_{X5}(u),\ \ldots,\ \mu_{X8}(u),\ \mu_{X10}(u),\ \mu_{X11}(u),\ \mu_{X12}(u),\ \mu_{X13}(u)\}$, where $M_4 = \{0.0368/a_1;\ 0.0053/a_2;\ 0.0071/a_3;\ 0.0028/a_4;\ 0.0016/a_5\}$;

$d_5$:   $\mu_{M5}(u) = \min\{\mu_{X1}(u),\ \mu_{X2}(u),\ 1 - \mu_{X3}(u),\ 1 - \mu_{X4}(u),\ \mu_{X5}(u),\ \ldots,\ \mu_{X10}(u),\ 1 - \mu_{X11}(u),\ \mu_{X12}(u),\ \mu_{X13}(u)\}$, where $M_5 = \{0.0368/a_1;\ 0.0053/a_2;\ 0.0071/a_3;\ 0.0028/a_4;\ 0.0016/a_5\}$;

$d_6$:   $\mu_{M6}(u) = \min\{1 - \mu_{X2}(u),\ 1 - \mu_{X10}(u),\ 1 - \mu_{X11}(u),\ 1 - \mu_{X12}(u)\}$, where $M_6 = \{0.0527/a_1;\ 0.0756/a_2;\ 0.2233/a_3;\ 0.2077/a_4;\ 0.0457/a_5\}$.

Then the rules $d_1 \div d_6$ can be rewritten in an even more compact form:

$d_1$:   "If $x = M_1$, then $y = F$"; $d_2$: "If $x = M_2$, then $y = MF$"; $d_3$: "If $x = M_3$, then $y = P$";

$d_4$:   "If $x = M_4$, then $y = VF$"; $d_5$: "If $x = M_5$, then $y = F$"; $d_6$: "If $x = M_6$, then $y = UF$".

The implication operation using, for example, the Lukasiewicz's formula [4]:

$$\mu_H(u,j) = \min\{1;\ 1 - u + j\},$$

transforms these rules into fuzzy matrix relations: $R_1, R_2, \ldots, R_6$, the intersection of which induces the final solution in the form of the following matrix:

$$
R = 
\begin{array}{c|ccccccccccc}
 & 0 & 0.1 & 0.2 & 0.3 & 0.4 & 0.5 & 0.6 & 0.7 & 0.8 & 0.9 & 1 \\
\hline
a_1 & 0.8937 & 0.9732 & 0.9920 & 0.9920 & 0.9920 & 0.9920 & 0.9920 & 0.9920 & 0.9920 & 0.9920 & 0.9473 \\
a_2 & 0.9947 & 0.9966 & 0.9966 & 0.9966 & 0.9966 & 0.9966 & 0.9966 & 0.9966 & 0.9966 & 0.9966 & 0.9244 \\
a_3 & 0.8129 & 0.9129 & 0.9929 & 0.9929 & 0.9929 & 0.9929 & 0.9929 & 0.9929 & 0.9767 & 0.8767 & 0.7767 \\
a_4 & 0.9972 & 0.9972 & 0.9972 & 0.9972 & 0.9972 & 0.9972 & 0.9972 & 0.9972 & 0.9923 & 0.8923 & 0.7923 \\
a_5 & 0.9984 & 0.9984 & 0.9984 & 0.9984 & 0.9984 & 0.9984 & 0.9984 & 0.9984 & 0.9984 & 0.9984 & 0.9543 \\
\end{array}
$$

According to [3], the fuzzy conclusion regarding the favorability of the ME for the $k$-th university ($k = 1 \div 5$) is reflected as the fuzzy set $E_k$ with the corresponding values of the its membership function from the $k$-th row of the matrix $R$. In particular, the conclusion relative to favorability of the ME for $a_1$ is interpreted in the type of fuzzy set: $E_1 = \{0.8937/0; 0.9732/0.1; 0.9920/0.2; 0.9920/0.3; 0.9920/0.4; 0.9920/0.5; 0.9920/0.6; 0.9920/0.7; 0.9920/0.8; 0.9920/0.9; 0.9473/1\}$. Then, setting the $\alpha$-level sets $E_{1\alpha}$ and finding their cardinal number as: $M(E_\alpha) = \sum_{j=1}^{n} u_j/n$ [3, 5], we have:

- for $0 < \alpha < 0.8937$: $\Delta\alpha = 0.8937$, $E_{1\alpha} = \{0; 0.1; 0.2; \ldots; 0.9; 1\}$, $M(E_{1\alpha}) = 0.5$;
- for $0.8937 < \alpha < 0.9473$: $\Delta\alpha = 0.0536$, $E_{1\alpha} = \{0.1; 0.2; \ldots; 0.9; 1\}$, $M(E_{1\alpha}) = 0.55$;
- for $0.9473 < \alpha < 0.9732$: $\Delta\alpha = 0.0259$, $E_{1\alpha} = \{0.1; 0.2; \ldots; 0.8; 0.9\}$, $M(E_{1\alpha}) = 0.50$;
- for $0.9732 < \alpha < 0.9920$: $\Delta\alpha = 0.0188$, $E_{1\alpha} = \{0.2; 0.3; \ldots; 0.8; 0.9\}$, $M(E_{1\alpha}) = 0.55$.

Then the numerical (defuzzified) estimate of $E_1$ is obtained as:

$$F(E_1) = \frac{1}{\alpha_{max}} \int_0^{\alpha_{max}} M(E_{1\alpha})d\alpha$$
$$= [0.5 \cdot 0.8937 + 0.55 \cdot 0.0536 + 0.5 \cdot 0.0259 + 0.55 \cdot 0.0188]/0.992 = 0.5036.$$

By similar actions It is possible to define the defuzzified values for the others outputs: $F(E_2) = 0.4965$; $F(E_3) = 0.4915$; $F(E_4) = 0.4842$; $F(E_5) = 0.4978$. From the point of view of competitiveness in the market of educational services among the considered alternatives, the best is the university $a_1$, which is characterized by the greatest estimation of its ME, i.e. by 0.5036. Further in descending order: $a_5$ (0.4978), $a_2$ (0.4965), $a_3$ (0.4915) and $a_4$ (0.4842).

## 4    The Choice of the Most Competitive University with the Application of Maximin Folding Technique

The characteristics of the ME $\{x_1, x_2, \ldots, x_{13}\}$ determine evaluation concepts, and the estimates of alternative universities $a_k$ ($k = 1 \div 5$) represent the degree of correspondence to these concepts (or criteria). According to the approach to the fuzzification of evaluation concepts proposed in the previous section, the evaluation of the competitiveness of universities by each $i$-th criterion is presented in the form of fuzzy sets $X_i$ ($i = 1 \div 13$). Then it is necessary to create the set of optimal alternatives in the form of

$$A = \bigcap_{i=1}^{13} X_i, \qquad (1)$$

where intersection operation "$\cap$" are realized as

$$\mu_A(a_k) = \min_i \{\mu_{X_i}(a_k)\}. \tag{2}$$

According to maximin folding technique [5], the alternative $a^*$ is the best, if it satisfy to equality

$$\mu_A(a^*) = \max_k \{\mu_A(a_k)\}. \tag{3}$$

Then the optimal alternative set is formed as follows:
$A = \{\min\{0.4498;\ 0.4111;\ 0.8924;\ 0.0080;\ 0.2645;\ 0.2551;\ 0.2873;\ 0.0835;\ 0.1063;\ 0.0368;\ 0.6973;\ 0.9473;\ 0.4219\},\ \min\{0.0053;\ 0.9244;\ 0.0224;\ 0.0034;\ 0.9587;\ 0.3121;\ 0.2373;\ 0.7981;\ 0.2280;\ 0.3570;\ 0.0163;\ 0.0460;\ 0.6837\},\ \min\{0.9052;\ 0.2406;\ 0.2287;\ 0.5729;\ 0.5669;\ 0.1871;\ 0.0071;\ 0.9655;\ 0.9709;\ 0.0655;\ 0.7767;\ 0.3807;\ 0.1032\},\ \min\{0.0028;\ 0.7923;\ 0.2414;\ 0.0211;\ 0.0035;\ 0.1183;\ 0.9948;\ 0.0028;\ 0.0039;\ 0.0289;\ 0.3890;\ 0.3181;\ 0.6449\},\ \min\{0.0016;\ 0.5378;\ 0.0356;\ 0.2292;\ 0.0728;\ 0.0018;\ 0.0262;\ 0.5519;\ 0.9588;\ 0.1137;\ 0.6742;\ 0.9543;\ 0.4139\}\}.$

In this case, the resulting priority vector for alternative universities is following:

$$\max\{\mu_A(a_k)\} = \max\{0.0080; 0.0034; 0.0071; 0.0028; 0.0016\} = 0.0080.$$

Thus, from the point of view of competitiveness in the market of educational services among the considered alternatives $a_1$ is the best university, which is characterized by value 0.0080. Further in descending order: $a_3$ (0.0071), $a_2$ (0.0034), $a_4$ (0.0028), $a_5$ (0.0016).

## 5   Conclusion

The approaches proposed in the article for establishing the most competitive university by aggregating consolidated expert estimates of key indicators of their ME envelop only arbitrarily chosen five universities, which predetermined the choice of the discrete universe $\{a_1, a_2, a_3, a_4, a_5\}$. On the base of this universe, the qualitative evaluation criteria are described by its fuzzy subsets. If more universities enveloped, then the quality of criteria declaration for estimating the indicators $X_i$ ($i = 1 \div 13$) by fuzzy sets can be noticeably improve, which will inevitably affect the adequacy of the subsequent ranking.

The ranking of the considered alternatives by fuzzy inference differs from the ranking by the maximin folding technique. This is explained by the fact that, in contrast to the method of fuzzy inference, where the construction of the chosen implicative rules is determined by the priority of the attributes $x_i$ ($i = 1 \div 13$), when using the maximin folding technique, the evaluation criteria $X_i$ ($i = 1 \div 13$) have equal orders of priorities. Nevertheless, in both cases of evaluation, the same university $a_1$ was chosen as the best alternative, which, as we hope, can be assign to the asset of our study.

**Acknowledgement.** The authors consider necessary to express their sincere appreciation to Professor R.R. Rzayev for the help that he rendered during the process of writing and preparing this paper.

# References

1. Asaul, A.N., Kaparov, B.M.: Management of a higher educational institution under national economy. Humanistic, St.-Petersburg (2007). (in Russian)
2. Zadeh, L.A.: The concept of a linguistic variable and its application to approximate reasoning. Inf. Sci. **8**(3), 199–249 (1965)
3. Rzayev, R.R.: Analytical Support for Decision-Making in Organizational Systems. Palmerium Academic Publishing, Saarbruchen (2016). (in Russian)
4. Lukasiewicz, J.: On Three-Valued Logic: Selected Works, pp. 87–88. NorthHolland Publishing Company, Amsterdam (1970). Ed. by, L. Borkowski
5. Andreichenkov, A.V, Andreichenkova, O.N.: Analysis, synthesis, planning decisions in the economy. Finance and Statistics, Moscow (2000). (in Russian)

# Approach to the Increasing the Validity of the Position-Binary Method for Cyclic Signal Recognition

Oktay Nusratov[1], Asker Almasov[1(✉)], and Hafiz Bayramov[2]

[1] Institute of Control Systems of ANAS, Baku AZ1141, Azerbaijan
askalmasov@gmail.com
[2] Azerbaijan State University of Economics, Baku AZ1001, Azerbaijan

**Abstract.** To increase the validity of the position-binary method of cyclic signals recognition a trivial approach is proposed on the base of a time-weighted comparison estimate. Real non-negative sigmoid and Gauss functions are used as the weight function, which in the recognition process differentiates the analyzed cyclic signals from the temporal characteristics of their position-binary components.

**Keywords:** Cyclic signal · Binary covering · Weight function

## 1 Introduction

Proposed in [1–3] the position-binary (PB) method for recognizing of cyclic signals is quite simple and, at the same time, having a high sensitivity to changes in the cyclic signal, is quite acceptable for solving a number of problems of monitoring, diagnostics and management. Here, as the informative features, the duration of the PB components (PBC) of the cyclic signals is used, and the closeness is determined by computing the numerical proximity parameters from the pairwise PBC coverings of the signals generated at each position in accordance with the following expression:

$$E = (\xi_{q(n-1)} + \eta_{q(n-1)}) \cdot q^{n-1} + (\xi_{q(n-2)} + \eta_{q(n-2)}) \cdot q^{n-2} + \ldots + (\xi_{q0} + \eta_{q0}) \cdot q^0,$$

$$(1)$$

where $\xi = \sum p_+$ is the total value of the PBS durations of the binary coverings of the recognizable and standard (etalon) signals generated by the transitions $(0 \rightarrow 1)$; $\eta = \sum p_-$ is the total value of the PBC durations of binary coatings formed by transitions $(1 \rightarrow 0)$; $n$ is the number of positions in the PBC-decomposition. In this case, the decision-making on the proximity of the signals is carried out in accordance with the relative values of the proximity estimates $E$ and the recognition threshold.

However, in some cases, in particular, when analyzing complex cyclic signals, the application of this method does not always guarantee the correctness of recognition. First, it is related that the PB-method based calculation of numerical estimates of proximity does not take into account the location of the PBC of cyclic signals along the time axis $t$. Therefore, over the past few years, we have undertaken studies to improve the PB-

© Springer Nature Switzerland AG 2019
R. A. Aliev et al. (Eds.): ICAFS-2018, AISC 896, pp. 551–558, 2019.
https://doi.org/10.1007/978-3-030-04164-9_73

method by involving in the numerical process an additional informative feature – the temporary location of the PBC. In particular, to improve the recognition score in [4] it is proposed to compare fuzzy interpretations of PBC of recognizable and reference cyclic signals on the base of their point estimates. As another alternative measure of comparison, a similarity measure of fuzzy sets for fuzzy interpretations of PBC was also used in [5].

In [6] to increase the validity of the PB-method for recognizing of cyclic signals fuzzy inference was used, which allows to involve the time characteristics of the PBC-location of $(0 \rightarrow 1/1 \rightarrow 0)$-coverings of the analyzed cyclic signals with the etalon ones into the computational process. Moreover, to establish the proximity measure the algorithm of point estimation of fuzzy relation was used, which interpret the position-binary covering of the recognized and reference signals. However, despite the encouraging results obtained in [6], it must be admitted that the fuzzy implicative rules used in the calculations and the Gaussian membership functions of fuzzy set that describe the inputs have not yet been optimized on the base of available set of the etalon cyclic signals. The last attempt to increase the validity of the PB-method was undertaken in [7], where the idea of using a Radial Basis Functional Neural Network was declared. By virtue of its logical basis is able to substantially differentiate the analyzed cyclic signals by the time characteristics of their PBC-location. As a result, the conducted studies have motivated authors to refer to another approach to solving the problem of increasing the validity of the PB-method for recognizing of cyclic signals based on numerical weighting of the PBC of $(0 \rightarrow 1/1 \rightarrow 0)$-coverings by applying weight functions.

## 2    Problem Definition

The etalon $e$ and recognizable $r_i$ ($i = 1 \div 3$) watt meters for the turnaround of the deep-pumping plant of the oil well, which characterize its technical state, are presented in Figs. 1 and 2, respectively. PBC of these signals formed at positions $q_5 \div q_0$ when $\Delta t = 50$ and amplitude 60 are also presented there.

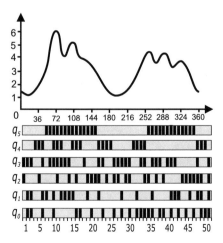

**Fig. 1.**  Etalon watt-meter and its PBC decomposition.

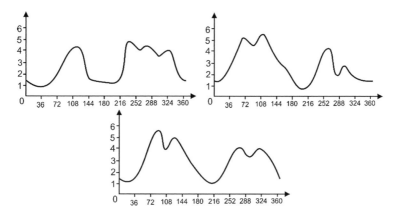

**Fig. 2.** Recognizable watt-meters and their PBC decomposition.

Taking into account the transitions $0 \rightarrow 1$ $(p_+)$ and $1 \rightarrow 0$ $(p_-)$ Fig. 3 shows the resultant sequences of PBC-covering of the compared pairs of signals $(e, r_j)$ $(j = 1 \div 3)$ for each position $q_k$ $(k = 0 \div 5)$ of PBC-decomposition.

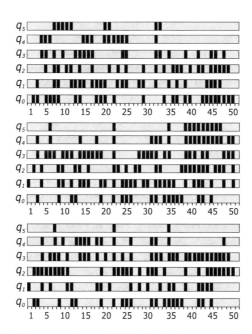

**Fig. 3.** Binary coverings of PBC of pairwise compared signals.

In addition to the PB-method for recognizing of cyclic signals it is necessary to find a mechanism capable consideration the time positions of PBC of $(0 \rightarrow 1/1 \rightarrow 0)$-coverings of recognizable cyclic signals $r_j$ $(j = 1 \div 3)$ with the etalon signal $e$ and, thereby, making the proximity estimates of the signals in a certain sense unique.

## 3   Weighing the PBC of Coverings in the Time Range: Weight Function

It is known that the weight function is used for summation, integration or averaging in order to give some elements a higher weight in the resultant value in comparison with other elements. In particular, the discrete weight function $w: A \rightarrow \Re^+$ is a positive function defined on a discrete set of values of $A$, which is usually finite and/or countable. The weight function $w(a) = 1$ corresponds to an unweighted situation, when all elements of the set $A$ have equal weights. For example, if the function $f: A \rightarrow \Re$ is defined on the real axis, then the unweighted sum $f$ on $A$ is defined as $\sum_{a \in A} f(a)$, i.e. as in the case of summation (1) within the framework of the PB-method applying. At the same time, by $w: A \rightarrow \Re^+$ weighted sum is defined as $\sum_{a \in A} w(a) \cdot f(a)$.

One of the most common applications of weighted sums is numerical integration and digital filtering. So, for example, if $B$ is a finite subset of the set $A$, then in the classical sense the cardinality of the set $|B|$ can be replaced by so-called weighted cardinality $\sum_{a \in A} w(a)$. If $A$ is a finite nonempty set, then the analog of the arithmetic mean $\{\sum_{a \in A} f(a)\}/|A|$ can be introduced in the form of following weighted average $\{\sum_{a \in A} w(a) \cdot f(a)\}/\{\sum_{a \in A} w(a)\}$. In particular, in the problems of multi-criteria choice of optimal alternatives for the transition from the set of particular values of quality criteria to the single universal integral criteria (for example, value criteria) weighted summation is used. So, for example, if the ranges of values of particular quality indicators differ by several orders, then before finding of the numerical value of the integral criterion $K$ the partial quality indicators $q_k$ are normalized, i.e. the range of variation $[\min\{q_k\}, \max\{q_k\}]$ of each of them is mapped to the segment $[0; 1]$ as follows:

$$q_k' = \frac{q_k - \min\{q_k\}}{\max\{q_k\} - \min\{q_k\}}. \tag{2}$$

In this case, integral criteria is calculated as $K = \sum_{k=1}^{n} q_k w_k$. As a result, the same effect of the particular criteria on the result is achieved with comparable values of the weights $w_1, w_2, \ldots, w_n$. In mathematical statistics, the weighted average is often used to compensate, so-called, *bias*. For the true value of $f$ measured as $f_t$ several times independently of each other with variances $\sigma_t^2$, the best approximation is obtained by averaging all measurement results with weights $w_t = 1/\sigma_t^2$: the resultant variance is less than each independent measurement $\sigma^2 = 1/(\sum w_t)$. In particular, in the maximum similitude method the differences are weighted by values similar to $w_t$.

On the base of these considerations to take into account the time characteristic of the PBC of $(0 \rightarrow 1/1 \rightarrow 0)$-coverings the real non-negative weight function $w(t)$ is introduced in (1) as an additional factor, which differentiates each PBC from any position of the binary covering. In other words, the proximity equality of cyclic signals can be represented as following weighted sum

$$E_w = \sum_{k=0}^{n-1} w(t)\left(\xi_{q(k-1)} + \eta_{q(k-1)}\right)q^k.$$

(3)

As weights $w(t)$ one can use various smooth curves described by real nonnegative functions. As a result of numerous experiments, the sigmoid and Gaussian weight functions was chosen. In respect to the time range under consideration, the sigmoid weight function (see Fig. 4) is given in the following form:

$$w(t) = \frac{1}{a + be^{-(t-t_0)}} + w_0$$

(4)

where $t_0 = 25$, $w_0 = 0.925$, $a = 20$, $b = 1$ are the parameters set by the heuristic way.

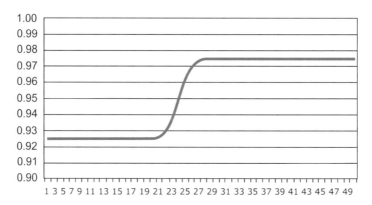

**Fig. 4.** Sigmoid weight function.

To demonstrate the proposed approach, the binary row of PBC, for example, located on the $q_5$-th position of the $(e, r_1)$-covering is chosen (see Fig. 5). The number of binary $0 \rightarrow 1/1 \rightarrow 0$-transitions considered in (1) is 9 here, and according to the PB-method each of these 9 transitions has a unit weight.

**Fig. 5.** PBC of $(e, r_1)$-covering located on the $q_5$-th position.

In the proposed formula (2) the weights of the PBC are already distinctive, i.e. differentiate by corresponding values of the sigmoid weight function (4). In particular, for each $(0 \rightarrow 1/1 \rightarrow 0)$-transition, respectively, we have:

- 7-th discrete step – the weight 0.925;
- 8-th discrete step – the weight 0.925;

- 9-th discrete step – the weight 0.925;
- 10-th discrete step – the weight 0.925;
- 11-th discrete step – the weight 0.925001;
- 19-th discrete step – the weight 0.925001;
- 20-th discrete step – the weight 0.930938;
- 31-th discrete step – the weight 0.974994;
- 32-th discrete step – the weight 0.974998.

If according to (1) the resulting value for the given position was 9 (the number of transitions weighted by the unit), then taking into account the sigmoidal weighting, the resulting value is the following number: $8.433292 = 1 \times 0.925 + 1 \times 0.925 + 1 \times 0.925 + 1 \times 0.925 + 1 \times 0.925001 + 1 \times 0.927362 + 1 \times 0.930938 + 1 \times 0.974994 + 1 \times 0.9749984$.

Now, as the weight function $w(t)$ in (3) let us define a Gaussian function (see Fig. 6) of the form:

$$w(t) = a \cdot \exp\{-(t - t_0)^2/\sigma^2\} + w_0 \qquad (5)$$

where $t_0 = 25$, $w_0 = 0.9$, $a = 0.1$, $\sigma = 25$ are the parameters set by the heuristic way.

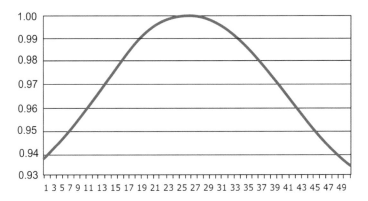

**Fig. 6.** Gaussian weight function.

Thus, applying the approach (3) we can obtain estimates of adjacency of pairwise compared cyclic signals $(e, r_j)$ $(j = 1 \div 3)$, but on the another scale. The obtained estimates are ordered in Table 1.

**Table 1.** Comparison of numerical estimates of binary coverings.

| Comparable pair of cyclic signals | Position weight | Evaluation of adjacency without considering the time characteristics of the PBC | | Evaluation of adjacency subject to the time characteristics of the PBC | | | |
| --- | --- | --- | --- | --- | --- | --- | --- |
| | | | | Sigmoid weight function | | Gaussian weight function | |
| | | The number $0 \rightarrow 1/1 \rightarrow 0$-transitions | Position-weighted value ($\times 2^k$) | Weighted number of transitions | Position-weighted value ($\times 2^k$) | Weighted number of transitions | Position-weighted value ($\times 2^k$) |
| $(e, r_1)$ | $q_5$ | 9 | 288 | 8.4333 | 269.8654 | 8.8091 | 281.8915 |
| | $q_4$ | 13 | 208 | 12.2023 | 195.2377 | 12.7938 | 204.7012 |
| | $q_3$ | 19 | 152 | 18.0557 | 144.4457 | 18.5108 | 148.0864 |
| | $q_2$ | 22 | 88 | 21.0749 | 84.2996 | 21.4343 | 85.7372 |
| | $q_1$ | 23 | 46 | 21.8341 | 43.6682 | 22.4999 | 44.9998 |
| | $q_0$ | 25 | 25 | 23.7919 | 23.7919 | 24.2167 | 24.2167 |
| Evaluation | | | 807 | | 761 | | 789 |
| $(e, r_2)$ | $q_5$ | 12 | 384 | 11.6134 | 371.6294 | 11.6078 | 371.4479 |
| | $q_4$ | 20 | 320 | 19.2637 | 308.2195 | 19.3892 | 310.2266 |
| | $q_3$ | 29 | 232 | 27.5893 | 220.7144 | 28.2733 | 226.1862 |
| | $q_2$ | 24 | 96 | 22.9872 | 91.9486 | 23.3202 | 93.2808 |
| | $q_1$ | 27 | 54 | 25.7688 | 51.5376 | 26.4533 | 52.9066 |
| | $q_0$ | 19 | 19 | 18.2174 | 18.2174 | 18.6487 | 18.6487 |
| Evaluation | | | 1105 | | 1062 | | 1072 |
| $(e, r_3)$ | $q_5$ | 4 | 128 | 3.8134 | 122.029 | 3.89096 | 124.5108 |
| | $q_4$ | 15 | 240 | 14.1373 | 226.1968 | 14.70329 | 235.2526 |
| | $q_3$ | 29 | 232 | 27.7253 | 221.8022 | 28.24695 | 225.9756 |
| | $q_2$ | 29 | 116 | 27.69199 | 110.768 | 28.16046 | 112.6418 |
| | $q_1$ | 19 | 38 | 18.1294 | 36.2588 | 18.55065 | 37.1013 |
| | $q_0$ | 22 | 22 | 21.0081 | 21.00813 | 21.53063 | 21.53063 |
| Total | | | 776 | | 738 | | 757 |

The nearest to the etalon is the signal, the covering with which is estimated by the least number. In all the cases considered, that is, when applying the PB-method and its two modification weighted by sigmoid and Gaussian functions, such signal is $r_3$.

## 4    Conclusion

Obviously, in terms of the above three cyclic signals, it would not be correct to assert that the recognition accuracy would increase due to modification of the PB-method by weight functions. It is necessary to verify experimentally on the base of tens and hundreds of cyclic signals. Nevertheless, the test results given in Table 1 allows to judge a significant increase in the validity of recognition of cyclic signals based on weighted summation, which establishes the evaluation of adjacency of cyclic signals.

**Acknowledgement.** The authors consider necessary to express their sincere appreciation to Professor R.R. Rzayev for the help that he rendered during the process of writing and preparing this article.

## References

1. Aliev, T.A., Nusratov, O.G.: Position-pulse-width method of diagnostics of cyclic processes. Theory Control Syst. **1**, 133–138 (1998). (in Russian)
2. Aliev, T.A., Nusratov, O.G.: Algorithms for the analysis of cyclic signals. Autom. Comput. Eng. **2**, 68–74 (1998). (in Russian)
3. Nusratov, O.G.: Position-pulse-width algorithm for the recognition of cyclic signals. Autom. Comput. Eng. **1**, 12–17 (2006). (in Russian)
4. Nusratov, O.G., Rzayev, R.R., Spevak, S.A.: Application of the apparatus of fuzzy sets to increase the validity of the position-binary method for identification of cyclic signals. Informatics and control problems. Trans. ANAS **29**(3), 76–81 (2009). (in Russian)
5. Nusratov, O.G., Almasov, A., Mamedova, A.M.: Positional-binary recognition of cyclic signals by fuzzy analyses of their informative attributes. Proc. Comput. Sci. **120**, 454–460 (2017)
6. Nusratov, O.G., Rzayev, R.R.: Positional-binary recognition of cyclic signals by fuzzy timing analysis of information indicators. In: 3rd International Conference "Problems of Cybernetics and Informatics", vol. 1, pp. 215–218. Elm, Baku (2010)
7. Aliev, T.A., Nusratov, O.G., Rzayev R.R.: Increase the validity of positional-binary recognition of cyclic signals with application of radial-basis functional neural network. In: 5th International Conference on Application of Information and Communication Technologies, pp. 701–703. Elm, Baku (2011)

# Weighted Estimate of Country Risk Using a Fuzzy Method of Maxmin Convolution

Sevinj Babayeva[1(✉)], Inara Rzayeva[2], and Tofig Babayev[1]

[1] Institute of Control Systems of ANAS, Baku AZ1141, Azerbaijan
babayevasevinj@yahoo.com
[2] Azerbaijan State University of Economics, Baku AZ1101, Azerbaijan

**Abstract.** Weighted attribute estimates and fuzzy method of maximin convolution based two approaches to evaluation the levels of country risk are considered. To obtain the final estimates of the country risk levels for an arbitrary set of alternatives these approaches are used on the base of expert conclusions regarding factors of country risk. The study is completed by comparative analysis of finale estimates of country risks.

**Keywords:** Country risk · Concordance coefficient · Fuzzy set

## 1 Introduction

Along with force majeure situations, country risks carry the dangers of political, legal, and socio-economic character. Therefore, to guarantee protection against such threats, it is necessary to take into account the economic and political situation in the aggregate (especially in emerging markets), which, in fact, predetermined the introduction of the concept of "country risk". Country risk (CR) is a multi-factor category characterized by a combined system of financial and economic, socio-political, and legal factors that distinguish the market of any country [1]. According to the degree of CR the countries are ranked by quantitative assessments. This ranking includes the following stages: (1) selection of financial and economic, socio-political and legal variables of CR; (2) identification of the weights of the selected variables of CR on the base of their relative influence on the CR-level; (3) expert estimation of CR factors using the established scale; (4) determination of weighted index reflecting the CR-level.

At present, many world rating agencies, and international institutions, such as Euromoney, Institutional Investor, Mood's Investor Service, the European Bank for Reconstruction and Development (EBRD), the World Bank (WB), etc., are currently ranking countries according to their CR-level. At the same time, existing approaches are conditioned by qualitative and/or quantitative, economic, combined, and structurally qualitative methods for estimating of CR.

© Springer Nature Switzerland AG 2019
R. A. Aliev et al. (Eds.): ICAFS-2018, AISC 896, pp. 559–567, 2019.
https://doi.org/10.1007/978-3-030-04164-9_74

## 2   Problem Definition

Well-known auditing firm Pricewaterhouse Coopers uses a limited set of variables to formulate the ratings of the investment attractiveness of states. These variables are formulated and denoted in the following form: $x_1$ – the level of corruption; $x_2$ – compliance of legislation; $x_3$ – the level of economic growth; $x_4$ – state policy on accounting and control; $x_5$ – state regulation [1]. On the base of above list of variables for the CR-aggregation it is necessary to conduct a preliminary expert analysis by conducting the comparative qualitative assessment of the risk factors (by simple ranking method on the base of expert preferences) and quantitative estimation of the weights of these factors (by applying the normalized values of the weights). Further, by determining the degree of consistency of the expert estimates relative to the priority $x_i$ ($i = 1 \div 5$) and their generalized weights it is necessary to compile the total index in the range from 0% to 100%.

Assuming the variables $x_i$ ($i = 1 \div 5$) as qualitative characteristics that exert relative effects on the CR-level, in addition to the above, it is necessary to carry out a multi-criteria evaluation of the alternative (hypothetical countries) relative to their SR-levels by a fuzzy maxmin convolution method.

## 3   Ranking of CR-Factors in the Orders of Experts' Preferences

Suppose that expert evaluations of the degrees of importance of CR-factors $x_i$ ($i = 1 \div 5$) are determined by independent questionnaire of 15 profile specialists. Each expert was asked to arrange the variable $x_i$ according to the principle: the most important variable should be designated by the number "1", followed the less important one by the number "2", and further in descending order of importance. The rank estimates obtained in this way are summarized in Table 1.

**Table 1.** Ranking of CR-factors in the orders of experts' preferences.

| Expert | Estimated factors and their ranks ($r_{ij}$) | | | | | Expert | Estimated factors and their ranks ($r_{ij}$) | | | | |
|---|---|---|---|---|---|---|---|---|---|---|---|
| | $x_1$ | $x_2$ | $x_3$ | $x_4$ | $x_5$ | | $x_1$ | $x_2$ | $x_3$ | $x_4$ | $x_5$ |
| 01 | 1 | 2 | 4 | 3 | 5 | 09 | 1 | 3 | 2 | 4 | 5 |
| 02 | 1 | 3 | 2 | 4 | 5 | 10 | 1 | 3 | 2 | 5 | 4 |
| 03 | 2 | 1 | 5 | 4 | 3 | 11 | 1 | 3 | 4 | 2 | 5 |
| 04 | 1 | 2 | 4 | 5 | 3 | 12 | 1 | 2 | 3 | 5 | 4 |
| 05 | 2 | 1 | 3 | 4 | 5 | 13 | 2 | 1 | 4 | 3 | 5 |
| 06 | 1 | 2 | 4 | 3 | 5 | 14 | 3 | 1 | 2 | 4 | 5 |
| 07 | 2 | 1 | 4 | 3 | 5 | 15 | 1 | 2 | 5 | 4 | 3 |
| 08 | 1 | 2 | 4 | 5 | 3 | $\sum r_{ij}$ | 21 | 29 | 52 | 55 | 65 |

To determine the degree of consistency of expert opinions, the Kendall concordance coefficient is applied, which demonstrates a multiple rank correlation of expert opinions. According to [2, 3], this coefficient is calculated by the formula:

$$W = \frac{12 \cdot S}{m^2(n^3 - n)},$$ (1)

where $m$ is the number of experts; $n$ is the number of CR-factor; $S$ is the deviation of expert conclusions from the average value of the ranking of the CR-factor, which is calculated, for example, by the formula (3):

$$S = \sum_{i=1}^{n} \sum_{j=1}^{m} [r_{ij} - m(n+1)/2]^2,$$ (2)

where $r_{ij} \in \{1, 2, 3, 4, 5\}$ is the rank of the $i$-th CR-factor, established by the $j$-th expert. In the case under consideration (see Table 1) the value of $S$ is 1450 and Kendall concordance coefficient is $W = 12 \cdot 1450/[15^2(5^3 - 5)] = 0.6444$. Condition $W > 0.6$ testifies the *strong* consistency of expert opinions on the importance of CR-factors.

## 4    Weight Identification of CR-Factors

Now, suppose that at the preliminary stage of the independent questionnaire, each expert was also instructed to establish the values of the normalized estimates of CR-factors weights. The results of this questionnaire are summarized in Table 2.

**Table 2.** The values of the normalized estimates of CR-factors weights.

| Expert | Normalized weights for factors $(\alpha_{ij})$ | | | | | Expert | Normalized weights for factors $(\alpha_{ij})$ | | | | |
|---|---|---|---|---|---|---|---|---|---|---|---|
| | $x_1$ | $x_2$ | $x_3$ | $x_4$ | $x_5$ | | $x_1$ | $x_2$ | $x_3$ | $x_4$ | $x_5$ |
| 01 | 0.300 | 0.250 | 0.150 | 0.225 | 0.075 | 09 | 0.275 | 0.175 | 0.200 | 0.100 | 0.250 |
| 02 | 0.350 | 0.175 | 0.200 | 0.150 | 0.125 | 10 | 0.300 | 0.200 | 0.250 | 0.100 | 0.150 |
| 03 | 0.225 | 0.250 | 0.150 | 0.175 | 0.200 | 11 | 0.300 | 0.175 | 0.150 | 0.250 | 0.125 |
| 04 | 0.275 | 0.250 | 0.175 | 0.100 | 0.200 | 12 | 0.300 | 0.250 | 0.200 | 0.100 | 0.150 |
| 05 | 0.250 | 0.275 | 0.200 | 0.175 | 0.100 | 13 | 0.225 | 0.250 | 0.175 | 0.200 | 0.150 |
| 06 | 0.300 | 0.250 | 0.150 | 0.200 | 0.100 | 14 | 0.200 | 0.300 | 0.250 | 0.150 | 0.100 |
| 07 | 0.200 | 0.375 | 0.150 | 0.175 | 0.100 | 15 | 0.300 | 0.250 | 0.125 | 0.150 | 0.175 |
| 08 | 0.325 | 0.300 | 0.150 | 0.025 | 0.200 | $\sum r_{ij}$ | 4.125 | 3.725 | 2.675 | 2.275 | 2.200 |

Starting from the data of Table 2 let us make preliminary calculations for the subsequent identification of the CR-factors weights. It is necessary to define the group estimates of the CR-factors and the numerical characteristics (degrees) of competence of all experts. To calculate the average value $\alpha_i$ from the $i$-th group of normalized estimates of the CR-factors weights, let us use the weighted degrees of expert competence by following difference equation:

$$\alpha_i(t+1) = \sum_{j=1}^{m} w_j(t)\alpha_{ij}, \tag{3}$$

where $w_j(t)$ is the weight characterizing the level of competency of the $j$-th expert $(j = 1 \div m)$ at time $t$. In this case, the process of finding group estimates of the normalized values has an iterative character, which is completed under following condition:

$$\max_i\{|\alpha_i(t+1) - \alpha_i(t)|\} \le \varepsilon \tag{4}$$

where $\varepsilon$ is the permissible accuracy of calculations, which is set in advance. In this case, let it be $\varepsilon = 0.0001$.

Let at initial stage $t = 0$ experts have the same levels of competence. Then, assuming for the general case $w_j(0) = 1/m$ as the initial value of the level of competence of the $j$-th expert, for the $i$-th group of normalized estimates of the CR-factors weights the average value in the first approximation is obtained from the partial equality:

$$\alpha_i(1) = \sum_{j=1}^{m} w_j(0)\alpha_{ij} = \frac{1}{m}\sum_{j=1}^{m} \alpha_{ij} \tag{5}$$

In accordance with (5), the averaged estimates of the CR-factors weights by groups in the first approximation are the following corresponding numbers: $\{\alpha_1(1); \alpha_2(1); \alpha_3(1); \alpha_4(1); \alpha_5(1)\} = \{0.27500; 0.24833; 0.17833; 0.15167; 0.14667\}$. It is not difficult to see that requirement (4) is not satisfied for the first approximation. Therefore, in order to proceed to the next stage, let us calculate the rating coefficient as: $\eta(1) = \sum_{i=1}^{5}\sum_{j=1}^{15} \alpha_i(1)\alpha_{ij} = 3.2042$. Then, according to the following equalities:

$$\begin{cases} w_j(1) = \dfrac{1}{\eta(1)}\sum_{i=1}^{5} \alpha_i(1) \cdot \alpha_{ij} \ (j = \overline{1, 14}), \\[4mm] w_{15}(1) = 1 - \sum_{j=1}^{14} w_j(1), \ \sum_{j=1}^{15} w_j(1) = 1, \end{cases} \tag{6}$$

where $w_{15}(1)$ is the competency indicator of the 15-th expert, let us calculate the of expert competence indicators in the first approximation as: $\{w_1(1); w_2(1); w_3(1); w_4(1); w_5(1); w_6(1); w_7(1); w_8(1); w_9(1); w_{10}(1); w_{11}(1); w_{12}(1); w_{13}(1); w_{14}(1); w_{15}(1)\} = \{0.0676; 0.0676; 0.0645; 0.0666; 0.0668; 0.0675; 0.0674; 0.0698; 0.0645; 0.0668; 0.0652; 0.0679; 0.0648; 0.0660; 0.0672\}$.

Now let us compute the average group estimate of the CR-factors in the second approximation by the formula (3), or more precisely from its particular expression: $\alpha_i(2) = \sum_{j=1}^{15} w_j(1)\alpha_{ij}$. In this case, the average estimates of the CR-factors for groups $i = 1 \div 5$ in the second approximation are the numbers: $\{\alpha_1(2); \alpha_2(2); \alpha_3(2); \alpha_4(2); \alpha_5(2)\} = \{0.27547; 0.24876; 0.17821; 0.15116; 0.14640\}$. Checking the obtained values for condition (4) and convincing that it is not satisfied again: $\max\{|\alpha_i(2) - {}_i(1)|\} = \max\{|0.2755 - 0.2750|; \ |0.2488 - 0.2483|; \ |0.1782 - 0.1783|; \ |0.1512 - 0.1517|; \ |0.1464 - 0.1467|\} = 0.0005 > \varepsilon$, it is necessary to calculate the rating coefficient as: $\eta(2) = \sum_{i=1}^{5}\sum_{j=1}^{15} \alpha_i(2)\alpha_{ij} = 3.2056$. Then the indicators of expert competence at the second approximation $w_j(2)$ $(j = 1 \div 15)$ will be following numbers: $\{w_1(2); w_2(2); w_3(2); w_4(2); w_5(2); w_6(2); w_7(2); w_8(2); w_9(2); w_{10}(2); w_{11}(2); w_{12}(2); w_{13}(2); w_{14}(2); w_{15}(2)\} = \{0.0676; 0.0676; 0.0645; 0.0666; 0.0668; 0.0675; 0.0674; 0.0699; 0.0645; 0.0668; 0.0652; 0.0679; 0.0647; 0.0660; 0.0672\}$.

The average group estimates of the CR-factors in the third approximation are obtained from the following particular expression of formula (3), namely: $\alpha_i(3) = \sum_{j=1}^{15} w_j(2)\alpha_{ij}$. In this case, the average values of the CR-factors for the groups $i = 1 \div 5$ in the third approximation are the following numbers: $\{\alpha_1(3); \alpha_2(3); \alpha_3(3); \alpha_4(3); \alpha_5(3)\} = \{0.27547; 0.24876; 0.17821; 0.15115; 0.14640\}$. The accuracy of the group estimates $x_i$ $(i = 1 \div 5)$ in the third approximation already satisfies the condition (4), that is, $\max\{|\alpha_i(3) - \alpha_i(2)|\} = \max\{|0.27547 - 0.27547|; \ |0.24876 - 0.24876|; \ |0.17821 - 0.17821|; \ |0.15115 - 0.15116|; \ |0.1464 - 0.1464|\} = 0.00001 < \varepsilon$, which is the basis for stopping calculations. Then $\{\alpha_1(3); \alpha_2(3); \alpha_3(3); \alpha_4(3); \alpha_5(3)\}$ are the summarized weights of CR-factors $x_i$ $(i = 1 \div 5)$.

## 5  Determination of the Weighted CR-Level

The method of expert evaluations supposes discussing the factors that influence to the CR-level by the group of especially involved specialists. Each of them is given a list of possible risks on the basis of variables $x_i$ $(i = 1 \div 5)$ and is offered to estimate of the probability of their occurrence in percentage terms on the base of the following five-point rating system: 5 – INSIGNIFICANT RISK; 4 – MOST PROBABLY THE RISK SITUATION DO NOT OCCUR; 3 – ABOUT THE POSSIBILITY OF RISK IT IS IMPOSSIBLE TO SAY ANYTHING DEFINITELY; 2 – THE RISK SITUATION WILL MOST PROBABLY COME; 1 – THE RISK SITUATION WILL MOST CERTAINLY COME. Further, expert judgements are analyzed for consistency by the rule: the maximum permissible difference between two expert opinions for any kind of risk with respect to $x_i$ $(i = 1 \div 5)$ should not exceed 3. This rule allows filter inadmissible deviations in expert judgements of the probability of occurrence of the risk for each CR-factor. The summary index, theoretically ranging from 0 to 100 can be calculate by following assessment criterion:

$$R = \frac{\displaystyle\sum_{i=1}^{5} \alpha_i e_i}{\displaystyle\max_i \sum_{i=1}^{5} \alpha_i e_i} \times 100, \tag{7}$$

where $\alpha_i$ is the weight of the importance of the $i$-th CR-factor; $e_i$ is the five-point evaluation system based expert judgement of the risk probability for $i$-th CR-factor. The minimum index symbolizes the maximum risk, and vice versa. CR-level is established on the base of the graduation of the resulting weighted estimates.

Suppose that the expert community is offered to test 10 alternative countries $a_k$ ($k = 1 \div 10$) by the five-point system: every expert need to assess the degree of influence of financial and economic, socio-political, and state-legal factors in these countries on their CR-level. So, estimates of the CR levels of these countries are obtained on the base of consolidated (averaged) expert opinions and application of the assessment criterion (7). Obtained estimates are summarized in the form of Table 3.

**Table 3.** Indexes of the CR-levels for alternative countries.

| Alternative countries | Weights of CR-factors | | | | | Index |
|---|---|---|---|---|---|---|
| | $\alpha_1$ | $\alpha_2$ | $\alpha_3$ | $\alpha_4$ | $\alpha_5$ | |
| $a_1$ | 4.50 | 4.75 | 4.5 | 4.75 | 4.25 | 91.27 |
| $a_2$ | 4.85 | 4.50 | 4.55 | 2.75 | 3.75 | 84.62 |
| $a_3$ | 3.75 | 4.00 | 3.25 | 3.85 | 3.25 | 73.30 |
| $a_4$ | 4.25 | 3.45 | 2.85 | 2.75 | 1.85 | 64.47 |
| $a_5$ | 4.00 | 2.55 | 3.00 | 2.25 | 1.85 | 57.64 |
| $a_6$ | 3.55 | 2.85 | 2.00 | 1.25 | 0.85 | 47.13 |
| $a_7$ | 2.25 | 1.75 | 1.25 | 1.85 | 1.50 | 35.54 |
| $a_8$ | 2.25 | 1.85 | 1.25 | 0.75 | 0.25 | 29.06 |
| $a_9$ | 5.00 | 4.75 | 4.85 | 4.85 | 4.75 | 97.04 |
| $a_{10}$ | 3.25 | 2.85 | 3.75 | 4.25 | 3.50 | 68.55 |

# 6   Ranking CR-Levels of the Countries Using the Fuzzy Method of Maxmin Convolution

The processing of expert judgements by the five-point system presented in Table 3 concerning the CR-factors for alternative $a_k$ ($k = 1 \div 10$) one can be carried out using the mathematical apparatus of the fuzzy sets theory by three stages.

Step 1. Construction of the membership function (fuzzification), which appropriates to the evaluation concept "NON-EXISTING RISK" [4]. In the case under consideration, this term can be reflected in the form of a fuzzy subset of the discrete finite set of estimated alternatives (in our case, countries) $\{a_1, a_2, ..., a_{10}\}$ in the following form: $A_i = \{\mu_{Ai}(a_1)/a_1; ...; \mu_{Ai}(a_{10})/a_{10}\}$, where $\mu_{Ai}(a_t)$ ($t = 1 \div 10$) is the value of the membership function of the fuzzy set $A_i$, which determines the ratio of the $t$-th country to the evaluation criterion $A_i$ = NON-EXISTING RISK. As the membership function it is possible to choose a Gaussian function of the form: $\mu_{Ai}(a_t) = exp\{-[e_i(a_t) - 5]^2/\sigma_i^2\}$, where $e_i(a_t)$ is the consolidated expert judgement for the country $a_t$ ($t = 1 \div 10$) obtained by five-point scale for compliance with the risk of the $i$-th factor as non-existent; $\sigma_i^2$ is the density of the location of the nearest elements, which is chosen as equal to 4 for all cases of the fuzzification [5].

Step 2. Determination of concrete values of the membership function $\mu_{Ai}(a_t)$ ($t = 1 \div 10$) according to the criteria $A_i$. In this case, assumed that $x_i$ ($i = 1 \div 5$) are linguistic variables, it is possible to represent one of their terms, namely: "NON-EXISTING RISK" by fuzzy subset $A_i$ of the discrete universe $U = \{a_1, a_2, ..., a_{10}\}$ as follows [4, 5]:

- $A_1 = \{0.9394/a_1; 0.9944/a_2; 0.6766/a_3; 0.8688/a_4; 0.7788/a_5; 0.5912/a_6; 0.1510/a_7; 0.1510/a_8; 1/a_9; 0.4650/a_{10}\}$;
- $A_2 = \{0.9845/a_1; 0.9394/a_2; 0.7788/a_3; 0.5485/a_4; 0.2230/a_5; 0.3149/a_6; 0.0713/a_7; 0.0837/a_8; 0.9845/a_9; 0.3149/a_{10}\}$;
- $A_3 = \{0.9394/a_1; 0.9506/a_2; 0.4650/a_3; 0.3149/a_4; 0.3679/a_5; 0.1054/a_6; 0.0297/a_7; 0.0297/a_8; 0.9944/a_9; 0.6766/a_{10}\}$;
- $A_4 = \{0.9845/a_1; 0.2821/a_2; 0.7185/a_3; 0.2821/a_4; 0.1510/a_5; 0.0297/a_6; 0.0837/a_7; 0.0109/a_8; 0.9944/a_9; 0.8688/a_{10}\}$;
- $A_5 = \{0.8688/a_1; 0.6766/a_2; 0.4650/a_3; 0.0837/a_4; 0.0837/a_5; 0.0135/a_6; 0.0468/a_7; 0.0036/a_8; 0.9845/a_9; 0.5698/a_{10}\}$.

Step 3. To identify the best alternative the convolution of available information. The set of optimal alternatives $A$ is determined by intersection of fuzzy sets containing estimates of alternatives according to the NON-EXISTING RISK criterion [4]. In this case, the rule for choosing the best alternative is

$$A = A_1 \cap A_2 \cap A_3 \cap A_4 \cap A_5. \tag{8}$$

Having the maximum value of the membership function of the fuzzy set $A$ alternative is considered optimal. According to [5], the intersection of fuzzy sets appropriates to the choice of the minimum value for the alternative $a_t$ ($t = 1 \div 10$) is

$$\mu_A(a_t) = \min_i\{\mu_{A_i}(a_t)\}. \tag{9}$$

According to (8) and (9) the set of optimal alternatives is formed as follows [5]:
$A = \{$min$\{0.9394; 0.9845; 0.9394; 0.9845; 0.8688\}$, min$\{0.9944; 0.9394; 0.9506; 0.2821; 0.6766\}$, min$\{0.6766; 0.7788; 0.4650; 0.7185; 0.4650\}$, min$\{0.8688; 0.5485; 0.3149; 0.2821; 0.0837\}$, min$\{0.7788; 0.2230; 0.3679; 0.1510; 0.0837\}$, min$\{0.5912; 0.3149; 0.1054; 0.0297; 0.0135\}$, min$\{0.1510; 0.0713; 0.0297; 0.0837; 0.0468\}$, min $\{0.1510; 0.0837; 0.0297; 0.0109; 0.0036\}$, min$\{1.0000; 0.9845; 0.9944; 0.9944; 0.9845\}$, min$\{0.4650; 0.3149; 0.6766; 0.8688; 0.5698\}\}$.

The resulting priority vector of alternatives is $\max_t\{\mu_A(a_t)\}$ = max$\{0.8688; 0.2821; 0.4650; 0.0837; 0.0135; 0.0297; 0.0036; 0.9845; 0.3149\}$.

Thus, from the point of view of the CR-level the best alternative is the country $a_9$, which corresponds to the value of 0.9845. Next in descending order: $a_1 \rightarrow 0.8688$,

$a_3 \rightarrow 0.4650$,    $a_{10} \rightarrow 0.3149$,    $a_2 \rightarrow 0.2821$,    $a_4 \rightarrow 0.0837$,    $a_5 \rightarrow 0.0837$,    $a_7 \rightarrow 0.0297$, $a_6 \rightarrow 0.0135$, $a_8 \rightarrow 0.0036$.

**Table 4.** The comparison of summarized results of CR-levels estimating.

| Alternative countries | Weight-counting technique | | Maxmin convolution method | |
|---|---|---|---|---|
| | Summarized estimate | Order | Summarized estimate | Order |
| $a_1$ | 91.27 | 2 | 0.8688 | 2 |
| $a_2$ | 84.62 | 3 | 0.2821 | 5 |
| $a_3$ | 73.30 | 4 | 0.4650 | 3 |
| $a_4$ | 64.47 | 6 | 0.0837 | 6 |
| $a_5$ | 57.64 | 7 | 0.0837 | 7 |
| $a_6$ | 47.13 | 8 | 0.0135 | 9 |
| $a_7$ | 35.54 | 9 | 0.0297 | 8 |
| $a_8$ | 29.06 | 10 | 0.0036 | 10 |
| $a_9$ | 97.04 | 1 | 0.9845 | 1 |
| $a_{10}$ | 68.55 | 5 | 0.3149 | 4 |

## 7 Conclusion

Within the framework of the first approach the generalized values of weights $x_i$ ($i = 1 \div 5$) are established on the base of the agreed expert judgements on the priority of the CR-factors. It becomes the basis for the reasoned formation of the final estimates of the CR-levels according to the established comparison test at the scale of the segment [0; 100]. The fuzzy maxmin convolution method, which is the essence of the second approach, solves the problem by using another way of aggregating of expert judgements of the CR-factors. A comparison of summarized results of CR-levels estimating of hypothetical alternatives (countries) $a_t$ ($t = 1 \div 10$) obtained by both methods is presented in Table 4, which shows that the orders of some estimates of the CR-levels do not coincide.

**Acknowledgement.** The authors consider necessary to express their sincere appreciation to Professor R.R. Rzayev for the help that he rendered during the process of writing and preparing this article.

## References

1. PwC Global Risk podcast series: https://www.pwc.com/gx/en/services/advisory/consulting/risk/cosoerm-framework/podcasts.html. Accessed 15 May 2018
2. Lin, A.S.: A note on the concordance correlation coefficient. Biometrics **56**, 324–325 (2012)
3. Lin, A.S., Wu, W.: Statistical Tools for Measuring Agreement. Springer, New York (2012). https://doi.org/10.1007/978-1-4614-0562-7

4. Zadeh, L.A.: The concept of a linguistic variable and its application to approximate reasoning. Inf. Sci. **8**(3), 199–249 (1965)
5. Rzayev, R.R.: Analytical Support for Decision-Making in Organizational Systems. Palmerium Academic Publishing, Saarbruchen (2016). (in Russian)

# Forecasting the Solvency of Legal Entities on the Base of Fuzzy Modeling of the Time Series of Their Financial Indicators

Elchin Aliyev[1($\boxtimes$)], Hajar Shikhaliyeva[1], Zaur Gaziyev[1],
and Adila Ali[2]

[1] Institute of Control Systems of ANAS, Baku AZ1141, Azerbaijan
elchin.aliyev@sinam.net, h.shixaliyeva@gmail.com
[2] University College London, London SE18YP, UK

**Abstract.** The methodology for assessing the prospective solvency of legal entities under uncertainty is considered on the base of fuzzy models predicting weakly structured time series of financial indicators and fuzzy method of multi-criteria choice of alternatives.

**Keywords:** Financial indicators of solvency · Fuzzy time series
Fuzzy set

## 1 Introduction

The conduct of credit operations is subject to multiple uncertainty factors associated, for instance, with the lack of comprehensive information on the financial position of the legal entity (LE) and its economic activity, the volatility of the insured mortgage market and the lack of necessary managerial qualities or proper experience of the enterprise manager. These and many other factors should be taken into account by the management of the commercial bank (CB) when conducting credit assessment of LE. Under these prerequisites it becomes evident that new methods of credit management oriented on compliance with the economic boundaries of credit, capable of preventing the unjustified credit investments and ensuring timely and full repayment of loans to significantly reduce the risk of default must be in put in place.

## 2 Problem Definition

The objects of research are LE – candidates for credit whose financial and economic activities are characterized by accounting reports for the last 12 months. The subject of the study is the assessment of their prospective solvencies. The foundation of the information and methodological base is the time series (TS) of financial indicators (FI) of solvency, formulas for calculating the main financial ratios (FR) and their normative values, as well as fuzzy methods of multi-criteria evaluation of alternatives.

The decision-making concerning the credit extension is reduced to a multi-criteria evaluation of the prospective solvencies of the LE. The assessment criterions are

© Springer Nature Switzerland AG 2019
R. A. Aliev et al. (Eds.): ICAFS-2018, AISC 896, pp. 568–576, 2019.
https://doi.org/10.1007/978-3-030-04164-9_75

chosen as following FR: $F_1$ – absolute liquidity ratio, $F_2$ – interim liquidity ratio, $F_3$ – general liquidity ratio, $F_4$ – equity ratio, $F_5$ – net profit ratio of liquidity. These ratios are calculated on the base of data of solvency indicators: $x_1$ – fund; $x_2$ – short-term financing investments; $x_3$ – receivables; $x_4$ – reserves and costs; $x_5$ – equity capital; $x_6$ – current liabilities; $x_7$ – result of the balance; $x_8$ – gross proceeds; $x_9$ – profit, using the following equalities [1]:

$$\begin{cases} F_1 = (x_1 + x_2)/x_6; \ F_2 = (x_1 + x_2 + x_3)/x_6; \\ F_3 = (x_1 + x_2 + x_3 + x_4)/x_6; \ F_4 = x_5/x_7; \ F_5 = x_9/x_8. \end{cases} \tag{1}$$

In order to apply the criteria (1) for the complex assessment of the perspective solvency of the LE it is necessary, first, to construct the TS-models of FI $x_i$ ($i = 1 \div 9$), second, to determine the forecasted values of $F_j$ ($j = 1 \div 5$) based on them, and, third, to apply one of the fuzzy methods of multi-criteria evaluation.

## 3  Forecasting of Financial Indicators of LE

Over the past two decades, the problem of predicting the fuzzy time series (FTS) is analyzed by numerous studies. Nevertheless, the basis of our calculations is the approach of S. Chen [2, 3], which provides the consistent implementation of the following procedures: (1) definition of the universe for historical data (HD) of TS and its partitioning into equal intervals; (2) definition of the evaluation criteria for the HD; (3) fazzification of the HD; (4) identification of internal fuzzy relations (IFR) and their grouping; (5) defuzzification of fuzzy outputs. In this case, HD should be considered as a weakly structured data, i.e. the one about which it is known to belong to a certain type. In particular, TS with the crisp HD cannot always adequately reflect information on the financial and economic behavior of LE. It is advisable to represent them by intervals, for example, in the form of the interval $x \in [x_{min}; x_{max}]$ or, better yet, describe the HD in the form of statement as "close to 7", i.e. by appropriate fuzzy set (FS) [4].

Suppose that the CB considers the requests of four LEs: $a_1$, $a_2$, $a_3$ and $a_4$, for a loan. Because the financial resources of the CB are limited, its management needs to select the best of LEs on the base of their FI TS enveloping 12 month (see Table 1).

According to the fuzzy approach, the forecasting of the FITS of the declared alternatives is carried out by performing following steps.

Step 1: *The definition of the universe and its division into equal intervals.* The universe $U$ for given TS is defined as $[D_{min} - D_1, D_{max} - D_2]$, where $D_{min}$ and $D_{max}$ are respectively the minimum and maximum values of HD, and the numbers $D_1 > 0$ and $D_2 > 0$ are selected by the user. In particular, for the TS "Fund" for alternative $a_1$ (see Table 1) we have: $D_{min} = 250$, $D_{max} = 289$. If to assume that $D_1 = 1$ and $D_2 = 2$, then the appropriate universe of this TS is the interval $U = [249; 291]$. As a rule, the universe is divided into equal parts according to the number of declared criteria for assessing the HD. Let it be 7. Then for TS "Fund" partial intervals will be: $u_1 = [249; 255]$, $u_2 = [255; 261]$, $u_3 = [261; 267]$, $u_4 = [267; 273]$, $u_5 = [273; 279]$, $u_6 = [279; 285]$, $u_7 = [285; 291]$.

**Table 1.**  TS of financial indicators of LE (in thousands of US dollars).

| LE | 1 | 2 | 3 | 4 | 5 | 6 | 7 | 8 | 9 | 10 | 11 | 12 |
|----|----|----|----|----|----|----|----|----|----|----|----|----|
| $x_1$ – Fund | | | | | | | | | | | | |
| $a_1$ | 250 | 259 | 286 | 266 | 271 | 279 | 263 | 277 | 264 | 257 | 289 | 276 |
| $a_2$ | 960 | 1007 | 984 | 983 | 961 | 993 | 981 | 990 | 978 | 965 | 966 | 1006 |
| $a_3$ | 955 | 973 | 992 | 967 | 991 | 1004 | 964 | 989 | 974 | 991 | 997 | 977 |
| $a_4$ | 1230 | 1236 | 1267 | 1271 | 1276 | 1242 | 1245 | 1243 | 1242 | 1253 | 1254 | 1246 |
| LE $x_2$ – Short-term financing investments | | | | | | | | | | | | |
| $a_1$ | 250 | 259 | 286 | 266 | 271 | 279 | 263 | 277 | 264 | 257 | 289 | 276 |
| $a_2$ | 960 | 1007 | 984 | 983 | 961 | 993 | 981 | 990 | 978 | 965 | 966 | 1006 |
| $a_3$ | 955 | 973 | 992 | 967 | 991 | 1004 | 964 | 989 | 974 | 991 | 997 | 977 |
| $a_4$ | 1230 | 1236 | 1267 | 1271 | 1276 | 1242 | 1245 | 1243 | 1242 | 1253 | 1254 | 1246 |
| LE $x_3$ – Receivables | | | | | | | | | | | | |
| $a_1$ | 250 | 259 | 286 | 266 | 271 | 279 | 263 | 277 | 264 | 257 | 289 | 276 |
| $a_2$ | 960 | 1007 | 984 | 983 | 961 | 993 | 981 | 990 | 978 | 965 | 966 | 1006 |
| $a_3$ | 955 | 973 | 992 | 967 | 991 | 1004 | 964 | 989 | 974 | 991 | 997 | 977 |
| $a_4$ | 1230 | 1236 | 1267 | 1271 | 1276 | 1242 | 1245 | 1243 | 1242 | 1253 | 1254 | 1246 |
| LE $x_4$ – Reserves and costs | | | | | | | | | | | | |
| $a_1$ | 250 | 259 | 286 | 266 | 271 | 279 | 263 | 277 | 264 | 257 | 289 | 276 |
| $a_2$ | 960 | 1007 | 984 | 983 | 961 | 993 | 981 | 990 | 978 | 965 | 966 | 1006 |
| $a_3$ | 955 | 973 | 992 | 967 | 991 | 1004 | 964 | 989 | 974 | 991 | 997 | 977 |
| $a_4$ | 1230 | 1236 | 1267 | 1271 | 1276 | 1242 | 1245 | 1243 | 1242 | 1253 | 1254 | 1246 |
| LE $x_5$ – Equity capital | | | | | | | | | | | | |
| $a_1$ | 250 | 259 | 286 | 266 | 271 | 279 | 263 | 277 | 264 | 257 | 289 | 276 |
| $a_2$ | 960 | 1007 | 984 | 983 | 961 | 993 | 981 | 990 | 978 | 965 | 966 | 1006 |
| $a_3$ | 955 | 973 | 992 | 967 | 991 | 1004 | 964 | 989 | 974 | 991 | 997 | 977 |
| $a_4$ | 1230 | 1236 | 1267 | 1271 | 1276 | 1242 | 1245 | 1243 | 1242 | 1253 | 1254 | 1246 |
| LE $x_6$ – Current liabilities | | | | | | | | | | | | |
| $a_1$ | 250 | 259 | 286 | 266 | 271 | 279 | 263 | 277 | 264 | 257 | 289 | 276 |
| $a_2$ | 960 | 1007 | 984 | 983 | 961 | 993 | 981 | 990 | 978 | 965 | 966 | 1006 |
| $a_3$ | 955 | 973 | 992 | 967 | 991 | 1004 | 964 | 989 | 974 | 991 | 997 | 977 |
| $a_4$ | 1230 | 1236 | 1267 | 1271 | 1276 | 1242 | 1245 | 1243 | 1242 | 1253 | 1254 | 1246 |
| LE $x_7$ – Result of the balance | | | | | | | | | | | | |
| $a_1$ | 250 | 259 | 286 | 266 | 271 | 279 | 263 | 277 | 264 | 257 | 289 | 276 |
| $a_2$ | 960 | 1007 | 984 | 983 | 961 | 993 | 981 | 990 | 978 | 965 | 966 | 1006 |
| $a_3$ | 955 | 973 | 992 | 967 | 991 | 1004 | 964 | 989 | 974 | 991 | 997 | 977 |
| $a_4$ | 1230 | 1236 | 1267 | 1271 | 1276 | 1242 | 1245 | 1243 | 1242 | 1253 | 1254 | 1246 |
| LE $x_8$ – Gross proceeds | | | | | | | | | | | | |
| $a_1$ | 250 | 259 | 286 | 266 | 271 | 279 | 263 | 277 | 264 | 257 | 289 | 276 |
| $a_2$ | 960 | 1007 | 984 | 983 | 961 | 993 | 981 | 990 | 978 | 965 | 966 | 1006 |
| $a_3$ | 955 | 973 | 992 | 967 | 991 | 1004 | 964 | 989 | 974 | 991 | 997 | 977 |
| $a_4$ | 1230 | 1236 | 1267 | 1271 | 1276 | 1242 | 1245 | 1243 | 1242 | 1253 | 1254 | 1246 |

(*continued*)

**Table 1.** (*continued*)

| LE | 1 | 2 | 3 | 4 | 5 | 6 | 7 | 8 | 9 | 10 | 11 | 12 |
|---|---|---|---|---|---|---|---|---|---|---|---|---|
|  | $x_1$ – Fund | | | | | | | | | | | |
| LE | $x_9$ – Profit | | | | | | | | | | | |
| $a_1$ | 250 | 259 | 286 | 266 | 271 | 279 | 263 | 277 | 264 | 257 | 289 | 276 |
| $a_2$ | 960 | 1007 | 984 | 983 | 961 | 993 | 981 | 990 | 978 | 965 | 966 | 1006 |
| $a_3$ | 955 | 973 | 992 | 967 | 991 | 1004 | 964 | 989 | 974 | 991 | 997 | 977 |
| $a_4$ | 1230 | 1236 | 1267 | 1271 | 1276 | 1242 | 1245 | 1243 | 1242 | 1253 | 1254 | 1246 |

Step 2: *The definition of the evaluation criteria.* Suppose that $A_1$, $A_2$, ..., $A_k$ are weakly structured evaluative concepts for qualitative interpretation of HD-values, which can be described by following FSs [5]:

- SMALL – $A_1 = 1/u_1 + 0.5/u_2 + 0/u_3 + 0/u_4 + 0/u_5 + 0/u_6 + 0/u_7$;
- NOT SUCH LARGE – $A_2 = 0.5/u_1 + 1/u_2 + 0.5/u_3 + 0/u_4 + 0/u_5 + 0/u_6 + 0/u_7$;
- LARGE – $A_3 = 0/u_1 + 0.5/u_2 + 1/u_3 + 0.5/u_4 + 0/u_5 + 0/u_6 + 0/u_7$;
- MORE THAN LARGE – $A_4 = 0/u_1 + 0/u_2 + 0.5/u_3 + 1/u_4 + 0.5/u_5 + 0/u_6 + 0/u_7$,
- ESSENTIALLY LARGE – $A_5 = 0/u_1 + 0/u_2 + 0/u_3 + 0.5/u_4 + 1/u_5 + 0.5/u_6 + 0/u_7$,
- VERY LARGE – $A_6 = 0/u_1 + 0/u_2 + 0/u_3 + 0/u_4 + 0.5/u_5 + 1/u_6 + 0.5/u_7$,
- TOO LARGE – $A_7 = 0/u_1 + 0/u_2 + 0/u_3 + 0/u_4 + 0/u_5 + 0.5/u_6 + 1/u_7$.

Step 3: *Fuzzification of the HD.* Fuzzification of the HD is carried out according to the following principle: if the value of the HD $F(t)$ gets into interval $u_i$ ($i = 1 \div 7$), then it is described as FS $A_i$, because the degree of belonging of $u_i$ to $A_i$ has the largest value. For example, for alternative $a_1$ the value 250 (HD for the 1st month of TS "Fund") gets into the interval $u_1$. Because $u_1$ with the highest degree belongs to FS $A_1$ ($\mu_{A1}(u_1)$) the given HD can be described, respectively, in the form of FS $A_1$. As a result, due to such fuzzification of the HD analogous FTS of FI for alternative LEs are constructed. For example, for the TS "Fund" the appropriate FTS is shown in Table 2.

Step 4: *Identify the IFRs and grouping them.* The IFRs are established between the fuzzified HDs. In particular, if as a result of fuzzyfication the crisp HD $F(t - 1)$ is replaced by FS $A_i$, and the HD $F(t)$ is replaced by FS $A_j$, then there is a connection between them that is called IFR of the form $A_i{\rightarrow}A_j$. In this case, $A_j$ reflects the current state of the FI, and $A_i$ - the previous one. So, for example, for the applicant $a_1$ the successive HDs of the FITS "Fund" for the 1st and 2nd months (see Table 2) are described respectively as $A_1$ and $A_2$, which are related by the implication "If HD is $A_1$, then next is $A_2$" or, simply, by fuzzy relation $A_1{\rightarrow}A_2$. In a similar manner, all IFRs into the considered FTS are identified. In particular, for the FITS "Fund" all IFRs are grouped and summarized in the form of Table 3. IFRs are grouped according to the principle: if the FS $A_3$ within the FTS (as in the case of "Fund") is associated with $A_2$, $A_4$ and $A_5$, then a local group $A_3{\rightarrow}A_2$, $A_4$, $A_5$ (or $A_{2, 4, 5}$) is formed concerning it.

Step 5: *Defuzzification of outputs of the fuzzy model of FITS.* To defuzzify the fuzzy outputs of the predictive model of FITS, the rules of S. Chen [2] were used, the essence of which consists in the following. Let the fuzzy analogue of the FI $F(t - 1)$ for the

**Table 2.** TS "Fund" for alternative LEs.

| Month | $a_1$ | | $a_3$ | | $a_3$ | | $a_4$ | |
|---|---|---|---|---|---|---|---|---|
| | HD | FS | HD | FS | HD | FS | HD | FS |
| 1 | 250 | $A_1$ | 960 | $A_1$ | 955 | $A_1$ | 1230 | $A_1$ |
| 2 | 259 | $A_2$ | 1007 | $A_7$ | 973 | $A_3$ | 1236 | $A_2$ |
| 3 | 286 | $A_7$ | 984 | $A_4$ | 992 | $A_6$ | 1267 | $A_6$ |
| 4 | 266 | $A_3$ | 983 | $A_4$ | 967 | $A_2$ | 1271 | $A_6$ |
| 5 | 271 | $A_4$ | 961 | $A_1$ | 991 | $A_5$ | 1276 | $A_7$ |
| 6 | 279 | $A_5$ | 993 | $A_5$ | 1004 | $A_7$ | 1242 | $A_3$ |
| 7 | 263 | $A_3$ | 981 | $A_4$ | 964 | $A_2$ | 1245 | $A_3$ |
| 8 | 277 | $A_5$ | 990 | $A_5$ | 989 | $A_5$ | 1243 | $A_3$ |
| 9 | 264 | $A_3$ | 978 | $A_3$ | 974 | $A_3$ | 1242 | $A_3$ |
| 10 | 257 | $A_2$ | 965 | $A_2$ | 991 | $A_5$ | 1253 | $A_4$ |
| 11 | 289 | $A_7$ | 966 | $A_2$ | 997 | $A_6$ | 1254 | $A_4$ |
| 12 | 276 | $A_5$ | 1006 | $A_7$ | 977 | $A_4$ | 1246 | $A_3$ |

**Table 3.** Groups of the IFRs for FITS "Fund".

| LE | Group 1 | Group 2 | Group 3 | Group 4 | Group 5 | Group 6 |
|---|---|---|---|---|---|---|
| $a_1$ | $A_1 \to A_2$ | $A_2 \to A_7$ | $A_3 \to A_{2,4,5}$ | $A_4 \to A_5$ | $A_5 \to A_3$ | $A_7 \to A_{3,5}$ |
| $a_2$ | $A_1 \to A_{5,7}$ | $A_2 \to A_{2,7}$ | $A_3 \to A_2$ | $A_4 \to A_{1,4,5}$ | $A_5 \to A_{3,4}$ | $A_7 \to A_4$ |
| $a_3$ | $A_1 \to A_3$ | $A_2 \to A_5$ | $A_3 \to A_{5,6}$ | $A_5 \to A_{3,6,7}$ | $A_6 \to A_{2,4}$ | $A_7 \to A_2$ |
| $a_4$ | $A_1 \to A_2$ | $A_2 \to A_6$ | $A_3 \to A_{3,4}$ | $A_4 \to A_{3,4}$ | $A_6 \to A_{6,7}$ | $A_7 \to A_3$ |

current month $(t - 1)$ is FS $A_i$. Then predict for the next month is determined in accordance with the following rules:

(1) if there is an unique relationship into the group localized relative to the $A_i$, for example, $A_i \to A_k$, where $A_k$ with the highest degree of belonging includes the interval $u_k$, then the predict is the bisecting point of the segment $u_k$;

(2) if FS $A_i$ is not related by any relationship, i.e. $A_i \to \varnothing$ and $A_i$ with the highest degree of belonging includes the interval $u_i$, then the predict is the bisecting point of the segment $u_i$;

(3) if there is a multi-valued relationship into the group localized relative to the $A_i$, for example, $A_i \to A_{1, 2, \dots, n}$, where the FS $A_1, A_2, \dots, A_n$ with the highest degrees of belonging include, respectively, the intervals $u_1, u_2, \dots, u_n$, then the appropriate predict is defined as $F(t) = (m_1 + m_2 + \dots + m_n)/n$, where $m_1, m_2, \dots, m_n$ are the bisecting points of the corresponding segments $u_1, u_2, \dots, u_n$.

For example, as can be seen from Table 2, the fuzzy analog of the FI "Fund" for 1st month is $A_1$, which forms only one relationship: $A_1 \to A_2$ (see Table 3). Therefore, applying the first rule, we get the appropriate predict as $258 = (255 + 261)/2$, implying that the interval $u_2 = [255; 261]$ belongs to $A_2$ with the greatest degree.

Let us consider another case. For the 12th month FI "Fund" of LE $a_3$ is described as FS $A_4$, which includes the interval $u_4 = [976; 984]$, but it is not a predicate, i.e. $A_4 \rightarrow \varnothing$. Therefore, the appropriate predict for the next month will be the bisecting point of the segment $u_4$, namely: $980 = (976 + 984)/2$.

Finally, let us consider the case of the application of the third rule. It is necessary to find predict for 10-th month of the FI "Fund" for LE $a_1$. Because fuzzy analogue of this indicator for $9^{th}$ month is $A_3$, concerning of which the group of relationships $A_3 \rightarrow A_2$, $A_4$, $A_5$ is established (see Table 3), then the predict for 10-th month will be arithmetical mean of the bisecting points of the segments $u_2 = [255; 261]$, $u_4 = [267; 273]$ and $u_5 = [273; 279]$, namely: $268 = [(255 + 261)/2 + (267 + 273)/2 + (273 + 279)/2]/3$.

Thus, applying above defuzzification rules to the all fuzzy outputs for all FTS the appropriate predictive models have been obtained for declared LEs. In particular, for the FITS "Fund" the obtained predicts are summarized in the form of Table 4.

Table 4. Forecasting the FI "Fund" for LEs.

| Month | $a_1$ | | $a_3$ | | $a_3$ | | $a_4$ | |
|---|---|---|---|---|---|---|---|---|
| | Fact | Predict | Fact | Predict | Fact | Predict | Fact | Predict |
| 1 | 250 | | 960 | | 955 | | 1230 | |
| 2 | 259 | 258 | 1007 | 1001 | 973 | 972 | 1236 | 1237 |
| 3 | 286 | 288 | 984 | 985 | 992 | 992 | 1267 | 1269 |
| 4 | 266 | 270 | 983 | 980 | 967 | 972 | 1271 | 1273 |
| 5 | 271 | 268 | 961 | 980 | 991 | 988 | 1276 | 1273 |
| 6 | 279 | 276 | 993 | 1001 | 1004 | 991 | 1242 | 1245 |
| 7 | 263 | 264 | 981 | 981 | 964 | 964 | 1245 | 1249 |
| 8 | 277 | 268 | 990 | 980 | 989 | 988 | 1243 | 1249 |
| 9 | 264 | 264 | 978 | 981 | 974 | 991 | 1242 | 1249 |
| 10 | 257 | 268 | 965 | 969 | 991 | 992 | 1253 | 1249 |
| 11 | 289 | 288 | 966 | 989 | 997 | 991 | 1254 | 1249 |
| 12 | 276 | 270 | 1006 | 989 | 977 | 972 | 1246 | 1249 |

## 4    Assessment of Current and Prospective Solvencies of LEs

For the current assessment of LE-solvency, the values of the FR for the 12th month are chosen as the basic HD, and for the prospective evaluation the corresponding predicts presented in Table 4 are applied. For this purpose in the beginning on the base of (1) it is necessary calculate the values of the FR $F_j$ ($j = 1 \div 5$), characterizing the solvencies of the LEs for the current 12th and predicted 13th months. The calculated and normative values for FR are summarized in Table 5.

The primary analysis of the calculated values of FRs relative to the corresponding normative indicators shows that all declared LEs can pretend to a loan, and, therefore, they form the acceptable alternatives set $\{a_1, a_2, a_3, a_4\}$. Multi-criterion evaluation of the current and prospective solvencies of the LEs $a_k$ ($k = 1 \div 4$) involves processing using the mathematical apparatus of the fuzzy sets theory under relevant information

**Table 5.** The values of FRs by results of FI "Fund" for LEs.

| FR | $a_1$ | | $a_2$ | | $a_3$ | | $a_4$ | | Normative values |
|---|---|---|---|---|---|---|---|---|---|
| | 12 | 13 | 12 | 13 | 12 | 13 | 12 | 13 | |
| $F_1$ | 0.174 | 0.171 | 0.107 | 0.108 | 0.086 | 0.089 | 0.133 | 0.133 | [0.1; 0.25] |
| $F_2$ | 1.334 | 1.329 | 0.713 | 0.715 | 0.601 | 0.603 | 0.5698 | 0.569 | [0.5; 1] |
| $F_3$ | 2.826 | 2.820 | 2.267 | 2.271 | 1.872 | 1.875 | 1.267 | 1.266 | [1; 2.5] |
| $F_4$ | 0.752 | 0.752 | 0.719 | 0.719 | 0.711 | 0.711 | 0.683 | 0.683 | 0.6 |
| $F_5$ | 0.285 | 0.285 | 0.116 | 0.116 | 1.500 | 1.4996 | 0.121 | 0.121 | [0; 2] |

obtained in the initial stage in the form of Table 5. Then evaluation of the current and prospective solvencies of the LEs $a_k$ can be realized by the fuzzy method of maxmin convolution in three stages [1].

Stage 1. *Construction of membership functions appropriating to evaluation concepts*: $F_1$ = PREFERABLE (absolute liquidity ratio), $F_2$ = DESIRABLE (interim liquidity ratio), $F_3$ = PERMISSIBLE (general liquidity ratio), $F_4$ = THE GREATEST (equity ratio), $F_5$ = THE BEST (net profit ratio of liquidity). The construction of such membership functions, as a rule, is carried out by experts. We will do otherwise. Assume that the intervals of normative values of FRs are universes. Then according to [5] the Gaussian functions $\mu_{Fj}(a) = \exp\{-(a - a_{j0})^2/\sigma_j^2\}$ can be choose as appropriate membership functions of fuzzy sets $F_j$ ($j = 1 \div 5$), where $a_{j0}$ and $\sigma_j^0$ are, respectively, the centers and densities of the distribution of the nearest elements. For each membership function of fuzzy sets $F_j$ ($j = 1 \div 5$) the appropriate parameters were obtained as following: $a_{10} = 0.175$, $\sigma_{10} = 0.025$; $a_{20} = 0.75$, $\sigma_{20} = 0.078$; $a_{30} = 1.75$, $\sigma_{30} = 0.23$; $a_{40} = 0.6$, $\sigma_{40} = 0.09$; $a_{50} = 2$, $\sigma_{50} = 0.625$.

Stage 2. *Determination of the concrete values of the membership functions according to the quality criteria $F_j$ ($j = 1 \div 5$)*. Starting from the current and predicted values of the FR (see Table 5) and the calculated corresponding values of the Gaussian membership functions $\mu_{Fj}(a)$ of the type of (2), let us form the FS to describe the qualitative criterion $F_j$ for the current 12th and predicted 13th months: $F_j(12)$ и $F_j(13)$. As subsets of the universe $\{a_1, a_2, a_3, a_4\}$ these FS are the following:

$F_1(12) = \{0.9993/a_1; 0.0005/a_2; 0.0000/a_3; 0.0561/a_4\}$,
$F_1(13) = \{0.9696/a_1; 0.0007/a_2; 0.0000/a_3; 0.0573/a_4\}$
$F_2(12) = \{0.0000/a_1; 0.7956/a_2; 0.0253/a_3; 0.0048/a_4\}$,
$F_2(13) = \{0.0000/a_1; 0.8155/a_2; 0.0282/a_3; 0.0046/a_4\}$
$F_3(12) = \{0.0000/a_1; 0.0063/a_2; 0.7553/a_3; 0.0121/a_4\}$,
$F_3(13) = \{0.0000/a_1; 0.0059/a_2; 0.7525/a_3; 0.0119/a_4\}$
$F_4(12) = \{0.0572/a_1; 0.1734/a_2; 0.2185/a_3; 0.4271/a_4\}$,
$F_4(13) = \{0.0578/a_1; 0.1724/a_2; 0.2189/a_3; 0.4246/a_4\}$
$F_5(12) = \{0.0005/a_1; 0.0001/a_2; 0.5275/a_3; 0.0001/a_4\}$,
$F_5(13) = \{0.0005/a_1; 0.0001/a_2; 0.5268/a_3; 0.0001/a_4\}$.

Stage 3. *The convolution of relevant information to identify the best alternative.* According to [1], the set of optimal alternatives $R$ is determined by the intersection of

FSs containing estimates of alternatives by qualitative criterions. Then, assuming that these criterions $F_j$ ($j = 1 \div 5$) for credit surveillance of considered LEs have the equal importance, the rule for choosing the best alternative can be present as: $R = F_1 \cap F_2 \cap F_3 \cap F_4 \cap F_5$, where $\mu_R(a_k) = \min\{\mu_{Fj}(a_k)\}$, $k = 1 \div 4$. In this case, the alternative with the maximum degree of belonging to the set $R$ is optimal. Then, for the current and prospective credit surveillances of LEs, the set of optimal alternatives is formed respectively as: $R(12) = \{\min\{0.9993; 0.0000; 0.000003; 0.0572; 0.0005\}$, $\min\{0.0005; 0.7956; 0.0063; 0.1734; 0.000113\}$, $\min\{0.0000; 0.0253; 0.7553; 0.2185;$ $0.5275\}$, $\min\{0.0561; 0.0048; 0.0121; 0.4271; 0.000119\}\}$, $R(13) = \{\min\{0.9696;$ $0.0000; 0.000007; 0.0578; 0.0005\}$, $\min\{0.0007; 0.8155; 0.0059; 0.1724; 0.000119\}$, $\min\{0.0000; 0.0282; 0.7525; 0.2189; 0.5268\}$, $\min\{0.0573; 0.0046; 0.0119; 0.4246;$ $0.0001\}\}$, or more specifically as:

$$R(12) = \{0/a_1; 0.000113/a_2; 0.000003/a_3; 0.000119/a_4\},$$
$$R(13) = \{0/a_1; 0.000113/a_2; 0.000007/a_3; 0.000119/a_4\}.$$

The resultant vectors, whose components determine the priority of alternatives for the current 12th and predicted 13th months, are following:

$$\max\mu_{R(12)}(a_k) = \max\{0; 0.000113; 0.000003; 0.000119\},$$
$$\max\mu_{R(13)}(a_k) = \max\{0; 0.000113; 0.000007; 0.000119\}.$$

The best alternative is the LE that is characterized by the largest value of the corresponding component of resultant vector. For the current month, it is $a_4 \rightarrow 0.000119$, and further in descending order: $a_2 \rightarrow 0.000113$; $a_3 \rightarrow 0.000003$ and $a_1 \rightarrow 0$. A similar scenario exists for the forecasted period as well.

# 5    Conclusion

Methodology for evaluating the current and prospective credit surveillances of LEs that was proposed in the current research is based on the application of fuzzy models of time series and fuzzy methods of multi-criteria evaluation under uncertainty. The proposed fuzzy model of TS is arithmetic, i.e. it is rather trivial in comparison with other existing analytical models of weakly structured time series. Nevertheless, this approach allows constructing short-term forecasts for the credit surveillances of LEs not only the next month, as shown in this research, but also for one year ahead. The obtained estimates do not claim to be optimal, as the applied Gaussian membership functions type of (2) were not optimized. The main purpose of this technique is the artificial formation of a temporary logical basis for the current and prospective comparison of alternatives in order to find the most solvent LE.

**Acknowledgement.** The authors consider necessary to express their appreciation to Professor R. R. Rzayev for the help that he rendered during the process of writing and preparing this article.

# References

1. Andreichenkov, A.V., Andreichenkova, O.N.: Analysis, Synthesis, Planning Decisions in the Economy. Finance and Statistics, Moscow (2000). (in Russian)
2. Chen, S.M.: Forecasting enrollments based on fuzzy time series. Fuzzy Sets Syst. **81**, 311–319 (1996)
3. Chen, S.M.: Forecasting enrollments based on fuzzy time series. Cybern. Syst.: Int. J. **33**, 1–16 (2002)
4. Zadeh, L.A.: The concept of a linguistic variable and its application to approximate reasoning. Inf. Sci. **8**(3), 199–249 (1965)
5. Rzayev, R.R.: Analytical Support for Decision-Making in Organizational Systems. Palmerium Academic Publishing, Saarbruchen (2016). (in Russian)

# Deep Parkinson Disease Diagnosis: Stacked Auto-encoder

Esam Al Shareef[1] and Dilber Uzun Ozsahin[1,2(✉)]

[1] Department of Biomedical Engineering, Near East University, Nicosia
North-Cyprus, Turkey
dilber.uzunozsahin@neu.edu.tr
[2] Gordon Center for Medical Imaging, Radiology Massachusetts General
Hospital and Harvard Medical School, Boston, USA

**Abstract.** In this work, we demonstrate the feasibility of deep learning based stacked auto-encoder for the Parkinson's Disease (PD) diagnosis. Features are extracted by the employed deep network from the input source. We transfer features learned from the SAE during pre-training to the fine-tuning phase in which each sample or patient's condition is labeled, which grants the network the time to learn and distinguish the healthy from the PD patients. The employed model is fine-tuned and tested on a small dataset in order to explore their generalization capabilities when trained using few data. Experimentally, the stacked auto-encoder showed a high accuracy and features extraction capability in diagnosing the Parkinson diseased patients where it achieved an accuracy of 89.5% which is considered as a promising result.

**Keywords:** Deep learning · Stacked auto-encoder · Parkinson's disease

## 1 Introduction

Parkinson's disease (PD) is classified as the second most common neurodegenerative condition [1]. Only 2% of men and 3 of women are ought to risk of developing Parkinson disease according to researchers from Olmsted County (Mayo Clinic) [2]. The success of deep learning in medicine is a main motivator of researchers to apply such intelligent systems in diagnosing the patient's conditions in an easier and more accurate manner. Thus, many researchers were conducted to diagnose diseases by training a deep network on a large of data which gives it the power of diagnosing new patients with small margins of error.

Doctors are still categorizing the patient conditions in a manual manner based on some visual examinations and laboratory tests [3]. Therefore, There is a need for an automatic and intelligent system that have the capability of accurate diagnosis of Parkinson Disease patients into healthy and PD [4]. Thus, in this work, we aim to use a powerful deep network in such diagnosis tasks. The deep network that is selected to be used in called the stacked auto-encoder which showed a great efficacy the different classification and diagnosis tasks in the medical field. Furthermore, the network is also examined on other medical applications [5–8], which aims to demonstrate the effects of medical image processing and enhancement on the performance a neural network.

© Springer Nature Switzerland AG 2019
R. A. Aliev et al. (Eds.): ICAFS-2018, AISC 896, pp. 577–585, 2019.
https://doi.org/10.1007/978-3-030-04164-9_76

Deep network named as stacked auto-encoder is expected to well perform since it has been applied to various medical classification tasks where it achieved high accuracies and because the data available are enough for training it and achieving a small error.

## 2    Stacked Auto-encoder

In unsupervised learning and deep architectures, auto-encoders play a fundamental role for transfer learning and other tasks. In [9], Pierre Baldi presented a general mathematical framework for the study of both linear and non-linear auto-encoders. The structure reveals insight into the various types of auto-encoders, their learning intricacy, vertical and horizontal composability in deep designs, basic focuses and crucial connections to information theory, Hebbian learning, and clustering.

By applying back-propagation, the auto-encoder attempts to reduce the error however much as could reasonably be expected amongst input and reconstruction by learning an encoder and a decoder [10].

## 3    Materials and Methods

This study is to investigate the generalization capability of a Stacked Auto-encoder (SAE) of two hidden layers in diagnosing the Parkinson disease. Hence, such SAE is must be first trained on a good number of data which grant it the power of diagnosing the cases of unseen and new patients. The following sections describe the data used in training and testing the SAE in addition to its training performance.

### 3.1    Database

The Deep network is known to be called as hungry-data systems; hence large number of data is required in order to obtain an effective trained deep network with high accuracy. Thus, in this work, we obtained the Parkinson database from a public data repository known as the University of California at Irvine (UCI) machine learning repository [7]. The database was developed collaboration between the University of Oxford and the National Centre for Voice and Speech, where the signals were recorded.

This dataset consists of a range of biomedical voice measurements from 31 people where 23 of them are with Parkinson's disease (PD). Each column of this dataset consists of 23 rows; each represents different parameters as shown in Table 1. Note that the parameter 'status' is to discriminate between the healthy and the Parkinson Diseased patients, in which it is set to 0 for healthy and 1 for PD. Therefore, this parameters 'status' was considered as output since the aim of this study is to identify or diagnose the patient's condition, whether it is healthy or with PD.

Table 2 shows the number of data in the database and the learning schemes used for training and testing the network. As seen in Table 2, the total number of data is 5,025; among 1593 are healthy patients while the rest is with PD. The data were split into training and testing sets in which 4174 and 850 samples are used for training and testing, respectively.

**Table 1.** Dataset splitting

| Attribute | Description |
|---|---|
| MDVP: Fo (Hz) | Average vocal fundamental frequency |
| MDVP: Fhi (Hz) | Maximum vocal fundamental frequency |
| MDVP: Flo (Hz) | Maximum vocal fundamental frequency |
| MDVP: Jitter (%)<br>MDVP: Jitter (Abs)<br>MDVP: RAP<br>MDVP: PPQ<br>Jitter: DDP | Several measures of variation in fundamental frequency |
| MDVP: Shimmer<br>MDVP: Shimmer (dB)<br>Shimmer: APQ3<br>Shimmer: APQ5<br>MDVP: APQ<br>Shimmer: DDA | Several measures |
| RPDE<br>D2 | Two nonlinear dynamical complexity measures |
| DFA | Signal fractal scaling exponent |
| Spread 1<br>Spread 2<br>PPE | Three nonlinear measures of fundamental frequency variation |
| NHR<br>HNR | Two measures of ratio of noise to tonal components in the voice |
| Concept class | Healthy, sick |

**Table 2.** Data set splitting

| Total number of data | Train | Test |
|---|---|---|
| 5025<br>Healthy: 1593<br>PD: 3432 | 4174 | 850 |

## 3.2  Training the Deep Models

Two auto-encoder networks were used to build the proposed stacked auto-encoder that is then used to be as the intelligent classifier of the Parkinson Disease data. The auto-encoder was first trained layer by layer using greedy layer wise training until a network of two hidden layer, one input, and one output network is formed. Therefore, these trained auto-encoders were all stacked together and the proposed stacked auto-encoder is formed.

As seen in Table 2, the SAE is trained on 4174 samples and tested on 850. Figure 2 shows the proposed architecture of the stacked auto-encoder designed for diagnosing the Parkinson Disease. As seen, the network consists of two hidden layers a different number of hidden neurons since two auto-encoders are used to build this network.

The input layer consists of 22 neurons as the input parameters are 22, and the output neurons are 2 because the network is designed to diagnose two classes: healthy or PD.

The output classes coding was considered as the following:

- Healthy output class: [1 0],
- PD output class: [0 1].

Note that the network is first pre-trained as it is deep network. Pre-training means that the networks are first trained layer by layer using Greedy-layer wise training (Hinton 2006). In this phase, there is no output labeling because the network is trained here to reconstruct its input from the extracted features in the hidden layer, which is why the number of output neurons is equal to the number of input neurons which is 22.

Once the networks finish pre-training, it is then fine-tuned using the conventional backpropagation algorithm. Here, the input data are labeled therefore, output neurons are two which means that the network is being trained to classify the samples into two classes: healthy and PD.

A sample of the data before normalization is shown in Table 3. The normalized data are shown in Table 4.

In this work, the two employed model is trained and tested using Matlab environment. The networks were simulated on a Windows 64-bit desktop computer with an Intel Core i7 4770 central processing unit (CPU) and 4 GB random access memory. Note that there was no graphical processing unit (GPU) available in the used desktop. The training and testing accuracies of the model were calculated as follows:

$$Accuracy = \frac{C}{T}$$

Where C is the number of correctly diagnosed data, and T is the total number of data used in training or testing.

Table 5 shows the training parameters values set during training the SAE.

The Fig. 1 shows the learning of the network. It shows the variation of error with the increase of number of iterations which were set as 1000 during the fine tuning of the network.

As seen in Fig. 1 the network was learning effectively until it reached a mean square error of 0.0461, and then it stopped. Note that this was achieved in 120 s and 1000 iterations as shown in Fig. 1.

Figure 1 shows the training performance evaluation of the network designed to diagnose the Parkinson disease. The SAE achieved a relatively good accuracy during training (92.5%), and therefore this may results eventually in a good generalization capability in diagnosing the unseen data or samples during testing.

## 4   Results Discussion

This paper presents a deep learning approach for the classification of patient medical data into healthy and PD. This work is based on a stacked auto-encoder that is trained to classify the medical data to learn the useful features that can distinguish both classes.

**Table 3.**  Training data before normalization

| 1 | 1 | 1 | 1 | 1 | 1 | 1 |
|---|---|---|---|---|---|---|
| 72 | 72 | 72 | 72 | 72 | 72 | 72 |
| 5.64310 | 12.6660 | 19.6810 | 25.6470 | 33.64200 | 40.6520 | 47.6490000 |
| 28.1990 | 28.4470 | 28.6950 | 28.9050 | 29.18700 | 29.4350 | 29.6820000 |
| 34.3980 | 34.8940 | 35.3890 | 35.8100 | 36.37500 | 36.8700 | 37.3630000 |
| 0.00662 | 0.00300 | 0.00481 | 0.00528 | 0.003350 | 0.00353 | 0.00422000 |
| 3.380000000000000e-02 | 1.680000000000000e-02 | 2.460000000000000e-02 | 2.660000000000000e-02 | 2.010000000000000e-02 | 2.290000000000000e-02 | 2.400000000000000e-02 |
| 0.00401 | 0.00132 | 0.00205 | 0.00191 | 0.00093 | 0.00119 | 0.00212 |
| 0.00317 | 0.00150 | 0.00208 | 0.00264 | 0.00130 | 0.00159 | 0.00221 |
| 0.01204 | 0.00395 | 0.00616 | 0.00573 | 0.00278 | 0.00357 | 0.00637 |

**Table 4.** Training data after normalization

| | | | | | | |
|---|---|---|---|---|---|---|
| 0.013888658 | 0.01388865 | 0.01388865 | 0.01388865 | 0.01388865 | 0.01388865 | 0.01388865 |
| 1 | 1 | 1 | 1 | 1 | 1 | 1 |
| 0.078376173 | 0.17591647 | 0.27334705 | 0.35620818 | 0.46724987 | 0.56461100 | 0.66179158 |
| 0.391652635 | 0.39509708 | 0.39854152 | 0.40145819 | 0.40537486 | 0.40881930 | 0.41224986 |
| 0.477749878 | 0.48463876 | 0.49151377 | 0.49736099 | 0.50520821 | 0.51208321 | 0.51893044 |
| 9.17111325103754e-05 | 4.14334300111134e-05 | 6.65722377557444e-05 | 7.31000170566706e-05 | 4.62944552464840e-05 | 4.87944558298175e-05 | 5.83777913992624e-05 |
| 2.36111166203717e-07 | 0 | 1.08333358611117e-07 | 1.36111142870378e-07 | 4.58333440277803e-08 | 8.47222419907454e-08 | 1.00000023333339e-07 |
| 5.54611240520400e-05 | 1.81000042233343e-05 | 2.82388954779645e-05 | 2.62944505798162e-05 | 1.26833362927785e-05 | 1.62944482464824e-05 | 2.92111179270386e-05 |
| 4.37944546631505e-05 | 2.06000048066678e-05 | 2.86555622418534e-05 | 3.64333418344464e-05 | 1.78222263807417e-05 | 2.18500050983345e-05 | 3.04611182187054e-05 |
| 0.0001669889278527972 | 5.46277905242622e-05 | 8.53222421307454e-05 | 7.93560185150043e-05 | 3.83777867325947e-05 | 4.93500115150027e-05 | 8.82389094779678e-05 |
| 0.0138886587 | 0.01388865 | 0.01388865 | 0.01388865 | 0.01388865 | 0.01388865 | 0.01388865 |

**Table 5.** Training network performance

| Learning parameters | Values (Fine-training) |
|---|---|
| Number of training samples | 4174 |
| Number of hidden layers of the network | 2 |
| Number of hidden neurons | 70, 50 |
| Learning rate | 0.23 |
| Maximum number of iterations | 1000\1000 |
| Transfer function | Sigmoid |

**Fig. 1.** Learning curve and MSE

This type of networks is trained using an algorithm called Greedy-layer wise training which is meant to train the network layer by layer in an unsupervised manner. This allows the network to gain the power of extracting of important features that will be used in the next training phase which is called Fine-tuning. Fine-tuning is to train pre-trained network to classify X-rays using the conventional backpropagation learning algorithm by using the weights obtained for the pre-training phase.

Upon training, the stacked auto-encoder was tested using 850 samples of the available data. Table 6 shows the performance of the network during testing. It is seen that the network achieved good diagnosis accuracy during testing which means that the network was capable of correctly diagnosing 87.2% of the samples used in testing. Moreover, the table shows the overall accuracy of the network which is the average accuracy of the training and testing phases.

**Table 6.** Testing results

| Deep networks | Number of testing images | Testing accuracy | Overall accuracy |
|---|---|---|---|
| SAE | 850 | 87.5% | 89.86% |

Figure 2 shows the regression plot of the network during testing. It shows the fitting between the desired and actual outputs of the network. It is seen that the actual output is not that far from the desired output, thus low error margins were obtained.

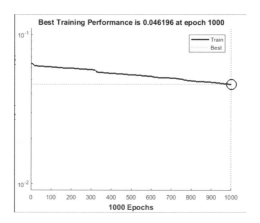

**Fig. 2.** Regression plot of the network

## 5 Conclusion

In this research, deep learning based neural networks were employed. Stacked auto-encoder of two hidden layers was used. Their features learned during pre-training are transferred to fine-tuning stage, in order to learn the classification of patient medical data into healthy and PD. We conclude that SAE, a well-designed and deeper architecture of sufficient complexity, was capable of achieving significantly satisfactory diagnosis accuracy when distinguishing between healthy and PD patients. Furthermore, it can be stated that the pre-training followed by conventional training of the network of deep neural network (SAE) to learn a new task can work accurately with a small margin of error, even when trained on a relatively small dataset. It is important to state that the depth of SAE network contributed to a better understanding of the input data by extracting different levels of abstractions, which contributed to achieving higher recognition rates; however, required a long time and iterations of 120 s and 1000 iterations, respectively.

Finally, our results can show that applying deep SAE model for the problem of Parkinson disease diagnosis is promising, in a way that similar or confusing Parkinson diseased patients can be diagnosed or correctly classified with good accuracies.

## References

1. Hughes, A., Daniel, S., Kilford, L., Lees, A.: Accuracy of clinical diagnosis of idiopathic Parkinson's disease: a clinico-pathological study of 100 cases. J. Neurol. Neurosurg. Psychiatry **55**, 181–184 (1992)

2. Elbaz, A., Bower, J., Maraganore, D., McDonnell, S., Peterson, B., Ahlskog, J., Schaid, D., Rocca, W.: Risk tables for parkinsonism and Parkinson's disease. J. Clin. Epidemiol. **55**, 25–31 (2002)
3. Little, M., McSharry, P., Hunter, E., Spielman, J., Ramig, L.: Suitability of dysphonia measurements for telemonitoring of Parkinson's disease. IEEE Trans. Biomed. Eng. **56**, 1015–1022 (2009)
4. Tolosa, E., Wenning, G., Poewe, W.: The diagnosis of Parkinson's disease. Lancet Neurol. **5**, 75–86 (2006)
5. Gil, D., Johnson, M.: Diagnosing Parkinson by using artificial neural networks and support vector machines. Global J. Comput. Sci. Technol. **9**, 63–71 (2009)
6. Helwan, A., Uzun Ozsahin, D.: Sliding window based machine learning system for the left ventricle localization in MR cardiac images. Appl. Comput. Intell. Soft Comput. **2017**, 1–9 (2017)
7. Helwan, A., Uzun, D., Abiyev, R., Bush, J.: One-year survival prediction of myocardial infarction. Int. J. Adv. Comput. Sci. Appl. **8**, 173–178 (2017)
8. Ozsahin, D., Isa, N., Uzun, B., Ozsahin, I.: Effective analysis of image reconstruction algorithms in nuclear medicine using fuzzy PROMETHEE. In: 2018 Advances in Science and Engineering Technology International Conferences (ASET) (2018)
9. Khemphila, A., Boonjing, V.: Parkinsons disease classification using neural network and feature selection. World Acad. Sci. Eng. Technol. Int. J. Math. Comput. Sci. **6**, 377–380 (2012)
10. Baldi, P.: Autoencoders, unsupervised learning, and deep architectures. In: Proceedings of the 2011 International Conference on Unsupervised and Transfer Learning Workshop, UTLW 2011, vol. 27, pp. 37–50 (2012)

# Machine Learning Comparative Analysis for Plant Classification

Elbrus Imanov[1($\boxtimes$)] and Abdallah Khaled Alzouhbi[2]

[1] Department of Computer Engineering, Near East University, Mersin 10,
North Cyprus, Turkey
elbrus.imanov@neu.edu.tr
[2] Department of Water Recycling, Machha, Akkar 1032, Lebanon
abdallazohbi@gmail.com

**Abstract.** Nowadays, digital image processing, artificial neural network and machine visualization have been pettishly progressing, and they cover a significant side of artificial cleverness and the rule among human beings and electro-mechanical devices. These technologies have been utilized in a wide range of agricultural operations, medicine and manufacturing. By this research the preparation of some functions has been conducted.

In this paper we introduce the classification of maize leaves from pictures that reveal many conditions, opening among pictures, by pre-processing, taking out, plant feature recognition, matching and training, and lastly getting the outcomes executed by Matlab, neural network pattern recognition application. These given features are separated to leaf maturity and picture interpretations, rotary motions and calibration, and they are calculated to develop an approach that gives us better classification algorithm results. A plant scientist may be introduced with a plant for recognition of its classes revealed in its natural home ground, to gather an in-depth recognition.

**Keywords:** Artificial neural network · Digital image processing
Machine visualization classification K-nearest neighbor
Support vector machine · Machine learning

## 1 Introduction

This assumption focuses on automatic identification throughout computer visualization and exploitation machine learning. All of the learning, compact knowledge base and operative processing can be reached through a neural network, even by using a simple traditional computer [1]. Artificial neural networks are popular with their capability to learn various tasks by using collected examples or training data. More importantly, ANNs are capable of making intelligent decisions on tasks for which they are trained. Machine learning suspects about the speculation behind artificial systems that extort information from pictures [2]. Integrating human knowledge expertise or skill with the desired functions of a computing machine used to-gather more correct decisions is the main function of the "learning algorithms" [3]. Machine learning and computer vision are the science and machinery of machines that have the power to check and

© Springer Nature Switzerland AG 2019
R. A. Aliev et al. (Eds.): ICAFS-2018, AISC 896, pp. 586–593, 2019.
https://doi.org/10.1007/978-3-030-04164-9_77

acknowledge. Snyder describes the expression computer vision as the method by a machine, usually a computing mechanism, mechanically processes a set of image and informs 'what is within the image'; it works on understanding the content of the image. Some of the important features which are usually detected include edges, corners, blobs, ridges, object parts, and etc. Common operations that are used to realize feature detectors in computer vision include delta function, Gaussian derivatives; Gabor filters coefficients [4]. Computer vision consists of activity of options, pattern classification confirming those options, and pattern recognition. Pattern classification is the association of patterns into sets of patterns having the same set of possession. Given a group of measurements of an unidentified item and the information of probable classes to which an item might belong to, a choice about to which class the unidentified object belongs can be made. Human being can reach the related desired solution for the pattern recognition and the obstacle avoidance; for example, within an uncertain environment by spending a little time period, however it would be more costly if the computer resources were implemented [3]. Pattern recognition is the procedure for classifying information or patterns grounded based on the information extracted from patterns. It is better to start by recognition of the objects and afterwards, category of the scene to be able to comprehend the bigger and complicated scene presented [5]. The sample to be recognized is most likely the sets of dimensions or interpretation presenting points in a suitable multidimensional room. This proposal was target-hunting to expand a structure that extracts entirely different options from a leaf picture and grouped different categories of leaves confirming the extracted options. The leaf image that can be easily transferred to the computer and its features can be extracted automatically by the computer by using image processing method. Moreover, this system uses the results of the category theme to spot the class of the latest leaf images [6]. Image processing technology is used to export a set of features which describe or symbolize the image. The amounts of these features give a brief demonstration about the information in such a figure.

This paper is extremely enthused by the real world implementation of machine learning using diverse sorts of classifiers. This proposal focuses on utilizing image processing and artificial neural network to mechanize categorization and perform plant detection based on the pictures of their leaves. Plant is everywhere we live and even places we are not living. Most of them carry important information about the human facts development [7]. Automatic plant categorization and identification can help botanists in their research, as well as assist laymen to classify and study plants more simply and more intensely. They additionally sorted the plants as trees, shrubs and herbs victimization complication classifier algorithms. But this work on creating an easy straightforward approach by simply considering leaf details victimization simple support vector machine classifier (SVM) and K-nearest neighbor (KNN). SVM works thoroughly on both well attributed datasets and the dataset that involves merely some cases available to conduct the training process. As a result of this advantageous feature, SVM classification becomes very useful [8]. For the division of the data points by planes, distant planes in space are used by the SVM classification [9]. By using the K-means algorithm, training data parts of each class are clustered one by one. Hence, an algorithm is offered forward, and besides, Support Vector Machine model is trained merely by exploitation of the cluster centers [10]. KNN that is used in classifying objects rely upon the feature space's closest training examples. KNN algorithm is a well-known simplest machine algorithm. The unlabeled query point that is involved in the classification process is basically associated with the label of its KNN [11]. K-nearest neighbor (KNN) rule is used for image classification without many complications.

## 2    Materials and Methods

The image database includes 200 paint picture divisions of maize plant usually shown in Mediterranean regions with 50 diverse kinds of shrubberies. Images were collected during several intervals during the diverse periods of a day. In accumulation, shrubberies with variable awning extent were designated to upsurge the trouble of the organizational problem. Support vector machine is a state of the art engine knowledge technique founded by the current arithmetic education concept. It has been magnificently useful in diverse classification difficulties. SVM achieves the organization by building a hyperactive smooth in such a way that the unraveling border among optimistic and undesirable samples is optimal. By using run length matrix the texture feature is extracted and then by using by ANN classifier these extracted features classification is conducted [12]. Extracting features from testing and training images, is followed by the conversion of vector represented patches into code words, which is realized by the process of k-means clustering on the whole vectors presented, that is a process to cluster n observation in our case. Extrication restless smooth then jobs as the choice superficial [13]. K-nearest neighbors, the k adjacent neighbor and do a popular elective.

## 3    Experimental Results

In addendum, as most of the image detection programs, a database of huckleberry or paper picture has to be done, in addition to knowledge way to elicit the advantages for the database, and a different technique to get back the top competition from the database. Put in data training, once the mark removal was total, two documents were achieved. Traineeship texture feature data and test texture feature data ranking using Support Vector Machine rooted in linear classifier. A software monotone was written in Matlab that would obtain jalousie documents that represent the exercise and check data, workouts the classifier using the train documents and then utilize the trial folder to do the arrangement task on the trial data. As a result, a system monotone would load all the data folders and make amendment to the data depends on to the suggested sample selected. In this part, process of printing pictures with neural network image processing classification is explained step by step. Main stages of the system are illustrated in the Fig. 1.

In the first step image data store function was used to automatically read all the given images. Second step counts each label function that is used in order to count and label as label maize (200), non maize (27). As a third step we have tried to display sampling of maize images using 'montage' function. Fourth step is the separation of

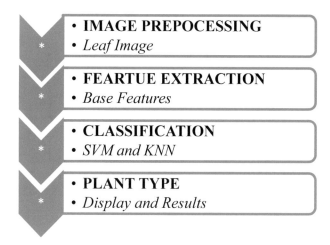

**Fig. 1.** Main stages of the system.

images into a training set and test set, feature extraction using bag of features, also known as bag of visual words is one method to extract features from images. To represent an image using this approach, an image can be treated as the document and occurrence of visual 'words' in images are used to generate a histogram that represents an image. The fifth step is creating visual vocabulary, to extract features from the given images that have used 'bag of features' function. These features are used as inputs to the SVM and K-NN classifiers to train them and classify them later. System offered a classification application which gives us a good opportunity to try as many different types of classifier as we want. The sixth step visualizes extracted features vectors, by using 'encode bag, image' and subplot functions. We create a table using the encoded features; using the new features to train a model and assess its performance by using different classifiers, the classification learner functions into the code or by choosing classifier learner icon from the tool strip. After starting session and selecting the required data that we have put into a table before, the original data will be displayed. Scene image data is shown in Fig. 2.

Blue points are for maize plant coordination, and orange ones are for non-maize plants coordination.

The sixth step involves many classifiers types available to choose and train the data-set. We run all the SVM and K-NN classifier types to pick up the one with highest accuracy which is the more suitable one for our dataset. Accuracies of KNN and SVM are shown in Fig. 3.

Randomly we chose linear SVM which have 96.7% accuracy, and export the model in order to use it in our code. Before exporting let's take a look on the matrix table of our model and see how it looks like. Confusion matrix of the system is given in Fig. 4.

Green areas are for maize feature data as we see in the image above, and the pink ones are for the non maize plants. For the white area this presents the common features between the two of them. Test out accuracy on test set to estimate the efficiency of the suggested technique is realized. The normal categorization rate is considered following

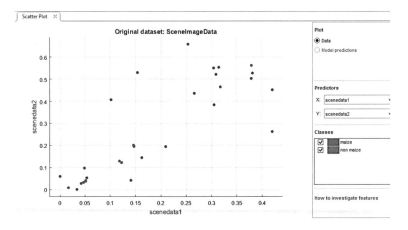

**Fig. 2.** Scene image data

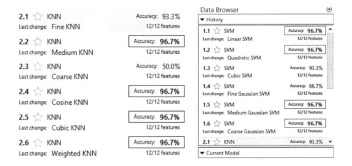

**Fig. 3.** KNN and SVM accuracies

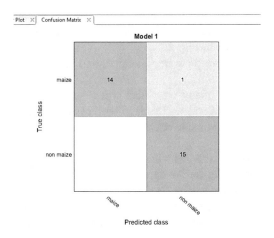

**Fig. 4.** Confusion matrix

to cycling the beyond procedure for nine or ten period [14]. As the examples of the experimenting set area unit unidentified to the classifier, the accomplishment ratio of classifying associate freelance testing dataset is mirrored by the forecasted accurateness acquired from this unidentified set. The categorization correctness of the native model operators will be prejudiced by regulating totally different parameters. For LBP, we have got utilizing totally different regulations for the parameters P and R. Equally, for LDP and LTP, the classification ratio was considered for varied k and t values, severally. Table 1 illustrates the classification rate of LBP, LTP, and LDP characteristic exemplification for various issue settings. Of these tests rotary motions invariant pattern mistreatment of each guide corresponding and support vector machine, severally. Classification rate by template corresponding to dissimilar factors setting is given in Table 1.

**Table 1.**  Classification ratio.

| Classification rate | Parameter setting | Operator |
|---|---|---|
| 82.8 | (P, R) = (8.1) | LBP |
| 83.3 | (P, R) = (16.2) | LBP |
| 83.8 | (P, R) = (8.1) | Uniform LBP |
| 85.0 | (P, R) = (16.2) | Uniform LBP |
| 87.0 | t = 5 | LTP |
| 86.3 | t = 10 | LTP |
| 85.5 | t = 15 | LTP |
| 87.0 | k = 2 | LDP |
| 89.3 | k = 3 | LDP |
| 87.8 | k = 4 | LDP |

It is determined that, LDP (k = 3) supplies the very best classification ratio between the native pattern factors exploitation each model corresponding and support vector machine. It's apprehensible that, the prevalence of LDP coding theme is as a result of the employment of strong edge restraint values in numerous guidelines for figuring the binary pattern, wherever the opposite strategies employ strength values of area environs. In these tests, support vector machine supplies advanced classification ratio corresponding to all native pattern operators. We will simply establish the best parameter intended for these operators. Optimal parameter setting for the local pattern operators is given in Table 2.

**Table 2.**  Optimal parameter setting for the local pattern operators.

| Parameter setting | Operator |
|---|---|
| (P, R) = (16.2) | LBP |
| (P, R) = (16.2) | Uniform LBP |
| t = 5 | LTP |
| k = 3 | LDP |

The act of the native prototype based mostly element illustration is additionally contrast with another presented plants categorization ways, particularly Gabor wavelets, and Haar wavelets remodel. Figure one displays the contrast between the popularity ratios of presented wavelets-based ways associate in Nursing LDP (k = 3). It will be seen that, LDP out-performs the opposite ways in conditions of classification accurateness. Returning to system, in order to test our test-set accuracy we used special function to extract our test-set accuracy, and we have got 85% accuracy which is considered as statistically good. Visualize how the classifier works, now, and as a final step let's see how our classifier behaves by showing some samples. Classifying maize plant results is shown in Fig. 5.

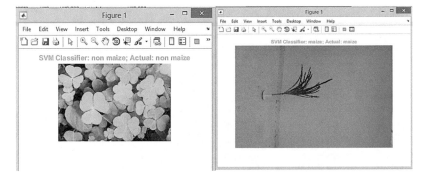

**Fig. 5.** Classifying maize plant results.

As we see in the pictures above our classifier behaves correctly and recognizes the maize and non maize plants. "SVM Classifier" case is for what the classifier expects and the "Actual" case is what the plant in reality is.

## 4   Conclusion

By this research article we studied on two distinct kinds of image and picture classifications for examination of the maize leaves. According to investigations with an accurate learning method, we can conclude that the SVM and KNN classifications have similar and almost same accuracies and performance. Hence, all the classifier square measures have been influential in classification and recognition of the plant used. We used the image database that involves 200 paint picture divisions of maize plant which is usually appeared in Mediterranean regions with 50 different kinds of shrubberies. Images were taken during several intervals in the course of the day. Lastly, after matching and training, outcomes were obtained by using Matlab's neural network pattern recognition application.

# References

1. Aliev, R.A., Fazlollahi, B., Aliev, R.R.: Soft Computing and its Application in Business and Economics, p. 388. Springer, Heidelberg (2004)
2. Gouk, H.G., Blake, A.M.: Fast sliding window classification with convolutional neural network. In: 2014 Proceedings of the 29th International Conference on Image and Vision Computing, Hamilton, New Zealand, pp. 114–118 (2014)
3. Karray, F.O., de Silva, C.: Soft Computing and Intelligent Systems Design, pp. 4–13, 223–224. Pearson Education Limited/British Library (2004)
4. Comaniciu, D., Ramesh, V., Meer, P.: Real time tracking of non-rigid objects using mean shift. In: 2000 Proceedings of the IEEE Conference on Computer Vision and Pattern Recognition, vol. 2, pp. 142–149 (2000)
5. Treisman, A.M., Gelade, G.: A feature integration theory of attention. Cogn. Psychol. **12**, 97–136 (1980)
6. Dallwitz, M.J.: A general system for coding taxonomic descriptions. Taxon **29**, 6 (1980)
7. Du, J.X., Wang, X.F., Zhang, G.J.: Leaf shape based plant species recognition. Appl. Math. Comput. **185**, 11 (2007)
8. Kim, J., Kim, B.S., Savarese, S.: Comparing image classification methods, K-nearest neighbor and support vector machines. In: 2012 Proceedings of the American Conference on Applied Mathematics, pp. 133–138 (2012)
9. Burgers, C.J.C.: A tutorial on supper vector machines for pattern recognition. Data Min. Knowl. Discov. **2**, 121–167 (1998)
10. Tran, Q.A., Zhang, Q.L., Li, X.: Reduce the number of support vectors by using clustering techniques. In: 2003 Proceedings of the Second International Conference on Machine Learning Cybernetics, Xi'an, pp. 1243–1248 (2003)
11. Bermmert, D., Demaine, E., Erickson, J., Longermans, S., Morin, P., Toussaint, G.: Output sensitive algorithms for computing nearest neighbors decision boundaries. Discrete Comput. Geom. **33**, 583–604 (2005)
12. Pujari, J.D., Yakkundimath, R., Byadgi, A.S.: Grading and classification of anthracnose fungal disease of fruits based on statistical texture features. Int. J. Adv. Sci. Technol. **52**, 121–132 (2013)
13. Ding, C., He, X.: K-means clustering via principal component analysis. In: 2004 Proceedings of International Conference on Machine Learning, pp. 225–232 (2004)
14. Alzouhbi, A.: Plant classification using SVM and KNN classifiers. Thesis Nicosia (2017)

# Exchange Market Algorithm for Selective Harmonic Elimination in Cascaded Multilevel Inverters

Ardavan MohammadHassani[1] and Ebrahim Babaei[1,2(✉)]

[1] Faculty of Electrical and Computer Engineering,
University of Tabriz, Tabriz, Iran
ardavan.mh@gmail.com, e-babaei@tabrizu.ac.ir
[2] Engineering Faculty, Near East University,
99138 Nicosia, Mersin 10, North Cyprus, Turkey

**Abstract.** Utilization of Exchange Market Algorithm (EMA) for selective harmonic elimination (SHE) in cascaded multilevel inverters (CMIs) is investigated in this paper. The nonlinear SHE equations are derived for a 7-level CMI based on the Fourier expansion of the output voltage waveform. A cost function is constructed according to the SHE equations and optimization is performed with EMA. In order to compare the results, the SHE equations are also solved with Genetic Algorithm (GA) with similar parameters as used for EMA. Optimization results prove that EMA is superior to GA in terms of convergence rate, and achieving exact global minima. For verifying the optimization results, simulations are performed on a 7-level CMI for different modulation indices.

**Keywords:** Cascaded multilevel inverters · Exchange Market Algorithm
Selective harmonic elimination

## 1 Introduction

Nowadays, cascaded multilevel inverters (CMIs) are widely used for various high-voltage and high-power applications. In such cases, the distortion of the output voltage is required to be as low as possible. One particular method for reducing the harmonic content of the output voltage is the selective harmonic elimination method (SHE). The main objective in SHE is to maintain the fundamental harmonic amplitude of the output voltage at its desired value and eliminate a certain number of low order harmonics. To do this, a certain number of equations are needed to be solved for a number of switching angles. However, the SHE equations are highly nonlinear and transcendental which makes them very difficult to solve.

Various methods have been introduced in the literature to solve the nonlinear SHE equations. Initial approaches were based on numerical methods. These methods include the well-known Newton's method [1]. The major drawback of these methods is that they require very accurate initial values for the switching angles. Although a number of methods have been introduced to achieve suitable initial values for 2-level voltage

© Springer Nature Switzerland AG 2019
R. A. Aliev et al. (Eds.): ICAFS-2018, AISC 896, pp. 594–601, 2019.
https://doi.org/10.1007/978-3-030-04164-9_78

waveforms [2], this is not straightforward for multilevel waveforms. In addition, these methods can only find one set of solutions.

Walsh functions were used for solving the SHE equations in [3]. In this method, the trigonometric SHE equations are transformed into a set of algebraic equations which are easier to solve. Also, this method is able to find more than one sets of solutions. However, an appropriate initial condition is still required for finding a solution. In addition, the computation time of this method is very high.

A method was proposed in [4] to solve the SHE equations based on symmetric polynomials and Groebner bases. However, the degree of the polynomials significantly increases with the number of levels which increases the computation time of this method.

Optimization methods have also attracted a large amount of attention for finding solutions to the nonlinear SHE equations. In these methods, the nonlinear SHE equations are first transformed into a cost function and then stochastic algorithms are used to search for all available sets of solutions. Genetic Algorithms (GA) have been used for SHE in [5]. Another promising algorithm is the Particle Swarm Optimization (PSO) [6]. Bee Algorithm (BA) was also used for SHE in [7]. Another algorithm for solving the SHE equations is the Ant Colony algorithm (AC) [8].

In this paper, application of the Exchange Market Algorithm (EMA) is explored for selective harmonic elimination in a 7-level CMI. Optimization results including running time and the probability of attaining a global solution are presented and compared with the same results obtained with GA. Results verify the superiority of EMA over GA. In order to further confirm the effectiveness of EMA, simulation results for a 7-level CMI are provided for three different modulation indices and the results are compared with the results obtained with GA.

The rest of this paper is structured as follows. In Sect. 2, the general topology and basic operation of a 7-level CMI is briefly presented. Then the nonlinear SHE equations are derived based on the Fourier expansion of the output voltage. EMA is briefly introduced in Sect. 3. Optimization problem and results are provided in Sect. 4. Section 5 presents the simulation results. Finally, conclusions are provided in Sect. 6.

## 2  Overview of the General Topology and Basic Operation of a 7-Level CMI

Cascaded multilevel inverters have several advantages over the diode-clamped, and flying-capacitor topologies. CMIs do not require extra clamping diodes or voltage balancing capacitors. In addition, the number of output voltage levels is very easy to adjust. CMIs consists of $N$ series-connected, single-phase full-bridge inverters which are called H-bridges. Each H-bridge can produce three different voltage outputs: $V_{dc}$, 0, and $-V_{dc}$. In order to generate a multilevel voltage, each H-bridge generates a quasi-square waveform by phase-shifting the switching timings of its positive and negative phase legs. The number of voltage levels in the output voltage of the CMI is $2N + 1$. Hence, a 7-level CMI consists of three H-bridges. The general topology for this CMI is shown in Fig. 1.

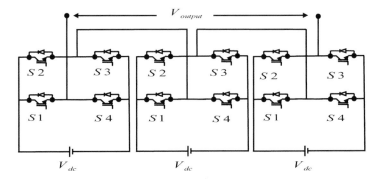

**Fig. 1.** General topology of a 7-level CMI

In order to synthetize a multilevel output voltage, each H-bridge is required to produce a step-wise voltage during its switching cycle. Hence, by connecting the H-bridges in series, a multilevel voltage can be produced. A sample multilevel voltage waveform alongside the output voltage of each H-bridge is shown in Fig. 2.

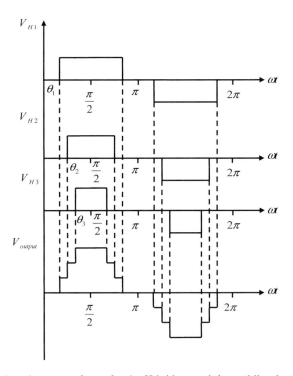

**Fig. 2.** Sample voltage waveforms for the H-bridges and the multilevel output voltage

The multilevel waveform shown in Fig. 2 has odd quarter-wave symmetry. Therefore, the even order harmonics in the Fourier expansion are zero. Hence, the Fourier expansion for this waveform can be written as

$$V_o(\omega t) = \sum_{h=3,5,7,\dots} \frac{4V_{dc}}{h\pi} \left( \sum_{i=1,2,3} \cos(h\theta_i) \right) \sin(h\omega t) \tag{1}$$

where, $h$ the harmonic is order, $V_{dc}$ is the input DC voltage for each H-bridge, and $\theta_i$ is the switching angle.

In this paper, the $5^{th}$ and $7^{th}$ harmonics are chosen to be eliminated while the fundamental harmonic is maintained at its desired value. Therefore, the nonlinear SHE equations can be written according to the switching angles as

$$\begin{aligned}
A_1 &= \frac{4V_{dc}}{\pi} \left( \cos(\theta_1) + \cos(\theta_2) + \cos(\theta_3) \right) = A_1^* \\
A_5 &= \frac{4V_{dc}}{5\pi} \left( \cos(5\theta_1) + \cos(5\theta_2) + \cos(5\theta_3) \right) = 0 \\
A_7 &= \frac{4V_{dc}}{7\pi} \left( \cos(7\theta_1) + \cos(7\theta_2) + \cos(7\theta_3) \right) = 0
\end{aligned} \tag{2}$$

where, $A_1^*$ is the reference value for the fundamental harmonic.

The reference value for the fundamental harmonic $A_1^*$ can be determined according to the modulation index $M$ as

$$A_1^* = \frac{12MV_{dc}}{\pi} \tag{3}$$

## 3 A Brief Introduction to EMA

Exchange Market Algorithm is a powerful and accurate optimization tool for solving various problems [9–11]. It has two effective searching and absorbent searching operators. Therefore, the limitations of other optimization algorithms are highly improved in EMA. EMA is based on the behavior of the members in a stock market. In EMA, the members are believed to operate similar to elite shareholders such as Warren Buffet. Two operating modes are considered in this Algorithm. In each mode the members are classified into three groups according to their share amounts. The first group are the elite members and do not change their shares. The second and third groups have less shares and try to buy and sell their shares so that they could move towards the first group. In the first mode, the stock prices face no oscillation and the members buy and sell shares in order to make the best possible benefit. In other words, the algorithm recruits members towards the elite members. On the other hand, in the second mode, the prices fall into oscillation and the members must take risks for achieving maximum benefit. This means that in this mode the algorithm searches for unknown points.

## 4  Implementation of SHE Using EMA

The optimization program is written with MATLAB and a study is conducted in order to derive the switching angles. The parameters for EMA are listed in Table 1.

**Table 1.** Optimization parameters for EMA.

| Parameter | Value |
|---|---|
| Initial population size | 100 |
| Maximum number of iterations | 500 |
| Common market risk | [0.1, 0.05] |
| Variable market risk | [0.1, 0.05] |
| Running time | 391.02 s |

The fitness function can be written as [7]

$$\text{cost} = \left(100\frac{A_1 - A_1^*}{A_1^*}\right)^4 + \sum_{h=3,5,7,\dots} \frac{1}{h}\left(50\frac{A_h}{A_1}\right)^2 \tag{4}$$

The switching angles are found by minimizing the cost function in Eq. (4). The switching angles are required to be within the region of $0 \le \theta_i \le \frac{\pi}{2}$, $i = 1, 2, 3$.

Figure 3 shows the harmonic conditions versus $M$. It can be clearly observed that the fundamental harmonic is near its reference for all values of $M$. This is due to the penalty term in Eq. (4). The low order harmonics are completely eliminated for a specific range of $M$. For other ranges, the harmonics are kept close to zero when the cost function is low.

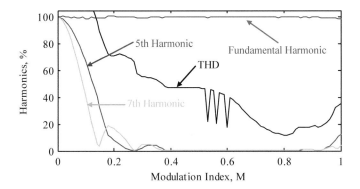

**Fig. 3.** Harmonic amplitudes versus modulation index

In order to further prove the efficiency of EMA, the optimization is also carried out with GA with similar parameters. The parameters employed in GA are listed in Table 2. By comparing the running times of both algorithms provided in Tables 1 and 2, it can be observed that EMA is superior over GA in reducing the computational burden.

**Table 2.** Optimization parameters for GA.

| Optimization parameter | Value |
| --- | --- |
| Initial population size | 100 |
| Maximum number of iterations | 500 |
| Mutation rate | 5% |
| Crossover rate | 85% |
| Running time | 737.66 s |

The cumulative distribution function (CDF) for EMA and GA are shown in Fig. 4. It can be seen that the probability of reaching global minima is higher in EMA. For example, the probability of EMA reaching $10^{-5}$ is 31% while with GA its 27%. Hence, EMA has a better performance in finding solutions than GA.

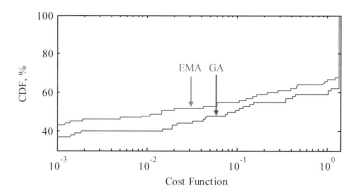

**Fig. 4.** CDF for EMA and GA

## 5   Simulation Results for a 7-Level CMI

A simulation model for the 7-level CMI shown in Fig. 1 is built in SIMULINK. The input DC sources have a voltage 12 V. An inductive load with resistance of $R = 5\,\Omega$ and an inductance of $L = 55\,\text{mH}$ is considered. Switching angles for $M = 0.5$ are derived using EMA. Simulated load voltage for the corresponding switching angles is shown in Fig. 5. As expected, the load voltage is a step-wise waveform. The FFT analysis for the simulated load voltage is shown in Fig. 6. It can be seen that EMA has efficiently eliminated the low order harmonics.

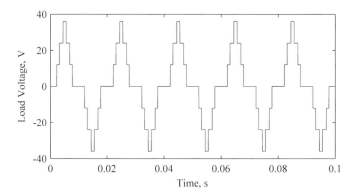

**Fig. 5.**  Simulated load voltage for $M = 0.5$

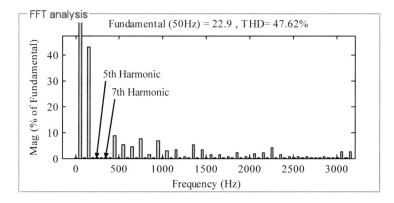

**Fig. 6.**  FFT report for simulated load voltage

In order to compare the efficiency of EMA with GA, the 5$^{th}$ and 7$^{th}$ harmonic amplitudes and the THD for the results provided by each algorithm for $M = 0.5, 0.6$, and 0.7 are listed in Table 3 for. By comparing the results, it can be seen that for $M = 0.5$ EMA is more efficient in eliminating the low order harmonics. However, the THD is lower with GA. For $M = 0.6$ EMA is more efficient in both eliminating the low order harmonics and reducing the THD. For $M = 0.7$ GA is more efficient than EMA in eliminating the low order harmonics. However, the THD is nearly equal for both algorithms.

**Table 3.**  Performance of EMA and GA in eliminating 5$^{th}$ and 7$^{th}$ harmonics and reducing the THD.

| Modulation index | EMA | | | GA | | |
|---|---|---|---|---|---|---|
| | h5% | h7% | THD% | h5% | h7% | THD% |
| $M = 0.5$ | 0.01 | 0.01 | 47.10 | 0.02 | 0.13 | 21.94 |
| $M = 0.6$ | 0 | 0 | 17.54 | 0.14 | 0.10 | 40.82 |
| $M = 0.7$ | 0.02 | 0.03 | 21.86 | 0 | 0 | 21.55 |

# 6   Conclusions

In this paper, EMA is employed for finding solutions to the SHE problem. Optimization results including the cost function, and harmonic amplitudes demonstrate that the algorithm is efficiently able to find solutions to the SHE problem. By comparing the CDF for both EMA and GA it is shown that the probability of finding global solutions is higher for EMA. Moreover, the running time of EMA is lower than GA. By running simulations on a 7-level CMI, it is shown that EMA is able to thoroughly eliminate the low order harmonics and efficiently control the fundamental harmonic. simulation results for other modulation indices are provided. Results demonstrate that for some values EMA is more effective in eliminating low order harmonics while GA is more efficient in reducing the THD.

# References

1. Enjeti, P., Lindsay, J.F.: Solving nonlinear equations of harmonic elimination PWM in power control. IEEE Electron. Lett. **23**(12), 656–657 (1987)
2. Kato, T.: Sequential homotopy-based computation of multiple solutions for selected harmonic elimination in PWM inverters. IEEE Trans. Circ. Syst. I Fundam. Theory Appl. **46**(5), 586–593 (1999). (1993–2003)
3. Ye, M., Song, P., Zhang, C.: Study of harmonic elimination technology for multilevel inverters. In: 3rd IEEE Conference on Industrial Electronics and Applications, ICIEA 2008, pp. 242–245. IEEE Press, New York (2008)
4. Yang, K., Zhang, Q., Yuan, R., Yu, W., Yuan, J., Wang, J.: Selective harmonic elimination with Groebner bases and symmetric polynomials. IEEE Trans. Power Electron. **31**(4), 2742–2752 (2016)
5. Dahidah, M.S.A., Agelidis, V.G., Rao, M.V.C.: Hybrid genetic algorithm approach for selective harmonic control. Energy Convers. Manag. **49**(2), 131–142 (2008)
6. Taghizadeh, H., Hagh, M.T.: Harmonic elimination of cascade multilevel inverters with nonequal DC sources using particle swarm optimization. IEEE Trans. Ind. Electron. **57**(11), 3678–3684 (2010)
7. Kavousi, A., Vahidi, B., Salehi, R., Bakhshizadeh, M.K., Farokhnia, N., Fathi, S.H.: Application of the bee algorithm for selective harmonic elimination strategy in multilevel inverters. IEEE Trans. Power Electron. **27**(4), 1689–1696 (2012)
8. Patil, S.D., Kadwane, S.G., Gawande, S.P.: Ant Colony Optimization applied to selective harmonic elimination in multilevel inverters. In: 2nd International Conference on Applied and Theoretical Computing and Communication Technology, pp. 637–640. IEEE Press, New York (2016)
9. Ghorbani, N., Babaei, E.: Exchange market algorithm. Appl. Soft Comput. **19**, 177–187 (2014)
10. Ghorbani, N., Babaei, E.: Exchange market algorithm for economic load dispatch. Int. J. Electr. Power Energy Syst. **75**, 19–27 (2016)
11. Ghorbani, N.: Combined heat and power economic dispatch using exchange market algorithm. Int. J. Electr. Power Energy Syst. **82**, 58–66 (2016)

# Fuzzy Portfolio Selection Model Using Linear Programming

Mustafa Menekay[✉]

Department of International Business, Near East University,
Nicosia, Mersin 10, TRNC, Turkey
mustafa.menekay@neu.edu.tr

**Abstract.** This paper development of crisp and fuzzy portfolio models using linear programming. Using Lagrange multiplier method the solution of linear programming is carried out. As a input data past gains of assets and values of expect gains are taken. The portfolio selection model, in the form of linear programming, based on the values of expected gains and the variance of securities is formulated. The gradient method is connected to discover weights values of assets. The $\alpha$ level method and interim number arithmetic is utilized to take care of fuzzy enhancement issue and locate the ideal fuzzy estimations of the securities.

**Keywords:** Fuzzy portfolio · Linear programming

## 1 Introduction

Outlining the right portfolio of benefits requires present day, capable and dependable scientific instruments and projects. In portfolio selection the expected gain and the risk measured by the variance are the two main components of a portfolio. The finding optimal values of these parameters, such as high gain and minimum risk are main problem of portfolio selection. Unfortunately asset having high gains usually have high risk. It is the point of the portfolio supervisor to discover a portfolio that augments expected increase under given hazard level or a portfolio that limits chance under given increase level. The scientific plan of portfolio analysis first was given by Prof. Harry Markowitz [1]. The main problem in portfolio selection is the distribution of assets in investment portfolio among different types of selection. Numerous portfolio determination models have been proposed. These models are optimization models. The probability hypothesis is one of primary apparatuses for investigating unknown in finance. But some of unknown factors differ from the random ones. For this reason probability theory cannot describe uncertainty completely. When portfolio optimization problems are characterized by some uncertainty, flexibility in problem constraints, the fuzzy sets are suitable representation for modeling this type of optimization problem. In this paper using Lagrange multiplier method the solving of crisp and fuzzy portfolio selection problems are considered.

R. A. Aliev et al. (Eds.): ICAFS-2018, AISC 896, pp. 602–608, 2019.
https://doi.org/10.1007/978-3-030-04164-9_79

## 2  Linear Programming

Linear programming can be viewed as optimization problem. How about we consider the accompanying linear programming issue with blended imperatives

$$f(x) = \sum_{j}^{n} c_j x_j \rightarrow \max \tag{1}$$

Subject to

$$\sum_{j-1}^{n} a_{uX_j} = B_i \text{ for } i = 1, 2, \ldots, p$$

$$\sum_{j-1}^{n} a_{ajX_j} \leq b_k \text{ for } q = p+1, p+2, \ldots, m \tag{2}$$

$$x_1 \geq 0, x_2 \geq 0, \ldots, x_n \geq 0$$

The problem is to maximize the objective under set of constraints. Here constraints are divided into two subsets. The first subset consists of p equality constraints, and the second subset has m-p inequality constraints. To solve this problem the Lagrange multiplier method is used. At first step the suitable energy function is formulated. Since the two equality and inequality sets are disjoint, the energy function is formulated as composite formula [2]. Energy function is constructed so that it penalizes every violation of the equality and inequality constraints.

$$E(x) = E_1(x) + E_2(x) + E_3(x) \tag{3}$$

Where

$$E_1(x) = c^T x \tag{4}$$

$$E_1(x) = \frac{K_1}{2} (A_p x - X_p)^T (A_p x - X_p) + \lambda_P^T (A_p x - b_p) - a_\lambda \lambda_p \tag{5}$$

$$E_3(x) = \frac{K_2}{2} \sum_{i-p+1}^{m} \Phi i(x)] \tag{6}$$

Here K1, K2, $\alpha$ are positive coefficients. Here E1(x) is energy function to be minimized. E2(x) and E3(x) are Lagrange multipliers that penalizes every violation of the equality and inequality constraints, correspondingly. Applying gradient approach the gradient of energy function with respect to x is calculated.

$$x(k+1) = x(k) + \mu \frac{\partial E}{\partial X} \qquad (7)$$

Then

$$\frac{\partial E}{\partial X} = \left( c + A_P^T \left[ K_1 r_p(k) - \lambda_p(k) \right] + K_2 \sum_{i-p+1}^{M} \Psi[r_i(x)] \right) \begin{Bmatrix} a_{i1} \\ a_{i2} \\ \ldots \\ a_{in} \end{Bmatrix} \qquad (8)$$

Here

$$r_p = A_P x - b_p \qquad (9)$$

$$\psi(v) = \frac{d\Phi(v)}{dv} \qquad (10)$$

$$\lambda_p(k+1) = \lambda_p(k) + v\left[ r_p(k) - a\lambda_p(k) \right] \qquad (11)$$

Where $\mu$ and $v$ are learning rates $\Psi(v)$ is differentiable piecewise function. Using formulas (8)–(11) the optimal values of unknown parameters x satisfying equality and inequality constraints are calculated.

## 3   Portfolio Modelling

In this section the upper described methodology is applied for solving portfolio selection. For portfolio selection Kanno and Yamazika [3] are proposed mean absolute deviation model. They use the absolute deviation risk function to replace the risk function in Markowitz's mean-variance model. The model can solve a linear programming problem instead of a quadratic programming. Based on mean absolute deviation model the semiabsolute deviation portfolio selection model was proposed [4]. The input data for absolute deviation model are past gains and expected gains in future. Expect that a portfolio director needs to designate his benefits among n hazardous securities based on recent past data or the organization's budgetary report and he want to maximize his portfolio gain under few given level of risk. Let xj are Aproportion of the sum investment given to the risky security j, $j = 1, 2, \ldots, n$. Assume that the data is obtained for the risky security j at period T. Obtained data are rate of gain of risky security j at duration t, where $t = 1, 2, \ldots, T$. The expected gain of portfolio $x = (x_1, x_2, \ldots, x_n)$ can be represented as

$$R(x) = \sum_{j-1}^{n} Er_j x_j \qquad (12)$$

Here $Er_j$ are expected gain of portfolio for j-th stock in time T. The expected gains of securities are evaluated by the deviation of the gain of portfolio. Also the sum of all portfolio weights must be 1. They are written as in the form of inequality and equality constraints.

$$\sum_{j-1}^{n}\left(Er_j - r_{jt}\right)x_j \le w, t = 1, 2, \ldots, T$$

$$\sum_{j-1}^{n} x_j = 1 \tag{13}$$

$$x_1 \ge 0, x_2 \ge 0, \ldots, x_n \ge 0$$

The issue is to discover such ideal weight estimations of xj under disparity and balance conditions (13), by utilizing them in target work (12) the estimation of portfolio gain would be most extreme. The gradient method is applied to solve problems (12) and (13). As an example the past estimations of gain rates for six stocks are taken. The value gains of the 6 stocks for 12 months as shown in Table 1.

**Table 1.** The value gains of stocks for 12 months

| | Stocks | | | | | |
|---|---|---|---|---|---|---|
| | 1 | 2 | 3 | 4 | 5 | 6 |
| Time | −0.0600 | −0.0700 | −0.1700 | 0.1200 | −0.0600 | −0.3400 |
| | −0.1000 | −0.0300 | 0.0900 | 0.0300 | 0.0300 | 0.1400 |
| | 0.0700 | 0.0800 | 0.0500 | 0.3400 | −0.0400 | −0.1000 |
| | 0.4000 | 0.1800 | 0.1700 | −0.1700 | 0.5600 | 0.0200 |
| | −0.1100 | −0.0800 | −0.1500 | −0.0300 | −0.0600 | 0.0300 |
| | −0.1100 | −0.0800 | 0.0200 | −0.0500 | −0.0300 | 0.2600 |
| | 0.0400 | 0.2700 | −0.0800 | −0.0100 | 0.2600 | −0.2000 |
| | 0.1300 | 0.1300 | 0.2900 | 0.3300 | 0.0800 | 0.2000 |
| | −0.0700 | −0.0300 | −0.0900 | 0.1900 | −0.1300 | 0.1200 |
| | 0.0400 | −0.0300 | 0.1100 | −0.0900 | 0.0900 | 0.5100 |
| | 0.0300 | 0.0800 | 0.0900 | −0.0200 | −0.0600 | −0.0600 |
| | 0.1100 | 0.0900 | 0.2000 | 0.1100 | 0.0800 | 0.0100 |

The value of expected gains for six stocks is estimated as (0.032, 0.045, 0.047, 0.065, 0.062, and 0.052) correspondingly. Using gradient method the optimal portfolio values of each stock for different values of tolerated risk level h are determined. In Fig. 1 the result of simulation when tolerated risk level h = 0.03 are given. The curves describe learning results of the trajectories of weight values for six stocks. The learning is carried out for 4000 iterations. At the result of learning the optimal value of stocks have been found x = (0.0, 0.0569, 0.0, 0.6109, 0.1774, 0.1580).

For different levels of risk the obtained values of stocks is given in Table 2. In the five alternatives, the alternative (0, 0.1075, 0, 0.5690, 0.1646, and 0.1548) is a better

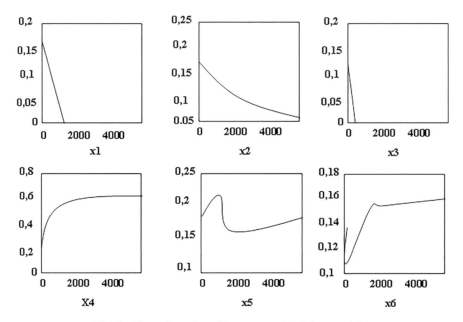

**Fig. 1.** The trajectories of investments for risky securities

investment strategy for the investors who don't like risky. The alternative (0.05, 0, 0, 0, 0.6762, 0.1983, 0.1330, and 0.0632) is better alternative for risky investors. By varying values of risk levels the different investment strategies can be generated.

**Table 2.** The levels of risk obtained values of stocks

| h | Stock | | | | | | Portfolio gain |
|---|---|---|---|---|---|---|---|
| | 1 | 2 | 3 | 4 | 5 | 6 | |
| 0.01 | 0 | 0.1075 | 0 | 0.5690 | 0.1646 | 0.1548 | 0.0601 |
| 0.02 | 0 | 0.0853 | 0 | 0.5892 | 0.1696 | 0.1558 | 0.0608 |
| 0.03 | 0 | 0.0569 | 0 | 0.6109 | 0.1774 | 0.1580 | 0.0615 |
| 0.04 | 0 | 0.0222 | 0 | 0.6387 | 0.1887 | 0.1561 | 0.0623 |
| 0.05 | 0 | 0 | 0 | 0.6762 | 0.1983 | 0.1330 | 0.0632 |

## 4   Fuzzy Portfolio Modeling

Fuzzy portfolio selection can be considered as general multiple objective multiple constraint fuzzy optimization. Fuzzy optimization is the collection of techniques that detail enhancement issues with adaptable, approximate or questionable imperatives and objectives. In general, fuzzy optimization can be divided into two categories: (1) To represent uncertainty in the constraints and the aim (objective functions), (2) To represent flexibility in the constraints and the aim. In this work the first type of fuzzy optimization problem is considered. Fuzzy optimization is formulated as

$$\widehat{R}(x) \sum_{j-1}^{n} \acute{C}_j x_j \rightarrow \text{fuzzy maximize} \tag{14}$$

Under set of fuzzy constraints.

$$\sum_{j-1}^{n} \bar{a}_{jt} x_j \leq \dot{b}, t = 1, 2, \ldots, T \tag{15}$$

$$x_i \geq 0, x_2 \geq 0, \ldots, x_n \geq 0$$

If we accept $\tilde{c}_j$ coefficients as fuzzy values of of expected gains of portfolio for j-th stock at time T and $\tilde{a}_{ij}$ coefficients as fuzzy values the deviation of the gain of portfolio we can get the fuzzy portfolio optimization model. Also the sum of all portfolio weights must be equal to one. The problem is to find such optimal weight values of $x_j$ under fuzzy inequalities and equality conditions (15), by using they in objective function (14) the value of portfolio return would be maximum. Some fuzzy optimization methods have been proposed in the literature in order to deal with different aspects of soft constraints. Zimmerman in [7] has considered the fuzzy optimization as a symmetric problem. In this formulation, fuzzy sets represent both the problem goals and the flexible (soft) constraints. In this work the first type of fuzzy optimization for portfolio selection problem is considered. In the work the $\alpha$ level procedure and interval arithmetic is used to solve fuzzy optimization problem and find the optimal fuzzy values of the securities. The values of expected returns for each stock are taken as fuzzy values that have trapezoid form. Then the fuzzy energy function for optimization is defined. Using interval calculus and $\alpha$-cuts the derivatives [5, 6] in (7) will be determined for adjusting the values of securities.

The values of expected gains for each stock are taken as fuzzy values that have trapezoid form. Then the fuzzy energy function for optimization is defined. Using interval calculus and $\alpha$-cuts the derivatives in (7) will be determined for adjusting the values of securities. If a trapezoid form is used for description of fuzzy variables, then any fuzzy number can be described by four parameters. $x = (x^1, x^2, x^3, x^4)$ [8, 9]. Then problem consists in determining the optimal values of four parameters. Then using (7) the updating formulas for these variables can be derived.

$$x^1(k+1) = x^1(k)\mu \frac{\partial E}{\partial x^1}$$

$$x^2(k+1) = x^2(k)\mu \frac{\partial E}{\partial x^2}$$

$$x^3(k+1) = x^3(k)\mu \frac{\partial E}{\partial x^3} \tag{16}$$

$$x^4(k+1) = x^4(k)\mu \frac{\partial E}{\partial x^4}$$

Using the $\alpha$ level procedure, the values of derivatives in (16) can be determined. $\alpha$-cut of the number of fuzzy is defined as

$$[x]\sigma = \left\lfloor [x]_\sigma^l [x]_\sigma^u \right\rfloor \quad [x]\sigma \leq \left\lfloor [x]_{\partial \leq} [x]_\sigma^u \right\rfloor$$

Here $[x]_\alpha^L$ and $[x]_\alpha^U$ are upper and lower limits of $[x]_\alpha$ correspondingly. To find values of unknown parameters the $\alpha$ level procedure is applied to energy function. Then the values of energy function for lower and upper limits would be determined. In the work the common energy function is determined as $E = (E^U + E^L)/2$ Using derivatives of energy function for $[x]_\alpha^L$ and $[x]_\alpha^U$, the derivatives of four parameters in (16) can be determined. The described procedure is applied to solve (14) and (15) under fuzzy values of expected gains and tolerated risk levels. As result of simulation, the fuzzy values of risky securities are defined.

# References

1. Markowitz, H.: Portfolio selection. J. Financ. **VII**(1), 77–91 (1952)
2. Cichocki, A., Unbehauen, R.: Neural Networks for Optimization and Signal Processing. Wiley, New York (1993)
3. Konno, H., Yamazaki, H.: Mean-absolute deviation portfolio optimization model and its applications to Tokyo stock market. Manag. Sci. **37**, 519–531 (1991)
4. Mansini, R., Speranza, M.G.: Heuristic algorithms for the portfolio selection problem with minimum transaction lots. Eur. J. Oper. Res. **114**, 219–233 (1999)
5. Budnick, F.S.: Applied Mathematics for Business, Economics, and the Social Sciences. McGraw-Hill International Edition. McGraw-Hill, New York (1993)
6. Lai, K.K., Wang, S.Y., Xu, J.P., Zhu, S.S., Fang, Y.: A class of linear interval programming problems and its application to portfolio selection. IEEE Trans. Fuzzy Syst. **10**(6), 698–703 (2002)
7. Zimmermann, H.J.: Description and optimization of fuzzy systems. Int. J. Gen Syst **2**, 209–215 (1976)
8. Ishibuchi, H., Morioka, K., Tanaka, H.: A fuzzy neural network with trapezoidal fuzzy weights. In: 3rd International Fuzzy Systems Conference Proceedings, Orlando, Florida, pp. 228–233 (1994)
9. Ishibuchi, H.: A fuzzy neural network. In: Pedrycz, W. (ed.) Fuzzy Modelling: Paradigms and Practice. Kluwer Academic Publisher, Boston (1996)

# Evaluation of Food Security in the Region Within the Framework of the FAO UN Cooperation Program Using the Fuzzy Inference

Elkhan Aliyev[1], Tarana Karimova[2], and Galib Hajiyev[2(✉)]

[1] Institute of Control Systems of ANAS, Baku AZ1141, Azerbaijan
Elkhan.Aliyev@fao.org
[2] Azerbaijan State University of Economics, Baku AZ1101, Azerbaijan
galib.haciyev@gmail.com

**Abstract.** It is proposed an approach for assessing the level of food security for the regions within the framework of the FAO Partnership Program. A mechanism of fuzzy inference is used to assess the influence on food security in the regions that form the basis for the Strategic Objectives approved in the FAO Strategic Framework Program for 2010–2019.

**Keywords:** Food security · Expert judgment · Fuzzy inference

## 1 Introduction

The main objective of the FAO Partnership Program (PP) with the various regional partner is to create a meaningful, financial and operational basis for active cooperation in the field of food security (FS) and poverty reduction in rural areas in the beneficiary countries. The activities of the FAO PP are determined both by current requirements and by the need to solve the most priority (acute) problems [1].

In [2] there was proposed the fuzzy cognitive model (FCM) for the integrated assessment of FS on the base of FAO reporting results. This model is designed to ensure the integrity, coordination and synergetic effect of activities supported by individual programs, projects and/or countries in accordance with FAO strategic goals (SG) for the five reasons: SG1 - hunger, problems of FS and poverty ($x_1$); SG2 - stable provision of goods and services from the agriculture, forestry and fisheries ($x_2$); SG3 - the extent of poverty in rural areas ($x_3$); SG4 - prerequisites for the creation of broader and more effective agricultural and food systems at the local, state, regional and international levels ($x_4$); SG5 - sustainability of livelihoods before threats and crises ($x_5$).

Due to this approach, transparent strategic management can be provided, the overall direction of the FAO activities can be determined, and general supervision and control of the FAO PP management can be carried out. However, most importantly, it is possible to ensure the initial technical analysis of projects to assess the level of FS in

R. A. Aliev et al. (Eds.): ICAFS-2018, AISC 896, pp. 609–618, 2019.
https://doi.org/10.1007/978-3-030-04164-9_80

the region, based on what to consider and more reasonably approve their funding under limited resources of FAO.

## 2 Problem Definition

So, in the context of the above, it is necessary to test the approach proposed in [2] by the simulation of the process of end-to-end assessment of the FS-level by way of example of several hypothetical regions that have undergone initial technical analysis and are characterized at the preliminary stage by expert judgments for the all factors which are the basis for the establishing of the FAO SG in the form of Table 1.

**Table 1.** Preliminary expert judgments of alternative projects.

| FAO SG determinants | Estimates by ten-point scale | | | | Assessment terms |
|---|---|---|---|---|---|
| | $a_1$ | $a_2$ | $a_3$ | $a_4$ | |
| $x_1$ – hunger, problems of FS and poverty | | | | | |
| $x_{11}$ – physical availability of resources | 7.76 | 2.17 | 8.95 | 3.24 | UNLIMITED |
| $x_{12}$ – economic availability of resources | 6.67 | 3.91 | 2.77 | 3.48 | UNLIMITED |
| $x_{13}$ – population's incomes | 9.29 | 4.11 | 9.78 | 1.00 | HIGH |
| $x_{14}$ – access to productive assets | 4.68 | 5.73 | 9.82 | 4.21 | UNLIMITED |
| $x_{15}$ – economic and other crises | 0.34 | 7.74 | 4.17 | 9.79 | INSENSIBLE |
| $x_{16}$ – political commitments | 5.74 | 2.22 | 5.83 | 1.34 | SUFFICIENT |
| $x_{17}$ – understanding of problems | 5.65 | 7.25 | 7.15 | 9.62 | FULL |
| $x_{18}$ – control and coordination means | 0.62 | 2.54 | 7.06 | 1.99 | EFFECTIVE |
| $x_{19}$ – consistency of strategies and investments | 5.05 | 1.66 | 7.31 | 8.59 | STRONG |
| $x_{1.10}$ – gender inequality | 3.33 | 8.56 | 0.31 | 3.96 | BE ABSENT |
| $x_2$ – stable provision of goods and services from the agriculture, forestry and fisheries | | | | | |
| $x_{21}$ – lack and degradation of the resource base | 0.19 | 1.01 | 3.21 | 6.01 | INSENSIBLE |
| $x_{22}$ – ecosystem services and biodiversity | 9.94 | 6.32 | 4.78 | 3.56 | WIDE |
| $x_{23}$ – climate fluctuation | 7.81 | 6.45 | 6.90 | 7.22 | INSIGNIFICANT |
| $x_{24}$ – migration | 6.91 | 4.01 | 4.89 | 8.18 | INESSENTIAL |
| $x_{25}$ – new threats | 5.48 | 7.71 | 7.73 | 3.49 | INESSENTIAL |
| $x_{26}$ – headaches in the management and policy | 0.62 | 4.40 | 1.38 | 4.80 | BE ABSENT |
| $x_{27}$ – sustainability of Eco development | 2.91 | 5.71 | 6.11 | 8.83 | STRONG |
| $x_{28}$ – sustainability of economic development | 0.49 | 5.30 | 9.73 | 9.54 | STRONG |
| $x_{29}$ – sustainability of social development | 6.76 | 6.39 | 1.70 | 2.47 | STRONG |
| $x_3$ – the extent of poverty in rural areas | | | | | |
| $x_{31}$ – policy orientation | 8.80 | 6.92 | 6.00 | 7.65 | AIMED |
| $x_{32}$ – access to productive assets and resources | 5.38 | 9.52 | 4.79 | 3.06 | SUFFICIENT |
| $x_{33}$ – access to services | 6.05 | 1.45 | 7.87 | 0.78 | SUFFICIENT |
| $x_{34}$ – public organizations (social capital) | 4.31 | 4.45 | 0.41 | 7.05 | STRONG |

*(continued)*

**Table 1.** (*continued*)

| FAO SG determinants | Estimates by ten-point scale | | | | Assessment terms |
|---|---|---|---|---|---|
| | $a_1$ | $a_2$ | $a_3$ | $a_4$ | |
| $x_{35}$ – employment opportunities | 0.17 | 4.54 | 4.49 | 9.40 | BOUNDED |
| $x_{36}$ – labor conditions | 8.17 | 9.13 | 1.10 | 4.48 | INSUFFICIENT |
| $x_{37}$ – social protection | 1.97 | 3.24 | 4.90 | 3.44 | INSUFFICIENT |
| $x_4$ – *prerequisites for the creation of broader and more effective agricultural - food systems* | | | | | |
| $x_{41}$ – favorability of the condition | 9.42 | 8.20 | 2.23 | 3.29 | SUFFICIENT |
| $x_{42}$ – potential (together with the private sector) | 7.86 | 2.53 | 3.36 | 4.95 | HIGH |
| $x_{43}$ – global market systems | 1.78 | 3.87 | 4.41 | 4.37 | PREFERRED |
| $x_5$ – *sustainability of livelihoods before threats and crises* | | | | | |
| $x_{51}$ – probability of natural disasters | 8.13 | 1.07 | 1.77 | 8.98 | LOW |
| $x_{52}$ – probability of juncture in food chains | 1.14 | 1.88 | 1.55 | 7.87 | LOW |
| $x_{53}$ – probability of socio-economic crises | 9.52 | 4.88 | 9.51 | 4.48 | LOW |
| $x_{54}$ – probability of armed conflicts | 5.51 | 5.97 | 1.60 | 6.84 | LOW |
| $x_{55}$ – probability of protracted crisis | 1.01 | 4.38 | 4.35 | 2.21 | LOW |
| $x_{56}$ – stability parameters (vulnerability) | 2.68 | 8.55 | 1.71 | 6.54 | WEAK |
| $x_{57}$ – needs | 9.67 | 9.06 | 5.88 | 7.69 | SATISFIED |
| $x_{58}$ – humanitarian protection | 7.99 | 9.22 | 1.55 | 5.26 | STRONG |
| $x_{59}$ – level of strategic partnership | 3.83 | 8.66 | 3.59 | 8.98 | HIGH |

# 3   Assessment of FS-Levels for Alternative Regions

Within the framework of the FCM proposed in [2] for analyzing the FS-levels in the region (country), fuzzy logical rules were formulated, which reflecting the cause-effect relations for estimation the foundation of the corresponding FAO SG. In particular, for the basis of SG-1 – *"The level of hunger, the problems of FS and poverty"* the case-effect relations relative to the terms from Table 1 are formed as follows:

$d_1$: "If $x_{11}$ = UNLIMITED and $x_{12}$ = UNLIMITED and $x_{13}$ = HIGH and $x_{14}$ = UNLIMITED and $x_{15}$ = INSENSIBLE and $x_{16}$ = SUFFICIENT and $x_{17}$ = FULL and $x_{18}$ = EFFECTIVE and $x_{19}$ = STRONG and $x_{1,10}$ = ABSENT, then $y_1$ = TOO INSENSITIVE";

$d_2$: "If $x_{11}$ = UNLIMITED and $x_{12}$ = UNLIMITED and $x_{13}$ = HIGH and $x_{16}$ = SUFFICIENT and $x_{17}$ = FULL and $x_{18}$ = EFFECTIVE and $x_{19}$ = STRONG and $x_{1,10}$ = ABSENT, then $y_1$ = VERY INSENSITIVE";

$d_3$: "If $x_{11}$ = UNLIMITED and $x_{12}$ = UNLIMITED and $x_{16}$ = SUFFICIENT and $x_{17}$ = FULL and $x_{18}$ = EFFECTIVE and $x_{19}$ = STRONG, then $y_1$ = MORE THAN INSENSITIVE";

$d_4$: "If $x_{11}$ = UNLIMITED and $x_{12}$ = UNLIMITED and $x_{13}$ = HIGH and $x_{14}$ = UNLIMITED, then $y_1$ = INSENSITIVE";

$d_5$: "If $x_{11}$ = LIMITED and $x_{12}$ = LIMITED and $x_{13}$ = LOW and $x_{14}$ = LIMITED, then $y_1$ = SENSITIVE";

$d_6$: "If $x_{11}$ = LIMITED and $x_{12}$ = LIMITED and $x_{16}$ = INSUFFICIENT and $x_{17}$ = NOT FULL and $x_{18}$ = NON-EFFECTIVE and $x_{19}$ = WEAK, then $y_1$ = MORE THAN SENSITIVE";

$d_7$: "If $x_{11}$ = LIMITED and $x_{12}$ = LIMITED and $x_{13}$ = LOW and $x_{16}$ = INSUFFICIENT and $x_{17}$ = NOT FULL and $x_{18}$ = NON-EFFECTIVE and $x_{19}$ = WEAK and $x_{1,10}$ = VISIBLE, then $y_1$ = VERY SENSITIVE";

$d_8$: "If $x_{11}$ = LIMITED and $x_{12}$ = LIMITED and $x_{13}$ = LOW and $x_{14}$ = LIMITED and $x_{15}$ = SENSIBLE and $x_{16}$ = INSUFFICIENT and $x_{17}$ = WEAK and $x_{18}$ = NON-EFFECTIVE and $x_{19}$ = WEAK and $x_{1,10}$ = VISIBLE, then $y_1$ = TOO SENSITIVE".

These rules form the fuzzy inference system (FIS) relative to satisfactory of alternatives for the indicator "The level of hunger and poverty". To simulate the FIS it is necessary to describe terms from the left and right parts of rules $d_1 - d_8$ in the form of fuzzy sets. According to [3] each term from the left-hand parts of rules can be described by fuzzy subset of the discrete universe $\{a_1, a_2, a_3, a_4\}$ in the form of $A_{ki} = \{m_{Aki}(a_1)/a_1; m_{Aki}(a_2)/a_2; m_{Aki}(a_3)/a_3; m_{Aki}(a_4)/a_4\}$, where $m_{Aki}(a_t)$ $(t = 1 \div 4)$ is the value of the Gaussian membership function in the type of: $m_{Aki}(a_t) = exp\{-[e_{ki}(a_t) - 10]^2/\sigma_{ki}^2\}$, where $e_{ki}(a_t)$ is the consolidated expert estimate of the alternative $a_t (t = 1 \div 4)$ (see Table 1). Then, according to above, we have:

- for $x_{11}$ $A_{\text{UNLIMITED}} = A_{11} = \{0.6325/a_1; 0.0038/a_2; 0.9039/a_3; 0.0158/a_4\}$,
- for $x_{12}$ $A_{\text{UNLIMITED}} = A_{12} = \{0.3640/a_1; 0.0340/a_2; 0.0086/a_3; 0.0210/a_4\}$,
- for $x_{13}$ $A_{\text{HIGH}} = A_{13} = \{0.9550/a_1; 0.0428/a_2; 0.9956/a_3; 0.0006/a_4\}$,
- for $x_{14}$ $A_{\text{UNLIMITED}} = A_{14} = \{0.0763/a_1; 0.1900/a_2; 0.9970/a_3; 0.0477/a_4\}$,
- for $x_{15}$ $A_{\text{INSENSIBLE}} = A_{15} = \{0.0002/a_1; 0.6283/a_2; 0.0456/a_3; 0.9959/a_4\}$,
- for $x_{16}$ $A_{\text{SUFFICIENT}} = A_{16} = \{0.1914/a_1; 0.0041/a_2; 0.2060/a_3; 0.0011/a_4\}$,
- for $x_{17}$ $A_{\text{FULL}} = A_{17} = \{0.1786/a_1; 0.5024/a_2; 0.4771/a_3; 0.9868/a_4\}$,
- for $x_{18}$ $A_{\text{EFFECTIVE}} = A_{18} = \{0.0003/a_1; 0.0060/a_2; 0.4560/a_3; 0.0029/a_4\}$,
- for $x_{19}$ $A_{\text{STRONG}} = A_{19} = \{0.1081/a_1; 0.0018/a_2; 0.5189/a_3; 0.8353/a_4\}$,
- for $x_{1,10}$ $A_{\text{ABSENT}} = A_{1,10} = \{0.0175/a_1; 0.8288/a_2; 0.0002/a_3; 0.0362/a_4\}$.

The terms from the right-hand parts as appropriate values of the output linguistic variable $y_1$ – the level of hunger, FS problems and poverty in the region can be described by appropriate membership functions of corresponding fuzzy subsets of discrete universe $J = \{0; 0.1; 0.2; \ldots; 1\}$. In particular, according to [4] $\forall j \in J$ we have: $TIS$ = TOO INSENSITIVE: $\mu_{TIS}(j) = 1$, if $j = 1$ and $\mu_{TIS}(j) = 0$, if $j < 1$; $VIS$ = VERY INSENSITIVE: $\mu_{VIS}(j) = j^2$; $MIS$ = MORE THAN INSENSITIVE: $\mu_{MIS}(j) = j^{1/2}$; $IS$ = INSENSITIVE: $\mu_{IS}(j) = j$; $S$ = SENSITIVE: $\mu_S(j) = 1 - j$; $MS$ = MORE THAN SENSITIVE: $\mu_{MIS}(j) = (1 - j)^{1/2}$; $VS$ = VERY SENSITIVE: $\mu_{VS}(j) = (1 - j)^2$; $TS$ = TOO SENSITIVE: $\mu_{TIS}(j) = 0$, if $j = 1$ and $\mu_{TIS}(j) = 1$, if $j < 1$. Then the implicative rules can be rewritten in the following symbolic form:

$d_1 : (x_{11} = A_{11}) \& (x_{12} = A_{12}) \& \ldots \& (x_{1,10} = A_{1,10},) \Rightarrow (y_1 = TIS);$

$d_2 : (x_{11} = A_{11}) \& (x_{12} = A_{12}) \& (x_{13} = A_{13}) \& (x_{16} = A_{16}) \& (x_{17} = A_{17}) \& \ldots \&$
$(x_{1,10} = A_{1,10}) \Rightarrow (y_1 = VIS);$

$d_3 : (x_{11} = A_{11}) \& (x_{12} = A_{12}) \& (x_{16} = A_{16}) \& (x_{17} = A_{17}) \& (x_{18} = A_{18}) \& (x_{19} = A_{19})$
$\Rightarrow (y_1 = MIS);$

$d_4 : (x_{11} = A_{11}) \& (x_{12} = A_{12}) \& (x_{13} = A_{13}) \& (x_{14} = A_{14}) \Rightarrow (y_1 = IS);$

$d_5 : (x_{11} = \neg A_{11}) \& (x_{12} = \neg A_{12}) \& (x_{13} = \neg A_{13}) \& (x_{14} = \neg A_{14}) \Rightarrow (y_1 = S);$

$d_6 : (x_{11} = \neg A_{11}) \& (x_{12} = \neg A_{12}) \& (x_{16} = \neg A_{16}) \& (x_{17} = \neg A_{17}) \& (x_{18} = \neg A_{18})$
$\& (x_{19} = \neg A_{19}) \Rightarrow (y_1 = MS);$

$d_7 : (x_{11} = \neg A_{11}) \& (x_{12} = \neg A_{12}) \& (x_{13} = \neg A_{13}) \& (x_{16} = \neg A_{16}) \& \ldots \& (x_{1,10} = \neg A_{1,10})$
$\Rightarrow (y_1 = VS);$

$d_8 : (x_{11} = \neg A_{11}) \& (x_{12} = \neg A_{12}) \& \ldots \& (x_{1,10} = \neg A_{1,10}) \Rightarrow (y_1 = TS).$

According to [3], in the discrete case, the logical operation "AND" between the fuzzy sets from the left-hand parts of the rules is realized by finding the minimum of the values of corresponding membership functions. Then the rules can be rewritten as:

$d_{11} : (x_1 = m_1) \Rightarrow (y_1 = TIS); d_{12} : (x_1 = m_2) \Rightarrow (y_1 = VIS); d_{13} : (x_1 = m) \Rightarrow (y_1 = MIS);$

$d_{14} : (x_1 = m_4) \Rightarrow (y_1 = IS); d_{15} : (x_1 = m_5) \Rightarrow (y_1 = S); d_{16} : (x_1 = m_6) \Rightarrow (y_1 = MS);$

$d_{17} : (x_1 = m_7) \Rightarrow (y_1 = VS); d_{18} : (x_1 = m_8) \Rightarrow (y_1 = TS),$

where: $M_1 = \{0.0002/a_1; 0.0018/a_2; 0.0002/a_3; 0.0006/a_4\};$ $M_2 = \{0.0003/a_1; 0.0018/a_2; 0.0002/a_3; 0.0006/a_4\}; M_3 = \{0.0003/a_1; 0.0018/a_2; 0.0086/a_3; 0.0011/a_4\}; M_4 = \{0.0763/a_1; 0.0038/a_2; 0.0086/a_3; 0.0006/a_4\}; M_5 = \{0.045/a_1; 0.8098/a_2; 0.0028//a_3; 0.0952/a_4\};$ $M_6 = \{0.3675/a_1; 0.4976/a_2; 0.0961/a_3; 0.0132/a_4\}; M_7 = \{0.045/a_1; 0.1712/a_2; 0.0044/a_3; 0.0132/a_4\};$ $M_8 = \{0.0450/\ a_1; 0.1712/a_2; 0.0028/a_3; 0.0041/a_4\}.$

As a result of the application of Lukasiewicz's implication: $\mu_H(u, j) = \min\{1; 1 - u + j\}$ [5], the corresponding fuzzy relations were obtained, the intersection of which induces the final general solution in the form of the following matrix:

$$R_1 = \begin{array}{c|ccccccccccc} & 0 & 0.1 & 0.2 & 0.3 & 0.4 & 0.5 & 0.6 & 0.7 & 0.8 & 0.9 & 1 \\ \hline a_1 & 0.9237 & 0.9998 & 0.9998 & 0.9998 & 0.9998 & 0.9998 & 0.9998 & 0.9998 & 0.9950 & 0.9487 & 0.6325 \\ a_2 & 0.9962 & 0.9982 & 0.9902 & 0.8902 & 0.7902 & 0.6902 & 0.5902 & 0.4902 & 0.3902 & 0.2902 & 0.1902 \\ a_3 & 0.9914 & 0.9998 & 0.9998 & 0.9998 & 0.9998 & 0.9998 & 0.9998 & 0.9998 & 0.9998 & 0.9998 & 0.9039 \\ a_4 & 0.9989 & 0.9477 & 0.8477 & 0.7477 & 0.6477 & 0.5477 & 0.4477 & 0.3477 & 0.2477 & 0.1477 & 0.0477 \end{array}.$$

The fuzzy conclusion relative to the level of hunger, the problem of PB and poverty in the region $a_k$ is reflected as the fuzzy subset $E_k$ of the discrete universe $J$ with the corresponding values of the appropriate membership function from the $k$-th row of the matrix $R_1$. For numerical evaluation of fuzzy conclusion, for example, for region $a_2$, which is interpreted as the fuzzy set: $E_2 = \{0.9962/0; 0.9982/0.1; 0.9902/0.2; 0.8902/0.3; 0.7902/0.4; 0.6902/0.5; 0.5902/0.6; 0.4902/0.7; 0.3902/0.8; 0.2902/0.9; 0.1902/1\}$, it is necessary to apply the defuzzification procedure in following manner:

- for $0<\alpha<0.1902: \Delta\underline{\alpha} = 0.1902, E_{2\alpha} = \{0;0.1;0.2;\ 0.3;\ldots;0.9;1\}, M(E_{2\alpha}) = 0.5;$
- for $0.1902<\alpha<0.2902: \Delta\alpha = 0.1, E_{2\alpha} = \{0;0.1;0.2;0.3;\ldots;0.8;0.9\}, M(E_{2\alpha}) = 0.45;$
- for $0.2902<\alpha<0.3902: \Delta\alpha = 0.1, E_{2\alpha} = \{0;0.1;0.2;0.3;\ldots;0.7;0.8\}, M(E_{2\alpha}) = 0.40;$
- for $0.3902<\alpha<0.4902: \Delta\alpha = 0.1, E_{2\alpha} = \{0;0.1;0.2;0.3;\ldots;0.6;0.7\}, M(E_{2\alpha}) = 0.35;$
- for $0.4902<\alpha<0.5902: \Delta\alpha = 0.1, E_{2\alpha} = \{0;0.1;0.2;0.3;0.4;0.5;0.6\}, M(E_{2\alpha}) = 0.30;$
- for $0.5902<\alpha<0.6902: \Delta\alpha = 0.1, E_{2\alpha} = \{0;0.1;0.2;0.3;0.4;0.5\}, M(E_{2\alpha}) = 0.25;$
- for $0.6902<\alpha<0.7902: \Delta\alpha = 0.1, E_{2\alpha} = \{0;0.1;0.2;0.3;0.4\}, M(E_{2\alpha}) = 0.20;$
- for $0.7902<\alpha<0.8902: \Delta\alpha = 0.1, E_{2\alpha} = \{0;0.1;0.2;0.3\}, M(E_{2\alpha}) = 0.15;$
- for $0.8902<\alpha<0.9902: \Delta\alpha = 0.1, E_{2\alpha} = \{0;0.1;0.2\}, M(E_{2\alpha}) = 0.10;$
- for $0.9902<\alpha<0.9962: \Delta\alpha = 0.006, E_{2\alpha} = \{0;0.1\}, M(E_{2\alpha}) = 0.05;$
- for $0.9962<\alpha<0.9982: \Delta\alpha = 0.002, E_{2\alpha} = \{0.1\}, M(E_{2\alpha}) = 0.10,$

where $M(E_\alpha)$ is cardinal number calculating by formula: $M(E_\alpha) = \sum_{j=1}^{n} i_j/n, i \in E_\alpha$. Then the numerical evaluation of $E_2$ can be obtained as following [4]:

$$F(E_2) = \frac{1}{\alpha_{max}} \int_0^{\alpha_{max}} M(E_{2\alpha})d\alpha = \frac{1}{0.9982} \int_0^{0.9982} M(E_{2\alpha})d\alpha$$

$$= (0.5 \cdot 0.1902 + 0.45 \cdot 0.1 + 0.40 \cdot 0.1 + 0.35$$
$$\times 0.1 + 0.3 \cdot 0.1 + 0.25 \cdot 0.1 + 0.2 \cdot 0.1 + 0.15 \cdot 0.1 + 0.10 \cdot 0.1 + 0.05 \cdot 0.006$$
$$+ 0.1 \cdot 0.002)/0.9982 = 0.3162.$$

The numerical estimates of fuzzy conclusions relative to levels of hunger, the problems of FS and poverty for other regions are obtained by similar actions as follows: $a_1 - F(E_1) = 0.4826; a_3 - F(E_3) = 0.4956; a_4 - F(E_4) = 0.2491.$

In a similar manner, estimates of alternative regions $a_k$ ($k = 1 \div 4$) relative to others FAO SGs $x_i$ ($i = 2 \div 5$) were obtained. All aggregated estimates are presented in Table 2.

## 4 Assessment of FS-Levels in the Regions Relative to FAO SG

To obtain final estimates of the FS-levels in the alternative regions $a_k$ ($k = 1 \div 4$), the following set of consistent judgements that determine the cause-effect relations between the bases of the FAO SG ($x_i$, $i = 1 \div 5$) and the FS-level ($y$) are applied:

Table 2. Aggregated estimates of alternative regions for all FAO SG.

| Strategy goals of FAO | $e(a_1)$ | $e(a_2)$ | $e(a_3)$ | $e(a_4)$ |
|---|---|---|---|---|
| FAO SG-1 | 0.4826 | 0.3162 | 0.4956 | 0.2491 |
| FAO SG-2 | 0.5000 | 0.2946 | 0.4013 | 0.4776 |
| FAO SG-3 | 0.4582 | 0.4985 | 0.3337 | 0.4077 |
| FAO SG-4 | 0.3550 | 0.4710 | 0.4950 | 0.4930 |
| FAO SG-5 | 0.4990 | 0.3314 | 0.3256 | 0.4992 |

$r_1$: "If the problem of hunger and poverty is sensitive and the provision of goods and services from the rural, forestry and fisheries sector is unstable, then the level of FS is low";

$r_2$: "If, in addition to the above, the prerequisites for the creation of agricultural and food systems are insignificant and the sustainability of livelihoods before threats and crises is unreliable, then the level of FS is more than low";

$r_3$: "If in addition to the conditions stipulated in $r_2$, it is known that the extent of poverty in rural areas is large, then the level of FS is too low";

$r_4$: "If the problem of hunger and poverty is sensitive, the scale of poverty in rural areas is large, the prerequisites for the creation of agricultural and food systems are insignificant, and the sustainability of livelihoods before threats and crises is unreliable, the level of FS is very low";

$r_5$: "If the problem of hunger and poverty is sensitive, the scale of poverty in rural areas is large, the prerequisites for the creation of agricultural and food systems are insignificant, but at the same time the sustainability of livelihoods before threats and crises is reliable, then the level of FS is still low";

$r_6$: "If the problem of hunger and poverty is insensitive, the provision of goods and services from the rural, forestry and fisheries sector is stable, and the extent of poverty in rural areas is not large, then the level of FS is high".

The analyses of these judgements allows forming a complete set of linguistic variables and rules for constructing the fuzzy inference system. For convenience, all variables are summarized in Table 3.

Then the judgements $r_1 \div r_6$ in symbolic forms can be written as following rules:

$r_1$: "If $x_1$ = SENSIBLE and $x_2$ = UNSTABLE, then $y$ = LOW";

$r_2$: "If $x_1$ = SENSIBLE and $x_2$ = UNSTABLE and $x_4$ = INSIGNIFICANT and $x_5$ = UNRELIABLE, then $y$ = MORE THAN LOW";

$r_3$: "If $x_1$ = SENSIBLE and $x_2$ = UNSTABLE and $x_3$ = LARGE and $x_4$ = INSIGNIFICANT and $x_5$ = UNRELIABLE, then $y$ = TOO LOW";

**Table 3.** Linguistic variables of fuzzy inference system for assessing the level of FS.

| Attribute | Input linguistic variables | | | | | Output |
|---|---|---|---|---|---|---|
| | $x_1$ | $x_2$ | $x_3$ | $x_4$ | $x_5$ | $y$ |
| Name | Problem of hunger and poverty | Provision of goods and services from the agriculture, forestry and fisheries | Extent of poverty in rural areas | Prerequisites for the creation of agricultural and food systems | Livelihoods before threats and crises | The level of FS |
| Terms | SENSIBLE, INSENSIBLE | STABLE, UNSTABLE | LARGE, NOT LARGE | SIGNIFICANT, INSIGNIFICANT | RELIABLE, UNRELIABLE | LOW, MORE THAN LOW, VERY LOW, TOO LOW, HIGH |
| Universe | [0; 1] | [0; 1] | [0; 1] | [0; 1] | [0; 1] | [0; 1] |

$r_4$: "If $x_1$ = SENSIBLE and $x_3$ = LARGE and $x_4$ = INSIGNIFICANT and $x_5$ = UNRELIABLE, then $y$ = VERY LOW";

$r_5$: "If $x_1$ = SENSIBLE and $x_3$ = LARGE and $x_4$ = INSIGNIFICANT and $x_5$ = RELIABLE, then $y$ = LOW";

$r_6$: "If $x_1$ = INSENSIBLE and $x_2$ = STABLE and $x_3$ = NOT LARGE, then $y$ = HIGH".

To fuzzificate the terms from the left-hand parts of these rules one can also use the Gaussian function in the form of: $\mu_{A1}(a_k) = \exp\{-[e_i(a_k)-1]^2/\sigma_i^2\}$, where $e_i(a_k)$ is the estimates of alternative regions $a_k$ ($k = 1 \div 4$) relative to FAO SGs $x_i$ ($i = 1 \div 5$) (see Table 2):

- for $x_1$ $A_{\text{SENSIBLE}} = A_1 = \{0.4275/a_1; 0.2266/a_2; 0.4459/a_3; 0.1670/a_4\}$,
- for $x_2$ $A_{\text{UNSTABLE}} = A_2 = \{0.4522/a_1; 0.2060/a_2; 0.3205/a_3; 0.4205/a_4\}$,
- for $x_3$ $A_{\text{LARGE}} = A_3 = \{0.3938/a_1; 0.4500/a_2; 0.2443/a_3; 0.3283/a_4\}$,
- for $x_4$ $A_{\text{INSIGNIFICANT}} = A_4 = \{0.2669/a_1; 0.4113/a_2; 0.4450/a_3; 0.4422/a_4\}$,
- for $x_5$ $A_{\text{UNRELIABLE}} = A_5 = \{0.4508/a_1; 0.2419/a_2; 0.2360/a_3; 0.4510/a_4\}$.

The terms from the right-hand parts can be described by appropriate fuzzy sets with the corresponding membership functions, namely $\forall j \in J = \{0; 0.1; 0.2; \ldots; 0.9; 1\}$ as [3, 4]: $TL$ = TOO LOW: $\mu_{TL}(j) = 0$, if $j = 1$ and $\mu_{TL}(j) = 1$, if $j < 1$; $VL$ = VERY LOW: $\mu_{VS}(j) = j^2$; $MS$ = MORE THAN STABLE: $\mu_{MS}(j) = j^{1/2}$; $S$ = STABLE: $\mu_S(j) = j$; $US$ = UNSTABLE: $\mu_{US}(j) = 1 - j$; $MUS$ = MORE THAN UNSTABLE: $\mu_{MUS}(j) = (1 - j)^{1/2}$; $VUS$ = VERY UNSTABLE: $\mu_{VUS}(j) = (1 - j)$; $TUS$ = TOO UNSTABLE: $\mu_{TUS}(j) = 0$, if $j = 1$ and $\mu_{TUS}(j) = 1$, if $j < 1$. Then the implicative rules $r_1 \div r_6$ can be rewritten in n the following more compact form:

$$r_1 : (x_1 = A_1)\&(x_2 = A_2) \Rightarrow (y = L);$$
$$r_2 : (x_1 = A_1)\&(x_2 = A_2)\&(x_4 = A_4)\&(x_5 = A_5) \Rightarrow (y = ML);$$
$$r_3 : (x_1 = A_1)\&(x_2 = A_2)(x_3 = A_3)\&(x_4 = A_4)\&(x_5 = A_5) \Rightarrow (y = TL);$$
$$r_4 : (x_1 = A_1)\&(x_3 = A_3)\&(x_4 = A_4)\&(x_5 = A_5) \Rightarrow (y = VL);$$
$$r_5 : (x_1 = A_1)\&(x_3 = A_3)\&(x_4 = A_4)\&(x_5 = \neg A_5) \Rightarrow (y = L);$$
$$r_6 : (x_1 = \neg A_1)\&(x_2 = \neg A_2)\&(x_3 = \neg A_3) \Rightarrow (y = H).$$

The transformation of these rules in the manner described above allows to find the general functional solution reflecting the cause-effect relations between the bases of the FAO SG ($x_i$, $i = 1 \div 5$) and the FS-level ($y$) for all regions in the form of matrix:

$$R_1 = \begin{bmatrix} & 0 & 0.1 & 0.2 & 0.3 & 0.4 & 0.5 & 0.6 & 0.7 & 0.8 & 0.9 & 1 \\ \hline a_1 & 0.4522 & 0.5522 & 0.6522 & 0.7522 & 0.8522 & 0.9522 & 0.8931 & 0.8231 & 0.7725 & 0.6725 & 0.5725 \\ a_2 & 0.4500 & 0.5500 & 0.6500 & 0.7500 & 0.8500 & 0.9500 & 0.9334 & 0.8634 & 0.8134 & 0.7834 & 0.7734 \\ a_3 & 0.4459 & 0.5459 & 0.6459 & 0.7459 & 0.8459 & 0.9459 & 0.9240 & 0.8540 & 0.8040 & 0.7740 & 0.6795 \\ a_4 & 0.4205 & 0.5205 & 0.6205 & 0.7205 & 0.8205 & 0.9205 & 0.9930 & 0.9230 & 0.8730 & 0.8430 & 0.8330 \end{bmatrix}.$$

As above, the fuzzy conclusion about the level of the AB in the region $a_k$ ($k = 1 \div 4$) is reflected as the fuzzy subset $E_k$ of the discrete universe $J$ with the corresponding values of the membership function from the $k$-th row of the matrix $R$. For numerical evaluation of these conclusions it necessary to apply the defuzzification procedure. In particular, for conclusion relative to level of FS in region $a_1$ described by fuzzy set $E_1 = \{0.4522/0; 0.5522/0.1; 0.6522/0.2; 0.7522/0.3; 0.8522/0.4; 0.9522/0.5; 0.8931/0.6; 0.8231/0.7; 0.7725/0.8; 0.6725/0.9; 0.5725/1\}$, we have follows:

- for $0 < \alpha < 0.4522 : \Delta\alpha = 0.4522$, $E_{1\alpha} = \{0; 0.1; 0.2; 0.3; \ldots; 0.8; 0.9; 1\}$, $M(E_{1\alpha}) = 0.50$;
- for $0.4522 < \alpha < 0.5522 : \Delta\alpha = 0.1$ $E_{1\alpha} = \{0.1; 0.2; 0.3; \ldots; 0.8; 0.9; 1\}$, $M(E_{1\alpha}) = 0.55$;
- for $0.5522 < \alpha < 0.5725 : \Delta\alpha = 0.0203$, $E_{1\alpha} = \{0.2; 0.3; \ldots; 0.8; 0.9; 1\}$, $M(E_{1\alpha}) = 0.60$;
- for $0.5725 < \alpha < 0.6522 : \Delta\alpha = 0.0797$, $E_{1\alpha} = \{0.2; 0.3; \ldots; 0.7; 0.8; 0.9\}$, $M(E_{1\alpha}) = 0.55$;
- for $0.6522 < \alpha < 0.6725 : \Delta\alpha = 0.0203$, $E_{1\alpha} = \{0.3; 0.4; \ldots; 0.7; 0.8; 0.9\}$, $M(E_{1\alpha}) = 0.60$;;
- for $0.6725 < \alpha < 0.7522 : \Delta\alpha = 0.0797$, $E_{1\alpha} = \{0.3; 0.4; 0.5; 0.6; 0.7; 0.8\}$, $M(E_{1\alpha}) = 0.55$;
- for $0.7522 < \alpha < 0.7725 : \Delta\alpha = 0.0203$, $E_{1\alpha} = \{0.4; 0.5; 0.6; 0.7; 0.8\}$ $M(E_{1\alpha}) = 0.60$;
- for $0.7725 < \alpha < 0.8231 : \Delta\alpha = 0.0505$, $E_{1\alpha} = \{0.4; 0.5; 0.6; 0.7\}$, $M(E_{1\alpha}) = 0.55$;
- for $0.8231 < \alpha < 0.8522 : \Delta\alpha = 0.0291$, $E_{1\alpha} = \{0.4; 0.5; 0.6\}$, $M(E_{1\alpha}) = 0.50$;
- for $0.8522 < \alpha < 0.8931 : \Delta\alpha = 0.0409$, $E_{1\alpha} = \{0.5; 0.6\}$, $M(E_{1\alpha}) = 0.55$;
- for $0.8931 < \alpha < 0.9522 : \Delta\alpha = 0.0591$, $E_{1\alpha} = \{0.5\}$, $M(E_{1\alpha}) = 0.50$.

Then numerical estimate is $F(E_1) = \frac{1}{0.9522} \int_0^{0.9522} M(E_{1\alpha})d\alpha = 0.5248$. Numerical estimates of fuzzy conclusions relative to FS-levels for other regions are calculated in the same way: $F(E_2) = 0.5482$; $F(E_3) = 0.5426$; $F(E_4) = 0.5723$. The region $a_4$ has the most estimate (0.5723), so it has the highest rating of FS-level among others. Further in descending order: $a_2(0.5482)$, $a_3(0.5426)$, $a_1(0.5248)$.

## 5 Conclusion

Numerical estimates of FS-levels in the regions are obtained relative to the basis of FAO SG. At the same time, both regions and preliminary expert judgements are chosen in an arbitrary order. Nevertheless, we hope that the proposed method of assessment significantly increases the degree of objectivity of the results relative to the level of FS for countries where the population has limited or insufficiently guaranteed regular access to high-quality food that necessary for an active and healthy lifestyle.

## References

1. The new strategic framework for FAO. http://www.fao.org/docrep/meeting/029/mi558e. Accessed 19 June 2018
2. Aliyev, E.T.: Monitoring the results and reporting of FAO UN on the base of application of fuzzy cognitive model. Math. Mach. Syst. **2**, 56–71 (2016). (in Russian)
3. Zadeh, L.A.: The concept of a linguistic variable and its application to approximate reasoning. Inf. Sci. **8**(3), 199–249 (1965)

4. Rzayev, R.R.: Analytical Support for Decision-Making in Organizational Systems. Palmerium Academic Publishing, Saarbruchen (2016). (in Russian)
5. Lukasiewicz, J.: On three-valued logic. In: Borkowski, L. (ed.) Selected Works, pp. 87–88. NorthHolland Publishing Company, Amsterdam (1970)

# A Study of Fuzzy and Boolean Logics for Resource Management in User-Centric Wireless Networks

Huseyin Haci[(✉)]

Electrical and Electronic Engineering, Near East University, Nicosia, Cyprus
huseyin.haci@neu.edu.tr

**Abstract.** The rapid growth of wireless technologies lead to the emergence of a wide range of collaborative wireless services that rely on Internet users and their devices, as part of the network operation. In these environments – user-centric networking environments – resource management has specific features which this paper addresses. This is a study concerning Fuzzy and Boolean logics based resource management aspects in user-centric networking.

**Keywords:** User-centric networks · Resource management · Self-organization

## 1 Introduction

Wireless Fidelity (Wi-Fi) became the de-facto technology solution for providing short-range Internet connectivity. Market growth forecasts (c.f. Wi-Fi Alliance http://www.wi-fi.org/) predict steady increase in demand for IEEE 802.11 based Wireless Local Area Networks (WLANs), as corroborated by the increasing number of innovative companies adopting 802.11 and investing in new applications for WLAN enabled handsets and consumer electronics. Such wide range of devices and the Wi-Fi pervasiveness is increasing the number of Wi-Fi hotspots, in a way that is not only becoming unmanageable, but also hardly controllable. For instance, every individual residential household has one Wi-Fi Access Point (AP) terminating Internet access. Such APs in average provide connectivity for a maximum of 3 devices (e.g. notebook, smartphones) during specific hours of the day. For the rest of the time, spectrum is unused and yet polluting the medium. In urban environments, this is a trend that tends to increase, which becomes a good environment to trigger User-centric Networking (UCN) [1]. Instead of being just simple consumer of Internet services, UCN aims to ignite the Internet user as an active stakeholder in the Internet connectivity and service distribution chain by sharing their resources and distributing services. In such self-organizing environments, adequate cooperation incentives are required to assist a faster adoption of the technology [2]. Trust models are the basis to assist in providing security in a scalable way. But above all, resource management (RM) is required to assist the self-organization nature of UCNs.

In the user-centric networking trend resource management is critical to ensure high performance. Although there have been a lot of research work on resource management, it has not been clear how the existing schemes can be adapted to user-centric

© Springer Nature Switzerland AG 2019
R. A. Aliev et al. (Eds.): ICAFS-2018, AISC 896, pp. 619–626, 2019.
https://doi.org/10.1007/978-3-030-04164-9_81

networks. This paper aims to provide a study of Fuzzy and Boolean logics for intelligent resource management in UCNs.

## 2  User-Centric Wireless Networks: Characterization and Challenges

The fundamental paradigm of UCNs relates to the notion of shared services. Such notion has stemmed from new service paradigms such as Peer-to-Peer (P2P), where users become active content providers instead of just being service consumers. Due to the pervasiveness and low-cost of technologies such as Wi-Fi, as well as to the notions of Software Defined Networking, sharing of resources is now also a characteristic that can be supported in the lower OSI Layers. In UCNs, one potential example is Internet access sharing. Internet users share their Internet access freely and transparently among themselves in a way that is legally and technically independent from infrastructure and access providers. Currently existing commercial examples of networks that make use of the UCN concept are FON, OpenSpark (c.f. https://open.sparknet.fi), and Whisher.

Internet access to UCNs are provided by one or more owners of devices that can directly (e.g. an AP) or indirectly (e.g. a smartphone with Internet access sharing applications) provide Internet access. We refer to these devices as micro-provider gateways (MPG). MPGs can be owned by end-users (e.g. smart phone, laptop, etc.) or by operators (e.g. AP, 4G/Wi-Fi station, etc.). To provide adequate operation levels and service robustness, UCNs should incorporate at least the following properties [2]: **connectivity sharing and relaying, cooperation, intermittent connectivity and multi-hop relaying, self-organization and trust**.

## 3  Wireless Medium Access Control

Since IEEE 802.11 based WLANs is seen as a promising fundamental infrastructure to provide connectivity in networks that deploy UCN technology, the initial step of any algorithm or technique design should be the careful study of legacy IEEE 802.11 protocols.

In [3] a straightforward summary of legacy 802.11 MAC protocol distributed coordination function (DCF) with its limitations to support quality of service (QoS) is given. Point coordination function (PCF) is used in legacy 802.11 to support QoS for time-bounded services. PCF is centrally coordinated by a station called Point Coordinator (PC), usually an AP, and provides prioritized access to the shared wireless medium. Among many others, unpredictable beacon delays and unknown transmission durations of polled stations are two major problems of PCF that can severely affect QoS.

Transmission Opportunity (TXOP) is a new attribute introduced in 802.11e [4] MAC to solve un-deterministic transmission time problem. It is an interval that a station can transmit once acquired the channel. TXOP is defined by its starting time and the duration it lasts. The maximum limit on this value is provided with a parameter called TXOP limit. On the other hand, the first problem of unpredictable beacon delays is solved by introducing a restriction that a station is only allowed to start a transmission if it can be completed before the upcoming beacon frame, i.e. TBTT. This restriction can significantly reduce the expected beacon delay.

Enhanced distributed channel access (EDCA) and hybrid coordination function controlled channel access (HCCA) are two channel access methods that respectively provides prioritized and parameterized QoS in IEEE 802.11e. In EDCA, there are four access categories (ACs) for voice, video, best effort and background traffic. In order to provide traffic prioritization for these ACs, the amount of channel sensing time (i.e. arbitration inter-frame space, AIFS), the size of the contention window (CW), and the duration that a station can transmit (TXOP) is varied by EDCA, as shown in Fig. 1.

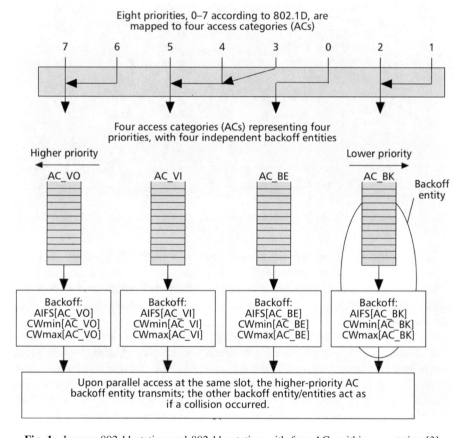

**Fig. 1.** Legacy 802.11 station and 802.11e station with four ACs within one station [3].

On the other hand, HCCA is based on a polling mechanism. Each station can have several traffic streams, which is characterized by a traffic specification (TSPEC), and the channel access is done according to the priority of the TSPEC. In order for a station to enter the channel, the AP (also called as the coordinator) should authorize it by sending a CFPoll frame to it. An example of this process is illustrated by Fig. 2, note that the coordinator is called Hybrid Coordinator (HC).

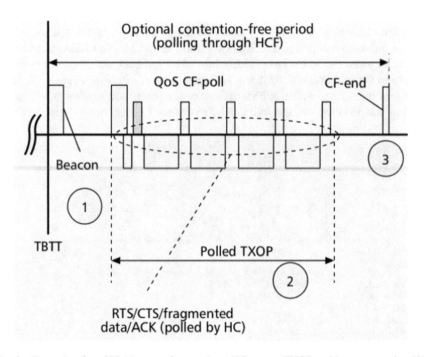

**Fig. 2.** Example of an 802.11e superframe where HC grants TXOPs without contention [3].

In [5] authors provided performance comparison of IEEE 802.11e QoS and power saving mechanisms with the legacy 802.11 MAC and power save mode. The proposed QoS mechanisms of IEEE 802.11e depend on predefined, fixed parameters (e.g. predefined ACs of EDCA), which can reduce the dynamism and effectiveness of the mechanisms.

In dynamic UCN scenarios it is very important to optimize the operation of the MAC protocol, according to the used spectrum provisioning schemes, which means that the MAC behavior may be different if focus is placed in: (i) spectrum management; (ii) radiation pattern schemes; (iii) adaptive modulation schemes; (iv) power management. However, existing multi-channel enabled protocols are restricted to the fixed channel-width spectrum distribution schemes, i.e. they do not benefit from the recent advances in spectrum management field that consists in variable channel widths [6]. In turn the recent proposals on variable channel widths concentrate all their research effort on the radio provisioning proposal and leave the medium access scheme task to CSMA/CA or variants, e.g. [7]. [8] proposed a novel method based on Fuzzy logic that can address the medium access control requirements for time critical applications. The proposed method uses a rule based Fuzzy logic system to set the arbitration threshold in wireless arbitration (WirArb) protocol. The study shows significant performance improvement over Boolean logic based systems, see Sect. 5.

## 4 Transmission Rate and Power Control

Frequent topology changes can be considered as one of the major challenges of UCN environments. Jointly adjusting the rate and power of transmission in wireless communications can be an efficient way to overcome this challenge. [9] proposed a cross-layer design called MRPC that exploits multi-rate transmissions with power control in ad hoc networks. At MRPC multiple, concurrent transmissions are allowed to happen, if they do not interfere with each other. The proposal comprises two main parts: MRPC-MAC and MRPC-Routing. MRPC-MAC is based on IEEE 802.11 MAC and uses the four-way handshake procedure (i.e. RTS/CTS/DATA/ACK) to carry out transmissions. Its support for multi-rate communications is based on Auto Rate Fallback (ARF) [10].

[11] considered a very interesting topic; potential negative effects of power control on WLAN performance. They listed these effects as follows; (1) make receiver-side interference (hidden terminal problem) and asymmetric channel access worse, (2) incorrectly lead to lowering the data rate of links and (3) make it difficult to differentiate above problems from mobility-induced channel variations at short timescales. To overcome these handicaps authors proposed a two-phase, synchronous rate and power control system, called Symphony. The key idea of Symphony is to synchronously use two phases – the reference phase (REF) and the operational phase (OPT) – to prevent adverse effects of dynamic power control. In the REF phase, the maximum transmit power is used and only rate adaptation [12] is performed. In this way the best achievable data rate for current channel conditions is found. On the other hand at the OPT phase both rate and power adaptation is performed. And next, system performances at REF and OPT phases are compared. If there is no performance degradation at OPT phase, it is shown that the power adaptation does not cause any negative effects and it can be safely employed besides rate adaptation. However, if there is performance degradation, the power adaptation is omitted and only rate adaptation is employed. Figure 3 illustrates the two-phase synchronous strategy applied.

Another practical solution to solve these problems is to realize single-channel and distributed approaches [13, 14]. Authors of [14] proposed a control scheme called Receiver Initiated power control Multi-Access (RIMA). To solve the hidden node problem, the receiver transmits ACK and CTS messages at full power and introduces an additional high-power pulsed in-band busy signal. For the exposed node problem, transmissions of RTS and data frames are done at low power. As shown in Fig. 4, this concept studies different reception, interference and carrier sense range and a signaling scheme that can improve location awareness and enable concurrent transmissions. For an example, refer to Fig. 4 and the relevant text from [14]. However, requirement for precise location information and lack of mobility support are two major drawbacks. [15] proposed a Fuzzy logic based power control technique. This technique introduces a combination of a time slot mechanism and an adaptive power control technique. It is shown that the proposed technique can outperform Boolean logic based techniques.

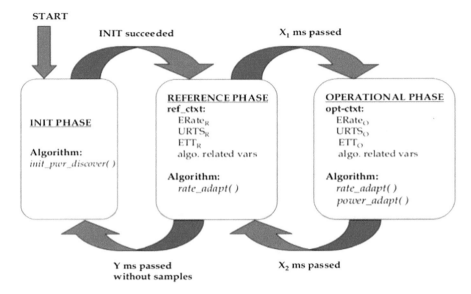

**Fig. 3.** Two-phase synchronous strategy of Symphony [11].

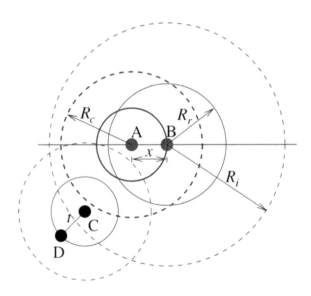

**Fig. 4.** Demonstration for exposed terminal problem solution [14].

## 5   Performance Results

Figure 5 compares the throughput performance versus number of users in the system for Boolean logic based CSMA/CA and WirArb techniques and Fuzzy logic based technique proposed by [8]. It can be seen that the Fuzzy logic based technique can significantly outperform the Boolean logic based techniques, especially when the number of users increases.

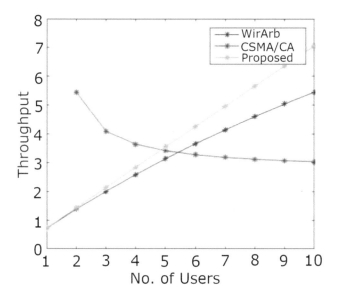

**Fig. 5.** Comparison between CSMA/CA, WirArb and proposed method [8].

Figure 6 shows the comparison of average energy per bit (milli Joules) versus packet generation per node (packet/minute) for Boolean logic based CSMA and CSMA/ACK and Fuzzy logic based technique proposed by [15]. The figure shows that the Fuzzy logic based technique can outperform the Boolean logic based techniques.

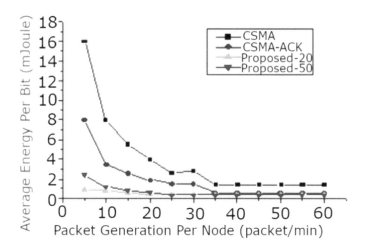

**Fig. 6.** Average energy per bit for different packet generation rate [15].

## 6   Conclusion

The fast pace growth of wireless technologies leads to the emergence of UCNs. This article aims to provide a comparative study of Fuzzy and Boolean logics for resource management in UCNs. Fuzzy logic based techniques have shown that they can outperform Boolean logic based techniques at providing higher throughput and lower average energy per bit.

## References

1. Lopes, L.A., Sofia, R., Osman, H., Haci, H.: A proposal for elastic spectrum management in wireless local area networks. In: 2014 IEEE Conference on Computer Communications Workshops, pp. 127–128. IEEE, Toronto (2014)
2. Lopes, L.A., Sofia, R., Haci, H., Zhu, H.: A proposal for dynamic frequency sharing in wireless networks. IEEE/ACM Trans. Netw. **24**, 2621–2633 (2016)
3. Mangold, S., Sunghyun, C., Hiertz, G.R., Klein, O., Walke, B.: Analysis of IEEE 802.11e for QoS support in wireless LANs'. IEEE Wirel. Commun. **10**, 40–50 (2003)
4. Gao, Y., Sun, X., Dai, L.: IEEE 802.11e EDCA networks: modeling, differentiation and optimization. IEEE Trans. Wirel. Commun. **13**, 3863–3879 (2014)
5. Perez-Costa, X., Camps-Mur, D.: IEEE 802.11E QoS and power saving features overview and analysis of combined performance. IEEE Wirel. Commun. **17**, 88–96 (2010)
6. Chen, X., Yuan, W., Cheng, W., Liu, W., Leung, H.: Access point selection under QoS requirements in variable channel-width WLANs. IEEE Wirel. Commun. Lett. **2**, 114–117 (2013)
7. Goldsmith, A.: Wireless Communications. Cambridge University Press, New York (2005)
8. Jacob, R.M., Sravan, M.S.: A novel method based on fuzzy logic to set the arbitration threshold in WirArb for time critical applications in wireless sensor network. In: 2017 International Conference on Networks & Advances in Computational Technologies, pp. 196–202. IEEE, Thiruvananthapuram (2017)
9. Shih, K.P., Chang, C.C., Chen, Y.D.: MRPC: a multi-rate supported power control MAC protocol for wireless ad hoc networks. In: 2009 IEEE Wireless Communications and Networking Conference, pp. 1–6. IEEE, Budapest (2009)
10. Maguolo, F., Lacage, M., Turletti, T.: Efficient collision detection for auto rate fallback algorithm. In: 2008 IEEE Symposium on Computers and Communications, pp. 25–30. IEEE, Marrakech (2008)
11. Ramachandran, K., Kokku, R., Zhang, H., Gruteser, M.: Symphony: synchronous two-phase rate and power control on 802.11 WLANs. IEEE/ACM Trans. Netw. **18**, 1289–1302 (2010)
12. Pefkianakis, I., Hu, Y., Lu, S.: History-aware rate adaptation in 802.11 wireless networks. In: 2011 IEEE Symposium on Computers and Communications, pp. 224–229. IEEE, Kerkyra (2011)
13. Li, X., Li, Q., Liu, K.: Efficient dual channel multiple access for multihop wireless networks. In: 2014 IEEE 17th International Conference on Computational Science and Engineering, pp. 855–858. IEEE, Chengdu (2014)
14. De, S., Egoh, K., Dosi, G.: A receiver initiated power control multi-access protocol in wireless ad hoc networks. In: 2007 IEEE Sarnoff Symposium, pp. 1–5. IEEE, Princeton (2007)
15. Caijun, C., Haiming, Y.: A lightweight MAC protocol with adaptive transmission power control in WSNs. In: 2011 Eighth International Conference on Fuzzy Systems and Knowledge Discovery, pp. 2208–2211. IEEE, Shanghai (2011)

# Application of Interval Analysis in Evaluation of Macroeconomic Impacts of Taxes

A. F. Musayev[1,2(✉)], R. I. Davudova[1,3,4], and A. A. Musayeva[3,4]

[1] ANAS Institute of Economy,
H. Javid Ave., 115, Baku AZ1143, Republic of Azerbaijan
akif.musayev@gmail.com, revanadavudova@gmail.com
[2] Near East University, Nicosia, Mersin 10, Northern Cyprus, Turkey
[3] ANAS Institute of Control Systems,
B. Vahabzadeh St., 9, Baku AZ1141, Republic of Azerbaijan
aygun.musayeva@gmail.com
[4] The Azerbaijan University,
J. Hajibeyli St., 71, Baku AZ1007, Republic of Azerbaijan

**Abstract.** The article is devoted to the study of the average tax burden and the relationship between the average tax burden and macroeconomic indicators of the balanced open economy. Based on the Samuelson model using interval analysis the interaction of the average tax burden with macroeconomic indicators was analyzed, as the norms of consumption, public expenditures, investments and rates of economic growth and are estimated on the interval values of the macroeconomic indicators of Azerbaijan Republic defined by the method Delphi during 2003–2017.

**Keywords:** Balanced open economy · Macroeconomic indicators
Average tax burden · Norm of consumption · Norm of public expenditures
Norm of investments · Rates of economic growth · Interval analysis

## 1 Introduction

Tax policy, being one of the main components of the state's socio-economic policy, plays an important role in shaping the revenues of the state budget, in ensuring macroeconomic stability and sustainable development of the country. It is also of great importance in stimulating economic activity. Tax compulsory burdens on the economy are examined in the study related to the assessment of the tax burden as a rule. However, in our opinion, in order to formulate a stimulating tax policy, it must be considered that the tax is an economic, social and legal category, and the consumption norms, economic growth rates, public expenditure norms, investment norms, and average tax burden and etc. factors that do not have the least impact, should be taken into consideration in the full assessment of its impact on the economic system. A number of studies have been conducted to study the tax burden, the integrated tax burden, the calculation and the impact of various factors. Thus, the influence of the shadow economy on the tax burden on the proportional growth model was investigated in [1]. Mamdani type fuzzy result was used to assess tax potential in [2, 3]. A Social

© Springer Nature Switzerland AG 2019
R. A. Aliev et al. (Eds.): ICAFS-2018, AISC 896, pp. 627–634, 2019.
https://doi.org/10.1007/978-3-030-04164-9_82

Accounting Matrix (SAM) has been established on the basis of "Delegate balance between the production and distribution of products and services" in the economic fields of the Republic of Azerbaijan, which is crucial to the Delphine method in [4]. Here is a combined scheme of SAM, taking into account the specific features of the economy of the Republic of Azerbaijan, the impact of changes in budget expenditure on different sectors of the economy. The Leontief's intersectoral balance model was studied by the application of interval analysis based on the matrix compiled by the authors in the calculations. In order to study the effect of government's administrative efforts on tax potential, an expert assessment of the impact of changes and additions made in tax administration on the tax potential was carried out in 2013, a fuzzy outcome method of the Mamdani type was used for this purpose [5, 6].

Analysis of the Balance Social Account shows that the normative values of the factors such as consumption, investment, public spending and so on, is not always possible to determine unequivocally. This is largely due to the fact that these norms are dependent on indefinite factors determined by a number social surveys or examinations typically at the time of listing, quantifiable and unquantifiable factors. In this case, accurate estimation of quantitative indicators can lead to loss of some features and signs of the system learned [7–9]. Interval estimates allow you to describe the real situations of the systems and processes investigated altogether [10–12]. Thus, interval estimation of economic system-specific indices and their interactions allows generating logical conclusions about the state of the economic system.

Let's consider the relationship between the tax burden and some macroeconomic indicators:

Public expenditure and tax burden. It is known that taxes serve the distribution of gross income and the tax burden does not help those who are less likely to fall, but this leads to an increase in revenues distributed among those with higher income and, ultimately, a decrease in government revenues, which, in turn, will result in the limitation of costs. Because the lower and middle income layers are more dependent on public spending, the additional income generated by the reduction of the tax burden will be less than the state support made to them.

Consumer rate and tax burden. Consumption norm is more complicated by tax burden. When analyzing the level of tax burden, assessment of the impact of consumption norms requires a special approach. In short, it should be taken into account more seriously as the consumption norm is affected by the accumulation, investment, demand and supply.

Economic growth and tax burden. The founder of the idea that lowering the tax burden would promote investment activity would have argued that tax revenues would increase as a result of the expansion of the tax base when the tax burden was lowered. However, the expansion of the tax base does not occur at the initial stage of the tax reform, which means that the reduction of the tax burden will result in a deficit in the state budget. If such changes occur frequently, foreign investors will not be able to invest in a country with such an unstable economy, even if the potential taxpayer's down payment can be more dangerous.

## 2  Analysis of the Macroeconomic Impacts of Tax Liability in the Open Economy

### 2.1  Methodology

The British economist, contributed a decisive contribution to the theory of economics of J. M. Keynes has shown that the dynamics of national income production and the employment level are determined not directly by the supply factors (the size of labor, capital, their productivity), but by the demand factors that ensure the realization these resources. Keynes was against the excessive mathematization of the economy, nuisance the perception of its layman.

Paul Samuelson once again demonstrated that mathematics is the only means of expression of the basic provisions of modern economic theories. His economic theory, based on graphic-analytical and mathematical analysis, was far from extreme ideological position, such as market economy or bureaucracy, public or private sector, fundamentalism or monetarism. He has given greater advantage to the idea of "golden middle" as a solid liberal.

The production capabilities of economic choice, alternatives of some macroeconomic indicators given in figures, production volumes and consumption tools are reviewed at the Samuelson [13] models in table and graphic form. The main feature of this model is that the problem of the balance of macroeconomic situations described by the main macroeconomic indicators is taken into consideration, the conditions for achieving balance are discussed and the reason why the macroeconomic system is not balanced is explained. In this case, the non-balanced economic system may be born of objective conditions - war, large social and economic reform, and cataclysm. Achieving balance in the real macroeconomic system is the most difficult financial and political task. To solve this, it becomes necessary to use public debt, tax exacerbation and money emissions.

The model takes into account the effectiveness of the choice of the public production and the extent to which the (extensive or intensive) impact on the development of the macroeconomic system, the rationale and suspicion of consumption norms and public economic efficiency, tax burden, public spending and investments, economic growth rates and overall macroeconomic equilibrium.

A direct dependence occurs between the additional charges reducing the efficiency and productivity of the economy by consumption requirements increasingly growing from this model.

Let's use the model given in [14] in order to evaluate the relationships between the key macroeconomic indicators. Here is a general mathematical model of the balanced open economy (BOE):

$$Y - C - \varepsilon B = G + I$$

So that, here: $Y$ – the volume of GDP forecasted (planned) by government agencies; $G$ – public expenditures (respectively, $g = G/Y$ – public expenditure norm); $I$ – investments (respectively, $s = I/Y$ – investing norm); $T$ – forecasted tax revenues by the state authorities (respectively, $\theta = T/Y = g/(1 - s)$ – average tax burden norm);

$C = (1 - s)(1 - \theta)Y$ – consumption volumes (respectively, $c = C/Y = (1 - s)(1 - \theta)$ consumption volumes norm); $\varepsilon$ – exchange rate; $B$ – the balance of the country's balance of payments. In the perfectly balanced open economy, where the balance of payments is zero, the following relationships of consumption norms $c_n$ with other macroeconomic norms are identified [14]:

It is unlikely that the rate of consumption of the ideal BOE is less than the rate of economic growth. The main reason for the production is consumption. Therefore, GDP growth rate cannot be less than the growth rate of consumption.

At the same time, the average tax burden in a balanced open economy is inextricably interconnected with the aforementioned macroeconomic indicators:

On the basis of historical experience it is possible to say that the average tax burden of macroeconomic systems will not be lower than 10%. At the same time, the ideal average tax burden can not exceed its maximum value in any case: $\theta_{max} = g_{max} = 1/\psi$. This case belongs to the administrative economy, more than 98% of the property of which is owned by the state-owned, and the investments (capital investment) are considered as the government spending here.

The following relationships can be used to evaluate the effectiveness of BOE:

Relevant simulations and assessments were made based on the P. Samuelson model [13] for the Open Economy using interval analysis to take into account the uncertainty related to the consumption rate, economic growth rate, public expenditure norm, investment norm and average tax burden using SAM, elements of which are interval numbers determined by Delphi Interrogation Methods.

## 2.2 Computational Experiments

We employed the interval mathematics platform and was created M-file in the MATLAB environment to evaluate the relationship of the main macroeconomic relations (Tables 1, 2, 3 and 4) based on the interval values of the macroeconomic indicators of Azerbaijan Republic defined by the method Delphi in 2003–2017 (Table 5).

**Table 1.** Relationships of consumption norms $c_n$ with macroeconomic norms

| № | Mathematical expression | Sign and name of the macroeconomic indicator |
|---|---|---|
| 1 | $c_{n1} = 1 - 1/\psi$ | $\psi$ – General index of structural effectiveness |
| 2 | $c_{n2} = (1 - \theta)(1 + \theta)$ | $\theta$ – Average tax burden |
| 3 | $c_{n3} = 1 - 2\sqrt{F}$ | $F$ – Economic growth rate |

**Table 2.** $\theta_n$ interconnection of macroeconomic indicators with average tax

| № | Mathematical expression | Sign and name of the macroeconomic indicator |
|---|---|---|
| 1. | $\theta_{n1} = (1 - c)(1 + c)$ | $c$ – Public consumption norm |
| 2. | $\theta_{n2} = \sqrt{F}/(1 - \sqrt{F})$ | $F$ – Economic growth rate |
| 3. | $\theta_{n3} = 1/(2\psi - 1)$ | $\psi$ – Total amount of public expenditure and investments |

**Table 3.** $F_n$ mutual interest of economic growth rates and macroeconomic indicators

| № | Mathematical expression | Sign and name of the macroeconomic indicator |
|---|---|---|
| 1. | $F_{n1} = 1/(4\psi^2)$ | $\psi$ – General index of structural effectiveness |
| 2. | $F_{n2} = \theta^2/(1 + \theta)^2$ | $\theta$ – Average tax burden |
| 3. | $F_{n3} = (1 - c)^2/4$ | $c$ – Public consumption norm |

**Table 4.** Effectiveness of BOE

| № | Mathematical expression | Sign and name of the macroeconomic indicator |
|---|---|---|
| 1. | $s = 1/\psi - g$ | $s$ – Investment norm |
| 2. | $F = \Delta Y/Y$ | $F$ – Economic development pace |
| 3. | $\psi = 1/(s + g)$ | $\psi$ – General index of structural effectiveness |
| 4. | $R = \Delta Y/I = F/S$ | $R$ – Social effectiveness index of investments |
| 5. | $S = \Delta Y/G = F/g$ | $S$ – Social effectiveness index of state expenditure |

**Table 5.** The interval values of the macroeconomic indicators norms

| Years | $c$ | | $\theta$ | | $F$ | |
|---|---|---|---|---|---|---|
| | Lower | Upper | Lower | Upper | Lower | Upper |
| 2003 | 0.16 | 0.20 | 0.13 | 0.26 | 0.16 | 0.20 |
| 2004 | 0.44 | 0.53 | 0.11 | 0.22 | 0.44 | 0.53 |
| 2005 | 0.17 | 0.21 | 0.12 | 0.25 | 0.17 | 0.21 |
| 2006 | 0.46 | 0.56 | 0.15 | 0.31 | 0.46 | 0.56 |
| 2007 | 0.47 | 0.57 | 0.16 | 0.32 | 0.47 | 0.57 |
| 2008 | 0.38 | 0.47 | 0.20 | 0.40 | 0.38 | 0.47 |
| 2009 | 0.10 | 0.13 | 0.22 | 0.44 | 0.10 | 0.13 |
| 2010 | 0.18 | 0.22 | 0.20 | 0.40 | 0.18 | 0.22 |
| 2011 | 0.21 | 0.25 | 0.23 | 0.45 | 0.21 | 0.25 |
| 2012 | 0.05 | 0.06 | 0.24 | 0.47 | 0.05 | 0.06 |
| 2013 | 0.06 | 0.07 | 0.25 | 0.50 | 0.06 | 0.07 |
| 2014 | 0.01 | 0.02 | 0.23 | 0.47 | 0.01 | 0.02 |
| 2015 | 0.15 | 0.18 | 0.19 | 0.38 | 0.15 | 0.18 |
| 2016 | 0.08 | 0.10 | 0.21 | 0.42 | 0.08 | 0.10 |
| 2017 | 0.00 | 0.01 | 0.20 | 0.40 | 0.00 | 0.01 |

Running results of M-file are presented in Tables 6, 7, 8 and 9.

Analyzing results it can be seen that during the whole period of observation, in 2003–2017, 2003–2008, 2004–2017 and 2012–2015 years the conditions are met, respectively:

$$\psi > c_2 > \theta > F_2; \quad \theta_2 \gg R \gg F \gg c_3; \quad S \gg \psi > R \gg s > F; \quad c > \theta_2 \gg R > F.$$

**Table 6.** Impact of public consumption norm $c$ on macroeconomic indicators

| Years | $c$ | | $\theta_{n1}$ | | $F_{n3}$ | |
|---|---|---|---|---|---|---|
| | Lower | Upper | Lower | Upper | Lower | Upper |
| 2003 | 0.16 | 0.20 | 0.19 | 0.47 | −0.46 | −0.32 |
| 2004 | 0.44 | 0.53 | 0.20 | 0.48 | 0.08 | 0.17 |
| 2005 | 0.17 | 0.21 | 0.06 | 0.37 | −0.49 | −0.35 |
| 2006 | 0.46 | 0.56 | 0.33 | 0.59 | −0.52 | −0.37 |
| 2007 | 0.47 | 0.57 | 0.45 | 0.67 | −0.36 | −0.24 |
| 2008 | 0.38 | 0.47 | 0.45 | 0.67 | 0.29 | 0.36 |
| 2009 | 0.10 | 0.13 | 0.14 | 0.43 | 0.07 | 0.16 |
| 2010 | 0.18 | 0.22 | 0.26 | 0.53 | −0.01 | 0.09 |
| 2011 | 0.21 | 0.25 | 0.33 | 0.58 | 0.52 | 0.57 |
| 2012 | 0.05 | 0.06 | 0.25 | 0.52 | 0.47 | 0.52 |
| 2013 | 0.06 | 0.07 | 0.16 | 0.45 | 0.75 | 0.77 |
| 2014 | 0.01 | 0.02 | 0.03 | 0.34 | 0.16 | 0.24 |
| 2015 | 0.15 | 0.18 | 0.04 | 0.35 | 0.38 | 0.43 |
| 2016 | 0.08 | 0.10 | 0.09 | 0.39 | 0.86 | 0.87 |
| 2017 | 0.00 | 0.01 | 0.02 | 0.33 | −0.46 | −0.32 |

**Table 7.** Influence of average tax burden $\theta$ on macroeconomic indicators

| Years | $\theta$ | | $F_{n2}$ | | $\psi$ | | $c_{n2}$ | |
|---|---|---|---|---|---|---|---|---|
| | Lower | Upper | Lower | Upper | Lower | Upper | Lower | Upper |
| 2003 | 0.13 | 0.26 | 0.087 | 0.230 | 2.45 | 4.40 | 0.04 | 0.41 |
| 2004 | 0.11 | 0.22 | 0.072 | 0.189 | 2.83 | 5.15 | 0.13 | 0.47 |
| 2005 | 0.12 | 0.25 | 0.074 | 0.196 | 2.53 | 4.56 | 0.11 | 0.45 |
| 2006 | 0.15 | 0.31 | 0.053 | 0.140 | 2.12 | 3.73 | 0.25 | 0.54 |
| 2007 | 0.16 | 0.32 | 0.050 | 0.131 | 2.07 | 3.65 | 0.28 | 0.55 |
| 2008 | 0.20 | 0.40 | 0.053 | 0.141 | 1.74 | 2.99 | 0.25 | 0.54 |
| 2009 | 0.22 | 0.44 | 0.051 | 0.136 | 1.65 | 2.80 | 0.26 | 0.55 |
| 2010 | 0.20 | 0.40 | 0.055 | 0.145 | 1.74 | 2.98 | 0.24 | 0.53 |
| 2011 | 0.23 | 0.45 | 0.057 | 0.152 | 1.61 | 2.71 | 0.22 | 0.52 |
| 2012 | 0.24 | 0.47 | 0.070 | 0.185 | 1.56 | 2.61 | 0.14 | 0.47 |
| 2013 | 0.25 | 0.50 | 0.072 | 0.190 | 1.49 | 2.49 | 0.13 | 0.46 |
| 2014 | 0.23 | 0.47 | 0.070 | 0.185 | 1.57 | 2.64 | 0.14 | 0.47 |
| 2015 | 0.19 | 0.38 | 0.045 | 0.120 | 1.80 | 3.11 | 0.31 | 0.57 |
| 2016 | 0.21 | 0.42 | 0.063 | 0.165 | 1.69 | 2.88 | 0.19 | 0.50 |
| 2017 | 0.20 | 0.40 | 0.054 | 0.143 | 1.75 | 3.01 | 0.24 | 0.53 |

**Table 8.**  Impact of economic growth $F$ on macroeconomic indicators

| Years | $F$ | | $c_{n3}$ | | $\theta_{n2}$ | |
|---|---|---|---|---|---|---|
| | Lower | Upper | Lower | Upper | Lower | Upper |
| 2003 | 0.164 | 0.200 | 0.10 | 0.19 | 0.18 | 0.48 |
| 2004 | 0.436 | 0.530 | −0.46 | −0.32 | 0.19 | 0.68 |
| 2005 | 0.174 | 0.212 | 0.08 | 0.17 | 0.07 | 0.35 |
| 2006 | 0.457 | 0.557 | −0.49 | −0.35 | 0.09 | 0.37 |
| 2007 | 0.472 | 0.574 | −0.52 | −0.37 | 0.19 | 0.48 |
| 2008 | 0.382 | 0.465 | −0.36 | −0.24 | 0.16 | 0.64 |
| 2009 | 0.104 | 0.127 | 0.29 | 0.36 | 0.14 | 0.43 |
| 2010 | 0.177 | 0.216 | 0.07 | 0.16 | 0.16 | 0.62 |
| 2011 | 0.208 | 0.254 | −0.01 | 0.09 | 0.18 | 0.68 |
| 2012 | 0.047 | 0.057 | 0.52 | 0.57 | 0.28 | 0.31 |
| 2013 | 0.058 | 0.070 | 0.47 | 0.52 | 0.32 | 0.36 |
| 2014 | 0.013 | 0.016 | 0.75 | 0.77 | 0.13 | 0.14 |
| 2015 | 0.146 | 0.178 | 0.16 | 0.24 | 0.62 | 0.73 |
| 2016 | 0.080 | 0.098 | 0.38 | 0.43 | 0.39 | 0.45 |
| 2017 | 0.004 | 0.005 | 0.86 | 0.87 | 0.07 | 0.08 |

**Table 9.**  Performance indicators of BOE

| Years | $s$ | | $F$ | | $\psi$ | | $R$ | | $S$ | |
|---|---|---|---|---|---|---|---|---|---|---|
| | L | U | L | U | L | U | L | U | L | U |
| 2003 | 0.51 | 0.71 | 0.164 | 0.200 | 1.04 | 1.69 | 0.11 | 0.25 | 0.79 | 1.45 |
| 2004 | 0.47 | 0.66 | 0.436 | 0.530 | 1.15 | 1.87 | 0.09 | 0.21 | 2.54 | 4.65 |
| 2005 | 0.46 | 0.65 | 0.174 | 0.212 | 1.13 | 1.83 | 0.11 | 0.25 | 0.85 | 1.55 |
| 2006 | 0.34 | 0.47 | 0.457 | 0.557 | 1.34 | 2.17 | 0.13 | 0.30 | 1.88 | 3.44 |
| 2007 | 0.31 | 0.44 | 0.472 | 0.574 | 1.38 | 2.24 | 0.14 | 0.31 | 1.83 | 3.35 |
| 2008 | 0.28 | 0.40 | 0.382 | 0.465 | 1.33 | 2.16 | 0.18 | 0.39 | 1.19 | 2.17 |
| 2009 | 0.25 | 0.35 | 0.104 | 0.127 | 1.36 | 2.21 | 0.19 | 0.43 | 0.29 | 0.54 |
| 2010 | 0.28 | 0.40 | 0.177 | 0.216 | 1.31 | 2.13 | 0.18 | 0.40 | 0.53 | 0.97 |
| 2011 | 0.28 | 0.39 | 0.208 | 0.254 | 1.28 | 2.09 | 0.19 | 0.43 | 0.59 | 1.07 |
| 2012 | 0.31 | 0.44 | 0.047 | 0.057 | 1.16 | 1.89 | 0.21 | 0.46 | 0.12 | 0.22 |
| 2013 | 0.31 | 0.44 | 0.058 | 0.070 | 1.15 | 1.86 | 0.22 | 0.48 | 0.15 | 0.27 |
| 2014 | 0.32 | 0.45 | 0.013 | 0.016 | 1.16 | 1.89 | 0.21 | 0.46 | 0.03 | 0.06 |
| 2015 | 0.51 | 0.71 | 0.164 | 0.200 | 1.04 | 1.69 | 0.11 | 0.25 | 0.79 | 1.45 |
| 2016 | 0.47 | 0.66 | 0.436 | 0.530 | 1.15 | 1.87 | 0.09 | 0.21 | 2.54 | 4.65 |
| 2017 | 0.46 | 0.65 | 0.174 | 0.212 | 1.13 | 1.83 | 0.11 | 0.25 | 0.85 | 1.55 |

*Note.* In Table 9 symbols L and U mean lower and upper bounds of the interval variables, respectively.

# 3 Conclusion

The article analyzes the macroeconomic impact of the tax burden on a balanced open economy, using the interval values of the macroeconomic indicators of Azerbaijan Republic defined by the method Delphi from 2003 to 2017, based on the model given in [14]. Based on computational experiments it is determined in 2003–2017, 2003–2008, 2004–2017 and 2012–2015 that the ratios of are satisfied, respectively:

$$\psi > c_2 > \theta > F_2; \quad \theta_2 \gg R \gg F \gg c_3; \quad S \gg \psi > R \gg s > F; \quad c > \theta_2 \gg R > F.$$

# References

1. Musayev A.F., Musayeva A.A.: A study of the impact of underground economy on integral tax burden in the proportional growth model under uncertainty. Adv. Fuzzy Systems. **2018**, article ID 6309787 (2018)
2. Musayev, A., Madatova, Sh., Rustamov, S.: Mamdani-type fuzzy inference system for evaluation of tax potential. In: Recent Developments and the New Direction in Soft-Computing Foundations and Applications, pp. 511–523. Springer, Cham (2018)
3. Musayev, A.F.: Tax potential and its assessment methods. Tax J. Azerbaijan Republic **5** (119), 75–86 (2014)
4. Musayev, A.F.: Opportunities for applying the Leontyev method to the budget system. Tax J. Azerbaijan Republic **3**(114), 79–102 (2013)
5. Musayev, A.F., Madatova, S.G., Rustamov, S.S.: Evolution of the impact of the tax legislation reforms on the tax potential by fuzzy inference method. In: 12th International Conference on Application of Fuzzy Systems and Soft Computing (ICAFS), pp. 507–514. Elsevier, Vienna (2016)
6. Musayev, A., Madatova, S.: Evaluation of the influence of government's administrative efforts on tax potential by the type of fuzzy conclusion method. Tax J. Azerbaijan Republic **3** (129), 83–96 (2016)
7. Rohn, J.: A Handbook of Results on Interval Linear Problems. Czech Academy of Sciences, Prague (2005)
8. Nguyen, H.T., Kreinovich, V., Wu, B., Xiang, G.: Computing Statistics under Interval and Fuzzy Uncertainty. Springer, Heidelberg (2012)
9. Mayer, G.: Interval Analysis and Automatic Result Verification. De Gruyter, Berlin, Boston (2017)
10. Alefeld, G., Herzberger, J.: Introduction to Interval Computations. Academic Press, New York (1983)
11. Moore, R.: Interval Analysis. Prentice-Hall, Upper Saddle River (1996)
12. Moore, R., Kearfott, R., Cloud, M.: Introduction to Interval Analysis. SIAM, Philadelphia (2009)
13. Samuelson, P., Nordhaus, W.: Economics, 16th edn. Mass Irwin/McGraw-Hill, Boston (1998)
14. Vladimirov, S.A.: On the macroeconomic essence of the strategic development objectives of effective, balanced macroeconomic systems. Scientific works of the North-West Institute of Management, branch of RANEPA, vol. 6, no. 4(21), pp. 105–116 (2015)

# Fuzzy Method of Assessing the Intensity of Agricultural Production on a Set of Criteria of the Level of Intensification and the Level of Economic Efficiency of Intensification

Tamara V. Alekseychik, Taras V. Bogachev, Denis N. Karasev,
Lyudmila V. Sakharova$^{(\boxtimes)}$, and Michael B. Stryukov

Rostov State University of Economics, B. Sadovaya street, 69,
344002 Rostov-on-Don, Russia
L_Sakharova@mail.ru

**Abstract.** The aim is a fuzzy-multiple work to develop a methodology to assess the dynamics of agricultural development in the region on the basis of a set of diverse indicators, as well as to compare (Rangers) agricultural facilities or regions. The technique is shown on the application of multi-level standard net [0, 1]–classification. It allows you to calculate complex numerical score for the level of intensification of agriculture on the criteria of the two groups for any number of years studied: level of intensification of production in the economy of the silk and the level of economic efficiency of intensification of production on the farm the silk, and also give practical recommendations for the further development of agriculture in the region. The proposed method has the following advantages: (1) a simple calculation scheme; (2) taking into account the large number of Razor estimates of significant indicators; (3) the use of only indicators that objectively reflect the effectiveness of the use of material and financial resources of agriculture; (4) the possibility of deviations vest dressed in the studied indicators in a complex assessment of the intensity of agricultural production in the region; (5) universality, which allows to apply it to the assessment of the intensity of not only agricultural but also industrial production.

**Keywords:** Methodology · Complex estimation
Intensity of agricultural production · Indicators · Theory of fuzzy sets

## 1 Introduction

The essence of the criterion of efficiency of agricultural production is the maximum production of agricultural products required by society at a given cost and amount of resources per unit of production, which ensures high quality products and rational use of labor, material, monetary and land resources [1]. In the aspect of optimization of agricultural production practical importance is the development of methods that allow to produce a comprehensive assessment of the intensity of agricultural production on the basis of a set of ranked indicators that objectively reflect the effectiveness of the use

© Springer Nature Switzerland AG 2019
R. A. Aliev et al. (Eds.): ICAFS-2018, AISC 896, pp. 635–642, 2019.
https://doi.org/10.1007/978-3-030-04164-9_83

of material and financial resources by agriculture, as well as the impact of agricultural production on the development of agriculture. Currently, there are methods for assessing the intensity of agricultural production on its individual indicators [2]. However, they do not allow to evaluate and rank agro-industrial enterprises, agricultural industries and entire regions on the basis of a comprehensive analysis of many indicators. For ranking in agriculture in practice, rating estimates are used, for example, the annual assessment of the club "Agro-300" [3]. The standard method of calculating the General economic rating provides for the use of only two indicators: revenue and profit from the sale of agricultural products [4]. Industry ratings are calculated quite difficult and are built using only three of the most important indicators for the industry: the volume of gross output of the industry, the cost of commodity products, profits from sales. The wide practical application of the above methods is complicated due to the following factors: (1) the complexity of the calculation of rating estimates; (2) the account in the construction of rating estimates of a small number of significant indicators; (3) use of indicators directly dependent on soil and weather-climatic conditions, as a result of which, for example, participants with the most favorable conditions of agricultural production do not require additional costs for agrotechnical measures usually get to the club "Agro-300". Thus, for a fair assessment of the dynamics of intensity of agricultural production, methods that objectively reflect the efficiency of material and financial resources of agriculture are necessary.

In this paper, a technique based on the methods of the theory of fuzzy sets and aimed at obtaining an objective comprehensive quantitative assessment of the intensity of agricultural production on a set of criteria of two groups: the level of intensification of production and the level of economic efficiency of intensification of production in agriculture. The novelty of the proposed method, as well as its difference from similar developments, is that integrated estimates are calculated for each of the indicators on the basis of time series of its values by means of normalizing formulas. The subsequent application to them of the standard five-level indistinct [0, 1] - classifier (previously used in the financial analysis and not used in the methods of estimation of intensity of production [5]) allows to calculate the normalized integrated assessment of the levels of intensification of production and economic efficiency of intensification of production in agriculture and to obtain a comprehensive assessment of the intensity of its production. Aggregation of the formed estimates on the basis of standard five–level fuzzy [0, 1] - classifiers allowed to receive the final complex assessment of intensity of agricultural production of the region on the example of the Rostov region.

## 2    General Principles of the Methodology for Assessing the Intensity of Agricultural Production

General principles of the author's method of assessing the intensity of agricultural production are described in detail in [6, 7]. The mathematical apparatus, which is the basis of the method and is a modification of the standard multi - level fuzzy [0, 1] - classifiers is disclosed in [8, 9]. The general principles of the method of evaluation of the intensity of agricultural production is reduced to the following algorithm.

Phase 1. Formation of a list of significant indicators of the level of intensification of production in agriculture for the period of n years (hereinafter: the first group of indicators), as well as significant indicators of the level of economic efficiency of intensification of production in agriculture for the same period (the second group of indicators).

Phase 2. Ranking of the importance of the studied indicators to assess the intensity of agriculture, the calculation of their weight coefficients based on expert assessments.

Phase 3. Calculation of normalized (that is, belonging to the segment [0, 1]) numerical values of the studied indicators of the first and second groups for the period of n years, for example, based on the formulas determined by the meaning of the problem.

Phase 4. The assignment of linguistic variables. Normalized values of indicators defined in Step 3 are numerical values of fuzzy variables with a universal set (carrier) in the form of a segment [0, 1]. They compare linguistic variables with terms-sets consisting of five terms: "very low level of the indicator"; "low level of the indicator"; "average level of the indicator"; "high level of the indicator"; "very high level of the indicator". The membership functions of linguistic variables are determined by trapezoidal functions.

In addition, we introduce into consideration the linguistic variables:

$\gamma$ = "integrated assessment of the intensity of agricultural production";
$\gamma_1$ = "assessment of the level of intensification of production in agriculture";
$\gamma_2$ = "assessment of the economic efficiency of the intensification of production in agriculture".

Universal set for each linguistic variable is a numerical segment [0, 1], and the set of values of all three variables $\gamma$, $\gamma_1$, $\gamma_2$ have term-set $G = \{G_1, G_2, G_3, G_4, G_5\}$, where

$G_1$ – "a stable tendency to reduce growth";
$G_2$ – "a tendency to reduce growth";
$G_3$ – "a tendency to stagnation";
$G_4$ – "a tendency to increase";
$G_5$ – "steady upward trend";

The functions of the accessory are also trapezoidal.

Phase 5. The transition from the numeric values of the metrics to numeric values assessments based on the common algorithm of the standard five-level indistinct [0, 1]–classifiers.

Phase 6. Linguistic recognition of the obtained numerical estimates in accordance with the definition of the term set, as well as the analysis of the obtained intensity estimates based on the numerical values of the indicators and recommendations for the correction of the situation.

## 3 Estimation of Intensity of Agricultural Production of the Rostov Region Taking into Account Positive and Negative Dynamics

The assessment of the level of intensity of agricultural production of the Rostov region is carried out on the basis of statistical data provided by the Ministry of agriculture of the Rostov region for 18 years, taking into account the positive and negative dynamics of their changes. The studied indicators of the level of intensification of production form four groups: the first group reflects the value of production assets on 1 hectare of agricultural land, the second group – energy resources; the third group – the characteristics of fixed production assets; the fourth group – current production costs. The role of total costs and the cost of fixed assets in the assessment of the level of intensification of production in agriculture is revealed through more detailed indicators: in the second group – through indicators of energy and energy supply and energy; in the third group – through the coefficients of updating agricultural machinery, the share of pedigree animals in the total population, as well as the density of cattle per 100 hectares of farmland (heads); in the third group – through indicators of the current production costs of crop and livestock. In General, to assess the level of intensification of production, there are statistical data on 14 indicators presented by the Ministry of agriculture and food of the Rostov region for 1996–2013.

Indicators of economic efficiency of intensification of production in agriculture of the Rostov region form six groups: the volume of gross income, the level of profitability, capital productivity, labor productivity, crop yields by groups, productivity of farm animals by groups.

Thus, the economic efficiency of the intensification of production in agriculture of the Rostov region should be evaluated by 10 indicators based on statistical data for 18 years. It is required on the basis of the obtained estimates of the level of intensification of production in agriculture of the Rostov region and the economic efficiency of the intensification of production in agriculture of the Rostov region to form a final comprehensive assessment of the intensification of production in agriculture of the Rostov region. The task is difficult to implement by the methods of classical mathematical modeling, primarily due to the need to take into account a significant amount of heterogeneous data. The source statistical material is a table of 24 rows and 18 columns; the contribution of each indicator to the final estimate is not equivalent. Indicators have different economic meaning, scale and dimension (for example, crop yields, measured in C/ha; capital productivity, rubles; level of profitability, %, etc.). In addition, there are currently no generally accepted standards for the indicators under review. "Positive" is the trend of constant positive growth of indicators;"negative" – zero or negative growth for each of the indicators.

Therefore, the calculation of the normalized values $x_i$ ($i = 1, 2, \ldots m_1$) of the studied indicators of the level of intensification of production in agriculture for the period of $N$ years is carried out on the basis of a scheme that integrates the time series of data for each of the indicators and takes into account the significance of different time periods due to weight coefficients:

$$x_i = 0,5 \left( 1 + \sum_{i=1}^{N-1} k_i I_i \right) \quad k_i = \frac{2(N-i)}{(N-1)N}, \tag{1}$$

where $k_i$ – the weight coefficients determined on the basis of the Fishburn rule, and the numbering of the time periods is in the reverse order (that is, in the example under consideration, the first period is the years 2012–2013, and the last, the 17th, the years 1996–1997). $I_i$ – integer functions defined in such a way that the value of "−1" corresponds to the negative increase in the $i$-th indicator; the value of "1"– a positive increase in the $i$-th indicator; the value of "0" – stagnation, zero increase. The analysis of the formula (1) shows that the scheme takes into account the time significance of each of the considered periods. If there is a positive increase in all periods, the amount in the bracket is equal to one, and the total numerical value of the indicator $x_i$ reaches its maximum and is equal to one. In the case of negative growth in all periods, the value of the indicator $x_i$ reaches a minimum and is equal to zero. The total weight of the time periods from 2008 to 2013 is 0.4902; from 2002 to 2007 – 0.3726; from 1996 to 2001 – only 0.1372. Thus, with a stable negative period in the last 5 years, the total numerical estimate for the I-th indicator is not higher than 0.5098 (which means "bad").

Similarly, the calculation of normalized values $y_i$ of the studied indicators of the level of economic efficiency of intensification of production in agriculture ($i = 1, 2, \ldots m_2$) is carried out. The calculation of complex assessments is carried out in accordance with the General scheme. The weights and the values of the membership functions for the first group of indicators are shown in the Table 1. For calculations based on the above schemes, a software package [10] was developed.

Based on Table 1 calculation and linguistic recognition $g_1$ = "estimates of the level of intensification of production in agriculture of the Rostov region": $g_1 = 0.6739$, $\mu(0.6739) = \mu_4(0.6739) = 1$. Therefore, $g_1$ = "assessment of the level of intensification of production in agriculture of the Rostov region" corresponds to the term $G_4$ – "growth trend". Based on the same Table 1 the critical indicators (those whose normalized value is less than 0.5) are determined. These are: (1) the density of cattle per 100 hectares of farmland (heads), $x_8 = 0.1831$; (2) the coefficient of renewal of agricultural machinery (%), combines of all types, $x_5 = 0.3297$; (3) the proportion of pedigree animals in the total population (%), sheep, $x_7 = 0.3636$; (4) energy, $x_2 = 0.4477$; (5) the proportion of pedigree animals in the total population (%), pigs, $x_6 = 0.5076$.

On the basis of the received estimates it is possible to draw a conclusion that in 1996–2013 in agriculture of the Rostov region the steady tendency to decrease in the following significant indicators characterizing level of intensification of agricultural production was observed: density of cattle on 100 hectares of farmland, specific weight of pedigree animals in the General livestock (sheep, pigs), coefficient of updating of agricultural machinery (combines of all types), power equipment.

Calculation and linguistic recognition $g_2$ = "evaluation of economic efficiency of intensification of production in agriculture of the Rostov region" is carried out similarly. It is established that $g_2 = 0.5861$, $\mu(0.5861) = \mu_4(0.5861) = 0.36$, $\mu(0.5861) = \mu_3(0.5861) = 0.64$.

**Table 1.** Weights and values of membership functions for the first group of indicators

| Indicators | Weights of indicators | Value of indicators | Terms of the linguistic variable of the first group | | | | |
|---|---|---|---|---|---|---|---|
| | | | $B_{i1}$ | $B_{i2}$ | $B_{i3}$ | $B_{i4}$ | $B_{i5}$ |
| 1. Cost of production. funds per 1 ha of agricultural land (ths. RUB./ha), $x_1$ | 1/4 | 0.96 | 0 | 0 | 0 | 0 | 1 |
| 2. Energy resources (HP/ha) | | | | | | | |
| 2.1. Energy security, $x_2$ | 1/8 | 0.45 | 0 | 0.02 | 0.98 | 0 | 0 |
| 2.2. Energy intensity, $x_3$ | 1/8 | 0.75 | 0 | 0 | 0 | 0.98 | 0.02 |
| 3. Characteristics of fixed assets | | | | | | | |
| 3.1. Coefficient of renewal of agricultural machinery (%) | | | | | | | |
| 3.1.1. Tractors, $x_4$ | 1/24 | 0.58 | 0 | 0 | 0.68 | 0.32 | 0 |
| 3.1.2. Combines of all types, $x_5$ | 1/24 | 0.33 | 0 | 1 | 0 | 0 | 0 |
| 3.2. The share of pedigree animals in the total population (%) | | | | | | | |
| 3.2.1. Pigs, $x_6$ | 1/24 | 0.51 | 0 | 0 | 1 | 0 | 0 |
| 3.2.2. Sheep, $x_7$ | 1/24 | 0.36 | 0 | 0.86 | 0.14 | 0 | 0 |
| 3.3. Density of cattle per 100 ha of farmland (heads), $x_8$ | 1/12 | 0.18 | 0.67 | 0.33 | 0 | 0 | 0 |
| 4. The amount of current production costs per 1 ha of agricultural land (ths. RUB/ha) | | | | | | | |
| 4.1. The size of the crop production costs per 1 ha of arable land (th. RUB/ha), $x_9$ | 1/8 | 0.91 | 0 | 0 | 0 | 0 | 1 |
| 4.2. The size of the current production costs of livestock for 1 goal. Farm animals (th. RUB per 1 head) | | | | | | | |
| 4.2.1. Dairy cattle, $x_{10}$ | 1/24 | 0.99 | 0 | 0 | 0 | 0 | 1 |
| 4.2.2. Beef cattle, $x_{11}$ | 1/24 | 0.79 | 0 | 0 | 0 | 0.56 | 0.44 |
| 4.2.3. Pigs, $x_{12}$ | 1/24 | 0.89 | 0 | 0 | 0 | 0 | 1 |
| Weights of terms of the linguistic variable of the first group | | | 0.06 | 0.11 | 0.20 | 0.16 | 0.48 |

Therefore, $g_2$ = "assessment of economic efficiency of intensification of production in agriculture of the Rostov region" corresponds to two terms $G_3$ – "the tendency to stagnation" and $G_4$ – "the tendency to growth", and the statement "there is a tendency to stagnation" is more true than the statement "there is a tendency to growth".

The indicators leading to the decrease in the final estimate, including: (1) the yield of calves per 100 main cows (heads), $y_9 = 0.2941$; (2) the yield of grain and leguminous without corn, $y_5 = 0.4443$; (3) the level of profitability (%), $y_2 = 0.4576$;

(4) the volume of production per ruble of invested capital (RUB.), $y_3 = 0.4902$. On the basis of the estimates we can conclude that in 1996–2013 in agriculture of the Rostov region there was a steady trend towards a decrease in the following significant indicators characterizing the level of economic efficiency of intensification of agricultural production: the yield of calves per 100 main cows, the yield of grain and legumes without maize, the level of profitability, the volume of production per ruble of invested capital.

Calculation $g$ = "complex assessment of intensification of production in agriculture of the region" is carried out. The term weights are defined as the arithmetic mean of the corresponding weights of the estimates $g_1$ and $g_2$: $g = 0.6301$, $\mu(0.6301) = \mu_4(0.6301) = 0,8$, $\mu(0.6301) = \mu_3(0.6301) = 0.2$.

Therefore, $g$ = "comprehensive assessment of the intensification of production in agriculture in the region" for 1996–2013 corresponds to two terms $G_3$ – "the tendency to stagnation" and $G_4$ – "the tendency to growth", the statement "there is a tendency to growth" is more true than the statement "there is a tendency to stagnation".

Thus, the proposed method allowed to perform a comprehensive analysis of the development of agriculture in the Rostov region in 1996–2013 on the basis of positive and negative dynamics of heterogeneous indicators of two groups: 14 indicators characterizing the intensification of production and 10 indicators – economic efficiency of intensification of production. The analysis was carried out taking into account the weight significance of indicators, as well as the ranking of the contribution of different time periods to the final assessment.

## 4  Conclusion

The method of estimation of ecological and economic efficiency of agricultural production in the region on the basis of standard five-level fuzzy [0, 1] –classifiers is offered. The practical significance of the method is that it allows to form a comprehensive assessment of the intensity of agricultural production in the region, as well as a comprehensive assessment of the impact of agricultural production on the ecology of the region on the basis of integrated estimates of time series of numerical statistical values of heterogeneous indicators, reflecting both the level and the rate of growth for the relevant periods; and the contribution of each of the indicators is estimated using a weight factor reflecting its importance.

Compared with existing estimation techniques, the proposed method of evaluation has a number of practically important advantages: (1) simple calculation; (2) into account when constructing estimates of a large number of diverse significant figures that allow variation depending on the available statistical material and features solve specific practical problems; (3) using only indicators objectively reflecting the efficiency of use of material and financial resources of agriculture and impacts of agricultural production on the ecology of the region; (4) the ability to vary the weight of the contribution of the studied indicators in the integrated assessment of the intensity of agricultural production in the region; (5) the adaptability and versatility of the proposed methodology, which allows to apply it to the assessment of the intensity of not only agricultural but also industrial production on a different scale; (6) the possibility of its use for ranking regions on a set of indicators, as well as forecasting their development, subject to the construction of trend models of the studied indicators.; (7) the ability to analyze on its basis the situation in the production under consideration and the formation of practical recommendations on the basis of the calculated integrated estimates of indicators.

# References

1. Bondarenko, L.: Ecological and economic efficiency and stability of grain production (based on the materials of the Krasnodar region). KubSU Publishing Center, Krasnodar (2000)
2. Minakov, I., Kulikov, I., Sokolov, V.: The Economics of Agriculture. Koloss, Moscow (2008)
3. Vartanyan, E.: Comparative analysis of methods for rating the effectiveness of agricultural enterprises. In: Interaction of business, state, science: a three-sided view on economic development, vol. 2, pp. 75–95 (2012)
4. Karminsky, A., Peresetsky, A., Petrov A.: Ratings in economy: methodology and practice. Finance and statistics, Moscow (2005)
5. Nedosekin, A.: Fuzzy Sets and Financial Management. AFA Library, Moscow (2003)
6. Stryukov, M., Sakharova, L., Alekseychik, T., Bogachev, T.: Methods of estimation of intensity of agricultural production on the basis of the theory of fuzzy sets. Int. Res. J. **7**(61), 123–129 (2017)
7. Sakharova, L., Stryukov, M., Akperov, I., Alekseychik, T., Chuvenkov A.: Application of fuzzy set theory in agro-meteorological models for yield estimation based on statistics. In: 9th International Conference on Theory and Application of Soft Computing, Computing with Words and Perception, Budapest, Hungary, pp. 820–829 (2017). Procedia Comput. Sc. **120**
8. Kramarov, S., Sakharova, L.: Management of complex economic systems by fuzzy classifiers method. Sci. Bull. South. Univ. Manag. **2**(18), 42–50 (2017)
9. Kramarov, S., Sakharova, L., Khramov, V.: Soft computing in management: management of complex multivariate systems based on fuzzy analog controllers. Sci. Bull. South. Univ. Manag. **3**(19), 42–51 (2017)
10. Albekov, A., Arapova, E., Karasev, D., Stryukov, M., Sakharova, L.: Program for evaluation of agricultural production intensity by means of fuzzy 5-point classifier. Certificate of registration of computer program № 2018613875. Federal service for intellectual property, Moscow (2018)

# Fuzzy-Logic Analysis of the Level of Comfort and Environmental Well-Being of the Urban Environment on the Example of Large Cities of Rostov Region

Elizabeth A. Arapova[1], Galina V. Lukyanova[1],
Lyudmila V. Sakharova[1(✉)], and Gurru I. Akperov[2]

[1] Rostov State University of Economics, B. Sadovaya str., 69,
344002 Rostov-on-Don, Russia
L_Sakharova@mail.ru
[2] Sothern University (IMBL), M. Nagibina pr., 33a/47,
344068 Rostov-on-Don, Russia

**Abstract.** The method of formation of the assessment of the level of comfort and ecological well-being in the region on the basis of five groups of heterogeneous indicators, such as: the state of the atmosphere in large cities, water quality, environmental management in the region, the impact of the environmental environment on the health of the population, the dynamics of disasters in the region. The time series of the corresponding statistical indicators for the Rostov region, as well as the data of remote sensing of the earth are used as initial data. The technique is based on the use of standard fuzzy multilevel [0, 1]- classifiers. The technique allows analyzing each sphere according to the complex of the most important heterogeneous indicators, as well as forming a numerical evaluation of the sphere based on the level of individual indicators and their dynamics. The system of fuzzy conclusions allows to aggregate the formed estimates into the final assessment of the region, which determines the level of ecological well-being of the region and allows to make an appropriate rating of the regions. The technique has versatility, allows you to change and Supplement the complexes of the studied indicators for each of the areas, as well as to change the weight coefficients of indicators in accordance with the opinion of environmental experts. The technique was tested on the example of the Rostov region.

**Keywords:** Standard fuzzy multi-level [0, 1] as a classifier
A comprehensive assessment · Indicators · Environmental

## 1 Introduction

At present, due to the increase of man-made loads on natural ecosystems, it is becoming increasingly important to systematically monitor the ecological state of the environment. The growth of megacities and the development of new industrial technologies inevitably leads to the violation of natural conditions that are comfortable for

© Springer Nature Switzerland AG 2019
R. A. Aliev et al. (Eds.): ICAFS-2018, AISC 896, pp. 643–650, 2019.
https://doi.org/10.1007/978-3-030-04164-9_84

humans. As a result, increased morbidity and mortality for environmental reasons, reduced quality of life [1].

The first stage on the way to environmental management is the monitoring of the state of the environment, currently being carried out in the Russian Federation at the state and municipal levels [2, 3]. The analysis presented in [2] and [3] provides comprehensive information about the ecological state of the region, but is difficult to imagine and does not allow to assess the level of comfort and environmental well-being in general, as well as to rank the regions by the indicated levels.

At present, there are a number of generally accepted methods that allow to estimate the degree of air pollution [4], soil [5] and water [6] on the basis of detailed criteria and scales on the basis of a set of specified indicators (chemical, microbiological, etc.). The methods of assessing chemical pollution of air, water and soil are reduced to comparing the recorded concentrations of chemical impurities with their maximum permissible concentrations in hazard classes, followed by the calculation of integrated indices. Such important environmental indicators as the volume of water used, emissions of pollutants into the atmosphere and water, waste disposal, morbidity of the population with environmentally-related diseases, indicators of environmental activities of municipalities are estimated, in general, on the basis of tables and charts.

There are attempts to develop methods that reflect the state of the region or municipality in the complex [7]; however, they take into account a small number of indicators and are based on a simple summation of points for several of the most important indicators (air pollution, water, soil). Therefore, one of the most important tasks of mathematical modeling at the present time is the creation of universal methods that allow to take into account a complex multi-level set of indicators that characterize the environmental state of the region, its dynamics, the impact of the environment on the population, as well as the effectiveness of the system of measures taken to improve the environmental situation.

This study proposes a method of comprehensive assessment of the level of comfort and environmental well-being in the region on the basis of five groups of heterogeneous indicators, such as: the state of the atmosphere in large cities, water quality, environmental management in the region, the impact of the environmental environment on the health of the population, the dynamics of disasters in the region. The time series of the corresponding indicators for the Rostov region were used as initial data. The method is based on the use of standard fuzzy multi-level [0, 1]- classifiers, which were used earlier in the financial analysis [8]. To analyze the ecology of the region, the author's developments in the field of assessing the state of complex systems based on the complexes of heterogeneous indicators were used [9, 10].

## 2    Methods of Water Quality Assessment in the Region

To assess the water quality in the region, statistical data on the proportion of water samples of water bodies that do not meet hygienic standards for chemical and microbiological indicators [1], Table 1.

**Table 1.** Application of pesticides in open ground of Rostov region-total, thousand hectares of crops

|  | 2012 | 2013 | 2014 | 2015 | 2016 |
|---|---|---|---|---|---|
| *Water bodies of the 1st category of water use* | | | | | |
| 1. By chemical indicators | 28.9 | 24.1 | 34.8 | 40.3 | 36.1 |
| 2. By microbiological indicators | 30.5 | 29.5 | 34.6 | 35.5 | 31.5 |
| *Decentralized drinking water supplies (wells, springs captai)* | | | | | |
| 3. By chemical indicators | 71.9 | 57.5 | 54.9 | 54.4 | 49.5 |
| 4. By microbiological indicators | 19.5 | 25.0 | 21.3 | 21.9 | 21.5 |
| *Water bodies of the 2nd category of water use (recreational water bodies)* | | | | | |
| 5. By chemical indicators | 47.3 | 35.4 | 50.4 | 41.4 | 42.2 |
| 6. By microbiological indicators | 42.7 | 41.2 | 47.3 | 50.1 | 42.1 |
| *Water bodies of the 2nd category of water use (sea)* | | | | | |
| 7. By chemical indicators | 83.2 | 64.8 | 89.6 | 98.5 | 100.0 |
| 8. By microbiological indicators | 86.1 | 80.6 | 56.7 | 37.7 | 59.8 |
| *Share of drinking water samples from the water supply network of settlements* | | | | | |
| 9. Towns by chemical indicators | 38.8 | 23.0 | 32.2 | 18.9 | 15.7 |
| 10. Towns by microbiological indicators | 1.8 | 1.2 | 0.6 | 1.6 | 1.2 |
| 11. Areas by chemical indicators | 51.5 | 37.8 | 42.0 | 37.5 | 35.2 |
| 12. Areas according to microbiological indicators | 6.4 | 4.4 | 4.3 | 3.9 | 4.0 |

## 2.1 Assessment of the Dynamics of Water Pollution in the Region

The linguistic variable: $g_v$ = "assessment of the dynamics of water pollution in the region" is introduced

A universal set for a linguistic variable is a numeric segment [0, 1]. The term set consists of five terms $G = \{G_1, G_2, G_3, G_4, G_5\}$, which conditionally assess the state of the system: $G_1$ – "steady tendency to improvement"; $G_2$ – "tendency to improvement"; $G_3$ – "stabilization"; $G_4$ – "tendency to deterioration"; $G_5$ – "steady tendency to deterioration".

Term membership functions are standard [9, 10]. The calculation of normalized values of the studied indicators for the period of $N$ years (5 years) is carried out on the basis of the scheme, taking into account the importance of different time periods due to weight coefficients:

$$x_i = 0.5\left(1 + \sum_{i=1}^{N-1} k_i I_i\right), \tag{1}$$

$$k_i = \frac{2(N-i)}{(N-1)N},$$

$k_i$–weight coefficients determined by the Fishburn rule; the numbering of time periods is carried out in the reverse order (that is, in the example under consideration, the first period is the years 2015–2016, and the last, the 4th, years 2012–2013). $I_i$ - integer functions defined in such a way that the value "1" corresponds to the increase in the I-th indicator (deterioration of the situation); the value "−1" – decrease in the i-th indicator; the value "0" – stabilization, lack of changes. Standard 5-point classifiers [9, 10] are used to form the estimate. It was found that for the Rostov region $g_v = 0.4196$, which corresponds to the term $G_3$ – "stabilization".

## 2.2 Assessment of Water Pollution in the Region

The linguistic variable: $g_u$ = "water pollution level in the region" is introduced. A universal set for a linguistic variable is a numeric segment [0, 1]. According to the standard classification [6], assessment of water pollution level, the term – set consists of three terms $G = \{G_1, G_2, G_3\}$: $G_1$ – "low pollution degree"; $G_2$ – "increased pollution degree"; $G_3$ – "high pollution degree". The membership functions correspond to their specification for the standard 3-point classifier [8, 9]. The calculation of normalized values of the studied indicators for the period of $N$ years (5 years) is carried out on the basis of the scheme, taking into account the importance of different time periods due to weight coefficients:

$$x = \sum_{i=1}^{N} k_i P_i, \tag{2}$$

$$k_i = \frac{2(N - i + 1)}{(N + 1)N},$$

$k_i$ – weight coefficients determined by the Fishburn rule; the numbering of time periods is carried out in the reverse order (that is, in the example under consideration, the first year is 2016, and the last is 2013); $P_i$ – the value of the indicator in the i–th year. It was found that for the Rostov region $g_u = 0.55$, which corresponds to the term $G_2$ – "increased degree of pollution".

## 2.3 Final Assessment of Water Quality in the Region

The numerical value of the resulting linguistic variable $g_2$ = "integrated assessment of water quality in the region" is calculated by aggregating the two estimates obtained in paragraphs 2.1 and 2.2: $g_v$ = "estimates of the dynamics of water pollution in the region" and $g_u$ = "estimates of the dynamics of water pollution in the region". The term set consists of five terms $G = \{G_1, G_2, G_3, G_4, G_5\}$, which conditionally assess the state of the system: $G_1$ – "the state of water in the region is excellent"; $G_2$ – "the state of water in the region is good"; $G_3$ – " the state of water in the region is satisfactory"; $G_4$ – "the state of water in the region is bad"; $G_5$ – "the state of water in the region is extremely bad". Standard 5-point classifiers are used to form the estimate. It was found that for the Rostov region $g_2 = 0.46$, which corresponds to the term $G_3$ – "the state of water in the region is satisfactory".

## 3    Methods of Assessment of the Ecological State of the Atmosphere in the Region

Time series of statistical data characterizing air pollution by the considered chemical impurities in the following major cities of the Rostov region: Rostov-on-Don, Azov, Taganrog, Tsimlyansk, the city of Shakhty (statistical data [1]) were used to form an assessment of the dynamics and level of air pollution. In addition, statistical data on emissions of pollutants into the air in the Rostov region for 2012–2016, thousands of tons from stationary sources, by economic activities were used. The numerical value of the final linguistic variable $g_1$ = "integrated assessment of the state of the atmosphere in the major cities of the region" is calculated by aggregation on the basis of standard five - point [0, 1]-classifiers of three estimates: $g_d$ = "assessment of the dynamics of air pollution in the region"; $g_s$ = "assessment of the level of air pollution in the region"; $g_z$ = "assessment of the dynamics of air pollution emissions". The methods of assessment are similar to those described above for assessing water pollution in the region. It was found that for the Rostov region $g_z$ = 0.463, which corresponds to the term $G_3$- "stabilization of the situation".

## 4    Environmental Management

### 4.1    Dynamics and Level of Production and Consumption Waste Utilization

Assessment of the dynamics of waste production and consumption is carried out on the basis of statistical data for 2012–2016 [2]: the number of generated production and consumption waste, tons; the amount of used production and consumption waste, tons; the number of neutralized production and consumption waste, tons; the share of used and neutralized production and consumption waste in the total volume of generated waste, %. The first three indicators reflect the dynamics of waste generation and disposal, and the fourth – their level. Parameters are considered to be equilibrium. The linguistic variable: $g_o$ = "assessment of dynamics and level of waste generation and utilization" is introduced. The formula (1) is used to calculate the normalized estimate of the 1st indicator); to aggregate the time series of the 2nd and 3rd indicators, a modified formula is used:

$$x_i = 0.5 \left( 1 - \sum_{i=1}^{N-1} k_i \, I_i \right), \tag{3}$$

$$k_i = \frac{2(N - i)}{(N - 1)N}.$$

The calculation of the normalized values of the 4-th indicator is realized on the basis of formula (2). Standard 5-point classifiers are used to form the estimate. It is established that for the Rostov region: $g_o = 0.453$, which corresponds to the term $G_3$ - "stabilization of the situation".

### 4.2   Dynamics of the Results of Control and Supervisory Activities

The analysis was carried out on the basis of statistical data for 2010–2016 [2], reflecting the results of control and supervisory activities: the number of inspections, identified and eliminated violations, issued and executed orders, the number of filed and collected fines, etc.. We introduce a linguistic variable: $g_n$ = "the evaluation of the dynamic results of inspection and oversight". It is revealed that for the Rostov region $g_n = 0.542$, which corresponds to the term $G_3$ – "stabilization of the situation".

### 4.3   Final Assessment of Environmental Management in the Region

The numerical value of the resulting linguistic variable $g_3$ = "integrated assessment of environmental management in the region" is calculated by aggregating the two assessments obtained in paragraphs 2.1 and 2.2: $g_o$ = "assessment of the dynamics and level of waste generation and disposal" and $g_n$ = "assessment of the dynamics of the results of supervisory activities". The universal set for a linguistic variable is a numeric segment [0, 1]. The term set consists of five terms that conditionally assess the state of the system $G = \{G_1, G_2, G_3, G_4, G_5\}$: $G_1$ – "excellent"; $G_2$ – "good"; $G_3$ – "satisfactory"; $G_4$ – "bad"; $G_5$ – "extremely bad". Standard 5-point classifiers are used to form the estimate. It was found that for the Rostov region $g_3 = 0.5$, which corresponds to the term $G_3$ – "satisfactory".

## 5   Environment and Health of the Population

A linguistic variable is introduced: $g_4$ = "assessment of the impact of the environmental environment on the health of the population". The assessment is calculated by aggregation using standard five-level [0, 1]- classifiers of estimates: $g_m$ = "assessment of mortality dynamics for environmentally caused reasons" and $g_p$ = "assessment of physical factors at facilities that do not meet sanitary and hygienic requirements". The term set consists of five terms $G = \{G_1, G_2, G_3, G_4, G_5\}$, which conditionally assess the state of the system: $G_1$ - "no negative impact of the environment on the health of the population has been revealed"; $G_2$ – "revealed mild negative impact on the health of the population"; $G_3$ – "revealed a medium significant negative influence of ecology on health of the population"; $G_4$ – "identified a clearly expressed negative influence on the health of the population"; $G_5$ – "identified a negative influence of ecology on health of the population". It was found that for the Rostov region $g_4 = 0.135$, which corresponds to the term $G_1$ - "the negative impact of the environment on the health of the population is not revealed".

## 6 Number of Emergency Situations of Natural, Man-Made and Biological-Social Character

The linguistic variable: $g_5$ = "assessment of the dynamics of disasters in the region" is introduced. Term-set consists of five terms $G = \{G_1, G_2, G_3, G_4, G_5\}$, conditionally assessing the state of the system: $G_1$ – "steady dynamics to reduce the number of catastrophes, excellent"; $G_2$ – "dynamics to reduce the number of catastrophes, good"; $G_3$ – "stabilization of the situation, satisfactory"; $G_4$ – "dynamics to increase the number of catastrophes, bad"; $G_5$ – "steady dynamics to increase the number of catastrophes, extremely bad". It is established that for the Rostov region $g_5 = 0,4$, which corresponds equally to two topics: $G_2$ – "the dynamics to reduce the number of disasters, good"; $G_3$ – "stabilizing the situation, satisfactory".

## 7 Aggregation of Data into a Final Integrated Assessment

The linguistic variable: $g$ = "assessment of the level of efficiency and environmental well-being in the region" is introduced. Term-set of five terms $G = \{G_1, G_2, G_3, G_4, G_5\}$, conditionally assessing the state of the system: $G_1$ – "the level of comfort and environmental well - being in the region is good (normal)"; $G_2$ – "the level of comfort and environmental well - being in the region is good (normal)"; $G_3$ – "the level of comfort and environmental well-being in the region is satisfactory (tense situation)"; $G_4$ –"the level of comfort and environmental well - being in the region is bad (crisis) "; $G_5$ – "the level of comfort and environmental well-being in the region is extremely bad (disaster)."

The integrated assessment is formed on the basis of the indicators obtained in paragraphs 1–4: $g_1$ = "integrated assessment of the state of the atmosphere in the major cities of the region"; $g_2$ = "integrated assessment of water quality in the region"; $g_3$ = "integrated assessment of environmental management in the region"; $g_4$ = "assessment of the impact of the environmental environment on public health"; $g_5$ = "assessment of the dynamics of disasters in the region". Normalized values of indicators correspond to the numerical values of linguistic variables. It is proposed to assume that the order of the parameters corresponds to the order of decreasing importance; as a result, it is proposed to calculate their weight coefficients based on the Fishburn rule. Based on the calculations it was found that for the Rostov region $g = 0.5333$, which corresponds to the term $G_3$ – "the level of comfort and environmental well-being in the region is satisfactory (tense situation)".

## 8 Conclusion

The method of formation of the level of comfort and environmental well-being in the region on the basis of five groups of heterogeneous indicators, including: the state of the atmosphere in large cities, water quality, environmental management in the region, the impact of the environmental environment on the health of the population, the dynamics of disasters in the region. The technique is tested using time series of statistical data for the Rostov region. The method is based on the use of standard fuzzy multi - level [0, 1] - classifiers, is universal and can be used to assess the environmental condition of other regions. The technique allows for modification of the complex of indicators for which the evaluation is performed, as well as the variation of the weight coefficients of the indicators. The advantage of the proposed method compared to existing methods of assessment is the integration of disparate heterogeneous indicators in the final assessment, which allows to assess the level of comfort and environmental well-being of the region as a whole.

# References

1. Ushakov, I., Volodin, A., Chikova, S., Zueva, T.: Medical aspects of protection of public health from the harmful effects of environmental factors. Hyg. Sanitation **6**, 29–34 (2005)
2. Environmental Bulletin of the Don "On the state of the environment and natural resources of the Rostov region in 2011/2016". Government-in the Rostov region, Rostov-on-Don. http://минприродыро.рф/state-of-the-environment/ekologicheskiy-vestnik/. Accessed 08 July 2018
3. Report on the state of sanitary and epidemiological welfare of the population in the Rostov region in 2016. Federal service for supervision of consumer rights protection and human welfare. Rostov-on-Don. http://61.rospotrebnadzor.ru/index.php?Itemid=116&catid=96:2009-12-30-08-03-55&id=6813u/index.php?Itemid=116&catid=96:2009-12-30-08-03-55&id=6813:-q-2016q&option=com_content&view=article. Accessed 30 Apr 2018
4. Bespamyatnov, G., Krotov, Y.: Maximum permissible concentrations of chemical substances in the environment, Chemistry, Leningrad (1985)
5. Ants, A., Karryev, B.: Evaluation of the Ecological State of the Soil. Krismas+, Moscow (2008)
6. Mishon, V.: Surface Water of the Earth: Resources, Use, Protection. VSU publishing House, Voronezh (1996)
7. Petrishchev, V., Dubovskaya, S.: Method of complex assessment of ecological state of urban areas. News Samara Sci. Cent. Russ. Acad. Sci. **15**(3), 234–238 (2013)
8. Nedosekin, A.: Fuzzy Sets and Financial Management. AFA Library, Moscow (2003)
9. Sakharova, L., Stryukov, M., Akperov, I., Alekseychik, T., Chuvenkov A.: Application of fuzzy set theory in agro-meteorological models for yield estimation based on statistics. In: 9th International Conference on Theory and Application of Soft Computing, Computing with Words and Perception, Budapest, Hungary, pp. 820–829 (2017). Procedia Comput. Sci. **120**
10. Kramarov, S., Smirnov, A., Sokolov, S.: System methods of analysis and synthesis of intelligent adaptive control. INFRA-M, Moscow (2016)

# Comparative Assessment of the Transport Systems of the Regions Using Fuzzy Modeling

Taras Bogachev[1]([✉]), Tamara Alekseychik[1],
and Viktor Bogachev[2]

[1] Rostov State University of Economics, Bolshaya Sadovaya Street 69,
344002 Rostov-on-Don, Russia
bogachev73@yandex.ru
[2] Rostov State Transport University (RSTU), Rostovskogo Strelkovogo Polka
Narodnogo Opolcheniya sq. 2, 344038 Rostov-on-Don, Russia

**Abstract.** A complex study of the state of the main transport systems of the subjects of the Southern Federal Region of the Russian Federation was carried out using fuzzy modeling. The information base for the study was the annual data on the main indicators of the characteristics of vehicles in the region for 2010–2016. The investigated indicators were considered in the form of relative indicators as a quotient of the division of their values into the sizes of the areas of the relevant subjects of the region. Then, estimates of the values of the indicators for 2016 were constructed as expected or desired in comparison with the largest values of the corresponding indicators per unit area of the subjects. Using the corresponding weight coefficients of the indicators and their expected estimates, fuzzy sets for the investigated indicators of each region's subject were determined. In order to identify the best subject of the region in terms of the state of the transport system in 2016, the maximin convolution method is applied. The received estimations have allowed to rank subjects of the region on a level of development of their transport systems. Also the analysis of transport systems of subjects and in the whole region is carried out taking into account the positive and negative dynamics of indicators based on their growth rates. The results of this analysis create the possibility of developing a strategy for the development of the region's transport system.

**Keywords:** Fuzzy set theory · Maximin convolution method
Analysis of transportation systems · Complex valuation

## 1 Introduction

Transport occupies one of the leading places in the economy of any region.

In this paper we propose a method for building a comprehensive assessment of the transport system of the region with the help of the fuzzy set theory, which allows not only to identify the level of the state of the transport complex, but also to rank the subjects of the region. This technique is a modification of the methods used to select the mode of transport for regional freight and passenger transport [1–3].

© Springer Nature Switzerland AG 2019
R. A. Aliev et al. (Eds.): ICAFS-2018, AISC 896, pp. 651–658, 2019.
https://doi.org/10.1007/978-3-030-04164-9_85

## 2  Purpose of the Study

Let us illustrate the proposed method by the South Federal Region (SFR) of the Russian Federation. In this task the subjects of the SFR are alternatives identified as follows: $a_1$ – Adigeya Republic, $a_2$ – Kalmikiya Rebublic, $a_3$ – Krasnodar Region, $a_4$ – Astrakhan Region, $a_5$ – Volgograd Region, $a_6$ – Rostov Region.

To analyze the main types of transport in the region: railroad and road. The information base for the study is the annual data on the main indicators SFR transport systems for 2010–2016 [4]:

$X_1$ – dispatch of goods by public railway transport (million tons)
$X_2$ – departures of passengers by railway transport for general use (thousand)
$X_3$ – railway density at the end of the year (km of tracks per 10000 $km^2$ of territory)
$X_4$ – carriage of goods by road transport organizations of all activities(million tons)
$X_5$ –turnover of motor transport organizations of all activities (million ton/km)
$X_6$ – transportation of passengers and passenger turnover of public buses (million people)
$X_7$ –passenger turnover of public buses (million passenger/km)

The purpose of the study is to build an assessment of the SFR transport system for 2016.

## 3  Creation a Comprehensive Assessment of the State of the SFR Transport System for 2016

Due to the fact that the subjects of the region have quite different areas, the proposed above indicators will be considered in the form of relative indicators as distinct from the division of their values into the areas of the relevant subjects of the region (Table 1).

**Table 1.** The estimated values of the indicators for 2016 for the subjects of the region

| Production figures | The values of the indicators for 2016 per unit area of the subject of the region per 10,000 square km | | | | | |
|---|---|---|---|---|---|---|
| | $a_1$ | $a_2$ | $a_3$ | $a_4$ | $a_5$ | $a_6$ |
| $X_1$ | 1,2821 | 0 | 4,2252 | 3,4490 | 1,4792 | 1,8119 |
| $X_2$ | 275,641 | 0,4016 | 2608,08 | 228,163 | 361,3818 | 1321,584 |
| $X_3$ | 262,8205 | 2,9451 | 37,4834 | 26,1224 | 12,66601 | 18,0198 |
| $X_4$ | 9,48718 | 0,08032 | 10,4371 | 0,8776 | 1,3640 | 5,9801 |
| $X_5$ | 562,8205 | 7,4966 | 999,6026 | 81,2245 | 176,3508 | 564,2574 |
| $X_6$ | 10,1282 | 4,5382 | 41,0728 | 15,8163 | 20,7174 | 31,6832 |
| $X_7$ | 158,9744 | 51,2718 | 571,6556 | 900,2041 | 394,0655 | 354,1584 |

In determining the assessments of the state of the transport system of the subjects of the region, the proposed indicators have different significance. In this connection it is offered to enter weight coefficients of indicators for each subject of the SFR as a share of values of indicators in the corresponding value of the indicator as a whole on the region (Table 2).

**Table 2.** Weight coefficients of indicators of the subjects of the region for 2016

| Production figures | Weights of indicators of subjects of the region for 2016 | | | | | |
|---|---|---|---|---|---|---|
| | $a_1$ | $a_2$ | $a_3$ | $a_4$ | $a_5$ | $a_6$ |
| $X_1$ | 0,104681 | 0 | 0,344989 | 0,281612 | 0,120777 | 0,147942 |
| $X_2$ | 0,057482 | 8,38E-05 | 0,543888 | 0,047581 | 0,075362 | 0,275603 |
| $X_3$ | 0,729941 | 0,00818 | 0,104104 | 0,072551 | 0,035178 | 0,050047 |
| $X_4$ | 0,33611 | 0,002846 | 0,369764 | 0,03109 | 0,048325 | 0,211866 |
| $X_5$ | 0,235317 | 0,003134 | 0,417937 | 0,03396 | 0,073733 | 0,235918 |
| $X_6$ | 0,081708 | 0,036611 | 0,33135 | 0,127596 | 0,167135 | 0,2556 |
| $X_7$ | 0,065413 | 0,021097 | 0,235217 | 0,370404 | 0,162145 | 0,145724 |

Then, let's build the estimates of the indicators $X_1$, $X_2$, $X_3$, $X_4$, $X_5$, $X_6$, $X_7$ as expected or desired in comparison with the highest values of the corresponding indicators per unit area of subjects for the period 2010–2016. For this purpose we will determine the proportion of values of indicators for 2016 per unit area compared to the highest values of the relevant indicators for the study period (Tables 3 and 4).

**Table 3.** The maximum values of the indicators of the region's subjects for 2010–2016

| Production figures | The maximum values indicators of subjects | | | | | |
|---|---|---|---|---|---|---|
| | $a_1$ | $a_2$ | $a_3$ | $a_4$ | $a_5$ | $a_6$ |
| $X_1$ | 2,820513 | 0,013387 | 5,298013 | 5,530612 | 1,514615 | 2,584158 |
| $X_2$ | 314,1026 | 1,606426 | 3650,331 | 327,7551 | 434,0124 | 1902,178 |
| $X_3$ | 262,8205 | 2,945114 | 37,48344 | 26,12245 | 12,66608 | 18,0198 |
| $X_4$ | 9,615385 | 0,254351 | 16,10596 | 1,285714 | 1,93977 | 6,079208 |
| $X_5$ | 676,9231 | 12,58367 | 1042,649 | 91,83673 | 181,7538 | 625,8416 |
| $X_6$ | 32,5641 | 4,819277 | 42,17219 | 21,38776 | 21,94863 | 40,55446 |
| $X_7$ | 335,8974 | 59,57162 | 599,2053 | 1098,98 | 394,0655 | 437,0297 |

**Table 4.** Expected assessments of indicators of subjects of the region for 2016

| Production figures | Expected assessments of indicators of subjects of the region | | | | | |
|---|---|---|---|---|---|---|
| | $a_1$ | $a_2$ | $a_3$ | $a_4$ | $a_5$ | $a_6$ |
| $X_1$ | 0,45454 | 0 | 0,7975 | 0,62362 | 0,97661 | 0,70115 |
| $X_2$ | 0,87755 | 0,25 | 0,71448 | 0,69614 | 0,83265 | 0,69477 |
| $X_3$ | 1 | 1 | 1 | 1 | 1 | 1 |
| $X_4$ | 0,98666 | 0,31579 | 0,64803 | 0,68254 | 0,70319 | 0,98371 |
| $X_5$ | 0,83144 | 0,59574 | 0,95871 | 0,88444 | 0,97027 | 0,90160 |
| $X_6$ | 0,31102 | 0,94167 | 0,97393 | 0,73950 | 0,94391 | 0,78125 |
| $X_7$ | 0,47328 | 0,86067 | 0,95402 | 0,81913 | 1 | 0,81038 |

Will process the received information using the tools of fuzzy set theory [2, 5]. Using the appropriate weighting factors of the indicators and their expected estimates, we define fuzzy sets for the studied indicators of each subject of the region:

$$\mu_{X1} = 0,0476/1,2821 + 0/0 + 0,2751/4,2252 \\ + 0,1756/3,4490 + 0,1180/1,4792 + 0,1037/1,1819;$$

$$\mu_{X2} = 0,0504/275,641 + 2,09.10^{-5}/0,4016 + 0,3886/2608,08 \\ + 0,0331/228,163 + 0,0628/361,382 + 0,1915/1321,584;$$

$$\mu_{X3} = 0,7299/262,821 + 0,0082/2,9451 + 0,1041/37,4834 \\ + 0,0726/26,1224 + 0,0352/12,6660 + 0,0500/18,0198;$$

$$\mu_{X4} = 0,3316/9,4872 + 0,0009/0,0803 + 0,2396/10,4371 \\ + 0,0212/0,8776 + 0,0340/1,3640 + 0,2084/5,9801;$$

$$\mu_{X5} = 0,1957/562,8205 + 0,0019/7,4966 + 0,4007/999,6026 \\ + 0,0300/81,2245 + 0,0715/176,3508 + 0,2127/564,2574;$$

$$\mu_{X6} = 0,0254/10,1282 + 0,0345/4,5382 + 0,3227/41,0728 \\ + 0,0944/15,8163 + 0,1578/20,7174 + 0,1997/31,6832;$$

$$\mu_{X7} = 0,0310/158,9744 + 0,0182/51,2718 + 0,2244/571,6556 \\ + 0,3034/900,2041 + 0,1621/394,0655 + 0,1181/354,1584.$$

In order to identify the best subject of the region in terms of the state of the transport system in 2016, we apply the method of maximum convolution of fuzzy set theory. The set of optimal alternatives, taking into account the varying degree of significance of the indicators, is defined as the intersection of fuzzy sets, which corresponds to the selection of a minimum value for each subject of the region:

$B = \{0{,}0254; 0; 0{,}1026; 0{,}02122; 0{,}0340; 0{,}0500\}$

Then we get

$\max\{0{,}0254; 0; 0{,}1026; 0{,}02122; 0{,}0340; 0{,}0500\} = 0{,}1026.$

A set of optimal alternatives can be considered as a complex assessment of the state of the transport system of the SFR in 2016 by the studied indicators. This set makes it possible to rank the subjects of the district by the level of the state of the transport system. The best state in the Krasnodar Region, then in the Rostov and Volgograd regions, Adigeya Republic, Astrakhan Region; the worst – in Kalmikiya Rebublic.

## 4   Assessment of Transport Systems of Constituent Entities of the SFR Given the Growth Rate over the 2010–2016

The analysis of the transport systems of the SFR and the whole district is carried out taking into account the positive and negative dynamics for 2010–2016 on the basis of the chain growth rates for each of the above main indicators of the characteristics of the transport systems of the region and the whole district.

Positive dynamics is considered to be subject to a positive increase in the indicator, negative-subject to a negative increase in the indicator. To take into account such dynamics, an integer function, $I_i$ – integer functions determined in such a way that the value "-1" corresponds to a negative increase in the $i$-th indicator; value "1" – a positive increase in the $i$-th indicator; value "0" – stagnation, zero increase.

For normalization of indicators of rates of increase of the probed indexes are introduced weighting coefficients according to the Fishburne rule:

$$k_i = \frac{2(n - i + 1)}{(n + 1)n},$$

moreover, the numbering of time periods is carried out in reverse order, that is, the first period is the period of 2015–2016, and the last period, 2010–2011, ($n = 6$).

Then, taking into account the significance of different time periods and the dynamics of the studied indicators, it is possible to obtain estimates of the state of transport systems of SFR subjects for 2010–2016:

$$Y_j = 0{,}5\left(1 + \sum_{i=1}^{6} K_i I_i\right) \cdot T_{\prod_j}$$

where $j = 1, \ldots, 6$ – number of the region subject, $T_{\prod_j}$ – growth rate of the corresponding indicator for 2016 in comparison with 2010 (Table 5).

**Table 5.** Estimates of the state of transport systems of the subjects of the SFR, taking into account the dynamics of their key indicators for 2010–2016.

| Production figures | Estimates of the state of the transport systems of subjects taking into account the dynamics of indicators | | | | | |
| --- | --- | --- | --- | --- | --- | --- |
| | $a_1$ | $a_2$ | $a_3$ | $a_4$ | $a_5$ | $a_6$ |
| $X_1$ | −0,104 | – | −0,06 | −0,144 | 0,017 | −0,08 |
| $X_2$ | 0,317 | −0,393 | 0,13 | −0,157 | −0,028 | −0,04 |
| $X_3$ | 0 | 0 | 0,01 | 0,02 | 0 | 0 |
| $X_4$ | 0,116 | −0,456 | −0,20 | −0,09 | 0,018 | 0,017 |
| $X_5$ | 0,354 | −0,212 | 0,001 | 0,117 | 0,365 | 0,217 |
| $X_6$ | 0 | 0 | 0 | 0 | 0 | 0 |
| $X_7$ | 0 | 0 | 0 | 0 | 0 | 0 |

On the basis of Table 5 we can draw the following conclusions.

1. For the indicator the dispatch of goods by public railway transport more stable growth is observed only in the Volgograd Region (the estimate is equal to 0,017), in Kalmikiya Rebublic this type of transport is practically not developed, in other subjects this indicator gradually decreases.
2. To measure departures of passengers by railway transport for general use in almost all regions there is a downward trend in this indicator, except for the Krasnodar Region.
3. Assessment for the indicator of transportation of goods by road transport of organizations of all types of activity is the highest in Adigeya Republic, but with a more detailed analysis of the growth rates, this is due to almost a sharp increase in the indicator in 2016; estimates in the Volgograd and Rostov regions are also positive and approximately the same, but insignificant. This suggests that in these regions first, there was a decline and then a slight increase in 2016. In other regions this indicator gradually decreased.
4. For the indicator of freight turnover of motor transport of organizations of all types of activity all estimates are positive, except for Kalmikiya Rebublic. This indicates the potential for the expansion of road transport in these regions, which is confirmed by a more detailed analysis of the growth rates for the period 2010-2016.
5. The estimates of the public bus passenger service indicator are very low, reflecting the absence of a trend of growth or decline in this indicator. This is confirmed by the analysis of the growth rate of this indicator for the period 2010–2016.
6. Similar to very low estimates for the passenger turnover of public buses, which reflects the absence of a trend of growth or decline in this indicator for the period 2010–2016.

Thus, the assessment of the state of the transport system of subjects with taking into account the growth rates of the studied indicators, it is possible to make a forecast of the existence or absence of a tendency of growth or reduction of indicators, which creates the possibility of developing a strategy for the development of the transport system of the region in an appropriate direction.

# 5   Conclusion

On the basis of the conducted researches the following algorithm of a technique of construction of a complex estimation of a state transport system of the region is proposed.

1. Select the subjects of the study region.
2. To carry out selection of the main indicators on which the integrated assessment of the condition of transport system of the region will be performed. Appropriate specialists are involved for this purpose.
3. Data collection on the main indicators characterizing various aspects of the transport system in the region.
4. Determination the weights of indicators.
5. Determination of shares of calculated values of indicators in comparison with the greatest (least) or desirable values of corresponding indicators for the investigated period.
6. The definition of fuzzy sets for the considered parameters.
7. Construction of a comprehensive assessment of the transport system of the region for the nearest period of development by the maximin convolution method.
8. Assessment of the state of transport systems of subjects of the region taking into account the growth rate during the study period.
9. Analysis of the conducted research and development of recommendations on the implementation of the necessary measures to improve the transport system of the region and the region as a whole.
10. The proposed methodology for analyzing the state of the transport system of the region allows to determine not only its comprehensive assessment, but also to make a forecast of trends in the studied indicators in order to develop strategies for the further development of both the transport system of the region and the region as a whole.

**Acknowledgements.** The research is conducted with a support of the Russian Foundation for Basic Research (#17-20-04236 ofi_m_RJD).

# References

1. Alekseychik, T., Bogachev, T., Bogachev, A., Bruhanova, V.: The choice of transport for freight and passenger traffic in the region, using econometric and fuzzy modeling. In: 9th International Conference on Theory and Application of Soft Computing, Computing with Words and Perception, ICSCCW 2017, 22–23 August 2017, Budapest, Hungary, vol. 120, pp. 830–834 (2017). Procedia Computer Science
2. Bogachev, T., Alekseychik, T., Stryukov, M., Gnedash, E.: Analysis of optimal transport vehicles for transporting in the Rostov Region. In: International Applied Science Conference « Russia and the EU: Development Paths and Potential » , pp. 788–792. RSUE, Rostov-On-Don (2016)

3. Alexeychik, T., Gnedash, E.: Decision-making by type of transport choice for passenger transportation in the region through econometric and fuzzy modeling. In: International Applied Science Conference « The role of innovations in the transformation of modern science » , vol. 1, pp. 17–20. Aeterna, Ufa (2017)
4. The ROSSTAT. Region of Russia. Socio-economic indicators. Statistical compendium. http://www.gks.ru/wps/wcm/connect/rosstat_main/rosstat/ru/statistics/publications/catalog/doc_1135087342078. Accessed 21 June 2018
5. Andreychikov, A., Andreichikova, O.: Analysis, Synthesis, Planning of Decisions in Economy. Finance and statistics, Moscow (2000)

# Bifurcation Analysis and Synergetic Management of the Dynamic System "Intermediary Activity"

Alexander V. Bratishchev[1], Galina A. Batishcheva[2(✉)], and Maria I. Zhuravleva[2]

[1] Don State Technical University, Gagarin square, 1, Rostov-on-Don, Russia
[2] Rostov State University of Economics,
Bolshaya Sadovaya Street, 69, 344002 Rostov-on-Don, Russia
gbati@mail.ru

**Abstract.** The full bifurcation analysis of the mathematical model of the dynamic system "Intermediary activity" proposed by V. P. Milovanov is carried out in the article. Using the poincaré transform, the behavior of trajectories at infinity is investigated. Two logically possible phase portraits of the system were obtained using theoretical analysis and numerical experiment in the Matlab package. The system of additive control of both cash and commodity flows is constructed by the method of analytical construction of aggregated regulators to achieve a given dynamic equilibrium from an arbitrary initial state. Dedicated class a valid reachable States. The stability of such state as a whole is proved. This model makes it possible to predict the development of the process for any given initial state of the system in advance, as well as to control the parameters of the system for designing a given dynamic equilibrium in advance.

**Keywords:** Cash and commodity flows · Autonomous system
Equilibrium state · Phase portrait · Parameter space · Asymptotic stability
Aggregated variable

## 1 Introduction

The presence of a mathematical model of interconnected economic processes makes it possible to involve methods of modern nonlinear dynamics and control theory for their analysis and control synthesis [1, 2]. In the framework of the theory of dynamical systems the full analysis of the system is its bifurcation analysis [3]. The results of the analysis naturally lead to the problem of managing this relationship. One of the modern methods of dynamic system control is the method of synergetic control developed, by Kolesnikov [4]. In the paper [5] considered the bifurcation analysis and synergetic control system, "Gross product - labour resource". In the present article the same mathematical apparatus is applied to the system of "Intermediary activity" [6]. Note that some dynamic systems of psychology are realized by the same mathematical model [7]. The mathematical model of the economic system "Intermediary activity" is given by the Autonomous system of differential equations:

© Springer Nature Switzerland AG 2019
R. A. Aliev et al. (Eds.): ICAFS-2018, AISC 896, pp. 659–667, 2019.
https://doi.org/10.1007/978-3-030-04164-9_86

$$\begin{cases} x'_f = a_1 - a_2 xy + a_3 xy^2 \\ y_t = b_1 - b_2 xy \end{cases}$$

Here: $x(t)$ - the amount of money at the disposal of the individual; $y(t)$ - the quantity of goods of type Y traded on the market; $a_1$ - entrepreneur's income not related to the sale of goods of type Y; $a_3 xy^2$ - is an income that an entrepreneur has by buying a type Y product on the market and organizing a supply chain for its resale; $a_2 xy$, $b_2 xy$ - competitive and exchange members, show how much money and goods type Y decreases from the market as a result of sales; $b_1$ - the value of the constant influx of product Y on the market per unit of time.

This system does not have an explicit solution, however for a specialist, answers to questions are important: how over time will change the value $x(t)$ – the amount of money and $y(t)$ – the amount of goods type y, if known their values at any fixed time; how you can control the flow of goods $y(t)$, regulating the amount of cash $x(t)$ from the entrepreneur and Vice versa.

## 2  Full Bifurcation Analysis of the System

State of equilibrium would still $S = \left( \dfrac{a_3 b_1^2}{b_2(a_2 b_1 - a_1 b_2)}, \dfrac{a_2 b_1 - a_1 b_2}{a_3 b_1} \right)$, if $a_2 b_1 - a_1 b_2 \neq 0$.
Under this condition, it is neither neutral nor multiple.

(1) If $a_2 b_1 - a_1 b_2 = 0$, then the final state of equilibrium system is missing, and all starting in the first-quarters trajectory go beyond its the limits.

Denote: $a = \dfrac{a_2 b_1 - a_1 b_2}{a_3 b_1} \Rightarrow S = \left( \dfrac{b_1}{b_2 a}, a \right)$.

The study of the nature of the equilibrium state with the help of Lyapunov's theorem gives such a result.

(2) If $a_2 b_1 - a_1 b_2 < 0$, that S is in the third quarter of the phase plane and is a saddle. Numerical experiment shows (Fig. 1) that all the trajectories of the first quarter goes on $(+\infty, 0)$.

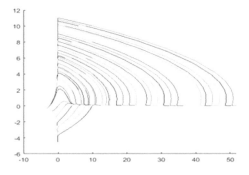

**Fig. 1.**  Phase portrait of the system (1) in the first quarter in case $a_2 b_1 - a_1 b_2 < 0$.

(3)  If $a_2b_1 - a_1b_2 > 0$, that S is in the first quarter of the phase plane and two variants are possible:

(a)  if $\left(\frac{a_1b_2a}{b_1} + \frac{b_1}{a}\right)^2 - 4a_3b_1a \geq 0$ that S – a stable node,

(b)  if $\left(\frac{a_1b_2a}{b_1} + \frac{b_1}{a}\right)^2 - 4a_3b_1a < 0$, that S- a steady focus.

The structure of the phase portrait in the first quarter is shown in Fig. 2:

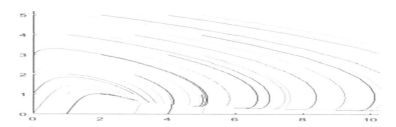

**Fig. 2.** Phase portrait of the system (1) in the first quarter in case $a_2b_1 - a_1b_2 > 0$.

In order to obtain a complete phase portrait, it is necessary to investigate the equilibrium States at infinity. To explore the vicinity of points $(0, \pm\infty)$ will do replacement variables in the system (1): $x = \frac{1}{u}, y = \frac{v}{u}$. Have

$$\begin{cases} u'_t = u(-a_1u^3 + a_2uv - a_3v^2) \\ v_t = b_1u^3 - b_2uv - a_1u^3v + a_2uv^2 - a_3v^3 \end{cases} \quad (1)$$

According to [8], the trajectories that enter the equilibrium state $S_1 = (0,0)$, must lie on the coordinate axes. By substituting, we make sure that they lie on the axis $u = 0$, that divides the phase plane into the right half-plane with two saddle sectors (Fig. 3(A)) and the left half-plane with two stable node sectors (Fig. 3(B)).

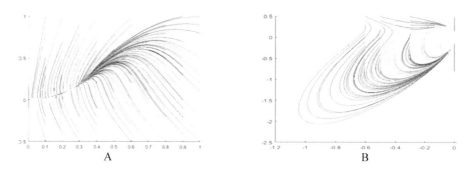

**Fig. 3.**  (A) Phase portrait of the system (2) in the right half-plane. (B) Phase portrait of the system (2) in the left half-plane.

These figures depict a phase portrait of the system (1) with parameters $a_1 = a_3 = b_2 = a_2 = 1, b_1 = 2$ (the node case) in the vicinity of the equilibrium state $S = (0,0)$.

Phase portrait of the system (1) with parameters $a_1 = a_3 = b_2 = 1, a_2 = b_1 = 2$ (focus case) in the vicinity of the equilibrium state $S_1 = (0,0)$ has a similar view (Fig. 4).

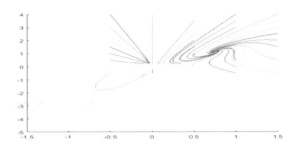

**Fig. 4.** Phase portrait of the system (2) in the case of focus.

Investigation of the behavior of trajectories at infinity in the vicinity of points $(0, \pm\infty)$. Make the change of variables in the system (1) $x = \frac{u}{v}, y = \frac{1}{v}$. Have

$$\begin{cases} u'_t = a_1 v^3 - a_2 uv + a_3 u - b_1 uv^3 + b_2 u^2 v \\ v_t = -b_1 v^4 + b_2 uv^2 \end{cases} \tag{2}$$

The theoretical analysis shows [5] that the equilibrium state $S_2 = (0,0)$ is a saddle-node. The trajectories lie on an axis $v = 0$ that divides the phase plane into the upper half-plane with two saddle sectors (Fig. 5(A)) and the lower half-plane with unstable nodes (Fig. 5(B)).

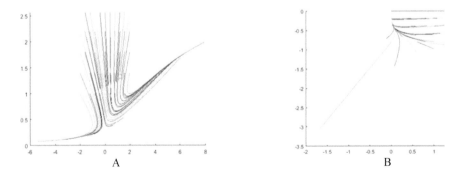

**Fig. 5.** (A) Phase portrait of the system (3) in the upper half-plane, (B) Phase portrait of the system (3) in the lower half-plane.

These figures depict a phase portrait of the system (2) with parameters $a_1 = a_2 = b_2 = a_2 = 1, b_1 = 2$ in the vicinity of the equilibrium state $S_1 = (0,0)$.

Thus, at infinity we have 4 equilibrium States. Schemes of phase portraits in the case of the node and in the case of focus are shown in Fig. 6(A) and (B), respectively.

**Conclusion.** Thus in both cases the phase plane is divided into three cells by trajectories connecting the point $(0, -\infty)$ with the points $(-\infty, 0)$ and $(0, +\infty)$ respectively.

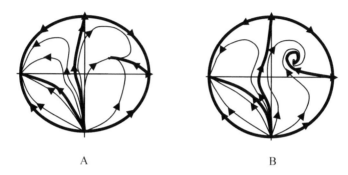

A                                    B

**Fig. 6.** (A) Scheme of the phase portrait of system (1) in the case of a node. (B) Scheme of the phase portrait of system (1) in the case of focus

## 3   Synergetic Management of the System

We apply the method of analytical design of aggregated regulators [4] to answer two questions:

(1)  how can you manage the amount of cash $x(t)$ from the entrepreneur, adjusting the flow of goods $y(t)$?
(2)  how can you manage the flow of goods $y(t)$ to regulate the amount of cash $x$ $(t)$ from the entrepreneur?

The control is sought as an additive component of the rate of change of the corresponding value and is a function of the current state of the system. The goal is to transfer an arbitrary initial state of the system to the preset state $(x_0, y_0)$.

3.1  $\begin{cases} x'_t = a_1 - a_2 x(t)y(t) + a_3 x(t)y^2(t) + u(t) \\ y_t = b_1 - b_2 x(t)y(t) \end{cases}$.

We look for a function (called an aggregated variable) $\psi_1(x, y)$, that defines an attracting invariant variety $\psi_1(x, y) = 0$ a controlled dynamical system (UDS) along which the paths of this the systems will be tightened to $(x_0, y_0)$. According to [4] the equilibrium state is the solution of the functional system:

$$\begin{cases} \psi_1(x,y) = 0 \\ b_1 - b_2 xy = 0 \end{cases},$$

and therefore lies on the hyperbola $y = \frac{b_1}{b_2 a}$. Its chart in the first quarter suggests such a choice of the aggregated variable $\psi_1 = x - a$ with parameter $a > 0$. The decision system has the form: $\begin{cases} x_0 = a \\ y_0 = \frac{b_1}{b_2 a} \end{cases}, a > 0.$

We verify the asymptotic stability condition [9]:

$$\left. f'_{2y} - \frac{\psi'_{1y}}{\psi'_{1x}} f'_{1x} \right|_{(x_0,y_0)} = -b_2 x_0 < 0.$$

That is $(x_0, y_0)$ steady. Since for any trajectory $\psi_1(x(t), y(t)) = x(t) - a \to 0, t \to +\infty$, that $\lim_{t \to \infty} x(t) = a$.

We will show that $\lim_{t \to \infty} y(t) = y_0 = \frac{b_1}{b_2 a}$.

From the property of an aggregated variable $\psi'_1(x(t), y(t)) = -T\psi_1(x(t), y(t))$ follows $x'_t = -T(x - a)$. Hence, the UDS has the form: $\begin{cases} x'_t = -T(x - a) \\ y'_t = b_1 - b_2 xy \end{cases}$, and management:

$$u(t) := -a_1 + a_2 x(t) y(t) - a_3 x(t) y^2(t) - T(x(t) - a)$$

Substituting the solution of the first equation $x_0(t) = Ce^{-Tt} + a$ in the second, we obtain a linear inhomogeneous equation of the first order $y'_t + b_2 x_0(t) y = b_1$. Its General solution has the form:

$$y(t) = \left( b_1 \int_0^t \exp\{b_2(\alpha\tau - C/Te^{-Tt} + C/T)\} d\tau + C_0 \right)$$
$$\cdot \left( \exp\{b_2(\alpha\tau - C/Te^{-Tt} + C/T)\} \right)^{-1}$$

Applying l'hospital's rule, find the limit:

$$\lim_{t \to \infty} y(t) = \lim_{t \to \infty} \frac{b_1}{b_2(a + Ce^{-Tt})} = \frac{b_1}{b_2 a}.$$

Thus, any UDS solution is drawn to a state of equilibrium. That is, the equilibrium is stable in General.

## Conclusion

(1) With fixed parameters $a_i, b_j$ the selected aggregated variable allows to synthesize the management of the amount of cash $x(t)$ in the entrepreneur to achieve only the States with coordinates $\left(a, \frac{b_1}{b_2 a}\right), a > 0$, lying on one branch of the hyperbola $y = \frac{b_1}{b_2 x}$.

(2) If we want to synthesize the control system to achieve an arbitrary preset state $(x_0, y_0), x_0, y_0 > 0$, it is necessary to change the parameters of the product $b_1, b_2$. Should be $b_1/b_2 = x_0 y_0$.

$$3.2 \quad \begin{cases} x_t' = a_1 - a_2 x(t)y(t) + a_3 x(t)y^2(t) \\ y_t' = b_1 - b_2 x(t)y(t) + u(t) \end{cases}.$$

We seek a function of the aggregated variable $\psi_2(x, y)$, that defines the attracting invariant manifold $\psi_2(x, y) = 0$ of UDS, along which trajectory the system will shrink $(x0, y0)$. The equilibrium state is the solution of the functional system:

$$\begin{cases} a_1 - a_2 xy + a_3 xy^2 = 0 \\ \psi_2(x, y) = 0 \end{cases}$$

and therefore lies on the cubic curve:

It follows from the graph (Fig. 7) that the equilibrium state lies on the branch of the curve located in the band $0 < y < \frac{a_2}{a_1}$. This prompts you to select an aggregated variable with a parameter $a \in \left(0, \frac{a_2}{a_3}\right)$.

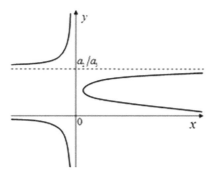

**Fig. 7.** Graph of the cubic curve.

The solution of the system looks like: $\begin{cases} x_0 = \frac{a_1}{a_2 a - a_3 a^2} \\ y_0 = a \end{cases}$, and lies in the first quarter when $a \in (0, a2/a3)$.

We verify the asymptotic stability condition [9]: $\left. f_{1x}' - \frac{\psi_{2x}'}{\psi_{2y}'} f_{1y}' \right|_{(x_0, y_0)} = -a_2 y_0 +$

$a_3 y_0^2 = -y_0 a_3 \left(\frac{a_2}{a_3} - a\right) < 0$

Since for any trajectory $\psi_2(x(t), y(t)) = y(t) - a \to 0, t \to \infty$, TO $\lim_{t\to\infty} y(t) = a$. We will show that $\lim_{t\to\infty} x(t) = x_0 = \frac{a_1}{a_2a - a_3a^2}$

Of $\psi_2'(x(t), y(t)) = -T\psi_2(x(t), y(t))$ follows $y_t' = -T(y - a)$. Hence, the UDS has the form:

$$\begin{cases} x_t' = a_1 - a_2x(t)y(t) + a_3x(t)y^2(t) \\ y_t' = -T(y - a) \end{cases}$$

a management : $u(t) := -b_1 + b_2x(t)y(t) - T(y(t) - a)$.

Substituting the solution of the second equation $y_0(t) = Ce^{-T\tau} + a$ in the first, and obtain a linear inhomogeneous equation of the first order:

$$x_t' + \left(a_2y_0(t) - a_3y_0^2(t)\right)x = a_1.$$

Its General solution has the form:

$$x(t) = \frac{a_1 \int_0^t exp\{\int_0^\tau (a_2y_0(v) - a_3y_0^2(v))dv\}d\tau + C_1}{exp\{\int_0^t a_2y_0(\tau) - a_3y_0^2(\tau))d\tau\}}.$$

Applying l'hospital's rule, find the limit:

$$\lim_{t\to\infty} x(t) = \lim_{t\to\infty} \frac{a_1}{a_2y_0(t) - a_3y_0^2(t)} = \frac{a_1}{a_2a - a_3a^2}.$$

Thus, any UDS solution is drawn to a state of equilibrium. That is, it is stable in General.

## Conclusion

1) with fixed parameters, the selected aggregated variable $a_i, b_j$ allows to synthesize the control to achieve only the States with coordinates:

$$\left(\frac{a_1}{a_2a - a_3a^2}, a\right), a \in \left(0, \frac{a_2}{a_3}\right).$$

2) If we want to synthesize the system to achieve an arbitrary preset state $(x_0, y_0), x_0, y_0 > 0$, we need to change the parameters of the entrepreneur $a_1, a_2, a_3$. Equality must be fulfilled: $y_0 = \frac{a_1}{a_2x_0 - a_3x_0^2}$.

## References

1. Pu, T.: Nonlinear Economic Dynamics. Udmurt University Publishing House, Izhevsk (2000)
2. Myasnikov, A.: Synergetic Effects in Modern Economy. 2 edn. Book house "LIBROKOM" (2013)

3. Bautin, N., Leontovich, E.: Methods of Qualitative Research of Dynamic Systems on a Plane, 2nd edn. Science, Hauppauge (1990)
4. Kolesnikov, A.: Synergetic Methods of Control of Complex Systems. The Theory of System Analysis. Komkniga, Moscow (2006)
5. Bratishchev, A., Zhuravleva, M.: Bifurcation analysis and synergetic control system, "gross product-labour resource". Bull. Rostov State Econ. Univ. (RINH) **2**(50), 147–155 (2015)
6. Milovanov, V.: Synergetics and Self-organization. Economy. Biophysics. Komkniga, Moscow (2005)
7. Milovanov, V.: Synergetics and Self-organization. General and Social Psychology. Komkniga, Moscow (2005)
8. Lefschetz, S.: Geometric Theory of Differential Equations. Publishing House/Foreign Literature, Moscow (1961)
9. Bratishchev, A.: The Mathematical Theory of Controlled Dynamical Systems. Introduction to Concepts and Methods. Publishing center DGTU, Rostov-on-don (2015)

# Patients Identification in Medical Organization Using Neural Network

V. Evdokimov Alexey[1], A. Kovalenko Vasiliy[1],
V. Kurbesov Alexandr[2]([✉]), and V. Shabanov Alexey[1]

[1] Electronic Medicine Ltd., 14 Line, 55, 344019 Rostov-on-Don, Russia
[2] Rostov State University of Economics, B. Sadovaya Str., 69,
344002 Rostov-on-Don, Russia
akurbesov@yandex.ru

**Abstract.** This article describes the method of patients identification by using neural network. Our complex filters increases recognition quality and greatly reduces computing resources needed for successful identification of a patient and can be applied to any neural network, used for identification. The resulting software is developed using only open source components and needs about 1 s to recognize a patient. Our experiment showed the effectiveness of recognition of patients not less than 99.9% with false positive ratio less than 0.00001 when using the software at the medical organization. Time interval was calculated from the first face recognition in video stream to successful identification with Euclidean metric less than 0.4. Frame rate of video stream was 25 frames per second. We measured the speed of the doctors' work before and after the introduction of the biometric identification system for patients. The patient's reception time is reduced by an average of 5 s. Based on 20.000 visits per month and 7 min per person on average, it is possible to take an additional 238 people a month without increasing the number of doctors.

**Keywords:** Optimization · Neural network · Identification · Image quality estimation

## 1 Introduction

Our main aim is to improve speed and quality of healthcare. By using neural network for patients identification we are reducing time of searching patient in medical software. Average improvement is about 15 s per patient.

We analyzed the existing software for biometric face recognition and found out that existing systems do not fit our conditions. Some systems have a high recognition quality even at a small size of the original image. However, as the image is enlarged, the quality increases insignificantly. In our case, however, it is possible to use an image of a sufficiently large size with relatively standard illumination conditions.

Other systems produce quite good results, but they work slowly and are not able to simultaneously recognize a large number of biometric images on rather weak computing power available in the medical organization.

R. A. Aliev et al. (Eds.): ICAFS-2018, AISC 896, pp. 668–675, 2019.
https://doi.org/10.1007/978-3-030-04164-9_87

Another problem with existing systems is that they use one picture to recognize, whereas in reality we can use a video stream. The main disadvantage of this approach is that in the initial video stream there are many frames of poor quality. If the input video stream processed frame-by-frame by the neural network, poor quality frames significantly increase the probability of false positive recognition of the patient and in addition dramatically increases the utilization of the computing resources of the system.

## 2  Analysis of Existing Systems

We analyzed the following existing face recognition systems: Based on open source systems, Based on OpenCV, Based on Dlib, With proprietary source code, VOCORD FaceControl 3D, Cascade-potok, Contur-potok, 3DiVi Face SDK, LUNA SDK, FindFace, Microsoft Face API.

Disadvantages of existing systems, leading to the inability to use them in our case. High overall cost of systems with a good quality of video stream recognition (since expensive cameras and a powerful hardware are required). Low speed. Focus on the use of a single frame, instead of the flow. Focus on the images of small size/poor quality (with increasing size and improving image quality, the quality of recognition increases slightly). Data storage in the cloud (which does not allow the customer to protect personal data, according to the law). High price. There is no possibility of integration with third-party systems.

The main aim was to develop a system that could be used in a medical organization and designed to produce a high percentage of successful recognition of a biometric template with a very low false positive identification ratio.

At the first stage, we estimated the dependence of the recognition quality on various factors: such as the size of the face on the image, the sharpness of the image, illumination, noise ratio and the structure of the neural network.

Our experiment showed that the quality of the original image is more important than the neural network structure. The maximum recognition ratio is reached at the size of the face 256*256 pixels. Further increase of face size only consumes additional resources without improving recognition ratio.

Table 1 shows result of our experiment. We use set of 10000 pictures for training and set of 250 pictures of each size for recognition. The system resulting recognition ratio is different (slightly higher) because it's measured on video stream (set of pictures). To identify person we need only one successful recognition, so our system is less dependant on false negative recognition ratio. But the false positive ratio is more important for our system.

To reduce computing resources we applied motion detection to the source video stream. It's made with simple and fast background subtractor which uses the 100 previous frames to determine the background. This method allows to reduce number of images for recognition by cutting off series of very similar images.

As our neural network convert face to vector generalizing the characteristics of a particular person, the efficiency of our neural network grows with increasing of the sharpness of the input image.

**Table 1.** Recognition time and ratio depending on the face size (single picture)

| Face size (pixels) | Face location (%) | Correct recognition (% of found faces) | Average time (ms) |
|---|---|---|---|
| 32 | 84 | 82 | 20 |
| 64 | 91 | 89,3 | 68 |
| 96 | 92 | 92.72 | 123 |
| 128 | 96 | 94.11 | 207 |
| 192 | 98 | 97.32 | 439 |
| 256 | 99.1 | 99.23 | 757 |
| 384 | 99.2 | 99.45 | 1649 |
| 512 | 99.4 | 99.51 | 3023 |

Using this idea we constructed fast filter for selecting best images to reduce neural network computational pipeline utilization. The filter drops about 90% of source frames without quality recognition loss. The filter computes the metric of image quality based on discrete cosine transform [1], applied to the image.

The correlation of DCT (discrete cosine transform) with a measure of image sharpness is based on the fact that the blurring of image reduces sharpness of borders. The blurred borders of the face details reduces overall quality of face recognition. Blur can be modelled as the overweight of low-frequency components in the spectral representation of the image. Or, a blur can be created by applying a low-pass filter - the corresponding terms of the DCT. This means that, for example, the JPEG format is such a filter. In this case, a logical approach to the problem of estimating image blur is to use the DCT coefficients.

This approach is applicable because of the high prevalence of the JPEG format, which is based on the DCT which means the availability of computationally effective implementations of the algorithm; high speed of calculation of DCT coefficients in contrast to a number of other methods for estimating image blur.

An image blur estimate, based on the analysis of DCT coefficients, can be used not only to determine the presence of blurred areas in the image, but also to determine the position of those areas [2].

The blur detection method is based on the observation that blur shifts the frequency content of blurred areas to DCT components with a lower frequency. In terms of the distribution of DCT coefficients, the blur can be marked as the region in the image where the DCT coefficients show smaller amplitude values.

In addition, the global blur introduced into the image has such a characteristic as the direction of the blur. This is reflected in the structure of the stored DCT coefficients. Thus, it is possible to introduce a blur measure for digital images, based on the analysis of the DCT coefficients.

The basic idea of constructing a measure of image sharpening is as follows: the image or the region of interest is divided into blocks of 8*8 pixels; each block stores the DCT coefficients. This block has a constant component (DC) and alternating components (AC). DC components make the main contribution to the definition of the hue for this pixel block, the ACs are responsible for the small details of this section and they are more sensitive to all kinds of artifacts in the image, in particular - blur. Having constructed a histogram of 64 columns corresponding to each coefficient in the DCT transformation, for all blocks of pixels 8*8 entering the image under study, the total number of components can be counted. The contribution of each component to the extent of blurring can be set by using a table of weights. The measure of the blurring will be the sum of only those values of the weights table, the values of the histogram of the distributions of the coefficients are 10 times smaller than the first component of the histogram (namely, the sum of the DC components of the entire image).

## 3   Effect of the Noise Level on Recognition Results

We also estimated the effect of the noise level on recognition results. To estimate the noise level, we used the ratio of the peak signal to noise - PSNR (Peak Signal-to-Noise Ratio). For the calculations, the following relations were used [3]:

$$MSE(i) = \frac{1}{WH} \sum_{x=0}^{W-1} \sum_{y=0}^{H-1} [Y_r(x, y, i) - Y_p(x, y, i)]^2 \tag{1}$$

where W and H are the frame width and height respectively in pixels, Yr and Yp are the luminance values of the reference and processed video frames respectively. The PSNR expressed in decibels for each pair i is then defined by:

$$PSNR(i) = 10lg \frac{I^2}{MSE(i)} \tag{2}$$

where I is the maximum pixel luminance value (e.g. 255 for 8-bit representation).

A frame obtained by "averaging" 100 frames was used as the reference frame. In the case of statistical independence of noise in the sample of images used, the noise level in the obtained frame is lower by 10 dB than in the original ones.

The studies were carried out under conditions of a real PSNR range of 20 … 35 dB. There was no significant dependence of the quality of recognition on the noise level in such conditions. This is due to the obtaining of values commensurable with the accuracy of the measured values.

We are using local binary pattern detection routine for fast face detection. We compared the detection quality of local binary pattern detection routine and the Haar cascade. As we are prefiltering images the detection quality is almost the same but the local binary pattern detection routine is much faster.

The quality of the automatic face recognition system depends on a variety of factors, presented to both the person and the scene as a whole. For example the recognition ratio decreases with increasing of the deviation of the face from vertical position. The main factors of image quality are head position (turn, tilt, deflection); requirements for the size of biometric features (eyes, lips, etc.); background, illumination level of the face; focus, blur and depth of field.

However, in real situations, the frames received by the biometric system are far from the desired quality.

To solve the listed problems associated with changing the biometric parameters of the template when changing the position of the face in the image, the following algorithm was used.

Find a face in the image using local binary patterns.

Find on the resulting image the positions of several face landmarks, such as the eyes and nose.

Affine image transformation to eliminate distortion.

Many modern systems, such as DeepFace, use the algorithm for constructing a 3D face pattern, but this leads to significant computational costs that did not fit in our case.

At the last stage we analyzed various configurations of neural network estimating the most effective correlation between the quality of recognition and computational costs. The most effective ratio is reached when we are using deep neural network that converts face image with size 256*256 pixels into 192 dimension vector.

## 4   Experiment

We cannot perform full set of experiment for every face size and different dimensions of output vector, due to the lack of computational resources. Every neural network training session takes more than 100 h on GeForce GTX 1080 and the time grows as we increase face image size and dimensions of the output vector.

We are using PyTorch to implement our neural network. The neural network consists of seven maxpool layers and one final fully connected layer and trained to group faces. Training goal was to reach the Euclidean metric value between any two 192 dimension vectors, produced from the photos of the same person less than 0.4. The Euclidean metric value between any two 192 dimension vectors, produced from the photos of different persons should be more than 0.75.

We are using Megaface dataset with additional of our patients faces to perform basic network training. The full training session takes 187 h on GeForce GTX 1080 (Table 2).

We are using modified facenet triplet loss algorithm [4] as the loss function.

Resulting topology of our neural network.

1. Convolution layer with input size 256*256*3, output size 128*128*64 and kernel size 7*7*3
2. Maxpool layer with input size 128*128*64, output size 64*64*64 and kernel size 3*3*64
3. Normalisation layer with size 64*64*64
4. Convolution layer with input size 64*64*64, output size 64*64*64 and kernel size 1*1*64
5. Convolution layer with input size 64*64*64, output size 64*64*192 and kernel size 3*3*64
6. Maxpool layer with input size 64*64*192, output size 32*32*192 and kernel size 3*3*192
7. Normalisation layer with size 32*32*192

8. Convolution layer with input size 32*32*192, output size 32*32*192 and kernel size 1*1*192
9. Convolution layer with input size 32*32*192, output size 32*32*384 and kernel size 3*3*192
10. Maxpool layer with input size 32*32*384, output size 16*16*384 and kernel size 3*3*384
11. Normalisation layer with size 16*16*384
12. Convolution layer with input size 16*16*384, output size 16*16*384 and kernel size 1*1*384
13. Convolution layer with input size 32*32*192, output size 16*16*384 and kernel size 3*3*192
14. Convolution layer with input size 16*16*384, output size 16*16*256 and kernel size 3*3*384
15. Convolution layer with input size 16*16*256, output size 16*16*256 and kernel size 3*3*256
16. Maxpool layer with input size 16*16*256, output size 8*8*256 and kernel size 3*3*256
17. Fully-connected layer with input size 8*8*256 and output size 1*32*192
18. Fully-connected layer with input size 1*32*192 and output size 1*192

**Table 2.** Successful identification ratio depending on vector size and neural network convolution layers (with kernel size > 1*1) number.

| Number of layers | Output vector length | Successful recognition ratio (%) |
|---|---|---|
| 5 | 96 | 84.1 |
| 6 | 96 | 85.4 |
| 7 | 96 | 87.6 |
| 8 | 96 | 87.7 |
| 5 | 128 | 91.1 |
| 6 | 128 | 94.2 |
| 7 | 128 | 98.9 |
| 8 | 128 | 99.0 |
| 5 | 192 | 94.3 |
| 6 | 192 | 96.7 |
| 7 | 192 | 99.1 |
| 8 | 192 | 99.1 |

On the one hand we would like to use small mini-batches as these tend to improve convergence during Stochastic Gradient Descent (SGD) [7]. On the other hand, computation speed is significantly higher then batches of tens to hundreds of exemplars are used. The main constraint with regards to the batch size, however, is the way we select hard relevant triplets from within the mini-batches. In most experiments we use a batch size of around 1.800 exemplars.

Another way to improve the quality of face recognition is to improve the biometric template.

Under real conditions, the deviation angles are significantly larger than those that allow us to obtain a high quality biometric template. Therefore, different approaches are used to extract as much information as is necessary to create a biometric template

The are two main method.

Use of several cameras located at different angles to the face. In consequence of redundancy in the number of video capture devices used, an acceptable number of biometric features, that are in the field of view of the camera, are required to build a set of anthropometric points, thereby increasing the accuracy of the created biometric template.

Using a software module that collects the resulting image from a set of disjoint fragments. A similar principle is used in the system of constructing three-dimensional images. It is also used in the new system of protection from Apple - Face ID. A user to register a new biometric template rotates his head, and an infrared camera collects a series of scattered images that are images of biometric features located at different angles.

Our software uses specially prepared biometric template. We are collecting biometric features from five cameras simultaneously using the same filter of source images as we use in recognition module. The resulting set of images is used to prepare set of biometric templates for patient.

## 5   Conclusions

Our complex filters increases recognition quality and greatly reduces computing resources needed for successful identification of a patient and can be applied to any neural network, used for identification.

The resulting software is developed using only open source components and needs about 1 s to recognize a patient. Our experiment showed the effectiveness of recognition of patients not less than 99.9% with false positive ratio less than 0.00001 when using the software at the medical organization.

Overall speed of identification described in Table 3.

**Table 3.** Overall speed of identification.

| Time interval (ms) | Successful identifications (per 10000 persons) | % |
|---|---|---|
| <100 | 224 | 2.24 |
| 100–200 | 674 | 6.74 |
| 200–400 | 928 | 9.28 |
| 400–600 | 1415 | 14.15 |
| 600–800 | 1874 | 18.74 |
| 800–1000 | 2012 | 20.12 |
| 1000–1200 | 1790 | 1.9 |
| 1200–1400 | 950 | 9.5 |
| >1400 | 132 | 1.32 |
| Not identified | 1 | 0.01 |

Time interval was calculated from the first face recognition in video stream to successful identification with Euclidean metric less than 0.4. Frame rate of video stream was 25 frames per second.

We measured the speed of the doctors' work before and after the introduction of the biometric identification system for patients. The patient's reception time is reduced by an average of 5 s. Based on 20,000 visits per month and 7 min per person on average, it is possible to take an additional 238 people a month without increasing the number of doctors.

# References

1. Marichal, X., Wei-Ying, M., HongJiang, Z.: Blur determination in the compressed domain using DCT information. In: Proceedings of 1999 International Conference on Image Processing, ICIP 1999, vol. 2 (1999)
2. Kalalembang, E., Usman, K., Gunawan, I.P.: DCT-based local motion blur detection. In: International Conference o Instrumentation, Communications, Information Technology, and Biomedical Engineering (ICICI-BME) (2009)
3. Huynh-Thu, Q., Ghanbari, M.: The accuracy of PSNR in predicting video quality for different video scenes and frame rates. Telecommun. Syst. **49**(1), 35–48 (2012)
4. Schroff, F., Kalenichenko, D., Philbin, J.: FaceNet: A unified embedding for face recognition and clustering. In: Proceedings of the IEEE Computer Society Conference on Computer Vision and Pattern Recognition (2015)
5. Chen, D., Ren, S., Wei, Y., Cao X., J. Sun J.: Joint cascade face detection and alignment. In: Proceedings of the IEEE Computer Society Conference on Computer Vision and Pattern Recognition (2014)
6. Weinberger, K.Q., Blitzer, J., Saul, L.K.: Distance metric learning for large margin nearest neighbor classification, pp. 2–3. In: NIPS. MIT Press (2006)
7. Wilson, D.R., Martinez, T.R.: The general inefficiency of batch training for gradient descent learning. Neural Netw. **16**(10), 1429–1451 (2003)

# Evaluation of Innovative Changes in the Structure of Economy of Regions on the Basis of Statistical Analysis and Theory of Fuzzy Sets

Aleksandr R. Groshev[✉]

Surgut State University, Lenin Ave. 1, Surgut 628412, Russia
maoovo@yandex.ru

**Abstract.** The study developed a model for determining the influence of macroeconomic factors on tax revenues, based on the application of the theory of algebra functions, elements of fuzzy sets and the analysis of proportional dependencies to predict income in the monoprophilic economy of the regions on the example of the Khanty-Mansiysk Autonomous Okrug – Ugra (KHMAO - Ugra) of Russia. An aggregated variable representing a nonlinear function of external (independent of the change in the structure of the district economy) forecast factors is introduced. The analysis indicates the possibility of using the fuzzy sets to predict budget revenues using aggregated variables.

**Keywords:** Fuzzy sets in the economy · Forecasting budget revenue
Aggregated variables

## 1 Introduction

The formation of a new competitive model of the Russian economy based on innovation is one of the main tools for improving the quality of life of the population. It is important that the current external conditions require special speed in solving new problems [1]. In this regard, the assessment and analysis of the ongoing innovative transformations of the economy is an important, but complex not only in practical but also in scientific terms.

Developers of programs of social and economic development of regions offered a variety of indicators and indicators that allow to judge the degree of achievement of the stated goals in the programs, but to build a system of evaluation, based on the analysis of statistics of dynamic series of values that adequately characterize the innovative changes in the structure of the regional economy, has not been yet succeeded.

One of the approaches to the assessment of changes in the structure of the economy is the analysis of changes in the structure of regional budget revenues. The relationship between the structure of the economy and the structure of budget revenues is quite obvious. This approach assumes that the transformation of the economic system of the region leads to changes in the structure of funds received in the budget revenues for each of the main sources (profit taxes, personal income taxes, property taxes, non-tax revenues).

© Springer Nature Switzerland AG 2019
R. A. Aliev et al. (Eds.): ICAFS-2018, AISC 896, pp. 676–680, 2019.
https://doi.org/10.1007/978-3-030-04164-9_88

At the federal level, the indicator of changes in the structure of budget revenues is "The share of oil and gas revenues in budget revenues in %". The same indicator, according to some experts, reflects the transformation of the country's economic system [2]. However, this indicator cannot be calculated and used to assess changes in the structure of income of the regions.

The use of number of dynamics for evaluation of changes revenue structure, based on data collected at present by official statistics, is impossible because of absence of relevant indicators. The inclusion of such indicators in the official reports of enterprises and organizations is also not possible due to the complexity of the methods of their possible accounting and description. As an example, "Recommendations for the collection and analysis of data on innovations" [3] is possible to consider. It seems appropriate to use indirect methods of constructing number of dynamics for analysis of changes in the structure of regional budget revenues, which are based on the processing of existing number of dynamics and the allocation of the necessary components.

It seems appropriate to use indirect methods of constructing time series to analyze changes in the structure of regional budget revenues, which are based on the processing of existing time series and the allocation of the necessary components.

## 2   Using Predictive Indicators

During the work on a long-term budget strategy for the period up to 2030, we have developed an approach and forecasting models based on the theory of function algebra, fuzzy sets and analysis of proportional dependencies of indicators [4]. In carrying out studies of the impact of macroeconomic factors on tax revenues, and the impact of tax incentives operating in the KHMAO - Ugra, an aggregate variable was introduced, which is a nonlinear function of external (independent of changes in the structure of the economy of the region) forecast factors [5, 6].

The revenues of the budget of the region were considered as forecast factors (see the Table 1).

The aggregate variable is a nonlinear function of three factors: (1) the average annual price of Brent crude oil in dollars USA; (2) the volume of oil production in the region; (3) the average annual exchange rate. The value of the aggregate variable was calculated based on the formula:

$$Xa_i = Pf1_i * Pf2_i * Pf3_i/M, \qquad (1)$$

where Xa - aggregate variable; Pf – forecast factors; M – scale coefficient. The Scale coefficient Mi is determined by the formula:

$$M_i = R(Pf1_i) + R(Pf2_i) * R(Pf3_i). \qquad (2)$$

Table 1 presents: (1) the average forecast of oil prices of specialists of PIRA EnergyGroup, the International energy Agency, The Economy Forecast Agency, the Agency of Economic Forecasting, the World Bank and the Central Bank of Russia; (2) the forecast of production volumes of specialists of the Autonomous institution of the KHMAO - Ugra scientific and analytical center for rational subsoil use of V. I. Shpilman; (3) forecast of the dollar of the Ministry of economic development.

**Table 1.** The scale factor, prognostic factors and the aggregated variable

| Year | Scale factor | Oil price | Extrac-tion volume | Dollar rate | Aggregate variable | Region revenue |
|------|-------------|-----------|--------------------|-------------|--------------------|----------------|
| 2004 | (1,1,1) = 3 | 38,3 | 255,6 | 28,81 | 94,01 | 100,6 |
| 2005 | (1,1,2) = 4 | 54,4 | 267,9 | 28,3 | 103,11 | 102,0 |
| 2006 | (1,3,1) = 5 | 65,4 | 275,6 | 27,17 | 97,94 | 97,5 |
| 2007 | (1,3,1) = 5 | 72,7 | 278,4 | 25,58 | 103,55 | 105,9 |
| 2008 | (2,1,2) = 5 | 97,7 | 277,6 | 24,86 | 134,85 | 139,7 |
| 2009 | (2,1,2) = 5 | 61,9 | 270,5 | 31,83 | 106,59 | 109,0 |
| 2010 | (1,2,2) = 5 | 79,6 | 266 | 30,36 | 128,57 | 121,1 |
| 2011 | (1,2,2) = 5 | 111 | 262,5 | 29,39 | 171,27 | 157,3 |
| 2012 | (2,2,2) = 6 | 121,4 | 259,9 | 31,08 | 163,44 | 157,8 |
| 2013 | (1,2,3) = 6 | 108,8 | 255,1 | 31,85 | 147,33 | 136,7 |
| 2014 | (1,2,2) = 5 | 98,9 | 250,3 | 38,61 | 191,16 | 197,0 |
| 2015 | (1,1,2) = 4 | 52,4 | 243,1 | 61,07 | 194,48 | 194,7 |
| 2016 | (1,1,1) = 3 | 41,9 | 239,2 | 67,19 | 224,47 | 197,9 |
| 2017 | (1,2,1) = 4 | 57,2 | 235,3 | 59,7 | 200,88 | 179,7 |
| 2018 | (1,2,2) = 5 | 60,5 | 232 | 68,7 | 192,85 | – |

The values of the forecast coefficient can also be seen in Table 1. To determine the R(Pfi), the coefficient of determination of the predicted factor Pf and the predicted indicator D on the interval i – Ki(Pf, D)is calculated.

$$R(Pf_i) = \begin{cases} 1, & if\ K_i(Pf,D) \leq 0,5 \\ 2, & if\ 0,5 < K_i(Pf,D) \leq 0,7 \\ 3, & if\ 0,75 < K_i(Pf,D) \leq 1 \end{cases} \tag{3}$$

Experts estimated the possible decrease in revenues in 2018 by 17.5 billion rubles, associated with the relocation of the management office of Gazprom from the KHMAO - Ugra to St. Petersburg. Thus, the revenue forecast for 2018 is as follows:

$$D_{2018} = Xa_{2018} - 17,5. \tag{4}$$

Statistical studies using the above-described forecasting models allowed us to hypothesize the possibility based on processing to distinguish from the existing series of dynamics necessary information for the analysis of innovative transformation of the economic system of the region.

Figure 1 shows the graphs of the dynamics of the indicator "budget revenues" and the aggregate variable of forecast factors [6].

The figure shows that the dynamics of the aggregate variable and the indicator under study coincide in all parts of the series except for 2004 and 2005.

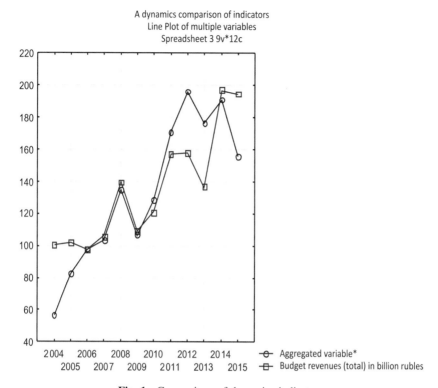

**Fig. 1.** Comparison of dynamics indicators

The deviations in the dynamics are explained by one-time transfers to the budget in the fourth quarter of 2004 and in the first quarter of 2005 of payment related to the repayment of tax debts of OJSC "Yukos Oil Company".

## 3   Model of Fuzzy Sets

To clarify the forecast, the elements of the fuzzy set theory were used, namely, the phasification of the constructed model, which allows to refine the forecast for the average, "optimistic" and "pessimistic" scenarios of the situation [7, 8].

To build the model, let's make assumptions: (1) the income of district a is a fuzzy triangular number; (2) for the fuzzy variable a the membership function $\mu(a)$ is defined, taking its greatest value, unit, with a = an, that is, the value corresponding to the average value of income; with a = amin (pessimistic forecast) and the maximum a = amax (optimistic forecast) values, the membership function turns to zero. Let's set the level that can be considered as the level of reliability of the results evaluation. Request: $\mu(a) \geq \alpha$. Hence, we find a range of blur values of district income:

$$\alpha a_n + (1 - \alpha)a_{\min} \le a \le \alpha a_n + (1 - \alpha)a_{\max}. \qquad (5)$$

Analysis of Table 1 shows that the standard deviation of the aggregate variable from the statistical value of income is 10.9 million rubles. Therefore, we can assume that $a_n$ = 192,85, $a_{min}$ = 181,95, $a_{max}$ = 203,75.

Substituting these values in the formula (5), we obtain:

$0,8 \cdot 192,85 + 0,2 \cdot 181,95 \le a \le 0,8 \cdot 192,85 + 0,2 \cdot 203,75,$
or $190,67 \le a \le 195,03$. Thus, with a probability of 0.8, it can be argued that the budget revenues will be from 190.67 to 195.03 million rubles.

## 4 Conclusion

The distinctive feature of the proposed method of indirect construction of series of dynamics is the ability to use for statistical analysis of changes in the structure of regional budget revenues official statistics. The analysis also points to the possibility of using the fuzzy sets mechanism to predict budget revenues using aggregated variables. The proposed method needs further formalization and has been tested only on the analysis of the data of the single-profile region, however, the results allow us to hope for the opportunity to further develop methods and procedures that allow us to build dynamic series that adequately characterize the innovative transformation of economic systems of different regions of the Russian Federation.

## References

1. Kudrin, A., Gurvich, E.: New model of growth for Russian economics. Quest. Econ. **12**, 4–36 (2014)
2. The main trends of socio-economic development of the Russian Federation in 2011–2015. http://economy.gov.ru/minec/about/collegium/2016250402. Accessed 21 Nov 2017
3. Recommendations for the collection and analysis of innovation data. (Oslo Manual). Joint OECD and Eurostat publication/official website of the Federal state statistics service. http://www.gks.ru/wps/wcm/connect/rosstat_main/rosstat/ru/statistics/science_and_innovations/science. Accessed 21 Nov 2017
4. Groshev, A.: Long-term budget planning in modern Russia (regional aspect): monograph. Research Center of SurGU, Surgut (2015)
5. Kazimova, G., et al.: Analysis of the effect of operating in the Khanty-Mansi Autonomous Region – Yugra tax privileges for the profit tax of the organizations for increase of budget revenues. Izv. Baykal State Univ. **25**(4), 630–636 (2015)
6. Karataeva, G., et al.: Evaluation of the influence of macroeconomic factors on tax revenues and the construction of a model of budget revenue forecasting. Innov. Dev. **3–1**(33), 42–73 (2016)
7. Kramarov, S., Sakharova, L.: Management of complex economic systems by fuzzy classifiers. Sci. Bull. South. Inst. Manag. **2**(18), 42–50 (2017)
8. Kramarov, S., Sakharova, L., Khramov, V.: Soft computing in management: management of complex multi-factor systems based on fuzzy analog controllers. Sci. Bull. South. Inst. Manag. **3**(19), 42–51 (2017)

# Assessment of the Sustainability of Agricultural Production in the Region on the Basis of Five-Level Fuzzy [0,1] – Classifier

Elizabeth A. Arapova[✉], Michael Yu Denisov, Elena A. Ivanova, and Yulia V. Kulikova

Rostov State University of Economics, B. Sadovaya str., 69, 344002 Rostov-on-Don, Russia
dist_edu@ntti.ru

**Abstract.** The technique of estimation of stability of agricultural production in the region on the basis of standard multilevel fuzzy [0,1] – classifiers is offered. The methodology makes it possible to form a comprehensive sustainability of agricultural production in the region based on the factors of three groups: economic, social and environmental sustainability. The resulting estimate is based on aggregation of estimates for each of the three listed areas. Evaluation of each area is based on the aggregation of the assessments of the mixed complex indicators. Aggregation is performed on the basis of time series of numerical values of heterogeneous indicators reflecting both the level and growth rates of indicators for the studied periods. The contribution of each of the indicators is estimated using a weighting factor reflecting its importance. The advantages of the proposed evaluation methodology in comparison with the existing methods consist in a simple scheme of calculations, taking into account a large number of heterogeneous significant indicators, allowing for variation depending on the available statistical material and features of a specific practical problem. It is adaptable and versatile, allowing it to be applied to assess the intensity of not only agricultural but also industrial production on various scales.

**Keywords:** Standard multilevel fuzzy [0,1] - classifier · Methods of evaluation Statistics · Integrated assessment

## 1 Introduction

The essence of the criterion of efficiency of agricultural production is the maximum production of agricultural products necessary for society at a given cost and the amount of resources per unit of production, providing high quality products and rational use of labor, material, monetary and land resources [1]. Currently, there is a deterioration in the environmental situation of natural and industrial systems. Therefore, the rationalization of agricultural production should be carried out not only from the standpoint of improving the efficiency of agricultural production, but also the reproduction of soil fertility, improvement of the ecological state of the natural environment. The system of

R. A. Aliev et al. (Eds.): ICAFS-2018, AISC 896, pp. 681–688, 2019.
https://doi.org/10.1007/978-3-030-04164-9_89

indicators of ecological and economic efficiency of agricultural production in the region should reflect not only economic, but environmental and social factors [2].

In the aspect of optimization of agricultural production on the basis of environmentally friendly technologies, the practical importance is the development of methods that allow for a comprehensive assessment of the intensity of agricultural production on the basis of a set of ranked indicators that objectively reflect the efficiency of material and financial resources of agricultural sectors, as well as the impact of agricultural production on the Currently, there are private methods of assessing the intensity of agricultural production on its individual indicators [3]. However, they do not allow to evaluate and rank agro-industrial enterprises, agricultural industries and entire regions on the basis of a comprehensive analysis of many indicators.

In this paper, a technique based on the methods of the theory of fuzzy sets and aimed at obtaining an objective comprehensive quantitative assessment of the intensity of agricultural production on a set of criteria of two groups: the level of intensification of production and the level of economic efficiency of intensification of production in agriculture. The novelty of the proposed method, as well as its difference from similar developments, is that integrated estimates are calculated for each of the indicators on the basis of time series of its values by means of normalizing formulas. The subsequent application to them of the standard five-level fuzzy [0,1]-classifier (used previously in financial analysis and not used in methods of production intensity assessment) allows to calculate the normalized comprehensive assessment of production intensification levels and economic efficiency of production intensification in agriculture, as well as to obtain a comprehensive assessment of the intensity of its production [4]. A similar method is used to assess the impact of agricultural production on the ecological state of the region. The time series of data on pesticides, soil fertility, air emissions of pollutants and water use by agriculture are used as a set of parameters on the basis of which the assessment is based. Aggregation of the formed estimates on the basis of standard five-level fuzzy [0,1]-classifiers allowed to obtain a final comprehensive assessment of the stability of agricultural production in the region on the example of the Rostov region.

## 2    Method of Estimation of Intensity of Agricultural Production

The author's method of assessing the intensity of agricultural production on the basis of standard five-level fuzzy [0,1]-classifiers is described in detail in [5, 6].

The first stage of the algorithm is aimed at forming a list of significant indicators of the level of intensification of production in agriculture for the period of n years (hereinafter: the first group of indicators), as well as significant indicators of the level of economic efficiency of intensification of production in agriculture for the same period (the second group of indicators). The second stage is aimed at ranking the importance of the studied indicators for assessing the intensity of agriculture, and the calculation of their weight coefficients on the basis of expert assessments. At the third stage, the calculation of normalized (i.e. belonging to the segment [0,1]) numerical values of the indicators of both groups is carried out on the basis of specially selected formulas. At the fourth stage, the task of linguistic variables corresponding to the studied indicators is performed, as well as a comprehensive assessment of the intensity of agricultural production", assessment of the level of intensification of production in agriculture" and evaluation of the economic efficiency of intensification of production in agriculture. A universal set for each linguistic variable is a numeric segment [0,1]. The set of values of all three estimates is the term-set $G = \{G_1, G_2, G_3, G_4, G_5\}$, where $G_1$ – "a stable tendency to decrease growth"; $G_2$ – "a tendency to decrease growth"; $G_3$ – "a tendency to stagnate"; $G_4$ – "a tendency to growth"; $G_5$ – "a stable tendency to growth". At the fifth stage, the transition from the numerical values of the indicators to the numerical values of the estimates is performed in accordance with the algorithm of standard five-level fuzzy [0,1]-classifiers [5, 6]. At the sixth stage there is a linguistic recognition of the obtained numerical estimates in accordance with the definition of the term set, as well as the analysis of the obtained intensity estimates based on the numerical values of the indicators and recommendations for the correction of the situation.

On the basis of the described algorithm, a complex numerical evaluation $g$ of the intensification of production in agriculture of the Rostov region is obtained [5, 6]:

$g = 0.6549$, $\mu(0.6549) = \mu_4(0.6549) = 1$.

Therefore, a comprehensive assessment of the intensification of production in agriculture of the Rostov region in accordance with the introduced above term-set, meets the term $G_4$ – "tendention to growth". Numerical estimates $g_1$ = "assessment of the level of intensification of production in agriculture of the Rostov region" and $g_2$ = "assessment of economic efficiency of intensification of production in agriculture of the Rostov region", calculated separately on the basis of formulas (4) and (5), respectively, are equal to: $g_1 = 0.6817$, $g_2 = 0.4536$. Therefore, "assessment of the level of intensification of production in agriculture of the Rostov region" corresponds to the term $G_4$ – "growth trend", and "assessment of economic efficiency of intensification of production in agriculture of the Rostov region" term $G_3$ – "trend to stagnation". Thus, the results of the quantitative analysis of the dynamics of agricultural production for 2014–2016 coincide with the results of the qualitative analysis for 1996–2013, and show that the overall development trends remain.

On the basis of the received estimates, as well as the calculation tables, the following practical recommendations were given for the further development of agriculture in the region: (1) to pay attention to the improvement of economic efficiency indicators of intensification of production in agriculture; (2) to pay special attention to increasing the yield of open ground vegetables; (3) to take the necessary measures to increase productivity; (4) to increase the percentage of calves per 100 main cows. The proposed algorithm can be generalized to any number of years, and also used to analyze the intensity of agricultural production in the regions as a whole (as well as their ranking), and individual industries (livestock, crop, etc.), as well as individual agricultural enterprises. The practical significance of the proposed method of evaluation is that it allows not only to calculate a comprehensive assessment of the state of agriculture, but also to study the impact on it of individual indicators and to isolate those whose values lead to a decrease in the final assessment, and therefore, need correction and additional research.

## 3  Assessment of the Impact of Agricultural Production on the Ecology of the Region

The method of formation of complex assessments described in paragraph 1 is used to assess the impact of agricultural production on the environmental condition of the region. The presented modification is a special case of the author's method of evaluation of complex systems based on complexes of heterogeneous indicators [7, 8]. As a set of indicators on the basis of which the assessment of the impact of agricultural production on the ecology of the region is formed, data on the introduction of pesticides in the soil, soil fertility, emissions of pollutants into the atmosphere by agricultural production in the region and water use by agriculture are used.

### 3.1  Assessment of Chemical Load Dynamics on the Ecological System of the Region

Calculations are made on the basis of time series reflecting the application of fertilizers in the open ground for 2012–2016 (Table 1, data from [9]).

**Table 1.** Application of pesticides in the open field of the Rostov region

| Name of indicator | 2012 year | 2013 year | 2014 year | 2015 year | 2016 year |
|---|---|---|---|---|---|
| *The volume of pesticides in the open field of the Rostov region-total, thousand tons* | | | | | |
| insecticides | 0.798 | 0.834 | 0.573 | 0.545 | 0.6 |
| fungicides | 0.587 | 0.736 | 0.408 | 0.404 | 0.65 |
| herbicides | 1.271 | 1.796 | 1.071 | 1.122 | 1.39 |
| *Application of pesticides in open ground of Rostov region-total, thousand hectares of crops* | | | | | |
| insecticides | 2401.42 | 2566.04 | 2334.640 | 2860.801 | 2393.76 |
| fungicides | 526 | 751.4 | 821.600 | 865.263 | 1975.25 |
| herbicides | 2157.06 | 2319 | 2517.600 | 2804.400 | 2399.363 |

The calculation of normalized values $x$ of the studied parameters for the period of $N$ years under consideration is based on the formula:

$$x_i = 0.5 \left( 1 + \sum_{i=1}^{N-1} k_i I_i \right), \quad k_i = \frac{2(N-i)}{(N-1)N}, \tag{1}$$

where $k_i$ are the weighting coefficients determined on the basis of rules Fishburnes, and the weighting factor for the period 2015–2016 is 0.4; for 2014–2015 is 0.3; for 2013–2014 is 0.3; for 2012–2013 is 0.1. $I_i$ there are integer functions, and their value "1" corresponds to an increase in the I-th indicator (deterioration of the situation); value "−1" – decrease in the $i$-th indicator; value "0" – stabilization, no changes.

The linguistic variable $g_1$ = "assessment of the dynamics of chemical load on the ecological system of the region", the universal set of which is a numerical segment [0,1], is introduced into consideration. Term-set consists of five terms $G = \{G_1, G_2, G_3, G_4, G_5\}$, conditionally assessing the state of the system: $G_1$ – "stable tendency to reduce the load"; $G_2$ – "tendency to reduce the load"; $G_3$ – "stabilization of the situation"; $G_4$ – "tendency to increase the load"; $G_5$ – "stable tendency to increase the load". Accessory functions – standard trapezoidal [8, 9]. Calculations show that for the Rostov region $g_1$ = 0.66, which means term $G_4$ – "the tendency to increase the load".

## 3.2   Evaluation of Soil Fertility

Evaluation is based on statistical data that reflects the content of humus in the soil by natural-agricultural zones, 1976–2013, by five year periods [9]. To assess the dynamics of humus content in the soil, a linguistic variable $g_2$ = "assessment of humus content dynamics in the soil" was introduced, the universal set of which is a numerical segment [0,1]. The term set consists of five terms $G = \{G_1, G_2, G_3, G_4, G_5\}$, which conditionally assess the state of the system: $G_1$ – "a stable tendency to increase the content of humus"; $G_2$ – "a tendency to increase the content of humus"; $G_3$ – "stabilization of the situation"; $G_4$ – "a tendency to decrease the content of humus"; $G_5$ – "a steady tendency to decrease the content of humus". The calculation of normalized values $x$ of the studied indicators for the period of $N$ years under consideration is based on the scheme:

$$x = 0.5 \left(1 - \sum_{i=1}^{N-1} k_i I_i \right), \quad k_i = \frac{2(N-i)}{(N-1)N}, \tag{2}$$

where $k_i$ – the weight coefficients determined on the basis of the Fishburn rule; for the period 2007–2013 the coefficient is the largest and is equal to 7/8. $I_i$ is integer functions. Calculations show that for the Rostov region $g_2$ = 0.63, which means belonging to two terms $G_3$ – "stabilization of the situation" and $G_4$ – "the tendency to decrease humus", the values of the functions of belonging are equal to 0.2 and 0.8, which means: the statement "there is a tendency to reduce the content of humus" is more true than the statement "there is a stabilization", which coincides with the qualitative assessment of [9].

To assess the content of humus in the soil introduced into consideration the linguistic variable: $g_3$ = "the level of humus in the soil". The term set consists of four terms $G = \{G_1, G_2, G_3, G_4, G_5\}$: $G_1$ – "norm"; $G_2$ – "environmental risk"; $G_3$ – "crisis zone"; $G_4$ – "disaster zone" (according to the generally accepted classification [10]). Normalized values of input parameters are calculated for each natural-agricultural zone as the ratio of indicators for 2013 to the value of the average humus content on virgin soil (4.2%). Standard 4-point classifiers [7, 8] are used. Calculations show that for the Rostov region $g_3$ = 0.385, which means belonging to the term $G_2$ – "environmental risk", which coincides with the estimates of [9].

### 3.3 Assessment of the Dynamics of Emissions of Pollutants into the Atmosphere

The analysis of the dynamics of emissions of pollutants into the atmosphere was carried out on the basis of data on emissions of pollutants into the atmosphere in the Rostov region, coming from stationary sources of agricultural production, thousands of tons [9]. The linguistic variable $g_4$ = "assessment of the dynamics of polluting emissions into the air" is introduced. Term-set consists of five terms $G = \{G_1, G_2, G_3, G_4, G_5\}$, which conditionally assess the state of the system: $G_1$ – "a stable trend to reduce emissions into the atmosphere"; $G_2$ – "a tendency to reduce emissions into the atmosphere"; $G_3$ – "stabilization of the situation"; $G_4$ – "tendency to increase emissions into the atmosphere"; $G_5$ – "steady tendency to increase emissions into the atmosphere". The time series of the wos aggregated parameters on the basis of formula (1). Standard 5-point classifiers are used to form the final estimate. Calculations show that for the Rostov region $g_4$ = 0.46, which means belonging to the term $G_3$ – "stabilization of the situation".

### 3.4 Assessment of the Dynamics of Water Use in Agriculture

The analysis was carried out on the basis of data on the volume of water used for 2007–2016 [9]. The final score is aggregated based on the formula (4). The linguistic variable $g_5$ = "assessment of the dynamics of the load on the water system of the region" is introduced. Term-set consists of five terms $G = \{G_1, G_2, G_3, G_4, G_5\}$: $G_1$ – "stable tendency to decrease the load on the water system of the region"; $G_2$ – "tendency to decrease the load on the water system of the region"; $G_3$ – "stabilization of the situation"; $G_4$ –"tendency to increase the load on the water system of the region"; $G_5$ – "a stable tendency to increase the load on the water system of the region". The calculations show that for the Rostov region $g_5$ = 0.5, which means equally belonging to two terms $G_1$ – "a stable tendency to reduce the load on the water system of the region"; $G_2$ – "a tendency to reduce the load on the water system of the region".

### 3.5 Comprehensive Assessment "Impact of Agricultural Production on the Ecology of the Region"

On the basis of standard 5-point classifiers, the estimates obtained in paragraphs 2.1–2.4 are aggregated: $g_1$ = "assessment of the dynamics of chemical load on the ecological system of the region"; $g_2$ = "assessment of the dynamics of humus content in the soil"; $g_3$ = "assessment of the level of humus content in the soil"; $g_4$ = "assessment of the dynamics of polluting emissions into the air; $g_5$ = "assessment of the dynamics of the load on the water system of the region".

The linguistic variable $g$ = "assessment of the negative impact of agricultural production on the ecological system of the region" is introduced. Term-set consists of five terms $G = \{G_1, G_2, G_3, G_4, G_5\}$, conditionally assessing the state of the system: $G_1$ – "extremely small"; $G_2$ – "minor"; $G_3$ – "average"; $G_4$ – "significant"; $G_5$ – "destructive". Membership functions have standard trapezoid forms. Calculations based on the results of the previous paragraphs show that for the Rostov region $g = 0.49$, which means equally belonging to the term $G_3$ – "the average negative impact of agricultural production on the environmental system of the region".

## 4  Comprehensive Assessment of Agricultural Production Sustainability in the Region

Assessment $\alpha$ = "integrated assessment of agricultural production sustainability "is formed by aggregation on the basis of standard five-point classifiers of the two assessments obtained in paragraphs 1 and 2: $\gamma$ = "integrated assessment of agricultural production intensity" and $\beta = 1-g$ where $g$ = "assessment of the negative impact of agricultural production on the ecological system of the region". The term set consists of five terms $G = \{G_1, G_2, G_3, G_4, G_5\}$: $G_1$ – "very bad"; $G_2$ – "bad"; $G_3$ – "satisfactory"; $G_2$ – "good"; $G_5$ – "excellent". Calculations based on the results of claim 1, 2 show that for the Rostov region $\alpha = 0.59$, which means equally belonging to the term $G_3$ – "satisfactory".

## 5  Conclusions

The method of stability of agricultural production in the region on the basis of standard five-level fuzzy [0,1]-classifiers is offered. The practical significance of the methodology is that it allows to assess the sustainability of agricultural production in the region based on the aggregation of estimates of the intensity of agricultural production in the region, as well as a comprehensive assessment of the impact of agricultural production on the environment of the region. The calculation of complex estimates is made on the basis of normalized values of heterogeneous indicators, formed on the basis of aggregation of time series of numerical statistical values of heterogeneous indicators, reflecting both the level and the rate of growth for the corresponding periods; and the contribution of each of the indicators is measured by a weight factor, reflecting its importance.

## References

1. Bondarenko, L.: Ecological and economic efficiency and stability of grain production (based on the materials of the Krasnodar region). KubSU Publishing Center, Krasnodar (2000)
2. Parkhomenko, N., Shchukina, L.: Methods of assessing the sustainability of agriculture in the region. Agric. Econ. **7**, 16–22 (2014)
3. Minakov, I., Kulikov, I., Sokolov, V.: The economics of agriculture. Koloss, Moscow (2008)
4. Nedosekin, A.: Fuzzy sets and financial management. AFA Library, Moscow (2003)
5. Stryukov, M., Sakharova, L., Alekseychik, T., Bogachev, T.: Methods of estimation of intensity of agricultural production on the basis of the theory of fuzzy sets. Int. Res. J. **7**(61), 123–129 (2017)

6. Sakharova, L., Stryukov, M., Akperov, I., Alekseychik, T., Chuvenkov A.: Application of fuzzy set theory in agro-meteorological models for yield estimation based on statistics. In: 9th International Conference on Theory and Application of Soft Computing, Computing with Words and Perception, Budapest, Hungary, vol. 120, pp. 820–829. Procedia Computer Science (2017)

7. Kramarov, S., Sakharova, L.: Management of complex economic systems by fuzzy classifiers method. Sci. Bull. South. Univ. Manag. 2(18), 42–50 (2017)

8. Kramarov, S., Sakharova, L., Khramov, V.: Soft computing in management: management of complex multivariate systems based on fuzzy analog controllers. Sci. Bull. South. Univ. Manag. 3(19), 42–51 (2017)

9. Environmental Bulletin of the Don "On the state of the environment and natural resources of the Rostov region in 2011/2016 year". Rostov-on-Don: Government-in the Rostov region. Rostov-on-Don. http://минприродыро.рф/state-of-the-environment/ekologicheskiy-vestnik/. Accessed 18 July 2018

10. Ants, A., Karryev, B.; Evaluation of the ecological state of the soil. Krismas+, Moscow (2008)

# Fuzzy-Multiple Analysis of Dependence Between Psychological Features of Human and Its Predispolence to Drug Addiction

E. Astapenko[(⊠)] and A. Maksimenko[(⊠)]

Southern University (Institute of Business and Law Management),
Rostov-on-Don, Russia
eugenia-a@yandex.ru, alina.maximencko@mail.ru

**Abstract.** The article deals with the problem of determination of chemical dependence using fuzzy mathematical sets. Psychological features of persons with chemical dependence are described.

**Keywords:** Character accentuation · Attitude to the disease · Coping strategy
Drug addiction

## 1   Introduction

The paper suggests an analysis between the psychological characteristics of a person, and his predisposition to dependence. The subject of the study was the psychological characteristics of people, such as character accentuations, the type of relationship to disease and coping mechanisms.

These studies and developed methods are designed for use by physicians psychiatrists, narcologists, social workers, psychotherapists, clinical psychologists, addiction specialists.

According to the estimation of the SAC (as of May 2015), about 8 million Russians (8% of the total Russian population) use drugs with varying degrees of frequency. The first in popularity is heroin, the second is marijuana, the third is synthetic mixtures - spices (the second most dangerous drug). The total amount spent by consumers on drugs is estimated at 4.5 million rubles daily or up to 1.5 trillion rubles annually. According to the head of the Federal Drug Control Service Viktor Ivanov, almost 70 thousand people die of drugs in Russia every year.

According to the Federal State Statistics Service, at the end of 2013, 308,300 drug addicts were registered in treatment and prevention organizations.

According to the National Scientific Center of Narcology of the Ministry of Health of the Russian Federation, in 2013 there were 96 narcological dispensaries in Russia (of which 88 were inpatient); 1845 institutions with out-patient narcological units (offices), including 252 adolescents; there were four independent rehabilitation centers (in the Kurgan and Sverdlovsk regions, North Ossetia, the Khanty-Mansi Autonomous Okrug).

© Springer Nature Switzerland AG 2019
R. A. Aliev et al. (Eds.): ICAFS-2018, AISC 896, pp. 689–693, 2019.
https://doi.org/10.1007/978-3-030-04164-9_90

In addition, there are non-state rehabilitation centers in Russia, their number as of December 1, 2014 was about 1 thousand (data of the Federal Drug Control Service), as well as over 60 rehabilitation centers of the Russian Orthodox Church.

From the above, it can be judged that the number of rehabilitation institutions in our country can not cope with the number of drug addicts. Moreover, it should be noted that to date, drug addiction is noticeably "younger", because of the easy availability of substances has become a more common problem in all social strata.

Reflected in numerous publications, the clinical experience of drug addiction treatment testifies that these diseases, especially such form as opioid dependence, in most cases are of low curative nature, showing high resistance to all known measures of therapeutic effect. To explain this situation, psychiatrists-narcologists rely on the process-biological mechanisms of the development of dependence on psychoactive substances (SAA), considering the problem of therapeutic resistance in the context of high progression of the disease, intensity and persistence of the pathological attraction to the drug (Ivanets and Vinnikova 2008). There is another line of reasoning that occurs in psychotherapeutically oriented specialists, which focuses on the personal factors of the pathogenesis of the narcological disease. In this case, the resistance of the disease to therapy is defined as the absence or lack of motivation for treatment in such patients, which in turn are the result of significant personality disorders. Changing personality (or addictive personality) generates a high degree of resistance in relation to medical measures, depriving it of conventionally beneficial aspects of drug use and the possibility, albeit pathological, but satisfaction of destructive personal tendencies (Sirota and Yaltonsky 2008; Belokrylov 2008; Dudko 2007; Valentik 2000) [3].

The topicality of this subject is that most rehabilitation programs do not presuppose a study of the personality of an addict individually, and are often set to recover from "old" drugs, and show their effectiveness precisely in this case. In recent times, the main type of dependence is dependence on synthetic drugs, such as spice and salt. Experience in the center shows that addicted to this type of drug are experiencing difficulties with the experience of an irresistible attraction to use and strong euphoric memories from the experience of use. That is why they must be considered separately and studied much deeper than those dependent on "classical" drugs and alcohol.

## 2    Description of the Study

The aim of the study was to study the psychological characteristics of people who suffer from chemical addiction and healthy people.

The subject of the study was the psychological characteristics of people, such as character accentuations, the type of relationship to disease and coping mechanisms.

The object of the study were 12 people who suffer from chemical dependence at the age of 26 to 38 years and 12 healthy people.

A technique has been developed that makes it possible to establish a correspondence between the psychological characteristics of a person and his predisposition to drug dependence, based on a fuzzy analysis of the statistical data of psychological testing using the ranking of fuzzy sets [1, 2].

Hypotheses are advanced.

1. Persons suffering from drug addiction have specific character accentuations.
2. Persons suffering from drug dependence are distinguished by specific reactions to difficult life situations (types of coping).

The results of testing of two groups of 12 people in five directions are compared:

1. character accentuation;
2. assessment of the behavior of the subject in a difficult life situation.

Based on the fuzzy ranking, the main features characteristic of a group of people suffering from drug addiction are highlighted. For the selected features a rating is constructed, on the basis of which the hypotheses put forward are checked.

## 3   The Results of the Study

### 3.1   Character Accentuation

Table 1 Functions of the fuzzy set belonging to the "Character accentuation", constructed on the basis of testing two groups of 12 respondents on the basis of Method 1; The 1st value corresponds to the sum of scores below the diagnostic number of the type being investigated, the 2nd value is higher than the diagnostic number of the type being investigated; The diagnostic number for each type is 12 points.

**Table 1.** Functions of the fuzzy set belonging to the "Character accentuation".

|    | Accentuation | Healthy people | Increase in excess standards | Sick people |
|----|--------------|----------------|------------------------------|-------------|
| 1. | Hypertensive | (9/12, 3/12) | → | (0, 1) |
| 2. | Excitable | (9/12, 3/12) | → | (7/12, 5/12) |
| 3. | Emotional | (8/12, 4/12) | → | (4/12, 8/12) |
| 4. | Pedantic | (11/12, 1/12) | → | (5/12, 7/12) |
| 5. | Troublesome | (1, 0) | → | (9/12, 3/12) |
| 6. | Cyclotum | (3/12, 9/12) | ← | (5/12, 7/12) |
| 7. | Demonstrative | (2/12, 10/12) | ← | (5/12, 7/12) |
| 8. | Unbalanced | (11/12, 1/12) | → | (1/12, 11/12) |
| 9. | Dysthymic | (5/12, 7/12) | ← | (11/12, 1/12) |
| 10. | Exalted | 4/12, 8/12) | → | (1/12, 11/12) |

We will denote Ai1, Ai2 fuzzy sets corresponding to healthy and sick people according to the i-th character of character accentuation. We consider that the center of the first subset (below the diagnostic number of the type being investigated) is 6 points; the second subset (above the diagnostic number of the type being investigated) is 18 points. Indices of ranking indicators are presented in the following table. Negative or zero ranking indexes have a cyclotimous, demonstrative and distinctive accentuation form. The other types can be ranked in descending order of the ranking factor as follows (Table 2):

1. disturbing;
2. pedantic;
3. emotive;
4. exalted;
5. is hypertensive;
6. unbalanced;
7. excitable.

**Table 2.** Calculation table "Character accentuation"

| | Accentuation | $H_+(A_{i1})$ | $H_+(A_{i2})$ | $H(A_{i2}, A_{i1})$ | Place in the rating |
|---|---|---|---|---|---|
| 1. | Hypertensive | 7,5 | 10,5 | 3 | 5 |
| 2. | Excitable | 7,5 | 8,5 | 1 | 7 |
| 3. | Emotional | 8 | 16 | 8 | 3 |
| 4. | Pedantic | 6,5 | 15,5 | 9 | 2 |
| 5. | Troublesome | 6 | 18 | 12 | 1 |
| 6. | Cyclotum | 16,5 | 15,5 | −1 | − |
| 7. | Demonstrative | 17 | 15,5 | −1,5 | − |
| 8. | Unbalanced | 17,5 | 11 | 2 | 6 |
| 9. | Dysthymic | 15,5 | 6,5 | −9 | − |
| 10. | Exalted | 16 | 17,5 | 7 | 4 |

Thus, the hypothesis is confirmed that a number of psychological accentuations accompany a narcological dependence.

### 3.2 Assessment of the Behavior of the Subject in a Difficult Life Situation

Table 3 Fuzzy set membership functions "Assessment of the subject's behavior in a difficult life situation", built on the basis of testing two groups of 19 respondents on the basis of Method 4; The 1st value corresponds to a low level of anxiety (0–6 points); 2nd - moderate (7–12 points); 3rd - high (13–18 points).

We denote Ci1, Ci2 fuzzy sets corresponding to healthy and sick people according to the i-th type of coping. We believe that the center of the first subset (low level of expression) is 3 points; the second subset (moderate level of expression) - 10 points; third subset (high level of expression) - 15 points. Indices of ranking indicators are presented in the following table. The negative ranking index has all kinds of copying, except for two (Table 4):

1. flight-avoidance;
2. acceptance of responsibility.

It should be noted that the most dramatically reduced, in comparison with the control group of healthy people, coping the planning of the solution to the problem, then self-monitoring, then the search for social support. Thus, the hypothesis that drug dependence is accompanied by certain types of coping is confirmed.

**Table 3.** Fuzzy set membership functions "Assessment of the subject's behavior in a difficult life situation"

| | Types of coping | Healthy people | Growth excess norms | Sick people |
|---|---|---|---|---|
| 1. | Confrontational | (3/12, 9/12, 0) | → | (3/12, 8/12, 1/12) |
| 2. | Distancing | (2/12, 8/12, 2/12) | = | (2/12, 8/12, 2/12) |
| 3. | Self-monitoring | (2/12, 3/12, 7/12) | ← | (2/12, 5/12, 5/12) |
| 4. | Finding social support | (1/12, 7/12, 4/12) | ← | (2/12, 8/12, 2/12) |
| 5. | Adoption of identity | (5/12, 7/12, 0) | = | (2/12, 10/12, 0) |
| 6. | Escape-avoidance | (0, 7/12, 5/12) | → | (3/12, 3/12, 6/12) |
| 7. | Planning a solution to a problem | (0, 9/12, 3/12) | = | (4/12, 5/12, 3/12) |
| 8. | Positive revaluation | (0, 7/12, 5/12) | ← | (2/12, 6/12, 4/12) |

**Table 4.** Calculation table "Assessment of the subject's behavior in a difficult life situation"

| | Types of coping | $H_+(D_{i1})$ | $H_+(D_{i2})$ | $H(D_{i2}, D_{i1})$ | Place in the rating |
|---|---|---|---|---|---|
| 1. | Confrontational | 9,13 | 9,04 | −0,09 | – |
| 2. | Distancing | 9,67 | 9,67 | 0 | – |
| 3. | Self-monitoring | 13,38 | 11,92 | −1,46 | – |
| 4. | Finding social support | 10,75 | 9,67 | −1,08 | – |
| 5. | Adoption of identity | 8,54 | 9,42 | 0,88 | 2 |
| 6. | Escape-avoidance | 11,04 | 12 | 1 | 1 |
| 7. | Planning a solution to a problem | 10,63 | 8,58 | −2,05 | – |
| 8. | Positive revaluation | 11,04 | 10,67 | −0,37 | – |

## 4  Conclusions

A technique has been developed that makes it possible to evaluate the degree of severity of the studied psychological signs in the studied group of people suffering from drug addiction in relation to a group of healthy people. The methodology is based on a fuzzy analysis of statistical data obtained using three standard methods.

## References

1. Blumin, S.L., Blyumin, S.L., Shuykova, I.A., Sarayev, P.V., Cherpakov, I.V.: Fuzzy Logic: Algebraic Bases and Applications: Monograph. LEGI, Lipetsk (2002)
2. Khaptahaeva, N.B., Dambaeva, S.V., Ayusheeva, N.N.: Introduction to the Theory of Fuzzy Sets: Textbook. Part I. Publishing house of the All-Russian State Technical University, Ulan-Ude (2004)
3. Tuchin, P.V., Valchuk, D.S.: Psychotherapeutic approaches to patients with opioid dependence. Moscow (2013)

# Fuzzy Models of the Results of the Mastering the Educational Programs in the Field of Information Security

Evgeniy Nikolaevich Tishchenko$^{(\boxtimes)}$, Elena Victorovna Zhilina,
Tatyana Nikolaevna Sharypova, and Galiya Nailevna Palyutina

Rostov State University of Economics,
Bolshaya Sadovaya Street, 69, Rostov-on-Don 344002, Russia
`celt@inbox.ru`

**Abstract.** The methodic of the evaluation of the results of the mastering the educational programs in the field of information security has been developed in the article. It is based on the theory of the fuzzy sets.

The method allows to make integral accounting both quantitative and qualitative factors of the adaptive testing within the framework of the intermediate certification of the student's mastering the discipline of the educational program in the field of information security. Setting the criterion of the significance of the confidence level of the membership functions of the input (output) variables' quality, you can change the final results depending on the group level of preparedness of students.

Within the framework of the accumulative point-rating system, the use of the theory of fuzzy sets allows to accumulate scores in a 100-point scale for all types of academic work and form the final score for each discipline of the educational program in the field of information security, depending on the maximum possible scores established for each volume work performed.

**Keywords:** Fuzzy model · Educational program · Linguistic variable
Membership function

## 1 Introduction

The essential feature of the higher education is the complexity of quantitative evaluation of the results of the mastering the educational programs. There is no uniquely understood list of quality indicators for training specialists in the field of information security, because there is no the clear ideas about what quantitatively measurable factors affect the learning process, what is the reliability of these indicators, etc. [5]. The methodic of the translation the summary total into the assessment is not usually considered in the works devoted to the tests.

The fuzziness of such a representation does not allow obsolete methods of mathematical modeling to receive adequate quantitative descriptions of the explored parameters, and therefore forces us to seek solutions to the classical problems of the educational process by non-classical methods.

R. A. Aliev et al. (Eds.): ICAFS-2018, AISC 896, pp. 694–701, 2019.
https://doi.org/10.1007/978-3-030-04164-9_91

The current and intermediate control of the student's mastering of each discipline of the educational program in the field of information security is proposed to be implemented within the accumulative point-rating system. The evaluation of the student's work is carried out in a 100-point scale for each supervised type of educational work, and for a specific module, discipline. The evaluation is set in the course of the current knowledge's control during the semester, and also in the intermediate control - passing credits and exams. The evaluation reflects the quality of mastering the educational material and the level of acquired knowledge and skills [1].

It is proposed to conduct intermediate certification by adaptive testing, adjusting to the student's level of training. At the same time, the change in the level of complexity of the next test issue occurs depending on the answer to the previous one: if the answer is yes, the level of complexity increases, with a negative one goes down. It should be noted that adaptive testing is possible if the difficulty of test tasks is predetermined, which presupposes their preliminary approbation and processing of results using the methods of analysis of variance and testing theory for determining reliability, validity and differentiating ability [2].

When integrating the results of the mastering the educational programs in the field of information security using linguistic variables, it is necessary to keep a record of quantitative and qualitative factors [4]. As a result of carrying out passive experiments, an evaluation of the explored characteristics is performed, for which limitations are set in the form of threshold values.

## 2    Main Part

When constructing a model for evaluating the results of the mastering the educational programs in the field of information security (OY), both quantitative factors (the number of questions, the number of correct answers, the total score) and the qualitative factors (level 2 (x1), level 3 (x2), 4 (x3), and 5 (x4)) are used.

The formation's model of the linguistic evaluation of the results of the mastering the educational programs in the field of information security (OY) has a general form (Fig. 1):

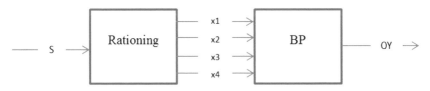

**Fig. 1.** The formation's model of the linguistic evaluation of the results of the mastering the educational programs in the field of information security in general.

where S - cumulative sum of points; x1 - level at 2; x2 - level at 3; x3 - level at 4; x4 - level at 5.

The method for the formation of a linguistic evaluation of the results of the mastering the educational programs in the field of information security is as follows

1. **To normalize the accumulated sum of points to the interval [0…100] by levels.**
   The variable $x_i$, $i = \overline{1,4}$ ("level by 2" (x1), "level by 3" (x2), "level by 4" (x3), "level by 5" (x4)) is normalized depending on the values of the auxiliary variable xi on the segment [0…100]:

$$
x_i = \begin{cases}
x_i \times 20, & if x_i \leq 1.2 \\
24 + (x_i - 1.2) \times 62.5, & if\ 1.2 < x_i \leq 1.6, \\
49 + (x_i - 1.2) \times 20, & if\ 1.6 < x_i \leq 2, \\
57 + (x_i - 1.6) \times 12.8, & if\ 2 < x_i \leq 2.7, \\
66 + (x_i - 2.7) \times 11.4, & if\ 2.7 < x_i \leq 3.4, \\
74 + (x_i - 3.4) \times 4.5, & if\ 3.4 < x_i \leq 5.4, \\
100 - 3^{(7.4-x_i)}, & if\ 5.4 < x_i < 7.4, \\
100, & if\ x_i \geq 7.4
\end{cases}
\tag{1}
$$

The formulae were compiled according to the compliance data of the 100-point scale used in Draft Regulations, 2011 for transferring the points to the traditional evaluation, and the numerical values $x'_i$, $i = \overline{1,4}$, obtained in the course of the experiment (the clarity of the calculations is shown in Fig. 2 [4]. For example, the value 7.4 corresponds to the maximum value of $x'_4$ when passing the adaptive test, starting with 1 difficulty level; values 3.4, 2 and 1.2 - the average values of $x'_3, x'_2, x'_1$; the values of 1.6, 2.7 and 5.4 are obtained by adding the left value and the difference between the adjacent values, divided in half (1.6 = 1.2 + (2-1.2)/2; 2.7 = 2 + (3.4 − 2)/2, etc.).

| 0 | 1,2 | 1,6 | 2,0 | 2,7 | 3,4 | 5,4 | 7,4 |
|---|---|---|---|---|---|---|---|
| | *20 | *62,5 | *20 | *12,8 | *11,4 | *4,5 | |
| 0 | 2,4 | 49 | 57 | 66 | 74 | 83 | 100 |
| unsatisfactory | | | satisfactory | | well | | excellent |

**Fig. 2.** The numerical values necessary to form the argument "x" of the quality-of-output function of the output variable

On each of the obtained segments there were coefficients of correspondence of numerical values to scores. For example, there are scores from 74 to 83 between [3.4, 5.4], therefore the numerical value of unity on this segment corresponds to 4.5 points ((83 − 74)/(5.4 − 3.4)).

2. **To set the classification scale to the linguistic variables $x_i$, $i = \overline{1,4}$.**
   For example, the linguistic variables "level by 2" (x1), "level by 3" (x2), "level by 4" (x3), "level by 5" (x4), we interpret as a term set with triar scale T2 = {DNM, MtS, CC}, where the value of DNM - does not match (2, 3, 4 or 5), MtS - matches slightly and CC - corresponds completely.

3. **To set the membership functions to variables $x_i$, $i = \overline{1,4}$.**

Each of the linguistic variables "level on 2 (3, 4, 5)" has one triangular membership curve and two T-shaped membership curves ($\mu_{x_i}^{DNM}$, $\mu_{x_i}^{MtS}$, $\mu_{x_i}^{CC}$, $i = \overline{1,4}$).

The membership function of the fuzzy term-sets to the linguistic variable "level on 2 (3, 4, 5)" (x1, x2, x3, x4) will have the following form:

(1) $\mu_{Tx1}^{CC}$ (x1, 0, 0, 15, 30); $\mu_{\Delta x1}^{DNM}$ (x1, 20, 35, 50); $\mu_{\Delta x1}^{MtS}$(x1, 40, 55, 100, 100);

(2) $\mu_{Tx2}^{DNM}$ (x2, 0, 0, 20, 40); $\mu_{\Delta x2}^{CC}$ (x2, 30, 50, 70); $\mu_{\Delta x2}^{MtS}$(x2, 60, 70, 100, 100);

(3) $\mu_{Tx3}^{HC}$ (x3, 0, 0, 50, 70); $\mu_{\Delta x3}^{CC}$ (x3, 55, 70, 85); $\mu_{\Delta x3}^{MtS}$(x3, 70, 90, 100, 100);

(4) $\mu_{Tx4}^{DNM}$ (x4, 0, 0, 70, 80); $\mu_{\Delta x4}^{MtS}$ (x4, 70, 80, 90); $\mu_{\Delta x4}^{CC}$(x4, 80, 90, 100, 100).

Figures 3, 4, 5 and 6 show graphs of the membership function of the fuzzy term-sets to the linguistic variable "level on 2 (3, 4, 5)" (x1, x2, x3, x4): input1 in the model corresponds to the linguistic variable "level on 2" - x1; input2 in the model corresponds to the linguistic variable "level on 3" - x2; input3 in the model corresponds to the linguistic variable "level on 4" - x3; input4 in the model corresponds to the linguistic variable "level on 5" - x4.

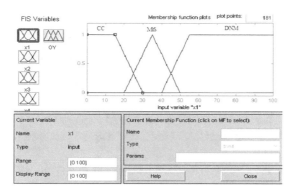

**Fig. 3.** Formation of input1- "level by 2" (x1).

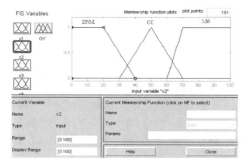

**Fig. 4.** Formation of input2 - "level by 3" (x2)

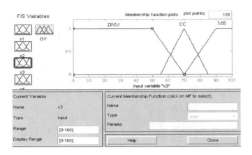

**Fig. 5.** Formation of input3 - "level for 4" (x3)

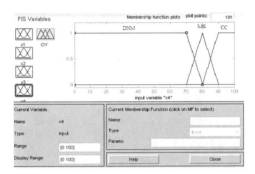

**Fig. 6.** Formation of input4 - "level by 5" (x4).

The simulation is carried out using a specialized package Fuzzy Logic Toolbox tool MATLAB [3]. The implementation of the fuzzy output is realized on the basis of the algorithm of Mamdani.

The values of parameters a, c, d and b can be refined in accordance with the experimental data. When classifiers (triar and tetra scales) are used on a carrier of a fuzzy set, the significance of linguistic variables is determined on the segment of the real axis [0,1].

4. **To set the classification scale of the explored parameter.**

   For example, the linguistic variable "Evaluation of the mastering the educational programs (OY) is interpreted as a term-set of values T1 = {two (H), three (Y), four (X), five (O)}.

5. **To set the quality's membership function of the explored parameter.**

   The linguistic variable "Evaluation of the mastering the educational programs" has two triangular membership curves and two T-shaped membership curves $(\mu_{OY}^H, \mu_{OY}^Y, \mu_{OY}^X, \mu_{OY}^O)$. The membership functions of fuzzy term-sets of the resultant variable OY will have the following form: $\mu_{OY}^H$(x, 0, 0, 25, 50); $\mu_{OY}^Y$(x, 40, 55, 70); $\mu_{OY}^X$(x, 60, 75, 90); $\mu_{OY}^O$(x, 80, 90, 100, 100). In Fig. 7 there is a graph of the membership functions of fuzzy term-sets in the linguistic variable "Evaluation of the mastering the educational programs", defined in MATLAB.

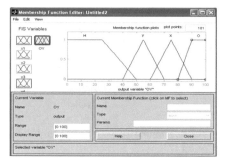

**Fig. 7.** Forming output1 - "Evaluation of the mastering the educational programs" (OY).

6. **To determine the linguistic meaning of the level of the explored factor.** A database of evaluation's rules has been formed (Table 1).

**Table 1.** Fuzzy productional rules.

| Rule | Kind of term | x1 | x2 | x3 | x4 | OY |
|------|-------------|----|----|----|----|-----|
| BR1 | DNM | + | | | | $OY^O$ |
| | MtS | | + | + | + | |
| | CC | | | | + | |
| BR2 | DNM | + | | | | $OY^X$ |
| | MtS | | + | | + | |
| | CC | | | + | | |
| BR3 | DNM | + | | | + | $OY^X$ |
| | MtS | | + | | | |
| | CC | | + | + | | |
| BR4 | DNM | | | | + | $OY^X$ |
| | MtS | + | + | + | | |
| | CC | | | | | |
| BR5 | DNM | + | | | + | $OY^Y$ |
| | MtS | + | | + | | |
| | CC | | + | | | |
| BR6 | DNM | + | | + | + | $OY^Y$ |
| | MtS | + | + | | | |
| | ПC | | | | | |
| BR7 | DNM | | + | + | + | $OY^H$ |
| | MtS | + | | | | |
| | CC | + | | | | |
| BR8 | DNM | | | | + | $OY^H$ |
| | MtS | | + | + | | |
| | CC | + | | | | |

7. **Accumulation of the conclusion** under all the rules was carried out using the max-disjunction operation. When dephasing, the center of gravity method was used for a discrete set of values of membership functions:

$$y' = \frac{\sum_{r=1}^{Y_{max}} y_r \mu_B(y_r)}{\sum_{r=1}^{Y_{max}} \mu_B(y_r)} \tag{2}$$

Where $Y_{max}$ is the number of elements $y_r$ in the region Y, which is quantized to calculate the "center of gravity".

Semester evaluation of the student's progress in each academic discipline of the educational program is deduced, based on the maximum sum of points, equal to 100 [1] (Table 2). If the linguistic estimates OY = (OY1... OYn) are known for each kind of study (F1... Fn) and the weighting factors p = ($p_1$,..., $p_n$) are defined, then the information aggregation operator is a weighted sum and is characterized by its linguistic evaluation, determined by the membership function on the 01-classifier.

$$\mu_{OY}(x) = \sum_{i=1}^{n} \mu_i(x)p_i$$

**Table 2.** Bulletin of scores on discipline.

| Full name | Name | | | | | | |
|-----------|------|----------|-------------|------|--------|------|--------|
| | Lectures | Practical lessons | Independent work | Test | Credit | Exam | The result |
| Information security | 8 | 8 | 10 | 12 | 12 | 50 | 100 |
| Remez M. S. | 97 | 79 | 83 | 82 | 87 | 84 | |

According to the Table 2 $\mu_Y(x) = 97 \times 0.08 + 79 \times 0.08 + 83 \times 0.1 + 82 \times 0.12 + 87 \times 0.12 + 84 \times 0.5 = 84.66 \approx 85$, which corresponds to the term of the linguistic variable OY "Evaluation of the mastering the educational programs in the field of information security" O - five with a confidence level of $\mu OYO = 0.5$. If you set the confidence level significance level KrZ = 0.05, the received value of the OY variable is accepted, so the student is graded "Excellent".

## 3  Conclusions

1. The method allows to make integral accounting both quantitative and qualitative factors of the adaptive testing within the framework of the intermediate certification of the student's mastering the discipline of the educational program in the field of information security. Setting the criterion of the significance of the confidence level of the membership functions of the input (output) variables' quality, you can change the final results depending on the group level of preparedness of students.

2. Within the framework of the accumulative point-rating system, the use of the theory of fuzzy sets allows to accumulate scores in a 100-point scale for all types of academic work and form the final score for each discipline of the educational program in the field of information security, depending on the maximum possible scores established for each volume work performed.

# References

1. Draft Regulations on the organization of the educational process at the Rostov State University of Economics with the use of credit units (credits) and a point-rating system, RSUE, Rostov-on-Don (2011)
2. Avanesov, V.: Composition of test tasks. In: Testing Center of the Ministry of Education of the Russian Federation, Moscow, p. 239 (2002)
3. Leonenkov, A.: Fuzzy modeling in the environment of MATLAB and fuzzyTECH. BHV-Petersburg, St. Petersburg (2005)
4. Zhilina, E.: Fuzzy models for assessing the success of mastering discipline by a student. Manage. Econ. Syst.: Electron. Sci. J. 35(11). http://www.uecs.ru. Accessed 21 Mar 2018
5. Tishchenko, E., Sharypova, T., Zhilina, E., Cherkezov, S.: Economic and mathematical modeling of complex cooperation of academic staff of educational cluster on the basis of fuzzy sets theory. J. Appl. Econ. Sci. (JAES) 5(11), 905–907 (2016)

# The Possibilities of Using Modern CASIO CG-50 Graphing Calculators for Volumetric and Complex Calculations, Including Fuzzy Calculations

Igor Vostroknutov[1]([✉]) and Yosuke Kaneda[2]

[1] CASIO Europe GmbX, Butyrskaya 77, 127015 Moscow, Russia
vostroknutov_i@mail.ru
[2] CASIO Europe GmbX, Casio-Platz 1, 22848 Norderstedt, Germany

**Abstract.** The general statement of a choice problem of the effective virtual business relations is formulated and possible ways of decision are defined basing on principles of fuzziness and the analytic hierarchy process. In the article, the computational capabilities of modern CASIO CG-50 graphical calculators are considered for solving complex mathematical problems. It is shown how the task of complex assessment of the state of an object can be solved using the CASIO CG-50 spreadsheets based on the aggregation of numerical values of a complex of indicators using a system of fuzzy inferences (fuzzy classifiers). In particular, we show: the calculation of the values of the membership functions of terms of the linguistic variable "indicator level" for each of the indicators from the initial formulas, the calculation of the numerical value of the linguistic variable, the complex evaluation of the state of the object in accordance with the parameters of the term sets.

**Keywords:** Fuzzy calculations · CASIO CG-50 graphic calculator
Calculation of weights of terms of linguistic variable

## 1 Introduction

Graphical calculators are widely used for teaching mathematical and economic disciplines in colleges and universities of most of the world's information developed countries. And they not only do not compete with computer and interactive technologies, but also successfully supplement them. Now the most popular calculators CASIO fx - CG20, fx-CG50 and CLASSPAD fx-CP400.

The computational capabilities of these CASIO graphical calculators make it possible to apply them not only in teaching mathematical, economic and technical subjects, but also in research and educational activities. As an example, let us consider the task of forming a complex assessment of the state of an object based on the aggregation of numerical values of a complex of indicators using a system of fuzzy-logical inferences (fuzzy [0,1] - classifiers) using the CASIO CG-50 calculator.

© Springer Nature Switzerland AG 2019
R. A. Aliev et al. (Eds.): ICAFS-2018, AISC 896, pp. 702–708, 2019.
https://doi.org/10.1007/978-3-030-04164-9_92

## 2  Work Algorithm

**1 step.** The task:
- the number of indicators N = 5,
- their weight coefficients, ki, i = 1,2,...,N, mandatory condition:

$$\sum_{i=1}^{N} k_i = 1$$

- their normalized numerical values xi (i = 1,2,...,N,), $0 \leq xi \leq 1$.

**2 step.** Calculation of the values of the membership functions of the terms of the linguistic variable "indicator level" for each of the indicators by the formulas given in Table 1.

**Table 1.** Functions of the membership terms.

| Terms $B_{il}$, $G_l$ | Function of belonging to fuzzy sets sets $B_{il}$, $G_l$ |
|---|---|
| $B_{i1}$ – «very low level of the indicator»; $G_1$ – «1-st term»; | $\mu_1(x) = \begin{cases} 1, & \text{if } 0 \leq x < 0.15 \\ 10(0.25 - x), & \text{if } 0.15 \leq x < 0.25 \\ 0, & \text{if } 0.25 \leq x < 1 \end{cases}$ |
| $B_{i2}$ – «low level of the indicator»; $G_2$ – «2-nd term»; | $\mu_2(x) = \begin{cases} 0, & \text{if } 0 \leq x < 0.15 \\ 10(x - 0.15), & \text{if } 0.15 \leq x < 0.25 \\ 1, & \text{if } 0.25 \leq x < 0.35 \\ 10(0.45 - x), & \text{if } 0.35 \leq x < 0.45 \\ 0, & \text{if } 0.45 \leq x \leq 1 \end{cases}$ |
| $B_{i3}$ – «average level of the indicator»; $G_3$ – «3-rd term»; | $\mu_3(x) = \begin{cases} 0, & \text{if } 0 \leq x < 0.35 \\ 10(x - 0.35), & \text{if } 0.35 \leq x < 0.45 \\ 1, & \text{if } 0.45 \leq x < 0.55 \\ 10(0.65 - x), & \text{if } 0.55 \leq x < 0.65 \\ 0, & \text{if } 0.65 \leq x \leq 1 \end{cases}$ |
| $B_{i4}$ – «high level of the indicator»; $G_4$ – «4-th term»; | $\mu_4(x) = \begin{cases} 0, & \text{if } 0 \leq x < 0.55 \\ 10(x - 0.55), & \text{if } 0.55 \leq x < 0.65 \\ 1, & \text{if } 0.65 \leq x < 0.75 \\ 10(0.85 - x), & \text{if } 0.75 \leq x < 0.85 \\ 0, & \text{if } 0.85 \leq x \leq 1 \end{cases}$ |
| $B_{i5}$ – «very high level of the indicator»; $G_5$ – «5-th term». | $\mu_5(x) = \begin{cases} 0, & \text{if } 0 \leq x < 0.75 \\ 10(x - 0.75), & \text{if } 0.75 \leq x < 0.85 \\ 1, & \text{if } 0.85 \leq x < 1 \end{cases}$ |

Each value is associated with a linguistic variable $B_i$, which term multiplicity $B_{il} = \{B_{i1}, B_{i2}, B_{i3}, B_{i4}, B_{i5}\}$, $i = 1,2,...,N$ consists of the following terms:

$B_{i1}$ – «very low level of the indicator» ;
$B_{i2}$ – «low level of the indicator» ;

$B_{i3}$ – «average level of the indicator» ;
$B_{i4}$ – «high level of the indicator» ;
$B_{i5}$ – «very high level of the indicator» .
**3 step.** Calculation of the weights of terms of a linguistic variable $\gamma$ – «evaluation of the object's condition» on the basis of formula:

$$p_i = \sum_{i=1}^{N} k_i \cdot \mu_{il}(x_i), l = 1, \dots, 5.$$

(the vector of weight coefficients is scalarly multiplied by the vector of values of the membership function of the corresponding term, see Fig. 1).

$\Gamma$ = "complex assessment of the state of the object", obtained by aggregating the numerical values $X = (X_1, X_2, \dots, X_N)$ of the complex from the $N$ indicators; here: $k_i$, $i = 1, 2, \dots, N$, - weight indicators; $\overline{g_l}$ – classifier nodes (fixed).

**4 step.** Calculation of the numerical value of a linguistic variable $\gamma$ on the basis of the formula:

$$\gamma = \sum_{l=1}^{5} p_l \cdot \overline{g_l}$$

where $\overline{g_l}$ – the middle of the intervals, which are the carriers of terms, $\overline{g_1} = 0.125$; $\overline{g_2} = 0.3$; $\overline{g_3} = 0.5$; $\overline{g_4} = 0.7$; $\overline{g_5} = 0.885$.

To solve this problem, use the cursor keys to select the Spreadsheet mode of the calculator main menu and enter it by pressing the (EXE) key. The data entry window opens (Fig. 2).

Let column A be used to enter parameter values Xi, column B for entering Bi parameter values. We introduce 5 arbitrary values in the cells A and B in the range from 0 to 1.

If we want to automatically recalculate all the values of the entered functions when changing the input parameters, then the following settings should be made: enter the SET UP mode (successively press the (SHIFT) and (SET UP) keys, in the opened settings window (Fig. 3) parameter Auto Calc switch to On mode by pressing the key (F1)). Then press (EXE) to go to the data entry screen.

We now turn to the calculation formulas. To do this, move the pointer to cell C1. Then (SHIFT) (=) enter the «=» sign in the formula string. The (SHIFT) key switches the keyboard to the yellow notation mode. The «=» sign is needed for the calculator to recalculate the entire expression using the formula entered, if one of the formula parameters changes. Then use the key (F4) to select the If mode. The CellIF operator is introduced. The CellIF – is a conditional operator. Its format is as follows:

CellIF(condition, value 1, value 2)

«Value 1» will be written to the cell, if the condition is met, «value 2» will be written, if not satisfied. For example, CellIF (A1 > 0,1,0) will write to cell 1, if cell A1 contains any number greater than 1, it will write zero if number $\leq$ 0.

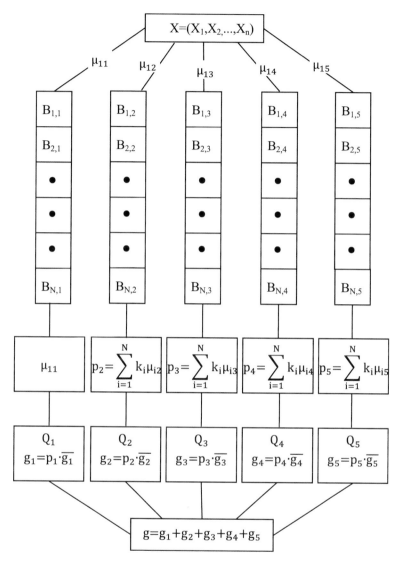

**Fig. 1.** Calculation of the numerical value of linguistic variables.

The calculation formula for cell C1 is as follows:
= CellIF(((A1 ≥ 0) And (A1 < 0.15)), 1, 0) + CellIF(((A1 ≥ 0.15) And (A1 < 0.25)), 10 × (0.25-A1), 0)

Here, the logical operator must be entered not by characters from the keyboard, but in the directory (SHIFT) (KATALOG). To do this, use the cursor control key to move the pointer down, highlight the AND operator and press (EXE) to insert it.

The calculation formula for cell C2:

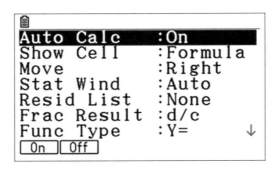

**Fig. 2.** Spreadsheet data entry window.

```
🔋
Auto Calc    :On
Show Cell    :Formula
Move         :Right
Stat Wind    :Auto
Resid List   :None
Frac Result  :d/c
Func Type    :Y=         ↓
[ On ][ Off ]
```

**Fig. 3.** Settings window.

= CellIF(((A2 ≥ 0) And (A2 < 0.15)), 0, 1) + CellIF(((A2 ≥ 0.15) And (A2 < 0.25)), 10 × (A2-0.15), 0) + CellIF(((A2 ≥ 0.25) And (A2 < 0.35)), 1, 0) + CellIF(((A2 ≥ 0.35) And (A2 < 0.45)), 10 × (0.45-A2), 0)

The calculation formula for cell C3:

= CellIF(((A3 ≥ 0) And (A3 < 0.35)), 0, 1) + CellIF(((A3 ≥ 0.35) And (A3 < 0.45)), 10 × (A3-0.35), 0) + CellIF(((A3 ≥ 0.45) And (A3 < 0.55)), 1, 0) + CellIF(((A3 ≥ 0.55) And (A3 < 0.65)), 10 × (0.65-A3), 0)

The calculation formula for cell C4:

= CellIF(((A4 ≥ 0) And (A4 < 0.55)), 0, 1) + CellIF(((A4 ≥ 0.55) And (A4 < 0.65)), 10 × (A4-0.55), 0) + CellIF(((A4 ≥ 0.65) And (A4 < 0.75)), 1, 0) + CellIF(((A4 ≥ 0.75) And (A4 < 0.85)), 10 × (0.85-A4), 0)

The calculation formula for cell C5:

= CellIF(((A5 ≥ 0) And (A5 < 0.75)), 0, 1) + CellIF(((A5 ≥ 0.75) And (A5 < 0.85)), 10 × (A5-0.75), 1)

Then, for certain values of $x_i$ (column A of the calculator) and $k_i$ (column B), calculate the value of $\mu_i$ in column C, as shown in Fig. 4.

Calculate the values of $p_i$: $p_i = \sum_{i=1}^{N} k_i \cdot \mu_{i1}(x_i)$.

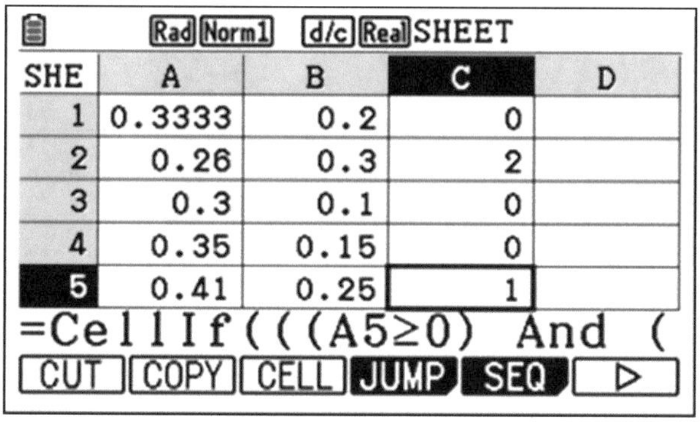

Rad Norm1 d/c Real SHEET

| SHE | A | B | C | D |
|---|---|---|---|---|
| 1 | 0.3333 | 0.2 | 0 | |
| 2 | 0.26 | 0.3 | 2 | |
| 3 | 0.3 | 0.1 | 0 | |
| 4 | 0.35 | 0.15 | 0 | |
| 5 | 0.41 | 0.25 | 1 | |

=CellIf(((A5≥0) And (

CUT COPY CELL JUMP SEQ ▷

**Fig. 4.** Calculation of the values $\mu_i$

In the general case this formula can be reduced to the form

$$p_i = \mu_l(x_l)\sum_{i=1}^{N} k_i$$

Then the formula for the cell D1: = B1 × CellSum(C1:C5). Correspondingly, for D2: = B2 × CellSum(C1:C5).

The formula can be copied from the first cell to all subsequent ones. To do this, place the cursor in cell D1 and press F2 (COPY). Then move the cursor to D2 and press F1 (PASTE). Then you need to correct the CellSum range (C1:C5). The values of the parameters $p_i$ in Fig. 5 (Fig. 6).

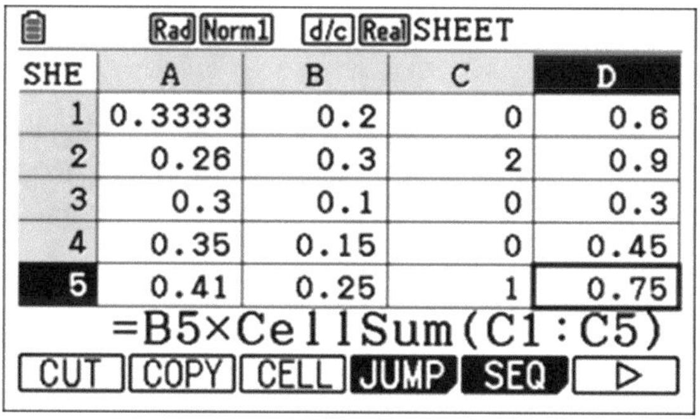

Rad Norm1 d/c Real SHEET

| SHE | A | B | C | D |
|---|---|---|---|---|
| 1 | 0.3333 | 0.2 | 0 | 0.6 |
| 2 | 0.26 | 0.3 | 2 | 0.9 |
| 3 | 0.3 | 0.1 | 0 | 0.3 |
| 4 | 0.35 | 0.15 | 0 | 0.45 |
| 5 | 0.41 | 0.25 | 1 | 0.75 |

=B5×CellSum(C1:C5)

CUT COPY CELL JUMP SEQ ▷

**Fig. 5.** Calculation of values $p_i$

Introduce into the column E the values $\overline{g_l}$ – of the middle of the intervals of the carriers of terms: $\overline{g_1} = 0.125$; $\overline{g_2} = 0..3$; $\overline{g_3} = 0.5$; $\overline{g_4} = 0.7$; $\overline{g_5} = 0.885$. It is easy to calculate the total numerical value of linguistic variables and their average value (Fig. 7.)

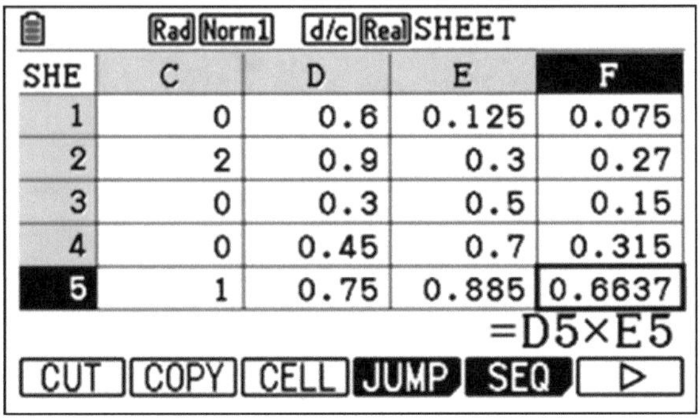

Rad Norm1 d/c Real SHEET

| SHE | C | D | E | F |
|---|---|---|---|---|
| 1 | 0 | 0.6 | 0.125 | 0.075 |
| 2 | 2 | 0.9 | 0.3 | 0.27 |
| 3 | 0 | 0.3 | 0.5 | 0.15 |
| 4 | 0 | 0.45 | 0.7 | 0.315 |
| 5 | 1 | 0.75 | 0.885 | 0.6637 |

=D5×E5

CUT COPY CELL JUMP SEQ ▷

**Fig. 6.** Calculation of values $g_i$

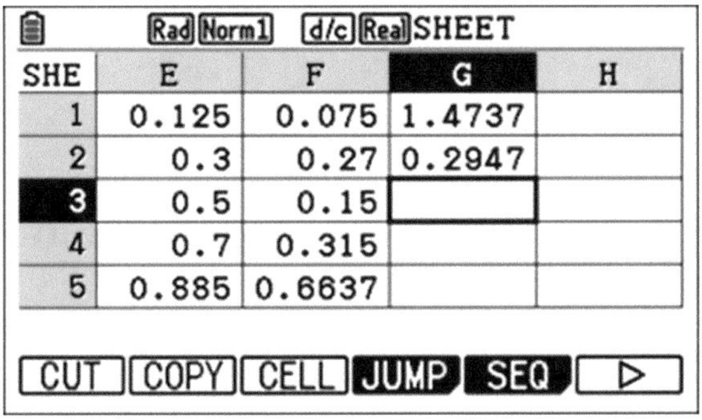

Rad Norm1 d/c Real SHEET

| SHE | E | F | G | H |
|---|---|---|---|---|
| 1 | 0.125 | 0.075 | 1.4737 | |
| 2 | 0.3 | 0.27 | 0.2947 | |
| 3 | 0.5 | 0.15 | | |
| 4 | 0.7 | 0.315 | | |
| 5 | 0.885 | 0.6637 | | |

CUT COPY CELL JUMP SEQ ▷

**Fig. 7.** The total numerical value of linguistic variables and their mean values

## 3 Conclusion

The convenience of using the Spreadsheet mode of the graphing calculator also consists in the fact that if we enter a new parameter group for linguistic variables, it automatically recalculates all the parameters according to the previously introduced formulas.

This example demonstrates that modern graphing calculators can be successfully applied for complex and volumetric calculations in teaching, teaching and research work.

Those who are interested in the capabilities of the CASIO CG-50 graphical calculators, we recommend that you read the user's manual [1].

## Reference

1. CG-50: User's Guide. http://support.casio.com/storage/en/manual/pdf/EN/004/fx-CG50_Soft_v311_EN.pdf

# Fuzzy-Logic Analysis of the State of the Atmosphere in Large Cities of the Industrial Region on the Example of Rostov Region

G. Vovchenko Natalia[1], B. Stryukov Michael[1],
V. Sakharova Lyudmila[1(✉)], and V. Domakur Olga[2]

[1] Rostov State University of Economics,
B. Sadovaya Str., 69, 344002 Rostov-on-Don, Russia
L_Sakharova@mail.ru
[2] Belarusian State Academy of Communication,
Skaryna St., 8/2, 220114 Minsk, Belarus

**Abstract.** The technique of forming a complex estimation of the state of the atmosphere in the industrial region, based on the aggregation of time series of statistical data through a system of fuzzy inference. The calculation of the estimate uses heterogeneous indicators of three groups: the dynamics of atmospheric pollution in the region, the level of atmospheric pollution in the region and the dynamics of polluting emissions into the air, with each group built its own comprehensive assessment. In turn, estimates of the dynamics of air pollution in the region and the level of air pollution in the region are formed by aggregation of the relevant estimates for the major cities of the region. The calculation of the assessment of the dynamics of the atmosphere pollution in a single city carried out on the basis of time series statistical data characterizing the pollution of air with various chemical contaminants (average concentration, standard index, the highest frequency of exceedance for each impurity). The estimation of the level of atmospheric pollution in a particular city is made on the basis of aggregation of statistical data on the total standard index, the highest repeatability and the index of atmospheric pollution for the current year. Statistical data on emissions of pollutants into the air from stationary sources by types of economic activity were used to estimate the dynamics of polluting emissions into the air. Standard fuzzy [0, 1] – classifiers are used for aggregation of indicators.

**Keywords:** Standard fuzzy multi-level [0, 1]-classifier
A comprehensive assessment · Indicators · Atmosphere

## 1 Introduction

The state of the atmosphere is the most important factor affecting the health, duration and quality of life of the population of the region [1]. In this regard, at the present time, the system monitoring of the ecological state of the atmosphere carried out in the Russian Federation at the state and municipal levels has become very important [2, 3].

© Springer Nature Switzerland AG 2019
R. A. Aliev et al. (Eds.): ICAFS-2018, AISC 896, pp. 709–715, 2019.
https://doi.org/10.1007/978-3-030-04164-9_93

For example, in the analytical annual reports of the Rostov region "Ecological Bulletin of the Don" on the state of the environment and natural resources of the Rostov region" the state of the atmosphere in large cities is studied in detail in order to inform specialists, scientists, designers, residents of the Rostov region [2].

The same aim is pursued by reports "On the state of sanitary and epidemiological welfare of the population in the Rostov region" [3]. The analysis presented in [2] and [3] provides comprehensive information on the state of the atmosphere in individual cities; at the same time, it is difficult to see and does not allow to form a final comprehensive assessment of the atmosphere in the region on the basis of a set of heterogeneous indicators.

At present, there are a number of generally accepted methods that allow to estimate the degree of air pollution in a particular locality on the basis of detailed criteria and scales [4]. Almost all of them are reduced to the comparison of the recorded concentrations of chemical impurities with their maximum permissible concentrations in hazard classes, followed by the calculation of the integral indices. There are no techniques that allow to evaluate not only the level of contamination of the atmosphere in the city with various chemical impurities, but also its dynamics, as well as the volume of pollutant emissions by sectors of economic activities, as well as integrated the results obtained in the final assessment of the state of the atmosphere in the region. In this study, a method of assessing the state of the atmosphere in the region based on aggregation and fuzzy-multiple analysis of time series of statistical data on air pollution by major chemical impurities, as well as emissions of pollutants into the air from stationary sources by economic activities is proposed. The technique is based on the use of standard fuzzy multi-level [0, 1]-classifiers [5] and tested on the example of the Rostov region.

## 2 Statistical Material and Weights of Indicators

To form the assessment of dynamics and level of air pollution time series of statistical data characterizing air pollution by the considered chemical admixtures for 2007–2016 in the following large cities of the Rostov region are used: Rostov-on-Don, Azov, Taganrog, Tsimlyansk, Mines (statistical data [2]). To create a method of complex assessment of the state of the atmosphere the authors used to assess the state of the systems based on the aggregation of numerical values of heterogeneous indicators by means of fuzzy multi-level [0, 1]-classifiers [6–9]. It is proposed to use two scales: (1) scale of assessment of pollution dynamics (classifiers of the second type); (2) scale of assessment of absolute value of pollution level (classifiers of the first type). For calculations, a software package is used that allows to evaluate a complex system based on the aggregation of a complex of heterogeneous indicators by means of fuzzy multi-level [0, 1]-classifiers [10].

Data on the content of impurities in the air are divided into four groups in accordance with the generally accepted scale of hazard classes of pollutants in the air [4]. It is proposed to calculate the weight coefficients of the indicators based on the existing weights of hazard classes by normalization. Standard coefficients of hazard classes of impurities are equal to:

k1 = 1.5 (1 hazard class); k2 = 1.3 (2 hazard class);
k3 = 1.0 (hazard class 3); k4 = 0.85 (hazard class 4).

For example, let the statistics show 7 impurities recorded, of which: (1) one belongs to the 1st hazard class (benzapyrene); (2) two to the 2nd hazard class (nitrogen dioxide, formaldehyde); (3) three to the third hazard class (dust, sulfur dioxide, nitrogen oxide); (4) one to the 4th hazard class (carbon dioxide).

Therefore, the sum of their standard coefficients is:

1 * 1.5 + 2 * 1.3 + 3 * 1.0 + 1 * 0.85 = 7.95. For each considered impurity of the indicator, the numerical series of three standard characteristics are given: CP, SI, NP (average concentration, standard index, the highest repeatability of exceeding the maximum permissible concentration). Therefore, taking into account the normalization, the weight coefficients of the indicator corresponding to each of the characteristics are equal:

1st class: 1.5/3/7.95 = 0.0628; 2-class: 1.3/3/7.95 = 0.0546;
3rd class: 1/3/7.95 = 0.0420; 4th class: 0.85/3/7.95 = 0.0356.

# 3    Formation of the Assessment of the Atmosphere on the Basis of the Dynamic Scale

The linguistic variable: gd(town) = "assessment of the dynamics of air pollution in the city (name)" is introduced. A universal set for a linguistic variable is a numeric segment [0,1]. The term set consists of five terms G = {G1, G2, G3, G4, G5}, which conditionally assess the state of the system: G1– "steady tendency to improvement"; G2– "tendency to improvement"; G3 – "stabilization"; G4 – "tendency to deterioration"; G5– "steady tendency to deterioration".

The functions of the term accessories have a standard trapezoidal shape [5, 8, 9] (at the same time, the increase in the term number corresponds to the increase in the numerical value of the variable: "0" is "a stable tendency to improve"; "1" is "a steady tendency to deteriorate"). The ranks of the numeric values Qcp., SI, NP of chemical impurities, in atmospheric air for 2007–2016 years in the city (name). The calculation of normalized values of the studied indicators for the period of $N$ years (10 years) is carried out on the basis of the following scheme, taking into account the importance of different time periods due to weight coefficients:

$$x_i = 0.5 \left( 1 + \sum_{i=1}^{N-1} k_i I_i \right), \tag{1}$$

$$k_i = \frac{2(N-i)}{(N-1)N},$$

$k_i$ - weight coefficients determined by the Fishburn rule; the numbering of time periods is carried out in the reverse order (that is, in the example under consideration, the first period is the years 2015–2016, and the last, the 9th, the years 2007–2008). $I_i$ – integer functions defined in such a way that the value "1" corresponds to the increase in the I-th indicator (deterioration of the situation); the value "−1" – decrease in the i-th indicator; the value "0" – stabilization, lack of changes. Each of the indicators is assigned a linguistic variable, a universal set, whose terms and functions coincide with similar characteristics of the final evaluation.

It is established that

$g_d(Azov) = 0.459$,     $g_d(Rostov\text{-}on\text{-}Don) = 0.522$,     $g_d(Taganrog) = 0.532$,     $g_d$ $(Tsimlyansk) = 0.513$, $g_d(Shakhty) = 0.475$, (term $G_3$ – "stabilization").

## 4   Assessment of Air Pollution Dynamics in the Region

The linguistic variable: $g_d$ = "assessment of the dynamics of air pollution in the region in large cities" is introduced. The same term-set and membership functions are used in the assessment of air pollution in the city. A set of indicators on the basis of which it is aggregated $g_d$ = "assessment of the dynamics of air pollution in the region in large cities": estimates of the dynamics of air pollution in large cities obtained in the previous stage. Weights of indicators are the proportion of the population of cities from the total population. Data aggregation based on fuzzy five-level [0,1]-classifiers shows that for the Rostov region $g_d = 0,500$, which corresponds to the term $G_3$ – "stabilization".

## 5   Assessment of the Level of Air Pollution in the Town

The linguistic variable $g_s(town)$ = "the level of air pollution in the town (name)" is introduced into consideration. A universal set for a linguistic variable is a numeric segment [0,1]. According to the standard classification of the assessment of the level of air pollution [4], the term – set consists of four terms $G = \{G_1, G_2, G_3, G_4\}$: $G_1$– "low air pollution in the city"; $G_2$ – "increased air pollution in the city"; $G_3$ – "high air pollution in the city"; $G_4$ – "very high air pollution in the city".

Accordingly, a 4-point fuzzy classifier with standard membership functions is applied [5, 8, 9]. A set of indicators on the basis of which the assessment is based: SI, NP, ISA in the city for 2016. Each of them corresponds to a linguistic variable, the terms of which are defined similarly to the terms of the final evaluation. Membership functions of terms each of the indicators obtained by fuzzification of the boundaries of the standard scale of assessment of the level of contamination, Table 1 [2].

The calculations show:

$g_s(Rostov\text{-}on\text{-}Don) = 0.533$,     $G_3$ – "high air pollution in the town";

$g_s(Azov) = 0.350$,     $G_2$ – "increased air pollution in the town";

$g_s(Taganrog) = 0.383$,     $G_2$ – "increased air pollution in the town";

$g_s(Tsimlyansk) = 0.200$,     $G_1$ – "low air pollution in the town";

$g_s(Shakhty) = 0.333$,     $G_2$ – "increased air pollution in the town".

**Table 1.** Membership functions of subsets of a term set $G$, here g = SI, NP, IZA

| $G_i$ | Fuzzy set membership function $G_i$ | | |
|---|---|---|---|
| | SI | NP | IZA |
| $G_1$ | $\mu_1 = \begin{cases} 1, & 0 \leq g < 1 \\ 2 - g, & 1 \leq g < 2 \end{cases}$ | $\mu_1 = \begin{cases} 1, & g = 0 \\ 1 - g, & 0 \leq g < 1 \end{cases}$ | $\mu_1 = \begin{cases} 1, & 0 \leq g < 4 \\ 5 - g, & 4 \leq g < 5 \end{cases}$ |
| $G_2$ | $\mu_2 = \begin{cases} g - 1, & 1 \leq g < 2 \\ 1, & 2 \leq g < 4 \\ 5 - g, & 4 \leq g < 5 \end{cases}$ | $\mu_2 = \begin{cases} g, & 0 \leq g < 1 \\ 1, & 1 \leq g < 19 \\ 20 - g, & 19 \leq g < 20 \end{cases}$ | $\mu_2 = \begin{cases} g - 5, & 4 \leq g < 5 \\ 1, & 5 \leq g < 6 \\ 7 - g, & 6 \leq g < 7 \end{cases}$ |
| $G_3$ | $\mu_3 = \begin{cases} g - 4, & 4 \leq g < 5 \\ 1, & 5 \leq g < 9 \\ 10 - g, & 9 \leq g < 10 \end{cases}$ | $\mu_3 = \begin{cases} g - 19, & 19 \leq g < 20 \\ 1, & 20 \leq g < 49 \\ 50 - g, & 49 \leq g < 50 \end{cases}$ | $\mu_3 = \begin{cases} g - 6, & 6 \leq g < 7 \\ 1, & 7 \leq g < 13 \\ 14 - g, & 13 \leq g < 14 \end{cases}$ |
| $G_4$ | $\mu_4 = \begin{cases} g - 9, & 9 \leq g < 10 \\ 1, & g \geq 10 \end{cases}$ | $\mu_4 = \begin{cases} g - 49, & 49 \leq g < 50 \\ 1, & g \geq 50 \end{cases}$ | $\mu_4 = \begin{cases} g - 14, & 13 \leq g < 14 \\ 1, & 0,8 \leq g \leq 1 \end{cases}$ |

## 6 Assessment of the Level of Air Pollution in the Region

We introduce a linguistic variable: $g_s$ = "the estimation of level of contamination of the atmosphere in the region". A universal set for a linguistic variable is a numeric segment [0,1]. Term-set consists of four terms $G = \{G_1, G_2, G_3, G_4\}$: $G_1$ – "low air pollution in the region"; $G_2$ – "increased air pollution in the region"; $G_3$ – "high air pollution in the region"; $G_4$ – "very high air pollution in the region".

The assessment is aggregated on the basis of a set of parameters: estimates of the level of air pollution in large cities of the region. The weights are the share of the urban population of the total population. To form an estimate, a standard 4-point classifier is used. Calculations show that for the Rostov region $g_s$ = 0.435, which corresponds to the term $G_2$ – "increased air pollution in the region".

## 7 Assessment of the Dynamics of Emissions of Pollutants in the Region

To assess the dynamics of emissions, statistical data on emissions of pollutants into the air in the Rostov region for 2012–2016, thousands of tons from stationary sources, by types of economic activity (statistical data [2]) were used. Different industries are sources of different pollutant complexes, so they can be ranked by hazard class, for example, in descending order of hazard as follows: (1) manufacturing; (2) production and distribution of electricity, gas and water; (3) mining; (4) transport and communications; (5) provision of other public, social and personal services; (6) other economic activities; (7) agriculture, hunting and forestry; (8) livestock. On the basis of ranking the weight coefficients are calculated according to the Fishburn rule (n = 8).

The linguistic variable: $g_z$ = "assessment of the dynamics of polluting emissions into the air" is introduced. A universal set for a linguistic variable is a numeric segment [0,1]. The term set consists of five standard terms $G = \{G_1, G_2, G_3, G_4, G_5\}$: $G_1$ – "sustainable trend towards reducing emissions to the atmosphere"; $G_2$ – "tendency to reduce emissions to the atmosphere"; $G_3$ – "stabilization of the situation";

$G_4$ – "tendency to increase emissions to the atmosphere"; $G_5$ – "sustainable tendency to increase emissions to the atmosphere".

Of membership function is standard trapezoid. A set of indicators on the basis of which the assessment is aggregated: emissions of pollutants from stationary sources by economic activities over five years. Time series of indicators are aggregated on the basis of the formula (1), in consequence of which normalized numerical values of indicators are formed. The numbering of time periods is carried out in the reverse order (that is, in the example under consideration, the first period is the years 2015–2016, and the last, the 4th, years 2012–2013). Standard 5-point classifiers are used for aggregation. It is established that for the Rostov region $g_z = 0.463$, that corresponds to the term "$G_3$ - stabilization of the situation".

## 8    Final Assessment of Air Pollution in the Region

The numerical value of the final linguistic variable $g_1$ = "integrated assessment of the state of the atmosphere in large cities of the region" is calculated by aggregation on the basis of standard five-point [0,1]- classifiers of the three estimates obtained above: $g_d$ = "assessment of the dynamics of pollution of the atmosphere in the region"; $g_s$ = "assessment of the level of air pollution in the region"; $g_z$ = "assessment of the dynamics of air pollution emissions". The universal set for a linguistic variable is a numeric segment [0,1]. Term-set consists of five terms $G = \{G_1, G_2, G_3, G_4, G_5\}$: $G_1$ – "the state of the atmosphere in the region is excellent (ecologically clean region on the state of the atmosphere)"; $G_2$ – "the state of the atmosphere in the region is good (normal)"; $G_3$ – "the state of the atmosphere in the region is satisfactory (tense situation)"; $G_4$ – "the state of the atmosphere in the region is bad (crisis)"; $G_5$ – "the state of the atmosphere in the region is extremely bad (disaster)". It was found that for the Rostov region: $g_1 = 0.468$, which corresponds to the term $G_3$ – "the state of the atmosphere in the region is satisfactory (tense situation)".

## 9    Conclusion

The method of formation of an assessment of the state of the atmosphere of fuzzy-multiple aggregation of data on air pollution by main chemical impurities, as well as emissions of pollutants into the air from stationary sources by economic activities is developed. The technique is tested using time series of statistical data for the Rostov region. The method is based on the use of standard fuzzy multi-level [0,1]- classifiers, is universal and can be used to assess the state of the atmosphere in other regions. The technique allows for modification of the complex of indicators for which the evaluation is performed, as well as the variation of the weight coefficients of the indicators. The advantage of the proposed method compared to the existing methods of assessment is the integration of disparate heterogeneous indicators in the final assessment, which allows to assess the state of the atmosphere in the major cities of the region as a whole.

# References

1. Ushakov, I.B.: Medical aspects of protection of public health from the harmful effects of environmental factors. Hyg. sanitat. **6**, 29–34 (2005)
2. Environmental Bulletin of the Don "On the state of the environment and natural resources of the Rostov region in 2011/2016". Government-in the Rostov region, Rostov-on-Don. http://минприродыро.рф/state-of-the-environment/ekologicheskiy-vestnik/. Accessed 18 July 2018
3. Report on the state of sanitary and epidemiological welfare of the population in the Rostov region in 2016. Federal service for supervision of consumer rights protection and human welfare, Rostov-on-Don. http://61.rospotrebnadzor.ru/index.php?Itemid=116&catid=96:2009-12-30-08-03-55&id=6813:-q-2016q&option=com_content&view=article. Accessed 30 Apr 2018
4. Bespamyatnov, G., Krotov, Y.: Maximum permissible concentrations of chemical substances in the environment. Chemistry, Leningrad (1985)
5. Nedosekin, A.: Fuzzy Sets and Financial Management. AFA Library, Moscow (2003)
6. Sakharova, L., Stryukov, M., Akperov, I., Alekseychik, T., Chuvenkov, A.: Application of fuzzy set theory in agro-meteorological models for yield estimation based on statistics. In: 9th International Conference on Theory and Application of Soft Computing, Computing with Words and Perception, Budapest, Hungary, pp. 820–829 (2017). Proc. Comput. Sci. 120
7. Kramarov, S., Temkin, I., Khramov, V.: The principles of formation of united geo-information space based on fuzzy triangulation. In: 9th International Conference on Theory and Application of Soft Computing, Computing with Words and Perception, 24–25 August 2017, Budapest, Hungary, pp. 835–843 (2017). Proc. Comput. Sci. 120
8. Kramarov, S., Sakharova, L.: Management of complex economic systems by fuzzy classifiers method. Sci. Bull. southern Univ. Manage. **2**(18), 42–50 (2017)
9. Kramarov, S., Sakharova, L., Khramov, V.: Soft computing in management: management of complex multivariate systems based on fuzzy analog controllers. Sci. Bull. southern Univ. Manage. **3**(19), 42–51 (2017)
10. Albekov, A., Arapova, E., Karasev, D., Stryukov, M., Sakharova, L.: Program for evaluation of agricultural production intensity by means of fuzzy 5-point classifier. Certificate of registration of computer program № 2018613875. Federal service for intellectual property, Moscow (2018)

# Application of Expected Utility to Business Decision Making Under U-Number Valued Information

Konul I. Jabbarova[✉]

Azerbaijan State Oil and Industry University,
Azadlyg Avenue, 20, Baku AZ1010, Azerbaijan
Konul.jabbarova@mail.ru

**Abstract.** Many real-world decision problems appear in some kind of "usual" situations. This is related to combination of fuzziness and probabilistic uncertainty of information, referred to as bimodal information. Aliev proposed the concept of U-number as a special case of Z-number, to form a new bridge to the concept of usuality of Zadeh. In this paper we consider application of expected utility to business decision making under information described by U-numbers. The obtained results correspond to human intuition-based decision making.

**Keywords:** Expected utility · Usuality · U-number · Business decision

## 1 Introduction

The expected utility is widely used for solving various decision making problems. In [1] testing of Expected Utility (EU) paradigm for a classic portfolio choice problem in a complete market space is considered. The assumptions of state independent preferences of consumers and a single commodity in each state are used. The authors also consider that demands depend on probabilities of states of nature as well as on prices and income. In [2] they analyze modeling of portfolio behaviour of banks in Jordan by using EU. The assumptions as homogeneity of the interest rate matrix are tested. It is shown, that availability of funds is more influential than interest rates for efficiency of portfolio decisions. It is needed to take into account that many real-world decision problems belong to activity in some kind "usual" situations. In order to formalize information relevant to such situations, Prof. Zadeh introduced the concept of usuality [3–5]. It is proposed to describe information in terms of linguistic evaluations of likelihood of events, such as "rarely", "often", "usually" etc. One can see that we deal with a combination of fuzziness and probabilistic uncertainty. real-world decision relevant information is characterized by fuzziness and partial reliability. In order to deal with this issue, Prof. Zadeh introduced the concept of a Z-number, as a pair of two fuzzy numbers $Z = (A, B)$ to describe information related to values of a random variable $X$ [6].

$A$ is a fuzzy constraint on values of $X$ and $B$ is a fuzzy reliability of $A$, and is considered as a value of probability measure of $A$. In [7, 8] a general and effective approach to computation with discrete Z-numbers is proposed. In [9, 10] Aliev

© Springer Nature Switzerland AG 2019
R. A. Aliev et al. (Eds.): ICAFS-2018, AISC 896, pp. 716–723, 2019.
https://doi.org/10.1007/978-3-030-04164-9_94

proposed to use the concept of a Z-number as a new bridge to the usuality concept. A special kind of Z-numbers, usual numbers (U-numbers) were introduced, where $B$ part takes values from the linguistic codebook of usuality. Let us note the state-of-the-art on application of EU under partially reliable information is scarce. [11] is devoted to the first application of EU in business decision making on the basis of direct computation with Z-numbers. The authors consider a benchmark decision problem where information on outcomes of alternatives and probabilities of states of nature are described by Z-numbers. The obtained Z-valued EUs of alternatives are compared by using an approach to human-like comparison of Z-numbers proposed in [12]. An analogous problem is considered in [13]. However, Z-number valued decision relevant information is reduced to fuzzy information to apply EU model. This naturally leads to loss of reliability information and obtained results may not be adequate for real-world problems. In this paper we consider application of EU paradigm to a typical decision making problem in a production firm. In the considered problem, information on outcomes of alternatives and probabilities of states of nature are described by U-numbers. The obtained results reflect human-like decision making because typical decision problems are naturally characterized by usual information.

The paper is structured as follows. In Sect. 2 preliminary information including a definition of a U-number, is given. In Sect. 3 we formulate and solve a decision problem under U-number valued information. Section 4 is conclusion.

## 2 Preliminaries

**U-numbers [9, 10].** Let $X$ be a random variable and $A$ be a fuzzy number playing a role of fuzzy constraint on values that the random variable may take: $X$ is $A$. The definition of a usual value of $X$ may be expressed in terms of the probability distribution of $X$ as follows [14]. If $p(x_i)$ is the probability of $X$ taking $x_i$ as its value, then [3–5]

$$usually(X \text{ is } A) = \mu_{usually}\left(\sum_i p(x_i)\mu_A(x_i)\right) \qquad (1)$$

or

$$usually(X \text{ is } A) = \mu_{most}\left(\sum_i p(x_i)\mu_A(x_i)\right) \qquad (2)$$

A usual number describing, "usually, professor's income is medium" is shown Fig. 1.

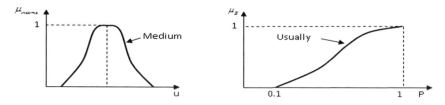

**Fig. 1.** An example of U-number

Formula (2) indicates that the probability that the event A occurs as the value for the variable $X$, is "*most*". As it was mentioned above, in [14] Zadeh provided an outline for the theory of usuality, however this topic requires further investigation. It is needed a more general approach for other usuality quantifiers. In this paper "usuality" will be a composite term characterized by fuzzy quantities as *always, usually, frequently/often, occasionally, seldom, almost never/rarely, never* (Fig. 2).

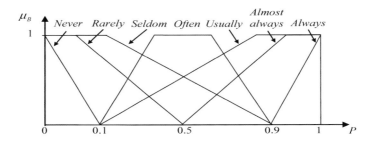

**Fig. 2.** The codebook of the fuzzy quantifiers of usuality

The framework for computations over U-numbers is given in [9, 10].

**Fuzzy Pareto Optimality (FPO) Principle Based Comparison of Z-numbers [12].**
Fuzzy Pareto optimality (FPO) principle allows to determine degrees of Pareto Optimality of multiattribute alternatives. We apply this principle to compare Z-numbers as multiattribute alternatives – one attribute measures value of a variable, the other one measures the associated reliability. According to this approach, by directly comparing Z-numbers $Z_1 = (A_1, B_1)$ and $Z_2 = (A_2, B_2)$ one arrives at total degrees of optimality of Z-numbers: $do(Z_1)$ and $do(Z_2)$. These degrees are determined on the basis of a number of components (the minimum is 0, the maximum is 2) with respect to which one Z-numbers dominates another one. $Z_1$ is considered higher than $Z_2$ if $do(Z_1) > do(Z_2)$.

## 3   Statement of the Problem

Consider business decision making for a production firm over a one-year horizon. Assume that production rate is much lower than current demand. The management considers the following decisions:

– Low increase of production capacity;
– Medium increase of production capacity by installing a new plant;
– High increase of production capacity by installing a new product line;
– Do nothing

The following states of nature (market conditions) are considered: low demand - $S_1$, medium demand - $S_2$, high demand - $S_3$, very high demand- $S_4$. Due to uncertainty of future demand, information on possible profit of the alternatives is characterized by fuzziness and partial reliability. In view of this, the decision problem is described in a form of the following U-number valued pay-off matrix:

The problem is to compute U-number valued EUs of the alternatives U(ai) and find the best alternative as an alternative with the maximal U:

Find $a^*$ such that $do(EU(a^*)) = \max\limits_{i=1,\dots,4} do(EU(a_i))$.

# 4  Solution of the Problem

Let us formalize the U-number valued information in Table 1. The codebooks for the used U-numbers are given in Tables 2 and 3.

**Table 1.**  U-number valued pay-off matrix.

|       | $S_1$   | $S_2$    | $S_3$    | $S_4$    |
|-------|---------|----------|----------|----------|
| $a_1$ | (A,U)   | (AA,U)   | (HA,U)   | (EA,S)   |
| $a_2$ | (L,U)   | (A,O)    | (EA,O)   | (VH,O)   |
| $a_3$ | (EL,S)  | (BA,U)   | (AA,U)   | (EH,U)   |
| $a_4$ | (A,U)   | (A,U)    | (A,U)    | (A,U)    |

**Table 2.**  The linguistic terms for A parts of U-numbers

| Linguistic value          | Fuzzy value                |
|---------------------------|----------------------------|
| Extremely Low (EL)        | $\{-92, -83, -67, -58\}$   |
| Very Low (VL)             | $(-67, -58, -42, -33)$     |
| Low (L)                   | $(-42, -33, -17, -8)$      |
| Low Average (LA)          | $(-17, -8, 8, 17)$         |
| Below Average (BA)        | $(8, 17, 33, 42)$          |
| Average (A)               | $(33, 42, 58, 67)$         |
| Above Average (AA)        | $(58, 67, 83, 92)$         |
| High Average (HA)         | $(83, 92, 108, 117)$       |
| Extremely Average (EA)    | $(108, 117, 133, 142)$     |
| High (H)                  | $(133, 142, 158, 167)$     |
| Very High (VH)            | $(158, 167, 183, 192)$     |
| Extremely High (EH)       | $(183, 192, 200)$          |

**Table 3.**  The linguistic terms for B parts of U-number-based criteria

| Linguistic value | Fuzzy value              |
|------------------|--------------------------|
| Rarely (R)       | $(0, 0, 0.1, 0.5)$       |
| Seldom (S)       | $(0, 0, 0.3, 0.9)$       |
| Often (O)        | $(0.1, 0.4, 0.6, 0.9)$   |
| Usually (U)      | $(0.1, 0.7, 1, 1)$       |

The information on probabilities of states of nature is also characterized by fuzziness and partial reliability. However, for simplicity of computations, we will not consider U-number valued probabilities of states of nature. We will conduct analysis for the cases of numeric and fuzzy probabilities.

Case 1. Let us consider the following numeric values of probabilities: $P(S_1) = 0.3$, $P(S_2) = 0.25$, $P(S_3) = 0.35$; $P(S_4) = 0.1$. In this case, the U-number valued EUs of the alternatives are found as follows:

$$EU(a_1) = P(S_1) \cdot (A, U) + P(S_2) \cdot (AA, U) + P(S_3) \cdot (HA, U) + P(S_4) \cdot (EA, S)$$
$$= (62.4, 69.6, 82.9, 90.1)(0, 0.0007, 0.299, 0.897);$$

$$EU(a_2) = P(S_1) \cdot (L, U) + P(S_2) \cdot (A, O) + P(S_3) \cdot (EA, O) + P(S_4) \cdot (VH, O)$$
$$= (48.95, 57.55, 72.95, 81.55)(0.0004, 0.089, 0.286, 0.74);$$

$$EU(a_3) = P(S_1) \cdot (EL, S) + P(S_2) \cdot (BA, U) + P(S_3) \cdot (AA, U) + P(S_4) \cdot (EH, U)$$
$$= (12.25, \ 20.25, \ 33.95, \ 41.05)(0.002, \ 0.226, \ 0.599, \ 0.897);$$

$$EU(a_4) = P(S_1) \cdot (A, U) + P(S_2) \cdot (A, U) + P(S_3) \cdot (A, U) + P(S_4) \cdot (A, U)$$
$$= (33, 42, 58, 67)(0.0004, 0.994, 0.994, 0.994);$$

In order to determine the best alternative, we will use the FPO principle based comparison of Z-numbers. The obtained results are as follows:

$$do(EU(a_1)) = 1, do(EU(a_2)) = 0.59, do(EU(a_4)) = 1, do(EU(a_3)) = 0,$$

$$do(EU(a_1)) = 1, do(EU(a_3)) = 0.49, do(EU(a_4)) = 1, do(EU(a_2)) = 0.67,$$

$$do(EU(a_1)) = 0.9, do(EU(a_4)) = 1, do(EU(a_2)) = 1, do(EU(a_3)) = 0.5.$$

As one can see, the best alternative is $a_4$.

Case 2. Let us consider the use of fuzzy probabilities. It is known, that fuzzy probabilities can be assigned only to $n - 1$ states of nature, whereas the remaining one should be computed [7]. Assume that $P(S_1) = (0.285, 0.3, 0.315)$; $P(S_2) = (0.2375, 0.25, 0.2625)$; $P(S_3) = (0.3325, 0.35, 0.3675)$; and $P(S_4)$ is unknown. We found that $P(S_4) = (0.055, 0.1, 0.145)$ [7]. Next, we calculated the U-valued EUs of all the alternatives:

$$EU(a_1) = P(S_1) \cdot (A, U) + P(S_2) \cdot (AA, U) + P(S_3) \cdot (HA, U) + P(S_4) \cdot (EA, S)$$
$$= (55.28, 69.6, 82.9, 99.4)(0, 0.0008, 0.299, 0.895);$$

$$EU(a_2) = P(S_1) \cdot (L, U) + P(S_2) \cdot (A, O) + P(S_3) \cdot (EA, O) + P(S_4) \cdot (VH, O)$$
$$= (40.3, 57.55, 72.95, 92.63)(0.0004, 0.089, 0.286, 0.59);$$

$$EU(a_3) = P(S_1) \cdot (EL, S) + P(S_2) \cdot (BA, U) + P(S_3) \cdot (AA, U) + P(S_4) \cdot (EH, U)$$
$$= (4.32, 20.25, 33.95, 51.1)(0.0001, 0.0002, 0.299, 0.89);$$

$$EU(a_4) = P(S_1) \cdot (A, U) + P(S_2) \cdot (A, U) + P(S_3) \cdot (A, U) + P(S_4) \cdot (A, U)$$
$$= (30.03, 42, 58, 73)(0.0004, 0.32, 0.99, 0.99);$$

The results on ranking of the obtained utility values are as follows:

$$do(EU(a_1)) = 1, do(EU(a_2)) = 0.6, do(EU(a_4)) = 1, do(EU(a_3)) = 0,$$

$$do(EU(a_1)) = 1, do(EU(a_3)) = 0, do(EU(a_4)) = 1, do(EU(a_2)) = 0.6,$$

$$do(EU(a_1)) = 0.9, do(EU(a_4)) = 1, do(EU(a_2)) = 1, do(EU(a_3)) = 0.$$

Let us conduct sensitivity analysis of the solution obtained in Case 2 with respect to usuality terms of outcomes of alternatives given in Table 1. We will conduct two experiments what happens when the level of usuality is decreased.

1. The level of usuality of outcomes is decreased from "usually" to "seldom" (Table 4).

Table 4. Pay-off table with the reduced level of usuality of the outcomes

|        | $S_1$   | $S_2$   | $S_3$   | $S_4$   |
|--------|---------|---------|---------|---------|
| $a_1$  | (A,S)   | (AA,S)  | (HA,S)  | (EA,S)  |
| $a_2$  | (L,S)   | (A,O)   | (EA,O)  | (VH,O)  |
| $a_3$  | (EL,S)  | (BA,S)  | (AA,S)  | (EH,S)  |
| $a_4$  | (A,S)   | (A,S)   | (A,S)   | (A,S)   |

The obtained values of expected utility:

$$EU(a_1) = P(S_1) \cdot (A, S) + P(S_2) \cdot (AA, S) + P(S_3) \cdot (HA, S) + P(S_4) \cdot (EA, S)$$
$$= (55.28, 69.6, 82.9, 99.4)(0, 0, 0.02, 0.7);$$

$$EU(a_2) = P(S_1) \cdot (L, S) + P(S_2) \cdot (A, O) + P(S_3) \cdot (EA, O) + P(S_4) \cdot (VH, O)$$
$$= (40.3, 57.55, 72.95, 92.63)(0.0002, 0.001, 0.115, 0.678);$$

$$EU(a_3) = P(S_1) \cdot (EL, S) + P(S_2) \cdot (BA, S) + P(S_3) \cdot (AA, S) + P(S_4) \cdot (EH, S)$$
$$= (4.32, 20.25, 33.95, 51.1)(0, 0, 0.02, 0.68);$$

$$EU(a_4) = P(S_1) \cdot (A, S) + P(S_2) \cdot (A, S) + P(S_3) \cdot (A, S) + P(S_4) \cdot (A, S)$$
$$= (30.03, 42, 58, 73)(0, 0, 0.02, 0.68);$$

The results of ranking:

$$do(EU(a_1)) = 0.5, do(EU(a_2)) = 1, do(EU(a_4)) = 1, do(EU(a_3)) = 0,$$

$$do(EU(a_1)) = 1, do(EU(a_3)) = 0, do(EU(a_4)) = 0, do(EU(a_2)) = 1,$$

$$do(EU(a_1)) = 1, do(EU(a_4)) = 0, do(EU(a_2)) = 1, do(EU(a_3)) = 0,$$

As one can see, the results are sensitive to change in usuality levels, $a_2$ becomes the best alternative instead of $a_4$. The reason is that this alternative is characterized by relatively higher reliability levels of outcomes.

2. The level of usuality of outcomes is decreased from "often" to "seldom". The computed values of expected utility:

$$EU(a_1) = P(S_1) \cdot (A, U) + P(S_2) \cdot (AA, U) + P(S_3) \cdot (HA, U) + P(S_4) \cdot (EA, S)$$
$$= (55.28, 69.6, 82.9, 99.4)(0, 0.0008, 0.299, 0.895);$$

$$EU(a_2) = P(S_1) \cdot (L, U) + P(S_2) \cdot (A, S) + P(S_3) \cdot (EA, S) + P(S_4) \cdot (VH, S)$$
$$= (40.3, 57.55, 72.95, 92.63)(0, 0, 0.05, 0.74);$$

$$EU(a_3) = P(S_1) \cdot (EL, S) + P(S_2) \cdot (BA, U) + P(S_3) \cdot (AA, U) + P(S_4) \cdot (EH, U)$$
$$= (4.32, 20.25, 33.95, 51.1)(0.0001, 0.0002, 0.299, 0.89);$$

$$EU(a_4) = P(S_1) \cdot (A, U) + P(S_2) \cdot (A, U) + P(S_3) \cdot (A, U) + P(S_4) \cdot (A, U)$$
$$= (30.03, 42, 58, 73)(0.0004, 0.32, 0.99, 0.99).$$

The results of ranking:

$$do(EU(a_1)) = 1, do(EU(a_2)) = 0, do(EU(a_4)) = 1, do(EU(a_3)) = 0,$$

$$do(EU(a_1)) = 1, do(EU(a_3)) = 0, do(EU(a_4)) = 1, do(EU(a_2)) = 0.4,$$

$$do(EU(a_1)) = 0.9, do(EU(a_4)) = 1, do(EU(a_2)) = 1, do(EU(a_3)) = 0.5,$$

As one can see, $a_4$ remains the best alternative. However, the optimality of $a_2$ is decreased due to the decrease of usuality level of its outcomes.

We can conclude that the considered solution is sensitive to the change in usuality levels of the outcomes. Indeed, change of usuality level of outcomes influences preferences. A DM would choose an alternative with higher reliability of outcomes if the reliability of initially chosen alternative is decreased.

## 5    Conclusion

In this paper we consider an application of the usuality concept in real-world business decision making problem characterized by bimodal information. Probabilities of states of nature and outcomes of alternatives are described by U-numbers. Solving of the problem is based on expected utility model which is applied by using computation and comparison of U-numbers. The obtained results are adequate to human-like decision making.

## References

1. Kubler, F., Selden, L., Wei, X.: Asset demand based tests of expected utility maximization. Am. Econ. Rev. **104**(11), 3459–3480 (2014)
2. Al-Tarawneh, A., Khataybeh, M.: Portfolio behaviour of commercial banks: the expected utility approach: evidence from jordan. Int. J. Econ. Financ. Issues **5**(2), 312–323 (2015)
3. Zadeh, L.A.: Fuzzy sets and commonsense reasoning. Institute of Cognitive Studies report 21, University of California, Berkeley (1984)
4. Zadeh, L.A.: Fuzzy sets as a basis for the management of uncertainty in expert systems. Fuzzy Sets Syst. **11**, 199–227 (1983)
5. Zadeh, L.A.: A computational theory of dispositions. In: Proceedings of 1984 International Conference on Computation Linguistics, Stanford, pp. 312–318 (1984)
6. Zadeh, L.A.: A note on Z-numbers. Inf. Sci. **181**, 2923–2932 (2011)
7. Zadeh, L.A.: Generalized theory of uncertainty (GTU) – principal concepts and ideas. Comput. Stat. Data Anal. **51**, 15–46 (2006)
8. Aliev, R.A., Alizadeh, A.V., Huseynov, O.H.: The arithmetic of discrete Z-numbers. Inf. Sci. **290**, 134–155 (2015)
9. Aliev, R.A.: Approximate arithmetic operations of U-numbers. Procedia Comput. Sci. **102**, 378–384 (2016)
10. Aliev, R.A.: Introduction to U-number calculus. Intell. Autom. Soft Comput. (2017). https://www.tandfonline.com/doi/full/10.1080/10798587.2017.1330311?scroll=top&needAccess=true
11. Aliev, R.R., Mraiziq, D.A.T., Huseynov, O.H.: Expected utility based decision making under z-information and its application. Comput. Intell. Neurosci. (2015). Article ID 364512. 11 p. http://dx.doi.org/10.1155/2015/364512
12. Aliev, R.A., Huseynov, O.H., Serdaroglu, R.: Ranking of z-numbers and its application in decision making. Int. J. Inf. Technol. Decis. Making **15**(6), 1503–1519 (2015)
13. Zeinalova, L.M.: Expected utility based decision making under z-information. Intell. Autom. Soft Comput. **20**(3), 419–431 (2014)
14. Zadeh, L.A.: Outline of a theory of usuality based on fuzzy logic. In: Fuzzy Sets, Fuzzy Logic, and Fuzzy Systems, pp. 694–712. World Scientific Publishing Co., Inc., River Edge (1996)

# Fuzzy-Based Failure Diagnostic Analysis in a Chemical Process Industry

Mohammad Yazdi[1]([✉]) and Mahlagha Darvishmotevali[2]

[1] Centre for Marine Technology and Ocean Engineering (CENTEC), Instituto Superior Técnico, Universidade de Lisboa, Lisbon 1049-001, Portugal
Mohammad.yazdi@tecnico.ulisboa.pt
[2] School of Tourism and Hotel Management, Near East University, Nicosia, Mersin 10, North Cyprus, Turkey
Mahlagha.darvish@neu.edu.tr

**Abstract.** Failure analysis is vital to prevent potential incidents in chemical process industries. The varieties of different failure analysis methods such as fault tree analysis (FTA) help assessors to optimize the amount of risk by providing corresponding corrective actions. However, such conventional failure analysis techniques still suffer from several shortages. As an example, availability of failure data in some cases is rare and besides they cannot be much more effective in dynamic structure. In this paper, a new framework based on probabilistic failure analysis using an integration of FTA and Petri-nets are proposed to provide ability in dynamic structure. Fuzzy logic is also used to deal with uncertainty conditions when there is a lack of information. A real case study of kick in chemical process industry is surveyed to show the effectiveness and efficiency of the proposed model.

**Keywords:** Failure analysis · Uncertainty · Aggregation

## 1 Introduction

Failure analysis in chemical process industries is a proper task which can extensively increase the reliability and safety performance of the system. Infrastructure of chemical process industries is associated with the high hazardous materials in risky conditions like as high temperature and pressure. Thus, it has high possibility for occurrence of accidents, mishaps, near misses, and incidents causes losses [1]. In this regards, attending to the failure analysis techniques in order to recognize potential failures of the studied system either are qualitative or quantitative is significant. Therefore, using improper and insufficient techniques for failure analysis is an invitation for future accidents.

The numerous fatal accidents have been occurred in recent years caused by explosion and fire and accordingly bring considerable asset losses, environmental pollutions, and several human losses. For example, on March 23, 2005, one the most disastrous accidents happened at BP Texas City Refinery which caused the 15 people were killed directly and more than 150 people were injured in accordance [2]. In this case, accident report emphasized that the utilized failure analysis technique to reduce

© Springer Nature Switzerland AG 2019
R. A. Aliev et al. (Eds.): ICAFS-2018, AISC 896, pp. 724–731, 2019.
https://doi.org/10.1007/978-3-030-04164-9_95

the risk measure was not reliable and therefore named as a main reason of the accident [3]. Thus, in recent decade the failure analysis methods have been broadly developed by several scholars to deal with shortages of conventional techniques and make them to be more effective and efficient to prevent such occurrence of accidents. Failure analysis techniques have been applied using hazard and operability study (HAZOP) [4], failure mode and effect critical analysis (FMECA) [5, 6], probabilistic risk assessment techniques like as fault tree analysis (FTA) [5], and in some cases Petri nets [7]. In this paper, a novel approach according to an integration of FTA and Petri nets is proposed as an extension to traditional failure analysis techniques under uncertainty circumstances. Fuzzy logic which was introduced to develop the conventional probability theory is therefore used to handle possible uncertainty during failure analysis.

## 2   Methods

The objective of the proposed method is to compute the probability of basic events (BEs), top event (TE), and accordingly to recognize the critical components under uncertainty conditions. The method is divided in three sections as shown in Fig. 1.

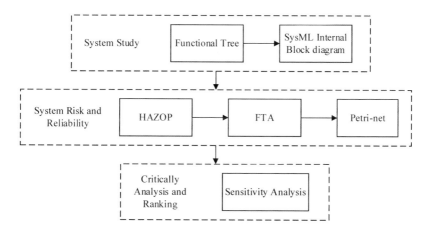

**Fig. 1.** The framework of proposed model

According to the proposed methodology, it is required that the system to be understood completely in terms of function, process, procedure, the relations between system, and subsystem. The aim of this phase is to evaluate how the system mechanisms fail and how it can be identified under uncertain conditions. the divisions of system and HAZOP study represents subsequently FTA as a deductive method identifies the causes of the deviation [8]. In FTA, the deviation is considered as an undesired event called TE and the root cause named as BEs. Once, FTA is structured, the probability of each BE events need to be computed. In much case, assessors can use reliability data handbooks whereas there are many conditions which caused that assessors are not able to use reliability data handbook like the specific components are not available in reliability data base. Using multi-experts knowledge as a proper alternative has enough capability to cope with shortages of data [4]. However, experts' knowledge brings uncertainty and ambiguity itself. To deal with this new state, fuzzy logic introduced by Prof. Zadeh can appropriately handle this part [9]. Thus, experts express their opinions related to possibility of each BE in qualitative expressions provided in [10]. Then, in the fuzzy environment the aggregation procedure is done and accordingly possibility of all BEs is transferred in to the probability.

According to intuitionistic fuzzy sets provided by Atanassov [11], the qualitative expressions are identified as intuitionistic fuzzy numbers (IFNs) which are provided in [10] due to the possibility of each BE. IFNs are developed to deal with the deficiencies of conventional fuzzy set numbers like as dealing with membership and non-membership functions.

Aggregate the expert's opinion using the intuitionistic fuzzy weighted averaging (IFWA) operator for any BEs, $BE_i = (i = 1, \ldots, m)$.

$$\alpha_{ij} = IFWA\left(\alpha_{ij}^1, \alpha_{ij}^2, \ldots, \alpha_{ij}^n\right) = \sum_{k=1}^{n} \lambda_k \alpha_{ij}^k \qquad (1)$$

where

$\alpha_{ij} = \left(\mu_{ij}, \nu_{ij}\right)$ is the final aggregated experts' opinions in expressions of IFN,

$\alpha_{ij}^k = \left(\mu_{ij}^k, \nu_{ij}^k\right)$ is the IFN that is reassigned by the corresponding qualitative expressions based on experts' opinion.

$\lambda_k$ is the given weight to each expert according to fuzzy analytical hierarchy process (FAHP) that signifies the significance of his/her opinion on $BE_i$, and satisfies $\lambda_k > 0$ $(k = 1, \ldots, n)$ and $\left(\sum_{k=1}^{n} \lambda_k = 1\right)$.

Boran et al. [12] indicated that Eq. 1 can be normalized to Eq. 2. Additionally, Anzilli and Facchinetti [13] represented that, Eq. 2 can be considered as a defuzzification IFNs which obtained by:

$$Val_S(x) = \frac{1}{2} \times (1 + \mu_S(x) - \nu_S(x)) \qquad (2)$$

where the set $(\mu_S(x), \nu_S(x))$ is named an IFN in intuitionistic fuzzy stand $\alpha = (\mu_S(x), \nu_S(x))$ simply signifies each IFN, where $\mu_\alpha \in [0, 1]$ and $\nu_\alpha \in [0, 1]$, and also satisfies $\mu_\alpha + \nu_\alpha \leq 1$. It should be noted that for an IFN $\alpha = (\mu_\alpha, \nu_\alpha)$, $\alpha^+ (1, 0)$ and $\alpha^- (0, 1)$ are nominated as the largest and smallest IFNs, respectively.

Once the crisp value as a possibility of an BE is computed then the possibility is converted to probability using Onisawa equations [14].

In order to calculate TE probability and identifying the critical BE, FTA is transferred to the Petri-nets. That is because, Petri-nets has many advantages to compare with FTA such as handling dynamic structure [7].

A typical PN is commonly utilized for examining qualitative or rational features of a system; however, on the other hand for quantitative performance examination, time notion is necessary to be taken into consideration in PN definitions. In this study, it is used the static part of the PN to analyze the behavioral of system reliability with consideration of the transition are not times. The mapping of logical FT into the PN, when all components are supposed to be completely practical at time 0 and the place representing the working state has a token is shown in Fig. 2.

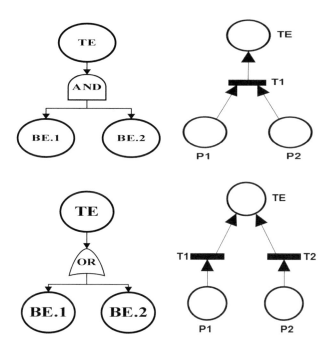

**Fig. 2.** Mapping of two input AND/OR logical Boolean gates into Petri-net.

Same as FTA, AND/OR logical Boolean gates can be computed in PN as well. The probability $\alpha_t$ which is involved to the token in the place $p_t$ is computed according to following Equations.

$$\alpha_{t_{AND}} = \prod_{i=1}^{n} \alpha_i \times \mu_i \tag{3}$$

$$\alpha_{t_{OR}} = 1 - \prod_{i=1}^{n} (1 - \alpha_i \times \mu_i) \tag{4}$$

where $\alpha_i$ is the probability of $p_i$ as $A = (\alpha_1, \alpha_2, \ldots, \alpha_n)$ and $\mu_i$ is the value of the certainty factor $t_i$ as $U : T \rightarrow [0, 1]$. $U = (\mu_1, \mu_2, \ldots, \mu_m)$.

The probability of a place has this possibility to be re-computed after its anterior transaction execute. Thus, the probability of follow-up places during the execution of the posterior transaction of place is updated. PN is required to be executed cylindrically till the probability in any place cannot be longer re-computed. It means that $A_k$ and $A_{k-1}$ are same where $k$ is the number of cycles. In order to obtain more details once can refer to the [15].

In the last step, sensitivity analysis (SA) is applied in order to identity the critical BEs. SA is capable to provide quantitative assessment to recognize weakest relationship between input events and proper design to substitute in a specific system. In addition, SA helps assessors to identify the important sources of variability as well as uncertainty during risk assessment procedure. In this paper, a basic significance according to the failure ranking of probabilities was utilized. The Risk Reduction Worth (RRW), as the maximum escalation in the system safety and reliability, is found as a ratio of the real probability of TE to the probability of TE, when $BE_i$ is substituted with BE which never happen [16]. This examination observes how the results of a calculation or model differ as separate expectations are altered. Besides, it can help the decision makers to comprehend the dynamics of the studied system.

## 3   Application of Study

Kick accidents is commonly occurred in oil and gas industry when there are working high pressure either governmental force to increase the amount of production or thermodynamic ones. According to several Kick reports, the fault tree is illustrated in Fig. 3 to represent the main causes of the accident. Four independent multi experts are employed in order to obtain the probability of each BE. Using fuzzy AHP explained in details by Yazdi and Kabir [17], the weight of the multi-employed experts are computed.

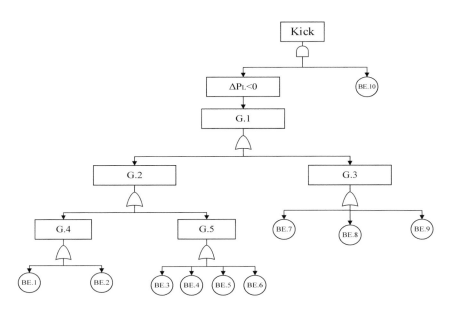

**Fig. 3.** Fault tree of Kick.

Then, it is asked from experts to express their opinions using Table 1 in order to estimate the occurrence possibility of each BE. The aggregation procedure under fuzzy environment has been done, and accordingly the possibility and probability of all BEs are computed. The results of computation are provided in Table 1.

**Table 1.** BEs description, corresponding qualitative terms and probability.

| BE Tag | Experts' opinions | | | | Aggregated IFN | Possibility | Probability | Rank |
|--------|------|------|------|------|----------------|-------------|-------------|------|
| | E1 | E2 | E3 | E4 | | | | |
| BE.1 | EL | VL | H | M | (0.732, 0.251) | 0.260 | 0.000546 | 1 |
| BE.2 | VH | H | M | M | (0.469, 0.498) | 0.514 | 0.005525 | 5 |
| BE.3 | FL | FL | M | VL | (0.559, 0.370) | 0.406 | 0.002439 | 3 |
| BE.4 | EL | VL | M | H | (0.723, 0.258) | 0.268 | 0.000605 | 2 |
| BE.5 | FL | M | H | L | (0.488, 0.430) | 0.471 | 0.004068 | 4 |
| BE.6 | EH | M | M | H | (0.304, 0.673) | 0.685 | 0.016706 | 9 |
| BE.7 | H | VH | M | FL | (0.341, 0.586) | 0.623 | 0.011286 | 8 |
| BE.8 | H | VL | H | FL | (0.426, 0.486) | 0.530 | 0.006171 | 6 |
| BE.9 | FL | M | M | H | (0.439, 0.500) | 0.530 | 0.006175 | 7 |
| BE.10 | FL | M | FL | VH | (0.439, 0.490) | 0.525 | 0.005963 | 10 |

Along with the associations between each BE, the petri-nets model for the kick analysis is developed and illustrated in Fig. 4.

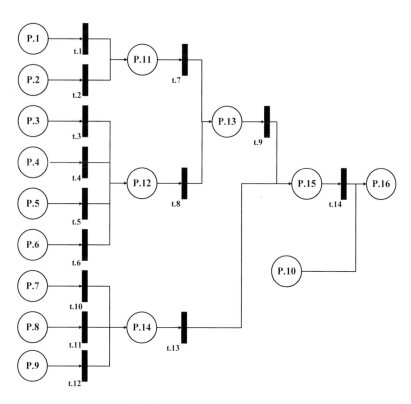

**Fig. 4.** Petri-net model of Kick.

According to the Petri-net model, the following matrices have this possibility clearly be computed. Once the input conditions based on AND/OR gates are fulfilled, the output event of the gates will occur. In this regards, the certainty factor value for each transitions of Petri-net model is assumed to be 1. Therefore,

$U = [1111111111111]$, primarily states:

$M_0 = [1111111111000000]$, and $A_0 = [0.000546 \quad 0.005525 \quad 0.002439 \quad 0.0006$ 05  0.004068  0.016706  0.011286  0.006171  0.006175  0.005963 0 0 0 0 0 0].

At this point, the set of transitions t.1, t.2, t.3, t.4, t.5, t.6, t.10, t.11, t.12, and t.14 are enabled. Subsequent to these transitions execute, it is obtained:

$M_1 = [1111111111111000]$ and $A_1 = [0.000546 \quad 0.005525 \quad 0.002439 \quad 0.0006$ 05  0.004068  0.016706  0.011286  0.006171  0.006175  0.005963  0.006068 0.023686 0 0.023454 0 0].

Then transition t.7, t.8, and t.13 are likewise enabled, Subsequent to the enabled transitions execute, the marking of the Petri-net and the probabilities of the places are re-computed.

$M_2 = [1111111111111100]$ and $A_2 = [0.000546 \quad 0.005525 \quad 0.002439 \quad 0.0006$ 05  0.004068  0.016706  0.011286  0.006171  0.006175  0.005963  0.006068 0.023686  0.029609  0.023454 0 0].

This process is continuing by enabling t.9 and t.13.

$M_3 = [1111111111111110]$ and $A_3 = [0.000546 \quad 0.005525 \quad 0.002439 \quad 0.00$ 0605  0.004068  0.016706  0.011286  0.006171  0.006175  0.005963  0.0060 68  0.023686  0.029609  0.023454  0.052370 0].

Now t.14 is also enabled and it is obtained:

$M_4 = [1111111111111111]$ and $A_4 = [0.000546 \quad 0.005525 \quad 0.002439 \quad 0.000$ 605  0.004068  0.016706  0.011286  0.006171  0.006175  0.005963  0.006068 0.023686  0.029609  0.023454  0.052370  0.000312].

As it can be seen the probability vector of $A_3 \neq A_4$. Thus execution process should be continued. Then in fifth iteration, it can be obtained:

$M_5 = [1111111111111111]$ and $A_5 = [0.000546 \quad 0.005525 \quad 0.002439 \quad 0.000$ 605  0.004068  0.016706  0.011286  0.006171  0.006175  0.005963  0.006068 0.023686  0.029609  0.023454  0.052370  0.000312].

As $A_4$ is equal to $A_5$, the execution process end, and $A_5$ or $A_4$ is the probability of Kick in the case study. Due to fact this case study is not dynamic and it is assumed to independent from time; thus, the result obtained by Petri-net is same as FTA. In order to recognize the critical BE, it is required to eliminate each transaction for once time and then continue Petri-net process to compute the probability of TE. The results of critical ranking for each BE are provided in Table 1 (last column).

## 4   Conclusion

Developing chemical process industries in recent decades have bring themselves several accidents which effect on several issues like as human fatality, environmental pollution, and assess loss. Therefore, it is vital using failure analysis techniques in order to prevent potential hazards in process industry. In this study, it is integrated two common failure analysis techniques (FTA and Petri-nets) under uncertainty conditions using intuitionistic fuzzy numbers. The efficiency and effectiveness of proposed model are examined by applying on real case study as kick in chemical process industry. At current present, we used static FTA for obtaining failure probability of each BE. As a direction for further study we plan to apply for dynamic FTA.

# References

1. Yazdi, M.: An extension of fuzzy improved risk graph and fuzzy analytical hierarchy process for determination of chemical complex safety integrity levels. Int. J. Occup. Saf. Ergon. 1–11 (2017). https://doi.org/10.1080/10803548.2017.1419654
2. Yazdi, M., Zarei, E.: Uncertainty handling in the safety risk analysis: an integrated approach based on fuzzy fault tree analysis. J. Fail. Anal. Prev. **18**(2), 392–404 (2018)
3. Kalantarnia, M., Khan, F., Hawboldt, K.: Modelling of BP Texas City refinery accident using dynamic risk assessment approach. Process Saf. Environ. Prot. **88**(3), 191–199 (2010)
4. Yazdi, M., Korhan, O., Daneshvar, S.: Application of fuzzy fault tree analysis based on modified fuzzy AHP and fuzzy TOPSIS for fire and explosion in process industry. Int. J. Occup. Saf. Ergon. 1–18 (2018). https://doi.org/10.1080/10803548.2018.1454636
5. Yazdi, M.: Improving failure mode and effect analysis (FMEA) with consideration of uncertainty handling as an interactive approach. Int. J. Interact. Des. Manuf. 1–18 (2018). https://doi.org/10.1007/s12008-018-0496-2
6. Yazdi, M., Daneshvar, S., Setareh, H.: An extension to fuzzy developed failure mode and effects analysis application for aircraft landing system. Saf. Sci. **98**(1), 113–123 (2017)
7. Kabir, S., Yazdi, M., Aizpurua, J.I., Papadopoulos, Y.: Uncertainty-aware dynamic reliability analysis framework for complex systems. IEEE Access **6**(1), 29499–29515 (2018). https://doi.org/10.1007/s12008-018-0496-2
8. Yazdi, M., Soltanali, H.: Knowledge acquisition development in failure diagnosis analysis as an interactive approach. Int. J. Interact. Des. Manuf. 1–18 (2018). https://doi.org/10.1007/s12008-018-0504-6
9. Zadeh, L.: Fuzzy sets. Inf. Control **8**(3), 338–353 (1965)
10. Yazdi, M.: Risk assessment based on novel intuitionistic fuzzy-hybrid-modified TOPSIS approach. Saf. Sci. 1–11(2018). https://doi.org/10.1016/j.ssci.2018.03.005
11. Atanassov, K.T.: On the concept of intuitionistic fuzzy sets. In: Studies in Fuzziness and Soft Computing. Springer, Heidelberg (2012)
12. Boran, F., Genç, S., Kurt, M., Akay, D.: A multi-criteria intuitionistic fuzzy decision making for supplier selection with TOPSIS method. Expert Syst. Appl. **36**(8), 11363–11368 (2009)
13. Anzilli, L., Facchinetti, G.: A new proposal of defuzzification of intuitionistic fuzzy quantities, vol. 5, no. 1, pp. 185–195. Springer, Cham (2016)
14. Yazdi, M.: Hybrid probabilistic risk assessment using fuzzy FTA and fuzzy AHP in a process industry. J. Fail. Anal. Prev. **17**(4), 756–764 (2017)
15. Zhou, K.Q., Zain, A.M.: Fuzzy Petri nets and industrial applications: a review. Artif. Intell. Rev. **45**(4), 405–446 (2016)
16. Dutuit, Y., Rauzy, A.: Efficient algorithms to assess component and gate importance in fault tree analysis. Reliab. Eng. Syst. Saf. **72**(2), 213–222 (2001)
17. Yazdi, M.: Footprint of knowledge acquisition improvement in failure diagnosis analysis. Qual. Reliab. Eng. Int. 1–18 (2018). https://doi.org/10.1002/qre.2408

# Fuzzy Differential Equations for Modeling and Control of Fuzzy Systems

Raheleh Jafari[1(✉)], Sina Razvarz[2], and Alexander Gegov[3]

[1] Centre for Artificial Intelligence Research (CAIR),
University of Agder, 4879 Grimstad, Norway
jafari3339@yahoo.com
[2] Departamento de Control Automático, CINVESTAV-IPN
(National Polytechnic Institute), 07360 Mexico City, Mexico
[3] School of Computing, University of Portsmouth, Buckingham Building,
Portsmouth PO13HE, UK

**Abstract.** A survey of the methodologies associated with the modeling and control of uncertain nonlinear systems has been given due importance in this paper. The basic criteria that highlights the work is relied on the various patterns of techniques incorporated for the solutions of fuzzy differential equations (FDEs) that corresponds to fuzzy controllability subject. The solutions which are generated by these equations are considered to be the controllers. Currently, numerical techniques have come out as superior techniques in order to solve these types of problems. The implementation of neural networks technique is contributed in the complex way of dealing the appropriate solutions of the fuzzy systems.

**Keywords:** Modeling · Fuzzy differential equation · Fuzzy system

## 1 Introduction

In recent days, many methods involving uncertainties have used fuzzy numbers [1–8], where the uncertainties of the system are represented by fuzzy coefficients. Fuzzy method is a highly favorable tool for uncertain nonlinear system modeling. The fuzzy models approximate uncertain nonlinear systems with several linear piecewise systems (Takagi-Sugeno method) [9]. Mamdani models use fuzzy rules to achieve a good level of approximation of uncertainties [10].

In comparison with the normal systems, FDEs are considered to be very non-complex. It is feasible for them to apply directly for nonlinear control. Fuzzy control through FDEs requires solution of the FDEs. Several approaches are incorporated. Some numerical approaches, such as Nystrom method [11] and Runge-Kutta method [12] can also be implemented for resolving FDEs. Laplace transform has been utilized for second-order FDE in [13]. The results of feedback control in refer to the wave equation has been illustrated in [14], whereas the open loop control in concerned to the wave equation has been demonstrated in [15].

Neural networks can also be implemented for resolving FDEs. [16] proposed a static neural network in order to resolve FDE. [17] illustrated that the solution of

© Springer Nature Switzerland AG 2019
R. A. Aliev et al. (Eds.): ICAFS-2018, AISC 896, pp. 732–740, 2019.
https://doi.org/10.1007/978-3-030-04164-9_96

ordinary differential equation (ODE) can be estimated with the help of neural network. [18] implemented neural approximations of ODEs to dynamic systems. [19] implemented dynamics neural networks for the approximation of the first-order ODE. In [20] a feed-forward neural network is suggested in order to resolve an elliptic PDE in 2D. In [21] by employing a feed forward neural network, controlled heat problem has been solved.

In this paper, a survey on the numerical solutions of the PDEs and FDEs is given. The solutions which are generated by these equations are considered to be the controllers. Here, it has been presented that the roots of the mentioned equations can be extracted with different methods. Studying of previous works by other researchers shows that no study has been done as a survey for the solutions of these equations, so that this survey will be a good beginning for those showing interest in the field of these kinds of equations.

## 2   Mathematical Preliminaries

The following definitions are used in this paper.

**Definition 1:** If $v$ is: (1) normal, there exists $\vartheta_0 \in \Re$ in such a manner $v(\vartheta_0) = 1$, (2) convex, $v(\gamma\vartheta + (1-\gamma)\theta) \geq \min\{v(\vartheta), v(\theta)\}, \forall\vartheta, \theta \in \Re, \forall\gamma \in [0,1]$, (3) upper semi-continuous on $\Re$, $v(\vartheta) \leq v(\vartheta_0) + \varepsilon, \forall\vartheta \in N(\vartheta_0), \forall\vartheta_0 \in \Re, \forall\varepsilon > 0, N(\vartheta_0)$ is a neighborhood, (4) $v^+ = \{\vartheta \in \Re, v(\vartheta) > 0\}$ is compact, then $v$ is a fuzzy variable, $v \in E : \Re \to [0,1]$.

**Definition 2:** The fuzzy number $v$ in association to the $\alpha$-level is illustrated as

$$[v]^\alpha = \{\vartheta \in \Re, v(\vartheta) \geq \alpha\} \tag{1}$$

where $0 < \alpha \leq 1, v \in E$.

Therefore $[v]^\alpha = v^+ = \{\vartheta \in \Re, v(\vartheta) > 0\}$. Since $\alpha \in [0,1]$, $[v]^\alpha$ is a bounded mentioned as $\underline{v}^\alpha \leq [v]^\alpha \leq \overline{v}^\alpha$. The $\alpha$-level of $v$ between $\underline{v}^\alpha$ and $\overline{v}^\alpha$ is explained as

$$[v]^\alpha = (\underline{v}^\alpha, \overline{v}^\alpha) \tag{2}$$

$\underline{v}^\alpha$ and $\overline{v}^\alpha$ signify the function of $\alpha$. We state $\underline{v}^\alpha = d_A(\alpha), \overline{v}^\alpha = d_B(\alpha), \alpha \in [0,1]$.

If $v_1, v_2 \in E$, the fuzzy operations are as follows

Sum,

$$[v_1 \oplus v_2]^\alpha = [v_1]^\alpha + [v_2]^\alpha = \left[\underline{v_1}^\alpha + \underline{v_2}^\alpha, \overline{v_1}^\alpha + \overline{v_2}^\alpha\right] \tag{3}$$

subtract,

$$[v_1 \ominus v_2]^\alpha = [v_1]^\alpha - [v_2]^\alpha = \left[\underline{v_1}^\alpha - \underline{v_2}^\alpha, \overline{v_1}^\alpha - \overline{v_2}^\alpha\right] \tag{4}$$

or multiply,

$$\underline{\omega}^{\alpha} \leq [v_1 \odot v_2]^{\alpha} \leq \overline{\omega}^{\alpha} \ or \ [v_1 \odot v_2]^{\alpha} = (\underline{\omega}^{\alpha}, \overline{\omega}^{\alpha}) \tag{5}$$

where, $\overline{\omega}^{\alpha} = \overline{v_1}^{\alpha}\overline{v_2}^{1} + \overline{v_1}^{1}\overline{v_2}^{\alpha} - \overline{v_1}^{1}\overline{v_2}^{1}, \underline{\omega}^{\alpha} = \underline{v_1}^{\alpha}\underline{v_2}^{1} + \underline{v_1}^{1}\underline{v_2}^{\alpha} - \underline{v_1}^{1}\underline{v_2}^{1}$ and $\alpha \in [0, 1]$.

# 3 Numerical Methods for Solving Partial and Fuzzy Differential Equations

## 3.1 Predictor-Corrector Method

The Predictor-corrector methodology is broadly utilized in order to resolve initial value problems. In [22], three numerical methodologies for resolving fuzzy ODEs are proposed. These methodologies are Adams-Bashforth, Adams-Moulton and predictor-corrector. Predictor-corrector is extracted by blending Adams-Bashforth and Adams-Moulton methodologies. Convergence and stability of the suggested methodologies are proved. Considering the convergence order of the Euler methodology which is $O(h)$ (as given in [23]), a higher order of convergency is achievable by utilizing the suggested methodologies in [22], to be mentioned that a predictor-corrector methodology of convergence order $O(h^m)$ is utilized where the Adams-Bashforth $m$-step methodology and Adams-Moulton $(m - 1)$ - step methodology are taken to be as predictor and corrector, respectively. By going with the ideas of [24], the suggested methodologies in [22] can resolve the stiff problems.

In [25] a numerical solution in concerned with hybrid FDE is researched. The improved predictor-corrector methodology is selected and altered in order to resolve the hybrid FDEs on the basis of the Hukuhara derivative. The symbolic systems associated with the computer to be mentioned as Maple and Mathematica are employed to carry out complex computations of algorithm. It is displayed that the solutions extracted using predictor-corrector methodology are more precise and well matched with the exact solutions.

In [26] an improved predictor-corrector method is presented in order to resolve FDE under generalized differentiability. The generalized characterization theorem is used for converting a FDE into two ODE systems. The significance of transforming a FDE to a system of ODEs is that any numerical technique which is suitable for ODEs can be applied. The improved predictor-corrector three-step methodology can be generated to improved predictor-corrector m-step methodologies of convergence order $O(h^m)$.

## 3.2 Adomian Decomposition Method

In [27] the Adomian decomposition method is used for finding the fuzzy solution of homogeneous fuzzy PDEs with specific fuzzy boundary and initial conditions. Seikkala derivative is utilized for resolving fuzzy heat equation with specific fuzzy boundary and initial conditions. The crisp form of heat equation is resolved by utilizing Adomian Decomposition method. After that the solution is extended in fuzzy form as a Seikkala solution.

In [28] the Adomian decomposition method is implemented for finding the numerical solution of hybrid FDEs. This methodology considers the approximate solution of a nonlinear equation as an infinite series which generally converges to the accurate solution. The comparison between the approximation solutions and the exact solutions shows that the convergency is quite close.

In [29] the convergence of the Adomian decomposition technique is proved for an initial-value problem. Convergence rates of the Adomian decomposition technique are studied in the context of the nonlinear Schrodinger equation.

### 3.3   Euler Method

In [23], the FDE is substituted by its parametric form. The classical Euler technique is implemented for resolving the novel system that contains two classical ODEs with initial conditions. The capability of technique is demonstrated by resolving several linear as well as nonlinear first-order FDEs.

In [30] two improvised Euler type methodologies to be mentioned as Max-Min improved Euler methodology and average improved Euler methodology are suggested for extracting numerical solution of linear as well as nonlinear ODEs at par with fuzzy initial condition. In this paper all the possible blends of lower as well as upper bounds in concerned with the variable are considered and then resolved by the suggested methodologies. Also, an exact method is laid down.

In [31] the numerical solution associated with linear, non-linear as well as system of ODEs with fuzzy initial condition is researched. Two Euler type methodologies namely Max-Min Euler methodology and average Euler methodology are laid down for extracting numerical solution related to the FDEs. Several investigators in their works have considered the left and right bounds of the variables in the differential equations. In this paper, the investigators constructed the methodologies by taking into account all possible combinations of lower as well as upper bounds of the variable. The solution extracted by Max-Min Euler methodology very closely matches with the outcomes extracted by [23] and exact solution.

### 3.4   Neural Network Method

In [32] a technique in order to resolve both ODEs and PDEs is presented and is dependent on the function approximation abilities of feedforward neural networks. This technique results in the development of solution presented in a differentiable and closed analytic form. This form applies a feedforward neural network as the basic estimation element that its parameters (weights and biases) are adjusted to diminish a suitable error function. In order to train the network, optimization methodologies have implemented, that need the calculation of the gradient error considering the network parameters. In the suggested methodology the model function is presented as the sum of two terms. The first term suffices the initial/boundary conditions, also does not include adjustable parameters. The second term includes a feedforward neural network to be trained in order to suffice the differential equation. The implementation of a neural architecture sums up several attractive features to the technique:

1. The implementation of neural networks supplies a solution with highly superior generalized attributes. Compared results with the finite element methodology which are depicted in this work describe this point vividly.
2. The technique is simple and can be implemented to ODEs, systems of ODEs and also to PDEs stated on orthogonal box boundaries. Furthermore, the process is in advancement to deal with the case of irregular (arbitrarily shaped) boundaries.
3. The technique can also be effectively imposed on parallel architectures.

This technique is simple and can be employed to both ODEs as well as PDEs by developing the suitable form of the trial solution.

In [33] a technique is introduced in order to resolve PDEs with boundary and initial conditions by utilizing neural networks. An evolutionary algorithm is employed for training the networks. The outcomes of implementing the methodology to a one-dimensional as well as two-dimensional problem are highly superior and convincing.

## 4 Comparison of Numerical Methods

In this section, an example of application has been laid down in order to compare the efficiency of the numerical methods to approximate the solution of FDEs.

**Example 1.** A tank with a heating system is shown in Fig. 1, where $R = 0.5$ and the thermal capacitance is $C = 2$. The temperature is $x$. The model is [34, 35],

$$\frac{d}{dt}x(t) = -\frac{1}{RC}x(t) \tag{6}$$

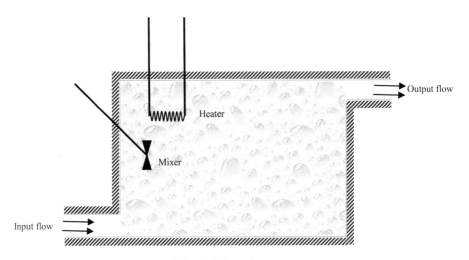

**Fig. 1.** Thermal system

where $t \in [0, 1]$ and $x$ is the amount of sinking in each moment. If the initial position is $x(0) = (\alpha - 1, 1 - \alpha)$ and $\alpha \in [0, 1]$, then the exact solutions of (6) are

$$x(t, \alpha) = [(\alpha - 1)e^t, (1 - \alpha)e^t] \tag{7}$$

To approximate the solution (7), we use four popular methods: Predictor-corrector method, Adomian decomposition method, Euler method, and Neural network method. The errors of these methods are shown in Table 1. The lower and upper bounds of absolute errors are shown in Figs. 2 and 3 respectively. The approximation errors of the neural network method is smaller than the other methods.

**Table 1.** Approximation errors

| $\alpha$ | Predictor-corrector | Adomian decomposition | Euler | Neural network |
|---|---|---|---|---|
| 0 | [0.2275, 0.4507] | [0.1196, 0.1623] | [0.0902, 0.1423] | [0.0423, 0.0812] |
| 0.2 | [0.2159, 0.4407] | [0.1006, 0.1553] | [0.0839, 0.1367] | [0.0381, 0.0702] |
| 0.4 | [0.2419, 0.4718] | [0.1228, 0.1713] | [0.0112, 0.1545] | [0.0519, 0.0901] |
| 0.6 | [0.2613, 0.4962] | [0.1486, 0.1923] | [0.1385, 0.1773] | [0.0734, 0.1578] |
| 0.8 | [0.2009, 0.4319] | [0.0933, 0.1441] | [0.0774, 0.1295] | [0.0201, 0.0635] |
| 1 | [0.2785, 0.2785] | [0.1633, 0.1633] | [0.1448, 0.1448] | [0.0801, 0.0801] |

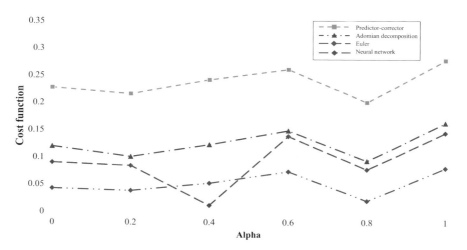

**Fig. 2.** The lower bounds of absolute errors

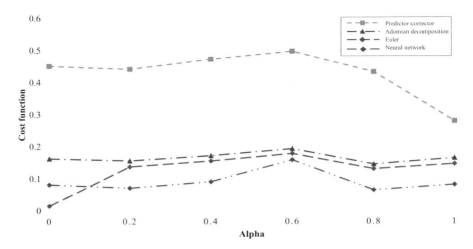

**Fig. 3.** The upper bounds of absolute errors

## 5    Conclusions

In this paper, some of numerical methodologies are demonstrated as a solution of PDEs and FDEs. This survey illustrates that the roots of the differential equation can be extracted with different algorithms. However, in few cases there exist no roots in differential equation. For obtaining the roots of system in a case that there is no exact solution, iteration methodologies can be utilized for estimating the solution. This survey supplies an input for those showing interest in the field of differential equations.

## References

1. Jafari, R., Razvarz, S.: Solution of fuzzy differential equations using fuzzy Sumudu transforms. In: IEEE International Conference on Innovations in Intelligent Systems and Applications, vol. 1, pp. 84–89 (2017)
2. Jafari, R., Yu, W.: Uncertainty nonlinear systems modeling with fuzzy equations. In: Proceedings of the 16th IEEE International Conference on Information Reuse and Integration, San Francisco, Calif, USA, 1 August, pp. 182–188 (2015)
3. Jafari, R., Yu, W.: Uncertainty nonlinear systems control with fuzzy equations. In: IEEE International Conference on Systems, Man, and Cybernetics, vol. 1, pp. 2885–2890 (2015)
4. Jafari, R., Yu, W., Li, X.: Numerical solution of fuzzy equations with Z-numbers using neural networks. Intell. Autom. Soft Comput. **1**, 1–7 (2017)
5. Jafari, R., Yu, W., Li, X., Razvarz, S.: Numerical solution of fuzzy differential equations with Z-numbers using Bernstein neural networks. Int. J. Comput. Intell. Syst. **10**, 1226–1237 (2017)
6. Razvarz, S., Jafari, R., Granmo, O.Ch., Gegov, A.: Solution of dual fuzzy equations using a new iterative method. In: Asian Conference on Intelligent Information and Database Systems, vol. 1, pp. 245–255 (2018)

7. Razvarz, S., Jafari, R., Yu, W.: Numerical solution of fuzzy differential equations with Z-numbers using fuzzy Sumudu transforms. Adv. Sci. Technol. Eng. Syst. J. (ASTESJ) **3**, 66–75 (2018)
8. Razvarz, S., Jafari, R., Yu, W., Khalili, A.: PSO and NN modeling for photocatalytic removal of pollution in wastewater. In: 14th International Conference on Electrical Engineering, Computing Science and Automatic Control (CCE) Electrical Engineering, vol. 1, pp. 1–6 (2017)
9. Takagi, T., Sugeno, M.: Fuzzy identification of systems and its applications to modeling and control. IEEE Trans. Syst. Man. Cybern. **15**, 116–132 (1985)
10. Mamdani, E.H.: Application of fuzzy algorithms for control of simple dynamic plant. IEE Proc. Control Theory Appl. **121**(12), 1585–1588 (1976)
11. Khastan, A., Ivaz, K.: Numerical solution of fuzzy differential equations by Nyström method. Chaos, Solitons Fractals **41**, 859–868 (2009)
12. Palligkinis, S.C., Papageorgiou, G., Famelis, I.T.: Runge-Kutta methods for fuzzy differential equations. Appl. Math. Comput. **209**, 97–105 (2009)
13. Ahmadi, M.B., Kiani, N.A., Mikaeilvand, N.: Laplace transform formula on fuzzy nth-order derivative and its application in fuzzy ordinary differential equations. Soft. Comput. **18**, 2461–2469 (2014)
14. Gibson, J.S.: An analysis of optimal modal regulation: convergence and stability. SIAM J. Control Optim. **19**, 686–707 (1981)
15. Kröner, A., Kunisch, K.: A minimum effort optimal control problem for the wave equation. Comput. Optim. Appl. **57**, 241–270 (2014)
16. Effati, S., Pakdaman, M.: Artificial neural network approach for solving fuzzy differential equations. Inform. Sci. **180**, 1434–1457 (2010)
17. Agatonovic-Kustrin, S., Beresford, R.: Basic concepts of artificial neural network (ANN) modeling and its application in pharmaceutical research. J. Pharm. Biomed. Anal. **22**, 717–727 (2000)
18. Yazdi, H.S., Pourreza, R.: Unsupervised adaptive neural-fuzzy inference system for solving differential equations. Appl. Soft Comput. **10**, 267–275 (2010)
19. Lee, H., Kang, I.S.: Neural algorithms for solving differential equations. J. Comput. Phys. **91**, 110–131 (1990)
20. Dissanayake, M.W.M.G., Phan-Thien, N.: Neural-network based approximations for solving partial differential equations. Commun. Numer. Methods Eng. **10**, 195–201 (2000)
21. Sukavanam, N., Panwar, V.: Computation of boundary control of controlled heat equation using artificial neural networks. Int. Commun. Heat Mass Transfer **30**, 1137–1146 (2003)
22. Allahviranloo, T., Ahmady, N., Ahmady, E.: Numerical solution of fuzzy differential equations by predictor-corrector method. Inform. Sci. **177**, 1633–1647 (2007)
23. Ma, M.A., Friedman, M., Kandel, A.: Numerical solutions of fuzzy differential equations. Fuzzy Sets Syst. **105**, 133–138 (1999)
24. Román-Flores, H., Rojas-Medar, M.: Embedding of level-continuous fuzzy sets on Banach spaces. Inform. Sci. **144**, 227–247 (2002)
25. Kim, H., Sakthivel, R.: Numerical solution of hybrid fuzzy differential equations using improved predictor–corrector method. Commun. Nonlin. Sci. Numer. Simul. **17**, 3788–3794 (2012)
26. Balooch Shahryari, M.R., Salahshour, S.: Improved predictor-corrector method for solving fuzzy differential equations under generalized differentiability. J. Fuzzy Set Valued Anal. **2012**, 16 (2012)
27. Pirzada, U.M., Vakaskar, D.C.: Solution of fuzzy heat equations using Adomian decomposition method. Int. J. Adv. Appl. Math. Mech. **3**, 87–91 (2015)

28. Paripour, M., Hajilou, E., Heidari, H.: Application of Adomian decomposition method to solve hybrid fuzzy differential equations. J. Taibah Univ. Sci. (2014). http://dx.doi.org/10.1016/j.jtusci.2014.06.002

29. Abdelrazec, A., Pelinovsky, D.: Convergence of the Adomian decomposition method for Initial-Value problems. Numer. Methods Partial Differ. Equ. **27**, 749–766 (2011)

30. Tapaswini, S., Chakraverty, S.: A new approach to fuzzy initial value problem by improved Euler method. Fuzzy Inf. Eng. **3**, 293–312 (2012)

31. Tapaswini, S., Chakraverty, S.: Euler-based new solution method for fuzzy initial value problems. Int. J. Artif. Intell. Soft. Comput. **4**, 58–79 (2014)

32. Lagaris, I.E., Likas, A., Fotiadis, D.I.: Artificial neural networks for solving ordinary and partial differential equations. IEEE Trans. Neural Netw. **9**, 987–1000 (1998)

33. Aarts, L.P., Van der Veer, P.: Neural network method for solving partial differential equations. Neural Process. Lett. **14**, 261–271 (2001)

34. Pletcher, R.H., Tannehill, J.C., Anderson, D.: Computational Fluid Mechanics and Heat Transfer. Taylor and Francis, Abingdon (1997)

35. Razvarz, S., Jafari, R.: Experimental study of $Al_2O_3$ nanofluids on the thermal efficiency of curved heat pipe at different tilt angle. J. Nanomater. Article ID 1591247, 1–7 (2018)

# Fuzzy Logic-Based Compositional Decision Making in Music

Javanshir Guliyev[1(✉)] and Konul Memmedova[2]

[1] Department of Acting, Near East University, Nicosia, North Cyprus, Turkey
javanshir55@gmail.com
[2] Department of Psychological Counselling and Guidance, Near East University,
Nicosia, North Cyprus, Turkey
konul.memmedova@neu.edu.tr

**Abstract.** The paper is dedicated to using Fuzzy Logic in compositional decision making processes. Fuzzy Logic framework enables us to take into account ambiguity in the music. The suggested in this paper method is based a "preference ordering" of the attributes of the given specific music.

**Keywords:** Fuzzy rules · Decision making · Music

## 1 Introduction

Research evidence in the area of application of Fuzzy Logic to music composition is very scarce. The conceptual use of Fuzzy Logic in music composition is presented in [1]. The author provides a brief overview of state-of-the-art of semiotics of music in automated composition systems. [2] uses Fuzzy Logic as the tool to advance computer music argument. The Author proposes an automated Fuzzy Logic system which could analyse each piece of music for its patterning. In [3] a-rhythmic compositions through the control of several sound synthesis parameters is considered. In [4] author has developed a chord creating mechanism. In [5] it is developed a Fuzzy Logic module Music program.

## 2 State of the Problem

The compositional decision problem in this study requires to construct relationship between controllable element (for example pitch, duration, amplitude etc.) and consequent evaluation. In this paper this relationship is described as set of IF … THEN rules with antecedents (input parameter) and consequent (output evaluation of created musical piece).

Each IF/THEN rule compares n attributes which produce a consequent. Attributes and consequent are described by fuzzy sets such as Poor, OK, Very Good etc.

For simplicity, without of loss of generality, assume that we deal with 2 attributes, such as "dynamic level" and "pitch" and consequent as "evaluation of quality".

© Springer Nature Switzerland AG 2019
R. A. Aliev et al. (Eds.): ICAFS-2018, AISC 896, pp. 741–745, 2019.
https://doi.org/10.1007/978-3-030-04164-9_97

Codebooks, which include linguistic terms of attributes and consequent of the created fuzzy *IF...THEN* model are shown in Tables 1, 2, 3 and Figs. 1, 2, 3.

**Table 1.** The encoded linguistic terms for the "dynamic level"

| Scale | Level of criteria | Linguistic value |
|-------|-------------------|------------------|
| 1.    | Poor (P)          | {0; 0.1; 0.2; 0.3} |
| 2.    | Not Soft (NS)     | {0.2; 0.3; 0.4; 0.5} |
| 3.    | Medium Soft (MS)  | {0.4; 0.5; 0.6; 0.7} |
| 4.    | Soft (S)          | {0.6; 0.7; 0.8; 0.9} |
| 5.    | Very Soft (VS)    | {0.8; 0.9; 1; 1} |

Fragment of the fuzzy IT-THEN model of music piece is given below [6]:

1. IF "dinamic level" is "very soft" and "pitch" is "very low" THEN "evaluation of quality" is "poor".
2. IF "dinamic level" is "soft" and "pitch" is "medium" THEN "evaluation of quality" is "poor"is OK.    (1)

Model (1) is contact depended and number of rules is identified bu composer. Given values of new attributes composer can canculate consequent evaluation by changing values of attributes one can get consequent evaluation as, for example, "very good".

Solution of the compositional decision problem is based on interpolation approach and consist of following steps:

Step 1. Given new values of attributes to calculate simularity degrees between current and existing in rule – base antesedent fuzzy values. Determine firing degrees of the rules by using the computed similarity degrees. We use simularity measure $S(A'_j, A_{ij})$ between current input $A'_j$ and antecedent $A_{ij}$ of *i-th* rule:

$$S(A'_j, A_{ij}) = \frac{1}{1 + D(A'_j, A_{ij})}$$    (2)

$D(A'_j, A_{ij})$ is a distance

$$D\left(A'_j, A_{ij}\right) = \frac{1}{n+1} \sum_{k=1}^{n} \left\{ \left| a'^L_{j\alpha_k} - a^L_{ij\alpha_k} \right| + \left| a'^R_{j\alpha_k} - a^R_{ij\alpha_k} \right| \right\}$$

where $a'^L_{j\alpha_k} = \min A'_j$, $a'^R_{j\alpha_k} = \max A'_j$.

**Table 2.** The encoded linguistic terms for the "pitch"

| Scale | Level of criteria | Linguistic value |
|---|---|---|
| 1. | Very Low (VL) | {0.05; 0.15; 0.25; 0.35} |
| 2. | Low (L) | {0.25; 0.35; 0.45; 0.55} |
| 3. | Medium (M) | {0.45; 0.55; 0.65; 0.75} |
| 4. | High (H) | {0.65; 0.75; 0.85; 0.95} |
| 5. | Very High (VH) | {0.85; 0.95; 1; 1} |

**Table 3.** The encoded linguistic terms for output

| Scale | Level of criteria | Linguistic value |
|---|---|---|
| 1. | Very Poor (VP) | {0; 0; 0.15; 0.25} |
| 2. | Poor (P) | {0.15; 0.25; 0.35; 0.45} |
| 3. | OK (O) | {0.35; 0.45; 0.55; 0.65} |
| 4. | Good (G) | {0.55; 0.65; 0.75; 0.85} |
| 5. | Very Good (VG) | {0.75; 0.85; 0.95; 1} |

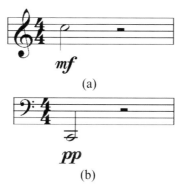

**Fig. 1.** Linguistic description of antecedent part of first (a) and second (b) rules.

The firing degree is determined as the minimal similarity of all the inputs:

$$\rho_i = \min_{j=1,\dots,n} S(A'_j, A_{ij}) \tag{3}$$

Step 2. On the base of the firing degrees we computed the coefficients of linear interpolation $w_i$, $i = 1,\dots,m$:

$$w_i = \frac{\rho_i}{\sum_{k=1}^{m} \rho_k} \tag{4}$$

Step 3. Then we computed current output $Y$:

$$Y = \sum_{i=1}^{m} w_i Y_i \qquad (5)$$

## 3  Numerical Example

Current attributes are: "dinamic level" is "not soft" and "pitch" is "medium".
What is consequent evaluation?
The obtained results of firing degrees on base of (3) are as follow:

$$\rho_1 = \min(0.4; 0.5) = 0.4$$

$$\rho_2 = \min(0.5; 1) = 0.5$$

Coefficients of linear interpolation are computed by using (4) are as follow:

$$w_1 = \frac{\rho_1}{\sum\limits_{i=1}^{10} \rho_i} = 0.8, \quad w_2 = \frac{\rho_2}{\sum\limits_{i=1}^{10} \rho_i} = 0.6,$$

The current output $Y$ is computed by using (5) as follows:

$$Y_1 = (0.26 \ \ 0.36 \ \ 0.46 \ \ 0.56).$$

According to the codebook in Table 3 this consequent evaluation is "OK".

## 4  Conclusion

This research contributes to production of music by using fuzzy concept. Use of this concept enables composer to take into account ambiguity in the music production. For solving compositional decision problem fuzzy IF...THEN model that describes relationship between compositional elements and overall performance is suggested.

Solution method of compositional decision making problem in music on base of suggested Fuzzy rule-base model by using interpolation approach to fuzzy reasoning is proposed. Computer simulation proofs validity of the suggested approach to music production.

## References

1. Whalley, I.: Towards a closed system automated composition engine: linking 'Kansei' and musical language recombinicity. In: ICMC2002, Goteborg, Sweden, pp. 200–203 (2002)
2. Milicevic, M.: Positive Emotion Learning through Music Listening. http://myweb.lmu.edu/mmilicevic/NEWpers/_PAPERS/papers.html

3. Cadiz, R.F.: Compositional control of computer music by fuzzy logic, Ph.D. thesis, Northwestern University (2006)
4. Elsea, P.: Fuzzy Logic and Musical Decisions. ftp://arts.ucsc.edu/pub/ems/FUZZY/Fuzzy_logic_and_Music.pdf
5. Sorensen, A., Brown, A.: jMusic: Music Composition in Java
6. http://jmusic.ci.qut.edu.au/index.html
7. Suiter, W.: The promise of fuzzy logic in generalised music composition, University of Wollongong, NSW, Australia

# Banknote Issuing Country Identification Using Image Processing and Neural Networks

Adnan Khashman[1,2(✉)], Waleed Ahmed[1,2], and Sadig Mammadli[3,4]

[1] European Centre for Research and Academic Affairs (ECRAA),
Nicosia, Mersin 10, Turkey
adnan.khashman@ecraa.com
[2] Final International University, Kyrenia, Mersin 10, Turkey
waleed.ahmed@final.edu.tr
[3] University of Kyrenia, Kyrenia, Mersin 10, Turkey
sadigm@gmail.com
[4] Odlar Yurdu University, AZ1008 Baku, Azerbaijan

**Abstract.** The work in this paper investigates developing an identification system for 21 countries using images of their banknotes and neural network classifiers. We consider the banknotes of 19 Asian countries, the European Union (EU), and the USA. Our motivation to investigate the Asian currencies is the increased global interaction in tourism and international trading with these countries where they have diverse and impressive banknote designs; thus making it difficult to identify by foreign visitors or traders. Our database comprises 504 original and pre-processed images of 6 banknotes of each of the 21 currencies. The investigated 19 Asian countries in this work are Afghanistan, Armenia, Azerbaijan, Bangladesh, Bhutan, Brunei, Burma, Cambodia, China, India, Kuwait, Maldives, Pakistan, Saudi Arabia, Sri Lanka, Syria, Tajikistan, Turkey, and United Arab Emirates. Most existing banknote identification systems aim to identify the currency value or decide whether a banknote is counterfeit. Our presented work is novel as it focuses on identifying the issuing country. Furthermore, we apply two pattern-averaging methods using $(5 \times 5)$ and $(10 \times 10)$ kernels, and follow two learning schemes to train and test the proposed neural identification models by using (50:50) and (75:25) training-to-validation data ratios. The obtained experimental results are considered as successful.

**Keywords:** Artificial intelligence · Image processing · Pattern averaging
Neural networks · Banknote identification · Currency recognition

## 1 Introduction

Over the past decade, rapid globalization and the availability of information technology means have made interacting with other countries much easier than we could have imagined. With the increase in electronic communication, international trading and tourism have become faster and cheaply available to all nations. Consequently, many countries; that were considered 'faraway' by the western hemisphere, have become within reach and very popular as traveling or trading destinations. The majority of these countries, which in many aspects remain mysterious to the foreign travelers, exist in Asia.

© Springer Nature Switzerland AG 2019
R. A. Aliev et al. (Eds.): ICAFS-2018, AISC 896, pp. 746–753, 2019.
https://doi.org/10.1007/978-3-030-04164-9_98

As interaction is increasing with such countries, both tourism and trade require financing which often takes place using the local currency of the destination country. Of course, international currencies such as the USA dollar and the EU Euro may well be accepted in these countries, but more often than not, each country opts to use its own currency [1, 2]. Due to the huge diversity in religions, cultures, and ethnicities of these Asian countries, each country reflects its own 'mark' on its currency; specially on its banknote designs. There is a tremendous variety of artistic designs on banknotes of many previously-considered mysterious countries. As a result of these varieties, recognizing the banknote issuing country requires a thorough visual inspection of the banknote which is a difficult task, thus, providing a fast solution to identifying a currency-issuing country by simply using banknote images would be advantageous.

Most of the previous works on processing banknotes focused on identifying the value of a banknote, or recognizing a counterfeit banknote from a real one [3–14]. Therefore, we can see that there is need for automatic fast system that identifies the countries that issue the currency or the banknote. In this paper, we propose such a novel system by using image processing and supervised neural network models based on the back propagation neural network (BPNN). Neural networks have been successfully used in many different applications with efficient results [14–24].

However, the use of artificially intelligent neural models requires training and machine learning; which in turn relies on using training and validation datasets (database); of course, the larger the dataset, the higher probability of a better learning. For this purpose, we constructed our own dataset of 504 banknote images representing 21 international currencies; comprising the US dollar, the Euro, and 19 Asian currencies of the following countries: Afghanistan, Armenia, Azerbaijan, Bangladesh, Bhutan, Brunei, Burma, Cambodia, China, India, Kuwait, Maldives, Pakistan, Saudi Arabia, Sri Lanka, Syria, Tajikistan, Turkey, and United Arab Emirates (UAE). The banknote images are gathered from a banknote museum that is freely available online [25].

The images in our database undergo pre-processing prior to using them in neural arbitration. In this work, we apply pattern averaging to all images using two different kernel sizes, i.e. $5 \times 5$ pixels and $10 \times 10$ pixels. The aim of averaging is to reduce the computational cost and the learning time. We also follow two schemes to train and test our neural network models. The schemes differ in how the dataset is divided during learning. In the first learning scheme, a training-to-validation data ratio of (50:50) is used; whereas, in the second learning scheme a ratio of (75:25) is used. Based on the obtained experimental results, we compare the neural models performance under the two learning schemes. Subsequent sections in the paper describe the banknote database, image processing phase, neural network design and implementation, results, and conclusion.

## 2  Banknote Database and Image Processing

The images of the 21 currencies in this database were collected from the freely available online banknote museum [25]. We consider six banknotes of each currency (see examples in Fig. 1). However, each considered banknote has two sides, which could also be viewed or scanned upsidedown. Therefore, for each banknote we further consider four images: front, back, front-inverted, and back-inverted as seen in an example in Fig. 2. As a result, there are 504 original color banknote images in our database determined as such:

$$21\, currencies \times 6\, banknotes \times 4\, banknotesides = 504\, banknoteimages$$

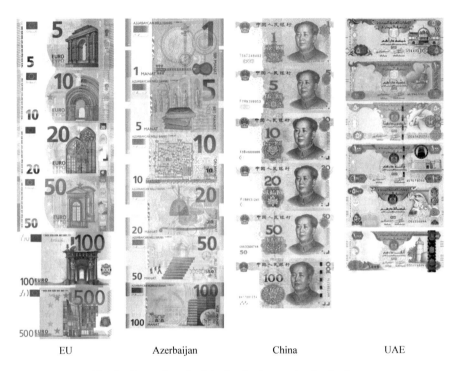

EU          Azerbaijan          China          UAE

**Fig. 1.** Examples of original color images in the database.

(a) front          (b) back          (c) front-inverted          (d) back-inverted

**Fig. 2.** The four sides of a Turkish banknote.

Each of the 504 color images in our dataset is pre-processed prior to using it with neural networks. The image processing phase comprises the following steps respectively:

$$[jpgcolorimage] \Rightarrow [jpggreyscaleimage] \Rightarrow$$
$$\Rightarrow [resized\ 200 \times 200\ pixels] \Rightarrow [rawimage]$$

As a result, we have 504 raw images of size $200 \times 200$ pixels each. Since these processed banknote images will be used as input data during neural network learning, their pixel values are fed to the neural network input layer using the one-pixel-per-node approach; consequently, the number of nodes (or input neurons) required would be very high (40,000 neurons) which is computationally expensive and causes excess time consumption. Therefore, another process is required to reduce the number of pixels of the input images.

This process is pattern averaging and it involves applying local kernels of defined sizes to each of the $200 \times 200$ size raw images, and then determine the average value of that kernel using Eq. (1).

$$PAV_i = \frac{1}{k_x k_y} \sum_{y=1}^{k_y} \sum_{x=1}^{k_x} p_i(x, y) \tag{1}$$

where $x$ is horizontal coordinate and $y$ is vertical coordinate of kernel $(k)$. Index $i$ indicates kernel number, $k_x$ and $k_y$ are width and height of kernel, $p_i(x, y)$ is image pixel value at $x$ and $y$ in kernel $i$, $PAV_i$ is kernel's average pattern value. The number of kernels in each image is equal to the number of neurons in the input layer and is defined as $i = \{0, 1, 2, ..., n\}$ and:

$$n = \left(\frac{X}{k_x}\right)\left(\frac{Y}{k_y}\right) \tag{2}$$

where $n$ is number of neurons in the input layer. In this work, we use two kernel sizes: $5 \times 5$ pixels ($k_x = k_y = 5$) and $10 \times 10$ pixels ($k_x = k_y = 10$), while the raw image size is $X \times Y = 200 \times 200$; therefore, we end up with two sets of images to use with the neural network model.

*First dataset* is obtained using local pattern averaging with $5 \times 5$ kernels. This results in ($n = 1600$). *Second dataset* is obtained using local pattern averaging with $10 \times 10$ kernels. This results in ($n = 400$). Thus, reducing the number of input layer neurons from 40000 to 1600 and 400 neurons, respectively.

The coding of the output of the neural network is dependent on the number of currency issuing countries in our dataset. Here we have a total of 21 issuing countries, therefore our neural network output will be represented by a $21 \times 21$ binary identity matrix. For example, using alphabetical order, the first country is Afghanistan, second is Armenia, … and so on. Having processed the images and prepared the input and output data for neural network learning, we then apply banknote images training and testing as will be described in the next section.

## 3   Neural Network Arbitration

In this work, two supervised neural networks that use error minimization and the back propagation learning algorithm are used. Considering the both image dataset with n = 1600 and n = 400 we have, respectively, the first neural model (ANN 1) with 1600 input neurons and the second neural model (ANN 2) with 400 input neurons; both receiving averaged pattern values of banknote images. The output layer in both models have 21 neurons according to the number of classifications in this work; i.e. 21 banknote issuing countries. The number of hidden layer neurons is usually determined during training and adjusted until optimum performance is achieved based on a predefined minimum error value. In our experiments, hidden neurons number was adjusted from one to hundred neurons. The activation function of the processing neurons in our neural networks is the sigmoid activation function. The important parameters in any supervised learning are the learning coefficient ($\eta$) and the momentum rate ($\alpha$). These parameters indicate how efficient a neural model learns, and are usually adjusted several times during learning to achieve optimum performance with minimal error and higher recognition rates. The architectural design of the ANN 1 neural network is shown in Fig. 3.

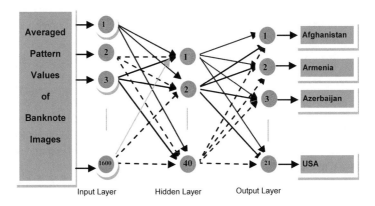

**Fig. 3.** Design of neural model ANN 1 (n = 1600) with learning scheme II (75:25)

## 4   Experimental Results

The neural networks simulation was completed using a 3.20 GHz PC with 4 GB of RAM, Windows 10 OS and Dev- C$^{++}$ compiler. Training the back propagation neural network models was carried out following two different learning schemes; where the ratio of training banknote images compared to testing images was differed. We adopted two learning schemes; namely, Scheme I with (50:50) training-to-validation data ratio, and Scheme II with (75:25) training-to-validation data ratio. The second scheme allows the neural model to 'see' more banknote images, thus allowing it to achieve higher correct identification ratios (CIR).

Table 1 shows a detailed listing of the experimental results for implementing both neural models ANN1 and ANN2; under the two learning schemes I and II. It can be seen from the results table that training neural network model ANN2 under scheme II achieved the highest results when considering the banknote issuing country correct identification rate (CIR). An overall CIR of 83.93% is considered as sufficient since the task of identifying 21 different issuing country based on showing banknote images only is a very difficult task indeed.

**Table 1.** Identification results and final training parameters of the neural network models.

| Neural model | ANN 1 | | ANN 2 | |
|---|---|---|---|---|
| Learning scheme | Scheme 1 (50:50) | Scheme II (75:25) | Scheme 1 (50:50) | Scheme II (75:25) |
| Kernel size (pixels) | $5 \times 5$ | $5 \times 5$ | $10 \times 10$ | $10 \times 10$ |
| Input neurons | 1600 | 1600 | 400 | 400 |
| Hidden neurons | 50 | 40 | 51 | 51 |
| Output neurons | 21 | 21 | 21 | 21 |
| Learning rate | 0.001 | 0.001 | 0.0001 | 0.0009 |
| Momentum rate | 0.20 | 0.2 | 0.33 | 0.37 |
| Obtained error | 0.0087 | 0.0095 | 0.0070 | 0.0069 |
| Iterations | 5000 | 5000 | 700 | 6895 |
| Training time (s)[a] | 1576 | 2445 | 50 | 1668 |
| CIR[b] – Training images | (248/252) 98.41% | (361/378) 95.50% | (245/252) 97.22% | (367/378) 97.09% |
| CIR[b] – Testing images | (103/252) 40.87% | (58/126) 46.03% | (106/252) 42.06% | (56/126) 44.44% |
| **CIR[b] – Overall** | **(351/504) 69.64%** | **(419/504) 83.13%** | **(351/504) 69.64%** | **(423/504) 83.93%** |

[a]using a 3.20 GHz PC with 4 GB of RAM, Windows 10 OS and Dev- C$^{++}$ compiler.
[b]CIR: Correct Identification Rate.

The performance of the other neural model ANN1 was equally impressive and slightly lagged behind ANN2 in terms of CIR value under learning scheme II. With similar CIR results, we must consider other factors when deciding upon the best performer; but first we can establish that learning scheme II (75:50) training-to-validation data ratio provides better learning to both models than scheme I. Now we consider other factors to decide whether ANN1 or ANN2 is the best performer. For this purpose, we examine Table 1 again and observe the error values and the training times. It is noticed that neural model ANN2 yielded less error value (0.0069) than ANN1 (0.0095). Also it is noticed that training the ANN2 required shorter time (1668 s) than training ANN1 (2445 s). Therefore, we can consider the best overall performer as ANN2 model.

## 5 Conclusions

This paper presented a novel application for using artificial neural networks. We proposed two neural network models; trained following two learning schemes, to identify the issuing countries of 21 currencies by only using their corresponding banknote images. The images undergo several preprocessing stages prior to presenting them to the neural models. The method of image pattern averaging is used to reduce the computational and time costs.

The banknote image dataset used in this work was built by the authors and comprises 504 color and processed images of 6 banknotes for each of the 21 currencies. There are 19 Asian currencies in addition to the currencies of USA and EU. The Asian countries in alphabetical order are Afghanistan, Armenia, Azerbaijan, Bangladesh, Bhutan, Brunei, Burma, Cambodia, China, India, Kuwait, Maldives, Pakistan, Saudi Arabia, Sri Lanka, Syria, Tajikistan, Turkey, and United Arab Emirates (UAE).

The obtained experimental results indicate that using a neural network classifier can be efficient in identifying a banknote issuing country despite the complexity of such task. Based on our implementations and experimental results, scheme II works better than scheme I, whereas the neural model ANN2 outperforms ANN1 in terms of correct identification rate (CIR), training time, and minimum error value. Future work will focus on improving the CIR results.

## References

1. Khashman, A., Sekeroglu, B.: Multi-banknote identification using a single neural network. In: Blanc-Talon, J., Philips, W., Popescu, D., Scheunders, P. (eds.) Advanced Concepts for Intelligent Vision Systems (ACIVS2005). LNCS, vol. 3708, pp. 123–129. Springer, Heidelberg (2005)
2. Khashman, A., Ahmed, W.: Intelligent banknote issuing country identification. In: 3rd International Conference on Computational Mathematics and Engineering Sciences (CMES-2018), 4–6 May 2018, Girne, N. Cyprus (2018)
3. Oyedotun, O.K., Khashman, A.: Banknote recognition: investigating processing and cognition framework using competitive neural network. Cogn. Neurodyn. **11**(1), 67–79 (2017)
4. Khashman, A., Sekeroglu, B.: A novel thresholding method for text separation and document enhancement. In: 11th Panhellenic Conference on Informatics, pp. 323–330, Greece (2007)
5. Yan, W.Q., Chambers, J., Garhwal, A.: An empirical approach for currency identification. Multimedia Tools Appl. **74**(13), 4723–4733 (2015)
6. Zlokazov, Y.E., Starikov, R.S., Odinokov, S.B.: Specificity of correlation pattern recognition methods application in security holograms identity control apparatus. Phys. Procedia **73**, 308–312 (2015)
7. Kamal, S., Chawla, S.S., Goel, N.: Feature extraction and identification of indian currency notes. In: 5th National Conference on Computer Vision, Pattern Recognition, Image Processing and Graphics (NCVPRIPG), Patna, India (2015)

8. Dominguez, R., Lara-Alvarez, C., Bayro-Corrochano, E.: Automated banknote identification method for the visually impaired. In: 19th Iberoamerican Congress on Pattern Recognition (CIARP). LNCS, vol. 8827, pp. 572–579. Springer, Heidelberg (2014)

9. Zhu, X., Ren, M.: A recognition method of RMB numbers based on character features. In: 2nd International Conference on Information, Electronics and Computer (ICIEAC), vol. 59, pp. 21–54 (2014)

10. Rahman, S., Banik, P., Naha, S.: LDA based paper currency recognition system using edge histogram descriptor. In: 17th International Conference on Computer and Information Technology (ICCIT), pp. 326–331 (2014)

11. Bruna, A., Farinella, G.M., Giuseppe, C.G.: Forgery detection and value identification of Euro banknotes. Sensors **13**(2), 2515–2529 (2013)

12. Gai, S., Liu, P., Liu, J.: Banknote image retrieval using rotated quaternion wavelet filters. Int. J. Comput. Intell. Syst. **4**(2), 268–276 (2011)

13. Jin, Y., Song, L., Tang, X.: A hierarchical approach for banknote image processing using homogeneity and FFD model. IEEE Signal Process. Lett. **15**, 425–428 (2008)

14. Khashman, A., Sekeroglu, B., Dimililer, K.: ICIS: a novel coin identification system. In: Intelligent Computing in Signal Processing and Pattern Recognition, pp. 913–918 (2006)

15. Khashman, A.: Intelligent face recognition: local versus global pattern averaging. In: Australasian Joint Conference on Artificial Intelligence, pp. 956–961 (2006)

16. Khashman, A., Dimililer, K.: Medical radiographs compression using neural networks and haar wavelet. In: EUROCON 2009, pp. 1448–1453. IEEE (2009)

17. Olaniyi, E.O., Khashman, A.: Onset diabetes diagnosis using artificial neural network. Int. J. Sci. Eng. Res. **5**(10), 754–759 (2014)

18. Khashman, A.: An emotional system with application to blood cell type identification. Trans. Inst. Meas. Control **34**(2–3), 125–147 (2012)

19. Oyedotun, O.K., Khashman, A.: Document segmentation using textural features summarization and feedforward neural network. Appl. Intell. **45**(1), 198–212 (2016)

20. Khashman, Z., Khashman, A.: Anticipation of political party voting using artificial intelligence. Procedia Comput. Sci. **102**, 611–616 (2016)

21. Oyedotun, O.K., Olaniyi, E.O., Khashman, A.: Deep learning in character recognition considering pattern invariance constraints. Int. J. Intell. Syst. Appl. **7**(7), 1–10 (2015)

22. Khashman, A.: Investigation of different neural models for blood cell type identification. Neural Comput. Appl. **21**(6), 1177–1183 (2012)

23. Khashman, Z., Khashman, A.: Modeling people's anticipation for Cyprus peace mediation outcome using a neural model. Procedia Comput. Sci. **120**, 734–741 (2017)

24. Olaniyi, E.O., Oyedotun, O.K., Khashman, A.: Heart diseases diagnosis using neural networks arbitration. Int. J. Intell. Syst. Appl. **7**(12), 72 (2015)

25. Banknote Museum. http://banknote.ws/. Accessed 14 Mar 2018

# Neural Network Modeling and Estimation of the Effectiveness of the Financing Policy Impact on the Socio-Economic Development of the Socio-Educational System

Alekperov Ramiz Balashirin and Ibrahimova Kyonul Akbar[✉]

Department of Computing Engineering, Odlar Yurdu University,
Baku AZ1008, Azerbaijan
ramiz62@rambler.ru, nazile003@mail.ru

**Abstract.** The questions concerning the usage of neural networks for estimation and forecasting of the influence of a financial policy on socio-economic development of socio-educational system within the conditions of information insufficiency are considered.

**Keywords:** Socio-educational system · Socio-economic development
Forecasting tasks · Neural network

## 1 Introduction

Researches confirm that the area that has the greatest reimbursement of outlays on economic development is a higher education, and that is the reason why the formation of a financial policy in the area of higher education is of utmost importance. Improving the quality of higher education and promoting its welfare in the economy, the socio-economic governance indicators will also improve.

In the economic literature there is no practice of assessing the effectiveness of the influence of financial policy on socio-economic development of higher education and the method of its calculation.

From the other hand, there are very few experimental data (information incompleteness), which does not allow full-fledged reasoning, assessment and forecasting of the economic efficiency of higher education development. This can be confirmed by data (Table 1), which we collected for the assessment of efficiency.

The economic effectiveness of higher education in the total regional product (TRP) for each year (Table 1) was calculated on the basis of the traditional economic efficiency assessment (see Column 4), where the total increase of TRP is determined by a change in the percentage of people with higher education involved in the region's economy (see Column 2).

As can be seen from the chart on the basis of a small number of experimental data it is difficult to solve the problem of extrapolation and interpolation according to efficiency forecasting.

© Springer Nature Switzerland AG 2019
R. A. Aliev et al. (Eds.): ICAFS-2018, AISC 896, pp. 754–759, 2019.
https://doi.org/10.1007/978-3-030-04164-9_99

**Table 1.** The economic effectiveness of financial policy's impact to socio-economic development in higher education sphere, in the total regional product (TRP)

| Years | Inputs (P) | | Outputs (T) |
|---|---|---|---|
| | Criteria | | The economic effectiveness of higher education in TRP, in % |
| | Average number of employed in the economy, in thousands people | Expenses related to the combined budget, in million AZN | |
| | $m_s$ | $z$ | $e$ |
| 2010 | 1382.9 | 1180.8 | 0.035 |
| 2012 | 1480.7 | 1453.2 | 0.03 |
| 2013 | 1514 | 1437.7 | 0.032 |
| 2014 | 1519.7 | 1553.9 | 0.029 |
| 2015 | 1502.5 | 1605.1 | 0.028 |
| 2016 | 1514.3 | 1754.4 | 0.026 |
| 2010 compared with 2016 (+, −) | 131.4 | 573.6 | −0.009 |

However, the problem of data incompleteness and forecasting is solved by various methods of interpolation and extrapolation [1], the analysis of which showed their low accuracy, the existence of hard requirements to the initial information, the complexity of implementation, which stipulates the development of methods based on new approaches, such as neural networks [2].

In this regard a methodology that is based on the use of neural networks for modeling and estimating economic efficiency with the dependence of the TRP (e) and the proportion of employed people (ms), and also taking into consideration the costs and expenses (z) allocated from the combined budget for socio-economic development of higher education, in conditions of incomplete information, is proposed.

The main idea of this approach is that between the input parameters z, ms and the output e, a functional dependency implemented through a neural network is established.

Neural networks can be taught to any functions represented by experimental data, as in our case, which allows avoiding the usage of a complex mathematical apparatus.

A network trained on a limited set of data is able to generalize the information received and to show good results on data that is not used during its training [2].

## 2  Statement of the Approximation Problem and Algorithm for Choosing the Type of Neural Network

In general, the task of estimating economic efficiency is formulated as the problem of approximating the function of two variables e = f (z, ms) on the basis of experimental data (Table 1).

First of all, using the method of least squares, we find the equation of economic efficiency approximation e (formula 1) (here, we have x = ms/Z):

$$e = 0.0299 * x + 1e - 06 \tag{1}$$

with the following parameters: 1. sample variances: S (X) = 0.25, S (e) = 0.000221, 2. covariance cov (x, e) = 0.00739.3, 3. The average error of approximation is 0.473%, 4. the empirical correlation ratio (correlation ratio) is 0.00221, 5. The determination coefficient R2 = 0.988428968, 6. The number of freedom degrees is 3, the value of Fkp (3, 3) = 10.1.

It follows that, on average, the calculated values deviate from the actual ones by 0.47%. As the error is less than 7%, this equation can be used as a regression. The values of determination coefficient show that 98.84% of cases of changes in x lead to a change in e. In other words, the accuracy of regression equation selecting is high. The remaining 1.16% change in e is due to factors not taken into consideration in the model. And also, according to the calculation results, F statistics, where F < Fkp = 10.1, the hypothesis of the absence of heteroscedasticity is accepted. According to the results of calculation of regression parameters, we can conclude that this regression equation approximates the functions of e by 98.84%.

From the graph (Fig. 1) of the change of the dynamics in economic efficiency difference from the calculated values (Formula 1) it can be seen that the real value of the regression coefficient varies proportionally with a decrease or increase in economic efficiency. That does not allow us to judge the adequacy of this equation to changes in model values beyond the range of the considered boundary parameters ms, z and e (Table 1). (Series1: regression coefficient, series2: economic efficiency).

However, the results of the approximation equation (Formula 1) can be used as a method of interpolation (within the experimental boundaries of the model parameters (Table 1) in order to gain the missing experimental data for preparing the neural network.

The main task of developing a neural network solution is to select the optimal neural network architecture. In this connection and for this task we will use the following types of neural networks [2]: radial basis (RBF), exact network with radial basis functions, and generalized regression (GRNN), which are most often used to solve the issue of approximating the experimental data.

On the other hand, the choice of neural network architecture, which one will be better and more practical, in most cases depends on the conditions of the issue.

An algorithm for obtaining a neural network model for estimating economic efficiency, which consists of the following blocks, which are modeled using the Neural Network Toolbox application package of the MATLAB system [3]:

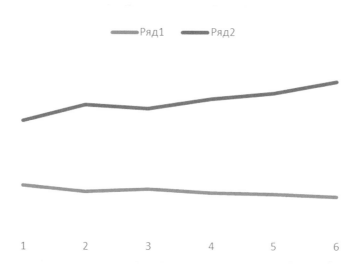

**Fig. 1.** Dynamics of changes of the correlation coefficient and economic efficiency.

1. Formation and input of experimental data,
   1.1. Initial data: P = {[1382.9; 1180.8], [1480.7; 1453.2], [1514; 1437.7], [1519.7; 1553.9], [1502.5; 1605.1], [1514.3; 1754.4]} – inputs of neural network. e = [0.035  0.03  0.32  0.029  0.028  0.026] – outputs of neural network. T = num2cell(e) – we convert a number group into cell array.
2. Creating neural networks: to create a generalized regression neural network net = newgrnn (p, t, spread); Accordingly, to construct a network with a radial basis, we use the function netrbf = newrb (p, t, goal, spread) and construct the exact network with radial basis functions using the function netrbe = newrbe (p, t, spread);
3. Set network parameters: goal = 1e−06; net.performFcn = 'sse';
4. Teaching the network on the basis of a pair (P, T) with teaching parameters net.-trainFcn = 'traingd'; net.trainParam.goal = 1e−06;
5. Adapt the networks with the parameter net.adaptFcn = 'adaptwb';
6. Let's simulate networks for comparing the results of outputs of all types of neural networks, and if two types are strongly divergent from the experimental values, we will form new interpolation points on the basis of formula 1 with a small number of partition points x−nx, and in y−ny. And now we turn to item 4. The process is repeated for reaching the acceptable results for all types of neural networks. As an assessment of acceptability, the determination coefficient (R2) is used.

## 3  Analysis of the Results Obtained

The results for the implementation of the developed algorithm are presented, in particular, the output values for T are presented in Table 2.

**Table 2.** Output values for T

| Used methods | Results (output T) | | | | | | Determination coefficient (R2) |
|---|---|---|---|---|---|---|---|
| Experimental data | 0.03500 | 0.03000 | 0.03200 | 0.02900 | 0.02800 | 0.02600 | |
| Approximation equation (MNR) | 0.03502 | 0.03047 | 0.03149 | 0.02924 | 0.02799 | 0.02581 | 0.9884 |
| Generalized regression neural network (GRNN) | 0.03500 | 0.03050 | 0.03160 | 0.02920 | 0.02800 | 0.02580 | 0.9902 |
| Radial basis neural network (RBF) | 0.03500 | 0.03030 | 0.03170 | 0.02920 | 0.02770 | 0.02570 | 0.9924 |
| Exact network with radial basis functions | 0.03500 | 0.03050 | 0.03150 | 0.02920 | 0.02800 | 0.02580 | 0.9883 |

Apparently, the values of the determination coefficients gained by neural networks are very close, which allows us to conclude that the use of neural networks for solving the approximation issues is acceptable because of their sensitivity to tracking even the smallest changes in the input data, thanks to the available in RBF and GRNN neuron networks of a hidden layer of neurons with nonlinear radial-basis activation functions. In this case, the coefficient of determination of the radial-basis and generalized regression neural networks is the highest and almost identical, respectively R2rbf = 0.9924 and R2grnn = 0.9902.

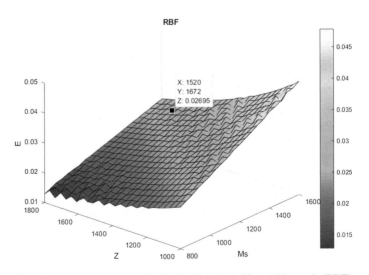

**Fig. 2.** Approximation results for Radial Basis Neural Network (RBF)

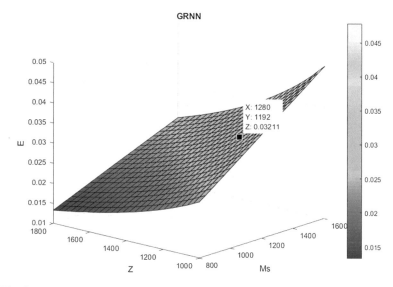

**Fig. 3.** Approximation results for Generalized Regression Neural Network (GRNN)

This proves the coincidence of the estimated and real values of economic efficiency. The results of the approximation on the basis of the given networks for nx = 21 and ny = 26 with intervals for ms, z respectively [800, 1600] [1000, 1800] are shown on Figs. 2 and 3.

## 4   Conclusion

As you can see from the pictures, the functions of approximation of economic efficiency are nonlinear, but at small intervals it seems that they are linear. It proves the fact that, for the experimental data under consideration, the values of the economic efficiency of neural networks are close to the results of the approximation equation gained through the Formula 1.

Thus, the analysis of the considered example of data approximation from known experimental data testifies the advisability of implementation of neural networks, which provides a high quality of approximation and can be used to assess and forecast the effectiveness of financial policy impact on the socio-economic development of the socio-educational system.

## References

1. Brezinski, C.: Interpolation and Extrapolation, vol. 2, 1st edn. North Holland, Amsterdam (2000)
2. Aliev, R.A., Fazlollahi, B., Aliev, R.R.: Soft Computing and Its Applications in Business and Economics. Springer, Heidelberg (2004)
3. Beale, M.H., Hagan, M.T., Demuth, H.B.: Neural Network Toolbox: User's Guide. Math Works, Inc., Natick (2014)

# Investigation of the "Input-Output" Model of the Azerbaijani Economy in Interval Information Conditions

V. J. Akhundov, S. K. Mammadova$^{(\boxtimes)}$, and A. M. Aliyev

Azerbaijan State Oil and Industry University, Azadlyg Avenue, 20,
AZ1010 Baku, Azerbaijan
`azeri46@mail.ru`, `akimovasamira9270@gmail.com`,
`azizaliyev@hotmail.com`

**Abstract.** In the proposed research study, Azerbaijan economy is analyzed by using the interval analysis method. The purpose of this article is to apply the method chosen by the method of interval direct expenditure coefficient matrix $A \in [A_{min}, A_{max}]$ to study the method of solution of the linear problem and to have a high computation of practical computation efficiency.

In the article by the method of Leontief's "Input-output" model interval analysis, the uncertainty of the parameters of the direct expenditure coefficient matrix $A \in [A_{min}, A_{max}]$, the total volume of product vector $X \in [X_{min}, X_{max}]$ and the output vector $Y \in [Y_{min}, Y_{max}]$ are described only as boundaries of possible values. The results of the researches show that, during the traditional interval estimates, the result range is usually great, and this reduces the importance of the answers. For this purpose, special methods have been developed to reduce the difference in result values.

**Keywords:** Input-output model · SAM · Interval number
Intermediate product · Sector of economy · Output

## 1 Introduction

In practice, economic issues are mainly provided in the form of inaccurate information. In this case, possible-statistical, fuzzy and interval models can be used as models describing indefinite data. In general, the interval analysis method has been developed to make calculations that are inaccurate and based on indefinite initial data. Uncertainty intervals allow us to describe changing or inaccurate initial information.

Usually, the assessment of the results of the economic events that may occur in the country's economy is carried out on the basis of the Social Accounts Matrix (SAM), based on Leontief's input-output model. However, in practice, accurate information about economic processes that may occur in the future is not available. This generates demands of the time to consider SAM as an interval. Considering SAM as an interval and the problem of interval analysis has been researched by Koparanova [1], Dymova [2], Moore [3], Lukashin [4] and others. However, in these studies, the problem of interval analysis has not been fully reflected in Leontief's direct expenditures coefficient model, at the same time changing the final demand and production level.

© Springer Nature Switzerland AG 2019
R. A. Aliev et al. (Eds.): ICAFS-2018, AISC 896, pp. 760–768, 2019.
https://doi.org/10.1007/978-3-030-04164-9_100

The purpose of the study is to analyze the results of the research on the structure of the SAM and the establishment of the SAM in the intermediate information environment of the Republic of Azerbaijan. In the research process, the Delphi interval method identified key national farming sectors and interpreted the results of the calculation of SAM based on the "Inter-sector balance of production and distribution of products and services" of the Republic of Azerbaijan. The article presents a coherent scheme of the SAM, taking into account the specifics of the country's economy, the impact of changes in budget expenditure on different sectors of the economy and the changes in the economy as a result of the transfer of budget expenditures to the economy. The calculations have been studied by interval analysis of Leontief's inter-sector balance method, taking into account the possible changes in the sectors of national economy, which, to our opinion, will be of great importance in the development of perspective plans of Azerbaijan's economy. The article estimates the interval forecasted value of the total volume of product by using the Differential Evolution optimization method during the possible changes in one or several sectors of the economic system.

## 2 Preliminaries

**Definition 1. Basic Arithmetic Operations Over Intervals.** Let $\bar{a} = [a_1, a_2]$ and $\bar{b} = [b_1, b_2]$ be intervals of the real line. The operation $\bar{a} * \bar{b}$, $* \in \{+, -, \cdot, /\}$, is defined as

$$\bar{a} * \bar{b} = [a_1, a_2] * [b_1, b_2] = [min(a_1 * b_1, a_1 * b_2, a_2 * b_1, a_2 * b_2),$$
$$max(a_1 * b_1, a_1 * b_2, a_2 * b_1, a_2 * b_2)].$$

Note that if $* = /$ is defined if and only if $0 \notin \bar{b}$.

**Definition 2. Differential Evolution Optimization.** Differential Evolution (DE) is a global search optimization method. DE method is of good convergence and flexibility and is not constrained by the properties of function to optimize.

DE is a stochastic method that uses uniform distribution-generated initial population, and operators of differential mutation, possibility crossover, and selection. The main idea of the algorithm is to extract distance and direction of information for search directly from the population: a new vector is generated by mutation which means randomly selecting from the population of 3 individuals– vectors $r_1 \neq r_2 \neq r_3$ and adding a weighted difference vector between two individuals to a third individual (population member). If the resulting vector yields ismore optimal value than a pre-determined population member, then it will replace the latter in the following generation.

## 3    Combined Scheme of Social Accounts Matrix of Azerbaijan

SAM is an accounting system covering common structural relationships between different sectors of the economy. SAM as the main law of the economy is a simple and effective way of representing the "basic cost of each income". For each country, it is important to take into account that SAM is a square matrix, and the columns and rows should match equally.

Table 1 presents an aggregate SAM in schematic form, which can be applied to the Azerbaijani economy [5]. In the basic formation of the proposed SAM, the country's economy is considered as four sectors:

1. The Production Sector; 2. The Household Sector; 3. The Government Sector; 4. The Foreign Sector.

Such an issue allows the study the effects of the fiscal policy change on the economy [6].

As can be seen in Table 1, production is an essential element of the SAM as the main economic activity. The production sector is divided into two parts - production activity and type of activity, which is crucial for modeling. Separation of the production sector into two parts provides the separation of intermediate products and total volume of product of the sector. Selection of the aggregated economics sectors by the Delphi method of production provides the analysis of the impact of changing the economic policy on the scale of consumption and the impact on households.

Households get income as a result of their activities. At the same time, they get income from the factor and get income from abroad, in the form of factor income and government payments. They share their gross income in the form of consumption and saving, which is reflected in the SAM [7].

The public sector receives its revenues from taxes, factors, and exports (mainly oil products). These revenues are distributed in the form of public consumption, transfers, accumulation and capital investment.

The foreign sector shows the connection between the country and the outside world. Access-to-output information has been expanded to disclose the balance of the external sector in the SAM scheme. The net export sector shows the difference between the prices of export and imports, which in turn makes it possible for SAM to be balanced in practice.

The purpose of our research is to analyze the results of experimental calculations for the Azerbaijani economy in the sector of production activity, which is the main sector of the SAM, as it is considered by Leontief's "input-output" interval analysis.

As we know, the equation of the classical Leontief "input-output" model is generally written as follows:

$$AX + Y = X \qquad (1)$$

As seen from the equation of the Leontief "input-output" model, the variables can be the total volume of production (X) or the output (Y). This allows to analysis four types of inter-sector balance:

**Table 1.** An aggregate social accounting matrix in a schematic form

| Receipts | Expenditures | | | | | | | |
|---|---|---|---|---|---|---|---|---|
| | 1 | 2 | 3 | 4 | 5 | 6 | 7 | 8 |
| | Production activities | Commodities | Factors | Households | Government | Capital | Rest of world | Totals |
| 1. Production Activities | | Domestic sales | | | | | Exports | Total sales |
| 2. Commodities | Intermediate inputs | | | Household consumption | Government consumption | Investment | | Aggregate demand |
| 3. Factors | Factor payments | | | | | | | Total factor income |
| 4. Households | | | Factor income | | Transfers | | Foreign remittances | Household income |
| 5. Government | Indirect taxes | Import taxes | Factor taxes | Direct taxes | | | | Government revenue |
| 6. Capital | | | | Savings | Savings | | Foreign savings | Total savings |
| 7. Rest of world | | Imports | | | Net export | | | Import |
| 8. Totals | Total costs | Aggregate supply | Total factor income | Household expenditures | Government expenditures | Total investment | Foreign exchange receipts | |

1. Direct material costs coefficients and the final product of all sectors are known: calculation of annual gross product output per sector;
2. All annual gross product output and designated direct costs of materials are known: calculation the total volume of product for each sector;
3. Direct material expenditure ratios, the total volume of product of a one part of the sector, and the amount of output other remained sectors are known: the total volume of product of the first sector and the total volume of product of the second sector;
4. Direct material expenditure ratios, the total volume of product of the national economy, and the volume of output of the national economy are known: the calibration of database based on statistical indicators and the establishment of the overall balance model, calculating the parameters that provide economic balance.

The practical application of the first three of the issues raised above is crucial and research has been made and certain results have been achieved. The problem of the fourth type of inter-sector balance, which is the logical continuation of the first three types of issues, is topical. Let's look at the Leontief "input-output" model in intermittent terms for research in this direction.

## 4   Leontief's "Input-Output" Model of Interval Analysis Methods

It should be taken into account that, in practice, changes in different diameters in almost all sectors of the economy are predicted in the forecast period. This implies the formation of prediction by interval method [8]. Inter-sector balance forecasting by the method of interval analysis mean formation of plans on the base of indexes in the form of interval [9].

The initial data required to compile the SAM forecasts can also be provided as the expected interval changes in both the final products and the intermediate products. In this case, we propose the general equation of the "input-output" interval forecast model of general product production:

$$[\text{Xmin}, \text{Xmax}] = [\text{Amin}, \text{Amax}] \, [\text{Xmin}, \text{Xmax}] + [\text{Ymin}, \text{Ymax}] \qquad (2)$$

Here, the minimum and maximum characterizes the possible level of changes in the country's economy.

We will consider solving (2) with respect to [Xmin, Xmax]: given interval-valued matrix [Amin, Amax] and interval-valued vector [Ymin, Ymax] one needs to compute [Xmin, Xmax]. Table 2 shows changing calculation interval taking into account approximate possibility of product output of sectors of economy on the base of "Inter-sector balance of production and distribution of product and services" for 2011.

In the research process, by the Delphi method [10] conducted an expert survey on 15 economic sectors that played a decisive role in the economic model of the Republic of Azerbaijan, the possible variation intervals of the A matrix were also defined (Table 3). The results show that statistical processing of discretionary information does not allow to take into account the ambiguity and uncertainty in the information [11].

**Table 2.** Interval number for output in case of possible changing

| Economic sectors | Y min | Y max |
|---|---|---|
| 1. Agriculture, hunting and forest products | 1326,46 | 1794,62 |
| 2. Fish and fishing products | 87,86 | 163,17 |
| 3. Products of the mining industry | 9610,74 | 10622,40 |
| 4. Manufacturing industry | 5473,65 | 7405,53 |
| 5. Electricity, gas and water | 345,70 | 367,09 |
| 6. Construction | 2486,25 | 3363,75 |
| 7. Trade services | 1052,15 | 1423,50 |
| 8. Services of hotels and restaurants | 34,93 | 47,25 |
| 9. Transport, postal and communication | 900,73 | 995,55 |
| 10. Financial, insurance and pension services | 218,16 | 236,34 |
| 11. Real estate, lease and other commercial services | 125,38 | 169,64 |
| 12. Services in public administration and defense, compulsory social insurance | 753,80 | 833,15 |
| 13. Educational services | 515,75 | 570,04 |
| 14. Health and social services | 377,71 | 417,47 |
| 15. Utility and other services | 339,83 | 415,35 |

**Table 3.** The changes in intervals of expenditure coefficient as a result of an expert survey, [aijmin, aijmax].

| | 1 | 2 | ... | 16 |
|---|---|---|---|---|
| 1 | [0.200, 0.203] | [0, 0.000] | ... | [0, 0.001] |
| 2 | 0 | [0.114, 0.116] | ... | 0 |
| 3 | [0, 0.000] | [0, 0.000] | ... | [0, 0.000] |
| 4 | [0.013, 0.015] | [0.001, 0.002] | ... | [0.008, 0.008] |
| 5 | [0.012, 0.015] | [0.001, 0.002] | ... | [0.006, 0.008] |
| 6 | [0.087, 0.093] | [0.006, 0.008] | ... | [0.008, 0.009] |
| 7 | [0.001, 0.002] | 0 | ... | [0, 0.000] |
| 8 | [0.011, 0.013] | 0 | ... | [0.003, 0.004] |
| 9 | [0.005, 0.007] | [0.004, 0.005] | ... | [0.005, 0.006] |
| 10 | [0.008, 0.010] | [0, 0.001] | ... | [0.011, 0.014] |
| 11 | [0.006, 0.007] | [0, 0.001] | ... | [0.002, 0.004] |
| 12 | [0.040, 0.041] | [0, 0.000] | ... | [0.009, 0.011] |
| 13 | 0 | 0 | ... | 0 |
| 14 | [0, 0.000] | [0, 0.000] | ... | [0, 0.000] |
| 15 | 0 | 0 | ... | [0, 0.000] |

The complexity of solving the intermittent equations system by analytical methods makes it a bit difficult for their practical use. For efficiency of calculations, we recommend using the Differential evolution optimization (DEO) method to solve system of interval linear Eq. (2) [12].

Using the DEO method, the estimated cost of the gross product is calculated in Table 4. In practice, the range of possible changes is lower, and therefore, the difference in interval between the results is minimal. It should be noted that for the purpose of checking the solution by the method of DEO, we have also taken the possible changes in the output (Ymin, Ymax) at 1% (in our example agricultural sector) at 10%. Thus, we have retained the current range of change of the Y vector (10% to 1326 ÷ 1795) at the level of 10% (1459 ÷ 1974) and the final product interval for other sectors. The main goal here is to evaluate the impact of the expected change in one sector of the national economy on other sectors.

**Table 4.** Calculation of the total product in the form of intervals using the DEO method

| Economic sectors | X min | X max |
|---|---|---|
| 1. Agriculture, hunting and forest products | 2334,20 | 2783,32 |
| 2. Fish and fishing products | 103,29 | 182,04 |
| 3. Products of the mining industry | 11115,78 | 12159,73 |
| 4. Manufacturing industry | 8741,77 | 10809,18 |
| 5. Electricity, gas and water | 719,43 | 733,98 |
| 6. Construction | 3363,42 | 4298,31 |
| 7. Trade services | 1718,65 | 2038,76 |
| 8. Services of hotels and restaurants | 156,89 | 173,98 |
| 9. Transport, postal and communication | 2774,82 | 2829,26 |
| 10. Financial, insurance and pension services | 337,43 | 355,74 |
| 11. Real estate, lease and other commercial services | 556,02 | 587,38 |
| 12. Services in public administration and defense, compulsory social insurance | 789,39 | 868,91 |
| 13. Educational services | 529,50 | 584,11 |
| 14. Health and social services | 577,28 | 614,24 |
| 15. Utility and other services | 454,08 | 528,63 |

The problem is solved by this method and the results are shown in Table 5. As can be seen from the results of the table, such a change in the Y vector on one sector (on the agricultural sector) has some influence on some of the national economy, but has almost no effect on other sectors. As seen from Tables 4 and 5, as a result of an increase in both the lower and upper limit of the output (Ymin, Ymax) as a result of an increase in the level of 10%, firstly, the gross agricultural output amounted to 2334.2–2783.3 million AZN the level of change interval is expected to be 2508.7–3000.8 million AZN.

The main goal here is to calculate how much of the intermediate product production (AX vector) of related sectors should be increased to provide a 10% increase in output agricultural product. This in turn ultimately leads to an increase in the total volume of production of these products.

As shown in Table 5, the increase in the total volume of agricultural production (ΔXmax = 3000.8-2783.3 = 217.5) is required to increase by 217.5 million AZN, in

**Table 5.** Influence of 10% increase in the final agricultural product on other sectors.

| Economic sectors | The value of the X-vector | |
| --- | --- | --- |
| | X min | X max |
| 1. Agriculture, hunting and forest products | 2508,67 | 3000,82 |
| 2. Fish and fishing products | 103,29 | 182,04 |
| 3. Products of the mining industry | 11119,45 | 12163,52 |
| 4. Manufacturing industry | 8755,15 | 10822,02 |
| 5. Electricity, gas and water | 725,24 | 739,63 |
| 6. Construction | 3364,32 | 4299,16 |
| 7. Trade services | 1721,05 | 2040,85 |
| 8. Services of hotels and restaurants | 157,10 | 174,18 |
| 9. Transport, postal and communication | 27789 | 2833,17 |
| 10. Financial, insurance and pension services | 337,75 | 356,10 |
| 11. Real estate, lease and other commercial services | 558,44 | 589,52 |
| 12. Services in public administration and defense, compulsory social insurance | 789,43 | 868,94 |
| 13. Educational services | 529,51 | 584,12 |
| 14. Health and social services | 577,46 | 614,40 |
| 15. Utility and other services | 454,21 | 528,76 |

the other branches of the national economy, the total volume of product - the manufacturing industry 12.8 million AZN, electricity, gas, water 6.3mln.AZN, 4million. AZN in transport, postal and communication sectors, 4 million AZN for mining products, 2 million AZN for trade services, and real estate-related services, rent and other commercial services - 2 million AZN. Finally, the calculations show that it is important to increase the total volume of product output of related sectors to 32.11 million AZN to increase the upper limit of total agricultural production (Xmax = 217.5) to 217.5 million AZN. As can be seen in Table 5, the total volume of product of related sectors should be increased to 33 million AZN, so as to ensure that the low level of agricultural production (Xmin = 174,47) is more than 174,47 million AZN. In agriculture, growth of total volume of product is not required in other sectors, which are almost non-affiliated, to ensure a 10% increase in production.

The proposed method allows take into account the ratio between the production sectors when designing farming plans at the national level, using the interval analysis method. The research work is relevant and practical because it allows the impact of changes in budget expenditures on different sectors of the economy.

# 5  Conclusion

The results of the research have shown that increasing or decreasing product volume in the form of interval in any sector of the national economy requires product amount of any sectors the country's economy to be increased or reduced in line with the range of production changes in other sectors. In the research, the output, intermediate consumption and direct expenditure coefficients for the design of optimal forecasts of total volume of production are examined in the form of intervals. This provides a more adequate forecast of the compiled national economy forecasts.

It has been clear from the research that it is possible to achieve more accurate results by making simple calculations and using the Differential Evolution Optimization method in simulation of one or more branches of the national economy.

## References

1. Koparanova, M.S., Koparanov, S.M.: Interval input-output model. In: IFAC 12th Triennial World Congress, Sydney, Australia, pp. 647–650 (1993)
2. Dymova, L., Sevastjanov, P., Pilarek, M.: A method for solving systems of linear interval equations applied to the Leontief input-output model of economics. Expert Syst. Appl.: Int. J. **40**(1), 222–230 (2013)
3. Moore, R., Kearfott, R., Cloud, M.: Introduction to Interval Analysis. Society for Industrial and Applied Mathematics, Philadelphia (2009)
4. Lukashin, Yu.P., Rakhlina, L.I.: Modern direction of statistical analysis of relationships and dependencies. Institute of World Economy and International Relations, Moscow, 54 (2012)
5. Karadag, M., Westaway, T.: A SAM-based computable general equilibrium model of the Turkish economy. Economic Research Paper 99/18, pp. 14–22 (1999)
6. Musayev, A.F., Uzelaltinbulat, S., Mammadova, S.K., Gardashova, L.A., Musayeva, A.A.: Estimation of impact of the changes made to the tax legislation to the tax receipts through fuzzy numbers. In: 9th International Conference on Theory and Application of Soft Computing, Computing with Words and Perception, August 2017, Budapest, Hungary, pp. 333–340 (2017)
7. Musayev, A.F.: Opportunities for applying the Leontief method to the budget system. Tax Mag. Azerbaijan **3**, 79–102 (2013)
8. Musayev, A.F., Madatova, S.G., Rustamov, S.S.: Mamdani-type fuzzy inference system for assessment of tax potential. In: Zadeh, L.A. (ed.) 6th World Conference on Computing dedicated to 50th Anniversary of Fuzzy Logic and Applications and 95th Anniversary, Berkley, USA (2016)
9. Musayev, A.F.: Mathematical modeling and forecasting in economics, Baku (1999)
10. Aliev, R.A., Aliev, R.R., Ahmedov, I.Z., Aliyeva, K.R.: Fuzzy delphi method. J. Knowl. **1**, 3–4 (2004)
11. Aliev, R.A., Fazlollahi, B., Aliev, R.R.: Soft Computing and Its Applications in Business and Economics. Springer, Berlin (2004)
12. Aliev, R.A., Pedrycz, W., Guirimov, B., Aliev, R.R., Ilhan, U., Babagil, M., Mammadli, S.: Type-2 fuzzy neural networks with fuzzy clustering and differential evolution optimization. Inf. Sci. **181**(9), 1591–1608 (2011)

# Forecasting Oil and Gas Reservoirs Properties Using of Fuzzy-Logic Based Methods

R. Y. Aliyarov, A. B. Hasanov[⊠], M. S. Ibrahimli, Z. E. Ismayilova, and A. J. Jabiyeva

Azerbaijan State Oil and Industry University, Azadlig Ave, 20, Baku, Azerbaijan
adalathasanov@yahoo.com

**Abstract.** As is known, the recognition of productive horizons within of oil and gas deposits is based on assessments of reservoir's filtration-capacitance properties, the values of which determine the nature of pore space and fluid saturation. Usually variability of reservoir's filtration-capacitance properties visualize by 2D and 3D petrophysical models which reflect results of well logging and core lab investigations. At the same time, considering that natural reservoirs are characterized by a multitude of parameters, such as tortuosity and the shape of the cross section of pore channels, the thickness of the layer of bound water, etc., which can not be strictly determined thru laboratory experiments so, petrophysical models based only on core researching can not be accepted as flawless. Another major problem in petrophysical modeling is the effect of scattering of experimental data in the phase space of parameters, which inexplicable as errors of calculations or measurements. As a try to avoid the above problems, should be considered fuzzy petrophysical models which is the most optimal, from the point of view of the representation of the polygons scattering experimental data and phase space parameters.

**Keywords:** Deep-lying sediments · Petrophysical models
Productive reservoirs · Fuzzy values · Predicting the reliability
Fuzzy logic-based method

## 1 Introduction

Fuzzy petrophysical models are a specific type of models realized by using fuzzy values described by the membership function, and a fuzzy relationship between systems of physic-geological parameters. The indistinctness of the parameters is determined by the heterogeneity of the geological objects (productive reservoirs) [1] under study with respect to the parameters under consideration and is expressed in the form of quantitative estimates of reliability. All this allows to eventually predicting the reliability of graphical images based on the distributions of the reliability of the initial data and the relationships between the parameters [2].

Accordingly, the forecast itself is realized in the form of a fuzzy model of the parameter distribution, in which for each spatial point the distribution of possible values is described, with an estimate of their reliability [3]. The negative side of this approach is the need to store and operate a large amount of information, as well as its

© Springer Nature Switzerland AG 2019
R. A. Aliev et al. (Eds.): ICAFS-2018, AISC 896, pp. 769–773, 2019.
https://doi.org/10.1007/978-3-030-04164-9_101

unconventionality. Such unconventionality is compensated by a high degree of visibility and the ability to reaching for each forecast parameter the distribution of its reliability by expected values [4, 5]. In result, the obtained fuzzy models allow to estimate the availability of the initial information of the implemented images and, thus, perform the expertise on the level of confidence, regardless of the type of geological models. The storage and processing of a large array of petrophysical data became possible due to created petrophysical database, organized using modern software tools.

The most suitable software for this is the Oracle platform, with the operating system Oracle Linux Server 7.4. This system allow to create, store, modify and archive huge data sets (Big Data), as well as cover standard methods of professional work with relational database management systems (Database Management Systems). Communications with the Oracle database are carried out using standard Open Database Connectivity (ODBC) methods and an appropriate API.

## 2    The Vector Time Series Method for Solution's and Prediction's Issues

In this article is explained about our experience by trying of predictions reservoirs properties using of fuzzy logic-based method. To this purpose, to being examined reservoirs properties in several oil wells using of vector fuzzy clustering method along well depth by the fuzzy "if-then" rules. Based on the rules of vector prediction were calculated the specific values of predicted reservoirs properties at high depths. But, since the issue we are dealing do not have a correct solving by a classic methods, so to predictions using vector rules, initially must build a fuzzy model of depending on the complexity of the behavior of the indicators that depend on one another and on precision of values. One of the solutions of the considered problem is the vector time series approaching, but the complex interrelations nature of the involved parameters, as well as the unclearness of the predictable values, demand building of a fuzzy model. Particularly, we considered the construction of Mamdani and Suggeno models using the Fuzzy C-means clustering method.

## 3    The Solution of the Problem

The solution of the problem consists of the following steps:

Step 1. Take interpolated data $\{Y_{t+j}\}, j = 1, \ldots, N * l$ on the basis of values $\{Y_{t+j}\}, j = 1, \ldots, l$

Step 2. To get fuzzy IF-THEN rules and set up fuzzy C-means clustering an interpolated $\{Y_{t+j}\}, j = 1, \ldots, N * l$ data should be developed as:

Rule $k$, $k = 1, \ldots, K$

$IF\ y_{1t-m}\ is\ A_{k1}^1\ and\ y_{1t-m+1}\ is\ A_{k2}^1\ and, \ldots, and\ y_{1t}\ is\ A_{km}^1$

$and\ y_{2t-m}\ is\ A_{k1}^2\ and\ y_{2t-m+1}\ is\ A_{k2}^2\ and, \ldots, and\ y_{2t}\ is\ A_{km}^2$

.

.

.

$and\ y_{nt-m}\ is\ A_{k1}^n\ and\ y_{nt-m+1}\ is\ A_{k2}^n\ and, \ldots, and\ y_{nt}\ is\ A_{nm}^n$

$THEN$

$y_{1t+1}\ is\ B_{n1}^1\ and\ y_{2t+1}\ is\ B_{n1}^2\ and, \ldots, and\ y_{nt+1}\ is\ B_{k1}^n$

Step 3. Checking the built-in fuzzy IF-THEN rules, and

Step 4. Prognostication

As an example of practical evaluations, let's initially observe an obtained earlier [1] prediction of quantitative parameters of productive reservoirs from Mesocaynozoiq sediments of the South Caspian area. Information involving for further evaluation reflected in the Table 1.

**Table 1.** Information regarding reservoirs properties of Mesocaynozoiq sediments of the South Caspian area

| Depth, m | Carbonate, % | Sand, % | Clay, % | Shale, % | Porosity, % |
|----------|-------------|---------|---------|----------|-------------|
| 170.1 | 5 | 35 | 34.5 | 30.5 | 22.7 |
| 327.5 | 4.6 | 59.9 | 18 | 22.1 | 30 |
| 342.5 | 9 | 16.1 | 62.7 | 21.2 | 26.2 |
| 361.5 | 12 | 1.3 | 64.7 | 34 | 21.6 |
| 414 | 14 | 26.6 | 38.8 | 34.6 | 21.4 |
| ... | ... | ... | ... | ... | ... |
| 5693 | 7.1 | 0.7 | 73.5 | 25.8 | 14 |
| 5693.5 | 6 | 30 | 53.1 | 16.9 | 10 |
| 5853 | 0 | 11.9 | 42.6 | 45.5 | 18.3 |

Using these data, we first interpolated the depth data in equal steps (100 m) then we predicted values to the next depth. But, for finding the optimal forecasting model, were calculated the predicted values and errors by different rules and parameter "m" (fuzzy indicator). Based on the first 2/3 of the given data, we built the model, but tested it based on the last 1/3 of the given data. The errors occurred during an installation and testing of the model are shown in Table 2.

In a similar way, were conducted calculations using the Sugeno model, as a result of which, the Mamdani model built on 5 clusters, was selected as the most optimal one.

Visualizations of reservoirs properties prediction results are presented by chart in Fig. 1.

**Table 2.** Errors admitted as a result of calculations (Mamdani model)

| Rules = 7, m = 2 | Training.er | 2.8488 | 16.146 | 11.443 | 7.7409 | 4.417 |
|---|---|---|---|---|---|---|
| | Testing.er | 3.9462 | 20.621 | 14.385 | 10.903 | 7.2658 |
| Rules = 5, m = 2 | Tr.e | 2.8132 | 15.797 | 11.342 | 7.7917 | 4.3365 |
| | Ts.e | 3.9674 | 19.917 | 13.964 | 10.791 | 7.188 |
| Rules = 7, m = 2.5 | Tr.e | 3.0533 | 15.864 | 12.489 | 9.0830 | 4.981 |
| | Ts.e | 3.9443 | 20.613 | 14.229 | 11.510 | 7.3897 |
| Rules = 5, m = 1.7 | Tr.e | 2.8167 | 15.926 | 11.566 | 7.7496 | 4.3270 |
| | Ts.e | 3.9480 | 19.924 | 14.169 | 10.759 | 7.2503 |
| Rules = 5, m = 1.5 | Tr.e | 2.8876 | 15.726 | 11.505 | 7.6771 | 4.556 |
| | Ts.e | 4.0545 | 19.266 | 13.527 | 10.888 | 7.3979 |

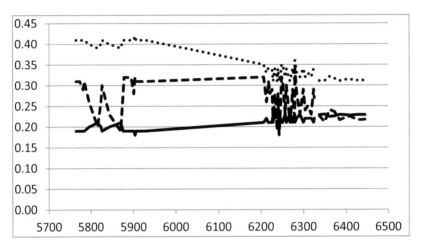

**Fig. 1.** Results of predictions some reservoirs properties: solid line – clay, dotted line – porosity, kneading line – water saturating

## 4   Conclusion

The forecasting of the reservoirs properties using of fuzzy logic-based method gives grounds to believe, that the values of porosity in deep-lying sediments confirm the probability of oil and gas presence in them. At the same time, it would be advisable, along with geophysical logs data, attract to evaluations also filter and capacitance characteristics of rocks in deep layers.

# References

1. Aliyarov, R.Y., Hasanov, A.B., et al.: Forecasting of qualitative characteristics of oil reservoirs. In: Materials of the Republican Scientific-Practical Conference devoted to the 95th Anniversary of H. Aliyev: Unity of science, education and production at the present stage of development, 7–8 May, pp. 22/31 (2018)
2. Latyshova, M.G.: Practical guidance on the interpretation of geophysical research of wells (in Russia), "Nedra", Moscow, 220p. (1991)
3. Aliyarov, R.Y., Ramazanov, R.A.: Prediction of multivariable properties of reservoir rocks by using fuzzy clustering. In: Proceedings of 12th International Conference on Application of Fuzzy Systems and Soft Computing, ICAFS 2016, 29–30 August 2016, Vienna, Austria, pp. 424–431, 28–30 August 2016
4. Wendelstein, B.Y., Rezvanov, R.A.: Geophysical methods for determining the parameters of oil and gas reservoirs: when calculating reserves and designing field development (in Russia), "Nedra", Moscow, 318p. (1978)
5. Dakhnov, V.N.: Petrophysics – the basis for the geological interpretation of the results of geophysical studies of wells, its state and immediate objectives (in Russia) Petrophysics of Oil and Gas Collectors. In: Papers of the Plekhanov Russian University of Economics and Gubkin Russian State University of Oil and Gas Institute, "Nedra", Moscow, Issue 115, p. 15 (1975)

# Application of the Combined State Concept to Behavioral Investment Decisions Under Interval-Valued Information

Khatira J. Dovlatova[1,2(✉)]

[1] Azerbaijan State Oil and Industry University, 20 Azadlig Avenue, Baku
AZ1010, Azerbaijan
xdovlatova@gmail.com
[2] Georgia State University, Atlanta, USA

**Abstract.** Investment decisions are characterized by behavioral issues and imperfect relevant information. It is needed to take into account behavior of decision maker that depends on states of economy in such probability. Application of the combined states approach for investment and sensitivity analysis can easily illustrate the opportunities of the proposed work. We used interval-valued probability in this paper.

**Keywords:** Behavioral decision making · Combined states
Investment problem · Interval-valued information · Choquet integral
Joint probability

## 1 Introduction

While improving the decision theories scientists try to take into consideration the features of human choices in formal decision models to make the latter closer to real behavioral decision making [1, 4, 5, 7, 9, 11, 17]. Risk issues were the first main behavioral incentives that become inevitable to take into consideration in the development of decision methods. Three main classifications of risk-related behaviors: risk neutrality, risk seeking and risk aversion cause to know. Gain- loss attitudes [1] and uncertainty attitudes are the other significant behavioral features.

The first theory developed for behavioral decision-making was the "Prospect Theory"; both the ratio of risk and the loss factor in the simple utility model included in it. Cumulative Prospect theory (CPT) [1], can be appropriate for both uncertainty and decision under risk, it is now one of the most effective decision theories. CPT has its origin in the use of Choquet integrals, as a consequence, can be viewed not only as an attitude to losses and risk, but also uncertain relationships. Choquet Expected Utility (CEU) [3] is an important specific model that can be used for both uncertain and risky situations. A wide sphere of investigations relates to mental - level models [8, 13, 14, 18] in which Decision Maker's (DM) behavior is composed of a group of situations. Each state delineates a Decision Maker's feasible decision- related condition and is mentioned to as a "mental state", a "state of mind", etc. In these models it is thought carefully about connection between situation of environment and mental state.

© Springer Nature Switzerland AG 2019
R. A. Aliev et al. (Eds.): ICAFS-2018, AISC 896, pp. 774–780, 2019.
https://doi.org/10.1007/978-3-030-04164-9_102

In "Behavioral decision making with combined states under imperfect information" (Rafik A. Aliev, Witold Pedrycz, Oleg H. Huseynov) the suggest a new approach to behavioral decision making based on a fundamental view to dependence of the Decision Maker behavior on objective conditions under imperfect information. Behavior of decision maker is described by states of decision maker. Which are possible behavioral conditions. A state of decision maker and a state of nature is consider as a combined state to describe dependence of subjective condition on objective conditions. The space of combined states is a framework for behavioral decision making modelling. In this paper we apply the combined states approach to solving of investment decision problem under interval-valued information.

## 2  Preliminaries

**Definition 1. [6]**
The intervals $p(s_i) = [\underline{p_i}\overline{p_i}]$ and $p(h_i) = [\underline{p_i}\overline{p_i}]$ i = 1, ...n are called the interval prob-

abilities of H and S if for any $p_i = [\underline{p_i}\overline{p_i}]$ there exist $p_i \in [\underline{p_i}\overline{p_i}]$ ..... $p_{i-1} \in \left[\underline{p_{i-1}}\overline{p_{i-1}}\right]$

..... $p_{i+1} \in \left[\underline{p_{i+1}}\overline{p_{i+1}}\right]$ ..... $p(n_i) = \left[\underline{p_n}\overline{p_n}\right]$ such that

$$\sum_{i=1}^{n} p_i = 1$$

Here $p_i$ denotes the basic probability, i.e. a numeric probability from an interval $p(s_i) = [\underline{p_i}\overline{p_i}]$ and $p(h_i) = [\underline{p_i}\overline{p_i}]$.

**Definition 2. Joint Interval Probability [2, 12]**
The joint probability is simply the probability that two events G and H will occur simultaneously.

For independent events Joint probability formula is below

$$P(G,H) = P(G) \times P(H)$$

To gauge a joint probability of two events G and H we require two types of information: marginal probabilities for G and H and information on a type of dependence between G and H directed to as a sign of dependence.

There are three main types of dependence: positive dependence, independence and negative dependence. Positive dependence is the process that G and H occur together. Negative dependence assumes that G and H don't commonly happen together. Independence implies that one existing process does not affect the happening of another. The special case of a positive dependence is directed to as an excellent dependence.

Positive dependence between G and H is described as:

$$P(H,G)[P_1(H,G), P_2(H,G)] = [P(H)P(G), min(P(H), P(G))]$$

Negative dependence between G and H is described as

$$P(H,G)[P1(H,G), P2(H,G)] = [max(P(H) + P(G) - 1, 0), P(H)P(G)]$$

### Definition 3. Comparison of Intervals. [10]

The degree to which $I = [\underline{I}, \overline{I}]$ is higher than $J = [\underline{J}, \overline{J}]$ is specified as follows.

$$d(I,J) = \begin{cases} \frac{\overline{I}-\overline{J}}{(\overline{I}-\overline{J})+(\underline{J}-\underline{I})}, \overline{I} > \overline{J}, \underline{J} \geq \underline{I} \\ 1. \overline{I} = \overline{J}, \underline{I} > \underline{J} \\ or\overline{I} > \overline{J}, \underline{I} \geq \underline{J} \\ or\overline{I} = \overline{J}, \underline{I} = \underline{J} \\ 1 - d(J,I) \\ otherwise \end{cases}$$

### Definition 4. Choquet Expected Utility [15]

Choquet Expected Utility is a utility model which is able to describe non-additivity of preferences. Imprecise information on probabilities is one of the fundamental reasons of non-additivity. A value of utility of alternative $f$ in CEU model for finite set of states of nature $\Omega = \{w_1, \ldots, w_n\}$ CEU is described as follows:

$$U(f_1(w_{11})) = \sum_{i=1}^{n} (u(f(w_{(i)})) - u(f(w_{(i+1)})))\eta(\{w_{(1)}, \ldots, w_{(i)}\}),$$

where $(i)$ in the index of the states $w$ implies that they are permuted such that $u(f(w_{(i)})) \geq u(f(w_{(i+1)}))$, and $u(f(w_{(n+1)})) = 0$ by convention.

### Definition 5. Lower Probability [16]

The lower probability $\eta$ is defined as follows.

$$\eta\{w_{(1)}, \ldots, w_{(j)}\} = p_{(1)} + \ldots + p_{(j)} \rightarrow min$$

$$\underline{P}_{(1)} \leq P_{(1)} \leq \overline{P}_{(1)},$$

$$\vdots$$

$$\underline{P}_{(j)} \leq P_{(j)} \leq \overline{P}_{(j)},$$
$$p_{(1)} + \ldots + p_{(j)} = 1$$

Here $p_i$ denotes possible numeric probability of $w_i$. As we can see, this problem is the problem of a linear programming.

## 3   State of Problem and Solution

Investment, is the process which is purchased with money that is anticipated to produce revenue or profit. Investments can be broken into three main groups: ownership, lending and cash equivalents. İnvestment is examined by approaches to behavioral problem and uncertainty of the information.

For ownership investment we have 3 alternatives:

1. Small business ($f_1$)
2. Tourism sector ($f_2$)
3. Transport ($f_3$)

We consider the following states in economy: $S_1$ - growth, $S_2$ - stagnation, $S_3$ - decline and 3 states of DM: $h_1$ - risk seeking, $h_2$ -risk aversion and $h_3$ - risk neutrality.

The states of combined states are shown in Table 1.

**Table 1.** Combined states space

|       | $S_1$ | $S_i$ | $S_3$ |
|-------|-------|-------|-------|
| $h_1$ | $(s_1, h_1)$ | $(s_i, h_1)$ | $(s_n, h_1)$ |
| $h_2$ | $(s_1, h_j)$ | $(s_i, h_j)$ | $(s_n, h_j)$ |
| $h_3$ | $(s_1, h_1)$ | $(s_i, h_m)$ | $(s_n, h_m)$ |

Assume probabilities of state of nature are as follows:

$$P(S_1) = [0.3, 0.4]$$

$$P(S_2) = [0.3, 0.5]$$

To compute $P(S_3)$ we use approach proposed in [10]

$$P(S_3) = [0.2, 0.3]$$

Assume probabilities of State of decision maker are as follows:

$$P(h_1) = [0.1, 0.3]$$

$$P(h_2) = [0.5, 0.6]$$

To compute $h_3$ we use approach proposed in [10]

$$P(h_3) = [0.2, 0.3]$$

For investment problem, we have different combinations of states of nature and states of decision maker.

We consider positive and negative dependence between S and H. For example, dependence between $h_1$ (risk averse) and $S_1, S_2, S_3$ is described as:

$$P(H_1, S_1) = [max(P_1(H_1) + P_1(S_1) - 1, 0), P_2(H_1)P_2(S_1)]$$
$$= [max((0.1 + 0.3) - 1, 0), 0.3 \times 0.4] = [0; 0.12]$$

$$P(H_1, S_2) = [P_1(H_1) \times P_1(S_2), min\, P_2(H_1)P_2(S_2)]$$
$$= [(0.1 \times 0.3)\, min(0.3; 0.5)] = [0.03; 0.3]$$

$$P(H_1, S_3) = [max(P_1(H_1) + P_1(S_3) - 1, 0), P_2(H_1)P_2(S_3)]$$
$$= [max((0.1 + 0.2) - 1, 0), (0.3 \times 0.3)] = [0; 0.06]$$

The obtained results are shown in Table 2.

**Table 2.** Define Joint probability

|  |  |  | Growth | | Stagnation | | Decline | |
|---|---|---|---|---|---|---|---|---|
|  |  |  | 0.3 | 0.4 | 0.3 | 0.5 | 0.2 | 0.3 |
| Risk averse | 0.1 | 0.3 | 0 | 0.12 | 0.03 | 0.3 | 0 | 0.06 |
| Risk seeking | 0.5 | 0.6 | 0.15 | 0.4 | 0 | 0.3 | 0.1 | 0.3 |
| Risk neutral | 0.2 | 0.3 | 0.06 | 0.3 | 0 | 0.15 | 0 | 0.09 |

**The values of utility of outcomes for combined states as shown in** Tables 3, 4 and 5.

**Table 3.** Utilities of outcomes for the first alternative

| Small business | Growth | Stagnation | Decline |
|---|---|---|---|
|  | (25% of yield) | (10% of yield) | (5% of yield) |
| Risk averse | $U(x) = \sqrt{10 \times 25\%} = 1.6$ | $U(x) = \sqrt{10 \times 10\%} = 1$ | $U(x) = \sqrt{10 \times 5\%} = 0.7$ |
| Risk seeking | $U(x) = (10 \times 25\%)^2 = 6.3$ | $U(x) = (10 \times 10\%)^2 = 1$ | $U(x) = (10 \times 5\%)^2 = 0.3$ |
| Risk neutral | $U(x) = (10 \times 25\%) = 2.5$ | $U(x) = (10 \times 10\%) = 1$ | $U(x) = (10 \times 5\%) = 0.5$ |

In this article we calculated utility for all business types with different percentage. Let us compute utility of alternatives by using Choquet Integral:

$$U(f) = (U_{(1)} - U_{(2)}) * \eta(\{w_{(1)}\}) + (U_{(2)} - U_{(3)}) * (\{w_{(1)}, w_{(2)}\}) + \ldots\ldots$$
$$+ (U_{(8)} - U_{(9)}) * (\{w_{(1)}, w_{(2)}, w_{(3)}, \ldots w_{(8)}\}) + (U_{(9)} - U_{(10)}) * (\{w_{(1)}, w_{(2)}, w_{(2)}, \ldots w_{(9)}\})$$

**Table 4.** Utilities of outcomes for the second alternative

| Tourism | Growth | Stagnation | Decline |
|---|---|---|---|
| | (23% of yield) | (15% of yield) | (3% of yield) |
| Risk averse | $U(x) = \sqrt{(10 \times 23\%)} = 1.5$ | $U(x) = \sqrt{(10 \times 15\%)} = 1.2$ | $U(x) = \sqrt{(10 \times 3\%)} = 0.5$ |
| Risk seeking | $U(x) = (10 \times 23\%)^2 = 5.3$ | $U(x) = (10 \times 15\%)^2 = 2.3$ | $U(x) = (10 \times 3\%)^2 = 0.1$ |
| Risk neutral | $U(x) = (10 \times 23\%) = 2.3$ | $U(x) = (10 \times 15\%) = 1.5$ | $U(x) = (10 \times 3\%) = 0.3$ |

**Table 5.** Utilities of outcomes for the third alternative

| Transport | Growth | Stagnation | Decline |
|---|---|---|---|
| | (20% of yield) | (12% of yield) | (10% of yield) |
| Risk averse | $U(x) = \sqrt{(10 \times 20\%)} = 1.4$ | $U(x) = \sqrt{(10 \times 12\%)} = 1.1$ | $U(x) = \sqrt{(10 \times 10\%)} = 1$ |
| Risk seeking | $U(x) = (10 \times 20\%)^2 = 4$ | $U(x) = (10 \times 12\%)^2 = 1.4$ | $U(x) = (10 \times 10\%)^2 = 1$ |
| Risk neutral | $U(x) = (10 \times 20\%) = 2$ | $U(x) = (10 \times 12\%) = 1.2$ | $U(x) = (10 \times 10\%) = 1$ |

1. $U(f_1) = [1.73; 2.78]$
2. $U(f_2) = [1.71; 2.7]$
3. $U(f_3) = [1.62; 2.18]$

By using Definition 3 The best alternative is $f_1$.

Let us conduct sensitivity analysis to consider influence of change probability of risk seeking to utility of alternatives. The obtained results are the following:

$P(H_2)$ is changed from [0.5; 0.6] to [0.4; 0.5]

The final result:

1. $U(f_1) = [1.73; 2.78]$
2. $U(f_2) = [1.71; 2.7]$
3. $U(f_3) = [1.62; 2.18]$

One we can see that some change of $P(h_2)$ doesn't influence the best alternative. Let us now change $P(H_1)$ from **[0.1; 0.3] to [0.5; 0.6],** The offered results:

1. $U(f_1) = [0.53; 0.72]$
2. $U(f_2) = [1.16; 1.9]$
3. $U(f_3) = [1.31; 1.69]$

As one can see significant change of $P(h_1)$ influence the choice of the best alternative. By using Definition 3 The best alternative is $f_3$.

## 4   Conclusion

In this paper we apply combined states approach to behavioral decision modelling in investment problem. Three possible states of decision maker as risk attitudes are considered. To model uncertainty related to human behavioral condition and state of economy interval-valued probability are used. Choquet integral based on utility is used to describe non-additivity of preferences under uncertainty. A sign of dependence decision maker and state of economy is taken into account sensitivity analysis of obtained solution illustrates validity of the proposed work.

## References

1. Tversky, A., Kahneman, D.: Advances in prospect theory: cumulative representation of uncertainty. J. Risk Uncertain. **5**(4), 297–323 (1992)
2. Wise, B.P., Henrion, M.: A framework for comparing uncertain inference systems to probability. In: Kanal, L.N., Lemmer, J.F. (eds.) Uncertainty in Artificial Intelligence, pp. 69–83. Elsevier Science Publishers, Amsterdam (1986)
3. Ergu, D., Kou, G.: Questionnaire design improvement and missing item scores estimation for rapid and efficient decision making. Ann. Oper. Res. **197**(1), 5–23 (2002)
4. Schmeidler, D.: Subjective probability and expected utility without additivity. Econometrica **57**(3), 571–587 (1989)
5. Kou, G., Lu, Y., Peng, Y., Shi, Y.: Evaluation of classification algorithms using MCDM and rank correlation. Int. J. Inf. Technol. Decis. Making **11**(1), 197–225 (2012)
6. Guo, P., Tanaka, H.: Decision making with interval probabilities. Eur. J. Oper. Res. **203**, 444–454 (2010)
7. von Neumann, J., Morgenstern, O.: Theory of Games and Economic Behaviour. Princeton University Press, Princeton (1944)
8. Compte, O., Postlewaite, A.: Mental processes and decision making, Working paper, Yale University, New Haven, USA (2009)
9. Ghirardato, P., Maccheroni, F., Marinacci, M.: Differentiating ambiguity and ambiguity attitude. J. Econ. Theory **118**, 133–173 (2004)
10. Aliev, R.A.: Uncertain Computation-Based on Decision Theory. World Scientific, Singapore (2018)
11. Aliev, R.A., Pedrycz, W., Fazlollahi, B., Huseynov, O.H., Alizadeh, A.V., Guirimov, B.G.: Fuzzy logic-based generalized decision theory with imperfect information. Inf. Sci. **189**, 18–42 (2011)
12. Williamson, R.C.: Probabilistic arithmetic. Ph.D. dissertation, University of Queensland, Australia (1989). http://theorem.anu.edu.au/williams/papers/thesis300dpi.ps
13. Brafman, R.I., Tennenholz, M.: Modeling agents as qualitative decision makers. Artif. Intell. **94**(1–2), 217–268 (1997)
14. El-Ghamrawy, S.M., Eldesouky, A.I.: An agent decision support model based on granular rough model. Int. J. Inf. Technol. Decis. Making **11**(4), 793–820 (2012)
15. Schmeidler, D.: Subjective probability and expected utility without additivity. Econometrita **57**(3), 571–587 (1989)
16. Wang, Z., Wang, W.: Extension of lower probabilities and coherence of belief measures. Lecture Notes in Computer Science, vol. 945, pp. 62–69 (1995)
17. Peng, Y., Kou, G., Wang, G., Wu, W.: Ensemble of software defect predictors: an AHP based evaluation method. Int. J. Inf. Technol. Decis. Making **10**(1), 187–206 (2011)
18. Shoham, Y., Cousins, S.B.: Logics of mental attitudes in AI: a very preliminary survey. In: Lakemeyer, G., Nebel, B. (eds.) Foundations of Knowledge Representation and Reasoning. Springer, Heidelberg, pp. 296–309 (1994)

# Prediction of Mechanical Lower Back Pain for Healthcare Workers Using ANN and Logistic Regression Models

Nuriye Sancar[1](✉), Mehtap Tinazli[1], and Sahar S. Tabrizi[2]

[1] Near East University, Nicosia, Northern Cyprus, Turkey
{nuriye.sancar,mehtap.tinazli}@neu.edu.tr
[2] University of Tabriz, 29 Bahman Blvd., 5166616471 Tabriz, Iran
tabrizi.s@tabrizu.ac.ir

**Abstract.** The aim of this study is the comparison of predictive capabilities of logistic regression (LR) model and artificial neural network (ANN) to predict chronic mechanical lower back pain (MLBP) for healthcare workers in North Cyprus. For this purpose, the dataset has been obtained from Near East University (NEU) Hospital healthcare employees after obtaining approval from the ethics committee. Since this work was defined as exploratory, stepwise regression methods were considered to be the most appropriate and therefore, the Forward Selection and Backward Elimination methods were compared to find the proper binary LR model by using Likelihood ratio test. In order to obtain accurate results, two ANN models (ANN_1 and ANN_2) were used in this study. The main different of these two models was the number of processing elements. In the both models the Levenberg–Marquardt (LM) algorithm, as one of the most common and fastest back-propagation training algorithms was used in this study. The predictive capabilities of the binary logistic regression and ANNs was evaluated by specificity, sensitivity, accuracy rates and area under the ROC curve. The comparison results show that ANN performs better than the logistic regression model for prediction of chronic MLBP for healthcare workers. However, two models are biologically acceptable, too.

**Keywords:** Low back pain · Risk factors · Healthcare-workers
ANN · Logistic regression

## 1 Introduction

Lower back pain is one of the most common complaints in the adult population and is a frequent cause of job losses [1]. It is considered that 70–90% of individuals will experience back pain during their lives [2]. Physically heavy work, frequent bending, load lifting, sudden movements and chronic traumas such as repetitive tasks can all cause lower back pain [3]. Lower back pain can be mechanical or non-mechanical in nature. It is known that most back pain is mechanical. Frequency of mechanical low back pain (MLBP) in healthcare workers has been reported between 46–65.8% in literature [4, 5]. It is important to distinguish whether MLBP is acute or chronic in order to determine the appropriate treatment. In North Cyprus society, there is no study

© Springer Nature Switzerland AG 2019
R. A. Aliev et al. (Eds.): ICAFS-2018, AISC 896, pp. 781–789, 2019.
https://doi.org/10.1007/978-3-030-04164-9_103

about risk factors for MLBP. Thus, the main purpose of this study is to develop a proper model that can predict chronic MLBP in healthcare workers in North Cyprus. As MLBP is a multifactor condition, each influencing factor only contributes a small amount of risk, which means that by addressing each individually, it will have a minimal impact on preventing the disease. Despite the development of advanced technology and techniques, specialists in the field have been unsuccessful in their attempts to diagnose and completely understand the underlying causes of MLBP [6, 7]. These deficiencies indicate that more extensive and in-depth research should be conducted along with the use of more accurate statistical processes to determine the specific risk factors associated with the disease.

Predictive models are utilized in various health care studies for prognosis and diagnostic purposes. The most commonly used predictive models in clinical trials are artificial neural networks (ANN) and logistic regression (LR). The origins of these two models are in various communities; computer science and statistics, however have many similarities [8]. In this paper, MLBP has been predicted by its associated prognostic risk factors using two predictive models: LR, which is a traditional statistical model that is frequently used for the prediction of binary data, and ANN, which is a more up-to-date and sophisticated model. Therefore, the aim of this study is to compare the predictive capabilities of the LR model and ANN in terms of the prediction of MLBP for healthcare workers in North Cyprus. Figure 1 demonstrates the block diagram of the study.

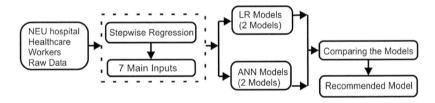

**Fig. 1.** The block diagram of the study

## 2   Materials and Methods

### 2.1   Data Collection

This study was conducted with NEU Hospital employees after obtaining approval permission of the ethics committee. The study participants included 140 healthcare employees who had experienced or were suffering from MLBP. The risk factors that caused the MLBP in the healthcare workers were identified and developed in a question form consisting of two separate sections. In the first part, which was composed of general questions, the demographic features, habits, history of chronic illness, drug use history and profession were questioned. In the second part, participants were asked questions regarding the characteristics of their back pain. Chronic lower back pain is defined as lower back pain that is experienced for over 12 weeks. Only those who had worked for at least a year or more in the health sector were included in the study.

## 2.2 Logistic Regression Analysis

The general form of logistic regression model is given in the form of generalized linear model as follow:

$$y_i = x_i\beta + \varepsilon_i, E(\varepsilon_i) = 0 \tag{1}$$

where $x_i = [1, x_{i1}, \ldots, x_{ip}]$ *is $i$-th row of* $X_{nx(p+1)}$ and $\beta' = [1, \beta_{i1}, \ldots, \beta_{ip}]$ is the vector of the parameters for i = 1, 2, ..., n. The distribution of each element $y_i$, given $x_{i1}, \ldots, x_{ip}$ is in an exponential family, in the sense that the probability density function (PDF) in Eq. (2) for each $y_i$ has the form in Eq. (2):

$$f(y_i, \eta_i, \varphi) = exp\left(\frac{y_i\eta_i - b(\eta_i)}{a(\varphi)} + h(y_i, \varphi)\right) \tag{2}$$

where $\eta_i = x_i\beta$ is a link function that maps the mean to the linear predictor; $y_i$ are jointly independent and $\varphi$ is the scale parameter. When y is a binary response (e.g., survived or died, infected or not infected, etc.), it has a Bernoulli distribution [9]. The PDF for Bernoulli distribution can be written as in Eq. (3):

$$f(y_i) = exp\left[y_i ln\left(\frac{\pi_i}{1 - \pi_i}\right) + ln(1 - \pi_i)\right] \tag{3}$$

with the probability $P(y_i = 1/x_i) = \pi_i$ *and* $y_i \sim Bernoulli(\pi_i)$. From Eq. (2), the link function is obtained as

$$\eta_i = x_i\beta = ln\left(\frac{\pi_i}{1 - \pi_i}\right) \tag{4}$$

In this form, the corresponding link function in Eq. (4) is called the logit function. By exponentiating both sides of the logit link function, the logistic regression model is obtained as

$$P(y_i = 1/x_i) = E(y_i) = \pi_i = \frac{e^{\eta_i}}{1 + e^{\eta_i}} \tag{5}$$

Logistic regression is generally used in areas such as modelling life probability in medical fields and evaluating disease risk factors. Binary logistic regression establishes the relationship between the binary dependent variable and the predictors using odds ratios. The odds ratio ($e^{\beta_k}$) is evaluated as the estimated increase in the probability of success associated with a one-unit change in the value of independent variable [10]. Unnecessary predictors will add noise to the estimation of other quantities in which we are interested. Since this work was defined as exploratory, stepwise regression methods were considered to be the most appropriate and therefore, the Forward Selection (FS) and Backward Elimination (BE) methods were compared using the Likelihood ratio test [11]. The likelihood-ratio test statistic $G^2$ is defined as in Eq. (6):

$$G^2 = -2log\left(\frac{l_0}{l_1}\right) = -2\left\{\log\left[l\left(\widetilde{\beta_0}, \widetilde{\beta_1}, \ldots, \widetilde{\beta_P}\right)|H_0\right] - \log\left[l\left(\widehat{\beta_0}, \widehat{\beta_1}, \ldots, \widehat{\beta_P}\right)|H_A\right]\right\}$$

(6)

for comparing the likelihoods ($l_0$ and $l_1$) of two models, which is comparing the log-likelihood under $H_0$ and the loglikelihood under $H_A$. The likelihood-ratio test is performed by comparing the fit of one model to the fit of the other.

### 2.3   Artificial Neural Network Architecture

One of the supervised learning algorithms, which is closely modelled upon the human brain's nervous behaviors and interconnecting, is called an Artificial Neural Network (ANN). A number of simple and highly interconnected processing components are used by ANNs [12]. ANNs use a number of interconnected processing components to process information. Various processing components which are highly interconnected and simple were used by an ANN to process information [12]. The main construction of the ANN consists of three main layers, called the Input, Hidden and Output. These layers consist of one or more processing elements (PEs), so-called neurons which represented by the small circles [13]. The received data by Input layer neurons was transmitted to the Hidden layer purely. This means that the Input layer PEs do not manipulate or modify data indicating that the so-called Input layer is a Passive layer. Other two layers, Hidden Layer and Output Layer, as an active layers play trainer and tester role on ANN [13].

## 3   Implementation

All the statistical analyses were performed using the SPSS version 23 (SPSS, Inc., Chicago, IL, USA). MATLAB ® R2016a Neural Network toolboxes were applied to develop the ANNs. Both models were constructed using a training set. A p value < 0.05 was considered statistically significant and 0.5 was considered as the cut point of the ANN. In this study, two logistic regression models based on Backward Elimination (BE) and Forward Selection (FS) and two different ANN models with various PEs numbers have been developed to establish the acceptable model by specifying the risk factors of chronic MLBP for the healthcare workers. Additionally, the sensitivity, specificity, accuracy rates and the area under the receiver operating characteristic curve (AUC), were evaluated as measures for classification and prediction performances of the models.

### 3.1   Logistic Regression Model Construction

The predictor variables (independent variables) are gender, job (doctor = 1, nurse = 2, patient consultant = 3, technician = 4, other = 5), age, education level, smoking, comorbid diseases, the event that started the low back pain (trauma, stress, infection, menstruation, heavy lifting, other), the onset of the low back pain (gradual = 1, sudden = 2), morning stiffness, improvement in lower back pain with exercise, improvement in the lower back pain condition with rest, gluteal pain, and night back pain. Age and education level were evaluated as the continuous variables. The remaining variables were evaluated as categorical variables. A dummy variable was used for each category of non-binary categorical variable, job and the event that started the MLBP. The predicted variable (dependent variable) was the condition of the MLBP, (suffering from chronic MLBP (coded 1) or suffering from acute MLBP (coded 0)). Out of the 140 healthcare workers considered in the study, 48.5% were male and 51.5% were female. In total, 38.6% of these people suffered from chronic MLBP, and 61.4% had acute MLBP. The results of the logistic regression analyses using the BE and FS methods are shown in Tables 1 and 2, respectively.

**Table 1.** Logistic regression analysis using backward elimination

| Logistic regression analysis with backward elimination (Step 9) | | | | | |
|---|---|---|---|---|---|
| Covariates | β | SE | Wald | p | Exp(β) |
| Trauma (T) | −2.115 | 0.911 | 5.390 | 0.020 | 0.121 |
| Stress (S) | 2.472 | 1.023 | 5.839 | 0.016 | 11.846 |
| Onset (sudden) (O) | 2.26 | 0.975 | 5.378 | 0.020 | 9.581 |
| Rest (R) | −2.174 | 0.981 | 4.911 | 0.027 | 0.114 |
| Exercise (E) | 1.976 | 0.749 | 6.959 | 0.008 | 7.213 |
| Nurse (N) | 3.017 | 1.246 | 5.863 | 0.015 | 20.43 |
| Night-back pain (NBP) | 1.844 | 0.743 | 6.159 | 0.013 | 6.323 |
| Constant | −2.899 | 1.103 | 6.907 | 0.009 | 0.055 |
| Model summary | β: logistic coefficient of each covariate in the model; Exp(β): Odds Ratio; −2 Log likelihood: 30.908; NagelKarke $R^2$: 0.886; Hosmer Lemeshow statistic: 9.571 (p = 0.214) | | | | |

**Table 2.** Logistic regression analysis using forward selection

| Logistic regression analysis with forward selection (Step 4) | | | | | |
|---|---|---|---|---|---|
| Covariates | B | SE | Wald | p | Exp(β) |
| Trauma | −2.257 | 0.811 | 7.745 | 0.005 | 0.105 |
| Onset (sudden) | 2.668 | 0.836 | 10.18 | 0.001 | 14.409 |
| Rest | −2.52 | 0.809 | 9.695 | 0.002 | 0.08 |
| Night-back pain | 2.076 | 0.861 | 5.815 | 0.016 | 7.974 |
| Constant | 0.326 | 0.771 | 0.179 | 0.673 | 1.385 |
| Model summary | −2 Log likelihood: 45.204; NagelKarke $R^2$: 0.788; Hosmer Lemeshow statistic: 3.759 (p = 0.585) | | | | |

The Hosmer-Lemeshow statistics were calculated for both models. Since $9.571 < \chi^2_{0.05,7} = 14.067$ and $3.759 < \chi^2_{0.05,5} = 11.070$, it was concluded that the models that were obtained by BE and FS methods were adequately fit to the data. The NagelKarke $R^2$ (0.886) for the model obtained by BE method is greater than the value for the model obtained by the FS method (0.788). The likelihood-ratio test statistic, $G^2$ for comparing the likelihoods of the two models is $G^2 = -2(L_{forward} - L_{backward}) = 45.204 - 30.908 = 14.296$ with $\Delta df = 2$. Since $\chi^2_{0.05,2} = 5.991 < 14.296$, it can be concluded that the model using the BE method produced a better fit than the FS method. The specifity, sensitivity and accuracy rates for the Backward LR stepwise method were 96.7%, 91.4% and 94.9%, respectively, and AUC is 0.923 as seen in Table 4.

## 3.2    ANN Training and Testing

In order to obtain accurate results, defining the testing and training data set, as well as determining the appropriate learning algorithm and the number of PEs are particularly crucial in ANN modelling [14]. Thus, the Levenberg–Marquardt (LM) algorithm, one of the most common and fastest back-propagation training algorithms [15, 16], was used in this study. The developed model used the same testing and training dataset that was used in the logistic regression model. Moreover, in order to obtain the appropriate numbers of PEs, a trial and error strategy was used in this study. In this regard, based on the training dataset, two different ANN models with two different PEs numbers were developed. Table 3 explains the details of the developed models.

**Table 3.** The developed ANNs' general details

| Architecture of ANNs | | |
|---|---|---|
| Type | Feed-forward back propagation | |
| Main parameters | Training method | Supervised |
| | Training algorithm | LMBP |
| | Number of data | 70% of all data (98) |
| | *Epochs* | 57 |
| | Cut-Point | 0.5 |
| Activation function | In output layer | Log-sigmoid |
| Neurons | Input | 7 (Trauma, Stress, Onset, Rest, Exercise, Nurse, Night-back Pain) |
| | Output | 1 |
| Number of PEs | ANN_1 | 15 |
| | ANN_2 | 10 |

# 4 Results

As can be seen in Table 4, the logistic regression based on the BE method and ANN_2 had the highest accuracy, specificity and sensitivity rates. Evidently, both developed models are capable of estimating chronic MLBP with high accuracy. Additionally, the AUC reveals that the Backward logistic regression model was lower than ANN_2 with 10 PEs for both types of datasets. The AUC for training was 0.923 and for testing it was 0.736 for the logistic regression model. On the other hand, the AUC for the ANN_2 model was 0.992 for training and 0.934 for testing. In conclusion, the developed ANN model has a better predictive capability than the constructed logistic regression model using BE in the terms of accuracy, specificity, sensitivity rates and the AUC. In last years, ANN has shown more satisfactory prediction performance than logistic model for medical outcomes in many studies [17–19]. However, although the ANN has better predictive capability than the logistic model, it is not able to explain which findings are more relevant in reaching the diagnosis. This can be regarded as a disadvantage for the ANN.

**Table 4.** Summary of the comparison of logistic and ANN models.

| Model | | Specificity | Sensitivity | Accuracy | AUC |
|---|---|---|---|---|---|
| Logistic models | | | | | |
| Backward elimination | Training | 96.7% | 91.4% | 94.9% | 0.923 |
| | Testing | 80% | 66.7% | 76.2% | 0.736 |
| Forward selection | Training | 93.1% | 88.9% | 91.8% | 0.873 |
| | Testing | 73.3% | 58.3% | 69.1% | 0.689 |
| ANN | | | | | |
| 15 neurons in the hidden layer | Training | 95.1% | 91.9% | 93.9% | 0.976 |
| | Testing | 83.3% | 75% | 81% | 0.925 |
| 10 neurons in the hidden layer | Training | 98.4% | 94.9% | 95.9% | 0.992 |
| | Testing | 86.6% | 83.3% | 85.7% | 0.934 |

The popularity of the logistic model may be attributed to the interpretability of model parameters and the ease of use. Commentaries of odds ratios of some predictors in logistic model with BE are as follows: Healthcare workers who are suffering from stress have 11.846 times more risk of experiencing chronic MLBP than those who are not suffering from stress. Moreover, for the healthcare workers suffering from MLBP who alleviate the condition with rest, the acute MLBP rates are 8.772 times more than who do not alleviate the MLBP with rest. Furthermore, for healthcare workers whose MLPB is relieved with activity, the chronic MLBP rates are 7.213 times greater than those who do not use activity to alleviate the condition. Healthcare workers who work as nurses have 20.43 times more risk of suffering from chronic MLBP than non-nurses. Finally, for the healthcare workers suffering from night back pain, the chronic MLBP rates are 6.323 times more than who do not experience night back pain.

## 5  Conclusion

This study was conducted to perform the predictive capabilities of two classification models, logistic regression and ANN for prediction of MLBP for health care workers. According to the high specificity, sensitivity, accuracy rates and the AUC, the ANN model with 10 neurons in the hidden layer has more accurate and feasible prediction capability than the logistic regression. Although ANN is more flexible and robust in identifying the complex relationship than logistic regression, there are some disadvantages of ANN: there are no particular methods for the variable selection procedure and mathematical formula between variables cannot be identified. The logistic regression model is a good tool for obtaining the mathematical relationship between the dependent and independent variables and for interpreting the model parameters. However, ANN has better prediction performance. Thus, these two methods should be considered as complementary methods. Furthermore, as a result of the present study, it has been found that MLBP is a common health problem for health care workers and it is highly prevalent among nurses, where trauma and stress are the most common etiologic causes in North Cyprus. Detecting and then preventing risk factors that may lead to chronic low back pain in the early stages will reduce both prevalence of chronic back pain, and job loss.

## References

1. Dıraçoğlu, D.: Sağlık personelinde kas-iskelet sistemi ağrıları. Turkiye Klinikleri J. Med. Sci. **26**, 132–139 (2006)
2. Rubin, D.: Epidemiology and risk factors for spine pain. Neurol. Clin. **25**(2), 353–371 (2007)
3. Domino, F.J.: The 5-Minute Clinical Consult 2011. Lippincott Williams & Wilkins, Philadelphia (2010)
4. Karahan, A., Kav, S., Abbasoglu, A., Dogan, N.: Low back pain: prevalence and associated risk factors among hospital staff. J. Adv. Nurs. **65**(3), 516–524 (2009)
5. Omokhodion, F.O., Umar, U.S., Ogunnowo, B.E.: Prevalence of low back pain among staff in a rural hospital in Nigeria. Occup. Med. **50**(2), 107–110 (2000)
6. Health Policy Institute. https://hpi.georgetown.edu/agingsociety/pubhtml/backpain/backpain.html. Accessed 21 June 2018
7. Chou, R., Huffman, L.H.: Nonpharmacologic therapies for acute and chronic low back pain: a review of the evidence for an American Pain Society/American College of Physicians clinical practice guideline. Ann. Intern. Med. **147**(7), 492–504 (2007)
8. Dreiseitl, S., Ohno-Machado, L.: Logistic regression and artificial neural network classification models: a methodology review. J. Biomed. Inform. **35**(5–6), 352–359 (2002)
9. Nelder, J.A., Baker, R.J.: Generalized Linear Models. Wiley, Hoboken (1972)
10. Montgomery, D.C., Peck, E.A., Vining, G.G.: Introduction to linear regression analysis, vol. 821. Wiley, Hoboken (2012)
11. Hosmer Jr., D.W., Lemeshow, S.: Applied Logistic Regression. Wiley, Hoboken (2004)
12. Rafiq, M.Y., Bugmann, G., Easterbrook, D.J.: Neural network design for engineering applications. Comput. Struct. **79**(17), 1541–1552 (2001)

13. Sobhani, J., Najimi, M., Pourkhorshidi, A.R., Parhizkar, T.: Prediction of the compressive strength of no-slump concrete: a comparative study of regression, neural network and ANFIS models. Constr. Build. Mater. **24**(5), 709–718 (2010)
14. Cannon, R.W.: Proportioning no-slump concrete for expanded applications. Concr. Int. **4**(8), 43–47 (1982)
15. Norouzi, J., Yadollahpour, A., Mirbagheri, S.A., Mazdeh, M.M., Hosseini, S.A.: Predicting renal failure progression in chronic kidney disease using integrated intelligent fuzzy expert system. Comput. Math. Methods Med. **2016** (2016)
16. Tabrizi, S.S., Sancar, N.: Prediction of body mass index: a comparative study of multiple linear regression, ANN and ANFIS models. In: International Conference on Theory and Application of Soft Computing, Computing with Words and Perception, ICSCCW, vol. 120, pp. 394–401. Procedia Computer Science, Budapest (2017)
17. Lin, C.C., Ou, Y.K., Chen, S.H., Liu, Y.C., Lin, J.: Comparison of artificial neural network and logistic regression models for predicting mortality in elderly patients with hip fracture. Injury **41**(8), 869–873 (2010)
18. Eftekhar, B., Mohammad, K., Ardebili, H.E., Ghodsi, M., Ketabchi, E.: Comparison of artificial neural network and logistic regression models for prediction of mortality in head trauma based on initial clinical data. BMC Med. Inform. Decis. Mak. **5**(1), 3 (2005)
19. Heydari, S.T., Ayatollahi, S.M.T., Zare, N.: Comparison of artificial neural networks with logistic regression for detection of obesity. J. Med. Syst. **36**(4), 2449–2454 (2012)

# Optimization of Agricultural Land Use on the Basis of Mathematical Methods of Financial Analysis and the Theory of Fuzzy Sets

V. Sakharova Lyudmila[1]([✉]), B. Stryukov Michael[2],
and I. Akperov Gurru[2]

[1] Rostov State University of Economics, B. Sadovaya Street 69,
344002 Rostov-on-Don, Russia
L_Sakharova@mail.ru
[2] Sothern University (IMBL), Prospekt M. Nagibina 33a/47,
344068 Rostov-on-Don, Russia

**Abstract.** The article presents the optimization model of unsustainable land use on the basis of data on crop acreage and yield for major agricultural crops grown in the Rostov region. The following mathematical apparatus used for modeling: methods of regression analysis and nonlinear optimization, commonly used in the mathematical methods of financial analysis to optimize portfolios of financial instruments. The model allows to calculate the optimal quantitative distribution of territories that ensures the maximum average yield with minimal average risk. Model modified on the basis of fuzzy set theory to apply in practical calculations, the overall estimated harvest of crop species.

**Keywords:** The optimization model · Agricultural land use
Yield of agricultural land · Fuzzy sets

## 1 Modification of Optimization Methods Financial Analysis for Agriculture

Optimization of unsustainable agricultural land use is one of the priority tasks of mathematical modeling in the agricultural sector. At the same time, the mathematical apparatus in this field is still insufficiently developed. Generally, traditional statistical and econometric methods used for modeling in agriculture. Their disadvantage is that they are suited to local conditions and do not have a sufficient level of universality. In this article we propose a simple model allowing to optimize land use on the basis of classical methods of financial analysis – the so-called "portfolio theory" [1, 2]. Consider the classic model of the securities market. At some time $t$ the investor can part of their funds to keep in a Bank account, and another part is spent on the purchase of securities traded on the market. A portfolio of financial instruments is a vector of values $X = (x_0, x_1, x_2, \ldots, x_N)^T$, where $x_0 \in R$ is the fraction of the investor's capital invested in risk-free asset; $x_k \in R$ is the fraction of the investor's capital invested in the action

© Springer Nature Switzerland AG 2019
R. A. Aliev et al. (Eds.): ICAFS-2018, AISC 896, pp. 790–798, 2019.
https://doi.org/10.1007/978-3-030-04164-9_104

number $k$ ($k = 1, 2, \ldots, N$). In this case, obviously, $x_0 + \sum\limits_{k=1}^{N} x_k = 1$. The number of $x_k$ can be both positive and negative (and zero). In the latter case, the investor takes money in debt from a Bank account or makes a short sale of shares. The effectiveness of the portfolio of financial instruments X is determined by the formula:

$$E_r = x_0 i + \sum_{k=1}^{N} x_k E_k$$

where $i$ is the efficiency of a Bank account (the interest rate), $E_k$—efficiency $k$-th stock (its profitability). The expected performance of the portfolio of financial instruments is called the mathematical expectation of its efficiency, and portfolio risk—standard deviation of its effectiveness. There is a task of portfolio optimization with the aim of obtaining maximum effectiveness with minimum risk, which can be solved in General form is impossible. The investor optimizes his portfolio should choose one of the criteria: either minimize risk $\sigma_n = \sqrt{DE_n}$ for a given level of efficiency of a portfolio; or to maximize expected efficiency $ME_n$ under a given level of portfolio risk.

Consider first the case, that is, we minimize risk for a given level of effectiveness $r$. Mathematical formulation of the problem has the form:

$$\sqrt{D\left(x_0 i + \sum_{k=1}^{N} x_k E_k\right)} \rightarrow \min, \quad M\left(x_0 i + \sum_{k=1}^{N} x_k E_k\right) = r, \quad x_0 + \sum_{k=1}^{N} x_k = 1,$$

$x_k \in R$ ($k = 0, 1, 2, \ldots, N$). Applying the properties of mathematical expectation and dispersion, the model can be rewritten in the following form:

$$\sum_{k=1}^{N} \sum_{i=1}^{N} \sigma_{ik} x_k x_i \rightarrow \min, \quad \sum_{k=1}^{N} r_k x_k = r, x_0 + \sum_{k=1}^{N} x_k = 1,$$

$x_K \in R$ ($k = 0, 1, 2, \ldots, N$), where $r_k = ME_k$—the expected efficiency of the $k$-th action ($k = 0, 1, 2, \ldots, N$), $\sigma_{kl} = \text{cov}(E_k, E_l)$ the covariance of the performance of the $k$-th and the $l$-th actions ($k = 0, 1, 2, \ldots, N, l = 0, 1, 2, \ldots, N$); obviously, $\sigma_{kk} = DE_k = \sigma_k^2$. Modify the considered model to optimize land use, providing a predetermined average expected yield while minimizing risk. Assume that the yield of each of the considered crops submitted to the statistical sample.

We introduce the notation: let there be optimization vector, $X = (x_1, x_2, \ldots, x_N)^T$ where $x_k$ is the proportion of the territory occupied by the $k$-th crop, $\sum\limits_{k=1}^{N} x_k = 1$. Obviously, in contrast to the problem of portfolio optimization valuable the BU-magician, in the task $x_k$ can only be positive. Then,

$$E_r = \sum_{k=1}^{N} x_k E_k$$

there is an average expected yield of a given agricultural area; $E_k$ is the average yield of the $k$-th crop (a random variable) defined on the basis of statistical sampling. Then, the average expected yield of agricultural areas will be called the expected value $E_x$, a risk is its standard deviation. Omitting detailed calculations, we formulate the optimization problem for agricultural areas:

$$\sum_{k=1}^{N}\sum_{i=1}^{N} \sigma_{ik} x_k x_i \rightarrow \min, \sum_{k=1}^{N} r_k x_k = r, \sum_{k=1}^{N} x_k = 1, x_K \in R \quad (k = 1, 2, \ldots, N),$$

where $r_k = ME_k$ is the proportion of the territory occupied by $k$-th crop, expected average yield-to-crops, computed as an average sample value $(k = 0, 1, 2, \ldots, N)$; $\sigma_{kl} = \mathrm{cov}(E_k, E_l)$—the covariance of yield, the $k$-th and the $l$-th crops $(k = 0, 1, 2, \ldots, N, l = 0, 1, 2, \ldots, N)$; obviously, $\sigma_{kk} = DE_k = \sigma_k^k$.

## 2 Optimizing Agricultural Areas of the Rostov Region

Optimization of agricultural territories in Rostov region will carry out in four stages in accordance with the quantitative and qualitative characteristics of statistical information on productivity [3].

*Stage 1.* Initially, we will consider the optimization of territories based on the distribution of agricultural products in four main types: grain crops, sunflower, potato, vegetables. The following indicators between 1990 and 2015 presented for each crop: sown area (thousand ha) gross agricultural output (thousand tonnes), yield (centners/ha). These statistics are in Tables 1 and 2. The average risk obtained on the basis of processing of statistical data, equal to 34%. The problem of optimization areas for the four crops as follows from the General formulation of the problem (1)–(2) has the form:

**Table 1.** Statistics for grain crops and sunflower: own area (thousand ha), gross harvest (thousand tons), yield (centners/ha).

| Year | Crops (№. 1) | | | Sunflower (№. 2) | | |
|------|----------|--------------|-------|----------|--------------|-------|
| | Own area | Gross harvest | Yield | Own area | Gross harvest | Yield |
| 1990 | 2940.7 | 9666.8 | 32.9 | 455.1 | 836.2 | 17.3 |
| 1996 | 2461.7 | 3199.3 | 13 | 694.4 | 522.9 | 7.5 |
| 1997 | 2483 | 4011.4 | 16.2 | 679.6 | 645.5 | 9.5 |
| 1998 | 2157.8 | 2719.4 | 12.6 | 809.9 | 607.9 | 7.5 |
| ... | | | | | | |
| 2014 | 3206.7 | 9506.9 | 29.6 | 524.6 | 761.2 | 14.5 |
| 2015 | 3296.0 | 9783.5 | 29.7 | 532.7 | 859.6 | 16.1 |

**Table 2.** Statistics for potato and vegetables: own area (thousand ha), gross harvest (thousand tons), yield (centners/ha).

| Year | Potato (№. 3) | | | Vegetables (№. 4) | | |
|------|----------|---------------|-------|----------|---------------|-------|
| | Own area | Gross harvest | Yield | Own area | Gross harvest | Yield |
| 1990 | 39.8 | 259.2 | 65.2 | 38 | 521.4 | 136.1 |
| 1996 | 77.7 | 297.3 | 383 | 48.5 | 233.6 | 48.1 |
| 1997 | 75.2 | 483.2 | 64.2 | 46.8 | 346.6 | 74 |
| 1998 | 72.1 | 351.2 | 48.7 | 43.5 | 312.9 | 71.9 |
| ... | | | | | | |
| 2014 | 33.4 | 397.2 | 119.0 | 33.5 | 701.8 | 210.7 |
| 2015 | 34.3 | 417.4 | 122.2 | 35.0 | 747.5 | 215.6 |

$$F(x_1, x_2, x_3, x_4) = \sigma_{11}x_1^2 + \sigma_{22}x_2^2 + \sigma_{33}x_3^2 + \sigma_{44}x_4^2 + 2\sigma_{12}x_1x_2 + 2\sigma_{13}x_1x_3$$
$$+ 2\sigma_{14}x_1x_4 + 2\sigma_{23}x_2x_3 + 2\sigma_{24}x_2x_4 + 2\sigma_{34}x_3x_4 \rightarrow \min \tag{1}$$

$$r_1x_1 + r_2x_2 + r_3x_3 + r_4x_4 = r \tag{2}$$

$$x_1 + x_2 + x_3 + x_4 = 1, x_1, x_2, x_3x_4 \in R \tag{3}$$

Solve the problem for different values of the average expected yield using the method of Lagrange multipliers and the conditions of Kuhn-Tucker [3, 4] (Table 3). Calculations show that the average expected yield r = 35 is unreachable because among the components of the solution are negative. The optimal allocation areas does not exceed 24% to 34%, with an average initial (statistical) distribution of agricultural territories. At the same time, the numerical experiment indicates that it is possible to achieve an average expected yield of the territory of 30 centners/ha at risk of 24% versus the statistical average yield 21,28 TL/ ha with an average risk of 34%.

**Table 3.** The results of optimization calculations for four main types of agricultural crops.

| Yield site | Average risk | Relative risk, % | Distribution of agricultural area under crops (fraction) | | | |
|------------|---------|-----------|----------------------|------------|----------|------------|
| | | | Crops, $X_1$ | Sunflower, $X_2$ | Potato, $X_3$ | Vegetables, $X_4$ |
| 20 | 4.47 | 22 | 0.4660 | 0.4948 | 0.0331 | 0.0061 |
| 25 | 5.86 | 23 | 0.6628 | 0.2700 | 0.04410 | 0.0231 |
| 30 | 7.30 | 24 | 0.8595 | 0.0452 | 0.0551 | 0.0402 |

*Stage 2.* In the second stage, we optimize the territory allocated for crops, four of their main types, greatly differing in agronomic characteristics: winter cereals, spring cereals, rice, bean. For each of the cultures represented by the following indicators for 2000–2015: sown area (thousand ha) gross harvest of agricultural products (thousand tonnes), yield (centners/ha). The statistical information is in Tables 4 and 5.

**Table 4.** Statistics for winter and spring cereals: own area (thousand ha), gross harvest (thousand tons), yield (centners/ha).

| Year | Winter cereals (№. 1) | | | Spring cereals (№. 2) | | |
|------|----------|---------------|-------|----------|---------------|-------|
|      | Own area | Gross harvest | Yield | Own area | Gross harvest | Yield |
| 2000 | 847.8 | 1780.5 | 21.0 | 1250.5 | 1668.8 | 13.3 |
| 2001 | 1262.9 | 3762.5 | 29.8 | 1189 | 1959.3 | 16.5 |
| 2002 | 1477.6 | 4409.4 | 29.8 | 1100.2 | 1785.8 | 16.2 |
| ... | | | | | | |
| 2015 | 2320.3 | 7622.6 | 32.9 | 803.7 | 1806.5 | 22.5 |

**Table 5.** Statistics for rice and bean: own area (thousand ha), gross harvest (th. tons), yield (centners/ha).

| Year | Rice (№. 3) | | | Bean (№. 4) | | |
|------|----------|---------------|-------|----------|---------------|-------|
|      | Own area | Gross harvest | Yield | Own area | Gross harvest | Yield |
| 2000 | 11.7 | 39.9 | 34.2 | 30.1 | 46.3 | 17.4 |
| 2001 | 10.8 | 33.6 | 31.1 | 44.5 | 107 | 25.9 |
| 2002 | 9.9 | 26 | 26.3 | 77.1 | 103.7 | 14.6 |
| 2003 | 10.2 | 29.6 | 29 | 86 | 61.2 | 7.8 |
| ... | | | | | | |
| 2015 | 15.8 | 88.9 | 56.2 | 82.9 | 165.1 | 19.4 |

The average yield is equal to: $r_s = 23.95$. As follows from the results of the 1st phase, the average risk of agricultural territories occupied by grain crops, equal to 6.13, that is, the relative risk agronomic areas equal to 25.6%. In accordance with the results of the regression analysis, there is a correlation between yields of winter and spring wheat and winter wheat and rice; yields of other crops are poorly correlated among themselves. The results are shown in Table 6.

**Table 6.** The results of optimization calculations for four main types of agricultural crops.

| The average expected yield site | Average risk | Relative risk, % | Distribution of agricultural area under crops (fraction) | | | |
|--------|--------|--------|--------|--------|--------|--------|
|        |        |        | Winter cereals, $X_1$ | Spring cereals, $X_2$ | Rice, $X_3$ | Bean, $X_4$ |
| 23.95 | 3.9 | 16 | 0.1447 | 0.3542 | 0.2408 | 0.2603 |
| 25 | 3.94 | 16 | 0.2108 | 0.2929 | 0.256 | 0.2403 |
| 27 | 4.03 | 15 | 0.336 | 0.1767 | 0.2848 | 0.2025 |
| 30 | 4.26 | 14 | 0.5237 | 0.00252 | 0.328 | 0.1458 |

Calculations show that the average expected yield r = 35 is unreachable because among the components of the solution are negative. As you can see from the table that the risk under the optimal allocation areas does not exceed 16% vs 25,6%, with an average initial (statistical) distribution of agricultural territories. At the same time, the numerical experiment indicates that it is possible to achieve an average expected yield of the territory of 30 centners/ ha at risk 14% against the statistical average yield 23,95 centners/ ha with an average risk of 25.6% (Tables 7 and 8).

**Table 7.** Area under winter and spring crops (thousand ha).

| Year | Winter wheat | Barley | The rye | Spring wheat | Spring barley | Corn for grain | Millet |
|------|------|------|------|------|------|------|------|
| № | 1 | 2 | 3 | 1 | 2 | 3 | 4 |
| 2000 | 813 | 11.6 | 23.2 | 15.9 | 890.8 | 109.7 | 234.1 |
| 2001 | 1205.7 | 24.1 | 33.1 | 13.5 | 878.8 | 121.3 | 175.4 |
| 2002 | 1414.9 | 33.1 | 29.6 | 10.5 | 924 | 78.2 | 87.5 |
| ... | | | | | | | |
| 2015 | 2280.8 | 34.6 | 4.9 | 15.1 | 415.3 | 233.6 | 139.7 |

**Table 8.** Average yields of varieties of winter and spring crops (centners/ha).

| Year | Winter wheat | Barley | The rye | Spring wheat | Spring barley | Corn for grain | Millet |
|------|------|------|------|------|------|------|------|
| № | 1 | 2 | 3 | 1 | 2 | 3 | 4 |
| 2000 | 21.9 | 25.1 | 17.9 | 14.8 | 14.7 | 16.4 | 9.4 |
| 2001 | 31 | 33.3 | 18.8 | 14.1 | 21.5 | 4.6 | 3.8 |
| ... | | | | | | | |
| 2013 | 23.9 | 30.0 | 10.5 | 11.2 | 15.1 | 27.8 | 12.9 |
| 2014 | 33.4 | 36.8 | 16.6 | 17.3 | 21.1 | 27.0 | 15.8 |
| 2015 | 32.5 | 37.0 | 13.9 | 19.6 | 21.9 | 28.3 | 16.0 |

*Stage 3.* In the third phase, optimize territories within crops. Distribution of winter crops will be optimized for three crops: wheat, barley, rye (respectively No. 1, 2 and 3 within the group of winter). Distribution of spring crops will do for four crops: wheat, barley, maize and millet (respectively, No. 1, 2, 3, 4 within a group of spring). The calculated data for the considered two groups are shown in Tables 9 and 10.

**Table 9.** The results of optimization calculations for four main types of agricultural crops.

| The average expected yield site | Average risk | Relative risk, % | Distribution of agricultural area under crops (fraction) | | |
|------|------|------|------|------|------|
| | | | Winter wheat, $X_1$ | Barley, $X_2$ | The rye, $X_3$ |
| 25 | 4.41 | 18 | 0.5878 | 0.1526 | 0.2596 |
| 27.74 | 5.06 | 18 | 0.8273 | 0.1387 | 0.03399 |
| 28 | 5.12 | 18 | 0.8501 | 0.1374 | 0.01258 |

The average yield of winter cereals can be improved by optimizing the distribution of territories to 28 centners/ha at risk 18% vs 27.74 centners/ha at the risk of 19%. From Table 10 it follows that the average yield of spring crops can be increased to 17.18 centners/ha at 24% vs 17.17 risk, with 26% of the risk. Further improvement of the results in these groups is impossible because of the uneven distribution of crops at the average yields.

**Table 10.** The results of optimization calculations for four main types of agricultural crops.

| The average expected yield site | Average risk | Relative risk, % | Distribution of agricultural area under crops (fraction) | | | |
|---|---|---|---|---|---|---|
| | | | Spring wheat, $X_1$ | Spring barley, $X_2$ | Corn for grain, $X_3$ | Millet, $X_4$ |
| 15 | 3.8 | 25 | 0.388 | 0.3462 | 0.05757 | 0.2082 |
| 17 | 4.14 | 24 | 0.3359 | 0.4977 | 0.1475 | 0.01892 |
| 17.18 | 4.17 | 24 | 0.3312 | 0.5113 | 0.1556 | 0.00189 |

**Table 11.** The results of optimization.

| The group of agricultural crops | Territory, thousand ha | The average yield (centners/ha) | Min. yield (centners/ha) | Maxim. yield (centners/ha) | The interval of the blur of productivity (th. TL) | The interval of the blur evaluation of the crop (th. TL) |
|---|---|---|---|---|---|---|
| Winter crops | 1800.48 | 27.74 | 17,6 | 35,3 | (27.71; 29,25) | (49891.30; 52664.04) |
| Spring grain crops | 8.66 | 17.17 | 7,1 | 26,1 | (15.16; 18,96) | (131.29; 164.19) |
| ... | | | | | | |
| Vegetables | 160.8 | 119.96 | 48,1 | 215,6 | (105,59; 139,09) | (16978.87; 22365.67) |

# 3   An Example of Optimization of Calculations the Agronomic Land by the Use of Fuzzy Set Theory

We will consider application of the obtained results for calculations of the distribution of farming areas. Suppose that the total agronomic territory of the Rostov region in 2018 of 4,000 ha. We map some results to the Table 10. The results of applying the proposed method indicate the feasibility of extending the acreage devoted to the growing of vegetables more than 6 times, potatoes – almost 7 times bean – 6, rice – almost 8 times; along with this, it is proposed to significantly reduce the planting of spring crops with relatively low yields (almost 10 times). In addition, the method indicates the need for optimization to cut the crops "light" sunflower almost 5 times. It is also proposed to cut the crops of winter grain crops in favor of more "weighty" vegetables and potatoes. It should be noted that the model was not laid restrictions (quotas) on the production of certain agricultural products. In addition, were not taken into account "weight" of different types of crops (e.g., sunflower and potatoes "weighed" in the model equally, which led to significant downsizing of the share of sunflower). The points allow us to conclude, that the model can be successfully applied to the "with equal weight" crops. However, in the calculations for heterogeneous agricultures must be attributed to the average optimum yield. In addition, the calculations of practical problems it is necessary to enter in the model of nonlinear optimization constraints, reflecting production quotas for specific agricultural products.

# 4  Evaluation of the Harvest of Agricultural Products, Based on the Theory of Fuzzy Numbers

As follows from the tables, the average value of crop yield may differ greatly from the maximum and minimum values (in the first place, because of favorable or unfavorable weather conditions). The difference between the estimated and the actual values will result in large errors of predicted and actual values of the total harvest. Therefore, to evaluate interval, which can be the value of the projected crop yield, we use the theory of fuzzy numbers.

To build the model will make assumptions: (1) the yield of each crop $a_i$ represent the fuzzy triangular number; (2) for each parameter $a_i$ defined by the membership function $a_i$ taking its maximum value, one, when $a = a_n$, that is, the value corresponding to average yield; the maximum $a = a_{max}$ and minimum $a = a_{min}$ values of a function becomes zero. Set the level $\alpha = 0, 8$, which can be considered as the confidence level evaluation. We require: $\mu(a) \geq \alpha$. Hence, using formula (3), we find a range of blur values for a given yield of crops:

$$\alpha a_n + (1 - \alpha)a_{min} \leq a \leq \alpha a_n + (1 - \alpha)a_{max}. \tag{4}$$

Formula (4) allows to estimate the total harvest of each group of agricultural products, at specified allotted under agronomic areas. Table 11 shows some values of defocus for the value of the harvest each of the crops (at the level of the reliability assessment results) for values of agronomic territories, calculated on the basis of the optimization model.

The mathematical model of optimization of land use based on portfolio theory, one of the main tools of portfolio optimization of securities in the classical mathematical methods of financial analysis. For calculations based on the model requires statistical sampling of a sufficient amount of yield of major crops in the optimized agricultural areas. The model, in General, allows to calculate the optimal distribution of areas for planting as the main groups of crops (cereals, sunflower, potatoes, vegetables, etc.) and species within groups (winter and spring cereals, rice, legumes) and also specific crops. The calculations show that the optimal distribution of agricultural territories based on the model, allows to increase the average expected yield of the territories with minimization of average risk. For the practical application of the model is appropriate to relate the yields of crop species to their optimal average values, to exclude the factor "difference weight" diverse agricultural crops (e.g., vegetables, and sunflower).

## References

1. **2**(5), 99–110 (2016)
2. Soloviev, V.I: Mathematical methods of risk management. Goo (2003)
3. Malykhin, V.I: Financial mathematics. YUNITI-DANA (2004)
4. Online calculator "Methods of optimization". http://math.semestr.ru
5. OnLine services for higher mathematics

# Deep Learning for Lung Lesion Detection

Ali Işın[1]([✉]) and Tazeen Sharif[2]

[1] Department of Biomedical Engineering, Near East University,
Nicosia, North Cyprus, Turkey
ali.isin@neu.edu.tr
[2] Department of Electrical Engineering, Ajman University,
Ajman, United Arab Emirates

**Abstract.** As the most fatal cancer type, early diagnosis of the lung cancer plays an important role for the survival of the patients. Diagnosis of the lung cancer involves screening the patients initially by Computed Tomography (CT) for the presence of lung lesions. This procedure requires expert radiologists which need to go over very large numbers of image slices manually in order to detect and diagnose lung lesions. Unfortunately this is a very time consuming process and its performance is very dependent on the performing radiologist. Thus assisting the radiologists by developing an automated computer aided detection (CAD) system is an interesting research goal. In this regard, as the aim of this paper a pre-trained AlexNet (deep learning) framework is transferred to develop and implement a robust CAD system for the classification of lung images depending on whether they bear a lung lesion or not. High performances of 98.72% sensitivity, 98.35% specificity and 98.48% accuracy are reported as a result.

**Keywords:** Deep learning · Lung lesion detection
Biomedical image processing · Transfer learning

## 1 Introduction

Lung cancer is the most fatal cancer type in the world. In regular medical routine, detection of lung tumors is carried out by clinicians, mainly radiologists, manually. First, doctor goes over series of Computed Tomography (CT) images of the patient slice by slice to accurately detect the presence and the location of the lung lesions. Then proper diagnosis is carried out based on the prior knowledge of the doctor and sometimes the additional information provided by other pathological examinations. Clinician's manual involvement in these procedures means they are time consuming, subjective and highly dependent on prior knowledge and expertise of the clinician performing the procedure. In this regard, results of such manual segmentations can include unwanted errors. Because of these concerns, development of robust computer aided automatic tumor detection and segmentation methods i.e. computer aided detection (CAD) systems, to enable objective and effective detection results, became a very exciting research area in all of medical imaging fields in the recent years [1].

For the special case of lung cancer detection, manual detection can be far more challenging. Small lung lesions can cover areas as small as 2 voxel diameter on the

© Springer Nature Switzerland AG 2019
R. A. Aliev et al. (Eds.): ICAFS-2018, AISC 896, pp. 799–806, 2019.
https://doi.org/10.1007/978-3-030-04164-9_105

images. Also their contrast levels can be very insignificant when compared to the contrast levels of the surrounding tissues. These make visual detection by the clinician a very tough process, resulting in missed tumors during detection. Therefore, automatic detection provides invaluable assistance to the clinicians.

In contrast, automatic detection methods do not require any human interaction. Mainly, almost all well-known image processing techniques along with machine learning methods can be implemented to carry out the automatic detections. Even in some methods previously gained knowledge is integrated to solve the problem. Automatic detection and segmentation methods can be mainly classified as discriminative or generative methods [1]. Discriminative methods require ground truth data to learn the relationship between input images containing the tumors with the ground truth to carry out decisions. Generally these methods involve extracting features from the images using different image processing techniques. Deciding which features to use is of great importance. In most cases final decision is made by using supervised machine learning techniques which, in order to perform well, require large image datasets with accurate ground truth data. In contrast, generative methods require prior knowledge, such as location and spatial extent of healthy tissues to generate probabilistic models, which carry out the final segmentation. Prior obtained maps of healthy tissues are implemented to segment the unknown tumors areas. Developing suitable probabilistic models by using prior knowledge is however a complicated task.

In recent years, an emerging technique of deep learning began to replace the aforementioned techniques. As the main deep learning method, convolutional neural networks (CNN) obtained state-of-the-art performances in many of the well-known object recognition challenges [2]. These marginal performances allowed deep learning methods to become highly recognized also in the field of medical image processing. In previous research, by obtaining record performances, application of the deep learning methods to the most complex medical tumor detection tasks proved to be very effective [1]. The main advantage of CNNs is that, due to their very deep computational layers, they learn highly representative complex features directly from the input images given to them. Oppositely, in traditional automatic classification applications, features representing the differences in tissue classes need to be extracted by hand using the aforementioned image processing techniques. Extracting highly representative features from the input images has the most powerful effect on the performance of computerized tumor detection applications. However, handcrafting these features requires high skill and knowledge. It is also very time-consuming, involving most of the work. Since CNNs automatically learn these complex representative features, the burden of feature handcrafting is eliminated and the performance of the classification is greatly enhanced.

Despite its clear improvements over traditional methods, implementing deep learning techniques also have some hassles. Training a deep convolutional neural network requires very large labeled training image dataset for improving performance by increasing the number of convolution layers. Also, increasing the network depth increases the computational cost due to an increase in the complex calculations carried out in the deep convolutional layers. In this regard, designing and training deep convolutional neural networks require powerful GPU powered computers.

An implementation of deep learning, namely transfer learning [3] presents an effective solution for this problem. In conditions where there are scarce training image data, not enough machine learning expertise and limited computational hardware available, researchers can use transfer learning as an efficient deep learning application. Basically, transfer learning means that, a pre-trained deep learning system is imported to be used as an efficient feature extractor for the desired application in question [4]. To be more explanative, a deep learning framework, such as a convolutional neural network, that is previously trained on a large annotated general image dataset where it has obtained high performance can be imported for a medical imaging application like lung cancer classification. This imported pre-trained CNN can be used as an automatic feature extractor for extracting highly representative features from the lung cancer images. Automatically extracted features are then delivered as an input into a more conventional, computationally more cost effective and easier to implement classifier. Then, this classifier carries out the final classification between normal healthy and cancerous lung tissues.

Implementation of such transfer learning technique to detect lung lesions in CT images is feasible for developing a robust and efficient lung lesion CAD system to provide automatic diagnosis assistance to the clinicians.

## 2   Method

In this part of the paper, methods for the development of a deep learning based computer aided detection (CAD) system for automatically detecting lung lesions are presented in detail.

The proposed system is a slight modification of the previous published work [4], which was a successful implementation of a transfer learning method for another medical task. The system can be considered in four main parts. First, a medical lung image dataset, in which some image slices contain lung lesions, is pulled from a public database. Then, that dataset is adjusted for the next step, which is feature extraction using transferred deep learning. After the features representing the lung image slices are extracted automatically, those features are used as an input into a conventional artificial neural network (ANN) based classifier for classifying the image slices into one of the output classes; not bearing or bearing lung lesion.

### 2.1   Lung Lesion Image Data

In this paper a CT based lung lesion image dataset is used for the development of the proposed CAD system. The Public Lung Database (PLD) [5] is selected for the proposed system. PLD dataset contains total of distinct 93 cases, where in each case there are varied number of CT image slices, ranging from 39 to 275 slices. Figure 1 shows some example CT image slices from the dataset.

From the whole dataset; 1,174 lung lesion bearing image slices are combined with 6,442 non-bearing slices to form the final training dataset of total 7,616 lung CT image slices. On the other hand, 782 lung lesion bearing and 4,294 non-bearing image slices are combined to form the final testing dataset that contains total of 5,076 lung CT images. Later, both datasets are randomly shuffled in themselves. In the end of this

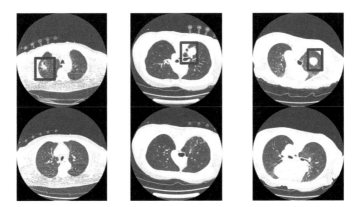

**Fig. 1.** Example CT slices from PLD dataset. Lung lesions are marked at lesion location.

dataset formation step, obtained training and testing datasets are ready for the next step which is feature extraction using transferred deep learning.

## 2.2 Transfer Deep Learning for Automatic Feature Extraction

After the lung CT images from the PLD dataset are selected, they need to be first prepared to be in line with the AlexNet [2] which is the chosen pre-trained CNN based deep learning framework for carrying out the automatic feature extraction for the proposed CAD system. AlexNet was trained on more than 1 million high resolution RGB images of the generic ImageNet dataset [6] and classifies those images into thousand different image classes with a very high performance. AlexNet's CNN architecture contains five convolutional layers and three fully-connected layers. Since those layers of the AlexNet are trained on a very large labeled dataset, connection weights of those layers can be generic enough to be easily transferred and implemented even for a complex medical image classification task. Accordingly, given the lung CT images from the PLD image dataset as inputs, outputs of the deepest AlexNet layers can be obtained as very representative features that provide valuable information regarding the presence of a lung lesion. Otherwise, without an automatic deep feature extractor like AlexNet, extracting similarly representative features from the complicated lung CT images using traditional feature extraction/image processing methods would be very difficult, adversely affecting the overall performance of the CAD system.

Images from the PLD image dataset are in DICOM format. First, those image slices are converted into $512 \times 512$ resolution.png format using a DICOM convertor. After the conversion, all images are adjusted into $227 \times 227$ resolution as accepted by the AlexNet. Another issue is that, AlexNet was designed for processing on RGB images. In this regard first convolutional layer of the network computes on a three dimensional input. To overcome this issue, our $227 \times 227$ sized lung CT images are reproduced three times for each image slice to imitate an RGB image. After the lung CT images of the PLD database are converted into $227 \times 227 \times 3$ format, they are fed as inputs into the transferred AlexNet. Features that are extracted automatically by the AlexNet from its 6$^{th}$ layer, which is the first fully connected layer (will be referred as FC6) and the 7$^{th}$ layer which is the second fully connected layer (will be referred as FC7) are used and compared. As a result, since both FC6 and FC7 layers contain 4096 neurons each, AlexNet extracts 4096 features from the every input lung CT image slice. These features are combined for all input image slices eventually forming a 4096 dimensional feature vector, i.e. since 4096 feature values are extracted from each input image slice, each input slice (or observation in general) is represented by 4096 variables/dimensions.

### 2.3    Lung Lesion Classification

Detection of lung lesions is carried out by classifying each lung CT image slice into one of the two output classes; not bearing lung lesion or bearing lung lesion. To carry out such classification, conventional and simple artificial neural networks with an architecture containing one input layer, one hidden layer and one output layer are designed and tested. All these ANNs are trained with scaled conjugate gradient back-propagation algorithm.

As explained in the previous section, transferred AlexNet method extracts 4096 features from each input image slice. In this regard, the dimension of the feature vector is 4096. This number is clearly too much for the simple architectures of our conventional ANNs which increases computational load heavily. Also it can cause over-fitting the training data. To solve this problem principal component analysis (PCA) is applied to select and use only the significant features as an input to the classifiers which in turn reduces the dimension of the feature vector, thus reducing computational cost and risk of over-fitting. By investigating the PCA output, we decided to use only the first 200 dimensions of the PCA transformed feature vector for the classification step. As a result feature vector dimension is reduced dramatically from 4096 to 200. Thus, PCA transformed feature matrix for training becomes 7616 × 200 sized and feature matrix for testing becomes 5076 × 200 sized.

After the dimension reduction is completed, a separate ANN is designed and trained one for FC6 and one for FC7 features. These two networks, namely N_FC6 and N_FC7, take their input feature vectors from the proposed automatic deep learning hierarchical feature extraction methods FC6 and FC7 respectively. Since output dimensions of all these feature extraction methods are reduced to 200 dimensions, number of input neurons in the input layer of all these networks is 200. Several different hidden layer neuron numbers are experimented and best performing numbers are chosen for each network. Output neuron number is two for both networks. When both output neuron activates (i.e. outputs 1) together it indicates a lesion bearing lung image. When none of them activate it indicates a non-lesion bearing image, in other words healthy lung. If only one neuron activates we consider that case as a un-classification. Output vectors for training of each network are same and prepared according to the input image class distribution provided by the ground truth information of the PLD dataset. Testing output vectors are also prepared similarly to compare the results of the testing and evaluate the detection performance of the classifier.

Training performances are recorded after the trainings of both networks are completed. After the training both networks are tested with the testing images. Again testing performance of each network is recorded for the final evaluation.

## 3   Results and Discussion

In this paper the lung CT image data from the PLD dataset is obtained, prepared and later fed into two different feature extraction methods. Dimensions of the features extracted from these methods are reduced to prevent over fitting by applying PCA. These features are then fed to their corresponding designed ANNs and trained to carry out classification of lung CT images into one of the possible classes; bearing or non-bearing a lesion.

After the training of the networks, they are tested using the testing dataset and the proposed deep learning based AlexNet feature extraction method (with its two variations) is evaluated. Performances of the methods are evaluated using three main metrics namely; Sensitivity (Sn), Accuracy (Ac) and Specificity (Sp).

$$Sn = \frac{T_p}{T_p + F_n + UN_p} \tag{1}$$

$$Sp = \frac{T_n}{T_n + F_p + UN_n} \tag{2}$$

$$Ac = \frac{T_p + T_n}{T_p + T_n + F_p + F_n + UN} \tag{3}$$

where $T_p$ gives the obtained true positive count, thus the correctly classified lung lesion bearing slices, $T_n$ gives the obtained true negative count, and thus the number of non-bearing slices that are correctly classified by the method, $F_p$ gives the obtained false positive count, thus non-bearing slices that are classified incorrectly as bearing slices. $F_n$ is the obtained false negative count, thus lung lesion bearing slices that are classified incorrectly as non-bearing and finally $UN$ is the total number of un-classifications, thus the slices that the classifier cannot assign a class label (where $UN_p$ and $UN_n$ are the un-classified lesion bearing and non-bearing image slices respectively). Basically, sensitivity provides performance measure regarding the detection of the lung lesion bearing images. On the other hand, specificity gives a performance measure regarding the detection of the lung images that do not contain lung lesions. Lastly, Accuracy provides performance measure regarding the detection of the both cases. Together, they give an indication regarding the overall performance of the designed lung lesion CAD system. Experiments using each different method are repeated several times. Best results obtained during the performed experiments regarding each different tested method are reported. Table 1 shows the classification counts that each method obtained after testing with the 5076 lung CT image slices of the testing dataset. Further, Table 2 shows the obtained sensitivity, specificity and accuracy measures for each method, which are calculated using the count values in Table 1.

**Table 1.** Comparison of the obtained $T_p$, $T_n$, $F_p$, $F_n$ and $UN$ counts for each method in their best performing runs.

| Network | $T_p$ | $F_n$ | $T_n$ | $F_p$ | UN | | Total test images |
|---------|-------|-------|-------|-------|-----|-----|------------------|
| | | | | | $UN_p$ | $UN_n$ | |
| N_FC6 | 765 | 12 | 4115 | 113 | 5 | 0 | 5076 |
| N_FC7 | 772 | 5 | 4223 | 71 | 5 | 4 | 5076 |

**Table 2.** Comparison of the results obtained after testing the proposed methods. Values are given as percentages.

| Network | Sn | Sp | Ac |
|---------|-------|-------|-------|
| N_FC6 | 97.83 | 95.83 | 96.14 |
| N_FC7 | 98.72 | 98.35 | 98.48 |

When the obtained results are investigated; it can be easily observed that the proposed methods based on transfer deep learning automatic feature extraction performed very well for the given lung lesion detection task, especially in terms of sensitivity. Among the two deep-learning based methods, N_FC7 which takes the automatic image features extracted by the 7th fully connected layer of the AlexNet performed better than the N_FC6. This proves the idea that the deeper layers of a deep learning framework trained on a huge and generic database (like the CNN based AlexNet) can be robust enough to be easily transferred and implemented even for a very complex medical imaging task, like lung lesion detection.

As the best performing method, N_FC7 miss-classified total of 5 lung lesion bearing images out of total 782 lung lesion bearing test images. Further it un-classified 5 more. Un-classifications are more tolerable since they are not presented to the radiologist as non-bearing results. So radiologist can re-check all un-classified images (which are very few in number) and give their own decisions. Additionally, un-classifications are based on our chosen network design and these can be corrected with little modifications on the design. In the end, N_FC7 outperformed the other method in all categories. With its very high performances of 98.72% sensitivity, 98.35% specificity and 98.48% accuracy, it proved to be an efficient method for developing a CAD system for assisting the radiologist in lung lesion detection tasks.

## 4 Conclusions

In current medical procedures, manually detecting lung lesions is a very time consuming process and its performance is very dependent on the performing radiologist. Thus assisting the radiologists by developing a CAD system is an interesting research goal. In this regard, we developed a transfer deep learning (AlexNet) based CAD system and implemented it for the automatic detection of the lung lesions.

First, lung CT images are obtained from the public PLD database. Then, proposed transfer deep learning based method is designed to extract representative features from the lung CT slices. After the feature extraction step is completed, high dimensions of the resulting feature vectors are reduced by applying the PCA method. As the final step, conventional ANN based classifiers are designed for each corresponding feature extraction method. These ANN classifiers are trained and tested using the images obtained from the PLD database. Obtained results are recorded and evaluated with respect to the metrics of accuracy, sensitivity and specificity.

The proposed AlexNet based automatic feature extraction method obtained final performances of 98.72% sensitivity, 98.35% specificity and 98.48% accuracy. Based on the obtained results, we can easily say that, an efficient CAD system which is based on an easily implementable state-of-the-art transfer deep learning automatic feature extraction method is successfully developed for the complex task of detection of lung lesions from CT images, which can provide invaluable help to the radiologists saving time and cost for the medical institutions.

Furthermore, since deep learning based methods are recently obtaining very high performances in most of the medical imaging tasks we can easily say that biomedical researchers are approaching towards developing an effective commercial CAD system. This gives us the ambition to develop deep learning based methods to solve other popular research problems in the wide field of biomedical imaging [7, 8].

# References

1. Işın, A., Direkoğlu, C., Şah, M.: Review of MRI-based brain tumor image segmentation using deep learning methods. Procedia Comput. Sci. **102**, 317–324 (2016)
2. Krizhevsky, A., Sutskever, I., Hinton, G.E.: Imagenet classification with deep convolutional neural networks. In: Advances in Neural Information Processing Systems, pp. 1097–1105 (2016)
3. Tajbakhsh, N., et al.: Convolutional neural networks for medical image analysis: Full training or fine tuning? IEEE Trans. Med. Imaging **35**(5), 1299–1312 (2016)
4. Işın, A., Ozdalili, S.: Cardiac arrhythmia detection using deep learning. Procedia Comput. Sci. **120**, 268–275 (2017)
5. Reeves, A.P., et al.: A public image database to support research in computer aided diagnosis. In: Annual International Conference of the IEEE Engineering in Medicine and Biology Society, EMBC 2009, pp. 3715–3718. IEEE, September 2009
6. Deng, J., Dong, W., Socher, R., Li, L.J., Li, K., Fei-Fei, L.: Imagenet: a large-scale hierarchical image database. In: IEEE Conference on Computer Vision and Pattern Recognition, CVPR 2009, pp. 248–255. IEEE, June 2009
7. Wang, G.: A perspective on deep learning. IEEE Access **4**, 8914–8924 (2016)
8. Işın, A., Ozsahin, D.U., Dutta, J., Haddani, S., El-Fakhri, G.: Monte carlo simulation of PET/MR scanner and assessment of motion correction strategies. J. Instrum. **12**(03), C03089 (2017)

# A Mobile Application to Teach Coding and Computational Thinking in Scratch Programming Language

M. A. Salahli[1]([⊠]), T. Gasimzadeh[2], F. Alasgarova[3], and A. Guliyev[3]

[1] Department of Computer and Instructional Technologies Education,
Çanakkale Onsekiz Mart University, Çanakkale, Turkey
msalahli@comu.edu.tr
[2] Department of Instrument Making Engineering,
Azerbaijan State Oil and Industry University, 20 Azadlig Avenue,
AZ 1010 Baku, Azerbaijan
tgasimzade@silkwaywest.com
[3] Department of Information Technology and Natural Sciences,
Azerbaijan Tourism and Management University, Koroglu Rehimov Street,
822/23, AZ 1072 Baku, Azerbaijan
flora.aleskerova@gmail.com, akber_guliyev@yahoo.com

**Abstract.** In this work, a mobile application developed on the Scratch programming language is described. Purpose of the application is to develop programming skills and computational thinking abilities of the students from secondary schools. The mobile application has been developed to solve two problems related to teaching Scratch programming language. The first problem related to educational aspects of the Scratch programming commonly used in secondary schools. The second problem is that Scratch does not have mobile applications for its learning.

The most important component of the mobile application - the activities are described in detail. The activities are designed to teach the Scratch programming language on the one hand and to develop the algorithmic, computational, and logical thinking skills of the students on the other. How to improve students' computational thinking skills by using activities is explained in an example.

**Keywords:** Scratch environment · Coding teaching · Mobile learning Computational thinking

## 1 Introduction

The role of the mobile application in programming education and the results of using this application are given in detail in (Salahli et al. 2017). In this submission, the activities component of the mobile application will be explained in more detail. In particular, the emphasis will be placed on the importance of activities in the development of students' computational skills of students. One of the important issues in education is the improvement of problem solving skills of students (Papadopoulos and Tegos 2012; Gülbahar 2014). The study of programming languages in secondary

R. A. Aliev et al. (Eds.): ICAFS-2018, AISC 896, pp. 807–814, 2019.
https://doi.org/10.1007/978-3-030-04164-9_106

schools plays an important role in the development of problem solving skills. So as programming education enhances students' high level skills in problem solving, critical thinking, logical thinking, and creative thinking (Siegle 2009; Calder 2010; Fesakis et al. 2009). In this sense, students who take the programming education at the required level and understand the programming logic can also succeed in other courses.

Research shows that, students often do not know exactly how to solve real world problems and how to develop the solution algorithms for these problems. In this context, by teaching the students how to solve such problems on the computer, in other words, by teaching them computational thinking, it is possible to solve the problem to a certain extent.

Various languages are used in programming education in secondary schools. Among of these languages Scratch is one of the most common. Scratch programming facilitates higher order thinking such as problem solving skills. Scratch offers a number of opportunities for students to create their algorithms and coding skills at a certain level.

On the other hand, considering the attention of students to mobile applications, widening of Scratch learning to mobile environment can give a new impetus to the programming learning.

This submittion describes the study to solve the above-mentioned problems related to Scratch learning.

In this work, a mobile application developed on the Scratch programming language is described. Purpose of the application is to develop programming skills and computational thinking abilities of the students from secondary schools.

## 2   Literature Review

Scratch is a programming language developed by MIT Media Lab to teach programming for children. Scratch offers an easier and more enjoyable environment for students. It also encourages users in advanced code writing. There are studies analyzing the impact of Scratch learning in the development of problem-solving skills of students. (Kob Siripat 2015) studied the creativity of students using Scratch. Author observed that Scratch positively influence the creativity of the students. Scratch using help students to develop their collaboration skills, logical thinking, and algorithmic thinking (Taylor et al. 2010). (Gülbahar and Kalelioğlu 2014; Shin and Park 2014) have investigated the influence of Scratch programming on problem solving skills of students. The results of these studies show that Scratch programming contributes positively to students' problem solving skills.

Nam et al. (2010) have studied the influence of scratch programming to programming achievement and problem solving skills of students. Sixteen students from the 6th grade participated in the study. The students were divided into experimental and control groups. The students in the experimental group were taught Scratch programming language. The students in the control group taught programming course were taught with classical methods. The results indicated that the students in the experimental group had a positive change in their programming and problem solving skills.

Wang et al. (2014) have investigated students' problem solving skills, their attitudes toward learning and their motivation by conducting a project-based study integrated with Scratch. A total of 91 students from middle schools participated in the research. Experimental results have led to significant advances in the success of the proposed learning approach.

Chen and Chung (2008) have developed a mobile learning system for English courses. Experimental results show that the proposed system positively affects the language learning performance of students in teaching English vocabulary.

Fernández-López et al. (2013) developed a mobile learning software aimed at improving the academic skills of special education students. The results show that the proposed has a positive effect on improving the basic skills of students with special needs.

Marcelino et al. (2018) and Shodiev (2015) describe the development of computer learning skills for students in the Scratch programming language.

## 3   The Mobile Learning Application

The Mobile Learning Application consists of 6 functional modules (Fig. 1).

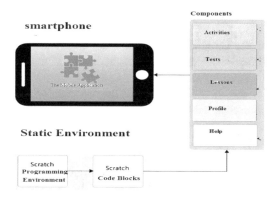

**Fig. 1.** Components of the mobile learning application

*User Login Module.* The purpose of the module is to record and tracking the activities of the students.

*Help Module.* The help module consists of 2 blocks. The information related to the Scratch software is saved in the first block. The other block is a guideline for the mobile learning application. The module allows easy use of the application without any external support.

*Student Profile Module.* This module processes and saves the statistics about of students' activities. All the activities data are kept in the database. When the student is entered to the system, he/she is informed about his/her programming skill level, and progress on the activities.

*Lessons Module.* Lessons to be taught divided into 8 blocks. These blocks contain totally 112 lesson units. The student can choose the lesson unit appropriate to their knowledge level and learning style.

*Test Module.* Test module has been developed so that students can evaluate their programming knowledge themselves. The component has easy, medium and difficult level tests. There are 10 questions in each level. After the test questions are answered, feedback on the test results is given. The feedback includes information such as the number of correct and incorrect answers, the success level of the student. Feedbacks provide information on the number of the correct and incorrect answers, performance levels, as well as recommendations on how to improve knowledge on test questions. A record of feedback is kept for each student. Students can access these records when they enter the module.

*Activities Module.* We will discuss the activities component in detail.

In this module, students are asked to use their Scratch programming skills to perform certain activities with specific goals. There are a total of 20 Scratch programming activities in the module. Events are listed from easy to difficult. This provides step-by-step learning of programming logic. In every activity, students have to reach a certain goal by using programming skills. Tips for reaching the goal are offered to students.

The contents of the activities are designed in accordance with the determined achievements. Activities were created using code blocks of the Scratch programming language. Students are given specific problems for solving each activity. These problems are presented as targets in the activities.

Drag-and-drop is used to perform activities. If the activity is performed correctly, students can go to next activities. In the case of the wrong activities, students are informed about the mistakes they have made, and what they need to perform the activity correctly.

The activities were created within specific acquisitions. Classification of the activities according to the achievements is given in Table 2. In the table, the number of the activity belonging to that acquisition is displayed against the achievements.

For more effective and efficient learning, certain teaching principles and methods have been used in the creation of activities. First, the creation of activities is based on the principle of simplicity to complexity. For example, in activity 1, students are taught only the step-by-step movement of the character in the scene. In activity 2, students learn how the character can move in the right direction and then return. There is a logical connection between all the activities. The use of this principle influences positively the student's self-confidence and his/her attitude towards programming. Teaching strategy through exploration has also been used in the creation of activities. Using their own cognitive skills, students solve the problems in the activities.

The contents of the activities are designed in accordance with the appropriate gains. Activities were created using code blocks of the Scratch programming language. Students are given specific problems for solving each activity. These problems are presented to the students as the targets of their activities. The targets and how they should be done, are presented on the screen for each activity.

When the student becomes active in terms of goals and actions, a three-part screen appears on the mobile phone (Fig. 2). In the first part, the visual state of the problem to be solved and the things to be done for the solution are expressed. The second part is where the Scratch programming language code blocks are used to solve the partition problem. The last section is the encoding section. The coding section specifies how many code blocks the student should use. The student must place the specified number of code blocks in the correct fields using the drag and drop method.

**Fig. 2.** Three – part screen for the activities component

After inserting the code blocks, the students can test whether the algorithm is correct by pressing the "Run" button. If there is an error in the algorithm for the solution of the problem, the application analyzes the error and the student provides clues to correct the errors. After reading the hints, students need to press the button "Let's try again" to create the correct algorithm.

When the students create the correct algorithm, they press the "Run" button again and the screen shows that they have successfully completed the activity (Fig. 3).

**Fig. 3.** The screen showing that the activity has been successfully completed

The development of computational thinking skills is considered one of the main objectives of the activities. Computational Thinking (CT) is a problem solving process. By learning of computational thinking, a student can better understand the relationship between real life concepts and events that the problem covers. As is known, the first step in the computational thinking process is the decomposition of the problem. The development of problem decomposition skills of students has been considered in some activities. Namely for this purpose, the activities, which oriented to solving problems by combining different code blocks (activities 9, 10, 14, 15, 19, 20), and to defining the sequence of algorithms for solving complex real world problems (activities 11, 12, 13, 14, 16, 17) have been created.

Let's continue our discussion on an example. We have been given a table that contains the names of students and their average scores. A list containing the names of the students who score "5" (highest score) should be making. In other words, we need to get the Table 2 from the data from the Table 1.

**Table 1.** Classification of activities according to coding learning achievements

| Student achievements | Activity numbers |
| --- | --- |
| The student finds the missing parts in the algorithm and ensures that it works correctly | 1, 2 |
| Student can use simple code blocks to solve problems | 3, 4 |
| Student can create algorithms that include loop and condition constructs. | 5, 6, 7, 10, 18, 20 |
| Student can use variables in problem solving | 8, 10, 12, 14, 20 |
| Student can solve problems by combining different code blocks | 9, 10, 14, 15, 19, 20 |
| Student can define the set of algorithms and their sequence for solving complex problems | 11, 12, 13, 14, 16, 17 |

In the first step we have to create Table 1. Does Scratch have the ability to create tables? Unfortunately, there is no option "create table". But the "Make list" there is able. At this stage, students are able to grasp, compare and abstract the patterns. Students are looking for answers to the question "How can we express a two-dimensional table by using list data type. The solution exists: fragmentation; the table will be converted to two lists. In this way, we are breaking down the problem into the following sub-problems:

1. Make the "Name" list
2. Make the "Scores" list
3. Compare the values in the "Score" list with the value "5"; In case of equality, the appropriate name from the "name" list is added to the "best_students" list (Fig. 4).

For the solution of the third sub- problem, the students have to know the conditions and loop operations.

**Table 2.** Student names and scores.

The names of the best students

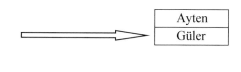

| Aytent | 5 |
|--------|---|
| Çiçek  | 4 |
| Barış  | 3 |
| Güler  | 5 |
| Fatma  | 4 |

| Ayten |
|-------|
| Güler |

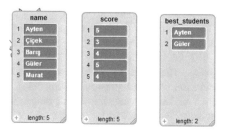

**Fig. 4.** "Name, score and best_students lists

The Scratch code fragment for solving the problem is shown in Fig. 5.

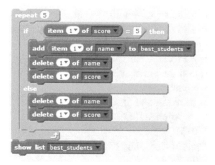

**Fig. 5.** The Scratch code fragment

## 4   Conclusion

In this study one mobile learning application for Scratch programming learning is discussed. The application was developed to support the programming education in the secondary schools. The influence of this application on the development of programming skills of students has been studied. Activities involving real-world problems has been created for the development of computational thinking skills of students.. The research results show that, proposed application positively contributes to the development of programming and computational thinking skills of students.

# References

Calder, N.: Using Scratch: an integrated problem-solving approach to mathematical thinking. Aust. Primary Math. Classroom **15**(4), 9–14 (2010)

Chen, G.D., Chang, C.K.: ve Wang, C.Y.: Ubiquitous learning website: scaffold learners by mobile devices with information-aware techniques. Comput. Educ. **50**(1), 77–90 (2008)

Fernandez-Lopez, A., Rodriguez-Fortiz, M.J., Rodriguez-Almendros, M.L., Martinez-Segura, M. J.: Mobile learning technology based on iOS devices to support students with special education needs. Comput. Educ. **61**, 77–90 (2013)

Fesakis, G., ve Serafeim, K.: Influence of the familiarization with Scratch on future teachers' opinions and attitudes about programming and ICT in education. In: ACM SIGCSE Bulletin, vol. 41, no. 3, pp. 258–262. ACM (2009)

Gülbahar, Y., ve Kalelioğlu, F.: The effects of teaching programming via Scratch on problem solving skills: a discussion from learners' perspective. Inf. Educ. **13**(1), 33–50 (2014)

Kob Siripat, W.: Effects of the media to promote the Scratch programming capabilities creativity of elementary school students. Procedia-Soc. Behav. Sci. **174**, 227–232 (2015)

Marcelino, M.J., et al.: Learning computational thinking and Scratch at distance. Comput. Hum. Behav. **80**, 470–477 (2018)

Nam, D., Kim, Y., ve Lee, T.: The effects of scaffolding-based courseware for the Scratch programming learning on student problem solving skill. In: Proceedings of the 18th International Conference on Computers in Education, Putrajaya, Malaysia. Asia-Pacific Society for Computers in Education (2010)

Papadopoulos, Y., Tegos, S.: Using microwords to introduce programming to services. In: Proceeding PCI 2012: Proceeding of the 2012 16th Panhellenic Conference on Informatics, Piraeus, Greece, pp. 180–185 (2012)

Salahli, M.A., et al.: One mobile application for the development of programming skills of secondary school students. Procedia Comput. Sci. **120**(2017), 502–508 (2017)

Shin, S., ve Park, P.: A study on the effect affecting problem solving ability of primary students through the Scratch programming. Adv. Sci. Technol. Lett. **59**, 117–120 (2014)

Shodiev, H.: Computational thinking and simulation in teaching science and mathematics. In: Interdisciplinary Topics in Applied Mathematics, Modeling and Computational Science, pp. 405–410. Springer (2015)

Siegle, D.: Developing student programming and problem-solving skills with visual basic. Gift. Child Today **32**(4), 24–29 (2009)

Taylor, M., Harlow, A., ve Forret, M.: Using a computer programming environment and an interactive whiteboard to investigate some mathematical thinking. Procedia-Soc. Behav. Sci. **8**, 561–570 (2010)

Wang, H.Y., Huang, I., ve Hwang, G.J.: Effects of an integrated Scratch and project-based learning approach on the learning achievements of gifted students in computer courses. In: 2014 IIAI 3rd International Conference on Advanced Applied Informatics (IIAIAAI), pp. 382–387 (2014)

# Study of the Uncertainty Heterogeneous Phase Equilibria Areas in the Binary YbTe-SnTe Alloy System

A. N. Mammadov[1], Z. S. Aliev[2,3(✉)], and M. B. Babanly[1]

[1] M. Nagiyev Institute of Catalysis and Inorganic Chemistry of ANAS,
Baku, Azerbaijan
asif.mammadov.47@mail.ru, babanlymb@gmail.com
[2] Azerbaijan State Oil and Industrial University, Baku, Azerbaijan
ziyasaliev@gmail.com
[3] Materials Science and Nanotechnology Department, Near East University,
99138 Mersin 10, Nicosia, North Cyprus, Turkey

**Abstract.** Using the Multipurpose Genetic Algorithm (MGA) the liquidus and solidus regions, binodal and spinodal boundaries of the solid solutions in the system YbTe-SnTe are optimized. The analysis correctly shows that the liquidus and solidus regions in the YbTe-SnTe system cannot be described within the framework of the ideal model of solutions. The use of functions, including the partial molar excess free Gibbs energies of all components, allowed for adequately taking into account deviations from the ideal model of the properties of equilibrium liquid and solid solutions. The analytical relationships between the variables and the phase transition parameters allowed us to estimate the sensitivity of the input data. It is established that the coordinates of the liquidus and solidus curves are insensitive to the melting enthalpy. At the same time, a high sensitivity of the liquidus and solidus coordinates to melting points of the YbTe and SnTe was observed. It was wound that, when melting point changes 10 K, the uncertainty areas for the liquidus and solidus vary in a wide range.

**Keywords:** Multipurpose genetic algorithm, uncertainty area, binary alloys system · Ytterbium telluride · Tin telluride

## 1 Introduction

Group IV metals binary chalcogenides and more complex thermodynamically stable and metastable alloys based on them are promising materials for the development of the devices such as high-sensitive IR detectors, thermoelectric converters [1, 2]. In the recent years, Group IV and V metals ternary chalcogenides, in particular, tellurides attract a lot attention as a novel class of functional materials called Topological Insulators (TIs) [3]. TIs are insulators in the bulk, but show metallic conduction at the surface due to the existence of well-defined topological surface states [4, 5]. The search and design of new phases such as TIs, requires plotting and revising of the phase diagrams using new approaches, in particular, through the principles of fuzzy logic. The search for the optimal model by selecting thermodynamic data for the

© Springer Nature Switzerland AG 2019
R. A. Aliev et al. (Eds.): ICAFS-2018, AISC 896, pp. 815–822, 2019.
https://doi.org/10.1007/978-3-030-04164-9_107

determination of phase equilibria is quite complex problem in materials science thanks to uncertainty of the input data. In the recent years, numerous research works have been developed on the optimization of the phase equilibria data using the Multipurpose Genetic Algorithm (MGA) in order to find more accurately limited and uncertain experimental thermodynamic data for the thermodynamic models of interest [6–8].

The subject of this work is the determination of the uncertainty heterogeneous phase equilibria areas in the YbTe-SnTe system. Information on the phase equilibria of this system is available in Refs. [9, 10]. The YbTe–SnTe system was found to have continuous series of cubic high-temperature solid solutions ($\alpha$), which undergo binodal phase decomposition at 950 K to $\alpha_1$- and $\alpha_2$-solid solutions based on the SnTe and YbTe, respectively. The integral thermodynamic functions of these solid solutions were calculated experimentally from electromotive force measurements (EMF) [11].

From this brief review it seems that the experimental phase equilibria and thermodynamic data for this system are very limited e.g., the thermal analysis for the investigated samples was carried out only up to 1300 K, whereas YbTe-rich alloys melt in within the temperature range from 1400 up to 2000 K [12]. The standard formation thermodynamic functions for the $Sn_xYb_{1-x}Te$ solid solutions were only determined in the temperature range from 298 K to 450 K [11].

Thus, the aim of this paper is to determine the uncertainty areas for the liquidus and solidus curves in the YbTe-SnTe system.

## 2  Thermodynamic Model

In this study, the binary system YbTe(1)-SnTe(2) have been accepted to have liquid and solid solutions [9]. From the condition that the chemical potentials of each component in the liquid and solid phases, which are in heterogeneous equilibrium at constant temperature and pressure, are equal, we can write [13]:

$$\mu_i^l(x_i^l) = \mu_i^s(x_i^s), \text{ where i} = 1, 2 \tag{1}$$

Here, $l$ and $s$ denote the liquid and solid phases, $x_i^l$ and $x_i^s$ are molar fractions in equilibrium liquid and solid solutions. The chemical potentials are given as:

$$\mu_i^l = \mu_i^{l,0} + RTlna_i^l, \ \mu_i^s = \mu_i^{s,0} + RTlna_i^s \tag{2}$$

From here we can write

$$\mu_i^{l,0} + RTlna_i^l = \mu_i^{s,0} + RTlna_i^s \tag{3}$$

and

$$\mu_i^{l,0} + RTlnx_i^l\gamma_i^l = \mu_i^{s,0} + RTlnx_i^s\gamma_i^s \tag{4}$$

In (2) and (3), $a_i^l$, $a_i^s$, $\gamma_i^l$ and $\gamma_i^s$ are thermodynamic activities and activity coefficients of the component $i$ in the liquid solution with respect to the pure liquid substance and in the solid solution with respect to the pure solid substance, respectively. From (4), we obtain:

$$\ln\left(\frac{x_i^l \gamma_i^l}{x_i^s \gamma_i^s}\right) = -\frac{\Delta\mu_i^{0(s\to l)}}{RT} \tag{5}$$

Here, $\Delta\mu_i^{0(s\to l)}$ is the change of chemical potential (Gibbs energy) on fusion of pure $i$. If the heat capacity of the liquid and the solid phases are assumed to be equal, the melting enthalpy does not depend on temperature. Then, we get:

$$\Delta\mu_i^{0(s\to l)} = \Delta G_i^{0,m} = \Delta H_i^m - T\Delta S_i^m \tag{6}$$

Here, $\Delta H_i^m$, $\Delta S_i^m = \Delta H_i^m / T_i^m$ are melting enthalpy and entropy, $T_i^m$ – is the melting temperature.

From Eqs. (5) and (6), we obtain:

$$\ln\left(\frac{x_i^l \gamma_i^l}{x_i^s \gamma_i^s}\right) = -\frac{\Delta H_i^m}{R}\left(\frac{1}{T} - \frac{1}{T_i^m}\right) = \frac{\Delta H_i^m}{R}\left(\frac{1}{T_i^m} - \frac{1}{T}\right) \tag{7}$$

If the liquid and solid solutions are ideal ($\gamma_i^l = \gamma_i^s = 1$) or pseudo-ideal ($\gamma_i^l/\gamma_i^s = 1$), from Eq. (7) for substances 1 and 2 we can get to:

$$x_1^l = x_1^s \exp\left(\frac{\Delta H_1^m}{R}\left(\frac{1}{T_1^m} - \frac{1}{T}\right)\right), \qquad x_2^l = x_2^s \exp\left(\frac{\Delta H_2^m}{R}\left(\frac{1}{T_2^m} - \frac{1}{T}\right)\right) \tag{8}$$

where, $R$ is the gas constant (8.314 J/mol K).

Equation (8) was used in [7] to elaboration the temperature and melting enthalpy of the basic components and to determine the liquidus and solidus regions in the UO2–PuO2 and UO2–BeO systems by using method of heuristic optimization and Bayesian statistics. The activity coefficients are measure of deviation of the solutions properties from the ideal state. For non-ideal solutions, $\gamma_i^l \neq 1$ and $\gamma_i^s \neq 1$. Based on their values, the partial molar excess free Gibbs energies are determined by using following equations:

$$\Delta\overline{G}_i^{exs,l} = RT\ln\gamma_i^l, \quad \Delta\overline{G}_i^{exs,s} = RT\ln\gamma_i^s \tag{9}$$

Substituting (9) into (7), we obtain an equation for describing the liquidus and solidus systems, taking into account the deviation of the properties of liquid and solid solutions from the ideal model:

$$ln\left(\frac{x_i^l}{x_i^s}\right) = -\frac{\Delta H_i^m}{R}\left(\frac{1}{T} - \frac{1}{T_i^m}\right) - \Delta\overline{G}_i^{exs,l/s}/RT \tag{10}$$

Here

$$\Delta\overline{G}_i^{exs,l/s} = \left(\Delta\overline{G}_i^{exs,l} - \Delta\overline{G}_i^{exs,s}\right) \tag{11}$$

Denoting

$$F_i(T) = \frac{\Delta H_i^m}{R}\left(\frac{1}{T} - \frac{1}{T_i^m}\right) + \Delta\overline{G}_i^{exs,l/s}/RT \tag{12}$$

from (10), we obtain equations for approximating the concentration of the liquidus and solidus mole fractions for each fixed temperature:

$$x_2^l = \frac{1 - \exp(F_1(T))}{\exp(F_2(T)) - \exp(F_1(T))} \tag{13}$$

$$x_2^s = x_2^l \cdot \exp(F_2) \tag{14}$$

Here $x_1^l = 1 - x_2^l;\quad x_1^s = 1 - x_2^s$

## 3    Methodology for Approximation of the Liquidus and Solidus of the System YbTe(1)-SnTe(2)

The values of the input parameters for the YbTe and SnTe are given in Table 1. These data contain a small uncertainty interval to check the sensitivity to the input data.

**Table 1.** The upper and lower limits of the variable search space [14, 15]

| Variable | Units | Lower value | Upper value |
|---|---|---|---|
| $\Delta H_{YbTe}^m$ | J/mol | 35600 | 36100 |
| $\Delta H_{SnTe}^m$ | J/mol | 45200 | 46000 |
| $T_{YbTe}^m$ | K | 1993 | 2003 |
| $\Delta T_{SnTe}^m$ | K | 1070 | 1080 |

The optimization of the phase diagram was carried out through fol-low scheme: initially, the search range for each parameter is defined for MGA on the basis of the obtained uncertainty or change in published parameter values. Next, the MGA changes the variable values depending on how well the values generate the solidus and liquidus curves, which correspond to the present experimental data, taking into account the uncertainties of the experimental solidus and liquidus curves.

For the YbTe-SnTe system the experimental data of the liquidus and solidus are available up to 1300 K. Therefore, in this stage, the coordinates of the liquidus and solidus curves were determined with the help of Eq. (8) of the ideal solution model (Fig. 1, curves 1, 2). Further, based on the initial experimental data available for a temperature interval of 1000–1300 K, the values of the partial excess free energies YbTe and SnTe were determined. These values were approximated in the form of functions:

$$\Delta \overline{G}_i^{exs,l/s}/RT = a + bT + cT^2 \tag{15}$$

In particular

$$\Delta \overline{G}_{YbTe}^{exs,l/s}/RT = -3,87867 + 0,00313T - 5,94727 \cdot 10^{-7}T^2 \tag{16}$$

$$\Delta \overline{G}_{SnTe}^{exs,l/s}/RT = -5,03255 + 0,00601T - 1,124474 \cdot 10^{-6}T^2 \tag{17}$$

The dependency parameters (16, 17) are optimized using the OriginLab computer program. Substituting (16, 17) in (12), and the function $F_i(T)$ in (13, 14), we can verify the sensitivity of the liquidus and solidus lines to the input data. For this study the initial uncertainty in concentration was accepted to be 0.01 and the uncertainty in the liquidus and solidus temperatures was about 10 K. Once the range of parameter values is optimized, values are sampled from this range in order to determine values ranges in the forward calculated solidus and liquidus curves. These results in fuzzy or uncertain bands those define the most probable positions of the curves given all data and the underlying model. These bands are displayed using error bars in Fig. 1. The uncertainty interval for the enthalpy of melting of tellurides of ytterbium and tin was about 1% of the reference value of these quantities.

## 4 Results and Discussion

The analysis shows that the liquidus and solidus regions in the YbTe-SnTe system cannot be described within the framework of the ideal model of solutions. Liquidus lines defined within the ideal or quasi ideal model (see Fig. 1, curves 1 and 2) are far away, especially the solidus (curve 2), from the heterogeneity equilibrium regions of the liquid solution-solid solution. The use of functions (12), including the partial molar excess free Gibbs energies for all components, allowed for taking into account deviations from the ideal model of the properties of equilibrium liquid and solid solutions. The analytical relationship between the variables and the parameters of the phase transition made it possible to estimate the sensitivity to the input data. It was found that the coordinates of the liquidus and solidus are insensitive to the enthalpy of melting (see Fig. 1, curves 4, 5 and 7, 8). At the same time, a high sensitivity of the liquidus and solidus coordinates to the values of the melting point of YbTe and SnTe was observed. Thus, a change in the melting point by 10 K leads to a wide band of uncertainty of the liquidus and solidus.

In the YbTe(1)-SnTe(2) system, below 950 K the α-solid solutions decompose into two α1 and α2-solid solutions [9, 10]. In the temperature range from 300 K to 400 K, the two-phase region is located 0.34–0.36 to 0.94–0.96 mol% YbTe. In this paper, MGA described in [6] was used to optimize the binodal curve (see Fig. 1). The optimization conditions were the following relations: $(x_{YbTe}^{\alpha_1}\gamma_{YbTe}^{\alpha_1} - x_{YbTe}^{\alpha_2}\gamma_{YbTe}^{\alpha_2})/\partial T \leq 0$, at a critical temperature $x_{YbTe}^{\alpha_1} = x_{YbTe}^{\alpha_2}$.

According to [11], for the part of the YbTe-rich binodal, the activity coefficient $\gamma_{YbTe}^{\alpha_1}$ increases from unity by decreasing the mole fraction of YbTe, and the $\gamma_{YbTe}^{\alpha_2}$ activity coefficient decreases from 2.6 for the part of the binodal with the increase in the fraction of YbTe. At a critical temperature, $\gamma_{YbTe}^{\alpha_1} = \gamma_{YbTe}^{\alpha_2}$. To determine the coordinates of the spinodal, we used the stability function $(\partial 2\Delta G0/\partial x2)$ P, T > 0. The analytical expression for the integral free mixing energy $\Delta G0$ of solid solutions of YbTe-SnTe is taken from work [16].

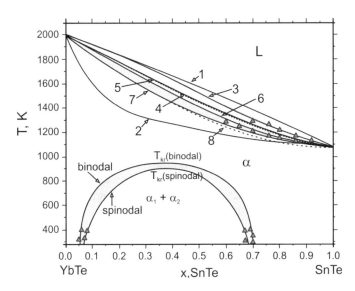

**Fig. 1.** Areas of uncertainty for the liquidus and solidus curves of YbTe-SnTe system. Experimental data taken from [9, 11]. 1, liquidus in the ideal approximation; 2, solidus in the ideal approximation; 3, liquidus taking into account the deviation from the ideal model ($\Delta H_{YbTe}^m = 35600, \Delta H_{SnTe}^m = 45200 \frac{J}{mol}; T_{YbTe}^m = 2003, T_{SnTe}^m = 1080\ K$); 4, Liquidus taking into account the deviation from the ideal model ($\Delta H_{YbTe}^m = 35600, \Delta H_{SnTe}^m = 45200 \frac{J}{mol}; T_{YbTe}^m = 1993, T_{SnTe}^m = 1070\ K$); 5, Liquidus (dashed) taking into account the deviation from the ideal model ($\Delta H_{YbTe}^m = 36100, \Delta H_{SnTe}^m = 4600 \frac{J}{mol}; T_{YbTe}^m = 1993, T_{SnTe}^m = 1070\ K$); 6, Solidus taking into account the deviation from the ideal model ($\Delta H_{YbTe}^m = 35600, \Delta H_{SnTe}^m = 45200 \frac{J}{mol}; T_{YbTe}^m = 2003, T_{SnTe}^m = 1080\ K$); 7, Solidus taking into account the deviation from the ideal model ($\Delta H_{YbTe}^m = 35600, \Delta H_{SnTe}^m = 45200 \frac{J}{mol}; T_{YbTe}^m = 1993, T_{SnTe}^m = 1070\ K$); 8, Solidus (dashed) taking into account the deviation from the ideal model ($\Delta H_{YbTe}^m = 36100, \Delta H_{SnTe}^m = 4600 \frac{J}{mol}; T_{YbTe}^m = 1993, T_{SnTe}^m = 1070\ K$)

# 5  Conclusions

The accurate analysis shows that the liquidus and solidus regions in the YbTe-SnTe system cannot be adequately described within the framework of the ideal model of solutions. The use of functions, including the partial molar excess free Gibbs energies of all components, allowed for taking into account deviations from the ideal model of the properties of equilibrium liquid and solid solutions. The analytical relationship between the variables and the parameters of the phase transition made it possible to estimate the sensitivity to the input data. It was found that the coordinates of the liquidus and solidus are insensitive to the standard enthalpy of melting. At the same time, a high sensitivity of the liquidus and solidus coordinates to the values of the melting point of YbTe and SnTe was observed. Thus, a change in the melting point by 10 K leads to a wide band of uncertainty of the liquidus and solidus. The Multipurpose Genetic Algorithm allowed optimizing the binodal and spinodal boundaries of solid solutions in the system YbTe-SnTe.

**Acknowledgments.** This work was performed in the frame of a scientific program of the international laboratory between the Institute of Catalysis and Inorganic Chemistry of the National Academy of Sciences of Azerbaijan (Azerbaijan) and Centro de Física de Materiales at Donostia (Spain).

# References

1. Kanatzidis, M.G.: The role of solid state chemistry in the discovery of new thermoelectric materials. Semicond. Semimet. **69**, 51–98 (2001)
2. Shevelkov, A.V.: Chemical aspects of thermoelectric materials engineering. Rus. Chem. Rev. **77**, 1–19 (2008)
3. Babanly, M.B., Chulkov, E.V., Aliev, Z.S., Shevelkov, A.V., Amiraslanov, I.R.: Phase diagrams in materials science of topological insulators based on metal chalcogenides. Russ. J. Inorg. Chem. **62**(13), 1703–1730 (2017)
4. Eremeev, S.V., Landolt, G., Menshchikova, T.V., Slomski, B., Koroteev, Y.M., Aliev, Z.S., Babanly, M.B., Henk, J., Ernst, A., Patthey, L., Eich, A., Khajetoorians, A.A., Hagemeister, J., Pietzsch, O., Wiebe, J., Wiesendanger, R., Echenique, P.M., Tsirkin, S.S., Amiraslanov, I. R., Dil, J.H., Chulkov, E.V.: Atom-specifik spin mapping and buried topological states in a homologous series of topolifical insulators. Nat. Commun. **3**, 635 (2012)
5. Franz, M.: Topological insulators: starting a new family. Nat. Mater. **9**, 536–537 (2010)
6. Preuss, M., Wessing, S., Rudolph, G., Sadowski. G.: Solving phase equilibrium problems by means of avoidance-based multiobjectivization. In: Springer Handbook of Computational Intelligence, pp. 1159–1169 (2015). Part E.58, Evol. Comput.
7. Stan, M., Reardon, B.J.: A Bayesian approach to evaluating the uncertainty of thermodynamic data and phase diagrams. CALPHAD **27**, 319–323 (2003)
8. Duong, T.C., Hackenberg, R.E., Landa, A., Honarmandi, A., Talapatra, A.: Revisiting thermodynamics and kinetic diffusivities of uranium–niobium with Bayesian uncertainty analysis. CALPHAD **55**, 219–230 (2016)
9. Aliev, Z.S., Ibadova, G.I., Tedenac, J.C., Babanly, M.B.: Study of the YbTe–SnTe–$Sb_2Te_3$ quasi-ternary system. J. Alloys Compd. **602**, 248–254 (2014)

10. Aliev, Z.S., Amiraslanov, I.R., Record, M.C., Tedenac, J.C., Babanly, M.B.: The YbTe-SnTe-Bi2Te3 system. J. Alloys Compd. **750**, 887–894 (2018)

11. Ibadova, G.I., Imamalieva, S.Z., Babanly, M.B.: Thermodynamic properties of solid solutions in the SnTe–YbTe system. Vestn. Bakinsk. Gos. Univ., Ser. Fiz.-Mat. Nauk 7–11 (2013)

12. Gamri, H., Djaballah, Y., Belgacem-Bouzida, A.: Thermodynamic modeling of the Eu-Te and Te-Ybsystems. J. Alloys Compd. **653**, 121–128 (2015)

13. Stolen, S., Grande, T.: Chemical Thermodynamics of Materials: Macroscopic and Microscopic Aspects. Wiley, Hoboken (2004)

14. Mamedov, A.N., Tagiev, E.R., Mashadiyeva, L.F., Babanly, M.B.: Thermodynamic calculation and 3D modeling of the liquidus surface of the YbTe–Sb$_2$Te$_3$– Bi$_2$Te$_3$ system. Int. Res. J. Pure and Appl. Chem. **10**(2), 1–5 (2016)

15. Iorish, V.S., Yungman, V.S.: BazadannykhTermicheskiekonstantyveshchestv [Database. - Thermal Constants of Substances] (2006, in Russ.). www.chem.msu.ru/cgi-bin/tkv.pl

16. Mamedov, A.N., Tagiev, E.R., Aliev, Z.S., Babanly, M.B.: Phase boundaries of the (YbTe)$_x$(PbTe)$_{1-x}$and (YbTe)$_x$(SnTe)$_{1-x}$solid solutions series. Russ. J. Inorg. Mater. **52**(6), 543–545 (2016)

# Decision Making in Investment Problem by Using Self-confidence Based Preference Relation

Aynur I. Jabbarova[✉]

Azerbaijan State Economic University, Baku, Azerbaijan
stat_aynur@mail.ru

**Abstract.** In this paper we use interval-valued preference relations with self-confidence for investment problem. For calculating priority vectors of this preference relations linear programming are used. We use TOPSIS method the same problem for check results first method.

**Keywords:** Linear programming · Self-confidence levels · Priority vector
TOPSIS method

## 1 Introduction

The priorities of real issues are indefinite. Different types of investigations have been conducted to determine this diversity in decision models. In general, these studies can be classified in the following areas. (1) incomplete priorities; (2) models of confidence; (3) fuzzy priority relations with partial trust.

Incomplete priority models can make obstacles to finding the right choices because of the lack of information and the complexity of alternatives. On the other hand, this indecision leads to failure of the group of decision-makers does not come to a conclusion. Formally, the alternatives $\mathcal{A}$, there exists $f, g \in \mathcal{A}$ such that neither $f \succeq g$ nor $g \succeq f$ is presumed [1]. For modelling of incomplete advantages, the vector advantages [1], indefinite probabilities [2–5], indefinite advantage and other approaches are offered [6, 7].

Fuzzy priority relations are used in cases where the decision-maker cannot prefer one of the alternatives due to complexity of alternatives, lack of knowledge and information and other factors. The advantages are of "distributed" nature to reflect that an alternative is better than the other one. Unlike the classical advantage relations, fuzzy priority relations (FPR) reflects that $\tilde{f}$ alternative is more advantageous than $\tilde{g}$ alternative in comparison of $\tilde{g}'$ alternative with $\tilde{f}'$ alternative.

In [8], new priority model is oferred. This model enables to define the priority degree given with the self-confidence level. The self-confidence level describes the confidence of the decision-maker in fuzzy priority. This approach is the best mean when it has the fuzzy priorities in [9–12] and indefinite priorities in [13–16].

R. A. Aliev et al. (Eds.): ICAFS-2018, AISC 896, pp. 823–829, 2019.
https://doi.org/10.1007/978-3-030-04164-9_108

## 2 Preliminaries

**Fuzzy Preferences with Self-confidence Level.** In real-world problems, a DM may not be completely sure in his preferences. In such cases, FPR is assigned by a self-confidence level described by a linguistic term form a predefined codebook. An FPR with self-confidence level proposed in [8] is described as follows.

**Definition 1 [8]. FPR with Self-confidence Level.** Let $R : \mathcal{A} \times \mathcal{A} \rightarrow T$ be a fuzzy preference relation with self-confidence based on a finite set of alternatives $\mathcal{A}$ shown as follows,

$$R = ((r_{ij}, s_{ij})) \tag{1}$$

where $r_{ij}$ denotes the degree or intensity of preference of alternative $\tilde{f}_i$ over alternative $\tilde{f}_j$, and $s_{ij}$ represents the self-confidence level on the preference value $r_{ij}$. It is assumed that $r_{ij} + r_{ji} = 1, s_{ij} = s_{ji}$ [8].

Consistency of an FPR with the self-confidence level is considered in terms of transitivity properties [8]. They consider weak stochastic transitivity, strong stochastic transitivity and additive transitivity at a confidence level $s$. These properties are considered as those of common FPR, but satisfied at some lowest possible self-confidence level.

The FPR with the self-confidence level [8] is a new step in development of a decision theory. It encompasses both a degree of preference and the related belief level. However, this approach is of two main shortcomings: the degree of preference is crisp and, what is more important, an essence of self-confidence level is not considered. However, a self-confidence level is naturally of a probabilistic character and may be considered as a fuzzy value of a probability measure of a fuzzy degree of preference. In this report, we propose a Z-valued preference relation as a more general preference model.

**Definition 2 [17]. Comparison of intervals.** The degree to which $[\underline{I}, \overline{I}]$ is higher than $[\underline{J}, \overline{J}]$ is defined as follows.

$$d(I, I) = \begin{cases} \dfrac{\overline{I} - \overline{J}}{(\overline{I} - \overline{J}) + (\underline{J} - \underline{I})}, & \overline{I} > \overline{J}, \quad \underline{J} \geq \underline{I} \\ 1, & \overline{I} = \overline{J}, \quad \underline{I} > \underline{J} \\ & or \quad \overline{I} > \overline{J}, \quad \underline{I} \geq \overline{J} \\ & or \quad \overline{I} = \overline{J}, \quad \underline{I} = \overline{J} \\ 1 - d(I, J), & otherwise \end{cases}$$

## 3   Statement of the Problem and a Solution Method

At first we applied self-confidence based preference relation method to our investment problem. A company is planning to make an investment in three sphere; A1-agriculture, A2-processing industry, A3-tourism sector/Each alternative is characterized by 3 criteria; C1-volume of income, C2-degree of risk, C3-enviromental impact.
The codebook for interval-valued level is given in Table 1.

**Table 1.**  The codebook for interval-valued confidence level

|              | Interval value |
| ------------ | -------------- |
| Medium       | $[0.4 \quad 0.6]$ |
| Medium high  | $[0.6 \quad 0.8]$ |
| High         | $[0.7 \quad 0.9]$ |
| Very high    | $[0.9 \quad 1]$ |

For calculating we comprised of intervals by using Definition.

$$d(VH, MH) = 1$$

$$d(VH, H) = 1$$

$$d(H, MH) = 1$$

Next, we offer $3 \times 3$ fuzzy preference relation with interval-valued self-confidence:

$$P = \begin{pmatrix} (0.5, VH) & (0.7, MH) & (0.9, H) \\ (0.3, MH) & (0.5, VH) & (0.7, H) \\ (0.1, H) & (0.3, H) & (0.5, VH) \end{pmatrix}.$$

We use the linear programming model for determine priority vector of $P$:
*Objective function*

$$\min z = z_{12} + z_{13} + z_{23}$$

subject to

$$\begin{cases} 0.5w_1 - 0.5w_2 - y_{12} = 0.2, \\ 0.5w_1 - 0.5w_3 - y_{13} = 0.4, \\ 0.5w_2 - 0.5w_3 - y_{23} = 0.2, \\ z_{12} - 2y_{12} \geq 0, \\ z_{12} + 2y_{12} \geq 0, \\ z_{13} - 3y_{13} \geq 0, \\ z_{13} + 3y_{13} \geq 0, \\ z_{23} - 3y_{23} \geq 0, \\ z_{23} + 3y_{23} \geq 0, \\ w_1 + w_2 + w_3 = 1, \\ w_i \geq 0, \quad i = 1,2,3 \\ z_{ij} \geq 0, \quad i,j = 1,2,3 \end{cases}$$

We solve this problem and find $z = 0.3$ and priority vector $w = (0.7, 0.3, 0)$. This results show that $1^{st}$ alternative is best alternative. Then we use TOPSIS method [17, 18] for solving this problem and compare with below method.

Importance weights of criteria: $w_1 = [0.4 - 0.5]$, $w_2 = [0.3 - 0.35]$, $w_3 = [0.15 - 0.3]$.

Decision matrix for investment problem is given in Table 2.

**Table 2.** Decision matrix

|  | $C_1$ | $C_2$ | $C_3$ |
|---|---|---|---|
| $A_1$ | 8 | 2 | 3 |
| $A_2$ | 6 | 5 | 4 |
| $A_3$ | 3 | 7 | 7 |

1. Calculate the normalized decision matrix by using following formula (Table 3):

**Table 3.** Normalized decision matrix

|  | $C_1$ | $C_2$ | $C_3$ |
|---|---|---|---|
| $A_1$ | 0.77 | 0.23 | 0.35 |
| $A_2$ | 0.57 | 0.57 | 0.47 |
| $A_3$ | 0.29 | 0.79 | 0.81 |

$$n_{ij} = \frac{c_{ij}}{\sqrt{\sum_{j=1}^{m} x_{ij}^2}}, \quad j = 1,\ldots,m, \quad i = 1,\ldots,n.$$

**Table 4.** Weighted normalized decision matrix

|       | $C_1$            | $C_2$             | $C_3$             |
|-------|------------------|-------------------|-------------------|
| $A_1$ | [0.308 0.385]    | [0.069 0.0805]    | [0.0525 0.105]    |
| $A_2$ | [0.228 0.285]    | [0.171 0.1995]    | [0.0705 0.141]    |
| $A_3$ | [0.116 0.145]    | [0.237 0.2765]    | [0.1215 0.243]    |

**Table 5.** Positive and negative ideal solutions

|                | $C_1$     | $C_2$      | $C_3$     |
|----------------|-----------|------------|-----------|
| Positive ideal | [0.385]   | [0.2765]   | [0.243]   |
| Negative ideal | [0.116]   | [0.069]    | [0.0525]  |

2. Calculate the weighted normalized decision matrix $r_{ij} = n_{ij} \cdot w_{i}$, where $j = 1,\ldots,m$, $i = 1,\ldots,n$ and $\sum\limits_{i=i}^{n} w_i = 1$ (Table 4).

3. Calculate the positive and the negative ideal solution by using following formula (Table 5):

$$A^+ = \{r_1^+,\ldots,r_n^+\} = \left\{ \left( \max_j r_{ij} \big| i \in I \right) \right\},$$

$$A^- = \{r_1^-,\ldots,r_n^-\} = \left\{ \left( \min_j r_{ij} \big| i \in I \right) \right\}.$$

4. Determine the separation measures, using n-dimensional Euclidean distance. The calculated separation of each alternative from the positive ideal solution by using following formula

$$d_j^+ = \left\{ \sum_{i=1}^{n} \overline{r}_{ij}^L - \overline{r}_i^- \right\}^{\frac{1}{2}}, \tag{3}$$

$A_1$   $(0.308 - 0.385)^2 + (0.069 - 0.2765)^2 + (0.0525 - 0.243)^2 = 0.077^2$
$+ 0.2075^2 + 0.1905^2 = 0.005929 + 0.043056 + 0.03629 = 0.085275;$

$A_2$   $(0.228 - 0.385)^2 + (0.171 - 0.2765)^2 + (0.0705 - 0.243)^2 =$
$0.157^2 + 0.1055^2 + 0.1725^2 = 0.024649 + 0.01113 + 0.029756 = 0.065535;$

$A_3$   $(0.116 - 0.385)^2(0.237 - 0.2765)^2 + (0.1215 - 0.243)^2 = 0.269^2 +$
$0.0395^2 + 0.1215^2 = 0.072361 + 0.00156 + 0.014762 = 0.088683;$

The calculated separation of each alternative from the negative ideal solution by following formula

$$d_j^- = \left\{ \sum_{i=1}^{n} \bar{r}_{ij}^U - \bar{r}_i^+ \right\}^{\frac{1}{2}}.$$  (4)

$A_1$  $(0.385 - 0.116)^2 + (0.0805 - 0.069)^2 + (0.105 - 0.0525)^2 =$
$0.269^2 + 0.0115^2 + 0.0525^2 = 0.072361 + 0.000132 + 0.002756 = 0.075249;$
$A_2$  $(0.285 - 0.116)^2 + (0.1995 - 0.069)^2 + (0.141 - 0.0525)^2 = 0.169^2$
$+ 0.1305^2 + 0.0885^2 = 0.028561 + 0.01703 + 0.007832 = 0.053423;$
$A_3$  $(0.145 - 0.116)^2 + (0.2765 - 0.069)^2 + (0.243 - 0.0525)^2 = 0.029^2$
$+ 0.2075^2 + + 0.1905^2 = 0.000841 + 0.043056 + 0.03629 = 0.080187$

5. The calculated the relative measures by using (5)

$$R_j = \frac{d_j^-}{(d_j^+ - d_j^-)}, j = 1, \ldots, m,$$  (5)

$$R_1 = \frac{0.075249}{0.075249 + 0.085275} = \frac{0.075249}{0.160524} \approx 0.469,$$

$$R_2 = \frac{0.053423}{0.065535 + 0.053423} = \frac{0.053423}{0.118958} \approx 0.45,$$

$$R_3 = \frac{0.080187}{0.088683 + 0.080187} = \frac{0.080187}{0.16887} \approx 0.475.$$

The ranking of relative measures the preference order are given in Table 6.

**Table 6.** The ranking of relative measure

| Alternatives | $R_j$ | Rank |
|---|---|---|
| $A_1$ | 0.469 | 3 |
| $A_2$ | 0.45 | 2 |
| $A_3$ | 0.475 | 1 |

The results represent that alternative $A_3$ is the best alternative. This result significantly differ from the result obtained by the self-confidence based preference relations. The reason is that information on DM's confidence level on assigned preference is disregarded. As one can see, this may lead to choice of a non-optimal alternative.

## 4 Conclusion

In this article, the issue of capital investment has been solved through a method based on interval-value fuzzy priority. This method is characterized by the self-confidence level that the decision maker has given to alternatives in advance. The issue has been solved through linear programming and has been assigned a priority vector. The issue has been solved through linear programming and has been assigned a priority vector.

Then this issue was solved by the TOPSIS algorithm and the best alternative was set. The results obtained through both methods have been analyzed and the results obtained by the first method have been shown to be more adequate.

# References

1. Ok, E.A.: Utility representation of an incomplete preference relation. J. Econ. Theory **104**, 429–449 (2002)
2. Nau, R.: The shape of incomplete preferences. Ann. Stat. **34**, 2430–2448 (2006)
3. Insua, D.R.: On the foundations of decision making under partial information. Theory Decis. **33**, 83–100 (1992)
4. Rigotti, R., Shannon, C.: Uncertainty and risk in financial markets. Econometrica **73**, 203–243 (2005)
5. Walley, P.: Statistical inferences based on a second-order possibility distribution. Int. J. Gen Syst **9**, 337–383 (1997)
6. Aliev, R.A.: Fuzzy Knowledge based Intelligent Robots. (Radio i Svyaz, Moscow) (in Russian) (1995)
7. Aliev, R.A., Huseynov, O.H.: A new approach to behavioral decision making with imperfect information. In: Proceedings of 6th International Conference on Soft Computing and, Computing with Words in System Analysis, Decision and Control, ICSCCW, pp. 227–237 (2011)
8. Liu, W., Dong, Y., Chiclana, F., Herrera-Viedma, E., Cabrerizo, F.J.: A new type of preference relations: fuzzy preference relations with self-confidence. In: IEEE International Conference on Fuzzy Systems (FUZZ-IEEE), pp. 1677–1684 (2016)
9. Blin, J.M., Whinston, A.B.: Fuzzy sets and social choice. J. Cybern. **3**(4), 17–22 (1973)
10. Dubois, D., Prade, H.: A review of fuzzy sets aggregation connectives. Inf. Sci. **36**, 85–121 (1985)
11. Herrera-Viedma, E., Herrera, F., Chiclana, F.: Multiperson decision-making based on multiplicative preference relations. Eur. J. Oper. Res. **129**(2), 372–385 (2001)
12. Herrera, F., Herrera-Viedma, E.: Aggregation operators for linguistic weighted information. IEEE Trans. Syst. Man Cybern. A **27**, 646–656 (1997)
13. De Baets, B., De Meyer, H.: Transitivity frameworks for reciprocal relations: cycle transitivity versus FG-transitivity. Fuzzy Sets Syst. **152**, 249–270 (2005)
14. Dubois, D.: The role of fuzzy sets in decision sciences: Old techniques and new directions. Fuzzy Sets Syst. **184**, 3–28 (2011)
15. McClellon, M.: Confidence models of incomplete preferences. Math. Soc. Sci. **83**, 30–34 (2016)
16. Chateauneuf, A., Faro, J.H.: On the confidence preferences model. Fuzzy Sets Syst. **188**(1), 1–15 (2012)
17. Aliev, R.A.: Uncertain Computation-Based Decision Theory, p. 540. World Scientific Publishing Co Pte Ltd. (2017)
18. Jahanshahloo, G.R., Hosseinzadeh Lotfi, F., Izadikhah, M.: An algorithmic method to extend TOPSIS for decision-making problems with interval data. Appl. Math. Comput. **175**, 1375–1384 (2006)

# Global Stability Analysis of HIV+ Model

Farouk Tijjani Saad[1], Tamer Sanlidag[2,4(✉)], Evren Hincal[1],
Murat Sayan[3,4], Isa Abdullahi Baba[1], and Bilgen Kaymakamzade[1]

[1] Department of Mathematics, Near East University, Nicosia, Cyprus
[2] Faculty of Medicine, Department of Medical Microbiology,
Celal Bayar University, Manisa, Turkey
tamer.sanlidag@neu.edu.tr
[3] Faculty of Medicine, Clinical Laboratory, PCR Unit,
Kocaeli University, Kocaeli, Turkey
[4] Research Center of Experimental Health Sciences,
Near East University, Nicosia, Cyprus

**Abstract.** We developed and studied a mathematical model of HIV+. Two equilibriums points were found, disease free and endemic equilibrium, and basic reproduction ratio $R_0$ was also calculated by the use of next generation matrix. Global stability analysis of the equilibria was carried out by the use of Lyapunov function, and it was shown that the stability of the equilibria depends on the magnitude of the basic reproduction ratio. When $R_0 < 1$, the disease free equilibrium is globally asymptotically stable, and disease dies out. On the other hand if $R_0 \geq 1$, the endemic equilibrium is globally asymptotically stable and epidemics occurs. Reported cases of 13646 HIV-1 positive were obtained in the year 2016 from Ministry of Health, Turkey (MOH). This data is used to present the numerical simulations, which supports the analytic result. $R_0$ was found to be 1.98998, which is bigger than 1, this shows the threat posed by HIV in Turkey.

**Keywords:** HIV · AIDS · Mathematical model · Global stability
Basic reproduction ratio · Turkey

## 1 Introduction

Historically, infectious diseases posed a real threat to the human population. Although they have been in human population all the time to some extent, the effects of epidemics are the most obvious and noticeable. Only in the 14th century Europe, around 25 million out of about 100 million individuals died from the Black Death [1]. Several diseases were discovered in America which included smallpox, measles, influenza, and typhus, whooping cough, the mumps, and diphtheria. Infectious disease was the main reason for the demise of the Indians [2]. In the early 1980's, some homosexual men in the United States were diagnosed with a type of fungal infection and a tumor called Candidiasis and Kaposi's sarcoma respectively. Paris Pasteur institute in 1984 detected a virus responsible for those diseases and called it the Human Immunodeficiency Virus (HIV) [3].

© Springer Nature Switzerland AG 2019
R. A. Aliev et al. (Eds.): ICAFS-2018, AISC 896, pp. 830–839, 2019.
https://doi.org/10.1007/978-3-030-04164-9_109

HIV is the virus that causes Acquired Immunodeficiency Syndrome (AIDS). The virus is responsible for attacking and destroying the immune systems mainly the CD4+ T-lymphocytes or T-cells [1]. On a normal basis, these CD4+ T-lymphocytes or T-cells detect foreign and infected cells, and attack, spread and kill them [1–3]. HIV is able to infect CD4+ T-lymphocytes and insert its genome to host genome. This integrated HIV genome may exist in 2 states. They can be either transcriptionally active generating new viruses that can infect other CD4+ T-lymphocytes or in latent state which may become activated later. In transcriptionally active stage, the infected CD4+ T-lymphocytes die due to cytopathic effect of the virus. As a result, the number of CD4+ T-lymphocytes, which are able to recognize foreign and infected cells, declines, and this decrement lead stoper manent and lasting damage to the immune system [3]. The immune system finally loses its ability to fight and kill infections due to the number of CD4+ T-lymphocytes count which is so small. When an individual reaches this stage, the person is said to have AIDS [4]. The time between getting infected with HIV and advancing to AIDS is, in general, five years but changes due to many factors. If the CD4+ T-lymphocytes cells count falls below $200\,\text{cells/mm}^3$, then the person is considered to be in the "AIDS phase", otherwise the person is said to be HIV infected [5]. The mode of transmission of HIV includes heterosexual intercourse, homosexual/bisexual intercourse, intravenous drug use, vertical transmission, and unknown reasons [6].

Since its discovery (HIV/AIDS), the extensive increase and epidemic continues around the globe. The greatest number in any one year was in 2003, where almost five million individuals became newly infected, with a total of 38 million HIV/AIDS patients, and almost three million people died from AIDS in the same year [7]. To continue its record of one of the most destructive epidemics in history, it killed at least 25 million people by 2005. Efforts to improve the use of antiretroviral treatment in some part of the world were still not enough to reduce a significant number of deaths, the HIV/AIDS epidemic claimed 3.1 million lives in 2005, of which about 570000 were children (UNAIDS/WHO [8]).

The region that suffered most is Africa, with Sub-Saharan Africa as the home of the epidemic. In 2010, 2.7 million people were newly infected, and 1.8 million patients died of AIDS-related causes across the globe. By the end of 2010, around 34 million individuals were victims of HIV/AIDS [9]. China stated that about 2.8 million died of AIDS-related causes in 2011, and there were about 7.8 million HIV-infected individuals by the end of 2011. In the year 1985, two patients were diagnosed with AIDS in Turkey, since then, the importance of AIDS started and still continue to be on the forefront. The number of new cases increases every year, with 34, 91, and 119 cases in 1990, 1995, and 1999 respectively [10].

In 2004, the Ministry of Health (MOH) published the total number of HIV patients 1802, of which 76% are between the ages of 15 to 49 years and sexually active. Among these patients, about 800 were AIDS patients. This is an official data from MOH, however, these numbers does not reflect the actual figures of HIV infected individual in Turkey, due to insufficiency of the registration system, and the lack awareness and phobia of the patients to attend health centers or hospitals [11].

At the end of 2011, the ministry stated that the total number of HIV/AIDS infected people was 5224. It also published in 2013 that at least 6000 individuals were infected with HIV, and the numbers of newly reported cases in 2010, 2011, and first six months of 2013 were 589, 1068, and 587 respectively. According to a report, poor knowledge of sexually transmitted diseases, poor socio-economic conditions, increase in number of unregistered sex workers, increase in number of homosexuals, and intravenous drug use contributed to the spread of HIV infection in Turkey. It was reported that, the main way of transmitting HIV in Turkey is via heterosexual sex (53%), then men having sex with men (MSM at 9%), and intravenous drug users (IDU at 3%) among the recorded cases. According to Positive Living Association Istanbul, personal communication, HIV/AIDS will become a major public health issue in Turkey in the coming years, as such; it must be regarded as a rising disease for Turkey [12].

The main ways in which HIV/AIDS is transmitted between individuals are now well understood, but the factors that contribute to the disparities in its prevalence and trends among populations remain an area of interest to scientific researchers. To understand these disparities, it is essential to understand the system, its components and its dynamics. Mathematical models of HIV/AIDS transmission dynamics are important research tool in this category [13, 14].

Primary purpose of any mathematical model of HIV transmission lies in using individual level inputs to project population level outcomes. Some of the important outcomes that can be examined with a model are; the incidence of infection, the prevalence of infection, or the doubling time of the epidemic. More important than these however, is simply the likelihood of an epidemic to occur that is whether there is sufficient transmission potential for a chain of infection to be sustained. This outcome is termed by a simple summary statistic: the reproduction number of the infectious process, $R_0$. In a susceptible population, $R_0$ represents the expected number of secondary infections generated by the first infected individual. If $R_0$ is equal or greater than 1 an epidemic is expected. If $R_0$ is less than 1, the infection is expected to die out [15].

The magnitude of $R_0$ is used to measure the risk of an epidemic or pandemic in any emerging infectious disease. It was used for understanding the outbreak and danger of SARS. $R_0$ was also used to characterize bovine spongiform encephalitis (BSE), foot and mouth disease (FMD), strains of influenza, and West Nile Virus [16–19]. The incidence and spread of dengue, Ebola, and scrapie have also been assessed by $R_0$ [20–22]. Tropical issues such as the risks of indoor airborne infection, bioterrorism, and computer viruses also depend on this important parameter [23–26].

In this paper, we shall first introduce the model involving systems of ODE, and then discuss the biological meaning of the parameters involved. We shall study the global stabilities of both disease free and endemic equilibria by the use of Lyapunov function. By the use of real data obtained from Turkey in 2016, we will conduct numerical simulations to support the analytic result.

The organization of the paper is as follows: In Sect. 2, the model is presented and the basic reproduction number is obtained. In Sect. 3, stability of the equilibria are investigated. In Sect. 4, results are obtained by numerical simulations of the real data obtained from Turkey in 2016. Finally Sect. 5 is the discussion and conclusion of the research.

## 2  Construction of the Model

The system of ordinary differential equations derived is for the whole Turkish population.

### 2.1  Susceptibles, S(t)

Consider the birth $\Lambda(t)$ to the susceptible population per unit time. Susceptibles individuals are removed through infection or through natural death. Let $\mu$ be the natural death rate for the whole population. The removal rate of susceptible individuals through infection is the number of new HIV infections per unit time. We use this rate in calculating HIV incidence which by definition is the number of new infected persons in a specified period of time divided by the number of uninfected persons that were exposed for this same time.

### 2.2  HIV Positives, H(t)

Let each susceptible have c contacts per unit time. Assume that a proportion H/N of these contacts are with infectives and at each of these contacts with infectives, a susceptible has a probability b of becoming infected. Let $\alpha$ the incidence rate, then the total probability of one susceptible getting HIV infected from any of their contacts per unit time is $\alpha H/N$. This is the expression for the force of infection. The force of infection is the probability that a susceptible will get an HIV infection per unit time. Therefore in a population of S susceptibles, the number of new HIV infections per unit time is given by $\alpha HS/N$. Infectives are recruited through new HIV infections described above and removed through death at rate v and through natural death at rate $\mu$. Hence, $1/v$ is the duration spent in the infective stage and $1/\mu$ is the life expectancy of the population. All these rates are assumed constant in the model.

### 2.3  Removed Phase, R(t)

Removed cases are recruited either through natural death $\mu$ or through deaths due to HIV at the rate v.

With these assumptions, we arrive at the following system of ODE.

$$\frac{dS}{dt} = \Lambda - \frac{\alpha SH}{N} - \mu S$$

$$\frac{dH}{dt} = \frac{\alpha SH}{N} - (v + \mu)H \tag{1}$$

$$\frac{dR}{dt} = (v + \mu)H + \mu S$$

$$S(t) > 0, \quad H(t) \geq 0, \quad and \quad R(t) \geq 0$$

Since $N = S + H + R$, then Eq. (1) can be reduced to

$$\frac{dS}{dt} = \Lambda - \frac{\alpha SH}{N} - \mu S$$

$$\frac{dH}{dt} = \frac{\alpha SH}{N} - (v + \mu)H \tag{2}$$

It follows from (1) that,

$$[S + H + R]' = \Lambda - \frac{\alpha SH}{N} - \mu S + \frac{\alpha SH}{N} - (v + \mu)H + \mu S + (v + \mu)H \leq \Lambda$$

Then

$$\lim_{t \to \infty} sup(S + H + R) \leq \Lambda$$

Thus the feasible region for (1) is

$$\pi = \{(S + H + R) : S + H + R \leq \Lambda, S(t) > 0, H(t) \geq 0, R(t) \geq 0\}$$

### 2.4    More Assumptions

For the simulation we make further assumptions as follows;

  (i)   S(0) is considered to be the whole population in 2016
  (ii)  We considered homogeneous mixing in the population
  (iii) $v = 0.01493$ is considered (That is averagely 67 years is the life span of HIV+ people)

### 2.5    Equilibrium Points

Equating the equations in (*) to zero and solving simultaneously we find two equilibrium points. Disease free and endemic equilibrium points.

$$E_0 = \left( S_0 = \frac{\Lambda}{\mu}, H_0 = 0 \right)$$

and

$$E_1 = \left( S_1 = \frac{v + \mu}{\alpha}, H_1 = \frac{\Lambda\alpha - \mu v - \mu^2}{\alpha(v + \mu)} \right)$$

## 2.6    Basic Reproduction Ratio

This is the number of secondary infections caused by a single infective individual in a completely susceptible population. It is denoted by $R_0$. Using the next generation of matrix (NGM) method we have,

$$F = [\alpha SH] \qquad\qquad V = [(v+\mu)H]$$
$$M = \partial F = [\alpha S] \qquad\qquad M(E_0) = [\alpha S_0]$$
$$K = \partial V = [v+\mu], \quad K^{-1} = \frac{1}{v+\mu}, \quad MK^{-1} = \frac{\alpha S_0}{v+\mu}.$$

The spectral radius, which is the dominant eigenvalue, is $\frac{\alpha S_0}{v+\mu}$

Hence the basic reproduction ration is;

$$R_0 = \frac{\alpha S_0}{v+\mu}$$

# 3    Global Stability Analysis of the Equilibria

In this section stability analysis of the two equilibrium points is obtained by the use of Lyapunov function. The conditions for the global stability of the equilibria in each case depends on the magnitude of the basic reproduction ratio $R_0$. Hence we have the following theorems and their proofs.

**Theorem 1:** The disease free equilibrium is globally asymptotically stable when $R_0 \leq 1$.

**Proof:** We construct the following Lyapunov function

$$V = (S - S_0 lnS) + H + C$$

Where $C = S_0 lnS_0 - S_0$

$$\dot{V} = \left(1 - \frac{S_0}{S}\right)\dot{S} + \dot{H}$$

$$= \left(1 - \frac{S_0}{S}\right)[\Lambda - \alpha SH - \mu S] + \alpha SH - (v+\mu)H$$

$$= \Lambda - \alpha SH - \mu S - \Lambda\frac{S_0}{S} + \alpha S_0 H + \mu S_0 + \alpha SH - (v+\mu)H$$

$$= \mu S_0\left(2 - \frac{S}{S_0} - \frac{S_0}{S}\right) - [(v+\mu) - \alpha S_0]H$$

$< 0$ by the relation between geometric and arithmetic means and if $\alpha S_0 \leq (v+\mu)$. This implies $\dot{V} \leq 0$ if $R_0 \leq 1$.

**Theorem 2:** The endemic equilibrium $E_1$ is globally asymptotically stable when $R_0 > 1$.

**Proof:** We construct the following Lyapunov function

$$V = (S - S_1 lnS) + (H - H_1 lnH) + C$$

Where

$$C = -[S_1 - S_1 lnS_1 + H_1 - H_1 lnH_1]$$

$$\dot{V} = \left(1 - \frac{S_1}{S}\right)\dot{S} + \left(1 - \frac{H_1}{H}\right)\dot{H}$$

$$= \left(1 - \frac{S_1}{S}\right)[\Lambda - \alpha SH - \mu S] + \left(1 - \frac{H_1}{H}\right)[\alpha SH - (v + \mu)H]$$

$$= \Lambda - \alpha SH - \mu S - \Lambda \frac{S_1}{S} + \alpha S_1 H + \mu S_1 + \alpha SH - (v + \mu)H - \alpha SH_1 + (v + \mu)H_1$$

$$= \mu S_1\left[2 - \frac{S}{S_1} - \frac{S_1}{S}\right] - \left[(v + \mu) - \alpha\left(\frac{v + \mu}{\alpha}\right)\right]H - [\alpha S - (v + \mu)]\left[\frac{\alpha \Lambda - \mu v - \mu^2}{\alpha(v + \mu)}\right]$$

$<0$ by the relation between arithmetic and geometric mean and if $\alpha \Lambda - \mu v - \mu^2 > 0$

This implies $\dot{V} \leq 0$ if $R_0 > 1$.

## 4   Results

In this section, results are calculated by simulating the model using the real data obtained from Turkey in 2016.

### 4.1   HIV in Turkey

Here we use the real data obtained from MOH, in which there were a total of 13646 HIV-1 positive reported cases in the year 2016, in the year 2016 to study and predict the dynamics of HIV in Turkey using our model. Table 1 presents the values of the parameters as calculated based on the data obtained.

#### 4.1.1   Equilibrium Points

$$E_0 = (s_0 = 2.15, h_0 = 0)$$
$$E_1 = (S_1 = 1.0804, H_1 = 1.0676)$$

**Table 1.** Model parameters as calculated from the data

| Parameters | Values |
|---|---|
| $s = \frac{S(t)}{N(t)}$ | 0.99992581 |
| $h = \frac{H(t)}{N(t)}$ | 0.00007419 |
| $\Lambda$ | 17.2 |
| $\mu$ | 8 |
| A | 7.4184 |
| V | 0.01493 |

### 4.1.2    Basic Reproduction Ratio

$$R_0 = \frac{\alpha S_0}{v + \mu} = 1.98998$$

Since $R_0 > 1$, the disease free equilibrium is not stable and the endemic equilibrium is stable. Hence, there is going to be epidemics.

### 4.1.3    Numerical Simulation
Simulating the above result, Fig. 1 shows the epidemic of HIV/AIDS in Turkey.

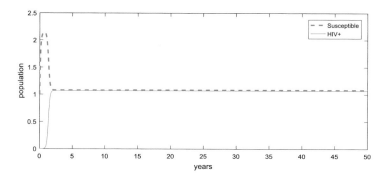

**Fig. 1.**  Dynamics of HIV/AIDS in Turkey

## 5    Discussion

A mathematical model for HIV+ is constructed and analyzed. Two equilibrium points (disease free and endemic) are found and stability of each of the equilibrium point was shown to depend on the magnitude of basic reproduction ratio, using Lyapunov function. It was shown that if $R_0 = \frac{\alpha S_0}{v + \mu}$ is less than one, the disease free equilibrium is globally asymptotically stable. Also, if the value is greater than or equals to one the endemic equilibrium is globally asymptotically stable.

The Turkish population in the year 2016 is 79814871, and the HIV positive population is 13646. The value of the basic reproduction ratio is 1.98998, which is bigger than one. This implies one HIV positive individual in Turkey can be able to transfer the disease to almost 2 individuals, hence there is going to be HIV epidemic in Turkey.

Numerical simulations were carried out and the results support the analytic findings. The results in the simulation shows that if appropriate measures are not taken, in 50 years to come, there will be more HIV positive individuals in Turkey as there are susceptible individuals.

The main limitation to our analysis may be that we did not account for the effect of behavioral change arising both from number of HIV cases in the community as well as awareness by governmental and nongovernmental organizations. Secondary, the possible effects of extensive use of antiretroviral drugs (ARVs) in terms of method of distributing drugs through public or private health institution or a combination of both could determine whether patients on ARVs revert back to the infective class. This together with reduced infectiousness due to lower viral loads for those on treatment was not accounted for.

Despite the limitations mentioned above, there are several implications of our findings to public health. First, the endemic equilibrium should be brought as low as possible especially during the first wave of the epidemic. This model suggests that this can be achieved by prolonging the lifetime of the HIV patients for as long as possible. Second, HIV prevalence at low prevalence levels become less sensitive to changes in the dynamics of HIV epidemic because it is overpowered by demographic changes especially the recruitment of susceptibles. At low prevalence levels, there is hence need to track trends in number of persons infected with HIV than tracking HIV prevalence.

# References

1. Anderson, R.M., Medley, G., May, F.R.M., Johnson, A.M.: A preliminary study of the transmission dynamics of the human immunodeficiency virus (HIV), the causative agent of AIDS. IMA J. Math. Appl. Med. Biol. **3**, 229–263 (1986)
2. Cai, L., Li, X., Ghosh, M., Guo, B.: Stability analysis of an HIV/Aids epidemic model with treatment. J. Comput. Appl. Math. **229**, 313–323 (2009)
3. Diekmann, O., Heesterbeek, J.A.P.: Mathematical Epidemiology of Infectious Diseases: Model Building, Analysis and Interpretation. Wiley, Hoboken (2000)
4. Al-Sheikh, S., Musali, F., Alsolami, M.: Stability analysis of an HIV/AIDS epidemic model with screening. Int. Math. Forum **6**(66), 3251–3273 (2011)
5. Li, Q., Cao, S., Chen, X., Sun, G., Liu, Y., Jia, Z.: Stability analysis of an HIV/AIDS dynamics model with drug resistance. Discrete Dyn. Nat. Soc. **2012**, 162527. https://doi.org/10.1155/2012/162527
6. Kockaya, G., Elbir, T.Z., Yenilmez, F.B., Dalgic, C., Malhan, S., Çerçi, P., Oksuz, E., Unal, S.: Analysis of the treatment costs of HIV/AIDS in Turkey. Farmeconomia Health Econ. Therapeutic Pathways **17**(1), 13–17 (2016). https://doi.org/10.7175/fe.V17i1.1219
7. Mukandavire, Z., Das, P., Chiyaka, C., Nyabadza, F.: Global analysis of an HIV/AIDS epidemic model. World J. Model. Siml. **6**(3), 231–240 (2010)

8. Sayan, M., Hıncal, E., Şanlıdağ, T., Kaymakamzade, B., Saad, F.T., Baba, İ.B.: Dynamics of HIV/AIDS in Turkey from 1985 to 2016. Qual. Quant. 1–13 (2017). https://doi.org/10.1007/s11135-017-0648-7

9. Ayranci, U.: Aids knowledge and attitudes in a Turkish population. An epidemiology study. BMC Public Health **5**, 95 (2005). https://doi.org/10.1186/1471-2458-5-95

10. Erbaydar, T., Eybaydar, N.P.: Status HIV/AIDS epidemic in Turkey. Acta Medica **1**, 19–24 (2012)

11. Alpsar, D., Agacfidan, A., Lübke, N., Verheyen, J., Eraksoy, H., Cagatay, A., Bozkaya, E., Kaiser, R., Akgül, B.: Molecular epidemiology of HIV in a cohort of men having sex with men from Istanbul. Med. Microbiol. Immunol. **202**, 251–255 (2013). https://doi.org/10.1007/500430-0120285-7

12. Cassels, S., Clark, S.J., Morris, M.: Mathematical models for HIV transmission dynamics: tools for social and behavioral science research. J. Acquir. Immune Defic. Syndr. **47**(1), 34–39 (2008)

13. Korenromp, E.L., Bakker, R., de Vlas, S.J., Gray, R.H., Wawer, M.J., Serwadda, D., Sewankambo, N.K., Habbema, J.D.: HIV dynamics and behaviour change as determinants of the impact of sexually transmitted disease treatment on HIV transmission in the context of the Rakai trial. AIDS **16**, 2209–2218 (2002)

14. White, R.G., Orroth., K.K., Korenromp, E.L., Bakker, R., Wambura, M., Sewankambo, N. K., Gray, R.H., Kamali, A., Whitworth, J.A., Grosskurth, H., Habbema, J.D., Hayes, R.J.: Can population differences explain the contrasting results of the Mwanza, Rakai, and Masaka HIV/sexually transmitted disease intervention trials: a modeling study. J. Acquir. Immune Defic. Syndr. **37**, 1500–1513 (2004)

15. Choi, B.C.K., Pak, A.W.P.: A simple approximate mathematical model to predict the number of severe acute respiratory syndrome cases and deaths. J. Epidemiol. Commun. **57**, 831–835 (2003)

16. Woolhouse, M.E.J., Anderson, R.M.: Understanding the epidemiology of BSE. Trends Microbiol. **5**, 421–424 (1997)

17. Ferguson, N.M., Donnelly, C.A., Anderson, R.M.: The foot – and – mouth epidemic in Great Britain: pattern of spread and impact of interventions. Science **292**, 1155–1160 (2001)

18. Mills, C.E., Robins, J.M., Lipsitch, M.: Transmissibility of 1918 pandemic influenza. Nature **432**, 904–906 (2004)

19. Wonham, M.J., de-Camino-Beek, T., Lewis, M.A.: An epidemiological model for West Nile virus: invasion analysis and control applications. Proc. R. Soc. B **271**, 501–507 (2004)

20. Luz, M.P., Cadeco, C.T., Massad, E., Struchiner, C.J.: Uncertainties regarding dengue modeling in Rio de Janeiro, Brazil. Mem. Inst. Oswaldo Cruz **98**, 871–878 (2003)

21. Hagmann, R., Charlwood, J.D., Gil, V., Conceicao, F., do-Rosario, V., Smith, T.A.: Malaria and its possible control on the Island of Principe. Malaria J. **2**, 15 (2003)

22. Chowell, G., Hengartner, N.W., Castillo-Chavez, C., Fenimore, P.W., Hyman, J.M.: The basic ratio number of Ebola and the effects of public health measures: the cases of Congo and Uganda. J. Theor. Biol. **229**, 119–126 (2004)

23. Gravenor, M.B., Papasozomenos, P., McLean, A.R., Neophytou, G.: A scrapie epidemic in cyprus. Epidemiol. Infect. **132**, 751–760 (2004)

24. Rudnick, S.N., Milton, D.K.: Risk of indoor airborne infection transmission estimated from carbon dioxide concentration. Indoor Air **13**, 237–245 (2003)

25. Kaplan, E.H., Craft, D.L., Wein, L.M.: Emergence response to small pox attack: the case for mass vaccination. Proc. Natl. Acad. Sci. **99**, 10935–10940 (2002)

26. Kaymakamzade, B., Şanlıdağ, T., Hıncal, E., Sayan, M., Saad, F.T., Baba, İ.B.: Role of awareness in controlling HIV/AIDS: a mathematical model. Qual. Quant. (2017). https://doi.org/10.1007/s11135-017-0640-2

# A New Compound Function-Based Z-number Valued Clustering

B. G. Guirimov[1] and O. H. Huseynov[2(✉)]

[1] SOCAR Midstream Operations, Baku, Azerbaijan
guirimov@hotmail.com
[2] Research Laboratory of Intelligent Control and Decision Making Systems
in Industry and Economics, Azerbaijan State Oil and Industry University,
Baku, Azerbaijan
oleg_huseynov@yahoo.com

**Abstract.** A large variety of hard and soft clustering methods exist including the deterministic, probabilistic and fuzzy clustering. However, these methods are devoted to handling different types of uncertainty. No works exist on clustering taking into account a confluence of probabilistic and fuzzy information. In such cases, reliability of extracted knowledge is one of the important issues to be studied. The concept of Z-number was introduced by Prof. Zadeh as a formal construct that express reliability of information under bimodal distribution. In this paper we suggest an approach to Z-number valued clustering of large data sets to describe reliability of data-driven knowledge. A numerical example is given that confirms validity of the proposed method.

**Keywords:** Data mining · Z-number · Clustering · Reliability

## 1 Introduction

Clustering is a well-known efficient data mining technique. A data set $X = \{x_1, x_2, \ldots, x_N\} \subset R^q$ is partitioned into C clusters according to data points similarity. Clustering is classified as hard clustering and soft clustering [1–3].

Hard clustering generates crisp sets of data points. In contrast, soft clustering separates data computing gradual membership of data points in clusters. In general, hard clustering may be viewed as a special case of soft clustering [4]. Two main types of soft clustering are probabilistic clustering [3, 5, 6] and fuzzy clustering [7, 8]. In probabilistic clustering, probability of data points in clusters is computed [3]. In fuzzy clustering, a membership degree of data point in a cluster is computed [3, 7]. The main probabilistic clustering methods include Expectation- Maximization method, probabilistic d-clustering, probabilistic exponential d-clustering etc.

Fuzzy clustering includes type-1 clustering [7, 9, 10], type-2 clustering [11–13], and intuitionistic clustering [14, 15].

Let us shortly overview a series of works devoted to application of both the fuzzy set theory-based and probabilistic approaches in clustering [4, 16–20]. [16] is devoted to application of fuzzy clustering and probabilistic clustering of very large (unloadable) data sets. A combination of fuzzy and probabilistic techniques is considered. In [4] they

© Springer Nature Switzerland AG 2019
R. A. Aliev et al. (Eds.): ICAFS-2018, AISC 896, pp. 840–847, 2019.
https://doi.org/10.1007/978-3-030-04164-9_110

uncover common features of fuzzy clustering and probabilistic clustering by using Renyi entropy measure. In [17] they consider fuzzy clustering of data which consist of normal probability distributions. The clustering is based on fuzzy C-means method with probabilistic distance structure. The advantage of this approach as opposed to classical FCM for the considered type of data is shown. [18] is devoted to a new approach for comparison of soft clustering methods. The approach is based on the use of the Earth Mover's distance and ordered weighted average. Systematic validation of proposed approach is provided. In [19] they conduct analysis of semi-supervised hard and soft clustering algorithms. [20] is devoted to integration of fuzzy, probabilistic and collaborative clustering [21] to partitioning of a mix of numeric and categorical data. For the purpose of a systematic view on the mixed data partition, the authors propose three algorithms. Two of them rely on the emphasis done solely to numeric and categorical data. The third one is applied to uncover the common features of structures of both types of data. A wide analysis of the approach is conducted.

There is a lack in general consideration of clustering of real-world data characterized by combination of fuzziness and probabilistic uncertainty. In such cases, reliability of clustering results becomes the one of key issues. However, reliability of information is not generally taken into account in all existing clustering methods. Reliability problems stem from a too small sample or uncertain, incomplete data. Indeed, real information in big data sets is characterized by partial reliability related to subjectivity of information sources etc.

Reliability of knowledge discovery by clustering is related mix of possibility and probability distributions referred to as bimodal distribution, e.g. [22–24]. The main representative of bimodal distributions is the Z-number concept. A Z-number, Z = (A, B). The first component, A, is a restriction on the values of a real-valued uncertain variable, X. The second component, B, is a measure of reliability of A.

The relationship between fuzziness and probability are rather collaborative than exclusive [2, 4]. Unfortunately, existing works do not deal with combination of fuzzy clustering and probabilistic clustering. The concept of Z-number joins probabilistic and fuzzy clustering to estimate reliability of knowledge discovery from data. In this paper the concept of Z-number is used for clustering of data with reliability measure.

The paper is structured as follows. Section 2 describes the preliminary information on bimodal distribution, Z-number, DEO etc. In Sect. 3 statement of the Z-clustering problem based on a new compound objective function. In Sect. 4 the solution method is given. In Sect. 5 a numerical example is presented. Section 6 is the conclusion.

## 2 Preliminaries

**Definition 1 Bimodal Distribution** [22]. Let X be a real-valued variable taking values in a finite set, U = {u1, …, un}. X can be associated with a possibility distribution, $\mu$, and a probability distribution, p, expressed as:

$$\mu = \mu_1/u_1 + \cdots + \mu_n/u_n,$$
$$p = p_1\backslash u_1 + \cdots + p_n\backslash u_n.$$

where $\mu_i/u_i$ implies that $\mu_i$, $i = 1, \ldots, n$, is the possibility that $X = u_i$. Similarly, $p_i\backslash u_i$ implies that $p_i$ is the probability that $X = u_i$.

The possibility distribution,$\mu$, may be combined with the probability distribution, p, through what is referred to as confluence:

$$\mu : p = (\mu_1, p_1)/u_1 + \cdots + (\mu_n, p_n)/u_n.$$

Distributions $\mu$ and p are compatible if their centroids are coincident, that is,

$$\sum_{i=1}^{n} u_i p_i = \frac{\sum_{i=1}^{n} u_i \mu_i}{\sum_{i=1}^{n} \mu_i}.$$

This condition is referred to as compatibility condition [22].

**Definition 2 A discrete Z-number** [23, 24]. A discrete Z-number is an ordered pair $Z = (A, B)$ where $A$ is a discrete fuzzy number playing a role of a fuzzy constraint on values that a random variable $X$ may take:

$$X \ is \ A,$$

and $B$ is a discrete fuzzy number with a membership function $\mu_B : \{b_1, \ldots, b_m\} \to [0, 1]$ $\{b_1, \ldots, b_m\} \subset [0, 1]$, playing the role of a fuzzy constraint on the probability measure of $A$:

$$P(A) \ is \ B.$$

**Definition 3 Differential Evolution Optimization.** Differential Evolution (DE) [25] is an effective numerical optimization method. In contrast to classical methods, it is a global search method not constrained by the properties of function to optimize. DE is of good convergence and flexibility to include different search strategies, constraints on variables, multiple-objectives, fuzzy processing and others. DE has been successfully applied to solve practical problems in image classification, clustering etc.

As a stochastic method, DE algorithm uses initial population randomly generated by uniform distribution, differential mutation, probability crossover, and selection operators. The main idea of the algorithm is to extract distance and direction of information for search directly from the population: in a standard strategy a new vector is generated by mutation which in DE means randomly selecting from the population 3 individuals– vectors $r_1 \neq r_2 \neq r_3$ and adding a weighted difference vector between two individuals to a third individual (population member). If the resulting vector yields a lower value of the objective function (assuming a lower value is a better one as is usually done in DE cost objective function) than a predetermined population member does, then in the following generation the newly generated vector will replace the vector with which it was compared.

## 3  Z-Clustering Using a New Compound Function

Assume data set $D = \{x_1, x_2, \ldots, x_N\} \subset R^q$ is given. Z-valued clustering problem is to partition D into C clusters described by Z-numbers: $Z^{c_j} = (A_j, B_j), j = 1, \ldots, C$. This implies that the membership of a data point $x$ in a cluster $Z^{c_j}$ is given as a membership degree and probability value. This is a multiobjective optimization problem [26]:

$$J = (J_m, J_p) = \left( \sum_{i=1}^{N} \sum_{j=1}^{C} U_{ij}^m \left\| x_i - c_j^m \right\|^2, \sum_{i=1}^{N} \sum_{j=1}^{C} (\left\| x_i - c_j^p \right\| p_j(x_i))^2 \right) \to \min \quad (1)$$

Where

$$U_{ij}^m = \frac{1}{\sum_{k=1}^{C} \left( \frac{\left\| x_i - c_j^m \right\|}{\left\| x_i - c_k^m \right\|} \right)^{\frac{2}{m-1}}} \quad (2)$$

$$c_j^m = \frac{\sum_{i=1}^{N} U_{ij}^m \cdot x_i}{\sum_{i=1}^{N} U_{ij}^m} \quad (3)$$

$$U_j(x_i) = \frac{(p_j(x_i))^2}{\left\| x_i - c_j^p \right\|}, \quad (4)$$

$$c_j^p = \sum_{i=1}^{N} \left( \frac{U_j(x_i)}{\sum_{i=1}^{N} U_j(x_i)} \right) \cdot x_i, \quad (5)$$

$$\sum_{i=1}^{N} p_j(x_i) = 1 \quad (6)$$

$$p_j(x_i) \geq 0, \quad (7)$$

$$\frac{\sum\limits_{i=1}^{N} U_{ij}^{m} \cdot x_i}{\sum\limits_{i=1}^{N} U_{ij}^{m}} = \sum\limits_{i=1}^{N} \left( \frac{U_j(x_i)}{\sum\limits_{i=1}^{N} U_j(x_i)} \right) \cdot x_i \qquad (8)$$

$c_j^m$- is a center of $j$-th fuzzy cluster, $c_j^p$- is a center of $j$-th probabilistic cluster, $U_{ij}^m$ – is a degree of membership of $i$-th data point to the $j$-th fuzzy cluster, $p_j(x_i)$ is a probability that $i$-th data point belongs to the $j$-th probabilistic cluster.

Objective function (1) is used to compound fuzzy clustering $J_m$ and probabilistic clustering $J_p$ criteria for construction of Z-clusters. Constraints (2) and (3) are typical constraints required to construct membership functions and to compute centers of fuzzy clusters. Constraint (5) is used to compute centers of probabilistic clusters. Constraint (8) is a compatibility condition used for closeness of fuzzy and probabilistic clusters. Solving of problem (1)–(8) provides membership degrees $U_{ij}^m(x_i)$ and probability distributions $p_j(x_i)$.

## 4   Solution Method

To solve complex problem (1)–(8), we propose to apply the DE Optimization approach. The reason is that gradient-based optimization algorithms (including the FCM) exhibit some disadvantages [13, 27]. In particular, they may not produce a global minimum but instead could get stuck in some local minimum. The other disadvantage is that standard FCM procedure becomes inefficient for values of m < 1.3 or m > 5 the FCM procedure often fails to find the minimum with required accuracy.

The DE optimization-based algorithm for solving problem (1)–(8) is as follows

1. Set a number of clusters $C$
2. Apply DE for minimization of objective function (1) within constraints (2)–(8). The solution to (1)–(8) provides:
   MF of each fuzzy cluster $U_j^m$, $j = 1, \ldots, C$;
   probability distribution for each probabilistic cluster $p_j$, $j = 1, \ldots, C$;
   fuzzy and probabilistic cluster centers $c_j^m$ and $c_j^p$.
3. Given the solution to (1)–(8) obtained at Step 1, consider a Z-valued cluster as a bimodal distribution $(U_j^m, p_j)$. Denote a fuzzy part of the bimodal distribution $A_j(x_i) = U_{ij}^m(x_i)$
4. Compute the reliability of $A_j$ as a value of probability measure:

$$P(A_j) = \sum\limits_{i=1}^{N} A_j(x_i) \cdot p_j(x_i) \qquad (9)$$

## 5 Numerical Example

Consider two-dimensional data (Fig. 1)

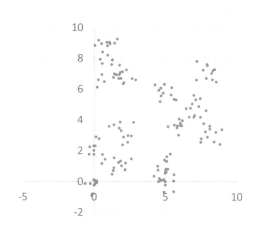

**Fig. 1.** Two-dimensional data.

Let us use the proposed approach to obtain Z-valued clusters for the data (fuzzifier $m = 2$). We have obtained 4 Z-valued clusters, each cluster is composed of membership function, $A$ and probability distributions (different distributions are obtained by relaxing constraint (8)). For each two-dimensional Z-clusters we have obtained two projections. For example, one of the projections is shown in Fig. 2($A$ part and probability distributions).

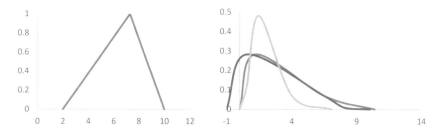

**Fig. 2.** Z-valued cluster: Membership function, A, and probability distributions, $p$

Thus, for each Z-valued cluster we obtained Z-numbers $Zi1 = (Ai1, Bi1)$ and $Zi2 = (Ai2, Bi2)$, $i = 1, 2$. Fuzzy numbers A1 and A2 are obtained as two projections, the related reliability parts B1 and B2 are computed by using (9). The obtained results for the considered cluster are (as triangular fuzzy numbers):

$$Z_{11} = ((2, 7.3, 10), (0.4, 0.7, 0.74)), \ Z_{12} = ((-0.9, 0.65, 4) \ (0.4, 0.7, 0.74))$$

Thus, the following if-then rule can be used to describe partial reliability in relationships within the considered data:

If $Z_{11} = ((2, 7.3, 10), (0.4, 0.7, 0.74))$ Then $Z_{12} = ((-0.9, 0.65, 4), (0.4, 0.7, 0.74))$

Analogously, We have obtained the Z-numbers for other clusters to form if-Then rules.

## 6   Conclusions

In this paper we propose an approach to Z-number valued clustering of large data set to account for reliability of knowledge discovery under bimodal information. The considered clustering problem formulation is based on a compound objective function that integrates fuzzy clustering and probabilistic clustering. The DE optimization-based method is used to solve the considered problem. A numerical example is given that confirms validity of the proposed method.

## References

1. Gosain, A., Dahiya, S.: Performance analysis of various fuzzy clustering algorithms: a review. Procedia Comput. Sci. Open Access J. **79**, 100–111 (2016)
2. Pedrycz, W.: Fuzzy equalization in the construction of fuzzy sets. Fuzzy Sets Syst. **119**(2), 329–335 (2001)
3. Borgelt, C.: Fuzzy and Probabilistic Clustering (2015). http://www.cost-ic0702.org/summercourse/files/clustering.pdf
4. Wang, S.T., Chung, K.F., Shen, H.B., Zhu, R.Q.: Note on the relationship between probabilistic and fuzzy clustering. Soft. Comput. **8**, 523–526 (2004)
5. Jain, A.K., Murty, M.N., Flynn, P.J.: Data clustering: a review. ACM Comput. Surv. (CSUR), Assoc. Comput. Mach. **31**(3), 264–323 (1999)
6. Wu, X., Kumar, V., Ross Quinlan, J.: Top 10 algorithms in data mining. Knowl. Inf. Syst. **14**(1), 1–37 (2008)
7. Bezdek, J.C.: Pattern Recognition with Fuzzy Objective Function Algorithm. Plenum, New York (1981)
8. Pal, N.R., Pal, K., Keller, J., Bezdek, J.C.: A possibilistic fuzzy c-means clustering algorithm. IEEE Trans. Fuzzy Syst. **13**(4), 517–530 (2005)
9. Kesavaraj, G., Sukumaran, S.: A study on classification techniques in data mining. In: 4th International Conference on Computing, Communications and Networking Technologies (ICCCNT), pp. 1–7. IEEE (2013)
10. Kaur, P., Soni, A.K., Gosain, A.: Robust kernelized approach to clustering by incorporating new distance measure. Eng. Appl. Artif. Intell. **26**(2), 833–847 (2013)
11. Patel, B.N., Prajapati, S.G., Lakhtaria, K.I.: Efficient classification of data using decision tree. Bonfring Int. J. Data Min. **2**(1), 6–12 (2012)

12. Kaur, P., Soni, A.K., Gosain, A.: Robust intuitionistic fuzzy c-means clustering for linearly and nonlinearly separable data. In: International Conference on Image Information Processing (ICIIP), pp. 1–6. IEEE (2011)
13. Aliev, R.A., Pedrycz, W., Guirimov, B.G., Aliev, R.R., Ilhan, U., Babagil, M., Mammadli, S.: Type-2 fuzzy neural networks with fuzzy clustering and differential evolution optimization. Inf. Sci. **181**, 1591–1608 (2011)
14. Chaira, T.: A novel intuitionistic fuzzy c means clustering algorithm and its application to medical images. Appl. Soft Comput. **11**, 1711–1717 (2011)
15. Zhang, H.M., Xu, Z.S., Chen, Q.: On clustering approach to intuitionistic fuzzy sets. Control Decis. **22**(8), 882–888 (2007)
16. Hathaway, R.J., Bezdek, J.C.: Extending fuzzy and probabilistic clustering to very large data sets. Comput. Stat. Data Anal. **51**(1), 215–234 (2006)
17. Nefti, S., Oussalah, M.: Probabilistic-fuzzy clustering algorithm. In: IEEE International Conference on Systems, Man and Cybernetics, pp. 4786–4791. IEEE (2004)
18. Anderson, D.T., Zare, A., Price, S.: Comparing fuzzy, probabilistic and possibilistic partitions using the earth mover's distance. IEEE Trans. Fuzzy Syst. **21**(4), 766–775 (2013)
19. Miyamoto, S., Obara N.: Algorithms of crisp, fuzzy, and probabilistic clustering with semi-supervision or pairwise constraints. In: IEEE International Conference on Granular Computing (GrC), IEEE Computer Society, pp. 225–230 (2013)
20. Pathak, A., Pal, N.R.: Clustering of mixed data by integrating fuzzy, probabilistic, and collaborative clustering framework. Int. J. Fuzzy Syst. **18**(3), 339–348 (2016)
21. Pedrycz, W.: Collaborative fuzzy clustering. Pattern Recogn. Lett. **23**(14), 1675–1686 (2002)
22. Zadeh, L.A.: A note on Z-numbers. Inf. Sci. **181**, 2923–2932 (2010)
23. Aliev, R.A., Huseynov, O.H., Aliyev, R.R., Alizadeh, A.V.: The Arithmetic of Z-numbers: Theory and Applications. World Scientific, Singapore (2015)
24. Aliev, R.A.: Uncertain Computation-Based Decision Theory. World Scientific, Singapore (2017)
25. Price, K., Storm, R., Lampinen, J.: Differential Evolution – A Practical Approach To Global Optimization. Springer, Berlin (2005)
26. Aliev, R.A., Guirimov, B.G., Huseynov, O.H.: Z-number based clustering for knowledge discovery with reliability measure of results. In: Proceedings of the International Conference on Information Society and Smart Cities (ISC 2018), Fitzwilliam College, University of Cambridge, Cambridge, United Kingdom, ISBN 978-1-912532-02-5 (2018)
27. Aliev, R.A., Guirimov, B.G.: Type-2 Fuzzy neural networks and their applications. Springer, Heidelberg (2014)

# Determination of Quality of Plastic Details Under Interval Uncertainty

Djahid A. Kerimov and Naila A. Gasanova[✉]

Azerbaijan State Oil and Industry University,
Azadlig 20, Nasimi, Baku, Azerbaijan
haciyevanaila64@gmail.com

**Abstract.** Application of the theory of planning of experiments has defined interrelation between an indicator of quality of ready details and their structure after production. It was established for the first time that by experimentally studying groups of mechanical parameters (hardness, purity of surface, shrinkage of surface elements), one can predict the nature of the change of mechanical uncontrollable parameters.

**Keywords:** Plastic materials · Quality indicators · ABS-plastic details
Polyamide tar (PT) · Polypropylene tar (PP)

## 1 Introduction

Joint action of relative humidity and temperature of ambient air is a real condition of operation of products out of plastic.

It should be noted that the most part of works devoted to studying of influence of these parameters on stability of properties of polymeric materials is carried out on polyamides. Systematic researches in this direction aren't conducted. Unfortunately, we have no opportunity for comparison of the received results as the technique of carrying out such researches is various.

Data on change of value of indicators of quality of thermoplastics according to the results of climatic tests are given in Tables 1, 2 and 3.

The analysis of data shows that properties of thermoplastics worsen with change of temperature and humidity of ambient air. Shrinkage size undergo to greatest change in particular at details from ABS. It is connected with ability of ABS plastic to moisture absorption where the quality of a surface of details from this material has improved.

As can be seen from the presented Tables 1, 2 and 3, the best stability of the values of quality indicators in climatic conditions is provided from ABS-plastic details.

Thus, in the operating conditions it is necessary to consider a possibility of structural changes of thermoplastic. The last in strong degree depend on initial extent of orientation and crystallinity. Therefore, stability of indicators of quality influence the manufacturing techniques of plastic details, because the technological parameters are very important in the formation of a particular structure.

© Springer Nature Switzerland AG 2019
R. A. Aliev et al. (Eds.): ICAFS-2018, AISC 896, pp. 848–851, 2019.
https://doi.org/10.1007/978-3-030-04164-9_111

**Table 1.** Values of indicators quality of details of polystyrene emulsion brands (ABS)-plastic from the first (I) and after (II) in the climatic trials

| Indicators of quality | Sb, % | ρ, kq/m3 | Ra, mkm | HB, MPa | σp, MPa |
|---|---|---|---|---|---|
| I | 0,74 | 1046,85 | 1,24 | 112,98 | 44,16 |
| II | 0,80 | 1044,62 | 1,11 | 106,50 | 41,67 |
| Change, % | 8,1 | 0,4 | 10,4 | 5,7 | 5,6 |

**Table 2.** Values of indicators quality of details of polyamide tar 68 (PT) from the first (I) and after (II) in the climatic trials

| Indicators of quality | Sb, % | ρ, kq/m3 | Ra, mkm | HB, MPa | σp, MPa |
|---|---|---|---|---|---|
| I | 2,36 | 1010,76 | 1,09 | 74,94 | 38,78 |
| II | 2,64 | 1013,10 | 0,89 | 71,04 | 36,55 |
| Change, % | 11,8 | 0,4 | 18,3 | 5,2 | 5,8 |

**Table 3.** Values of indicators quality of details of polypropylene tar (PP) from the first (I) and after (II) in the climatic trials

| Indicators of quality | Sb, % | ρ, kq/m3 | Ra, mkm | HB, MPa | σp, MPa |
|---|---|---|---|---|---|
| I | 2,41 | 911,46 | 1,13 | 72,95 | 34,46 |
| II | 2,62 | 908,02 | 0,99 | 69,07 | 32,02 |
| Change, % | 8,9 | 0,4 | 12,3 | 5,3 | 7,1 |

## 2  Statement at the Problem and Solution

It is known that determination of durability of a finished product is almost impossible without his direct destruction that, in turn, leads to impossibility of his operation.

The offered method of determination of durability of products without any violation of its structure and properties takes into account the heterogeneity of the distribution of material properties, and also provides an opportunity to assess the quality of the product as a whole. It is especially expedient to apply this method at production of plastic details in the adjusted technological process where ready details after their production go on the assembling.

Results of a research on establishment of multicomponent connection of indicators of quality of plastic details are given in the real stage. It should be noted also that the indicators of quality chosen by us are characterized by the relatively low laboriousness of their determination directly on the detail itself. The possibility of using serial equipment, the simplicity of the method of determination, the high sensitivity to the change in the quality of the detail and a low measurement error of their assessment.

Preliminary results of the research showed that the empirical equation establishing the interrelation of the studied indicators has the following form:

$$\sigma_p = S_b^{n_1} R_a^{n_2} HB^{n_3} K^{n_4}, \tag{1}$$

$K-$ volume coefficient, $K = \frac{V_g}{V_\Phi} = \frac{M_g}{\rho_g V_\Phi}$; $V_g, M_g, \rho_g-$ respectively volume, weight and density of a detail; $V_\Phi-$ form cavity volume.

Equation (2) in log scale will take a form

$$\sigma_p' = n_1 S_b' + n_2 R_a' + n_3 HB' + n_4 K'. \tag{2}$$

The solution of system of the equations was carried out by using DEO methods. Results of calculation for specially made program are given in Table 4.

**Table 4.** Values of equation coefficients (3) for details from (ABS)-plastic, polyamide (PA) and polypropylene tar (PP)

| Material of details | Equation coefficients (3) | | | |
|---|---|---|---|---|
| | n1 | n2 | n3 | n4 |
| ABS | 0,0738835 | –0,1437933 | 0,0593783 | 0,4139313 |
| PA | 0,0313837 | –0,0914700 | 0,4013026 | 0,2029645 |
| PP | 0,0531273 | –0,0762046 | 0,4762118 | 0,9572402 |

Substituting numerical values of coefficients in the Eq. (3)

$$\sigma_p'' = n_1 S_b'' + n_2 R_a'' + n_3 HB'' + n_4 K'', \tag{3}$$

we will receive for ABS

$$\sigma_p'' = 0,07388 S_b'' - 0,14379 R_a'' + 0,05937 HB'' + 0,41393 K''; \tag{4}$$

for PA

$$\sigma_p'' = 0,03138 S_b'' - 0,09147 R_a'' + 0,40130 HB'' + 0,20296 K''; \tag{5}$$

for PP

$$\sigma_p'' = 0,05312 S_b'' - 0,07620 R_a'' + 0,47621 HB'' + 0,95724 K''. \tag{6}$$

To determine the adequacy of the obtained dependence, the mean square deviations of the observed values σobs were calculated. From the calculated σcal predicted by Eqs. (4–6) is given in Table 5.

**Table 5.** Values of $D_{\sigma_p}$ for details from (ABS)-plastic, polyamide (PA) and polypropylene tar (PP)

| Material of details | $D_{\sigma_p}$ |
|---|---|
| ABS | $=1,734321$ |
| PA | 1,483756 |
| PP | 1,152331 |

## 3  Conclusions

The received results show that durability of a plastic detail is directly proportional to sizes of shrinkage, hardness and density and is inversely proportional to the size of roughness of a surface. The greatest impact on durability of details is exerted by change of volume characteristics (elements of surfaces).

The obtained dependence has, in our opinion, of great practical importance for quality control of plastic parts of oil equipment. The received empirical equation is not only mathematical communication between separate parameters, but also one of the main estimates of quality of a ready detail of plastic.

## References

1. Kerimov, D.A.: Fundamentals and practical methods of optimization of quality of plastic details of the oil-field equipment. Dissertation of Doctor of Technology Sciences, Baku (1985)
2. Mitropolsky, A.K.: Technology of statistical calculations. Mathematical Statistics (1971)
3. Nalimov, V.V., Chernova, N.A.: Statistical methods of planning of extreme experiments. Mathematical Statistics (1965)

# Application of Cloud Technologies for Optimization of Complex Processes of Industrial Enterprises

N. R. Yusupbekov[1], F. R. Abdurasulov[2], F. T. Adilov[3($\boxtimes$)], and A. I. Ivanyan[3]

[1] Tashkent State Technical University, Tashkent, Uzbekistan
dodabek@mail.ru
[2] "Oltin Yo'l GTL" Company, Tashkent City, Republic of Uzbekistan
info@oltinyolgtl.com
[3] LLC "XIMAVTOMATIKA", Tashkent City, Republic of Uzbekistan
{Farukh.Adilov,Arsen.Ivanyan}@himavtomatika.uz

**Abstract.** This article introduces the specialists of manufacturers with the achievements of science in the field of cloud technologies of the Industrial Internet of things for specific purposes to improve the reliability and efficiency of the process, as well as the optimization of complex technological process in the framework of the changing conditions of its flow. The article presents the joint developments of UOP and HPS divisions of Honeywell Company.

**Keywords:** "Cloud" technologies · Model · Honeywell Connected Plant Consultant · "Cloud" infrastructure · Optimization · Process · Simulator "Digital twin"

## 1 Introduction

In the modern world, the level of technological effectiveness of industrial enterprises is constantly increasing, more complex and sophisticated production technologies are used, which are almost impossible to repeat and perfectly implement in different geographical conditions of the planet. For a successful result, firstly, the maximum technical thoroughness is required in the design of an industrial enterprise with the mandatory involvement of the licensor of this technology, and, secondly, advanced tools for modeling the design decisions, checking the operation of these models and operational regulation of the implemented process in order to its constant optimization by setting up the model and using the forecast data obtained from the analysis of the model.

## 2 Description of Technical Solutions

The task of the maximum level of detail information about the technological process and the industrial enterprise as a whole is solved by creating IT-model that stores this information and uses it on demand. The solution of this problem is already largely the

© Springer Nature Switzerland AG 2019
R. A. Aliev et al. (Eds.): ICAFS-2018, AISC 896, pp. 852–858, 2019.
https://doi.org/10.1007/978-3-030-04164-9_112

key to the success of the implementation of the conceived idea, because many modern enterprises still cannot solve the problem of creating a centralized information resource. Many articles are devoted to this problem [1, 2], and in this article we will not stop on it in more detail.

After solving the problem of creating a single centralized information resource, the issue of competent and effective use of a huge source of accumulated information to achieve specific results of an industrial enterprise arises. Moreover, here unexpected and unexplored scientific horizons are opened by "cloud" technologies [3, 4].

In this article, as it was mentioned earlier, we would like to stay on the specific joint developments of the two divisions of Honeywell Company - UOP, which is one of the main world developers-licensors of production technologies in the oil and gas and chemical industries, and HPS - developer of advanced automation and information technologies in the industry.

These developments are based on the concepts of Honeywell Connected Plant and Honeywell Connected Performance Services (CPS) and are essentially based on a 24-h online connection to the "cloud" of the industrial enterprise on the one hand and expert services of Honeywell on the other hand. Honeywell's achievements in the field of software and information-communication technologies make it possible to implement a number of computational and analytical tasks directly in the cloud and to practice the behavior of enterprise models with the development of specific forecasts and recommendations [5].

Let us dwell on these achievements, arranged in the form of specific scientific and technical solutions.

## 2.1 Process Reliability Advisor

Software-technical solution Reliability Process Advisor analyzes the performance of the operated process unit to detect problems in the process, using key process parameters embedded in the process model created by the UOP as process technology licensor with appropriate subject matter expertise. This software and hardware solution also provides the ability to detect events early to diagnose and fix problems through built-in cause and effect models. This analysis allows increasing the reliability of the process, its performance, equipment performance and overall efficiency of the process plant.

Figure 1 below illustrates the basic concept of a "cloud" Process Reliability Advisor software-technical solution.

This basic concept is established on three key aspects:

- **Fundamental calculations:** a set of mathematical models and calculations is configured in detailed accordance with the configuration of the process unit to ensure strict simulation. In addition to the models of the technological processes and units developed by the licensor, a set of failure scenarios for various technologies has also been developed and implemented, which ensures early detection of events related to the problems of the plant performance, and the issuance of recommendations for the elimination of these events and their consequences. The concept of an operational corridor is implemented, taking into account the configurable

## Process Reliability Advisor
*Allows to Detect and Solve Potential Problems Before They Happen*

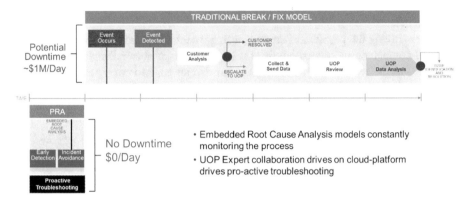

**Fig. 1.** Basic concept of Process Reliability Advisor "cloud" software-technical solution.

parameters of the process unit and equipment, which is an additional convenient tool for the operational and dispatching staff, allowing to control the process in conditions as close as possible to the limits of operational readiness and reliability.

– **User interface:** the CPS dashboard is based on UOP recommendations, but also takes into account the specific experience of end users. As failures occur, a dedicated troubleshooting guide provides the user with additional tools to quickly identify the root cause of the failure. The human-machine interface of the CPS dashboard also includes a set of intuitive associative hints, compiled based on many years of experience of the UOP company.

– **Mode of continuous maintenance of the unit:** throughout the period of unit's lifecycle using CPS technology dedicated UOP specialist works with enterprise operation team on continuous basis, and provides additional expert help in monitoring, elimination and prevention of failures, which finally results to increase of unit's performance. This specialist also provides regular recommendations for updating of configuration of model of enterprise and real process (for example, in case of catalyst reload and plant modernization) and performs a number of operations by himself in remote mode via "cloud" infrastructure.

Complexity and specifics of each exact process stipulates necessity of development of dedicated models for every licensed process. Today there are already software-technical solutions and models available for following processes and units:

(1)  Production of complex of aromatic hydrocarbons
(2)  Platforming - catalytic reforming with continuous catalyst regeneration
(3)  Oleflex - obtaining of light olefins by dehydrogenation of propane
(4)  Catalytic cracking on fluidized bed of catalyst
(5)  Hydrocracking
(6)  Gas fractionation units
(7)  Penex - isomerization of pentane-hexane fraction

(8)  Residue products upgrading
(9)  General refinery configuration.

## 2.2   Process Optimization Advisor

Process Optimization Advisor software-technical solution provides an arsenal of opportunities to develop operational recommendations for the management of process areas and assets. Using the built-in models and tools for managing these models, configured for a specific technological process taking into account its technical features, economic indicators and problem areas, the software-technical solution Process Optimization Advisor allows you to carry out detailed online simulation of the model directly in the cloud and to develop in the prompt mode optimal settings for operational management of the technological process. Experts from the licensor-developer, thus providing an unprecedented level of technical support and optimization for a running industrial plant, check these recommended settings.

The Concept of the "Cloud" Software-Hardware Solution Process Optimization Advisor Consists of the Following Key Elements [6]:

– **Continuous data processing:** software and network component of Process Optimization Advisor continuously processes the data received from the Cloud historian, which, in its turn, carries out continuous data exchange with the RTDB server (real-time database) and the laboratory information management system (LIMS) of the operated industrial enterprise. Continuous data collection is carried out in automatic mode in real time basis.

– **High fidelity basic models:** as in the case of Process Reliability Advisor software-technical solution, the mathematical engine of Process Optimization Advisor software-technical solution is based on the basic dynamic models configured in the environment of the specialized Honeywell UniSim$^{TM}$ Design software, which already includes the built-in optimization tool (Optimizer) for each process unit taking into account its operating limits and calculated technical and economic indicators. This aspect will be considered in more details below.

– **Constant feedback:** the results of calculated optimization recommendations are displayed through the CPS dashboard, similar to the panel described for the Process Reliability Advisor solution. Interface of CPS dashboard for the Process Optimization Advisor solution involves the provision of detailed information on the economic indicators of the technological process, changes in operating costs and potential options to increase profitability of the unit and technological process in general. An example of a detailed display of the CPS dashboard for the Process Optimization Advisor software-hardware solution is shown in Fig. 2.

– **Ease of maintenance:** cloud infrastructure enables remote configuration of the mathematical calculations underlying the work of the Process Optimization Advisor, remote modification of the model, migration and software platform updates. All these activities can be done online by a Honeywell UOP specialist.

Process Optimization Advisor technology also considers development of specialized software-technical solution for each process unit and for each technological process.

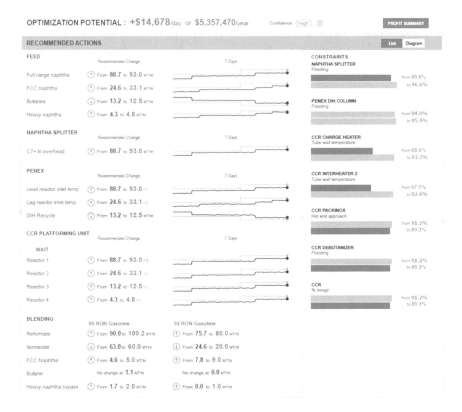

**Fig. 2.** Example of detailed display of CPS dashboard for Process Optimization Advisor solution.

## 2.3    Creation of "Digital Twin" - Model of Enterprise Based on Unisim Design Software Solution

The successful implementation of the above-described Process Optimization Advisor and Process Reliability Advisor is impossible without the presence of a well-designed and proven by the example of numerous projects the software-hardware platform for mathematical modeling of technological processes and units. There are only several developers of such platforms in the world and one of these companies is the developer company Honeywell UOP. Namely based on the Unisim technology from Honeywell UOP the software package Unisim Design is developed.

Unisim Design is a software-hardware solution used for the design and optimization of industrial processes, equipped with the functionality of the subsequent dynamic simulation of these processes.

The software package Unisim Design has been successfully used for 30 years by many industrial enterprises, in the first place, to create training simulators for operational and dispatching personnel of enterprises. The training model is based on high-precision mathematical modeling of real processes, describing in detail the technological processes by differential equations of energy transfer, material balance, hydraulics and chemical kinetics. The modeling is carried out taking into account the thermodynamic properties of flows and materials of equipment, mechanical dynamic characteristics of valves, pumps and tanks, linear dimensions of devices [2].

The innovative idea of Honeywell UOP, covered in this article, is the creation on the basis of Unisim Design software-hardware solution "cloud" models of technological units and combining them into a single integrated "cloud" model of the production complex, called "digital twin" which in fact represents a virtual projection of the industrial facility into the "cloud", which can be used in solving both theoretical and practical problems of a huge wide range.

The simplified architecture of the cloud solution for creating a digital twin based on the Unisim technology is shown in Fig. 3 below.

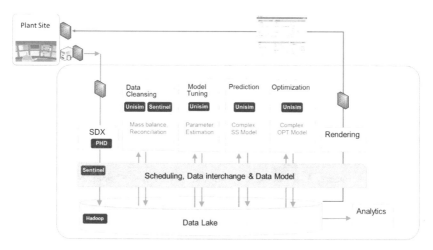

**Fig. 3.** Simplified architecture of "cloud" solution for creation of "digital twin" on the basis of Unisim technology.

The importance of the issue of creating " digital twins" is confirmed by the increasing number of publications on this topic ([7–10]), but all of them are related with the creation of models of individual units of equipment, devices or compact package units. Creating a full-fledged twin of a complex technological process, which involves large-scale industrial installations, involves the creation of a mathematical model and associated software and network "cloud" infrastructure of an incredible level of complexity. And in this regard, Honeywell UOP's development and experience are truly unique.

Without any doubt, cloud technologies in the industry are the next inevitable step for the industrial enterprise to enter a new stage of productivity improvement in today's conditions. The use of standard approaches of autonomous enterprise management does not provide opportunities for further development and, moreover, in the conditions of constant complication of the technological process scheme, caused by the requirement to extract maximum benefit from the same range of raw materials, can often lead to wrong production decisions, associated with the lack of full information about the current operational situation and, as a consequence, to decrease in the productivity of the enterprise in the long-term aspect.

# References

1. Yusupbekov, N.R., Sattarov, Sh.B., Adilov, F.T., Ivanyan, A.I.: integrated informational-control system for group of gas-condensate fields. In: 8th International Conference on Soft Computing, Computing with Words and Perceptions in System Analysis, Decision and Control, ICSCCW 2015, Antalya, Turkey (2015)
2. Yusupbekov, N.R., Abdurasulov, F.R., Sattarov, Sh.B., Hwa, J.J., Adilov, F.T., Ivanyan, A. I.: Application of innovative IT-solutions in complex high-technology production on the example of Uzbekistan Gas to Liquid (UzGTL) project. In: Proceedings of the International Conference on Integrated Innovative development of Zarafshan region: achievements, challenges and prospects, vol. II, pp. 10–15 (2017)
3. Yusupbekov, N.R., Sattarov, Sh.B., Ivanyan, A.I.: Industrial internet of things – new level in evolution of automation systems. Chemical technology control and management. J. Korea Multimed. Soc. Special ISSUE, 21–25 (2016)
4. Yusupbekov, N.R., Sattarov, Sh.B., Doudin, I.S., Adilov, F.T., Ivanyan, A.I.: Application of solutions of connection of production cluster to analytical cloud in chemical industry. In: Proceedings of the International Conference on Integrated Innovative Development of Zarafshan Region: Achievements, Challenges and Prospects, vol. I, pp. 246–252 (2017)
5. Data's Big Impact on Manufacturing: A Study of Executive Opinions. Study by Honeywell and KRC Research Inc. (2016)
6. Jones, K., Romatier, C.: Leveraging the Cloud to Drive Complex-wide Profitability. Honeywell UOP, 25 East Algonquin Rd, Des Plaines, IL 60016 (2017)
7. Patrakhin, V.A.: Proactive maintenance of equipment as practical implementation of GE Digital Twin concept. World of Automation, №. 2 (2017)
8. Frankel, A., Larsson, J.: A better way: digital twin will improve efficiency of engineering design and manufacturing processes. Mechanical engineering and related industries, CAD/CAM/CAE Observer #3 (103), (2016)
9. Grieves, M.: Digital Twin: Manufacturing Excellence through Virtual Factory Replication. Digital Twin White Paper (2014)
10. Glaessgen, E.H., Stargel, D.S.: The Digital Twin Paradigm for Future NASA and U.S. Air Force Vehicles (2017)

# Decision-Making on Restriction of Water Inflows into Oil Wells in Dependence on the Type of Initial Information

B. N. Koilybayev[1], A. S. Strekov[2], K. T. Bissembayeva[1],
P. Z. Mammadov[3], D. A. Akhmetov[1], and O. G. Kirisenko[2(✉)]

[1] Caspian State University of Technology and Engineering named after
Sh. Yessenov, Mangistau Region, 32 Microdistricts, 130003 Aktau,
Republic of Kazakhstan
nomad_bk@bk.ru, d_akhmetov@KBM.KZ
[2] Oil and Gas Institute, Azerbaijan National Academy of Sciences, F.Amirov 9,
AZ1000 Baku, Azerbaijan
a.s.strekov@mail.ru, oleg.kirisenko@gmail.com
[3] Azerbaijan State Oil and Industry University, Azadliq ave. 34,
AZ1010 Baku, Azerbaijan
parviz08@list.ru

**Abstract.** The report is devoted to the problem of limiting water inflows into oil wells. A large number of studies have been devoted to this subject, theoretical and experimental work has been carried out, but the inadequacy of the decision-making process on the choice of the bottomhole zone treatment method significantly reduces the effectiveness of the problem solution. Depending on the type of initial information, various methods were used at different times, in particular statistical methods known from the theory of fuzzy sets and others. Investigations and analysis of the processes of water inflows show that a number of factors make an influence on efficiency of this process. In this regard, in this paper, the collected data about influence of the of various factors on effectiveness are subjected to fuzzy cluster analysis with the establishment of fuzzy rules that express the influence of the selected factors on the effect duration and the amount of additional extracted oil. Four clusters were obtained with the further formulation of fuzzy rules according to the principle "if …, then …" as a result of the application of fuzzy cluster-analysis program.

**Keywords:** Well · Water inflow · Bottomhole zone · Fuzzy sets
Fuzzy cluster-analysis · Oil recovery · Permeability · Well rate
Polymer solution

## 1 Introduction

When assessing the effectiveness of waterproofing operations in production wells and making decisions, there are difficulties due to uncertainties of a different nature. In this connection, depending on the type of initial information, different methods were used at different times, in particular statistical methods known from the theory of fuzzy sets,

© Springer Nature Switzerland AG 2019
R. A. Aliev et al. (Eds.): ICAFS-2018, AISC 896, pp. 859–864, 2019.
https://doi.org/10.1007/978-3-030-04164-9_113

etc. Investigation and analysis of the processes of limiting water inflows show that the effectiveness of this process is affected by the whole a number of factors. These are factors that characterize the geological and physical, technical conditions for wells exploitation, in particular: the well rate on fluid before the treatment of the bottomhole well zone (BWZ) with polymer solution in $m^3$/day, depression on layer in MPa, formation permeability in $mkm^2$, the dismemberment of bed, the coefficient of oil recovery from the reservoir, the length of filter in m, the amount of polymer solution per 1 m of perforated thickness of the formation in kg, the coverage polymer solution of BWZ area with shear rates equal or having the same order, and the shear rates of shear (0.1-1 s-1), in%. Efficiency of treatment is characterized by continuation of effect of limitation of water inflows, expressed in months, and the amount of additional produced oil.

Often there are situations when, in the presence of the same data, in the end, fundamentally different results are obtained. In order to find specific expressions for these dependencies and the parameters of their characteristics, in particular, the methods of mathematical statistics are used in the processing of data and, in the final, the actual experimental material or the results of field observations are replaced by the obtained laws and some integral, that is, in general, an estimate of the tightness of communication. In accordance with the adopted technology, the found law is transferred to the researched object in the form of coupling equations between the influencing factors and performance indicators. This way is often the source of erroneous conclusions, since in most cases, the formulation of goals and constraints in making decisions to improve the effectiveness of water inflow restriction in the presence of multifactory, multicriteriality are unclear, in short, all this requires the use of the appropriate apparatus.

In this regard, change of effect duration of restricting water inflows, in depend on geological-physical, technical conditions of operation and technological factors of Azerbaijan deposits, was considered in [1]. In these works, statistical and linguistic models were obtained, respectively, expressing the dependence of the effectiveness of the polymer effect on pointed factors.

However, these works do not show the ways of decisions-making with use of obtained models, taking into account multicriteria. They consider one criterion, which greatly simplifies the formulation and solution of the problem.

## 2    Problem Formulation

To solve the problem of choosing the optimal combination of application condition and methods of increasing oil recovery, the most suitable method is the odd modeling method, which is based on the fuzzy cluster analysis procedure. To conduct this procedure, it is necessary to justify the selection and collect data characterizing the geological, physical, technical and technological conditions of bottomhole well zone and the criteria characterizing the effectiveness of the process of isolating water inflows by polymer solutions.

Based on the analysis of experimental studies results and the experience of field application of polymer solutions for isolation of water inflows in production wells, geological, technical, technological data and results of isolation of water inflows with polymer solutions at various sites were collected and processed as factors and criteria. The data were subjected to cluster analysis using the corresponding program [2–5].

## 3 Results of the Implementation of Fuzzy Clustering Algorithm

With the change of geological-physical, technical and technological factors characterizing the bottom-hole zone and treatment technology, effect duration of limiting water inflows increases on the one hand, and on the other hand it decreases. For example, an increase of formation permeability, amount of extracted oil (AEO), the amount of polymer per 1 m of filter, the coverage of the BWZ by the polymer solution leads to the growth of effect duration of limiting water inflows, and an increase in the dismemberment of formation, the liquid flow rate, the depression per layer, the length of filter reduce the effect duration of limiting water inflows. Obviously, all this makes it difficult to make effective decisions when applying the technology of BWZ processing with polymer solution. Therefore, as a criterion, we also accepted additional oil production.

With the application of noted program, the initial massif, sorted by the age-values of effect duration of limiting water flow by polymer solutions and additionally extracted oil, was divided into four fuzzy clusters by fuzzy clusterization (Table 1). Y values belonging to each fuzzy class will be described by the following linguistic variables: small, medium, good and very good effect. As an example, in Fig. 1 we show the distribution of effect value of limiting water flow on four fuzzy clusters and the corresponding values of membership function, where $\mu_1$, $\mu_2$, $\mu_3$, $\mu_4$ - function values of accessories to classes "small, medium, good and very good effect". The obtained values of membership function $\mu_1$, $\mu_2$, $\mu_3$, $\mu_4$ for Y are assigned to each from eight initial variables. As a result, we get a training massif from sorted ascending Y eight initial variables and values of membership function for each of classes.

Using data shown in Fig. 1, corresponding values of Y and variables $x_1$–$x_8$ for each of four fuzzy clusters were determined. The obtained values are summarized in Table 1.

Based on obtained fuzzy clusters (Table 1), fuzzy linguistic rules are formulated in the form "if …., then … ", with the help of which fuzzy forecasting will be carried out.

**Rule 1.** If well flow rate is from average to very high, depression is from average to very high, permeability is from low to average, dismemberment is from average to very high, COR is small to average, filter length is from large to very large, the amount of injected polymer per 1 m of filter is small, the coverage percentage of BWZ is small, the duration of effect is small and the volume of additionally produced oil is low.

**Rule 2.** If well flow rate is from average to high, depression on formation is from average to very high, the permeability of reservoir is from small to high, dismemberment is from small to high, COR is from small to high, the length of filter from average to very large, the amount of injected polymer per 1 m of filter is from small to

**Table 1.** Intervals of variables change and duration of effect by classes

| If | | | | | | | | Then | |
|---|---|---|---|---|---|---|---|---|---|
| Interval of factors change | | | | | | | | Interval of criteria change | |
| $X_1$, m³/day | $X_2$, MPa | $X_3$, mkm² | $X_4$ | $X_5$ | $X_6$, m | $X_7$, kg | $X_8$, % | $Y_1$, months | $Y_2$,t. |
| >50 very high | >2,6 very high | 0,068–0,26 very small | >14 very high | 0,16–0,25 small | >50 very high | 51,8–96 small | >96 small | 0,07–6,6 small | 17–80 low |
| 40–50 high | 1,5–2,6 high | 0,26–0,5 average | >10–14 high | >0,25–0,45 average | >20–50 high | >96–272 average | 96–100 average | 6,67–13,83 average | 80–800 average |
| 20–40 average | 0,07–1,5 average | 0,5–0,7 high | >7–10 average | >0,45–0,75 high | >10–20 average | >272–500 high | 100 high | 13,97–21,63 good | 800–2500 good |
| 4–20 Small | <0,07 small | >0,7 very high | 1–7 small | >0,75 high | 5–10 small | >500 very high | 100 very high | 22–33,33 very good | >2500< very good |

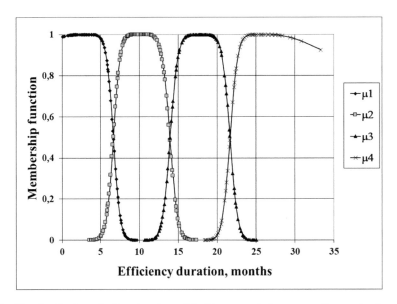

**Fig. 1.** The distribution of effect duration of limiting water inflows according to the degrees of its belonging to four distinguished fuzzy clusters: $\mu_1$- small; $\mu_2$ - average; $\mu_3$ - good; $\mu_4$ - very good.

large, the coverage percentage of BWZ is from average to large, the duration of effect is average and the volume of additional produced oil is average.

**Rule 3.** If well flow rate is small to average, depression on formation is from small to average, the permeability of reservoir is small to high, dismemberment is from small to average, the COR is from average to very high, the length of the filter is from small to average, the amount of polymer per 1 m of filter is from average to large, the coverage percentage of BWZ is from average to large, the duration of the effect is good and the volume of additional produced oil is good.

**Rule 4.** If well flow rate is small to average, depression on formation is from low to average, the permeability of reservoir is from average to very high, the dismemberment is from small to average, the COR is from average to very high, the length of filter is from small to average, the amount of polymer per 1 m of filter is from average to very large, the coverage percentage of COR is from high to very high, the duration of effect is very good and the volume of the additional produced oil is very good.

These fuzzy rules have been tested at one of the fields during the implementation of polymer flooding. The conditions of the object correspond to the input part of rules 3, 4. It should be noted good coherence between results and the above rules.

Applying rule 3, 4 and using the training massif, we'll obtain the following values of membership function to class "the duration of the effect is good" and the duration of effect is very good. The belonging of variables $x_1$–$x_8$ to each of classes will be equal to the minimum of the computed ones. In view of the fact that the minimum $\mu_4 > \mu_3$ we attribute this set of variables to the class "effect duration is very good".

Dephasing is performed in the following way.

We select from the training massif the data on effect duration (Y) and the membership function ($\mu_4$) belonging to the fourth class (you can also use the data shown in the figure and sort them in ascending order $\mu_4$). The desired effect duration (Y) is interpolated to calculated membership to the fourth class $\mu_4 = 0{,}9806$. As a result we get that the predicted effect duration is 26,6 months. In the examining massif effect duration of restricting water inflows was 28 months. Such calculations were also made for another wells from considering massif with good result.

Thus, use of the method of fuzzy modeling allowed, on the basis of complex data on geological, physical, technical and technological conditions of wells operation at Azerbaijan deposits, to develop linguistic models in the form of appropriate rules "if …, then …. ". These rules allow in the conditions of insufficient initial information to predict the effect duration of limiting water inflows during treatment of BWZ using polymer solution.

# References

1. Abasov, M.T., Djafarova, N.M., Strekov, A.S., Efendiyev, G.M., Manafov, G.R.: Prediction of time of water inflows restriction with polymer solutions in producing wells. In: Proceedings of the Azerbaijan National Academy of Sciences, The Sciences of Earth, №. 2, pp. 97–102 (2001)
2. Bezdek, J.C., Ehrlich, R., Full, W.E.: The fuzzy C-means clustering algorithm. Comput. Geosci. **10**, 191–203 (1984)
3. Aliev, R.A., Guirimov, B.G.: Type-2 Fuzzy Neural Networks and Their Applications. Springer, Switzerland (2014). http://www.springer.com/us/book/9783319090719

4. Turksen, I.B.: Full type 2 to type n fuzzy system models. In: Seventh International Conference on Soft Computing, Computing with Words and Perceptions in System Analysis, Decision and Control, Izmir, Turkey, vol. 2013, p. 21 (2013)
5. Efendiyev, G.M., Mammadov, P.Z., Piriverdiyev, I.A., Mammadov, V.N.: Clustering of geological objects using FCM-algorithm and evaluation of the rate of lost circulation. In: 12th International Conference on Application of Fuzzy Systems and Soft Computing, 29–30 August 2016, Vienna, Austria (2016). Procedia Comput. Sci. **102**, 159–162 (2016)

# Classification of Hard-to-Recover Hydrocarbon Reserves of Kazakhstan with the Use of Fuzzy Cluster-Analysis

D. A. Akhmetov[1], G. M. Efendiyev[2($\boxtimes$)],
M. K. Karazhanova[1], and B. N. Koylibaev[1]

[1] Yessenov University, 32 Microdistricts, 130003 Aktau,
Mangistau Region, Kazakhstan
[2] Institute of Oil and Gas, 9 F. Amirov Street, AZ1000 Baku, Azerbaijan
galib_2000@yahoo.com

**Abstract.** This report is devoted to the classification of hard-to-recover oil reserves. The analysis of existing classifications has been carried out preliminary and the necessity of using a method that takes into account the whole range of characteristics allowing to classify oil and conditions of occurrence to a particular class has been shown. In this connection, we applied the method of fuzzy cluster analysis. The tasks of cluster-analysis have been widely used in economics, sociology, medicine, geology, oilfield practice and other industries, i.e. wherever there are sets of objects of an arbitrary nature, described in the form of vectors $x = \{x_1, x_2, \ldots, x_N\}$, which must be automatically divided into groups of homogeneous objects according to the similarity within the homogeneous object (cluster) and the difference between these objects. A considerable amount of literature has accumulated in this direction. As noted in the literature, there are more than one hundred different clustering algorithms, among them hierarchical and non-hierarchical cluster-analyzes, fuzzy clustering.

In order to classify hard-to-recover reserves, we performed clustering using the fuzzy cluster-analysis algorithm. For this purpose, data were collected on the viscosity, oil density and permeability of oil conditions from the oilfields of Kazakhstan. As a result, 4 classes were obtained, each of which characterizes the difficulty of extracting oil: the layer is permeable, highly viscous and very heavy oil; medium permeability layer, viscous and heavy oil; high-permeability reservoir, medium viscosity oil and medium-density oil; low-permeability reservoir, low viscosity oil, light oil.

**Keywords:** Permeability · Density · Viscosity · Hard-to-recover reserves
Fuzzy clustering

## 1 Introduction

As is known, nowadays the share of hard-to-recover oils is more than half of the world's explored reserves and is considered by experts as the main resource base for the development of oil production in the 21st century. The possibility of applying this or that development technology is determined by the geological structure and the

© Springer Nature Switzerland AG 2019
R. A. Aliev et al. (Eds.): ICAFS-2018, AISC 896, pp. 865–872, 2019.
https://doi.org/10.1007/978-3-030-04164-9_114

conditions of occurrence, the physico-chemical properties of the reservoir fluid, the state and reserves of hydrocarbon raw materials, geographical conditions, etc.

Adopting the right technological solutions requires a detailed study of the properties of oil, the conditions of its occurrence, the methods of development.

Currently, due to the depletion of oil fields being developed, oil and gas companies are increasingly paying attention to the improvement of methods for development of deposits with hard-to-recover reserves of heavy oils and natural bitumen [1]. Improving the technologies for heavy oil and natural bitumen production is becoming increasingly important, as the reserves of these resources already exceed the reserves of conventional (light) oil (Fig. 1), and due to the continued increase in production and associated decrease in light oil reserves, the share of heavy oil in the structure reserves of hydrocarbons will increase steadily.

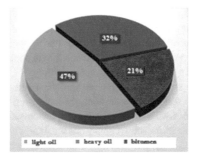

**Fig. 1.** The ratio of the explored reserves of light oil, heavy oil and natural bitumen.

The problem of development of heavy oil deposits constantly attracts specialists' attention as important and relevant, especially in Kazakhstan, especially in the Western oil and gas producing regions (Mangistau and Atyrau).

**Analysis and Brief Review of Scientific Literature**

To the problem of increasing the efficiency of the development of deposits of high-viscosity oils is devoted to the works of many researchers who made a great contribution to the development of the scientific foundations of this direction, who carried out interesting theoretical and experimental studies. A characteristic feature of the development of oil deposits in recent years is a change in the structure of oil reserves in the direction of increasing the share of hard-to-recover oil, which is due to a deterioration in the quality of reservoirs, an increase of watercut of oil produced, unfavorable geological and physical characteristics and conditions of occurrence of oil. Naturally, the efficiency indicators, in particular, the extraction rates of oil, the coefficients of oil recovery, the economic efficiency of the development of deposits of hard-to-recover oils are generally much lower than those for deposits with normal oil [2].

Nowadays, classifications of hard-to-recover oil reserves have been proposed, in particular, according to GOST R 51858-2002, the classification is based on the content of paraffins: oil with low-paraffin, in which the paraffin content is up to 1.5%; waxy, 1.5–6%; high paraffin content when the content of paraffin is more than 6%. Asphaltene and resin content: low-tar, tar content up to 5%; resinous, 5–15%; highly resinous, more than 15%. The density of oil is classified as follows: density up to 834.5 (at 15 °C) and up to 830 (at 20 °C) is characteristic for particularly light oils; 834.5–854.4 and 830–850 - light oil; 854.4–874.4 and 850–870 - average density; 874.4–899.3 and 870–895 - heavy oil and oil with a density of more than 899.3 and 895 at 15 °C and at 20 °C, respectively - bituminous.

On the viscosity of oil: $\leq 5$ mPa s - insignificant viscosity; 5–10 - low viscosity; 10–30 - increased viscosity and, finally, more than 30 mPa s - high viscosity. However, other limits of this value were also upheld - 10.40 and 50 mPa s [3–5].

It is proposed in [6] to divide high-viscosity oils (HVO) into three groups: the first group is characterized by a change in viscosity in the range from 30–100 mPa s and density of 834–929 kg/m$^3$, the second group is characterized by a viscosity change in the range from 100 to 500 mPa s and a change in density in the range of 882–955 kg/m$^3$, and the third group is characterized by a viscosity change above 500 mPa s and density $\geq$ 934 kg/m$^3$. According to [7], oil reserves with a viscosity of more than 50 mPa s are usually referred to as hard-to-recover. In general, liquid hydrocarbons, that exist in nature, devide: low-viscosity (0.4–10 mPa s), medium viscosity (10–50 mPa s), high viscosity (50–1500 mPa s), heavy oil (more than 1500 mPa s), and bitumen (more than (20–25) * 10$^3$ mPa s) [6]. The largest deposits of high-viscosity oils in the territory of the former CIS in Kazakhstan are: Karazhanbas - 230 million tons; North Buzachi - 195 million tons; Kenkiyak - 72 million tons; in Azerbaijan: Balakhany-Sabunchi-Novels - 114 million tons.

In Russia, hard-to-recover oil is considered to be an alternative source of hydrocarbon raw [8], since they differ from conventional oils not only by increased density, but also by complex composition.

At present, the distinctive feature of the oil production process is the increase in the global structure of hydrocarbon resources in the share of hard-to-recover reserves, which are mainly heavy and highly viscous oil [9]. Reserves of such oils, as noted in the literature, significantly exceed the reserves of light and low-viscosity oils. According to experts, the reserves of hard-to-recover oil are at least 1 trillion and are considered not so much as a reserve of oil production, but rather as a major development base for the coming years [9].

Due to the fact that oil is a very complex natural object, researchers study it in various aspects. So, they study the genesis of oil and the formation of oil deposits, the issues of their search and exploration, the chemical composition of oils and the development of ways for their processing. The solution of all these and related issues is possible with a rational classification, which allows a fairly complete and clear characterization of oil from both a scientific and a practical point of view. This will help to choose a main method of stimulation on the reservoir. Along with this, the creation of such a classification is a very difficult task, which, according to a large number of researchers, has not found a satisfactory solution nowadays [10–14].

As the main reasons for the difficulties in classification, the authors of the works note [10, 12]: the complexity and diversity of the chemical composition of the oils (the composition of not only different oilfields, but also oil extracted from different wells of the same field); insufficient knowledge of the composition of oil and factors affecting the formation and transformation of petroleum in the bowels of the earth, as well as the significant development of methods for studying the composition and properties of oils and the resulting extremely rapid change in their level of study; the need to choose from a variety of physical, physico-chemical, structural and other characteristics of oils and the conditions of their occurrence the classification characteristics with maximum informativeness; the need to take into account and predict the possible directions for the development of chemistry and geochemistry of oil in the future.

Taking into account the mentioned reasons will allow to justify the choice of the stimulation method on the layer and to achieve an increase in the efficiency of its application.

It should be noted that by now a lot of work has accumulated, in which attempts have been made to classify oils, of which three main groups can be distinguished: chemical, geochemical (genetic), technological (commercial, commodity). However, in the opinion of geologists and developers, since the properties of oil are associated with its composition, and the composition in its turn is a function of geological and geochemical history, the division of existing classifications into the specified groups is very conditional [15].

On the basis of the analysis and generalization of the literature data, the authors [10] present the typing of hard-to-recover oils and their classification according to the proposed criterion, called the quality index. As the authors note, the expansion in recent years of the volumes of exploration and development of hard-to-recover oils necessitates the study of their qualitative features. The analysis of the physico-chemical properties of hard-to-recover oils and their occurrence conditions was carried out in the work [16]. The wide range of different oils in composition and properties is one of the main reasons for the inadequate research, which in turn makes it difficult to solve the problems of both oil production and transportation of hydrocarbon raw materials, as well as technological problems in petrochemicals and oil refining. This explains the researchers' attention to this problem. In [10], in order to develop a classification of the quality of oil and its application in the problems of analyzing the quality indicators of various types of hard-to-recover oils, appropriate studies were carried out. The authors note, that on the basis of the proposed by E.Khalimov, E.M. in 1987, definitions of the concept of hard-to-recover reserves in [11] formulated the basic principles and criteria for recognition, put in other words the attribution of oil reserves to hard-to-recover. Different papers give different definitions to hard-to-recover oil reserves, hence the difficulties associated with their classification.

Thus, in most cases, for example, according to [11, 12, 17, 18], oil reserves are represented by low-mobility oil (in particular, with high viscosity or density and high content of solid paraffins), oils with high (more than 500 $m^3$/t) or low (less than 200 $m^3$/t) gas saturation, or in the presence of aggressive components (hydrogen sulphide, carbon dioxide) in dissolved and/or free gas in quantities that require special equipment for drilling and oil production. According to the literature data, in particular [12], the second group of hard-to-recover oils consists of oils with complicated occurrence conditions (in geologically complex layers and deposits, in oil-and-water and oil-and-gas zones, in low permeability and low-porosity reservoirs, with an abnormally high or anomalously low reservoir temperature and others), as well as oil located on the territory of permafrost and on the shelves of the seas [18]. Based on the generalization of the criteria for classifying petroleum reserves as hard-to-recover, a list of the main types of hard-recovering oils are presented in the form of a table in work

[10], compiled from the results of research in a large number of scientific publications [11, 12, 17, 18].

A complex index of oil quality was proposed in [19], taking into account such parameters as total sulfur content in oil, concentration of chloride salts, $\rho$ - oil density, fraction content at different temperatures and regression equation for calculating this index. Later in work [20] it was suggested to use this complex indicator for estimating the qualitative characteristics of oils in oil deposits of oilfields and oil-bearing basins [10]. This approach was applied in the works of the authors [14, 21, 22] in the evaluation and analysis of the regularities of spatial changes in the quality parameters of oils in the oil-bearing territories of different regions, countries and continents. The oil quality is determined in the form of a value inverse to the indicator noted above, proved to be quite convenient for analyzing the qualitative features of different types of hard-to-recover oils [10]. At the same time, the increase in the values of the oil quality index corresponds to the rise of the quality of oil, and the decrease to a decline, which is convenient when comparing different types of hard-to-recover oil in terms of quality.

The results of researches of optical properties of oil are given in work [23]. The results of laboratory studies of density, viscosity, optical density and coefficients of light absorption of oil samples from various oil and gas bearing regions of Russia are analyzed. The work describes the international and Russian oil classification, according to which oil is distributed with increased viscosity and medium density. The typification of oil was carried out according to the nature of the approximation of the density and the root-mean-square value of Ksp, three classes of oil were distinguished with a rather high value of the reliability of the approximation: light (with density $\rho < 870$ kg/m$^3$), medium (with density $870 < \rho < 920$ kg/m$^3$) and heavy density $\rho > 920$ kg/m$^3$). According to the nature of the viscosity approximation, three classes of oil are distinguished: low-viscosity (with a viscosity of $1 < \mu < 5$ mPa s), with an increased viscosity (with a viscosity of $5 < \mu < 30$ mPa s) and high viscosity (with a viscosity $\mu > 30$ mPa s) [23].

At the 11th World Petroleum Congress in Houston (USA, 1987), a classification was proposed in which were divided into oil and bitumen (conditionally "classification 1" [23]). To bitumens, it was suggested to refer hydrocarbons with a density $\rho$ above 1000 kg/m$^3$ and with dynamic viscosity $\mu$ more than 10,000 mPa s in the reservoir conditions, and oil was divided into four classes: light, medium, heavy and extra heavy.

In 2012, Russia introduced the classification of oil and gas reserves and resources in a new version to integrate the Russian classification into the most common international classifications.

As noted in the literature, in international classifications consider a greater variety of oil in density ("classification 2") [23]. According to this classification, the oil density is taken into account at a temperature of 15.56 °C, and five classes of oil are distinguished: extra light, light, medium, heavy, bituminous. In some cases, very light (less than 800 kg/m$^3$), light (800–840 kg/m$^3$), medium (840–880 kg/m$^3$), heavy (880–920 kg/m$^3$) and very heavy (over 920 kg/m$^3$) of oil (according to "classification 3"). It is known that deposits with hard-to-recover oil are characterized by low and unstable well rates, for the operation of which it is necessary to create and introduce modern and expensive technologies.

From the above brief overview it also follows that oil is hard-to-recover due to the quality of the raw materials: heavy (density more than 0.92 g/cm$^3$);high viscosity (more than 30 mPa s under normal conditions) and according to the conditions of occurrence: very low permeability of reservoirs (less than 0.05 μm$^2$) [24].

Thus, hard-to-recover reserves are deposits (oilfields, objects of development) that are relatively unfavorable for extraction by geological conditions of oil occurrence and (or) physical properties, the development of which is economically ineffective with existing technologies in the current tax system.

The very concept of "hard-to-recover reserves" is a relative term, since stocks can be different in terms of the difficulty of extracting them. Therefore, their classification is necessary with an assessment of the difficulty of extraction. There are many methods for classifying different objects.

The tasks of cluster-analysis, or, as they are sometimes called, the tasks of automatic classification, in recent years have been widely used in economics, sociology, medicine, geology, oilfield practice and other industries, i.e. wherever there are sets of objects of an arbitrary nature, described in the form of vectors x = {x$_1$, x$_2$,..., x$_N$}, which must be automatically divided into groups of homogeneous objects according to the similarity within the homogeneous object (cluster) and the difference between these objects. A considerable amount of literature has accumulated in this direction [25–30].

As noted in the literature, there are more than one hundred different clustering algorithms, among them hierarchical and non-hierarchical cluster-analyzes, fuzzy clustering. In recent years, these methods are widely used in problems of information mining, data collection and processing (Data mining) [27–30]. In order to classify hard-to-recover reserves, we performed clustering using the fuzzy cluster-analysis algorithm. For this purpose, data were collected on the viscosity, oil density and permeability of oil conditions from the oilfields of Kazakhstan. As a result, four classes were obtained, each of which characterizes the difficulty of extracting oil:

- the layer is permeable, highly viscous and very heavy oil;
- medium permeability layer, viscous and heavy oil;
- high-permeability reservoir, medium viscosity oil and medium-density oil;
- low-permeability reservoir, low viscosity oil, light oil.

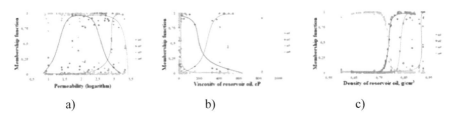

a)                    b)                    c)

**Fig. 2.** Term-sets of different levels of permeability (a), viscosity of reservoir oil (b), density of reservoir oil (c).

## 2   Conclusions

As a result of fuzzy cluster-analysis, four clusters are obtained, each of which represents a certain category of oil and conditions of occurrence due to extraction difficulties. The graphs of membership functions for each factor are constructed, which allow to separate the values of factors into separate term-sets. In this case, light oil is oil with a density $(700-815)$ kg/m$^3$; average density - $(800-890)$ kg/m$^3$; heavy $(892-910)$ kg/m$^3$, very heavy - $(910-933)$ kg/m$^3$ (Fig. 2c).

On the viscosity: low viscosity - $(0.17-6.9)$ mPa s; the average viscosity is $(8.6-55)$ mPa s; Viscous - $(55-160)$ mPa s; 160 mPa s and higher (Fig. 2b). Permeability: high permeability $(840-2180)$ mD; permeable $(480-800)$ mD; average permeability $(130-690)$ mD; low-permeability - (less than 130) mD (Fig. 2a).

## References

1. Nikolin, I.V.: Methods for the development of heavy oils and natural bitumen. In: Science – The Foundation for Solving the Technological Problems of Development of Russia, vol. 2, pp. 54–68 (2007)
2. Company News. In: Oil Industry, vol. 1, p. 9 (2009)
3. Vakhitov, G.G., Morozov, V.D., Safiullin, R.K.H.: Problems of well development of high-viscosity oil and natural bitumen deposits abroad. In: Review, Inform. "Oilfield Business", vol. 19, no. 126, p. 49 (1986)
4. Verevkin, K.I., Diyashev, R.N.: Classification of hydrocarbons in the selection of methods for their extraction. In: Oil Industry, vol. 3, pp. 31–34 (1982)
5. Skorovarov, Y.N., Trebin, G.F., Kapyryp, Y.V.: Properties of high-viscosity oil deposits of the USSR. In: Geology of Oil and Gas, vol. 2, pp. 24–27 (1985)
6. Halimov, E.M., Klimushin, I.M., Ferdman, L.N.: Geology of high-viscosity oil deposits of the USSR. Nedra, Reference Book, 174 p. (1987)
7. Steam-thermal effect on the reservoir. http://proofoil.ru/Oilproduction/Steamaction.html
8. Makarevich, V.N., Iskritskaya, N.I., Bogoslovsky, S.A.: Resource potential of heavy oil deposits in the European part of the Russian Federation. In: Oil and Gas Geology. Theory and Practice, vol. 3, p. 7 (2012). http://www.ngtp.ru/rub/6/43_2012.pdf
9. Nugiev, M.A.: About none-quilibrium reological properties of high-viscosity oils of some deposits of Western Kazakhstan. http://kaznipi.kz, https://refdb.ru/look/3007729.html
10. Yashchenko, I., Polishchuk, Y., Kozin, E.: Hard-to-recover oil: classification and analysis of qualitative features. Oil & Gas J. Russia. Geol. Geophys. **11**, 64–70 (2015)
11. Lisovsky, N.N., Halimov, E.M.: On the classification of hard-to-recover reserves. In: Nedra, Vestnik, vol. 6, pp. 33–35 (2009)
12. Purtova, I.P., Varichenko, A.I., Shpurov, I.V.: Hard-to-recover oil reserves. Terminology. Problems and state of development in Russia. In: Science and Fuel and Energy, vol. 6, pp. 21–26 (2011)
13. Polishchuk, Y.M., Yashchenko, I.G.: Comparative analysis of the quality of Russian oil. In: Technologies TEK, vol. 3, pp. 51–56 (2003)
14. Yashchenko, I.G., Polishchuk, Y.M.: Comparative analysis of the quality of hard-to-recover oils. In: The Gas Industry, vol. 5, no. 722, pp. 18–23 (2015)
15. Classification of oils. https://studfiles.net/preview/1772355/page:2

16. Maksutov, R., Orlov, G., Osipov, A.: Development of high-viscosity oil reserves in Russia. In: Technologies of the Fuel and Energy Sector, vol. 6, pp. 36–40 (2005)
17. Yashchenko, I.G., Polishchuk, Y.M.: Hard-to-recover oil: physical and chemical properties and regularities of placement. In: Novikov, A.A. (ed.) Tomsk, B-Spectrum, p. 154 (2014)
18. Shpurov, I.V., Rastrogin, A.E., Bratkova, V.G.: On the problem of development of hard-to-recover oil reserves in Western Siberia. In: Oil Industry, vol. 12, pp. 95–97 (2014)
19. Degtyarev, V.N.: About the bank of quality of oil. In: Oil Industry, vol. 3, pp. 62–63 (1997)
20. Polishchuk, Y.M., Yashchenko, I.G.: Analysis of eurasian oils quality. In: Oil industry, vol. 1, pp. 66–68 (2002)
21. Kritskaya, E.B., Chizh, D.V.: Study of changes in the physico-chemical parameters of the Ciscaucasian oil. In: Voronezh State University. Series: Chemistry. Biology. Pharmacy. Vestnik, vol. 1, pp. 21–23 (2013)
22. Klubkov, S.: Stimulation of TRIZ development will help support the level of oil production in Russia. Oil & Gas J. Russia 7(95), 6–11 (2015)
23. Raupov, I.R., Kondrasheva, N.K., Burkhanov, R.N.: Development of a mobile device for measuring the optical properties of oil in solving geological and fishing problems. Electron. Sci. J. Oil Gas Bus. **3**, 17–32 (2014). http://ogbus.ru/issues/3_2014/ogbus_3_2014_p17-32_RaupovIR_en.pdf
24. Kluvert, N.-B., Savenok, O.V.: Hard-to-recover hydrocarbon reserves, important resources in the territory of the Federal Republic of Nigeria. In: The Twenty-First International Scientific and Practical Conference, Current State of Natural and Technical Sciences, Moscow, 46 (2015). https://cyberleninka.ru/article/n/ekonomicheskaya-znachimost-razrabotki-osvoeniya-i-dobychi-bituma-iz-bituminoznogo-peska-i-tyazheloy-nefti-v-nigerii
25. Krylov, V.Y., Ostryakova, T.V.: Mathematical methods of data processing in psychological studies: new methods of cluster-analysis based on the psychological theory of L.S. Vygotsky's concept development. Psychol. J. **1**, 16 (1995)
26. Savchenko, T.N.: The application of methods of cluster-analysis for the processing of data of psychological research. In: Experimental Psychology, vol. 2 (2010)
27. Bezdek, J.C., Ehrlich, R., Full, W.: The fuzzy c-means clustering algorithm. Comput. Geosci. 10, 191–203 (1984)
28. Aliev, R.A., Guirimov, B.G.: Type-2 Fuzzy Neural Networks and Their Applications. http://www.springer.com/us/book/9783319090719
29. Turksen, I.B.: Full Type 2 to Type n Fuzzy System Models. In: Seventh International Conference on Soft Computing, Computing with Words and Perceptions in System Analysis, Decision and Control. Turkey, Izmir, p. 21 (2013)
30. Efendiyev, G.M., Mammadov, P.Z., Piriverdiyev, I.A., Mammadov, V.N.: Clustering of geological objects using FCM-algorithm and evaluation of the rate of lost circulation. In: 12th International Conference on Application of Fuzzy Systems and Soft Computing, ICAFS 2016, Vienna, Austria, 29–30 August. Procedia Comput. Sci. **102**, 159–162 (2016)

# Knowledge-Based Planning for Industrial Automation Systems: The Way to Support Decision Making

N. R. Yusupbekov[1], Sh. M. Gulyamov[1], S. S. Kasimov[2], and N. B. Usmanova[2(✉)]

[1] Tashkent State Technical University,
2, University Street, 100095 Tashkent, Uzbekistan
dodabek@mail.ru
[2] Tashkent University of Information Technologies, 108, Amir Temur Street,
100200 Tashkent, Uzbekistan
nargizausm@mail.ru

**Abstract.** The issue of adaptation of industrial automation system is considered in line with paradigm of multi-agent systems as a method of distributed artificial intelligence, wherein subsequent agents can be integrated into systems that jointly solve complex problems. In this regard we propose to use technique based on associative interrelation between the terms within multi-agents when piece of knowledge is represented as a function that maps a domain of clauses. While using trie-based structures (as possible representation technique) for good reasoning and decision making we can properly integrate metric information about the environment and semantic information provided by the user. In other words, we achieve another, higher level of representation of the environment when necessary knowledge is properly directed to actually reasoning (knowledge-based reasoning) and decision making.

**Keywords:** Industrial automation system · Associative interrelation
Trie-based structure · Knowledge-based reasoning

## 1 Introduction

The recent trends in information technology together with appropriate advances in industrial automation make manufacturing companies more competitive, bringing on research and development arena of enterprises the new challenges. The needs to satisfy increasing customer demands, with ever increasing volume of newly implemented products and services, are becoming the key success factors for product manufacturers.

Derived from the necessity of shaping the product development process in efficient, interactive, collaborative way, there multidisciplinary task of bringing the new paradigms of product development and product lifecycle management with product design

© Springer Nature Switzerland AG 2019
R. A. Aliev et al. (Eds.): ICAFS-2018, AISC 896, pp. 873–879, 2019.
https://doi.org/10.1007/978-3-030-04164-9_115

data or specific knowledge attracting the difficulties and 'right site' implementation, in a way overall collaborative process serve proper product development.

We refer to the issue of adaptation of industrial automation system while there is a difference between product lifecycle which is getting shorter and lifecycle of industrial automation systems: in most cases we do not know strict requirements for the industrial automation systems due to market change, so they are often unknown at the beginning of the planning process, and demands to product quality should be 'wrapped' into the process of planning. We propose an approach to support the adaptation of industrial automation systems in line with changes of requirements in planning with the help of multi-agent system wherein software agents reflect the environment changes through information metrics formed via associative trie-based structures being the basis for decision making.

## 2    Background and Motivation of Research

Multi-agent systems (MAS) represent the new paradigm for distributed systems of artificial intelligence, while emergence and active development of methods and technologies of distributed artificial intelligence, trends in hardware and software support of the concept of distribution and openness, brought to research filed the realization of the important fact that agents can be integrated into systems that jointly solve complex problems.

According to theoretical insights, the multi-agent system is viewed as a set of intelligent agents distributed across the network, migrating through and searching of relevant data, knowledge and procedures thus cooperating in the decision making process [1–3]. Some researchers call this way of collaborative actions as a community of software robots, aimed to satisfy the various information and computing needs of users [4]. The scope of research in the field of multi-agent systems is now very broad and comparable to the breadth of research in the field of artificial intelligence. This is due to integration of conceptual statements, system approach, the complexity of the architecture and the variety of components of each individual agent, as well as variety of mathematical and software tools in describing and developing. We can add to count also structural complexity and existence of different options agent interact each other, the complexity and diversity of components of the external environment wherein agents function.

Agent-based modeling is a computational method that enables a researcher to create, analyze, and experiment with models composed of agents interacting within an environment. The use of multi-agent technologies allows solving the actual tasks of automated resource planning in real time on the basis of new algorithms for automated production planning. Multi-agent resource planning technologies are based on a new approach to the description and modeling of complex systems. In particular, in MAS the enterprise is modeled as a dynamic network of software agents of needs and opportunities. In such a network there can be different units, production orders and specific resources [5, 6].

Nowadays there are task-oriented distributed systems used in complex industrial enterprises, wherein knowledge and resources are distributed among sufficiently independent agents, but a common control unit taking decisions in critical situations was still exist. The paradigm of completely decentralized systems, in which management occurs only through local interactions between agents can be seen as further step in this regard. At the same time, the agent gradually began its functional orientation towards the solution in a way to universal integrity (autonomy).

It is worth to note that software agents serve as an integral part of the system within essential feature of multi-agent technologies to be a fundamentally new method for solving problems: the solution is obtained automatically as a result of the interaction of a set of independent, purposeful program modules [6].

Thus, it is within particular interest to determine how those agents get information for decision making, in a challenging issues of enterprise management, while in an unpredictable dynamic environment of modern business and related requirements. Solving such problem is subject to real life demands, i.e. taking into account the life cycle of industrial production.

# 3 Methodology

The concept of a mobile software agent constitutes the basis of the multi-agent approach that is implemented and functions as an independent specialized computer program or an element of artificial intelligence.

The management task within life cycle of modern manufacturing enterprise is related with the need to solve many complex tasks of organizing production processes and preparing production, planning and organizing production management, managing production resources and personnel [7]. Taking into account modern requirements for automation of all stages of the product life cycle, including the manufacturing process, automated control and decision support systems are widely used in the management of modern enterprises.

To ensure effective automation of production management, automation of production resource planning in real time is necessary. To do this, we need new algorithms that allow us to find the optimal solution for a limited time.

The main task of such a system is to build and maintain a balance of interests of all participants in the production process. For such systems, the transition from centralized solutions to distributed ones becomes characteristic. A number of principles of modern design provide more flexibility and effectiveness of management decisions. In such schemes, agents of needs and opportunities interact in a special way: orders and resources can come into direct connection with each other and initiate a process of reviewing and agreeing plans as the anticipated or unforeseen events occur with each of these elements. Due to such a dynamic network organization, the system being developed at any time can reconsider the links between these elements and reconcile plans in a coordinated manner. In other words, automatic real-time planning of enterprise resources is provided within the framework of distributed planning solutions.

Another important property (as a result of distributed planning) is adaptability. Taking into account the capabilities of multi-agent production planning systems and the mechanisms of multi-agent algorithms, as well as the specifics of their application, it is advisable to consider the decision-making problems based on knowledge, taking into account the availability of information at different levels and layers of the production cycle as the basis for intellectual analysis in the systems of automated design of production processes.

According to the provisions of automation of production processes, different levels of automation of the enterprise are distinguished: at the management level - ERP system (corporate information system for automation of accounting and management); at the level of a process engineer - process control system (software and hardware for the management of process equipment); at the level of designers - the system of automation of design works (organizational and technical system for the performance of design activities with the use of computer technology).

Management execution systems, for example appear to be of particular interest from the standpoint of implementing activity management functions, such as status monitoring and resource allocation, operational planning, dispatching production, document management, data collection and storage, human resources management, quality management, production process management and several others. As one can note, this is a rather complex set of functions, the implementation of which requires the management system to respond at the right time, in the right situation.

To discover the proper activity behavior for agents on planning process we are giving herein with some considerations regarding the agents' activity. In general, when dealing with the term 'planning' we informally refer to the generation of the sequence of actions to solve a complex problem. Furthermore, knowledge can be seen as an expert system represented in various ways (e.g. by well-known techniques for representation of knowledge like Production Systems, Logical Calculus and Structured Models).

Referring to the terminology of artificial intelligence, we apply rules within production systems as one of the well-known techniques of knowledge representation [8, 9]. A production system includes a knowledge base, represented by production rules, a working memory to hold the matching patterns of data that causes the rules to fire and an interpreter, also called the inference engine, that decides which rule to fire.

The addition to and deletion from working memory depends on the consequent part of the fired rule. Adding of new elements to the working memory is required to maintain firing of the subsequent rules.

The product model of knowledge - a model based on rules, allows to represent knowledge in the form of sentences of the type 'If (condition), then (action)'. Being the fragments of the semantic network, production model is based on temporary relationships between the states of the objects. The production model has the disadvantage that when a sufficiently large number of products are accumulated, they begin to contradict each other due to the irreversibility of disjunctions. Several modifications are made for this case by the developers: to complicate the system, including the modules of fuzzy inference or other means of conflict resolution, e.g. priority rules, depth rules, heuristic exclusion, return mechanisms, etc. In this regard we propose to use technique based of associative interrelation between the terms when piece of knowledge is represented as a function that maps a domain of clauses. The function may take algebraic or relational form depending on the type of applications. For this we use trie-based structures which have good reasoning and decision making possibilities. Using such representation technique we can properly integrate metric information about the environment and semantic information provided by the user. In other words, we achieve another, higher level of representation of the environment when necessary knowledge is properly directed to actually reasoning (knowledge-based reasoning) and decision making.

## 4  Examples to Demonstrate the Methodology

Although concise, we present two examples to show how metric information can serve to achieve more clear understanding the context. First example refers to the sorting array of strings (or words) using trie-based representation, while characters in set of words are basis for searching the symbols (Example 1 given below). Another example deals with the task of trie-based searching the words, when the value represents the area in memory space to which corresponding request will be directed (Example 2).

Both examples are somehow the practical demonstration of using the metric information. Several other codes and implementations (more examples in java can be found at https://stackoverflow.com) with respect to finding out the relations between initial and metric-related information are feasible, both in developing the software architecture for industrial automation, and considering the managing procedures. Thus any changes in existing technical process can be taken into consideration along with layers, e.g. process, field, group/process control, production (lifecycle), management level.

By properly integrate metric information about the environment and semantic information provided by the user, higher level of representation of the environment can be achieved when necessary knowledge is properly directed to actually reasoning and decision making.

Example 1:

```
* Adds a string to the trie
*
* @param s String to add to the trie
*/
public void add(String s) {
HashMap<Character, HashMap> curr_node =
root;
for (int i = 0, n = s.length(); i < n; i++) {
Character c = s.charAt(i);
if (curr_node.containsKey(c))
curr_node = curr_node.get(c);
else {
curr_node.put(c, new HashMap<Character,
HashMap>());
curr_node = curr_node.get(c);
}
}
curr_node.put('\0', new HashMap<Character,
HashMap>(0)); // term
}
```

```
public boolean contains(String s) {
HashMap<Character, HashMap> curr_node =
root;
for (int i = 0, n = s.length(); i < n; i++) {
Character c = s.charAt(i);
if (curr_node.containsKey(c))
curr_node = curr_node.get(c);
else
return false;
}
if (curr_node.containsKey('\0'))
return true;
else
return false;
}
public static void main(String[] args) {
TrieExample t = new TrieExample();
t.add("TEMP");
t.add("TEMPLATE");
t.add("TEMPERATURE");
System.out.println(t.contains("PRE") + " " +
false);
System.out.println(t.contains("TEMP") + " " +
false);
System.out.println(t.contains("TEMPL") + " "
+ false);
System.out.println(t.contains("TEMPLES") + "
" + true);
}
}
Output:

false false
false false
false false
true true
```

Example 2:

```
class Suffix_tree{
SuffixTrieNode root = new SuffixTrieNode();
Suffix_tree(String txt) {

    for (int i = 0; i < txt.length(); i++)
        root.insertSuffix(txt.substring(i), i);
}
void search_tree(String pat) {
    List<Integer> result = root.search(pat);
        System.out.println("Pattern not found");
    else {
        int patLen = pat.length();
        for (Integer i : result)
          System.out.println("Pattern found at position " +
                (i - patLen));


}
    }
```

## 5  Conclusion and Further Work

In this paper we have discussed the possibility of using associative interconnections as a background information for trie-based structures, serving for good reasoning and decision making for industrial automation.

It is obvious within such task statement, that appropriate use case(s) should be provided, including software architecture development (in the form of platform in best case) which objective is to integrate information at different levels of production lifecycle, performing consistency and uniformity of processing. In addition, theoretical justification will be required in further steps of research as acquisition and presentation of information are the main motivating aspects to develop 'smart' industrial automation system, wherein processes are performed in efficient and harmonized way, to respond the current and future needs.

**Acknowledgement.** The authors would like to acknowledge Sh. Narzullaev, Master of Telecommunication Engineering specialty, Tashkent University of Information Technologies for providing support in getting software codes examples according to task statement.

## References

1. Wooldridge, M.: An Introduction to Multiagent Systems. Wiley, Hoboken (2002)
2. Weiss, G.: Multiagent Systems: A Modern Approach to Distributed Artificial Intelligence. The MIT Press, Cambridge (1999)
3. Usmanova, N.: Agent-based modeling of distributed environment: assigning properties to agent. In: Eighth World Conference on Intelligent Systems for Industrial Automation, Proceedings, WCIS 2014, Tashkent (2014)
4. Nissim, R., Brafman, R., Domshlak, C.: A general, fully distributed multi-agent planning algorithm. In: ISF Research/2007, William Frankel Center for Computer Science (2007)
5. Obitko, M., Marik, V.: Ontologies for multi-agent systems in manufacturing domain. In: DEXA 2002: Proceedings of the 13th International Workshop on Database and Expert Systems Applications, p. 597. IEEE Computer Society, Washington, DC (2002)
6. Cheung, C.F., Wang, W.M., Kwok, S.K.: Knowledge-based inventory management in production logistics: a multi-agent approach. Proc. Inst. Mech. Eng. Part B J. Eng. Manuf. **219**(3), 299–308 (2005)
7. http://w3.siemens.com/mcms/plant-engineering-software/en/pages/default.aspx
8. Melik-Merkumians, M., Moser, T., Schatten, A., Zoitl, A., Biffl, S.: Knowledge-based Runtime Failure Detection for Industrial Automation Systems. http://ceur-ws.org/Vol-641/paper_06.pdf
9. Moser, T., Mordinyi, R., Winkler, D.: Efficient automation systems engineering process support based on semantic integration of engineering knowledge. https://doi.org/10.1109/etfa.2011.6059098

# Logical-Linguistic Model of Functioning of Computer Systems' Software

H. Z. Igamberdiev[1], A. N. Yusupbekov[1], D. A. Mirzaev[2(✉)], and N. A. Kabulov[1]

[1] Tashkent State Technical University,
2, University Street, 100095 Tashkent, Uzbekistan
dodabek@mail.ru
[2] Tashkent University of Information Technologies,
108, Amir Temur Street, 100200 Tashkent, Uzbekistan
mdilshod@mail.ru

**Abstract.** The paper considers logical - linguistic model of software functioning process of the data control systems in the mode of functional failure formation. In addition, the structure of the process model of a functional failure formation is studied and illustrated as a fuzzy mapping of functional failures to the set of input sets of investigation object.

**Keywords:** Software · Information and control systems · Cause - effect model Process of functional failure formation

## 1 Introduction

The most important factor that determines the quality of modern software of computer systems, networks and complexes is the reliability of its functioning. This problem is addressed by many researchers and developers of software complexes and products, because the failures that lead to the stoppage of whole production or customer service, the mistakes of designers, profligate leaked confidential The financial and reputational risks. Under these conditions, the decision on the possibility of operating computer systems software must be sufficiently reliable.

The theory of reliability was originally born in the field of technical sciences and the term "reliability" characterized property of the object to perform the given functions, keeping in time and space established performance indicators in the specified Ranges.

The issues of improving the reliability of information and control systems software should be considered using the principles and methods of system analysis and synthesis of complex systems – such as principles of integrity, flexibility, determination, the necessary diversity, openness, etc. [1–4]. In accordance with the principle of integrity with increasing the size and complexity of the original system increases the likelihood of acquiring the latest new special characteristics and properties, without inheriting the properties characteristic of compound elements of complex System. Realization of the principle of necessary in solving problems of reliability improvement of software of computational systems is reduced to search of means and programming methods

R. A. Aliev et al. (Eds.): ICAFS-2018, AISC 896, pp. 880–885, 2019.
https://doi.org/10.1007/978-3-030-04164-9_116

possessing necessary potential possibilities. The principle of software flexibility implies the ability of the system to adapt or adapt to changes in the external and internal parameters of the object. The principle of determination or purpose of conditionality implies the presence of a clearly defined goal, the achievement of which ensures the control of the correct functioning of the software of the computer system. Finally, the principle of external complement requires the availability of the necessary resources, which contribute to compensation unaccounted disturbing external and internal disturbances [5]. Under a complex or large system we understand a set of interconnected constituent elements constituting a single whole, which is endowed with traits absent from its individual elements [6, 7]. Computer software is not just a system as a result of simple addition (summation) of individual elements, possessing additional properties that are absent from the constituent elements of the system. It has additional characteristics that are absent from composite elements and which arise in the process of interaction of the latter. The program system consisting of various subroutines and functional blocks, is capable to execute the given task in full volume, at the same time, that each its integral element realizes the certain functions and can function separately, being part of a single whole and bringing its useful effect to the functioning of the system as a whole [1].

## 2   Formulation of the Problem

The evaluation of fault tolerance of the software of computational systems is carried out mainly in terms of probabilistic character [8, 9]. It is stated in the paper that the probabilistic characteristics of software products before the first refusal in the execution of programs in some forms can be considered as a criterion of their reliability. In addition, it is necessary to know the characteristics of the operation of the software system in the process of recovery [10].

In work [9] the model of process of functioning of the software at formation of functional refusal which is understood as refusal is proposed, the consequence of refusal is full or partial failure of performance of functions which are entrusted to programs. This model is an imitation of the functioning of a program product. It is based on the use of a relationship that more accurately reflects the process under consideration and takes into account the causal relationships that take place in the formation of a functional failure.

Let us turn to Fig. 1, given in [9] and from which it follows that the set of input sets (IS) can be represented in the form of the following four subsets (A ÷ C):

- X1: non-standard input set => X2 (case A);
- Failure situation => X3 (case B);
- Operation of means to prevent the transition of a failing situation in a functional failure => X4 (case C);
- Functional failure or normal operation (Case D).

Explanation in more details: X1 => X4 means insignificant move the input set into a functional failure; X1 => X2 => X4 – input set => fault situations => functional

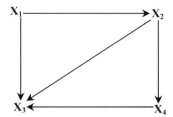

**Fig. 1.** The logical model of the software in the mode of formation of functional failure

failure: X1 => X2 => X3 => X4 – input set => fault situations => working means of compensation transition of fault situations into functional failure => functional failure.

The possibility of developing a computational process for some branches can be described by a system of statements using linguistic, subjective or certain probabilities. If the terms can be replaced with certain numbers, it is possible to come to the traditional statistical conclusions. Unfortunately, an unambiguous definition of these and similar terms are not feasible, although in some cases this fact to ignore, the law of probability distribution is assigned (because the rigorously prove the applicability of a law distribution is virtually impossible), and reliable statistical outcomes are obtained of these conditions [11].

## 3   Results

The advantage of the proposed scheme is that it allows to get additional information, accounting of the linguistic probabilities (frequencies) of the input set, fault situations, operation of the compensation means, functional failures. The analysis shows that the causal models provide a more complete picture of the reliability of software compared to models that take into account only failures without fulfilling their causes. Figure 2 considers the area of input sets of software functioning parameters of information-control systems. If in case of "a" failure in the implementing of the program the failure takes place at input vector belonging to a subset of D, and in the case of "b" in input set of a subset of C, then the estimations of reliability cannot be compared. In the first case described by input vector subjective (linguistic) probability <VERY LIKELY>, the second <NOT VERY LIKELY> or <ALMOST IMPOSSIBLE>.

Quality of SW functioning based on a scenario is described by a system of statements, for example: (if the frequency of the INPUT VECTOR set B, $\omega_{IV} = HIGH$ and frequency of functional failures $\omega_{FF} = LESS\ \omega_{IV}$, then the quality of the functioning is HIGH). To describe fuzzy type approval is used, i.e. frequency $\omega_{XI} = GREAT,\ldots,$ fuzzy relations $R_j$ between the elements of the sets $X_i$ for example, the absolute difference of frequencies $\omega_{X4}^{(f)} = LARGE$ in output of $j$- type and in functional refusal of "$f$" type, and so on.

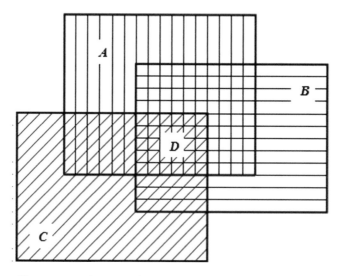

**Fig. 2.** Area of input vectors data: A region of tested input vectors data; B area of input vectors that were specified in the technical assignment; C area of real input vectors; D of real input vectors included in the technical assignment in which the software was tested.

Fuzzy approval and fuzzy relations in the general case are presented in the form of fuzzy sets:

$$\omega_{xj} = \bigcup_i \omega_{xj}^{(i)} / \mu_{\omega_{xj}}^{(i)}, j = 1, 2, \ldots;$$

where $\omega_{xj}^{(i)}$ - the possible value of the parameter $\omega_{xj}$; $\mu_{\omega_{xj}}^{(i)}$ - membership function, which determines the degree of probability on which the value $\omega_{xj}^{(i)}$ relates to the class, $\mu_{\omega}^{(i)}(x) \rightarrow [0, 1]$.

It should be noted that the possibility of assigning the frequency of events (e.g., functional refusal) to a particular class of frequencies can increase the information content of the measurement, since as additional information can be viewed all the properties held by the class [10].

Note that a similar approach (use of statements to evaluate the software) is used in [12] for a description of the two systems: {best, with high quality and worst with low quality software.

Based on the scenario, we can formulate a system of statements for software that determine the quality of his functioning. By taking similar states as references on the basis of the principle of reference [13] it is possible to determine the system preferences on the basis of which to assess the reliability of the components of the software functioning quality. Application of the reference principle to the evaluation of the software functioning quality is discussed in [14].

The process of the software functioning in accordance with cause and consequence model will be considered as a mapping of the set input vector functional refusal, i.e. $F$:

$X1 \Rightarrow X4$. INPUT VECTOR ranked by membership function from a position of assignment to a particular class of probabilistic linguistic terms described. Similarly, we can rank the set of FUNCTIONAL REFUSALS, selecting a subset of FR, which took place at the <VERY LIKELY>, <LIKELY>, etc. Accordingly, X1 and X4 have the form

$$X1 = \bigcup_i x_1^{(j)}/\mu_{X1}^j, \quad j = 1, 2, \ldots;$$

$$X4 = \bigcup_i x_4^j/\mu_{X4}^j, \quad j = 1, 2, \ldots;$$

Since the values of fuzzy or linguistic variables (LP) $x_1^{(j)} \in X1, x_4^j \in X4$, fuzzy subsets correspond to the membership functions $\mu_{X1} \in F(X1), \mu_{X4} \in F(X4)$, then the mapping $F : F(X1) = > F(X4)$ can be generally assumed fuzzy. It can be obtained as a fuzzy match for all functional failures and input vectors:

$$F = \bigcup_{j \in J} \mu_{X1}^j \times \mu_{X4}^j,$$

$$\mu_{X1}^j \in F(X1), \mu_{X4}^j \in F(X4),$$

Thus, we can state the problem of determining (derivability) fuzzy vector values $x_4^j \in X4$ with a new set of fuzzy vector values $x_1^j \in X1, j = 1, 2, \ldots$, e.g.: $\mu_{X4} = \mu_{X1} \otimes F \Leftrightarrow X4 = X1 * F$.

In [8] the possibility of developing a model of a fuzzy system based on logical structure is shown. This model includes the definition of the system $F$ with some operator $*$ such that: (a) $X4_i = X1_i * F$ for a given pair of "input-output" $X1_i \Rightarrow X4_i$; (b) the logical structure defined by, for example, models; $M_1 M_1$ - if $X1$ (LARGE); that $X4$ (GREAT); $M_2$ if $X1$ (GREAT), the $X4$ (SMALL). Applicability of this approach to solving problems is limited essentially by the fact that for the fuzzy output (if $X1^*X1^*$, then the $X4^*M_1X4^*M_1$) we must prove that $M_1M_1$ is (between) $X1^j$ and $X1^{j+1}$ (for $M_1$ or $M_2$). In this case, defining $X4^j = X1^j * F$ and $X4^{j+1} = X1^{j+1} * F$, it can be argued that the $X4^*$ is (between), $X4^j$ and $X4^{j+1}$ and fairly $X4^* = X1^* * F$.

## 4   Results

Nowadays the software reliability analysis is mainly based on the theoretical concepts inherent to the reliability of technical systems. This is due to the fact that among the entire set of functional failures there is a rather large subset of those that appear by hardware. For such failures traditional reliability assessment seems justified, although in this case insufficient statistics are significantly related. However, ambiguous functional failures should not be ignored, which though are much smaller (compared to the hardware type) subsets but nevertheless can be fatal.

Vital features using the classical notion of security with respective attitude towards software are: fundamentally different physical nature of reliability for software and technical systems: while the 'aging' is fundamentally absent in software, so the 'time to failure' is used as the main and essentially only indicator of software reliability, requiring so far a separate study; functional failure is subjective, that leads to dialectical contradiction - it is necessary to collect 'objective' statistics comparably to subjective factor and inability to obtain exact distributions of probabilistic parameters the formation of functional failure.

The software is too complex product of technical purposes, the quality of its operation cannot be described by only one scalar- indicator of what is, for example, the time to failure. Obviously, synthesis of a vector indicator's quality would be one component of the indicator.

# References

1. Stepanovich-Tsvetkova, G.: Increase of software reliability: system-functional approach. Internet Mag. "Medieval" **7**(5) (2015). http://naukovedenie.ru/index.php?p=vol7-5
2. O'Connor, J., McDermott, I.: The Art of Systems Thinking: Essential Skills for Creativity and Problem Solving. Thorsons, Toronto (1997)
3. Blauberg, I., Sadovskiy, V., Yudin, E.: Systemic approach in modern science. Problems of System Research Methodology, pp. 7–48 (1970)
4. Voskoboynikov, A.: System studies: basic concepts, principles and methodology. Knowl. Underst. Skill (6) (2013). zpu-journal.ru/ezpu/2013/6/Voskoboinikov_Systems-Research/
5. Beer, S.: Cybernetics and Management, 2nd edn. English Universities Press, London (2006)
6. Lipaev, V.: Software Quality. Yanus-K, Moscow (2002)
7. Fatrell, R., Shafer, D., Shafer, L.: Program Project Management. Achieve Optimal Quality with Minimum Costs. Vil'yams (2003)
8. Minaev, Y.: To the issue of analysis and selection of the reliability index of computer software. Cybern. Syst. Anal. **2**, 46–60 (1992)
9. Lipaev, V.: Software reliability (overview of the concepts). Autom. Mech. **10**, 5–32 (1986)
10. Myers, G.: Reliability Software Securing. Wiley, Hoboken (1980)
11. Yusupbekov, N., Gulyamov, Sh., Temerbekova, B., Mirzaev, D.: Linguistic cause and effect model of functioning of formation of functional refusal. Int. J. Emerg. Technol. Adv. Eng. **6**(1), 205–211 (2016)
12. Longbottom, R.: Reliability Computing Systems. Energoizdat (1985)
13. Kuzmin, V.: Construction of Group Decisions in the Spaces Clear and Fuzzy Binary Relativeness. Nauka (1982)
14. Mirzaev, D.: Indicators of the reliability of software information management systems. In: Problems of Information and Telecommunication Technologies, Scientific and Technical Conference, 3-Section, Collection of Reports, pp. 351–353 (2015)

# A Novel Technique to Solve Fully Fuzzy Nonlinear Matrix Equations

Raheleh Jafari[1](✉), Sina Razvarz[2], and Alexander Gegov[3]

[1] Centre for Artificial Intelligence Research (CAIR), University of Agder,
Grimstad, Norway
jafari3339@yahoo.com
[2] Departamento de Control Automatico,
CINVESTAV-IPN (National Polytechnic Institute), Mexico City, Mexico
[3] School of Computing, University of Portsmouth, Buckingham Building,
Portsmouth PO1 3HE, UK

**Abstract.** Several techniques are suggested in order to generate estimated solutions of fuzzy nonlinear programming problems. This work is an attempt in order to suggest a novel technique to obtain the fuzzy optimal solution related to the fuzzy nonlinear problems. The major concept is on the basis of the employing nonlinear system with equality constraints in order to generate nonnegative fuzzy number matrixes $\widetilde{\gamma}, \widetilde{\gamma}^2, \ldots, \widetilde{\gamma}^n$ that satisfies $\widetilde{D}\widetilde{\gamma} + \widetilde{G}\widetilde{\gamma}^2 + \ldots + \widetilde{P}\widetilde{\gamma}^n = \widetilde{Q}$ in which $\widetilde{D}, \widetilde{G}, \ldots, \widetilde{P}$ and $\widetilde{Q}$ are taken to be fuzzy number matrices. An example is demonstrated in order to show the capability of the proposed model. The outcomes show that the suggested technique is simple to use for resolving fully fuzzy nonlinear system (FFNS).

**Keywords:** Fuzzy solution · Fuzzy numbers · Fully fuzzy nonlinear system
Fully fuzzy matrix equations

## 1 Introduction

An area of applied mathematics which contains many applications in different fields of science is resolving fuzzy nonlinear systems [1–8]. In [9] a numerical method is proposed for solving fuzzy systems. Theoretical aspects related to the fuzzy linear system are investigated in [10]. In [11] the Jacobi as well as Gauss Seidel techniques are suggested in order to find the solution of fuzzy linear system. In [12] the Conjugate gradient approach is suggested in order to resolve fuzzy symmetric positive definite system of linear equation. In [13] an iterative algorithm in order to resolve dual linear systems is proposed. In [14] LU decomposition technique is applied in order to solve fuzzy system of linear equation. In [15] a certain decomposition technique is applied in order to resolve fully fuzzy linear system of equations.

Generally, there is no approach on the basis of matrices which yields fuzzy solutions for FFNS. In this paper, a novel method is proposed in order to resolve the fully fuzzy nonlinear matrix equations (FFNME), $\widetilde{D}\widetilde{Y} + \widetilde{G}\widetilde{Y}^2 + \ldots + \widetilde{P}\widetilde{Y}^n = \widetilde{Q}$, in which $\widetilde{D}, \widetilde{G}, \ldots, \widetilde{P}$ are $n \times n$ arbitrary triangular fuzzy number matrices, $\widetilde{Q}$ is a $n \times 1$

© Springer Nature Switzerland AG 2019
R. A. Aliev et al. (Eds.): ICAFS-2018, AISC 896, pp. 886–892, 2019.
https://doi.org/10.1007/978-3-030-04164-9_117

arbitrary triangular fuzzy number matrix, also the unknown $\tilde{Y}, \tilde{Y}^2, \ldots, \tilde{Y}^n$ are matrices having $n$ positive fuzzy numbers. The fuzzy matrices $\tilde{Y}^2, \ldots, \tilde{Y}^n$ are defined with following elements:

$$\text{If } \tilde{Y} = \begin{bmatrix} \tilde{y}_{1,1} \\ \tilde{y}_{2,1} \\ \vdots \\ \tilde{y}_{n,1} \end{bmatrix}, \text{ then } \tilde{Y}^2 = \begin{bmatrix} \tilde{y}_{1,1}^2 \\ \tilde{y}_{2,1}^2 \\ \vdots \\ \tilde{y}_{n,1}^2 \end{bmatrix}, \ldots, \tilde{Y}^n = \begin{bmatrix} \tilde{y}_{1,1}^n \\ \tilde{y}_{2,1}^n \\ \vdots \\ \tilde{y}_{n,1}^n \end{bmatrix}.$$

A nonlinear system with equality constraints is applied in order to obtain non-negative fuzzy number matrix $\tilde{Y}, \tilde{Y}^2, \ldots, \tilde{Y}^n$ that satisfies $\tilde{D}\tilde{Y} + \tilde{G}\tilde{Y}^2 + \ldots + \tilde{P}\tilde{Y}^n = \tilde{Q}$.

This paper is organized as follows: Some basic definitions are given in Sect. 2. In Sect. 3, a novel technique in order to resolve FFNS is suggested with numerical example. Conclusion is given in Sect. 4.

## 2 Basic Definitions and Notations

In this section the essential notations utilized in fuzzy operations are given.

**Definition 1.** A fuzzy number is a fuzzy set $\tilde{z} : \mathbb{R}^1 \rightarrow [0, 1]$ such that

   i. $\tilde{z}$ is upper semi-continuous.
  ii. $\tilde{z}(x) = 0$ outside some interval $[k, l]$.
 iii. There exist real numbers $l$ and $m$, $k \leq l \leq m \leq n$, where
  1. $\tilde{z}(x)$ is monotonically increasing on $[k, l]$,
  2. $\tilde{z}(x)$ is monotonically decreasing on $[m, n]$,
  3. $\tilde{z}(x) = 1, l \leq x \leq m$.

The set of all fuzzy numbers is displayed by $E^1$ [16, 17].

**Definition 2.** $\tilde{C} = (\tilde{c}_{ij})$ is named a fuzzy number matrix, if each element of $\tilde{C}$ be a fuzzy number. $\tilde{C}$ is named a positive (negative) fuzzy matrix, also is displayed by $\tilde{C} > 0 \, (\tilde{C} < 0)$ if each element of $\tilde{C}$ be positive (negative). $\tilde{C}$ is named non-positive (non-negative), also displayed by $\tilde{C} \leq 0 \, (\tilde{C} \geq 0)$ if each element of $\tilde{C}$ is non-positive (non-negative).

**Definition 3.** Suppose $\tilde{p} = (p_m, p_l, p_u)$ as well as $\tilde{q} = (q_m, q_l, q_u)$ be two triangular fuzzy numbers. Hence:

1. $\tilde{p} \oplus \tilde{q} = (p_m + q_m, p_l + q_l, p_u + q_u)$,
2. $-\tilde{p} = (-p_u, -p_l, -p_m)$,
3. $\tilde{p} \ominus \tilde{q} = (p_m - q_u, p_l - q_l, p_u - q_m)$.

The fuzzy multiplication is displayed by $\hat{*}$ [18]. It is performed with the below mentioned equation:

$$\tilde{p} \mathbin{\widehat{*}} \tilde{q} = (s_m, s_l, s_u),$$

where

$$s_l = p_l \cdot q_l,$$
$$s_m = \min(p_m.q_m, p_m.q_u, p_u.q_m, p_u.q_u),$$
$$s_u = \max(p_m.q_m, p_m.q_u, p_u.q_m, p_u.q_u).$$

In a case that $\tilde{p}$ be a triangular fuzzy number as well as $\tilde{q}$ be a non-negative one, the following is concluded:

$$\tilde{p} \mathbin{\widehat{*}} \tilde{q} = \begin{cases} (p_m.q_m, p_l.q_l, p_u.q_u), & p_m \geq 0, \\ (p_m.q_u, p_l.q_l, p_u.q_u), & p_m < 0, p_u \geq 0, \\ (p_m.q_m, p_l.q_l, p_u.q_m), & p_m < 0, p_u < 0. \end{cases}$$

## 3   Fully Fuzzy Nonlinear Matrix Equation

Consider the below mentioned FFNME:

$$\begin{bmatrix} \tilde{d}_{11} & \tilde{d}_{12} & \cdots & \tilde{d}_{1n} \\ \tilde{d}_{21} & \tilde{d}_{22} & \cdots & \tilde{d}_{2n} \\ \vdots & \vdots & \vdots & \vdots \\ \tilde{d}_{n1} & \tilde{d}_{n2} & \cdots & \tilde{d}_{nn} \end{bmatrix} \begin{bmatrix} \tilde{y}_{11} \\ \tilde{y}_{21} \\ \vdots \\ \tilde{y}_{n1} \end{bmatrix} + \begin{bmatrix} \tilde{g}_{11} & \tilde{g}_{12} & \cdots & \tilde{g}_{1n} \\ \tilde{g}_{21} & \tilde{g}_{22} & \cdots & \tilde{g}_{2n} \\ \vdots & \vdots & \vdots & \vdots \\ \tilde{g}_{n1} & \tilde{g}_{n2} & \cdots & \tilde{g}_{nn} \end{bmatrix} \begin{bmatrix} \tilde{y}_{11}^2 \\ \tilde{y}_{21}^2 \\ \vdots \\ \tilde{y}_{n1}^2 \end{bmatrix} + \cdots + \begin{bmatrix} \tilde{p}_{11} & \tilde{p}_{12} & \cdots & \tilde{p}_{1n} \\ \tilde{p}_{21} & \tilde{p}_{22} & \cdots & \tilde{p}_{2n} \\ \vdots & \vdots & \vdots & \vdots \\ \tilde{p}_{n1} & \tilde{p}_{n2} & \cdots & \tilde{p}_{nn} \end{bmatrix} \begin{bmatrix} \tilde{y}_{11}^n \\ \tilde{y}_{21}^n \\ \vdots \\ \tilde{y}_{n1}^n \end{bmatrix} = \begin{bmatrix} \tilde{q}_{11} \\ \tilde{q}_{21} \\ \vdots \\ \tilde{q}_{n1} \end{bmatrix}$$

In which $\tilde{d}_{ij}$, $\tilde{g}_{ij}$ as well as $\tilde{p}_{ij} (for\, 1 \leq i,\, j \leq n)$, are arbitrary triangular fuzzy numbers, the elements $\tilde{q}_{i1}$ as well as the unknown elements $\tilde{y}_{i1}$ are nonnegative fuzzy numbers. Utilizing matrix notation, the following is extracted

$$\tilde{D} \mathbin{\widehat{*}} \tilde{Y} + \tilde{G} \mathbin{\widehat{*}} \tilde{Y}^2 + \ldots + \tilde{P} \mathbin{\widehat{*}} \tilde{Y}^n = \tilde{Q}. \tag{1}$$

The fuzzy number matrices $\tilde{Y} = (\tilde{y}_1, \tilde{y}_2, \ldots, \tilde{y}_n)^T$, $\tilde{Y}^2 = (\tilde{y}_1^2, \tilde{y}_2^2, \ldots, \tilde{y}_n^2)^T, \ldots, \tilde{Y}^n = (\tilde{y}_1^n, \tilde{y}_2^n, \ldots, \tilde{y}_n^n)^T$ demonstrated by $\tilde{y}_i = (\tilde{u}_{i1}, \tilde{y}_{i1}, \tilde{v}_{i1})$, $\tilde{y}_i^2 = (\tilde{u}_{i1}^2, \tilde{y}_{i1}^2, \tilde{v}_{i1}^2), \ldots, \tilde{y}_i^n = (\tilde{u}_{i1}^n, \tilde{y}_{i1}^n, \tilde{v}_{i1}^n)$, $(for\, 1 \leq i \leq n)$, are the solutions of the fuzzy matrix system Eq. (1) if

$$\tilde{d}_i \mathbin{\widehat{*}} \tilde{Y} + \tilde{g}_i \mathbin{\widehat{*}} \tilde{Y}^2 + \ldots + \tilde{p}_i \mathbin{\widehat{*}} \tilde{Y}^n = \tilde{q}_i, \, 1 \leq i \leq n, \tag{2}$$

where

$$\tilde{q}_i = (\tilde{a}_{i1}, \tilde{q}_{i1}, \tilde{c}_{i1}),$$
$$\tilde{d}_i = ((\tilde{b}_{i1}, \tilde{d}_{i1}, \tilde{e}_{i1}), (\tilde{b}_{i2}, \tilde{d}_{i2}, \tilde{e}_{i2}), \ldots, (\tilde{b}_{in}, \tilde{d}_{in}, \tilde{e}_{in})),$$
$$\tilde{g}_i = ((\tilde{f}_{i1}, \tilde{g}_{i1}, \tilde{h}_{i1}), (\tilde{f}_{i2}, \tilde{g}_{i2}, \tilde{h}_{i2}), \ldots, (\tilde{f}_{in}, \tilde{g}_{in}, \tilde{h}_{in})),$$
$$\tilde{p}_i = ((\tilde{r}_{i1}, \tilde{p}_{i1}, \tilde{s}_{i1}), (\tilde{r}_{i2}, \tilde{p}_{i2}, \tilde{s}_{i2}), \ldots, (\tilde{r}_{in}, \tilde{p}_{in}, \tilde{s}_{in})).$$

**Definition 4.** In the nonnegative FFNME Eq. (1), with new notations $\tilde{D} = (B, D, E)$, $\tilde{G} = (F, G, H)$, ..., $\tilde{P} = (R, P, S)$ in which $B, D, E, F, G, H, ...., R, P, S$ are crisp matrices, we say that $\tilde{Y}, \tilde{Y}^2, ..., \tilde{Y}^n$ are the solutions if:

$$\begin{cases} BU + FU^2 + \ldots + RU^n = A, \\ DY + GY^2 + \ldots + PY^n = Q, \\ EV + HV^2 + \ldots + SV^n = C. \end{cases} \tag{3}$$

Moreover, if $U \geq 0, Y - U \geq 0, V - Y \geq 0, Y^2 - U^2 \geq 0, V^2 - Y^2 \geq 0, ..., Y^n - U^n \geq 0, V^n - Y^n \geq 0$, so it can be denoted that $\tilde{Y}, \tilde{Y}^2, ..., \tilde{Y}^n$ are consistent solutions of the nonnegative FFNME.

## 3.1   The Proposed Technique

Here, a novel technique in order to obtain fuzzy solutions of an FFNME is suggested. Take into consideration the FFNME Eq. (2) in which all the parameters $\tilde{d}_{ij}, \tilde{g}_{ij}, ..., \tilde{p}_{ij}, \tilde{y}_{i1}$ as well as $\tilde{q}_{i1}$ are demonstrated as $(b_{ij}, d_{ij}, e_{ij}), (f_{ij}, g_{ij}, h_{ij}), ...,$ $(r_{ij}, p_{ij}, s_{ij}), (u_{i1}, y_{i1}, v_{i1})$ and $(a_{i1}, q_{i1}, c_{i1})$ respectively. Hence the FFNME is written as below

$$(B, D, E)(U, Y, V) + (F, G, H)\left(U^2, Y^2, V^2\right) + \ldots + (R, P, S)(U^n, Y^n, V^n) = (A, Q, C), \tag{4}$$

Assuming $(b_{ik}, d_{ik}, e_{ik})\widehat{*}(u_{k1}, y_{k1}, v_{k1}) + (f_{ik}, g_{ik}, h_{ik})\widehat{*}\left(u_{k1}^2, y_{k1}^2, v_{k1}^2\right) + \ldots + (r_{ik}, p_{ik}, s_{ik})\widehat{*}\left(u_{k1}^n, y_{k1}^n, v_{k1}^n\right) = \left(k_{k1}^{(j)}, o_{k1}^{(j)}, x_{k1}^{(j)}\right), 1 \leq i, j, k \leq n$, in which each $(u_{k1}, y_{k1}, v_{k1})$ is a nonnegative triangular fuzzy number. The FFNME (2) can be displayed as below:

$$\sum_{k=1}^{n} \left(k_{k1}^{(j)}, o_{k1}^{(j)}, x_{k1}^{(j)}\right) = (a_{i1}, q_{i1}, c_{i1}), \quad 1 \leq i \leq n. \tag{5}$$

Utilizing arithmetic operations, described in Sect. 2, the following nonlinear programming is obtained in which, the artificial variables $r_i, i = 1, 2, ..., n^2$ is added. Minimize $r_1 + r_2 + \ldots + r_n^2$,

$$subject\ to \begin{cases} \sum_{k=1}^{n} w_{k1}^{(1)} + r_1 = d_{11} \\ \sum_{k=1}^{n} w_{k1}^{(2)} + r_2 = d_{21} \\ \qquad \vdots \\ \sum_{k=1}^{n} w_{k1}^{(n)} + r_n = d_{n1}, \\ \sum_{k=1}^{n} q_{k1}^{(1)} + r_{n+1} = b_{11}, \\ \qquad \vdots \\ \sum_{k=1}^{n} u_{k1}^{(n)} + r_{3n} = f_{n1}. \end{cases}$$

## 4 Example

To illustrate the technique proposed in this paper, consider the following examples.

**Example 4.1.** Consider the following FFNME:

$$\begin{bmatrix} (2,3,5) & (2,4,5) \\ (1,2,3) & (3,4,6) \end{bmatrix} \begin{bmatrix} \tilde{x}_{11} \\ \tilde{x}_{21} \end{bmatrix} + \begin{bmatrix} (1,2,3) & (3,5,6) \\ (3,4,5) & (1,3,4) \end{bmatrix} \begin{bmatrix} \tilde{x}_{11}^2 \\ \tilde{x}_{21}^2 \end{bmatrix} = \begin{bmatrix} (19,140,467) \\ (14,136,436) \end{bmatrix},$$

where $\tilde{x}_{11}, \tilde{x}_{21}, \tilde{x}_{11}^2, \tilde{x}_{21}^2$ are triangular fuzzy numbers.

Assuming $\tilde{x}_{11} = (y_{11}, x_{11}, z_{11}), \tilde{x}_{21} = (y_{21}, x_{21}, z_{21}), \tilde{x}_{11}^2 = \left(y_{11}^2, x_{11}^2, z_{11}^2\right)$ and $\tilde{x}_{21}^2 = \left(y_{21}^2, x_{21}^2, z_{21}^2\right)$. The given FFNME is written as follows:

$$\begin{cases} (2,3,5)\widehat{*}(y_{11}, x_{11}, z_{11}) + (2,4,5)\widehat{*}(y_{21}, x_{21}, z_{21}) + (1,2,3)\left(y_{11}^2, x_{11}^2, z_{11}^2\right) + \\ \qquad (3,5,6)\left(y_{21}^2, x_{21}^2, z_{21}^2\right) = (19,140,467), \\ (1,2,3)\widehat{*}(y_{11}, x_{11}, z_{11}) + (3,4,6)\widehat{*}(y_{21}, x_{21}, z_{21}) + (3,4,5)\left(y_{11}^2, x_{11}^2, z_{11}^2\right) + \\ \qquad (1,3,4)\left(y_{21}^2, x_{21}^2, z_{21}^2\right) = (14,136,436). \end{cases}$$

Wherein

$$\begin{cases} \left(2y_{11} + 2y_{21} + y_{11}^2 + 3y_{21}^2, 3x_{11} + 4x_{21} + 2x_{11}^2 + 5x_{21}^2, 5z_{11} + 5z_{21} + 3z_{11}^2 + 6z_{21}^2\right) = (19,140,467), \\ \left(y_{11} + 3y_{21} + 3y_{11}^2 + y_{21}^2, 2x_{11} + 4x_{21} + 4x_{11}^2 + 3x_{21}^2, 3z_{11} + 6z_{21} + 5z_{11}^2 + 4z_{21}^2\right) = (14,136,436). \end{cases}$$

Applying the proposed technique, the above FFNME is converted into the following crisp system:

$$\begin{cases} 2y_{11} + 2y_{21} + y_{11}^2 + 3y_{21}^2 = 19, \\ 3x_{11} + 4x_{21} + 2x_{11}^2 + 5x_{21}^2 = 140, \\ 5z_{11} + 5z_{21} + 3z_{11}^2 + 6z_{21}^2 = 467, \\ y_{11} + 3y_{21} + 3y_{11}^2 + y_{21}^2 = 14, \\ 2x_{11} + 4x_{21} + 4x_{11}^2 + 3x_{21}^2 = 136, \\ 3z_{11} + 6z_{21} + 5z_{11}^2 + 4z_{21}^2 = 436. \end{cases}$$

Minimize $r_1 + r_2 + \ldots + r_6$

$$\begin{cases} 2y_{11} + 2y_{21} + y_{11}^2 + 3y_{21}^2 + r_1 = 19, \\ 3x_{11} + 4x_{21} + 2x_{11}^2 + 5x_{21}^2 + r_2 = 140, \\ 5z_{11} + 5z_{21} + 3z_{11}^2 + 6z_{21}^2 + r_3 = 467, \\ y_{11} + 3y_{21} + 3y_{11}^2 + y_{21}^2 + r_4 = 14, \\ 2x_{11} + 4x_{21} + 4x_{11}^2 + 3x_{21}^2 + r_5 = 136, \\ 3z_{11} + 6z_{21} + 5z_{11}^2 + 4z_{21}^2 + r_6 = 436. \end{cases}$$

where $r_1 + r_2 + \ldots + r_6 \geq 0$. The optimal solution is $y_{11} = 1, y_{21} = 2, y_{11}^2 = 1, y_{21}^2 = 4, x_{11} = 4, x_{21} = 4, x_{11}^2 = 16, x_{21}^2 = 16, z_{11} = 6, z_{21} = 7, z_{11}^2 = 36, z_{21}^2 = 49$. Hence the fuzzy solution is given by $\tilde{x}_{11} = (1,4,6), \tilde{x}_{21} = (2,4,7), \tilde{x}_{11}^2 = (1,16,36)$ and $\tilde{x}_{21}^2 = (4,16,49)$.

## 5  Concluding Remarks

The fuzzy nonlinear systems are extremely significant in numerical analysis. In this paper, a novel technique in order to extract the nonnegative fuzzy optimal solutions of FFNME, $\tilde{D}\tilde{Y} + \tilde{G}\tilde{Y}^2 + \ldots + \tilde{P}\tilde{Y}^n = \tilde{Q}$ is suggested, in which $\tilde{D}, \tilde{G}, \ldots, \tilde{P}$ are $n \times n$ arbitrary triangular fuzzy number matrices, $\tilde{Q}$ is a $n \times 1$ arbitrary triangular fuzzy number matrix, also the unknown $\tilde{Y}, \tilde{Y}^2, \ldots, \tilde{Y}^n$ are matrices having $n$ positive fuzzy numbers. A nonlinear system with equality constraints to FFNME is utilized in order to resolve FFNME. The suggested technique is validated with a numerical example.

## References

1. Jafari, R., Razvarz, S.: Solution of fuzzy differential equations using fuzzy Sumudu transforms. In: IEEE International Conference on Innovations in Intelligent Systems and Applications, vol. 1, pp. 84–89 (2017)
2. Jafari, R., Yu, W.: Uncertainty nonlinear systems modeling with fuzzy equations. In: Proceedings of the 16th IEEE International Conference on Information Reuse and Integration, San Francisco, California, USA, 1 August 2015, pp. 182–188 (2015)
3. Jafari, R., Yu, W.: Uncertainty nonlinear systems control with fuzzy equations. In: IEEE International Conference on Systems, Man, and Cybernetics, vol. 1, pp. 2885–2890 (2015)
4. Jafari, R., Yu, W., Li, X.: Numerical solution of fuzzy equations with Z-numbers using neural networks. Intell. Autom. Soft Comput. 1, 1–7 (2017)
5. Jafari, R., Yu, W., Li, X., Razvarz, S.: Numerical solution of fuzzy differential equations with Z-numbers using Bernstein neural networks. Int. J. Comput. Intell. Syst. 10, 1226–1237 (2017)
6. Razvarz, S., Jafari, R., Granmo, O.Ch., Gegov, A.: Solution of dual fuzzy equations using a new iterative method. In: Asian Conference on Intelligent Information and Database Systems, vol. 1, pp. 245–255 (2018)
7. Razvarz, S., Jafari, R., Yu, W.: Numerical solution of fuzzy differential equations with Z-numbers using fuzzy Sumudu transforms. Adv. Sci. Technol. Eng. Syst. J. (ASTESJ) 3, 66–75 (2018)
8. Razvarz, S., Jafari, R., Yu, W., Khalili, A.: PSO and NN modeling for photocatalytic removal of pollution in wastewater. In: 14th International Conference on Electrical Engineering, Computing Science and Automatic Control (CCE) Electrical Engineering, vol. 1, pp. 1–6 (2017)
9. Takagi, T., Sugeno, M.: Structure identifical of systems and its application to modelling and control. IEEE Trans. Syst. Man. Cybern. 15, 116–132 (1985)
10. Dubois, D., Prade, H.: Systems of linear fuzzy constraints. Fuzzy Sets Syst. 3, 37–48 (1980)
11. Allahviranloo, T.: Numerical methods for fuzzy system of linear equations. Appl. Math. Comput. 155, 493–502 (2004)
12. Abbasbandy, S., Jafarian, A., Ezzati, R.: Conjugate gradient method for fuzzy symmetric positive definite system of linear equations. Appl. Math. Comput. 171, 1184–1191 (2005)
13. Wang, X., Zhong, Z., Ha, M.: Iteration algorithms for solving a system of fuzzy linear equations. Fuzzy Sets Syst. 119, 121–128 (2001)
14. Abbasbandy, S., Ezzati, R., Jafarian, A.: LU decomposition method for solving fuzzy system of linear equations. Appl. Math. Comput. 172, 633–643 (2006)

15. Nasseri, S.H., Sohrabi, M., Ardil, E.: Solving fully fuzzy linear systems by use of a certain decomposition of the coefficient matrix. Int. J. Comput. Math. Sci. **2**, 140–142 (2008)
16. Goetschel, R., Voxman, W.: Elementary calculus. Fuzzy Sets Syst. **18**, 31–43 (1986)
17. Nguyen, H.T.: A note on the extension principle for fuzzy sets. J. Math. Anal. Appl. **64**, 369–380 (1978)
18. Feuring, T.H., Lippe, W.M.: Fuzzy neural networks are universal approximators. In: IFSA World Congress, vol. 2, pp. 659–662 (1995)

# Algebraic Properties of Z-Numbers Under Additive Arithmetic Operations

Akif V. Alizadeh[1], Rashad R. Aliyev[2], and Oleg H. Huseynov[3(✉)]

[1] Department of Control and Systems Engineering, Azerbaijan State Oil and Industry University, 20 Azadlig Ave., Baku AZ1010, Azerbaijan
akifoder@yahoo.com, a.alizade@asoiu.edu.az
[2] Department of Mathematics, Eastern Mediterranean University, Famagusta, North Cyprus, Turkey
rashadaliyev@yahoo.com
[3] Research Laboratory of Intelligent Control and Decision Making Systems in Industry and Economics, Azerbaijan State Oil and Industry University, 20 Azadlig Ave., Baku AZ1010, Azerbaijan
oleg_huseynov@yahoo.com

**Abstract.** Prof. L.A. Zadeh introduced the concept of a Z-number for description of real-world information. A Z-number is an ordered pair $Z = (A, B)$ of fuzzy numbers $A$ and $B$ used to describe a value of a random variable $X$. $A$ is an imprecise estimation of a value of $X$ and $B$ is an imprecise estimation of reliability of $A$. A series of important works on computations with Z-numbers and applications were published. However, no study exists on properties of operation of Z-numbers. Such theoretical study is necessary to formulate the basics of the theory of Z-numbers. In this paper we prove that Z-numbers exhibit fundamental properties under additive arithmetic operations.

**Keywords:** Fuzzy arithmetic · Probabilistic arithmetic · Associativity law Commutativity law · Z-number

## 1 Introduction

Real-world information is often characterized by imprecision and partial reliability. Indeed, imprecision actually takes place when values of interest are estimated by a human being using natural language. At the same time, these estimations are often partially reliable as human knowledge and experience are not perfect.

To develop a basis for dealing with imprecision and partial reliability, Zadeh [12] introduced the concept of a Z-number as an ordered pair $Z = (A, B)$ of fuzzy numbers used to describe a value of a random variable $X$, where $A$ is a fuzzy constraint on values of $X$ and $B$ is a reliability considered as a value of probability measure of $A$.

Several studies exist on computation with Z-numbers, which utilize the classical fuzzy arithmetic and probabilistic arithmetic. Zadeh suggests a general framework of computation of a Z-number-valued function on the basis of the Zadeh's extension principle [12]. However, the proposed approach is characterized by high complexity, it requires to deal with variational problems.

© Springer Nature Switzerland AG 2019
R. A. Aliev et al. (Eds.): ICAFS-2018, AISC 896, pp. 893–900, 2019.
https://doi.org/10.1007/978-3-030-04164-9_118

Kang et al. [4] proposed to dealing with Z-numbers which naturally arise in the areas of control, decision making, modeling and others. The approach is based on converting a Z-number to a fuzzy number on the basis of an expectation of a fuzzy set. However, converting Z-numbers to fuzzy numbers leads to loss of original information reducing the benefit of using original Z-number-based information. The work of Zadeh [14] is devoted to computation over continuous Z-numbers and several important practical problems. The suggested investigation is based on the use of normal probability density functions for modeling random variables. Aliev and colleagues [1–3] suggested a general and computationally effective approach to computation with Z-numbers. The approach is applied to computation of arithmetic and algebraic operations, t-norms and s-norms, and construction of typical functions. However, this approach is also based on classical fuzzy arithmetic.

The main disadvantage of classical interval arithmetic and fuzzy arithmetic is that fundamental properties of arithmetic operations over real numbers are lost [5–10]. This creates problems with solving fuzzy equations, defining derivatives of fuzzy functions etc. In order to resolve this problem, in [7–9] they introduced basics of a new approach to fuzzy arithmetic which relies on the use of so-called horizontal membership functions (HMFs).

Thus, there is a need for investigation of properties of operations over Z-numbers. This would provide a strong basis for the theory of Z-numbers and its further development. In this paper, we study properties of Z-numbers under additive arithmetic operations. Validity of the study is illustrated in examples.

The paper is organized as follows. Section 2 includes basic concepts used in the paper. In Sect. 3 we prove basic laws of arithmetic operations over Z-numbers and provide examples. Section 4 concludes.

## 2   Preliminaries

**Definition 1. Arithmetic operations of random variables** [11]. Let $X_1$ and $X_2$ be two independent continuous random variables with probability distributions $p_1$ and $p_2$. A probability distribution $p_{12}$ of $X_{12} = X_1 * X_2$, $* \in \{+, -, \cdot, /\}$ is referred to as a convolution (a resulting probability distribution of an arithmetic operation) of $p_1$ and $p_2$ and is defined as follows.

$$X_{12} = X_1 + X_2 : \quad p_{12}(x) = \int_{-\infty}^{\infty} p_1(x_1) p_2(x - x_1) dx_1,$$

$$X_{12} = X_1 - X_2 : \quad p_{12}(x) = \int_{-\infty}^{\infty} p_1(x_1) p_2(x_1 - x) dx_1,$$

$$X_{12} = X_1 X_2 : \quad p_{12}(x) = \int_{-\infty}^{\infty} p_1(x_1) p_2(x/x_1) \frac{1}{|x_1|} dx_1,$$

$$X_{12} = X_1/X_2 : \quad p_{12}(x) = \int_{-\infty}^{\infty} |x_1| p_1(x_1) p_2(x_1/x) dx_1.$$

Let $X_1$ and $X_2$ be two independent discrete random variables with the corresponding outcome spaces $X_1 = \{x_{11}, \ldots, x_{1i}, \ldots, x_{1n_1}\}$ and $X_2 = \{x_{21}, \ldots, x_{2i}, \ldots, x_{2n_2}\}$ and the corresponding discrete probability distributions $p_1$ and $p_2$. The probability distribution of $X_1 * X_2$ is the convolution $p_{12} = p_1 \circ p_2$ of $p_1$ and $p_2$ which is determined as follows:

$$p_{12}(x) = \sum_{x=x_1 * x_2} p_1(x_1) p_2(x_2),$$

for any $x \in \{x_1 * x_2 | x_1 \in X_1, x_2 \in X_2\}$, $x_1 \in X_1, x_2 \in X_2$.

**Definition 2. Probability measure of a continuous fuzzy number** [13]. Let $X$ be a continuous random variable with pdf $p$. Let $A$ be a continuous fuzzy number describing a possibilistic restriction on values of $X$. A probability measure of $A$ denoted $P(A)$ is defined as

$$P(A) = \int_{\mathcal{R}} \mu_A(x) p(x) dx.$$

**Definition 3. A Z-number** [12]. A Z-number is an ordered pair $Z = (A, B)$, where $A$ is a fuzzy number playing a role of a fuzzy constraint on values that a random variable $X$ may take:

$$X \text{ is } A,$$

and $B$ is a fuzzy number with a membership function $\mu_B : [0, 1] \rightarrow [0, 1]$, playing a role of a fuzzy constraint on the probability measure of $A$:

$$P(A) \text{ is } B.$$

## 3 Basic Laws of Additive Arithmetic Operations over Z-Numbers

### 3.1 Commutative Law

Let us verify if the commutative law for sum of Z-numbers is true:

$$Z_{12} = Z_1 + Z_2 = Z_2 + Z_1 = Z_{21}. \tag{1}$$

That is,

$$(A_1, B_1) + (A_2, B_2) = (A_2, B_2) + (A_1, B_1)$$

**Proof.** At first, we have to consider operation over fuzzy numbers:

$$A_{12} = A_1 + A_2 = A_2 + A_1 = A_{21}. \tag{2}$$

Commutativity (2) was proved in [6–8] in terms of HMFs. Now we need to prove equality of fuzzy restrictions over values of probability measures of $A_{12}$ and $A_{21}$:

$$B_{12} = B_{21}, \tag{3}$$

where the basic values are defined as

$$b_{12} = \int_R \mu_{A_{12}}(u) p_{12}(u) du, \tag{4}$$

$$b_{21} = \int_R \mu_{A_{21}}(u) p_{21}(u) du. \tag{5}$$

Taking into account that (2) holds, we have $\mu_{A_{12}} = \mu_{A_{21}}$. At the same time, it is known that commutativity law holds for convolution of sum $\circ_+$ of random variables:

$$p_{12} = p_1 \circ_+ p_2 = p_2 \circ_+ p_1 = p_{21}. \tag{6}$$

Thus, $B_{12} = B_{21}$. Thus, the commutativity law holds for Z-numbers:

$$Z_1 + Z_2 = Z_2 + Z_1.$$

**Example.** Consider Z-numbers $Z_1 = (A_1, B_1)$ and $Z_2 = (A_2, B_2)$, the components of which are trapezoidal fuzzy numbers (TFNs):

$$A_1 = (10, 14, 14, 20), \ B_1 = (0.6, 0.8, 0.8, 0.9);$$

$$A_2 = (8, 16, 16, 18), \ B_2 = (0.4, 0.6, 0.6, 0.8).$$

Let us compute $Z_{12}$ and $Z_{21}$, and verify if (1) holds. At first, consider $Z_{12}$. We have obtained $A_{12}$

$$A_{12} = (18, 30, 30, 38).$$

The fuzzy reliability value $B_{12}$ is obtained as

$$B_{12} = (0.5322, 0.6725, 0.6725, 0.7776).$$

For $Z_{21}$ we have obtained the following results:

$$A_{21} = (18, 30, 30, 38), \quad B_{21} = (0.5322, 0.6725, 0.6725, 0.7776).$$

Thus, $A_{12} = A_{21}$ and $B_{12} = B_{21}$. Therefore, commutativity law holds:

$$Z_1 + Z_2 = Z_2 + Z_1.$$

## 3.2 Associative Law

Let us study whether associative law for sum of Z-numbers holds:

$$Z_{231} = Z_1 + (Z_2 + Z_3) = (Z_1 + Z_2) + Z_3 = Z_{123}. \tag{7}$$

**Proof.** At first, consider associativity of operations over fuzzy numbers:

$$A_{231} = A_1 + (A_2 + A_3) = (A_1 + A_2) + A_3 = A_{123}. \tag{8}$$

In [6–8] they proved that the associativity holds. Let us now prove that

$$B_{231} = B_{123}, \tag{9}$$

where

$$b_{231} = \int_R \mu_{A_{231}}(u) \cdot p_{231}(u) du, \tag{10}$$

$$b_{123} = \int_R \mu_{A_{123}}(u) \cdot p_{123}(u) du. \tag{11}$$

Taking into account that (8) holds, we have $\mu_{A_{231}} = \mu_{A_{123}}$. At the same time, it is known that associativity law holds for convolution of sum $\circ_+$ of random variables:

$$p_{231} = p_1 \circ_+ p_{23} = p_1 \circ_+ (p_2 \circ_+ p_3) = (p_1 \circ_+ p_2) \circ_+ p_3 = p_{123}. \tag{12}$$

Then $B_{231} = B_{123}$.
Thus, the associativity law holds for Z-numbers:

$$Z_{231} = Z_1 + (Z_2 + Z_3) = (Z_1 + Z_2) + Z_3 = Z_{123}.$$

**Example.** Consider Z-numbers $Z_1 = (A_1, B_1)$, $Z_2 = (A_2, B_2)$, and $Z_3 = (A_3, B_3)$ the components of which are trapezoidal fuzzy numbers (TFNs):

$$A_1 = (10, 14, 14, 20), \ B_1 = (0.6, 0.8, 0.8, 0.9);$$

$$A_2 = (8, 16, 16, 18), \ B_2 = (0.4, 0.6, 0.6, 0.8);$$

$$A_3 = (16, 18, 18, 21), \ B_2 = (0.1, 0.3, 0.3, 0.6).$$

Let us verify if associativity (7) holds. At first, let us consider the left hand side of (7). For $A_{231}$ we have obtained:

$$A_{23} = (24, 34, 34, 39)$$

and

$$A_{231} = A_{23} + A_1 = (34, 48, 48, 59).$$

For the value of fuzzy reliability $B_{123}$ we obtained:

$$B_{23} = (0.3824, 0.5175, 0.5175, 0.6892),$$

and

$$B_{231} = (0.3909, \ 0.4704, \ 0.4704, \ 0.6417).$$

Consider the right hand side of (7). The fuzzy numbers are found as

$$A_{12} = (18, 30, 30, 38),$$

and

$$A_{123} = A_{12} + A_2 = (34, 48, 48, 59).$$

The value of fuzzy reliability $B_{123}$ has been obtained as follows:

$$B_{12} = (0.5322, 0.6725, 0.6725, 0.7776),$$

and

$$B_{123} = (0.3909, \ 0.4704, \ 0.4704, \ 0.6417).$$

Thus, $A_{231} = A_{123}$ and $B_{231} = B_{123}$. One has that associativity law (7) holds.

### 3.3    Additive Inverse Element $-Z_1$ of Element $Z_1$

Let us prove existence of an additive inverse element of a Z-number:

$$Z_1 + (-Z_1) = Z_{11} = (0, 1), \tag{13}$$

where $Z_{11} = (A_1, B_1) - (A_1, B_1) = (A_{11}, B_{11}).$

**Proof.** The fuzzy $A_{11}$ is the singleton $A_{11} = 0$ as it is shown in [6–8]. Let us consider fuzzy reliability $B_{11}$. As $\mu_{A_{11}}(x) = 1$ iff x = 0, one has for any basic value

$$b_{11} = \int_R 1 \cdot p_{11}(u)du = 1, \tag{14}$$

Thus,

$$Z_1 + (-Z_1) = (0, 1) = 0_Z. \tag{15}$$

### 3.4 Cancellation Law

Let us investigate cancellation law for sum of Z-numbers:

$$Z_1 + Z_3 = Z_2 + Z_3 \Rightarrow Z_1 = Z_2, \tag{16}$$

**Proof.** In order to prove that the cancelation law holds for Z-numbers, we will use inverse element of a Z-number. Adding an inverse interval $-Z_3$ to the both sides of (16) we get

$$Z_1 + Z_3 + (-Z_3) = Z_2 + Z_3 + (-Z_3). \tag{17}$$

By applying (15), one has

$$Z_1 + 0_Z = Z_2 + 0_Z, \tag{18}$$

and

$$Z_1 = Z_2. \tag{19}$$

Thus, cancellation law holds for Z-numbers.

## 4 Conclusion

It is proved that the basic laws of additive arithmetic operations holds for Z-numbers. The proofs are based on the analogous properties of fuzzy arithmetic and probabilistic arithmetic. The obtained results are necessary for strong formulation of such important concepts as equations of Z-numbers, differentiation and integration of Z-number valued functions and other concepts.

# References

1. Aliev, R.A., Alizadeh, A.V., Huseynov, O.H.: The arithmetic of discrete Z-numbers. Inform. Sci. **290**, 134–155 (2015)
2. Aliev, R.A., Alizadeh, A.V., Huseynov, O.H.: The arithmetic of continuous Z-numbers. Inform. Sci. **373**, 441–460 (2016)
3. Aliev, R.A., Alizadeh, A.V., Huseynov, O.H., Jabbarova, K.I.: Z-number based linear programming. Int. J. Intell. Syst. **30**, 563–589 (2015)
4. Kang, B., Wei, D., Li, Y., Deng, Y.: A method of converting Z-number to classical fuzzy number. J. Inf. Comput. Sci. **9**, 703–709 (2012)
5. Piegat, A., Plucinski, M.: Computing with words with the use of Inverse RDM models of membership functions. Appl. Math. Comput. Sci. **25**(3), 675–688 (2015)
6. Piegat, A., Plucinski, M.: Fuzzy number addition with the application of horizontal membership functions. Sci. World J. **2015**, 16 (2015). Article ID 367214
7. Piegat, A., Landowski, M.: Is the conventional interval-arithmetic correct? Journal of Theoretical and Applied Computer Science **6**(2), 27–44 (2012)
8. Piegat, A., Landowski, M.: Multidimensional approach to interval uncertainty calculations. In: Atanassov, K.T., et al. (eds.) New Trends in Fuzzy Sets, Intuitionistic: Fuzzy Sets, Generalized Nets and Related Topics, vol, Volume II: Applications, IBS PAN-SRI PAS, Warsaw, Poland, pp. 137-151 (2013)
9. Piegat, A., Landowski, M.: Two interpretations of multidimensional RDM interval arithmetic - multiplication and division. Int. J. Fuzzy Syst. **15**, 488–496 (2013)
10. Piegat, A., Plucinski, M.: Some advantages of the RDM-arithmetic of intervally-precisiated values. Int. J. Comput. Intell. Syst. **8**(6), 1192–1209 (2015)
11. Williamson, R.C., Downs, T.: Probabilistic arithmetic. I. Numerical methods for calculating convolutions and dependency bounds. Int. J. Approx. Reason. **4**(2), 89–158 (1990)
12. Zadeh, L.A.: A note on Z-numbers. Inform. Sci. **181**, 2923–2932 (2011)
13. Zadeh, L.A.: Probability measures of fuzzy events. J. Math. Anal. Appl. **23**(2), 421–427 (1968)
14. Zadeh, L.A.: Methods and systems for applications with Z-numbers. United States Patent, Patent No.: US 8,311,973 B1, Date of Patent: 13 Nov 2012

# Construction of Fuzzy Control System Rule-Base with Predefined Specificity

Nigar E. Adilova[1,2(✉)]

[1] Joint MBA Program, Azerbaijan State Oil and Industry University,
20 Azadlig Ave., Baku AZ1010, Azerbaijan
adilovanigarr@gmail.com
[2] Georgia State University, Atlanta, USA

**Abstract.** Fuzzy rule-base very frequently is used for function approximation, modeling and identification of systems and objects. It is known that main characteristic of fuzzy rule base arte specificity, granularity and interpretability. Research on trade-off between specificity and granularity of fuzzy rule-bases is very scars. In this paper we develop an approach to construct fuzzy rule-base with predefined specificity degree. The problem is to find such granularity of domain which gives desirable degree of specificity of designed fuzzy rule-base.

**Keywords:** Fuzzy rule-base · Specificity · Granularity · Interpretability

## 1 Introduction

Specificity of fuzzy rule-base determines degree of informativeness of results of reasoning made by using existing rules in knowledge-base. The minimal specificity is related with the size, decomposition of the imput domain of rules. High level of specificity requires low level of granularity of input space. But it can lead to loss of interpretability of constructed fuzzy rule-base.

The principle of minimal specificity is considered in [1]. In this paper possibilistic approach to investigation of specificity is considered. It is shown that principle of minimal specificity leads to certainty rule.

$$\eta(x, y) = \mu_A(u) \to \mu_B(v)$$
$$a \to b = \max(1 - a, b) \,(\text{Dienes implication})$$

Trade-off between specificity and granularity is investigated in [2].

The specificity of fuzzy sets measures the degree to which the set designates a unique element of set. The sets of the rule is the specificity of the fuzzy sets in the consequent of the rule. A lower specificity indicates that the consequent provides a wider range of positive value. In particular when the output domain is partitioned by triangular FSs with peak points $\{c_1, c_{2,}, \ldots, c_n\}$ the specificity of the rule "IF X is $A_i$ then Z is $C_k$". Is the distance $[C_{k-1} - C_{k+1}]$ between neighboring peak points of the output domain decomposition.

© Springer Nature Switzerland AG 2019
R. A. Aliev et al. (Eds.): ICAFS-2018, AISC 896, pp. 901–904, 2019.
https://doi.org/10.1007/978-3-030-04164-9_119

Input domain decomposition by using expert selection of linguistic terms for control systems is considered in [1, 2].

In [2] properties of fuzzy models are investigated. It is shown that learning methods of construction of fuzzy models dominates on expert-driven models on different characteristic of fuzzy rules, but interpretability is not so good in comparison with expert opinion-based models. In this paper we investigate specificity problem of fuzzy rule-bases to achieve predefined specificity of models by using of granularity of input domain of fuzzy IF-THEN rules.

In [3] a systematic study of specificity of fuzzy IF ... THEN rules is proposed. A new measure of specificity is introduced on the basis of the Yager's idea.

The paper is structured as follows. Section 2 describes the preliminary information on fuzzy number specificity measure of fuzzy number, fuzzy IF ... THEN rules etc. In Sect. 3 statement of the fuzzy modeling of control system by using fuzzy IF ... THEN rules is given. In Sect. 4 investigation of total specificity of fuzzy rule-base of the considered control system is presented Sect. 5 is the conclusion.

## 2    Preliminaries

***Definition 1:*** A fuzzy number is a set $A$ on $R$ which possesses the following properties: (a) $A$ is a normal fuzzy set; (b) $A$ is a convex fuzzy set; (c) $\alpha-$ cut of $A$, $A^{\alpha}$ is a closed interval for every $\alpha \in (0, 1]$; (d) the support of $A$, $A^{+0}$ is bounded [4].

***Definition 2:*** The specificity measure defined on the basis of cardinality concept is as follows [5]:

$$Sp(A) = \int\limits_0^{hgt(A)} \frac{1}{|A^{\mu}|} d\mu \qquad (1)$$

where $|A^{\mu}|$ is a cardinality of $\mu$-cut of $A$.

## 3    Statement of the Problem

Assume that fuzzy control system includes following 7 rules.

1. If the error $e$ is negative big (NB) THEN the control action $u$ is negative big (NB).
2. If the error $e$ is negative medium (NM) THEN the control action $u$ is negative medium (NM).
3. If the error $e$ is negative small (NS) THEN the control action $u$ is negative small.
4. If the error $e$ is zero (ZE) THEN the control action $u$ is zero.
5. If the error $e$ is positive small (PS) THEN the control action $u$ is positive small (PS).
6. If the error $e$ is positive medium (PM) THEN the control action $u$ is positive medium (PM).
7. If the error $e$ is positive big (PB) THEN the control action $u$ is positive big (PB).

Membership functions for error $e$ and control action $u$ are shown in Fig. 1.

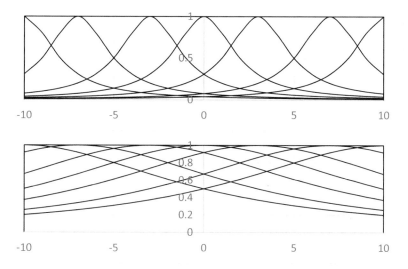

**Fig. 1.** Membership functions of the error and the control action

The composed fuzzy relation matrix is as follows.

| 1.00 | 0.92 | 0.73 | 0.67 | 0.50 | 0.34 | 0.34 |
|------|------|------|------|------|------|------|
| 0.91 | 1.00 | 0.86 | 0.67 | 0.61 | 0.50 | 0.37 |
| 0.61 | 0.86 | 1.00 | 0.92 | 0.74 | 0.67 | 0.50 |
| 0.61 | 0.73 | 0.92 | 1.00 | 0.92 | 0.73 | 0.67 |
| 0.50 | 0.67 | 0.74 | 0.92 | 1.00 | 0.86 | 0.67 |
| 0.37 | 0.50 | 0.61 | 0.67 | 0.86 | 1.00 | 0.91 |
| 0.34 | 0.34 | 0.50 | 0.67 | 0.73 | 0.92 | 1.00 |

The problem is to investigate specificity of a fuzzy output of the control system given specificity of a current fuzzy input.

## 4  Experimental Investigation

Let's us investigate dependence of specificity of fuzzy output on specificity of fuzzy current input. The fuzzy output is completed as $\mu_U = \max_e \min[\mu_E(e_{SC}^c), \mu_R(u, e)]$.

The obtained results are:

| 0   | 1    | 0,92 | 0,73 | 0,67 | 0,5  | 0,34 | 0,34 |
|-----|------|------|------|------|------|------|------|
| 0   | 0,91 | 1    | 0,86 | 0,67 | 0,61 | 0,5  | 0,37 |
| 0,2 | 0,61 | 0,86 | 1    | 0,92 | 0,74 | 0,67 | 0,5  |
| 1   | 0,61 | 0,73 | 0,92 | 1    | 0,92 | 0,73 | 0,67 |
| 0,2 | 0,5  | 0,67 | 0,74 | 0,92 | 1    | 0,86 | 0,67 |
| 0   | 0,37 | 0,5  | 0,61 | 0,67 | 0,86 | 1    | 0,91 |
| 0   | 0,34 | 0,34 | 0,5  | 0,67 | 0,73 | 0,92 | 1    |

|     | 0.61 | 0.73 | 0.92 | 1 | 0.92 | 0.73 | 0.67 |
|-----|------|------|------|---|------|------|------|

Sp(e) = 0,866667
Sp(u) = 0,252476

| 1   | 1    | 0,92 | 0,73 | 0,67 | 0,5  | 0,34 | 0,34 |
|-----|------|------|------|------|------|------|------|
| 0,2 | 0,91 | 1    | 0,86 | 0,67 | 0,61 | 0,5  | 0,37 |
| 0   | 0,61 | 0,86 | 1    | 0,92 | 0,74 | 0,67 | 0,5  |
| 0   | 0,61 | 0,73 | 0,92 | 1    | 0,92 | 0,73 | 0,67 |
| 0   | 0,5  | 0,67 | 0,74 | 0,92 | 1    | 0,86 | 0,67 |
| 0   | 0,37 | 0,5  | 0,61 | 0,67 | 0,86 | 1    | 0,91 |
| 0   | 0,34 | 0,34 | 0,5  | 0,67 | 0,73 | 0,92 | 1    |
|     | 1    | 0,92 | 0,73 | 0,67 | 0,5  | 0,34 | 0,34 |

Sp(e) = 0,9
Sp(u) = 0,318071

The obtained results imply that rule-base for considered control system is not perfect. We have to update rule-base to achieve predefined specificity.

## 5  Conclusion

In this paper specificity of rule-base of fuzzy control system is analyzed. The analysis has shown that specificity level of the rule base isn't sufficiently high, the specificity of fuzzy output is quite low, even when specificity of input is high. To achieve sufficiently high specificity it is needed to attain more optimal granularity of fuzzy rule base.

## References

1. Dubois, D., Prade, H.: Possibility theory: qualitative and quantitative aspects, vol. 1, pp. 169–226. Kluwer Academic Publishers (1998)
2. Sudkamp, T.: Granularity and specificity in fuzzy rule-based systems, vol. 7, pp. 257–274. Springer, Heidelberg (2001)
3. Kacprzyk, J.: On measuring specificity of if-then rules. Int. J. Approx. Reason. **11**(1), 29–53 (1994)
4. Aliev, R.A.: Fundamentals of the Fuzzy Logic-Based Generalized Theory of Decisions. Springer, Heidelberg (2013)
5. Yager, R.R.: Measuring tranquility and anxiety in decision making: an application of fuzzy sets Internat. Int. J. Gen Syst **8**, 139–146 (1982)

# Application of Neural Networks for Segmentation of Catering Services Market Within the Overall System of Consumer Market on the Model of Restaurant Business with the Aim to Advance the Efficiency of Marketing Policy

Alekperov Ramiz Balashirin[1] and Iskenderli Ilhama Tarlan[2(✉)]

[1] Department of Computing Engineering, Odlar Yurdu University,
Baku AZ1008, Azerbaijan
ramiz62@rambler.ru
[2] Department of Economy, International Relationships and Social Sciences,
Odlar Yurdu University, Baku AZ1008, Azerbaijan
ilhamaiskandarova@gmail.com

**Abstract.** The subject of classification of catering services market from the standpoint of prospecting of the best market positioning is considered with the aim to identify the means of effective organization of marketing policy.

**Keywords:** Segmentation of catering services market · Consumer market
Neural network · Neural network with Kohonen's layer

## 1 Introduction

The principal task of marketing in restaurant business is the attraction of clients and the increase of sales' volume, so it demands to carry out the marketing policy in restaurant properly and competently. This marketing policy permits to identify the current problems, to conduct the necessary changes to adapt the business to the altered operation conditions, to work out the optimal advertisement campaign, to alter the attitude of customers towards the restaurant, to design the effective means of clients' retention. In order to arrive this task it is necessary to explore the existing market and to solve the problem of catering services' market segmentation.

Bear in mind that the fact of expansion of catering services' market suggests about the grow of population's standard of life in connection with the increase of their incomes and salaries. On the other hand, by segmenting the market it is possible to determine the more favorable locations for placement of restaurant business.

Embracing the mentioned, the problem of segmentation of catering services' market is stated from the standpoint of their expansion on the basis of the territorial units (responders) with the application of artificial neural networks (ANN) [1].

© Springer Nature Switzerland AG 2019
R. A. Aliev et al. (Eds.): ICAFS-2018, AISC 896, pp. 905–913, 2019.
https://doi.org/10.1007/978-3-030-04164-9_120

ANN are capable to process the vast amount of data, to analyze the similarity between respondents, to identify the patterns and to forecast the future. It can advise to the managers of restaurant business how to organize their business with the consideration of population's income level and with the comparison to restaurants being in the most favorable position on catering services' market.

Neural networks are demonstrating good results in devising the forecasts due to their practice to be trained. And, as opposed to traditional approaches to the solution of forecasting and classification task, the predictive analysis in the issue of ANN is easily adapted to the alterations of behavior – when new data arrive the prediction becomes better.

## 2   Modelling the Segmentation of Catering Services' Market

Let's formulate the problem in terms of cluster analysis. Let the set of T entities of some market is given. Every entity is being characterized by the definite values of classification marks (both quantitative and qualitative ones) of the same bank. The manner to determine the proximity between the objects is specified. It is required to breakdown the set of objects into disjoint subsets (classes, clusters, groups) in the way that the objects resembling to each other by a variety of marks are gathered into one class.

In general to solve the given problem the Kohonen's networks are used. They are self-organized artificial neural network (ANN) [2]. On the basis of Kohonen's layer the neural network (utilizing a non-controlled training) groups respondents (T) on the evidence of their similar marks (x) and have a capability to ignore noise data. Under such a training the train set consists of only the input values for x variables, there is no the comparison of neuron's outputs with the etalon values.

The essence of the neural network with the Kohonen's layer summarizes in the estimation of difference – d (distance) of weight vector w of each neuron in the network from the vector of input pattern xp and in search for neuron which weight coefficient (w) has the minimal standoff distance from x. For every j-th neuron and i-th mark of p pattern of input values the minimal standoff distance usually is calculated according to the formula 1 [2]:

$$d_j = \sum_{l=1}^{m} \left( w_i^j - x_i^p \right)^2 \qquad (1)$$

where $j = 1, \ldots, m$, m stands for total amount of neurons. The neuron with most minimal difference d is considered a winner and subsequently it's weights are adjusted according to the formula 2:

$$w_{new}^j = w_{old}^j (1 - \alpha) + \alpha x^p \qquad (2)$$

where $\alpha : [0, \ldots, 1]$ – training coefficient.

The segmentation of catering services' sphere is carried out with a glance of the categorization of entrepreneur subjects and turnover of catering services within the economic regions of Azerbaijan Republic (see Table 1). The software implementation of segmentation task has been carried out with the application of artificial neural network with Kohonen's layer on the basis of the Neural Network Toolbox application software package of the MATLAB system [3].

The general algorithm of segmentation program consists of the following steps:

1. Input of marks' values for every p-th pattern (object) of the set of $x^p = \left[ x_1^p, x_2^p, \ldots, x_k^p \right]$:

$$
\begin{aligned}
P = [&3201, 710, 910, 680, 514, 260, 1509, 46, 200, 665 \\
&3317, 754, 995, 728, 546, 315, 1544, 41, 205, 672; \\
&3438, 763, 1041, 752, 559, 380, 1561, 42, 274, 692; \\
&4083, 902, 1166, 867, 680, 423, 1792, 47, 321, 672; \\
&5458, 1096, 1294, 994, 803, 749, 2216, 134, 390, 790; \\
&6258, 1244, 1429, 1063, 912, 865, 242, 137, 445, 820; \\
&140.9, 20.9, 22.6, 13.9, 12.4, 6.9, 42, 1.4, 7.7, 20.6; \\
&150, 22.5, 24.8, 15, 13, 8.7, 42.3, 1.3, 9, 20.8; \\
&165.4, 23.7, 24.8, 15.6, 12.8, 9.8, 42.3, 1.3, 9.7, 21.4; \\
&196.4, 27.6, 27.8, 17.9, 15.5, 11, 45.4, 1.4, 11.5, 24.2; \\
&283.8, 33.5, 30.5, 19.8, 18.3, 20.1, 55, 5.2, 13.2, 25; \\
&333.8, 40.3, 32.7, 22.3, 20.5, 23, 58.6, 4.4, 14.1, 26; \\
&231.1, 20.8, 34.5, 11.1, 17.5, 13.3, 56.3, 2.2, 8.3, 22.6; \\
&320.2, 28.1, 45.8, 14.5, 23.3, 18.3, 71.8, 2.9, 10.5, 30.1; \\
&405.4, 31.2, 49, 17.5, 25.1, 21.9, 77.8, 3, 12.2, 36.7; \\
&499, 33.5, 51.9, 18.5, 26.3, 23.4, 85.9, 3.2, 13.8, 44.5; \\
&627.6, 38.7, 55.3, 20.3, 28.1, 26.4, 93.8, 3.4, 15.3, 47.1; \\
&752.4, 44.8, 58.9, 22.9, 30.2, 30, 102, 3.6, 16.2, 50.2; \\
&55.3, 5, 8.2, 2.7, 4.2, 3.2, 13.5, 0.5, 2, 5.4; \\
&56.6, 5, 8.1, 2.6, 4.1, 3.2, 12.7, 0.5, 1.9, 5.3; \\
&59.6, 4.6, 7.2, 2.6, 3.7, 3.2, 11.4, 0.45, 1.8, 5.4; \\
&62.3, 4.2, 6.5, 2.3, 3.3, 2.9, 10.7, 0.4, 1.7, 5.6; \\
&65.6, 4, 5.8, 2.1, 2.9, 2.8, 9.8, 0.35, 1.6, 4.9; \\
&67.7, 4, 5.3, 2.1, 2.7, 2.7, 9.2, 0.32, 1.5, 4.5];
\end{aligned}
$$

2. Determination of border values of p:

$$\text{for } i = 1 : \text{size}(P, 1) \ \max(P(i, :)); P(i, :) = P(i, :)/\max(P(i, :)); \text{end}$$

3. Determination of confidence coefficient (Cef_dov)
4. Determination of a number of clusters (num)

**Table 1.** Categorization of entrepreneur catering subjects and of catering services turnover within the economic regions of Azerbaijan Republic

| Factors | Years | | Economic regions | | | | | | | | | |
|---|---|---|---|---|---|---|---|---|---|---|---|---|
| | | | Baku city | Absheron | Gandja-Kazakh | Shaki-Zakatala | Lenkoran | Quba-Khachmaz | Arran | Upper-Karabakh | Mountainous-Shirvan | Nakhichevan AR |
| | | | P1 | P2 | P3 | P4 | P5 | P6 | P7 | P8 | P9 | P10 |
| The number of entrepreneur catering subjects (in units) (A) and the number of seating places (in 1000 places) (B) | 2010 | A x1 | 3201 | 710 | 910 | 680 | 514 | 260 | 1509 | 46 | 200 | 665 |
| | 2011 | x2 | 3317 | 754 | 995 | 728 | 546 | 315 | 1544 | 41 | 205 | 672 |
| | 2012 | x3 | 3438 | 763 | 1041 | 752 | 559 | 380 | 1561 | 42 | 274 | 692 |
| | 2013 | x4 | 4083 | 902 | 1166 | 867 | 680 | 423 | 1792 | 47 | 321 | 672 |
| | 2014 | x5 | 5458 | 1096 | 1294 | 994 | 803 | 749 | 2216 | 134 | 390 | 790 |
| | 2015 | x6 | 6258 | 1244 | 1429 | 1063 | 912 | 865 | 242 | 137 | 445 | 820 |
| | 2010 | B x7 | 140.9 | 20.9 | 22.6 | 13.9 | 12.4 | 6.9 | 42 | 1.4 | 7.7 | 20.6 |
| | 2011 | x8 | 150 | 22.5 | 24.8 | 15 | 13 | 8.7 | 42.3 | 1.3 | 9 | 20.8 |
| | 2012 | x9 | 165.4 | 23.7 | 24.8 | 15.6 | 12.8 | 9.8 | 42.3 | 1.3 | 9.7 | 21.4 |
| | 2013 | x10 | 196.4 | 27.6 | 27.8 | 17.9 | 15.5 | 11 | 45.4 | 1.4 | 11.5 | 24.2 |
| | 2014 | x11 | 283.8 | 33.5 | 30.5 | 19.8 | 18.3 | 20.1 | 55 | 5.2 | 13.2 | 25 |
| | 2015 | x12 | 333.8 | 40.3 | 32.7 | 22.3 | 20.5 | 23 | 58.6 | 4.4 | 14.1 | 26 |
| The catering service turnover (in min. manat) (C) and their share (in %) (D) | 2010 | C x13 | 231.1 | 20.8 | 34.5 | 11.1 | 17.5 | 13.3 | 56.3 | 2.2 | 8.3 | 22.6 |
| | 2011 | x14 | 320.2 | 28.1 | 45.8 | 14.5 | 23.3 | 18.3 | 71.8 | 2.9 | 10.5 | 30.1 |
| | 2012 | x15 | 405.4 | 31.2 | 49 | 17.5 | 25.1 | 21.9 | 77.8 | 3 | 12.2 | 36.7 |
| | 2013 | x16 | 499 | 33.5 | 51.9 | 18.5 | 26.3 | 23.4 | 85.9 | 3.2 | 13.8 | 44.5 |
| | 2014 | x17 | 627.6 | 38.7 | 55.3 | 20.3 | 28.1 | 26.4 | 93.8 | 3.4 | 15.3 | 47.1 |
| | 2015 | x18 | 752.4 | 44.8 | 58.9 | 22.9 | 30.2 | 30 | 102 | 3.6 | 16.2 | 50.2 |

(continued)

**Table 1.** (*continued*)

| Factors | Years | | Economic regions | | | | | | | | | |
| | | | Baku city | Absheron | Gandja-Kazakh | Shaki-Zakatala | Lenkoran | Quba-Khachmaz | Arran | Upper-Karabakh | Mountainous-Shirvan | Nakhichevan AR |
| | | | P1 | P2 | P3 | P4 | P5 | P6 | P7 | P8 | P9 | P10 |
| D | 2010 | x19 | 55.3 | 5 | 8.2 | 2.7 | 4.2 | 3.2 | 13.5 | 0.5 | 2 | 5.4 |
| | 2011 | x20 | 56.6 | 5 | 8.1 | 2.6 | 4.1 | 3.2 | 12.7 | 0.5 | 1.9 | 5.3 |
| | 2012 | x21 | 59.6 | 4.6 | 7.2 | 2.6 | 3.7 | 3.2 | 11.4 | 0.45 | 1.8 | 5.4 |
| | 2013 | x22 | 62.3 | 4.2 | 6.5 | 2.3 | 3.3 | 2.9 | 10.7 | 0.4 | 1.7 | 5.6 |
| | 2014 | x23 | 65.6 | 4 | 5.8 | 2.1 | 2.9 | 2.8 | 9.8 | 0.35 | 1.6 | 4.9 |
| | 2015 | x24 | 67.7 | 4 | 5.3 | 2.1 | 2.7 | 2.7 | 9.2 | 0.32 | 1.5 | 4.5 |

5. Determination of training speed parameter (kohonen)
6. Dimensioning of an empty array 'assignment_buffer' which indicate the belonging of objects to classes
7. Formation of Kohonen's neural network:

$$
\begin{aligned}
\text{net} = \text{newc}([&\min(P(1,:))\max(P(1,:)); \; \min(P(2,:))\max(P(2,:)); \; \min(P(3,:))\max(P(3,:)); \\
&\min(P(4,:))\max(P(4,:)); \; \min(P(5,:))\max(P(5,:)); \; \min(P(6,:))\max(P(6,:)); \\
&\min(P(7,:))\ldots \\
&\max(P(7,:)); \; \min(P(8,:))\max(P(8,:)); \; \min(P(9,:))\max(P(9,:)); \ldots \\
&\min(P(10,:))\max(P(10,:)); \; \min(P(11,:))\max(P(11,:)); \\
&\min(P(12,:))\max(P(12,:)); \ldots \\
&\min(P(13,:))\max(P(13,:)); \; \min(P(14,:))\max(P(14,:)); \ldots \\
&\min(P(15,:))\max(P(15,:)); \; \min(P(16,:))\max(P(16,:)); \ldots \\
&\min(P(17,:))\max(P(17,:)); \; \min(P(18,:))\max(P(18,:)); \ldots \\
&\min(P(19,:))\max(P(19,:)); \; \min(P(20,:))\max(P(20,:)); \ldots \\
&\min(P(21,:))\max(P(21,:)); \; \min(P(22,:))\max(P(22,:)); \ldots \\
&\min(P(13,:))\max(P(23,:)); \; \min(P(14,:))\max(P(24,:))], \\
&\text{num,kohonen,Cef\_dov});
\end{aligned}
$$

8. Determination of initial number of training epochs (epochs)
9. Start the training epoch.

$$\text{net} = \text{train}(\text{net}, P);$$

10. Determination of weight values w = net.IW{1};
11. Calculation and determination of clusters: X = P'; count = zeros(1,num);

```
for i = 1:size(X,1)
for j = 1:(size(w,1))
for l = 1:size(X,2)
    distance(l) = (X(i,l) − w(j,l))²;
end
    distances(j) = sqrt(sum(distance));
end
    [min_distance(i), assignment(i)] = min(distances);
    count(assignment(i)) = count(assignment(i)) + 1;
end
```

12. If the end of training epoch is not reached then go to step 9
13. Match the 'assignment' with the 'assignment_buffer'. If there is no changes in the grouping of objects, by other words if the objects belonging to one cluster didn't change their memberships then go to the end of algorithm (step 16), else go to step 14.
14. Let 'assignment_buffer' = 'assignment';
15. Increase the number of training epochs and go to step 8.
16. The end of algorithm.

## 3 Result Analysis

For the input data $x = [x_1, x_2, \ldots, x_{24}]$ (see Table 1) of all over 10 territorial regions of Azerbaijan Republic (objects) the 4 clusterization centers were selected. The results with the most important characteristics were obtained and shown on dynamic graphs on Fig. 1. The graphs on Fig. 1 are demonstrating the followings:

(a) the first graph demonstrates all objects and corresponding centers for the selected criteria such as number of entrepreneur subjects ($x_1$, $x_6$) and nutrition turnover ($x_{18}$);
(b) the second graph demonstrates the values of particular criteria for different centroids;
(c) the third graph presents the weights of the clusters (i.e. the number of objects assigned to each centroid);
(d) the fourth graph demonstrates the change's dynamic of overall standoff of particular objects from the correspoding centroids.

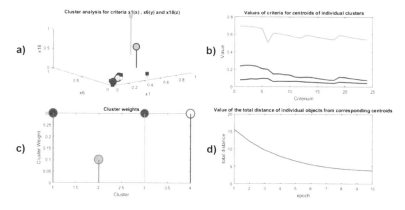

**Fig. 1.** The cluster analysis of catering services' sphere

The maximum number of training epochs was determined and is equal to 12. After mentioned number of training epochs at the training speed equal to 0.01 the number of objects assigned to each centroid were not change. The training lasted 15.7107 sec., overall standoff of objects from centroid was 4.7550.

The results of allocation of objects in line with the relevant centroids – clusters – are indicated on Table 2. The analysis of those clusters displays that Baku city economic region greatly distinguishes from other regions by the all of factors. The close regions to the Baku region by the level of economy development with the consideration of all the factors are:

- Absheron, Nakhichevan AR, Gandja-Kazakh and Arran regions – the $3^{rd}$ segment;
- Lenkoran, Quba-Khachmaz, Upper-Karabakh and Mountainous-Shirvan regions – the $4^{th}$ segment;
- Shaki-Zakatala region – the $1^{st}$ segment;

**Table 2.** Allocation of economic regions into classification clusters.

| Economic regions | Cluster # by the number of entrepreneur catering subjects factor | Cluster # by the catering service turnover factor | Cluster # with consideration of all factors |
|---|---|---|---|
| Baku city | 2 | 2 | 2 |
| Absheron | 4 | 1 | 3 |
| Gandja-Kazakh | 4 | 3 | 3 |
| Shaki-Zakatala | 4 | 1 | 1 |
| Lenkoran | 3 | 1 | 4 |
| Quba-Khachmaz | 3 | 1 | 4 |
| Arran | 4 | 3 | 3 |
| Upper-Karabakh | 3 | 1 | 4 |
| Mountainous-Shirvan | 3 | 1 | 4 |
| Nakhichevan AR | 3 | 1 | 3 |

But at the same time the results of analysis indicates that individually on the number of entrepreneur catering subjects factor and on the catering service turnover factor the economic regions are classified into 3 clusters, though the classification was expected to be divided into 4 clusters.

## 4    Conclusion

Application of neural network with the Kohonen's layer allots the possibility to organize effectively the marketing policy from the standpoint of search for better market position, which are the most promising for the expansion of restaurant business and the conduct of advertisement campaign of a restaurant.

## References

1. Aliev, R.A., Fazlollahi, B., Aliev, R.R.: Soft Computing and its Applications in Business and Economics. Springer, Heidelberg (2004)
2. Kohonen, T.: Self-Organizing Maps, 3rd Extended edn. Springer, New York (2001). ISBN 978-3-540-67921-9
3. Beale, M.H., Hagan, M.T., Demuth, H.B.: Neural Network Toolbox. User's Guide. Math Works, Inc., Natick (2014)

# Detecting Adverse Events in an Active Theater of War Using Advanced Computational Intelligence Techniques

Jozef Zurada[1,2(✉)], Donghui Shi[3], Waldemar Karwowski[4],
Jian Guan[1], and Erman Çakıt[5]

[1] Department of Computer Information Systems, College of Business,
University of Louisville, Louisville, KY 40292, USA
{jozef.zurada, jeff.guan}@louisville.edu
[2] WSB Gdansk, Gdansk, Poland
[3] Department of Computer Engineering, Anhui Jianzhu University,
230601 Hefei, China
sdonghui@gmail.com
[4] Department of Industrial Engineering and Management Systems,
University of Central Florida, Orlando, FL 32816, USA
wkar@ucf.edu
[5] Department of Industrial Engineering, Aksaray University,
68100 Aksaray, Turkey
ermancakit@aksaray.edu.tr

**Abstract.** This study investigates the effectiveness of advanced computational intelligence techniques in detecting adverse events in Afghanistan. The study first applies feature reduction techniques to identify significant variables. Then it uses five cost-sensitive classification methods. Finally, the study reports the resulting classification accuracy rates and areas under the receiver operating characteristics charts for adverse events for each method for the entire country and its seven regions. It appears that when analysis is performed for the entire country, there is little correlation between adverse events and project types and the number of projects. However, the same type of analysis performed for each of its seven regions shows a connection between adverse events and the infrastructure budget and the number of projects allocated for the specific regions and times. Among the five classifiers, the C4.5 decision tree and *k*-nearest neighbor seem to be the best in terms of global performance.

**Keywords:** Detecting adverse events · Active war theater
Computational intelligence · Soft computing

## 1 Introduction

The U.S. Department of Defense (DoD) defines irregular warfare as a non-conventional warfare which includes non-proportional force to subdue, hassle, and coerce civilians in the region(s) in which opposite forces are not large and effective [1]. In such warfare, conducted in overseas countries such as Afghanistan, one of the main tasks of the

© Springer Nature Switzerland AG 2019
R. A. Aliev et al. (Eds.): ICAFS-2018, AISC 896, pp. 914–921, 2019.
https://doi.org/10.1007/978-3-030-04164-9_121

military is to protect the civilian population from terrorist attacks. To better understand the dynamic of irregular warfare and to address its military and non-military challenges, the DoD modified its force structure and created the Human Social Culture Behavior (HSCB) modeling program [2]. The military uses HSCB models to understand the behavior and structure of organizational units at the macro level (i.e., health, economics, and security) and at the micro level (i.e., terrorist networks, tribes, and customs) [3]. These HSCB models are very complex as they exhibit non-linear and fuzzy behavior and are often ill defined with respect to their socio-economic-cultural factors.

Some studies used data-driven models and statistical approaches to predict terrorists' behavior. Since terrorist attacks are not random in space and time, it is possible to discover representative patterns in adverse activity over time and space by analyzing the geospatial intelligence on reported incidents [4]. Additionally, Geographic Information Systems (GIS) can provide crucial information about the spatial patterns of adverse events using incident-based data [5]. Reed et al. [6] used incident-based data and a time-correlation approach to predict patterns and trends in future behaviors of terrorists. These trends could help decision-makers to allocate more resources and personnel to the places that are more likely to be attacked. Inyaem et al. [7] showed that adaptive neuro-fuzzy inference system (ANFIS) outperformed fuzzy inference system (FISs) in adverse event classification and incident analysis.

The studies by [8–10] used the data set provided by the HSCB program management of the U.S. DoD. They applied linear regression, neural networks (NNs), ANFIS, FIS as well as fuzzy C-means and subtractive clustering for predicting four categories of events, i.e., the number of killed, the number of wounded, the number of hijacked, and the number of events based on infrastructure development spending, the number of projects, and other variables. These four categories of events are collectively called "adverse events". The four dependent/output variables representing adverse events were measured on a continuous scale. The studies used mean absolute percentage error (MAPE) and mean absolute error (MAE) to evaluate the prediction results. The reported MAPE and MAE values were large and they varied among the four predicted variables, the regions, and periods.

This study uses the same data set provided by HSCB program and proposes a different, more feasible, and novel approach through recoding of the dependent variables. We present the following rationale for our approach. Table 1 shows the descriptive statistics for the four output features for the entire data set. These are killed, wounded, hijacked, and events at month t. The mean values for the four attributes are within the [0.08, 0.52] range. The maximum number of people killed, wounded, hijacked, and the number of events in a given month could occasionally be as high as 103, 261, 156, and 38, respectively. One can see that each of these four variables is a sparse vector/column where from 87.2% to 98.1% of their values are 0's, with a 0 representing lack of events. As a result, the four output variables are highly unbalanced. The descriptive statistics calculated for seven regions (Central, Eastern, Northeastern, Southeastern, Western, Northwestern, and Southwestern) show very similar patterns. Thus when the four variables are measured on the continuous scale we deal with the regression problem and the accurate prediction of the number of killed, wounded, hijacked, and events is an extremely challenging task regardless of the techniques used in [8–10]. When the values of MAE for the four adverse events are several times larger than their mean values, as shown in Table 1, MAE could be considered large.

**Table 1.** The descriptive statistics for four output variables for the entire data set

| Variables | Min | Max | Mean | Std Dev | % of 0's | MAE [10] |
|---|---|---|---|---|---|---|
| Dead(t) | 0 | 103 | 0.36 | 2.15 | 92.4 | 0.66 |
| Wounded(t) | 0 | 261 | 0.52 | 4.08 | 93.1 | 1.00 |
| Hijacked(t) | 0 | 156 | 0.08 | 1.24 | 98.1 | 0.20 |
| Event(t) | 0 | 38 | 0.28 | 1.09 | 87.2 | 0.66 |

This paper proposes a different approach by transforming/recoding the dependent variables. We recoded the values of the four output variables as follows. If a number of people killed, wounded, hijacked, or events is $\geq 1$, it is represented as 1 or "Yes", and if the number is 0 it remains 0 or "No." In other words, we reduced the regression task investigated by [8–10] to a classification task. We detect/predict if someone was killed or not, wounded or not, hijacked or not, or if an event happened or did not happen. We evaluate the effectiveness of five advanced computational intelligence techniques: NNs, $k$-nearest neighbor ($k$NN), support vector machines with sequential minimal optimization (SMO), C4.5 algorithm for decision trees (C4.5), and random forest (RF) in detecting adverse events based on infrastructure development spending (the project types), the number of projects, and other variables in an active theater of war in Afghanistan using the data set provided by the HSCB program management (2002–2010) of the U.S. DoD. These five classifiers are meta-classifiers, i.e., cost-sensitive classifiers. The study first uses feature reduction techniques to identify significant variables, then applies the above classification methods, and finally reports correct classification accuracy rates and areas under the receiver operating characteristics curves (AUROC) for the adverse events for each method for the entire country and its seven regions. Due to space constraints, the rates for C4.5 for seven regions are shown only.

## 2 Methods

Weka was used for computer simulation. Since NN, $k$NN, and C4.5 are commonly used methods for classification, we only provide a brief and informal description of cost-sensitive classifiers, SMO, and RF.

Cost-sensitive classifier: As the four output variables in the data set are highly unbalanced, we used a cost-sensitive learning approach to address the class imbalance issue and achieve better performance. A 2 by 2 cost matrix can be used to stand for the different cost of each class of misclassification. Weka provides a cost-sensitive learning algorithm that reweighs training instances according to the total cost assigned to each of the two classes: (1 – "Yes") and (0 – "No") [11]. It is a meta-classifier that makes its base classifier cost-sensitive and it is used for wrapping the base classifiers: NN, $k$NN, SMO, C4.5, and RF. The "best" cost ratio to use for learning is determined empirically

[12]. In the cost-sensitive learning algorithms, the cost matrix $[C_{00}, C_{01}; C_{10}, C_{11}]$ is manually adjusted to $[0, 1; C_{10}, 0]$. In the study, $C_{01}$ is set to 1 and the values of $C_{10}$ are defined as (the number of "No" cases)/(the number of "Yes" cases) for globally better performance in the overall classification accuracy rates as well as the rates for the "Yes" and "No" categories. $C_{10}$ is the cost when an actual "Yes" case is incorrectly taken to be a "No" case. In order to obtain a better result, varied cost values represented by $C_{10}$ were set for each of the four output variables.

SMO: Support vector machines (SVMs) are supervised learning models. They can be used in classification and regression using the concept of a maximum margin hyperplane. SVM algorithms produce a model that can usually be expressed in terms of a few support vectors and can be applied to nonlinear problems using kernel functions. SMO is an algorithm used in Weka that applies John Platt's sequential minimal optimization algorithm for training a support vector classifier [13].

Radom Forest (RF): It builds a randomized decision tree in each iteration of the bagging algorithm, and often produces excellent predictors [11]. RF is an ensemble learning method for classification and regression. The algorithm constructs a multitude of decision trees at training time and outputs the class, which is equal to the mean prediction of the individual trees.

## 3 Data Set

The original data set provided by HSCB program contains 101 attributes and 33,600 records representing years 2002 through 2010. It contains the following input variables: the year, month, province, district, and region for the allocated aid; the budget amount in [\$] for fourteen different project types $B1(t-2)$ through $B14(t-2)$ at years 2002–2008; $B1(t-1)$ through $B14(t-1)$ at years 2003–2009; $B1(t)$ through $B14(t)$ at years 2004–2010; the number for each of the fourteen project/aid types $A1(t-2)$ through $A14(t-2)$, $A1(t-1)$ through $A14(t-1)$, and $A1(t)$ through $A14(t)$; urban and rural male and female population densities; and the number of killed, wounded, hijacked, and the number of events at month $t-1$. The four output variables are the number of killed, wounded, hijacked, and events at month t. In this study, the four output variables have been recoded to 1's and 0's. The fourteen project types (denoted by B) and their number (denoted by A) include: B1, A1: Commerce and Industry; B2, A2: Community Development; B3, A3: Education; B4, A4: Emergency Assistance; B5, A5: Energy; B6, A6: Environment; B7, A7: Gender; B8, A8: Governance; B9, A9: Health; B10, A10: Security; B11, A11: Transport; B12, A12: Water and Sanitation; B13, A13: Agriculture; and B14, A14: Capacity Building.

## 4 Computer Simulation and Discussion of the Results

The data set was divided into four subsets, each with one output variable representing the event killed, wounded, hijacked, or the presence of an event, each taking the value of 1 "Yes" or 0 "No". As the original data set contains 101 variables (97 input variables and 4 output variables), we used attribute selection tools for the data set representing the entire country and its seven regions to obtain the best subsets of input variables. The process of attribute selection is done in two parts. First, the attribute evaluator is used to identify and assess attribute subsets and then the best-first search method is applied to retain attributes. Table 2 shows that only 5 or 6 attributes were retained from 97 input attributes.

**Table 2.** Significant variables for four output variables for the entire country

| Dead | Wounded | Hijacked | Events |
|------|---------|----------|--------|
| 1. A1(t − 2) − Commerce and Industry | 1. Urban Female Population Density | 1. B1(t − 1) − Commerce and Industry | 1. A2(t − 1) − Community Development |
| 2. Urban Male Population Density | 2. Dead(t − 1) | 2. B6(t) − Environment | 2. Urban Male Population Density |
| 3. Dead(t − 1) | 3. Wounded(t − 1) | 3. Dead(t − 1) | 3. Dead(t − 1) |
| 4. Wounded(t − 1) | 4. Hijacked(t − 1) | 4. Hijacked(t − 1) | 4. Wounded(t − 1) |
| 5. Hijacked(t − 1) | 5. Event_Nu (t − 1) | 5. Event_Nu(t − 1) | 5. Hijacked(t − 1) |
| 6. Event_Nu(t − 1) | | | 6. Event_Nu(t − 1) |

Table 3 shows the rates for five classifiers for the entire country. One can see that most correct classification accuracy rates for "Yes" cases are low. The reason is the data are unbalanced, i.e., too few "Yes" cases in the four data sets. The "Yes" cases are within the [47.6, 74.9] range for all four adverse events. More specifically, for dead, wounded, hijacked, and events the "Yes" rates are within the [60.6, 74.9], [60.7, 65.5], [47.6, 64.9], and [54.1, 61.9] range, respectively. As expected the overall rates and the rates for the "No" cases are much higher and they fall within the [63.9, 89.2] and [62.9, 93.4] ranges, respectively, for four adverse events The AUROC values, which testify to the global performance of the five methods and are measured on the [0, 1] scale, fall into the range of [.688, .805]. In terms of the AUROC values, C4.5 and $k$NN seem to produce the highest performance. We also performed analysis for the seven different regions to identify significant variables. Depending on the region and the category of the adverse event, from three to fifteen variables are retained. One can also see more regularity and patterns in these results. For example, for most regions the number of killed, wounded, hijacked, and events at time t − 1, one month before the adverse event, appear to be significant as well as the population densities. Among the project types: Governance, Energy, Emergency Assistance, Security, and Education, in this order, at different periods are the most significant. Among the number of projects: Emergency Assistance, Education, Security, and Capacity Building, in this order, at different periods appear to be significant. Surprisingly, investments in Community Development, Transport, Water and Sanitation, and Agriculture at different periods are very rarely significant in detecting the four adverse events. However, for region Central none of the project types or their number has been identified as significant. One can observe similar patterns in regions Southwestern and Northwestern where only one variable representing project types Security and Energy, respectively, seem to be significant in detecting dead and wounded, respectively.

**Table 3.** The correct classification accuracy rates in [%] and AUROC for five classifiers for the whole country

| Classifier | Overall | Yes | No | AUROC | $C_{10}$ | Overall | Yes | No | AUROC | $C_{10}$ |
|---|---|---|---|---|---|---|---|---|---|---|
| | *Dead* | | | | | *Wounded* | | | | |
| NN | 88.6 | 60.6 | 90.9 | 0.766 | 12 | 88.8 | 60.7 | 90.9 | 0.765 | 14 |
| *k*NN | 79.5 | 70.0 | 80.2 | 0.797 | 22 | 87.3 | 65.1 | 88.9 | 0.787 | 14 |
| SMO | 63.9 | 74.9 | 62.9 | 0.689 | 22 | 88.8 | 60.9 | 90.8 | 0.758 | 22 |
| C4.5 | 78.7 | 71.7 | 79.3 | 0.797 | 25 | 87.0 | 65.5 | 88.5 | 0.778 | 14 |
| RF | 79.7 | 62.6 | 81.1 | 0.747 | 25 | 86.1 | 61.1 | 87.9 | 0.751 | 22 |
| | *Hijacked* | | | | | *Events* | | | | |
| NN | 75.0 | 56.7 | 75.3 | 0.694 | 52 | 88.4 | 54.1 | 93.4 | 0.743 | 7 |
| *k*NN | 75.4 | 63.9 | 75.6 | 0.769 | 100 | 84.6 | 61.6 | 88.0 | 0.804 | 10 |
| SMO | 89.2 | 64.9 | 89.7 | 0.773 | 80 | 88.3 | 54.2 | 93.3 | 0.738 | 10 |
| C4.5 | 76.3 | 60.4 | 76.3 | 0.731 | 100 | 86.2 | 61.9 | 89.7 | 0.805 | 7 |
| RF | 77.8 | 47.6 | 78.4 | 0.688 | 100 | 85.5 | 58.1 | 89.5 | 0.788 | 10 |

Table 4 presents the rates for seven regions for four adverse events generated by the C4.5 classifier that exhibits the best global performance measured by the AUROC values. For the four adverse events, the rates for "Yes" cases are within the range [51.7, 84.9], where 51.7% is the rate for hijacked in region Southwestern and 84.9% is the rate for events in region Southeastern. One can see the AUROC values for hijacked are the lowest within [.571, .765] and $C_{10}$ values are the highest within [124, 800]. The AUROC values for events are the highest within [.668, .893] and $C_{10}$ values are the lowest within [4, 14]. The higher $C_{10}$ values show that when there are fewer "Yes" cases in the data, a bigger cost value needs to be set in order to obtain a balanced classification accuracy rates. Most of the AUROC values with higher $C_{10}$ are lower than the AUROC values with lower $C_{10}$.

For dead, the highest and the lowest AUROC values are .851 and .718, for the Southwestern region and the Eastern region, respectively. For wounded, the AUROC value for region Southwestern is .786 (the highest) and for Western region is .714 (the lowest). For hijacked, the AUROC value is the highest (.765) for the Northeastern region and for the Southwestern region is the lowest (.571). For events, the highest AUROC value is .893 for the Southeastern region and the lowest is .668 for the Eastern region. For all four adverse events, the highest AUROC value is .893 representing events in the Southeastern region and the lowest is .571 representing hijacked in the Southwestern region.

**Table 4.** The correct classification accuracy rates in [%] and AUROC using C4.5 for seven regions

| Region | Overall | Yes | No | AUROC | $C_{10}$ | Overall | Yes | No | AUROC | $C_{10}$ |
|---|---|---|---|---|---|---|---|---|---|---|
| | *Dead* | | | | | *Wounded* | | | | |
| Central | 89.2 | 67.3 | 90.5 | 0.789 | 17 | 91.7 | 65.3 | 93.2 | 0.764 | 17 |
| Eastern | 69.8 | 69.2 | 69.9 | 0.718 | 50 | 73.6 | 65.9 | 74.3 | 0.722 | 50 |
| Northeastern | 85.5 | 66.1 | 86.1 | 0.752 | 80 | 84.1 | 65.8 | 84.7 | 0.731 | 60 |
| Southeastern | 83.6 | 74.2 | 84.6 | 0.823 | 10 | 89.3 | 63.5 | 91.7 | 0.783 | 11 |
| Western | 78.0 | 79.9 | 77.8 | 0.799 | 20 | 86.4 | 60.0 | 88.1 | 0.714 | 16 |
| Northwestern | 83.4 | 68.6 | 83.9 | 0.780 | 80 | 75.0 | 68.9 | 75.1 | 0.750 | 37 |
| Southwestern | 80.3 | 80.0 | 80.4 | 0.851 | 5 | 83.6 | 71.3 | 85.6 | 0.786 | 6 |
| | *Hijacked* | | | | | *Events* | | | | |
| Central | 87.5 | 60.2 | 88.0 | 0.748 | 150 | 80.9 | 71.4 | 81.9 | 0.816 | 13 |
| Eastern | 74.2 | 61.6 | 74.5 | 0.701 | 800 | 69.9 | 63.7 | 71.0 | 0.668 | 6 |
| Northeastern | 78.9 | 64.5 | 79.3 | 0.765 | 170 | 82.7 | 75.5 | 83.3 | 0.829 | 14 |
| Southeastern | 78.5 | 62.0 | 78.9 | 0.750 | 200 | 81.9 | 84.9 | 81.3 | 0.893 | 6 |
| Western | 74.0 | 67.2 | 74.2 | 0.740 | 140 | 80.4 | 75.0 | 81.2 | 0.797 | 7 |
| Northwestern | 94.2 | 64.9 | 94.6 | 0.649 | 124 | 74.8 | 66.0 | 74.8 | 0.737 | 13 |
| Southwestern | 56.2 | 51.7 | 56.3 | 0.571 | 180 | 80.6 | 84.1 | 79.5 | 0.874 | 4 |

## 5   Conclusion

Several previous studies used FIS, ANFIS, neural networks, and linear regression models to predict the number of four adverse events (dead, wounded, hijacked, and events) based on infrastructure development spending and other variables. Because the predictive models generated large errors, this study proposed a different and more feasible approach. This study investigates the effectiveness of five cost-classification methods for detecting adverse events using the data set provided by the HSCB program management of the U.S. DoD. It appears that when analysis is performed for the entire country, there is little correlation between adverse events and fourteen project types and the number of projects. However, the same type of analysis performed for each of its seven regions shows a relationship between adverse events and the infrastructure budget and the number of projects allocated for the specific regions and periods. Among the five classifiers, C4.5 and *k*-NN seem to be the best in terms of the global performance.

The results shows that the AUROC values for events are generally higher than the AUROC values for dead, wounded and hijacked; and that the AUROC values for hijacked are generally lower than the AUROC values for dead, wounded and events. When there are fewer "Yes" cases in a data set, we have to increase the cost values to obtain a balanced accuracy rates for "Yes" cases. Even so, the performance with a bigger cost value is still lower than that with a smaller cost value under most circumstances. Our results show that it is easier to obtain better results with a more balanced data set than to achieve better results through adjustment of cost in an unbalanced data set.

The models presented in this study might support decision makers who analyze historical economic data on how regional funds allocation can best help minimize adverse events. Though the models used Afghanistan data, they may be applicable for other countries, such as Iraq and Somalia, where the threat of terrorist and military activities is present.

**Acknowledgment.** This study was supported in part by Grant no. 10523339, Complex Systems Engineering for Rapid Computational Socio-Cultural Network Analysis, from the Office of Naval Research awarded to Dr. W. Karwowski at the University of Central Florida.

# References

1. Clancy, J., Crossett, C.: Measuring effectiveness in irregular warfare. Parameters **37**(2), 88 (2007)
2. HSCB Modeling Program (2009). http://www.dtic.mil/biosys/docs/HSCBnews-spring-2009.pdf
3. Stanton, J.: Evolutionary cognitive neuroscience: dual use discipline for understanding & managing complexity and altering warfare. In: International Studies Association Conference, Portugal (2007). SSRN. http://ssrn.com/abstract=1946864
4. Open Source Center (OSC): Afghanistan-Geospatial Analysis Reveal Patterns in Terrorist Incidents 2004–2008, 30 April 2009. http://www.fas.org/irp/dni/osc/afghan-geospat.pdf. Accessed 3 May 2012
5. Berrebi, C., Lakdawalla, D.: How does terrorism risk vary across space and time? An analysis based on the israeli experience. Defense Peace Econ. **18**(2), 113–131 (2007)
6. Reed, G.S., Colley, W.N., Aviles, S.M.: Analyzing behavior signatures for terrorist attack forecasting. J. Defense Model. Simul.: Appl. Methodol. Technol. **10**, 1–12 (2011)
7. Inyaem, U., Meesad, P., Haruechaiyasak, C., Tran, D.: Terrorism event classification using fuzzy inference systems. Int. J. Comput. Sci. Inf. Secur. (IJCSIS) **7**(3), 243–256 (2010)
8. Çakıt, E., Karwowski, W.: Assessing the relationship between economic factors and adverse events in an active war theater using fuzzy inference system approach. Int. J. Mach. Learn. Comput. **5**(3), 252–257 (2015)
9. Çakıt, E., Karwowski, W.: Fuzzy inference modelling with the help of fuzzy clustering for predicting the occurrence of adverse events in an active theater of war. Appl. Artif. Intell. **29**, 945–961 (2015)
10. Çakıt, E., Karwowski, W.: Understanding the social and economic factors affecting adverse events in an active theater of war: a neural network approach. In: Advances in Cross-Cultural Decision Making. Advances in Intelligent Systems and Computing, vol. 610, pp. 215–223. Springer (2017)
11. Witten, I.H., Frank, E., Hall, M.A.: Data Mining: Practical Machine Learning Tools and Techniques, 3rd edn. Morgan Kaufmann, Burlington (2011)
12. Elkan, C.: The Foundations of cost-sensitive learning. In: International Joint Conference on Artificial Intelligence, vol. 17, no. 1, pp. 973–978. Lawrence Erlbaum Associates Ltd. (2001)
13. Platt, J.: Fast Training of support vector machines using sequential minimal optimization. In: Schoelkopf, B., Burges, C., Smola, A. (eds). Advances in Kernel Methods - Support Vector Learning (1998)

# Country Selection for Business Location Under Imperfect Information

R. R. Aliyev[(⊠)]

Warwick Business School, WBS, The University of Warwick, Coventry, UK
rafig.aliyev@hotmail.com

**Abstract.** One of important problems in international business is evaluation of the relative priority of countries to enter for archiving sustainable business success. The objective of this study is to develop country selection model in imperfect situation, which is characterized with uncertain, imprecise and partially reliable decision-relevant information. The suggested approach is based on use of fuzzy AHP method, that is expected to provide decision makers tool to select more suitable candidate countries for sustainable business success.

**Keywords:** Sustainable international business · Country selection model
Imperfect information · Fuzzy AHP

## 1 Introduction

Country selection problem fundamentally matters to international business [3, 4]. In [1, 2] authors investigate impact of economic, social and political environments to advantage when entering a new business location. Global business strategies in terms of internationalization are considered in [5]. The research in [6] highlights risk factors and project reward factor for construction country selection model for sustainable business. These authors use hybrid of objective and subjective information for creation of country selection model. A multi-criteria decision model to estimation priorities for foreign investment is given in [7]. The suggested methodology combines AHP, TOPSIS and multi-period multi-attribute decision making (MP-MADNI) methods.

In [8] authors develop method of selecting the country to maximize fit degree between the retailer criteria and country characteristics. Ethical, cultural, economic and other factors are considered as necessary attribute of the internationalization decision.

[9] investigate country selection as a function of geographical or cultural distance.

In this study we develop fuzzy AHP-based country selection for international business location under imperfect information.

The paper is structured as follows. In Sect. 2, we present some prerequisite material including fuzzy number, fuzzy AHP, etc. In Sect. 3 we give statement of the country selection problem. In Sect. 4 we illustrate application of fuzzy AHP to real-world country selection problem. Section 5 offers conclusions.

© Springer Nature Switzerland AG 2019
R. A. Aliev et al. (Eds.): ICAFS-2018, AISC 896, pp. 922–928, 2019.
https://doi.org/10.1007/978-3-030-04164-9_122

## 2   Preliminaries

*Fuzzy Number* **[10]**

A fuzzy number is a set $A$ on $R$ which possesses the following properties: (a) $A$ is a normal fuzzy set; (b) $A$ is a convex fuzzy set; (c) $\alpha$ – cut of $A$, $A^{\alpha}$ is a closed interval for every $\alpha \in (0, 1]$; (d) the support of $A$, $A^{+0}$ is bounded.

*Triangular Fuzzy Numbers* **[10]**

A fuzzy number $A = (a, b, c)$ is called triangular if its membership function is described as

$$\mu_A(x) = \begin{cases} 0, & x < a \\ \frac{x-a}{b-a}, & a \leq x \leq b \\ \frac{c-x}{c-b}, & b \leq x \leq c \\ 1, & x > c \end{cases}$$

**Fuzzy AHP**

Fuzzy AHP applied in this study includes the following steps [6].

**Step 1.** Comparison of criteria and alternatives via linguistic terms (see Table 1). Construction of pair wise comparison matrice.

**Table 1.** The linguistic terms and the corresponding triangular fuzzy numbers

| Saaty scale | Definition | TFN |
|---|---|---|
| 1 | Equally important (Eq. Imp.) | (1, 1, 1) |
| 3 | Weakly important (W. Imp.) | (2, 3, 4) |
| 5 | Fairly important (F. Imp.) | (4, 5, 6) |
| 7 | Strongly important (S. Imp.) | (6, 7, 8) |
| 9 | Absolutely important (A. Imp.) | (9, 9, 9) |
| 2 | The intermittent values between adjacent scales | (1, 2, 3) |
| 4 | | (3, 4, 5) |
| 6 | | (5, 6, 7) |
| 8 | | (7, 8, 9) |

**Step 2.** Calculation of mean of fuzzy comparison values of each criterian.
**Step 3.** Calculation of fuzzy weights of each criterian.
**Step 4.** Defuzzification of obtained fuzzy weights.
**Step 5.** Normalization of defuzzified weights.
**Step 6.** Calculation of the scores for each alternative.
**Step 7.** Ranking of alternatives.

## 3  Statement of the Problem

Using factor analysis authors in [6] extracted main factors affecting to country selection problem: business environment (BE), market opportunity (MO), project success (PS), and market experience (ME). Using these factors the country risk level is identified as average of BE and MO. Project reword level is determined as average of PS and ME.

In this study as criteria for selection of candidate country we use market stability $(C_1)$; market competition $(C_2)$; quality of national governance $(C_3)$; ease of doing business $(C_4)$; average profit rate $(C_5)$; stability of profit performance $(C_6)$. As candidate countries for selection 4 countries are considered: $A_1$(Brazil); $A_2$ (Romania); $A_3$(Germany); $A_4$(UK).

The problem is to choose best country $(A_n^*)$ that fits all criteria $C_j, \ j = 1, 6$.

## 4  Solution of the Problem by Using Fuzzy AHP

For solution of the country selection problem stated in Sect. 3 we use fuzzy AHP method. As fuzzy number triangular fuzzy number is used.

In accordance with fuzzy-AHP application procedure described in Sect. 2, first we construct pair wise comparisons of criteria $C_i, i = \overline{1,6}$ (see Table 2).

**Table 2.** Pairwise comparison matrise

| C | C1 | | | C2 | | | C3 | | | C4 | | | C5 | | | C6 | | |
|---|---|---|---|---|---|---|---|---|---|---|---|---|---|---|---|---|---|---|
| C1 | 1 | 1 | 1 | 4 | 5 | 6 | 1 | 1 | 1 | 1 | 1 | 1 | 1 | 1 | 1 | 0.2 | 0.2 | 0.3 |
| C2 | 0.2 | 0.2 | 0.3 | 1 | 1 | 1 | 0.13 | 0.14 | 0.17 | 1 | 1 | 1 | 1 | 1 | 1 | 1 | 1 | 1 |
| C3 | 1 | 1 | 1 | 6 | 7 | 8 | 1 | 1 | 1 | 0.2 | 0.2 | 0.3 | 0.2 | 0.2 | 0.3 | 1 | 1 | 1 |
| C4 | 1 | 1 | 1 | 1 | 1 | 1 | 4 | 5 | 6 | 1 | 1 | 1 | 1 | 1 | 1 | 1 | 1 | 1 |
| C5 | 1 | 1 | 1 | 1 | 1 | 1 | 4 | 5 | 6 | 1 | 1 | 1 | 1 | 1 | 1 | 1 | 1 | 1 |
| C6 | 4 | 5 | 6 | 1 | 1 | 1 | 1 | 1 | 1 | 1 | 1 | 1 | 1 | 1 | 1 | 1 | 1 | 1 |

Then we calculate geometric means of fuzzy comparison values as

$$r_i = \left( \prod_{j=1}^{n} d_{ij} \right)^{1/n} , i = 1, 2, \ldots, n$$

Results of calculation is given in Table 3.

**Table 3.** Geometric means of fuzzy comparison values

| Criteria | Ri | | |
|---|---|---|---|
| C1 | 0.9347 | 0.9701 | 1.0699 |
| C2 | 0.5246 | 0.5529 | 0.5888 |
| C3 | 0.7418 | 0.8088 | 0.8909 |
| C4 | 1.2599 | 1.3077 | 1.3480 |
| C5 | 1.2599 | 1.3077 | 1.3480 |
| C6 | 1.2599 | 1.3077 | 1.3480 |
| Total | 5.9808 | 6.2548 | 6.5936 |
| Reverse (power of-1) | 0.1672 | 0.1599 | 0.1517 |
| Increasing order | 0.1517 | 0.1599 | 0.1672 |

Next relative fuzzy weight of each criterian are defined as

$$w_i = r_i \otimes (r_1 \otimes r_2 \otimes \ldots \otimes r_n)^{-1}$$

Results are shown in Table 4.

**Table 4.** Fuzzy weights of each criterian

| Criteria | $w_i$ | | | Sum | Mi |
|---|---|---|---|---|---|
| C1 | 0.1418 | 0.1551 | 0.1789 | 0.4758 | 0.1586 |
| C2 | 0.0796 | 0.0884 | 0.0984 | 0.2664 | 0.0888 |
| C3 | 0.1125 | 0.1293 | 0.1490 | 0.3908 | 0.1303 |
| C4 | 0.1911 | 0.2091 | 0.2254 | 0.6256 | 0.2085 |
| C5 | 0.1911 | 0.2091 | 0.2254 | 0.6256 | 0.2085 |
| C6 | 0.1911 | 0.2091 | 0.2254 | 0.6256 | 0.2085 |
| Total | 0.9073 | 1.0001 | 1.1025 | 3.0099 | 1.0033 |

Defuzzifacation of fuzzy weights and their normalization is performed and results are shown in Tables 4 and 5.

**Table 5.** Normalized and defuzzified weights of criteria

| Criteria | Mi | Ni |
|---|---|---|
| C1 | 0.1586 | 0.1581 |
| C2 | 0.0888 | 0.0885 |
| C3 | 0.1303 | 0.1298 |
| C4 | 0.2085 | 0.2079 |
| C5 | 0.2085 | 0.2079 |
| C6 | 0.2085 | 0.2079 |

Now we proceed to calculation of weights of alternatives $A_i$, $i = 1, \ldots, 4$ with respect to criteria $C_j$, $j = 1, \ldots, 6$. The same procedures as it was used above are applied for calculation respective values for alternatives.

Let's to illustrate this calculation for criterian $C_1$.

Pair waise comparision of $A_i$, $i = 1, \ldots, 4$ with respect to $C_1$ is shown in Table 6.

**Table 6.** Pair wise comparisions $A_I$, $i = 1,..,4$ with respect $C_1$

| C1 | A1 | | | | | | A2 | | | A3 | | | A4 | | |
|----|----|----|----|----|----|----|------|------|------|------|------|------|------|------|------|
| A1 | 1 | 1 | 1 | 1 | 1 | 1 | 0.17 | 0.20 | 0.25 | 0.17 | 0.20 | 0.25 | | | |
| A2 | 1 | 1 | 1 | 1 | 1 | 1 | 0.25 | 0.33 | 0.50 | 0.17 | 0.20 | 0.25 | | | |
| A3 | 4 | 5 | 6 | 2 | 3 | 4 | 1 | 1 | 1 | 0.25 | 0.33 | 0.50 | | | |
| A4 | 4 | 5 | 6 | 4 | 5 | 6 | 2 | 3 | 4 | 1 | 1 | 1 | | | |

Similar to criterion calculation prodecduries we determine geometric means (Table 7) and fuzzy weights (Tables 8) of alternatives.

**Table 7.** Geometric means of fuzzy comparison values

| Criteria | Ri | | |
|----------|--------|--------|--------|
| A1 | 0.4082 | 0.4472 | 0.5000 |
| A2 | 0.4518 | 0.5081 | 0.5946 |
| A3 | 1.1892 | 1.4953 | 1.8612 |
| A4 | 2.3784 | 2.9428 | 3.4641 |
| Total | 4.4277 | 5.3935 | 6.4199 |
| Reverse (power of-1) | 0.2259 | 0.1854 | 0.1558 |
| Increasing order | 0.1558 | 0.1854 | 0.2259 |

Defuzzified and normalized weights of $A_i$, $i = 1, \ldots, 4$ with respect to $C_1$ is given in Table 9.

Analogously we calculated defuzzified and normalized relative weights of each alternative with respect $C_j$, $j = 1, \ldots, 6$.

Consequently we have aggregated results for each alternative $A_i$, $i = 1, \ldots, 4$ with respect to each criterion (Table 10).

Ranking of total scores of alternatives shows that the best candidate country is $A_4$ (UK). This results coinside with existing study on country selection for international business location. In [6] it is shown that UK belongs to group of countries where business climate is characterized with low-risk and high-reward.

**Table 8.** Fuzzy weights.

| Criteria | $w_i$ | | | Sum | $M_i$ |
|---|---|---|---|---|---|
| A1 | 0.0636 | 0.0829 | 0.1130 | 0.2595 | 0.0865 |
| A2 | 0.0704 | 0.0942 | 0.1343 | 0.2989 | 0.0996 |
| A3 | 0.1853 | 0.2772 | 0.4204 | 0.883 | 0.2943 |
| A4 | 0.3706 | 0.5456 | 0.7825 | 1.6987 | 0.5662 |
| Total | 0.6898 | 1.0000 | 1.4503 | 3.1400 | 1.0467 |

**Table 9.** weights of $A_i$, i = 1, ..., 4 with respect $C_1$

| Criteria | Mi | Ni |
|---|---|---|
| A1 | 0.0865 | 0.0826 |
| A2 | 0.0996 | 0.0952 |
| A3 | 0.2943 | 0.2812 |
| A4 | 0.5662 | 0.541 |

**Table 10.** Aggregated values for each alternative via to each criteria

| Criteria | | Scores of Alternatives with respect to related Criterion | | | |
|---|---|---|---|---|---|
| | Weights | A1 | A2 | A3 | A4 |
| C1 | 0.160 | 0.083 | 0.095 | 0.281 | 0.541 |
| C2 | 0.170 | 0.083 | 0.095 | 0.281 | 0.541 |
| C3 | 0.070 | 0.083 | 0.095 | 0.281 | 0.541 |
| C4 | 0.200 | 0.083 | 0.095 | 0.281 | 0.541 |
| C5 | 0.200 | 0.091 | 0.394 | 0.121 | 0.395 |
| C6 | 0.200 | 0.091 | 0.394 | 0.121 | 0.395 |
| Total | | 0.09 | 0.10 | 0.20 | 0.70 |

## 5   Conclusion

Country selection problem in international business is challenging problem to archive sustainable business success. This scientific problem usually is accompanied with business environment characterized with uncertain imprecise and partially reliable decision model of country selection for business location. Solution results of real-world problem coinside with existing practical situation in country selection, also with existing literature on the problem and approves validity and applicability of the suggested approach.

# References

1. Han, S.H., Kim, D.Y., Jang, H.S., Choi, S.: Strategies for contractors to sustain growth in the global construction market. Habitat Int. **34**, 1–10 (2010)
2. Teo, E.A.L., Chan, S.L., Tan, P.H.: Empirical investigation into factors affecting exporting construction services in SMEs in Singapore. J. Constr. Eng. Manag. **133**, 582–591 (2007)
3. Chen, C.: Entry mode selection for international construction markets: the influence of host country related factors. Constr. Manag. Econ. **26**, 303–314 (2008)
4. Chen, C., Wang, Q., Martek, I., Li, H.: International market selection model for large Chinese contractors. J. Constr. Eng. Manag. **142**, 04016044 (2016)
5. Jung, W., Han, S.H., Lee, K.W.: Country portfolio solutions for global market uncertainties. J. Manag. Eng. **28**, 372–381 (2012)
6. Lee, K.W., Han, S.H.: Quantitative analysis for country classification in the construction industry. J. Manag. Eng. (2017)
7. Blonigen, B.A., Davies, R.B., Waddel, G.R., Naughton, H.T.: FDI in space: spatial autoregressive relationships in foreign direct investment. Eur. Econ. Rev. **51**(5), 1303–1325 (2007). https://doi.org/10.1016/j.euroecorev.2006.08.006
8. Ghauri, P., Elg, U., Sinkovics, R.: Foreign direct investment-location attractiveness for retailing forms in the European Union. In: Ghauri, P.N., Oxelheim, L. (eds.) European Union and the Race for Foreign Direct Investment in Europe, pp. 407–428. Oxford, Pergamon (2004)
9. Vida, I.: An empirical inquiry into international expansions of US retailers. Int. Mark. Rev. **17**(4/5), 454–475 (2000)

# Material Selection Methods: A Review

M. B. Babanli[1]($\boxtimes$), F. Prima[2], P. Vermaut[2], L. D. Demchenko[3],
A. N. Titenko[4], S. S. Huseynov[1], R. J. Hajiyev[1],
and V. M. Huseynov[1]

[1] Azerbaijan State Oil and Industry University, AZ1010 Baku, Azerbaijan
mustafababanli@yahoo.com
[2] Chimie ParisTech, UMR CNRS 7045, 11 rue Pierre et Marie Curie,
75005 Paris, France
[3] National Technical University of Ukraine "KPI", Kiev 03056, Ukraine
[4] Institute of Magnetism under NAS and MES of Ukraine, Kiev 03142, Ukraine

**Abstract.** Material selection is an important problem attracting theoretical and practical interest. Nowadays, a lot of materials and alloys are designed. In most alloys some properties are good and in compliance with the requirements, but some of them are not acceptable. Generally, for material selection methods it is necessary to have unique synergy of theoretical knowledge and practical experiences data. Scientists used and developed some selection methods due to all of these.

**Keywords:** Material selection method
Multi-criteria decision making (MCDM) · Fuzzy approach methods
Z-number

## 1 Introduction

Material selection includes three stages: initial screening, development and comparison of alternatives, and final one is determination of best solution [1, 2]. A systematic study of material selection in the mechanic field is provided in Ref. [2]. The author describe the main categories of materials, their properties, and in general, evolution of material selection study. The considered approaches mainly fall within the classical (analytical equations based) optimization techniques, such that property limits, geometric restrictions, material indexes (e.g. structural index), cost (cost and performance relation) and other criteria are considered. The book includes a series of important case studies. A systematic study on quantitative approaches to material selection, including analytical and some computer-aided approaches is also provided in [1]. In [1–3] it is mentioned that two main kinds of information are relevant for selection: (i) Screening and ranking information and, (ii) supporting information. The former one is commonly related to shifting through the database due to the technical and economical requirements of design, whereas the latter one supporting information is based on knowledge about microstructure, performance in specific environment and other issues [25]. In this review, the main categories of material selection methods are compiled and classifed.

© Springer Nature Switzerland AG 2019
R. A. Aliev et al. (Eds.): ICAFS-2018, AISC 896, pp. 929–936, 2019.
https://doi.org/10.1007/978-3-030-04164-9_123

## 2   Multi-criteria Decision-Making (MCDM) Approaches and the Related Techniques

One can say that using multi-criteria decision-making (MCDM) for selection of materials is considered for new materials with complex application, and when each material has a competitive advantage in performance criteria [4]. MCDM methods include multiobjective optimization and multiattribute decision making (MADM). MADM strategies can be fitted to material plan and improvement. MADM methods are used to compare materials by using set of characteristics.

One of the approaches for initial screening is the Ashby method [2]. This method is based on material selection charts where the axes for density and strength are used. Various lines describing fixed ratios of these properties are drawn. Materials placed on the line exhibit approximately the same ratio, those which are upper a line perform better, and those that are below – perform worse. An application of the Ashby method to initial screening in the framework of multiobjective optimisation is considered in [5].

Review of methods of material screening and choosing is provided in [6]. Authors discuss the existing literature in the field, including potential application and various approaches. Also, advantages and drawbacks of screening and choosing methods in material selection is discussed in the paper. In [1] the simplest method for comparing candidate materials referred to as weigthed properties method is described. According to this method, each material characteristic is assigned a weight measuring the importance for an alloy in a considered applied filed. Values of attributes are multiplied by corresponding weigths and then summed to produce the overall performance of alloy. Finally, an alloy with the highest overall performance is chosen as the best option. In realm of MADM such approach is referred to as simple additive weighting (SAW). This method is simple, but includes several well known disadvantages (mainly, difficulty of assigning weights and additive aggregation). In order to problem of adequate determination of weigths, a digital logic method (DLM) is used [1]. Weigths are determined by using pairwise comparison of material properties. Only two properties are considered at a time and provided that all possible pairs are analyzed, the correct weight is found by using relative scalling.

As it is written in [7] many methods suffer from a lack of the support to the selection of proper engineering criteria or parameters which are desired especially for the new designers and those who have little knowledge about the operations in the selection process. As it is mentioned in [29], one of the shortcomings of the existing tools is that there is a need for defining a weighting method which is both user-friendly and can clearly represent the project's requirements. In this article, it is underlined importance of ranking and selection of the optimal material during engineering design process. The authors indicate tending to focus on cost and benefit criteria as minuses for proposed methods of ranking in materials selection. According to the authors, due to these financial aspects, some technical or biological properties are missing. In response to this perceived gap, Technique for Order Preference by Similarity to Ideal Solution (TOPSIS) method and objective weighting were used in this paper. Accordingly, in this work the TOPSIS method was modified and updated. Further, the system for objective weighting was developed for cases with target values of attributes, and its drawbacks are addressed. Also it has been shown that the offered normalization technique is not only able to account for the impact of criteria, but also for the criteria goal values. TOPSIS method in this work was applied for several cases for validation of the proposed model results. One of the most famous MADM methods is AHP method [8]. In this approach a decision problem is considered within three-level hierarchy: the aim of choice, criteria, and alternatives. Criteria evaluations and weights are determined by using pairwise comparison matrices. This comparison increases reliability of computation results but complicates the decision procedures. AHP method is widely applied for material selection purposes. In [8] they apply this method for analysis of Al/SiC composite material properties and making decisions. Paper [9] is devoted to application of AHP for semiconductor switching devices. In [10] an aluminum alloy was chosen by using AHP with the aim to minimize environmental impact in screw manufacturing. The paper [11] suggests that the designers can be able to work without preparing material selection decision matrices and can mainly stress on finding the most important criterion controlling the entire selection process. In the article, five material selection problems are examined from different point of views. VIKOR, TOPSIS and PROMETHEE methods are used. The comprehension of the matter hints at the impact of the criteria having maximum priority weight on the working efficiency of VIKOR, TOPSIS and PROMETHEE techniques. [12] is devoted to material selection for flywheel. The application of a flywheel can be hindered by sudden failure. In order to ensure continuous work of a flywheel it is needed to delay the time to failure which is due to fatigue and/or brittle structure. Performance index in this particular case will be the ratio of fatigue limit over the material density [13]:

If the failure occurs due to the brittle structure of the material, then the fracture toughness (Kic) of the material will be the most important parameter to consider. The fragmentability (F) is also very important parameter while considering time to failure because if it is possible to break the flywheel into small parts such that the risk of the hazards can be minimized [14]. It is demonstrated that both methods VIKOR and ELECTRE II have determined the same optimal material for the flywheel design [13]. These methods have a great potential to solve such kind of material selection problems (Table 1).

**Table 1.** Quantitative data of material selection for flywheel [13].

| Sl. № | $\sigma_{limit}/\rho$ | $K_{iC}/\rho$ | C/m | F |
|---|---|---|---|---|
| 300M ($A_1$) | 100 | 8.6125 | 4200 | 3 |
| 2024-T3 ($A_2$) | 49.6454 | 13.4752 | 2100 | 3 |
| 7050-T73651($A_3$) | 78.0142 | 12.5532 | 2100 | 3 |
| Ti-6Al-4 V ($A_4$) | 108.8795 | 26.0042 | 10500 | 3 |
| E glass-epoxy FRP ($A_5$) | 70 | 10 | 2735 | 9 |
| S glass-epoxy FRP ($A_6$) | 165 | 25 | 4095 | 9 |
| Carbon-epoxy FRP ($A_7$) | 440.2516 | 22.0126 | 35470 | 7 |
| Kevlar 29-epoxy FRP ($A_8$) | 242.8571 | 28.5714 | 11000 | 7 |
| Kevlar 49-epoxy FRP ($A_9$) | 616.4384 | 34.2466 | 25000 | 7 |

In [15] 10 MCDM methods are compared in solving material selection problems for 3 various purposes. Among the considered methods, PROMETHEE II (EXPROM2), complex proportional of with gray relations (COPRAS-G), ORESTE (Organization, Rangement Et Synthese De DonnesRelationnelles), VIKOR and operational competitiveness rating analysis (OCRA) are used. All these methods indicate selected materials which are best and worst. It is shown that for selection of any type of material with various criteria and number of alternatives the named methods can be efficiently used. The VIKOR method is proven to be the best method amongst these methods because of relatively better performance.

However, in [16] it is argued that though VIKOR method is quite a comprehensive tool for solving the problems of material selection it still does not account for constraints and goals of design with continuous variables. [17] is devoted to comparison of VIKOR and ELECTRE methods in selection of materials for engineering purpose. VIKOR is a compromise ranking method and ELECTRE is an outranking method. At the same time both of them are used to compare materials with under considered requirements.

One of the novel methods of material selection, referred to as Z-transformation method is applied in [18]. In the process of designing of every product the selection of materials is considered very important. Different researchers have offered and applied approaches in selection of various materials with the quantitative and qualitative properties. But some of these methods are complex. That is why modified digital logic (MDL) method is suggested to normalize the criteria. Three cases in mechanical design and lightweight naval structures have been carried out to compare capabilities of this new technique with MDL. Advantages of Z-transformation methods are illustrated to be compared with MDL method where it is difficult to identify any superiority of material properties when their selecting. Designers evaluated this advantage of the Z-transformation method because trades try to make use of the lower ranked materials. Despite a wide range of studies on materials selection using different MCDM methods exist, Indian scientists suggested a new approach of MCDM [19]. As it is mentioned in this paper, applicability and capability of new methods (for a given engineering application): (a) evaluation of mixed data (EVAMIX), and (b) complex proportional assessment (COPRAS) methods. Also, statistic information for two illustrative examples given this work could be used which proves that these two methods can be successfully applied to solve parctical problems in material selection. The main advantage of EVAMIX, is that it different mathematical models are used to deal with the ordinal and cardinal decision. On the other hand, COPRAS method can be easily applied to evaluate the alternatives and select the optimal material, even if the physical meaning of the decision making process is not known. Two illustrative examples and comparison of MCDM methods provided in the work to test the capability, accuracy and applicability of EVAMIX and COPRAS for complex material selection problems with ordinal and cardinal criteria. The authors of the article [20] use EXPROM2, COPRAS-G, ORESTE, OCRA methods in gear material selection. They have considered 9 different materials as the candidates for the gear design and amongst them were 5 selection criteria.

It was observed that the results obtained with COPRAS-G and EXPROM 2 are very similar to that obtained by the two most popular MCDM methods VIKOR and PROMETHEE II [15].

## 3  Fuzzy Logic and Soft Computing Based Approaches

The methods based on fuzzy logic and soft computing have a capability to deal with complexity and uncertainty of real-world problems in material selection. In the present study, we will consider some of these works to reflect the findings in the field.

In the research [21] they used fuzzy VIKOR as a tool to select alternative material for instrument panel of electric car. The fuzzy VIKOR is used to deal with linguistic evaluations provided by a decision maker. In [22] they applied fuzzy logic and genetic algorithm to develop shape memory alloy (SMA) actuators with the ability to take the required shape while heating and having a lot of application fields. Playing the main role in the system design actuators are based on hydraulic, electric, and pneumatic technology and have driving mechanism property. Physical changes of the actuators result in "memorization" of a specific shape displaying significant hysteresis. Delays and inaccuracy of the motion control of actuators are the result of hysteresis behavior. The authors have presented geometrical implementation based numerical Preisach model for the hysteresis in SMA. This model is used in a PID control strategy. The genetic algorithm is used to derive optimal values of PID parameters in computer simulation.

In [23] it is mentioned that engineers and designers have to make right materials choice meeting product requirements, such as weight saving, higher product performance, and cost reduction. The goal of the research is to suggest a new, interval-valued intuitionistic fuzzy sets (IVIFSs) and multi attributive border approximation area comparison (MABAC) based approach. It is used to settle material selecting problems with incomplete weight data. The authors offer that polypropylene is the best for the automotive instrument panel and Co–Cr alloys-wrought alloy is the optimal variant for hip prosthesis. IVIF-MABAC and other relevant representative methods have been compared and it was concluded that alternative materials are in a good agreement with the before derived ones. This new approach can be used to solve other material selection problems with robustness and efficiency. In [24] for selecting materials of an automotive instrument panel Fuzzy PROMETHEE (Preference Ranking Organization Method for Enrichment Evaluation) and PROMETHEE II (EXPROM2) methods are used. These methods are based on trapezoidal fuzzy interval numbers. The comparison of fuzzy PROMETHEE method with other three various fuzzy MCDM methods (fuzzy VIKOR, fuzzy TOPSIS, and fuzzy ELECTRE) is shown. The fuzzy methods including fuzzy multicriteria approaches, fuzzy inference, fuzzy expert systems and other methods are successively used for such problems as automotive component material selection [21], piezoelectric material selection [2], material selection in the electronics industry [3], material selection in concurrent product design [25] and other problems.

[26] is devoted to fuzzy logic based material selection in an applied field. The author reviews the main methods for material selection including classical MADM methods, MODM methods, expert systems and justify use of fuzzy logic. The proposed

approach is based on expert-driven fuzzy If-Then rules with inputs describing important characteristics as price, corrosion resistance, yield strength, toughness and others. The author uses some simplifying techniques to reduce number of fuzzy rules. Approximately 20–40 candidate alloys are compared to choose the best alternative in each case study. For validation of the proposed fuzzy logic based approach, the comparison with classical SAW method is used. The comparison is based on analysis of minimal, maximal and average performance indexes produced by the classical and fuzzy approaches for each case study. The results of comparison show that two methods provide close results In [27] Zadeh proposed a concept of Z-number to account for fuzziness and partial reliability of real-world decision relevant information. The first approach for material selection problem under Z-number valued information is proposed in [28]. The problem of material selection is formulated as a MADM problem. Criteria values and weights measuring properties of candidate alloys and the related importance are characterized by partial reliability. The solution approach is based on aggregation of Z-number-valued information and Z-number ranking procedures. Application of the proposed approach to selection of an optimal titanium alloy on the basis of three criteria is considered.

## 4   Conclusion

Based on the comprehensive review on the material selection subsection, fuzzy approach is a very good choice in material selection. Applying linguistic values by using fuzzy logic gives this method an advantage in comparison with other methods.

However, the proposed newest method, solving the problem of material selection in the Z-environment with Z-numbers by taking into account the partial reliability of the relevant information. As mentioned above the main advantage of this method is formulation of material selection problem as a MADM problem. Criteria values and weights measuring properties of candidate alloys and the related importance are characterized by partial reliability. The solution approach is based on aggregation of Z-number-valued information and Z-number ranking procedures.

## References

1. Frang, M.: Quantitative Methods of Material Selection. Handbook of Material Selection (2002)
2. Ashby, M.: Materials Selection in Mechanical Design. Butterworth-Heinemann, Oxford (2010)
3. Cebon, D., Ashby, M.: Data systems for optimal material selection. Adv. Mat. Process. **161** (6), 51–54 (2003)
4. Jahan, A., Edwards, K.L.: Multi-criteria Decision Analysis for Supporting the Selection of Engineering Materials in Product Design. Butterworth-Heinemann, Oxford (2016)
5. Ashby, M.: Multi-objective optimization in material design and selection. Acta Materilia **48**, 359–369 (2000)
6. Jahan, A., Ismail, M.Y., Sapuan, S.M., Mustapha, F.: Material screening and choosing methods – a review. Mater. Des. **31**, 696–705 (2010)

7. Cavallini, C., Giorgetti, A., Citti, P., Nicolaie, F.: Integral aided method for material selection based on quality function deployment and comprehensive VIKOR algorithm. Mater. Des. **47**, 27–34 (2013)

8. Zafarani, H.R., Hassani, A., Bagherpour, E.: Achieving a desirable combination of strength and workability in Al/SiC composites by AHP selection method. J. Alloy. Compd. **589**, 295–300 (2014)

9. Shimin, V.V., Shah, V.A., Lokhande, M.M.: Material selection for semiconductor switching devices in electric vehicles using Analytic Hierarchy Process (AHP) method. In: IEEE International Conference on Intelligent Control and Energy Systems (ICPEICES) (2016)

10. Kiong, S.C., et al.: Decision making with the Analytical Hierarchy Process (AHP) for material selection in screw manufacturing for minimizing environmental impacts. Appl. Mech. Mater. **315**, 57–62 (2013)

11. Athawale, V.M., Chakraborty, S.: Material selection using multi-criteria decision-making methods: a comparative study. In: Proceedings of Institution of Mechanical Engineers, Part L, vol. 226, no. 4, pp. 267–286 (2012). Journal of Materials: Design and Applications

12. Flywheels move from steam age technology to Formula 1: Jon Stewart (2012)

13. Jee, D.-H., Kang, K.-J.: A method for optimal material selection aided with decision making theory. Mater. Des. **21**(3), 199–206 (2000)

14. Rai, D., Jha, G.K., Chatterjee, P., Chakraborty, S.: Material selection in manufacturing environment using compromise ranking and regret theory-based compromise ranking methods: a comparative study. Univ. J. Mater. Sci. **1**(2), 69–77 (2013)

15. Chatterjee, P., Chakraborty, S.: Material selection using preferential ranking methods. Mater. Des. **35**, 384–393 (2012)

16. Jahan, A., Bahraminasab, M., Edwards, K.L.: A target-based normalization technique for materials selection. Mater. Des. **35**, 647–654 (2012)

17. Kl, E.: Selecting materials for optimum use in engineering components. Mater. Des. **26**, 469–474 (2005)

18. Fayazbakhsh, K., Abedian, A., Manshadi, B.D., Khabbaz, R.S.: Introducing a novel method for materials selection in mechanical design using Z-transformation in statistics for normalization of material properties. Mater. Des. **30**, 4396–4404 (2009)

19. Chatterjee, P., Athawale, V.M., Chakraborty, S.: Materials selection using complex proportional assessment and evaluation of mixed data methods. Mater. Des. **32**, 851–860 (2011)

20. Milani, A.S., Shanian, A., Madoliat, R., Nemes, J.A.: The effect of normalization norms in multiple attribute decision making methods: a case study in gear material selection. Struct. Multidisc. Optim. **29**, 312–318 (2005)

21. Jeya Girubha, R., Vinodh, S.: Application of fuzzy VIKOR and environmental impact analysis for material selection of an automotive component. Mater. Des. **37**, 478–486 (2012)

22. Ahn, K.K., Kha, N.B.: Modeling and control of shape memory alloy actuators using Preisach model, genetic algorithm and fuzzy logic. Mechatronics **18**, 141–152 (2008)

23. Xue, Y.-X., You, J.-X., Lai, X.-D., Liu, H.-C.: An interval-valued intuitionistic fuzzy MABAC approach for materialselection with incomplete weight information. Appl. Soft Comput. **38**, 703–713 (2016)

24. Gul, M., Celik, E., Gumus, A.T., Guneri, A.F.: A fuzzy logic based PROMETHEE method for material selection problems. Beni-Suef Univ. J. Basic Appl. Sci. **7**, 68–79 (2018)

25. Zhu, X.F.: A web-based advisory system for process and material selection in concurrent product design for a manufacturing environment. Adv. Manuf. Technol. **25**, 233–243 (2005)

26. Welling, D.A.: A fuzzy logic material selection methodology for renewable ocean energy applications by proquest, Umi Dissertation Publishing (2011)

27. Zadeh, L.A.: A note on Z-numbers. Inf. Sci. **181**, 2923–2932 (2011)

28. Babanli, M.B., Huseynov, V.M.: Z-number-based alloy selection problem. In: 12th International Conference on Application of Fuzzy Systems and Soft Computing, ICAFS 2016, Vienna, Austria, vol. 102, pp. 183–189 (2016). Procedia Computer Science
29. Jahan, A., Ismail, M.Y., Shuib, S., Norfazidah, D., Edwards, K.L.: An aggregation technique for optimal decision-making in materials selection. Mater. Des. **32**, 4918–4924 (2011)

# Review on the New Materials Design Methods

M. B. Babanli[1($\boxtimes$)], F. Prima[2], P. Vermaut[2], L. D. Demchenko[3],
A. N. Titenko[4], S. S. Huseynov[1], R. J. Hajiyev[1],
and V. M. Huseynov[1]

[1] Azerbaijan State Oil and Industry University, 1010 Baku, Azerbaijan
mustafababanli@yahoo.com
[2] Chimie ParisTech, UMR CNRS 7045, 11 rue Pierre et Marie Curie,
75005 Paris, France
[3] National Technical University of Ukraine "KPI", Kiev 03056, Ukraine
[4] Institute of Magnetism Under NAS and MES of Ukraine, Kiev 03142, Ukraine

**Abstract.** For a long time experimental approach was main method for material design. However, experimental approach has many drawbacks. With the development of the computing sciences, a new era of synthesis of alloys or different materials began. Scientists proposed and developed various approaches for the synthesis of new alloys which relies on phase diagrams, Thermo-Calc, machine learning, neural network and fuzzy concepts.

**Keywords:** Materials design · Alloys · Neural network · Fuzzy logic
Z-number theory

## 1 Introduction

Developing of machinery constantly requires the creation of new materials with desired properties. This, in turn, is impossible without the involvement of new multicomponent systems and the development of appropriate methods for the production of alloys on their basis. The latter is especially important when it comes to the creation of new special-purpose materials, such as antifriction, wear-resistant, high-damping, electrical, magnetic, radiation-resistant, etc. materials.

A characteristic feature of the modern period of development of materials and technologies is a significant complication of both the compositions of alloys and the processes of obtaining from them parts with special properties. The problem is that the achieved doping complexity has led to the fact that their further development and optimization of promising compositions are continuously associated with ensuring simultaneous consideration of a significant number of factors directly affecting their performance.

In this paper, a detailed review of an alloy modeling technologies is presented. Its implementation led to the creation of new high-temperature cast nickel alloys for cooled blades of gas turbine plants, including those operating under the active influence of sea salt corrosion.

© Springer Nature Switzerland AG 2019
R. A. Aliev et al. (Eds.): ICAFS-2018, AISC 896, pp. 937–944, 2019.
https://doi.org/10.1007/978-3-030-04164-9_124

## 2    Classical Approaches

The Ref. [1] is aimed to report experimental data on binary Ti-Al and ternary Ti-Al-X systems. The Ti-Al binary phase diagram was reported based on the experimental data whereas the Ti-Al-X (X = Cr, Mo and W) phase diagrams have been plotted using the Thermo-Calc software. Results of the high temperature X-ray diffraction examination clearly show that the α-phase is stable near the melting point. Despite some equivocal points still remain (composition and morphology may changes at very high temperatures during observations), from the aftereffects of the high temperature X-ray diffraction, the Ti-Al binary phase diagram is reproduced qualitatively. Thermo-Calc Software developments were presented in [2]. From this paper it is clear that Thermo-Calc is a very useful software and database package. Generally, this database package includes all kinds of phase transformation calculations, phase diagram, phase equilibrium, and thermodynamic assessments. With Thermo-Calc application-oriented interface, many types of process simulations for metallurgy, materials science, alloy design and development, geochemistry, any thermodynamic system in the fields of chemistry etc. could be performed.

Other modern material design method is FLAPW (full-potential linearized augmented plane wave) which also involves a collaboration experimental material science and computational approach. FLAPW is one such method that provides the requisite level of numerical accuracy, despite of complexity. Thus, this method accurately could predict new materials properties, describe the physics of the experiments, and be applicable to new and complex structures [3]. In [4] they use computer simulations to overcome difficulties in analysis of the designed materials that often restrict the use of physical experiments (e.g. restrictions on time and cost, and conditions of experiments).

## 3    Modern Computational Approaches

Despite that computer experiments help to facilitate materials design approaches, they still are restricted much when are based on hard computational schemes. In view of this modern approaches including machine learning, big data and soft computing approaches become desired tools.

### 3.1    D-Electron Design Approach

One of new calculation method were proposed by Japanese scientists in [49] for titanium alloys which alloyed with Mo, Nb, Ta, Sn and Zr and has a body centered cubic (bcc) crystal structure. This method is based on electronic structures of molecular orbitals. By using this new method, electronic structures of bcc were calculated for titanium alloys which alloyed above mentioned elements, also theoretically were determined two alloying parameters. First parameter is the bond order (Bo), second parameter is (Md) the metal d-orbital energy level. Bo is covalent bond strength which shows bond between alloying element and titanium, Md is that energy level which is correlated with the metallic radius of elements and electro negativity.

## 3.2  Machine Learning Approaches

Japanese scientists made calculations and machine learning to predict material synthesis and design [5]. Density functional theory based material data considering every possible element combinations was constructed and used to support vector machines. The properties of predicted material are corresponded with experimental data. Desired material properties based on material combinations are able to be predicted too. Flow between the material database and designing materials has become the bridge. This approach makes it possible to reveal undiscovered desired materials and targeted material mining using big data. Targeted material synthesis is carried out experimentally.

Up until now, candidate molecules for energetic materials have been screened using predictions from expensive quantum simulations and thermochemical codes. In [6] it is demonstrated that machine learning techniques can be used to predict the properties of CNOHF energetic molecules by using their molecular structures. In this paper were challenged the assumption about importance of huge set of dates for machine learning. It is known that for finding new materials with best properties decreasing of the cost and time consumption and error experiments is one of the outstanding factor in materials science. In carrying out experiments and calculations on the materials with good properties data-driven machine learning tools have been used recently. It is required to find material to lower dissipation feature using Landau model for shape memory alloys in order to minimize the number of experiments [7].

## 3.3  Big Data Approaches

Let us provide a short introduction about Big Data over Material science. In 2007, Turing Prize laureate Jim Gray celebrated four scientific paradigms [8]: empirical; theoretical; computational; data exploration or eScience (Big data).

The object of research in this area is large data (BIG DATA) and focused on extracting generalized knowledge from data. This concept reflects the idea of using, storing, analyzing, and retrieving data from a great deal of data collected at great speeds and from different sources. Big data is making significant changes to all areas of science and engineering and is improving interaction among researchers. In paper by Ashley A. White the author demonstrates the future impact of big data on materials science [9]. To speed up materials discovery, researches are presently using computers more widely. Melissae Fellet noted, that Big Data Analytics deliver materials science insights. Analysis of big data of materials patterns and structure help researchers to mine the data for hidden relationships search [10]. Design of new materials is complex process based on fusion of serendipity and difficult methodical work. Researchers perform time-consuming synthesis of new materials by using chemical knowledge and intuition to infer material performance. The materials were designed for powerful batteries, lightweight aircraft components, tough body armour etc.

In [11] they discuss the challenges and opportunities associated with materials research. An information about 4 specific efforts of material science: Materials Project [12], Open Quantum Materials Database [13], expert database at the University of California, Santa Barbara and The University of Utah [14], and the Citrination platform [15].

These bases allow to aggregate, analyze, and visualize large amount of research data for free. The data may be used for machine learning and improvement of new materials discovery.

Material property databases built from literature data, and methods of data aggregation from literature are considered in [16]. Here authors consider manual aggregation of data for forming interactive databases to support interactive visualization of the original experimental data and additional metadata. The described databases include materials for thermoelectric energy conversion, and for Li-ion batteries electrodes.

There are very good open access databases in the crystallography field. One of them is the well-known Inorganic Crystal Structure Database (ICSD), managed by FIZ Karlsruhe. It includes over 180 000 entries on the crystal structures of metals, minerals, and other inorganic compounds [17]. The Cambridge Structural Database of the Cambridge Crystallography Data Centre is also a popular database. It includes small molecule organic and metal-organic crystal structures (more than 800 000 entries) [18]. The Crystallography Open Database (COD) include 120 000 entries of structures and a search infrastructure [19, 20]. Pearson's Crystal Structure Database (274 000 entries) [21] and the Protein Databank for nucleic acids, and other complex materials [22]. The Novel Materials Discovery (NOMAD) Laboratory [23] manages the largest open-access database of all important codes of computational materials science. It can build several Big-Data Services to support materials science and engineering. Extracting hidden information from repositories of computational materials science, it is possible by using NOMAD. Data sharing and the collaborative databases role is discussed in work [24]. The authors consider the topic of data reuse in the Materials Genome Initiative (MGI). Especially, they consider the role of 3 computational databases (that rely on the density functional theory methods) for researchers. They also propose recommendations on data reuse technical aspects, discuss future fundamental challenges, including those of data sharing in MGI perspective [25–30]. They discuss perspective form of data sharing: the use of density functional theory databases formed by experimental groups and theory for wide applications of materials design. Note that, it is possible to compute material properties by using density functional theory, to analyze the electronic structure of a material on the basis of approximate solutions to Schrodinger's equation [31].

## 4   Fuzzy Logic and Soft Computing Approaches

The paradigms of soft computing include fuzzy logic [32, 33], artificial neural networks [34], evolutionary computing, chaos theory and other algorithms. Each paradigm, has its own advantages and disadvantages. The use of fuzzy logic for modeling and prediction of properties allows to describe development process and interpret results better [35–37].

Papers [38–40] show the necessity to account for non-linearity and uncertainty factors that characterize modeling of material design problems. This requires searching for new ways in formalization of systematic approaches to material design. These papers are devoted to the application of soft computing to deal with these factors.

The use of hybrid soft computing approaches [41, 42] allows achieving better results for material engineering due to fusion of advantages of different paradigms. The strategies applied are ANN-fuzzy models (57% works on material engineering, fuzzy-genetic and neuro-genetic (18%) and neuro-fuzzy-genetic (7%) approaches [43].

Nowadays, a huge amount of works devoted to application of fuzzy logic and soft computing exist. A systematic review of the use of fuzzy logic and soft computing approaches for material engineering is conducted in [43, 44]. In [45] they have developed a rule-based fuzzy logic model for predicting shear strength of Ni–Ti alloys specimens which were produced using powder metallurgy method. As input variables the authors selected processing time and temperature and designed a fuzzy model with two inputs and one output variable. Model accuracy is assessed by four statistical parameters. The results of this model and the artificial neural network (ANN) model have been compared and it was concluded that fuzzy rule-based model possesses better predicting capability. Less than 33% experimental data are required for the developed fuzzy model. The presented fuzzy model has higher accuracy and more economical performance and it can be used powerful tool in predicting shear strength of Ni–Ti alloys in powder metallurgy. From literature it is known that traditional approach to deal with material synthesis based on experimental outcomes is used by scientists and practitioners during a long time. However, classical approach has several drawbacks and for eliminating of these flaws different methods were suggested. One of these methods was investigated in [46]. In this paper they applied the fuzzy set theory to knowledge mining from big data on material characteristics. Author proposes fuzzy clustering-generated If-Then rules as a basis for computer synthesis of new materials. For this approach proposed fuzzy If-Then rules based model to predict properties of new materials. This model is constructed on the basis of fuzzy clustering of big data on dependence between material composition and related properties. The motivation to use fuzzy model is inspired by necessity to construct an intuitively well-interpretable development strategy from imperfect and complex data. Validity of the proposed approach is verified on an example of prediction properties of Ti-Ni alloy and Computer experiments of the proposed fuzzy model show its better performance as compared to physical experiments based analysis [46]. In [47] a method of design of knowledge base of fuzzy logic controller using GA is proposed. To compare the effectiveness of approach with that of a previous GA-fuzzy hybrid approach, a combination of prediction of power requirement and surface finish in grinding is used. The results show advantage of the proposed method. In [48] fuzzy logic and GA are used to find optimal solution to the multi objective problem of recyclable materials selection. Comparison of proposed method with Sustainability Express Solid Work is also presented. The proposed method can assist product designers to design a high recyclability product without ignoring technical perspectives. Some problems of the results, the limitations and sensitivity of the applied method and the accuracy of the electron elastic cross-sections are discussed.

## 5   Conclusion

Analyzing a wide diversity of approaches to material selection and synthesis, one can observe a tendency to shift research efforts from physical experiments to systematic analysis based on mathematical models and computational schemes. The latter, in turn, evaluates from traditional analytical methods and computational schemes to modern approaches that are based on collaboration of fuzzy logic and soft computing, machine learning, big data and other new methods. The aim to apply fuzzy logic and soft computing methods is to improve research using the advantage of dealing with: imprecision of experimental data; partial reliability of experimental data, prediction results and expert opinions; uncertainty of material properties stemming from complex relationship between material components; a necessity to analyze, summarize, and reason with large amount of information of various types (numeric data, linguistic information, graphical information, geometric information etc.).

Fuzzy logic, Z-number theory and Soft computing methods have a good capability to effectively capture and process imprecise experimental data, that is interpret, classify, learn, and compute with them. Z-number theory has a promising capability to account for fuzzy and partially reliable information due to ability to fuse fuzzy computation and probabilistic arithmetic. Indeed, variability of experimental conditions, complex content and structure of materials, imperfect expert knowledge demand to consider reliability of information on material behavior as restricted.

Uncertainty of material properties requires combining FL and efficient learning methods as ANNs, evolutionary algorithms and others to more adequately model and predict possible material behavior. Fuzzy logic, Z-number theory and Soft computing may help to improve abilities of big data principles to deal with huge amount and variety of information. In this realm, fuzzy clustering, Neuro-fuzzy inference systems, intelligent databases, soft CBR, computational intelligence based KBs and information search algorithms provide bridge between complexity, imperfectness, qualitative nature of information and research techniques. Particularly, this may help to get intuitive general interpretation of material science results obtained by various techniques, and ways to get practical results would be then more evident.

## References

1. Hashimoto, K., Kimura, M., Mizuhara, Y.: Alloy design of gammatitanium aluminides based on phase diagrams. Intermetallics **6**, 667–672 (1998)
2. Andersson, J.O., Helander, T., Höglund, L., Shi, P., Sundman, B.: Thermo-Calc & DICTRA, computational tools for materials science. Calphad **26**, 273–312 (2002)
3. Weinert, M., Schneider, G., Podloucky, R., Redinger, J.: FLAPW: applications and implementations. J. Phys.: Condens. Matter **21**(8), 084201 (2009)
4. Abreu, M.P.: On the development of computational tools for the design of beam assemblies for Boron neutron capture therapy. J. Comput. Aided Mater. Des. **14**, 235–251 (2007)
5. Takahashi, K., Tanaka, Y.: Material synthesis and design from first principle calculations and machine learning. Comput. Mater. Sci. **112**, 364–367 (2016)

6. Elton, D.C., Boukouvalas, Z., Butrico, M.S., Fuge, M.D., Chung, P.W.: Applying machine learning techniques to predict the properties of energetic materials (2018)

7. Dehghannasiri, R., Xue, D., Balachandran, P.V., Yousefi, M.R., Dalton, L.A., Lookman, T., Dougherty, E.R.: Optimal experimental design for materials discovery. Comput. Mater. Sci. **129**, 311–322 (2017)

8. Hey, T., Tansley, S., Tolle, K. (eds.): The Fourth Paradigm: Data-Intensive Scientific Discovery. Microsoft Corporation, p. 287 (2009)

9. White, A.A.: Big data are shaping the future of materials science. MRS Bull. **38**, 594–595 (2013)

10. Fellet, M.: Big Data Analytics Deliver Materials Science Insights (2017). http://www.lindau-nobel.org/blog-big-data-analytics-deliver-materials-science-insights/

11. Hill, J., Mulholland, G., Persson, K., Seshadri, R., Wolverton, C., Meredig, B.: Materials science with large-scale data and informatics: unlocking new opportunities. MRS Bull. **41**, 399–409 (2016)

12. http://www.materialsproject.org

13. http://oqmd.org

14. www.mrl.ucsb.edu:8080/datamine/thermoelectric.jsp

15. https://citrination.com

16. Seshadri, R., Sparks, T.D.: Perspective: interactive material property databases through aggregation of literature data. APL Mater. **4**(5), 053206 (2016)

17. Belsky, A., Hellenbrandt, M., Karen, V.L., Luksch, P.: Acta Crystallogr. Sect. B **58**, 364 (2002)

18. Allen, F.H.: Acta Crystallogr. Sect. B **58**, 380 (2002)

19. Downs, R.T., Hall-Wallace, M.: Am. Miner. **88**, 247 (2003)

20. Gražulis, S., Chateigner, D., Downs, R.T., Yokochi, A.F.T., Quirós, M., Lutterotti, L., Manakova, E., Butkus, J., Moeck, P., Le Bail, A.: J. Appl. Crystallogr. **42**, 726 (2009)

21. Villars, P.: Pearson's Crystal Data: Crystal Structure Database for Inorganic Compounds (2007)

22. Berman, H.M., Westbrook, J., Feng, Z., Gilliland, G., Bhat, T., Weissig, H., Shindyalov, I. N., Bourne, P.E.: Nucleic Acid Res. **28**, 235 (2000)

23. https://nomad-coe.eu/

24. Jain, A., Persson, K.A., Ceder, G.: The materials genome initiative: data sharing and the impact of collaborative ab initio databases. J. APL Mater. **4**(5), 1–14 (2016)

25. Sumpter, B.G., Vasudevan, R.K., Potok, T., Kalinin, S.V.: A bridge for accelerating materials by design. NPJ Comput. Mater. **1**, 15008 (2015)

26. Christodoulou, J.A.: Integrated computational materials engineering and materials genome initiative: accelerating materials innovation. Adv. Mater. Process. **171**(3), 28–31 (2013)

27. White, A.A.: Universities prepare next-generation workforce to benefit from the materials genome initiative. MRS Bull. **38**, 673–674 (2013)

28. Olson, G.B., Kuehmann, C.J.: Materials genomics: from CALPHAD to flight. Scr. Mater. **70**, 25–30 (2014)

29. White, A.A.: Interdisciplinary collaboration, robust funding cited as key to success of materials genome initiative program. MRS Bull. **38**, 894–896 (2013)

30. White, A.: Workshop makes recommendations to increase diversity in materials science and engineering. MRS Bull. **38**, 120–122 (2013)

31. Ceder, G., Hautier, G., Jain, A., Ong, S.P.: Recharging lithium battery research with first-principles methods. MRS Bull. **36**, 185–191 (2011)

32. Zadeh, L.A.: Fuzzy logic = computing with words. IEEE Trans. Fuzzy Syst. **4**(2), 103–111 (1996)

33. Aliev, R.A., Aliev, R.R.: Soft Computing and Its Application. World Scientific, New Jersey (2001)

34. Pedrycz, W., Peters, J.F.: Computational Intelligence in Software Engineering. Advances in Fuzzy Systems, Applications and Theory, vol. 16. World Scientific, Singapoure (1998)

35. Babanli, M.B., Huseynov, V.M.: Z-number-based alloy selection problem. In: 12th International Conference on Application of Fuzzy Systems and Soft Computing, ICAFS 2016, Vienna, Austria. Procedia Comput. Sci. **102**, 183–189 (2016)

36. Chen, S.-M.: A new method for tool steel materials selection under fuzzy environment. Fuzzy Sets Syst. **92**, 265–274 (1997)

37. Cheng, J., Feng, Y., Tan, J., Wei, W.: Optimization of injection mold based on fuzzy moldability evaluation. J. Mater. Process. Technol. **21**, 222–228 (2008)

38. Lee, Y.-H., Kopp, R.: Application of fuzzy control for a hydraulicforging machine. Fuzzy Sets Syst. **99**, 99–108 (2001)

39. Elishakoff, I., Ferracuti, B.: Fuzzy sets based interpretation of the safety factor. Fuzzy Sets Syst. **157**, 2495–2512 (2006)

40. Rao, H.S., Mukherjee, A.: Artificial neural networks for predicting the macromechanical behaviour of ceramic-matrix composites. Comput. Mater. Sci. **5**, 307–322 (1996)

41. Hancheng, Q., Bocai, X., Shangzheng, L., Fagen, W.: Fuzzy neural network modeling of material properties. J. Mater. Process. Technol. **122**, 196–200 (2002)

42. Chen, D., Li, M., Wu, S.: Modeling of microstructure and constitutive relation during super plastic deformation by fuzzy-neural network. J. Mater. Process. Technol. **142**, 197–202 (2003)

43. Odejobi, O.A., Umoru, L.E.: Applications of soft computing techniques in materials engineering: a review. Afr. J. Math. Comput. Sci. Res. **2**(7), 104–131 (2009)

44. Datta, S., Chattopadhyay, P.P.: Soft computing techniques in advancement of structural metals. Int. Mater. Rev. **58**, 475–504 (2013)

45. Tajdari, M., Mehraban, A.G., Khoogar, A.R.: Shear strength prediction of Ni–Ti alloys manufactured by powder metallurgy using fuzzy rule-based model. Mater. Des. **31**, 1180–1185 (2010)

46. Babanli, M.B.: Synthesis of new materials by using fuzzy and big data concepts. Procedia Comput. Sci. **120**, 104–111 (2017)

47. Nandi, A.K., Pratihar, D.K.: Automatic design of fuzzy logic controller using a genetic algorithm-to predict power requirement and surface finish in grinding. J. Mater. Process. Technol. **148**, 288–300 (2004)

48. Sakundarini, N., Taha, Z., Abdul-Rashid, S.H., Ghazilla, R.A.R.: Incorporation of high recyclability material selection in computer aided design. Mater. Des. **56**, 740–749 (2014)

49. Morinaga, M., Kato, M., Kamimura, T., Fukumoto, M., Harada, I., Kubo, K.: Theoretical design of b-type titanium alloys. In: Titanium 1992, Science and Technology, Proceedings of 7th International Conference on Titanium, San Diego, CA, USA, pp. 276–283 (1992)

# Z-Number Based Diagnostics of Parkinson's Diseases

Elcin Huseyn[✉]

Research Laboratory of Intelligent Control and Decision Making Systems
in Industry and Economics, Azerbaijan State Oil and Industry University,
20 Azadlig Avenue, Baku 1010, Azerbaijan
e.huseyn@vitroel.com

**Abstract.** Parkinson's disease is a neuro-degenerative movement disorder that causes voice and speech disorders. Therefore, the disease can be diagnosed as a dysfunctional disease. In this study, Computational Intelligence Methods, a new method for classifying Parkinson's disease, have been used. This method has been tested with two data sets and compared with classical methods. According to the obtained results, this method yielded better results than the classical methods.

**Keywords:** Parkinson's disease · Z-number · Artificial neural network
Computational intelligence methods · Neuro-Fuzzy IS · Type-2 FIS
Z-FIS

## 1 Introduction

Every year in the world, thousands of people are suffering from Parkinson's disease. Parkinson's disease is a disease with different stages, usually seen in the elderly, affecting the whole body [1]. Diagnosis of the disease can be easily done by specialist physicians with highly technological devices. However, it can be very difficult and troublesome to apply this type of diagnosis and diagnostic methods for patients to each individual. The most common negativity in individuals suffering from this disorder is speaking disorders as a result. Therefore, the diagnosis of the disease by analyzing the voice signals of the person is emerging as an alternative diagnostic method. The great advantage of this method is that it is possible to diagnose the person even without leaving the person's home or even to determine the stages of the disease. In addition, the course of the disease can be monitored from a distance. Diagnosis of the disease and the degree of Parkinson's disease in the patient can be done with the help of voice record samples [2]. Classification is performed by extracting various attributes from the sound recordings of the test. This grading process can be done with many grading methods. In this study, the classification process was done with a method that was not used before. Computational Intelligence Methods have been used in deep learning constructs for bounce reshaping. Deep learning techniques are used in many fields of science and they are very successful according to the known classical methods [3]. In this study, based on a number of symptoms from deep nervous network structures, the aim is to apply a series of computational intelligence methods including artificial neural

© Springer Nature Switzerland AG 2019
R. A. Aliev et al. (Eds.): ICAFS-2018, AISC 896, pp. 945–950, 2019.
https://doi.org/10.1007/978-3-030-04164-9_125

network, Neuro-Fuzzy IS, Type-2 FIS and Z-FIS to detect Parkinson's disease in patients. The data used to train the systems and test the performance of the system is taken from real patients who are sick and are being treated.

## 2  Preliminaries

The work intends to apply a number of computational intelligence methods including artificial neural network, Neuro-Fuzzy IS, Type-2 FIS, and Z-FIS to detect Parkinson Decease in patients on the basis of a number of symptoms.

The data to train the systems and test the system's performance are taken from real patients suffering the decease and taking treatment.

Two versions of detection systems are considered:

1. Detection on the basis of patient's behavioral characteristics
2. Detection on the basis of patient's voice features

Input Data for version 1 System:

- memory, hallucinations, mood, motivation, speech, saliva, swallowing, handwriting, cutting_food, dressing, hygiene, turning_in_bed, falling, freezing, walking, tremors, numbness.

Output Data for version 1 System

- Hoehn and Yahr stage (hoehn_yahr)

Fragment of data:

| Input data | Memory | 2 | 1 | 0 | 2 | 1 | 0 | 1 | 0 | 0 | 1 | ... |
|---|---|---|---|---|---|---|---|---|---|---|---|---|
| | Hallucinations | 2 | 1 | 2 | 1 | 0 | 0 | 1 | 1 | 1 | 0 | ... |
| | Mood | 4 | 1 | 0 | 2 | 1 | 0 | 0 | 0 | 1 | 0 | ... |
| | Motivation | 4 | 0 | 1 | 4 | 2 | 1 | 0 | 1 | 0 | 3 | ... |
| | Speech | 2 | 2 | 2 | 1 | 0 | 2 | 1 | 1 | 0 | 2 | ... |
| | Saliva | 0 | 1 | 2 | 0 | 0 | 1 | 0 | 1 | 0 | 2 | ... |
| | Swallowing | 0 | 1 | 1 | 1 | 0 | 0 | 1 | 0 | 1 | 2 | ... |
| | Handwriting | 2 | 1 | 3 | 2 | 1 | 1 | 1 | 1 | 0 | 1 | ... |
| | Cutting_food | 1 | 1 | 2 | 1 | 1 | 1 | 0 | 0 | 1 | 0 | ... |
| | Dressing | 1 | 1 | 2 | 1 | 1 | 1 | 1 | 1 | 1 | 0 | ... |
| | Hygiene | 1 | 1 | 1 | 1 | 1 | 1 | 0 | 1 | 1 | 0 | ... |
| | Turning_in_bed | 1 | 2 | 2 | 1 | 1 | 1 | 1 | 1 | 1 | 0 | ... |
| | Falling | 1 | 2 | 1 | 1 | 0 | 1 | 0 | 1 | 0 | 0 | ... |
| | Freezing | 1 | 1 | 0 | 0 | 0 | 0 | 0 | 0 | 0 | 0 | ... |
| | Walking | 1 | 2 | 1 | 2 | 0 | 1 | 1 | 1 | 2 | 1 | ... |
| | Tremors | 3 | 1 | 2 | 1 | 0 | 0 | 2 | 1 | 0 | 1 | ... |
| | Numbness | 2 | 1 | 2 | 1 | 1 | 0 | 1 | 0 | 0 | 0 | ... |
| Output data | Hoehn_yahr | 3 | 3 | 3 | 2 | 3 | 3 | 1 | 2 | 1 | 2 | ... |

## 3   Artificial Neural Network Based Systems

Let us consider application of Artificial Neural Network (ANN) to diagnostics of Parkinson' disease. Two versions of ANNs are used.

<u>Version 1</u>

3 Layered Feed-Forward NN with
Sigmoidal hidden and linear output layer neurons
Number of inputs: 19
Number of hidden neurons: 10
Number of outputs: 1
Data are scaled for faster learning of network
Number of training input-output data pairs: 350
Number of test input-output data pairs: 39
Training based on Differential Evolution algorithm
Several training experiments done. The results are shown in Figs. 1, 2, 3 and 4.
RMSE after training (on train data): 0.63

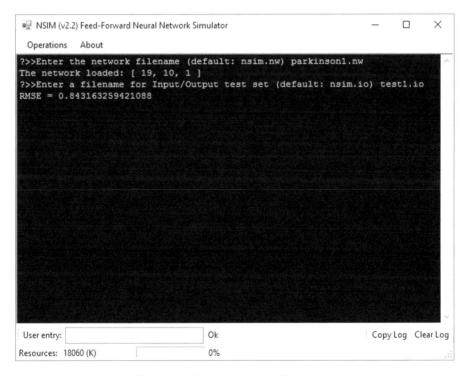

**Fig. 1.** ANN training results (Version 1)

Version 2
3 Layered Feed-Forward NN with
Sigmoidal hidden and linear output layer neurons
Number of inputs: 38
Number of hidden neurons: 16
Number of outputs: 1
Data are not scaled
Input Data are 38 measured voice features.
Output is Hoehn and Yahr stage (hoehn_yahr).
Number of training input-output data pairs: 340
Number of test input-output data pairs: 34
Training based on Differential Evolution algorithm

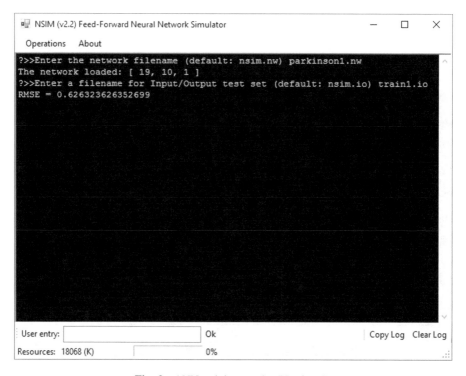

**Fig. 2.** ANN training results (Version 1)

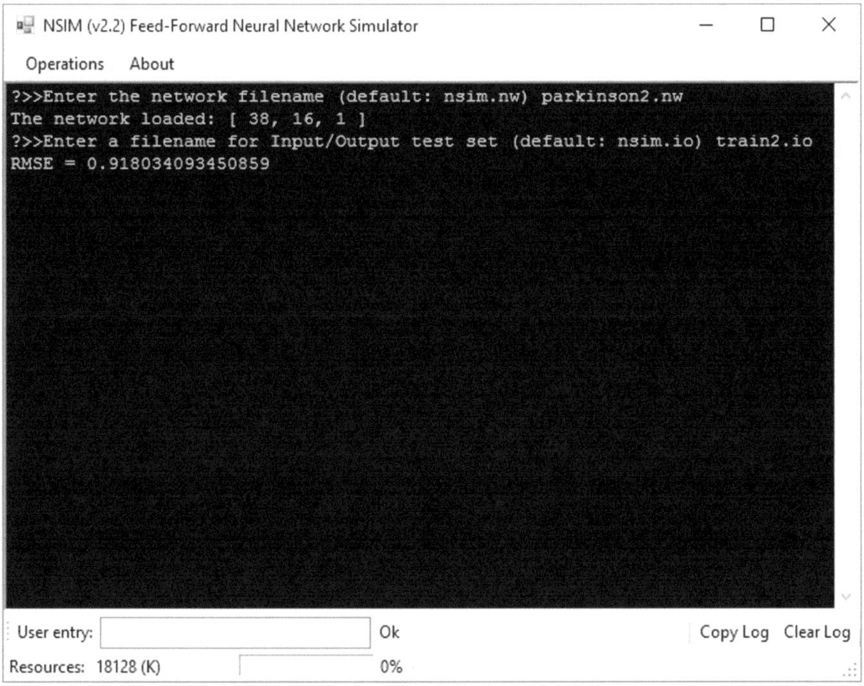

**Fig. 3.** ANN training results (Version 2)

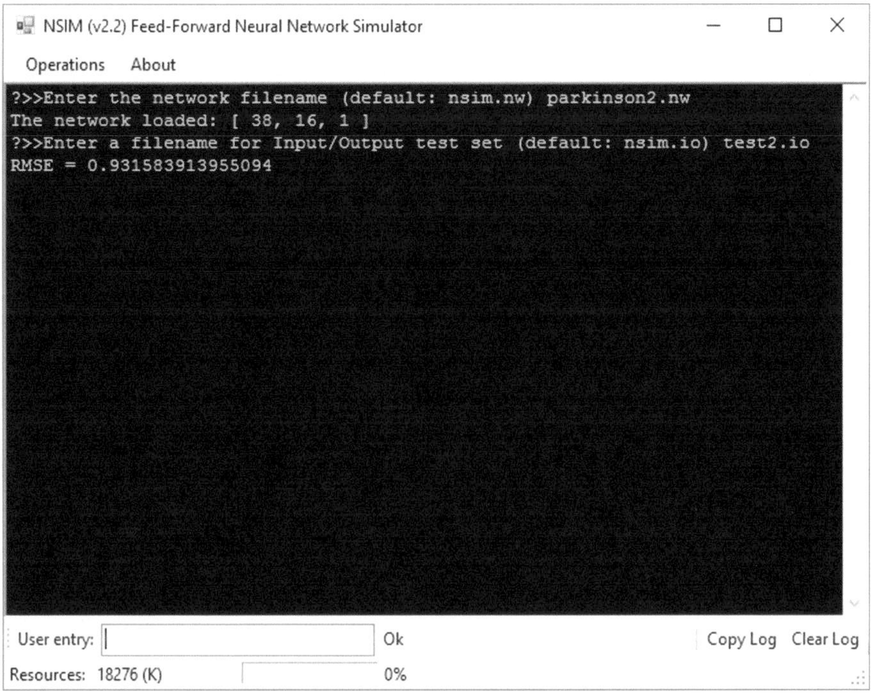

**Fig. 4.** ANN training results (Version 2)

## 4  Conclusion

Parkinson's disease can be diagnosed from dysphonias by using a number of features extracted from the voice signals from patients in the class recall. It is a new approach to diagnose Parkinson's disease using the artificial neural network, Neuro-Fuzzy IS, Type-2 FIS and Z-FIS. artificial neural network, Neuro-Fuzzy IS, Type-2 FIS and Z-FIS structures were measured with two datasets and produced more successful results than other classical methods commonly used in the literature. These successful results are an indication that deep learning structures are an important tool for grading.

## References

1. Ogut, F., Yavuzere, U.M., Akyurekli, O., Kalaycı, T.: Evaluation criteria of voice change in Parkinson's disease. Ege Tıp Dergisi **39**(1) (2000)
2. Little, M.A., McSharry, P.E., Roberts, S.J., Costello, D.A., Moroz, I.M.: Exploiting nonlinear recurrence and fractal scaling properties for voice disorder detection. Biomed. Eng. Online Skin **6**, 23 (2007)
3. LeCun, Y., Bengio, Y., Hinton, G.: Deep learning. Nature **521**, 436–444 (2015)

# An Integrated MCDM Approach
# to the Problem of New Graduate Job Selection
# Under Fuzzy Environment

Metin Dagdeviren[1] and Aylin Adem[2(✉)]

[1] Industrial Engineering Department, Gazi University, 06570 Ankara, Turkey
[2] TUSAŞ-Kazan Vocational School, Department of Administration and
Organization, Gazi University, Ankara, Turkey
aylinadem@gazi.edu.tr

**Abstract.** The job s/he chooses as a starting point for her/his career for a young workforce graduated from college with certain qualifications is crucial in ensuring the correct placement and sustainability of the workforce. From this point of view, along with the human resources department of a large-scale business, to determine the tendency of newly graduated students to select a business area, a study was conducted with four human resource specialists. In this context, three alternative working areas were selected for the new graduate student. It was included in the selection problem by determining the criteria that would be effective in selecting a job and providing job satisfaction. These selection criteria were derived from factors included in Herzberg's Dual Factor Theory. In the solution of the selection problem, a two-stage integrated structure was used. Analytical Hierarchy Process (AHP) method was used to gain criteria's weight and Hesitant Fuzzy VIKOR method was used for determining the order of the alternatives which are the next step after obtaining the weights. The reason for using AHP method in achieving the criterion weights can be expressed as that AHP is a simple method and gives consistent results. In the same way, the HF-VIKOR solution approach is used for ordering the alternatives, with the thought that they will act hesitantly when making this decision which will affect the lives of the new graduates.

**Keywords:** Hesitant fuzzy sets · VIKOR · Job selection
AHP · MCDM

## 1 Introduction

The choice of human resource and its placement in the right job both affects the efficiency to be taken from the worker and is important in terms of the use of business resources. There are many studies about selection of personnel in the literature [1–4]. The more important it is for a business to hire the right person, in the same way, the more important it is for the person to work the right job. In particular, this situation becomes even more important for candidates who have recently graduated from college and have not yet faced the complicated structure of business life. A wrong decision at the beginning of the career of the person can lead to both missed opportunities,

© Springer Nature Switzerland AG 2019
R. A. Aliev et al. (Eds.): ICAFS-2018, AISC 896, pp. 951–957, 2019.
https://doi.org/10.1007/978-3-030-04164-9_126

unhappiness, insecurity for business, and similar problems for other jobs that s/he will enter later. For this reason, in this study, the problem of personnel selection is considered from the perspective of personnel. Within the scope of current study, it was tried to identified the tendency of the graduated young labour force candidate to choose the line of business (academicianship, entrepreneurship, government office). The study first started with a request from Human Resource (HR) department of a large-scale business. The objective of this business is to evaluate the criterion and the importance of the criteria by determining the criteria's weight that the newly graduated students pay attention to in their job selection. From this point of view, the business wants to unequivocally reveal what the students who graduated from university with degrees, or the young people they want to hire to train, want from the business. Job satisfaction criteria in Herzberg's dual factor theory have been re-adapted by the authors and experts who have been working in HR for a long time in the country according to the requirements of our country and the century [5]. Six main criteria have been identified and four new graduates are asked to evaluate these criteria using AHP method AHP method was developed by Saaty to gain proportion scales from both discrete and continuous binary comparisons. Binary comparisons are fundamental in the use of the AHP and this method is widely used in the MCDM literature. Saaty's 1–9 scale was used as a scale for evaluation [6]. With the Expert Choice program, the AHP method was applied and the criteria weights were determined. Then, the method of HF-VIKOR was adopted and it was determined which of the three alternative occupation groups came to the fore. HF-VIKOR solution approach was used for ordering the alternatives, with the thought that they will act hesitantly when making this decision which will affect the lives of the new graduates. This work is structured as follows: The methods used in the study are given in the second section, the application of the proposed integrated method is discussed in the third part of the study, the results of the study and the discussions about what can be done in the future are given in the fourth section.

## 2    HF - VIKOR Method

Opricovic brought a new sight to MCDM literature with the VIKOR method, to decipher MCDM with contradictory and incommensurable criteria [7]. This method aims to prioritize determined alternatives. "$A_i$ ($i = 1, 2, ..., m$)", regarding to the criteria $C_j$ ($j = 1, 2, ..., n$) and determines reconciliation solution for a decision making problem. It demonstrates the multi-criteria ordering index based on closeness to the ideal solution. [8, 9]. Please see [7] to get more information about VIKOR method. Within the context of this study, Extended VIKOR Method with a Hesitant Fuzzy Set (HF-VIKOR) are introduced.

### 2.1    Preliminaries

In this part of our study, a concise information about HFS preliminaries is exhibited.

**Definition 1.** "Let X be a constant hesitant fuzzy set (HFS) on X is in terms of a function that when operative to X turnovers a subset of [0, 1]".

The HFS is expressed by a mathematical symbol as follows [9]:

$$\text{``}E = \{< x, h_E(x) >|\ x \in X\}\text{''} \tag{1}$$

where is a set of some values in [0, 1], indicating the probable membership degrees of the component to the set $E$.

$h_1$ and $h_2$ is two HFEs. The representation of the mathematical operations in hesitant fuzzy sets, which are used is as follows. The former demonstrate union of sets, the latter shows intersection sets:

$$\text{``(1)}\ \tilde{h}_1 \cup \tilde{h}_2 = \cup_{\gamma_1 \in \tilde{h}_1, \gamma_2 \in \tilde{h}_2} \max\{\gamma_1, \gamma_2\}\text{''}$$

$$\text{``(2)}\ \tilde{h}_1 \cap \tilde{h}_2 = \cup_{\gamma_1 \in \tilde{h}_1, \gamma_2 \in \tilde{h}_2} \min\{\gamma_1, \gamma_2\}\text{''}$$

**Definition 2.** $h_1$ and $h_2$ is two HFSs. The hesitant normalized Hamming distance measure between h1 and h2 is pointed out in Eq. (2) [9];

$$\text{``}\| h_1 - h_2 \| = \frac{1}{l} \sum_{j=1}^{l} | h_{1\sigma(j)} - h_{2\sigma(j)} |\text{''} \tag{2}$$

l(h) is the number of the component in the $h$.

To compute properly, there are some methods used to equalize their length when comparing them. Pessimistic decision-maker(DM) generally hope adverse consequence of the decision-making process; on the contrary, optimistic DM mostly expect admirable results. According to the decision maker's attitude, either the minimum value or the maximum value can be added to hesitant fuzzy sets [9]. In our study, experts expressed their manner as pessimistic. The similar approach is seen in many existing papers in the related literature [9–11].

## 2.2  Extended VIKOR Method with a Hesitant Fuzzy Set

In this study, Zhang and Wei's HF-VIKOR Method is utilized [9]

1. Detecting the positive and negative ideal solution:

$M^* = \{h_1^*, \ldots, h_n^*\}$, where

$$\text{``}h_j^* = \cup_{i=1}^{m} h_{ij} = \cup_{\gamma_{1j} \in h_{1j}, \ldots, \gamma_{mj} \in h_{mj}} \max\{\gamma_{1j}, \ldots, \gamma_{mj}\},\ j = 1, 2, \ldots n\text{''} \tag{3}$$

$M^- = \{h_1^-, \ldots, h_n^-\}$ where

$$\text{``}h_j^- = \cap_{i=1}^{m} h_{ij} = \cup_{\gamma_{1j} \in h_{1j}, \ldots, \gamma_{mj} \in h_{mj}} \min\{\gamma_{1j}, \ldots, \gamma_{mj}\},\ j = 1, 2, \ldots n\text{''} \tag{4}$$

2. Calculation $S_i$ and $R_i$:

$$\text{“} S_i = \sum_{j=1}^{n} w_j \parallel h_j^* - h_{ij} \parallel / \parallel h_j^* - h_j^- \parallel, \quad i = 1, 2, \ldots, m \text{”} \tag{5}$$

$$\text{“} R_i = \max_j w_j \parallel h_j^* - h_{ij} \parallel / \parallel h_j^* - h_j^- \parallel, \quad i = 1, 2, \ldots, m \text{”} \tag{6}$$

min $S_i$ refers to a maximum group utility, likewise, min $R_i$ shows a minimum individual regret.

3. Compute the values $Q_i$ using the formulation below, which can be found in paper [7]:

$$\text{“} Q_i = v(S_i - S^*) / (S^- - S^*) + (1 - v)(R_i - R^*) / (R^- - R^*) \text{”} \tag{7}$$

where

$$\text{“} S^* = \min_i S_i, \quad S^- = \max_i S_i, \text{”} \tag{8}$$

$$\text{“} R^* = \min_i R_i, \quad R^- = \max_i R_i, \text{”} \tag{9}$$

4. Sequence the alternatives according to the $S$, $R$ and $Q$ values.
5. Determine the best solution or compromise solutions via traditional VIKOR's acceptable advantage and acceptable stability conditions.

    "$DQ = 1/(m - 1)$"; $m$ is the number of alternatives. If $m$ is smaller than 4, $DQ$ value is taken 0.25.

## 3  Application

In this part of the present study, a real-world problem is solved to exemplify the application of the suggested method (see Fig. 1). There are four new graduates to evaluate six main criteria that introduced in this study. There are three alternatives (A1: Academicianship, A2: Entrepreneurship, A3: Government officer) to be assessed with respect to these six criteria: C1: Wage and salary increases, C2: Status in the organization, C3: Company policies, C4: Competence of supervision, C5: Interpersonal relations with colleagues at work and C6: Working conditions and job security. As seen in Fig. 1; this study starts with gaining criteria weights, which were obtained using AHP with Expert Choice program: $w = (0.297, 0.260, 0.078, 0.068, 0.137, 0.160)^{\mathrm{T}}$. In the second stage, Hesitant Fuzzy VIKOR method was used for determining sequencing determined profession areas after obtaining the weights. Hesitant fuzzy decision matrix $H = (h_{ij})_{mxn}$ is given in Table 1, where $h_{ij}$ ($i = 1, 2, 3, j = 1, 2, 3,$

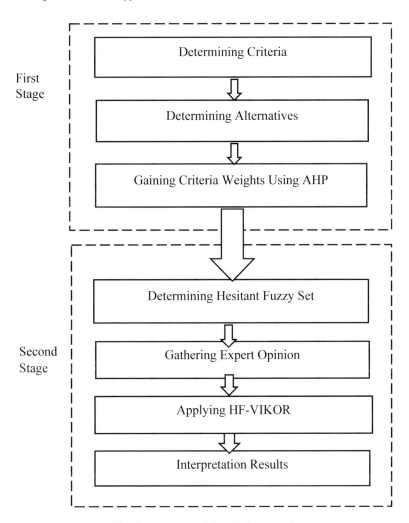

**Fig. 1.** Two-stage integrated approach

**Table 1.** Hesitant fuzzy decision matrix

|    | C1 | C2 | C3 | C4 | C5 | C6 |
|----|----|----|----|----|----|----|
| A1 | (0.5; 0.6; 0.7) | (0.5; 0.8; 0.9) | (0.4; 0.6) | (0.3; 0.4; 0.7; 0.9) | (0.5; 0.8) | (0.6; 0.7; 0.8; 0.9) |
| A2 | (0.7; 0.8) | (0.6; 0.7) | (0.5; 0.6; 0.7) | (0.3; 0.4; 0.5; 0.7) | (0.3; 0.4; 0.5) | (0.2; 0.3) |
| A3 | (0.6; 0.7; 0.8) | (0.5; 0.7; 0.8) | (0.2; 0.4; 0.5; 0.6) | (0.8; 0.9) | (0.2; 0.3; 0.4; 0.5) | (0.4; 0.5) |

*4, 5, 6)* are in the form of HFEs. To solve this problem was used HF VIKOR method according to the algorithm introduced in Sect. 2:

1. Determine the Positive and Negative Ideal Solution (PIS and NIS):

$$M^* = \{h_1^*, h_2^*, h_3^*, h_4^*, h_5^*, h_6^*\} = \{0.800; 0.900; 0.700; 0.900; 0.800; 0.900\}$$
$$M^- = \{h_1^-, h_2^-, h_3^-, h_4^-, h_5^-, h_6^-\} = \{0.500; 0.500; 0.200; 0.300; 0.200; 0.200\}$$

2. Calculate S and R values:

$$S_1 = \frac{w_1\|h_1^*-h_{11}\|}{\|h_1^*-h_1^-\|} + \frac{w_2\|h_2^*-h_{12}\|}{\|h_2^*-h_2^-\|} + \frac{w_3\|h_3^*-h_{13}\|}{\|h_3^*-h_3^-\|} + \frac{w_4\|h_4^*-h_{14}\|}{\|h_4^*-h_4^-\|}$$
$$+ \frac{w_5\|h_5^*-h_{15}\|}{\|h_5^*-h_5^-\|} + \frac{w_6\|h_6^*-h_{16}\|}{\|h_6^*-h_6^-\|}$$

$S_1 = 0.436$; $S_2 = 0.511$; $S_3 = 0.500$

$$R_1 = \max\left\{ \begin{array}{c} \frac{w_1\|h_1^*-h_{11}\|}{\|h_1^*-h_1^-\|}, \frac{w_2\|h_2^*-h_{12}\|}{\|h_2^*-h_2^-\|}, \frac{w_3\|h_3^*-h_{13}\|}{\|h_3^*-h_3^-\|}, \frac{w_4\|h_4^*-h_{14}\|}{\|h_4^*-h_4^-\|}, \\ \frac{w_5\|h_5^*-h_{15}\|}{\|h_5^*-h_5^-\|}, \frac{w_6\|h_6^*-h_{16}\|}{\|h_6^*-h_6^-\|} \end{array} \right\}$$

$R_1 = 0.19$; $R_2 = 0.16$; $R_3 = 0.15$

3. Calculate Q values:

Let $v = 0.25, 0.5$ and $0.75$ and calculate the $Q_i$ values ($i = 1, 2, 3$):
For $v = 0.25$, $Q_i$ values: $Q_1 = 0.750$, $Q_2 = 0.437$, $Q_3 = 0.213$.
For $v = 0.5$, $Q_i$ values: $Q_1 = 0.500$, $Q_2 = 0.625$, $Q_3 = 0.426$.
For $v = 0.75$, $Q_i$ values: $Q_1 = 0.250$, $Q_2 = 0.812$, $Q_3 = 0.640$.
Step 4 and 5. Order the alternatives with respect to S, R and Q. Derive the compromise solution:

All calculation and their consequences are seen in Table 2.

**Table 2.** Alternative ranking and compromise solution

|  | A1 | A2 | A3 | Ranking | Compromise solution |
|---|---|---|---|---|---|
| S | 0.436 | 0.511 | 0.500 | 1-3-2 | *Academicianship* |
| R | 0.190 | 0.160 | 0.150 | 3-2-1 | *Government officer* |
| Q (v = 0.25) | 0.750 | 0.437 | 0.213 | 3-2-1 | *Government officer-entrepreneurship* |
| Q (v = 0.50) | 0.500 | 0.625 | 0.426 | 3-1-2 | *Government officer-entrepreneurship-academicianship* |
| Q (v = 0.75) | 0.250 | 0.812 | 0.640 | 1-3-2 | *Academicianship* |

## 4   Results and Conclusion

In this study, a two-stage integrated method was used to solve a real life selection problem. Criteria weights were obtained with AHP, and then alternatives were ranked with HF-VIKOR after using obtained criteria weights. Because of the advantage of HFS in portraying the vague situations, the traditional VIKOR method is extended to use the MCDM problems with uncertainty. The problem was solved using several $v$ values. For $v = 0.25$, alternatives A3 and A2 are chosen compromise solution, for $v = 0.5$ all alternatives are in the compromise solution cluster and for $v = 0.75$, alternative A1 is chosen the best solution. In this paper, only focused on one of the traditional distance measures: Hamming distance. In future study, researchers can solve HF-VIKOR using different distance measures and do analysis to see differences between these measurements methods. In this study, it was also accepted that decision maker' are pessimist. In the literature there are several assumptions such as optimism. Researchers can investigate different decision maker assumptions to do sensitivity analysis.

## References

1. Tunacan, S., Çetin, C.: Lise öğretmenlerinin iş doyumunu etkileyen faktörlerin tespitine ilişkin bir araştırma. M.Ü. Atatürk Eğitim Fakültesi Eğitim Bilimleri Dergisi **29**, 155–172 (2009)
2. Koç, S., Bardak, A., Yılmaz, K.: Identifying workplace expectations of final-year nursing students. Hemşirelikte Eğitim ve Araştırma Dergisi **11**(3), 43–50 (2014)
3. Bozkurt, Ö., Bozkurt, İ.: A field study on job satisfaction effecting internal factors in education sector. Doğuş Üniversitesi Dergisi **9**(1), 1–18 (2008)
4. Akkaya, G., Turanoğlu, B., Öztaş, S.: An integrated fuzzy AHP and fuzzy MOORA approach to the problem of industrial engineering sector choosing. Expert Syst. Appl. **42** (24), 9565–9573 (2015)
5. Groover, M.P.: Work Systems and the Methods, Measurement, and the Management of Work. Pearson Prentice Hall, Upper Saddle River (2007)
6. Saaty, T.L.: The Analytic Hierarchy Process. Mcgraw-Hill, New York (1980)
7. Opricovic, S., Tzeng, G.H.: Compromise solution by MCDM methods: a comparative analysis of VIKOR and TOPSIS. Eur. J. Oper. Res. **156**, 445–455 (2004)
8. Liao, H., Xu, Z., Zeng, X.: Hesitant fuzzy linguistic VIKOR method and its application in qualitative multiple criteria decision making. IEEE Trans. Fuzzy Syst. **23**(5), 1343–1355 (2015)
9. Zhang, N., Wei, G.: Extension of VIKOR method for decision making problem based on hesitant fuzzy set. Appl. Math. Model. **37**, 4938–4947 (2013)
10. Liu, H.W., Wang, G.J.: Multi-criteria decision-making methods based on intuitionistic fuzzy sets. Eur. J. Oper. Res. **179**, 220–233 (2007)
11. Merigó, J.M., Casanovas, M.: Induced aggregation operators in decision making with the Dempster-Shafer belief structure. Int. J. Intell. Syst. **24**, 934–954 (2009)

# Neutrosophic Fuzzy Analytic Hierarchy Process Approach for Safe Cities Evaluation Criteria

Serhat Aydın[1], Ahmet Aktas[2], and Mehmet Kabak[2(✉)]

[1] Air Force Academy, National Defense University, 34334 Istanbul, Turkey
[2] Department of Industrial Engineering, Gazi University, 06570 Ankara, Turkey
mkabak@gazi.edu.tr

**Abstract.** Many decisions in real life applications require taking different criteria into consideration simultaneously. To find a moderate solution to these kind of decisions, multiple criteria decision making approaches are used. Furthermore, crisp logic is unable to express the vagueness and uncertainty in evaluation of criteria. Hence, multiple criteria decision making approaches are commonly integrated with fuzzy logic. In recent years, ordinary fuzzy sets have been extended to new types. Neutrosophic set is one of the extensions of fuzzy sets and introduces a new component called "indeterminacy", and carry more information than fuzzy sets. In this study, Neutrosophic Fuzzy Analytic Hierarchy Process (NF-AHP) method is presented. NF-AHP is used to construct a ranking model for cities in views of safety under four main criteria and 21 sub-criteria. The obtained results show that the most important criterion of city safety is personal security.

**Keywords:** Neutrosophic fuzzy sets · Analytic Hierarchy Process
Safe cities

## 1 Introduction

Decision-making is a daily activity that is done frequently by people such as selecting the object to be bought, determining the direction to walk, choosing the house to rent, etc. [1]. There must be at least two alternatives to speak about making a decision. If the decision maker makes his/her decision by considering one criterion and the alternative values are crisp, it is easy to decide. In this case, the alternative, which is close to the desired value, can be determined as the decision. However, in many cases the decision problem is in a more complex structure. A number of criteria should be taken into consideration and none of the alternatives is dominant in views of all criteria. For example, let us consider a car selection decision. Some criteria such as comfort, price, fuel consumption, service costs, luggage volume, etc. can be considered in the selection process. Cheaper cars are usually worse on comfort criterion or cars with higher luggage volume are generally more expensive. In such decisions, decision maker have to find a compromise alternative by considering all criteria simultaneously.

© Springer Nature Switzerland AG 2019
R. A. Aliev et al. (Eds.): ICAFS-2018, AISC 896, pp. 958–965, 2019.
https://doi.org/10.1007/978-3-030-04164-9_127

Multiple criteria decision making (MCDM) techniques are very useful at finding a moderate solution in complex decision problems consisting a finite number of alternatives [2]. Decision makers commonly use these techniques because they are very flexible that can be utilized on almost all decision problems. Some of the most common MCDM techniques are Analytic Hierarchy Process (AHP), Analytic Network Process (ANP) and Technique of Order Preference by Similarity to Ideal Solution (TOPSIS).

Among different MCDM methods, AHP method is the most popular method for researchers. This method is firstly introduced by Saaty in 1980 [3]. Because of the ability to model tangible and intangible judgments in an aggregated manner, AHP studies are confronted in the literature in a wide application area. Some examples of decision problems that utilize AHP can be listed as healthcare service quality evaluation [4], weapon selection [5], quality award assessment [6], 3D printer selection [7], etc.

Fuzzy logic is introduced into the literature by Zadeh [8]. Expression of linguistic and uncertain statements gets possible by utilizing fuzzy logic. Fuzzy logic is firstly integrated into decision making literature in 1970 by Zadeh and Bellman [9].

By that time, fuzzy applications of decision making is also popular for researchers. Study numbers for the fuzzy extensions of the MCDM methods mentioned above are about 20% and 50% of total studies.

First studies for fuzzy extensions of AHP are developed by Van Laarhoven and Pedrycz in 1983 [10] and by Buckley in 1985 [11]. While Van Laarhoven and Pedrycz only consider triangular membership functions, Buckley demonstrated usability of different types of membership functions in AHP. In 1996, Chang proposed extent analysis for Fuzzy Analytic Hierarchy Process [12]. This study presents the steps for determination of weights when triangular fuzzy numbers are used for pair-wise comparison. By the progress in fuzzy concept, new extensions of fuzzy AHP are introduced into the literature. For example, Kahraman et al. proposed interval type-2 fuzzy extension of AHP [13] and Xu and Liao extended AHP into intuitionistic fuzzy environment [14].

The main purpose of this study is to present a neutrosophic fuzzy extension of AHP in order to construct a ranking model for safe cities. Differ from the other fuzzy sets; a neutrosophic set is expressed by three parameters, which are called Truthiness, Indeterminacy and Falsity. Neutrosophic logic was developed by Smarandache [15] as a generalization of fuzzy logic. A neutrosophic set <T, I, F> is composed of three parameters which are a degree of truth (T), a degree of indeterminacy (I), and a degree of falsity (F), where T, I, and F $\in\, ]^-0, 1^+[$ and $^-0 \leq \sup T_A(x) + \sup I_A(x) + \sup F_A(x) \leq 3^+$. It is difficult to apply the neutrosophic set operators in the real application. Therefore, Wang et al. [16] proposed a single valued neutrosophic sets, which is an extension of neutrosophic sets. For each point x in X, we have $T_A(x), I_A(x), F_A(x) \in [0, 1]$ and $0 \leq T_A(x), I_A(x), F_A(x) \leq 3$.

Classic AHP uses integers in computing importance scale. However, real-world problems involve substantial vagueness and uncertainty, which necessitates using fuzzy numbers [17]. Therefore, AHP and neutrosophic set were combined and transformed into an integrated model called Neutrosophic Fuzzy AHP (NF-AHP). In this study, NF-AHP is used to construct a ranking model for cities in views of safety.

Since world population grows every day, food and other resources needed for survive diminishes with increasing rate. Safety in cities emerges as an important and risky issue and safe cities indicators help people in this regard. The Economist Intelligence Unit regularly publish Safe Cities Reports. They present rank of cities in views of each safety aspects by considering different weight values of safety indicators. To construct a scientific basis for safe city assessment, importance degree of safety indicators should be determined by using analytical approaches. Hence, the main aim in this study is to determine the importance degree of safety indicators for cities by using NF-AHP method. By considering the aforementioned advantages, the method gives people the ability of making an assessment between cities by considering truth, indeterminacy and falsity of judgments of decision makers.

The rest of the paper goes as follows: in the second part, Neutrosophic Fuzzy AHP methodology is explained. The third part consists the application steps for constructing the ranking model for safe cities. This study is concluded in the fourth part by summarizing the obtained results and highlighting the future research directions.

## 2 Neutrosophic Fuzzy AHP

Steps of NF-AHP method is given as follows. Due to the page limits, readers can refer to related cited references in steps for details.

**Step 1: Structure hierarchy.** Define the problem and identify the criteria, sub criteria and alternatives of the decision making problem. Then organize the problem as a hierarchy.

**Step 2: Establish pair wise comparison matrix of the factors.** Experts compare the factors in pairwise matrices in each level using the scale by Radwan et al. [18] in Table 1. The results of the pair-wise comparison on n factors can be summarized in an n x n matrix R.

**Step 3: Aggregate individual neutrosophic evaluation into group neutrosophic evaluation.** Experts' evaluation is captured in Step 2. Then, neutrosophic weighted arithmetic average aggregation operator [19] is used in order to get group Neutrosophic evaluation.

**Step 4: Check the consistency of pairwise comparison matrix.** Xu and Liao [14] proposed a method in order to construct consistent preference relation. Hence we used this method to construct a perfect consistent neutrosophic preference relation.

**Step 5. Determine neutrosophic relative weight of each preference relation.** After checking consistency, we continue the calculations with neutrosophic matrices. Eigenvector procedure is performed to get the neutrosophic weights vector $w = \{w_1, w_2, \ldots w_n\}$. In this procedure firstly, normalization procedure is applied. Then the weights are computed.

**Step 6. Calculate final weight.** The rating of each alternative is multiplied by the weights of the sub-criteria and aggregated to get local ratings with respect to each main criterion. The local ratings are then multiplied by the weights of the main criteria and aggregated to get global ratings.

**Table 1.** Linguistic variables and importance weight based on neutrosophic values.

| Linguistic term | Neutrosophic set | Linguistic term | Neutrosophic set |
|---|---|---|---|
| Extremely highly preferred | $\langle 0.90, 0.10, 0.10 \rangle$ | Mildly lowly preferred | $\langle 0.10, 0.90, 0.90 \rangle$ |
| Extremely preferred | $\langle 0.85, 0.20, 0.15 \rangle$ | Mildly preferred | $\langle 0.15, 0.80, 0.85 \rangle$ |
| Very strongly to extremely preferred | $\langle 0.85, 0.25, 0.20 \rangle$ | Mildly preferred to very lowly preferred | $\langle 0.20, 0.75, 0.80 \rangle$ |
| Very strongly preferred | $\langle 0.75, 0.20, 0.20 \rangle$ | Very lowly preferred | $\langle 0.25, 0.75, 0.75 \rangle$ |
| Strongly preferred | $\langle 0.70, 0.30, 0.30 \rangle$ | Lowly preferred | $\langle 0.30, 0.70, 0.70 \rangle$ |
| Moderately highly to strongly preferred | $\langle 0.65, 0.30, 0.35 \rangle$ | Moderately lowly preferred to lowly preferred | $\langle 0.35, 0.70, 0.65 \rangle$ |
| Moderately highly preferred | $\langle 0.60, 0.35, 0.40 \rangle$ | Moderately lowly preferred | $\langle 0.40, 0.60, 0.60 \rangle$ |
| Equally to moderately preferred | $\langle 0.55, 0.40, 0.45 \rangle$ | Moderately to equally preferred | $\langle 0.45, 0.60, 0.55 \rangle$ |
| Equally preferred | $\langle 0.50, 0.50, 0.50 \rangle$ | Equally preferred | $\langle 0.50, 0.50, 0.50 \rangle$ |

**Step 7. Convert neutrosophic weights in to the crisp numbers.** After get overall neutrosophic weights, they can be converted into crisp numbers in order to ranking criteria [18].

# 3   Application

In this section, safe city evaluation criteria weights are determined. In this study, the alternative cities will be ranked according to the safety criteria. The evaluation criteria can be seen on Table 2. In order to show the steps of the methodology, main criteria weights calculations will be explained in detail and sub-criteria weights values will be given.

**Step 1.** Evaluation criteria are determined as they are given in Table 2.

**Step 2.** Experts compare the factors in pairwise matrices in each level using the scale in Radwan et al. [19]. The experts' weights are appointed 0.4, 0.3, 0.3, respectively. Pairwise comparison matrices are given in Table 3.

**Step 3.** Then, simplified neutrosophic weighted arithmetic average aggregation operator is used in order to get group Neutrosophic evaluation. Table 4 shows the aggregated individual neutrosophic evaluation. For example, the aggregation of $\tilde{r}_{...21}$ is calculated by the following equation:

**Table 2.** Safe city evaluation criteria.

| Main criteria | Sub criteria |
|---|---|
| Digital security (C1) | Frequency of identity theft (C11) |
| | Percentage of computers infected (C12) |
| | Percentage with Internet access (C13) |
| Health security (C2) | Access to healthcare (C21) |
| | No. of beds per 1,000 (C22) |
| | No. of doctors per 1,000 (C23) |
| | Quality of health services (C24) |
| | Life expectancy years (C25) |
| | Infant mortality (C26) |
| Infrastructure security (C3) | Quality of road infrastructure (C31) |
| | Quality of electricity infrastructure (C32) |
| | Disaster management/business continuity plan (C33) |
| | Frequency of vehicular accidents (C34) |
| | Frequency of pedestrian deaths (C35) |
| Personal security (C4) | Prevalence of petty crime (C41) |
| | Prevalence of violent crime (C42) |
| | Organized crime (C43) |
| | Frequency of terrorist attacks (C44) |
| | Severity of terrorist attacks (C45) |
| | Gender safety (C46) |
| | Perceptions of safety (C47) |

**Table 3.** Comparison matrix of main criteria by experts

| | | C1 | C2 | C3 | C4 |
|---|---|---|---|---|---|
| C1 | E1 | $\langle 0.50, 0.50, 0.50 \rangle$ | $\langle 0.35, 0.70, 0.65 \rangle$ | $\langle 0.40, 0.65, 0.60 \rangle$ | $\langle 0.15, 0.80, 0.85 \rangle$ |
| | E2 | $\langle 0.50, 0.50, 0.50 \rangle$ | $\langle 0.25, 0.70, 0.70 \rangle$ | $\langle 0.35, 0.70, 0.65 \rangle$ | $\langle 0.30, 0.70, 0.70 \rangle$ |
| | E3 | $\langle 0.50, 0.50, 0.50 \rangle$ | $\langle 0.45, 0.60, 0.55 \rangle$ | $\langle 0.35, 0.70, 0.65 \rangle$ | $\langle 0.15, 0.80, 0.85 \rangle$ |
| C2 | E1 | $\langle 0.65, 0.30, 0.35 \rangle$ | $\langle 0.50, 0.50, 0.50 \rangle$ | $\langle 0.60, 0.35, 0.40 \rangle$ | $\langle 0.40, 0.65, 0.60 \rangle$ |
| | E2 | $\langle 0.75, 0.25, 0.25 \rangle$ | $\langle 0.50, 0.50, 0.50 \rangle$ | $\langle 0.60, 0.35, 0.40 \rangle$ | $\langle 0.45, 0.60, 0.55 \rangle$ |
| | E3 | $\langle 0.55, 0.40, 0.45 \rangle$ | $\langle 0.50, 0.50, 0.50 \rangle$ | $\langle 0.65, 0.30, 0.35 \rangle$ | $\langle 0.20, 0.75, 0.80 \rangle$ |
| C3 | E1 | $\langle 0.60, 0.35, 0.40 \rangle$ | $\langle 0.40, 0.65, 0.60 \rangle$ | $\langle 0.50, 0.50, 0.50 \rangle$ | $\langle 0.25, 0.75, 0.75 \rangle$ |
| | E2 | $\langle 0.65, 0.30, 0.35 \rangle$ | $\langle 0.40, 0.65, 0.60 \rangle$ | $\langle 0.50, 0.50, 0.50 \rangle$ | $\langle 0.30, 0.70, 0.70 \rangle$ |
| | E3 | $\langle 0.65, 0.30, 0.35 \rangle$ | $\langle 0.35, 0.70, 0.65 \rangle$ | $\langle 0.50, 0.50, 0.50 \rangle$ | $\langle 0.40, 0.65, 0.60 \rangle$ |
| C4 | E1 | $\langle 0.85, 0.20, 0.15 \rangle$ | $\langle 0.60, 0.35, 0.40 \rangle$ | $\langle 0.75, 0.25, 0.25 \rangle$ | $\langle 0.50, 0.50, 0.50 \rangle$ |
| | E2 | $\langle 0.70, 0.30, 0.30 \rangle$ | $\langle 0.55, 0.40, 0.45 \rangle$ | $\langle 0.70, 0.30, 0.30 \rangle$ | $\langle 0.50, 0.50, 0.50 \rangle$ |
| | E3 | $\langle 0.85, 0.20, 0.15 \rangle$ | $\langle 0.80, 0.25, 0.20 \rangle$ | $\langle 0.60, 0.35, 0.40 \rangle$ | $\langle 0.50, 0.50, 0.50 \rangle$ |

**Table 4.** Aggregated individual evaluation into group neutrosophic evaluation

|    | C1 | C2 | C3 | C4 |
|----|----|----|----|----|
| C1 | $\langle 0.50, 0.50, 0.50 \rangle$ | $\langle 0.35, 0.69, 0.66 \rangle$ | $\langle 0.37, 0.68, 0.63 \rangle$ | $\langle 0.20, 0.77, 0.83 \rangle$ |
| C2 | $\langle 0.48, 0.31, 0.35 \rangle$ | $\langle 0.50, 0.50, 0.50 \rangle$ | $\langle 0.62, 0.34, 0.38 \rangle$ | $\langle 0.36, 0.67, 0.66 \rangle$ |
| C3 | $\langle 0.63, 0.32, 0.37 \rangle$ | $\langle 0.39, 0.67, 0.62 \rangle$ | $\langle 0.50, 0.50, 0.50 \rangle$ | $\langle 0.31, 0.71, 0.70 \rangle$ |
| C4 | $\langle 0.81, 0.23, 0.20 \rangle$ | $\langle 0.66, 0.34, 0.36 \rangle$ | $\langle 0.70, 0.30, 0.31 \rangle$ | $\langle 0.50, 0.50, 0.50 \rangle$ |

$$\tilde{r}_{...21} = \left\langle \begin{array}{c} 1 - (1 - 0.65)^{0.4} \times (1 - 0.75)^{0.3} \times (1 - 0.55)^{0.3}, \\ 1 - (1 - 0.30)^{0.4} \times (1 - 0.25)^{0.3} \times (1 - 0.40)^{0.3}, \\ 1 - (1 - 0.35)^{0.4} \times (1 - 0.25)^{0.3} \times (1 - 0.45)^{0.3} \end{array} \right\rangle \tag{1}$$

$$\tilde{r}_{...21} = \langle 0.48, 0.31, 0.35 \rangle \tag{2}$$

**Step 4.** The consistency of neutrosophic pairwise comparison matrix with respect to the goal is constructed as shown in Table 5

**Table 5.** Consistency pairwise comparison matrix with respect to the goal

|    | C1 | C2 | C3 | C4 |
|----|----|----|----|----|
| C1 | $\langle 0.50, 0.50, 0.50 \rangle$ | $\langle 0.35, 0.69, 0.66 \rangle$ | $\langle 0.47, 0.53, 0.55 \rangle$ | $\langle 0.22, 0.83, 0.79 \rangle$ |
| C2 | $\langle 0.66, 0.68, 0.35 \rangle$ | $\langle 0.50, 0.50, 0.50 \rangle$ | $\langle 0.62, 0.34, 0.39 \rangle$ | $\langle 0.42, 0.53, 0.55 \rangle$ |
| C3 | $\langle 0.63, 0.68, 0.37 \rangle$ | $\langle 0.39, 0.33, 0.62 \rangle$ | $\langle 0.50, 0.50, 0.50 \rangle$ | $\langle 0.31, 0.71, 0.70 \rangle$ |
| C4 | $\langle 0.82, 0.23, 0.20 \rangle$ | $\langle 0.66, 0.34, 0.36 \rangle$ | $\langle 0.70, 0.30, 0.31 \rangle$ | $\langle 0.50, 0.50, 0.50 \rangle$ |

**Step 5.** In order to determine local neutrosophic relative weights of each preference relation eigenvector procedure is performed. For example, relative weight of "Digital Security" is calculated as follows;

$$W_{Digital\ Security} = \{0.42, 0.54, 0.53\} \tag{3}$$

**Step 6.** In the sixth step, the local ratings are then multiplied by the weights of the main criteria and aggregated to get global ratings. Table 6 represents the global ratings.

**Step 7.** After get all global ratings of criteria and sub-criteria, they can be convert into crisp numbers in order to get ranking. For example, weight of digital security;

**Table 6.** The global ratings of the criteria and sub-criteria

| Main criteria | Criteria weight | Sub criteria | Sub criteria local weight | Global weight |
|---|---|---|---|---|
| C1 | $\langle 0.42, 0.54, 0.53 \rangle$ | C11 | $\langle 0.33, 0.64, 0.67 \rangle$ | $\langle 0.14, 0.83, 0.84 \rangle$ |
| | | C12 | $\langle 0.67, 0.36, 0.32 \rangle$ | $\langle 0.28, 0.71, 0.68 \rangle$ |
| | | C13 | $\langle 0.76, 0.24, 0.23 \rangle$ | $\langle 0.32, 0.65, 0.64 \rangle$ |
| C2 | $\langle 0.64, 0.34, 0.34 \rangle$ | C21 | $\langle 0.56, 0.42, 0.41 \rangle$ | $\langle 0.36, 0.61, 0.61 \rangle$ |
| | | C22 | $\langle 0.35, 0.31, 0.61 \rangle$ | $\langle 0.22, 0.54, 0.74 \rangle$ |
| | | C23 | $\langle 0.37, 0.29, 0.58 \rangle$ | $\langle 0.24, 0.53, 0.72 \rangle$ |
| | | C24 | $\langle 0.64, 0.33, 0.33 \rangle$ | $\langle 0.41, 0.55, 0.56 \rangle$ |
| | | C25 | $\langle 0.51, 0.43, 0.45 \rangle$ | $\langle 0.32, 0.62, 0.64 \rangle$ |
| | | C26 | $\langle 0.75, 0.21, 0.22 \rangle$ | $\langle 0.48, 0.48, 0.49 \rangle$ |
| C3 | $\langle 0.54, 0.42, 0.41 \rangle$ | C31 | $\langle 0.52, 0.49, 0.49 \rangle$ | $\langle 0.28, 0.71, 0.70 \rangle$ |
| | | C32 | $\langle 0.35, 0.68, 0.66 \rangle$ | $\langle 0.19, 0.81, 0.80 \rangle$ |
| | | C33 | $\langle 0.37, 0.63, 0.63 \rangle$ | $\langle 0.20, 0.79, 0.78 \rangle$ |
| | | C34 | $\langle 0.62, 0.40, 0.40 \rangle$ | $\langle 0.33, 0.65, 0.65 \rangle$ |
| | | C35 | $\langle 0.49, 0.50, 0.53 \rangle$ | $\langle 0.27, 0.71, 0.72 \rangle$ |
| C4 | $\langle 0.80, 0.21, 0.20 \rangle$ | C41 | $\langle 0.38, 0.65, 0.62 \rangle$ | $\langle 0.30, 0.72, 0.70 \rangle$ |
| | | C42 | $\langle 0.42, 0.59, 0.58 \rangle$ | $\langle 0.34, 0.68, 0.66 \rangle$ |
| | | C43 | $\langle 0.52, 0.49, 0.47 \rangle$ | $\langle 0.42, 0.60, 0.58 \rangle$ |
| | | C44 | $\langle 0.52, 0.47, 0.48 \rangle$ | $\langle 0.41, 0.58, 0.58 \rangle$ |
| | | C45 | $\langle 0.55, 0.43, 0.45 \rangle$ | $\langle 0.44, 0.55, 0.56 \rangle$ |
| | | C46 | $\langle 0.60, 0.37, 0.40 \rangle$ | $\langle 0.48, 0.51, 0.59 \rangle$ |
| | | C47 | $\langle 0.59, 0.39, 0.41 \rangle$ | $\langle 0.47, 0.52, 0.53 \rangle$ |

$$CW_{Digital\ Security} = (3 + (0.4231) - 2(0.5467) - (0.5336))/4 = 0.44 \qquad (4)$$

The other main criteria weights are; 0.65 for health security, 0.57 for infrastructure security, and 0.79 for personal security.

## 4   Conclusion

Safety is one of the most important aspects for assessment of suitability of a region for living. Safe city indicators help people to understand livability of a city in views of a specific aspect. However, different indicators should be simultaneously considered for determining safety of a city. To develop an analytic tool for safe city assessment, we utilize multiple criteria decision making, since MCDM allows decision makers to take different tangible and intangible aspects of decision problems in an aggregated manner.

In this study we propose a NF-AHP method for safe city assessment. The method uses the neutrosophic scale in order to establish pair wise comparison matrix. In this method, many experts can be utilized in order to get comparison matrix. And neutrosophic weighted arithmetic average aggregation operator is used in order to get group Neutrosophic evaluation. Then consistency of pairwise comparison matrix is checked. And the Eigenvector procedure is performed to get the neutrosophic weights vector. Finally global weights are calculated and they converted into crisp values in order to get ranking results.

# References

1. Kirkwood, C.W.: Strategic Decision Making: Multiobjective Decision Analysis with Spreadsheets. Duxbury Press, Belmont, California (1997)
2. Yoon, K.P., Hwang, C.L.: Multiple Attribute Decision Making: An Introduction. SAGE Publications, California (2005)
3. Saaty, T.: The Analytic Hierarchy Process. McGraw-Hill, New York (1980)
4. Aktas, A., Cebi, S., Temiz, I.: A new evaluation model for service quality of healthcare systems based on AHP and information axiom. J. Intell. Fuzzy Syst. **28**(3), 1009–1021 (2015)
5. Dagdeviren, M., Yavuz, S., Kilinci, N.: Weapon selection using the AHP and TOPSIS methods under fuzzy environment. Expert Syst. Appl. **36**, 8143–8151 (2009)
6. Aydın, S., Kahraman, C., Kaya, İ.: A new fuzzy multicriteria decision making approach: an application for European quality award assessment. Knowl. Based Syst. **32**, 37–46 (2012)
7. Çetinkaya, C., Kabak, M., Özceylan, E.: 3d printer selection by using fuzzy analytic hierarchy process and PROMETHEE. Int. J. Inf. Technol. **10**(4), 371–380 (2017)
8. Zadeh, L.A.: Fuzzy sets. Inf. Control **8**(3), 338–353 (1965)
9. Bellman, R., Zadeh, L.: Decision-making in a fuzzy environment. Manage. Sci. **17**(4), 141–164 (1970)
10. Van Laarhoven, P.J.M., Pedrycz, W.: A fuzzy extension of Saaty's priority theory. Fuzzy Sets Syst. **11**(1–3), 229–241 (1983)
11. Buckley, J.J.: Fuzzy hierarchical analysis. Fuzzy Sets Syst. **17**(3), 233–247 (1985)
12. Chang, D.Y.: Applications of the extent analysis method on fuzzy AHP. Eur. J. Oper. Res. **95**(3), 649–655 (1996)
13. Kahraman, C., Öztayşi, B., Sarı, İ.U., Turanoğlu, E.: Fuzzy analytic hierarchy process with interval type-2 fuzzy sets. Knowl. Based Syst. **59**, 48–57 (2014)
14. Xu, Z., Liao, H.: Intuitionistic fuzzy analytical hierarchy process. IEEE Trans. Fuzzy Syst. **22**(4), 749–761 (2014)
15. Samarandache, F.: A Unifying Field in Logics. Neutrosophy: Neutrosophic Probability, Set and Logic. American Research Press, Rehoboth (1998)
16. Wang, H., Smarandache, F., Zhang, Y., Sunderraman, R.: Single valued neutrosophic sets. In: Multispace and Multistructure, vol. 4, pp. 410–413 (2010)
17. Aydın, S., Kahraman, C.: Multiattribute supplier selection using fuzzy analytic hierarchy process. Int. J. Comput. Intell. Syst. **3**(5), 553–565 (2010)
18. Radwan, N.M., Senousy, M.B., Riad, A.E.D.M.: Neutrosophic AHP multi criteria decision making method applied on the selection of learning management system. Int. J. Adv. Comput. Technol. **8**, 95–105 (2016)
19. Ye, J.: A multicriteria decision-making method using aggregation operators for simplified neutrosophic sets. J. Intell. Fuzzy Syst. **26**, 2459–2466 (2014)

# Author Index

Printed in the United States
By Bookmasters